NEUROBIOLOGY OF THE PARENTAL BRAIN

NEUROBIOLOGY OF THE PARENTAL BRAIN

ROBERT S. BRIDGES

ELSEVIER

AMSTERDAM • BOSTON • HEIDELBERG • LONDON • NEW YORK • OXFORD
PARIS • SAN DIEGO • SAN FRANCISCO • SINGAPORE • SYDNEY • TOKYO
Academic Press is an imprint of Elsevier

AP

Academic Press is an imprint of Elsevier
30 Corporate Drive, Suite 400, Burlington, MA 01803, USA
525 B Street, Suite 1900, San Diego, California 92101-4495, USA
32 Jamestown Road, London NW1 7BY, UK

British Library Cataloguing in Publication Data
A catalogue record for this book is available from the British Library

Library of Congress Cataloguing in Publication Data
A catalogue record for this book is available from the Library of Congress

ISBN 978-0-12-374285-8

For information on all Academic Press publications
visit our website at www.elsevierdirect.com

Typeset by Charon Tec Ltd., A Macmillan Company. (www.macmillansolutions.com)

Printed and bound in the USA

08 09 10 11 10 9 8 7 6 5 4 3 2 1

I dedicate this book to my sons, Samuel Erskine Bridges, and Mark Stafford Bridges, who have provided me with the wonderful rewards and occasional challenges of parenting.

CONTENTS

III NEUROENDOCRINE ADAPTATIONS OF PARENTING: PREGNANCY, LACTATION, AND OFFSPRING 201

33 Neuroendocrine and Behavioral Adaptations Following Reproductive Experience in the Female Rat 509

ELIZABETH M. BYRNES, BENJAMIN C. NEPHEW AND ROBERT S. BRIDGES

34 Plasticity in the Maternal Neural Circuit: Experience, Dopamine, and Mothering 519

ALISON S. FLEMING, ANDREA GONZALEZ, VERONICA M. AFONSO AND VEDRAN LOVIC

ACKNOWLEDGMENTS

This publication is an outgrowth of the "Parental Brain Conference: Parenting and the Brain" that was held in Boston, Massachusetts in June 2007. This conference was generously supported by contributions from The Heinz Family Philanthropies, Tufts University and Tufts Cummings School of Veterinary Medicine, the University of Richmond, Boston College, the 2003 Mother–Infant Conference – Douglas Hospital, Montreal, Canada, the British Society for Neuroendocrinology, and the Center for Behavioral Neuroscience at Georgia State University.

Major support for the conference and the resultant publication was received from the National Institutes of Health. The NIH issues the following disclaimer regarding this publication: "This project was supported (in part) by NIH Research Conference Grant 1R13MH080562-01 funded by the National Institute of Mental Health, the National Institute on Drug Abuse, the National Institute of Child Health and Human Development, and the Office of Research on Women's Health through the Office of the Director, National Institutes of Health. The views expressed in this publication do not necessarily reflect the official policies of the Department of Health and Human Services; nor does mention by trade names, commercial practices, or organizations imply endorsement by the U.S. Government."

Finally, I would like to thank my staff assistant, Janine Stuczko, Tufts Cummings School of Veterinary Medicine's Continuing Education Associate Director, Susan Brogan, and my colleagues, Drs. Elizabeth Byrnes and Phyllis Mann, for their support and assistance in making the conference a success.

LIST OF CONTRIBUTORS

Numbers in parentheses indicate the pages on which the authors' contributions begin.

Veronica M. Afonso (519) Department of Psychology, Erindale College, University of Toronto, Toronto, Canada, L5L1C6

Rachael A. Augustine (249) Department of Anatomy and Structural Biology, Centre for Neuroendocrinology, University of Otago, P.O. Box 913, Dunedin, New Zealand

Karen L. Bales (417) Department of Psychology, University of California-Davis, 1 Sheilds Avenue, Davis, CA 95616, USA

Elizabeth A. Becker (435) Department of Psychology, University of Wisconsin-Madison, Madison, WI, USA

Sarah L. Berga (175) Departments of Gynecology and Obstetrics, Psychiatry and Behavioral Sciences, Emory University School of Medicine, Atlanta, GA, USA

Richard Bertram (235) Department of Mathematics and Program Molecular Biophysics, The Florida State University, 107 Chieftan Way, P.O. Box 3064370, Tallahassee, FL 32306-4370, USA

Janet K. Bester-Meredith (435) Biology Department, Seattle Pacific University, 3307 3rd Avenue, West, Seattle, WA 98119, USA

Maria L. Boccia (377) Frank Porter Graham Child Development Institute, CB# 8185, The University of North Carolina at Chapel Hill, Chapel Hill, NC 27599-8185, USA

Ericka M. Boone (417) Department of Psychiatry, Brain Body Center, University of Illinois at Chicago, Chicago, IL 60612, USA

Oliver J. Bosch (115) Department of Behavioural Endocrinology, Institute of Zoology, University of Regensburg, Universitätsstrasse 31 93053 Regensburg, Germany

Robert S. Bridges (509) Department of Biomedical Sciences, Tufts University – Cummings School of Veterinary Medicine, 200 Westboro Road, North Grafton, MA 01536, USA

Paula J. Brunton (203) Laboratory of Neuroendocrinology, Centre for Integrative Physiology, School of Biomedical Sciences, College of Medicine and Veterinary Medicine, University of Edinburgh, Hugh Robson Building, George Square, Edinburgh EH8 9XD, UK

John D. Buntin (269) Department of Biological Sciences, P.O. Box 413, University of Wisconsin, Milwaukee, WI 53217, USA

Elizabeth M. Byrnes (509) Department of Biomedical Sciences, Tufts University – Cummings School of Veterinary Medicine, 200 Westboro Road, North Grafton, MA 01536, USA

Newton S. Canteras (75) Faculdade de Medicina Veterinária e Zootecnia – Instituto de Ciências Biomédicas, Universidade de São Paulo, São Paulo, Brasil

C. Sue Carter (417) Department of Psychiatry, Brain Body Center, University of Illinois at Chicago, Chicago, IL 60612, USA

Frances A. Champagne (307) Department of Psychology, Columbia University, 406 Schermerhorn Hall, 1190 Amsterdam Avenue, New York, NY 10027, USA

James P. Curley (319) Department of Psychology, Columbia University, 406 Schermerhorn Hall, 1190 Amsterdam Avenue, New York, NY 10017, USA

Kimberly L. D'Anna (103) Department of Zoology, University of Wisconsin, 1117 West Johnson Street, Madison, WI 53706, USA

Esterina d'Asti (293) Department of Psychiatry, Douglas Hospital Research Center, McGill University, Montreal, 6875 Lasalle Boulevard, Verdun, QC, Canada H4H 1R3

Alison J. Douglas (225) Centre for Integrative Physiology, College of Medicine and Veterinary Medicine, University of Edinburgh, CIP, School of Biomedical Science, Hugh Robson Building, George Square, Edinburgh, EH8 9XD, UK

Marcel Egli (235) Space Biology Group, Swiss Federal Institute of Technology, Zurich (ETHZ), Technoparkstrasse 1, ETH-Technopark, Zurich, Switzerland CH-8005, UK

Marcelo Febo (61) Department of Psychology, Northeastern University, 360 Huntington Avenue, 125NI, Boston, MA 02115, USA

Luciano F. Felício (75) Faculdade de Medicina Veterinária e Zootecnia – Instituto de Ciências Biomédicas, Universidade de São Paulo, São Paulo, Brasil

Craig F. Ferris (61) Department of Psychology, Northeastern University, 360 Huntington Avenue, 125NI, Boston, MA 02115, USA

Jeffrey E. Fite (461) U.S. Army Research Institute, 121 Morande Street, Fort Knox, KY 40121-4141, USA

Alison S. Fleming (519) Department of Psychology, Erindale College, University of Toronto, Toronto, Canada, L5L1C6

Marc E. Freeman (235) Department of Biological Science and Program in Neuroscience, The Florida State University, 107 Chieftan Way, P.O. Box 3064370, Tallahassee, FL 32306-4370, USA

Jeffrey A. French (461) Departments of Psychology and Biology, University of Nebraska at Omaha, Omaha, NE 68182-0274, USA

Liisa A. M. Galea (493) Program in Neuroscience, Department of Psychology and Brain Research Center, University of British Columbia, 2136 West Mall, Vancouver, BC, V6T 1Z4, USA

Stephen C. Gammie (103) Department of Zoology, University of Wisconsin, 1117 West Johnson Street, Madison, WI 53706, USA

Jody M. Ganiban (391) Department of Psychology, NW The George Washington University, 2125 G St., Washington, DC 20057, USA

Erin D. Gleason (435) Department of Psychology, University of Wisconsin, 1202 West Johnson Street, Madison, WI 53706, USA

Kyle L. Gobrogge (347) Department of Psychology, Florida State University, 209 Copeland Avenue, Tallahassee, FL 32306-1270, USA

Andrea Gonzalez (519) Department of Psychology, Erindale College, University of Toronto, Toronto, Canada, L5L1C6

David R. Grattan (249) Department of Anatomy and Structural Biology, Centre for Neuroendocrinology, University of Otago, P.O. Box 913, Dunedin, New Zealand

Minako Hashii (361) Kanazawa University, 21st Century Center for Excellence Program on Innovation Brain Science on Development, Learning and Memory, 13-1 Takara-machi, Kanazawa, Ishikawa 920-8640, Japan

Kenshi Hayashi (361) Kanazawa University, 21st Century Center for Excellence Program on Innovation Brain Science on Development, Learning and Memory, 13-1 Takara-machi, Kanazawa, Ishikawa 920-8640, Japan

Haruhiro Higashida (361) Kanazawa University, 21st Century Center for Excellence Program on Innovation Brain Science on Development, Learning and Memory, 13-1 Takara-machi, Kanazawa, Ishikawa 920-8640, Japan

Teresa H. Horton (449) Department of Neurobiology and Physiology, Northwestern University, 2205 Tech Drive, Room 2-160 Hogan Hall Evanston, IL 60208, USA

Sarah Blaffer Hrdy (407) Department of Anthropology University of California, Davis, CA 95616, USA

Md. Saharul Islam (361) Kanazawa University, 21st Century Center for Excellence Program on Innovation Brain Science on Development, Learning and Memory, 13-1 Takara-machi, Kanazawa, Ishikawa 920-8640, Japan

Marianna A. Jimenez (449) Department of Neurobiology and Physiology, Northwestern University, 2205 Tech Drive, Room 2-160 Hogan Hall Evanston, IL 60208, USA

Duo Jin (361) Kanazawa University, 21st Century Center for Excellence Program on Innovation Brain Science on Development, Learning and Memory, 13-1 Takara-machi, Kanazawa, Ishikawa 920-8640, Japan

Craig H. Kinsley (481) Department of Psychology, University of Richmond, Richmond, VA 23173, USA

Keita Koizumi (361) Kanazawa University, 21st Century Center for Excellence Program on Innovation Brain Science on Development, Learning and Memory, 13-1 Takara-machi, Kanazawa, Ishikawa 920-8640, Japan

Sharon R. Ladyman (249) Department of Anatomy and Structural Biology, Centre for Neuroendocrinology, University of Otago, P.O. Box 913, Dunedin, New Zealand

Kelly G. Lambert (481) Department of Psychology, Copley Science Center, Room 133, Randolph-Macon College, Ashland, VA 23005, USA

Grace Lee (103) Neuroscience Training Program, University of Wisconsin, 1117 West Johnson Street, Madison, WI 53706, USA

Leslie D. Leve (391) Oregon Social Learning Center, 10 Shelton McMurphy Boulevard, Eugene, OR 97401, USA

Jon E. Levine (449) Department of Neurobiology and Physiology, Northwestern University, 2205 Tech Drive, Room 2-160, Hogan Hall Evanston, IL 60208, USA

Frederick Lévy (23) INRA, UMR85 Physiologie de la Reproduction et des Comportements, F-37380, Nouzilly, France

Hong-Xiang Liu (361) Kanazawa University, 21st Century Center for Excellence Program on Innovation Brain Science on Development, Learning and Memory, 13-1 Takara-machi, Kanazawa, Ishikawa 920-8640, Japan

Yan Liu (347) Department of Psychology, Florida State University, 209 Copeland Avenue, Tallahassee, FL 32306-1270, USA

Hong Long (293) Department of Psychiatry, Douglas Hospital Research Center, McGill University, Montreal, 6875 Lasalle Boulevard, Verdun, QC, Canada H4H 1R3

Joseph S. Lonstein (145) Program in Neuroscience and Department of Psychology, Giltner Hall, Michigan State University, East Lansing, Michigan 48823, USA

Olga Lopatina (361) Kanazawa University, 21st Century Center for Excellence Program on Innovation Brain Science on Development, Learning and Memory, 13-1 Takara-machi, Kanazawa, Ishikawa 920-8640, Japan

Jeffrey P. Lorberbaum (83) Psychiatry Department, Penn State University, Hershey Medical Center, Hershey, PA, USA

Vedran Lovic (519) Department of Psychology, Erindale College, University of Toronto, Toronto, Canada, L5L1C6

Dario Maestripieri (163) Department of Comparative Human Development, The University of Chicago, Chicago, IL 60637, USA

Catherine A. Marler (435) Departments of Psychology and Zoology, University of Wisconsin, 1202 West Johnson Street, Madison, WI 53706, USA

De'Nise T. Mckee (235) Department of Biological Science and Program in Neuroscience, The Florida State University, 107 Chieftan Way, P.O. Box 3064370, Tallahassee, FL 32306-4370, USA

Simone L. Meddle (225) Centre for Integrative Physiology, College of Medicine and Veterinary Medicine, University of Edinburgh, CIP, School of Biomedical Science, Hugh Robson Building, George Square, Edinburgh, EH8 9XD, UK

Carolyn C. Meltzer (175) Departments of Radiology, Neurology, and Psychiatry and Behavioral Sciences, Emory University School of Medicine, Atlanta, GA, USA

Stephanie M. Miller (147) Department of Psychology, Giltner Hall, Michigan State University, East Lansing, Michigan 48823, USA

Ginger A. Moore (391) Department of Psychology, Moore Building, The Pennsylvania State University, University Park, PA 16802, USA

Joan I. Morrell (39) Center for Molecular and Behavioral Neuroscience, Rutgers, The State University of New Jersey, 197 University Avenue, Newark, NJ 07102, USA

Eydie L. Moses-Kolko (175) Department of Psychiatry, Women's Behavioral Health Care, Western Psychiatric Institute and Clinic, University of Pittsburg School of Medicine, 410 Oxford Building, 3811 O'Hara Street, Pittsburg, PA 15213, USA

Toshio Munesue (361) Kanazawa University, 21st Century Center for Excellence Program on Innovation Brain Science on Development, Learning and Memory, 13-1 Takara-machi, Kanazawa, Ishikawa 920-8640, Japan

Lindsay Naef (247, 293) Department of Psychiatry, Douglas Hospital Research Center, McGill University, Montreal, 6875 Lasalle Boulevard, Verdun, QC, Canada H4H 1R3

Jenae M. Neiderhiser (391) Department of Psychology, Moore Building, The Pennsylvania State University, University Park, PA 16802, USA

Benjamin C. Nephew (509) Department of Biomedical Sciences, Tufts University – Cummings School of Veterinary Medicine, 200 Westboro Road, North Grafton, MA 01536, USA

Inga D. Neumann (115) Department of Behavioural Endocrinology, Institute of Zoology, University of Regensburg, Universitätsstrasse 31 93053 Regensburg, Germany

Michael Numan (3) Department of Psychology, Boston College, 140 Commonwealth Avenue, Chestnut Hill, MA 02467, USA

Daniel E. Olazábal (333) Laboratory of Neuroscience, School of Sciences, Universidad de la República, Montevideo, Uruguay

Jodi L. Pawluski (493) Program in Neuroscience, Department of Psychology and Brain Research Center, University of British Columbia, 2136 West Mall, Vancouver, BC, V6T 1Z4, USA

Cort A. Pedersen (377) The Department of Psychiatry, CB# 7160, The University of North Carolina at Chapel Hill, Chapel Hill, NC 27599-7160, USA

Mariana Periera (39) Center for Molecular and Behavioral Neuroscience Rutgers, The State University of New Jersey, 197 University Avenue, Newark, NJ 07102, USA

Selvakumar Ramakrishnan (269) Department of Biological Sciences, P.O. Box 413, University of Wisconsin, Milwaukee, WI 53217, USA

Corinna N. Ross (461) Barshop Institute for Longevity and Aging Studies, University of Texas Health Science Center at San Antonio, 15355 Lambda Drive San Antonio, TX 78245, USA

John A. Russell (203) Laboratory of Neuroendocrinology, Centre for Integrative Physiology, School of Biomedical Sciences, College of Medicine and Veterinary Medicine, University of Edinburgh, Hugh Robson Building, George Square, Edinburgh EH8 9XD, UK

Johanna Schneider (449) Department of Biochemistry and Molecular Genetics, University of Virginia, Jordan Hall Room 1229, P.O. Box 800733, Charlottesville, VA 22908, USA

Katherine Seip (39) Center for Molecular and Behavioral Neuroscience Rutgers, The State University of New Jersey, 197 University Avenue, Newark, NJ 07102, USA

Sharon A. Stevenson (103) Department of Zoology, University of Wisconsin, 1117 West Johnson Street, Madison, WI 53706, USA

Danielle S. Stolzenberg (3) Department of Psychology, Boston College, 140 Commonwealth Avenue, Chestnut Hill, MA 02467, USA

April D. Strader (269) Department of Physiology, Southern Illinois University School of Medicine, University of Wisconsin, Milwaukee, WI 53217, USA

James E. Swain (83) Child Study Center, Yale University School of Medicine, New Haven, CT, USA

Luz Torner (131) Instituto Mexicano del Seguro Social, Centro de Investigacion, Biomedica de Michoacan, Ventura Puente esq, Salvador Carrillo, S/N, Col. Cuauhtemoc, Michoacan, Mexico

Brian C. Trainor (435) Department of Psychology, University of California-Davis, 1 Sheilds Avenue, Davis, CA 95616, USA

Claire-Dominique Walker (293) Department of Psychiatry, Douglas Hospital Research Center, McGill University, Montreal, 6875 Lasalle Boulevard, Verdun, QC, Canada H4H 1R3

Zuoxin Wang (347) Department of Psychology, Florida State University, 209 Copeland Avenue, Tallahassee, FL 32306-1270, USA

Katherine L. Wisner (175) Departments of Psychiatry, Obstetrics and Gynecology, and Epidemiology, University of Pittsburgh, School of Medicine, Pittsburg, PA, USA

Barbara Woodside (249) Department of Psychology, Center for Studies in Neurobiology, Concordia University, Montreal, Quebec, Canada

Zhifang Xu (293) Department of Psychiatry, Douglas Hospital Research Center, McGill University, Montreal, 6875 Lasalle Boulevard, Verdun, QC, Canada H4H 1R3

Shigeru Yokoyama (361) Kanazawa University, 21st Century Center for Excellence Program on Innovation Brain Science on Development, Learning and Memory, 13-1 Takara-machi, Kanazawa, Ishikawa 920-8640, Japan

Larry J. Young (333) Department of Psychiatry and Behavioral Sciences, Center for Behavioral Neuroscience, Yerkes National Primate Research Center, Emory University School of Medicine, Atlanta, Georgia 30329, USA

PARENTING AND THE BRAIN: AN OVERVIEW

ROBERT S. BRIDGES

Department of Biomedical Sciences, Tufts University – Cummings School of Veterinary Medicine, North Grafton, MA, USA.

INTRODUCTION

It seems a truism to state that the mother's physiological and mental states impact the survival and reproductive success of her offspring. Whereas a given level of parental input is crucial for raising a healthy baby or offspring, diminished parental provisioning and care can adversely affect the outcome of the young. One aspect of parental care that certainly affects reproductive outcome is the mental state of the mother and/or father. It has been reported, for example, that approximately 50% of women display some form of postpartum "blues," while anywhere from 5% to 15% of mothers experience postpartum depression (PPD). Both of these conditions have the potential to adversely affect the functioning of the family unit.

A key factor in regulating parental care is the brain. The brain is the main integrator of sensory and chemical information which is channeled into behavioral, physiological, and emotional outputs. Given the primary role of the brain in maternal, as well as paternal care, a group of internationally renowned researchers gathered in Boston, Massachusetts in June, 2007 to share their research findings and perspectives on the involvement of the central nervous system in the regulation of parental behavior. This book is an outgrowth of this "Parenting and the Brain Conference," and includes contributions from conference speakers. The aim of this book is to present research of basic and clinical scientists who use state of the art scientific approaches to examine the role of the central nervous system in both maternal and paternal care. Both normal parental behavior and maladaptive responses during the critical periods associated with raising offspring are discussed. Basic underlying mechanisms that regulate the adaptations of the maternal and paternal brain as well as translational aspects of this research are presented. Issues related to postpartum mood disorders, such as PPD, anxiety and aggression in women, and inadequate parental bonding to infants, are discussed in an effort to identify novel linkages between the basic and clinical sciences on these crucial topics and to advance our understanding of women's and men's health issues. The objective of this exchange of ideas is to promote new research initiatives that can positively impact the mental health of parents.

A NEURO-DEVELOPMENTAL MODEL

The operations of the parental mind may best be understood when conceptualized within what can be termed a "Neuro-Developmental" framework or model (see Figure I.1). This model incorporates both physiological and experiential input throughout the lifespan of the parent from prenatal and postnatal life through the peripubertal period and adulthood. Various inputs act on the central nervous system and an underlying genetic potential to modify the parental brain and the responses of the mother and father toward their young. These factors lead to both behavioral and physiological outputs which may be transferred to subsequent generations via genetic, epigenetic, and non-genetic mechanisms.

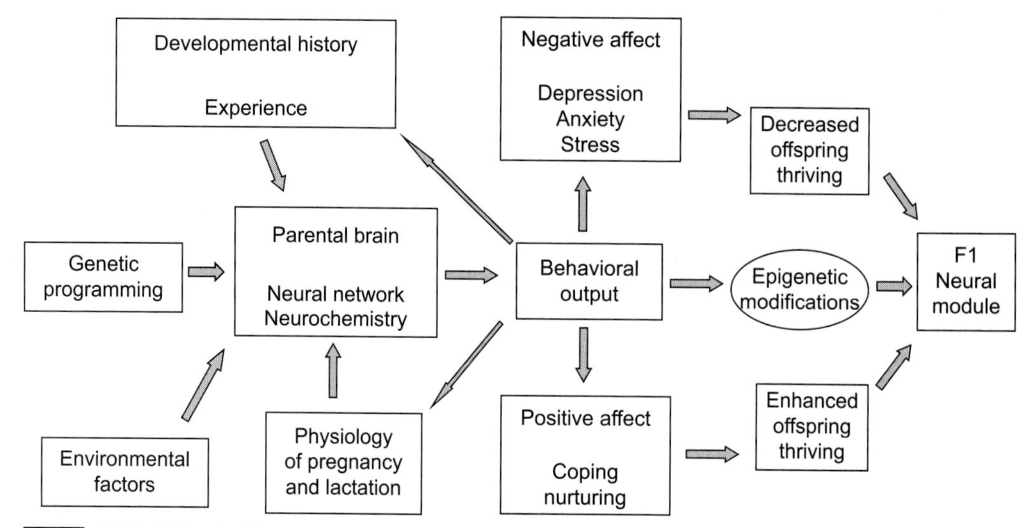

FIGURE I.1 This "Neuro-Developmental Model" depicts the relationships among genetic, environmental, and experiential factors over the course of development in directing the activity of the parental brain. The neurobiology of the parental brain is impacted by early as well as adult inputs that can have long-lasting actions which are transmitted to future generations via epigenetic, genetic, and environmental factors. Likewise, the next F1 generation receives a similar set of inputs which are transmitted to subsequent generations, that is, F2, F3, etc.

SECTION 1: THE NEUROANATOMICAL BASIS OF MATERNAL BEHAVIOR

Perhaps the best starting point to gain an understanding of this model is the parental brain itself. In Section 1 of this book, contributions focus on the neuroanatomical basis of maternal behavior. During the past three decades, our understanding of the neural network of maternal and paternal behavior has grown extensively. The early seminal research in rats of Dr. Michael Numan and his colleagues and, more recently, the studies in sheep of the Dr. Frederic Levy and collaborators have identified key neural sites and pathways that serve as the anatomical substrates of parental care. Whereas the identification of a complete neural circuitry underlying parental care will likely require extensive additional studies, good progress has been made in laying out what appear to be key neural loci and pathways that regulate this evolutionarily critical set of behaviors. Based on the results of these anatomical studies, Dr. Joan Morrell and colleagues have examined the neuroanatomical basis of maternal motivation and the relative reward properties of young vs. a substance of abuse, cocaine. They have shown how the reward properties of these factors shift across lactation; young are more salient stimuli early in lactation and cocaine, which acts through what may be a similar dopaminergic reward system, has greater saliency after the middle of lactation. Drs Febo and Ferris present work that uses functional magnetic resonance imagery (fMRI) to identify regions of the brain that are activated when mothers are exposed to suckling stimuli from young as well as selected hormonal and drug manipulations. Their studies demonstrate a much broader activation of the sensory cortex to suckling than previously realized and raise new questions regarding the involvement of cortical structures in parental care.

Drs Felicio and Canteras explore the neurochemical and neuroanatomical basis of maternal choice when a mother is given the opportunity to either care for her young or hunt for food. Their research identifies a role for opioids, acting in the midbrain at the level of the periaqueductal grey, in stimulating hunting. This work indicates that neural mechanisms exist within the mother, which allow her to cope with conflicting behavioral tasks or demands. Activation of the neuroanatomical substrates associated with human parenting is presented in the chapter by Drs. Swain and Lorberbaum who utilize fMRI technology to identify areas of the brain that respond to infant cues in mothers and fathers. Together, these research teams present

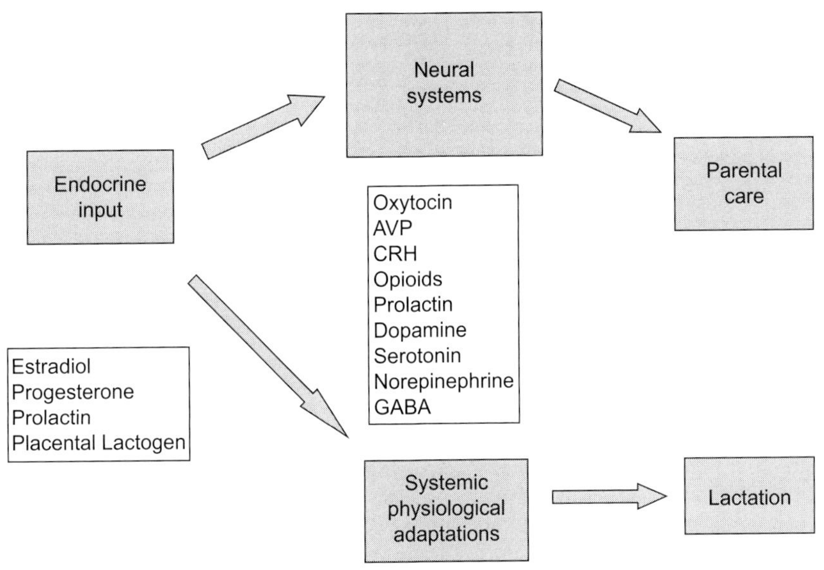

FIGURE I.2 Biochemical inputs regulating neural systems involved in parental care and the physiological adaptations of parenting. The hormones, peptides, and proteins listed have identified roles in mediating the behavioral and physiological actions of the parental brain.

an up to date perspective of our understanding of the neuroanatomical substrates that underlie the expression of parental care. Many of the subsequent chapters draw on these underpinnings in their examinations of the biochemical and neurochemical regulation of parental care across a range of mammals from rodents to non-human primates to humans.

SECTION II: ADAPTIVE AND MALADAPTIVE PARENTING

The behavioral output associated with parental care and its regulation are the topics of Section 2. During the postpartum period, mothers display a range of behavioral adaptations which function to protect and promote the growth and survival of their young. Among these alterations are increases in the levels of protective behavior, that is, maternal aggression, reductions in anxiety, attenuations in maternal stress responsivity, and increased food consumption or hyperphagia (see Section 3). A number of molecules that include hormones and neurochemicals have been demonstrated to regulate parental care (see Figure I.2). The studies of Dr. Stephen Gammie and colleagues focus on the role of one of these factors, the peptide corticotropin releasing factor (CRF), in the control of maternal aggression. Their studies suggest that decreased activation of the central CRF system facilitates the display of maternal aggression, possibly by lowering fear and anxiety. The work of Neumann and Bosch explores the relationships between stress responsiveness, anxiety and the actions of the neuropeptide oxytocin and the pituitary/neural protein prolactin. Rats bred for high levels of anxiety-related behavior display both increased aggression and higher levels of central oxytocin release as a function of maternal defense. Prenatal stress, as well as other environmental and genetic factors, modify the normal levels of stress responsivity which may affect peripartum adaptations and increase the risk of mood disorders. Dr. Luz Torner's work focuses on the involvement of neural prolactin as a mediator of stress responsiveness during lactation. Neural-derived prolactin is reported to be released within the brain where it acts as an anti-stress agent to reduce anxiety and the reactivity of the hypothalamo–pituitary–adrenal axis. It is noteworthy that, in addition to prolactin's anti-stress actions, prior studies in rats have shown that prolactin can stimulate the onset of maternal behavior through its actions on a key maternal network locus, the medial preoptic area. The chapter by Joseph Lonstein elegantly demonstrates the importance of sensory stimuli received from the offspring in modifying the level of anxiety in the mother. It is proposed that sensory induced alterations in neurochemicals, that is, GABA, norepinephrine, and oxytocin, act as part of a neural network that involves projections from the midbrain periaqueductal grey to the bed nucleus of the stria terminalis to potentiate anxiolytic-like responses in mothers which helps the mother cope more effectively with the demands of motherhood.

Behaviors that are less adaptive to the species include rejection of the young as well as the development of postpartum mood disorders. In female rhesus monkeys maternal rejection and infant abuse can be transmitted across generations. Dr. Dario Maestripieri's studies of abusive rhesus mothers indicate that early experience produces long-term shifts in serotonergic functions that then play an important role in the transmission of normal and abnormal parental care. Interplay among genotype, experience, and behavior are evident as reflected by reductions in cerebrospinal fluid levels of the serotonin metabolite 5-HIAA, in abused females who themselves then become abusive mothers. This research is particularly important, since it examines neurochemical correlates of maternal abuse using non-human primates. The next step along the evolutionary walkway focuses on parental affect in women. Dr. Moses-Kolko and colleagues provide a lucid overview of postpartum mood disorders in women and the possible mediators of these states. Work discussed includes the incidence and treatment of postpartum "blues," PPD, postpartum anxiety disorders, and postpartum psychosis. Positron emission tomography (PET) studies are presented that examine the possible role of the serotonin-1A and dopamine D2 receptors in depression within the context of a neuroendocrine model of PPD. These studies are set within a context of maternal adaptation and multifactorial regulators.

SECTION III: NEUROENDOCRINE ADAPTATIONS OF PARENTING: PREGNANCY, LACTATION, AND OFFSPRING

Section 3 focuses on the neuroendocrine adaptations of parenting. These chapters examine changes accompanying the states of pregnancy and lactation in the mother as well as in the offspring. Drs Russell and Brunton take a broad approach, focusing on the health implications of fetal programming on the hypothalamo–pituitary–adrenal axis and stress responsivity in adulthood. Their work in rodents and pigs is placed within the context of a range of human conditions: infertility, drug-taking behavior, and domestic violence. Their work illustrates how environmental input can influence neuroendocrine development which, in turn, may impact parental care. Dr. Douglas's chapter examines the role of oxytocin during the periparturitional period, and focuses on the involvement of central oxytocin and its processing by oxytocinase, the enzyme that degrades oxytocin, in parturition and perinatal behaviors. A mathematical model that accounts for the secretion and control of oxytocin and prolactin during early pregnancy is described by Dr. Marc Freeman and colleagues. Mating in the rat induces a neuroendocrine memory that results in twice daily surges of circulating prolactin which are crucial for pregnancy maintenance. Use of this novel approach which merges mathematics with neurobiology provides researchers with a powerful means to define the temporal relationships among a set of biochemical markers. It is proposed that this approach can be used to decipher causal relationships between key physiological processes involved in pregnancy maintenance together with lactation and the neurobiology of parental care.

Along a slightly different line of study, Dr. Woodside and colleagues focus on changes in metabolic systems that enable females to meet the energetic demands of fetal growth and lactation. One adaptive mechanism proposed that may mediate the effects of hormones on enhanced food intake during pregnancy and lactation is a state of leptin resistance which appears to be induced by activation of central prolactin receptors. The similarities in the hormonal and neurochemical mechanisms involved in meeting the energy demands of parenting are presented in work in pigeons and doves by Dr. John Buntin's research team. Using a non-mammalian ring dove model, they found that, during the reproductive cycle of raising young, prolactin-stimulated hyperphagia is mediated by prolactin's actions on two appetite-stimulating agents, neuropeptide Y and agouti-related peptide. The value of utilizing comparative approaches is that fundamental evolutionary adaptations within the parental brain that occur across both mammalian and non-mammalian species can be established. The research in the final chapter in this section by Dr. Dominique-Walker and colleagues examines the process of elevated maternal dietary fat intake during late pregnancy and lactation and its consequences on dopamine function when the offspring of these mothers become adults. Elucidation of how nutritional "programming" alters motivational and reward systems can aid in understanding the neurobiology of drug abuse which appears to involve similar neurochemical systems.

SECTION IV: MATERNAL CARE: FROM GENES TO ENVIRONMENT

Section 4 of the book addresses the role of genes and the environment and their interactions in the expression of maternal behavior. Dr. Frances Champagne, whose earlier work with Dr. Michael Meaney, helped promote the concept of the epigenetic transmission of maternal care, discusses evidence that supports the idea that throughout the pre-conceptual, prenatal, and postnatal periods, maternal factors can alter the reproductive behavior of the female offspring. One aspect of this transgenerational modification of behavior appears to involve the degree of DNA methylation of the estrogen receptor within select brain regions. These studies may help explain phenotypic similarities between mothers and daughters. Dr. James Curley then explores the contributions of paternal and maternal genotypes to parental care using reciprocal hybrid breeding of mice strains that differ in the levels of parental care. "Parent-of-Origin" effects may account for the differential influences of maternal and paternal genes on the behavior of the offspring.

Studies using prairie voles have provided some of the most convincing evidence for an important role for oxytocin in affiliative and parental care. Drs Olazábal and Young report that individual variations in oxytocin receptor density in the brain contribute to individual variations in parental care. One area where a positive correlation exists between receptor density and parental care is the nucleus accumbens, whereas a negative correlation is found in the lateral septum. Comparative analyses in studies across other rodent species support the argument that the level of oxytocin activity plays an important role in accounting for the natural variations in maternal care in mammals. The neurochemical regulation of partner preference is then presented in a chapter by Dr. Wang and his colleagues. Their focus is on the role of dopamine in this process. A vole model is one that can be used to study issues related to mental health, including conditions such as autism and drugs of abuse in humans. One attribute of this model is that it allows for the study of the interactions between social and drug rewards. Drs Pedersen and Boccia provide a detailed analysis of the involvement of oxytocin in the development of maternal care. Their studies in rats demonstrate that the behavioral input that the young receive from the mother alter their behaviors in adulthood through an oxytocinergic mechanism. This provides a new perspective for understanding the origins of human behavioral and emotional problems that may arise as a result of the young receiving inadequate nurturing during early development. The molecular basis of social memory and maternal care is further explored in the chapter by Dr. Higashida and co-workers. Using an array of null mutations, they demonstrate that CD38, a transmembrane glycoprotein, plays a crucial role in maternal nurturing and social behavior through its regulation of oxytocin release. Thus, the CD38 gene may be an important factor in a range of neuro-developmental disorders, possibly including autism.

Related studies in humans have approached the genetic involvement in parenting through the study of twins and co-twins. Dr. Ganiban and colleagues examine genetic and environmental influences on parenting within the context of genotype-environment correlations (rGE). The inputs of the characteristics from the parent and child within the framework of contextual factors, such as the environment, provide an integrative approach to assess social and genetic influences on human parenting.

SECTION V: THE NEUROBIOLOGY OF PATERNAL CARE

This section explores the neurobiological basis of the flip side of parental care, paternal care. The renowned anthropologist, Dr. Sarah Blaffer Hrdy, provides an evolutionary context for the role of fathers in human parental care, comparing paternal care in non-human primates with that found in various human societies. Dr. Hrdy notes the scarcity of biparental care in primates, including the great apes, and the limited involvement of men in parenting within certain human groupings. She emphasizes the important role that alloparenting plays in child rearing, primarily by female cohorts, that is, grandmothers, aunts, etc., and stresses the limited understanding of the neurobiological underpinnings of male nurturing.

The work of Dr. Carter and colleagues on the developmental consequences of early experience in the prairie vole on alloparental and biparental care is then presented. The actions of oxytocin, especially in male prairie voles during early development, on later parental care and bond formation are highlighted. Of interest is the finding that manipulations of the mother during the early postpartum period are capable of altering behaviors of her offspring later in life through actions on oxytocin and arginine vasopressin (AVP) systems. Males appear to be more sensitive to alterations in AVP or the AVP-1a receptor. As proposed by these authors, identifying differences between males and females in their early life experiences may help us understand more about sex differences in the vulnerability to disorders that are associated with atypical social behaviors, including parental care.

Our understanding the role of the male in parental care in recent years has also been advanced by research in the biparental California mouse, *Peromyscus californicus*. Field studies have demonstrated that the male California mouse actively participates in raising his young. In the chapter by Dr. Marler and colleagues, the relationships between paternal pup retrieval and the level of resident-intruder aggression of the offspring as adults are discussed. A particular focus is on the mechanisms that underlie the actions of testosterone and vasopressin in this process. Their data emphasize the long-term developmental consequences of the experience of the young on their own behavior as adults. These effects of early behavioral experience illustrate once again how behavioral experience can be transferred between generations, ultimately generating an organism (F1 module, see Figure I.1) which itself provides a neural pallet for parental care that is subject to modifications by experience, genes, and their interactions.

One common endocrine change characteristic of pregnancy in mammals is the significant elevation in circulating progesterone. In addition to its crucial role in pregnancy maintenance, progesterone has an important behavioral role. It both primes the female during pregnancy to respond maternally at birth and controls the timing of the onset of maternal care. Until recently, a possible role for progesterone in paternal care was unexplored. Using mice in which the progesterone receptor (PR) has been deleted or "knocked out," Dr. Horton and colleagues present a series of studies which support the idea that the actions of progesterone through its receptor in the brain promote "bad" parenting behavior in males. Male mice lacking the PR display decreased pup killing and enhanced parental care. The broader reproductive effects of the PR, including the effects of "knocking out" the PR on male sexual behavior are also presented. These studies are particularly intriguing, since they identify a factor, that is, a hormone, which may reduce reproductive success and suggest certain neurochemical commonalities in the regulation of maternal and paternal care.

Involvement of the father in parental care in primates is perhaps best demonstrated in marmosets where fathers display extensive care of the young. Dr. Jeffrey French and his colleagues discuss sources of variation of paternal care, focusing on the possible roles of circulating testosterone and genetic relatedness. The extent of paternal care in male marmosets is inversely correlated with circulating testosterone levels in the fathers and positively correlated with the genetic relatedness between the infants and the father. Marmosets also have a unique embryonic development in which a shared blood supply results in the exchange of stem cells between fraternal twins. Studies in genetic chimeras found that fathers who have a greater genetic relatedness to the offspring display altered levels of paternal care toward the chimeric young. This results in changes in key developmental parameters, including puberty and adult stress responsivity. Thus, the marmoset provides a rich source for studying the interrelationships among early experience, hormones, and genes over the course of development in the male.

SECTION VI: REPRODUCTIVE EXPERIENCE: MODIFICATIONS IN BRAIN AND BEHAVIOR

The final section of this book focuses on how reproductive experience alters the neurobiological processes in female mammals.

The overriding theme of this section is that the maternal brain changes as a function of reproductive experience which leads to adaptations that help to promote reproductive success of the mother and the offspring. The initial chapter by Drs Lambert and Kinsley place the maternal–young interactions within a unique context of neuroeconomics to examine how the reproductive state and experience of the female may alter reproductive success. A set of behavioral responses associated with the maternal state, including learning, foraging strategies, and risk assessment are set on a background reproductive experience. They report how repeated maternal experience strengthens the neurobiological infrastructure that supports the maternal state and ultimately offspring survival. Contextual appropriate problem solving during the postpartum period, for example, appears to be a key evolutionary adaptation that has been incorporated into the operation of the maternal brain.

Drs Pawluski and Galea summarize their research on the role of reproductive experience on hippocampal plasticity and morphology and hippocampal dependent spatial memory in the mother. Reproductive experience results in a long-term enhancement of working and reference memory that persists beyond the cessation of lactation and enhanced neurogenesis in the hippocampus. These findings indicate that neural regions, such as the hippocampus, that are historically not considered part of the "maternal circuit" may mediate other aspects of maternal performance which are altered by reproductive experience. Dr. Byrnes and colleagues explore a range of behavioral and endocrine changes associated with a single or multiple reproductive experiences in rats. Long-term reductions in both prolactin and anxiety are present in females that have previously given birth and raised young. Prior experience also results in changes in activation of the immediate early gene, c-Fos, in select nuclei of the amygdala when experienced females are presented with non-tactile pup stimuli. This indicates that the function of this older "olfactory brain" which mediates maternal care in numerous species is altered, and perhaps, improved as a result of reproductive experience. Finally, it is reported that the level of maternal aggression increases as a result of repeated pregnancies and lactations, with multiparous lactating rats being almost twice as aggressive as age-matched primiparous dams. These long-term adaptations in neuroendocrine function and behavioral state likely enhance the reproductive competency of the female and increase her chances of future reproductive success.

The final chapter by Dr. Fleming and colleagues highlights the principles and parallel findings in the animal and human literature regarding the involvement of hormones, neurochemical systems, and experience underlying the regulation of maternal behavior. Their presentation is framed within two phases of maternal behavior: the "wanting" or motivational phase and the "doing" phase. Within this context, an important focus is on the role of the dopamine system in maternal motivation emphasizing behavioral changes mediated through dopaminergic projection sites, that is, the nucleus accumbens and the medial prefrontal cortex. The importance of the prefrontal cortex in women in planning, cognition, and attention suggests a greater role for this brain region in maternal care in women than in rodents. Overall, the prefrontal cortex likely plays a greater role in the organization of mothering in humans, a contention supported by imaging studies. The authors also discuss how parental experience alters subsequent responsiveness toward offspring. The concept of once a mother, always a mother, emerges from this work.

FUTURE DIRECTIONS IN THE STUDY OF THE NEUROBIOLOGY OF THE PARENTAL BRAIN

What will be the hot topics or focal research areas in the field of the neurobiology of the parental brain over the next decade? Based on our existing knowledge base and technological developments in the sciences, major advances should occur in at least three broad areas of basic and clinical research. These areas include: (1) the interplay of genes and environment across development, (2) elucidation of key neurochemical systems that mediate various actions of the parental brain, and (3) determination of fundamental similarities and differences in the neurobiological basis of parental behavior between females and males.

Perhaps the research area that has gained the greatest impetus recently is the interplay between genes and experience across development, impacting the functions and capacity of the parental brain. One important illustration of this interplay emerges from the research of

Dr. Michael Meaney. His team has elegantly demonstrated underlying mechanisms for the epigenetic transfer of maternal behavior in rats. The effects of gene modifications by DNA methylation and histone acetylation across early as well as adult development in the regulation of parental behavior and reproductive processes have only begun to be explored. Moreover, the long-term effects in adulthood of reproductive and behavioral experience on the brain have just touched the surface on how experience can modify the parental brain and the physiological set points of the mother. Another area of study that will help foster growth in this area is elucidation of specific genes that may alter the actions of the parental brain or which direct its activities. Although it seems unlikely that there is a specific gene that drives "parental behavior," select aspects of behavior may be controlled or modulated by a constellation of genes that regulate factors such as anxiety, arousal, learning, attention, and olfaction, for example, all of which combine to affect the parent's responses.

It is also likely that, during the next few years, our understanding of the neurochemical control of parental care will reach a greater understanding and move beyond focusing on single chemical and endocrine factors. A better understanding of the complexity of the interactions and relationships among various neurochemical systems is likely to emerge. For example, how oxytocin, prolactin, and dopamine, all key players in maternal care, interact within the context of early development and during pregnancy and lactation should become better understood. In addition, the specific context of involvement of a range of established and new neuropeptides should emerge. The neurochemical control of the maternal brain will be better understood, both at the time of establishment of care at parturition and throughout the long-lactational period. These findings will potentially lead to novel treatments of a range of postpartum mood disorders, most notably PPD.

The third area of research on the parental brain that has vast potential for increasing our understanding of the parental brain is that of sex similarities and differences in the neurobiology of the parental mind. As emphasized by Dr. Sarah Blaffer Hrdy, our understanding of the biological basis of paternal care is rudimentary. There is great potential for expanding our understanding of the paternal brain within ecological and evolutionary contexts. Research in animals such as marmosets offers an excellent model to expand on our understanding of male parenting. Functional MRI and PET studies of changes in mother and father brains in response to babies also provide avenues to identify regions of the brain involved in processing stimuli received from infants and neurochemical/receptor systems that are altered in parents with mood disorders.

One *added value* of animal and human models used to study parental care is that these models not only are useful to study how the parental brain works, but these animal models can also be used to study disorders associated with other atypical social behaviors, that is, autism, depression. Moreover, results of studies that examine the formation of social bonds between parents and offspring, as well as between adults, can serve as the basis for understanding the neurobiological basis of bonding. Moreover, such studies may lead to a better understanding of addictive disorders, which may involve similar neurochemical systems as those involved in the emotional attachment of parent to young.

Finally, the objective of research in the field of the neurobiology of parental care is to stimulate new ideas and findings that translate into improvements in human and animal health and the human condition. After all, the study of the parental brain is relevant to just about everyone, since each of us has been parented and many of us will be parents in our lifetimes.

I

THE NEUROANATOMICAL BASIS OF MATERNAL BEHAVIOR

1

HYPOTHALAMIC INTERACTION WITH THE MESOLIMBIC DOPAMINE SYSTEM AND THE REGULATION OF MATERNAL RESPONSIVENESS

MICHAEL NUMAN AND DANIELLE S. STOLZENBERG

Department of Psychology, Boston College, 140 Commonwealth Avenue, Chestnut Hill, MA 02467, USA

INTRODUCTION

This chapter will review research from our laboratory dealing with the neural basis of maternal behavior in rats. The major components of maternal behavior in this species are retrieval behavior, nest building, nursing behavior, and pup grooming (Numan & Insel, 2003). Retrieval behavior, which occurs when the postpartum female carries individual pups in her mouth, serves to transport the altricial pups from one location to another, for example, to a new nest site, or to bring displaced pups back to the current nest site. Nests keep the poikilothermic pups warm in the mother's absence, and nursing behavior, where the female crouches over her young to expose her mammary region, is necessary to feed the young. Licking and grooming the pups is important in aiding pup urination and defecation, and it also has profound effects on the emotional development of the young. Postpartum rats do not form selective attachments to their own young; they will care for their own young and those from another mother (foster pups).

Several investigators have differentiated retrieval behavior from nursing behavior. Terkel *et al.* (1979) have classified retrieving as an active maternal response because it is initiated by the female, while they have classified nursing as a passive maternal response that is primarily initiated by the pups. Similarly, Hansen *et al.* (1991a) have referred to retrieval as an appetitive maternal response and nursing as a consummatory maternal response, which corresponds to Stern's (1996) reference to retrieval as a pronurturant response and nursing as a nurturant response. Following along these lines, we would like to classify pup-seeking behaviors and retrieval behavior as voluntary proactive maternal responses, while nursing/crouching behavior is more of a reflexive maternal response that is closely tied to proximal pup stimulation. Indeed, Stern (1991) has shown that when male rats are immobilized through drug treatment and placed over pups, they will show the reflexive crouching/nursing posture in response to nuzzling pups. Of course, if the males were not immobilized they would not show any parental behavior on their initial exposure to pups.

At parturition, the primiparous female rat shows the full complement of maternal behaviors on her first exposure to her own pups or to foster pups (Numan & Insel, 2003). In contrast, the naïve virgin (nulliparous) estrous cycling female rat will not care for foster pups upon her initial exposure to them; she avoids such pups and may even attack them (Numan & Insel, 2003). This difference is very important because it shows that infant stimuli do not automatically elicit maternal responses in female rats. Important internal changes take place in the female who has just given birth which increase maternal responsiveness or maternal motivation. Clearly, the brain circuits which are operative in the first-time mother must be different from those in the naïve virgin female rat.

An important question is whether research on the neural mechanisms of maternal behavior in rodents is relevant to an understanding of the brain control of human maternal behavior. For us, it makes sense that such research should be relevant. Since maternal behavior is a defining characteristic of mammals, one should expect that evolutionarily ancient core neural circuits underlying the behavior are present across all mammals. A research program and theoretical framework that are directed at uncovering neural circuits which control maternal responsiveness to infant stimuli, rather than only focusing on the neural control of species-typic maternal responses, should be able to tap into a core neural circuitry regulating maternal motivation.

VIRGIN SENSITIZATION AND THE HORMONAL BASIS OF RODENT MATERNAL BEHAVIOR

Although the naïve virgin female rat does not show maternal behavior when initially presented with pups, if one cohabitates her with pups over a series of days some dramatic behavioral changes are observed. Initially, the virgin female avoids the pups, but after about 3–4 days she tolerates their proximity, and then, beginning about 7 days from the time of initial exposure, she begins to care for the pups: she retrieves them to a single location, builds a nest around them, grooms them and hovers over them in a nursing posture even though she cannot lactate (Rosenblatt, 1967; Fleming & Luebke, 1981; Stern, 1997). This pup-induced maternal behavior has been designated sensitized maternal behavior, and the process has been called sensitization. The number of days of pup exposure that is required before maternal behavior is initiated is referred to as the female's sensitization latency. In the sensitization procedure, the virgin female is presented on a daily basis with freshly nourished pups that are provided by a group of postpartum females. This method is necessitated by the fact that initially maternal behavior does not occur, and even when it is induced, the female does not lactate.

The primiparous parturient female is immediately responsive to pups, and does not require a period of cohabitation, because her brain has been affected by the endocrine events associated with pregnancy and pregnancy termination (Numan & Insel, 2003). These endocrine changes include rising blood levels of estradiol and lactogens (pituitary prolactin and placental lactogens) superimposed on a major drop in progesterone levels. As initially shown by Moltz *et al.* (1970), if one treats virgin female rats with a hormone regimen which mimics these endocrine changes, one can reduce sensitization latencies from the typical 7 days shown by control nulliparae to 1–2 days in the hormone treated females (see also Bridges, 1984; Bridges & Ronsheim, 1990).

Another model, referred to as the pregnancy termination model, has been developed by Rosenblatt's group (Rosenblatt & Siegel, 1975; Siegel & Rosenblatt, 1975) to explore the hormonal basis of maternal behavior in rats. Briefly, on days 15–17 of pregnancy, primigravid female rats are generally unresponsive to pups. However, if one terminates pregnancy via hysterectomy (removal of the uterus, placentas, and pups) on day 15 of a 22-day pregnancy, and then presents the pregnancy-terminated females with pups 48 h later, a sensitization latency of about 1 day is observed. Interestingly, if females are similarly treated except that they are also *ovariectomized* on day 15, sensitization latencies increase to about 2–3 days. Finally, if females are hysterectomized and ovariectomized and treated with estradiol at the time of surgery, then when they are presented with pups 48 h later, most females show maternal behavior on their first day of pup exposure. Since hysterectomy results in a decline in serum progesterone levels, these results conform with the view that declining progesterone and rising estradiol contribute to the immediate onset of maternal behavior at parturition. Hysterectomy-induced maternal behavior occurs because when the surgery is performed around day 15 of pregnancy it prematurely activates those stimulatory hormonal events which would naturally occur closer to the normal time of parturition (see Numan and Insel (2003) for mechanistic details).

Table 1.1 summarizes the work we have just described. For later reference, it is worth comparing the average sensitization latencies of the following groups: virgin females (7 days); 15-day pregnant hysterectomized and ovariectomized females (15HO: 2–3 days); 15-day

TABLE 1.1 Sensitization latencies for female rats exposed to different treatments

Treatment	Average sensitization latency (days)
Naïve virgins	7
15-Day pregnant HO	2
15-Day pregnant HO + E	0

Note: For pregnant females, pregnancy is terminated on day 15 of a 22-day pregnancy via hysterectomy (H) and ovariectomy (O). The 15-day pregnant HO + E females receive 20 μg/kg of estradiol benzoate (sc) at the time of surgery. Pups are presented to females 48 h post-surgery.

pregnant hysterectomized and ovariectomized females treated with estradiol (15HO + E: 0 days). In a variety of studies, research has shown that the sensitization latencies of these groups differ significantly from one another. Clearly, the third group approximates the maternal responsiveness of the naturally parturient female and this is the group that is composed of females whose brains have been exposed to declining progesterone and rising estradiol and lactogen levels (see Numan & Insel, 2003). However, note that the 15HO females show significantly shorter sensitization latencies than the virgin females and significantly longer latencies than the 15HO + E females. Therefore, some stimulation of maternal responsiveness occurs even if an estradiol rise does not follow pregnancy termination. We have referred to the 15HO preparation as one which gives rise to suboptimal or partial hormonal priming of maternal responsiveness, and as will be shown, this preparation has helped us explore the neurochemical basis of maternal behavior in rats.

THE MEDIAL PREOPTIC AREA AND MATERNAL MOTIVATION

The results reviewed above clearly show that the endocrine events associated with the end of pregnancy increase maternal motivation or maternal responsiveness. We use the term maternal motivation in a very mechanistic way to refer to the fact that pregnancy hormones potentiate the ability of infant-related stimuli to evoke proactive voluntary maternal responses. Our goal in this paper is to describe what we know about such central neural mechanisms which underpin this increase in maternal responsiveness. As outlined below, the medial preoptic area (MPOA) of the rostral hypothalamus and the adjoining ventral part of the bed nucleus of the stria terminalis (vBST, located in the telencephalon) are among the sites where pregnancy hormones act to prime/stimulate neural circuits which regulate proactive voluntary maternal responses to infant-related cues.

Figure 1.1 shows frontal and sagittal sections of the rat brain indicating the location of the MPOA in relation to other neural structures. For later reference, note the locations of the nucleus accumbens (NA) and ventral pallidum (VP), both of which are located in the telencephalon, and the ventral tegmental area (VTA), located in the midbrain. It is now well established that an intact MPOA is necessary for normal maternal behavior in rats, other rodents, and sheep (see Numan (2006) for a review). Electrical lesions or knife cuts which sever the lateral connections of the MPOA disrupt maternal behavior in rats as do axon sparing excitotoxic amino acid lesions of the MPOA (Numan *et al.*, 1988). These lesions and knife cuts also typically damage the vBST. The results indicate that the lateral efferents of MPOA/vBST neurons are particularly important for maternal behavior.

What is the nature of the maternal behavior deficit that occurs after MPOA damage? All studies agree that retrieving behavior and nest building are abolished or severely disrupted. Studies differ with respect to findings on nursing behavior (Numan & Insel, 2003). Many studies report a severe, near total elimination of all components of maternal behavior, including nursing

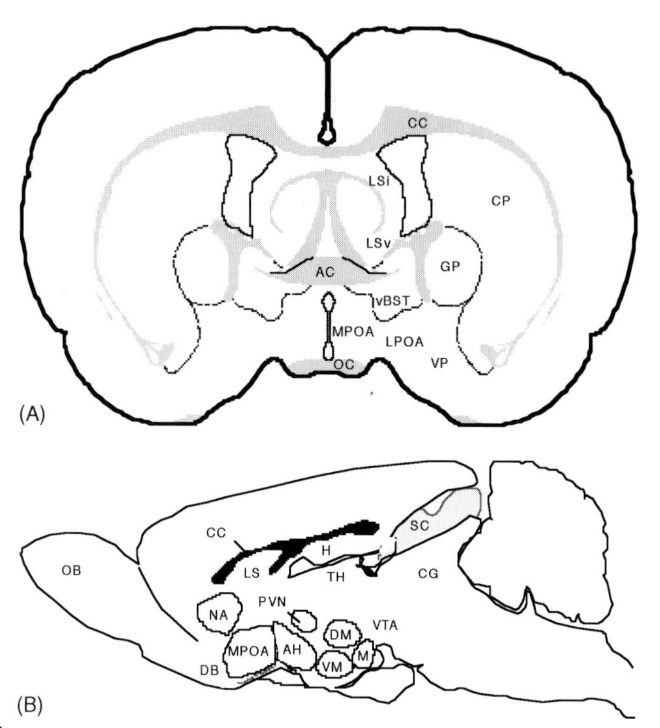

FIGURE 1.1 Frontal (A) and sagittal (B) sections of the rat brain at the level of the MPOA. Abbreviations: AC, anterior commissure; AH, anterior hypothalamic nucleus; CC, corpus callosum; CG, central gray (periaqueductal gray); CP, caudate-putamen; DB, nucleus of the diagonal band of Broca; DM, dorsomedial hypothalamic nucleus; GP, globus pallidus; H, hippocampus; LPOA, lateral preoptic area; LS, lateral septum; LSi, intermediate nucleus of lateral septum; LSv, ventral nucleus of lateral septum; M, mammillary bodies; MPOA, medial preoptic area; NA, nucleus accumbens; OB, olfactory bulb; OC, optic chiasm; PVN, paraventricular hypothalamic nucleus; SC, superior colliculus; TH, thalamus; vBST, ventral bed nucleus of stria terminalis; VM, ventromedial hypothalamic nucleus; VP, ventral pallidum; VTA, ventral tegmental area. (*Source*: Adapted from Swanson's (1992) rat brain atlas. Reproduced from Numan and Insel (2003; fig.5.7, p. 130). Copyright 2003 by Springer-Verlag, with the kind permission of Springer Science and Business Media.)

(Numan, 1974; Numan *et al.*, 1988), while other studies report that some nursing occurs in MPOA damaged females, although the duration of the behavior is significantly depressed in comparison to control females (Terkel *et al.*, 1979; Numan & Callahan, 1980). These results suggest that the MPOA is most concerned with the regulation of proactive voluntary maternal responses and that reflexive nursing postures can occur in preoptic damaged females in response to proximal pup stimuli (ventral trunk stimulation caused by pup nuzzling and suckling).

It is important to note that the effects of MPOA damage on maternal behavior in rats are relatively specific (Numan & Insel, 2003). Although such females do not care properly for their young, they show normal levels of sexual behavior, locomotor activity, body weight regulation, and hoarding behavior. This last point is particularly relevant because it shows that MPOA damage does not cause an oral motor deficit, in that candy can be transported by the mouth although pups are not retrieved (Numan & Corodimas, 1985).

Given that MPOA damage depresses maternal behavior, one can ask whether hormones act at this site to stimulate maternal behavior. MPOA neurons contain estrogen receptors (Shughrue *et al.*, 1997), progesterone receptors (Numan *et al.*, 1999), and prolactin receptors (Bakowska & Morrell, 1997), making it a likely site where hormones might act to stimulate maternal behavior. In fact, a series of studies have shown that discrete implants of either estradiol (Numan *et al.*, 1977; Fahrbach & Pfaff, 1986; Felton *et al.*, 1998) or lactogenic hormones (Bridges *et al.*, 1990, 1996) into the MPOA can promote the onset of full maternal behavior. In the Numan *et al.* (1977) study, a pregnancy termination model was employed. Primigravid rats were hysterectomized and ovariectomized on day 15 of pregnancy and presented with pups 48h later. Those female with estradiol implants into the MPOA showed maternal behavior on their first day of pup exposure (sensitization latency of 0 days), while females that received cholesterol implants into the MPOA or estradiol implants into other hypothalamic regions showed a delayed onset of maternal behavior, with sensitization latencies averaging around 2 days.

Oxytocin (OT), a hormone that is released by the neural lobe of the pituitary, is closely related to the physiological events of parturition and milk ejection, but it has poor penetrance across the blood–brain barrier (see Numan & Insel, 2003; Numan, 2006) and therefore, as a hormone, it is unlikely to influence maternal behavior. However, OT also serves as a neurotransmitter or neuromodulator within central neural circuits, and in this role OT has been found to facilitate the onset of maternal behavior in rats and other species (see Numan & Insel, 2003; Numan, 2006). Importantly, the MPOA is one of the sites where OT acts to stimulate maternal behavior since Pedersen *et al.* (1994) have shown that direct injections of an OT receptor antagonist into the MPOA disrupt the onset of maternal behavior in parturient rats.

With respect to the stimulation of the onset of maternal behavior, it is possible that there is an interaction between estrogenic effects and oxytocinergic systems at the level of the MPOA. Since estradiol is involved in the induction of OT receptor expression in MPOA (Pedersen *et al.*, 1994; Champagne *et al.*, 2001), one aspect of estrogen action may be to prime MPOA neurons to receive oxytocinergic input, and this, coupled with lactogenic effects on MPOA neurons, could result in functional alterations of MPOA neurons which allow them to participate in neural circuits regulating maternal responsiveness to infant cues.

Finally, it is noteworthy that the expression of Fos proteins (both c-Fos and Fos B) is activated in MPOA and vBST neurons during maternal behavior (Numan & Insel, 2003). Since Fos expression indicates that neurons have been affected by inputs, these results support the view that MPOA neurons participate in neural circuits regulating maternal behavior. In further support, Brown *et al.* (1996) have reported that a transgenic mouse line with a null mutation of the Fos B gene shows an absence of maternal behavior. Importantly, we have not only shown that mother–pup interactions activate Fos expression in MPOA neurons, but that the hormonal events associated with pregnancy termination also have this effect (Sheehan & Numan, 2002). Since Fos proteins are transcription factors which serve to activate specific genes, it is possible that Fos production during all aspects of the maternal state is influencing the phenotype of MPOA neurons so that they can function in neural circuits relevant to maternal behavior. A simple hypothesis follows: pregnancy hormones act on MPOA neurons to modify the phenotype of these neurons, and one mechanism through which this may be achieved is through the mediation of Fos proteins. The altered phenotype of MPOA neurons could include a change in the concentration of neurotransmitter/neuromodulator receptors that these neurons contain, such as an increase in OT receptors, as well as a change in the neurotransmitter/neuromodulator content of MPOA neurons. Since the MPOA is capable of receiving olfactory and tactile sensory inputs from pups (see Numan & Insel, 2003), these functional changes in MPOA neurons may allow them to be activated by pup stimuli and to release neurotransmitters or neuromodulators within circuits critical for maternal behavior. After the hormonal events associated with pregnancy termination have waned, the continued occurrence of maternal behavior during the postpartum period may be dependent on the continuance of Fos expression which may then maintain the functional integrity of MPOA neurons (Stack & Numan, 2000).

Support for the view that Fos proteins are expressed in MPOA neurons that participate in neural circuits underlying maternal behavior comes for the work of Stack *et al.* (2002). This research is based on the fact that although bilateral knife cuts which sever the lateral connections of the MPOA eliminate the proactive voluntary components of maternal behavior, unilateral cuts are ineffective. In other words, normal maternal behavior can occur as long as the MPOA in one hemisphere is functioning. Using this knowledge, Stack *et al.* (2002) found that unilateral knife cuts of the lateral MPOA/vBST connections disrupted both c-Fos and Fos B expressions in MPOA/vBST neurons on the

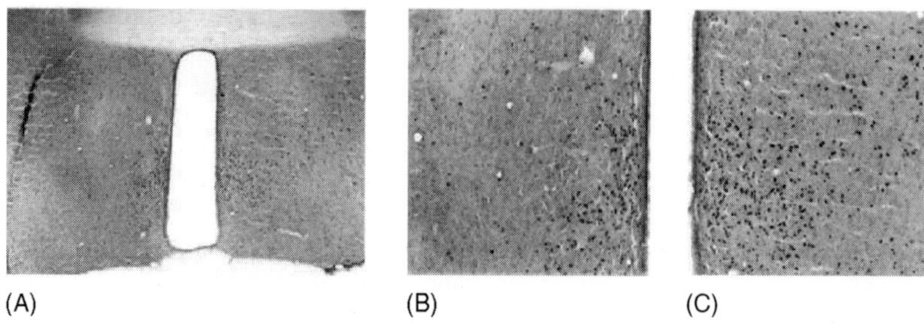

FIGURE 1.2 Photomicrographs showing the effect of a unilateral knife cut severing the lateral connections of the MPOA and adjoining ventral bed nucleus of the stria terminalis (vBST) on the expression of c-Fos within cells in the ipsilateral and contralateral MPOA and vBST of maternal rats. (A) A low power magnification showing both the ipsilateral and contralateral regions. The knife cut is visible on the left side, the anterior commissure is located dorsally, and the third ventricle is in the middle of the photomicrograph. (B) and (C) show high power magnifications of the ipsilateral and contralateral regions, respectively. The knife cut clearly decreases c-Fos expression on the ipsilateral side. (*Source*: Modified from Stack *et al.* (2002). Reproduced from Numan and Insel (2003; fig. 5.15, p. 158). Copyright 2003 by Springer-Verlag, with the kind permission of Springer Science and Business Media.)

side of the brain ipsilateral to the cut. These results, shown in Figure 1.2, suggest that Fos expression during postpartum maternal behavior in rats is marking MPOA/vBST neurons that contribute to neural circuits which control voluntary proactive maternal responses.

In a recent study using the conditioned place preference paradigm, Mattson and Morrell (2005) found that the MPOA also expresses Fos when postpartum females are searching for pups in an environment where pups had previously been located but are no longer present. Such MPOA Fos expression was substantially greater than that which occurred when females were searching in an environment previously associated with cocaine administration. These results are consistent with the view that MPOA neurons participate in neural circuits that regulate pup-seeking and other proactive voluntary maternal responses. In a related finding, Lee *et al.* (2000) found that MPOA lesions disrupted an operant bar pressing response in postpartum females when pups were used as the reinforcing stimulus but not, importantly, when food was the reinforcer.

A PROPOSED NEURAL MODEL

Given that the MPOA is essential for the normal expression of maternal responsiveness to infant stimuli, what is the mechanism through which it exerts its effects? The main purpose of this chapter is to present the evidence in support of the view that one of the ways in which MPOA output circuits affect maternal behavior is through their interaction with the mesolimbic dopamine (DA) system. Therefore, we need to simply summarize the anatomy and function of the mesolimbic DA system. The reader is referred to several important papers from which the following analysis was derived: Swanson (1982), Alexander *et al.* (1986), Mogenson (1987), Nicola *et al.* (2000), Voorn *et al.* (2004), Heimer and Van Hoesen (2006), and Numan (2006). The midbrain gives rise to several major ascending DA systems which terminate in the telencephalon, and two of these systems are the nigrostriatal DA system and the mesolimbic DA system. Although there is overlap in the organization and function of these two systems, some significant differences exist. As shown in Figure 1.3, the nigrostriatal system, which originates in the substantia nigra pars compacta (SNc) and terminates in the caudate nucleus and putamen (dorsal striatum; part of the extrapyramidal motor system), functions to facilitate an organism's motor responsiveness to sensory stimuli that are primarily of neocortical origin. The mesolimbic DA system originates from neurons in the VTA and one of its major sites of termination is the NA (ventral striatum; part of the extrapyramidal motor system). In contrast to the nigrostriatal system, one of the functions of the mesolimbic system is to facilitate an organism's motor responsiveness to sensory stimuli that are primarily being relayed to the NA from

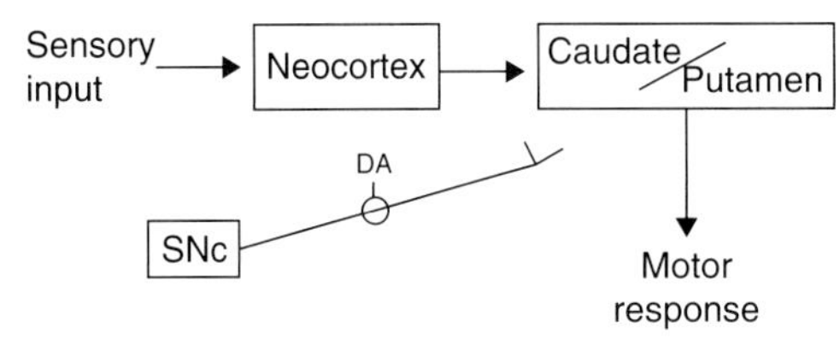

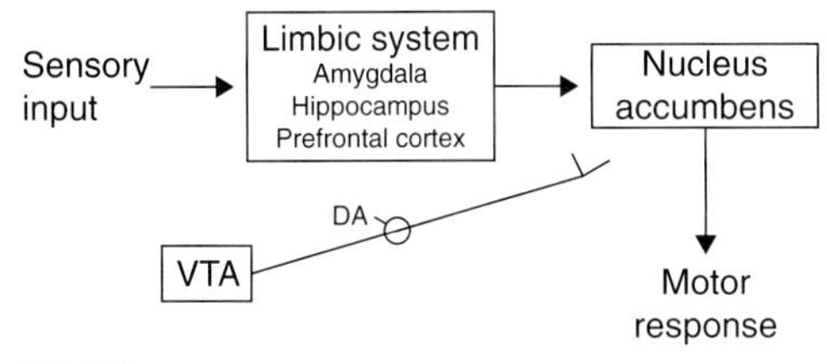

FIGURE 1.3 Diagrammatic representations of aspects of the functional neuroanatomy of the nigrostriatal and mesolimbic DA systems. The nigrostriatal system originates from DA neurons in the SNc of the midbrain with ascending projections to the caudate-putamen of the telencephalon. The caudate-putamen also receives sensory and other inputs from the neocortex. DA release into caudate-putamen facilitates an organism's motor responsiveness to sensory stimuli that are primarily of neocortical origin. The mesolimbic system originates from DA neurons in the VTA, and one of its targets in the telencephalon is the NA, which also receives afferents from the limbic system (amygdala, hippocampus, prefrontal cortex). DA release into NA facilitates an organism's motor responsiveness to sensory stimuli that are primarily of limbic system origin.

the limbic system. Therefore, the mesolimbic DA system can be viewed as part of the limbic motor system (Mogenson, 1987), which makes it our main concern with respect to maternal behavior regulation.

Figure 1.4 presents a neural model of mesolimbic DA function in the context of maternal motivation or maternal responsiveness to infant stimuli. Infant stimuli gain access to NA via afferents from the limbic system (amygdala, hippocampus, prefrontal cortex), but such input is not appropriately processed to allow for voluntary proactive maternal responses when VTA activity is low and DA is not being strongly released into NA. However, when the VTA is active and DA is released into NA, the organism is more likely to show appropriate maternal responsiveness.

In further development of this model, we can ask what determines when VTA-DA neurons become active. That depends on the motivational state of the organism. In the case of maternal behavior, we propose that after the MPOA has been appropriately primed by hormones, it becomes responsive to pup stimuli, which are probably tactile and olfactory in nature (see Numan & Insel, 2003). As shown in Figure 1.4, pup stimulation-induced activation of the MPOA excitatory efferents to the VTA results in DA release into NA, opening a gate which allows pup stimuli to be processed by the limbic motor system. Importantly, note the dual role of pup stimuli: they enter the system at two levels, the MPOA and NA. The NA will be involved in gating sensory inputs from pups only if the MPOA also responds to pup stimuli. In the non-maternal virgin female, therefore, where the MPOA is not fully operative because it has not been hormonally primed, pup stimuli will not fully activate the mesolimbic DA system and maternal behavior will not occur.

The remainder of this chapter will present further elaborations of the model presented in Figure 1.4 as well as the evidence in support of it.

ANATOMICAL EVIDENCE FOR AN MPOA-TO-VTA CONNECTION RELEVANT TO MATERNAL BEHAVIOR

The following pieces of anatomical evidence support the view that MPOA efferents relevant to maternal behavior are capable of influencing the function of the VTA.

1. With the use of anterograde and retrograde tracers, neuroanatomical tract tracing studies have shown that MPOA and vBST neurons project strongly to the VTA (Simerly & Swanson, 1988; Numan & Numan, 1996, 1997; Geisler & Zahm, 2005).

2. Neuroanatomical tract tracing studies undoubtedly label a functionally heterogeneous population of MPOA neurons that project to VTA. However, using a double labeling immunohistochemical procedure, Numan and Numan (1997) showed that MPOA and vBST neurons which express Fos during maternal behavior project to the VTA. Therefore, to the extent that Fos is marking MPOA neurons that are part of a maternal circuit, one of the termination sites of such a circuit is the VTA.

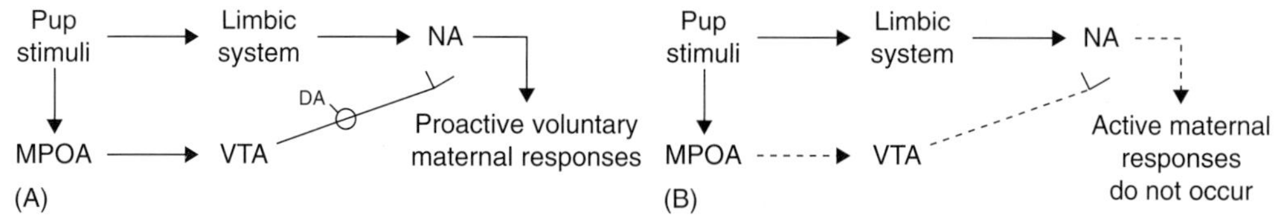

FIGURE 1.4 Neural model to explain how the interaction of MPOA with the mesolimbic DA system might regulate maternal responsiveness. Neural circuits indicated by dashed lines are not active, while circuits shown in solid lines are neurally active. (A) In a postpartum maternal rat, pup stimuli reach both the MPOA and the NA. Since the MPOA has been appropriately primed by pregnancy hormones, it responds to such afferent input and, in turn, activates VTA DA neurons. DA release into NA opens a gate which allows the limbic motor system to respond appropriately to pup stimuli. The details of this gating mechanism are described in Figure 1.6. (B) In the non-maternal virgin female, the MPOA does not respond to pup stimuli because it has not been primed by hormones. VTA-DA neurons are not activated and the limbic motor system does not process pup stimuli, which results in a lack of maternal responsiveness.

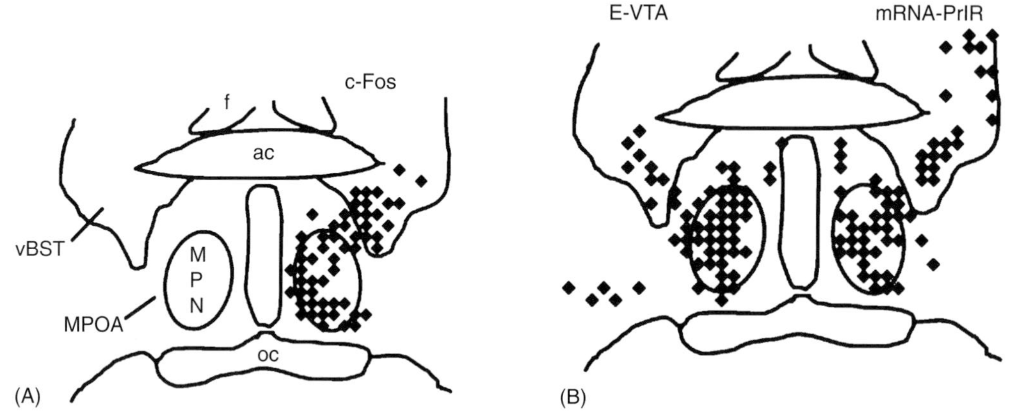

FIGURE 1.5 Representative frontal sections through the level of the MPOA on which is plotted the approximate location of cells: which express c-Fos during maternal behavior (right side of panel (A); derived from Stack and Numan, 2000); which both bind estradiol and project to the VTA (E-VTA: left side of panel (B); derived from Fahrbach *et al.*, (1986)); which express mRNA for the long form of the prolactin receptor (mRNA-PrlR: right side of panel (B); derived from Bakowska and Morrell (1997)). ac, anterior commissure; f, fornix; MPN, medial preoptic nucleus; oc, optic chiasm; vBST, ventral part of the bed nucleus of the stria terminalis.

3. Fahrbach *et al.* (1986) have shown that estradiol-binding neurons in the MPOA and vBST project to the VTA. These results fit with the idea that the hormonal events of late pregnancy prime MPOA neurons so that they become responsive to pup stimuli, which then enables activation of MPOA efferents to the VTA. In line with this perspective, Lonstein *et al.* (2000) have reported that a large proportion of MPOA and vBST neurons which express Fos during maternal behavior also contain intracellular estrogen receptors. Figure 1.5 shows the anatomical distribution of (i) c-Fos in the MPOA/vBST during maternal behavior (Stack & Numan, 2000), (ii) estradiol-binding MPOA/vBST neurons which project to VTA (Fahrbach *et al.*, 1986), and (iii) mRNA for the long form of the prolactin receptor in MPOA/vBST (Bakowska & Morrell, 1997). The overlap is remarkable, supporting the view that one of the neural circuits affected by hormones which facilitate maternal behavior is the MPOA/vBST-to-VTA circuit.

4. The above evidence clearly shows that MPOA/vBST neurons relevant to maternal behavior project to the VTA. There is also some evidence that such efferents ultimately influence VTA-DA projections to NA. First, Fos expression increases in the medial NA during maternal behavior (Stack *et al.*, 2002), supporting the view that a modification of NA function occurs during maternal behavior. Second, DA action on the D1 type DA receptor is capable of activating Fos expression in NA (Robertson & Jian, 1995;

TABLE 1.2 Some evidence supporting the involvement of the mesolimbic dopamine system in the maternal behavior of rats

Evidence	Literature reference
1. DA is released into NA during maternal behavior.	Champagne *et al.* (2004) and Hansen *et al.* (1993)
2. Electrical lesions of the VTA disrupt maternal behavior.	Gaffori and Le Moal (1979) and Numan and Smith (1984)
3. 6-Hydroxydopamine lesions of VTA or NA disrupt maternal behavior.	Hansen *et al.* (1991a,b)
4. Microinjection of DA receptor antagonists into NA disrupts maternal behavior.	Keer and Stern (1999) and Numan *et al.* (2005a)

DA, dopamine; NA, nucleus accumbens; VTA, ventral tegmental area.

Moratalla *et al.*, 1996; Hunt & McGregor, 2002). In the context of these findings, Stack *et al.* (2002) presented data on the question of whether unilateral damage to MPOA/vBST neurons would disrupt Fos activation in the ipsilateral NA during maternal behavior. Fully maternal primiparous rats received unilateral excitotoxic amino acid lesions of the MPOA/vBST or sham lesions. The lesions, because they were unilateral, did not disrupt maternal behavior. For the sham-lesioned females, during maternal behavior Fos expression increased in the NA of both hemispheres when compared to the level of Fos expression in females that were not exposed to pups. In contrast, for females with unilateral MPOA/vBST damage, Fos expression during maternal behavior only increased in the NA contralateral to the MPOA lesion. The NA in the hemisphere that was ipsilateral to the MPOA lesion exhibited a level of Fos expression which matched that of females that were not exposed to pups. These results coincide with anatomical results which show that MPOA projections to VTA and VTA projections to NA are to a large extent ipsilateral or uncrossed (Swanson, 1982; Simerly & Swanson, 1988; Numan & Numan, 1996; Geisler & Zahm, 2005).

THE MESOLIMBIC DA SYSTEM AND MATERNAL BEHAVIOR

Table 1.2 summarizes the evidence which indicates that DA action at the level of NA is necessary for normal maternal behavior in rats. As indicated, DA is released into NA during maternal behavior and interference with DA function at the level of NA or disruption of VTA-DA neurons causes deficits in maternal behavior. Importantly, downregulation of mesolimbic DA function primarily disrupts the proactive voluntary components of maternal behavior, while nursing behavior is much less affected (Hansen *et al.*, 1991a, b; Keer & Stern, 1999). These effects match those which occur after damage to MPOA circuits, bolstering the view that the two systems interact to influence maternal responsiveness.

There are two main types of DA receptors, both of which are located in NA: the D1 type and the D2 type receptor (Robertson & Jian, 1995; Missale *et al.*, 1998). Given that 6-hydroxydopamine lesions of mesolimbic DA neurons (Hansen *et al.*, 1991a, b) or microinjections of a mixed D1–D2 antagonist into NA (Keer & Stern, 1999) disrupt the proactive voluntary components of maternal behavior in rats, it is important to determine whether D1, D2, or both receptor types are involved. Some evidence suggests that D1 receptors might be more important. As already stated, Fos expression increases in NA during maternal behavior and activation of D1 receptors is capable of increasing Fos expression in NA. Second, Champagne *et al.* (2004) have reported that postpartum female rats which show high levels of active maternal responses (licking/grooming pups) have more D1 type DA receptors in NA than do those females which exhibit a lower level of pup grooming. In a recent study (Numan *et al.*, 2005a), we confirmed that DA–D1 receptors in NA are more importantly involved in the regulation of the proactive voluntary components of maternal behavior (retrieval behavior) than are D2 receptors. Postpartum female rats were injected with various doses (0, 1, and 3μg) of either a standard D1 antagonist (SCH 23390) or a standard D2 antagonist (eticlopride) into the medial NA (referred to as the shell region of NA: NAs), and the effects on maternal behavior were observed. Although nursing behavior was relatively intact after each treatment, the D1 antagonist, but not the D2 antagonist, disrupted retrieval behavior in a dose-dependent fashion. It should be noted that all females, irrespective of treatment, promptly approached their pups when they were placed outside the nest area. In addition, the D1 females retrieved some of their pups, indicating that they could perform the response. Their main deficit was an inability to complete the retrieval of all pups to the nest. They would retrieve one or two pups to the nest and then begin nursing without returning for the other pups, or returning only after a long delay (a litter consisted of six pups). One interpretation of these results is that D1 receptor antagonism in NA makes pups stimuli less salient in the sense that they are less likely to stimulate proactive voluntary maternal responses.

In Figure 1.4 we proposed that DA release into NA opens a gate which allows pup stimuli to be processed by the limbic motor system. We can now specify that DA action on D1 receptors in NA has this effect. But what is the mechanism that mediates this gating function? This question is actually quite controversial and research findings have proposed more than one answer (Pennartz *et al.*, 1994; Nicola *et al.*, 2000; Numan, 2006). As we have already indicated, the NA receives inputs from limbic structures, and this input is primarily glutamatergic (Pennartz *et al.*, 1994). The major output neurons (referred to as medium spiny neurons) of NA are GABAergic and one of their major sites of termination is the VP (Pennartz *et al.*, 1994). The outputs of VP are capable of influencing behavioral reactivity through efferents which affect brainstem and cortical motor systems (Pennartz *et al.*, 1994; Zahm, 2006). The most prevalent view of the function of the NA-ventral pallidal circuit is that espoused by Nicola *et al.* (2000). They summarize evidence which supports the view that DA action at the level of NA acts to suppress weak inputs from the limbic system while enhancing the effects of strong inputs. In other words, the neuromodulatory effects of DA allow only strong inputs to the NA to be processed through to the VP and

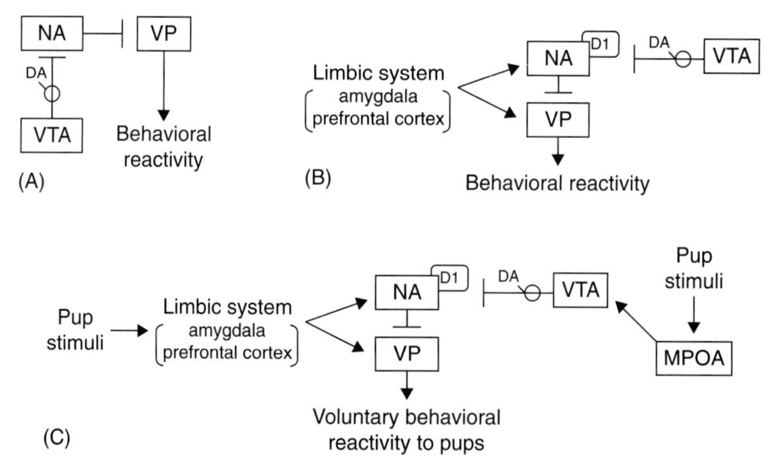

FIGURE 1.6 Neural models of mesolimbic DA function (A and B) and the interaction of this system with the MPOA (C). Axons ending in an arrow signify excitation and those ending in a bar represent an inhibitory process. (A) Mogenson's (1987) view of the function of DA within the NA-VP circuit. NA output neurons are shown as inhibiting VP. DA release into NA depresses the output of NA, which disinhibits VP, causing behavioral reactivity. (B) A modification of the Mogenson view based on the work of Numan *et al.* (2005a). Afferent input from the limbic system reaches both the NA and VP. In the absence of VTA DA activity, NA activation by limbic afferents antagonizes the ability of VP to respond to these afferents. However, when the VTA is active, DA action on D1 receptors in NA depresses the responsiveness of NA to limbic afferents, and this process opens a gate which allows the VP to respond to these afferents, causing behavioral reactivity. (C) A neural model showing how the MPOA might interact with the system shown in panel B. MPOA activation by pup stimuli causes DA release on to D1 receptors in NA which disinhibits VP. This process allows VP to respond to pup stimuli so that proactive voluntary maternal responses can occur.

other NA output stations. Therefore, *stimulation* of NA efferents by certain limbic inputs is deemed critical for behavioral responsiveness. Earlier pioneering research by Mogenson (1987) suggested just the opposite type of process. He suggested that DA acts to inhibit excitatory transmission in NA and that this effect releases VP from inhibition by NA GABAergic efferents (see Figure 1.6(A)). He argued that the consequent increase in VP activity stimulates behavioral reactivity. In other words, stimulation of VP efferents was deemed essential for behavioral responsiveness. As we will describe below, our research on maternal behavior coincides with Mogenson's model rather than that of Nicola *et al.* Therefore, based on Mogenson's model, we have developed a more elaborate model of the sensory–motor integration which occurs in the NA-ventral pallidal circuit under the influence of DA input (Numan *et al.*, 2005a). This model is based on the findings that VP, like NA, receives glutamatergic excitatory inputs from both the amygdala and the prefrontal cortex. Our model is shown in Figure 1.6(B), where we propose that external stimuli derived from limbic afferents are capable of activating both NA and VP. This dual activation is a crucial aspect of the model. Under baseline conditions, these two forces would oppose one another and VP would not show a strong response to sensory stimulation. However, when VTA-DA neurons are active, DA action on D1 receptors suppresses the response of NA to sensory input. Consequently, NA inhibition of VP would be decreased and this action would therefore allow the VP to respond to sensory input, giving rise to behavioral reactivity.

Just what types of biologically significant external stimuli an organism responds to depends on the stimuli that activate VTA dopaminergic input to NA. Comparing Figure 1.4 with Figure 1.6, we suggest that when MPOA neurons activate VTA-DA input to NA D1 receptors the disinhibitory process described above opens a gate which allows the VP to respond to pup-related stimuli and that VP efferent activity allows for the occurrence of proactive voluntary maternal responses (see Figure 1.6(C)).

The models described in Figure 1.6 propose that DA action on D1 receptors inhibits NA output to the VP. With respect to the regulation of maternal behavior, what is the evidence in support of this view? In the context of the fact that D1 antagonism in NA disrupts maternal behavior in postpartum rats, note that NA lesions do not have such a disruptive effect (Numan *et al.*, 2005b). These findings fit with our model that during postpartum maternal behavior DA may be depressing NA output and therefore NA lesions would only be adding to the proposed depressant effect of DA rather than counteracting the DA effect. In fact, our model predicts that depression of VP activity should inhibit maternal behavior, which is exactly what Numan *et al.* (2005b) found in an additional experiment. We microinjected various doses of muscimol into either NA or VP of postpartum rats. Muscimol is a GABAA receptor agonist which causes a temporary neural inhibition at its site of action. Muscimol injections into NA did not disrupt maternal behavior, while all doses (5, 15, or 25 ng) injected into VP disrupted retrieval behavior and the highest doses also depressed nursing behavior. Note that muscimol injections into VP would simulate activity of GABAergic NA efferents to VP, just those inputs that our model proposes should be inhibited during maternal behavior. Overall, therefore, this evidence suggests that the output of the VP, but not that of NA, is essential for maternal behavior. Since the mesolimbic DA system with an action on DA–D1 receptors in NA is also essential for maternal behavior, the best conclusion is that DA functions to depress NA activity which releases the VP from NA inhibition.

An interesting issue arises at this point. If NA output is postulated to be depressed during maternal behavior, why should Fos expression increase in NA during maternal behavior? It is important to realize that Fos expression within neurons should not be taken as evidence that the neurons are firing action potentials (see Hoffman & Lyo, 2002). Fos expression within neurons simply means that the neurons involved have been exposed to events which can trigger Fos expression. DA action on D1 receptors on NA neurons is best conceived as exerting neuromodulatory effects, and our view is that such effects render certain NAs neurons *less* responsive to glutamatergic inputs from the amygdala and prefrontal cortex (Harvey & Lacey, 1997; Charara & Grace, 2003; Maeda *et al.*, 2004). Since DA action on D1 receptors can activate the cAMP-protein kinase A cascade (as well as other signal transduction cascades) and since the fos gene contains a calcium/cAMP response element, D1 activation should also increase Fos expression (Simpson & Morris, 1995; Hoffman & Lyo, 2002). Whether such

Fos expression functions to maintain a reduced responsiveness of NA projection neurons to limbic inputs remains to be determined.

MPOA INTERACTION WITH THE VTA-TO-NA-TO-VP CIRCUIT AND MATERNAL BEHAVIOR

We have presented anatomical evidence in favor of an MPOA-to-VTA connection relevant to maternal behavior and we have presented behavioral evidence that the mesolimbic DA system is involved in maternal behavior. Here, we present neurobehavioral evidence supporting a linkage between MPOA efferents and the VTA-to-NA-to-VP circuit in maternal behavior control. In the first published study on this topic, Numan and Smith (1984) showed that bilateral damage to a neural system which extends between the MPOA and VTA disrupts maternal behavior in postpartum rats. They found that a unilateral knife cut of the lateral MPOA connections paired with a contralateral electrical lesion of the VTA disrupted maternal behavior to a much greater extent than did a variety of control lesions. As one would expect, retrieving and nest building were more severely disrupted by the effective lesions than was nursing behavior. Although these results support a linkage between MPOA and VTA neurons in the control of maternal behavior, because non-specific lesioning procedures were used, which

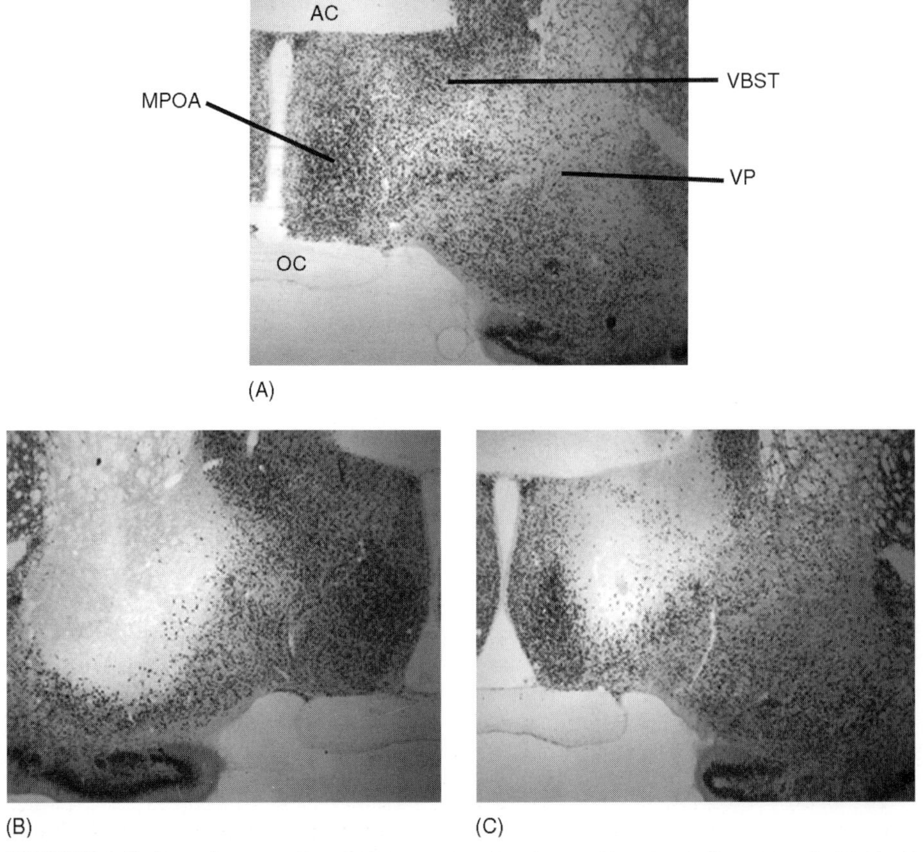

FIGURE 1.7 Frontal sections through the preoptic region immunohistochemically processed with NeuN primary antibody. (A) Normal morphology of the MPOA, vBST, and VP. (B) A unilateral NMDA (*N*-methyl-D-aspartic acid, an exctitotoxic amino acid) lesion aimed at the VP. (C) A unilateral NMDA lesion aimed at the MPOA/vBST region. The NeuN antibody recognizes a neuronal specific protein. If neurons are destroyed, the tissue appears white. AC, anterior commissure; OC, optic chiasm. (*Source*: Reproduced with the permission of Elsevier, Inc., from Numan *et al.* (2005b). Copyright 2005 by Elsevier.)

would also damage axons of passage through the MPOA and VTA, these results do not provide definitive evidence.

If MPOA activation of VTA-DA input to NA releases the VP from NA inhibition during maternal behavior, then we should be able to show that an MPOA lesion on one side of the brain paired with a VP lesion on the other side of the brain would disrupt the proactive voluntary aspects of maternal behavior. In a recent study (Numan *et al.*, 2005b) we provided evidence for this prediction through the use of excitotoxic amino acid lesions which would spare axons of passage. Fully maternal postpartum rats received an excitotoxic amino acid lesion of the MPOA on one side of the brain and an excitotoxic amino acid lesion of VP on the contralateral side (see Figure 1.7). Control females received the two lesions on the same side of the brain, or received contralateral sham lesions. The contralateral lesions were more effective than the ipsilateral lesions in disrupting retrieving behavior, while nursing behavior was much less affected (see Figure 1.8).

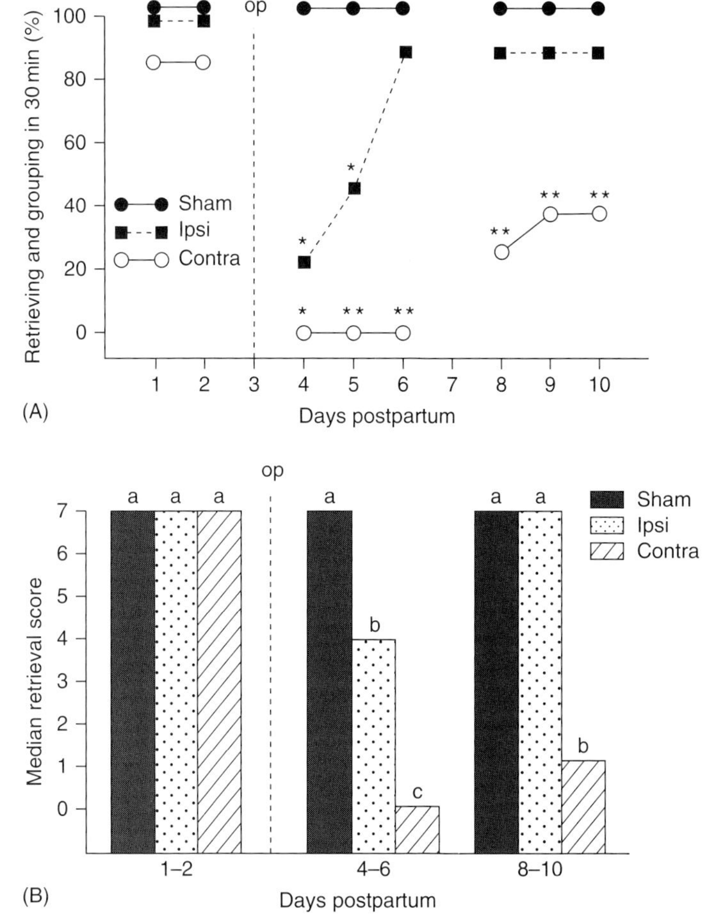

FIGURE 1.8 (A) Percentage of females retrieving all of their pups to a single nest site within 30 min of pup presentation pre- and post-operatively, during the postpartum period. Retrieval tests were not administered on day 7 postpartum. Females received one of the following: unilateral excitotoxic amino acid lesions of the MPOA and VP that were either contralateral or ipsilateral to one another, or sham lesions. The lesions were performed on day 3 postpartum (op). *Significantly different from Sham group. **Significantly different from Sham and Ipsi groups. (B) For the same groups, median retrieval scores averaged over days 1–2 pre-operatively, and 4–6 and 8–10 post-operatively. Within each time interval, groups that do not share a common letter differ significantly from one another. Retrieval rating scale: all pups retrieved to nest in 5 min = 7; in 10 min = 6; in 15 min = 5; in 30 min = 4; in 45 min = 3; in 1 h = 2; in 2 h = 1; not all pups retrieved and grouped in 2 h = 0. (*Source*: Reproduced with the permission of Elsevier, Inc., from Numan *et al.* (2005b). Copyright 2005 by Elsevier.)

Importantly, the contralateral lesions disrupted retrieval behavior in a relatively specific manner. Such females showed normal locomotor activity and they were also able to pick up candy in their mouths and carry the candy to other parts of the cage. The size and weight of the candy approximated that of the pups.

Although these two studies did not specifically manipulate VTA-DA neurons, their findings, in conjunction with the work previously reviewed, adds support to the proposal that MPOA interaction with the mesolimbic DA system regulates proactive voluntary maternal responses.

ACTIVATION OF DOPAMINERGIC NEURAL SYSTEMS STIMULATES THE ONSET OF MATERNAL BEHAVIOR IN RATS

We have reviewed the evidence that interference with the mesolimbic DA system disrupts maternal behavior. We have recently initiated a research program to investigate whether upregulation of the mesolimbic DA system can stimulate the onset of maternal behavior in rats (this research forms part of Danielle Stolzenberg's doctoral dissertation thesis). To do this research we employed the 15HO pregnancy termination model because, as indicated in Table 1.1, the maternal responsiveness of these females is partially or suboptimally primed. Such females show sensitization latencies of 2–3 days instead of the typical 7 days observed in naïve virgins. In the first experiment of this series, Stolzenberg *et al.* (2007) examined whether microinjection of a D1–DA receptor agonist, SKF 38393, into the medial NA could reduce the sensitization latencies of 15HO females. Recall that our neural model proposes that the hormonal events of late pregnancy prime the MPOA so that it becomes fully responsive to pup stimulation, which allows MPOA activation of mesolimbic DA input to NA. In the partially primed 15HO female, MPOA efferents to VTA may not be fully activated by pup stimuli. However, if we were to add DA–D1 stimulation to NA we might be able to mimic a fully active MPOA and produce full maternal behavior.

Primigravid female rats were implanted with bilateral cannulas into the medial NA. On day 15 of pregnancy the females were hysterectomized and ovariectomized and 48 h later (day 0 of testing) they were presented with pups and their maternal behavior was tested. On days 0–1–2 of testing they received bilateral NA injections of either sterile water, 0.2 μg, or 0.5 μg of SKF 38393. Freshly nourished pups were presented daily until full maternal behavior occurred (retrieval of all pups to a common nest site, hovering/crouching over pups, and grooming pups) on two consecutive days or until 5 days had elapsed.

The injection of 0.5 μg of the D1 agonist into the medial NA resulted in a dramatic facilitation of maternal behavior, with 90% of the females showing full maternal behavior after only 24 h of pup exposure. In other words, these females behaved like 15HO + E females (see Table 1.1). In this study we also determined that once maternal behavior occurred in the 0.5 μg group, it was fully maintained in the absence of continued microinjections. Since D1 stimulation of NA promotes the onset of maternal behavior, while D1 antagonism in NA disrupts postpartum maternal behavior (Numan *et al.*, 2005a), we conclude that the maintenance of maternal behavior in the current experiment after the termination of SKF 38393 injections is the result of *endogenous* DA activity at D1 receptors in NA. One possibility to explain these results, which is related to the concept of maternal memory (Numan & Insel, 2003; Numan, 2006; Numan *et al.*, 2006), is that once maternal behavior occurs in first-time mothers, the MPOA is reorganized in some way to allow it to be fully activated by pup stimuli in the absence of continued hormone priming, or in the current case, in the absence of continued microinjections of SKF 38393 into NA.

In additional experiments in this study (Stolzenberg *et al.*, 2007), it was found that microinjections of 0.5 μg of SKF 38393 into the caudate-putamen, which would mimic activation of the nigrostriatal system, did not stimulate maternal behavior in 15HO females. Furthermore, emphasizing the importance of D1 receptors in NA, injections of various doses of quinpirole, a D2 agonist, into NA also did not have a stimulatory effect on maternal behavior. These results substantiate the anatomical and chemical specificity of the SKF 38393 stimulatory effect in NA.

Because of the importance of the MPOA for maternal behavior, and because the MPOA receives DA input from diencephalic sources

(Simerly *et al.*, 1986) and expresses both D1 and D2–DA receptors (Bakowska & Morrell, 1995), Stolzenberg *et al.* (2007) explored the effects of microinjection of 0.5μg of SKF 38393 into MPOA on the onset of maternal behavior in 15HO rats. We also attempted to replicate the NA stimulatory effect. The results are shown in Figure 1.9, where it can be seen that D1 agonist stimulation of either NA or MPOA was equally effective in facilitating the onset of maternal behavior in 15HO females when compared to females who received injections of sterile water into these regions. One interesting possibility is that DA action on D1 receptors in MPOA contributes to the process which allows pup stimuli to activate MPOA efferents to the VTA, and that D1 activation of NA opens the gate which allows the VP to respond to pup stimuli. Therefore, in the partially primed female, based on this model, D1 activation at either the level of the MPOA or the NA would allow for the initiation of full maternal behavior.

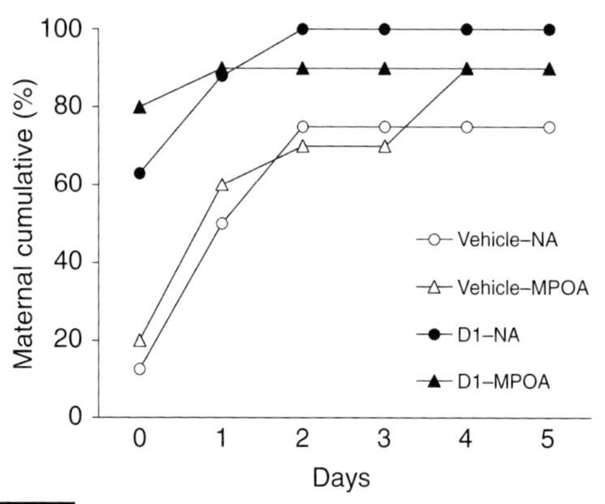

FIGURE 1.9 Cumulative percentage of female rats showing full maternal behavior on each test day. Females were hysterectomized and ovariectomized on day 15 of pregnancy and maternal behavior tests commenced 48h later (day 0 of testing). Females received bilateral microinjections of either 0μg (vehicle-NA) or 0.5μg (D1-NA) of SKF 38393 (n = 8, 8 respectively) into the NA on days 0, 1, and 2 of testing, or 0 (vehicle-MPOA) or 0.5 (D1-MPOA) μg of SKF 38393 (n = 10, 10 respectively) into the MPOA. Females that received SKF 38393 injections into either the NA or MPOA became fully maternal significantly faster than did the vehicle-treated females. (*Source*: Reproduced with the permission of the American Psychological Association, from Stolzenberg *et al.* (2007). Copyright 2007 by the American Psychological Association.)

Barbara Woodside's laboratory has uncovered some findings which are very relevant to a role for DA action on the MPOA with respect to maternal behavior. Service and Woodside (2007) have reported that microinjection of L-NAME into MPOA disrupts retrieval behavior in postpartum rats, suggesting a role for MPOA nitric oxide (NO) in maternal behavior since L-NAME is a nitric oxide synthase (NOS) inhibitor. Furthermore, coinjection of SKF 38393 with L-NAME counteracted L-NAME's inhibitory effect (Service & Woodside, 2006). These results suggest an interaction between D1–DA receptor stimulation and NO within MPOA in the regulation of maternal behavior, and the effects of DA acting on MPOA D1 receptors may be downstream from the NO effect. These results, since they were done in postpartum rats, suggest the D1–DA receptors in MPOA may not only contribute to the onset of maternal behavior, but may also be involved in its maintenance (see Miller & Lonstein, 2005; Numan *et al.*, 2005a). Interestingly, an interaction between NO and DA within the MPOA is also involved in the regulation of male sexual behavior in rats (Hull & Dominguez, 2006). It has long been recognized that there are many similarities between MPOA regulation of maternal behavior and male sexual behavior in rats (Numan, 1974; Numan, 1985; Numan & Insel, 2003). The meaning of these similarities with respect to the overall neural circuitry regulating maternal behavior and male sexual behavior remains to be determined (see Newman, 1999; Numan & Insel, 2003).

Figure 1.10 presents a final neural model which attempts to integrate the various findings we have presented in this paper. The MPOA is shown as containing DA neurons, NOS neurons, and output neurons which project to VTA. All of these MPOA neurons are shown as containing glutamate receptors (GluR: this possibility was derived from the work of Hull and Dominguez (2006)), and the output neurons are shown as also containing D1 receptors. When the MPOA is fully primed by pregnancy hormones, pup stimuli are capable of activating this integrated MPOA network so that the output neurons projecting to the VTA are stimulated. The resultant release of DA into NA, through an action on D1 receptors, in turn functions to disinhibit the VP, allowing the VP to process sensory input from pups so that voluntary

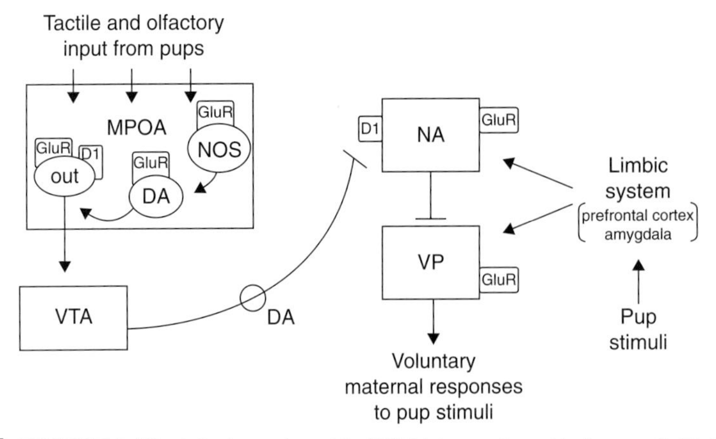

FIGURE 1.10 A final neural model of MPOA interaction with the mesolimbic DA system in the control of maternal behavior. See text for details. D1, D1 dopamine receptor; DA, dopamine; GluR, glutamate receptor; MPOA, medial preoptic area; NA, nucleus accumbens; NOS, nitric oxide synthase containing neuron; out, output or projection neuron from MPOA; VP, ventral pallidum; VTA, ventral tegmental area.

proactive maternal responses, such as pup retrieval, can occur. In contrast, if a female were only partially primed to show maternal behavior, then through the experimental addition of D1–DA receptor stimulation at either MPOA or NA one could compensate for the missing hormones to activate full maternal behavior. In this model, DA–D1 effects on MPOA facilitate the responsiveness of MPOA output neurons to pup stimuli, while at the level of NA DA–D1 activation is proposed to depress the responsiveness of NA output neurons to pup stimuli. If this is indeed the case, the details of the intercellular and intracelluar effects of DA–D1 activation within each nucleus would have to be worked out to provide a proper explanation for these different outcomes.

OXYTOCIN AND THE MPOA-TO-VTA-TO-NA-TO-VP CIRCUIT

As we have indicated previously, oxytocin (OT) as a hormone is involved in parturition and milk ejection, two physiological conditions closely tied to the maternal state. In addition, we noted that OT also serves as a neurotransmitter/neuromodulator within the central nervous system. The primary source of OT neurons is the paraventricular hypothalamic nucleus (PVN: Numan & Insel, 2003), and research has clearly shown that the central release of OT at diverse neural sites is involved in stimulating the onset of maternal behavior at parturition in rats and other species (Numan & Insel, 2003). We have already mentioned that the MPOA is one site of OT action (Pedersen *et al.*, 1994) in its stimulation of the onset of maternal behavior in rats. These same authors also found that microinjection of an OT receptor antagonist into VTA similarly disrupts the onset of maternal behavior in parturient rats. Finally, in recent work on prairie voles (Olazabel & Young, 2006), evidence has been presented favoring a role for OT action at the level of NA in maternal behavior. Given that OT receptors are located in MPOA, vBST, VTA, and NA (Pedersen *et al.*, 1994; Kremarik *et al.*, 1995; Veinante & Freund-Mercier, 1997; Champagne *et al.*, 2001), it is possible that the coordinated release of OT at each of these sites in the parturient mammal plays a critical role in potentiating the linkages within the MPOA-to-VTA-to-NA-to-VP circuit which we have described as being critical for maternal responsiveness (see Figure 1.11). Since OT neural systems appear to play a greater role in the onset of maternal behavior at parturition and a lesser role in the continuance of the behavior, once it has become established (see Numan, 2006), it is interesting to speculate that OT plays a role in consolidating the function of the maternal circuit so that it can continue to operate in the absence of further oxytocinergic mediation. Just how OT might interact with DA neural systems, Fos expression, and with NO remains to be determined. However, the OT work clearly

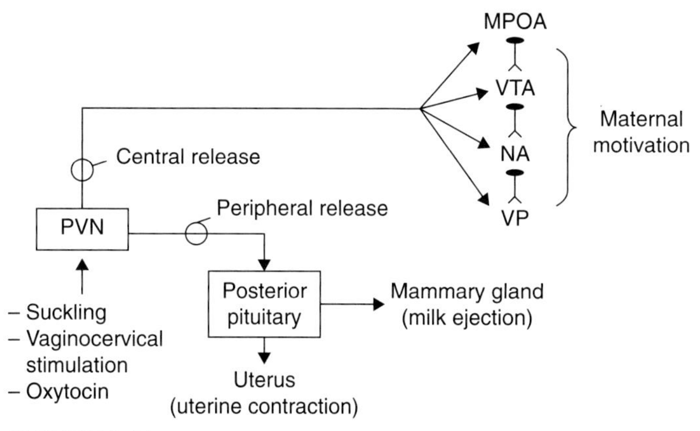

FIGURE 1.11 OT systems and maternal behavior. OT is released from the PVN as a hormone via projections to the posterior pituitary. OT is also released into the brain, where its actions at the level of the MPOA, VTA, and perhaps NA, facilitate the onset of maternal behavior in rats. VP, ventral pallidum.

supports the idea that the MPOA interacts with the mesolimbic DA system in the regulation of maternal responsiveness to infant stimuli.

ACKNOWLEDGEMENTS

Michael Numan and Danielle S. Stolzenberg are supported by NSF grant IOB 0312380.

REFERENCES

Alexander, G. E., DeLong, M. R., and Strick, P. L. (1986). Parallel organization of functionally segregated circuits linking basal ganglia and cortex. *Annu. Rev. Neurosci.* 9, 357–381.

Bakowska, J. C., and Morrell, J. I. (1995). Quantitative autoradiographic analysis of D1 and D2 receptors in rat brain in early and late pregnancy. *Brain Res.* 703, 191–200.

Bakowska, J. C., and Morrell, J. I. (1997). Atlas of the neurons that express mRNA for the long form of the prolactin receptor in the forebrain of the female rat. *J. Comp. Neurol.* 386, 161–177.

Bridges, R. S. (1984). A quantitative analysis of the roles of dosage, sequence, and duration of estradiol and progesterone exposure in the regulation of maternal behavior in the rat. *Endocrinology* 114, 930–940.

Bridges, R. S., and Ronsheim, P. M. (1990). Prolactin (PRL) regulation of maternal behavior in rats: Bromocriptine treatment delays and PRL promotes the rapid onset of behavior. *Endocrinology* 126, 837–848.

Bridges, R. S., Numan, M., Ronsheim, P. M., Mann, P. E., and Lupini, C. E. (1990). Central prolactin infusions stimulate maternal behavior in steroid-treated, nulliparous female rats. *Proc. Natl Acad. Sci. USA* 87, 8003–8007.

Bridges, R. S., Robertson, M. C., Shiu, R. P. C., Friesen, H. G., Stuer, A. M., and Mann, P. E. (1996). Endocrine communication between conceptus and mother: Placental lactogen stimulation of maternal behavior. *Neuroendocrinology* 64, 57–64.

Brown, J. R., Ye, H., Bronson, R. T., Dikkes, P., and Greenberg, M. E. (1996). A defect in nurturing in mice lacking the immediate early gene fosB. *Cell* 86, 297–309.

Champagne, F., Diorio, J., Sharma, S., and Meaney, M. J. (2001). Naturally occurring variations in maternal behavior in the rat are associated with differences in estrogen-inducible central oxytocin receptors. *Proc. Natl Acad. Sci. USA* 98, 12736–12741.

Champagne, F. A., Chretien, P., Stevenson, C. W., Zhang, T. Y., Gratton, A., and Meaney, M. J. (2004). Variations in nucleus accumbens dopamine associated with individual differences in maternal behavior in the rat. *J. Neurosci.* 24, 4113–4123.

Charara, A., and Grace, A. A. (2003). Dopamine receptor subtypes selectively modulate excitatory afferents from hippocampus and amygdala in rat nucleus accumbens. *Neuropsychopharmacology* 28, 1412–1421.

Fahrbach, S. E., and Pfaff, D. W. (1986). Effects of preoptic region implants of dilute estradiol on the maternal behavior of ovariectomized, nulliparous rats. *Horm. Behav.* 20, 354–363.

Fahrbach, S. E., Morrell, J. I., and Pfaff, D. W. (1986). Identification of medial preoptic neurons that concentrate estradiol and project to the midbrain in the rat. *J. Comp. Neurol.* 247, 364–382.

Felton, T. M., Linton, L. N., Rosenblatt, J. S., and Morrell, J. I. (1998). Estrogen implants in the lateral habenular nucleus do not stimulate the onset of maternal behavior in female rats. *Horm. Behav.* 35, 71–80.

Fleming, A. S., and Luebke, C. (1981). Timidity prevents the nulliparous female from being a good mother. *Physiol. Behav.* 27, 863–868.

Gaffori, O., and Le Moal, M. (1979). Disruption of maternal behavior and the appearance of cannibalism after ventral mesencephalic tegmentum lesions. *Physiol. Behav.* 23, 317–323.

Geisler, S., and Zahm, D. S. (2005). Afferents to the ventral tegmental area in the rat-anatomical substratum for integrative functions. *J. Comp. Neurol.* 490, 270–294.

Hansen, S., Harthon, C., Wallin, E., Lofberg, L., and Svensson, K. (1991a). Mesotelencephalic dopamine system and reproductive behavior in the female rat: Effects of ventral tegmental 6-hydroxydopamine lesions on maternal and sexual responsiveness. *Behav. Neurosci.* 105, 588–598.

Hansen, S., Harthon, C., Wallin, E., Lofberg, L., and Svensson, K. (1991b). The effects of 6-OHDA-induced dopamine depletions in the ventral and dorsal striatum on maternal and sexual behavior in the female rat. *Pharmacol. Biochem. Behav.* 39, 71–77.

Hansen, S., Bergvall, A. H., and Nyiredi, S. (1993). Interaction with pups enhances dopamine release in the ventral striatum of maternal rats: a microdialysis study. *Pharmacol. Biochem. Behav.* 45, 673–676.

Harvey, J., and Lacey, M. G. (1997). A postsynaptic interaction between dopamine D1 and NMDA receptors promotes presynaptic inhibition in the rat nucleus accumbens via adenosine release. *J. Neurosci.* 17, 5271–5280.

Heimer, L., and Van Hoesen, G. W. (2006). The limbic lobe and its output channels: Implications for emotional functions and adaptive behavior. *Neurosci. Biobehav. Rev.* 30, 126–147.

Hoffman, G. E., and Lyo, D. (2002). Anatomical markers of activity in neuroendocrine systems: Are we all 'Fos-ed out'?. *J. Neuroendocrinol.* 14, 259–268.

Hull, E. M., and Dominguez, J. M. (2006). Getting his act together: Roles of glutamate, nitric oxide, and dopamine in the medial preoptic area. *Brain Res.* 1126, 66–75.

Hunt, G. E., and McGregor, I. S. (2002). Contrasting effects of dopamine antagonists and frequency induction of Fos expression induced by lateral hypothalamic stimulation. *Behav. Brain Res.* 132, 187–201.

Keer, S. E., and Stern, J. M. (1999). Dopamine receptor blockade in the nucleus accumbens inhibits maternal retrieving and licking, but enhances nursing behavior in lactating rats. *Physiol. Behav.* 67, 659–669.

Kremarik, P., Freund-Mercier, M. J., and Stoeckel, M. E. (1995). Oxytocin and vasopressin binding sites in the hypothalamus of the rat: Histoautoradiographic detection. *Brain Res. Bull.* 36, 195–203.

Lee, A., Clancy, S., and Fleming, A. S. (2000). Mother rats bar-press for pups: Effects of lesions of the MPOA and limbic sites on maternal behavior and operant responding for pup-reinforcement. *Behav. Brain Res.* 108, 215–231.

Lonstein, J. S., Greco, B., De Vries, G. J., Stern, J. M., and Blaustein, J. D. (2000). Maternal behavior stimulates c-fos activity within estrogen receptor alpha-containing neurons in lactating rats. *Neuroendocrinology* 72, 91–101.

Maeda, T., Fukazawa, Y., Shimizu, N., Ozaki, M., Yamamoto, H., and Kishioka, S. (2004). Electrophysiological characteristic of corticoaccumbens synapses in rat mesolimbic system reconstructed using organotypic slice cultures. *Brain Res.* 1015, 34–40.

Mattson, B. J., and Morrell, J. I. (2005). Preference for cocaine- versus pup-associated cues differentially activates neurons expressing either Fos or cocaine- and amphetamine-regulated transcript in lactating maternal rodents. *Neuroscience* 135, 315–328.

Miller, S. M., and Lonstein, J. S. (2005). Dopamine D1 and D2 receptor antagonism in the preoptic area produces different effects on maternal behavior in lactating rats. *Behav. Neurosci.* 119, 1072–1083.

Missale, C., Nash, S. R., Robinson, S. W., Jaber, M., and Caron, M. G. (1998). Dopamine receptors: From structure to function. *Physiol. Rev.* 78, 189–225.

Mogenson, G. J. (1987). Limbic-motor integration. *Prog. Psychobiol. Physiol. Psychol.* 12, 117–167.

Moltz, H., Lubin, M., Leon, M., and Numan, M. (1970). Hormonal induction of maternal behavior in the ovariectomized nulliparous rat. *Physiol. Behav.* 5, 1373–1377.

Moratalla, R., Xu, M., Tonegawa, S., and Graybiel, A. M. (1996). Cellular responses to psychomotor stimulant and neuroleptic drugs are abnormal in mice lacking the D1 dopamine receptor. *Proc. Natl Acad. Sci. USA* 93, 14928–14933.

Newman, S. W. (1999). The medial extended amygdala in male reproductive behavior. *Ann. NY Acad. Sci.* 877, 242–257.

Nicola, S. M., Surmeier, D. J., and Malenka, R. C. (2000). Dopaminergic modulation of neuronal excitability in the striatum and nucleus accumbens. *Annu. Rev. Neurosci.* 23, 185–215.

Numan, M. (1974). Medial preoptic area and maternal behavior in the female rat. *J. Comp. Physiol. Psychol.* 87, 746–759.

Numan, M. (1985). Brain mechanisms and parental behavior. In *Handbook of Behavioral Neurobiology* (N. Adler, D. Pfaff, and R. W. Goy, Eds.), Vol. 7, pp. 537–605. Plenum Press, New York.

Numan, M. (2006). Hypothalamic neural circuits regulating maternal responsiveness toward infants. *Behav. Cog. Neurosci. Rev.* 5, 163–190.

Numan, M., and Callahan, E. C. (1980). The connections of the medial preoptic region and maternal behavior in the rat. *Physiol. Behav.* 25, 653–665.

Numan, M., and Smith, H. G. (1984). Maternal behavior in rats: Evidence for the involvement of preoptic projections to the ventral tegmental area. *Behav. Neurosci.* 98, 712–727.

Numan, M., and Corodimas, K. P. (1985). The effects of paraventricular hypothalamic lesions on maternal behavior in rats. *Physiol. Behav.* 35, 417–425.

Numan, M., and Numan, M. J. (1996). A lesion and neuroanatomical tract-tracing study of the role of the bed nucleus of the stria terminalis in retrieval behavior and other aspects of maternal responsiveness in rats. *Dev. Psychobiol.* 29, 23–51.

Numan, M., and Numan, M. J. (1997). Projection sites of medial preoptic and ventral bed nucleus of the stria terminalis neurons that express Fos during maternal behavior in female rats. *J. Neuroendocrinol.* 9, 369–384.

Numan, M., and Insel, T. R. (2003). *The Neurobiology of Parental Behavior*. Springer-Verlag, New York.

Numan, M., Rosenblatt, J. S., and Komisaruk, B. R. (1977). Medial preoptic area and onset of maternal behavior in the rat. *J. Comp. Physiol. Psychol.* 91, 146–164.

Numan, M., Corodimas, K. P., Numan, M. J., Factor, E. M., and Piers, W. D. (1988). Axon-sparing lesions of the preoptic region and substantia innominata disrupt maternal behavior in rats. *Behav. Neurosci.* 102, 381–396.

Numan, M., Roach, J. K., del Cerro, M. C. R., Guillamon, A., Segovia, S., Sheehan, T. P., and Numan, M. J. (1999). Expression of intracellular progesterone receptors in the rat brain during different reproductive states, and involvement in maternal behavior. *Brain Res.* 830, 358–371.

Numan, M., Numan, M. J., Pliakou, N., Stolzenberg, D. S., Mullins, O. J., Murphy, J. M., and Smith, C. D. (2005a). The effects of D1 and D2 dopamine receptor antagonism in the medial preoptic area, ventral pallidum, or nucleus accumbens on the maternal retrieval response and other aspects of maternal behavior in rats. *Behav. Neurosci.* 119, 1588–1604.

Numan, M., Numan, M. J., Schwarz, J. M., Neuner, C. M., Flood, T. F., and Smith, C. D. (2005b). Medial preoptic area interactions with the nucleus accumbens-ventral pallidum circuit and maternal behavior in rats. *Behav. Brain Res.* 158, 53–68.

Numan, M., Fleming, A. S., and Levy, F. (2006). Maternal behavior. In *Knobil & Neill's Physiology of Reproduction* (J. D. Neill, Ed.), pp. 1921–1993. Elsevier Inc., New York.

Olazabel, D. E., and Young, L. J. (2006). Oxytocin receptors in the nucleus accumbens facilitate "spontaneous" maternal behavior in adult prairie voles. *Neuroscience* 141, 559–568.

Pedersen, C. A., Caldwell, J. D., Walker, C., Ayers, G., and Mason, G. A. (1994). Oxytocin activates the postpartum onset of rat maternal behavior in the ventral tegmental area and medial preoptic area. *Behav. Neurosci.* 108, 1163–1171.

Pennartz, C. M. A., Groenewegen, H. J., and Lopes Da Silva, F. H. (1994). The nucleus accumbens as a complex of functionally distinct neuronal ensembles: An integration of behavioural, electrophysiological and anatomical data. *Prog. Neurobiol.* 42, 719–761.

Robertson, G. S., and Jian, M. (1995). D1 and D2 dopamine receptors differentially increase Fos-like immunoreactivity in accumbal projections to the ventral pallidum and midbrain. *Neuroscience* 64, 1019–1034.

Rosenblatt, J. S. (1967). Nonhormonal basis of maternal behavior in the rat. *Science* 156, 1512–1514.

Rosenblatt, J. S., and Siegel, H. I. (1975). Hysterectomy-induced maternal behavior during pregnancy in the rat. *J. Comp. Physiol. Psych.* 89, 685–700.

Service, G., and Woodside, B. (2006). The effects of bilateral infusions of L-NAME into the medial preoptic area on pup retrieval is reversed by simultaneous administration of SKF 38393 in postpartum rats. Program No. 577.9. *2006 Abstract Viewer/Itinerary Planner*. Society for Neuroscience, Washington, DC.

Service, G., and Woodside, B. (2007). Inhibition of nitric oxide synthase within the medial preoptic area impairs pup retrieval in lactating rats. *Behav. Neurosci.* 121, 140–147.

Sheehan, T. P., and Numan, M. (2002). Estrogen, progesterone, and pregnancy termination alter neural activity in brain regions that control maternal behavior in rats. *Neuroendocrinology* 75, 12–23.

Shughrue, P. J., Lane, M. V., and Merchenthaler, I. (1997). Comparative distribution of estrogen receptor-α and-β mRNA in the rat central nervous system. *J. Comp. Neurol.* 388, 507–525.

Siegel, H. I., and Rosenblatt, J. S. (1975). Hormonal basis of hysterectomy-induced maternal behavior during pregnancy in the rat. *Horm. Behav.* 6, 211–222.

Simerly, R. B., and Swanson, L. W. (1988). Projections of the medial preoptic nucleus: A Phaseolus vulgaris leucoagglutinin anterograde tract-tracing study in the rat. *J. Comp. Neurol.* 270, 209–242.

Simerly, R. B., Gorski, R. A., and Swanson, L. W. (1986). Neurotransmitter specificity of cells and fibers in the medial preoptic nucleus: An immunohistochemical study in the rat. *J. Comp. Neurol.* 246, 343–363.

Simpson, C. S., and Morris, B. J. (1995). Induction of c-fos and zif/268 gene expression in rat striatal neurons, following stimulation of D1-like dopamine receptors, involves protein kinase A and protein kinase C. *Neuroscience* 68, 97–106.

Stack, E. C., and Numan, M. (2000). The temporal course of expression of c-Fos and Fos B within the medial preoptic area and other brain regions of postpartum female rats during prolonged mother–young interactions. *Behav. Neurosci.* 114, 609–622.

Stack, E. C., Balakrishnan, R., Numan, M. J., and Numan, M. (2002). A functional neuroanatomical investigation of the role of the medial preoptic area in neural circuits regulating maternal behavior. *Behav. Brain Res.* 131, 17–36.

Stern, J. M. (1991). Nursing posture is elicited rapidly in maternally naïve, haloperidol-treated female and male rats in response to ventral trunk stimulation from active pups. *Horm. Behav.* 25, 504–517.

Stern, J. M. (1996). Somatosensation and maternal care in Norway rats. In *Advances in the Study of Behavior* (J. S. Rosenblatt and C. T. Snowdon, Eds.), Vol. 25, pp. 243–294. Academic Press, San Diego, CA.

Stern, J. M. (1997). Offspring-induced nurturance: Animal–human parallels. *Dev. Psychobiol.* 31, 19–37.

Stolzenberg, D. S., McKenna, J. B., Keough, S., Hancock, R., Numan, M. J., and Numan, M. (2007). Dopamine D1 receptor stimulation of the nucleus accumbens or the medial preoptic area promotes the onset of maternal behavior in pregnancy-terminated rats. *Behav. Neurosci.* 121, 907–919.

Swanson, L. W. (1982). The projections of the ventral tegmental area and adjacent regions: A combined retrograde fluorescent and immunoflourescence study in the rat. *Brain Res. Bull.* 9, 321–353.

Swanson, L. W. (1992). *Brain Maps: Structure of the Rat Brain.* Elsevier Inc., Amsterdam.

Terkel, J., Bridges, R. S., and Sawyer, C. H. (1979). Effects of transecting the lateral neural connections of the medial preoptic area on maternal behavior in the rat: Nest building, pup retrieval, and prolactin secretion. *Brain Res.* 169, 369–380.

Veinante, P., and Freund-Mercier, M. (1997). Distribution of oxytocin- and vasopressin-binding sites in the rat extended amygdala: A histoautoradiographic study. *J. Comp. Neurol.* 383, 305–325.

Voorn, P., Vanderschuren, L. J. M. J., Groenewegen, H. J., Robbins, T. W., and Pennartz, C. M. A. (2004). Putting a spin on the dorsal–ventral divide of the striatum. *Trends Neurosci.* 27, 468–474.

Zahm, D. S. (2006). The evolving theory of basal forebrain functional–anatomical "macrosystems". *Neurosci. Biobehav. Rev.* 30, 148–172.

2

NEURAL SUBSTRATES INVOLVED IN THE ONSET OF MATERNAL RESPONSIVENESS AND SELECTIVITY IN SHEEP

FREDERIC LÉVY

UMR 6175 INRA-CNRS-Universite de Tours-Haras Nationaux, PRC, INRA 37380 Nouzilly, France

Maternal behavior displayed by the mother in response to the young assumes a wide variety of patterns, some of which are common to most placental mammals, while others depend primarily on the maturity of the young at birth. The species-characteristic pattern of maternal behavior also depends on the social structure of the species and their ecological setting.

For most mammals, maternal behavior usually emerges at or close to parturition. Just after birth, the female shows a very rapid interest in the newborn. Cleaning of the neonate and the consumption of amniotic fluid and placenta are a widespread behavior among mammalian orders. Mothers of many mammals also emit characteristic vocalizations in response to their young and show retrieval, gathering, herding, or carrying behaviors. As well, most new mothers protect their young from predators and conspecifics. Nursing is the most important and common pattern of maternal behavior in mammals, which occurs within the first hours after parturition. However, besides these common patterns, differences in mothering style can be influenced by the developmental condition of the young at birth, the habitat, and the social organization of the species.

In so-called altricial species (most rodents, canids, felids), the mother builds a nest in which she gives birth to a large litter of young that are not fully developed and have limited sensory and locomotor abilities. In most of these altricial species, the newborn young stay huddled together within the nest for a number of days. Hence there is little need for the mother to recognize individual members of her litter. The physiological and neural control of the onset of maternal behavior has been studied most in altricial mammals, in particular the rat (see other chapters in this book). The other main type of maternal behavior is exhibited by precocial mammals. These species (most ungulates) have a small litter of fully developed young capable of following the mother shortly after birth. Indeed, these species are constantly on the move in search of food, and it is vital that young are born with mature sensory and motor systems to follow rapidly their mothers. Because own and alien young can potentially co-occur in the same flock, nursing ungulate females potentially risk having their limited milk supply usurped by young that are not their offspring. Mothers of these species, therefore, develop discriminative maternal care favoring their own young, allowing them to suckle while rejecting any alien young that may approach the udder. In this respect, the establishment of a selective bond within the first few hours after parturition represents one of the essential characteristics of maternal behavior in precocial species. Individual recognition of young has been demonstrated in sheep, goats, cattle, and horses (Herscher *et al.*, 1963; Klopfer *et al.*, 1964; Smith *et al.*, 1966; Hudson & Mullord, 1977; Maletinska *et al.*, 2002). This characteristic is different from maternal responsiveness which reflects the interest toward any newborn and occurs immediately at birth in both altricial and precocial species. Hence, precocial species offer the unique opportunity to understand mechanisms involved not only in the immediate receptivity and display of maternal care, but also in the exclusive care for the young, that is, the maternal bond.

The aim of this chapter is to review the data on the neural and neurochemical mechanisms of these two characteristics of maternal behavior, that is, maternal responsiveness and selectivity, and to consider the possible relations between the neural networks controlling these two facets of maternal attachment in precocial species. Thus, we first introduce the main physiological and sensory determinants of maternal responsiveness and selectivity and, secondly, the different neural substrates that mediate these responses. Finally, we will examine how previous maternal experience provided by interacting with the young during the first parental cycle results in changes in the underlying neural mechanisms controlling maternal care. In precocial mammals, most work investigating these issues has been conducted in sheep, and this species, therefore, is the focus of this review.

HORMONAL AND SENSORY DETERMINANTS OF MATERNAL RESPONSIVENESS AND SELECTIVITY

Maternal Responsiveness

High levels of maternal responsiveness are contingent on parturition and do not occur outside the peripartum period. At that time, mothers accept any neonate indicating that they are in a state of responsiveness to cues common to any neonate (Poindron & Le Neindre, 1980; Poindron *et al.*, 1980). They show a strong attraction to amniotic fluid and immediately after expulsion of the neonate, the mother licks her newborn until the young is cleaned. This behavior is associated with the emission of low-pitched bleats. If the young is removed, the mother responds with high-pitched bleats. As soon as the lamb can stand, it searches for the udder and the mother responds by arching her body, thus facilitating access to the udder. Suckling usually occurs within the first 2 h after parturition. From these behavioral observations, a series of criteria have been defined to assess the acceptance and the rejection of the newborn by a female. Licking behavior, emission of maternal bleats, udder acceptance, and suckling are indicative of maternal responsiveness, whereas aggressive behavior, emission of high bleats, and udder refusal are indicative of maternal rejection.

The state of high maternal responsiveness is restricted to a limited period called the sensitive period (Poindron *et al.*, 2007). If the newborn lamb is removed at birth, without having any contact with the mother, maternal responsiveness fades rapidly. After 12 h of mother–young separation ewes are unable to display maternal responsiveness. The fading of maternal responsiveness depends also on the time at which this separation occurred. For example, 12–24 h of separation have little consequence on maternal response in ewes which had 4 h of contact before separation (Lévy *et al.*, 1991). Therefore, the mother needs to experience a limited amount of contact with a neonate just after parturition following which maternal behavior can be maintained.

The existence of a limited period of maternal receptivity is caused by several physiological changes that occur at parturition. However, two factors which act in synergy are essential: the prepartum rise in estrogen and the vaginocervical stimulation (VCS) induced by the expulsion of the neonate. In non-pregnant, but multiparous, ewes primed with progesterone and estradiol, the induction of maternal behavior in response to a neonate occurs in the majority of cases when ewes have received 5 min of VCS. By contrast, few females responded maternally without this stimulation (Keverne *et al.*, 1983). These results have been further confirmed by administering low doses of estradiol and even in ewes at estrus (Poindron *et al.*, 1988; Kendrick & Keverne, 1991; Kendrick *et al.*, 1992a). This stimulation not only increased the number of maternal ewes, but also induced the full complement of maternal responsiveness, licking, maternal bleats and acceptance at the udder. On the other hand, VCS without a pre-treatment of steroids was ineffective. This confirms the importance of steroid priming (Poindron *et al.*, 1988; Kendrick & Keverne,

1991) in this process. This facilitatory action of VCS has also been demonstrated in parturient ewes. Five minutes of VCS 1 h after parturition reinduced maternal responses that normally occur only at birth, especially licking toward a newborn lamb (Keverne *et al.*, 1983). In contrast, if stimulation of the genital tract was prevented at the time of parturition by peridural anesthesia, maternal behavior was disrupted (Krehbiel *et al.*, 1987).

The neural mechanisms by which VCS initiates the onset of maternal responsiveness appears to be linked to the rise in oxytocin (OT) levels in cerebrospinal fluid at parturition and following artificial VCS (Kendrick *et al.*, 1986; Lévy *et al.*, 1992). These results led to the hypothesis that OT was involved in stimulating maternal behavior. This hypothesis was first supported by a study in non-pregnant ewes in which intracerebroventricular injection of OT induced maternal responses within a minute in steroid-primed ewes (Kendrick *et al.*, 1987; Keverne & Kendrick, 1991). Central infusion of a receptor antagonist partially blocked the induction of maternal care by OT, and the use of a pharmacologic agonist was as efficient as OT itself (Kendrick, 2000). Interestingly, OT, like VCS, was ineffective when given without estradiol priming. The crucial role of OT has also been demonstrated in parturient females. Inhibition of genital stimulation feedback with peridural anesthetic just before parturition prevented both central OT release and maternal behavior. Central infusion of OT reversed these behavioral effects (Lévy *et al.*, 1992).

These physiological factors modify the mother's responsiveness to infant cues and specifically amniotic fluid which coats the neonate. It is established that the ewe develops an olfactory attraction to amniotic fluid at parturition which did not exist prior to birth (Lévy *et al.*, 1983, 1995b). The simultaneous occurrence of attraction to amniotic fluid and the occurrence of the sensitive period was an indication that the same factors controlled maternal olfactory functioning and maternal behavior. Indeed, various experiments indicate that VCS facilitates attraction to amniotic fluid as it does for maternal responsiveness (Lévy *et al.*, 1990b). Peridural anesthesia which blocks vaginal sensation in parturient ewes also induces a loss of preference for amniotic fluid. Conversely, attraction to amniotic fluid once it has disappeared in post-parturient ewes can be restored by artificial VCS up to 4 h postpartum. This effect of VCS appears to be mediated in part by OT. Intracerebroventricular infusion of OT in peridural anesthetized ewes also restored the preference for amniotic fluid (Lévy *et al.*, 1990b).

Attraction to amniotic fluid is a necessary step for the development of maternal responsiveness. Washing the neonate to remove amniotic fluids greatly disrupts the onset of maternal care (Lévy & Poindron, 1987). Amniotic fluid itself is also sufficient to induce maternal responsiveness in a context wherein females typically reject young. That is, parturient experienced ewes accepted 1-day-old lambs whose coats were treated with amniotic fluid. The origin of amniotic fluid had no reliable effect on maternal acceptance which suggests that amniotic fluid would only contain cues responsible for general attractiveness, but not for individual recognition. However, contrary to the privation of olfactory cues provided by the amniotic fluid, prepartum lesion of the main olfactory mucosa had fewer consequences on the display of maternal care suggesting that some compensation by other sensory cues can occur (Lévy *et al.*, 1995b).

Maternal Selectivity

During the early phase of maternal responsiveness, ewes progressively learn the characteristics of their offspring, so that they are able to recognize them and only to accept them at the udder. The development of selectivity of ewes toward lambs was systematically studied by assessing ewes' willingness to let own, but not alien, lambs suckle at the udder. Own and alien lambs were presented consecutively at different postpartum periods ranging from birth to 4 h postpartum (Keller *et al.*, 2003). A relative high proportion of ewes showed selectivity at suckling as early as 30 min after parturition and after 4 h of contact the majority of ewes were selective. Thus, recognition at suckling is a very rapid learning process that takes place at parturition. The requisite duration of contact between ewe and lamb necessary to produce long-term retention of the recognition learning was also studied. A short exposure to the lamb for 4 h just after birth, which produces good selectivity, was not adequate to sustain selectivity after a 24 or 36 h mother–young separation (Lévy *et al.*, 1991; Keller *et al.*, 2005). However, if ewes were permitted to interact with their lambs for 7 days, they retained their selectivity over 36 h separation, but not over a 3 day separation (Keller *et al.*, 2005). Interestingly, unlike the long-term retention of responsiveness found with maternal experience in the rat studies (for review, see Lévy & Fleming, 2006), in sheep, the olfactory recognition memory is quite short-lived, and does not persist for very long. However, a recent study suggests that olfactory recognition memory of the lamb could be strengthened through a process labeled reconsolidation, dependent on protein synthesis and induced by a short reunion period with the lamb during the separation period. Indeed, if after 7 days of mother–young contact and 8 h of separation, ewes are subcutaneously injected with a protein synthesis inhibitor during 10 min of reunion, then the mother's recognition of the lamb during a subsequent test for long-term selectivity is disrupted (Perrin *et al.*, 2007a). These data suggest that in normal conditions, reunion with the lamb triggers reconsolidation processes in the mother's brain that could strengthen lamb recognition memory, thus rendering it more resistant to the passage of time.

As for maternal responsiveness, the key factor responsible for the formation of maternal selectivity is based on the expulsion of the neonate. The importance of VCS for the formation of the maternal bond was investigated by studying the ability of VCS to induce adoption of an alien lamb after the selective bond with the familiar lambs had been formed (Kendrick *et al.*, 1991). Five minutes of mechanical VCS was effective in inducing complete acceptance of an alien lamb at 1-day postpartum. A similar result was reported in goats (Romeyer *et al.*, 1994).

Numerous studies consistently indicate that the sense of smell plays a primary role in ewes' selective acceptance of lambs for nursing (Poindron & Le Neindre, 1980; Poindron *et al.*, 1993; Porter *et al.*, 1994; Lévy *et al.*, 1996). When ewes were rendered hyposmic by irrigating the olfactory mucosa with zinc sulfate solution to destroy the olfactory receptors (Poindron, 1976; Lévy *et al.*, 1995b), they subsequently showed no selective preference for their own young, but nursed alien young as well as their own. Sectioning the nerves of the accessory olfactory system was without effect (Lévy *et al.*, 1995b). In related experiments, manipulation of the lambs' sensory cues, rather than of the mothers' sensory modality produced similar effects (Poindron & Le Neindre, 1980). Ewes that were exposed for 12 h to their lambs in a confined, double-walled mesh cage (receiving olfactory, but not physical, contact) developed selective bonds with that lambs. In contrast, ewes that were exposed to lambs in an airtight transparent box (thereby eliminating olfactory cues) did not develop a selective bond. Thus, contrary to what happens for maternal responsiveness, selectivity at suckling depends primarily on olfactory cues, and no compensatory mechanism emerges after lesioning of the olfactory mucosa. However, a learning based on auditory and visual cues for recognition of the young at longer distances can occur several hours later (Keller *et al.*, 2003).

NEURAL NETWORKS INVOLVED IN MATERNAL RESPONSIVENESS AND SELECTIVITY

As the preceding section reveals, the establishment of maternal responsiveness and of maternal selectivity does not seem to constitute a sole and unique behavioral process but rather two different events sharing some common features, namely, the importance of VCS (for a more comprehensive review on that matter, see Poindron *et al.* (2007)). The following discussion of the related neural circuitry will give a better understanding of these two aspects of maternal behavior.

Maternal Responsiveness

The use of immediate early genes as markers of neuronal activation has allowed the investigation of the neuronal networks involved in the activation of maternal responsiveness in the ewe. In one study, neural activation resulting from a brief exposure to lambs after parturition (30 min) was compared with that found in females receiving a treatment to induce maternal receptivity in the absence of any exposure to lambs (Da Costa *et al.*, 1997). In another study, neural activation was measured in mothers rendered anosmic before parturition. This allows for the display of maternal responsiveness, but prevents the establishment of bonding (Keller *et al.*, 2004a). Both studies revealed that the induction of maternal responsiveness involves activation of an extensive neural circuitry including various limbic and hypothalamic areas. Indeed, the pattern of gene expression observed between ewes displaying maternal behavior (i.e., maternally responsive and selective) and ewes receiving VCS or rendered anosmic before parturition (i.e., maternally responsive, but not selective), were remarkably similar in hypothalamic regions, particularly the medial preoptic area (MPOA) and the paraventricular nucleus (PVN), and also in the bed nucleus of the stria terminalis (BNST). These findings help identify neuroanatomical substrates, likely involved in the onset of maternal responsiveness.

The functional involvement of some of these structures was explored using various pharmacological approaches. The effects of MPOA and BNST inactivation were investigated in primiparous ewes by infusing an anesthetic into these regions before parturition and during the first 2 h postpartum (Perrin *et al.*, in press). MPOA inactivation greatly impaired the whole repertoire of maternal behavior at parturition, whereas inactivation of the BNST or of adjacent sites (septum or diagonal band of Broca) or infusion of cerebrospinal fluid did not. However, when interest for the lamb was challenged by a separation/reunion lamb test performed at 2 h postpartum, ewes with MPOA inactivation exhibited little reaction after separation of their lambs and did not show any motivation to reunite with them. Ewes with BNST inactivation were also less motivated to join their lamb. These findings led to reassess the minor role previously assigned to the MPOA (Kendrick *et al.*, 1997) and to consider this brain region as a key structure for the control of maternal responsiveness as it has been demonstrated in the rat (Numan & Insel, 2003). Furthermore, whereas the BNST does not seem to be involved in the different components of maternal responsiveness at parturition, this structure could regulate approach behavior when the young are away. Interestingly, in the rat neurochemical lesions that specifically destroy the BNST without damaging the MPOA primarily disrupt retrieval behavior, although other components of maternal behavior were unaffected. (Numan & Numan, 1996).

The functional involvement of the PVN in the onset of maternal responsiveness is thought to operate through its OT release. Indeed, release of OT at birth and/or in response to VCS occurs primarily in the PVN, the main source of OT release in the brain (Da Costa *et al.*, 1996). Furthermore, levels of OT immunoreactivity and mRNA expression are increased in the PVN at parturition suggesting an enhancement of OT synthesis and storage at this time (Broad *et al.*, 1993a). Parturition also increases OT receptor mRNA expression in the cell bodies of the PVN (Broad *et al.*, 1999). The extensive activation of the PVN OT system appears to facilitate maternal responsiveness. In fact, when OT is infused into the PVN by retrodialysis in non-pregnant animals primed with a steroid treatment, maternal interest toward lambs is stimulated (Da Costa *et al.*, 1996). Interestingly, this treatment induced the full maternal repertoire comparable to that reported following central infusion of OT and VCS (Keverne *et al.*, 1983; Kendrick *et al.*, 1987). Altogether, these findings indicate that VCS occurring at parturition induces OT release largely in the PVN and activation of the OT system is promoted by OT itself through autoreceptors (Da Costa *et al.*, 1996; Kendrick *et al.*, 1997). However, OT release also occurs in the BNST, the MPOA, and the main olfactory bulb (MOB) during parturition and/or following VCS (Kendrick *et al.*, 1992b; Da Costa *et al.*, 1996). Hence, OT may stimulate maternal receptivity or at least some of its components within these brain regions. For example, in steroid-treated, non-pregnant ewes, infusions of OT by retrodialysis in the MPOA or in the MOB reduced aggression toward lambs, but had no effects on acceptance behaviors (Kendrick *et al.*, 1997). Part of the effects of OT in inducing maternal responsiveness could be mediated through its modulation of transmitter release (Lévy *et al.*, 1995a; Kendrick *et al.*, 1997). For instance, noradrenaline is released together with OT during parturition at a number of different brain regions including the MPOA, the BNST, the PVN, and the MOB (Kendrick *et al.*, 1992b, 1997; Lévy *et al.*, 1995a). Moreover, retrodialysis infusions of OT increased noradrenaline release in the MPOA (Kendrick *et al.*, 1992b) and in the MOB (Lévy *et al.*, 1995a). Therefore, it is possible that OT acts on pre-synaptic noradrenergic terminals and in this way a restricted pattern of potentiated noradrenaline release could occur at sites controlling maternal responsiveness. It would be of interest to evaluate, for instance, the effects of adrenergic antagonists directly injected into the MPOA to see whether this treatment would reduce maternal responsiveness at parturition.

Along with OT, the opioid, pre-proenkephalin, and corticotrophin-releasing hormone are also synthesized in the PVN and their mRNA expression increases at birth (Broad *et al.*, 1993b, 1995). These neuropeptides have a modulatory role in contributing to the induction of maternal behavior induced by OT. Central administration of the opioid agonist, morphine, to steroid-treated ewes facilitated the induction of maternal responsiveness by VCS and the concomitant release of OT in the cerebrospinal fluid, while its antagonist, naltrexone, injected either peripherally (Caba *et al.*, 1995) or centrally had the opposite effects (Keverne & Kendrick, 1991). Similar results were also obtained after a central infusion of CRH (Keverne & Kendrick, 1991).

In rats, a co-localization of the immediate early gene, c-fos, with OT and CRH peptides has been found in PVN neurons (Verbalis *et al.*, 1991). Kendrick's laboratory tested the hypothesis that this immediate early gene which is upregulated in PVN cells following birth (Da Costa *et al.*, 1997; Keller *et al.*, 2004a), modulates gene expression for these neuropeptides and consequently participates in the induction of maternal responsiveness. However, while c-fos/c-jun antisense infusions into the PVN reduced the birth-induced increase in OT concentration and upregulation of CRH and preproenkephalin mRNA expression in the PVN, the animals were fully maternal and only displayed a slight deficit in maternal bleats (Da Costa *et al.*, 1999). Hence this study failed to unravel the links between these peptides and the onset of maternal responsiveness.

Maternal Selectivity

A clear differentiation between neural substrates controlling maternal responsiveness and those involved in the olfactory memory process associated with selectivity was suggested by the Fos studies reported above (see previous section). The identification of activated brain regions during olfactory bonding was undertaken using two experimental strategies. One strategy consisted of comparing c-fos mRNA expression in mothers that were exposed to a lamb after parturition and in non-pregnant animals that received an artificial VCS without having any contact with a lamb (Da Costa *et al.*, 1997). The other study compared Fos immunoreactivity after 2h of exposure to a lamb in intact maternal and selective mothers and in anosmic mothers displaying maternal responsiveness, but not individual lamb recognition (Keller *et al.*, 2004a). Both studies reported changes in Fos expression in animals which were undergoing olfactory memory formation and were mainly restricted to the main olfactory processing regions, that is the MOB, the piriform cortex, the frontal medial cortex, and the orbitofrontal cortex. In addition, in intact ewes as compared with anosmic ewes, significant increases of Fos labeling were found at parturition in the cortical nucleus of the amygdala, in the entorhinal cortex, and also in the hippocampal formation, specifically the subiculum (Keller *et al.*, 2004a). Hence, both studies reach the conclusion that the olfactory memory formation associated with selectivity calls on a specific neural network different from the one involved in maternal responsiveness.

Which of these structures is involved in the memorization of lamb olfactory cues was the subject of numerous pharmacological, neurochemical, and electrophysiological studies. Since the MOB plays a critical role in olfactory information processing, its function in learning of the scent of the familiar lamb has been extensively studied using a combination of methodological approaches. First, the hypothesis that parturition specifically modifies the processing of lamb odors was tested. Electrophysiological recordings were performed in awake ewes before and after birth from mitral cells in the olfactory bulb. These cells receive and transmit olfactory signals (Kendrick *et al.*, 1992c). In recordings made during pregnancy, none of these cells responded preferentially to lamb or amniotic fluid odors. Instead, the majority of cells responded preferentially to food odors. After birth, there was a dramatic increase in the number of cells that responded preferentially to lamb odors, supporting the idea that the change in salience of the lamb odor that occurs at the time of parturition is mediated by a shift in olfactory cell responsivity. However, in relation to the issue of the recognition of the familiar lamb, while the majority of cells did not differentiate between lamb odors, a proportion of the cells did respond preferentially to the odor of the ewe's own lamb. Thus, during the first anatomical relay processing olfactory information, a coding for the familiar lamb odor is established when recognition of lamb odors is a priority.

These shifts in electrical properties of mitral cells are reflected in concurrent parturitional changes in the release of two peptides, the inhibitory gamma-amino-butyric-acid (GABA) and the excitatory glutamate, within the MOB. In comparison to before birth, after parturition, when ewes have established a selective bond with their lambs, the odors of these lambs, but not those of alien ones, increase the release of both peptides. The increase of GABA after birth was significantly greater than that of glutamate (Kendrick *et al.*, 1992c) which may reflect an experience-based increase in the inhibition of activity of certain mitral cells.

Other experimental evidence supports these correlations. Infusion of a GABAa receptor antagonist, bicuculline, in the MOB prevented lamb recognition after the selective bonding had been formed (Kendrick, 1994). The general increase of GABA would refine the olfactory signal by inhibiting mitral cells except those activated by the odor of the familiar lamb.

The selective increase in reactivity of the MOB to familiar lamb odors also results from noradrenergic input from the locus coeruleus (Lévy *et al.*, 1999). A dramatic increase of noradrenaline release occurred during the first 2 h after parturition, that is the learning phase of lamb odor (Lévy *et al.*, 1993). Lesions of noradrenergic projections to the MOB or direct infusions of β-adrenergic antagonist reduces the number of ewes developing the olfactory memory without affecting maternal responsiveness and odor perception (Pissonnier *et al.*, 1985; Lévy *et al.*, 1990a). In contrast to the inhibitory effects of GABA on mitral cell activity, the parturitional release of noradrenaline causes the disinhibition of mitral cells, permitting potentiation of the glutamate system by the retrograde messenger, nitric oxide (Kendrick *et al.*, 1997).

OT release within the MOB may also act to facilitate lamb odor memory because it modulates noradrenaline release. Infusion of OT within the MOB caused a significant increase in noradrenaline concentrations in the MOB (Lévy *et al.*, 1995a), as also demonstrated in the rat (Dluzen *et al.*, 2000). Further evidence that OT can modulate olfactory processing directly at the level of the MOB comes from electrophysiological studies showing that OT decreases mitral cell activity (Yu *et al.*, 1996). Also, the functional relevance of the interaction between noradrenaline and OT systems of the MOB in social memory has been demonstrated in male rats. Phentolamine, an α-adrenoceptor antagonist, abolished social recognition responses after OT-induced elevation of noradrenaline levels within the MOB (Dluzen *et al.*, 2000). Overall, OT in the MOB may play a dual role in the maternal behavior of sheep. It would facilitate maternal responsiveness by mediating olfactory attraction to amniotic fluid (Lévy *et al.*, 1990b) and maternal selectivity by facilitating olfactory memory formation induced by noradrenaline release.

Outside the MOB, the role of neural regions, some of which show Fos activation during formation of olfactory lamb memory, were studied by reversible inactivation or pharmacological manipulation. Infusion of an anesthetic, tetracaine, within the piriform cortex had no effect on the establishment of olfactory selectivity (Broad *et al.*, 2002b). It only reduced rejection of a strange lamb from the same breed, but not from a different one. This could indicate an involvement of the piriform cortex in the fine discrimination of similar lamb odors, as it has been suggested by the authors. However, the assumption that lambs of the same breed share some olfactory characteristics remains to be demonstrated. Numerous studies suggest an involvement of the piriform cortex in olfactory memory. Fos activation in this structure has been reported in olfactory fear conditioning in rats (Schettino & Otto, 2001), in olfactory social recognition in mice (Ferguson *et al.*, 2001), and following exposure to sexually conditioned odor in rats (Kippin *et al.*, 2003). Synaptic changes within the piriform cortex have also been shown to occur after olfactory learning (Litaudon *et al.*, 1997; Mouly *et al.*, 2001; Mouly & Gervais, 2002). A functional dissociation between anterior and posterior piriform cortex has also been reported, the posterior part together with the entorhinal cortex being more engaged in olfactory memorization (Mouly *et al.*, 2001; Mouly & Gervais, 2002). Therefore, it would be of interest to more specifically inactivate the posterior part of the piriform cortex to see whether it has a real effect on olfactory lamb memory. Lastly, inactivation of the medial frontal cortex did not prevent the formation of an olfactory memory for the familiar lamb, but it did inhibit the aggressive rejection directed to strange lambs (Broad *et al.*, 2002a).

In contrast, recent studies of maternal selectivity have demonstrated a pre-eminent role for both the medial nucleus of the amygdala (MeA) and the cortical nucleus of the amygdala (CoA) structures that receive olfactory input from the MOB (Keller *et al.*, 2004b). Infusion of the anesthetic, lidocaine, for the first 8 h postpartum in either of these nuclei prevented animals from learning to discriminate their own lamb from an alien lamb and, hence, both were permitted to suckle. This effect did not result from a disturbance of maternal acceptance, because ewes displayed the full repertoire of maternal responsiveness toward the lamb at parturition. Also, lidocaine-induced impairment did not appear to be mediated by preventing memory retrieval. The termination of lidocaine infusions in non-selective mothers did not restore selectivity 2 h later, after the lidocaïne had worn off. Moreover, lidocaine infusions in the CoA in selective mothers at 48 h postpartum did not impair selectivity. These results also indicate that the effects of lidocaine are not the result of blocking the display of rejection behavior. Taken together, these results favor the hypothesis that the CoA and the MeA form a hub in the network specifically controlling maternal selectivity.

Interestingly, the MeA was also found to be essential for individual recognition of a conspecific in mice. Infusion of an OT antagonist in this structure prevents the recognition of a familiar over an unfamiliar animal (Ferguson *et al.*, 2001). In addition, OT infusion in the MeA of OT knock-out mice restores normal olfactory recognition of a familiar conspecific (Ferguson *et al.*, 2002). OT receptor activation in the MeA may also be involved in lamb olfactory memory, since an increase of OT receptors in the MeA at the time of parturition was observed (Broad *et al.*, 1999b). However, our preliminary result failed to prevent the formation of a lamb olfactory memory by infusion of a specific OT antagonist within the MeA.

The cholinergic system appears to be of particular importance in the formation of selectivity. The cholinergic system is activated at parturition and during the immediate postpartum period (Kendrick *et al.*, 1986; Lévy *et al.*, 1993). Moreover, pharmacological studies indicate that the centrally acting muscarinic antagonist, scopolamine, impairs olfactory lamb recognition (Lévy *et al.*, 1997; Ferreira *et al.*, 1999). Using a specific cholinergic neurotoxin, IgG-saporin, to specifically destroy the basal forebrain cholinergic system, we found a severe impairment of olfactory lamb recognition without any evidence for deficits of maternal responsiveness (Ferreira *et al.*, 2001a, b). Importantly, impairment of selectivity was observed in animals for which loss of cholinergic neurons in each basal forebrain nucleus and their respective limbic, olfactory, and cortical targets was higher than 75%. However, which of its projection sites is important for the impaired lamb recognition remains to be determined.

The neural network involved in olfactory recognition of the lamb probably includes more brain structures. Entorhinal cortex is also activated during lamb odor memory formation (Da Costa *et al.*, 1997; Keller *et al.*, 2004a). Lesions of the perirhinal–entorhinal cortex were found to be critical in olfactory recognition memory of conspecifics in rats (Bannerman *et al.*, 2002), of a sexual partner in male hamsters (Petrulis & Eichenbaum, 2003) and of familiar conspecific odors in female hamsters (Petrulis *et al.*, 2000). These data suggest that the perirhinal–entorhinal cortex may be important for the formation of olfactory memories, and increased Fos expression in the entorhinal cortex of intact ewes, but not of anosmic ewes, might reflect such an involvement.

While the neural network controlling the establishment of lamb olfactory recognition has been explored, studies about brain regions supporting memory consolidation remain sparse. During lamb memory formation, extensive immediately early genes activation was found throughout the olfactory processing network. In contrast, only a few brain structures are engaged in retrieval of lamb memory once consolidated. During the early stages of memory consolidation (4 h postpartum), piriform and entorhinal cortices showed significant expression of activation markers, while retrieval of more consolidated memory (7 days postpartum) led to an enhanced activation of frontal and orbitofrontal cortices (Keller *et al.*, 2004a, 2005; Sanchez-Andrade *et al.*, 2005). Thus, consolidation processes induce time-dependent reorganization in the network engaged in lamb recognition. As the memory consolidates, the sites engaged by retrieval shift partly to frontal regions that are thought to be one of the final storage sites of consolidated memories (Frankland & Bontempi, 2005).

INFLUENCE OF MATERNAL EXPERIENCE ON NEURAL NETWORKS INVOLVED IN MATERNAL RESPONSIVENESS AND SELECTIVITY

The influence of maternal experience in the development of maternal behavior has been abundantly documented in mammals (for review, see Fleming & Li, 2002; González-Mariscal & Poindron, 2002). Previous maternal experience is a requisite for optimal expression of maternal behavior at parturition in ungulates, rabbits, and primates. In sheep, inexperienced mothers display temporary delays in the expression of maternal responsiveness or some behavioral disturbances (Poindron & Le Neindre, 1980; Dwyer & Lawrence, 2000). Primiparous mothers are slower to begin licking their lambs after parturition, are more likely to show rejection behaviors, and to refuse the lamb at the udder (O'Connor *et al.*, 1992; Dwyer & Lawrence, 2000). These deficits result in higher mortality in lambs of primiparous ewes (Putu *et al.*, 1986). If the role of experience in maternal responsiveness has been clearly established, its possible influence on the olfactory recognition of the lamb remains contradictory. Unpublished observations indicate a slower latency to bond by 2 h postpartum in primiparous ewes (Kendrick, 1994), while in another study primiparous mothers were as efficient as their biparous or multiparous counterparts at 2 h postpartum (Keller *et al.*, 2003). One plausible explanation for this discrepancy is that in the former study the deficit in selectivity was due to a slower onset of maternal responsiveness, whereas in the latter study ewes showing absence of maternal interest were excluded. However, a positive influence of maternal experience was found in selectivity based on the learning of visual and auditory cues of the lamb (Keller *et al.*, 2003).

The physiological and neurobiological mechanisms underlying maternal responsiveness and maternal selectivity could serve as the basis for parity differences in behavior. Indeed, maternal experience greatly influences the action of the physiological factors identified. In inexperienced, non-pregnant ewes, neither steroid priming nor VCS induced acceptance of the lamb (Le Neindre *et al.*, 1979; Keverne & Kendrick, 1991). Also, estrogen-primed nulliparous ewes were unresponsive to other artificial means of inducing maternal responsiveness, either with OT, opiates, or corticotrophin-releasing factor (Keverne & Kendrick, 1991). One of the effects of maternal experience could be to increase the brain's sensitivity to hormonal factors involved in maternal responsiveness. For instance, parallel to the lower capacity of inexperienced females to respond to estrogen administration, a difference of estrogen receptors-alpha (ERα) in hypothalamic regions was found between primiparous and multiparous ewes during late pregnancy (Meurisse *et al.*, 2005a). Interestingly, previous maternal experience was associated with a higher density of ERα in the PVN and the MPOA that are involved in maternal responsiveness. Furthermore, maternal experience also enhanced expression of OT receptor mRNA in the PVN which could account for theses difference in maternal responsiveness. The OT receptors of the PVN are probably autoreceptors and their activation has a positive feedback action facilitating OT release in a number of terminals sites (Freund-Mercier & Richard, 1984). Their increased expression in experienced ewes would allow for a much more effective positive feedback action to facilitate the co-ordinated release of OT at these sites. This hypothesis is supported by the finding showing that OT release was higher in the MOB of multiparous than that of primiparous ewes following parturition (Lévy *et al.*, 1995a). Thus, a lower OT activity at different sites may partially explain the more frequent display of maternal disturbances found in primiparous ewes.

Maternal experience also influences the release of neurotransmitters. There was more of noradrenaline, acetylcholine, GABA glutamate, and OT release in the MOB in experienced mothers (Keverne *et al.*, 1993; Lévy *et al.*, 1993, 1995a). Moreover, an influence of maternal experience on the interactions between neurotransmitters and neuropeptide has been reported (Lévy *et al.*, 1995a). For instance, retrodialysis infusion of OT induced an increase in noradrenaline concentrations in the MOB of multiparous, but not in primiparous females. The differences in transmitter release had disappeared 6 h after parturition based on the fact that experimental VCS, an event that mimics parturition, was almost equipotent in multiparous and primiparous ewes in causing neurotransmitter release (Keverne *et al.*, 1993; Lévy *et al.*, 1993). Thus, the first parturition appears to induce a neural maturation of the MOB within the first 6 h postpartum.

Maternal experience is also able to modify the contribution of a neural region in regulating the onset of maternal behavior. Recently, we found that inactivation of the MPOA resulted in less impairment of maternal responsiveness in multiparous mothers than in their primiparous counterparts (Perrin *et al.*, 2007b). This result indicates that maternal behavior relies less on MPOA influence in maternally experienced ewes. Thus, one could hypothesize that the neural network involved in maternal behavior is changed by maternal experience so that other brain structures could compensate for MPOA inactivation. The PVN and the BNST are good candidates for this, since they are connected with the MPOA (Tillet *et al.*, 1993; Scott *et al.*, 2003), and involved in maternal behavior (Da Costa *et al.*, 1996, 1999; Perrin *et al.*, in press). Whereas these brain regions are key components in a maternal neural circuit, this circuit likely includes other brain structures which will be revealed by future study.

CONCLUSION

Across mammalian species, sheep are unique in showing maternal care together with a specialized form of rapid learning producing recognition of individual young. These two behavioral processes which lead to maternal attachment are synchronized at parturition by VCS induced by the expulsion of the neonate, although the mechanisms controlling these two processes differ (Table 2.1). Concerning the sensory inputs, olfactory cues from amniotic fluid are a requisite for the initial approach to the neonate, even though anosmic ewes can still be maternal. On the other hand, selectivity is strictly dependent on the learning of individual olfactory signature. At a physiological level, OT, supported by the opioid peptides and CRH, is the key factor for maternal responsiveness, while its role in lamb olfactory memory is not determined yet. At the neural level, the control of maternal responsiveness is mainly hypothalamic (PVN, MPOA, and BNST) and has little in common with the circuitry involved in selectivity which mainly concerns olfactory processing regions (MOB, CoA, MeA, and frontal cortices) and the cholinergic and noradrenergic systems.

How the two neural networks are functionally interconnected is currently unknown. Using anterograde and retrograde tracers, we observed an interconnection not only between the MOB and the CoA, but also between the BNST and the CoA and between the MPOA/BNST and the MeA (Lévy *et al.*, 1999; Meurisse *et al.*, 2005b). Both olfactory nuclei were interconnected with a more intense projection from the CoA to the MeA. Based on these anatomical studies and the neurobiological data reported in this review, a hypothetical neural model is presented to explain how the two sets of brain regions come into play around parturition to control maternal responsiveness and selectivity (Figures 2.1 and 2.2). At parturition, VCS induces activation of the parvocellular OT neurons in the PVN via neurotransmitter pathways probably, noradrenergic, serotoninergic, and dopaminergic (Figure 2.1). This activation co-ordinates release of OT in a number of brain regions and more specifically in the MPOA/BNST. From these two

TABLE 2.1 Comparison between maternal responsiveness and selectivity in sheep

	Maternal responsiveness	Selectivity at suckling
Behavior	Acceptance of any neonate Licking, maternal bleats, nursing Occurs within the first 12 h postpartum	Acceptance of the familiar lamb; refusal of any alien lamb Established within 4 h of contact
Physiological determinants	E2, VCS, OT, CRH, opioids	VCS, OT (?)
Sensorial determinants	Amniotic fluid	Individual olfactory signature
Neural structures	MPOA, BNST, PVN	MOB, CoA, MeA, frontal cortices, LC, CBF
Influence of experience	Determined E2 and OT receptors Importance of the MPOA	Contradictory NA, Ach, GABA, Glu and OT release in the MOB

E2, estradiol; VCS, vaginocervical stimulation; OT, oxytocin; CRH, corticotrophin-releasing hormone; MPOA, medial preoptic area; BNST, bed nucleus of the stria terminalis; PVN, paraventricular nucleus; MOB, main olfactory bulb; CoA, cortical amygdala; MeA, medial amygdala; LC, locus coeruleus; CBF, cholinergic basal forebrain; NA, noradrenaline; Ach, acetylcholine; GABA, gamma-amino-butyric-acid; Glu, glutamate.

structures, which brain regions are activated to promote motor responses necessary for the performance of maternal acceptance behaviors (licking, bleating, nursing) is not known and remains to be identified. Based on findings in rat studies, the ventral tegmental area, the nucleus accumbens, and its major efferent projections the ventral pallidum may be good candidates for participation in a maternal neural network (see Chapter 1; Numan & Insel, 2003). Also, at parturition VCS induces activation of the cholinergic system of the basal forebrain and the locus coeruleus which project to the MOB (Figure 2.2). The noradrenergic activation induces neural changes within the MOB which results in an enhancement of pattern activity in response to the familiar lamb odor favoring its discrimination. The changes of mitral cell activity induced by the familiar lamb odor would activate the CoA and the MeA. Thus, these nuclei would respond more strongly to odor from the familiar lamb than the alien lamb. This is consistent with the findings showing a discriminative response of the MeA to urine from familiar and unfamiliar males (Binns & Brennan, 2005). In turn, this olfactory network stimulated by the familiar odor would activate the MPOA/BNST/PVN network to promote maternal acceptance. Concurrently, nursing would activate the MPOA/BNST through the activation of the oxytocinergic system in the PVN. This would result in a tighter coupling between the MOB/CoA/MeA and the MPOA/BNST/PVN networks such that after a few hours postpartum, learned odors systematically evoke maternal acceptance. On the other hand, stimulation of the MOB by alien lamb odor would result in an inhibition or a lack of activation of the MPOA/BNST/PVN network. Concomitantly, unfamiliar lamb odor would activate brain regions that regulate rejection behavior, like the medial frontal cortex of which inactivation impairs the aggressive motor rejection response (Broad *et al.*, 2002a). It is worth noting that this is a working neural model

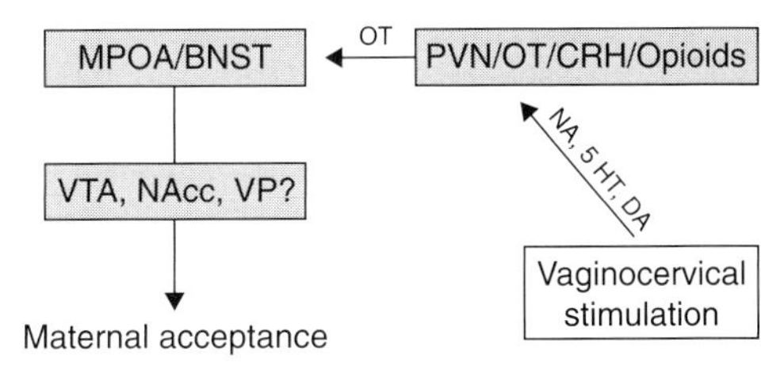

FIGURE 2.1 A neural model of the regulation of maternal responsiveness established at parturition in sheep. OT, oxytocin; CRH, corticotrophin-releasing hormone; MPOA, medial preoptic area; BNST, bed nucleus of the stria terminalis; PVN, paraventricular nucleus; NA, noradrenaline; 5 HT, serotonine; DA, dopamine; VTA, ventral tegmental area; NAcc, nucleus accumbens; VP, ventral pallidum.

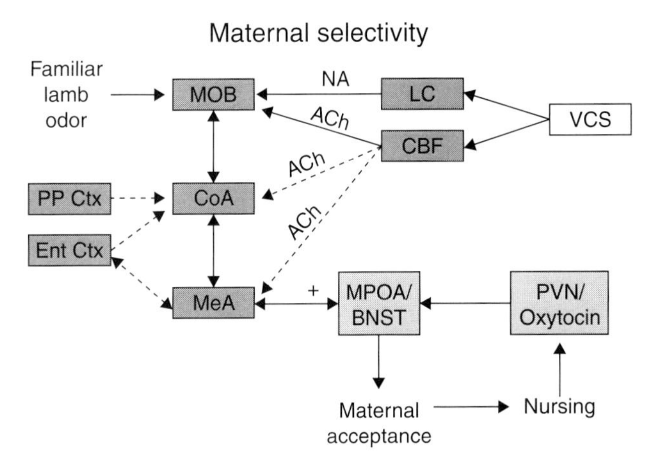

FIGURE 2.2 A neural model of the regulation of maternal selectivity established at parturition in sheep. VCS, vaginocervical stimulation; MPOA, medial preoptic area; BNST, bed nucleus of the stria terminalis; PVN, paraventricular nucleus; MOB, main olfactory bulb; CoA, cortical amygdala; MeA, medial amygdala; LC, locus coeruleus; CBF, cholinergic basal forebrain; NA, noradrenaline; Ach, acetylcholine; PP Ctx, posterior piriform cortex; Ent Ctx, entorhinal cortex. Stripe arrow = hypothetical arrow.

and that additional studies are needed to unveil which other neural structures control acceptance behavior or learning of lamb olfactory cues.

REFERENCES

Bannerman, D. M., Lemaire, M., Yee, B. K., Iversen, S. D., Oswald, C. J., Good, M. A., and Rawlins, J. N. (2002). Selective cytotoxic lesions of the retrohippocampal region produce a mild deficit in social recognition memory. *Exp. Brain Res.* 142, 395–401.

Binns, K. E., and Brennan, P. A. (2005). Changes in electrophysiological activity in the accessory olfactory bulb and medial amygdala associated with mate recognition in mice. *Eur. J. Neurosci.* 21, 2529–2537.

Broad, K. D., Kendrick, K. M., Sirinathsinghji, D. J., and Keverne, E. B. (1993a). Changes in oxytocin immunoreactivity and mRNA expression in the sheep brain during pregnancy, parturition and lactation and in response to oestrogen and progesterone. *J. Neuroendocrinol.* 5, 435–444.

Broad, K. D., Kendrick, K. M., Sirinathsinghji, D. J., and Keverne, E. B. (1993b). Changes in pro-opiomelanocortin and pre-proenkephalin mRNA levels in the ovine brain during pregnancy, parturition and lactation and in response to oestrogen and progesterone. *J. Neuroendocrinol.* 5, 711–719.

Broad, K. D., Keverne, E. B., and Kendrick, K. M. (1995). Corticotrophin releasing factor mRNA expression in the sheep brain during pregnancy, parturition and lactation and following exogenous progesterone and oestrogen treatment. *Mol. Brain Res.* 29, 310–316.

Broad, K. D., Lévy, F., Evans, G., Kimura, T., Keverne, E. B., and Kendrick, K. M. (1999). Previous maternal experience potentiates the effect of parturition on oxytocin receptor mRNA expression in the paraventricular nucleus. *Eur. J. Neurosci.* 11, 3725–3737.

Broad, K. D., Hinton, M. R., Keverne, E. B., and Kendrick, K. M. (2002a). Involvement of the medial prefrontal cortex in mediating behavioural responses to odour cues rather than olfactory recognition memory. *Neuroscience* 114, 715–729.

Broad, K. D., Mimmack, M. L., Keverne, E. B., and Kendrick, K. M. (2002b). Increased BDNF and trk-B mRNA expression in cortical and limbic regions following formation of a social recognition memory. *Eur. J. Neurosci.* 16, 2166–2174.

Caba, M., Poindron, P., Krehbiel, D., Lévy, F., Romeyer, A., and Vénier, G. (1995). Naltrexone delays the onset of maternal behavior in primiparous parturient ewes. *Pharmacol. Biochem. Behav.* 52, 743–748.

Da Costa, A. P., Guevara-Guzman, R. G., Ohkura, S., Goode, J. A., and Kendrick, K. M. (1996). The role of oxytocin release in the paraventricular nucleus in the control of maternal behaviour in the sheep. *J. Neuroendocrinol.* 8, 163–177.

Da Costa, A. P., Broad, K. D., and Kendrick, K. M. (1997). Olfactory memory and maternal behaviour-induced changes in c-fos and zif/268 mRNA expression in the sheep brain. *Mol. Brain Res.* 46, 63–76.

Da Costa, A. P., De La Riva, C., Guevara-Guzman, R., and Kendrick, K. M. (1999). C-fos and c-jun in the paraventricular nucleus play a role in regulating peptide gene expression, oxytocin and glutamate release, and maternal behaviour. *Eur. J. Neurosci.* 11, 2199–2210.

Dluzen, D. E., Muraoka, S., Engelmann, M., Ebner, K., and Landgraf, R. (2000). Oxytocin induces preservation of social recognition in male rats by activating α-adrenoceptors of the olfactory bulb. *Eur. J. Neurosci.* 12, 760–766.

Dwyer, C. M., and Lawrence, A. B. (2000). Maternal behaviour in domestic sheep (*Ovis aries*): Constancy and change with maternal experience. *Behaviour* 137, 1391–1413.

Ferguson, J. N., Aldag, J. M., Insel, T. R., and Young, L. J. (2001). Oxytocin in the medial amygdala is essential for social recognition in the mouse. *J. Neurosci.* 21, 8278–8285.

Ferguson, J. N., Young, L. J., and Insel, T. R. (2002). The neuroendocrine basis of social recognition. *Front. Neuroendocrinol.* 23, 200–224.

Ferreira, G., Gervais, R., Durkin, T. P., and Lévy, F. (1999). Postacquisition scopolamine treatments reveal the time course for the formation of lamb odor recognition memory in parturient ewes. *Behav. Neurosci.* 113, 136–142.

Ferreira, G., Meurisse, M., Gervais, R., Ravel, N., and Levy, F. (2001a). Extensive immunolesions of basal forebrain cholinergic system impair offspring recognition in sheep. *Neuroscience* 106, 103–115.

Ferreira, G., Meurisse, M., Tillet, Y., and Levy, F. (2001b). Distribution and co-localization of choline acetyltransferase and p75 neurotrophin receptors in the sheep basal forebrain: Implications for the use of a specific cholinergic immunotoxin. *Neuroscience* 104, 419–439.

Fleming, A. S., and Li, M. (2002). Psychobiology of maternal behavior and its early determinants in nonhuman mammals. In *Handbook of Parenting* (M. H. Bornstein, Ed.), Vol. 2, pp. 61–97. Lawrence Erlbaum Associates, Mahwah, NJ.

Frankland, P. W., and Bontempi, B. (2005). The organization of recent and remote memories. *Nat. Rev. Neurosci.* 6, 119–130.

Freund-Mercier, M. J., and Richard, P. (1984). Electrophysiological evidence for facilitatory control of oxytocin neurones by oxytocin during suckling in the rat. *J. Physiol.* 352, 447–466.

González-Mariscal, G., and Poindron, P. (2002). Parental care in mammals: Immediate internal and sensory factors of control. In *Hormones, Brain and Behavior* (D. W. Pfaff, A. P. Arnold, A. M. Etgen, S. E. Fahrfbach, and R. T. Rubin, Eds.), Vol. 1, pp. 215–298. Academic Press, New York.

Herscher, L., Richmond, J. B., and Moore, A. U. (1963). Maternal behavior in sheep and goats. In *Maternal Behavior in Mammals* (H. L. Rheingold, Ed.), pp. 203–232. John Wiley and Sons Inc., New York.

Hudson, S. J., and Mullord, M. M. (1977). Investigations on maternal bonding in dairy cattle. *Appl. Anim. Ethol.* 3, 271–276.

Keller, M., Meurisse, M., Poindron, P., Nowak, R., Shayit, M., Ferreira, G., and Lévy, F. (2003). Maternal experience influences the establishment of visual/auditory, but not of olfactory recognition of the newborn baby lamb by ewes at parturition. *Dev. Psychobiol.* 43, 167–176.

Keller, M., Meurisse, M., and Lévy, F. (2004a). Mapping the neural substrates involved in maternal responsiveness and lamb olfactory memory in parturient ewes using Fos imaging. *Behav. Neurosci.* 118, 1274–1284.

Keller, M., Perrin, G., Meurisse, M., Ferreira, G., and Levy, F. (2004b). Cortical and medial amygdala are both involved in the formation of olfactory offspring memory in sheep. *Eur. J. Neurosci.* 20, 3433–3441.

Keller, M., Meurisse, M., and Levy, F. (2005). Mapping of brain networks involved in consolidation of lamb recognition memory. *Neuroscience* 133, 359–369.

Kendrick, K. M. (1994). Neurobiological correlates of visual and olfactory recognition in sheep. *Behav. Process.* 33, 89–112.

Kendrick, K. M. (2000). Oxytocin, motherhood and bonding. *Exp. Physiol.* 85, 111–124.

Kendrick, K. M., and Keverne, E. B. (1991). Importance of progesterone and estrogen priming for the induction of maternal behavior by vaginocervical stimulation in sheep: Effects of maternal experience. *Physiol. Behav.* 49, 745–750.

Kendrick, K. M., Keverne, E. B., Baldwin, B. A., and Sharman, D. F. (1986). Cerebrospinal fluid levels of acetylcholinesterase, monoamines and oxytocin during labour, parturition, vaginocervical stimulation, lamb separation and suckling in sheep. *Neuroendocrinology* 44, 149–156.

Kendrick, K. M., Keverne, E. B., and Baldwin, B. A. (1987). Intracerebroventricular oxytocin stimulates maternal behaviour in the sheep. *Neuroendocrinology* 46, 56–61.

Kendrick, K. M., Lévy, F., and Keverne, E. B. (1991). Importance of vaginocervical stimulation for the formation of maternal bonding in primiparous and multiparous parturient ewes. *Physiol. Behav.* 50, 595–600.

Kendrick, K. M., da Costa, A. P., Hinton, M. R., and Keverne, E. B. (1992a). A simple method for fostering lambs using anoestrus ewes with artificially induced lactation and maternal behaviour. *Appl. Anim. Behav. Sci.* 34, 345–357.

Kendrick, K. M., Keverne, E. B., Hinton, M. R., and Goode, J. A. (1992b). Oxytocin, amino acid and monoamine release in the region of the medial preoptic area and bed nucleus of the stria terminalis of the sheep during parturition and suckling. *Brain Res.* 569, 199–209.

Kendrick, K. M., Lévy, F., and Keverne, E. B. (1992c). Changes in the sensory processing of olfactory signals induced by birth in sheep. *Science* 256, 833–836.

Kendrick, K. M., Da Costa, A. P. C., Broad, K. D., Ohkura, S., Guevara, R., Lévy, F., and Keverne, E. B. (1997). Neural control of maternal behaviour and olfactory recognition of offspring. *Brain Res. Bull.* 44, 383–395.

Keverne, E. B., and Kendrick, K. M. (1991). Morphine and corticotrophin-releasing factor potentiate maternal acceptance in multiparous ewes after vaginocervical stimulation. *Brain Res.* 540, 55–62.

Keverne, E. B., Lévy, F., Poindron, P., and Lindsay, D. R. (1983). Vaginal stimulation: An important determinant of maternal bonding in sheep. *Science* 219, 81–83.

Keverne, E. B., Lévy, F., Guevara-Guzman, R., and Kendrick, K. M. (1993). Influence of birth and maternal experience on olfactory bulb neurotransmitter release. *Neuroscience* 56, 557–565.

Kippin, T. E., Cain, S. W., and Pfaus, J. G. (2003). Estrous odors and sexually conditioned neutral odors activate separate neural pathways in the male rat. *Neuroscience* 117, 971–979.

Klopfer, P. H., Adams, D. K., and Klopfer, M. S. (1964). Maternal imprinting in goats. *Proc. Natl. Acad. Sci. USA* 52, 911–914.

Krehbiel, D., Poindron, P., Lévy, F., and Prud'Homme, M. J. (1987). Peridural anesthesia disturbs maternal behavior in primiparous and multiparous parturient ewes. *Physiol. Behav.* 40, 463–472.

Le Neindre, P., Poindron, P., and Delouis, C. (1979). Hormonal induction of maternal behavior in non-pregnant ewes. *Physiol. Behav.* 22, 731–734.

Lévy, F., and Poindron, P. (1987). The importance of amniotic fluids for the establishment of maternal behavior in experienced and non-experienced ewes. *Anim. Behav.* 35, 1188–1192.

Lévy, F., and Fleming, A. (2006). The neurobiology of maternal behavior in mammals. In *The Development of Social Engagement* (P. J. Marshall and N. A. Fox, Eds.), p. 427. Oxford University Press, Oxford.

Lévy, F., Poindron, P., and Le Neindre, P. (1983). Attraction and repulsion by amniotic fluids and their olfactory control in the ewe around parturition. *Physiol. Behav.* 31, 687–692.

Lévy, F., Gervais, R., Kinderman, U., and Orgeur, P. (1990a). Importance of β-noradrenergic receptors in the olfactory bulb of sheep for recognition of lambs. *Behav. Neurosci.* 104, 464–469.

Lévy, F., Keverne, E. B., Piketty, V., and Poindron, P. (1990b). Physiological determinism of olfactory attraction for amniotic fluids in sheep. In *Chemical Signals in Vertebrates* (D. W. MacDonald, D. Müller-Schwarze, and S. E. Natynczuck, Eds.), Vol. 5, pp. 162–165. Oxford University Press, Oxford.

Lévy, F., Gervais, R., Kindermann, U., Litterio, M., Poindron, P., and Porter, R. (1991). Effects of early post-partum separation on maintenance of maternal responsiveness and selectivity in parturient ewes. *Appl. Anim. Behav. Sci.* 31, 101–110.

Lévy, F., Kendrick, K. M., Keverne, E. B., Piketty, V., and Poindron, P. (1992). Intracerebral oxytocin is important for the onset of maternal behavior in inexperienced ewes delivered under peridural anesthesia. *Behav. Neurosci.* 106, 427–432.

Lévy, F., Guevara-Guzman, R., Hinton, M. R., Kendrick, K. M., and Keverne, E. B. (1993). Effects of parturition and maternal experience on noradrenaline and acetylcholine release in the olfactory bulb of sheep. *Behav. Neurosci.* 107, 662–668.

Lévy, F., Kendrick, K. M., Goode, J. A., Guevara-Guzman, R., and Keverne, E. B. (1995a). Oxytocin and vasopressin release in the olfactory bulb of parturient ewes: Changes with maternal experience and effects on acetylcholine, gamma-aminobutyric acid, glutamate and noradrenaline release. *Brain Res.* 669, 197–206.

Lévy, F., Locatelli, A., Piketty, V., Tillet, Y., and Poindron, P. (1995b). Involvement of the main but not the accessory olfactory system in maternal behavior of primiparous and multiparous ewes. *Physiol. Behav.* 57, 97–104.

Lévy, F., Kendrick, K. M., Keverne, E. B., Proter, R. H., and Romeyer, A. (1996). Physiological, sensory, and experiential factors of prenatal care in sheep. In *Advances in the Study of Behavior* (J. S. Rosenblatt and C. T. Snowdon, Eds.), Vol. 25, pp. 385–416. Academic Press, San Diego, CA.

Lévy, F., Richard, P., Meurisse, M., and Ravel, N. (1997). Scopolamine impairs the ability of parturient ewes to learn to recognise their lambs. *Psychopharmacology* 129, 85–90.

Lévy, F., Meurisse, M., Ferreira, G., Thibault, J., and Tillet, Y. (1999). Afferents to the rostral olfactory bulb in sheep with special emphasis on the cholinergic, noradrenergic and serotonergic connections. *J. Chem. Neuroanat.* 16, 245–263.

Litaudon, P., Mouly, A. M., Sullivan, R., Gervais, R., and Cattarelli, M. (1997). Learning-induced changes in rat piriform cortex activity mapped using multisite recording with voltage sensitive dye. *Eur. J. Neurosci.* 9, 1593–1602.

Maletinska, J., Spinka, M., Vichova, J., and Stehulova, I. (2002). Individual recognition of piglets by sows in the early post-partum period. *Behaviour* 139, 975–991.

Meurisse, M., Gonzalez, A., Delsol, G., Caba, M., Levy, F., and Poindron, P. (2005a). Estradiol receptor-alpha expression in hypothalamic and limbic regions of ewes is influenced by physiological state and maternal experience. *Horm. Behav.* 48, 34–43.

Meurisse, M., Perrin, G., Keller, M., and Lévy, F. (2005b). Neuroanatomical connections of the cortical and medial amygdala in sheep. In *7° Colloque de la Société des Neurosciences*. Lille, France.

Mouly, A. M., and Gervais, R. (2002). Polysynaptic potentiation at different levels of rat olfactory pathways following learning. *Learn. Mem.* 9, 66–75.

Mouly, A. M., Fort, A., Ben-Boutayab, N., and Gervais, R. (2001). Olfactory learning induces differential long-lasting changes in rat central olfactory pathways. *Neuroscience* 102, 11–21.

Numan, M., and Numan, M. (1996). A lesion and neuroanatomical tract-tracing analysis of the role of the bed nucleus of the stria terminalis in retrieval behavior and other aspects of maternal responsiveness in rats. *Dev. Psychobiol.* 29, 23–51.

Numan, M., and Insel, T. R. (2003). *The Neurobiology of Parental Behavior*. Springer-Verlag, New York.

O'Connor, C. E., Lawrence, A. B., and Wood-Gush, D. G. M. (1992). Influence of litter size and parity on maternal behaviour at parturition on Scottish Blackface sheep. *Appl. Anim. Behav. Sci.* 33, 345–355.

Perrin, G., Ferreira, G., Meurisse, M., Verdin, S., Mouly, A. M., and Levy, F. (2007a). Social recognition memory requires protein synthesis after reactivation. *Behav. Neurosci.* 121, 148–155.

Perrin, G., Meurisse, M., and Lévy, F. (2007b). Inactivation of the medial preoptic area or the bed nucleus of the stria terminalis differentially disrupts maternal behavior in sheep. *Horm. Behav.* 52, 461–73.

Petrulis, A., and Eichenbaum, H. (2003). The perirhinal–entorhinal cortex, but not the hippocampus, is critical for expression of individual recognition in the context of the Coolidge effect. *Neuroscience* 122, 599–607.

Petrulis, A., Peng, M., and Johnston, R. E. (2000). The role of the hippocampal system in social odor discrimination and scent-marking in female golden hamsters (*Mesocricetus auratus*). *Behav. Neurosci.* 114, 184–195.

Pissonnier, D., Theiry, J. C., Fabre-Nys, P., Poindron, P., and Keverne, E. B. (1985). The importance of olfactory bulb noradrenalin for maternal recognition in sheep. *Physiol. Behav.* 35, 361–363.

Poindron, P. (1976). Mother–young relationships in intact or anosmic ewes at the time of suckling. *Biol. Behav.* 2, 161–177.

Poindron, P., and Le Neindre, P. (1980). Endocrine and sensory regulation of maternal behavior in the ewe. In *Advances in the Study of Behavior* (J. S. Rosenblatt, R. A. Hinde, and C. Beer, Eds.), Vol. 11, pp. 75–119. Academic Press, New York.

Poindron, P., Le Neindre, P., Raksanyi, I., Trillat, G., and Orgeur, P. (1980). Importance of the characteristics of the young in the manifestation and establishment of maternal behaviour in sheep. *Reprod. Nutr. Dev.* 20, 817–826.

Poindron, P., Lévy, F., and Krehbiel, D. (1988). Genital, olfactory, and endocrine interactions in the development of maternal behaviour in the parturient ewe. *Psychoneuroendocrinology* 13, 99–125.

Poindron, P., Nowak, R., Lévy, F., Porter, R. H., and Schaal, B. (1993). Development of exclusive mother–young bonding in sheep and goats. *Oxford Rev. Reprod. Biol.* 15, 311–364.

Poindron, P., Lévy, F., and Keller, M. (2007). Maternal responsiveness and maternal selectivity in domestic sheep and goats: the two facets of maternal attachment. *Dev. Psychobiol.* 49, 54–70.

Porter, R. H., Lévy, F., Nowak, R., Orgeur, P., and Schaal, B. (1994). Lambs' individual odor signatures: Mosaic hypothesis. *Adv. Biosci.* 93, 233–238.

Putu, I. G., Poindron, P., Oldham, C. M., Gray, S. J., and Ballard, M. (1986). Lamb desertion in primiparous and multiparous Merino ewes induced to lamb with dexamethasone. *Proc. Aust. Soc. Anim. Prod.* 16, 315–318.

Romeyer, A., Poindron, P., Porter, R. H., Lévy, F., and Orgeur, P. (1994). Establishment of maternal bonding and its mediation by vaginocervical stimulation in goats. *Physiol. Behav.* 55, 395–400.

Sanchez-Andrade, G., James, B. M., and Kendrick, K. M. (2005). Neural encoding of olfactory recognition memory. *J. Reprod. Develop.* 51, 547–558.

Schettino, L. F., and Otto, T. (2001). Patterns of Fos expression in the amygdala and ventral perirhinal cortex induced by training in an olfactory fear conditioning paradigm. *Behav. Neurosci.* 115, 1257–1272.

Scott, C. J., Clarke, I. J., and Tilbrook, A. J. (2003). Neuronal inputs from the hypothalamus and brain stem to the medial preoptic area of the ram: Neurochemical correlates and comparison to the ewe. *Biol. Reprod.* 68, 1119–1133.

Smith, F. V., Van-Toller, C., and Boyes, T. (1966). The "critical period" in the attachment of lambs and ewes. *Anim. Behav.* 14, 120–125.

Tillet, Y., Batailler, M., and Thibault, J. (1993). Neuronal projections to the medial preoptic area of the sheep, with special reference to monoaminergic afferents: Immunohistochemical and retrograde tract tracing studies. *J. Comp. Neurol.* 330, 195–220.

Verbalis, J. G., Stricker, E. M., Robinson, A. G., and Hoffman, G. E. (1991). Cholecystokinin Activates c-fos expression in hypothalamic oxytocin and corticotropin-releasing hormone neurons. *J. Neuroendocrinol.* 3, 205–213.

Yu, G. Z., Kaba, H., Okutani, F., Takahashi, S., Higuchi, T., and Seto, K. (1996). The action of oxytocin originating in the hypothalamic paraventricular nucles on mitral and granule cells in the rat main olfactory bulb. *Nauroscience* 72, 1073–82.

3

MATERNAL MOTIVATION AND ITS NEURAL SUBSTRATE ACROSS THE POSTPARTUM PERIOD

MARIANA PEREIRA, KATHARINE M. SEIP AND JOAN I. MORRELL

Center for Molecular and Behavioral Neuroscience Rutgers, The State University of New Jersey, Newark Campus, New Jersey, USA

In rats and other mammals, the maternal condition is associated with a constellation of profound behavioral adaptations mediated by underlying physiological and neuroendocrine substrates that ultimately allow the postpartum female to effectively care for her developing offspring under a variety of challenging environmental conditions. Clearly the brain is readied to mediate this behavioral pattern by the physiological events of pregnancy and parturition, since the newly parturient female rat shows immediate maternal responsiveness on her first exposure to the novel stimulation of newborn pups. During the 3–4 weeks following parturition, the postpartum female rat is almost entirely dedicated to caring for her pups until they are weaned. She will spend most of her time caring for the litter (Grota & Ader, 1969, 1974; Leon *et al.*, 1978; Pereira *et al.*, 2007a), will protect and defend her pups by attacking intruder conspecifics that approach her nest and pups (Erskine *et al.*, 1978a, b; Ostermeyer, 1983; Ferreira & Hansen, 1986; Lonstein & Gammie, 2002) and will overcome stressful and risky situations in order to bring her in close proximity to pups (Nissen, 1930; Fleming & Luebke, 1981; Fahrbach & Pfaff, 1982; Hård & Hansen, 1985; Ferreira *et al.*, 1989; Bitran *et al.*, 1991; Pereira *et al.*, 2005; Lonstein & Morrell, 2007). This highly consuming and vitally important task of pup care requires an underlying process to assure maternal attraction toward and interest in pup-related stimuli continues across the postpartum period. Thus, the importance of maternal motivation as the initiator and sustainer of these care giving sequences is paramount.

The hormones associated with late pregnancy and parturition, including estrogen, progesterone, prolactin, and oxytocin (Bridges, 1990; Insel, 1990; Numan & Insel, 2003) act in specific temporal concert both peripherally and in discrete brain regions to initiate the immediate maternal responsiveness toward pups. Whereas these neuroendocrine events importantly coordinate its onset, they are not necessary to maintain maternal responsiveness across the postpartum period (Numan & Insel, 2003). Continued expression of maternal responsiveness instead relies on the continued somatosensory and chemosensory experiences acquired by the mother while interacting with her pups (Magnusson & Fleming, 1995). During this maintenance period, there is currently no evidence that the neuroendocrine events of postpartum are necessary to support the female's motivation to nurture her offspring.

Among the factors regulating the maternal responsiveness of the postpartum female (genetic, environmental, experiential), the particular needs and behavioral capacities of the developing pups are of major importance. We posit that the interaction of the female–pup dyad tailors the responses of the female to the needs of the developing pups such that the changing characteristics of the pups furnishes signals regulating the maternal state and the behavioral responses of the female. Thus, in particular we posit that the interaction of the female–pup dyad regulates maternal motivation to crucially tailor the range of her responses. Her care giving behaviors can then be adjusted with considerable plasticity to the behavioral

Neurobiology of the Parental Brain

capacities and physiological needs of the pups. In this chapter we present our current data on several aspects of maternal motivation as the postpartum progresses, with particular attention to how some of the pup–female interactions signal the tailoring of maternal motivation.

DEFINITIONS AND MEASURES OF MATERNAL MOTIVATION

Traditionally, motivated behaviors have been divided into appetitive and consummatory phases (Hinde, 1982). The appetitive phase consists of flexible seeking behaviors that bring the animal into close contact with a desired stimulus or goal, thus allowing the consummatory interaction with the stimulus; together they are a continuum that compose a complete behavioral sequence. Consummatory responses depend on the outcome of the appetitive phase, and generally consist of more rigid, reflexive behaviors that are performed once the goal is achieved. Consummatory behaviors directed toward the desired stimulus change the nature of the stimulus, alter the physiological status of the individual engaged in the consummatory behaviors, and typically decrease the incentive value of the stimulus for the individual. Subsequent appetitive behaviors are therefore affected by these consummatory responses. The two most common methods used to measure the appetitive components of motivated behaviors are classical operant conditioning procedures and conditioned place preference (CPP) (Lonstein & Morrell, 2007).

In the case of maternal behavior specifically, pup-seeking behaviors (Fleming *et al.*, 1994; Lee *et al.*, 1999; Mattson *et al.*, 2001, 2003) and certain active maternal behaviors (Hansen *et al.*, 1991a, b; Stern, 1996; Numan & Insel, 2003; Numan, 2007) have both been considered appetitive maternal responses. Initial responses to chemosensory and other cues from pups, and retrieval can be considered components of the appetitive phase of maternal behavior. Once pups are grouped, nursing behavior, often viewed as the major consummatory response and of more reflexive quality, can be initiated by the ventral somatosensory stimulation from pups (Hansen *et al.*, 1991a, b; Stern, 1996; Numan & Insel, 2003; Numan, 2007). We posit that it is formally possible to view even responses to chemosensory cues and pup retrieval as consummatory components of the behavioral sequence, since contact with these cues from a maternal female leads to a fixed action pattern following the concentration gradient, and finding of pups again stimulates a fixed action pattern of retrieval that is repeatedly executed. While both views are valid, it is our premise that use of CPP as a measure of motivated appetitive phase of the behavioral sequence is uniquely independent of any of the stimulus properties of the pups, which we view as the most initial phase of the appetitive component of the behavior.

The expression of maternal behavior relies on the reinforcing stimulation of the pups. That pups and their associated sensory stimuli are powerful reinforcers to the maternally behaving female (Hauser & Gandelman, 1985; Fleming *et al.*, 1989, 1994; Lee *et al.*, 1999; Mattson *et al.*, 2001) is known from both operant and CPP procedures. Postpartum females will bar press for hours, almost insatiably, in order to gain access to and retrieve pups (Wilsoncroft, 1969; Lee *et al.*, 1999) and will prefer to spend time in an environment that was previously paired with pups over the environment not paired with pups (Fleming *et al.*, 1994). This is in contrast to the non-maternal virgin female rat, which when first presented with newborn pups initially actively avoids them (Fleming & Luebke, 1981; Rosenblatt & Mayer, 1995). However, after continuous exposure to pups, nulliparous females can be induced into a maternal state and begin to exhibit maternal (Cosnier & Couturier 1966; Rosenblatt 1967) and associated affective behaviors (Agrati *et al.*, 2007; Ferreira *et al.*, 2002; Mayer and Rosenblatt, 1993; Pereira *et al.*, 2005) and to find pups reinforcing (Fleming *et al.*, 1994; Seip & Morrell, 2008a). This is further evidence of the power of the chemosensory and somatosensory stimuli provided by the pups.

It is useful to consider the differences in the appetitive-consummatory continuum that are measured by operant conditioning vs. CPP. In operant procedures, the female presses the bar to obtain pups, which are then invariably retrieved and subsequently nursed. Thus, every operant response after the initial pup delivery is affected by the previous maternal interaction with the pup. In CPP procedures, on the other hand, the appetitive phase of the motivated behavior is measured after conditioning in which pups are repeatedly paired with a unique cue-associated chamber, and then preference for this pup-associated chamber is subsequently tested in the absence of pups. Thus the CPP measures the initial appetitive responses prior to any aspect of interaction with the pups and provides a measure similar to a break point analysis in a bar-pressing paradigm, all prior to achieving the stimulus. Together these two methodological approaches provide mutually enriching data points. In both cases the stimulation provided by the pups has been shown to be important in driving motivation. If females are prevented from interacting with the pups in the operant chamber, subsequent bar-pressing behavior extinguishes almost immediately (Lee *et al.*, 1999). Females' exposure to important somatosensory and chemosensory stimuli associated with pups (Magnusson & Fleming, 1995) and the duration of this exposure (Fleming *et al.*, 1994) are factors necessary for the subject to establish pup-associated chamber preference.

PORTRAIT OF MATERNAL MOTIVATION DURING THE PROGRESSION OF THE POSTPARTUM PERIOD

Examination of early and late postpartum females reveals a fundamental change in level of motivation. Prior to our work, the majority of the studies on maternal motivation of females rodents were performed during the first postpartum week. Our work expands the analyses of maternal motivation by examining several time points across the postpartum period. Females in either early (postpartum days: PPDs 4–8) or late (PPDs 12–16) postpartum were conditioned to associate pups with a uniquely cued chamber, and to associated various non-pup stimuli with the opposite chamber of a CPP apparatus. Preference for a particular stimulus-associated chamber was determined, in the absence of the unconditioned stimuli; time spent in the stimulus-associated chamber is the measure of preference, that is, motivation for the stimulus. Chamber preference is defined by a conservative quantitative criterion that we have developed (Mattson *et al.*, 2001). Chamber preference of an individual can be identified if the subject spends at least 50% of the test time in that chamber, and that this is at least 25% more than in any other chamber. This criterion clarifies the motivational categorization of the subjects and avoids categorizing individuals based on very small differences in the time spend in one associated chamber over another. This phase of our work has used contemporaneously aged pups that are developmentally age-matched to the female by the stage of the postpartum period.

In the first use of CPP for pup-associated stimuli, female rats in early postpartum (up to day 8 postpartum) consistently preferred the pup-associated chamber over the alternative, neutral chamber (Fleming *et al.*, 1994). Our work has included the late postpartum period. We find that preference for the pup-associated chamber dramatically declines as the postpartum period progresses. Thus, early in the postpartum period, most females preferred the pup-associated chamber, whereas later in the postpartum period most females preferred the non-pup-associated chamber (Wansaw *et al.*, 2008).

Pup deprivation reveals additional features of postpartum motivation. Separation of pups from the maternal female is a commonly used procedure used to facilitate the testing of maternal behavior. Presumably, this mimics the natural situation when the female must leave the nest area to forage for sustenance. We suggest that the measure of pup-seeking in the absence of the pup stimulus determined with CPP is uniquely suited to measuring the motivation of the female returning to the nest voluntarily due to her physiology and/or CNS function. This initiation of pup-seeking is an important aspect of the behavioral sequence for motivated behavior, as without it, no further stage of appetitive or consummatory process would occur.

One way to look at the period of pup deprivation is that this tool may reveal the nature of maternal motivation in the postpartum female that is given a presumably common choice of continuing to forage in the environment vs. returning to an environment where the pup stimulus is reliably found, the maternal nest. Much of foraging consists of examining objects or environmental features that may be relatively neutral in reward value for the rat and can be represented by neutral objects or an empty neutral space as the alternative choice in a CPP paradigm.

To investigate maternal motivation throughout the postpartum period using this strategy, we separated different groups of early and late postpartum females from their pups for times ranging from 15 min to 22 h (Wansaw *et al.*, 2008) before place preference conditioning and testing. Pups and females were first reunited in the cued chamber in which its association to the pup was being conditioned. Females were also conditioned to associate a different chamber to neutral pup-sized object, from which they were deprived for the same time. Pups used for these conditioning phases were deprived of maternal care for a maximum of 2–3 h, so they experienced deprivation matched to time of maternal deprivation for points up to 2 h but not for the longer time points. As can be seen in Figure 3.1, early postpartum females expressed material pup-associated place preference regardless of their experience of pup deprivation prior to conditioning, but the picture was very different in the late postpartum female. The late postpartum females did not express a pup-associated place preference unless they were deprived of pups for 22 h prior to conditioning. These females then expressed pup-associated place preference that was as strong as in early postpartum (Wansaw *et al.*, 2008).

These data concur with the original place preference findings of Fleming *et al.* (1994), showing that a substantial majority of early postpartum females deprived of pups for 23 h before conditioning and testing preferred the pup-associated chamber. Our data also extend knowledge of maternal motivation to include information on the state of the late postpartum female, which requires longer deprivation periods from their litter to alter their motivational state such that they had a response equal to that of the early postpartum females. We posit that it is unlikely that this response is due to the relief of milk-engorged mammary glands experienced by the females with long pup deprivation upon being reunited with pups, since non-lactating maternal virgin females also express stronger pup-associated chamber preference when deprived of pups for 23 h (Fleming *et al.*, 1994).

Choice between pup- and cocaine-associated place reveals early postpartum motivational strength. The profound strength of maternal motivation in early postpartum became evident when pups were contrasted with the highly reinforcing but categorically different stimulus cocaine, a psychomotor stimulant with

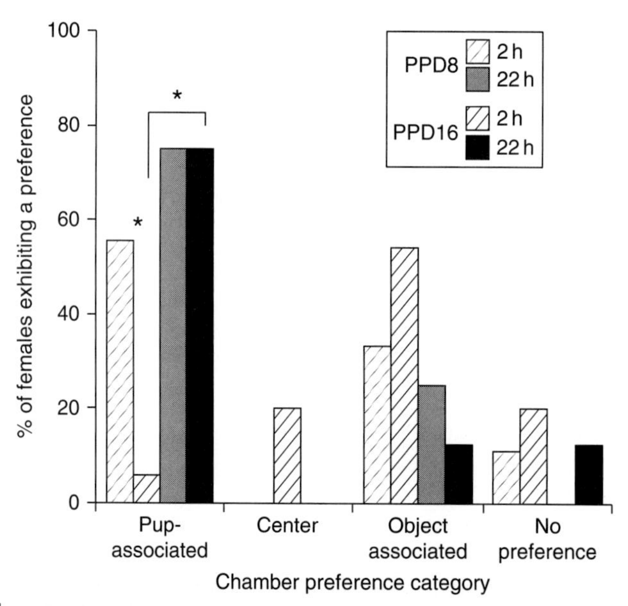

FIGURE 3.1 Percentage of early (grey bars) and late (black bars) postpartum dams in each preference category after 2 h (striped bars) and 22 h (solid bars) of pup deprivation prior to conditioning and testing. Postpartum females were conditioned to associate neutral context cues with either age-matched pups to the postpartum stage or pup-sized objects. Early postpartum females expressed strong pup-associated chamber preference regardless of the length of pup deprivation used. However, late postpartum females would not prefer the pup-associated chamber preference unless deprived of pups for 22 h, at which point most females expressed maternal motivation that was as strong as in early postpartum. $*p < 0.05$ designate the most relevant significant differences between groups in each figure.

well-documented reinforcing properties. Using a dual-choice CPP procedure, postpartum females were conditioned to associate pups with one chamber and cocaine with the alternative, non-pup chamber. Given the choice between cues associated with pups vs. those associated with cocaine, early postpartum dams (PPD 8) continued to express strong preference for the pup-associated chamber over moderately rewarding cocaine (Mattson *et al.*, 2001) and maintained notable pup-associated chamber preference even when cocaine's reward value was maximized (Seip & Morrell, 2007b). In comparison, almost all late postpartum females (PPD 16) preferred the cocaine-associated chamber (Mattson *et al.*, 2001; Seip & Morrell, 2007b) (Figure 3.2).

These striking postpartum differences in maternal motivation cannot be attributed to variations in early and late postpartum females' responsivity to cocaine. When contrasted with saline-associated chambers, preference for cocaine-associated chambers remains remarkably similar at these two postpartum times across a range of doses and administration routes (Seip *et al.*, 2008a). Moreover, plasma levels of cocaine and its metabolites, as well as their respective time courses, do not change substantially across postpartum, nor do the effects of cocaine on locomotor activity (Vernotica & Morrell, 1998; Wansaw *et al.*, 2005).

The picture that emerges from these data sets is that the highest levels of maternal motivation are found in the early postpartum period, with diminished levels in the late postpartum period. However, the surprising stimulation of pup-associated preference in the late postpartum females after lengthy separation from her pups makes very clear that, although the level of maternal motivation is changing as postpartum progresses, it remains a powerful force that can be called upon by the environmental situation the female and her offspring meet. Together these experiments suggest that a variety of changes in the postpartum period might be at work to adjust the motivation of the female and the nature of the responses she provides from her care giving repertoire. The experiments in the next two sections ask questions about how the expression of care-giving fits with our CPP

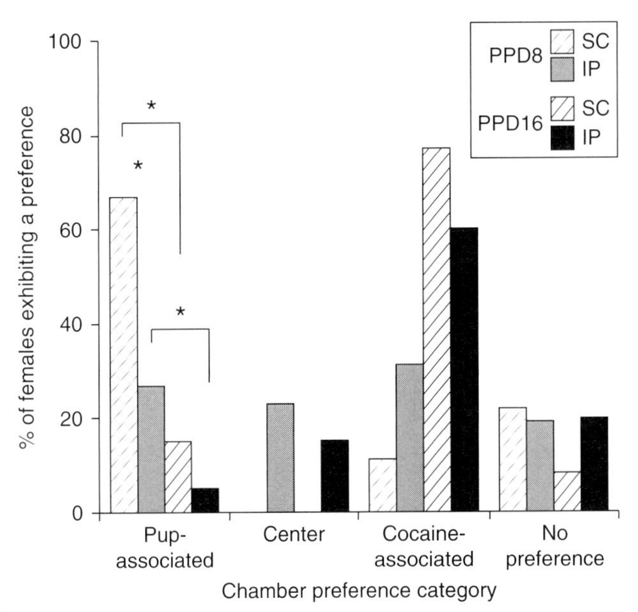

FIGURE 3.2 Percentage of early (grey bars) and late (black bars) postpartum females in each preference category after place preference conditioning sessions with age-matched pups to the postpartum stage and either 10.0 mg/kg SC (striped bars) or IP (solid bars) cocaine. Given the choice between cues associated with pups vs. those associated with cocaine, early postpartum dams continued to express strong preference for the pup-associated chamber over moderately rewarding cocaine (10.0 mg/kg SC) and maintained notable pup-associated chamber preference even when cocaine's reward value was maximized (10.0 mg/kg IP). In comparison, almost all late postpartum females preferred the cocaine-associated chamber.

data on motivational changes, and whether the neuroendocrine tone set by prolactin or the neurotransmitter tone in the dopamine (DA) system of the female might underlie the altered maternal motivation and expression of pup-care behaviors found as the postpartum period progresses.

COORDINATING CHANGES IN MATERNAL MOTIVATION AND BEHAVIOR ACROSS THE POSTPARTUM PERIOD

A notable reduction in the expression of maternal (Grota & Ader, 1969, 1974; Reisbick *et al.*, 1975; Rosenblatt, 1975; Pereira *et al.*, 2007a, b) and associated affective behaviors, such as maternal aggression (Flannelly & Flannelly, 1987; Mayer *et al.*, 1987; Giovenardi *et al.*, 2000) and anxiety-like behaviors (Neumann, 2001; Lonstein, 2005) occurs beginning around PPD 12. In the laboratory, the lactating female and her litter are usually confined in a single cage and hence measures of maternal behavior are generally recorded in the inescapable presence of the pups. Particularly in the case of developing pups during the 10 days prior to weaning, the female cannot escape her pup's demands for contact and nursing. Prior to our ongoing work, only one study (Grota & Ader, 1969, 1974) examined expression of maternal behavior using a procedure that provides the female an alternative pup-free environment, allowing the female to choose and to spend only voluntary time in pup-care activities. We have used a two- or four-chambered home cage designed such that the female and her pups have access to a chamber with a secure nest box, while only the female has additional access to alternative pup-free chambers. Using this apparatus we, similar to Grota & Ader (1969, 1974), found that early postpartum dams voluntarily spend approximately 80% of their time with the pups (Pereira *et al.*, 2007a). Furthermore, the majority of this time is consumed by care-taking behaviors (Grota & Ader 1969, 1974; Leon *et al.*, 1978; Pereira *et al.*, 2007a) (Figure 3.3). As postpartum progresses, the females reduce their time in contact with pups to around 50–60%. Thus, the pups are alone more frequently and for longer durations; females also decrease their care-giving behaviors when with the pups (Grota & Ader 1969, 1974; Pereira *et al.*, 2007a).

Less voluntary time with the pups, however, does not mean that the pups have freedom to roam unattended by the mother. We and others have found that females actively barricade the pups into the nest with bedding material, preventing the pups from leaving the nest in her absence, until very late in the postpartum period (Leon *et al.*, 1978; Pereira *et al.*, 2007a). The female stocks the nest with appropriate food for initial sampling by pups. Much of the initial exploration of the environment outside the maternal nest area is always in the presence of the mother which guides the food and exploration choices (Galef & Clark, 1972).

An ongoing experiment provides a further example of the behavioral plasticity of postpartum females. Subjects were housed with pups in a four-chambered apparatus, with two enclosed (dark) nest boxes in two of the chambers, designed such that only females could move from chamber to chamber. Typically

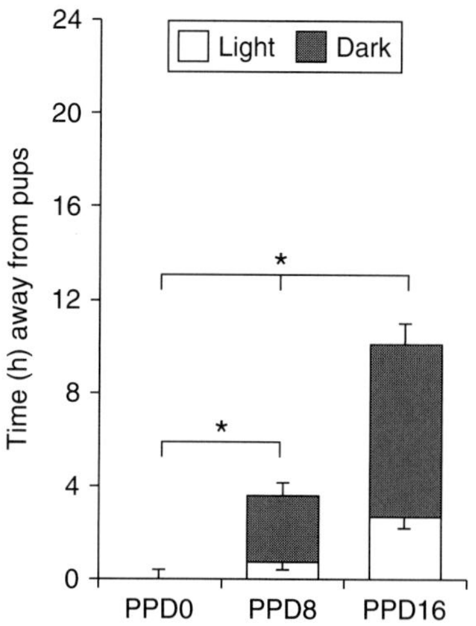

FIGURE 3.3 Use of a two-chambered maternal behavior allowed measure of the voluntary time (Median ± median absolute deviation (MAD)) that postpartum females are away from pups across PPDs 0, 8, and 16 during the light (white bars) and dark (black bars) phases of the 12-h light/dark cycle. Early postpartum females spent more than 80% of their time in close contact with the pups; the expression of maternal behavior was high, with much nursing. The time voluntarily spent with pups progressively decreased and by PPD 12–16 females spent less than 57% in contact. This was accompanied by a significant decrease in maternal responses, such as pup retrieval, pup licking, nest building, and time spent nursing.

one nest box was chosen by the female for pup-nest. As mentioned, maternal responses declined across postpartum. However, when raising large litters, females in the late postpartum period moved some pups to the second nest box. Thus, the late postpartum female will readily move up to 300 g in pup weight (3 times the mass of an entire early litter) over several meters, which represents a considerable work and can be seen as a measure of motivation. Thus, motivation to care for the pups at the later postpartum stage could be seen as not in fact diminished, but different and likely driving a tailored component of the pup-care activities. Older pups survive the absence of the mother due to their developmental progress and the protective measures of the female, appropriate to their age and needs.

POSSIBLE FACTORS UNDERLYING CHANGES IN MATERNAL STATE AS THE POSTPARTUM PERIOD PROGRESSES

The Nature of the Pup Stimulus: We posit that the changing behavioral capacities and physiological demands of the pups as they develop are significant determinants of the postpartum female's evolving maternal motivation and care-giving. Up to this point in our use of the place preference conditioning, all the females had home cage litters and were conditioned with pups that were age-matched to their postpartum stage. A key piece of data supporting our supposition concerning pup development and maternal motivation was found when we provided late postpartum females with young (4–7 days old) for conditioning, instead of the 12–15 day-old pups used in the prior experiments (above). Conditioning with younger pups resulted in an increased number of late postpartum females that preferred pup-associated chambers vs. those conditioned with pups age-matched to their PPD (Wansaw *et al.*, 2008) (Figure 3.4). We also found that these younger pups elicit early postpartum maternal-like care-giving behaviors from late postpartum females, and that this continues for more days in the late postpartum period than when the females care for older pups (Figure 3.5). Thus, late postpartum females retrieved and grouped 4–7 day-old pups, expressed more licking and nest building behaviors, and adopted nursing

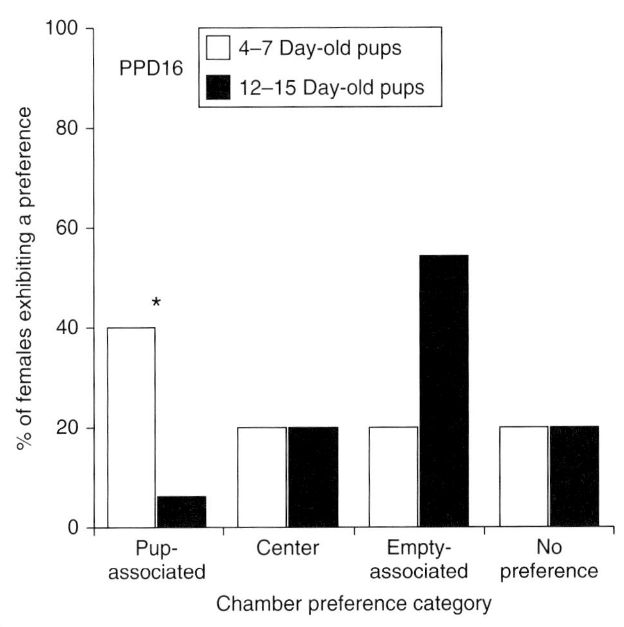

FIGURE 3.4 Percentage of late postpartum females in each preference category after place preference conditioning sessions with either 4–7 (white bars) or 12–15 (black bars) day-old pups vs. empty chamber. Late postpartum females expressed strong pup-associated place preference after conditioning with 4–7 day-old pups instead of pups age-matched to their PPD.

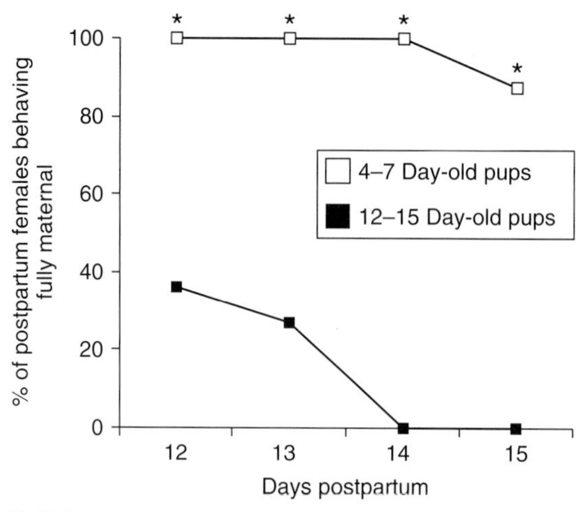

FIGURE 3.5 Percentage of postpartum females behaving fully maternally (retrieving and grouping all pups into the nest and adopting a nursing position over pups during the 30-min maternal behavior test) over PPDs 12–15 toward 4–7 day-old pups (white squares) compared to 12–15 day-old pups (black squares). Younger pups greatly increased % of fully maternal late postpartum females. Thus, late postpartum females readily retrieved and grouped 4–7 day-old pups, expressed more licking and nest building behaviors, and adopted nursing postures faster and for longer durations compared to 12–15 day-old pups.

postures faster and for longer durations, all similar to those levels found in early postpartum females with pups of similar age (Pereira *et al.*, 2007b).

Our data accord well with prior findings that the normal decline in maternal behavior (Reisbick *et al.*, 1975; Mayer & Rosenblatt, 1980, 1984) and associated affective behavioral changes (Flannelly & Flannelly, 1987; Lonstein, 2005) are considerably delayed if late postpartum females are maintained with 1–8 day-old pups (Reisbick *et al.*, 1975; Giovenardi *et al.*, 2000). Our data extends these findings to demonstrate that there is, in addition to diminished maternal behavior, a diminished initial appetitive response, suggesting that there is indeed a significant appropriate linking of changing appetitive state and changing pup-directed maternal response. The capacity for plasticity in the behavioral repertoire of postpartum dams to adapt their maternal responses according to the particular characteristics and demands of their pups is similarly evident when she is raising overlapping litters. If a female rat conceives during the postpartum estrus, approximately 6–15 h following parturition (Connor & Davis, 1980a, b; Gilbert *et al.*, 1980), she will gestate a second litter while simultaneously nurturing her recently delivered one. It is likely that, following the birth of the new litter, older offspring remain and share the nest with their younger siblings, in which case postpartum females will preferentially care for the junior litter (Gilbert *et al.*, 1983; Uriarte *et al.*, 2007).

Rodents bear extremely altricial young that, undergo rapid and dramatic developmental changes (Rosenblatt *et al.*, 1985) over the course of the postpartum period until weaning. This development encompasses changes in the nature of the sensory, motoric, thermoregulatory, and behavioral capacity of the pups (Rosenblatt *et al.*, 1985) and includes marked increases in pup body weight and milk demand (Stern & Keer, 2002). It is important to consider how the changing characteristics of the developing pups "inform" the mother about the pups' particular needs for care. Future experiments can clarify which key features of the stimulation provided by the pup to the female change across their natural course of development to influence the exact nature of a female's maternal motivation and care giving responses.

Does Prolactin Support the Ongoing Maternal State, Particularly Maternal Motivation? One particularly salient example of how changes in the needs of the developing pups mediate maternal care-giving is evident in nursing behavior. Across development, changes in the duration or intensity of pups' suckling stimulation (i.e., changes in the ingestive motivation of

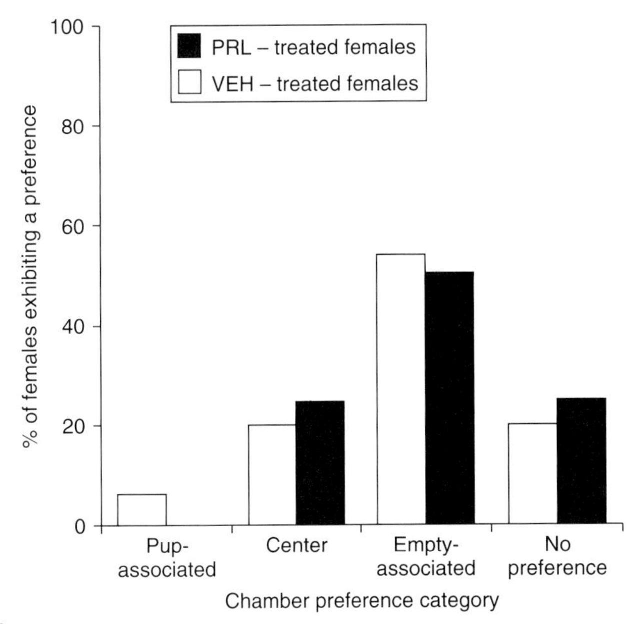

FIGURE 3.6 Percentage of prolactin-treated late postpartum females in each preference category after place preference conditioning sessions with age-matched pups to the postpartum stage and empty chamber. Enhanced levels of circulating prolactin did not promoted pup-associated chamber preference in late postpartum dams.

the pups themselves) occur and these changes are thought to importantly determine the patterns of prolactin secretion and of nursing (Taya & Greenwald, 1982a, b; Stern & Keer, 2002). In the first 10 days after parturition, the total duration and number of nursing bouts, and the circulating prolactin levels are considerably higher than in the second half of the postpartum period (Smith *et al.*, 1975; Taya & Sasamoto, 1981; Taya & Greenwald, 1982a, b; Mattheij *et al.*, 1985; Freeman, 1988). This profile of higher plasma prolactin in early postpartum, lower in late postpartum correlates with the profile of higher numbers of early postpartum females with a pup-associated place preference, and much lower numbers of late postpartum females with pup preference. Further, plasma prolactin levels and nursing behavior are also influenced by the age of the pups (Giovenardi *et al.*, 2000), the number of pups suckling (Taya & Greenwald, 1982b), and the hunger of the pups (Stern & Keer, 2002).

In the context of these data, it is intriguing to consider the extent to which prolactin levels attributable to the ventral stimulation received from pups coincide with changes in maternal motivation strength in early and late postpartum females. Remarkably, prolactin release can be stimulated by such ventral stimulation even in the absence of ability to nurse due to surgical intervention or in the model of a virgin with pup-induced maternal behavior (Marinari & Moltz, 1978; Woodside & Popeski, 1999). Thus, it is possible that pup-induced changes in circulating prolactin levels could be informative of the pups' developmental characteristics and consequently regulate maternal motivation and behavior.

We posited that maintaining a higher level of circulating prolactin in late postpartum would increase the number of females with pup-associated place preference in this period. This possibility was examined by administering ovine prolactin (0.5 mg/0.2 ml/rat, sc) to late postpartum females twice daily between PPDs 9 and 15. Our experiment is based on the paradigm used in Bridges *et al.* (1985). In late postpartum, our data show that enhanced levels of circulating prolactin did not promote pup-associated chamber preference (Figure 3.6; Pereira *et al.*, 2007b). It is worth noting that prolactin-treated mothers spent more time nursing the litter (Figure 3.7), and this was reflected in a substantial increase in pup weight gain

(7% more, a remarkable 20 g total, than pups raised by vehicle-treated females, $p < 0.05$), implying that better milk production was a peripheral effect of the treatment. These females also engaged in more corporal and anogenital pup licking than control females (Figure 3.7).

While pup place preference was not stimulated with our peripheral treatment, it remains an open question as to whether brain region specific or if stimulation by prolactin together with other agents might demonstrate specificity in the role of prolactin in maternal motivation. Our experiment deals exclusively with the challenge of reinstating chamber preference for pups in the late postpartum female. So, it also remains an open question as to whether initiation of maternal motivation just prior to the onset of maternal expression might be supported by prolactin, in a manner analogous to the findings demonstrating the facilitatory effect of prolactin in the onset expression of maternal behavior (Moltz *et al.*, 1970; Bridges *et al.*, 1985).

Does DA support the maternal state, particularly maternal motivation? The mesolimbic DA system is critically involved in the regulation of most motivated behaviors (Robbins & Everitt, 1996; Berridge & Robinson, 1998; Ikemoto & Panksepp, 1999; Salamone *et al.*, 2007). A number of studies have manipulated the mesolimbic DA system and demonstrated its role in the behavioral repertoire of the maternal state including maternal motivation. For example, disrupting DA transmission during conditioning attenuates pup-associated chamber preference (Fleming *et al.*, 1994) and selectively disrupts most forms of maternal behaviors (retrieval and grouping of the pups at the nest site, pup licking, and nest building), while leaving nursing behavior relatively intact in early postpartum females (Gaffori & Le Moal, 1979; Giordano *et al.*, 1990; Hansen *et al.*, 1991a, b; Stern & Taylor, 1991; Keer & Stern, 1999; Stern & Keer, 1999; Ferreira *et al.*, 2000; Silva *et al.*, 2001, 2003; Byrnes *et al.*, 2002; Numan *et al.*, 2005a, b). Interestingly, these maternal deficits occur only if mothers have recently interacted with pups but not if they have experienced a period of separation from them (Hansen, 1994; Keer & Stern, 1999), suggesting that blocking

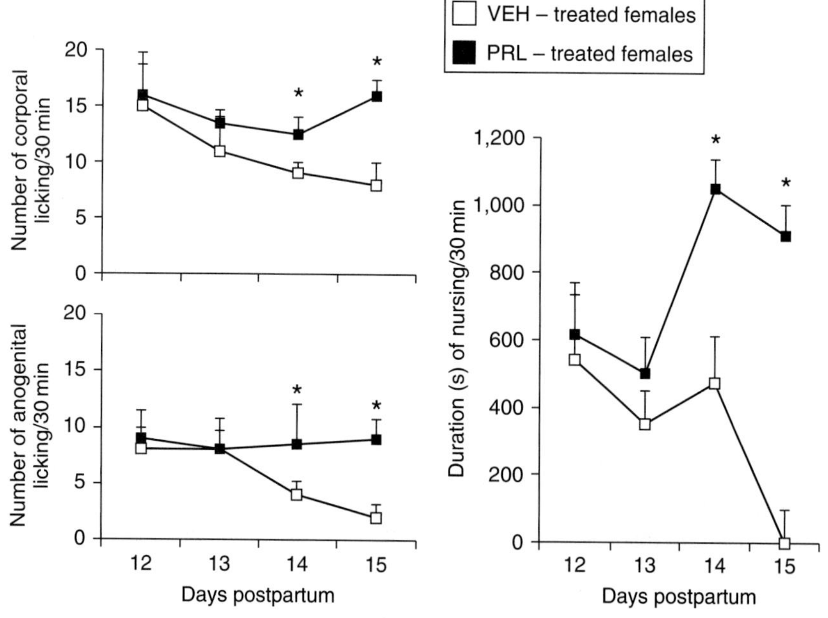

FIGURE 3.7 Number of pup licks and duration of nursing behavior in a 30-min maternal behavior test of late postpartum females treated with prolactin (black squares) over PPDs 12–15 compared to vehicle-(white squares) treated females. In the last 2 days of treatment, prolactin-treated postpartum females exhibited increase corporal and anogenital pup licking and spent more time nursing the litters than control postpartum females. Data is expressed as medians ± MAD.

brain DA primarily affects motivation to care the pups and not the motor mechanisms of caring. Also, that these behavioral impairments reflect motivational deficits is reflected in similar effects of DA antagonists on other types of "approach" behaviors. For example, if postpartum rats are muzzled so they cannot retrieve pups, they will substitute for retrieval considerable time nudging and pushing at the pups with their snouts and handling them with their paws (Stern & Keer, 1999). Blocking mesolimbic DA transmission also severely reduced those compensatory behaviors of pushing pups with the snout and moving them with the paws (Stern & Keer, 1999). It has been posited that the mesolimbic DA system modulate maternal motivation via interconnections within the maternal circuit, especially with the medial preoptic area (MPOA) and the ventral bed nucleus of the stria terminalis (vBNST) (Numan, 2006, 2007).

The importance of the dopaminergic activity for maternal state has been further examined by measuring DA response when pups are removed from and then reunited with their mothers after varying lengths of pup deprivation. Hansen originally showed that reunion with pups after a period of separation increases extracellular DA concentrations in the nucleus accumbens (NA) (Hansen *et al.*, 1993) and active maternal responses (Hansen, 1994). Moreover, reunion with pups after a 3–12 h separation period restores, in a manner that is dependent on length of mother–litter separation, the behavioral deficits in pup-caring activities observed in postpartum females treated with low subcateleptic doses of DA antagonists (Stern & Taylor, 1991; Keer & Stern, 1999) or following 6-OH-DA lesions in the ventral tegmental area or NA (Hansen *et al.*, 1991a, b; Hansen, 1994).

Some investigators have considered the physiological condition of the pups as an independent variable. While early postpartum females normally care avidly for pups on being reunited with them after a brief separation period, this response can be further increased with use of demanding pups, pups that have undergone an extended period of maternal separation (Pereira & Ferreira, 2006). It is worth mentioning that in this case, postpartum females were with a group of pups, separated from them for only a brief 15-min period, and then provided with demanding pups. This procedure reveals that demanding pups overrode the maternal behavior deficits caused by treatment with subcataleptic doses of DA antagonists (Pereira & Ferreira, 2006), restoring these behaviors to levels exhibited by control lactating females. Based on these results, Pereira & Ferreira (2006) posit that demanding pups elicited more DA release, which counteracted the effects of DA antagonist. Thus, the motivational power of the pup's condition, separate from the motivational state of the female, suggests that both components of the dyad have the capacity to adjust the motivational valence of the interaction and that dopaminergic tone is key for both components.

In the context of all the data suggesting that DA has a significant role in maternal state, we posited two things. First, perhaps the progressive decline in pup-associated place preference may be due to a progressive decrease in dopaminergic tone in response to the developing characteristics of the pups. Secondly, we posit that maternal motivation in late postpartum females should improve with increased dopaminergic activity. To test this hypothesis, late postpartum females received systemic injections of the selective D1 DA receptor agonist SKF38393, twice a day from PPDs 9–15. The pup-associated place preference and the maternal expression of these females were taken as measures of the maternal state of the females. These females were deprived of pups for only 2 h before CPP conditioning and testing.

The vast majority of these late postpartum SKF-treated females showed increased expression of maternal behaviors and continued to fully express all components of maternal behavior over 11–16 days postpartum, whereas vehicle-treated females expressed a marked decline in maternal behavior (Figure 3.8) (Pereira *et al.*, 2007b). An additional novel finding was that a significant subset of SKF-treated females preferred the pup-associated chamber over the alternative, empty chamber, whereas vehicle-treated females, as expected from our prior work, substantially preferred the alternative option (Figure 3.9) (Pereira *et al.*, 2007b).

These results accord with and extend the findings of others in early postpartum females showing that the onset of the expression of maternal behavior in hormonally primed females is facilitated by D1 receptor stimulation (Stolzenberg *et al.*, 2007). Our studies demonstrate that dopaminergic action on D1 receptors is necessary to maintain maternal motivation to

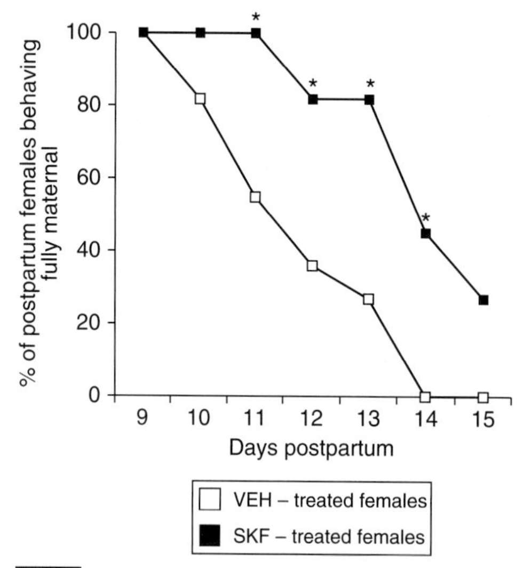

FIGURE 3.8 Percentage of SKF-treated postpartum females behaving fully maternally (retrieving and grouping all pups into the nest and adopting a nursing position over pups during the 30-min maternal behavior test) over days 9–15 postpartum compared to vehicle. D1 DA receptor stimulation greatly increased % of fully maternal late postpartum females over PPDs 11–15 compared with vehicle-treated females which had a marked natural decline in all maternal behaviors.

seek and care for the pups in the second half of the postpartum period. These findings also suggest that increases in dopaminergic response after a prolonged period of deprivation of the maternal female from her pups before conditioning and testing may be an important underpinning of the robust pup place preference responses of early and late postpartum females deprived of pups for long periods (Fleming *et al.*, 1994; Wansaw *et al.*, 2008).

Collectively these data suggest that offspring provide mothers with stimuli that promote their own care and nurturance as a function of their physiological needs, including promoting the very initial stages of the appetitive components of maternal behavior. Further, it may be that a decreasing capacity of pups' incentive properties diminishes their impact on the dopaminergic system and contributes to waning initial appetitive responses in late postpartum. However, the system remains plastic in that stimulus properties such as pup age and deprivation time from pups or pup neediness can sufficiently stimulate even the late postpartum system, such that these pups can stimulate the initial appetitive responses and be fully cared for by behaviors appropriate to their developmental stage.

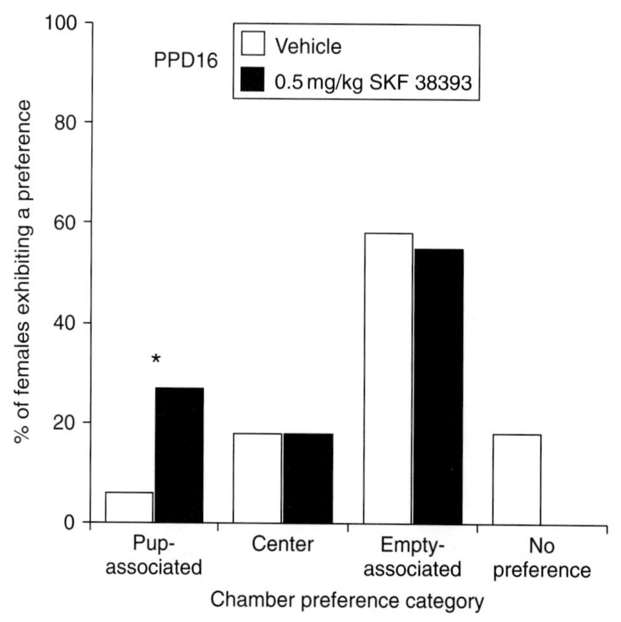

FIGURE 3.9 Percentage of SKF-treated late postpartum females in each preference category after place preference conditioning sessions with pups age-matched to the postpartum stage vs. empty chamber. A significant subset of SKF-treated females preferred the pup-associated chamber over the alternative, empty chamber, whereas vehicle-treated females substantially preferred the empty chamber.

NEURAL SUBSTRATE OF MATERNAL MOTIVATION

We (Mattson & Morrell, 2005) recently identified subsets of neurons within specific neuroanatomical regions that are activated when females on PPD 10 express preference for chambers associated with pups or with cocaine. In the brains of these postpartum females, neurons activated by preference for each specific stimulus-associated chamber were identified using immunocytochemical procedures to visualize c-Fos, the protein product of an immediate early gene associated with neuronal activation, and to cocaine- and amphetamine-related transcript (CART). As preference testing occurred in the absence of the stimuli themselves, preference-associated neuronal activation could be directly attributed to the female's underlying motivation rather than a stimulus-response contingency.

More c-Fos-immunoreactive (c-Fos-ir) neurons were identified in the medial prefrontal cortex (mPFC), nucleus accumbens (NA), and basolateral nucleus of amygdala (blAMYG) of females that preferred the cocaine-associated chamber than those preferring the pup-associated chamber or no chamber (controls). Females preferring the pup-associated chamber also exhibited substantial activation in these regions, though this activation was relatively less than that of cocaine-preferring females. While these regions were also associated with elevated CART-ir in cocaine-preferring females, the NA was the exclusive site of these three areas in which CART-ir was elevated in females preferring the pup-associated chamber. All of these regions are key components within a general activating circuitry supporting motivation and goal-directed behavior.

Outside of these three brain regions, only the MPOA contained more c-Fos-ir and CART-ir neurons in females preferring the pup-associated chamber compared to females preferring the cocaine-associated chamber. These activated neurons comprised a novel component of a circuit that may underpin motivated, goal-directed behavior directed specifically at biologically relevant, natural stimuli. While motivation directed at either pups or cocaine may result from the concerted activity of a hierarchically organized and regionally distributed network of neurons in mPFC, NA, and blAMYG, activity within regions such as MPOA may participate importantly in preference responses directed at specific stimuli. It is also likely that this specificity of preference responses may be attributed to the activation of particular subsets of neurons (Carelli, 2002) within each region showing distinct patterns of increased activation associated with contrasting stimulus-associated chamber preference. It is likely that these regions associated with preference-specific responses are initially activated by the mPFC, a higher-order cortical region associated with executive decision-making and judgment based on relative contrast. The mPFC may then activate components of the general motivational circuitry and specific stimulus-directed circuitry (e.g., MPOA). This feed-forward process provides a likely mechanism by which the "set point" or excitability of these downstream, non-cortical regions might be modulated to enable stimulus-specific motivation. In the unique case of maternal motivation, this feed-forward process is likely modulated by endocrinological and environmental factors of the maternally responsive female rat.

The MPOA is one likely site of this important integration. Lesions to the MPOA nearly eliminate strong bar-pressing for pups in an operant procedure (Lee *et al.*, 1999), whereas MPOA stimulation promotes pup-associated chamber preference in maternally responsive females (Morgan *et al.*, 1999). Together with the increased levels of neuronal activation associated with pup-associated chamber preference (Mattson & Morrell 2005), the MPOA comprises an integral component of the neural circuitry supporting maternal motivation. In a further analysis we have examined the pattern of activated neurons in the MPOA, using immunocytochemistry for c-Fos in early and late postpartum females that are conditioned to associate a cued chamber with either pups or with no stimulus. Brains of females that preferred the chamber associated with pups were compared to controls, which were parturient females exposed to the CPP procedure but that were exclusively exposed to chambers with no stimuli for association conditioning.

We performed a detailed anatomical analysis of activated neurons in the MPOA and (vBNST), in these early or late postpartum females preferring a pup-associated chamber (Morrell *et al.*, 2006; Seip *et al.*, 2007b). In early postpartum females, more activated neurons were found in the caudal portion of the MPOA (cMPOA) and vBNST of pup-preferring females compared to their controls

(Morrell *et al.*, 2006; Seip *et al.*, 2007b) (Figure 3.10). Our preliminary analysis suggests that the same trend was not observed in pup-preferring females in late postpartum, which did not differ from controls. In addition it appears that the overall number of neurons activated in both females with a conditioned pup place preference and controls from the late postpartum period is globally reduced compared to early postpartum females regardless of their pup preference condition. Possibly the activation of the same neural substrates change globally across the postpartum period, as well as with respect to the MPOA/vBNST

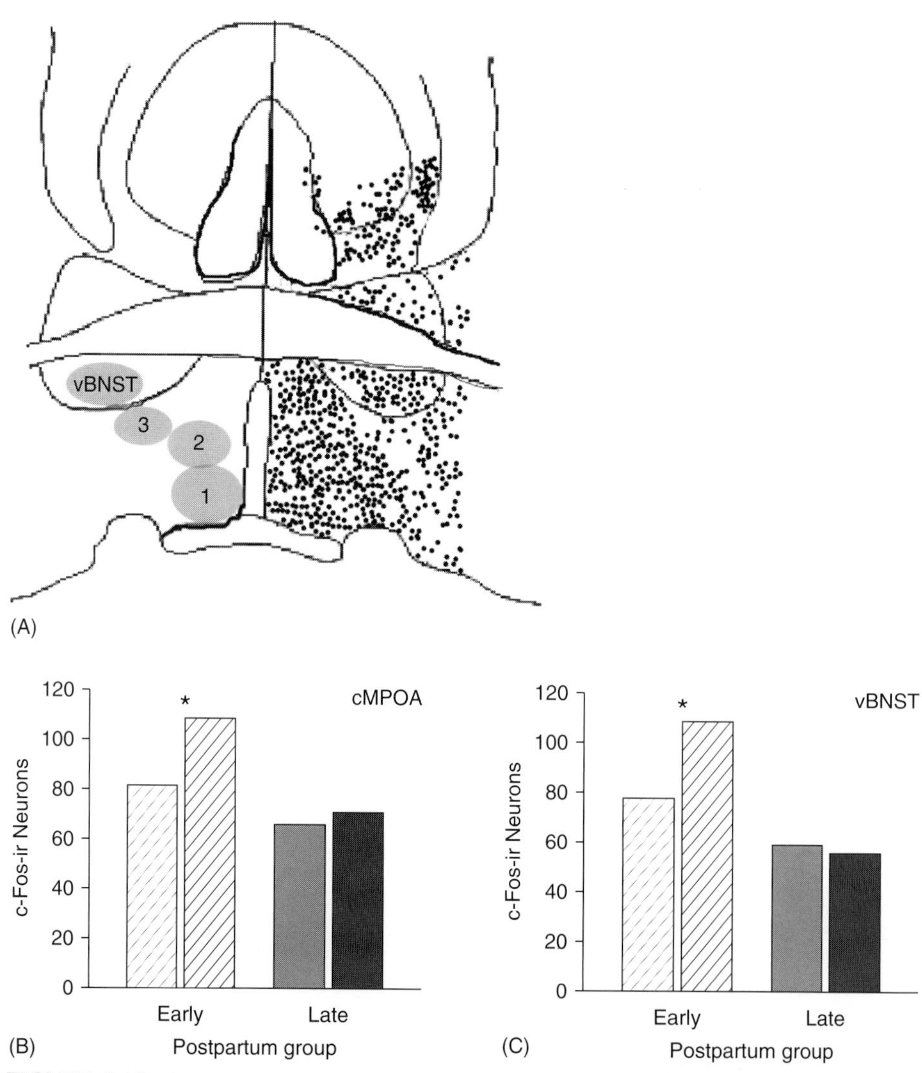

FIGURE 3.10 (A) Drawing is a representation of cross sections including the MPOA/BNST of female rats during early postpartum. Each black dot represents one c-Fos-ir neuron. Previous analyses of c-Fos-ir neurons associated with the performance of maternal behavior (Numan & Insel, 2003) are delineated by grey circles; circles 1–3 are components of the MPOA while the fourth circle, labeled vBNST, contains a representative component of the vBNST, defined dorsally by the anterior commissure. (B–C) Mean numbers of c-Fos-ir neurons identified in pup-preferring (black bars) and control (grey bars) females during early postpartum (stripes) and late postpartum (solid) in the caudal MPOA (cMPOA; B) and ventral BNST (vBNST; C). Asterisks designate significant differences between preference groups, where $p < 0.05$.

regions which might mediate maternal motivation more particularly. This is an open question requiring further experimentation.

Subregion-specific analyses were subsequently performed on brain sections taken from early postpartum females to identify whether discrete subnuclei within either MPOA or BNST played a particularly meaningful role in maternal motivation of early postpartum females. While a rostrocaudal difference in activated neurons of the MPOA neurons was confirmed, c-Fos-ir neurons did not appear to be localized in discrete subnuclei (Seip *et al.*, 2007b) (Figure 3.11), suggesting that the integrity of the entire MPOA and BNST may be necessary to support maternal motivation.

As MPOA neurons activated by the expression of maternal behavior probably synthesize and release gamma-aminobutyric acid (GABA) (Lonstein & de Vries, 2000), we explored whether motivationally activated neurons in the MPOA may similarly synthesize and release GABA. Given the technical limitations of available antibodies to GABA and its precursor enzyme, glutamic acid decarboxylase, we measured the extent to which maternal

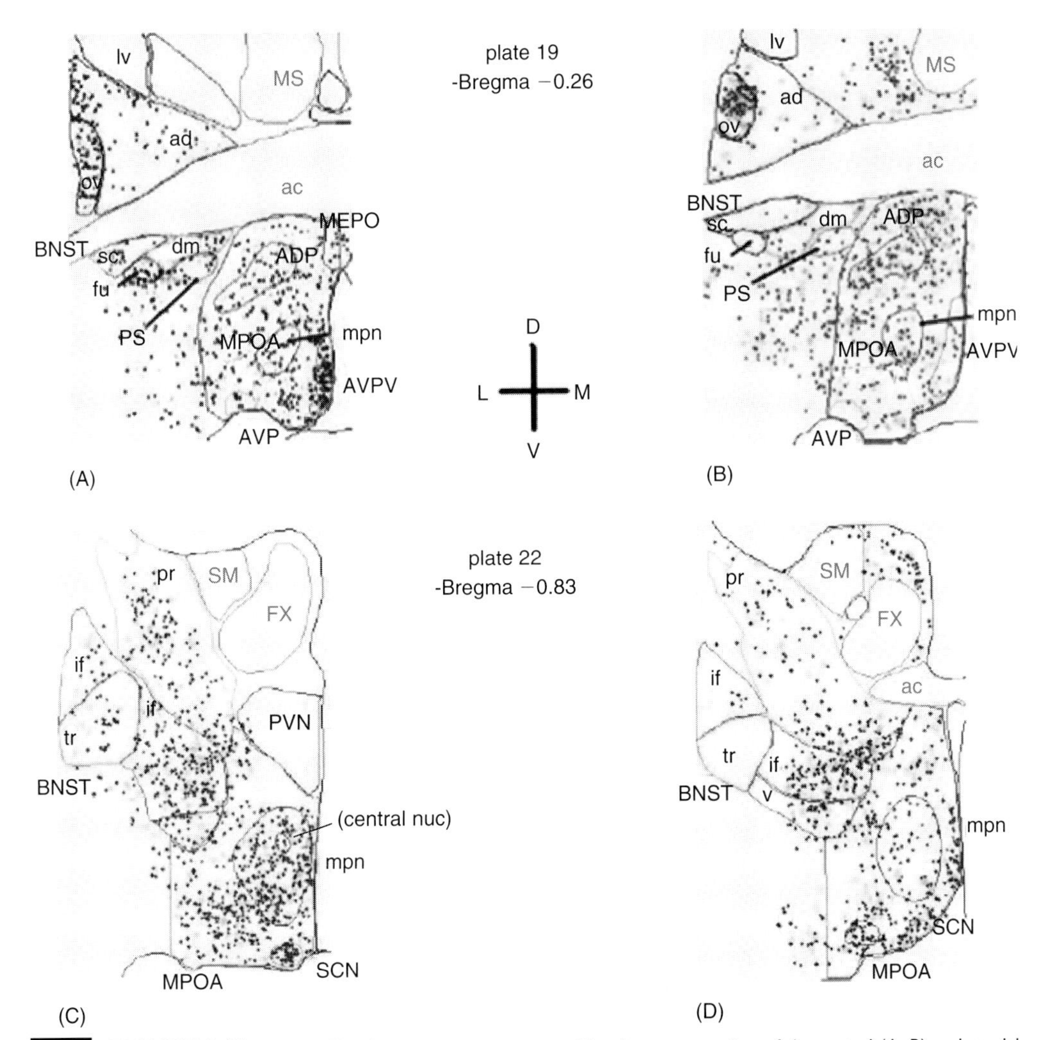

FIGURE 3.11 (A–B) The drawings represent one side of a cross section of the rostral (A–B) and caudal (C–D) portions of the MPOA/BNST of the female rat during early postpartum. More c-Fos-ir neurons (black dots) were found in females that preferred the pup-associated chamber (A, C) vs. controls which preferred an empty chamber (B, D). Subnuclei are abbreviated and plates are defined as in Swanson (1992).

motivation-activated (c-Fos-ir) neurons in pup-preferring females expressed various calcium-(Ca^{2+})-binding proteins, including parvalbumin, calretinin, and calbindin (Seip *et al.*, 2007b). Despite the well-documented colocalization of GABA with parvalbumin and even calretinin, very few parvalbumin-ir and calretinin-ir cells were found anywhere in the hypothalamus; while a moderate number of calbindin-ir cells were identified in MPOA/vBNST, this calbindin-ir population did not overlap substantially with the c-Fos-ir population (Seip *et al.*, 2007b). As calbindin is known to colocalize in both glutamatergic and GABAergic neurons, depending on the discrete brain region, it would have been difficult to infer the neurochemical identity of c-Fos-ir MPOA/vBNST neurons, even if there had been substantial overlap. This question remains unanswered.

Overall Summary: In this chapter we present our current data on the nature of maternal motivation as the postpartum progresses, with particular attention to building a portrait that includes the late postpartum period. Using CPP as a measure of the initial phases of the appetitive responses of postpartum females, we have demonstrated that preference for the pup-associated chamber dramatically declines as the postpartum period progresses if females are expressing preference for pups age-matched to their postpartum stage, and with pup deprivation mimicking only the length of time a female would be voluntarily away from her pups. Thus, early in the postpartum period, most females preferred the pup-associated chamber, whereas later in the postpartum period most females preferred the non-pup-associated chamber.

The strength of maternal motivation in early postpartum was further probed when pups were contrasted with the highly reinforcing but categorically different stimulus cocaine, a psychomotor stimulant with well-documented reinforcing properties. The uniform preference of late postpartum females for the non-pup stimulus (either a neutral stimulus or cocaine) after relatively short deprivation from their pups was strikingly different from the pup-preferring response of the early females. This pattern of diminished initial pup-seeking motivation was in general accord with our data showing that as the postpartum period progresses the parturient female dramatically reduces the voluntarily time she spends with her pups, and tailors the nature of her care giving behaviors to maturing pups.

Nonetheless, late postpartum females could express surprising levels of motivation to seek their pups if they were separated from their pups for prolonged periods before conditioning, or if they were conditioned with pups with more altricial characteristics. Furthermore if the increased work of moving mature pups were taken as a measure of motivation late postpartum females readily engaged in increasing work to move older pups to location they preferred, and assiduously confined pups to nest areas whether they were present or not. Thus, they were motivated to tailor their care giving to the needs of the maturing pups, while at the same time the initial phases of their motivational responses to seek pups and to be relentlessly involved with the pups was diminishing.

Together these experiments suggest that a variety of changes in the postpartum period may mediate changing motivational patterns. While prolactin does not independently restore late postpartum motivation to levels seen in the early postpartum female, it does modify some components of care giving behaviors of the late postpartum females such that they are more similar to those of early postpartum females or late postpartum females provided with more altricial pups. Interestingly we did find that globally increased dopaminergic tone in the late postpartum female can modestly increase pup-associated place preference, even when contemporaneous pups are the stimuli. This suggests that possibly several factors mediate the response of the female and that DA is one of them. This hypothesis accords well with the positive place preference impact of the younger stimulus pups or prolonged pup deprivation prior to conditioning on the number of late postpartum females that prefer pup-associated chambers, both of which may increase the dopaminergic tone of the female. Together these data suggest that DA is a significant neurotransmitter in the CNS circuits of the postpartum female over which the nature of the pup stimulus might signal the developmental state or physiological needs of the pups. Thus, the offspring provide mothers with stimulation that promotes the appropriate state of maternal motivation and the appropriate expression of a particular subset of care giving behaviors. Presumably dopaminergic tone normally decreases as the postpartum period progresses but the system remains plastic such that pup age or neediness can sufficiently stimulate even the late postpartum system and promote the survival of the pups.

The motivationally activated neurons that we have demonstrated, using c-Fos and CART immunoreactivity, are found within structures (the mPFC, NA, and blAMYG) well understood to be part of the neural circuit that supports motivation and responses to stimuli with incentive salience. Novel finding in this work include our data on the importance of the mPFC and the MPOA for the initial and purely appetitive phase of maternal motivation. We posit that the neurons important for maternal motivation are arranged in a hierarchically activated distributed network that includes subsets of neurons in all these structures. It is further likely that dopaminergic input to most of these structures may be involved in the regulation of the motivational state of the parturient female as she progresses through the postpartum period. Our data suggesting that it may be a globally reduced activation of these neurons in the late postpartum female may reveal CNS events important for the changes in the maternal state of the female as she progresses through the postpartum period, including changes in some CNS processes outside those regulating motivation.

REFERENCES

Agrati, D., Zuluaga, M. J., Fernández-Guasti, A., and Ferreira, A. (2007). Maternal condition modifies the behavioral, but not the endocrine stress response of virgin female rats. *Parental Brain Conference*, US, 2007.

Berridge, K. C., and Robinson, T. E. (1998). What is the role of dopamine in reward: Hedonic impact, reward learning, or incentive salience?. *Brain Res. Rev.* 28, 309–369.

Bitran, D., Hilvers, R. J., and Kellogg, C. K. (1991). Ovarian endocrine status modulates the anxiolytic potency of diazepam and the efficacy of gamma-aminobutyric acid-benzodiazepine receptor-mediated chloride ion transport. *Behav. Neurosci.* 105(5), 653–662.

Bridges, R. S. (1990). Endocrine regulation of parental behavior in rodents. In *Mammalian Parenting: Biochemical, Neurobiological, and Behavioral Determinants* (N. A. Krasnegor and R. S. Bridges, Eds.), pp. 93–117. Oxford Press, New York.

Bridges, R. S., DiBiase, R., Loundes, D. D., and Doherty, P. C. (1985). Prolactin stimulation of maternal behavior in female rats. *Science* 227(4688), 782–784.

Byrnes, E. M., Rigero, B. A., and Bridges, R. S. (2002). Dopamine antagonists during parturition disrupt maternal care and the retention of maternal behavior in rats. *Pharmacol. Biochem. Behav.* 73, 869–875.

Carelli, R. M. (2002). Nucleus accumbens cell firing during goal-directed behaviors for cocaine vs 'natural' reinforcement. *Physiol. Behav.* 76, 379–387.

Connor, J. R., and Davis, H. R. (1980a). Postpartum estrus in Norway rats. *I. Behavior. Biol. Reprod.* 23(5), 994–999.

Connor, J. R., and Davis, H. N. (1980b). Postpartum estrus in Norway rats. II. Physiology. *Biol. Reprod.* 23(5), 1000–1006.

Cosnier, J., and Couturier, C. (1966). Comportement maternal provoqu´e chez les rates adultes castrees. *C. R. Seances Soc. Biol. Ses. Fil.* 160, 789–791.

Erskine, M. S., Denenberg, V. H., and Goldman, B. D. (1978a). Aggression in the lactating rat: Effects of intruder age and test arena. *Behav. Biol.* 23(1), 52–66.

Erskine, M. S., Barfield, R. J., and Goldman, B. D. (1978b). Intraspecific fighting during late pregnancy and lactation in rats and effects of litter removal. *Behav. Biol.* 23(2), 206–218.

Fahrbach, S. E., and Pfaff, D. W. (1982). Hormonal and neural mechanisms underlying maternal behavior in the rat. In *The Physiological Mechanisms of Motivation* (D. W. Pfaff, Ed.), pp. 253–285. Springer-Verlag, New York.

Ferreira, A., and Hansen, S. (1986). Sensory control of maternal aggression in *Rattus norvegicus*. *J. Comp. Psychol.* 100(2), 173–177.

Ferreira, A., Hansen, S., Nielsen, M., Archer, T., and Minor, B. G. (1989). Behavior of mother rats in conflict tests sensitive to antianxiety agents. *Behav. Neurosci.* 103(1), 193–201.

Ferreira, A., Picazo, O., Uriarte, N., Pereira, M., and Fernández-Guasti, A. (2000). Inhibitory effect of buspirone and diazepam, but not 8-OH-DPAT, on maternal behavior and aggression. *Pharmacol. Biochem. Behav.* 66, 389–396.

Ferreira, A., Pereira, M., Agrati, D., Uriarte, N., and Fernández-Guasti, A. (2002). Role of maternal behavior on aggression, fear and anxiety. *Physiol. Behav.* 77(2–3), 197–204.

Flannelly, K. J., and Flannelly, L. (1987). Time course of postpartum aggression in rats (*Rattus norvegicus*). *J. Comp. Psychol.* 101, 101–103.

Fleming, A. S., and Luebke, C. (1981). Timidity prevents the virgin female rat from being a good mother: Emotionality differences between nulliparous and parturient females. *Physiol. Behav.* 27(5), 863–868.

Fleming, A. S., Cheung, U., Myhal, N., and Kessler, Z. (1989). Effects of maternal hormones on 'timidity' and attraction to pup-related odors in female rats. *Physiol. Behav.* 46(3), 449–453.

Fleming, A. S., Korsmit, M., and Deller, M. (1994). Rat pups are potent reinforcers to the maternal animal: Effects of experience, parity, hormones, and dopamine function. *Psychobiology* 22, 44–53.

Freeman, M. E. (1988). The ovarian cycle in the rat. In *The Physiology of Reproduction* (E. Knobil and J. D. Neill, Eds.), Vol. 1, pp. 1893–1928. Raven Press, New York.

Gaffori, O., and Le Moal, L. E. (1979). Disruption of maternal behavior and appearance of cannibalism after ventral mesencephalic tegmentum lesions. *Physiol. Behav.* 23, 317–323.

Galef, B. G., and Clark, M. M. (1972). Mother's milk and adult presence: Two factors determining initial dietary selection by weanling rats. *J. Comp. Physiol. Psychol.* 78, 220–225.

Gilbert, A. N., Pelchat, R. J., and Adler, N. T. (1980). Postpartum copulatory and maternal behaviour in Norway rats under seminatural conditions. *Anim. Behav.* 28(4), 989–995.

Gilbert, A. N., Burgoon, D. A., Sullivan, K. A., and Adler, N. T. (1983). Mother-weanling interactions in Norway rats in the presence of a successive litter produced by postpartum mating. *Physiol. Behav.* 30, 267–271.

Giordano, A. L., Johnson, A. E., and Rosenblatt, J. S. (1990). Haloperidol-induced disruption of retrieval behavior and reversal with apomorphine in lactating rats. *Physiol. Behav.* 48, 211–214.

Giovenardi, M., Consiglio, A. R., Barros, H. M., and Lucion, A. B. (2000). Pup age and aggressive behavior in lactating rats. *Braz. J. Med. Biol. Res.* 33(9), 1083–1088.

Grota, L. J., and Ader, R. (1969). Continuous recording of maternal behavior in *Rattus norvegicus*. *Anim. Behav.* 17, 722–729.

Grota, L. J., and Ader, R. (1974). Behavior of lactating rats in a dual-chambered maternity cage. *Horm. Behav.* 5(4), 275–282.

Hansen, S. (1994). Maternal behavior of female rats with 6-OHDA lesions in the ventral striatum: Characterization of the pup retrieval deficit. *Physiol. Behav.* 55(4), 615–620.

Hansen, S., Bergvall, A. H., and Nyiredi, S. (1993). Interaction with pups enhances dopamine release in the ventral striatum of maternal rats: A microdialysis study. *Pharmacol. Biochem. Behav.* 45(3), 673–676.

Hansen, S., Harthon, C., Wallin, E., Löfberg, L., and Svensson, K. (1991a). Mesotelencephalic dopamine system and reproductive behavior in the female rat: Effects of ventral tegmental 6-hydroxydopamine lesions on maternal and sexual responsiveness. *Behav. Neurosci.* 105(4), 588–598.

Hansen, S., Harthon, C., Wallin, E., Löfberg, L., and Svensson, K. (1991b). The effects of 6-OHDA-induced dopamine depletions in the ventral or dorsal striatum on maternal and sexual behavior in the female rat. *Pharmacol. Biochem. Behav.* 39(1), 71–77.

Hård, E., and Hansen, S. (1985). Reduced fearfulness in the lactating rat. *Physiol. Behav.* 35(4), 641–643.

Hauser, H., and Gandelman, R. (1985). Lever pressing for pups: Evidence for hormonal influence upon maternal behavior of mice. *Horm. Behav.* 19(4), 454–468.

Hinde, R. A. (1982). *Ethology*. Fontana Paperbacks: Glasgow, United Kingdom.

Ikemoto, S., and Panksepp, J. (1999). The role of nucleus accumbens dopamine in motivated behavior: A unifying interpretation with special reference to reward-seeking. *Brain Res. Rev.* 31, 6–341.

Insel, T. R. (1990). Regional changes in brain oxytocin receptor postpartum: Time course and relationship to maternal behaviour. *J. Neuroendocrinol.* 2, 539–545.

Keer, S. E., and Stern, J. M. (1999). Dopamine receptor blockade in the nucleus accumbens inhibits maternal retrieval and licking, but enhances nursing behavior in lactating rats. *Physiol. Behav.* 67, 659–669.

Lee, A., Clancy, S., and Fleming, A. S. (1999). Mother rats bar-press for pups: Effects of lesions of the mpoa and limbic sites on maternal behavior and operant responding for pup-reinforcement. *Behav. Brain Res.* 100(1–2), 15–31. Corrected and republished in: *Behav. Brain Res.*, 2000, 108(2), 215–231.

Leon, M., Croskerry, P. G., and Smith, G. (1978). Thermal control of mother-young contact in rats. *Physiol. Behav.* 21, 793–811.

Lonstein, J. S. (2005). Reduced anxiety in postpartum rats requires recent physical interactions with pups, but is independent of suckling and peripheral sources of hormones. *Horm. Behav.* 47(3), 241–255.

Lonstein, J. S., and de Vries, G. J. (2000). Maternal behaviour in lactating rats stimulates c-fos in glutamate decarboxylase-synthesizing neurons of the medial preoptic area, ventral bed nucleus of the stria terminalis, and ventrocaudal periaqueductal gray. *Neuroscience* 100(3), 557–568.

Lonstein, J. S., and Gammie, S. C. (2002). Sensory, hormonal, and neural control of maternal aggression in laboratory rodents. *Neurosci. Biobehav. Rev.* 26(8), 869–888.

Lonstein, J. S., and Morrell, J. I. (2007). Neuroendocrinology and neurochemistry of maternal motivation and behavior. In *Handbook of Neurochemistry and Molecular Neurobiology. Behavioral Neurochemistry, Neuroendocrinology and Molecular Neurobiology* (A. Lajtha and J. D. Blaustein, Eds.), pp. 195–245. Springer Reference.

Magnusson, J. E., and Fleming, A. S. (1995). Rat pups are reinforcing to the maternal rat: Role of sensory cues. *Psychobiology* 23, 69–75.

Marinari, K. T., and Moltz, H. (1978). Serum prolactin levels and vaginal cyclicity in concaveated and lactating female rats. *Physiol. Behav.* 21(4), 525–528.

Mattheij, J. A., Swarts, H. J., and van Mourik, S. (1985). Plasma prolactin in the rat during suckling without prior separation from pups. *Acta Endocrinol. (Copenh)* 108(4), 468–474.

Mattson, B. J., and Morrell, J. I. (2005). Preference for cocaine- versus pup-associated cues differentially activates neurons expressing either Fos or cocaine- and amphetamine-regulated transcript in lactating, maternal rodents. *Neuroscience* 135(2), 315–328.

Mattson, B. J., Williams, S., Rosenblatt, J. S., and Morrell, J. I. (2001). Comparison of two positive reinforcing stimuli: Pups and cocaine throughout the postpartum period. *Behav. Neurosci.* 115(3), 683–694.

Mattson, B. J., Williams, S. E., Rosenblatt, J. S., and Morrell, J. I. (2003). Preferences for cocaine- or pup-associated chambers differentiates otherwise behaviorally identical postpartum maternal rats. *Psychopharmacol.* 167(1), 1–8.

Mayer, A. D., and Rosenblatt, J. S. (1980). Hormonal interaction with stimulus and situational factors in the initiation of maternal behavior in non-pregnant rats. *J. Comp. Physiol. Psychol.* 94(6), 1040–1059.

Mayer, A. D., and Rosenblatt, J. S. (1984). Prepartum changes in maternal responsiveness and nest defense in *Rattus norvegicus*. *J. Comp. Psychol.* 98(2), 177–188.

Mayer, A. D., and Rosenblatt, J. S. (1993). Persistent effects on maternal aggression of pregnancy but not of estrogen/progesterone treatment of non-pregnant ovariectomized rats revealed when initiation of maternal behavior is delayed. *Horm. Behav.* 27, 132–155.

Mayer, A. D., Reisbick, S., Siegel, H. I., and Rosenblatt, J. S. (1987). Maternal aggression in rats: Changes over pregnancy and lactation in a Sprague-Dawley strain. *Aggress. Behav.* 13, 29–43.

Moltz, H., Lubin, M., Leon, M., and Numan, M. (1970). Hormonal induction of maternal behavior in the ovariectomized nulliparous rat. *Physiol. Behav.* 5(12), 1373–1377.

Morgan, H. D., Watchus, J. A., Milgram, N. W., and Fleming, A. S. (1999). The long lasting effects of electrical simulation of the medial preoptic area and medial amygdala on maternal behavior in female rats. *Behav. Brain Res.* 99(1), 61–73.

Morrell, J. I., Mattson, B. J., Wansaw, M. P., Smith, K. S., Seip, K. M. (2006). Postpartum neuroendocrine and endocrine changes may participate in postpartum phases of maternal motivation as measured by neuronal activation during pup-associated place preference. *International Congress of Neuroendocrinology*, US.

Neumann, I. D. (2001). Alterations in behavior and neuroendocrine stress coping strategies in pregnant, parturient, and lactating rats. *Prog. Brain Res.* 133, 143–152.

Nissen, H. W. (1930). A study of maternal behavior in the white rat by means of the obstruction method. *J. Genetic. Psychol.* 37, 377–393.

Numan, M. (2006). Hypothalamic neural circuits regulating maternal responsiveness toward infants. *Behav. Cognit. Neurosci. Rev.* 5(4), 163–190.

Numan, M. (2007). Motivational systems and the neural circuitry of maternal behavior in the rat. *Dev. Psychobiol.* 49(1), 12–21.

Numan, M., and Insel, T. R. (2003). *The Neurobiology of Parental Behavior*. Springer-Verlag, New York.

Numan, M., Numan, M. J., Schwarz, J. M., Neuner, C. M., Flood, T. F., and Smith, C. D. (2005a). Medial preoptic area interactions with the nucleus accumbens-ventral pallidum circuit and maternal behavior in rats. *Behav. Brain Res.* 158(1), 53–68.

Numan, M., Numan, M. J., Pliakou, N., Stolzenberg, D. S., Mullins, O. J., Murphy, J. M., and Smith, C. D. (2005b). The effects of D1 or D2 dopamine receptor antagonism in the medial preoptic area, ventral pallidum, or nucleus accumbens on the maternal retrieval response and other aspects of maternal behavior in rats. *Behav. Neurosci.* 119(6), 1588–1604.

Ostermeyer, M. C. (1983). Maternal aggression. In *Parental Behavior in Rodents* (R. W. Elwood, Ed.), pp. 151–179. Wiley, London.

Pereira, M., and Ferreira, A. (2006). Demanding pups improve maternal behavioral impairments in sensitized and haloperidol-treated lactating female rats. *Behav. Brain Res.* 175(1), 139–148.

Pereira, M., Uriarte, N., Agrati, D., Zuluaga, M. J., and Ferreira, A. (2005). Motivational aspects of maternal anxiolysis in lactating rats. *Psychopharmacol.* 180(2), 241–248.

Pereira, M., Dziopa, E. I., and Morrell, J. I. (2007a). Expression of maternal behavior during the late postpartum period: Effect of D1 dopamine receptor stimulation. *Parental Brain Conference*, US.

Pereira, M., Dziopa, E. I., and Morrell, J. I. (2007b). Identifying factors that underlie maternal motivation during the late postpartum period. *SfN's 37th Annual Meeting*, US.

Reisbick, S., Rosenblatt, J. S., and Mayer, A. D. (1975). Decline of maternal behavior in the virgin and lactating rat. *J. Comp. Physiol. Psychol.* 89(7), 722–732.

Rosenblatt, J. S. (1967). Non-hormonal basis of maternal behavior in the rat. *Science* 156, 1512–1514.

Rosenblatt, J. S. (1975). Prepartum and postpartum regulation of maternal behavior in the rat. *Ciba Found. Symp.* 33, 17–32.

Rosenblatt, J. S., and Mayer, A. D. (1995). An analysis of approach/withdrawal processes in the initiation of maternal behavior in the laboratory rat. In *Behavioral Development, Concepts of Approach/Withdrawal and Integrative Levels* (K. E. Hood, E. Greenberg, and E. Tobach, Eds.), pp. 177–230. Garland Press, New York.

Rosenblatt, J. S., Mayer, A. D., and Siegel, H. I. (1985). Maternal behavior among nonprimate mammals. In *Handbook of Behavioral Neurobiology* (N. Adler, D. Pfaff, and R. W. Goy, Eds.), Vol. 7, pp. 229–298. Plenum Press, New York.

Robbins, T. W., and Everitt, B. J. (1996). Neurobehavioural mechanisms of reward and motivation. *Curr. Opin. Neurobiol.* 6, 228–236.

Salamone, J. D., Correa, M., Farrar, A. M., and Mingote, S. M. (2007). Effort-related functions of nucleus accumbens dopamine and associated forebrain circuits. *Psychopharmacol.* 191, 461–482.

Silva, M. R. P., Bernardi, M. M., and Felicio, L. F. (2001). Effects of dopamine receptor antagonists on ongoing maternal behavior in rats. *Pharmacol. Biochem. Behav.* 68, 461–468.

Silva, M. R. P., Bernardi, M. M., Cruz-Casallas, P. E., and Feliciom, L. F. (2003). Pimozide injections into the nucleus accumbens disrupt maternal behaviour in lactating rats. *Pharmacol. Toxicol.* 93, 42–47.

Seip, K. M., and Morrell, J. I. Strong maternal motivation in nulliparous female rats emerges after prolonged pup-exposure. *Physiol. Behav.*, In revision 2008a.

Seip, K. M., and Morrell, J. I. (2007b). Increasing the incentive salience of cocaine challenges preference for pup- over cocaine-associated stimuli during early postpartum: Place preference and locomotor analyses in the lactating female rat. *Psychopharmacol.* 194(3), 309–319.

Seip, K. M., Pereira, M., Wansaw, M. P., Dziopa, E. I., Reiss, J. I., and Morrell, J. I. Incentive salience of cocaine across the postpartum period of the female rat. *Psychopharmacol.*, In press, 2008a.

Seip, K. M., Dziopa, E. I., Wansaw, M. P., and Morrell, J. I. (2007b). Localization and cellular characterization of neural substrates contributing to pup-associated preference in the lactating female rat. *Parental Brain Conference*, US.

Smith, M. S., Freeman, M. E., and Neill, J. D. (1975). The control of progesterone secretion during the estrous cycle and early pseudopregnancy in the rat: Prolactin, gonadotrophin, and steroid levels associated with rescue of the corpus luteum of pseudopregnancy. *Endocrinology* 96, 219–226.

Stern, J. M. (1996). Somatosensation and maternal care in Norway rats. In *Advances in the Study of Behavior. Parental care – Evolution, Mechanisms, and Adaptive Significance* (J. S. Rosenblatt and C. T. Snowdon, Eds.), Vol. 25, pp. 243–294. Academic Press, New York.

Stern, J. M., and Taylor, L. A. (1991). Haloperidol inhibits maternal retrieval and licking, but facilitates nursing behavior and milk ejection in lactating rats. *J. Neuroendocrinol.* 3, 591–596.

Stern, J. M., and Keer, S. E. (1999). Maternal motivation of lactating rats is disrupted by low dosages of haloperidol. *Behav. Brain Res.* 99(2), 231–239.

Stern, J. M., and Keer, S. E. (2002). Acute hunger of rat pups elicits increased kyphotic nursing and shorter intervals between nursing bouts: Implications for changes in nursing with time postpartum. *J. Comp. Psychol.* 116, 83–92.

Stolzenberg, D. S., Numan, M. J., and Numan, M. (2007). Dopamine D1 receptor stimulation of the nucleus accumbens or the medial preoptic area promotes the onset of maternal behavior in pregnancy-terminated rats. *Parental Brain Conference*, US.

Swanson, L. W. (1992). *Brain maps: Structure of the rat brain.* Elsevier, New York.

Taya, K., and Sasamoto, S. (1981). Changes in FSH, LH and prolactin secretion and ovarian follicular development during lactation in the rat. *Endocrinol. Jpn.* 28(2), 187–196.

Taya, K., and Greenwald, G. S. (1982a). Peripheral blood and ovarian levels of sex steroids in the lactating rat. *Endocrinol. Jpn.* 29(4), 453–459.

Taya, K., and Greenwald, G. S. (1982b). Mechanisms of suppression of ovarian follicular development during lactation in the rat. *Biol. Reprod.* 27(5), 1090–1101.

Uriarte, N., Ferreira, A., Rosa, X. F., and Lucion, A. B. (2007). Maternal behavior in lactating rats: Effects of concurrent pregnancy and of overlapping litters. *Parental Brain Conference*, US.

Vernotica, E. M., and Morrell, J. I. (1998). Plasma cocaine levels and locomotor activity after systemic injection in virgin and in lactating maternal female rats. *Physiol. Behav.* 64(3), 399–407.

Wansaw, M. P., Lin, S. N., and Morrell, J. I. (2005). Plasma cocaine levels, metabolites, and locomotor activity after subcutaneous cocaine injection are stable across the postpartum period in rats. *Pharmacol. Biochem. Behav.* 82(1), 55–66.

Wansaw, M. P., Pereira, M., and Morrell, J. I. Characterization of maternal motivation in the lactating rat: Contrasts between early and late postpartum responses. *Horm. Behav.*, In press 2008.

Wilsoncroft, W. E. (1969). Babies by bar-press: Maternal behavior in the rat. *Behav. Res. Meth. Instrum.* 1, 229–230.

Woodside, B., and Popeski, N. (1999). The contribution of changes in milk delivery to the prolongation of lactational infertility induced by food restriction or increased litter size. *Physiol. Behav.* 65(4–5), 711–715.

4

IMAGING THE MATERNAL RAT BRAIN

MARCELO FEBO AND CRAIG F. FERRIS

Department of Psychology, Northeastern University, Boston, MA, USA

INTRODUCTION

Breastfeeding has played a critical role in the survival of mammals. Lactation is an intermittent event in which mothers forage for sustenance while simultaneously converting food into nutrient-rich milk for their young. During the reproductive period defined by lactation, a bond is formed between a mother and her offspring that favors their protection and social development. In rodents, the initiation and maintenance of maternal behavior progresses through a complex interaction between the endocrine status of the dam prior to and following parturition and the continued interaction and stimulation from pups until weaning (Numan, 1994). Hormones combine with the physiology of parturition to foster maternal behavior in first time mothers. However, from postpartum day (PPD) 4 through postpartum day 20 the maintenance of maternal behavior is more strongly influenced by learning and the tactile, auditory and odor stimuli coming from the pups.

Perhaps one of the greatest advantages realized in the evolution of breastfeeding is the potential for a mother to determine the social development of her young (Fleming *et al.*, 2002). A prime example of the relationship between motherhood and offspring is the discovery that variations in maternal care can lead to non-genomic transmission of individual behavioral differences across generations (Francis *et al.*, 1999). Rat dams that spend more time licking and grooming their pups during the first week of lactation raise offspring that are less responsive to stress as adults and are more likely to provide adequate attendance to their own pups (Weaver *et al.*, 2004). Thus, high-nurturing mothers raise high-nurturing daughters, a behavioral phenotype that can be cross-fostered.

How does the brain respond to pup suckling, this unique environmental stimulus that has such important consequences? More specifically what neural circuits are activated in the dam's brain with pup suckling that would contribute to maternal behavior and the facilitation of mother–infant bonding? Site-specific lesions of the medial preoptic area (MPOA), bed nucleus of the stria terminalis, lateral septum, amygdala, cingulate cortex, nucleus accumbens and substantia nigra have demonstrated their roles in either the onset and/or maintenance of maternal behaviors in lactating rats (Slotnick, 1967; Slotnick & Nigrosh, 1975; Numan & Corodimas, 1985; Stack *et al.*, 2002; Li & Fleming, 2003). Immunostaining for immediate early gene proteins as cellular markers of neuronal activity has revealed a more intricate neuronal network comprising a number of cortical, thalamic, hypothalamic, limbic and brainstem sites (Kendrick *et al.*, 1997; Lin *et al.*, 1998; Lonstein *et al.*, 1998; Lee *et al.*, 1999b; Li *et al.*, 1999). Unfortunately, the temporal window for these cellular markers is up to 2 h after stimulus presentation leaving in doubt the precise onset and location of neuronal activity associated with the start of pup suckling. With the advent of newer imaging modalities like functional magnetic resonance imaging (fMRI) with the blood oxygen level-dependent (BOLD) technique these issues can be resolved. Recent work in our laboratory using BOLD fMRI has unveiled an interesting view of the maternal brain, where distributed *realtime* activity across higher cortical centers in conjunction with classically known regions of the hypothalamus, forebrain and midbrain is elicited selectively by pup suckling.

Functional MRI is a non-invasive technology with exceptional temporal and spatial resolution making it possible to map in seconds functionally relevant neural networks activated by a variety of environmental and chemical stimuli (Lahti *et al.*, 1999; Boyett-Anderson *et al.*, 2003; Fize *et al.*, 2003; Tenney *et al.*, 2004). Increased neuronal activity is accompanied by an increase in metabolism concomitant with changes in cerebral blood flow and blood volume in the area of elevated neural activity. BOLD fMRI is a technique sensitive to the oxygenation status of hemoglobin (Ogawa *et al.*, 1990). While fMRI has neither the cellular spatial resolution of immunostaining, nor the millisecond temporal resolution of electrophysiology, it does show synchronized changes in neuronal activity across multiple brain areas, providing a unique insight into functional neuroanatomical circuits coordinating the thoughts, memories and emotions for particular behavioral states. By using fMRI to gain an understanding of the neural regions associated with key maternal activities, that is, nursing, windows are created to elucidate the basic neurobiological underpinnings of motherhood. In the present chapter we discuss the technology and methods for performing imaging studies on awake dams during pup suckling. This unique experimental model allowed us to address three questions: (1) Does pup suckling activate reward circuitry? (2) Where is the cortical representation of nipple stimulation? (3) What role does oxytocin (OT) play in suckling-induced brain activation?

IMAGING THE NEURAL RESPONSE TO PUP SUCKLING

Our laboratory has developed the technology and methods for imaging changes in brain activity in fully conscious rat dams in response to the onset of pup suckling (Ferris *et al.*, 2005). Studies are performed with a multi-concentric dual-coil, small animal restrainer (Insight Neuroimaging Systems, LLC, Worcester MA). The basic configuration used for rat MR imaging is shown Figure 4.1. The BOLD fMRI signal (T2 or T2* weighted signal in technical jargon) is detected through the manipulation of precessing protons in the various brain tissue compartments when a subject is placed within

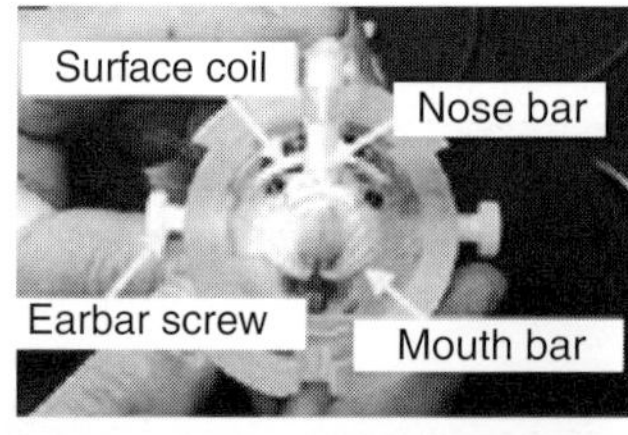

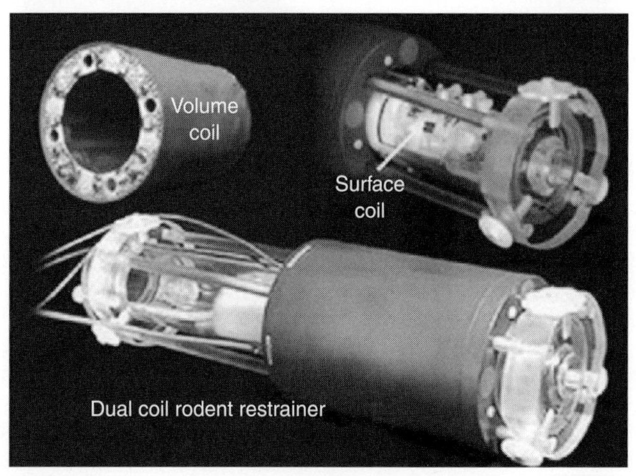

FIGURE 4.1 Restrainer components and Rf electronics for functional imaging of awake rats.

an external magnetic field (B_0 field provided by the MR scanner). Two radiofrequency (Rf) coils, one for transmission and one for reception, facilitate the manipulation and detection of changes in proton resonant frequency, as an indirect measurement of the fMRI signal *in vivo*. Prior to imaging studies, dams are anesthetized with 2–4% isoflurane. A topical anesthetic of 10% lidocaine cream is applied to the skin and soft tissue around the ear canals and over the bridge of the nose. A plastic semicircular headpiece with blunted ear supports that fit into the ear canals are positioned over the ears (Figure 4.1). The head is placed into a cylindrical head holder with the animal's canines secured over a bite bar and ears positioned inside of the head holder with adjustable screws fitted into lateral sleeves (Figure 4.1). An adjustable Rf surface coil built into the head holder is pressed firmly on the head and locked into place. This Rf coil facilitates the detection of changes in resonant frequency signals from brain tissues and works in conjunction with the overlying volume Rf coil that transmits Rf pulses into the brain (Ludwig *et al.*, 2004). The body of the animal is placed into a body restrainer. The body restrainer "floats" down the center of the chassis connecting at the front and rear end-plates and buffered by rubber gaskets. The headpiece locks into a mounting post on the front of the chassis. This design isolates all of the body movements from the head restrainer and minimizes motion artifact. Once the animal is positioned in the body holder, a volume Rf coil is slid over the head restrainer and locked into position.

Prior to imaging, virgin females are routinely acclimated to the restrainer and the imaging protocol. Animals are anesthetized with isoflurane as described above. When fully conscious, the restraining unit is placed into a black opaque tube "mock scanner" with a tape-recording of an MRI pulse sequence for 90 min in order to simulate the bore of the magnet and an imaging protocol. This procedure is repeated every other day for 4 days. With this procedure, rats show a significant decline in respiration, heart rate, motor movements and plasma corticosterone when comparing the first to the last acclimation periods (King *et al.*, 2005). The reduction in autonomic and somatic measures of arousal and stress improve the signal resolution and quality of the MR images. After this acclimation procedure females are pair housed with male breeders for mating.

Experiments are conducted in a Bruker Biospec 4.7-T/40-cm horizontal magnet (Oxford Instrument, Oxford, UK) equipped with a Biospec Bruker console (Bruker, Billerica, MA USA). Functional images are acquired using a T2 weighted multi-slice fast spin echo sequence. A single data acquisition acquires 12, 1.2 mm slices in 6–8 s. The imaging sessions usually consist of 3–5 min of baseline data followed by 5–10 min of stimulation data. At the end of each imaging session a high-resolution anatomical data set that exactly matches the geometry of the functional scan is collected.

Lactating dams are used between PPDs 4 and 8. The adapted configuration used for maternal studies is shown in Figure 4.2. For suckling studies, the hind limbs of the dams are loosely tethered and raised just above the floor of the body tube. This provides a visual inspection of the ventrum from outside of the magnet and prevents the dam from kicking and injuring the pups during suckling. A cradle containing four to six pups is positioned under the dam in the magnet. The body tube has a window exposing the dam's ventrum to the pups. A thin plastic shield separates the pups from the mother. When the shield is pulled away the pups are exposed to the six hind-limb nipples and begin suckling. We are able to visually confirm when pups came onto the most caudal teats but are not able to determine whether all teats are suckled during the stimulation period. In all pup stimulation studies, suckling usually occurs within seconds of removing the shield. To promote suckling in the magnet, dams are prevented from having physical contact with their pups for 2 h prior to imaging. This is accomplished by inverting a shallow perforated Plexiglas box over the huddled pups.

IMAGING THE REINFORCING NATURE OF PUP SUCKLING

Pups have strong reinforcing properties that compete with cocaine self-administration and place preference paradigms (Hecht *et al.*, 1999; Mattson *et al.*, 2001). Indeed, dams will even train to lever press for pups in an operant response paradigm (Lee *et al.*, 1999a). Activation of the dopaminergic mesocorticolimbic system or "reward pathway" is hypothesized to be part of a central mechanism contributing to the reinforcing property of pups. The origin

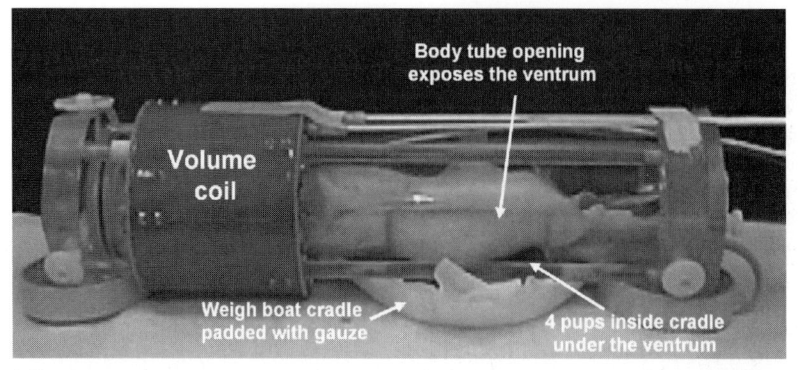

(A)

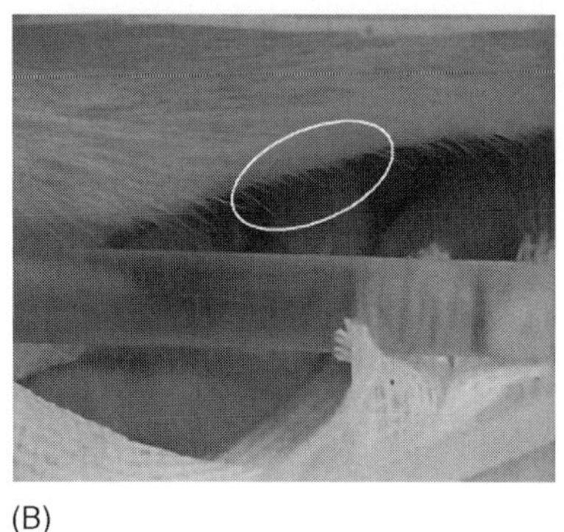

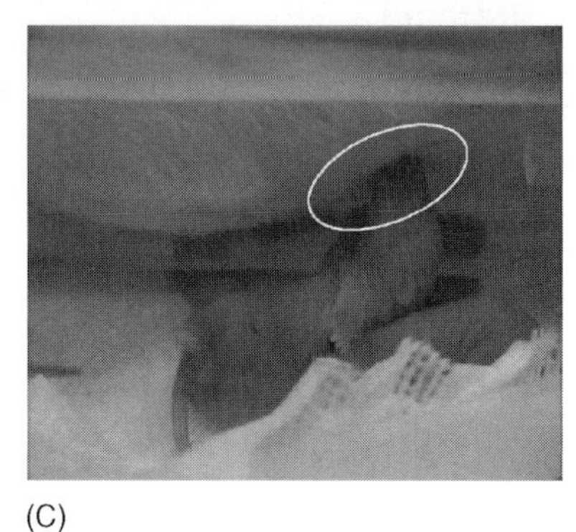

(B) (C)

FIGURE 4.2 Adaptation of the imaging setup for imaging awake lactating dams. Shown in (A) is a PPD 4–8 dam held in the body tune with an opening toward the ventrum that allows access to teats (shown in B and C). (*Source*: Febo *et al.*, 2005b, with permission from the Society for Neuroscience.)

of the dopaminergic innervations to the major components of the reward system, nucleus accumbens and prefrontal cortex, is the ventral tegmental area (VTA). Activity within the VTA–accumbens–prefrontal cortex pathway can thus be expected to play a part in accentuating the rewarding nature of pups thereby reinforcing maternal behaviors. Lesioning the VTA reduces maternal behavior (Numan & Smith, 1984). Depleting dopamine levels in this VTA/accumbens projection with a neurotoxin reduces pup retrieval (Hansen *et al.*, 1991a, b) and direct electrolytic lesion of the accumbens alters maternal memory over the postpartum period (Lee *et al.*, 1999b).

Given the strong association between the mesocorticolimbic dopaminergic system and the reinforcing properties of pup suckling and cocaine, our lab designed studies to compare activation of the reward pathway in response to both stimuli using fMRI (Ferris *et al.*, 2005). Mothers were imaged in response to pups or intracerebroventricular (ICV) cocaine while a control group of virgin females were imaged in response to cocaine alone. Three-dimensional (3D) models and contiguous two-dimensional activation maps comprising the mesocorticolimbic and nigrostriatal dopamine systems are shown for each stimulus condition (Figure 4.3). Pup suckling causes areas of positive BOLD activation localized to both dorsal (caudate–putamen) and ventral (accumbens) striatum and prefrontal cortex (left column). These areas of activation are very similar to that observed with cocaine stimulation in virgin females (middle column) but not cocaine stimulation in lactating dams (right column). The volume of activation of the primary forebrain areas comprising the reward system (i.e., accumbens and prefrontal cortex) can be viewed in a 3D model for each stimulus condition. The most obvious difference among the three stimulus conditions in terms of positive BOLD signal is the absence of activation in

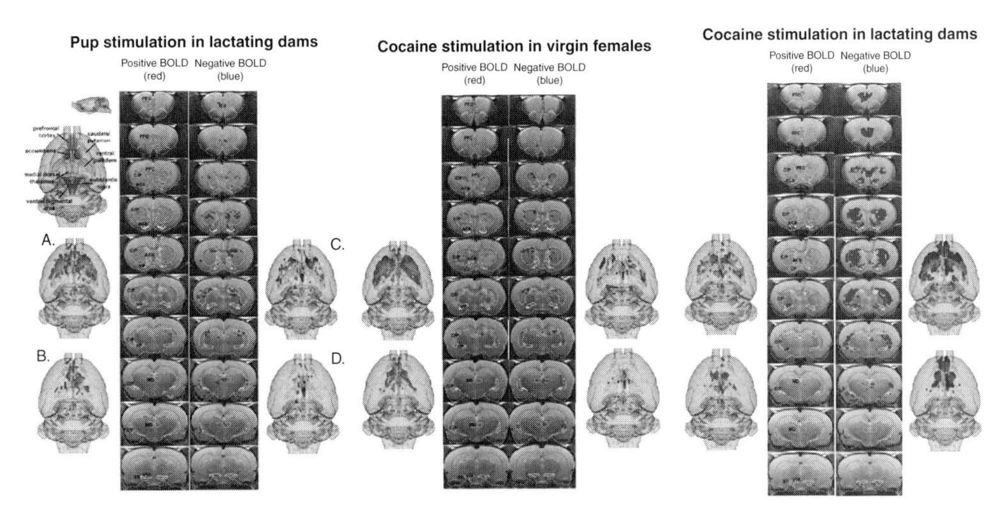

FIGURE 4.3 Composite brain maps of positive and negative BOLD responses to suckling pups and cocaine. Shown are data for virgin rats and PPD 4–8 dams given cocaine or dams exposed to suckling pups. The three columns of coronal anatomical scans surrounded by 3D glass brains indicate areas of the mesocorticolimbic and nigrostriatal dopamine systems that showed increases in BOLD responses (red-positive BOLD) or decreases (blue-negative BOLD) for the three conditions. The 3D brain in the upper far left corner highlights the anatomical subregions that were studied. (*Source*: Adapted from Ferris *et al.*, 2005.) (See Color Plate)

the prefrontal cortex in lactating dams exposed to cocaine. Although the activation of the accumbens appears to be comparable between lactating dams stimulated with pups and cocaine in the 3D presentation, careful investigation of the contiguous coronal sections shows that the cocaine stimulation is limited to the most caudal–dorsal accumbens (putative core), whereas pups stimulation is much broader, including rostral–ventral accumbens (putative shell). Pup stimulation also causes modest but significant changes in negative BOLD signal (blue). These negative BOLD data were very similar to cocaine stimulation in virgin females (middle column). One of the more striking differences among stimulation conditions was the robust negative BOLD signal change observed throughout the mesocorticolimbic and nigrostriatal dopaminergic systems in lactating dams exposed to cocaine.

These fMRI studies show pup suckling is a robust stimulus for activating the mesocorticolimbic dopaminergic system. The simultaneous activation of the nigrostriatal and mesocorticolimbic systems is not surprising since cocaine blocks monoamine reuptake and enhances dopaminergic activity across these areas. However, the coordinated activation of parallel dopaminergic systems with pup suckling was unexpected and may reflect the natural pattern of brain activation with reinforcing stimuli. Indeed, dorsal and ventral striatal pathways comprise portions of the basal ganglia through which thalamic and cortical afferent connections facilitate or inhibit expression of repetitive motor sequences, as well as cognitive and motivational processes (Voorn *et al.*, 2004).

The mesocorticolimbic dopaminergic system appears to be critical neurochemical pathways in anticipatory and consummatory aspects of maternal behavior. Dams between PPDs 4 and 10, show elevated extracellular dopamine levels in the nucleus accumbens in response to pup suckling and in anticipation of pup interactions (Hansen *et al.*, 1993; Pruessner *et al.*, 2004). Lesioning the VTA/accumbens circuit with dopamine-depleting 6-hydroxydopamine reduces pup retrieval (Hansen *et al.*, 1991a, b). Lactating dams treated with different dopamine receptor antagonists in the nucleus accumbens show a reduction in several measures of maternal behavior (Keer & Stern, 1999; Numan *et al.*, 2005). Collectively these findings support the notion that pup activation of the dopamine system is an important reinforcer in the anticipatory aspect of dams seeking pups and in the consummatory aspect of maternal behavior such as pup retrieval and pup licking and grooming. There are a handful of clinical imaging studies of recent mothers responding to sensory stimuli from newborn babies. A variety of sensory stimuli have been presented to mothers inside of the MRI machine, including pictures of their own vs. an unknown child (Bartels & Zeki, 2004; Leibenluft *et al.*, 2004; Ranote *et al.*, 2004), pictures of newborns under various emotive states (Seifritz *et al.*, 2003) and baby cry sounds (Lorberbaum *et al.*, 2002). Areas comprising the ventral striatum, caudate–putamen, anterior cingulate, insular cortex, amygdala, midbrain regions, prefrontal cortex were all cited among the areas responding to infants. Therefore, neuronal activity within a mother's brain reward circuitry may in fact generalize to a variety of infant-specific sensory stimuli and not just the tactile sensation of suckling. Consistent throughout the human clinical work was the activation (or deactivation) of the prefrontal cortex, particular limbic regions of the frontal cortex such as the orbitofrontal (Nitschke *et al.*, 2004), insular (Seifritz *et al.*, 2003), anterior cingulate (Bartels & Zeki, 2004) and mesial prefrontal areas (Lorberbaum *et al.*, 2002). This supports an additional role for limbic subregions of the prefrontal cortex in maternal responding and perhaps more specifically in maternal motivation to care for their newborns. Recently we designed a study in which virgin rats were treated with cocaine for 14 days. Females were then withdrawn from further treatment and pair housed with males for breeding. After pregnancy they were imaged for their response to suckling between PPD 4 and 8. We observed significantly less positive BOLD activation in the orbital subregion of the medial prefrontal cortex of dams pretreated with cocaine as compared to vehicle control dams. The reduced positive BOLD response to pups may reflect an underlying change in the dam's motivational state (Febo & Ferris, 2007).

Pup suckling stimulates the same mesocorticolimbic dopamine systems as cocaine in virgin females, evidence that both reinforcing stimuli elicit similar brain mechanisms regarding motivation and reward. However, the reproductive condition of the female appears to be a critical determinant in predicting cocaine's effect on brain activation. Virgin females of the same age as the maternal, lactating females show a robust positive BOLD signal change in the reward system in response to cocaine. However, the same dose of cocaine given to lactating dams causes a predominately negative BOLD response in the reward pathway possibly reflecting a suppression of synaptic and neuronal activity. The difference is probably due to neuroadaptive changes over the reproductive period of birth and early lactation that foster mother–infant bonding, in part, by making suckling a rewarding experience. Our previous work shows that in virgins, estrogen influences the BOLD response to cocaine (Febo *et al.*, 2005a). Therefore one contributing factor to the underlying neuroadaptive changes determining the differential sensitivity to pups and cocaine may be sex steroids, particularly estrogen (Febo *et al.*, 2003). It is important to add that virgin females trained to self-administer cocaine and subsequently breed, show a reduction in self-administration during early lactation (Hecht *et al.*, 1999). As noted previously, the reinforcing properties of pups as measured with conditioned place preference exceeds that of cocaine in early lacta tion (Mattson *et al.*, 2001, 2003; Mattson & Morrell, 2005). Both of these behavioral tests for cocaine seeking are also modulated by estrogen administration in females (Becker, 1999).

The pronounced negative BOLD over most of the mesocorticolimbic dopaminergic system in lactating dams exposed to cocaine is noteworthy. The robust negative BOLD signal observed in these studies when lactating dams are exposed to cocaine is interpreted as brain deactivation. This raises the critical question of why cocaine stimulates the reward system and is an effective reinforcer in virgin rats but less effective, even inhibitory, in dams during early lactation? If pups and cocaine activate a common neural substrate as the present findings would indicate, then the response of one would affect the other depending on the salience and magnitude of the reinforcing stimuli (Mattson *et al.*, 2001). Introduction of a highly addictive drug like cocaine into a critical reproductive period of social bonding might be expected to usurp or disrupt the natural stimulus/reward relationship between pups and dams (Panksepp *et al.*, 2002). Instead, evolution may have provided dams with a resistance against competing hedonic stimuli during early lactation assuring the salience of pup-seeking over other rewards (Mattson *et al.*, 2001).

SENSORY CORTEX REPRESENTATION OF THE MATERNAL VENTRUM AND NIPPLE

As noted above, human imaging studies show that breastfeeding is more than just simple tactile stimulation but seems to involve multiple sensory modalities. Given the importance of breastfeeding on mother–infant bonding, it is surprising how little is known about the areas of the cerebrum that represent the sensory maps associated with suckling and lactation. The sensory system is organized such that information coming from the landscape of peripheral receptors (i.e., body surface, cochlea and retina) is relayed to the cortex through the spinothalamic pathway and topographically represented in the cerebrum. In the case of the peripheral fibers arising from the mammillae, additional primary afferent projections to the dorsal column and lateral cervical nuclei are relayed to the paraventricular hypothalamus, and are believed to be involved in the milk-ejection reflex and nursing behaviors (Tsingotjidou & Papadopoulos, 1996; Lonstein & Stern, 1997; Li *et al.*, 1999). It is important to underscore the fact that the cortical representation of afferent signals coming from the glabrous skin of the nipples and areolae of mammals is unknown (Xerri *et al.*, 1994). Tactile or suckling stimulation of the nipple does not evoke neuronal activity in the somatosensory cortex of multiparous dams even though the same cortical area is highly responsive to deflection of the hair immediately surrounding the areolar skin (Xerri *et al.*, 1994). Again, fMRI in conscious dams is ideally suited for providing insight into this question of cortical representation of nipple stimulation during lactation.

To address this issue, we designed an imaging study using four stimulation conditions. In one condition, lactating dams were exposed to their pups for a 5-min stimulation period. In the second condition, a separate group of lactating dams were artificially suckled in the absence of pups. In the third condition, a separate group of lactating dams were exposed to gentle rubbing of the ventrum around the nipples with a flat edged wooden ruler. In the fourth condition, virgin females were exposed to this rubbing stimulus. Figure 4.4 summarizes our major finding. As can be observed, wide areas of the cerebrum exhibit an increase in brain activity during suckling stimulation, suggesting that neural activity is modified over extended regions of the cerebral cortex in response to a highly specific stimulus. Therefore, although auditory, olfactory and non-suckling tactile stimulation from pups can contribute to cortical activity, the isolated suckling stimulus is fully capable of causing a widespread cortical response.

This finding suggests that during lactation incoming sensory information from a single modality is sufficient to drive activity of others. Enhanced sensitivity of the cortical map during breastfeeding may help the dam to perceive, process and remember external stimuli critical to the care and protection of her young. Indeed, ventral stimulation from nuzzling and suckling pups influences the following maternal functions in rodents: nursing behavior (Stern & Johnson, 1990); ongoing maternal motivation and the development of maternal memory (Morgan *et al.*, 1992); the formation of a conditioned place preference where the mother learns to return to the specific site where she received suckling stimulation (Fleming *et al.*, 1994; Walsh *et al.*, 1996); maternal aggression (Stern & Kolunie, 1993).

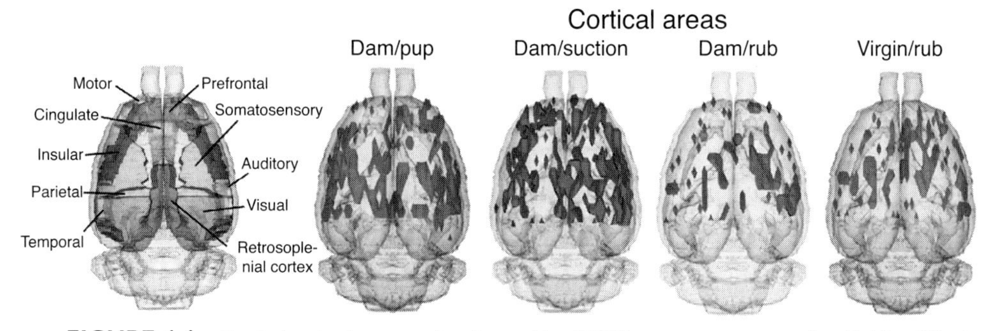

FIGURE 4.4 Cortical activation maps showing positive BOLD responses to pup and artificial suckling in PPD 4–8 dams or ventrum rubbing in dams and virgin rats. The 3D glass brains indicate areas of the cortical mantle that showed increases in BOLD responses (red-positive BOLD). The 3D brain in the far left highlights the various major subdivisions of the rat cortex. (*Source*: Adapted from Febo *et al.*, 2008, in press.) (See Color Plate)

Maternal motivation and maternal memory are likely to be interrelated since maternal memory involves a modification of maternal responsiveness. When a primiparous female rat gives birth and is exposed to pups for the first time, her maternal responsiveness is highly dependent on the endocrine changes associated with pregnancy termination, and a critical process involved in this hormonal induction of maternal behavior is a shift in the valence of pups odors from negative (or aversive) to positive (or attractive) (Numan & Insel, 2003). In the naïve nulliparous female rat, pup odors are primarily aversive, leading to avoidance behavior. Interestingly, if the primiparous female is allowed only a few hours of maternal contact with pups, her maternal responsiveness is permanently modified: the pups can be removed and when young pups are returned several weeks later the female will care for them rather than avoid them, even though she is not under the influence of the endocrine changes associated with the end of pregnancy (Bridges, 1975; Orpen *et al.*, 1987). One interpretation of these results is that the initial maternal experience modifies the brain so that pup odors become permanently attractive and activate approach behavior instead of avoidance behavior, even in the absence of hormonal stimulation. Importantly, primiparous females who are unable to receive *both* ventral tactile stimulation (which includes suckling stimulation) and perioral tactile stimulation from pups show very little maternal behavior when first exposed to pups postpartum and they also do not develop maternal memory or the long-term retention of maternal responsiveness (Morgan *et al.*, 1992). Therefore, it is possible that suckling-induced activation of the olfactory cortex may contribute to causing a relatively permanent shift in the valence of pup odors from negative to positive. Perhaps, there are reciprocal pathways to the olfactory bulbs that are activated when a female is suckling, hence cuing her senses more into the odors of young.

COMPARISON OF SUCKLING VS OXYTOCIN STIMULATED BRAIN ACTIVITY

OT synthesis mainly occurs in neurons of the paraventricular (PVN) and supraoptic nucleus (SON) of the hypothalamus. It is transported to and released from nerve terminals in the posterior pituitary and various regions of the brain. Suckling stimulates the release of OT simultaneously into the bloodstream and central nervous system (CNS) of postpartum rats (Neumann *et al.*, 1993b). Systemically, this neurohypophyseal hormone enhances smooth muscle contractility, which is important for milk "let-down" during nursing. There is evidence to suggest that the release of OT in the CNS during parturition initiates the onset of maternal behaviors. Indeed, the expression of maternal behaviors in sheep and rats is delayed following blockade of OT receptors during parturition (van Leengoed *et al.*, 1987; Levy *et al.*, 1992). OT release in response to suckling has been measured using microdialysis in the substantia nigra, olfactory bulbs, mediobasal hypothalamus, bed nucleus of the stria terminalis, MPOA and septum of parturient sheep (Kendrick *et al.*, 1997), as well as in sites of origin, the PVN and SON (Neumann *et al.*, 1993a, b). The release of OT within the latter two hypothalamic nuclei can be blocked by administration of an oxytocin receptor antagonist (OTA), suggesting a positive feedback mechanism controlling its own release (Neumann *et al.*, 1994). The specific effect of OT on brain activity following parturition, particularly during breastfeeding remains unclear. Here we tested whether OT modulates suckling-stimulated brain activity in postpartum dams.

With fMRI we observed that continuous suckling from pups during a 10-min period resulted in robust activation of brain areas involved in olfactory, emotional and reward processing. Data from these studies are shown in Figure 4.5. Suckling-stimulated brain activation was paralleled by OT administration (sc) and was partly blocked by an OTA. Our present results provide evidence that OT modulates brain activity during quiescent, or motorically inactive suckling. The best evidence is provided for the following regions: PVN, olfactory tubercle, anterior olfactory nucleus (AON), insular cortex, piriform cortex, cortical amygdala, MPOA and prefrontal cortex. For each of these regions, suckling and ICV OT administration increased the BOLD signal. In addition, OTA administration blocked the suckling-induced increase in the BOLD signal. Several of these areas have been associated with the olfactory system, either indirectly or by direct projections received from the olfactory bulbs (Price, 1973). This suggests that OT release during nursing contributes to olfactory-related neural activity.

The results for the PVN are not surprising since the PVN is involved in the milk-ejection reflex and OT exerts a positive feedback effect on its own release from PVN neurons (Neumann *et al.*, 1994). Importantly, the activated PVN is likely to be releasing OT into central neural

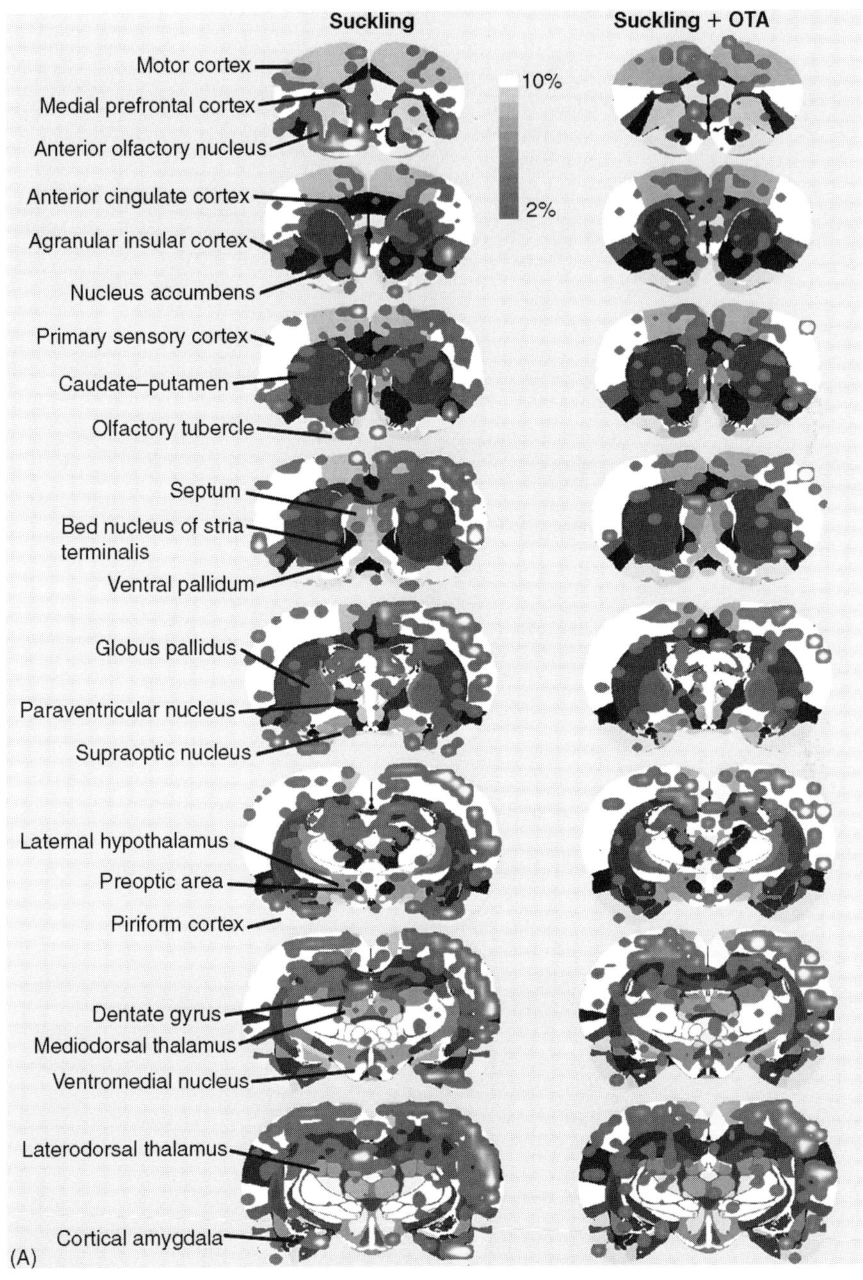

FIGURE 4.5 Composite brain maps of positive BOLD activity in response to suckling, suckling after OT receptor blockade (A) and in response to OT administration (B) in PPD 4–8 dams. Colored areas in (A) indicate increases in BOLD activity. BOLD activation maps in (B) were compared to anatomical maps of OT receptor binding. (*Source*: Adapted from Febo *et al.*, 2005b.) (See Color Plate)

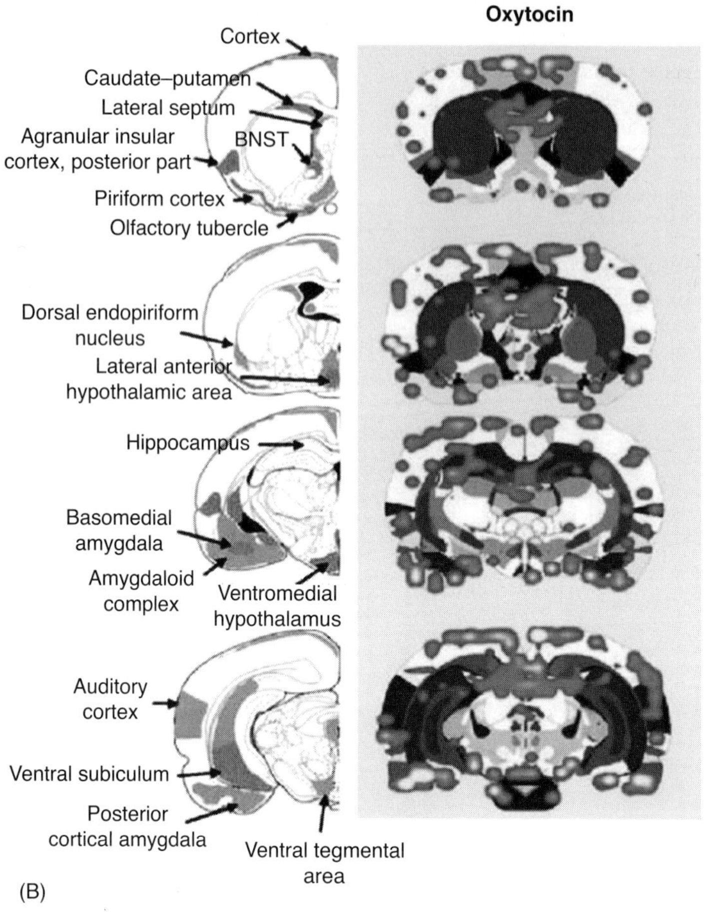

FIGURE 4.5 (Continued)

sites as well as into the periphery via the posterior pituitary. In the rat, destruction of the PVN delays the onset of maternal behaviors (Insel & Harbaugh, 1989). Similar effects have been observed by administering OTA (van Leengoed *et al.*, 1987). Our present data provides evidence that OT continues to be released within this area in fully maternal postpartum rats. With respect to the olfactory-related areas (Olf Tub, AON, insular, etc.), the increased BOLD signal in these areas may be related to aspects of maternal behavior control. Naïve virgin female rats are not maternally responsive but instead avoid pups and find their odors aversive, while postpartum rats are attracted to pups and their odors (Numan & Insel, 2003). Perhaps the increased BOLD signal observed in these regions as a result of suckling-induced OT release into the brain maintains the postpartum functional change in the olfactory system so that pup odors continue to be attractive. Importantly, although *acute* ICV injection of OTA does not cause a major disruption of maternal behavior, perhaps a long-term disruption of central oxytocinergic effects would be much more disruptive (Pedersen *et al.*, 1995; Pedersen & Boccia, 2003).

The increased BOLD signal in the MPOA as a result of suckling-induced OT release into the brain may also be related to maternal behavior. An intact MPOA is essential for maternal behavior, and estradiol and prolactin act on the MPOA to stimulate maternal behavior (Numan *et al.*, 1977; Bridges *et al.*, 1990; Numan & Insel, 2003). In addition, direct OTA administration into the MPOA blocks the onset of maternal behavior in parturient rats (Pedersen *et al.*, 1994).

We observed activation of the accumbens with suckling which probably relates to the release of dopamine (Hansen *et al.*, 1993). The fact that we also observed accumbens activation with OT could suggest that this neuropeptide somehow induces the release of dopamine within this region, and perhaps the prefrontal cortex. Since there is no evidence of OT receptor density in the accumbens (Tribollet *et al.*, 1988; Yoshimura *et al.*, 1993; Veinante & Freund-Mercier, 1997; Vaccari *et al.*, 1998) or OT release within this region (Kendrick *et al.*, 1997), it is possible that OT's actions are mediated via the VTA and/or MPOA (Numan & Smith, 1984).

FINAL SUMMARY

To move beyond the rodent data and to make some general statements, the results of the current series of experiments suggest that suckling stimulation during nursing may have dramatic influences on mammalian maternal behavior; it appears to influence maternal motivation and the mother–infant bond, maternal aggressiveness and learning and memory mechanisms which probably enable the mother to recognize environments which are safe places to care for infants. Activation of wide areas of the cortex by suckling appears to be involved in these effects, and some of these processes may be influenced by suckling-induced release of OT in selected neural sites (Febo *et al.*, 2005b). The neuroanatomical substrates activated by OT administration in the lactating rat closely paralleled that observed with suckling stimulation alone. Moreover, blockade of OT receptors with a specific antagonist selectively reduced brain activity in many of these common areas, evidence that endogenous OT has a role in the neurobiology of nursing. There were non-overlapping brain regions, namely the caudate–putamen, septum and the thalamus, that could involve other neuropeptide and neurotransmitter systems independent of OT release. These may include vasopressin, opioids and dopamine. Understanding the neurobiology of maternal–infant bonding may improve the future clinical treatment of mothers afflicted by psychiatric disease states such as depression, anxiety and drug addiction. Novel sensory experiences, particularly those associated with epochal developmental events like lactation, can alter cortical representation, affecting memory, perception and behavior.

REFERENCES

Bartels, A., and Zeki, S. (2004). The neural correlates of maternal and romantic love. *Neuroimage* 21, 1155–1166.

Becker, J. B. (1999). Gender differences in dopaminergic function in striatum and nucleus accumbens. *Pharmacol. Biochem. Behav.* 64, 803–812.

Boyett-Anderson, J. M., Lyons, D. M., Reiss, A. L., Schatzberg, A. F., and Menon, V. (2003). Functional brain imaging of olfactory processing in monkeys. *Neuroimage* 20, 257–264.

Bridges, R. S. (1975). Long-term effects of pregnancy and parturition upon maternal responsiveness in the rat. *Physiol. Behav.* 14, 245–249.

Bridges, R. S., Numan, M., Ronsheim, P. M., Mann, P. E., and Lupini, C. E. (1990). Central prolactin infusions stimulate maternal behavior in steroid-treated, nulliparous female rats. *Proc. Natl. Acad. Sci. USA* 87, 8003–8007.

Febo, M., and Ferris, C. F. (2007). Development of cocaine sensitization before pregnancy affects subsequent maternal retrieval of pups and prefrontal cortical activity during nursing. *Neuroscience* 148, 400–412.

Febo, M., Gonzalez-Rodriguez, L. A., Capo-Ramos, D. E., Gonzalez-Segarra, N. Y., and Segarra, A. C. (2003). Estrogen-dependent alterations in D2/D3-induced G protein activation in cocaine-sensitized female rats. *J. Neurochem.* 86, 405–412.

Febo, M., Ferris, C. F., and Segarra, A. C. (2005a). Estrogen influences cocaine-induced blood oxygen level-dependent signal changes in female rats. *J. Neurosci.* 25, 1132–1136.

Febo, M., Numan, M., and Ferris, C. F. (2005b). Functional magnetic resonance imaging shows oxytocin activates brain regions associated with mother–pup bonding during suckling. *J. Neurosci.* 25, 11637–11644.

Febo, M., Stolberg, T. L., Numan, M., Bridges, R. S., Kulkarni, P., and Ferris, C. F. (2008). Nursing stimulation is more than tactile sensation: it is a multisensory experience. Accepted for publication in *Horm Behav.*

Ferris, C. F., Kulkarni, P., Sullivan, J. M. Jr., Harder, J. A., Messenger, T. L., and Febo, M. (2005). Pup suckling is more rewarding than cocaine: Evidence from functional magnetic resonance imaging and three-dimensional computational analysis. *J. Neurosci.* 25, 149–156.

Fize, D., Vanduffel, W., Nelissen, K., Denys, K., Chef d'Hotel, C., Faugeras, O., and Orban, G. A. (2003). The retinotopic organization of primate dorsal V4 and surrounding areas: A functional magnetic resonance imaging study in awake monkeys. *J. Neurosci.* 23, 7395–7406.

Fleming, A. S., Suh, E. J., Korsmit, M., and Rusak, B. (1994). Activation of Fos-like immunoreactivity in the medial preoptic area and limbic structures by maternal and social interactions in rats. *Behav. Neurosci.* 108, 724–734.

Fleming, A. S., Kraemer, G. W., Gonzalez, A., Lovic, V., Rees, S., and Melo, A. (2002). Mothering begets mothering: The transmission of behavior and its neurobiology across generations. *Pharmacol. Biochem. Behav.* 73, 61–75.

Francis, D., Diorio, J., Liu, D., and Meaney, M. J. (1999). Nongenomic transmission across generations of maternal behavior and stress responses in the rat. *Science* 286, 1155–1158.

Hansen, S., Harthon, C., Wallin, E., Lofberg, L., and Svensson, K. (1991a). The effects of 6-OHDA-induced dopamine depletions in the ventral or dorsal striatum on maternal and sexual behavior in the female rat. *Pharmacol. Biochem. Behav.* 39, 71–77.

Hansen, S., Harthon, C., Wallin, E., Lofberg, L., and Svensson, K. (1991b). Mesotelencephalic dopamine system and reproductive behavior in the female rat: Effects of ventral tegmental 6-hydroxydopamine lesions on maternal and sexual responsiveness. *Behav. Neurosci.* 105, 588–598.

Hansen, S., Bergvall, A. H., and Nyiredi, S. (1993). Interaction with pups enhances dopamine release in the ventral striatum of maternal rats: A microdialysis study. *Pharmacol. Biochem. Behav.* 45, 673–676.

Hecht, G. S., Spear, N. E., and Spear, L. P. (1999). Changes in progressive ratio responding for intravenous cocaine throughout the reproductive process in female rats. *Dev. Psychobiol.* 35, 136–145.

Insel, T. R., and Harbaugh, C. R. (1989). Lesions of the hypothalamic paraventricular nucleus disrupt the initiation of maternal behavior. *Physiol. Behav.* 45, 1033–1041.

Keer, S. E., and Stern, J. M. (1999). Dopamine receptor blockade in the nucleus accumbens inhibits maternal retrieval and licking, but enhances nursing behavior in lactating rats. *Physiol. Behav.* 67, 659–669.

Kendrick, K. M., Da Costa, A. P., Broad, K. D., Ohkura, S., Guevara, R., Levy, F., and Keverne, E. B. (1997). Neural control of maternal behaviour and olfactory recognition of offspring. *Brain Res. Bull.* 44, 383–395.

King, J., Garelick, T. S., Brevard, M. E., Chen, W., Messenger, T L., Duong, T. Q., and Ferris, C. F. (2005). Procedure for minimizing stress for fMRI studies in conscious rats. *J. Neurosci. Methods* 148, 154–160.

Lahti, K. M., Ferris, C. F., Li, F., Sotak, C. H., and King, J. A. (1999). Comparison of evoked cortical activity in conscious and propofol-anesthetized rats using functional MRI. *Magn. Reson. Med.* 41, 412–416.

Lee, A., Clancy, S., and Fleming, A. S. (1999a). Mother rats bar-press for pups: Effects of lesions of the mpoa and limbic sites on maternal behavior and operant responding for pup-reinforcement. *Behav. Brain Res.* 100, 15–31.

Lee, A., Li, M., Watchus, J., and Fleming, A. S. (1999b). Neuroanatomical basis of maternal memory in postpartum rats: Selective role for the nucleus accumbens. *Behav. Neurosci.* 113, 523–538.

Leibenluft, E., Gobbini, M. I., Harrison, T., and Haxby, J. V. (2004). Mothers' neural activation in response to pictures of their children and other children. *Biol. Psychiatry* 56, 225–232.

Levy, F., Kendrick, K. M., Keverne, E. B., Piketty, V., and Poindron, P. (1992). Intracerebral oxytocin is important for the onset of maternal behavior in inexperienced ewes delivered under peridural anesthesia. *Behav. Neurosci.* 106, 427–432.

Li, C., Chen, P., and Smith, M. S. (1999). Neural populations in the rat forebrain and brainstem activated by the suckling stimulus as demonstrated by cFos expression. *Neuroscience* 94, 117–129.

Li, M., and Fleming, A. S. (2003). Differential involvement of nucleus accumbens shell and core subregions in maternal memory in postpartum female rats. *Behav. Neurosci.* 117, 426–445.

Lin, S. H., Miyata, S., Matsunaga, W., Kawarabayashi, T., Nakashima, T., and Kiyohara, T. (1998). Metabolic mapping of the brain in pregnant, parturient and lactating rats using fos immunohistochemistry. *Brain Res.* 787, 226–236.

Lonstein, J. S., and Stern, J. M. (1997). Role of the midbrain periaqueductal gray in maternal nurturance and aggression: c-fos and electrolytic lesion studies in lactating rats. *J. Neurosci.* 17, 3364–3378.

Lonstein, J. S., Simmons, D. A., Swann, J. M., and Stern, J. M. (1998). Forebrain expression of c-fos due to active maternal behaviour in lactating rats. *Neuroscience* 82, 267–281.

Lorberbaum, J. P., Newman, J. D., Horwitz, A. R., Dubno, J. R., Lydiard, R. B., Hamner, M. B., Bohning, D. E., and George, M. S. (2002). A potential role for thalamocingulate circuitry in human maternal behavior. *Biol. Psychiatry* 51, 431–445.

Ludwig, R., Bodgdanov, G., King, J., Allard, A., and Ferris, C. F. (2004). A dual RF resonator system for high-field functional magnetic resonance imaging of small animals. *J. Neurosci. Methods* 132, 125–135.

Mattson, B. J., and Morrell, J. I. (2005). Preference for cocaine- versus pup-associated cues differentially activates neurons expressing either Fos or cocaine- and amphetamine-regulated transcript in lactating, maternal rodents. *Neuroscience* 135, 315–328.

Mattson, B. J., Williams, S., Rosenblatt, J. S., and Morrell, J. I. (2001). Comparison of two positive reinforcing stimuli: Pups and cocaine throughout the postpartum period. *Behav. Neurosci.* 115, 683–694.

Mattson, B. J., Williams, S. E., Rosenblatt, J. S., and Morrell, J. I. (2003). Preferences for cocaine- or pup-associated chambers differentiates otherwise behaviorally identical postpartum maternal rats. *Psychopharmacology (Berl)* 167, 1–8.

Morgan, H. D., Fleming, A. S., and Stern, J. M. (1992). Somatosensory control of the onset and retention of maternal responsiveness in primiparous Sprague-Dawley rats. *Physiol. Behav.* 51, 549–555.

Neumann, I., Ludwig, M., Engelmann, M., Pittman, Q. J., and Landgraf, R. (1993a). Simultaneous microdialysis in blood and brain: Oxytocin and vasopressin release in response to central and peripheral osmotic stimulation and suckling in the rat. *Neuroendocrinology* 58, 637–645.

Neumann, I., Russell, J. A., and Landgraf, R. (1993b). Oxytocin and vasopressin release within the supraoptic and paraventricular nuclei of pregnant, parturient and lactating rats: A microdialysis study. *Neuroscience* 53, 65–75.

Neumann, I., Koehler, E., Landgraf, R., and Summy-Long, J. (1994). An oxytocin receptor antagonist infused into the supraoptic nucleus attenuates intranuclear and peripheral release of oxytocin during suckling in conscious rats. *Endocrinology* 134, 141–148.

Nitschke, J. B., Nelson, E. E., Rusch, B. D., Fox, A. S., Oakes, T. R., and Davidson, R. J. (2004). Orbitofrontal cortex tracks positive mood in mothers viewing pictures of their newborn infants. *Neuroimage* 21, 583–592.

Numan, M. (1994). A neural circuitry analysis of maternal behavior in the rat. *Acta Paediatr. Suppl.* 397, 19–28.

Numan, M., and Smith, H. G. (1984). Maternal behavior in rats: Evidence for the involvement of preoptic projections to the ventral tegmental area. *Behav. Neurosci.* 98, 712–727.

Numan, M., and Corodimas, K. P. (1985). The effects of paraventricular hypothalamic lesions on maternal behavior in rats. *Physiol. Behav.* 35, 417–425.

Numan, M., and Insel, T. R. (2003). *The Neurobiology of Parental Behavior (Hormones, Brain, and Behavior Series)*. Springer-Verlag, New York.

Numan, M., Rosenblatt, J. S., and Komisaruk, B. R. (1977). Medial preoptic area and onset of maternal behavior in the rat. *J. Comp. Physiol. Psychol.* 91, 146–164.

Numan, M., Numan, M. J., Pliakou, N., Stolzenberg, D. S., Mullins, O. J., Murphy, J. M., and Smith, C. D. (2005). The effects of D1 or D2 dopamine receptor antagonism in the medial preoptic area, ventral pallidum, or nucleus accumbens on the maternal retrieval response and other aspects of maternal behavior in rats. *Behav. Neurosci.* 119, 1588–1604.

Ogawa, S., Lee, T. M., Kay, A. R., and Tank, D. W. (1990). Brain magnetic resonance imaging with contrast dependent on blood oxygenation. *Proc. Natl. Acad. Sci. USA* 87, 9868–9872.

Orpen, B. G., Furman, N., Wong, P. Y., and Fleming, A. S. (1987). Hormonal influences on the duration of postpartum maternal responsiveness in the rat. *Physiol. Behav.* 40, 307–315.

Panksepp, J., Knutson, B., and Burgdorf, J. (2002). The role of brain emotional systems in addictions: A neuro-evolutionary perspective and new 'self-report' animal model. *Addiction* 97, 459–469.

Pedersen, C. A., and Boccia, M. L. (2003). Oxytocin antagonism alters rat dams' oral grooming and upright posturing over pups. *Physiol. Behav.* 80, 233–241.

Pedersen, C. A., Caldwell, J. D., Walker, C., Ayers, G., and Mason, G. A. (1994). Oxytocin activates the postpartum onset of rat maternal behavior in the ventral tegmental and medial preoptic areas. *Behav. Neurosci.* 108, 1163–1171.

Pedersen, C. A., Johns, J. M., Musiol, I., Perez-Delgado, M., Ayers, G., Faggin, B., and Caldwell, J. D. (1995). Interfering with somatosensory stimulation from pups sensitizes experienced, postpartum rat mothers to oxytocin antagonist inhibition of maternal behavior. *Behav. Neurosci.* 109, 980–990.

Price, J. L. (1973). An autoradiographic study of complementary laminar patterns of termination of afferent fibers to the olfactory cortex. *J. Comp. Neurol.* 150, 87–108.

Pruessner, J. C., Champagne, F., Meaney, M. J., and Dagher, A. (2004). Dopamine release in response to a psychological stress in humans and its relationship to early life maternal care: A positron emission tomography study using [11C]raclopride. *J. Neurosci.* 24, 2825–2831.

Ranote, S., Elliott, R., Abel, K. M., Mitchell, R., Deakin, J. F., and Appleby, L. (2004). The neural basis of maternal responsiveness to infants: An fMRI study. *Neuroreport* 15, 1825–1829.

Seifritz, E., Esposito, F., Neuhoff, J. G., Luthi, A., Mustovic, H., Dammann, G., von Bardeleben, U., Radue, E. W., Cirillo, S., Tedeschi, G., and Di Salle, F. (2003). Differential sex-independent amygdala response to infant crying and laughing in parents versus nonparents. *Biol. Psychiatry* 54, 1367–1375.

Slotnick, B. M. (1967). Disturbances of maternal behavior in the rat following lesions of the cingulate cortex. *Behaviour* 29, 204–236.

Slotnick, B. M., and Nigrosh, B. J. (1975). Maternal behavior of mice with cingulate cortical, amygdala, or septal lesions. *J. Comp. Physiol. Psychol.* 88, 118–127.

Stack, E. C., Balakrishnan, R., Numan, M. J., and Numan, M. (2002). A functional neuroanatomical investigation of the role of the medial preoptic area in neural circuits regulating maternal behavior. *Behav. Brain Res.* 131, 17–36.

Stern, J. M., and Johnson, S. K. (1990). Ventral somatosensory determinants of nursing behavior in Norway rats. I. Effects of variations in the quality and quantity of pup stimuli. *Physiol. Behav.* 47, 993–1011.

Stern, J. M., and Kolunie, J. M. (1993). Maternal aggression of rats is impaired by cutaneous anesthesia of the ventral trunk, but not by nipple removal. *Physiol. Behav.* 54, 861–868.

Tenney, J. R., Duong, T. Q., King, J. A., and Ferris, C. F. (2004). FMRI of brain activation in a genetic rat model of absence seizures. *Epilepsia* 45, 576–582.

Tribollet, E., Barberis, C., Dreifuss, J. J., and Jard, S. (1988). Autoradiographic localization of vasopressin and oxytocin binding sites in rat kidney. *Kidney Int.* 33, 959–965.

Tsingotjidou, A., and Papadopoulos, G. C. (1996). Neuronal expression of Fos-like protein along the afferent pathway of the milk-ejection reflex in the sheep. *Brain Res.* 741, 309–313.

Vaccari, C., Lolait, S. J., and Ostrowski, N. L. (1998). Comparative distribution of vasopressin V1b and oxytocin receptor messenger ribonucleic acids in brain. *Endocrinology* 139, 5015–5033.

van Leengoed, E., Kerker, E., and Swanson, H. H. (1987). Inhibition of post-partum maternal behaviour in the rat by injecting an oxytocin antagonist into the cerebral ventricles. *J. Endocrinol.* 112, 275–282.

Veinante, P., and Freund-Mercier, M. J. (1997). Distribution of oxytocin- and vasopressin-binding sites in the rat extended amygdala: A histoautoradiographic study. *J. Comp. Neurol.* 383, 305–325.

Voorn, P., Vanderschuren, L. J., Groenewegen, H. J., Robbins, T. W., and Pennartz, C. M. (2004). Putting a spin on the dorsal–ventral divide of the striatum. *Trends Neurosci.* 27, 468–474.

Walsh, C. J., Fleming, A. S., Lee, A., and Magnusson, J. E. (1996). The effects of olfactory and somatosensory desensitization on Fos-like immunoreactivity in the brains of pup-exposed postpartum rats. *Behav. Neurosci.* 110, 134–153.

Weaver, I. C., Cervoni, N., Champagne, F. A., D'Alessio, A. C., Sharma, S., Seckl, J. R., Dymov, S., Szyf, M., and Meaney, M. J. (2004). Epigenetic programming by maternal behavior. *Nat. Neurosci.* 7, 847–854.

Xerri, C., Stern, J. M., and Merzenich, M. M. (1994). Alterations of the cortical representation of the rat ventrum induced by nursing behavior. *J. Neurosci.* 14, 1710–1721.

Yoshimura, R., Kiyama, H., Kimura, T., Araki, T., Maeno, H., Tanizawa, O., and Tohyama, M. (1993). Localization of oxytocin receptor messenger ribonucleic acid in the rat brain. *Endocrinology* 133, 1239–1246.

5 MATERNAL CHOICES: NEURAL MEDIATION – CARING FOR YOUNG OR HUNTING?

LUCIANO F. FELÍCIO AND NEWTON S. CANTERAS

Faculdade de Medicina Veterinária e Zootecnia, Instituto de Ciências Biomédicas, Universidade de São Paulo, São Paulo, Brasil

BALANCING MATERNAL CARE AND FORAGING DURING LACTATION

Every mother has to maintain a balance between means of survival and reproduction. In other words, all mothers balance tradeoffs between subsistence and reproduction (Blaffer-Hrdy, 1999). For mammals, balancing subsistence and reproduction during the postpartum period is critical. This demands a great deal of adaptability and versatility. Thus, a successful strategy for the maintenance of a certain species might depend on selecting the most adaptive behavior. During the postpartum period, dams spend great amounts of energy lactating, nursing, and caring for their young. Meanwhile, supply availability may vary. The female brain is equipped with mechanisms that allow the mother to switch from nursing to hunting. Such feature might be important for achieving the best possible level of successful reproductive experience. Spending the optimum amount of time and energy on foraging and hunting may depend on the circumstances established by environmental constraints. If there is plenty of food, a lactating female may spend more time taking care of her litter. In contrast, a food-lacking environment will probably force her to use more of her working time looking for nutrients to meet her and her litter needs. Thus, such mechanisms would have to be plastic and adaptive as well. Processing environmental clues during pregnancy, anticipating the postpartum situation in terms of food supply availability may play an important adaptive role. Such ability may allow a female to prepare herself for the postpartum environmental challenges, or even better, anticipate her adjustment to forthcoming demands. Thus, at the proper time, her endogenous means for tradeoff performances can express an optimum adjustment. Our studies have investigated these switching mechanisms and their neuroanatomical basis. Recent data from our laboratory examining the neural mechanisms underlying the morphine-induced maternal behavior inhibition shed some light on key neural sites involved in the switching from nursing to hunting. Those data strongly suggest that the periaqueductal gray (PAG) plays a major role in such behavioral selection. In addition, by the use of biochemical and pharmacological approaches, we have been able to reveal some adaptive features of this switching mechanism.

THE ROLE OF OPIOIDS

The combined action of steroid hormones such as estradiol and progesterone results in changes in opioid binding in the brain. In addition, opioid stimulation on opioid receptors modulates the action of genes that encode for those receptors (Teodorov *et al.*, 2006). Both binding and sensitivity to opioids decrease from pregnancy to lactation. Studies on the influence of opioids on maternal behavior usually take advantage of pharmacological tools. Morphine, a classical opioid agonist, stimulates various types of opioid receptors, that is, the mu (μ), kappa (κ), and delta (δ) opiate receptors. Morphine is a non-peptide, differing in its chemical nature from the endogenous opioids that are peptides. In order to investigate the role of the endogenous opioids some studies have utilized opioid peptide molecules such as β-endorphin that are endogenous substances. Since those molecules do not cross the blood–brain barrier in a large extent, they have either to be infused directly into specific brain areas such as medial preoptic area (MPOA) or into the central ventricular system to be distributed throughout the brain. Another elegant pharmacological approach to investigate the role of endogenous opioids on their receptors is to use opioid antagonists. Thus, if a drug has its effect reversed by the antagonist, this is interpreted that the drug action was due to its binding to that specific receptor. In addition, if a function is compromised by the antagonist itself, this suggests that role was being played by one or more endogenous opioids (Sukikara *et al.*, 2007). Other techniques including measurements of gene expression, gene knockout, and lesions have been used to investigate the possible opioid role in maternal behavior. It has been well established that opioids play a role in maternal behavior. Treatment of postpartum female rats with morphine and other opioidergic agonists inhibit maternal behavior (Bridges & Grimm, 1982; Rubin & Bridges, 1984; Mann *et al.*, 1990, 1991; Felicio *et al.*, 1991). Lactating animals also undergo functionally relevant changes in opioid regulation of pain sensitivity and maternal behavior (Kinsley & Bridges, 1988). A decrease of opioid peptide activity is important for the expression of certain postpartum behaviors, particularly those related to the female approaching and becoming attached to the pups. Since the opioid antagonist naloxone reverses the morphine-induced inhibition of maternal behavior, it has been suggested that these morphine actions are due to its stimulation of opioid receptors (Bridges & Grimm, 1982; Mann & Bridges, 1992). A number of studies examining the distribution of Fos protein have shown many areas of the lactating rat brain mobilized in response to maternal behavior during physical interaction between dams and pups, including the MPOA and the ventral bed nucleus of the stria terminalis (Stafisso-Sandoz *et al.*, 1998). Since, the MPOA is an opioid-sensitive brain region (Gulledge *et al.*, 2000), which is essential for the expression of maternal behavior (Numan, 1988, 1994), studies have focused on the role of opioids in maternal behavior in this brain region. Both morphine and β-endorphin block maternal behavior in females when placed into the MPOA (Rubin & Bridges, 1984; Mann *et al.*, 1995). Thus, most of the classical studies on the opioid inhibitory role on maternal behavior have focused on this neural region (Numan, 1988; Stafisso-Sandoz *et al.*, 1998).

Since opioid systems are plastic and adaptive, long-term treatment with morphine may alter maternal behavior sensitivity to opioids as well. We and others have demonstrated that long-term treatment with morphine during late pregnancy makes lactating rats more sensitive to morphine-induced inhibition of maternal behavior (Miranda-Paiva *et al.*, 2001, 2003; Slamberová *et al.*, 2001; Sukikara *et al.*, 2006; Yim *et al.*, 2006). Of particular relevance, in morphine-experienced dams those low doses of morphine inhibited maternal behavior by acting on the lateral PAG, and not in the MPOA. Thus, low doses of morphine induce a striking activation of the lateral PAG and an intense maternal inhibition in morphine-treated dams. Naloxone injections were able to re-establish maternal behavior in these dams.

The PAG has an important role in the organization and modulation of various motivated behaviors, that is defensive, sexual, and maternal behaviors. There is, however, a scarceness of research on the role of the PAG on maternal behavior. Some reports focused on the somatosensory aspects of the PAG's role in the physiological control of intense nursing or crouching posture (Lonstein & Stern, 1997a, b). The opioidergic-induced lateral PAG activation that inhibits maternal behavior also stimulates the expression of foraging and predatory behaviors (Comoli *et al.*, 2003, 2005, Sukikara *et al.*, 2006). Since the behavioral repertoire during lactation has to be versatile and adaptive, neural systems with these characteristics may be useful for the animals to respond adequately to the demands of the environment. Repeated treatment with morphine induces different kinds of adaptive mechanisms. Both tolerance, for its sedative effects, and reversal tolerance or sensitization, for its excitatory and addictive effects have been well described (Post, 1980; Honkanen *et al.*, 1994; Felicio *et al.*, 2001;

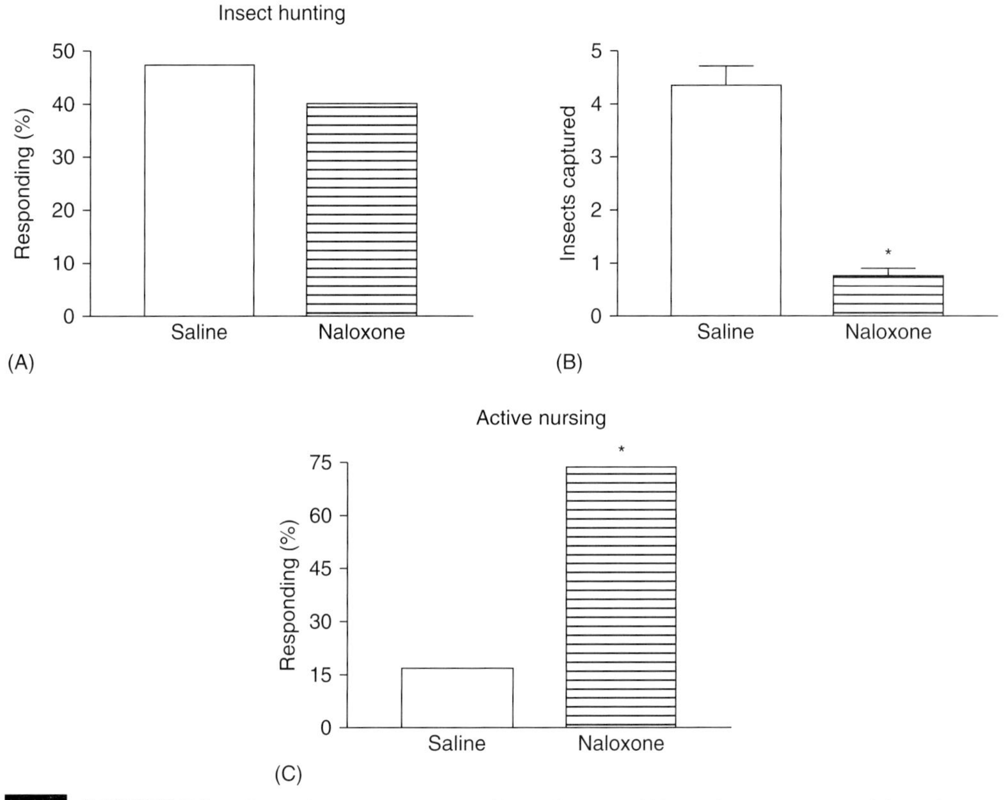

FIGURE 5.1 Effects of naloxone treatment (0.1 mg/kg; group Naloxone) on percentage of dams showing full maternal behavior (A), number of insects captured by each group (B), and percentage of dams displaying hunting (C). Number of insects captured by each group expressed as mean ± SEM. *$p < 0.05$ compared with the saline group (*Source*: From Sukikara *et al.*, 2007).

Yim *et al.*, 2006). Consistently, morphine pretreatment during late pregnancy increases opioid inhibitory effects on maternal behavior (Miranda-Paiva *et al.*, 2001). Thus, plasticity in opioidergic transmission may provide a substrate for the versatility and adaptations required for a dam during the postpartum period (i.e., inhibits maternal behavior and stimulates foraging). Specifically, opioids may underlie decision-making in lactating rats. Thus opioidergic transmission plays a role in the physiological state that may eventually underline decision-making in lactating rats. In order to address this possibility, lactating rats were tested simultaneously for maternal behavior and hunting. By exposing lactating dams simultaneously to pups and insects, it was found that: (1) opioidergic stimulation reduces maternal behavior by motivating predatory hunting in lactating females; (2) the opioid antagonist naloxone re-establishes maternal behavior by reducing hunting in lactating rats treated with morphine; and (3) dams treated only with naloxone increase active nursing while decreasing insect hunting (Figure 5.1). Treatment with the opioid antagonist naloxone also decreased the number of insects captured and increased the percentage of animals displaying nursing behavior. These results suggest that opioidergic transmission plays a key role in the regulation of behavioral selection during lactation with endogenous opioids stimulating hunting at the expense of maternal behavior during lactation (Sukikara *et al.*, 2007).

What brain region regulates this opioid response? It is established that the rostral lateral PAG is functionally important for the inhibition of maternal behavior during lactation. Interestingly, lesions in the rostral lateral PAG in lactating dams promote a switch back from hunting to maternal behavior (see Figure 5.2).

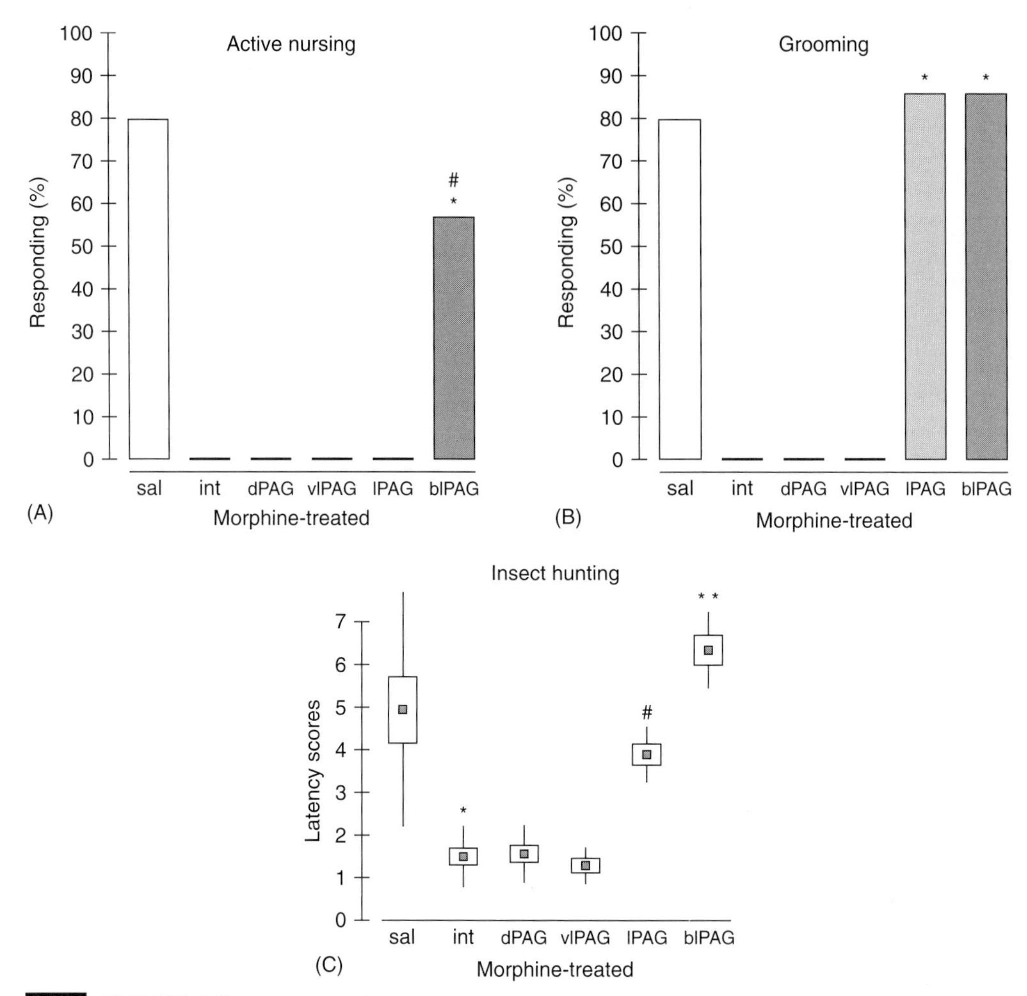

FIGURE 5.2 Percentage of animals expressing active nursing (A) and grooming the pups (B), as well as the latency scores to start capturing the roaches (C), for the saline-treated dams (sal); the intact morphine-treated animals (int); the morphine-treated rats with unilateral lesions in the dorsal (dPAG), ventrolateral (vlPAG), or lateral (lPAG) PAG; and the morphine-treated rats with bilateral lesions in the lateral PAG (blPAG). (C) shows the mean time taken to capture the first prey, averaged over each experimental group, after being transformed into rank-order data (1, 1.0–300 s; 2, 301–600 s; 3, 601–900 s; 4, 901–1,200 s; 5, 1,201–1,500 s; 6, 1,501–1,800 s; 7, >1,801 s); for illustration purposes only, these categorized data are expressed as arithmetic mean ± SE (boxes) and arithmetic mean ± SD (whiskers). (A) $*p < 0.005$ compared with the lPAG group, $\#p = 0.289$ compared with the saline group; (B) $*p < 0.005$ compared with the morphine-treated intact, dPAG, or vlPAG groups; (C) $*p < 0.0085$ compared with the saline group, $\#p < 0.002$ compared with the morphine-treated intact, dPAG, or vlPAG groups, $**p < 0.002$ compared with the morphine-treated lPAG group (*Source*: From Sukikara *et al.*, 2006).

Moreover, in morphine-treated dams the lateral PAG influences switching from maternal to hunting behavior. These findings support a previously unsuspected role for the PAG in the selection of adaptive behavioral responses. The PAG is in an advantaged position in the brain to play such a role, since the PAG receives projections from the prefrontal cortex that mediates behavioral planning. The PAG also receives projections from limbic structures that influence emotional state (Beitz, 1995). A role for the PAG is further supported by studies that

FIGURE 5.3 (A) Opioid administration stimulates hunting, which correlates with neuronal activation in the lateral PAG as reflected by an increase in the Fos-labeled neurons in the PAG. (B) When no conflict is evident and maternal behavior is observed, Fos-labeled neurons in the lateral PAG are less activated.

demonstrate that bilateral lesions in the lateral PAG produce intense direct maternal behavior even in morphine-treated animals exposed to insects. Moreover, hunting and Fos activity in the PAG are increased after morphine treatment, whereas maternal behavior is associated with low Fos expression (see Figure 5.3).

More investigation is needed on the neurobiological basis of the physiological states in which two adaptive behaviors compete with each other. How do two behaviors compete to an optimum time schedule strategy within the specific environmental context? Are there neurobiological mechanisms that help anticipating postpartum situations that would be challenging for the mother? It is possible that such female neural mechanisms play a role in the evolution of the species.

HORMONAL MEDIATION

Opioids stimulate prolactin release, a hormone known to stimulate maternal behavior. The influence of opioids on maternal behavior can be both direct and indirect (Bridges *et al.*, 1993). Endogenous changes in these molecules influence the sensitivity of sensory systems. This is particularly true with respect to olfactory clues (Kinsley & Bridges, 1990). Hormonal changes during late pregnancy induce the postpartum expression of maternal behavior. In general, progesterone serum levels that are kept high during pregnancy start declining before parturition. Estrogen and prolactin serum levels increase before parturition. Other hormones such as oxytocin, opioids, and sulfated cholecystokinin-octapeptide (CCK) also participate in the maintenance of maternal behavior (Felicio *et al.*, 1991; Mann *et al.*, 1995). Recently, we have reported that morphine treatment during late pregnancy increases serum progesterone levels without affecting corticosterone, estradiol, or prolactin levels (Felicio *et al.*, 2007). Since progesterone stimulates feeding behavior, it may activate hyperphagia during pregnancy. It is possible that the elevated levels of progesterone have a programming effect that continues to induce appetitive signals postpartum. This, in turn, could prompt foraging behavior. One can only suspect that the tardive effects described for this morphine treatment would be mediated by progesterone. This finding seems to fit in a more general theory. Specifically, endogenous increasing appetite clues during late pregnancy would make the organism more willing to display hunting and foraging behavior during the postpartum period.

INTERACTION OF CCK AND OPIOIDS CONTROLLING MATERNAL BEHAVIOR

Opioids and CCK seem to play functionally competitive roles that may have implications for maternal behavior in lactating rats (Felicio *et al.*, 1991, 2001; Miranda-Paiva & Felicio, 1999; Miranda-Paiva *et al.*, 2002, 2007). One of the excitatory effects of morphine is increasing appetite (Mucha & Iversen, 1986; Pomonis *et al.*, 2000). CCK, a gut peptide that is also present in the brain, is considered a classical endogenous satiety signal. Also, this peptide reverses morphine actions in pain threshold. Consistently, CCK reverses opioid inhibitory effects on maternal behavior (Felicio *et al.*, 1991; Mann *et al.*, 1995). It has been found that infusions of β-endorphin, an endogenous opioid, into the ventricular system of lactating rats block normal maternal behavior, while CCK can have effects opposite to those of opioids. The first study on this topic evaluated whether intracerebroventricular administration of CCK would be able to antagonize the inhibitory effect of β-endorphin on maternal behavior. The results of this study demonstrated that CCK prevented the β-endorphin-induced increase in latencies to retrieve the first pup, retrieve all pups, and to group and crouch over rat pups. In addition, reductions in the percentage of rats retrieving all pups and displaying full maternal behavior were prevented by CCK. These data suggest that CCK can act as an opioid antagonist in neural systems that control maternal behavior (Felicio *et al.*, 1991). Other studies tested the site-specificity of the β-endorphin–CCK interaction. Direct infusions of CCK into the MPOA blocked the disruptive effects of β-endorphin on the maintenance of maternal behavior in postpartum lactating rats. In addition, proglumide, a CCK receptor antagonist, disrupted maternal behavior in postpartum lactating rats by increasing latencies to retrieve and crouch over the young (Mann *et al.*, 1995). These results support the concept of a dual peptidergic control in the MPOA. More recently, given evidence that morphine-induced activation of the rostral lateral PAG is required to inhibit maternal behavior in lactating rats, we tested whether CCK would act similarly in the PAG. This hypothesis was confirmed. Morphine's inhibitory effect on maternal responsiveness was blocked by CCK injections into the rostral PAG, but not in nearby regions of the mesencephalic reticular nucleus (Miranda-Paiva *et al.*, 2007). CCK has two receptor subtypes, CCK1 and CCK2. In order to evaluate if both receptors were involved in this opioid–CCK interaction in the control of maternal behavior, CCK1 and CCK2 antagonists were tested in lactating rats treated with morphine. Peripheral injections of antagonists of the CCK1 receptor (lorglumide) and the CCK2 receptor (L-365,260) intensified the effects of morphine on maternal behavior. These results suggest that CCK antagonism of opioid-induced disruption of maternal behavior occurs due to the action of CCK on both CCK1 and CCK2 receptor subtypes (Miranda-Paiva & Felicio, 1999). It is possible that blocking CCK receptors increases opioidergic tone, producing the possible adaptive consequences. Curiously, pretreatment with a CCK antagonist during late pregnancy increases opioid inhibitory effects on maternal behavior (Miranda-Paiva *et al.*, 2002). These data suggest that long-term interactions between CCK and opioids might influence expression of maternal behavior and perhaps hunting as well.

It has been proposed, hedonistic perception and social interaction are modulated by opioids as well (Panksepp *et al.*, 1980). Thus, during pregnancy and lactation the opioid systems are involved in a dynamic combination of molecular, hormonal, and multipeptidergic actions that result in sensory and motivational changes that create the physiological condition needed to generate adaptive responses. Thus, competent neural mechanisms exist to cope with the adaptability and versatility required within the maternal environment. Mechanisms involved in these processes are starting to be identified.

ACKNOWLEDGMENTS

The authors would like to thank Ms. Cássia Cristiane Medea for drawing the mother rats. The authors' work is supported by FAPESP, CAPES, and CNPq. This chapter is dedicated to Lucilia de Freitas Felicio (in memoriam).

REFERENCES

Beitz, A. J. (1995). Periaqueductal gray. In *The Rat Nervous System* (G. Paxinos, Ed.), pp. 173–182. Academic Press, Sidney.

Blaffer-Hrdy, S. (1999). *Mother Nature: Maternal Instincts and How They Shape the Human Species*. Ballentine Publishing Group, New York.

Bridges, R. S., Felicio, L. F., Pellerin, L. J., Stuer, A. M., and Mann, P. E. (1993). Prior parity reduces post-coital diurnal and nocturnal prolactin surges in rats. *Life Sci.* 53, 439–445.

Bridges, R. S., and Grimm, C. T. (1982). Reversal of morphine disruption of maternal behavior by concurrent treatment with the opiate antagonist naloxone. Science. 218, 166–168.

Comoli, E., Ribeiro-Barbosa, E. R., and Canteras, N. S. (2003). Predatory hunting and exposure to a live predator induce opposite patterns of Fos immunoreactivity in the PAG. *Behav. Brain Res.* 138, 17–28.

Comoli, E., Ribeiro-Barbosa, E. R., Negrao, N., Goto, M., and Canteras, N. S. (2005). Functional mapping of the prosencephalic systems involved in organizing predatory behavior in rats. *Neuroscience* 130, 1055–1067.

Felicio, L. F., Mann, P. E., and Bridges, R. S. (1991). Intracerebroventricular cholecystokinin infusions block beta-endorphin-induced disruption of maternal behavior. *Pharm. Biochem. Behav.* 39, 201–204.

Felicio, L. F., Mazzini, B. K., Cacheiro, R. G., Cruz, T. N., Florio, J. C., and Nasello, A. G. (2001). Stimulation of either cholecystokinin receptor subtype reduces while antagonists potentiate or sensitize a morphine-induced excitatory response. *Peptides* 22, 1299–1304.

Felicio, L. F., Sukikara, M. H., Felippe, E. C. G., Anselmo-Franci, J. A., and Oliveira, C. A. (2007). Endocrine aspects of the opioidergic stimulation in pregnant rats. *Abstract Int. Behav. Neurosci. Soc.* 16, 96.

Gulledge, C. C., Mann, P. E., Bridges, R., Bialos, M., and Hammer, R. P., Jr. (2000). Expression of mu-opioid receptor mRNA in the medial preoptic area of juvenile rats. *Brain Res. Dev. Brain Res.* 119, 269–276.

Honkanen, A., Piepponen, T. P., and Ahtee, L. (1994). Morphine-stimulated metabolism of striatal and limbic dopamine is dissimilarly sensitized in rats upon withdrawal from chronic morphine treatment. *Neurosci. Lett.* 180, 119–122.

Kinsley, C. H., and Bridges, R. S. (1988). Parity-associated reductions in behavioral sensitivity to opiates. *Biol. Reprod.* 39, 270–278.

Kinsley, C. H., and Bridges, R. S. (1990). Morphine treatment and reproductive condition alter olfactory preferences for pup and adult male odors in female rats. *Dev. Psychobiol.* 23, 331–347.

Lonstein, J. S., and Stern, J. M. (1997a). Role of the midbrain periaqueductal gray in maternal nurturance and aggression: c-Fos and electrolytic lesion studies in lactating rats. *J. Neurosci.* 17, 3364–3378.

Lonstein, J. S., and Stern, J. M. (1997b). Somatosensory contributions to c-fos activation within the caudal periaqueductal gray of lactating rats: Effects of perioral, rooting, and suckling stimuli from pups. *Horm. Behav.* 32, 155–166.

Mann, P. E., and Bridges, R. S. (1992). Neural and endocrine sensitivities to opioids decline as a function of multiparity in the rat. *Brain Res.* 580, 241–248.

Mann, P. E., Pasternak, G. W., and Bridges, R. S. (1990). Mu 1 opioid receptor involvement in maternal behavior. *Physiol. Behav.* 47, 133–138.

Mann, P. E., Kinsley, C. H., and Bridges, R. S. (1991). Opioid receptor subtype involvement in maternal behavior in lactating rats. *Neuroendocrinology* 53, 487–492.

Mann, P. E., Felicio, L. F., and Bridges, R. S. (1995). Investigation into the role of cholecystokinin (CCK) in the induction and maintenance of maternal behavior in rats. *Horm. Behav.* 29, 392–406.

Miranda-Paiva, C. M., and Felicio, L. F. (1999). Differential role of cholecystokinin receptor subtypes in opioid modulation of ongoing maternal behavior. *Pharmacol. Biochem. Behav.* 64, 165–169.

Miranda-Paiva, C. M., Nasello, A. G., Yin, A. J., and Felicio, L. F. (2001). Morphine pretreatment increases opioid inhibitory effects on maternal behavior. *Brain Res. Bull.* 55, 501–505.

Miranda-Paiva, C. M., Nasello, A. G., Yim, A. J., and Felicio, L. F. (2002). Puerperal blockade of cholecystokinin (CCK_1) receptors disrupts maternal behavior in lactating rats. *J. Mol. Neurosci.* 18, 97–104.

Miranda-Paiva, C. M., Ribeiro-Barbosa, E. R., Canteras, N. S., and Felicio, L. F. (2003). A role for the periaqueductal grey in opioidergic inhibition of maternal behaviour. *Eur. J. Neurosci.* 18, 667–674.

Miranda-Paiva, C. M., Canteras, N. S., Sukikara, M. H., Nasello, A. G., Mackowiak, I. I., and Felicio, L. F. (2007). Periaqueductal gray cholecystokinin infusions block morphine-induced disruption of maternal behavior. *Peptides* 28, 657–662.

Mucha, R. F., and Iversen, S. D. (1986). Increased food intake after opioid microinjections into nucleus accumbens and ventral tegmental area of rat. *Brain Res.* 397, 214–224.

Numan, M. (1988). Neural basis of maternal behavior in the rat. *Psychoneuroendocrinology* 13, 47–62.

Numan, M. (1994). Maternal behavior. In *The Physiology of Reproduction* (E. Knobil and J. D. Neill, Eds.), 2nd edn., pp. 221–302. Raven Press, New York.

Panksepp, J., Herman, B. H., Vilberg, T., Bishop, P., and DeEskinazi, F. G. (1980). Endogenous opioids and social behavior. *Neurosci. Biobehav. Rev.* 4, 473–487.

Pomonis, J. D., Jewett, D. C., Kotz, C. M., Briggs, J. E., Billington, C. J., and Levine, A. S. (2000). Sucrose consumption increases naloxone-induced c-Fos immunoreactivity in limbic forebrain. *Am. J. Physiol. Regul. Integr. Comp. Physiol.* 278, R712–R719.

Post, R. (1980). Intermittent versus continuous stimulation: Effect of time interval on the development of sensitization or tolerance. *Life Sci.* 26, 1275–1282.

Rubin, B. S., and Bridges, R. S. (1984). Disruption of ongoing maternal responsiveness in rats by central administration of morphine sulfate. *Brain Res.* 307, 91–97.

Slamberová, R., Szilagyi, B., and Vathy, I. (2001). Repeated morphine administration during pregnancy attenuates maternal behavior. *Psychoneuroendocrinology* 26, 565–576.

Stafisso-Sandoz, G., Polley, D., Holt, E., Lambert, K. G., and Kinsley, C. H. (1998). Opiate disruption of maternal behavior: Morphine reduces, and naloxone restores, c-fos activity in the medial preoptic area of lactating rats. *Brain Res. Bull.* 45, 307–313.

Sukikara, M. H., Mota-Ortiz, S. R., Baldo, M. V., Felicio, L. F., and Canteras, N. S. (2006). A role for the periaqueductal gray in switching adaptive behavioral responses. *J. Neurosci.* 26, 2583–2589.

Sukikara, M. H., Platero, M. D., Canteras, N. S., and Felicio, L. F. (2007). Opiate regulation of behavioral selection during lactation. *Pharmacol. Biochem. Behav.* 87, 315–320.

Teodorov, E., Modena, C. C., Sukikara, M. H., and Felicio, L. F. (2006). Preliminary study of the effects of morphine treatment on opioid receptor gene expression in brain structures of the female rat. *Neuroscience* 141, 1225–1231.

Yim, A. J., Miranda-Paiva, C. M., Florio, J. C., Oliveira, C. A., Nasello, A. G., and Felicio, L. F. (2006). A comparative study of morphine treatment regimen prior to mating and during late pregnancy. *Brain Res. Bull.* 68, 384–391.

6 IMAGING THE HUMAN PARENTAL BRAIN

JAMES E. SWAIN[1] AND JEFFREY P. LORBERBAUM[2]

[1] *Child Study Center, Yale University School of Medicine, New Haven, CT, USA*
[2] *Psychiatry Department, Penn State University, Hershey Medical Center, Hershey, PA, USA*

BRAIN IMAGING OF HUMAN PARENT–INFANT RELATIONSHIPS

In this chapter, we present data on the brain basis of human parental behavior and thoughts. We limit our review to studies which used the high resolution and non-invasive brain imaging technique of blood-oxygen-dependant functional magnetic resonance imaging (fMRI) to measure regional brain activity in response to infant auditory, visual and related stimuli. fMRI assays brain activity by indirectly measuring changes in regional blood oxygenation. The differences between a region's oxygenated and deoxygenated hemoglobin, between states of action vs. inaction for instance, provide characteristic magnetic signals localized to millimeters that are detected by scanners positioned around each subject's head. An important caveat throughout the interpretation of parenting fMRI studies, however, is that these brain activity measurements are indirect and represent an integration of instantaneous electrical brain activity over at least seconds. Furthermore, fMRI related blood flow change lags behind synaptic events over 3–6 seconds.

Furthermore, experimental design captures brain activity over periods of a few seconds or tens of seconds. On the one hand, short blocks or events may capture briefly held mental states, but miss bigger changes such as sustained emotion, while on the other hand longer blocks may capture more complex brain responses, but also average them out making subtle responses more difficult to detect. Brain activity during these blocks may then be measured and compared between periods of attending to stimuli of interest and control stimuli to generate maps of the brain indicating differences in brain activity that may be important for one set of perceptions and thoughts vs. another. So far, a small group of studies have been published in which parents have been the subjects and infant cries and pictures have been stimuli. In such studies, for example, comparisons of brain activity measured during baby cry vs. control sound experience may yield significant differences in certain brain regions that relate to the parental experience of a baby cry, and so the associated parenting thoughts and behaviors. Some of these studies had small subject numbers and fixed effects analyses that do not take inter-subject variability into account while other studies use random effects analyses that account for inter-subject variability and permit generalization of findings. However, before we describe the parenting studies, we begin with a brief review of recent literature on social thoughts and empathy that are instructive as to how interpret human parenting brain imaging experiments, and to tease apart aspects of human parenting brain circuitry.

THE NEUROBIOLOGY OF HEALTHY EMPATHY AND PARENTING

Empathy, defined as appropriate perception, experience, and response to another's emotion, is especially relevant to parenting in which infants needs are great, yet most communication is non-verbal. The growing field of cognitive neuroscience, propelled by modern brain imaging techniques and interest in deficits of empathy in such disorders as autism, has revealed networks of brain activity relating to empathy and emotional mirroring (Gallese *et al.*, 2004; Uddin *et al.*, 2007). These empathy systems seem to overlap significantly with brain responses of parents to infant stimuli reviewed in this paper. Two of these overlapping regions are the cingulate and insular cortices.

In one fascinating study, focusing on the neuroanatomy of empathy using fMRI techniques, Singer and colleagues measured brain activity while volunteers experienced a painful stimulus or observed a signal indicating that their loved one ("other"), present in the same room, had received a similar pain stimulus (Singer *et al.*, 2004). They found a separation of circuits responding to the sensory-discriminative components of pain from the autonomic-affective aspects. Specifically, both experiencing pain and observing and empathically experiencing the pain of a loved one activated the insula and anterior cingulate with the latter activating relatively more anterior regions of these structures. The experience of one's own pain oneself also activated the brainstem, cerebellum, and sensorimotor cortex. Such decoupled yet parallel representations of empathy in cortical structures such as the insula and observing and empathically anterior cingulate have been postulated to be necessary for our empathic abilities to mentalize, that is, to understand the thoughts, beliefs, and intentions of others in relation to ourselves (Frith & Frith, 2003). It may well be that humans use separate circuits to decouple representations of the external vs. internal information to understand physical properties and assess personal emotional values. This framework may be of great importance to those studying the brain substrates of relationships, including the interface of external experiences with the internal representations that pervade our mood and allow planning.

In another wonderful study of the cingulate in mediating the brain basis of social behavior, Eisenberger and colleagues utilized virtual reality to simulate feelings of social isolation. In this study (Eisenberger *et al.*, 2003), the subject is involved in a virtual game of Cyberball which includes three players. Suddenly, the subject player is excluded from the virtual game and there is a rapid change in the anterior cingulate cortex activity. Perhaps the cingulate mediates a social loss/separation/attachment/grief system, which may be so important to parenting as well as to critical developments in each individual. Thus, in addition to registering physical pain, the anterior cingulate may also be an important circuit in thinking about a range of emotional signals (including social pain such as in witnessing the pain of a loved one, social rejection, or stimuli of one's child or romantic love), all involving coordinated shifts in attention, decision-making, memory, mood regulation, and directed behavior.

The insula has also been raised as an important center for integrating emotional information (Carr *et al.*, 2003) with connections to cortical mirror areas in the posterior parietal, inferior frontal, and superior temporal cortical regions also of interest. In one study subjects were shown pictures of standard emotional faces (happy, sad, angry, surprised, disgusted, and afraid) and fMRI was used to measure responses to two behavioral tasks: (i) mere observation and (ii) observation as well as internal simulation of the emotion observed.

As expected, imitation produced greater activity in frontotemporal areas in the mirror network, including the premotor face area, the dorsal pars opercularis of the inferior frontal cortex, and the superior temporal sulcus. Imitation also produced greater activity in the right anterior insula and right amygdala. This is particularly intriguing in light of evidence that the anterior insula responds to pleasant "caress-like" touch (Olausson *et al.*, 2002) and that the insula plays a crucial role in emotional and interpersonal interaction in health and mental illness such as autism (Dapretto *et al.*, 2006). A further confirmation of the insula's role in emotion recognition comes from the study of patients with strokes. Stroke patients with insular lesions showed a significantly greater deficit in emotion recognition than other stroke patients (Bodini *et al.*, 2004). We speculate that cingulate and insula will continue to emerge as key areas of importance during the social recognition of others important to the personal transformations that are typical in the initial formation of a new family as well as other relationships. Further research on the brain basis of thinking about other minds (mentalization) and empathy is also beginning to dissect the brain basis of complex social emotional thinking

(Pelphrey *et al.*, 2005; Saxe, 2006b).This research suggests that specific regions in the medial prefrontal cortex and temporal cortex mediate empathic thoughts and collaborative behaviors. Perhaps studies of high-risk families will fail to show these patterns of activation. Furthermore, early intervention programs shown to have beneficial long-term effects on child development (Olds *et al.*, 2004, 2007), may be associated with "corrections" in the activation patterns seen in these limbic cortical regions in response to infant stimuli.

In another innovative approach to explicitly study the biological bases of adult attachment, line drawings were presented to subjects during brain imaging which were meant to activate the attachment system with depictions of illness, solitude, separation, and abuse (Buchheim *et al.*, 2006). Subjects with organized compared to disorganized attachment patterns showed decreased activity in the right amygdala, left hippocampus, and right inferior frontal gyrus sub regions, which are considered to be important to fear and diminished social bonding and attachment. In the following section, we describe attempts to specifically understand the brain basis of parental attachment by presenting emotionally charged infant stimuli during brain imaging. We hypothesize that such "parenting" brain circuits share much in common with those that regulate other social attachments.

The experiments to date on parents using baby sound and visual stimuli with brain fMRI are summarized in Tables 6.1 and 6.2, respectively. These inclusive reference tables are intended to suggest patterns of response across all studies and stimuli at a glance and to stimulate future studies. Parent brain areas of increased activity with baby stimuli are indicated in these tables with "ACT," while areas of decreased activity are indicated by "DEACT." It is important to keep in mind that human brains are awash with blood at all times and we are just able to measure differences between different periods – not yet absolute or nuanced mental states. Also indicated are the number of subjects, age of infants at time of scan, type of study (magnet strength and block or event design), and stimuli used in each study. Statistical methods vary across studies, but all findings satisfy the criteria of fixed effects at $p < 0.001$, or random effects at $p < 0.05$. Each of these studies along with closely related research is detailed in the following sections.

Parental Brains and Baby Cry Stimuli

The first experiments using the pioneering approach of studying brain activity in mothers while they listen to infant cries were conducted by Lorberbaum and colleagues. Building on the thalamocingulate theory of maternal behavior in animals (MacLean, 1990), they initially predicted that baby cries would selectively activate cingulate and thalamus in mothers (ranging from 3 weeks to 3.5 years postpartum) exposed to an audio taped 30 s standard baby cry, not from their own infant (Lorberbaum *et al.*, 1999). They later expanded their hypotheses to include the basal forebrain's MPOA/ventral BNST and its rich reciprocal connections as being critical to parental behaviors (Lorberbaum *et al.*, 2002). These include the descending connections to modulate more basic reflexive caring behaviors such as nursing, licking, grooming and carrying reflexes in rodent studies, and ascending connections such as the mesolimbic and mesocortical dopamine systems for more general motivation and flexible responses to tend a crying infant or prepare for a threat (Numan and Insel, 2003; Lorberbaum *et al.*, 2002). In their first study (Lorberbaum *et al.*, 1999), a group of four mothers were studied for their response to 30 s of a standard cry compared with 30 s of a control sound consisting of white noise that was shaped to the temporal pattern and amplitude of the cry. With cry vs. control sound, the four mothers (3 weeks to 3 years postpartum) showed increased activity in the subgenual anterior cingulate and right inferior mesial prefrontal/orbitofrontal cortex to infant cries using a fixed effects data analysis. In a methodologically more stringent follow-up study that used a larger sample size and a random effects analysis, brain activity was measured in 10 healthy, breastfeeding, first-time mothers with infants 1–2 months old. Subjects listened to standard infant cry recordings (not from their own baby) compared to white noise-shaped control sounds varying in the same pattern as the cry in the same fashion as their previous study (Lorberbaum *et al.*, 1999). Brain regions posterior to the brainstem were not imaged. Activated regions included the anterior and posterior cingulate, thalamus, midbrain, hypothalamus, septal region, dorsal and ventral striatum, medial prefrontal cortex, right orbitofrontal/insula/temporal polar cortex region, and right lateral temporal cortex and fusiform gyrus. Additionally, when cry response was compared with the inter-stimulus rest periods, instead of the control sound (which some mothers judged to be aversive), the amygdala was active. Activation of the fusiform gyrus is interesting because this structure has been implicated in human face and voice recognition along with related social cognitions that might be impaired in autism (Schultz, 2005).

These initial studies suggest human parent activations to baby cry fit with regions that were thought to be involved in animal parenting behavior (Numan & Insel, 2003) even though the cry stimuli did not originate from the parent's own infant and the white-noise shaped control sounds were emotionally negative (sounded like harsh static on the television). Certainly, attention is important in parenting, but it is unclear whether parent brain activations from paying attention to infants is simply another instance of attention in general, or if some regions are peculiar to parenting attention *per se*. Indeed, it has been shown by auditory event-related brain potential studies that auditory attention does require anterior cingulate and temporal cortices (Tzourio *et al.*, 1997). However, in another study using similar techniques, women responded significantly more to a baby cry than to an emotionally neutral vocalization in these regions (Purhonen *et al.*, 2001b); in a third study, mothers responded more than control women to infant cries (Purhonen *et al.*, 2001a). These results suggest a general increase in the mother's alertness and arousal toward baby signals which may function to assist the mother by her ability to be continuously alert or attuned to the infant's needs. Support for this view might be found in studying parents who are having difficulty sustaining or appropriately modulating their attention and arousal in response to infant cries, that is those suffering postpartum depression or who mistreat or neglect their children. In one such physiological study of parents who maltreat their children (Frodi & Lamb, 1980), audiovisual infant stimuli elicited exaggerated physiological responses. Indeed, infant crying is a proximate risk factor for infanticide (Soltis, 2004), perhaps due to parents' failure to regulate their arousal and maintain a caring stance. Future work may shed light on this critical question: What is going on in a healthy parent's brain and parent–infant relationship (Swain *et al.*, 2004b) compared to a parent at risk for neglect and abuse? One might think that healthy parents would attend to infant cues and respond appropriately, but not be so aroused as to make an impulsive, disinhibited decision.

Hypothesizing that gender and experience would also influence neural responses to baby sounds such as baby cry and laughter, Seifritz *et al.* (2003) studied four groups: mothers and fathers of children under age 3, and non-parent males and females, with 10 subjects in each group. They used a more event-related fMRI design, measuring brain responses to brief 6 s sounds. Over the entire sample, intensity-matched baby sounds of crying and laughing compared to "neutral" sounds (white noise pulsed at 5-Hz with an averaged frequency spectrum similar to the infant vocalizations) produced more brain activity in bilateral temporal regions. These regions might be important for hearing processes (Heshyl's gyrus and temporal poles), processing human vocalizations, and empathic emotion processing. They also reported that women as a group, including parents and non-parents (but not males), had a decrease in activity in response to both baby cry and laughter in the subgenual anterior cingulate cortex. This finding is, however, contrary to the other studies (Lorberbaum *et al.*, 1999, 2002; Swain *et al.*, 2003, 2004a, 2005) which highlights the importance of the choice of stimuli in these experiments as well as not viewing the anterior cingulate as one structure without subdivisions. Perhaps also, 6 s vs. 30 s stimuli have very different meanings to new parents and there may be non-linear or multiphasic anterior cingulate responses. Finally, within-group analyses showed that parents activated more to infant crying than laughing in the right amygdala, while non-parent response was greater for infant laughing then crying (Seifritz *et al.*, 2003). These within-group results suggest a potential change in amygdala function with being a parent, although there was no direct comparison of parents to non-parents. Inclusion of psychological measures of parenting parameters will make future studies more practically insightful. These data do, however, represent the first attempts to extend the previous work on parental brain circuits to include gender and experience-dependant aspects of human parenting.

Relevant to parent responses to infant sounds, other fMRI research has been exploring brain responses to emotionally laden human vocalizations, such as having non-parents listen to adult cries and laughter. Some of the brain responses overlap with those found in the parent–infant studies. To reveal emotion circuits, subjects were asked to self-induce happy or sad emotions to correspond with the laughing or crying stimuli, respectively. For pitch detection, subjects were asked to detect pitch shifts. Both conditions led to bilateral activation of the amygdala, insula, and auditory cortex with a right-hemisphere advantage in the amygdala, and larger activation during laughing than crying in the auditory cortex with a slight right-hemisphere advantage for laughing. Both of these responses are likely due to acoustic stimulus features. These results suggest that certain brain regions, including the amygdala, activate to emotionally meaningful sounds like laughing and crying independent of the emotional involvement, suggesting the pattern recognition aspect of these sounds is crucial for this activation and that emotional valence might be represented elsewhere in the brain (Sander *et al.*, 2003). Frontal areas may be good candidates for altering emotional valence as suggested by a more recent work by Sander and colleagues in which they found a correlation between activity in the orbitofrontal cortex (OFC) in response to angry utterances and an emotional sensitivity scale across a group of young adults (Sander *et al.*, 2005). Sander and colleagues also report a striking gender effect with infant laughing and crying stimuli vs. a control sound activating the amygdala and anterior cingulate of women, while the control stimuli elicited stronger activations in men. Independent of listeners' gender, auditory cortex and posterior cingulate were more strongly activated by the control stimuli than by infant laughing or crying (Sander *et al.*, 2007). Perhaps the gender-dependent correlates of neural activity in amygdala and anterior cingulate reflect neural predispositions in women for responses to preverbal infant vocalizations, whereas the gender-independent similarity of activation patterns in posterior and anterior cingulate reflects the relatively sensory-based and cognitive levels of neural processing. The question of whether parent brains are actually differentially sensitive to infant stimuli across gender will require a direct contrast of men vs. women in parents and non-parents.

In an attempt to further this research on the neurocircuitry underlying emotionally laden parenting thoughts, behaviors, and parent–infant attachment, and based in part on the work of Lorberbaum *et al.* (2002), Swain and colleagues have been gathering data sets on groups of new parents across a range of experience, temperament, and parent–infant interaction styles using each parent's own baby cries and including comprehensive interviews and self-reports (Swain *et al.*, 2003). In this design, parents underwent brain fMRI during 30 s blocks of infant cries generated by their own infant as well as a "standard" cry and control noises matched for pattern and intensity. In addition, they added a longitudinal component with scans and interviews done at two time-points: 2–4 weeks and 12–16 weeks postpartum. These times were chosen to coincide with the transforming experience of having a baby known to be associated with increased tendency for parents to be highly preoccupied in the early postpartum (Leckman *et al.*, 1999). Swain and colleagues (2003) hypothesized that parental responses to own baby cries would include specific activations in thalamo–cortico–basal ganglia circuits based on research on potentially related human habitual, ritualistic and obsessive-compulsive thoughts and behaviors (Baxter, 2003; Leckman *et al.*, 2004). They also reasoned that emotional alarm, arousal, and salience detection centers including amygdala, hippocampus, and insula (LeDoux, 2003; Britton *et al.*, 2006) would be activated by baby cry stimuli. The experimental block design was used in order to give parents a chance to reflect on their experience of parenting and, according to our hypothesis, become more preoccupied with their infants' well being and safety. In a group of first-time mothers ($n = 9$) at 2–4 weeks postpartum, regions that were relatively more active with parent's own baby cry stimuli compared with other baby cry included the midbrain, basal ganglia, cingulate, amygdala, and insula (Swain *et al.*, 2003). Activation of these regions may reflect an increase in arousal, obsessive/anxiety circuits that may be normally temporarily more active for parents and persistently sensitive with some mental illnesses (Swain *et al.*, 2007). Preliminary analysis of the parenting interview data shows that mothers were significantly more preoccupied than fathers, which was consistent with the relative activations for mothers compared with fathers in the amygdala and basal ganglia (Swain *et al.*, 2004a). In the group of primiparous mothers, given the same stimuli at 3–4 months postpartum, amygdala and insula activations were not evident; instead activity had shifted to medial prefrontal cortical and hypothalamic (hormonal control) regions (Swain *et al.*, 2004a). This may reflect a change in regional brain responses as the parent–infant relationship develops, and the mothers learn to associate their infant cries more with more flexible social behaviors and more mature attachment and less with alarm and anxiety.

TABLE 6.1 Human parent brain responses to own and other infant cries. Anatomical brain regions with increased activity (ACT) and areas of decreased activity (DEACT) are indicated. Dashes indicate no significant changes in brain activity with exposure to baby sounds

Authors (year):	Lorberbaum *et al.* (1999)	Lorberbaum *et al.* (2002)	Seifritz *et al.* (2003)	Swain *et al.* (2003, 2004)		Swain *et al.* (2003, 2004)		Sander *et al.* (2007)
Number of subjects:	$n = 4$	$n = 10$	$n = 20$	$n = 7–14$		$n = 7–8$		$n = 18$
Age of infants at the time of scan:	3 weeks to 3.5 years	1–2 months	<3 years	Time 1: 2–4 weeks Time 2: 3–4 months		Time 1: 2–4 weeks Time 2: 3–4 months		N/A
Parental groups:	Mothers only	Mothers only	Mothers + fathers + non-parents	Novice + multiparous mothers		Novice + multiparous fathers		Non-parents
Type of study:	1.5 T, 30 s blocks, fixed effects	1.5 T, 30 s blocks random effects	1.5 T, 6 s events random effects	3 T, 30 s blocks fixed effects		3 T, 30 s blocks fixed effects		3 T blocks
Baby cry used:	Other cry > white noise	Other cry > control noise	Other cry + laugh	Own cry > control	Other cry > control	Own cry > control	Other cry > control	Laughing, crying ♀ > ♂
Septal regions (MPOA/VBNST/ caudate head)	–	ACT	–	ACT	ACT	–	–	–
Hypothalamus	–	ACT	–	ACT	–	ACT	–	–
Thalamus	–	ACT	–	–	ACT	–	ACT	–
Striatum/putamen/ nucleus accumbens	–	ACT	–	ACT	ACT	–	ACT	–
Anterior cingulate	ACT	ACT	DEACT	ACT	ACT	ACT	ACT	ACT
Middle cingulate	–	ACT	ACT	–	–	–	–	–
Posterior cingulate	–	ACT	–	–	–	–	–	ACT
Amygdala	–	ACT (cry-rest)	ACT	ACT	–	–	–	ACT

Lentiform nucleus globus pallidus	–	ACT	–	ACT	ACT	–	–	–
Hippocampus	–	–	–	ACT	–	ACT	–	–
Midbrain	–	ACT	–	ACT	–	ACT	ACT	–
Insula	–	ACT	ACT	ACT	ACT	–	ACT	–
Orbitofrontal/ inferior frontal gyri	ACT	ACT	–	ACT	ACT	–	ACT	–
Medial frontal gyrus	ACT	ACT	DEACT	DEACT	ACT	–	–	–
Ventral prefrontal cortex	ACT	ACT	ACT	–	–	–	–	–
Temporoparietal cortex	–	ACT	ACT	ACT	ACT	ACT	ACT	–
Parahippocamal/ limbic lobe	–	–	–	ACT	–	ACT	–	–
Occipital cortex	Not examined	Not examined	–	ACT	ACT	–	ACT	–
Fusiform gyrus	–	ACT	–	ACT	ACT	–	ACT	–
Temporal/auditory cortex	–	ACT	–	ACT	–	–	ACT	ACT
Cerebellum	Not examined	Not examined	–	ACT	ACT	–	ACT	–

Activations and deactivations, measured by functional magnetic resonance imaging, satisfied significance criteria of random effects analysis at $p < 0.05$ or fixed effects analysis at $p < 0.001$ at a minimum; T, Tesla (unit of magnetic field strength); blocks, periods of stimulus exposure and fMRI data acquisition; events, brief exposures to infant stimuli during fMRI experiments; other cry, cry of an unfamiliar baby; own cry, cry of the subject's own baby; MPOA, medial preoptic area; BNST, bed nucleus of the stria terminalis.

Parental Brains and Baby Visual Stimuli

Several groups are using baby visual stimuli to activate parental brain circuits (Swain *et al.*, 2003, 2006; Bartels & Zeki, 2004b; Leibenluft *et al.*, 2004; Nitschke *et al.*, 2004; Ranote *et al.*, 2004; Strathearn *et al.*, 2005; Noriuchi *et al.*, 2008) with a variety of designs, parent populations, and infant age.

We begin with the work of Swain and colleagues who used own and other baby photographs (taken 0–2 weeks postpartum) shown to groups of mothers and fathers with similar block design for pictures as was used for cries (Swain *et al.*, 2003, 2006). Photographs were chosen by the parents themselves in order to provide the most potent and ethologically appropriate signals to evoke their own parenting emotions involving motivation and reward (i.e. loving feelings). In these studies, own vs. other baby picture contrasts revealed activations in frontal and thalamo–cortical circuits at 2–4 weeks postpartum. Specific characterization of these regions according to differences by gender, experience, and postpartum time of assessment are underway. Correlations between parent response to own vs. other baby picture and parent behaviors attained from videotapes of interactions with their infant reveals activations in superior temporal lobe, OFC, and ventral tegmental areas which might be key to regulating the motivation and reward associated with empathy, approach, and caring behaviors as well as social bonding.

With the idea that parental love may make use of the same reward and emotion circuits as romantic love (Bartels & Zeki, 2000), Bartels and Zeki used photographs of own, familiar, and unfamiliar infants (9 months to 3.5 years of age) as stimuli for parent brains (Bartels & Zeki, 2004b). They measured brain activity in 20 healthy mothers, while viewing still face photographs of their own child compared to age-matched photographs of other children. There was increased activity in the midbrain (periaqueductal gray and substantia nigra regions), dorsal and ventral striatum, thalamus, left insula, OFC, sub-, pre-, and supragenual anterior cingulate, and superior medial prefrontal cortex. There were also increases in the cerebellum, left fusiform, and left occipital cortex, but decreases in the left amygdala, posterior cingulate, medial frontal gyrus, and temporoparietal regions potentially important in critically evaluating others. Bartels and Zeki also compared mother brain responses of own child vs. familiar child to the best friend vs. familiar friend in order to control for familiarity and positive affect, and they argue that responses were unique to the own child stimuli. They suggested that parent–infant attachment may be regulated by a push–pull mechanism that selectively activates motivation and reward systems, while at the same time suppressing critical social assessment and negative emotion systems (Bartels & Zeki, 2004b).

Using a similar approach, but focusing on early stage romantic love, attachment, and mate selection (Fisher *et al.*, 2002a, b), Aron, Fisher and colleagues conducted fMRI studies of brain response to photographs of beloved and familiar individuals (Aron *et al.*, 2005; Fisher *et al.*, 2005). They replicated the findings of Bartels and Zeki's romantic love findings (Bartels & Zeki, 2000). They also reported that activations specific to the beloved in dopamine-rich areas associated with mammalian reward and motivation areas of the midbrain (right ventral tegmental area) and caudate nucleus (right postero-dorsal body and medial parts) correlated with facial attractiveness scores. Furthermore, activation in the right anteromedial caudate was correlated with questionnaire scores that quantified intensity of romantic passion for the individuals whose photographs were used as stimuli. Finally, activity in the left insula–putamen–globus pallidus correlated with trait affect intensity, whereas activity in limbic and cortical regions, including insula, cingulate parietal, inferior temporal, and middle temporal cortex was correlated with the length of time in love. Taken together, these studies suggest that romantic love uses subcortical motivation and reward systems to focus thoughts and behaviors on a specific individual, while limbic cortical regions process individual emotional factors. In a related study of romantic relationships using aversive stimuli, Najib and colleagues investigated women whose romantic relationship had ended within the 4 months preceding the experiment. They found that acute grief related to the loss of a romantic attachment figure modulated activity in some of the same areas implicated in social attachment and parenting when they recalled sad thoughts about their loved one (Najib *et al.*, 2004). Decreases in activity in this case were associated with sad thoughts about the loss of the romance compared with neutral thoughts about someone else. This included deactivations in temporal cortex, insula, and anterior cingulate/prefrontal cortex. In contrast to the romance-studies which found activations in the anterior cingulate, they also found that romantic grief was consistently associated with deactivations in this region. Finally, they found that deactivation in the anterior cingulate, insula, and amygdala were positively correlated with the level of a subject's grief.

Returning to a study of parent–child relationship using photographs of much older children (5–12 years old), mothers viewed pictures of their own vs. other children's faces during brain fMRI measurements and were asked to press a button to indicate identity (Leibenluft *et al.*, 2004) – but not feelings. Some social cognition regions that were not activated in the Bartels and Zeki (2004b) study were significantly activated in this study, including the anterior paracingulate, posterior cingulate, and the superior temporal sulcus which are important for empathy (Saxe, 2006a). This may be explained by the use of much older children, which might involve a different set of circuits relevant to those particular relationships. It may also be that the cognitive task interacts with affective responses to face images in some way (Gray, 2001). In addition, some of the more subcortical activations from Bartels and Zeki (2004b) were not activated in the study by Leibenluft *et al.* (2004) because of the smaller sample size and the lack of focus on loving thoughts. Differences in child photo affective facial expressions (happy vs. neutral vs. sad) may also constitute a confounding factor. Finally, differences between studies may emerge because of differences in sample populations that need to be controlled. Although all of the studies were of "normative" parent populations, most studies only screened for clinical psychiatric disease. Perhaps different populations may process infant cues in different ways. Studies involving more specific tasks and correlations between brain activations and relationship-specific variables will be able to tease apart the particular roles of different brain regions in different aspects of those relationships.

In another study focusing on parents' brains using visual stimuli, Nitschke *et al.* (2004) studied six healthy, primiparous mothers' brains at 2–4 months postpartum as they viewed smiling pictures of their own and unfamiliar infants. They reported OFC activations that correlated positively with pleasant mood ratings. In contrast, areas of visual cortex that also discriminated between own and unfamiliar infants were unrelated to mood ratings (Nitschke *et al.*, 2004). Perhaps, as they suggest, activity in the OFC – which may vary across individuals – is involved with high order dimensions of maternal attachment. Perhaps the complex aspects of parenting may be quantified using fMRI of frontal brain areas to help predict the risks of mood problems in parents.

With the innovative and perhaps more realistic and ethologically appropriate use of videotaped infant stimuli, Ranote and colleagues conducted a similar experiment (Ranote *et al.*, 2004). In their study, 10 healthy mothers viewed alternating 40 s blocks of their own infant's video, a neutral video, and an unknown infant. For these women, there was significant activation in the "own" vs. "unknown" infant comparison in the left amygdala and temporal pole. They interpreted this circuit as regulating emotion and theory-of-mind regions relating to the ability to predict and explain other people's behaviors. Certainly, this fits with fMRI experiments on biological motion, which activate similar regions (Morris *et al.*, 2005). It is important to note that all of these visual paradigms used to examine differences between one's own infant and unfamiliar infants employ a complex set of brain systems necessary for sensory perception, identification, and emotional response. Yet, it now appears from a number of studies that despite the multisensory complexities of audiovisual stimuli, meaningful analysis of fMRI data is possible. For example, there seems to be a striking inter-subject synchronization among emotions regulating brain areas responding to audiovisual cues during observation of the same scenes of an emotionally powerful movie (Hasson *et al.*, 2004). Also, the intensity with which subjects perceive different features in a movie (color, faces, language, and human bodies) was correlated with activity in separate brain areas (Bartels & Zeki, 2004a). Finally, regional activity between brain areas that are known to be anatomically connected have been shown to be simultaneously active during movie viewing (Bartels & Zeki, 2005). Using infant stimuli, it may be possible to delineate parental emotion regulation under stress. Thinking about the contribution of the infant's affect to maternal brain function prompted a recent study by Noriuchi and colleagues (Noriuchi *et al.*, 2007) which used silent video clips of their own and other infants in play or separation situations. First, they confirmed the increased activity, associated with recognizing own baby pictures, in certain brain regions including cortical orbitofrontal, anterior insula, and

TABLE 6.2 Human parent brain responses to infant pictures. Anatomical brain regions with increased activity (ACT) during infant cry and areas of decreased activity (DEACT) are indicated. Dashes indicate no significant changes in brain activity with exposure to baby pictures

Authors (year):	Bartels and Zeki (2004a, b)	Nitschke *et al.* 2004	Ranote *et al.* 2004	Strathearn and Montague (2005)	Swain *et al.* 2003, 2004a, b, 2005		Swain *et al.* 2003, 2004a, b, 2005		Noriuchi *et al.* 2008
Number of subjects:	*n* = 19	*n* = 6	*n* = 10	*n* = 8	*n* = 9–14		*n* = 4–9		*n* = 13
Age of infants at the time of scan:	9 months to 3.5 years	2–4 months	4–8 months	3–18 months	Time 1: 2–4 weeks Time 2: 3–4 months		Time 1: 2–4 weeks Time 2: 3–4 months		16.5 months
Parental groups:	Mothers only	Mothers only	Mothers only	Mothers only	Novice + multiparous *mothers*		Novice + multiparous *fathers*		Mothers only
Type of study: (all silent visuals)	2 T, 15 s blocks random effects	1.5 T, 30 s blocks fixed effects	1.5 T, 20–40 s videos random effects	3 T, 6 s events fixed effects	3 T, 30 s blocks fixed effects		3 T, 30 s blocks fixed effects		1.5 T, 32 s video blocks random effects
Baby visual:	Own > other	Own > other	Own > other	Own > other	Own > other	Baby > house	Own > other	Baby > house	Own > other
Septal regions (MPOA/VBNST/ caudate head)	–	–	–	–	ACT	–	–	–	–
Hypothalamus	–	–	–	–	ACT	–	–	–	ACT
Thalamus	ACT	–	ACT	ACT	ACT	ACT	ACT	ACT	ACT
Striatum/putamen/ nucleus accumbens	ACT	–	ACT	ACT	–	ACT	–	–	ACT

Anterior cingulate	ACT	–	–	–	ACT	ACT	ACT	ACT	–
Middle cingulate	–	–	–	–	ACT	ACT	ACT	ACT	–
Posterior cingulate	DEACT	–	–	–	–	–	–	–	ACT
Amygdala	DEACT	–	–	–	–	ACT	–	–	–
Lentiform Nucleus/globus pallidus	–	–	ACT	ACT	–	ACT	–	–	–
Hippocampus	–	–	ACT	ACT	–	–	–	–	–
Midbrain	ACT	–	–	–	ACT	ACT	ACT	ACT	ACT
Insula	ACT	–	–	–	–	–	–	–	ACT
Orbitofrontal/ inferior frontal	ACT	ACT	–	–	ACT	ACT	ACT	ACT	ACT
Medial frontal gyrus	DEACT	–	–	–	DEACT	ACT	–	ACT	ACT
Ventral prefrontal	ACT	–	–	–	–	–	–	–	–
Temporoparietal	DEACT	–	–	–	ACT	ACT	ACT	ACT	ACT
Parahippocamal/ limbic lobe	–	–	–	–	–	ACT	–	–	–
Occipital cortex	ACT	–	ACT	ACT	ACT	ACT	ACT	ACT	–
Fusiform Gyrus	ACT	–	ACT	ACT	ACT	ACT	ACT	ACT	ACT
Temporal/auditory cortex	–	ACT	–	–	–	–	–	–	ACT
Cerebellum	–	ACT	ACT	ACT	ACT	ACT	–	–	–

See Table 6.1 footnote.

precuneus cortical areas as well as subcortical regions which include the periaqueductal gray and putamen. These are areas active in arousal and reward learning. Furthermore, they found strong and specific differential responses of mother's brain to her own infant's distress in substantia nigra, caudate nucleus, thalamus, posterior and superior temporal sulcus, anterior cingulate, dorsal regions of OFC, right inferior frontal gyrus, and dorsomedial prefrontal cortex. They interpreted OFC and related activations as part of circuits required for the execution of well-learned movements. This could also be interpreted as activation in emotion regulation and habitual behavioral response systems that are active in a range of normal and abnormal emotion-control states including obsessive-compulsive disorder (Leckman & Mayes, 1999; Feygin *et al.*, 2006; Swain *et al.*, 2007). They also found correlations in OFC between own baby response and happiness as well as to their own distressed baby response in the superior temporal regions. This is consistent with the emerging importance of these areas in social behaviors. Indeed, it seems that movies and other more complex stimuli may be used to stimulate parents' brains in ways that may be measured and combined with behavioral measures to better understand the functional architecture of parenting brain systems.

Finally, Strathearn and colleagues have also been studying healthy mother–infant dyads using fMRI to examine maternal brain regions activated in response to visual infant facial cues of varying affect (smiling, neutral, and crying). They have completed a pilot study of eight healthy right-handed mothers, without a history of psychiatric impairment or child maltreatment along with their infants aged between 3 and 8 months. They assessed serum oxytocin levels sequentially from the mothers during a standardized period of mother–infant interaction, during which they acquired infants' facial expression videotapes. Maternal brain activity was then assayed with fMRI in response to 6 s exposures to the facial images of their own infant compared with familiar and unknown infant facial images (Strathearn, 2002). Areas of significant activation (uncorrected $p < 0.005$) unique to own infant viewing included brain reward areas with dopaminergic projections (ventral striatum, thalamus, and nucleus accumbens), areas containing oxytocin projections (amygdala, bed nucleus of the stria terminalis, and hippocampus), the fusiform gyrus (involved in face processing), and bilateral hippocampi (involved in episodic memory processing). Further, a positive, but non-significant trend in this small sample was seen in serum oxytocin concentrations before and after mother–infant interaction (prior to scanning), suggesting a possible correlation between brain activation and peripheral affiliative hormone production. A further study, which was limited to the presentation of crying infant faces, revealed activation of the anterior cingulate and insula bilaterally (Strathearn *et al.*, 2005).

Careful use of a variety of baby stimuli to activate parent brains, along with correlations of parental brain activity with psychometric parameters will help in the understanding of these circuits. It may also be helpful to include comprehensive measurements of parent physiology during infant response. In addition to understanding normal parental behavior, this field promises to elucidate abnormalities of parental circuitry that may be manifest in postpartum depression and anxiety. Such understanding may suggest optimal detection and treatment strategies for these conditions that have profound deleterious effects on the quality of parent–infant interactions, and the subsequent long-term health risks and resiliencies of infants. These studies will also enhance our understanding of social circuits important for empathy across a range of relationships.

SPECIAL PARENT POPULATIONS FOR IMAGING

In addition to understanding normal human parenting in order to optimize health outcomes, research on parents with health issues from substance abuse and mood disorders to different conditions of birth and infant feeding is needed to improve the recognition and treatment, of compromised parenting circumstances.

Recently published follow-up data on the offspring of depressed and anxious mothers who have increased mental health risks (Brown *et al.*, 1987; Kendler *et al.*, 1993; Heim *et al.*, 1997; Sroufe *et al.*, 1999), underscores the significance of work in this area. Clearly, parental wellness (and/or the presence of other attuned care giving adults) has long-term positive effects on resiliency and emotional well-being of children as they grow up and for decades later. Indeed, longitudinal studies of high-risk infants suggest that secure attachment in the perinatal period is associated with a degree of resiliency and protection against the development of psychopathology later in life (Werner, 2004).

Parental mental health problems in the postpartum, such as depression and anxiety, are common and contribute significantly to parent–infant attachment problems. Postpartum depression follows 10–15% of all deliveries (Caplan *et al.*, 1989) and more than 60% of patients have an onset of symptoms within the first 6 weeks postpartum (Stowe & Nemeroff, 1995). These common problems, including preterm delivery, postpartum depression and anxiety and substance abuse have received much less investigative attention and not a single fMRI study (Squire & Stein, 2003). A growing body of evidence from naturalistic longitudinal studies attests to an adverse impact of postpartum depression, with depressed mothers less sensitively attuned to their infants, less affirming, and more negative in describing their infant. These disturbances in early mother–infant interactions were found to predict poorer infant cognitive outcome at 18 months (Murray & Cooper, 2003) and at later time-points, namely 7 years of age (Kim-Cohen *et al.*, 2005).

However, a recent study showed that maternal remission from depression within 3 months was associated with significant decreases in the mood symptoms of their children at 7–17 years of age (Weissman *et al.*, 2006). We would predict an even more dramatic effect in younger children. In efforts to understand the underlying physiology, brain imaging studies are currently underway (Mayes *et al.*, 2005) with parents at risk for postpartum depression. Such work will outline future opportunities to identify families at risk for pathological attachment, assess treatments and improve parent–child attachment.

Among normal parents, variations in breastfeeding may provide a natural way to study variations in oxytocin. To this end, groups of mothers who breastfed or exclusively formula fed their infant are being studied. Preliminary analysis of these data (Kim *et al.*, 2007) showed that at 3 weeks postpartum, there were greater activations in limbic–hypothalamic–midbrain circuits in response to own baby cry and picture among breastfeeding mothers vs. formula feeding mothers. This suggests that at early postpartum times, breastfeeding and perhaps the associated oxytocin facilitates parent brain sensitivity to own baby cry.

Another question being addressed is whether the mode of delivery affects parent brain function. A controversial literature suggests a link between cesarean section and depression and we may be able to discern the neural correlates of this risk in empathy circuits (Swain *et al.*, 2008) and design prevention measures accordingly.

SUMMARY AND MODEL

Functional MRI experiments on parenting using baby stimuli are just beginning to make a meaningful contribution to our understanding of the parental brain. This selective review of the physiology of parenting identifies many brain areas that are likely important in regulating human parenting. For this review, virtually all of the studies involving infant stimuli to study parent brains with fMRI are summarized and contrasted in Table 6.1 (baby cry stimuli), and Table 6.2 (baby picture stimuli). So far, it appears that a set of brain circuits of parental response to baby stimuli, whether picture or cry, is emerging. This appears to center on the feedback loops involving hypothalamus, midbrain, basal ganglia regions, anterior cingulate, prefrontal cortex, and thalamus – all requiring motivation and reward. More complex planning and social emotional/empathy responses may involve frontal, insular, fusiform, and occipital areas. Other important aspects of parenting may be regulated by context and memory processing regions including the hippocampus, parahippocampus, and amygdala. Clearly, baby pictures and cries can be used to selectively activate brain circuits related to arousal, mood, and social and habitual behaviors. However, different groups have used a mixture of stimuli including baby cries, laughter, and child pictures of very different ages and different facial affect and experience. A clearer picture of the specificity of different brain areas may emerge as brain responses in these areas are linked to specific aspects of parenting by adding sophisticated interviews, naturalistic assessments of parent–infant interaction, and bonding.

Based on brain imaging of parents, informed by animal work (Swain *et al.*, 2007), we suggest the following crude model for further discussion (Figure 6.1) to account for parenting behaviors. First, key parenting sensory signals that include cry, visuals, as well as touch and smell (A) must be organized in sensory cortical regions which appraise the input and interact with subcortical memory and motivation structures (B). Ultimately, sufficient motivation will activate corticolimbic modules (C). We have delineated as (1) reflexive caring impulses, such as those studied in preclinical animal models and requiring little or no cortical input such as licking, grooming, and nursing in which the hypothalamus, MPOA, and other limbic and thalamocingulate circuits are of primary importance (Numan & Insel, 2003). In humans, this might be a caring endophenotype (Panksepp, 2006) and amenable to further study. In addition, (2) cognitive and emotional circuits would be brought online, including those that regulate "mirroring," empathy, planning, and further cognitive flexibility, including the inferior frontal, insular, and superior-temporoparietal cortical regions. These regions might allow accurate modeling of the baby's mind to predict their needs and plan behavior (Baron-Cohen & Wheelwright, 2004; Pelphrey *et al.*, 2004; Decety & Grezes, 2006; Saxe, 2006a). Finally, other alarm/emotion-preoccupation anxiety systems might be activated (3) to increase arousal and regulate parental worries and habitual responses in coordination with memory systems. Such arousal or alarm could help or hinder parental behavior. These emotional circuits might include the ventral tegmental area, striatum, amygdala, insula, cingulate cortex, and OFC (Mayes *et al.*, 2005; Swain *et al.*, 2007). Working together, these reflexive, cognitive, and emotional modules would interact with each other in the experience of parental love and attachment formation, and work together to generate coordinated hormonal, autonomic, and behavioral output required for parenting (D) via motor cortex and hypothalamus. The output would also feed back to sensory systems (B) during dynamic interactions with the infant to generate new input (1).

In summary, this review is an attempt to synthesize our current understanding of human

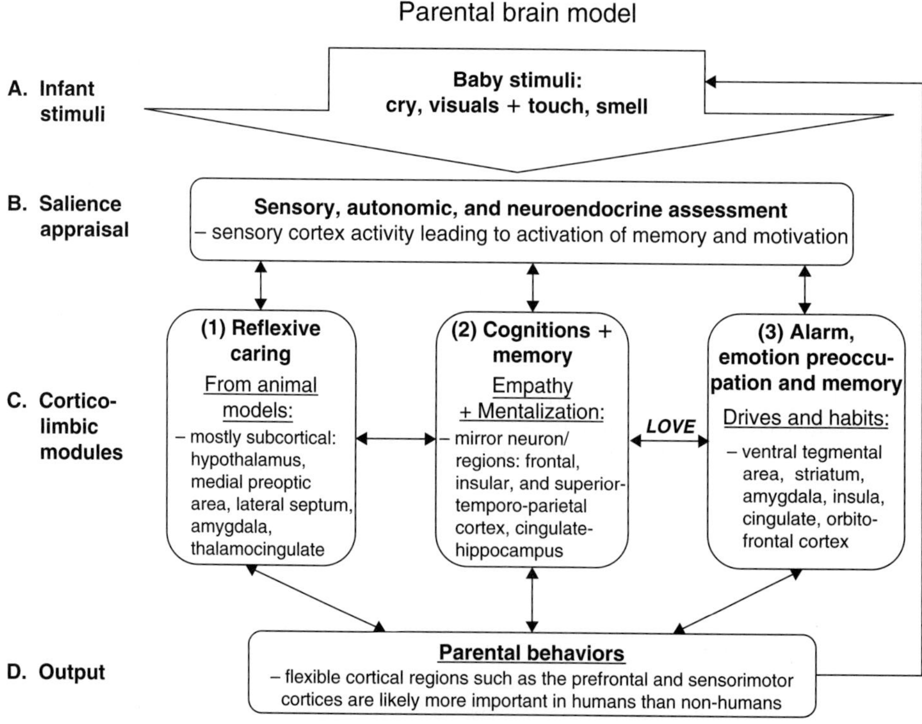

FIGURE 6.1 Human parental circuits. Brain regions expected to be important to human parenting. This is based on human and animal studies.

parent–infant bonding, largely from the perspective of the parent's brain physiology. The parent–infant bond, so central to the human condition, may also determine risks for mood and anxiety disorders, and the potential for resiliency and protection against the development of psychopathology later in life, not to mention the far-reaching aspects of human attachment across individual behaviors and between cultures. Efforts to characterize this reciprocal interaction between caregiver and infant and to assess its impact have provided a powerful theoretical and empirical framework in the fields of social and emotional development.

FUTURE DIRECTIONS

Future studies will likely focus on the use of more movie stimuli and different sensory systems, such as the olfactory system, to expand and refine our understanding of the parental brain. This approach will require careful consideration and study of how these patterns of brain activation may differ between attachment groups. Do mothers with insecure patterns of attachment respond differently to their infant cues? Are neglecting mothers unresponsive to these cues or do they fail to receive reward signals in the brain? Longitudinal research designs may help in this regard. In addition, it will be important to clarify the role of different neuroendocrine pathways and different genetic variations in mediating parenting brain activations.

A helpful approach to these questions will include systematic studies of well-characterized, but different, populations of parents using a range of infant stimuli paradigms and psychometric tools. As in other areas of cognitive neuroscience, there will be debates about whether to use more ethologically correct, but poorly controlled, stimuli vs. more tightly controlled, but less natural stimuli. Both types of experiments will be needed to tease apart the basic apparatus of baby responsiveness and bond formations as well as the parts of the circuit that are actually at work in normal day-to-day parenting. This work will also require the study of both parents *and* infants to understand how their interactions contribute to their bond and infant outcomes.

In the near future, we expect that differences in parental response patterns will be reported in specific clinical populations, such as those with postpartum depression and substance abuse. This may lead to future assessments of parent mental health risk and resilience profiles using standardized imaging techniques and to improvements in the detection, treatment, and prevention of mental illness that interferes with parenting.

ACKNOWLEDGMENTS

The authors would like to acknowledge the generous support of colleagues, research assistants, and research participants at our respective institutions:

JES was supported by grants from the Institute for Research on Unlimited Love (unlimitedlove institute.org), the National Alliance for Research on Schizophrenia and Depression (narsad.org), the Yale Center for Risk, Resilience and Recovery, and Associates of the Yale Child Study Center. Dr Swain would especially like to acknowledge the mentorship of Drs James F. Leckman, Linda C. Mayes, Robert T. Schultz, Ruth Feldman, and Robert T. Constable. Helpful suggestions were also provided by Kim Pilyoung and Virginia Eicher.

JPL was supported by the child neglect consortium RFA grant, NINDS R01 NS40259-01, NICHD R-03 HD49422-01, the Center for Advanced Imaging Research (CAIR) at the Medical University of South Carolina, the National Alliance for Research on Schizophrenia and Depression (narsad.org), and Penn State's Child Youth and Family Consortium. Dr Lorberbaum would especially like to acknowledge the ideas of Drs Paul Maclean and Mark S. George in getting started in this line of research as well as the collaborative research efforts of Samet Kose and Muhammed M. Tariq in performing this research.

REFERENCES

Aron, A., Fisher, H., Mashek, D. J., Strong, G., Li, H., and Brown, L. L. (2005). Reward, motivation, and emotion systems associated with early-stage intense romantic love. *J. Neurophysiol.* 94, 327–337.

Baron-Cohen, S., and Wheelwright, S. (2004). The empathy quotient: An investigation of adults with Asperger syndrome or high functioning autism, and normal sex differences. *J. Autism Dev. Disord.* 34, 163–175.

Bartels, A., and Zeki, S. (2000). The neural basis of romantic love. *Neuroreport* 11, 3829–3834.

Bartels, A., and Zeki, S. (2004a). Functional brain mapping during free viewing of natural scenes. *Hum. Brain Mapp.* 21, 75–85.

Bartels, A., and Zeki, S. (2004b). The neural correlates of maternal and romantic love. *Neuroimage* 21, 1155–1166.

Bartels, A., and Zeki, S. (2005). Brain dynamics during natural viewing conditions – a new guide for mapping connectivity *in vivo*. *Neuroimage* 24, 339–349.

Baxter, L. R. Jr., (2003). Basal ganglia systems in ritualistic social displays: Reptiles and humans; function and illness. *Physiol. Behav.* 79, 451–460.

Bodini, B., Iacoboni, M., and Lenzi, G. L. (2004). Acute stroke effects on emotions: An interpretation through the mirror system. *Curr. Opin. Neurol.* 17, 55–60.

Britton, J. C., Phan, K. L., Taylor, S. F., Welsh, R. C., Berridge, K. C., and Liberzon, I. (2006). Neural correlates of social and nonsocial emotions: An fMRI study. *Neuroimage* 31, 397–409.

Brown, G. W., Bifulco, A., and Harris, T. O. (1987). Life events, vulnerability and onset of depression: Some refinements. *Br. J. Psychiatr.* 150, 30–42.

Buchheim, A., Erk, S., George, C., Kachele, H., Ruchsow, M., Spitzer, M., Kircher, T., and Walter, H. (2006). Measuring attachment representation in an fMRI environment: A pilot study. *Psychopathology* 39, 144–152.

Caplan, H. L., Cogill, S. R., Alexandra, H., Robson, K. M., Katz, R., and Kumar, R. (1989). Maternal depression and the emotional development of the child. *Br. J. Psychiatr.* 154, 818–822.

Carr, L., Iacoboni, M., Dubeau, M. C., Mazziotta, J. C., and Lenzi, G. L. (2003). Neural mechanisms of empathy in humans: A relay from neural systems for imitation to limbic areas. *Proc. Natl Acad. Sci. USA* 100, 5497–5502.

Dapretto, M., Davies, M. S., Pfeifer, J. H., Scott, A. A., Sigman, M., Bookheimer, S. Y., and Iacoboni, M. (2006). Understanding emotions in others: Mirror neuron dysfunction in children with autism spectrum disorders. *Nat. Neurosci.* 9, 28–30.

Decety, J., and Grezes, J. (2006). The power of simulation: Imagining one's own and other's behavior. *Brain Res.* 1079, 4–14.

Eisenberger, N. I., Lieberman, M. D., and Williams, K. D. (2003). Does rejection hurt? An FMRI study of social exclusion. *Science* 302, 290–292.

Feygin, D. L., Swain, J. E., and Leckman, J. F. (2006). The normalcy of neurosis: Evolutionary origins of obsessive-compulsive disorder and related behaviors. *Prog. Neuropsychopharmacol. Biol. Psychiatr.* 30, 854–864.

Fisher, H., Aron, A., Mashek, D., Li, H., Strong, G., and Brown, L. L. (2002a). The neural mechanisms of mate choice: A hypothesis. *Neuroendocrinol. Lett.* 23(suppl 4), 92–97.

Fisher, H. E., Aron, A., Mashek, D., Li, H., and Brown, L. L. (2002b). Defining the brain systems of lust, romantic attraction, and attachment. *Arch. Sex. Behav.* 31, 413–419.

Fisher, H., Aron, A., and Brown, L. L. (2005). Romantic love: An fMRI study of a neural mechanism for mate choice. *J. Comp. Neurol.* 493, 58–62.

Frith, U., and Frith, C. D. (2003). Development and neurophysiology of mentalizing. *Philos. Trans. R. Soc. Lond. B Biol. Sci.* 358, 459–473.

Frodi, A. M., and Lamb, M. E. (1980). Child abusers' responses to infant smiles and cries. *Child Dev.* 51, 238–241.

Gallese, V., Keysers, C., and Rizzolatti, G. (2004). A unifying view of the basis of social cognition. *Trends Cogn. Sci.* 8, 396–403.

Gray, J. R. (2001). Emotional modulation of cognitive control: Approach-withdrawal states double-dissociate spatial from verbal two-back task performance. *J. Exp. Psychol. Gen.* 130, 436–452.

Hasson, U., Nir, Y., Levy, I., Fuhrmann, G., and Malach, R. (2004). Intersubject synchronization of cortical activity during natural vision. *Science* 303, 1634–1640.

Heim, C., Owens, M. J., Plotsky, P. M., and Nemeroff, C. B. (1997). The role of early adverse life events in the etiology of depression and posttraumatic stress disorder. Focus on corticotropin-releasing factor. *Ann. NY Acad. Sci.* 821, 194–207.

Kendler, K. S., Kessler, R. C., Neale, M. C., Heath, A. C., and Eaves, L. J. (1993). The prediction of major depression in women: Toward an integrated etiologic model. *Am. J. Psychiatr.* 150, 1139–1148.

Kim, P., Leckman, J. F., Mayes, L. C., Feldman, R., and Swain, J. E. (2007). Longitudinal study of breastfeeding vs. formula-feeding mothers' brain response to infant stimuli during early postpartum period. *Proceedings of the Society for Neuroscience Annual Meeting* 748, 11.

Kim-Cohen, J., Moffitt, T. E., Taylor, A., Pawlby, S. J., and Caspi, A. (2005). Maternal depression and children's antisocial behavior: Nature and nurture effects. *Arch. Gen. Psychiatr.* 62, 173–181.

Leckman, J. F., and Mayes, L. C. (1999). Preoccupations and behaviors associated with romantic and parental love. Perspectives on the origin of obsessive-compulsive disorder. *Child Adolesc. Psychiatr. Clin. N. Am.* 8, 635–865.

Leckman, J. F., Mayes, L. C., Feldman, R., Evans, D. W., King, R. A., and Cohen, D. J. (1999). Early parental preoccupations and behaviors and their possible relationship to the symptoms of obsessive-compulsive disorder. *Acta Psychiatr. Scand. Suppl.* 396, 1–26.

Leckman, J. F., Feldman, R., Swain, J. E., Eicher, V., Thompson, N., and Mayes, L. C. (2004). Primary parental preoccupation: Circuits, genes, and the crucial role of the environment. *J. Neural Transm.* 111, 753–771.

LeDoux, J. (2003). The emotional brain, fear, and the amygdala. *Cell. Mol. Neurobiol.* 23, 727–738.

Leibenluft, E., Gobbini, M. I., Harrison, T., and Haxby, J. V. (2004). Mothers' neural activation in response to pictures of their children and other children. *Biol. Psychiatr.* 56, 225–232.

Lorberbaum, J. P., Newman, J. D., Dubno, J. R., Horwitz, A. R., Nahas, Z., Teneback, C. C., Bloomer, C. W., Bohning, D. E., Vincent, D., Johnson, M. R., Emmanuel, N., Brawman-Mintzer, O., Book, S. W., Lydiard, R. B., Ballenger, J. C., and George, M. S. (1999). Feasibility of using fMRI to study mothers responding to infant cries. *Depress. Anxiety* 10, 99–104.

Lorberbaum, J. P., Newman, J. D., Horwitz, A. R., Dubno, J. R., Lydiard, R. B., Hamner, M. B., Bohning, D. E., and George, M. S. (2002). A potential role for thalamocingulate circuitry in human maternal behavior. *Biol. Psychiatr.* 51, 431–445.

MacLean, P. D. (1990). *The Triune Brain in Evolution: Role in Paleocerebral Functions*. Plenum Press, New York.

Mayes, L. C., Swain, J. E., and Leckman, J. F. (2005). Parental attachment systems: Neural circuits, genes, and experiential contributions to parental engagement. *Clin. Neurosci. Res.* 4, 301–313.

Morris, J. P., Pelphrey, K. A., and McCarthy, G. (2005). Regional brain activation evoked when approaching a virtual human on a virtual walk. *J. Cogn. Neurosci.* 17, 1744–1752.

Murray, L., and Cooper, P. J. (2003). The impact of postpartum depression on child development. In *Aetiological Mechanisms in Developmental Psychopathology* (I. Goodyer, Ed.). Oxford University Press, Oxford.

Najib, A., Lorberbaum, J. P., Kose, S., Bohning, D. E., and George, M. S. (2004). Regional brain activity in women grieving a romantic relationship breakup. *Am. J. Psychiatr.* 161, 2245–2256.

Nitschke, J. B., Nelson, E. E., Rusch, B. D., Fox, A. S., Oakes, T. R., and Davidson, R. J. (2004). Orbitofrontal cortex tracks positive mood in mothers viewing pictures of their newborn infants. *Neuroimage* 21, 583–592.

Noriuchi, M., Kikuchi, Y., and Senoo, A. (2008). The functional neuroanatomy of maternal love: mother's response to infant's attachment behaviors. *Biol. Psychiatr.* 63, 415–423.

Numan, M., and Insel, T. R. (2003). *The Neurobiology of Parental Behavior*. Springer-Verlag, New York.

Olausson, H., Lamarre, Y., Backlund, H., Morin, C., Wallin, B. G., Starck, G., Ekholm, S., Strigo, I., Worsley, K., Vallbo, A. B., and Bushnell, M. C. (2002). Unmyelinated tactile afferents signal touch and project to insular cortex. *Nat. Neurosci.* 5, 900–904.

Olds, D. L., Kitzman, H., Cole, R., Robinson, J., Sidora, K., Luckey, D. W., Henderson, C. R. Jr., Hanks, C., Bondy, J., and Holmberg, J. (2004). Effects of nurse home-visiting on maternal life course and child development: Age 6 follow-up results of a randomized trial. *Pediatrics* 114, 1550–1559.

Olds, D. L., Sadler, L., and Kitzman, H. (2007). Programs for parents of infants and toddlers: Recent evidence from randomized trials. *J. Child Psychol. Psychiatr.* 48, 355–391.

Panksepp, J. (2006). Emotional endophenotypes in evolutionary psychiatry. *Prog. Neuropsychopharmacol. Biol. Psychiatr.* 30, 774–784.

Pelphrey, K. A., Morris, J. P., and McCarthy, G. (2004). Grasping the intentions of others: the perceived intentionality of an action influences activity in the superior temporal sulcus during social perception. *J Cogn Neurosci* 16, 1706–1716.

Pelphrey, K. A., Morris, J. P., Michelich, C. R., Allison, T., and McCarthy, G. (2005). Functional anatomy of biological motion perception in posterior temporal cortex: An FMRI study of eye, mouth and hand movements. *Cereb. Cortex* 15, 1866–1876.

Purhonen, M., Kilpelainen-Lees, R., Paakkonen, A., Ypparila, H., Lehtonen, J., and Karhu, J. (2001a). Effects of maternity on auditory event-related potentials to human sound. *Neuroreport* 12, 2975–2979.

Purhonen, M., Paakkonen, A., Ypparila, H., Lehtonen, J., and Karhu, J. (2001b). Dynamic behavior of the auditory N100 elicited by a baby's cry. *Int. J. Psychophysiol.* 41, 271–278.

Ranote, S., Elliott, R., Abel, K. M., Mitchell, R., Deakin, J. F., and Appleby, L. (2004). The neural basis of maternal responsiveness to infants: An fMRI study. *Neuroreport* 15, 1825–1829.

Sander, D., Grandjean, D., Pourtois, G., Schwartz, S., Seghier, M. L., Scherer, K. R., and Vuilleumier, P. (2005). Emotion and attention interactions in social cognition: Brain regions involved in processing anger prosody. *Neuroimage* 28, 848–858.

Sander, K., Brechmann, A., and Scheich, H. (2003). Audition of laughing and crying leads to right amygdala activation in a low-noise fMRI setting. *Brain Res. Protoc.* 11, 81–91.

Sander, K., Frome, Y., and Scheich, H. (2007). FMRI activations of amygdala, cingulate cortex, and auditory cortex by infant laughing and crying. *Hum. Brain Mapp.* 28, 1007–1022.

Saxe, R. (2006a). Uniquely human social cognition. *Curr. Opin. Neurobiol.* 16, 235–239.

Saxe, R. (2006b). Why and how to study theory of mind with fMRI. *Brain Res.* 1079, 57–65.

Schultz, R. T. (2005). Developmental deficits in social perception in autism: The role of the amygdala and fusiform face area. *Int. J. Dev. Neurosci.* 23, 125–141.

Seifritz, E., Esposito, F., Neuhoff, J. G., Luthi, A., Mustovic, H., Dammann, G., von Bardeleben, U., Radue, E. W., Cirillo, S., Tedeschi, G., and Di Salle, F. (2003). Differential sex-independent amygdala response to infant crying and laughing in parents versus nonparents. *Biol. Psychiatr.* 54, 1367–1375.

Singer, T., Seymour, B., O'Doherty, J., Kaube, H., Dolan, R. J., and Frith, C. D. (2004). Empathy for pain involves the affective but not sensory components of pain. *Science* 303, 1157–1162.

Soltis, J. (2004). The signal functions of early infant crying. *Behav. Brain Sci.* 27, 443–458. discussion 459–490

Squire, S., and Stein, A. (2003). Functional MRI and parental responsiveness: A new avenue into parental psychopathology and early parent–child interactions?. *Br. J. Psychiatr.* 183, 481–483.

Sroufe, L. A., Carlson, E. A., Levy, A. K., and Egeland, B. (1999). Implications of attachment theory for developmental psychopathology. *Dev. Psychopathol.* 11, 1–13.

Stowe, Z. N., and Nemeroff, C. B. (1995). Women at risk for postpartum-onset major depression. *Am. J. Obstet. Gynecol.* 173, 639–645.

Strathearn, L. (2002). A 14-year longitudinal study of child neglect: Cognitive development and head growth. In *14th International Congress on Child Abuse and Neglect*, Denver, CO.

Strathearn, L., Li, J., and Montague, P. R. (2005). An fMRI study of maternal mentalization: Having the baby's mind in mind. *Neuroimage* 26, S25.

Swain, J. E., Leckman, J. F., Mayes, L. C., Feldman, R., Constable, R. T., and Schultz, R. T. (2003). The neural circuitry of parent–infant attachment in the early postpartum. In *American College of Neuropsychopharmacology*, Puerto Rico.

Swain, J. E., Leckman, J. F., Mayes, L. C., Feldman, R., Constable, R. T., and Schultz, R. T. (2004a). Neural substrates and psychology of human parent–infant attachment in the postpartum. *Biol. Psychiatr.* 55, 153S.

Swain, J. E., Mayes, L. C., and Leckman, J. F. (2004b). The development of parent–infant attachment through dynamic and interactive signaling loops of care and cry. *Behav. Brain Sci.* 27, 472–473.

Swain, J. E., Leckman, J. F., Mayes, L. C., Feldman, R., and Schultz, R. T. (2005). Early human parent–infant bond development: fMRI, thoughts and behaviors. *Biol. Psychiatr.* 57, 112S.

Swain, J. E., Leckman, J. F., Mayes, L. C., Feldman, R., and Schultz, R. T. (2006). Own baby pictures induce parental brain activations according to psychology, experience and postpartum timing. *Biol. Psychiatr.* 59, 126S.

Swain, J. E., Lorberbaum, J. P., Kose, S., and Strathearn, L. (2007). Brain basis of early parent–infant interactions: Psychology, physiology, and *in vivo* functional neuroimaging studies. *J. Child Psychol. Psychiatr.* 48, 262–287.

Swain, J. E., Tasgin, E., Mayes, L. C., Feldman, R., Constable, R. T., and Leckman, J. F. (2008). Cesarean delivery affects maternal brain response to own baby cry. *J. Child Psychol. Psychiatr.* submitted

Tzourio, N., Massioui, F. E., Crivello, F., Joliot, M., Renault, B., and Mazoyer, B. (1997). Functional anatomy of human auditory attention studied with PET. *Neuroimage* 5, 63–77.

Uddin, L. Q., Iacoboni, M., Lange, C., and Keenan, J. P. (2007). The self and social cognition: The role of cortical midline structures and mirror neurons. *Trends Cogn. Sci.* 11, 153–157.

Weissman, M. M., Pilowsky, D. J., Wickramaratne, P. J., Talati, A., Wisniewski, S. R., Fava, M., Hughes, C. W., Garber, J., Malloy, E., King, C. A., Cerda, G., Sood, A. B., Alpert, J. E., Trivedi, M. H., and Rush, A. J. (2006). Remissions in maternal depression and child psychopathology: A STAR*D-child report. *JAMA* 295, 1389–1398.

Werner, E. E. (2004). Journeys from childhood to midlife: Risk, resilience, and recovery. *Pediatrics* 114, 492.

ADAPTIVE AND MALADAPTIVE PARENTING

7

ROLE OF CORTICOTROPIN RELEASING FACTOR-RELATED PEPTIDES IN THE NEURAL REGULATION OF MATERNAL DEFENSE

STEPHEN C. GAMMIE[1,2], KIMBERLY L. D'ANNA[1], GRACE LEE[1] AND SHARON A. STEVENSON[1]

[1] *Department of Zoology, University of Wisconsin, Madison, WI 53706, USA*
[2] *Neuroscience Training Program, University of Wisconsin, Madison, WI 53706, USA*

BACKGROUND ON MATERNAL DEFENSE

Lactating females of most mammalian species actively protect their offspring when they are young and defenseless. This protective behavior is termed maternal defense or maternal aggression and in some species the females actively attack a given threat to their young. Because maternal defense is highly conserved among mammals it is thought to increase the fitness of the offspring by defending them from harm. Although this hypothesis is difficult to test directly, work in voles and hamsters suggests that heightened defense leads to increased offspring survival (Giordano *et al.*, 1984; Heise & Lippke, 1997). Further, maternal defense can bring harm to the mother, so the conserved expression of an increased risk to the individual needs to be linked to some benefit and offspring survival is the most parsimonious explanation. Although almost no studies on maternal defense in humans have been conducted, examples of the protective mother permeate popular culture. For example, the basis of the Terminator movie series involves a mother's protection of her son. The basis of the animated feature, Finding Nemo, is that an otherwise skittish father must overcome his fears to find and rescue his son Nemo. Interestingly, for the prolog of this movie, Nemo's mother dies while attempting to protect him from a barracuda attack. She was successful in her protection, but she dies in the act and this highlights the dangers of maternal protection. Although the ability to turn on and off this complex social behavior of offspring protection provides an excellent opportunity for examining the control of aggression, maternal aggression has received relatively little attention.

When they are rearing and protecting pups, lactating rodents fiercely attack intruders (Lonstein & Gammie, 2002). Female house mice exhibit fierce aggression toward intruders when they are protecting pups and this ability to turn on and off maternal aggression in rodents provides an excellent opportunity for examining the neural control of aggression. The advantage of studying maternal defense in mice (or rats) is that the dependent measure is aggression, which is easily quantifiable. Thus, one can conduct manipulations and examine effects on a robust measure. Maternal aggression differs from male aggression in a number of critical ways. (1) Expression of male aggression is facilitated by testosterone, or its aromatization to estradiol (Compaan *et al.*, 1994), in most rodent species, while the expression of maternal aggression is facilitated by estradiol,

Neurobiology of the Parental Brain

progesterone, and prolactin released during pregnancy and lactation (Mann *et al.*, 1984; Stern & McDonald, 1989; Bridges, 1996; Lonstein & Gammie, 2002). (2) Sensory input to the lactating females from the suckling by pups plays a role in maintaining the expression of maternal aggression in both rats and mice (Svare & Gandelman, 1976; Stern & Kolunie, 1993; Lonstein & Gammie, 2002). (3) A number of neuromodulators have different effects on maternal and male aggression. For example, low levels of a serotonin agonist eliminate intermale aggression and female–female aggression (a mild territorial aggression between female rodents), but have no effect on maternal aggression (Parmigiani *et al.*, 1998). Site-specific injections of a serotonin agonist can even enhance maternal aggression (De Almeida & Lucion, 1997). Additionally, knockout studies indicate differential actions of a variety of genes on maternal vs. intermale aggression (Gammie & Nelson, 1999; Gammie *et al.*, 2000; Del Punta *et al.*, 2002; Gammie *et al.*, 2005). Because females only express maternal aggression in association with rearing pups, it is reasonable that females developed a specialized mechanism for the control of maternal aggression that differs from other forms of aggression. Relative to intermale aggression, though, maternal defense has been greatly understudied. Basic research into the control of maternal aggression may provide critical information for developing sex-specific interventions for aggressive behavior problems in humans. Also, an understanding of maternal aggression will provide general information on rapidly induced, short-term defense aggression.

WHY STUDY CORTICOTROPIN RELEASING FACTOR (CRF)-RELATED PEPTIDES? POSSIBLE COMMON ROLES IN ANXIETY AND DEFENSE CHANGES DURING LACTATION

In association with lactation and the production of maternal aggression, separate studies have shown that dams exhibit decreased indices of fearfulness and anxiety in a number of experimental paradigms. For example, relative to non-lactating females, lactating females show diminished fear/anxiety using the acoustic startle paradigm, the open field, the elevated plus maze, the defensive burying paradigm, the punished drinking paradigm, and the light/dark choice test (Fleming & Luebke, 1981; Hansen *et al.*, 1985; Hard & Hansen, 1985; Ferreira *et al.*, 1989; Bitran *et al.*, 1991; Maestripieri & D'Amato, 1991; Lonstein *et al.*, 1998; Fernandez-Guasti *et al.*, 2001; Lonstein & Gammie, 2002). One line of evidence that the regulation of the decreased fear and heightened aggression is similar is that sensory input from pups is required for both changes (Erskine *et al.*, 1978; Hard & Hansen, 1985; Stern & Kolunie, 1993). Thus, it is likely that common underlying neuromodulators regulate both fear/anxiety and maternal aggression, but this raises the question of the benefits and costs of linking these two systems. One possible benefit is that adjusting fear/anxiety systems alters the reaction to a potentially threatening and normally fear-evoking stimulus, such that an attack, as opposed to other options (e.g., flight), is produced.

The CRF-related peptide system is a promising candidate for this lactation-associated reduction in fearfulness. In addition to its critical role in activation of the HPA axis (Vale *et al.*, 1981), CRF released within the CNS mediates stress-induced behaviors (Smagin *et al.*, 2001), including an ability to induce fear and anxiety (Berridge & Dunn, 1989; Spadaro *et al.*, 1990; Liang *et al.*, 1992; Owens & Nemeroff, 1993; Stenzel-Poore *et al.*, 1994; Hammack *et al.*, 2002; Spina *et al.*, 2002). During lactation the CNS responsiveness to CRF and stress is suppressed. For example, intracerebroventricular (icv)-injected CRF is less potent at stimulating increases in neuronal activity as assessed by c-Fos activation in lactating relative to non-lactating females (da Costa *et al.*, 1997). CRF normally enhances response to an acoustic startle, but during lactation this enhancement is diminished (Walker *et al.*, 2003). Further, restraint stress is less potent at stimulating increases in neuronal activity during lactation (da Costa *et al.*, 1996). Thus, during lactation the CNS dampens responsiveness to CRF and stress.

The primary receptor for CRF is the G-protein coupled receptor, CRFR1 (Smith *et al.*, 1998; Weninger *et al.*, 1999; Bittencourt & Sawchenko, 2000), but CRF can also bind CRFR2 (Kishimoto *et al.*, 1995) with 10-fold less affinity than for CRFR1 (Vaughan *et al.*, 1995). Ucn 1, 2, and 3 are peptides related to CRF and these bind CRFR2 with high affinity; Ucn 1 can also bind to CRFR1 with equal efficacy as CRF (Vaughan *et al.*, 1995; Lewis *et al.*, 2001; Reyes *et al.*, 2001). In terms of localization, CRFR1 and CRFR2 are found in mostly distinct regions of the CNS with co-expression in only a few regions, including BNST, paraventricular nucleus of the hypothalamus, and portions of the septal region (Smagin & Dunn, 2000; Van Pett *et al.*, 2000).

EFFECTS OF CRF-RELATED PEPTIDES ON MATERNAL DEFENSE

A primary interest of our work is the central actions of CRF-related peptides. CRF is a potent activator of the HPA axis (Vale *et al.*, 1981), a peripheral effect, and recent work suggests Ucn 1 and 3 can also trigger HPA responses (de Groote *et al.*, 2005). Importantly, the end product of HPA activation, elevated gluco-corticoids (corticosterone), though, does not affect maternal aggression output (Al-Maliki, 1980). In fact, during maternal aggression, glucocorticoids are elevated (Neumann *et al.*, 2001; Deschamps *et al.*, 2003), which is not surprising given that the primary function of corticosterone is glucose mobilization and fighting involves high metabolic output. Although stress reactivity in terms of HPA responses to some stressors are muted during lactation (Walker *et al.*, 1995; da Costa *et al.*, 1996), in our studies we are concerned with altered central stress reactivity that underlies the behavioral responses to stressors.

Because CRF neurotransmission is decreased during lactation, we were interested in testing the hypothesis that CRF levels are inversely linked to levels of maternal aggression. Our first approach was to use icv injections CRF to test whether this peptide could decrease defense as predicted. As seen in Figure 7.1(A), we found CRF to dose-dependently impair maternal aggression (Gammie *et al.*, 2004). In this study we did not examine effects of CRF on other maternal behaviors, but did observe that all treated mice were nursing within 5 min of being reunited with pups following the aggression test. We did not find an effect of a CRF receptor antagonist on aggression and interpreted this finding as reflecting a "floor effect." In other words, CRF neurotransmission may already be low in key areas during lactation and keeping it low does not affect intensity. This interpretation suggests some other modulators are involved in positively regulating levels of aggression.

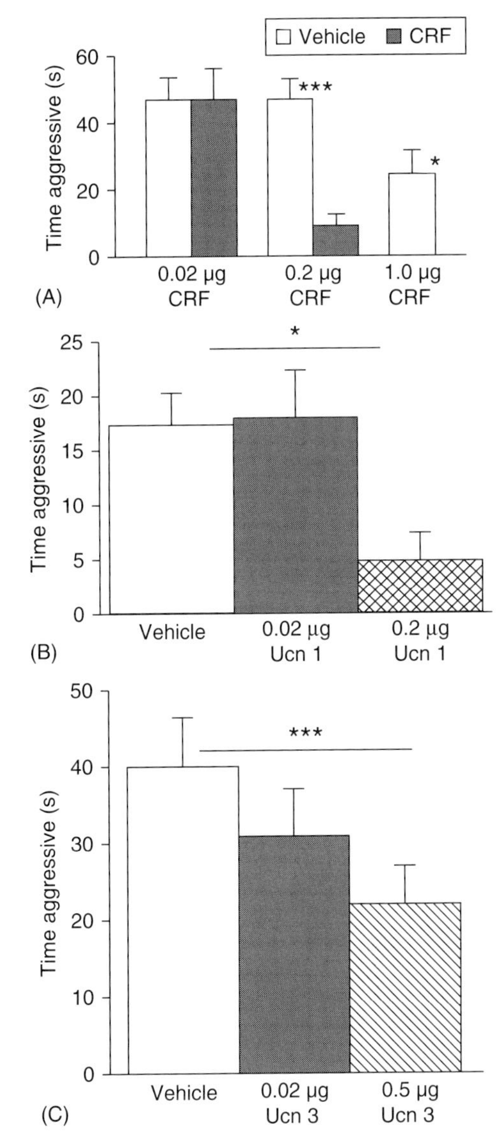

FIGURE 7.1 CRF-related peptides (CRF, Ucn 1, and Ucn 3) impair maternal aggression. Icv injection of CRF dose-dependently impairs total time aggressive at 0.2 and 1.0 μg, but not at 0.02 μg relative to vehicle injections (A). Icv Ucn 1 decreases aggression at 0.2 μg, but not 0.02 μg relative to vehicle injections (B). Icv Ucn 3 impairs aggression at 0.5 μg, but not 0.2 μg relative to vehicle injections (C). (*Source*: Data redrawn from D'Anna *et al.* (2005) and Gammie *et al.* (2004)). $^{*}p < 0.05$; $^{**}p < 0.001$.

In a second study, we examined the effects of Ucn 1 and 3 (peptides related to CRF; see above) on maternal aggression and found these also

c-Fos increases with icv injections

CRF = ↑ in LS & BNST, + 12 other regions

Ucn 1 = ↑ in LS & BNST, + 1 other region

Ucn 3 = ↑ in LS & BNST, + 4 other regions

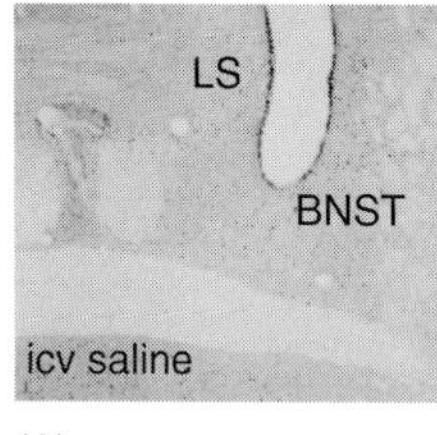

(A)

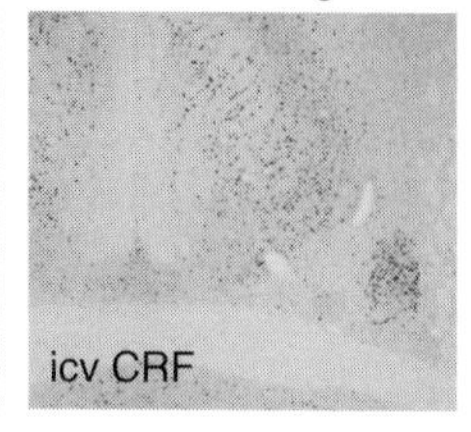

(B)

FIGURE 7.2 Following icv injection of CRF-related peptides (CRF, Ucn 1, or Ucn 3), significant increases in c-Fos were observed in both LS and BNST. In addition to LS and BNST, each peptide activated other regions, but none of these were common for all three peptides (Gammie *et al.*, 2004; D'Anna *et al.*, 2005). An example of elevated c-Fos with 1 µg CRF (B) relative to saline (A) in LS and BNST is shown in the lower panel.

to impair aggression in a dose-dependent manner (Figures 7.1(B),(C)) (D'Anna *et al.*, 2005). An important new feature in this study was our examination of pup retrieval immediately following the aggression test. Here, we found no effect of Ucn 1 or 3 on retrieval yet a significant impairment of aggression. This finding suggests that the peptide actions are specific to maternal defense and not generally to a range of maternal behaviors.

For this study, as for the CRF study, we examined levels of c-Fos in various brain regions following injection of peptide that impaired aggression. Interestingly, at doses where all three peptides impair aggression, significant increases in c-Fos activity were noted in only two common brain regions, lateral septum (LS) and BNST dorsal (BNSTd) (Figure 7.2). Additional evidence suggests these regions may be important for maternal defense regulation. For example, the normal c-Fos increases induced by central injection of CRF are blunted in LS and BNSTd in lactating vs. virgin females (da Costa *et al.*, 1997). Also, c-Fos increases that occur with stress are suppressed in LS in lactating vs. virgin females (da Costa *et al.*, 1996). CRF enhancement of the acoustic startle is diminished during lactation (Walker *et al.*, 2003) and a role for BNSTd in this suppressed startle response is possible. Because of these results, we are now very interested in both LS and BNSTd as possible critical sites for the negative regulation of maternal aggression.

As a follow-up to these studies, we have now begun studies injecting CRF and Ucn 3 directly into LS and have found that this treatment alone is sufficient to impair maternal aggression (D'Anna & Gammie, unpublished observations). It appears that activation of CRFR2 alone is sufficient to impair aggression, as evidenced by Ucn 3 being able to impair aggression when injected into LS. LS is enriched with CRFR2 and receives projections of Ucn 3 containing neurons that closely match the sites of high CRFR2 (Li *et al.*, 2002). Ucn 1-positive neurons and projections are also found in LS (Kozicz *et al.*, 1998). Several lines of research suggest a functional role for CRFR1 in LS. For example, activation of CRFR1 in LS triggers increases in glutamatergic neurotransmission (Liu *et al.*, 2004). Further, CRFR1 immunoreactivity in LS is found in mice (Chen *et al.*, 2000) and *in situ* probes for CRFR1 developed by the Allen Brain Atlas (http://www.brain-map.org/welcome.do) show detectable levels of CRFR1 in LS. Whether each CRF-related peptide is equally effective in these regions, which receptors are involved in this response, and what are the downstream targets of this action are not known.

Our findings that both CRF and related peptides, and that certain stressors applied postpartum impair aggression (described below), suggest that decreased CRF neurotransmission plays a critical, permissive role in the expression of maternal aggression. The link between elevated CRF and decreased maternal aggression also provides an explanation for why maternal aggression and fear/anxiety are often inversely correlated. As indicated below, for many signaling molecules examined, an inverse relationship of maternal defense and fear/anxiety has been found.

MATERNAL AGGRESSION IN CRFR1 AND CRFR2 KNOCKOUT MICE

One approach to examining the role for CRF-related peptides in maternal defense is to evaluate aggression in association with loss of either receptor. For studies on the CRFR1 knockout (KO) mice, we wanted to improve rates of reproduction and elevate overall maternal care and bred the deletion into mice we had selectively bred to show elevated maternal defense that show high levels of maternal care (see below). We also mated the test female mice with outbred sires (thus producing heterozygote offspring) to avoid deficits that can be observed in KO pups of KO dams (Smith *et al.*, 1998). We identified a trend toward decreased maternal defense in CRFR1 KO mice, but this did not reach significance (Gammie *et al.*, 2007). Importantly, we identified significant deficiencies in the KO mice both in nursing levels and in pup weight. Because suckling positively regulates maternal defense, one explanation for a trend for lower aggression in the KO mice was decreased contact with pups. Indeed, when we run pup weight (a measure of pup contact and suckling) as a covariate, no measure of maternal aggression is close to differing from wild-type (WT). Our finding is also consistent with our previous work that a general CRF receptor antagonist does not overtly alter maternal defense (Gammie *et al.*, 2004). Again, the interpretation is that the findings represent a floor effect whereby action on CRFR1 is relatively low during lactation and keeping this low does not enhance aggression. Thus, other modulators are thought to positively enhance the behavior.

For CRFR2 KO mice, we found significantly lower levels of maternal aggression relative to WT mice (Gammie *et al.*, 2005). Our interpretation is that KO mice have impaired aggression because they overproduce CRF and they have an intact CRFR1 (they have previously been shown to have high levels of anxiety) (Bale *et al.*, 2000). A recent alternate explanation of high anxiety in these mice is altered GABAergic output from LS to central amygdala (Henry *et al.*, 2006). Recent experiments suggest that the CRFR2 KO mice show a heightened sensitivity to the inhibitory effects of certain stressors (D'Anna & Gammie, unpublished observations). Interestingly, we did not find a difference in intermale aggression in either CRFR2 or CRFR1 KO mice (Gammie *et al.*, 2005; Gammie & Stevenson, 2006b), suggesting CRF does not play identical roles in the control of these two forms of aggression among these knockouts.

EFFECT OF STRESS ON MATERNAL DEFENSE

We recently found that restraint stress impairs maternal aggression (Figure 7.3) (Gammie & Stevenson, 2006a). This finding is interesting because a number of previous studies

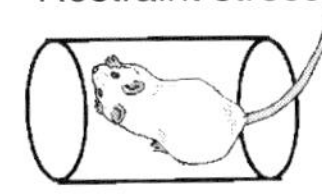

FIGURE 7.3 Recent work in our lab has shown that applying an acute stressor, such as restraint, during lactation is sufficient to impair maternal defense (top panel) (Gammie & Stevenson, 2006a). More recent pilot studies indicate that both forced swimming and exposure to fox odor will also impair defense. Given that the central release of CRF has been implicated in some of the behavioral responses to stress, we are interested in whether these effects can be mitigated using CRF antagonists.

have suggested that CRF is released centrally in association with the behavioral response to stress (Smagin *et al.*, 2001) and these results indicate CRF-related peptides may indeed be naturally released under high levels of stress as a natural modulator of maternal defense. More recently, we found that a brief swim stress or exposure to fox odor also impairs maternal defense (Stevenson & Gammie, unpublished observations) (Figure 7.3). Interestingly, when we examined light/dark box performance immediately following an aggression test either with or without a swim stress, we found low anxiety to be significantly linked with high maternal defense. Thus, we see a nice experimental link between elevated fear/anxiety and decreased maternal defense. Whether CRF-related peptides are involved in this association is unknown, but CRF-related peptide release could easily explain the findings. Altered central stress reactivity during lactation via altered CRF-related peptide signaling may heighten the expression of maternal defense by minimizing the extent to which fear- or anxiety-provoking stimuli can inhibit a mother's protection of her offspring. However, this does not suggest that lactating mothers are completely immune to the effects of stress (as suggested by our stress studies) and it may be adaptive to allow some stressors to modify behavior. For example, an environment with high stressors may not be optimal for pup rearing and hence protection of offspring may not be worth the risk or investment.

GENE ARRAY STUDIES ON HIGH MATERNAL DEFENSE MICE

We recently employed a powerful approach for examining the genetic basis of maternal aggression, namely to select for high levels of this trait using outbred mice. We have now selected for high maternal aggression for over 14 generations (Gammie *et al.*, 2006b). Using the selected high maternal defense and non-selected control mice, we then examined gene expression differences to understand which genes may contribute to differences in aggression. Although no differences in CRF were detected (a non-significant 10% decrease was noted in aggressive mice), significant increases in CRF binding protein were found in aggressive relative to unselected control mice (Gammie *et al.*, 2006a). Real-time PCR also suggested heightened CRF binding protein in selected mice ($p = 0.051$). CRF binding protein is a soluble non-membrane bound protein that can act in the CNS by binding either CRF or Ucn 1 (Westphal & Seasholtz, 2006), both of which we have found to impair aggression. Given that both CRF and Ucn 1 are anxiogenic, it is not surprising that CRF binding protein-deficient mice show heightened fear and anxiety (Karolyi *et al.*, 1999). CRF binding protein is expressed in the CNS, including in LS and BNSTd, as well as a number of cortical and subcortical regions (Potter *et al.*, 1992). Our finding of elevated CRF binding protein in high maternal defense mice is consistent with the inhibitory roles for CRF and Ucn 1 in aggression and suggests modulating the binding protein may be a way to modulate aggression levels. Our array analysis also identified additional candidate genes for regulating maternal defense, such as neurotensin which was lower in high aggressive mice. Interestingly, we recently found neurotensin to be a potent inhibitor of maternal defense (Stevenson & Gammie, unpublished observations) and we have recent evidence that neurotensin neurons could be a target of CRF-related peptides acting in LS.

ASSOCIATION OF FEAR AND ANXIETY PATHWAYS WITH MATERNAL DEFENSE REGULATION

Our work with CRF-related peptides indicates a mechanism whereby elevating anxiety may be associated with reduced defense. However, we recently completed a study which examined the opposite association, namely decreased fear and anxiety and elevated aggression. For this study, we used the benzodiazepine, chlordiazepoxide (CDP), which elevates endogenous GABA signaling, and replicated previous work that found benzodiazepines to elevate aggression. We found CDP to elevate levels of aggression by an average of 20 s (Lee & Gammie, 2007). A main focus of this study was to identify which brain regions show altered brain activity with application of benzodiazepines that enhance maternal defense. We found c-Fos activity of LS (and a few other regions, including caudal periaqueductal gray) to be significantly reduced by benzodiazepines at the dose that elevates aggression. This result again suggests LS as a critical site in maternal defense regulation but also confirms that decreased anxiety can be associated with heightened aggression. Interestingly, at higher doses of CDP, which would also be expected to be anxiolytic, we found aggression to be impaired and this may be due to sedative effects of drug. Thus, for a given modulator, such as GABA, we can find examples of fear and anxiety both being inversely and positively associated with maternal defense. This result highlights that although common signaling molecules underlie both fear and anxiety and maternal defense, the effects or links between the two can be complex.

A common counterexample to the link of low anxiety to high aggression is found in rats bred for high and low anxiety (HAB and LAB, respectively). In this case, HAB rats have higher maternal defense than LAB rats (Bosch *et al.*, 2005). However, a number of factors should be included in assessing this model. (1) These two populations are now highly inbred (having been separated by over 10 years) and genetic drift (whereby a trait becomes fixed by chance in an inbred population) could easily account for these differences. For example, we identified extremely high maternal aggression in one line of mice selected for high wheel-running, but this trait was not found in three other lines selected for this trait (Gammie *et al.*, 2003). Thus, high maternal aggression was fixed in this line by chance. In fact, if one just randomly compares a given control line (out of 4) with a given selected line (out of 4), in one out of every three comparisons a significant difference is found, but each of those finding can be considered spurious when the overall model is examined and no effect of selection is found on aggression. (2) The levels of aggression between HAB and LAB rats become equivalent when both lines are exposed to low stressors during pregnancy (Neumann *et al.*, 2005), indicating the aggression levels are statistically equivalent when examined in an altered environment. Under these conditions, LAB aggression is 20% higher than for HAB rats. (3) LAB rats show poor maternal care, indicating general deficits in maternal behavior. In our lab, poor maternal care is always associated with low maternal defense and LAB rats may have suppressed aggression due to general deficits from genetic drift. Even if the results from the HAB and LAB rats are taken at face value, these still produce an *N* of 1 in terms of linking elevated anxiety with heightened maternal aggression. Further, ongoing work using these rats has consistently suggested that underlying modulators, such as oxytocin, that decrease anxiety (Rosenzweig-Lipson *et al.*, 2004; Ring *et al.*, 2006), elevate maternal defense (Bosch *et al.*, 2005), again supporting an inverse relationship between these two traits. When we examined anxiety in our mice selected for high maternal aggression, we saw a higher number of middle square and closed arm entries but no difference in open or closed times (Gammie *et al.*, 2006b). Anxiety was examined in Generation 5 and it would be interesting to see whether anxiety differences exist now that additional selection has occurred. It is worth pointing out that we only have one selected line of high aggression mice and thus the caveats that apply for trying to examine a correlated trait without replication apply here as well. In other words, if one really wants to look for a correlation between maternal defense and fear and anxiety using a selection model, one needs to select for one of the traits across multiple lines (at least four) and one must also maintain unselected control lines (at least four as well).

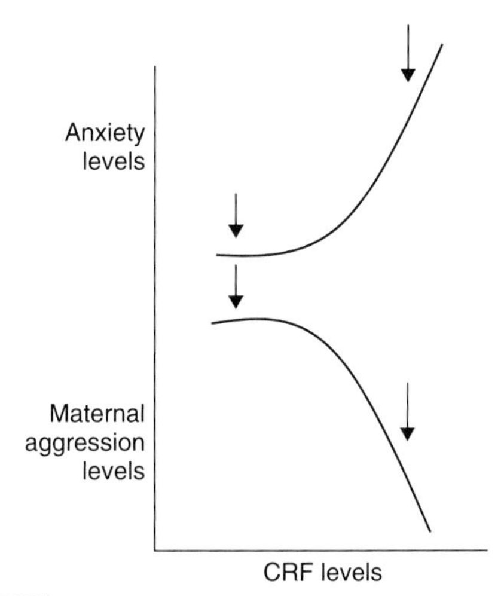

FIGURE 7.4 Overview of how of CRF regulates aggression and fear/anxiety. Cumulative results from our work with CRF (and its related peptides) suggest that although some small baseline levels of CRF signaling are needed for full maternal aggression expression (assuming all other systems are active), over most levels CRF is inhibitory to maternal defense (lower half of graph). Interestingly, for these same levels as seen in numerous other studies in mice and rats, one finds CRF to elevate anxiety (upper half of graph). Thus, for this peptide, one can see why an inverse association of maternal aggression and fear/anxiety can be found. As shown in Figure 7.5, though, a number of other signaling molecules can contribute to both aggression and fear/anxiety making this relationship much more complex.

ADVANTAGES OF LINKING CENTRAL STRESS-RELATED PATHWAYS WITH MATERNAL DEFENSE

When broadly surveyed, research indicates that decreases in anxiety measures are most commonly associated with elevations of maternal aggression, but many exceptions exist. Because the final regulation of fear and anxiety vs. maternal defense is different, a full concordance between these traits should not be expected. When examining specific signaling molecules that can affect both fear/anxiety and maternal aggression systems, an inverse regulation of the two systems is often seen (see Figure 7.4 and 7.5). Thus, when examining the underlying neuronal systems, a biological explanation exists for why an inverse association can sometimes be found between fear/anxiety and heightened defense. As shown for GABA signaling above, though, complex outputs can occur which can alter or reverse this relationship. To go beyond general inverse correlations between fear/anxiety and maternal aggression, it is important to continue to elucidate more precisely where and how CRF-related peptides

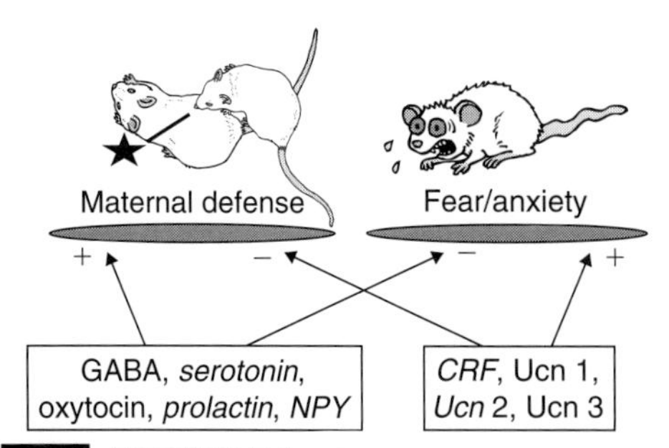

FIGURE 7.5 One approach for examining how and why maternal defense and fear/anxiety are commonly inversely associated is to examine how underlying neuromodulators affect both systems. As shown in this schematic, a subset of signaling molecules, such as CRF-related peptides, activates (+) fear/anxiety pathways, yet also inhibits (−) maternal defense. Further, for a different set of signaling molecules, such as GABA, serotonin, and oxytocin, evidence exists for these to activate maternal aggression (+) while decreasing (−) fear/anxiety. Prolactin and NPY can both decrease fear/anxiety, but only indirect evidence suggests they elevate maternal defense. An important qualifier for this framework is that actions of signaling molecules can differ across brain regions and signaling molecules can have an inverted U dose–response curve, such as we have seen with GABA, whereby this inverse association of behavioral traits can become a positive association, depending on levels of signaling molecules.

Altered response with lactation
Freeze → Fight!
Flight
Fight
Fright (give up/ go limp)
Default behavioral response

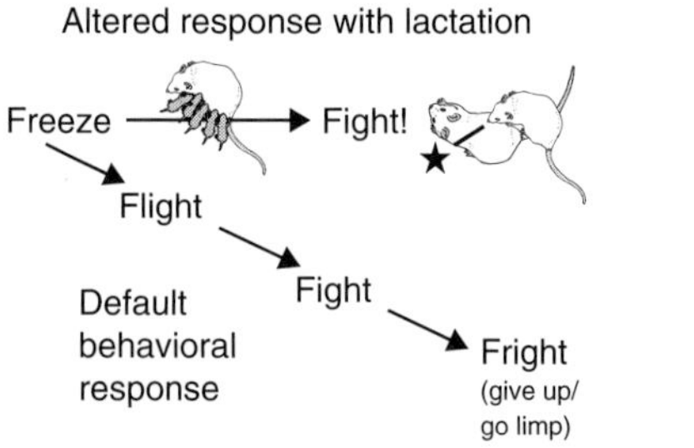

FIGURE 7.6 One framework for the expression of maternal aggression is an alteration in the underlying default behavioral response to stress. As shown, a typical default behavioral response to stress is freezing behavior, followed by flight, followed by fighting, and then followed by fright, which can involve an animal going limp or giving up (Bracha *et al.*, 2004). Each of these responses can be adaptive. For example, freezing allows an animal to evaluate a potential unexpected problem and can help an animal avoid predator detection. Flight can be adaptive because removing oneself from a potentially dangerous situation decreases likelihood of receiving injury. Fighting can be adaptive when an animal is preserving a resource or if one cannot escape a potential interaction. Finally, fright can be adaptive because if an animal cannot flee or defeat another animal, then giving up may decrease possible harm. One model for maternal aggression, then, is that this default pathway is altered such that the animal will quickly transition from freeze to fight with flight being superseded. A value of this switch is that it allows lactating females to quickly defend offspring and also helps remove flight (which would not be advantageous to offspring) as an option.

and other anxiety-related signaling molecules alter maternal defense and anxiety.

Because females of most species provide a primary role in care of offspring, the way females respond to stress has been hypothesized to be fundamentally different than for males (Taylor *et al.*, 2000). Protective and nurturing behavior by lactating females is critical for the survival of most mammalian offspring. Females that remain with and defend offspring when faced with an attacker, then, do not have the same options (fight or flight) as males (Taylor *et al.*, 2000). More recently, it has been suggested that the default behavioral response to stress progresses from freeze to flight to fight to fright (which involves going limp or giving up) (Bracha *et al.*, 2004). With lactation, an important shift in the default behavioral response may be seen whereby fight precedes flight and allows for a quick protection of offspring (Figure 7.6). An understanding of the neural basis of maternal aggression may shed insights into what may be the fundamental sex differences in the response to stress and how those change with lactation. A critical value of studying maternal aggression is that its production is linked to alterations during lactation of neuromodulators (e.g., CRF-related peptides) that are key players in fear, anxiety, and depression (including postpartum). Changes in central stress reactivity during lactation leave females particularly vulnerable to mood disorders, including postpartum depression. Thus, understanding how and why central stress reactivity is altered during lactation and how this integrates with the defense/protection of offspring, could provide unique insights into mood disorders that occur during the postpartum period and provide the basis for future translational research.

FUTURE DIRECTIONS

CRF-related peptides are among many signaling molecules involved in the regulation of maternal defense. For future studies it will be important to understand in what brain regions these peptides are acting and what are the downstream targets. This includes an understanding of how CRF-related peptides interact with other signaling molecules that regulate maternal defense.

As part of a larger picture, it will be valuable to ascertain whether certain CRF-related peptide manipulations (e.g., site-directed injections) are also simultaneously altering fear/anxiety measures. With this information, one can develop a more complex understanding of how fear and anxiety and maternal defense pathways intersect. Another important framework for studies in maternal defense is understanding how the circuitries are activated. One of the most fascinating aspects of maternal defense is its relative brief temporal appearance across the life-history of the animal. For example, understanding how CRF-related peptide signaling (or other signaling molecules, such as oxytocin) is modulated during lactation and how that specifically contributes to defense will be a critical aspect in developing an understanding of maternal defense.

ACKNOWLEDGMENTS

This work was supported by National Institutes of Health grants R01 MH066086 to S.C.G and an American Psychological Association Diversity Program in Neuroscience Fellowship to K.L.D.

REFERENCES

Al-Maliki, S. (1980). Influences of stress-related hormones on a variety of attack behaviour in laboratory mice. In *Adaptive Capabilities of the Nervous System* (P. McConnell, Ed.), Vol. 53, pp. 421–426. Elsevier, Amsterdam.

Bale, T. L., Contarino, A., Smith, G. W., Chan, R., Gold, L. H., Sawchenko, P. E., Koob, G. F., Vale, W. W., and Lee, K. F. (2000). Mice deficient for corticotropin-releasing hormone receptor-2 display anxiety-like behaviour and are hypersensitive to stress. *Nat. Genet.* 24, 410–414.

Berridge, C. W., and Dunn, A. J. (1989). CRF and restraint-stress decrease exploratory behavior in hypophysectomized mice. *Pharmacol. Biochem. Behav.* 34, 517–519.

Bitran, D., Hilvers, R. J., and Kellogg, C. K. (1991). Ovarian endocrine status modulates the anxiolytic potency of diazepam and the efficacy of gamma-aminobutyric acid-benzodiazepine receptor-mediated chloride ion transport. *Behav. Neurosci.* 105, 653–662.

Bittencourt, J. C., and Sawchenko, P. E. (2000). Do centrally administered neuropeptides access cognate receptors?: An analysis in the central corticotropin-releasing factor system. *J. Neurosci.* 20, 1142–1156.

Bosch, O. J., Meddle, S. L., Beiderbeck, D. I., Douglas, A. J., and Neumann, I. D. (2005). Brain oxytocin correlates with maternal aggression: Link to anxiety. *J. Neurosci.* 25, 6807–6815.

Bracha, H. S., Ralston, T. C., Matsukawa, J. M., Williams, A. E., and Bracha, A. S. (2004). Does "fight or flight" need updating?. *Psychosomatics* 45, 448–449.

Bridges, R. S. (1996). Biochemical basis of parental behavior in the rat. In *Parental Care: Evolution, Mechanisms, and Adaptive Significance* (J. S. Rosenblatt and C. T. Snowden, Eds.), pp. 215–237. Academic Press, San Diego, CA.

Chen, Y., Brunson, K. L., Muller, M. B., Cariaga, W., and Baram, T. Z. (2000). Immunocytochemical distribution of corticotropin-releasing hormone receptor type-1 (CRF(1))-like immunoreactivity in the mouse brain: Light microscopy analysis using an antibody directed against the C-terminus. *J. Comp. Neurol.* 420, 305–323.

Compaan, J. C., Wozniak, A., De Ruiter, A. J., Koolhaas, J. M., and Hutchison, J. B. (1994). Aromatase activity in the preoptic area differs between aggressive and nonaggressive male house mice. *Brain Res. Bull.* 35, 1–7.

da Costa, A. P., Wood, S., Ingram, C. D., and Lightman, S. L. (1996). Region-specific reduction in stress-induced c-fos mRNA expression during pregnancy and lactation. *Brain Res.* 742, 177–184.

da Costa, A. P., Kampa, R. J., Windle, R. J., Ingram, C. D., and Lightman, S. L. (1997). Region-specific immediate-early gene expression following the administration of corticotropin-releasing hormone in virgin and lactating rats. *Brain Res.* 770, 151–162.

D'Anna, K. D., Stevenson, S. A., and Gammie, S. C. (2005). Urocortin 1 and 3 impair maternal defense behavior in mice. *Behav. Neurosci.* 119, 161–171.

De Almeida, R. M., and Lucion, A. B. (1997). 8-OH-DPAT in the median raphe, dorsal periaqueductal gray and corticomedial amygdala nucleus decreases, but in the medial septal area it can increase maternal aggressive behavior in rats. *Psychopharmacology* 134, 392–400.

de Groote, L., Penalva, R. G., Flachskamm, C., Reul, J. M. H. M., and Linthorst, A. C. E. (2005). Differential monoaminergic, neuroendocrine and behavioural responses after central administration of corticotropin-releasing factor receptor type 1 and type 2 agonists. *J. Neurochem.* 94, 45–56.

Del Punta, K., Leinders-Zufall, T., Rodriguez, I., Jukam, D., Wysocki, C. J., Ogawa, S., Zufall, F., and Mombaerts, P. (2002). Deficient pheromone responses in mice lacking a cluster of vomeronasal receptor genes. *Nature* 419, 70–74.

Deschamps, S., Woodside, B., and Walker, C. D. (2003). Pups presence eliminates the stress hyporesponsiveness of early lactating females to a psychological stress representing a threat to the pups. *J. Neuroendocrinol.* 15, 486–497.

Erskine, M. S., Barfield, R. J., and Goldman, B. D. (1978). Intraspecific fighting during late pregnancy and lactation in rats and effects of litter removal. *Behav. Biol.* 23, 206–218.

Fernandez-Guasti, A., Ferreira, A., and Picazo, O. (2001). Diazepam, but not buspirone, induces similar anxiolytic-like actions in lactating and ovariectomized Wistar rats. *Pharmacol. Biochem. Behav.* 70, 85–93.

Ferreira, A., Hansen, S., Nielsen, M., Archer, T., and Minor, B. G. (1989). Behavior of mother rats in conflict tests sensitive to antianxiety agents. *Behav. Neurosci.* 103, 193–201.

Fleming, A. S., and Luebke, C. (1981). Timidity prevents the virgin female rat from being a good mother: Emotionality differences between nulliparous and parturient females. *Physiol. Behav.* 27, 863–868.

Gammie, S. C., and Nelson, R. J. (1999). Maternal aggression is reduced in neuronal nitric oxide synthase-deficient mice. *J. Neurosci.* 19, 8027–8035.

Gammie, S. C., and Stevenson, S. A. (2006a). Effects of daily and acute restraint stress during lactation on maternal aggression and behavior in mice. *Stress* 9, 171–180.

Gammie, S. C., and Stevenson, S. A. (2006b). Intermale aggression in corticotropin-releasing factor receptor 1 deficient mice. *Behav. Brain Res.* 171, 63–69.

Gammie, S. C., Huang, P. L., and Nelson, R. J. (2000). Maternal aggression in endothelial nitric oxide synthase-deficient mice. *Horm. Behav.* 38, 13–20.

Gammie, S. C., Hasen, N. S., Rhodes, J. S., Girard, I., and Garland, T., Jr. (2003). Predatory aggression, but not maternal or intermale aggression, is associated with high voluntary wheel-running behavior in mice. *Horm. Behav.* 44, 209–221.

Gammie, S. C., Negron, A., Newman, S. M., and Rhodes, J. S. (2004). Corticotropin-releasing factor inhibits maternal aggression in mice. *Behav. Neurosci.* 118, 805–814.

Gammie, S. C., Hasen, N. S., Stevenson, S. A., Bale, T. L., and D'Anna, K. D. (2005). Elevated stress sensitivity in corticotropin-releasing factor receptor 2 deficient mice decreases maternal, but not intermale aggression. *Behav. Brain Res.* 160, 169–177.

Gammie, S. C., Auger, A. P., Jessen, H. M., Vanzo, R. J., Awad, T. A., and Stevenson, S. A. (2006a). Altered gene expression in mice selected for high maternal aggression. *Genes Brain Behav.* 36, 713–722.

Gammie, S. C., Garland, T., and Stevenson, S. A. (2006b). Artificial selection for increased maternal defense behavior in mice. *Behav. Genet.* 36, 713–722.

Gammie, S. C., Bethea, E. D., and Stevenson, S. A. (2007). Altered maternal profiles in corticotropin-releasing factor receptor 1 deficient mice. *BMC Neurosci.* 8, 17.

Giordano, A. L., Siegel, H. I., and Rosenblatt, J. S. (1984). Effects of mother-litter separation and reunion on maternal aggression and pup mortality in lactating hamsters. *Physiol. Behav.* 33, 903–906.

Hammack, S. E., Richey, K. J., Schmid, M. J., LoPresti, M. L., Watkins, L. R., and Maier, S. F. (2002). The role of corticotropin-releasing hormone in the dorsal raphe nucleus in mediating the behavioral consequences of uncontrollable stress. *J. Neurosci.* 22, 1020–1026.

Hansen, S., Ferreira, A., and Selart, M. E. (1985). Behavioural similarities between mother rats and benzodiazepine-treated non-maternal animals. *Psychopharmacology* 86, 344–347.

Hard, E., and Hansen, S. (1985). Reduced fearfulness in the lactating rat. *Physiol. Behav.* 35, 641–643.

Heise, S., and Lippke, J. (1997). Role of female aggression in prevention of infanticidal behavior in male common voles, *Microtus arvalus* (Pallas, 1779). *Aggressive Behav.* 23, 293–298.

Henry, B., Vale, W., and Markou, A. (2006). The effect of lateral septum corticotropin-releasing factor receptor 2 activation on anxiety is modulated by stress. *J. Neurosci.* 26, 9142–9152.

Karolyi, I. J., Burrows, H. L., Ramesh, T. M., Nakajima, M., Lesh, J. S., Seong, E., Camper, S. A., and Seasholtz, A. F. (1999). Altered anxiety and weight gain in corticotropin-releasing hormone-binding protein-deficient mice. *Proc. Natl Acad. Sci. USA* 96, 11595–11600.

Kishimoto, T., Pearse, R. V., II, Lin, C. R., and Rosenfeld, M. G. (1995). A sauvagine/corticotropin-releasing factor receptor expressed in heart and skeletal muscle. *Proc. Natl Acad. Sci. USA* 92, 1108–1112.

Kozicz, T., Yanaihara, H., and Arimura, A. (1998). Distribution of urocortin-like immunoreactivity in the central nervous system of the rat. *J. Comp. Neurol.* 391, 1–10.

Lee, G., and Gammie, S. C. (2007). GABA enhancement of maternal defense in mice: Possible neural correlates. *Pharmacol. Biochem. Behav.* 86, 176–187.

Lewis, K., Li, C., Perrin, M. H., Blount, A., Kunitake, K., Donaldson, C., Vaughan, J., Reyes, T. M., Gulyas, J., Fischer, W. et al. (2001). Identification of urocortin III, an additional member of the corticotropin-releasing factor (CRF) family with high affinity for the CRF2 receptor. *Proc. Natl Acad. Sci. USA* 98, 7570–7575.

Li, C., Vaughan, J., Sawchenko, P. E., and Vale, W. W. (2002). Urocortin III-immunoreactive projections in rat brain: Partial overlap with sites of type 2 corticotrophin-releasing factor receptor expression. *J. Neurosci.* 22, 991–1001.

Liang, K. C., Melia, K. R., Miserendino, M. J., Falls, W. A., Campeau, S., and Davis, M. (1992). Corticotropin-releasing factor: Long-lasting facilitation of the acoustic startle reflex. *J. Neurosci.* 12, 2303–2312.

Liu, J., Yu, B. J., Neugebauer, V., Grigoriadis, D. E., Rivier, J., Vale, W. W., Shinnick-Gallagher, P., and Gallagher, J. P. (2004). Corticotropin-releasing factor and urocortin I modulate excitatory glutamatergic synaptic transmission. *J. Neurosci.* 24, 4020–4029.

Lonstein, J. S., and Gammie, S. C. (2002). Sensory, hormonal, and neural control of maternal aggression in laboratory rodents. *Neurosci. Biobehav. Rev.* 26, 869–888.

Lonstein, J. S., Simmons, D. A., and Stern, J. M. (1998). Functions of the caudal periaqueductal gray in lactating rats: Kyphosis, lordosis, maternal aggression, and fearfulness. *Behav. Neurosci.* 112, 1502–1518.

Maestripieri, D., and D'Amato, F. R. (1991). Anxiety and maternal aggression in house mice (*Mus musculus*): A look at interindividual variability. *J. Comp. Psychol.* 105, 295–301.

Mann, M. A., Konen, C., and Svare, B. (1984). The role of progesterone in pregnancy-induced aggression in mice. *Horm. Behav.* 18, 140–160.

Neumann, I. D., Toschi, N., Ohl, F., Torner, L., and Kromer, S. A. (2001). Maternal defence as an emotional stressor in female rats: Correlation of neuroendocrine and behavioural parameters and involvement of brain oxytocin. *Eur. J. Neurosci.* 13, 1016–1024.

Neumann, I. D., Kromer, S. A., and Bosch, O. J. (2005). Effects of psycho-social stress during pregnancy on neuroendocrine and behavioural parameters in lactation depend on the genetically determined stress vulnerability. *Psychoneuroendocrinology* 30, 791–806.

Owens, M. J., and Nemeroff, C. B. (1993). The role of corticotropin-releasing factor in the pathophysiology of affective and anxiety disorders: Laboratory and clinical studies. *Ciba Found. Symp.* 172, 296–308.

Parmigiani, S., Ferrari, P. F., and Palanza, P. (1998). An evolutionary approach to behavioral pharmacology: Using drugs to understand proximate and ultimate mechanisms of different forms of aggression in mice. *Neurosci. Biobehav. Rev.* 23, 143–153.

Potter, E., Behan, D. P., Linton, E. A., Lowry, P. J., Sawchenko, P. E., and Vale, W. W. (1992). The central distribution of a corticotropin-releasing factor (Crf)-binding protein predicts multiple sites and modes of interaction with Crf. *Proc. Natl Acad. Sci. USA* 89, 4192–4196.

Reyes, T. M., Lewis, K., Perrin, M. H., Kunitake, K. S., Vaughan, J., Arias, C. A., Hogenesch, J. B., Gulyas, J., Rivier, J., Vale, W. W. et al. (2001). Urocortin II: A member of the corticotropin-releasing factor (CRF) neuropeptide family that is selectively bound by type 2 CRF receptors. *Proc. Natl Acad. Sci. USA* 98, 2843–2848.

Ring, R. H., Malberg, J. E., Potestio, L., Ping, J., Boikess, S., Luo, B., Schechter, L. E., Rizzo, S., Rahman, Z., and Rosenzweig-Lipson, S. (2006). Anxiolytic-like activity of oxytocin in male mice: Behavioral and autonomic evidence, therapeutic implications. *Psychopharmacology* 185, 218–225.

Rosenzweig-Lipson, S. J., Boikess, S., Li, J. P., Luo, B., Malberg, J. E., Platt, B., Potestio, L. M., and Ring, R. H. (2004). Oxytocin produces anxiolytic-like effects in multiple animal models of anxiety in rodents. *FASEB J.* 18, A959–A959.

Smagin, G. N., and Dunn, A. J. (2000). The role of CRF receptor subtypes in stress-induced behavioural responses. *Eur. J. Pharmacol.* 405, 199–206.

Smagin, G. N., Heinrichs, S. C., and Dunn, A. J. (2001). The role of CRH in behavioral responses to stress. *Peptides* 22, 713–724.

Smith, G. W., Aubry, J. M., Dellu, F., Contarino, A., Bilezikjian, L. M., Gold, L. H., Chen, R., Marchuk, Y., Hauser, C., Bentley, C. A. et al.

(1998). Corticotropin releasing factor receptor 1-deficient mice display decreased anxiety, impaired stress response, and aberrant neuroendocrine development. *Neuron* 20, 1093–1102.

Spadaro, F., Berridge, C. W., Baldwin, H. A., and Dunn, A. J. (1990). Corticotropin-releasing factor acts via a third ventricle site to reduce exploratory behavior in rats. *Pharmacol. Biochem. Behav.* 36, 305–309.

Spina, G., Merlo-Pich, E., Akwa, Y., Balducci, C., Basso, M., Zorrilla, P., Britton, T., Rivier, J., Vale, W., and Koob, F. (2002). Time-dependent induction of anxiogenic-like effects after central infusion of urocortin or corticotropin-releasing factor in the rat. *Psychopharmacology* 160, 113–121.

Stenzel-Poore, M. P., Heinrichs, S. C., Rivest, S., Koob, G. F., and Vale, W. W. (1994). Overproduction of corticotropin-releasing factor in transgenic mice: A genetic model of anxiogenic behavior. *J. Neurosci.* 14, 2579–2584.

Stern, J. M., and McDonald, C. (1989). Ovarian hormone-induced short-latency maternal behavior in ovariectomized virgin Long-Evans rats. *Horm. Behav.* 23, 157–172.

Stern, J. M., and Kolunie, J. M. (1993). Maternal aggression of rats is impaired by cutaneous anesthesia of the ventral trunk, but not by nipple removal. *Physiol. Behav.* 54, 861–868.

Svare, B., and Gandelman, R. (1976). Postpartum aggression in mice: The influence of suckling stimulation. *Horm. Behav.* 7, 407–416.

Taylor, S. E., Klein, L. C., Lewis, B. P., Gruenewald, T. L., Gurung, R. A., and Updegraff, J. A. (2000). Biobehavioral responses to stress in females: Tend-and-befriend, not fight-or-flight. *Psychol. Rev.* 107, 411–429.

Vale, W., Spiess, J., Rivier, C., and Rivier, J. (1981). Characterization of a 41-residue ovine hypothalamic peptide that stimulates secretion of corticotropin and beta-endorphin. *Science* 213, 1394–1397.

Van Pett, K., Viau, V., Bittencourt, J. C., Chan, R. K., Li, H. Y., Arias, C., Prins, G. S., Perrin, M., Vale, W., and Sawchenko, P. E. (2000). Distribution of mRNAs encoding CRF receptors in brain and pituitary of rat and mouse. *J. Comp. Neurol.* 428, 191–212.

Vaughan, J., Donaldson, C., Bittencourt, J., Perrin, M. H., Lewis, K., Sutton, S., Chan, R., Turnbull, A. V., Lovejoy, D., and Rivier, C. (1995). Urocortin, a mammalian neuropeptide related to fish urotensin I and to corticotropin-releasing factor. *Nature* 378, 287–292.

Walker, C. D., Trottier, G., Rochford, J., and Lavallee, D. (1995). Dissociation between behavioral and hormonal responses to the forced swim stress in lactating rats. *J. Neuroendocrinol.* 7, 615–622.

Walker, D. L., Toufexis, D. J., and Davis, M. (2003). Role of the bed nucleus of the stria terminalis versus the amygdala in fear, stress, and anxiety. *Eur. J. Pharmacol.* 463, 199–216.

Weninger, S. C., Dunn, A. J., Muglia, L. J., Dikkes, P., Miczek, K. A., Swiergiel, A. H., Berridge, C. W., and Majzoub, J. A. (1999). Stress-induced behaviors require the corticotropin-releasing hormone (CRH) receptor, but not CRH. *Proc. Natl Acad. Sci. USA* 96, 8283–8288.

Westphal, N. J., and Seasholtz, A. F. (2006). CRH-BP: The regulation and function of a phylogenetically conserved binding protein. *Front. Biosci.* 11, 1878–1891.

8

MATERNAL STRESS ADAPTATIONS PERIPARTUM: MOM'S INNATE ANXIETY DETERMINES MATERNAL CARE AND AGGRESSION

INGA D. NEUMANN AND OLIVER J. BOSCH

Department of Behavioural Neuroendocrinology, University of Regensburg, Regensburg, Germany, UK

Remarkable physiological and behavioral changes have been extensively described in the mammalian maternal brain in the peripartum period. These profound adaptations mainly start as complex and direct consequences of hormonal signals arising from the fetus. They continue in lactation as a result of close interactions between mother and offspring, for example, during suckling.

Perhaps the main alterations are related to the changing demands of the developing fetus or newborn offspring, from intrauterine provision of sufficient nutrients to postpartum care and protection including the intrauterine provision of sufficient nutrients, a stable hormonal and biochemical environment, safe birth, and an immediate postpartum care including lactation; maternal behavior, and protection of the offspring. In this respect, alterations in the activity of numerous neuroendocrine systems are critically involved in the promotion of these complex maternal adaptations. Altered synthetic and secretory activity of the oxytocin (OXT) and prolactin (PRL) systems within the maternal brain are likely to contribute to the fascinating switch in physiological, but also in a variety of behavioral parameters including emotionality, cognition, and, most importantly, social behaviors such as maternal care and aggression.

In this chapter we aim to summarize mechanisms underlying the neuroendocrine and emotional changes observed peripartum. We will mainly focus on the increased activity of the neuropeptides OXT and PRL during this period and their involvement in the blunted stress responsiveness and anxiety. Moreover, brain OXT plays an important role in the regulation not only of maternal behavior, but more specifically also of maternal aggression. Here, in a rat model for genetically-determined high (HAB) and low (LAB) levels of anxiety-related behavior, could be found a clear link between the innate level of anxiety and the intensity of maternal behavior including maternal aggressive behavior. Importantly, we demonstrate that differences in local OXT release patterns, for example, within the central amygdala (CeA) and the hypothalamic paraventricular nucleus (PVN) are responsible for the differing styles of maternal protective behavior.

Moreover, we will discuss data demonstrating that adverse chronic life experiences, such as chronic stress in pregnancy or even prenatal stress, will affect maternal adaptations in lactation.

CHANGES IN HPA AXIS RESPONSIVENESS PERIPARTUM

Profound alterations in the activity of the hypothalamo–pituitary–adrenal (HPA) axis, both under basal and stimulated conditions, have been consistently demonstrated to occur at the end of pregnancy and in lactation (Slattery & Neumann, 2008). The fine-tuned regulation of the HPA activity has a particular significance to pregnancy and lactation, times at which the metabolic demands on the mother are increased and, simultaneously, her immune competence is challenged.

Studies in lactating and pregnant rats have demonstrated a state of basal hypercorticalism (Stern *et al.*, 1973; Walker *et al.*, 1992; Fischer *et al.*, 1995; Windle *et al.*, 1997a). Sensory stimuli by the pups is required to maintain this state and it appears to be important to meet the metabolic demands of the offspring, especially in lactation (Lightman *et al.*, 2001). However, in pregnancy excessive levels of circulating glucocorticoids have to be avoided, as high glucocorticoid levels exert complex adverse effects on the fetal development *in utero* (for review see Weinstock, 2001; Welberg & Seckl, 2001).

In this context, it has been reported in several species including humans that the responsiveness of the HPA axis to the majority of stimuli studied so far is severely attenuated peripartum, both at the end of pregnancy and in lactation. Consequently, the rise in plasma adrenocorticotropic hormone (ACTH), and corticosterone (rodents) or cortisol (humans) in response to exposure to a given psychological stressor (e.g., restraint: da Costa *et al.*, 1996; noise: Windle *et al.*, 1997b; novel environment: Neumann *et al.*, 1998a), swimming (Walker *et al.*, 1995; Toufexis *et al.*, 1998; Neumann *et al.*, 1998b), foot shock (Stern *et al.*, 1973), ether vaporation (Banky *et al.*, 1994), intraperitoneal injection of NaCl (Lightman & Young, 1989), or to immunological stressors (lipopolysaccharide: Shanks *et al.*, 1999; interleukin: Brunton *et al.*, 2005) is attenuated. Also, in response to an ethologically relevant stimulus, that is, defense of the offspring by the dam during the maternal defense test and display of aggressive behavior, ACTH and corticosterone secretion was found to be relatively low (Neumann *et al.*, 2001). However, there is also good evidence that the presence of the pups and the relevance of the stressor for the safety of the pups modulate the maternal stress response (Deschamps *et al.*, 2003).

MECHANISMS OF BLUNTED HPA AXIS RESPONSE

Lack of Excitatory Noradrenergic and Opioid Neurotransmission

Various brain mechanisms have been revealed to underlie the blunted stress responsiveness in the peripartum period (for review see Lightman *et al.*, 2001; Neumann, 2001; Brunton & Russell 2008; Walker *et al.*, 2001; Slattery & Neumann, 2008). These include, for example, a reduced excitatory noradrenergic activity within the PVN (Toufexis *et al.*, 1998; Douglas *et al.*, 2005). Noradrenaline from brain stem nuclei is a major excitatory input to the PVN and mediates the stress response of parvocellular PVN neurons via postsynaptic α1-, α2-, and β-adrenergic receptors (Plotsky, 1987; Itoi *et al.*, 1994; Han *et al.*, 2002). On day 20 of pregnancy in rats, local intra-PVN release of noradrenaline in response to swim stress was found to be almost completely abolished. This was accompanied by a reduced expression of α1A adrenoreceptor mRNA in the parvo- and magnocellular PVN and reduced sensitivity of the HPA axis to noradrenergic neurotransmission (Douglas *et al.*, 2005).

There is also lack of another excitatory influence on HPA axis responses. Endogenous opioids are important to trigger an appropriate HPA axis response in virgin rats, but this excitatory input is absent in late pregnancy (Douglas *et al.*, 1998). Furthermore, the opioidergic input is actually reversed into an inhibitory tone during parturition (Wigger *et al.*, 1999) as revealed by intravenous infusion of naloxone prior to stressor exposure. Thus, the activity of two mainly excitatory inputs to the hypothalamic PVN, noradrenaline and endogenous opioids, is significantly blunted in late pregnancy and lactation, contributing to the attenuated stress responsiveness.

Other mechanisms of HPA axis hyporesponsiveness include blunted stress-induced expression of immediate early genes at various relevant brain sites (da Costa *et al.*, 1996; Woodside & Amir, 1997; Shanks *et al.*, 1999; Amico *et al.*, 2004), reduced activity of corticotropin-releasing hormone (CRH) neurons in the PVN (Douglas & Russell, 1994; Toufexis *et al.*, 1998; Johnstone *et al.*, 2000) and reduced CRH receptor binding and CRH receptor signaling at the adenohypophysis resulting in attenuated corticotroph cell responses to CRH (Neumann *et al.*, 1998b; Toufexis *et al.*, 1999).

Involvement of Intracerebral OXT and PRL

The neuropeptides OXT and PRL have emerged as maternal brain factors regulating the blunted stress response peripartum. Both OXT and PRL are main reproduction-related hormones promoting labor, lactogenesis, and milk-ejection. Increased hormone synthesis in magnocellular hypothalamic neurons and in lactotroph cells, respectively, and increased secretion into blood are required around birth to fulfill these reproduction-related physiological demands. In addition, OXT (Kendrick *et al.*, 1988; Moos *et al.*, 1989; Moos *et al.*, 1991; Neumann *et al.*, 1993) and PRL (Torner *et al.*, 2004) are released in response to reproduction-related stimuli, such as birth or suckling, within distinct brain regions including the PVN, supraoptic nucleus, septum, hippocampus, and olfactory bulb. Additionally, PRL release in response to suckling was found within the PVN and medial preoptic area (MPOA) (Torner *et al.*, 2004). In the context of central regulation of the stress response at central levels it is important to mention that local somato-dendritic OXT release has also been demonstrated in response to several psychological and physical stressors (for review see Landgraf & Neumann, 2004). Similarly, local release of PRL within the PVN and the MPOA occurs in response to restraint stress, both in male and female rats (Torner *et al.*, 2004). This, together with the finding of elevated PRL gene expression within the hypothalamus in pregnancy and lactation (Torner & Neumann, 2002) and elevated receptor binding in several brain regions in lactation (Pi & Grattan, 1999; Augustine *et al.*, 2003) makes PRL a novel, potentially important, candidate regulating peripartum stress hyporesponsiveness.

Indeed, down-regulation of the long form of PRL receptors within the brain of lactating dams by use of antisense oligonucleotides disinhibited the HPA axis response to a given stressor (Torner *et al.*, 2002). In support, chronic up-regulation of brain PRL by intracerebroventricular (ICV) infusion in ovariectomized virgin female rats via osmotic minipumps, thus mimicking the increased brain neuropeptide availability peripartum, reduced the neuronal, hormonal, and behavioral HPA axis responses to stress (Donner *et al.*, 2007). In more detail, five days of ICV PRL treatment reduced the stress-induced neuronal activation within several relevant brain regions, and the restraint stress-induced expression of c-Fos and CRH mRNA within the PVN (Figure 8.1). These results indicate a significant involvement of PRL in the attenuated HPA axis response peripartum.

Chronic OXT treatment exerts a similar effect in virgin rats (Windle *et al.*, 1997a, 2004). However, acute ICV infusion of an OXT receptor antagonist prior to stressor exposure resulted in disinhibition of ACTH and corticosterone secretion only in virgin and male, but not in pregnant and lactating rats (Neumann *et al.*, 2000a, b).

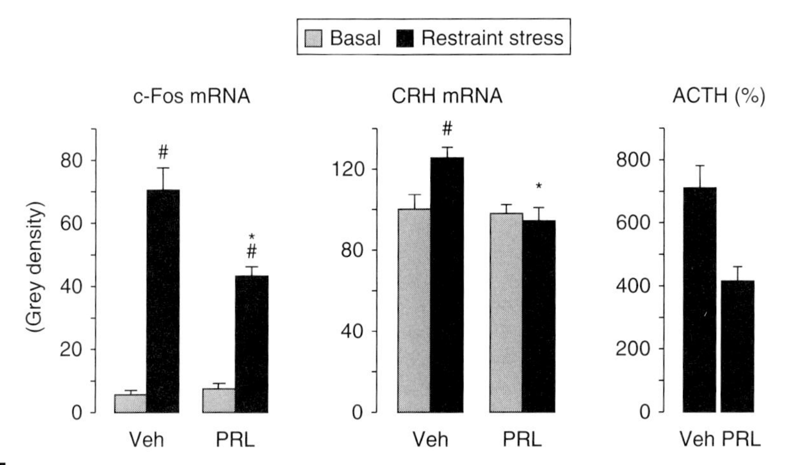

FIGURE 8.1 Effects of chronic central PRL (1.0 μg/0.5 μl/h via osmotic mini-pumps) on c-Fos mRNA (left) and CRH mRNA (middle) expression in the PVN of non-stressed control animals (grey bars) and in response to 30-min restraint (black bars). The percent increase of plasma ACTH 30 min after restraint is shown in the right graph. Restraint increased c-Fos and CRH mRNA expression as well as plasma ACTH, whereas PRL treatment reduced these stress responses. Data are means + SEM. $^{\#}p < 0.01$ vs. respective basal. $^{*}p < 0.01$ vs. corresponding vehicle group. (*Source*: Data from Donner *et al.*, 2007. Copyright *Eur. J. Neurosci.*)

This may indicate that other inhibitory brain mechanisms such as those mentioned above are involved and act in concert with OXT to attenuate HPA axis responses peripartum.

ALTERATIONS IN EMOTIONALITY PERIPARTUM: INVOLVEMENT OF OXT AND PRL

In addition to physiological (particularly neuroendocrine) adaptations complex behavioral alterations have been related to the peripartum period; above all the appearance of intense maternal behavior and maternal aggression for protection of the offspring. Both OXT and PRL are importantly involved in the promotion of maternal behavior (Pedersen & Prange, 1979; Bridges *et al.*, 1984; for review see Numan & Insel, 2003; Pedersen *et al.*, 2006).

In humans, it has been reported that nursing mothers are more likely to describe positive mood states, be less anxious (Fleming *et al.*, 1990; Altshuler *et al.*, 2000; Groer, 2005; Breitkopf *et al.*, 2006; Lonstein, 2007) and show increased calmness compared with bottle-feeding mothers (Carter & Altemus, 1997; Heinrichs *et al.*, 2001). These findings are complemented by the description of altered emotionality in lactating rats including reduced anxiety (Hard & Hansen, 1985; Windle *et al.*, 1997a; Neumann, 2001), and a reduced behavioral response to white noise (Windle *et al.*, 1997b). As summarized by Lonstein (Lonstein, 2007), sensory stimuli arising from the pups or the baby during close physical contact with the mother, rather than nursing itself, is important for maintaining the reduced level of anxiety in lactation. This is in contrast to the blunted HPA axis responses in lactation, which requires the suckling stimulus (Tu *et al.*, 2005).

Up-regulation of the activity of both the brain OXT (Windle *et al.*, 1997a; Neumann *et al.*, 2000a; Lonstein, 2007) and PRL (Torner *et al.*, 2002; Donner *et al.*, 2007) systems have been directly linked to the attenuation of emotional responses and to the reduced anxiety level found in lactation. For example, chronic ICV infusion of OXT (Windle *et al.*, 1997a) or PRL (Donner *et al.*, 2007) into virgin rats over 5 days results in reduced anxiety-related behavior. Also, blockade of central OXT receptors increased anxiety in pregnant and lactating dams, but not in virgin rats, demonstrating the involvement of endogenous OXT in the low anxiety levels seen peripartum (Neumann *et al.*, 2000a). However, the involvement of other neurochemical systems, which show alterations in the postpartum period, and which are important regulators of emotionality including sexual steroids (Laconi *et al.*, 2001; Frye & Walf, 2004; Wenzel *et al.*, 2005; Toufexis *et al.*, 2006), noradrenaline (Tanaka *et al.*, 2000; Fendt *et al.*, 2005), GABA (Crestani *et al.*, 1999; Low *et al.*, 2000; for review see Nemeroff, 2003), or CRH (for review see Bakshi & Kalin, 2000; Keck & Holsboer, 2001), is less clear in this context. As mentioned above, in pregnant and lactating rats, a reduced CRH neuronal stress response in the PVN and the CeA have been described (Johnstone *et al.*, 2000; Lightman *et al.*, 2001; Walker *et al.*, 2001; Bosch *et al.*, 2007), brain regions important for both HPA axis regulation and emotionality (Davis &Whalen, 2001; Wigger *et al.*, 2004; Phelps & LeDoux, 2005). Since CRH exerts anxiogenic effects (Bakshi *et al.*, 2002), the low activity of the brain CRH system is likely to contribute to the reduced emotional response of the dam. Moreover, a low CRH activity has been linked to enhanced maternal behavior (for review see Pedersen *et al.*, 1991) and maternal aggression when protecting the offspring (Gammie *et al.*, 2004).

MATERNAL BEHAVIOR AND AGGRESSION: LINK TO MOM'S ANXIETY

In humans, substantial inter-individual differences exist, for example, with respect to the quality of maternal care, the degree of emotional changes or stress vulnerability peripartum, all of which are regulated in a complex manner involving biological and social factors. From animal studies it has become clear that differences in maternal care are dependent on the genetic predisposition (Takayanagi *et al.*, 2005; Neumann *et al.*, 2005a; Pedersen *et al.*, 2006), as well as on early life experiences (for review see Champagne & Meaney, 2001, 2006; Bosch *et al.*, 2007), in particular the quality of maternal behavior received as a neonate (Kaffman & Meaney, 2007). Furthermore, adverse life events like prenatal stress (Bosch *et al.*, 2007) or chronic stress during pregnancy (Neumann *et al.*, 2005a; Leonhardt *et al.*, 2007) are likely to, at least partly, prevent the physiological and behavioral adaptations of the maternal brain.

Are High Anxiety Dams (HAB) Better Mothers? Differences in Maternal Care

Recently, we could demonstrate differences in maternal behavior between rat dams selectively bred for high (HAB) or low (LAB) anxiety-related behavior (Figure 8.2; Neumann *et al.*, 2005a; Bosch *et al.*, 2006). Robust behavioral differences between HAB and LAB rats, mainly characterised in males, are seen in various tests of emotionality including the open field, the plus-maze, and the light-and-dark box, (Henniger *et al.*, 2000; Ohl *et al.*, 2001). Moreover, HAB and LAB rats differ in the stress-coping style during forced swimming (Liebsch *et al.*, 1998; Neumann *et al.*, 1998c; Keck *et al.*, 2005) and in male aggression (Veenema *et al.*, 2007). Furthermore, the HAB and LAB rats are characterised by differences in their brain arginine vasopressin (AVP) system; AVP expression and release is elevated in HAB compared with LAB rats due to a single nucleotide polymorphism (SNP) in the promoter region of the AVP gene (Murgatroyd *et al.*, 2004). Crucially, the elevated AVP system has been shown to directly contribute to the high anxiety level observed in HABs (for review see Landgraf & Wigger, 2003). Importantly, in female HAB and LAB rats, the differences in inborn anxiety persist during pregnancy (Neumann *et al.*, 1998c) and in lactation (Neumann *et al.*, 2005a).

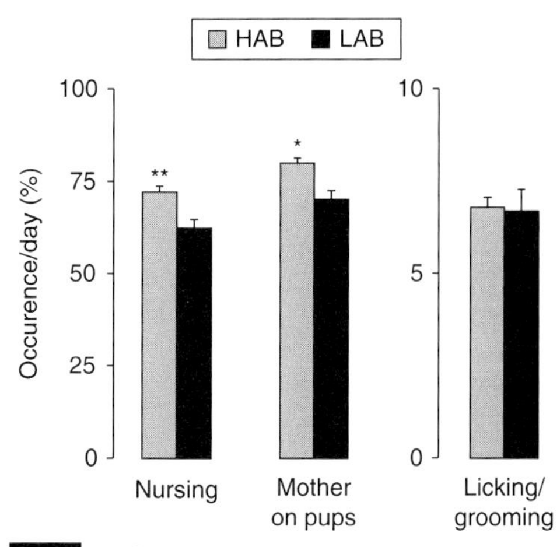

FIGURE 8.2 Maternal care in the home cage of HAB and LAB dams on lactation day 1. HAB dams show more nursing behavior and spend more time in contact with their offspring (*mother on pups*: nursing, carrying the pups, licking/grooming) compared with LABs. Interestingly, we found no difference in the parameter *licking/grooming* between HAB and LAB dams. Data are percent occurrence per day + SEM. $*p < 0.05$, $**p < 0.01$ vs. LAB. (*Source*: Data from Bosch *et al.*, 2006. Copyright *Eur. J. Neurosci.*)

According to the recent literature regarding the profound effects of maternal factors early in life on adult emotionality and stress vulnerability (Plotsky & Meaney, 1993; Ogawa *et al.*, 1994; Suchecki *et al.*, 1995; Caldji *et al.*, 1998; Francis & Meaney, 1999; Wigger & Neumann, 1999; Brake *et al.*, 2004; Ladd *et al.*, 2004; Macri *et al.*, 2004; Milde *et al.*, 2004; Weaver *et al.*, 2004; for review see Kaffman & Meaney, 2007), the possibility exists that the behavioral differences between HAB and LAB rats are not only genetically-determined but also, at least partly, due to differences in maternal behavior received. Therefore, we extensively monitored and compared the maternal behavior of HAB and LAB dams in their home cage. Whereas HAB and LAB dams delivered a virtually identical number of pups, the delivery process was significantly slower in HABs (Neumann *et al.*, 2005b). To our initial surprise, all the data accumulated to date clearly indicate that HAB dams spend more time on the litter, leave the nest less often and, remarkably, display more nursing behavior (Figure 8.2; Neumann *et al.*, 2005a; Bosch *et al.*, 2006). The differences in maternal care between HAB and LAB dams are robust, in as much as they are (i) independent of the time of the day, (ii) constant over at least the first 5 days postpartum, and (iii) also seen in dams which were separated from their litter for 3 h daily (Neumann *et al.*, 2005b). However, no difference in licking and grooming behavior has ever been found between the rat lines (Figure 8.2); probably due to the fact that licking and grooming behavior is rarely shown by Wistar rat dams. From these results we conclude that HAB offspring develop a high anxiety phenotype, which is accompanied by high stress susceptibility despite the reception of intense maternal care as neonates; further emphasizing the strong genetic determination of their behavior.

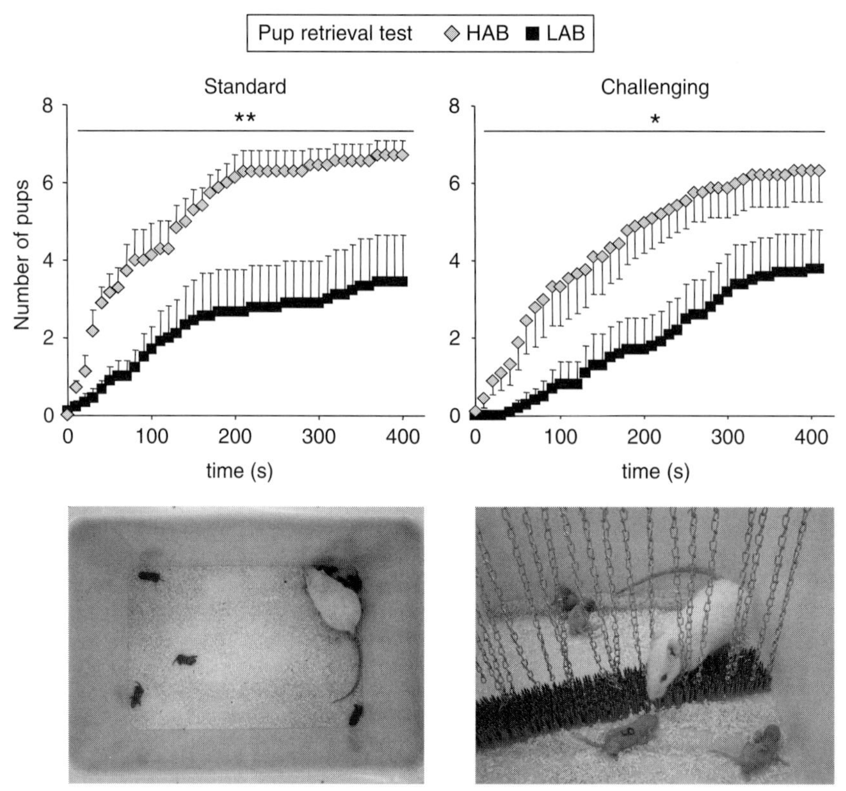

FIGURE 8.3 Maternal behavior of HAB and LAB dams in a novel environment, i.e. during the pup retrieval test. Under standard conditions (left; day 3 of lactation; van Leengoed *et al.*, 1987) HAB dams retrieved their pups faster despite their high level of anxiety. Surprisingly, even in a more challenging environment (see text for further details) HAB dams retrieved their pups faster compared with LABs on day 4 of lactation. Data are the means ± SEM; $^{*}p < 0.05$; $^{**}p < 0.01$ vs. LAB. (*Source*: Data from Neumann *et al.*, 2005a. Copyright *Psychoneuroendocrinology*.)

HAB Dams Care More – Even Under Stressful Conditions

In addition to the differences in maternal care between HAB and LAB dams in the home cage, we determined whether the high level of maternal behavior could also be confirmed in a novel environment despite the inherent increase in anxiety provoked. Therefore, we performed the pup retrieval test, an established test for maternal behavior and motivation (van Leengoed *et al.*, 1987). In this test, dams have to retrieve their pups into a corner of the novel arena (35 cm × 55 cm). Lactating HAB dams clearly won the competition for retrieving their pups: HABs started to collect their pups earlier and retrieved more pups within a given time period (Figure 8.3, left panel). This indicates that their high level of anxiety does not prevent them from being highly protective even in an unknown and potentially dangerous environment.

In order to further corroborate this finding, we created a challenging arena for the pup retrieval test, which should be highly stressful for the dam (Figure 8.3, right panel; Neumann *et al.*, 2005a). Thus, eight pups were placed into the outer two compartments of a cage arena (35 cm × 55 cm) divided into three compartments by two vertical fronts of small iron chains. Movement of the iron chains while passing them induces ultrasonic noise. Moreover, below the iron chains was an unpleasant floor surface (stripes of plastic shoe cleaner), which represented a further deterrent for the rats. This device was clearly avoided by virgin female rats (Neumann *et al.*, 2005a). Even in this challenging situation, HAB dams retrieved their pups faster indicating that the high anxiety level of HAB dams is linked to a protective mothering style combined with a high maternal motivation. We have shown that this involves the presence of the pups as lactating HAB rats are more anxious compared to LABs on the elevated plus-maze (Neumann *et al.*, 2005a). Thus, these results show that maternal drive can overcome innate levels of high anxiety.

HAB Dams Are More Aggressive During Maternal Defense

Maternal aggression is an important aspect of the complex patterns of maternal behavior in most mammals (Erskine *et al.*, 1978; Rosenblatt *et al.*, 1994; Lonstein & Gammie, 2002; Numan & Insel, 2003). In order to further test our hypothesis that HAB dams display high levels of pup protection, we compared maternal aggression between HAB and LAB dams during the maternal defense test performed in their home cage (Neumann *et al.*, 2001; Bosch *et al.*, 2005). Although more anxious, HAB dams were more offensive toward a virgin female intruder, as they displayed more attacks with a reduced attack latency compared with LABs (Figure 8.4). The high level of aggression was efficient to impress the virgin female intruders as they were significantly more anxious on the elevated plus-maze than intruders exposed to LABs, indicating that the high level of aggression of HAB dams is indeed behaviorally relevant and protective.

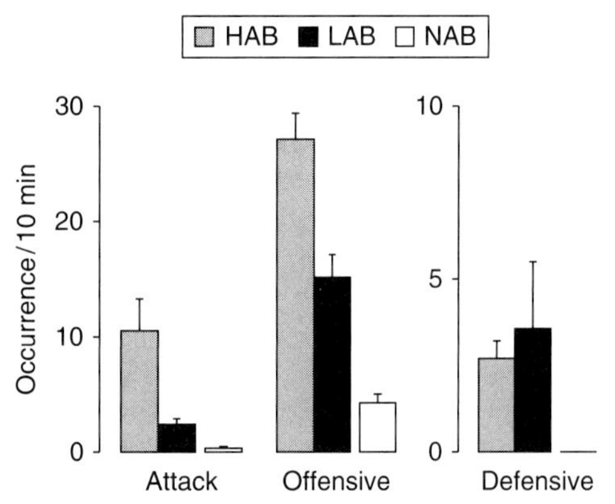

FIGURE 8.4 Maternal aggression of lactating HAB (grey bars) and LAB (black bars) residents as well as of residents unselected for anxiety (normal anxiety-related behavior: NAB; white bars) during the 10-min maternal defense test on day 3 of lactation. HAB residents show more attacks and total offensive behavior, whereas LABs and NABs are less aggressive. Data are means + SEM. (*Source*: Data from Bosch *et al.*, 2004. Copyright *Neuroscience*; Bosch *et al.*, 2005. Copyright *J. Neurosci.*)

Interestingly, in male rodents innate aggression toward a male conspecific is inversely related to the level of anxiety (mice: Nyberg *et al.*, 2003; HAB/LAB rats: Veenema *et al.*, 2007). Moreover, there are also reports from unselected and genetically similar rat dams demonstrating that pharmacological manipulations, which increased their emotionality, consequently reduced their maternal aggressive behavior (Johns *et al.*, 1994; Lonstein *et al.*, 1998). Therefore, we conclude that the high level of maternal aggression in HAB dams is specifically linked to a highly protective mothering style determined by their innate hyper-anxiety.

In humans, anxious mothers who are highly protective toward their children cause increased shyness in their young (Rubin & Burgess, 2002; Coplan *et al.*, 2007). Such shy children are known to be at increased risk of developing social, emotional, and adjustment difficulties (Rubin *et al.*, 2002). This intense maternal care has recently been named “helicopter parenting” and is indicative of parents trying to keep extensively “in touch” with their children. Worryingly, long-term consequences of “helicopter parenting” have not been described yet. Thus, in this context, using our model of HAB and LAB dams, it would be of interest to monitor long-term consequences of increased or reduced maternal care in HAB and LAB offspring, respectively. Indeed, daily separation from the mother during the first 14 days of life was found to reduce the level of anxiety in male HAB offspring, whereas the same treatment increased the level of anxiety in LAB offspring (Neumann *et al.*, 2005b). This substantiates the saying “Too much is too much” – also for maternal care.

Brain OXT Mediates Maternal Aggression in HAB Dams

As discussed above, there is evidence from studies performed in sheep (Kendrick *et al.*, 1988) and rats (Moos *et al.*, 1989; Neumann & Landgraf, 1989; Neumann *et al.*, 1993) for OXT release within distinct brain regions both during parturition and suckling. Besides various regulatory effects on neuroendocrine functions peripartum (Moos *et al.*, 1984; Neumann *et al.*, 1994, 1996), brain OXT plays an important role in regulating maternal care (Pedersen & Prange, 1979; van Leengoed *et al.*, 1987; Numan & Insel, 2003; Pedersen & Boccia, 2003; Pedersen *et al.*, 2006). However, the neuropeptidergic regulation of maternal aggression is less clear, and even partially controversial (Consiglio & Lucion, 1996; Giovenardi *et al.*, 1998; Elliott *et al.*, 2001; Lubin *et al.*, 2003). An increased OXT release was found within the PVN of unselected Wistar dams during the maternal defense test (Bosch *et al.*, 2004). As HAB and LAB dams differ robustly in their level of maternal aggression, we used this model to study underlying neurochemical mechanisms of high vs. low maternal aggression focussing on the brain OXT system. OXT release was monitored within both the CeA (Figure 8.5) and the hypothalamic PVN using intracerebral microdialysis before, during, and after maternal defense against a virgin female intruder (Bosch *et al.*, 2005). In HAB dams, OXT release within both the CeA and the PVN was highly elevated during maternal defense. In contrast, in LAB dams, local neuropeptide release was significantly reduced within the PVN and remained at a relatively low level within the CeA. Importantly, the amount of locally released OXT was positively correlated with the level of aggressive behavior displayed by the dam.

In order to reveal the behavioral relevance of such locally released OXT in HAB dams, we infused an OXT receptor antagonist via bilateral retrodialysis into the CeA or the PVN (Bosch *et al.*, 2005). In both regions, blockade of OXT receptor-mediated actions reduced the amount of offensive behavior, specifically the number of attacks against the intruder. In contrast, and supporting the role of endogenous OXT in maternal aggression, local infusion of synthetic OXT into the PVN of LAB dams tended to increase some aspects of offensive behavior (lateral threats).

As OXT receptor expression and binding was not found to differ between HAB and LAB dams, it is likely that differences in local release patterns rather than differences in neuropeptide binding underlie the differences in offensive behavior during lactation.

One possible mechanism underlying maternal aggression is via inhibition of CRH neurons. CRH has been shown to reduce maternal

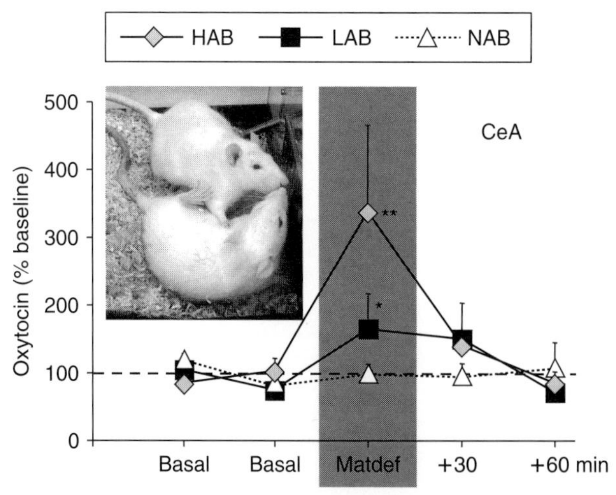

FIGURE 8.5 Different oxytocin release patterns within the CeA of lactating HAB (grey symbol), LAB (black symbol), and NAB (white symbol) residents in response to maternal defense. The increased oxytocin release during maternal defense corresponds with the amount of aggression displayed by the respective lactating residents (see Figure 8.4). For microdialysis, 30-min dialysates were sampled under basal conditions (samples 1 and 2). During the third dialysis sampling period, a virgin intruder rat was placed into the cage of the lactating resident for 10 min (maternal defense). Also shown is a picture of a typical aggressive attack of the dam. Data are percentage of baseline + SEM. $^{*}p < 0.05$, $^{**}p < 0.01$ vs. basal. (*Source*: Data from Bosch *et al.*, 2004. Copyright *Neuroscience*; Bosch *et al.*, 2005. Copyright *J. Neurosci.*)

aggression in mice, and inhibition of brain CRH activity seems to be a prerequisite for the expression of maternal aggression (Gammie *et al.*, 2004). Indeed, chronic ICV OXT treatment of ovariectomized virgin female rats reduced the CRH neuronal responsiveness (Windle *et al.*, 2004). Therefore, the hypothesis exists that OXT released within the PVN during the maternal defense test exerts a local inhibitory effect on CRH neurons, thus further promoting aggressive behavior.

Within the CeA, OXT could contribute to the modulation of local neuronal circuitries involved in stress recognition (Ebner *et al.*, 2005; Huber *et al.*, 2005; Kirsch *et al.*, 2005), which may allow the aggressive response of HAB dams.

IMPORTANCE OF MATERNAL ADAPTATIONS PERIPARTUM FOR MATERNAL MENTAL HEALTH

Robust adaptations in maternal responsiveness to a variety of stressors are a marker of the peripartum period. These adaptations, specifically the general attenuation of the stress responsiveness of the HPA axis, and reduced emotional (e.g., anxiety) and neuronal (e.g., hypothalamic expression of CRH) responses, are important not only for protection of the offspring against excessive glucocorticoid levels, but probably also for the maternal mental health.

In humans, the postpartum period is a time of increased vulnerability to mood disorders, which are found in about 20–30% of women within the first 6 weeks of postpartum. Thus, mood disorders are the most common serious medical complication of the puerperium (O'Hara & Swain, 1996; Llewellyn *et al.*, 1997; Pedersen, 1999; Mastorakos & Ilias, 2000). Although neurobiological mechanisms of, for example, postpartum depression are largely unknown, maladaptations of the stress circuitries and neuroendocrine systems described in this review are likely to contribute to postpartum mood disorders. Such maladaptations may include insufficient activation of the brain OXT and PRL systems, lack of inhibition of hypothalamic noradrenaline release, and, consequently, elevated activity of the brain CRH system increasing the risk for mood disturbances and postpartum depression.

Chronic Pregnancy Stress: Effects on Adaptations in Lactation

Indeed, chronic stress experienced during pregnancy has been demonstrated to compromise the performance of maternal behavior and alter emotional parameters in lactation (Muir *et al.*, 1986; Pardon *et al.*, 2000; Meek *et al.*, 2001; Patin *et al.*, 2002; Darnaudery *et al.*, 2004; Neumann *et al.*, 2005a). Specifically, pregnancy stress prevented the attenuation of the HPA axis responsiveness in lactation (Neumann *et al.*, 2005a). In humans, an unstable social environment, and in particular stressful life events experienced during pregnancy have been shown to increase the risk of postpartum depression (Llewellyn *et al.*, 1997; Pedersen, 1999; Federenko & Wadhwa, 2004). Importantly, a genetic predisposition to high emotionality and prior history of depression disorders should interact with stressful life events in pregnancy to increase the subsequent risk of developing a postpartum mood disorder (O'Hara & Swain, 1996; Rini *et al.*, 1999; Lobel *et al.*, 2000). Indeed, this could be confirmed in an animal study (Neumann *et al.*, 2005a). Chronic pregnancy stress specifically compromised the adaptations of the HPA axis in lactating HAB dams, whereas no effect was found in LAB dams. Thus, high stress susceptibility linked with hyperanxiety in HAB rats makes them more vulnerable to the adverse effects of chronic stress exposure in pregnancy with respect to postpartum adaptations.

Early Life Stress: Effects on Lactation-Associated Adaptations in Female Offspring

Another factor that probably contributes to postpartum mood disorders is early life stress, which is believed to constitute a risk factor for the development of adult psychopathologies including depression (Kofman, 2002). In a recent study we could show that prenatal stress exerts long-lasting behavioral and neuroendocrine effects in the female offspring. These adaptations become more pronounced during the lactation period (Bosch *et al.*, 2007). Thus, lactating dams which were prenatally stressed, that is, their mothers were stressed during pregnancy, showed impaired maternal behavior in their home cage. Moreover, prenatal stress exposure prevented the attenuation of the HPA axis response to a mild stressor, as reflected by higher ACTH and corticosterone plasma concentrations measured 5–15 min after exposure to a novel environment. We could identify two possible neuronal mechanisms underlying the effects of prenatal stress on lactation: (i) lack of peripartum adaptations of the brain CRH system, and (ii) maladaptations of the brain AVP system. In detail, whereas low levels of CRH mRNA expression were found within the hypothalamic PVN in non-stressed control dams compared with virgin rats, CRH mRNA expression was significantly higher in prenatally stressed lactating dams. With respect to the AVP system, we found a general up-regulation of neuropeptide synthesis both in the magno- and parvocellular parts of the PVN of lactating dams, supporting earlier reports (Walker *et al.*, 2001). However, in prenatally stressed dams a further rise in AVP synthesis was found specifically within the parvocellular PVN (Bosch *et al.*, 2007). AVP, like CRH (Stenzel-Poore *et al.*, 1994; Bakshi *et al.*, 2002; Heinrichs & Koob, 2004), is a prominent anxiogenic factor, and a regulator of neuroendocrine stress responses (Landgraf & Neumann, 2004; Wigger *et al.*, 2004). Importantly, and adding further credence to this hypothesis, hyperactivity of CRH and AVP systems are likely to be involved in behavioral and neuroendocrine aberrations related to psychopathologies, including depression (Nemeroff, 1996; Keck & Holsboer, 2001).

Therefore, genetic and environmental stress factors and their interactions which have been generally linked to the development of adult psychopathologies, may also play an important role in specifically preventing physiological adaptations in the postpartum period, thus contributing to peripartum mood disorders.

REFERENCES

Altshuler, L. L., Hendrick, V., and Cohen, L. S. (2000). An update on mood and anxiety disorders during pregnancy and the postpartum period. *Prim. Care Companion J. Clin. Psychiatr.* 2, 217–222.

Amico, J. A., Mantella, R. C., Vollmer, R. R., and Li, X. (2004). Anxiety and stress responses in female oxytocin deficient mice. *J. Neuroendocrinol.* 16, 319–324.

Augustine, R. A., Kokay, I. C., Andrews, Z. B., Ladyman, S. R.., and Grattan, D. R. (2003). Quantitation of prolactin receptor mRNA in the maternal rat brain during pregnancy and lactation. *J. Mol. Endocrinol.* 31, 221–32.

Bakshi, V. P., and Kalin, N. H. (2000). Corticotropin-releasing hormone and animal models of anxiety: Gene–environment interactions. *Biol. Psychiatr.* 48, 1175–1198.

Bakshi, V. P., Smith-Roe, S., Newman, S. M., Grigoriadis, D. E., and Kalin, N. H. (2002). Reduction of stress-induced behavior by antagonism of corticotropin-releasing hormone 2 (CRH2) receptors in lateral septum or CRH1 receptors in amygdala. *J. Neurosci.* 22, 2926–2935.

Banky, Z., Nagy, G. M., and Halasz, B. (1994). Analysis of pituitary prolactin and adrenocortical response to ether, formalin or restraint in lactating rats: Rise in corticosterone, but no increase in plasma prolactin levels after exposure to stress. *Neuroendocrinology* 59, 63–71.

Bosch, O. J., Kromer, S. A., Brunton, P. J., and Neumann, I. D. (2004). Release of oxytocin in the hypothalamic paraventricular nucleus, but not central amygdala or lateral septum in lactating residents and virgin intruders during maternal defence. *Neuroscience* 124, 439–448.

Bosch, O. J., Meddle, S. L., Beiderbeck, D. I., Douglas, A. J., and Neumann, I. D. (2005). Brain oxytocin correlates with maternal aggression: Link to anxiety. *J. Neurosci.* 25, 6807–6815.

Bosch, O. J., Kromer, S. A., and Neumann, I. D. (2006). Prenatal stress: Opposite effects on anxiety and hypothalamic expression of vasopressin and corticotropin-releasing hormone in rats selectively bred for high and low anxiety. *Eur. J. Neurosci.* 23, 541–551.

Bosch, O. J., Musch, W., Bredewold, R., Slattery, D. A., and Neumann, I. D. (2007). Prenatal stress increases HPA axis activity and impairs maternal care in lactating female offspring: Implications for postpartum mood disorder. *Psychoneuroendocrinology* 32, 267–278.

Brake, W. G., Zhang, T. Y., Diorio, J., Meaney, M. J., and Gratton, A. (2004). Influence of early postnatal rearing conditions on mesocorticolimbic dopamine and behavioural responses to psychostimulants and stressors in adult rats. *Eur. J. Neurosci.* 19, 1863–1874.

Breitkopf, C. R., Primeau, L. A., Levine, R. E., Olson, G. L., Wu, Z. H., and Berenson, A. B. (2006). Anxiety symptoms during pregnancy and postpartum. *J. Psychosom. Obstet. Gynaecol.* 27, 157–162.

Bridges, R. S., DiBiase, R., Loundes, D. D., and Doherty, P. C. (1984). Prolactin stimulation of maternal behavior in female rats. *Science* 227, 782–784.

Brunton, P. J., Meddle, S. L., Ma, S., Ochedalski, T., Douglas, A. J., and Russell, J. A. (2005).

Endogenous opioids and attenuated hypothalamic–pituitary–adrenal axis responses to immune challenge in pregnant rats. *J. Neurosci.* 25, 5117–5126.

Brunton, P. J., and Russell, J. A. (2008). The expectant brain: adapting for motherhood. *Nat. Rev. Neurosci.* 9, 11–25.

Caldji, C., Tannenbaum, B., Sharma, S., Francis, D., Plotsky, P. M., and Meaney, M. J. (1998). Maternal care during infancy regulates the development of neural systems mediating the expression of fearfulness in the rat. *Proc. Natl. Acad. Sci. USA* 95, 5335–5340.

Carter, C. S., and Altemus, M. (1997). Integrative functions of lactational hormones in social behavior and stress management. *Ann. NY Acad. Sci.* 807, 164–174.

Champagne, F., and Meaney, M. J. (2001). Like mother, like daughter: Evidence for non-genomic transmission of parental behavior and stress responsivity. *Prog. Brain Res.* 133, 287–302.

Champagne, F. A., and Meaney, M. J. (2006). Stress during gestation alters postpartum maternal care and the development of the offspring in a rodent model. *Biol. Psychiatr.* 59, 1227–1235.

Consiglio, A. R., and Lucion, A. B. (1996). Lesion of hypothalamic paraventricular nucleus and maternal aggressive behavior in female rats. *Physiol. Behav.* 59, 591–596.

Coplan, R. J., Arbeau, K. A., and Armer, M. (2007). Don't fret, be supportive! Maternal characteristics linking child shyness to psychosocial and school adjustment in Kindergarten. *J. Abnorm. Child Psychol.* doi:10.1007/s10802-007-9183-7.

Crestani, F., Lorez, M., Baer, K., Essrich, C., Benke, D., Laurent, J. P., Belzung, C., Fritschy, J. M., Luscher, B., and Mohler, H. (1999). Decreased GABAA-receptor clustering results in enhanced anxiety and a bias for threat cues. *Nat. Neurosci.* 2, 833–839.

da Costa, A. P., Wood, S., Ingram, C. D., and Lightman, S. L. (1996). Region-specific reduction in stress-induced c-fos mRNA expression during pregnancy and lactation. *Brain Res.* 742, 177–184.

Darnaudery, M., Dutriez, I., Viltart, O., Morley-Fletcher, S., and Maccari, S. (2004). Stress during gestation induces lasting effects on emotional reactivity of the dam rat. *Behav. Brain Res.* 153, 211–216.

Davis, M., and Whalen, P. J. (2001). The amygdala: Vigilance and emotion. *Mol. Psychiatr.* 6, 13–34.

Deschamps, S., Woodside, B., and Walker, C. D. (2003). Pups presence eliminates the stress hyporesponsiveness of early lactating females to a psychological stress representing a threat to the pups. *J. Neuroendocrinol.* 15, 486–497.

Donner, N., Bredewold, R., Maloumby, R., and Neumann, I. D. (2007). Chronic intracerebral prolactin attenuates neuronal stress circuitries in virgin rats. *Eur. J. Neurosci.* 25, 1804–1814.

Douglas, A. J., and Russell, J. A. (1994). Corticotrophin-releasing hormone, proenkephalin A and oxytocin mRNA's in the paraventricular nucleus during pregnancy and parturition in the rat. *Gene Ther.* 1(suppl 1), S85.

Douglas, A. J., Johnstone, H. A., Wigger, A., Landgraf, R., Russell, J. A., and Neumann, I. D. (1998). The role of endogenous opioids in neurohypophysial and hypothalamo–pituitary–adrenal axis hormone secretory responses to stress in pregnant rats. *J. Endocrinol.* 158, 285–293.

Douglas, A. J., Meddle, S. L., Toschi, N., Bosch, O. J., and Neumann, I. D. (2005). Reduced activity of the noradrenergic system in the paraventricular nucleus at the end of pregnancy: Implications for stress hyporesponsiveness. *J. Neuroendocrinol.* 17, 40–48.

Ebner, K., Bosch, O. J., Kromer, S. A., Singewald, N., and Neumann, I. D. (2005). Release of oxytocin in the rat central amygdala modulates stress-coping behavior and the release of excitatory amino acids. *Neuropsychopharmacology* 30, 223–230.

Elliott, J. C., Lubin, D. A., Walker, C. H., and Johns, J. M. (2001). Acute cocaine alters oxytocin levels in the medial preoptic area and amygdala in lactating rat dams: Implications for cocaine-induced changes in maternal behavior and maternal aggression. *Neuropeptides* 35, 127–134.

Erskine, M. S., Barfield, R. J., and Goldman, B. D. (1978). Intraspecific fighting during late pregnancy and lactation in rats and effects of litter removal. *Behav. Biol.* 23, 206–218.

Federenko, I. S., and Wadhwa, P. D. (2004). Women's mental health during pregnancy influences fetal and infant developmental and health outcomes. *CNS Spectr.* 9, 198–206.

Fendt, M., Siegl, S., and Steiniger-Brach, B. (2005). Noradrenaline transmission within the ventral bed nucleus of the stria terminalis is critical for fear behavior induced by trimethylthiazoline, a component of fox odor. *J. Neurosci.* 25, 5998–6004.

Fischer, D., Patchev, V. K., Hellbach, S., Hassan, A. H., and Almeida, O. F. (1995). Lactation as a model for naturally reversible hypercorticalism plasticity in the mechanisms governing hypothalamo–pituitary–adrenocortical activity in rats. *J. Clin. Invest.* 96, 1208–1215.

Fleming, A., Ruble, D., Flett, G., and Van Wagner, V. (1990). Adjustment in first-time mothers: Changes in mood and mood content during the early postpartum months. *Dev. Psychol.* 26, 137–143.

Francis, D. D., and Meaney, M. J. (1999). Maternal care and the development of stress responses. *Curr. Opin. Neurobiol.* 9, 128–134.

Frye, C. A., and Walf, A. A. (2004). Estrogen and/ or progesterone administered systemically or to the amygdala can have anxiety-, fear-, and pain-reducing effects in ovariectomized rats. *Behav. Neurosci.* 118, 306–313.

Gammie, S. C., Negron, A., Newman, S. M., and Rhodes, J. S. (2004). Corticotropin-releasing factor inhibits maternal aggression in mice. *Behav. Neurosci.* 118, 805–814.

Giovenardi, M., Padoin, M. J., Cadore, L. P., and Lucion, A. B. (1998). Hypothalamic paraventricular nucleus modulates maternal aggression in rats: Effects of ibotenic acid lesion and oxytocin antisense. *Physiol. Behav.* 63, 351–359.

Groer, M. W. (2005). Differences between exclusive breastfeeders, formula-feeders, and controls: A study of stress, mood, and endocrine variables. *Biol. Res. Nurs.* 7, 106–117.

Han, S. K., Chong, W., Li, L. H., Lee, I. S., Murase, K., and Ryu, P. D. (2002). Noradrenaline excites and inhibits GABAergic transmission in parvocellular neurons of rat hypothalamic paraventricular nucleus. *J. Neurophysiol.* 87, 2287–2296.

Hard, E., and Hansen, S. (1985). Reduced fearfulness in the lactating rat. *Physiol. Behav.* 35, 641–643.

Heinrichs, M., Meinlschmidt, G., Neumann, I., Wagner, S., Kirschbaum, C., Ehlert, U., and Hellhammer, D. H. (2001). Effects of suckling on hypothalamic–pituitary–adrenal axis responses to psychosocial stress in postpartum lactating women. *J. Clin. Endocrinol. Metab.* 86, 4798–4804.

Heinrichs, S. C., and Koob, G. F. (2004). Corticotropin-releasing factor in brain: A role in activation, arousal, and affect regulation. *J. Pharmacol. Exp. Ther.* 311, 427–440.

Henniger, M. S., Ohl, F., Holter, S. M., Weissenbacher, P., Toschi, N., Lorscher, P., Wigger, A., Spanagel, R., and Landgraf, R. (2000). Unconditioned anxiety and social behaviour in two rat lines selectively bred for high and low anxiety-related behaviour. *Behav. Brain Res.* 111, 153–163.

Huber, D., Veinante, P., and Stoop, R. (2005). Vasopressin and oxytocin excite distinct neuronal populations in the central amygdala. *Science* 308, 245–248.

Itoi, K., Suda, T., Tozawa, F., Dobashi, I., Ohmori, N., Sakai, Y., Abe, K., and Demura, H. (1994). Microinjection of norepinephrine into the paraventricular nucleus of the hypothalamus stimulates corticotropin-releasing factor gene expression in conscious rats. *Endocrinology* 135, 2177–2182.

Johns, J. M., Noonan, L. R., Zimmerman, L. I., Li, L., and Pedersen, C. A. (1994). Effects of chronic and acute cocaine treatment on the onset of maternal behavior and aggression in Sprague-Dawley rats. *Behav. Neurosci.* 108, 107–112.

Johnstone, H. A., Wigger, A., Douglas, A. J., Neumann, I. D., Landgraf, R., Seckl, J. R., and Russell, J. A. (2000). Attenuation of hypothalamic–pituitary–adrenal axis stress responses in late pregnancy: Changes in feedforward and feedback mechanisms. *J. Neuroendocrinol.* 12, 811–822.

Kaffman, A., and Meaney, M. J. (2007). Neurodevelopmental sequelae of postnatal maternal care in rodents: Clinical and research implications of molecular insights. *J. Child Psychol. Psychiatr.* 48, 224–244.

Keck, M. E., and Holsboer, F. (2001). Hyperactivity of CRH neuronal circuits as a target for therapeutic interventions in affective disorders. *Peptides* 22, 835–844.

Keck, M. E., Sartori, S. B., Welt, T., Muller, M. B., Ohl, F., Holsboer, F., Landgraf, R., and Singewald, N. (2005). Differences in serotonergic neurotransmission between rats displaying high or low anxiety/depression-like behaviour: Effects of chronic paroxetine treatment. *J. Neurochem.* 92, 1170–1179.

Kendrick, K. M., Keverne, E. B., Chapman, C., and Baldwin, B. A. (1988). Intracranial dialysis measurement of oxytocin, monoamine and uric acid release from the olfactory bulb and substantia nigra of sheep during parturition, suckling, separation from lambs and eating. *Brain Res.* 439, 1–10.

Kirsch, P., Esslinger, C., Chen, Q., Mier, D., Lis, S., Siddhanti, S., Gruppe, H., Mattay, V. S., Gallhofer, B., and Meyer-Lindenberg, A. (2005). Oxytocin modulates neural circuitry for social cognition and fear in humans. *J. Neurosci.* 25, 11489–11493.

Kofman, O. (2002). The role of prenatal stress in the etiology of developmental behavioural disorders. *Neurosci. Biobehav. Rev.* 26, 457–470.

Laconi, M. R., Casteller, G., Gargiulo, P. A., Bregonzio, C., and Cabrera, R. J. (2001). The anxiolytic effect of allopregnanolone is associated with gonadal hormonal status in female rats. *Eur. J. Pharmacol.* 417, 111–116.

Ladd, C. O., Huot, R. L., Thrivikraman, K. V., Nemeroff, C. B., and Plotsky, P. M. (2004). Long-term adaptations in glucocorticoid receptor and mineralocorticoid receptor mRNA and negative feedback on the hypothalamo–pituitary–adrenal axis following neonatal maternal separation. *Biol. Psychiatr.* 55, 367–375.

Landgraf, R., and Wigger, A. (2003). Born to be anxious: Neuroendocrine and genetic correlates of trait anxiety in HAB rats. *Stress* 6, 111–119.

Landgraf, R., and Neumann, I. D. (2004). Vasopressin and oxytocin release within the brain: A dynamic concept of multiple and variable modes of neuropeptide communication. *Front. Neuroendocrinol.* 25, 150–176.

Leonhardt, M., Matthews, S. G., Meaney, M. J., and Walker, C. D. (2007). Psychological stressors as a model of maternal adversity: Diurnal modulation of corticosterone responses and changes in maternal behavior. *Horm. Behav.* 51, 77–88.

Liebsch, G., Montkowski, A., Holsboer, F., and Landgraf, R. (1998). Behavioural profiles of two Wistar rat lines selectively bred for high or low anxiety-related behaviour. *Behav. Brain Res.* 94, 301–310.

Lightman, S. L., and Young, W. S., 3rd. (1989). Lactation inhibits stress-mediated secretion of corticosterone and oxytocin and hypothalamic accumulation of corticotropin-releasing factor and enkephalin messenger ribonucleic acids. *Endocrinology* 124, 2358–2364.

Lightman, S. L., Windle, R. J., Wood, S. A., Kershaw, Y. M., Shanks, N., and Ingram, C. D. (2001). Peripartum plasticity within the hypothalamo–pituitary–adrenal axis. *Prog. Brain Res.* 133, 111–129.

Llewellyn, A. M., Stowe, Z. N., and Nemeroff, C. B. (1997). Depression during pregnancy and the puerperium. *J. Clin. Psychiatr.* 58(suppl 15), 26–32.

Lobel, M., DeVincent, C. J., Kaminer, A., and Meyer, B. A. (2000). The impact of prenatal maternal stress and optimistic disposition on birth outcomes in medically high-risk women. *Health Psychol.* 19, 544–553.

Lonstein, J. S. (2007). Regulation of anxiety during the postpartum period. *Front. Neuroendocrinol.* 28, 115–141.

Lonstein, J. S., and Gammie, S. C. (2002). Sensory, hormonal, and neural control of maternal aggression in laboratory rodents. *Neurosci. Biobehav. Rev.* 26, 869–888.

Lonstein, J. S., Simmons, D. A., and Stern, J. M. (1998). Functions of the caudal periaqueductal gray in lactating rats: Kyphosis, lordosis, maternal aggression, and fearfulness. *Behav. Neurosci.* 112, 1502–1518.

Low, K., Crestani, F., Keist, R., Benke, D., Brunig, I., Benson, J. A., Fritschy, J. M., Rulicke, T., Bluethmann, H., Mohler, H., and Rudolph, U. (2000). Molecular and neuronal substrate for the selective attenuation of anxiety. *Science* 290, 131–134.

Lubin, D. A., Elliott, J. C., Black, M. C., and Johns, J. M. (2003). An oxytocin antagonist infused into the central nucleus of the amygdala increases maternal aggressive behavior. *Behav. Neurosci.* 117, 195–201.

Macri, S., Mason, G. J., and Wurbel, H. (2004). Dissociation in the effects of neonatal maternal separations on maternal care and the offspring's HPA and fear responses in rats. *Eur. J. Neurosci.* 20, 1017–1024.

Mastorakos, G., and Ilias, I. (2000). Maternal hypothalamic–pituitary–adrenal axis in pregnancy and the postpartum period. Postpartum-related disorders. *Ann. NY Acad. Sci.* 900, 95–106.

Meek, L. R., Dittel, P. L., Sheehan, M. C., Chan, J. Y., and Kjolhaug, S. R. (2001). Effects of stress during pregnancy on maternal behavior in mice. *Physiol. Behav.* 72, 473–479.

Milde, A. M., Enger, O., and Murison, R. (2004). The effects of postnatal maternal separation on stress responsivity and experimentally induced colitis in adult rats. *Physiol. Behav.* 81, 71–84.

Moos, F., Freund-Mercier, M. J., Guerne, Y., Guerne, J. M., Stoeckel, M. E., and Richard, P. (1984). Release of oxytocin and vasopressin by magnocellular nuclei *in vitro*: Specific facilitatory effect of oxytocin on its own release. *J. Endocrinol.* 102, 63–72.

Moos, F., Poulain, D. A., Rodriguez, F., Guerne, Y., Vincent, J. D., and Richard, P. (1989). Release of oxytocin within the supraoptic nucleus during the milk ejection reflex in rats. *Exp. Brain Res.* 76, 593–602.

Moos, F., Ingram, C. D., Wakerley, J. B., Guerne, Y., Freund-Mercier, M. J., and Richard, P. (1991). Oxytocin in the bed nucleus of the stria terminalis and lateral septum facilitates bursting of hypothalamic oxytocin neurons in suckled rats. *J. Neuroendocrinol.* 3, 163–171.

Muir, J. L., Brown, R., and Pfister, H. P. (1986). A possible role for oxytocin in the response to a psychological stressor. *Pharmacol. Biochem. Behav.* 25, 107–110.

Murgatroyd, C., Wigger, A., Frank, E., Singewald, N., Bunck, M., Holsboer, F., Landgraf, R., and Spengler, D. (2004). Impaired repression at a vasopressin promoter polymorphism underlies overexpression of vasopressin in a rat model of trait anxiety. *J. Neurosci.* 24, 7762–7770.

Nemeroff, C. B. (1996). The corticotropin-releasing factor (CRF) hypothesis of depression: New findings and new directions. *Mol. Psychiatr.* 1, 336–342.

Nemeroff, C. B. (2003). The role of GABA in the pathophysiology and treatment of anxiety disorders. *Psychopharmacol. Bull.* 37, 133–146.

Neumann, I., and Landgraf, R. (1989). Septal and hippocampal release of oxytocin, but not vasopressin, in the conscious lactating rat during suckling. *J. Neuroendocrinol.* 1, 305.

Neumann, I., Russell, J. A., and Landgraf, R. (1993). Oxytocin and vasopressin release within the supraoptic and paraventricular nuclei of pregnant, parturient and lactating rats: a microdialysis study. *Neuroscience* 53, 65–75.

Neumann, I., Koehler, E., Landgraf, R., and Summy-Long, J. (1994). An oxytocin receptor antagonist infused into the supraoptic nucleus attenuates intranuclear and peripheral release of oxytocin during suckling in conscious rats. *Endocrinology* 134, 141–148.

Neumann, I., Douglas, A. J., Pittman, Q. J., Russell, J. A., and Landgraf, R. (1996). Oxytocin released within the supraoptic nucleus of the rat brain by positive feedback action is involved in parturition-related events. *J. Neuroendocrinol.* 8, 227–233.

Neumann, I. D. (2001). Alterations in behavioral and neuroendocrine stress coping strategies in pregnant, parturient and lactating rats. *Prog. Brain Res.* 133, 143–152.

Neumann, I. D., Johnstone, H., Hatzinger, M., Landgraf, R., Russell, J. A., and Douglas, A. (1998a). Neuroendocrine adaptations of the hypothalamo–pituitary–adrenal (HPA) axis throughout pregnancy. *J. Physiol.* 508, 289–300.

Neumann, I. D., Wigger, A., Liebsch, G., Holsboer, F., and Landgraf, R. (1998b). Increased basal activity of the hypothalamo–pituitary–adrenal axis during pregnancy in rats bred for high anxiety-related behaviour. *Psychoneuroendocrinology* 23, 449–463.

Neumann, I. D., Torner, L., and Wigger, A. (2000a). Brain oxytocin: Differential inhibition of neuroendocrine stress responses and anxiety-related behaviour in virgin, pregnant and lactating rats. *Neuroscience* 95, 567–575.

Neumann, I. D., Wigger, A., Torner, L., Holsboer, F., and Landgraf, R. (2000b). Brain oxytocin inhibits basal and stress-induced activity of the hypothalamo–pituitary–adrenal axis in male and female rats: Partial action within the paraventricular nucleus. *J. Neuroendocrinol.* 12, 235–243.

Neumann, I. D., Toschi, N., Ohl, F., Torner, L., and Kromer, S. A. (2001). Maternal defence as an emotional stressor in female rats: Correlation of neuroendocrine and behavioural parameters and involvement of brain oxytocin. *Eur. J. Neurosci.* 13, 1016–1024.

Neumann, I. D., Kromer, S. A., and Bosch, O. J. (2005a). Effects of psycho-social stress during pregnancy on neuroendocrine and behavioural parameters in lactation depend on the genetically-determined stress vulnerability. *Psychoneuroendocrinology* 30, 791–806.

Neumann, I. D., Wigger, A., Kromer, S., Frank, E., Landgraf, R., and Bosch, O. J. (2005b). Differential effects of periodic maternal separation on adult stress coping in a rat model of extremes in trait anxiety. *Neuroscience* 132, 867–877.

Numan, M., and Insel, T. R. (2003). The neurobiology of parental behaviour. In *Hormones, Brain, and Behavior Series* (G. F. Ball, J. Balthazart, and R. J. Nelson, Eds.). Springer, New York.

Nyberg, J. M., Vekovischeva, O., and Sandnabba, N. K. (2003). Anxiety profiles of mice selectively bred for intermale aggression. *Behav. Genet.* 33, 503–511.

O'Hara, M. W., and Swain, A. M. (1996). Rates and risks of postpartum depression – A meta-analysis. *Int. Rev. Psychiatr.* 8, 37–54.

Ogawa, T., Mikuni, M., Kuroda, Y., Muneoka, K., Mori, K. J., and Takahashi, K. (1994). Periodic maternal deprivation alters stress response in adult offspring: Potentiates the negative feedback regulation of restraint stress-induced adrenocortical response and reduces the frequencies of open field-induced behaviors. *Pharmacol. Biochem. Behav.* 49, 961–967.

Ohl, F., Toschi, N., Wigger, A., Henniger, M. S., and Landgraf, R. (2001). Dimensions of emotionality in a rat model of innate anxiety. *Behav. Neurosci.* 115, 429–436.

Pardon, M., Gerardin, P., Joubert, C., Perez-Diaz, F., and Cohen-Salmon, C. (2000). Influence of prepartum chronic ultramild stress on maternal pup care behavior in mice. *Biol. Psychiatr.* 47, 858–863.

Patin, V., Lordi, B., Vincent, A., Thoumas, J. L., Vaudry, H., and Caston, J. (2002). Effects of prenatal stress on maternal behavior in the rat. *Brain Res. Dev. Brain Res.* 139, 1–8.

Pedersen, C. A. (1999). Postpartum mood and anxiety disorders: A guide for the nonpsychiatric clinician with an aside on thyroid associations with postpartum mood. *Thyroid* 9, 691–697.

Pedersen, C. A., and Prange, A. J., Jr. (1979). Induction of maternal behavior in virgin rats after intracerebroventricular administration of oxytocin. *Proc. Natl. Acad. Sci. USA* 76, 6661–6665.

Pedersen, C. A., and Boccia, M. L. (2003). Oxytocin antagonism alters rat dams' oral grooming and upright posturing over pups. *Physiol. Behav.* 80, 233–241.

Pedersen, C. A., Caldwell, J. D., McGuire, M., and Evans, D. L. (1991). Corticotropin-releasing hormone inhibits maternal behavior and induces pup-killing. *Life Sci.* 48, 1537–1546.

Pedersen, C. A., Vadlamudi, S. V., Boccia, M. L., and Amico, J. A. (2006). Maternal behavior deficits in nulliparous oxytocin knockout mice. *Genes Brain Behav.* 5, 274–281.

Phelps, E. A., and LeDoux, J. E. (2005). Contributions of the amygdala to emotion processing: From animal models to human behavior. *Neuron* 48, 175–187.

Pi, X. J., and Grattan, D. R. (1999). Increased expression of both short and long forms of prolactin receptor mRNA in hypothalamic nuclei of lactating rats. *J. Mol. Endocrinol.* 23, 13–22.

Plotsky, P. M. (1987). Facilitation of immuno reactive corticotropin-releasing factor secretion into the hypophysial-portal circulation after activation of catecholaminergic pathways or central norepinephrine injection. *Endocrinology* 121, 924–930.

Plotsky, P. M., and Meaney, M. J. (1993). Early, postnatal experience alters hypothalamic corticotropin-releasing factor (CRF) mRNA, median eminence

CRF content and stress-induced release in adult rats. *Brain Res. Mol. Brain Res.* 18, 195–200.

Rini, C. K., Dunkel-Schetter, C., Wadhwa, P. D., and Sandman, C. A. (1999). Psychological adaptation and birth outcomes: The role of personal resources, stress, and sociocultural context in pregnancy. *Health Psychol.* 18, 333–345.

Rosenblatt, J. S., Factor, E. M., and Mayer, A. D. (1994). Relationship between maternal aggression and maternal care in the rat. *Aggres. Behav.* 20, 243–255.

Rubin, K., and Burgess, K. (2002). Parents of aggressive and withdrawn children. In *Handbook of Parenting* (M. Bornstein, Ed.), Vol. 1, pp. 383–418. Lawrence Erlbaum Associates, Hillsdale, NJ.

Rubin, K., Burgess, K., and Coplan, R. (2002). Social withdrawal and shyness. In *Blackwell Handbook of Childhood Social Development. Blackwell Handbooks of Developmental Psychology* (P. Smith and C. Hart, Eds.), pp. 330–352. Blackwell Publishers, Malden, MA.

Shanks, N., Windle, R. J., Perks, P., Wood, S., Ingram, C. D., and Lightman, S. L. (1999). The hypothalamic–pituitary–adrenal axis response to endotoxin is attenuated during lactation. *J. Neuroendocrinol.* 11, 857–865.

Slattery, D. A., and Neumann, I. D. (2008). No stress please! Mechanisms of stress hyporesponsiveness of the maternal brain. *J. Physiol.* 586, 377–385.

Stenzel-Poore, M. P., Heinrichs, S. C., Rivest, S., Koob, G. F., and Vale, W. W. (1994). Overproduction of corticotropin-releasing factor in transgenic mice: A genetic model of anxiogenic behavior. *J. Neurosci.* 14, 2579–2584.

Stern, J. M., Goldman, L., and Levine, S. (1973). Pituitary–adrenal responsiveness during lactation in rats. *Neuroendocrinology* 12, 179–191.

Suchecki, D., Nelson, D. Y., Van Oers, H., and Levine, S. (1995). Activation and inhibition of the hypothalamic–pituitary–adrenal axis of the neonatal rat: effects of maternal deprivation. *Psychoneuroendocrinology* 20, 169–182.

Takayanagi, Y., Yoshida, M., Bielsky, I. F., Ross, H. E., Kawamata, M., Onaka, T., Yanagisawa, T., Kimura, T., Matzuk, M. M., Young, L. J., and Nishimori, K. (2005). Pervasive social deficits, but normal parturition, in oxytocin receptor-deficient mice. *Proc. Natl. Acad. Sci. USA* 102, 16096–16101.

Tanaka, M., Yoshida, M., Emoto, H., and Ishii, H. (2000). Noradrenaline systems in the hypothalamus, amygdala and locus coeruleus are involved in the provocation of anxiety: Basic studies. *Eur. J. Pharmacol.* 405, 397–406.

Torner, L., and Neumann, I. D. (2002). The brain prolactin system: Involvement in stress response adaptations in lactation. *Stress* 5, 249–257.

Torner, L., Toschi, N., Nava, G., Clapp, C., and Neumann, I. D. (2002). Increased hypothalamic expression of prolactin in lactation: Involvement in behavioural and neuroendocrine stress responses. *Eur. J. Neurosci.* 15, 1381–1389.

Torner, L., Maloumby, R., Nava, G., Aranda, J., Clapp, C., and Neumann, I. D. (2004). *In vivo* release and gene upregulation of brain prolactin in response to physiological stimuli. *Eur. J. Neurosci.* 19, 1601–1608.

Toufexis, D. J., Thrivikraman, K. V., Plotsky, P. M., Morilak, D. A., Huang, N., and Walker, C. D. (1998). Reduced noradrenergic tone to the hypothalamic paraventricular nucleus contributes to the stress hyporesponsiveness of lactation. *J. Neuroendocrinol.* 10, 417–427.

Toufexis, D. J., Tesolin, S., Huang, N., and Walker, C. D. (1999). Altered pituitary sensitivity to corticotropin-releasing factor and arginine vasopressin participates in the stress hyporesponsiveness of lactation in the rat. *J. Neuroendocrinol.* 11, 757–764.

Toufexis, D. J., Myers, K. M., and Davis, M. (2006). The effect of gonadal hormones and gender on anxiety and emotional learning. *Horm. Behav.* 50, 539–549.

Tu, M. T., Lupien, S. J., and Walker, C. D. (2005). Measuring stress responses in postpartum mothers: Perspectives from studies in human and animal populations. *Stress* 8, 19–34.

van Leengoed, E., Kerker, E., and Swanson, H. H. (1987). Inhibition of post-partum maternal behaviour in the rat by injecting an oxytocin antagonist into the cerebral ventricles. *J. Endocrinol.* 112, 275–282.

Veenema, A. H., Torner, L., Blume, A., Beiderbeck, D. I., and Neumann, I. D. (2007). Low inborn anxiety correlates with high intermale aggression: Link to ACTH response and neuronal activation of the hypothalamic paraventricular nucleus. *Horm. Behav.* 51, 11–19.

Walker, C. D., Lightman, S. L., Steele, M. K., and Dallman, M. F. (1992). Suckling is a persistent stimulus to the adrenocortical system of the rat. *Endocrinology* 130, 115–125.

Walker, C. D., Trottier, G., Rochford, J., and Lavallee, D. (1995). Dissociation between behavioral and hormonal responses to the forced swim stress in lactating rats. *J. Neuroendocrinol.* 7, 615–622.

Walker, C. D., Toufexis, D. J., and Burlet, A. (2001). Hypothalamic and limbic expression of CRF and vasopressin during lactation: Implications for the control of ACTH secretion and stress hyporesponsiveness. *Prog. Brain Res.* 133, 99–110.

Weaver, I. C., Cervoni, N., Champagne, F. A., D'Alessio, A. C., Sharma, S., Seckl, J. R., Dymov, S., Szyf, M., and Meaney, M. J. (2004). Epigenetic programming by maternal behavior. *Nat. Neurosci.* 7, 847–854.

Weinstock, M. (2001). Alterations induced by gestational stress in brain morphology and behaviour of the offspring. *Prog. Neurobiol.* 65, 427–451.

Welberg, L. A., and Seckl, J. R. (2001). Prenatal stress, glucocorticoids and the programming of the brain. *J. Neuroendocrinol.* 13, 113–128.

Wenzel, A., Haugen, E. N., Jackson, L. C., and Brendle, J. R. (2005). Anxiety symptoms and disorders at eight weeks postpartum. *J. Anxiety Disord.* 19, 295–311.

Wigger, A., and Neumann, I. D. (1999). Periodic maternal deprivation induces gender-dependent alterations in behavioral and neuroendocrine responses to emotional stress in adult rats. *Physiol. Behav.* 66, 293–302.

Wigger, A., Lorscher, P., Oehler, I., Keck, M. E., Naruo, T., and Neumann, I. D. (1999). Non-responsiveness of the rat hypothalamo–pituitary–adrenocortical axis to parturition-related events: Inhibitory action of endogenous opioids. *Endocrinology* 140, 2843–2849.

Wigger, A., Sanchez, M. M., Mathys, K. C., Ebner, K., Frank, E., Liu, D., Kresse, A., Neumann, I. D., Holsboer, F., Plotsky, P. M., and Landgraf, R. (2004). Alterations in central neuropeptide expression, release, and receptor binding in rats bred for high anxiety: Critical role of vasopressin. *Neuropsychopharmacology* 29, 1–14.

Windle, R. J., Shanks, N., Lightman, S. L., and Ingram, C. D. (1997a). Central oxytocin administration reduces stress-induced corticosterone release and anxiety behavior in rats. *Endocrinology* 138, 2829–2834.

Windle, R. J., Wood, S., Shanks, N., Perks, P., Conde, G. L., da Costa, A. P., Ingram, C. D., and Lightman, S. L. (1997b). Endocrine and behavioural responses to noise stress: Comparison of virgin and lactating female rats during non-disrupted maternal activity. *J. Neuroendocrinol.* 9, 407–414.

Windle, R. J., Kershaw, Y. M., Shanks, N., Wood, S. A., Lightman, S. L., and Ingram, C. D. (2004). Oxytocin attenuates stress-induced c-fos mRNA expression in specific forebrain regions associated with modulation of hypothalamo–pituitary–adrenal activity. *J. Neurosci.* 24, 2974–2982.

Woodside, B., and Amir, S. (1997). Lactation reduces Fos induction in the paraventricular and supraoptic nuclei of the hypothalamus after urethane administration in rats. *Brain Res.* 752, 319–323.

9 ROLE OF PROLACTIN IN THE BEHAVIORAL AND NEUROENDOCRINE STRESS ADAPTATIONS DURING LACTATION

LUZ TORNER

Centro de Investigación Biomédica de Michoacán,
Instituto Mexicano del Seguro Social,
Morelia, Michoacán, Mexico

INTRODUCTION

Prolactin (PRL) exerts a wide array of physiological functions (i.e., lactogenesis, maternal behavior, food ingestion, immunocompetence (Bole-Feysot *et al.*, 1998; Freeman *et al.*, 2000). Some of these functions are especially relevant in the peripartum period by promoting survival of the young. In fact, PRL is one of the hormones required to adapt the female organism to meet her role as a mother. During pregnancy PRL promotes lobuloalveolar development and after birth PRL stimulates lactogenesis in the mammary gland and the display of maternal behavior (Bridges *et al.*, 1985; Bole-Feysot *et al.*, 1998; Freeman *et al.*, 2000). Besides these reproductive adaptations that directly ensure nutrition and care of the offspring, PRL promotes other changes that support reproductive success in an indirect fashion.

A variety of behavioral changes including a decrease in anxiety and behavioral responses to stress take place during lactation (Hard & Hansen, 1984; Asher *et al.*, 1995; Walker *et al.*, 1995; Windle *et al.*, 1997; Toufexis *et al.*, 1999; Heinrichs *et al.*, 2001; Neumann, 2001). For instance, around parturition the responsiveness to a stressor is significantly attenuated, as revealed by reduced secretory responses of the hypothalamo–pituitary–adrenal (HPA) axis and a reduction of anxiety-related behavior. These physiological adaptations belong to a complex pattern of maternal behaviors and are partially driven by alterations of excitatory and inhibitory inputs to the hypothalamic corticotropin releasing hormone (CRH) system, the main regulator of the HPA axis. Similarly, the concerted actions of neuropeptides like prolactin (PRL) and oxytocin (OXT) that are significantly involved in a broad set of reproductive functions also exert effects on the reactivity of the HPA axis and on anxiety-related behavior. These actions constitute part of the dynamic adaptations in stress responsiveness observed peripartum that support reproductive function and favor the survival of the young.

PRL SOURCES AND PRL RECEPTORS

Pituitary PRL

The major synthesis of PRL, a 23-kilodalton protein, takes place in the lactotrophic cells of the anterior pituitary or adenohypophysis (Freeman *et al.*, 2000), but is not restricted to this tissue. Several sources of PRL have been identified, including the placenta and the brain (Ben-Jonathan *et al.*, 1996; Freeman *et al.*, 2000).

Placental PRL and Placental Lactogens

During pregnancy, both the maternal and fetal components of the placenta produce PRL-like substances (Ben-Jonathan *et al.*, 1996). For instance, the decidua produces a PRL-like molecule, which is apparently identical in the human to pituitary PRL based on chemical, immunological, and biological criteria (Andersen, 1990), but is somewhat dissimilar to that in the rat (Gu *et al.*, 1994). Additionally, a family of placental lactogens (Yamaguchi *et al.*, 1994) is produced by the placenta during pregnancy. These PRL-like molecules are able to bind to PRL-receptors (PRL-Rs), although with different potencies (Shiu *et al.*, 1973).

Brain PRL

PRL produced in the brain and its cerebral receptors will be referred here as the brain PRL system, to differentiate it from the pituitary PRL.

PRL-synthesizing neurons have been observed in several brain regions in both male and female mammals, with greater concentrations in the female than the male rat brain (DeVito, 1988). PRL-synthesizing neurons were initially shown by immunocytochemistry (Fuxe *et al.*, 1977). A local synthesis of PRL was indicated as PRL-like immunoreactivity was shown to persist in the brain several weeks after hypophysectomy (DeVito, 1989; Paut-Pagano *et al.*, 1993; Torner *et al.*, 1995). PRL cell bodies are localized in the lateral hypothalamic area, (Fuxe *et al.*, 1977; Hansen *et al.*, 1982; Harlan *et al.*, 1989; Siaud *et al.*, 1989; Paut-Pagano *et al.*, 1993), and specifically in the dorsomedial, ventromedial (Griffond *et al.*, 1994), arcuate, supraoptic (SON), and paraventricular (PVN) nuclei (Hansen *et al.*, 1982). In particular, the magnocellular neurons of the PVN and the SON, the major sites of OXT and vasopressin synthesis, contain immunoreacting positive PRL neurons (Clapp *et al.*, 1994; Torner *et al.*, 1995; Mejia *et al.*, 1997).

In addition, PRL-like immunoreactive fibers profusely innervate several areas of the brain, including the parvocellular portion of the PVN (Paut-Pagano *et al.*, 1993). This area is a main component of the stress response circuit, as CRH-synthesizing neurons are localized in the PVN (Merchenthaler *et al.*, 1982). Other regions involved in the regulation of anxiety-related behavior and HPA axis activity (Charney *et al.*, 1998) include extrahypothalamic limbic brain regions such as the bed nucleus of the stria terminalis, the amyg-dala, and the locus coeruleus which are also innervated by PRL immunoreactive fibers (Paut-Pagano *et al.*, 1993).

Amplification of PRL mRNA by the polymerase chain reaction (PCR) has confirmed the expression of PRL in the brain (DeVito *et al.*, 1992; Emanuele *et al.*, 1992; Clapp *et al.*, 1994; Torner *et al.*, 1999). PRL gene expression appears to be localized in the hypothalamus. The nuclei screened by PCR to date include the PVN and the SON (Clapp *et al.*, 1994; Torner *et al.*, 1999). Studies using double immunofluorescence labeling followed by confocal microscopy indicate the coexistence of PRL and vasopressin-related antigens within the same neurons of the PVN and SON (Mejia *et al.*, 1997). Expression of hypothalamic PRL is upregulated by vasointestinal peptide (Bredow *et al.*, 1994) and by estrogen (DeVito *et al.*, 1992) in a manner similar to that found in the pituitary. This last finding was confirmed by semi-quantitative PCR that detected higher PRL mRNA expression in the PVN and the SON during the estrous phase (Torner *et al.*, 1999).

Presence of PRL-Rs in the Brain

PRL-Rs are membrane-bound proteins belonging to the class 1 cytokine-hematopoietin receptor superfamily (Kelly *et al.*, 1991; Bole-Feysot *et al.*, 1998). Several isoforms have been recognized in each vertebrate species, where the main difference is the length and composition of the cytoplasmic domain (Kelly *et al.*, 1991; Bole-Feysot *et al.*, 1998), conferring different properties and signal transduction actions to the receptor isoforms. Long (591 amino acids) and short (291 amino acids) isoforms have been identified in the rat brain and in other species. In general, activation of the receptors involves ligand-induced sequential receptor dimerization. Formation of long form homodimers leads to signal transduction via phosphorylation of a tyrosine kinase, Janus kinase 2 (JAK2), while long and short form heterodimers prevent PRL signaling (Perrot-Applanat *et al.*, 1997).

In the female rat brain, both long and short form PRL-Rs have been mapped in hypothalamic regions that include the PVN, SON, arcuate, and ventromedial nuclei, and the medial preoptic area (MPOA); and in non-hypothalamic regions such as the amygdala and the cerebral cortex (Bakowska & Morrell, 1997; Pi & Grattan, 1998, 1999a). Noteworthy, the highest density of PRL-Rs, with predominance of the long form, is located in choroid plexus cells (Crumeyrolle-Arias *et al.*, 1993; Chiu & Wise,

1994; Bakowska & Morrell, 1997). Areas where the long form is significantly expressed include the bed nucleus of the stria terminalis, the central gray of the midbrain, thalamus, and olfactory bulb (Bakowska & Morrell, 1997). As mentioned above, some of these areas (e.g., PVN, amygdala) are involved in the regulation of neuroendocrine and behavioral stress responses.

Recent studies using *in situ* hybridization revealed that in both the SON and the PVN, the mRNA of the long form of the PRL-R is predominantly colocalized (up to 80%) with OXT neurons, but in very few vasopressin neurons. Interestingly, the proportion of PRL-R positive OXT neurons increases significantly during pregnancy and lactation (Kokay *et al.*, 2006). Since OXT is known to stimulate PRL release within the PVN (Arey & Freeman, 1992), it is hypothesized that OXT might regulate its own release via alterations in PRL secretion.

ACTIONS OF PRL IN THE BRAIN

Numerous reports on behavioral effects of PRL acting centrally have been documented. Experimental strategies have included application of exogenous PRL to the brain to mimic a hyperprolactinemic state and enhance brain PRL-R expression (Mangurian *et al.*, 1992; Sugiyama *et al.*, 1994). Also, infusion of PRL-R antisense probes or PRL antagonists as well as the deletion of the PRL or the PRL-R gene in mice has helped elucidate central actions of PRL (for reviews see Bern & Nicoll, 1968; Bole-Feysot *et al.*, 1998; Freeman *et al.*, 2000; Goffin *et al.*, 2002).

Behavioral actions of PRL include the regulation of grooming (Drago *et al.*, 1983; Drago & Lissandrello, 2000), food intake (Noel & Woodside, 1993; Li *et al.*, 1995), the sleep–wake cycle (Roky *et al.*, 1994), and exerting dose-dependent facilitatory or inhibitory effects on sexual behavior (Drago & Lissandrello, 2000). The latter effect may involve PRL actions on the CRH system (Kooy *et al.*, 1990; Calogero *et al.* 1996), as well as on gonadotropin-releasing hormone (GnRH) cells (Grattan *et al.*, 2007). Perhaps the most studied function of PRL in the brain is the regulation of maternal behavior (Bridges *et al.*, 1985, 1990; for review see Mann & Bridges, 2001). Promotion of nurturing behavior is shortened in virgin female rats which are not spontaneously maternal (Rosenblatt, 1967) by administering PRL into the ventricular system of the brain (Voci & Carlson, 1973; Bridges *et al.*, 1990; 2001), or directly into the MPOA (Bridges *et al.*, 1990; 2001; Mann & Bridges, 2001). Studies with PRL-R knockout mice confirmed the impairment of maternal behavior associated with the lack of brain PRL-Rs (Lucas *et al.*, 1998). In contrast, PRL knockout mice are still maternal when confronted with foster pups (Horseman *et al.*, 1997), suggesting involvement of other neuropeptide systems and adaptive processes not yet fully understood. More recently, downregulation of brain PRL-R expression by antisense infusion into the cerebral ventricles of lactating rats resulted in impaired maternal behavior, giving further support to the involvement of the brain PRL system in maternal care (Torner *et al.*, 2002).

It is also important to note that peripheral PRL enters the central nervous system through a receptor-mediated active uptake by the choroid plexus cells of the brain ventricles (Walsh *et al.*, 1987). Accordingly, it has been shown that PRL upregulates the expression of its own receptors (Sugiyama *et al.*, 1994), which is hypothesized to activate its target areas.

CONTRIBUTION OF THE BRAIN PRL SYSTEM TO STRESS-RELATED ADAPTATIONS IN THE PERIPARTUM PERIOD

Activation of the Brain PRL System Peripartum

PRL blood levels in the female rat rise during early and late pregnancy, but become low at midpregnancy, when placental lactogens, proteins belonging also to the PRL family, begin to rise (Tonkowicz & Voogt, 1983). During lactation, PRL levels rise within 1–3 min of nursing initiation, and fall when nursing is terminated (Grosvenor *et al.*, 1986; Freeman *et al.*, 2000). Accordingly, increased PRL-R expression in the brain has been reported during pregnancy and lactation (Sugiyama *et al.*, 1994; Bakowska & Morrell, 1997; Pi & Grattan, 1999a, b). For instance, PRL-R mRNA expression and PRL-R density are significantly increased in the choroid plexus, the MPOA, and in the arcuate hypothalamic nuclei of lactating compared to diestrous rats (for review see Grattan *et al.*,

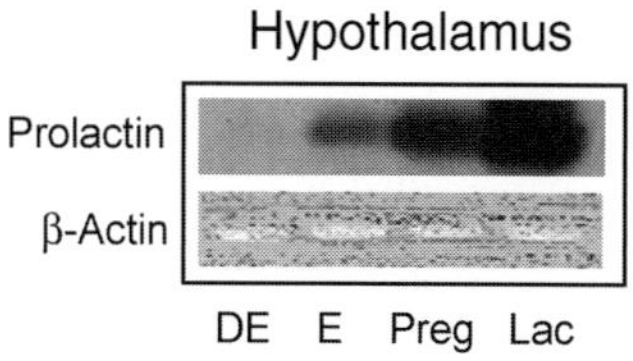

FIGURE 9.1 *PRL mRNA expression in the hypothalamus is increased in pregnancy, and becomes maximal during lactation.* PRL mRNA from isolated hypothalami of virgin rats in diestrus (DE) and estrus (E), and from pregnant (preg) and lactating (lac) rats was reverse-transcribed, and the corresponding cDNA was amplified by PCR. A southern blot representative of five independent experiments is shown, containing an oligonucleotide band of 388 bp corresponding to the amplified cDNA fragment. (*Source*: Data modified from Torner *et al.*, 2002.)

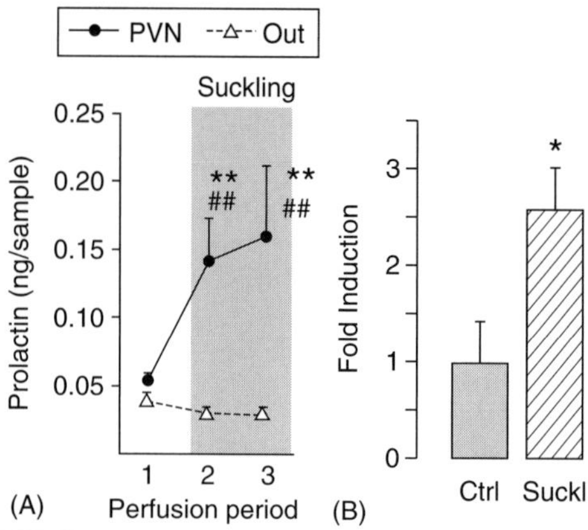

FIGURE 9.2 *Local release and synthesis of brain PRL in response to suckling.* (A) PRL content in three consecutive 30-min PP perfusates collected within the PVN (black circles) or outside of this area (triangles), of conscious lactating female rats before and during suckling. The pups were separated from the dams overnight, were returned to the home cage before starting the second perfusion interval, and were allowed to suckle during the second and third perfusion periods (shaded area). Data are expressed as ng/sample. * $p < 0.05$, ** $p < 0.01$ vs. basal; ## $p < 0.01$ vs. outsider group. (B) PRL mRNA quantification by real-time PCR in hypothalami from non-suckled (ctrl) and suckled (suckl) lactating rats previously left overnight without their pups but one. The pups were either returned to the home cage and allowed to suckle the dam or left separated. The dams were sacrificed immediately after 60 min suckling or without being suckled (ctrl). Determination of fold induction of PRL mRNA was calculated as a ratio. The results are expressed as means + SEM. * $p < 0.05$ and ** $p < 0.01$ vs. control group. (*Source*: Data modified from Torner *et al.*, 2004.)

2001). Moreover, in various hypothalamic nuclei, including the PVN and the SON, PRL-R mRNA expression and PRL-R immunoreactivity is more evident in lactating animals (Pi & Grattan, 1999a, b). The increased activity of the brain PRL system peripartum is also reflected by an increase in PRL gene transcription in the brain (Torner *et al.*, 2002). PCR amplification revealed a huge increase of PRL mRNA in the hypothalamus during pregnancy, which became maximal in lactation, in comparison to non-pregnant virgin female rats (see Figure 9.1). Thus, an overall activation of the brain PRL system occurs in the peripartum period that serves to sustain its reproductive and non-reproductive functions in the central nervous system.

Endogenous Release of Brain PRL

Since ascribing a role for PRL of neural origin has been difficult, it has been largely assumed that exclusively the peripheral PRL promotes the above mentioned functions in the brain. However, recent demonstrations of dynamic changes in local PRL release and expression support a role for brain PRL as a neurotransmitter/neuromodulator.

PRL release was first observed in hypothalamic explants (DeVito *et al.*, 1991; Torner *et al.*, 1995). Later, the *in vivo* release of PRL was monitored using push–pull (PP) perfusion (Torner *et al.*, 2004) within the PVN and MPOA of urethane-anesthetized female and male rats in response to a depolarizing medium containing 56 mM K^+, a procedure used to characterize the release from excitable neuronal structures. Brain PRL release was also monitored in conscious, lactating rats before and during nursing. An initial 30-min period of PP perfusion samples were collected from the PVN (Figure 9.2(A)) of lactating rats without the pups (except one to prevent behavioral changes). Before the start of the second sampling period, all pups were returned to the dam. Suckling of the young triggered a significant and selective increase in PRL release within the PVN (Figure 9.2(A)), but not outside of the targeted region. A similar release of PRL was observed within the MPOA in response to suckling. Identity of PRL in the perfusates was confirmed by western blot, radioimmunoassay, and the specific Nb2 cell bioassay (Torner *et al.*, 2004). In addition, using real-time PCR techniques, a significant increase in PRL mRNA in response to suckling was observed in the hypothalamus of lactating rats 60 min after the start of suckling (Figure 9.2(B)). These results demonstrate dynamic, stimulus-induced changes of PRL synthesis in the brain (Torner *et al.*, 2004).

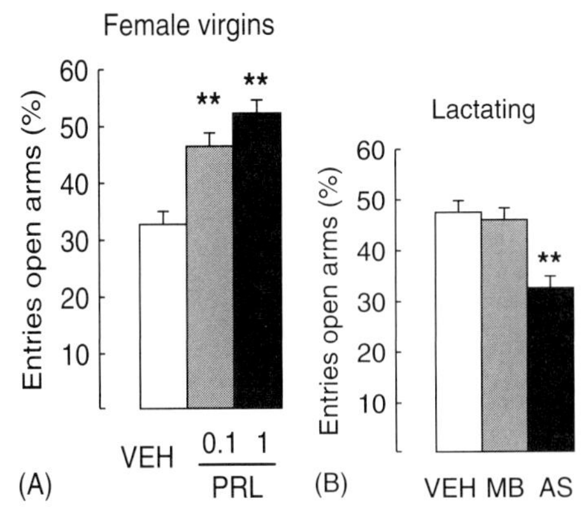

FIGURE 9.3 *Anxiolytic effect of* PRL *in virgin and lactating rats on the EPM.* (A) An acute ICV infusion of ovine PRL (0.1 and 1 μg/5 μl) 10 min before EPM exposure decreased the anxiety-related behavior in virgin female rats as shown by an increase in the percentage of entries into the open arms of the maze. (B) 5 days of chronic ICV infusion of PRL-R antisense oligodeoxynucleotide (AS) in order to downregulate brain PRL-R expression increased the anxiety-related behavior of lactating rats on the plus-maze, as indicated by a decrease in the percentage of entries into the open arms, compared to vehicle (Veh)- or mixed bases (MB)-treated animals. Data are the mean ± SEM. ** $p < 0.01$. (*Source*: Data modified from Torner *et al.*, 2001, 2002.)

Involvement of PRL in Stress Regulation

Pituitary PRL is released into the blood in response to stress exposure (Neill, 1970; Noel *et al.*, 1972; Seggie & Brown, 1975; Meyerhoff *et al.*, 1988) and as such is considered a marker of stress. An involvement of PRL in stress response mechanisms was initially suggested by studies showing that PRL administration intracerebroventricular (ICV) prevented stress-induced hyperthermia (Drago & Amir, 1984) and gastric ulcer formation (Drago *et al.*, 1985). Experimentally induced hyperprolactinemia using pituitary grafts caused antidepressive-like effects in animals subjected to forced swimming tests (Drago *et al.*, 1990). Finally, chronic restraint stress induced the expression of the long form of the PRL-R in choroid plexus cells, probably due to stress-induced PRL secretion (Fujikawa *et al.*, 1995).

Regulation of Anxiety by PRL in Virgin Females

We recently showed that PRL acts as an endogenous anxiolytic substance interacting with brain PRL-Rs. Acute administrations of 0.1 or 1 μg of PRL ICV to virgin female rats increased, in a dose-dependent fashion, the females exploration of the open arms of the elevated plus-maze (EPM) – the standard test validated for the detection of emotional responses to anxiolytic or anxiogenic substances in rodents (Pellow *et al.*, 1985) – thus indicating an anxiolytic effect of PRL (Figure 9.3(A)). Similar results were observed in male rats and were validated by comparing the actions of PRL with the established anxiolytic drug diazepam (Torner *et al.*, 2001). The anxiolytic action of PRL was further confirmed after downregulation of the PRL-R expression in the brain of virgin rats by antisense targeting, using a computer-modeled oligodeoxynucleotide probe that decreases autoradiographic receptor binding up to 72% (Torner *et al.*, 2001). Reduction of central PRL availability by antisense targeting treatment significantly increased the anxiety-related behavior of virgin female rats on the EPM. Of interest, an intravenous administration of ovine PRL before plus-maze exposure at doses resembling stress-induced plasma PRL concentrations (Neill, 1970) also resulted in reduced anxiety (Torner *et al.*, 2001). These findings suggest coordinated anxiolytic action of brain and pituitary PRL, the latter entering the brain through active and selective uptake by choroid plexus cells (Walsh *et al.*, 1987).

PRL Effects on Anxiety in Lactating Females

Blockade of PRL-R expression in the brain likewise induced a significant increase in the anxiety-related behavior of lactating rats displayed on the EPM (Torner *et al.*, 2002), that is, a reduction in the number of entries into the open arms of the maze (Figure 9.3(B)). These results support a role for PRL as an endogenous anxiolytic with an especially relevant role during the peripartum period.

A reduction in stress perception may function to reduce the mother's fearfulness and neophobia to the pups during lactation (Fleming *et al.*, 1989) and to increase aggressive behavior to protect the offspring (Erskine *et al.*, 1978; Neumann *et al.*, 2001). Recently, it was shown that chronic elevation of PRL levels within the brain resulted in reduced neuronal activation within the hypothalamus, specifically within the PVN, in response to an acute stressor (Donner *et al.*, 2007). Therefore, we suggest that PRL also contributes to the reduction in stress perception described around peripartum.

Effects of PRL on Neuroendocrine Stress Responses

During the peripartum period, the responsiveness of the HPA axis to a variety of stressors is reduced, as characterized by blunted adrenocorticotropic hormone (ACTH) and corticosterone responses (Stern *et al.*, 1973; Lightman & Young, 1989; Walker *et al.*, 1995, Neumann *et al.*, 1998; Johnstone *et al.*, 2000). Although the current evidence suggests that reproduction-related changes underlie the hyporesponsive state, the mechanisms involved are still poorly understood. Proposed mechanisms include alterations in the perception of the stressor reduction in brainstem excitatory inputs to the hypothalamic CRH/vasopressin cells (the main ACTH secretagogues), reduced synthetic activity of CRH and vasopressin neurons, and alterations at the level of the pituitary corticotrophs (Da Costa *et al.*, 1996; Toufexis & Walker, 1996; Douglas *et al.*, 1998; Neumann *et al.*, 1998; Johnstone *et al.*, 2000).

In addition to these possibilities, previous studies have shown that neuroendocrine stress responses are attenuated in experimentally induced hyperprolactinemia (Schlein *et al.*, 1974; Carter & Lightman, 1987). This raises the possibility that PRL, either released from the pituitary into the blood and taken up by choroid plexus cells into the brain or released within the brain itself from neuronal structures may be a significant factor in the blunted HPA response that occurs during lactation.

Inhibitory Action of PRL on the HPA Axis Reactivity in Virgin Female Rats

Chronic ICV ovine PRL infusions in virgin female rats causes a shift in HPA axis activity toward an attenuated ACTH secretory response following exposure to a novel environment (Torner *et al.*, 2001). In contrast, ICV infusion of antisense oligos against the PRL-R and down regulation of PRL-R expression in the brain significantly enhanced ACTH secretion, that is, causing a further elevation of the stress-induced ACTH secretion (Torner *et al.*, 2001).

Effect of PRL on the HPA Axis Reactivity in Lactating Females

In other studies, using osmotic minipumps, we infused the PRL-R antisense probe or control substances (mixed base oligonucleotides or vehicle) into the lateral cerebral ventricle of lactating rats to assess the role of PRL in regulating the HPA axis during lactation (Torner *et al.*, 2002). After 5 days of antisense treatment, basal levels of ACTH were not altered, whereas the stress-induced rise in ACTH secretion was significantly more pronounced compared to that found in both groups of control rats, thus revealing a disinhibition of the HPA axis (Figure 9.4(A)). It is noted that ACTH responses in antisense-treated lactating rats were very similar compared to those observed in virgin rats (vehicle treatment), indicating reversal of the attenuated ACTH response when PRL-R functions are blocked (or reduced) during lactation. Thus, the attenuated HPA axis responses seen in lactation are at least partly due to an inhibitory effect of brain and/or pituitary PRL acting at brain PRL-Rs.

These results are an initial demonstration of a pharmacological disinhibition of the attenuated HPA axis response. Other pharmacological manipulations employing selective antagonists to OXT (Neumann *et al.*, 2000) or endogenous opioids (Douglas *et al.*, 1998) to identify inhibitory factors that may also affect ACTH/corticosterone secretion failed to show such disinhibition. Furthermore, the lack of change

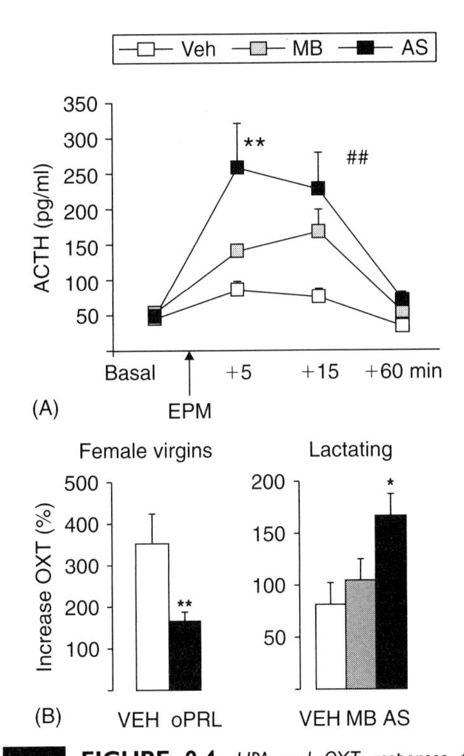

FIGURE 9.4 *HPA and OXT responses to stress are partially reduced by* PRL *during lactation.* (A) Plasma ACTH concentrations under basal conditions and 5, 15, and 60 min after exposure to a novel environment (EPM) of lactating rats treated chronically with either vehicle (veh), a mixture of oligonucleotides (MB), or PRL-R antisense (AS) during 4 days via osmotic minipumps (1.0 μg/0.5 μl/h). Vehicle-treated dams show a blunted ACTH release in response to stress, as it is characteristic during lactation. PRL-R AS treatment significantly increases the stress-induced ACTH release. Data are means + SEM. (*Source*: Data modified from Torner *et al.*, 2002). (B) Plasma OXT concentrations in virgin female rats chronically treated with ICV vehicle or ovine PRL or lactating rats chronically treated with ICV vehicle, MB or PRL-R AS. Shown is the percentage of the increase of OXT released after exposure to forced swimming (60 s, 19°C). Data are means + SEM. Numbers in parentheses indicate group size. *$p < 0.05$ vs. veh- and MB-treated rats; #$p < 0.05$ vs. veh-treated rats; **$p < 0.01$ vs. vehicle-treated rats. (*Source*: Data modified from Torner *et al.*, 2002.)

in the HPA axis responsiveness of lactating rats after lesions of noradrenergic inputs to the PVN is thought to reflect a reduced noradrenergic excitation in the peripartum period (Toufexis & Walker, 1996). Hence, the suppression of ACTH and corticosterone secretion in pregnancy and lactation may be a consequence of several inhibitory factors. It is therefore even more remarkable that blockade of brain PRL-Rs results in disinhibition of the HPA axis response in lactation, These findings support the idea that brain PRL plays a major role in modification of the HPA axis responsivity.

Effects on OXT System Reactivity

Similar to the HPA axis system, the OXT secretory responses to non-reproduction-related stimuli are reduced at the end of pregnancy (Neumann *et al.*, 1993, 1998; Douglas *et al.*, 1995), during parturition (Neumann *et al.*, 2003), and in lactation (Carter & Lightman, 1987; Lightman & Young, 1989; Neumann *et al.*, 1994, 1995; Douglas *et al.*, 1998). A stimulatory effect of PRL on OXT secretion at a hypothalamic level was first suggested based on *in vitro* studies with hypothalamic explants (Ghosh & Sladek, 1995a, b). Increased basal OXT plasma concentrations have been described during lactation and in hyperprolactinemic animals (Carter & Lightman 1987). In agreement, we found that basal OXT secretion is increased after chronic ICV infusion of PRL in virgin rats (Torner *et al.*, 2002; Figure 9.4(B)). In contrast, brain PRL exerts an inhibitory effect on the responsiveness of the OXT system to stress. This is shown by attenuated stress-induced OXT secretion after induction of hyperprolactinemia in virgin female rats in a similar manner to that observed in lactation (Carter & Lightman, 1987; Torner *et al.*, 2002). Further support for a PRL-mediated inhibition of OXT is that antisense targeting of the brain PRL-R in lactating rats results in a disinhibition of the OXT response to forced swimming, (Figure 9.4(B); Torner *et al.*, 2002). Thus, we can conclude that the blunted neuroendocrine responsiveness to various stressors, including the HPA axis and OXT system may, at least in part, be due to an inhibition by centrally acting PRL.

Clinical studies provide parallel findings to our animal studies. As in our studies, reduced anxiety scores have been reported in women after nursing (Asher *et al.*, 1995; Heinrichs *et al.*, 2001; Mezzacappa & Katlin, 2002), and reduced concentrations of PRL in plasma are present in women suffering from postpartum depression (Abou-Saleh *et al.*, 1998). These

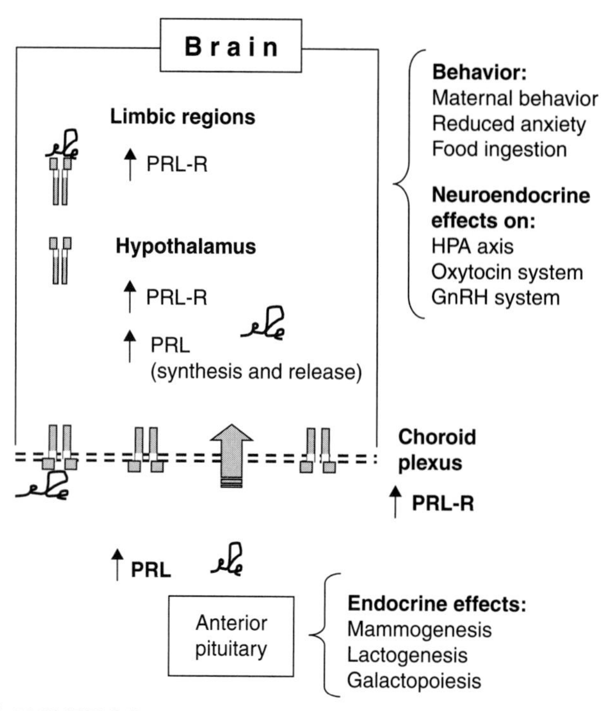

FIGURE 9.5 *Schematic view of the PRL system in the maternal brain.* PRL expression increases both in the pituitary and the hypothalamus. PRL-R expression also increases in several brain areas including the PVN, SON, and choroid plexus cells. Pituitary PRL taken up by choroid plexus cells and brain PRL intracerebrally released affect HPA axis and OXT system responses to stress and modulate maternal behavior, anxiety, and aggression during lactation.

observations point to a relationship between PRL and mood (Groer & Davis, 2006). However, it is important to mention that pathological hyperprolactinemic states in humans are often accompanied by an altered psychological profile with an increased incidence of depression and anxiety, positively modified by the administration of dopaminergic drugs (Thienhaus & Hartford, 1986; Rocco *et al.*, 1993; Reavley *et al.*, 1997). At present, it is unclear what accounts for the discrepancy between the findings of an anxiolytic effect of PRL and the rise in anxiety shown in hyperprolactinemic patients. Explanations could include alterations in brain dopaminergic activity or in serotonergic system influencing pituitary PRL secretion. Alternatively, the possibility of either a selective downregulation of PRL-Rs in specific brain regions as a consequence of high, long-term exposure to PRL (Pi & Voogt, 2001), or a dysregulation of second messenger coupling mechanisms need to be considered. Overall, possible anxiety-related effects of PRL in humans in the context of the pathological or physiological states of hyperprolactinemia need further clarification.

CONCLUSIONS

In summary, our findings of significant actions of centrally acting PRL on anxiety-related and maternal behaviors and on neuroendocrine responses to stressors in lactation are consistent with the demonstration of PRL enhanced brain PRL activity. Locally released PRL in the PVN could act in a paracrine way to affect CRH and OXT neuronal activity, or interact with PRL-Rs in discrete brain areas to affect emotionality (Figure 9.5). In addition central actions of pituitary PRL taken up into the brain compartment also need to be considered. Synergistic actions of the brain PRL and OXT systems, both activated to meet reproduction-related demands, support neuroendocrine and emotional adaptations which promote reproductive events.

ACKNOWLEDGMENTS

I wish to thank Professor Inga D. Neumann for her supervision and contribution to the studies here reported. These were performed at the Max Planck Institute (MPI) of Psychiatry and at the Institute of Zoology of the University of Regensburg, in Germany. I thank Professor Rainer Landgraf from the MPI and Professor Carmen Clapp from the National University of Mexico for their contribution to some of the studies. I thank also my coworkers Nicola Toschi, Agnes Pohlinger, Rodrigue Maloumby, Gabriel Nava, and Jorge Aranda. LT is currently supported by a grant from FOFOI, of the Instituto Mexicano del Seguro Social (N.R. 2005/1/I/049). Some fragments were extracted and modified after Torner and Neumann, (2002), with permission of Informaworld. (http://www.informaworld.com).

REFERENCES

Abou-Saleh, M. T., Ghubash, R., Karim, L., Krymski, M., and Bhai, I. (1998). Hormonal aspects of postpartum depression. *Psychoneuroendocrinology* 23, 465–475.

Andersen, J. R. (1990). Decidual prolactin studies of decidual and amniotic prolactin in normal and pathological pregnancy. *Dan. Med. Bull.* 37, 154–165.

Arey, B. J., and Freeman, M. E. (1992). Activity of oxytocinergic neurons in the paraventricular nucleus mirrors the periodicity of the endogenous stimulatory rhythm regulating prolactin secretion. *Endocrinology* 130(1), 126–132.

Asher, I., Kaplan, B., Modai, I., Neri, A., Valevski, A., and Weizman, A. (1995). Mood and hormonal changes during late pregnancy and puerperium. *Clin. Exp. Obstet. Gynecol.* 22, 321–325.

Bakowska, J. C., and Morrell, J. I. (1997). Atlas of the neurons that express mRNA for the long form of the prolactin receptor in the forebrain of the female rat. *J. Comp. Neurol.* 386, 161–177.

Ben-Jonathan, N., Mershon, J. L., Allen, D. L., and Steinmetz, R. W. (1996). Extrapituitary prolactin: Distribution, regulation, functions, and clinical aspects. *Endocr. Rev.* 17, 639–669.

Bern, H. A., and Nicoll, C. A. (1968). The comparative endocrinology of prolactin. *Recent Progr. Horm. Res.* 24, 681–720.

Bole-Feysot, C., Goffin, V., Edery, M., Binart, N., and Kelly, P. A. (1998). Prolactin (PRL) and its receptor: Actions, signal transduction pathways and phenotypes observed in PRL receptor knockout mice. *Endocr. Rev.* 19, 225–268.

Bridges, R. S., DiBiase, R., Loundes, D. D., and Doherty, P. C. (1985). Prolactin stimulation of maternal behavior in female rats. *Science* 227, 782–784.

Bridges, R. S., Numan, M., Ronsheim, P. M., Mann, P. E., and Lupini, C. E. (1990). Central prolactin infusions stimulate maternal behavior in steroid-treated nulliparous female rats. *Proc. Natl Acad. Sci. USA* 87, 8003–8007.

Bridges, R. S., Rigero, B., Byrnes, E. M., Yang, L., and Walker, A. M. (2001). Central infusions of the recombinant human prolactin antagonist, S179D-PRL, delay the onset of maternal behavior in steroid-primed, nulliparous female rats. *Endocrinology* 142, 730–739.

Bredow, S., Kacsoh, B., Obal, F., Jr., Fang, J., and Krueger, J. M. (1994). Increase of prolactin mRNA in the rat hypothalamus after intracerebroventricular injection of VIP or PACAP. *Brain Res.* 660, 301–308.

Calogero, A. E., Burrello, N., Ossino, A. M., Weber, R. F., and D'Agat, R. (1996). Interaction between prolactin and catecholamines on hypothalamic GnRH release *in vitro*. *J. Endocrinol.* 151, 269–275.

Carter, D. A., and Lightman, S. L. (1987). Oxytocin responses to stress in lactating and hyperprolactinaemic rats. *Neuroendocrinology* 46, 532–537.

Charney, D. S., Grillon, C. C. G., and Bremner, J. D. (1998). The neurobiological basis of anxiety and fear: Circuits, mechanisms, and neurochemical interactions (Part II). *Neuroscientist* 4, 122–132.

Chiu, S., and Wise, P. M. (1994). Prolactin receptor mRNA localization in the hypothalamus by *in situ* hybridization. *J. Neuroendocrinol.* 6, 191–199.

Clapp, C., Torner, L., Gutierrez-Ospina, G., Alcantara, E., Lopez-Gomez, F. J., Nagano, M., Kelly, P. A., Mejia, S., Morales, M. A., and Martinez de la Escalera, G. (1994). The prolactin gene is expressed in the hypothalamic–neurohypophyseal system and the protein is processed into a 14-kDa fragment with activity like 16-kDa prolactin. *Proc. Natl Acad. Sci. USA* 91, 10384–10388.

Crumeyrolle-Arias, M., Latouche, J., Jammes, H., Djiane, J., Kelly, P. A., Reymond, H. J., and Haour, F. (1993). Prolactin receptors in the rat hypothalamus: Autoradiographic localization and characterization. *Neuroendocrinology* 57, 457–466.

Da Costa, A. P., Wood, S., Ingram, C. D., and Lightman, S. L. (1996). Region-specific reduction in stress-induced c-fos mRNA expression during pregnancy and lactation. *Brain Res.* 742, 177–184.

DeVito, W. J. (1988). Distribution of immunoreactive prolactin in the male and female rat brain: Effects of hypophysectomy and intraventricular administration of colchicine. *Neuroendocrinology* 47, 284–289.

DeVito, W. J. (1989). Immunoreactive prolactin in the hypothalamus and cerebrospinal fluid of male and female rats. *Neuroendocrinology* 50, 182–186.

DeVito, W. J., Stone, S., and Avakian, C. (1991). Stimulation of hypothalamic prolactin release by veratridine and angiotensin II in the female rat: Effect of ovariectomy and estradiol administration. *Neuroendocrinology* 54, 391–398.

DeVito, W. J., Avakian, C., Stone, S., and Ace, C. I. (1992). Estradiol increases prolactin synthesis and prolactin messenger ribonucleic acid in selected brain regions in the hypophysectomized female rat. *Endocrinology* 131, 2154–2160.

Donner, N., Bredewold, R., Maloumby, R., and Neumann, I. D. (2007). Chronic intracerebral prolactin attenuates neuronal stress circuitries in virgin rats. *Eur. J. Neurosci.* 25(6), 1804–1814.

Douglas, A. J., Neumann, I., Meeren, H. K., Leng, G., Johnstone, L. E., Munro, G., and Russell, J. A. (1995). Central endogenous opioid inhibition of supraoptic oxytocin neurons in pregnant rats. *J. Neurosci.* 15, 5049–5057.

Douglas, A. J., Johnstone, H., Wigger, A., Landgraf, R., Russell, J. A., and Neumann, I. D. (1998). The role of endogenous opioids in neurohypophysial and hypothalamic–pituitary–adrenal axis hormone secretory responses to stress in pregnant rats. *J. Endocrinol.* 158, 285–293.

Drago, F., and Amir, S. (1984). Effects of hyperprolactinemia on core temperature of the rat. *Brain Res. Bull.* 12, 355–358.

Drago, F., and Lissandrello, CO. (2000). The "low-dose" concept and the paradoxical effects of prolactin on grooming and sexual behaviour. *Eur. J. Pharm.* 405, 131–137.

Drago, F., Bohus, B., Gispen, W. H., Scapagnini, U., and DeWied, D. (1983). Prolactin-enhanced grooming behavior: Interaction with ACTH. *Brain Res.* 263, 277–282.

Drago, F., Continella, G., Conforto, G., and Scapagnini, U. (1985). Prolactin inhibits the development of stress-induced ulcers in the rat. *Life Sci.* 36, 191–197.

Drago, F., Pulvirenti, L., Spadaro, F., and Pennsini, G. (1990). Effects of TRH and prolactin in the behavioral despair (swim) model of depression in rats. *Psychoneuroendocrinology* 15, 349–356.

Emanuele, N. V., Jurgens, J. K., Halloran, M. M., Tentler, J. J., Lawrence, A. M., and Kelley, M. R. (1992). The rat prolactin gene is expressed in brain tissue: Detection of normal and alternatively spliced prolactin messenger RNA. *Mol. Endocrinol.* 6, 35–42.

Erskine, M., Barfield, R. J., and Goldman, B. D. (1978). Intraspecific fighting during late pregnancy and lactation in rats and effects of litter removal. *Behav. Biol.* 23, 206–213.

Fleming, A. S., Cheung, U. S., Myhal, N., and Kessler, Z. (1989). Effects of maternal hormones on "timidity" and attraction to pup-related odors in females. *Physiol. Behav.* 46, 449–453.

Freeman, M. E., Kanyicska, B., Lerant, A., and Nagy, G. (2000). Prolactin: Structure, function, and regulation of secretion. *Physiol. Rev.* 80, 1523–1631.

Fujikawa, T., Soya, H., Yoshizato, H., Sakaguchi, K., Doh-Ura, K., Tanaka, M., and Nakasima, K. (1995). Restraint stress enhances the gene expression of prolactin receptor long form at the choroid plexus. *Endocrinology* 136, 5608–5613.

Fuxe, K., Hökfelt, T., Eneroth, P., Gustafsson, J. A., and Skett, P. (1977). Prolactin-like immunoreactivity: Localization in nerve terminals of rat hypothalamus. *Science* 196, 899–900.

Goffin, V., Binart, N., Touraine, P., and Kelly, P. A. (2002). Prolactin: The new biology of an old hormone. *Annu. Rev. Physiol.* 64, 47–67.

Ghosh, R., and Sladek, C. D. (1995a). Role of prolactin and gonadal steroids in regulation of oxytocin mRNA during lactation. *Am. J. Physiol.* 269, E76–E84.

Ghosh, R., and Sladek, C. D. (1995b). Prolactin modulates oxytocin mRNA during lactation by its action on the hypothalamo–neurohypophyseal axis. *Brain Res.* 672, 24–28.

Grattan, D. R., Pi, X. J., Andrews, Z. B., Augustine, R. A., Kokay, I. C., Summerfield, M. R., Todd, B., and Bunn, S. J. (2001). Prolactin receptors in the brain during pregnancy and lactation: Implications for behavior. *Horm. Behav.* 40, 115–124.

Grattan, D. R., Jasoni, C. L., Liu, X., Anderson, G. M., and Herbison, A. E. (2007). Prolactin regulation of gonadotropin-releasing hormone neurons to suppress luteinizing hormone secretion in mice. *Endocrinology* 48(9), 4344–4351.

Griffond, B., Colard, C., Deray, A., Fellman, D., and Bugnon, C. (1994). Evidence for the expression of dynorphin gene in the prolactin-immunoreactive neurons of the rat lateral hypothalamus. *Neurosci. Lett.* 165, 89–92.

Groer, M. W., and Davis, M. W. (2006). Cytokines, infections, stress, and dysphoric moods in breastfeeders and formula feeders. *J. Obstet. Gynecol. Neonatal Nurs.* 35(5), 599–607.

Grosvenor, C. E., Shyr, S. W., Goodman, G. T., and Mena, F. (1986). Comparison of plasma profiles of oxytocin and prolactin following suckling in the rat. *Neuroendocrinology* 43, 679–685.

Gu, Y., Soares, M. J., Srivastava, R. K., and Gibori, G. (1994). Expression of decidual prolactin-related protein in the rat decidua. *Endocrinology* 135, 1422–1427.

Hansen, B. L., Hansen, G. N., and Hagen, C. (1982). Immunoreactive material resembling ovine prolactin in perikarya and nerve terminals of the rat hypothalamus. *Cell Tissue Res.* 226, 121–131.

Hard, E., and Hansen, S. (1984). Reduced fearfulness in the lactating rat. *Physiol. Behav.* 35, 641–643.

Harlan, R. E., Shivers, B. D., Fox, S. R., Kaplove, K. A., Schachter, B. S., and Pfaff, D. W. (1989). Distribution and partial characterization of immunoreactive prolactin in the rat brain. *Neuroendocrinology* 49, 7–22.

Heinrichs, M., Meinlschmidt, G., Neumann, I. D., Wagner, S., Kirschbaum, C., Ehlert, U., and Hellhammer, D. H. (2001). Effects of suckling on hypothalamic–pituitary–adrenal axis responses to psychosocial stress in postpartum lactating women. *J. Clin. Endocrinol. Metabol.* 86, 4804–4978.

Horseman, N. D., Zhao, W., Montecino-Rodriguez, E., Tanaka, M., Nakashima, K., Engle, S. J., Smith, F., Markoff, E., and Dorshkind, K. (1997). Defective mammopoiesis, but normal hematopoiesis, in mice with a targeted disruption of the prolactin gene. *EMBO J.* 16, 6926–6935.

Johnstone, H. A., Wigger, A., Douglas, A. J., Neumann, I. D., Landgraf, R., Seckl, J. R., and Russell, J. A. (2000). Attenuation of hypothalamic–pituitary–adrenal axis stress responses in late pregnancy: changes in feedforward and feedback mechanisms. *J. Neuroendocrinol.* 12, 811–822.

Kelly, P. A., Djiane, J., Postel-Vinay, M. C., and Edery, M. (1991). The prolactin/growth hormone receptor family. *Endocr. Rev.* 12, 235–251.

Kokay, I. C., Bull, P. M., Davis, R. L., Ludwig, M., and Grattan, D. R. (2006). Expression of the long form of the prolactin receptor in magnocellular oxytocin neurons is associated with specific prolactin regulation of oxytocin neurons. *Am. J. Physiol. Regul. Integr. Comp. Physiol.* 290, R1216–R1225.

Kooy, A., De Greef, W. J., Vreeburg, J. T., Hackeng, W. H., Ooms, M. P., Lamberts, S. W., and Weber, R. F. (1990). Evidence for the involvement of corticotropin-releasing factor in the inhibition of gonadotropin release induced by hyperprolactinemia. *Neuroendocrinology* 51, 261–266.

Li, C., Kelly, P. A., and Buntin, J. D. (1995). Inhibitory effects of anti-prolactin receptor antibodies on prolactin binding in brain and prolactin-induced feeding behavior in ring doves. *Neuroendocrinology* 61, 125–135.

Lightman, S. L., and Young, W. S., III (1989). Lactation inhibits stress-mediated secretion of corticosterone and oxytocin and hypothalamic accumulation of corticotropin-releasing factor and enkephalin messenger ribonucleic acids. *Endocrinology* 124, 2358–2364.

Lucas, B. K., Ormandy, C. J., Binart, N., Bridges, R. S., and Kelly, P. A. (1998). Null mutation of the prolactin receptor gene produces a defect in maternal behavior. *Endocrinology* 139, 4102–4107.

Mann, P. E., and Bridges, R. S. (2001). Lactogenic hormone regulation of maternal behavior. *Prog. Brain Res.* 133, 251–262.

Mangurian, L. P., Walsh, R. J., and Posner, B. I. (1992). Prolactin enhancement of its own uptake at the choroid plexus. *Endocrinology* 131, 698–702.

Mejia, S., Morales, M. A., Zetina, M. E., Martinez de la Escalera, G., and Clapp, C. (1997). Immunoreactive prolactin forms colocalize with vasopressin in neurons of the hypothalamic paraventricular and supraoptic nuclei. *Neuroendocrinology* 66, 151–159.

Merchenthaler, I., Vigh, S., Petrusz, P., and Schally, A. (1982). Immunocytochemical localization of corticotropin releasing factors (CRF) in the rat brain. *Am. J. Anat.* 165, 385–396.

Meyerhoff, J. L., Oleshansky, M. A., and Mougey, E. H. (1988). Psychologic stress increases plasma levels of prolactin, cortisol, and POMC-derived peptides in man. *Psychosom. Med.* 50, 295–303.

Mezzacappa, E. S., and Katlin, E. S. (2002). Breastfeeding is associated with reduced perceived stress and negative mood in mothers. *Health Psychol.* 21(2), 187–193.

Neill, J. D. (1970). Effects of "stress" on serum prolactin and luteinizing hormone levels during the estrous cycle of the rat. *Endocrinology* 87, 1192–1197.

Neumann, I., Russell, J. A., and Landgraf, R. (1993). Oxytocin and vasopressin release within the supraoptic and paraventricular nuclei of pregnant, parturient and lactating rats: A microdialysis study. *Neuroscience* 53, 65–75.

Neumann, I., Landgraf, R., Takahashi, Y., Pittman, Q. J., and Russell, J. A. (1994). Stimulation of oxytocin release within the supraoptic nucleus and into blood by CCK-8. *Am. J. Physiol.* 267, R1626–R1631.

Neumann, I., Landgraf, R., Bauce, L., and Pittman, Q. J. (1995). Osmotic responsiveness and crosstalk involving oxytocin, but not vasopressin or aminoacids, between the supraoptic nuclei in virgin and lactating rats. *J. Neurosci.* 15, 3408–3417.

Neumann, I. D. (2001). Alterations in behavioural and neuroendocrine stress coping strategies in pregnant, parturient and lactating rats. *Progr. Brain Res.* 133, 143–152.

Neumann, I. D., Johnstone, H. A., Hatzinger, M., Liebsch, G., Shipston, M., Russell, J. A., Landgraf, R., and Douglas, A. J. (1998). Attenuated neuroendocrine responses to emotional and physical stressors in pregnant rats involve adenohypophyseal changes. *J. Physiol.* 508, 289–300.

Neumann, I. D., Torner, L., and Wigger, A. (2000). Brain oxytocin: Differential inhibition of neuroendocrine stress responses and anxiety-related behaviour in virgin, pregnant and lactating rats. *Neuroscience* 95, 567–575.

Neumann, I. D., Toschi, N., Ohl, F., Torner, L., and Krömer, S. A. (2001). Maternal defense as an emotional stressor in female rats: Correlation of neuroendocrine and behavioral parameters and involvement of brain oxytocin. *Eur. J. Neurosci.* 13, 1016–1024.

Neumann, I. D., Bosch, O. J., Toschi, N., Torner, L., and Douglas, A. J. (2003). No stress response of the hypothalamo–pituitary–adrenal axis in parturient rats: Lack of involvement of brain oxytocin. *Endocrinology* 144, 2473–2479.

Noel, G. L., Suh, H. K., Stone, J. G., and Frantz, A. G. (1972). Human prolactin and growth hormone release during surgery and other conditions of stress. *J. Clin. Endocrinol. Metab.* 35, 840–851.

Noel, M. B., and Woodside, B. (1993). Effects of systemic and central prolactin injections on food intake, weight gain, and estrous cyclicity in female rats. *Physiol. Behav.* 54, 151–154.

Paut-Pagano, L., Roky, R., Valatx, J., Kitahama, K., and Jouvet, M. (1993). Anatomical distribution of prolactin-like immunoreactivity in the rat brain. *Neuroendocrinology* 58, 682–695.

Pellow, S., Chopin, P., File, S. E., and Briley, M. (1985). Validation of open:closed arms entries in an elevated plus-maze as a measure of anxiety in the rat. *J. Neurosci. Meth.* 14, 149–167.

Perrot-Applanat, M., Gualillo, O., Pezet, A., Vincent, V., Edery, M., and Kelly, P. A. (1997). Dominant negative and cooperative effects of mutant forms of prolactin receptor. *Mol. Endocrinol.* 11, 1020–1032.

Pi, X. J., and Grattan, D. R. (1998). Differential expression of the two forms of prolactin receptor mRNA within microdissected hypothalamic nuclei of the rat. *Brain Res. Mol. Brain Res.* 59, 1–12.

Pi, X. J., and Grattan, D. R. (1999a). Increased expression of both short and long forms of prolactin receptor mRNA in hypothalamic nuclei of lactating rats. *J. Mol. Endocrinol.* 23, 13–22.

Pi, X. J., and Grattan, D. R. (1999b). Increased prolactin receptor immunoreactivity in the hypothalamus of lactating rats. *J. Neuroendocrinol.* 11, 693–705.

Pi, X. J., and Voogt, J. L. (2001). Mechanisms for suckling-induced changes in expression of prolactin receptor in the hypothalamus of the lactating rat. *Brain Res.* 891, 197–205.

Reavley, S., Fisher, A. D., Owen, D., Creed, F. H., and Davis, J. R. E. (1997). Psychological distress in patients with hyperprolactinemia. *Clin. Endocrinol.* 47, 343–348.

Rocco, A., Mori, F., Baldelli, R., Aversa, A., Munizzi, M. R., Nardone, M. R., Fabbrini, A., and Falaschi, P. (1993). Effect of chronic bromocriptine treatment on psychological profile of patients with PRL-secreting pituitary adenomas. *Psychoneuroendocrinology* 18, 57–66.

Roky, R., Valatx, J. L., Paut-Pagano, L., and Jouvet, M. (1994). Hypothalamic injection of prolactin or ist antibody alters the rat sleep–wake cycle. *Physiol. Behav.* 55, 1015–1019.

Rosenblatt, J. S. (1967). Nonhormonal basis of maternal behaviour in the rat. *Science* 156, 1512–1514.

Schlein, P. A., Zarrow, M. X., and Denenberg, V. H. (1974). The role of prolactin in the depressed or "buffered" adrenocorticosteroid response of the rat. *J. Endocrinol.* 62, 93–99.

Seggie, J. A., and Brown, G. M. (1975). Stress response patterns of plasma corticosterone, prolactin, and growth hormone in the rat, following handling or exposure to novel environment. *Can. J. Physiol. Pharm.* 53, 629–637.

Shiu, R. P., Kelly, P. A., and Friesen, H.G. (1973). Radioreceptor assay for prolactin and other lactogenic hormones. *Science* 180(89), 968–971.

Siaud, P., Manzoni, O., Balmefrezol, M., Barbanel, G., Assenmacher, I., and Alonso, G. (1989). The organization of prolactin-like-immunoreactive neurons in the rat central nervous system. Light- and electron-microscopic immunocytochemical studies. *Cell Tissue Res.* 255, 107–115.

Stern, J. M., Goldman, L., and Levine, S. (1973). Pituitary–adrenal responses during lactation in rats. *Neuroendocrinology* 12, 179–191.

Sugiyama, T., Minoura, H., Kawabe, N., Tanaka, M., and Nakashima, K. (1994). Preferential expression of long form prolactin receptor mRNA in the rat brain during the estrous cycle, pregnancy and lactation: Hormones involved in its gene expression. *J. Endocrinol.* 141, 325–333.

Thienhaus, M. D., and Hartford, J. T. (1986). Depression in hyperprolactinemia. *Psychosomatics* 27, 663–664.

Tonkowicz, P. A., and Voogt, J. L. (1983). Termination of the prolactin surges with development of placental lactogen secretion in the rat. *Endocrinology* 113, 1314–1318.

Torner, L., Mejia, S., Lopez-Gomez, F. J., Quintanar, A., Martinez de la Escalera, G., and Clapp, C. (1995). A 14-kilodalton prolactin-like fragment is secreted by the hypothalamo–neurohypophyseal system of the rat. *Endocrinology* 136, 5454–5460.

Torner, L., Nava, G., Dueñas, Z., Corbacho, A., Mejia, S., Lopez, F., Cajero, M., Martinez de la Escalera, G., and Clapp, C. (1999). Changes in the expression of neurohypophyseal prolactins during the estrous cycle and after estrogen treatment. *J. Endocrinol.* 161, 423–432.

Torner, L., Toschi, N., Pohlinger, A., Landgraf, R., and Neumann, I. D. (2001). Anxiolytic and anti-stress effects of brain prolactin: Improved efficacy of antisense targeting of the prolactin receptor by molecular modeling. *J. Neurosci.* 21, 3207–3214.

Torner, L., Toschi, N., Nava, G., Clapp, C., and Neumann, I. D. (2002). Increased hypothalamic expression in lactation: Involvement in behavioral and neuroendocrine stress responses. *Eur. J. Neurosci.* 15, 1381–1389.

Torner, L., Maloumby, R., Nava, G., Aranda, J., Clapp, C., and Neumann, I. D. (2004). *In vivo* release and gene upregulation of brain prolactin in response to physiological stimuli. *Eur. J. Neurosci.* 19, 1601–1608.

Torner, L., and Neumann, I. D. (2002). The brain prolactin system: involvement in stress response adaptations in lacation. *Stress.* 5(4): 249–257.

Toufexis, D. J., and Walker, C. D. (1996). Noradrenergic facilitation of the adrenocorticotropin response to stress is absent during lactation in the rat. *Brain Res.* 737, 71–77.

Toufexis, D. J., Rochford, J., and Walker, C. D. (1999). Lactation-induced reduction in rats' acoustic startle is associated with changes in noradrenergic neurotransmission. *Behav. Neurosci.* 113, 176–184.

Voci, V. E., and Carlson, N. R. (1973). Enhancement of maternal behaviour and nest building following systemic and diencephalic administration of prolactin and progesterone in the mouse. *J. Comp. Physiol. Psychol.* 88, 388–393.

Walker, C. D., Trottier, G., Rochford, J., and Lavallée, D. (1995). Dissociation between behavioral and hormonal responses to the forced swim stress in lactating rats. *J. Neuroendocrinol.* 7, 615–622.

Walsh, R. J., Slaby, F. J., and Posner, B. I. (1987). A receptor-mediated mechanism for the transport of prolactin from blood to cerebrospinal fluid. *Endocrinology* 120, 1846–1850.

Windle, R. J., Wood, S., Shanks, N., Perks, P., Conde, G. L., da Costa, A. P. C., Ingram, C. D., and Lightman, S. L. (1997). Endocrine and behavioural responses to noise stress: Comparison of virgin and lactating female rats during non-disrupted maternal activity. *J. Neuroendocrinol.* 9, 407–414.

Yamaguchi, M., Ogren, L., Endo, H., Soares, M. J., and Talamantes, F. (1994). Co-localization of placental lactogen-I, placental lactogen-II, and proliferin in the mouse placenta at midpregnancy. *Biol. Reprod.* 51, 1188–1192.

10

MOTHER–INFANT TOUCH, NEUROCHEMISTRY, AND POSTPARTUM ANXIETY

JOSEPH S. LONSTEIN[1,2] AND STEPHANIE M. MILLER[2]

[1]*Neuroscience Program, Giltner Hall, Michigan State University, East Lansing, MI 48824, USA*

[2]*Department of Psychology, Giltner Hall, Michigan State University, East Lansing, MI 48824, USA*

Infants provide their mothers with a rich array of sensory experiences that uniquely influence maternal physiology and behavior. The continual physical contact between mothers and their infants is particularly salient and has widespread effects. Indeed, somatosensory inputs provided by neonates are required for maintaining lactation (Wakerley *et al.*, 1994), recalibrating food intake and energy balance (Johnstone & Higuchi, 2001; Lederman, 2004; Woodside, this volume), suspending ovarian cyclity (McNeilly, 2001; Smith & Grove, 2002), suppressing hypothalamic–pituitary–adrenal (HPA) responses (Heinrichs *et al.*, 2001, 2002; Neumann, 2001; Tu *et al.*, 2005), as well as establishing and maintaining maternal behavior and aggression (Fleming *et al.*, 1996; Stern, 1996; Lonstein & Gammie, 2002; Fleming, this volume).

Somatosensory cues from infants are also responsible for the remarkable changes in emotional state occurring after females give birth. Compared to non-mothers, Postpartum laboratory rodents display reduced anxiety-related behaviors in many paradigms (for review, see Lonstein, 2007). This has been suggested to be necessary for the onset of maternal behavior despite the novel and anxiety-generating sensory cues that neonates emit (Fleming & Luebke, 1981), and for dams' heightened aggression toward threatening and potentially infanticidal intruders (Hard & Hansen, 1985). Although these explanations are certainly reasonable and attractive, they are complicated by data demonstrating that anxiety is actually highest at the end of pregnancy when maternal behavior is emerging (Rosenblatt & Siegel, 1975; Neumann *et al.*, 1998), and that anxiety in dams is sometimes positively correlated with their maternal aggression (Bosch *et al.*, 2005). It may be that, rather than being outright permissive for her nurturing behaviors, a mother's level of anxiety within a normal range has more subtle influences on her maternal style and, consequently, offspring development (Champagne & Meaney, 2001; Champagne & Curley, 2005).

Most human mothers also experience a reduction in anxiety, which is part of a well-documented general improvement in mood during the postpartum period compared to before or during pregnancy (Kumar & Robson, 1984; Kendell *et al.*, 1987; Cowley & Roy-Byrne 1989; Engle *et al.*, 1990; Fleming *et al.*, 1990; Altshuler *et al.*, 2000; Sjogren *et al.*, 2000; Breitkopf *et al.*, 2006). The benefits of low maternal anxiety are numerous, and include reduced infant abuse (Whipple & Webster-Stratton, 1991; Nayak & Milner, 1998; De Bellis *et al.*, 2001), increased likelihood of breastfeeding (Sjolin *et al.*, 1977; Barnett & Parker, 1986; Galler *et al.*, 1999; Clifford *et al.*, 2006; Dennis, 2006; Forster *et al.*, 2006), enhanced physical (Barnett & Parker, 1986; O'Brien *et al.*, 2004) and cognitive (Cogill *et al.*, 1986; Galler *et al.*, 2000) development of infants, and enhanced mother infant emotional attachment (Barnett & Parker,

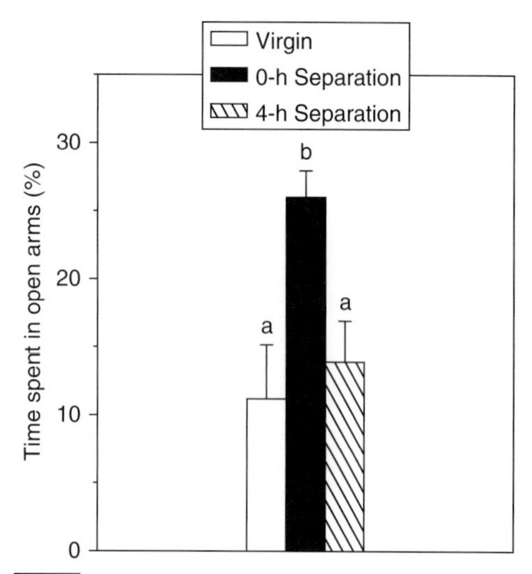

FIGURE 10.1 Effects of removing the litter for 0 or 4 hours before testing on the percentage of time dams spent in the open arms of an elevated plus-maze on day 7 postpartum. Diestrous virgins were included as a high-anxiety control group. Different letters above bars indicate significant differences between groups, $p < 0.05$. (*Source*: Figure modified from Lonstein (2005)).

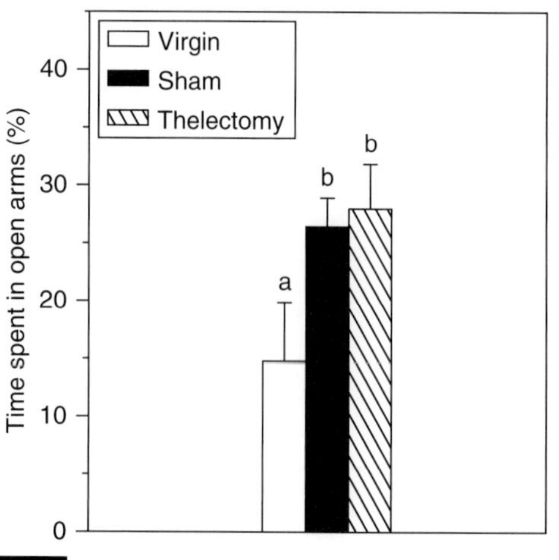

FIGURE 10.2 Effects of pre-mating thelectomy on the later percentage of time dams spent in the open arms of an elevated plus-maze on day 7 postpartum. Diestrous virgins were included as a high-anxiety control group. Different letters above bars indicate significant differences between groups, $p < 0.05$. (*Source*: Figure modified from Lonstein (2005)).

1986; Manassis *et al.*, 1994; Turner *et al.*, 2002; Woodruff-Borden *et al.*, 2002; Adam *et al.*, 2004; Moore *et al.*, 2004; Zelkowitz & Papageorgiou, 2005).

In both rodents and humans, recent somatosensory inputs from infants are critical for reduced maternal anxiety. Removing most or all of the pups for as little as 2–4 h causes anxiety-related behaviors in lactating rats to revert back to that found in virgins (Neumann, 2003; Lonstein, 2005; Figure 10.1). Furthermore, placing the pups in a wire mesh cage through which dams could see, smell, and hear the pups, but not touch them, does not maintain mothers' low anxiety (Lonstein, 2005). This anxiolytic effect of direct physical contact with infants is not exclusive to laboratory rats. Women with recent physical contact with their infants also show lower anxiety than postpartum women without such contact (Heinrichs *et al.*, 2001).

The specific somatosensory cues from infants that produce anxiolytic effects in their mothers do not involve suckling in rats, and may not involve suckling in humans. Surgical removal of the teats in rats does not prevent physical contact with infants from reducing anxiety (Lonstein, 2005; Figure 10.2), similar to the lack of effects of nipple removal on the overall display of maternal behavior or aggression in dams who can otherwise touch their infants (Mayer *et al.*, 1987; Stern & Kolunie, 1993). In women, we believe it is unclear if suckling is absolutely necessary for the anxiolytic effect of infant contact. Women who recently breastfed their infants have been reported to be less anxious, but anxiety is similarly low in women who simply held their infants on their laps (Heinrichs *et al.*, 2001). This can be contrasted to the requirement for recent suckling for suppressed HPA responsiveness in women (Heinrichs *et al.*, 2001; Tu *et al.*, 2005). The literature suggesting that breastfeeding itself is critical for mothers' reduced anxiety is fraught with complicating methodological factors, as was recently discussed elsewhere (Lonstein, 2007). Further support that even non-breastfeeding contact with infants is adequate for reduced anxiety is obtained from data showing that the direction of contact does not need to be from infant to mother, let alone very close contact and suckling, because mother-initiated, infant-directed contact in the form of maternal massage of the infant can also reduce mother's anxiety (Feijo *et al.*, 2006) and produce other improvements in maternal mood (Fujita *et al.*, 2006).

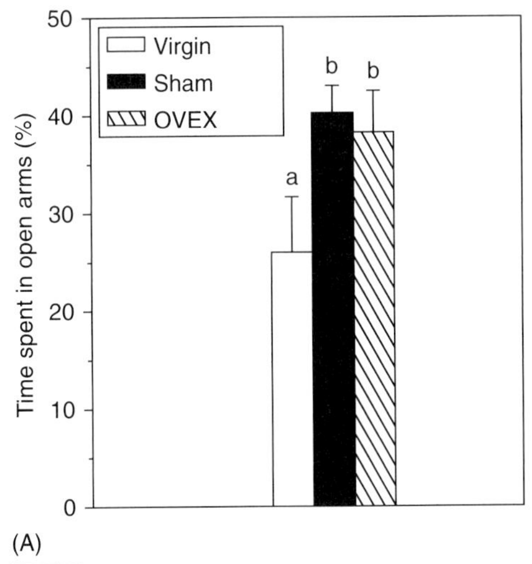

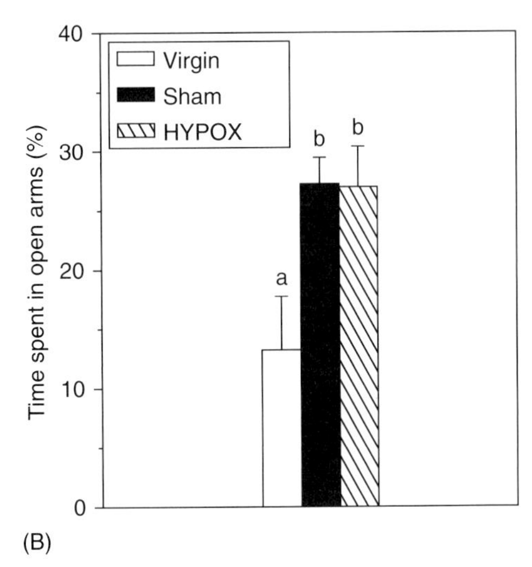

FIGURE 10.3 Effects of postpartum ovariectomy (A), or prepartum hypophysectomy (B), on the percentage of time dams spent in the open arms of an elevated plus-maze on day 7 postpartum. Diestrous virgins were included as a high-anxiety control group. Different letters above bars indicate significant differences between groups, $p < 0.05$. (*Source*: Figure modified from Lonstein (2005)).

The neurochemicals released in response to infant touch to produce this anxiolytic effect on mothers are numerous. Given the dramatic fluctuations in hormones across pregnancy, parturition, and lactation (Bridges, 1996; Numan & Insel, 2003), it is conceivable that mothers' reduced anxiety would be a consequence of their endocrine state. Nonetheless, work from our laboratory and others has demonstrated that neither ovariectomy, adrenalectomy, nor hypophysectomy prevent mother rats from displaying lower anxiety-related behaviors compared to non-mothers (Hansen, 1990; Lonstein, 2005; Figure 10.3). Furthermore, ovariectomized virgin female rats induced to become maternal through repeated exposure to neonates (i.e., maternally sensitized) show somewhat reduced anxiety (Ferreira *et al.*, 2002; Pereira *et al.*, 2005). These data suggest that neurochemicals other than those released peripherally that can then cross into the brain are largely responsible for reducing anxiety in female rats. This does not mean that hormones and peptides from peripheral origin have no role, but instead of directly mediating the reduction in anxiety, they establish a state of high maternal motivation (see Lonstein & Morrell, 2006; Morrell, this volume) that leads to infant-contacting behaviors. These behaviors permit mothers' receipt of somatosensory inputs from neonates, which then more directly influence mothers' anxiety. Our laboratory has recently been focusing on intracerebral activity of three neurochemicals – oxytocin, GABA, and norepinephrine – as potential substrates for how somatosensory inputs from pups reduce anxiety in their mothers.

OXYTOCIN

The neuropeptide, oxytocin (OT), has received a great deal of attention for modulating anxiety in mammals. High central OTergic activity reduces anxiety in laboratory rats and mice (McCarthy *et al.*, 1996; Windle *et al.*, 1997a, b, 2006; Bale *et al.*, 2001; Ring *et al.*, 2006), while animals without the ability to synthesize OT show elevated anxiety (Mantella *et al.*, 2003; Amico *et al.*, 2004). OT is also anxiolytic in humans, including after intranasal infusion of OT in men (Heinrichs *et al.*, 2003).

Central OTergic activity is upregulated during early lactation when mothers interact with infants. For example, increased OT release is found in the preoptic area and bed nucleus of the stria terminalis (BST) while mothers interact with infants (Kendrick *et al.*, 1992, 1986; Neumann *et al.*, 1993). This could influence anxiety, but may instead be more related to maternal behavior because such increases are not observed in the other sites implicated in anxiety, such as the paraventricular hypothalamus, septum, or amygdala (Bosch *et al.*, 2004). At the level of the OT receptor, there is transient upregulation at or very soon after parturition in the ventromedial hypothalamus, lateral septum, central and medial amygdala, and dorsal BST (Insel, 1986, 1990; Young *et al.*, 1997; Meddle *et al.*, 2007), but OT receptor expression thereafter does not differ between lactating and virgin rats in these or other sites traditionally involved in generating anxiety (Insel, 1986, 1990; Young *et al.*, 1997; Francis *et al.*, 2000). Furthermore, it is unclear if parturition and interactions with pups can modulate Fos expression in OTergic cells of the parvocellular PVN (Fenelon *et al.*, 1993; Meddle *et al.*, 2007), which might be expected to be the source of intracerebral OTergic that reduces anxiety. Nonetheless, increased OTergic activity contributes to low anxiety in mother rats. Neumann *et al.* (2000) demonstrated that intracerebroventricular infusion of a highly specific OT receptor antagonist decreases the percentage of time spent in the open arms of an elevated plus-maze, indicating increased anxiety. In contrast, OT antagonism has no effects on the plus-maze behavior of virgin females (Neumann *et al.*, 2000).

Areas of the brain where OT receptor activity produces anxiolytic effects in lactating rats are probably numerous, but our laboratory has demonstrated that the midbrain periaqueductal gray (PAG) is one influential site. The PAG is necessary for a myriad of behavioral and physiological processes (Depaulis & Bandler, 1991), including acting as a supraspinal "final common pathway" for anxiety- and fear-related behaviors in animals (Brandao *et al.*, 2003). Electrical stimulation of the PAG elicits feelings of anxiety and panic in rodents and humans (Nashold *et al.*, 1969; Jenck *et al.*, 1989), and functional MRI reveals that the PAG is activated when people feel anxious (Dunckley *et al.*, 2005). Conversely, lesions of the PAG are anxiolytic (LeDoux *et al.*, 1988; Kim *et al.*, 1993; Vianna *et al.*, 2001), and manipulation of many neurochemical systems in the PAG reduce or increase anxiety (Menard & Treit, 1999; Brandao *et al.*, 2003). Additionally, its cells respond to exogenous OT (Ogawa *et al.*, 1992), and many regions of the PAG contain OTergic terminals (Buijs *et al.*, 1983) and express OT receptors (Yoshimura *et al.*, 1993).

In lactating female rats, the ventrocaudal PAG (cPAGv) is particularly sensitive to somatosensory inputs from pups (Lonstein & Stern, 1997a, b). This sensitivity is relevant for dams' ability to nurse the pups in the upright crouched (i.e., kyphotic) posture. It is also relevant for dams' anxiety-related behavior because lesioning the cPAGv produces a further decrease in their anxiety (Lonstein *et al.*, 1998). The ability of suppressed activity in the cPAGv to further reduce anxiety might reflect the normal state of this site when mothers interact with pups, possibly the result of increased OTergic neurotransmission. We have found that cPAGv infusion of the same highly specific OT receptor antagonist used by Neumann *et al.*, (2000) increased anxiety in dams to the levels found in virgin females, while OT receptor antagonism in the cPAGv had no effects in virgin females (Figuera *et al.*, in preparation; Figure 10.4). Conversely, in mothers that are separated from their pups before testing (which increases anxiety), infusion of 2 ng of OT into each hemisphere of the cPAGv reduced anxiety back to low levels. This effect was also specific to the postpartum state, because OT infused into the cPAGv of virgins did not reduce anxiety (Figueira *et al.*, in press; Figure 10.4).

Pups regularly root and probe in their dam's ventrum in search of warmth and the opportunity to feed. Even in the absence of subsequent suckling, this stimulation has neural and neuroendocrine consequences for the dam (e.g., Stern & Siegel, 1978; Lonstein & Stern, 1997b), perhaps including OT release in the cPAGv. Interestingly, any effect of gentle touch on OT content in the PAG is not exclusive to the mother–infant dyad, as repeated stroking of the vental skin (but not dorsal skin) of male rats increases OT content in their PAG (Lund *et al.*, 2002), and may eventually be found to reduce their anxiety. The source of OTergic input to the cPAGv has not yet been determined with retrograde tracing, but one would expect that it arises from the parvocellular PVN. However, cell-body lesions encompassing much of the parvocellular PVN affect neither exploration of or defecation in a novel chamber, nor freezing in response to an acoustic startle stimulus (Olazabal & Ferreira, 1997). This suggests that cells outside the PVN are actually the source of OTergic input to the cPAGv (Jirikowski *et al.*, 1989), but the location of such cells remains to be determined.

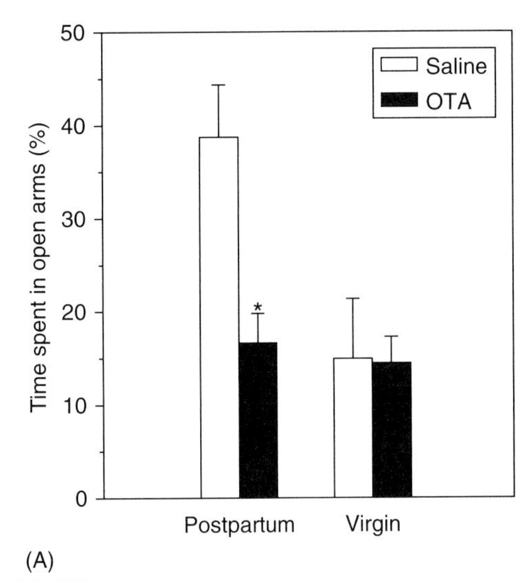

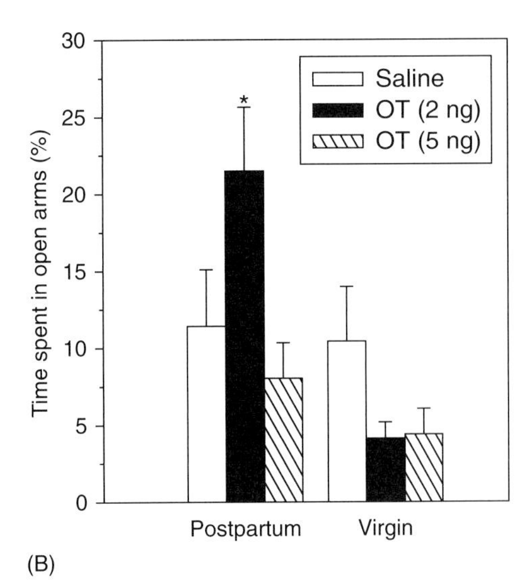

FIGURE 10.4 Effects of cPAGv infusion of 125 ng/hemisphere of an OT receptor antagonist (A) or 2 or 5 ng of OT (B) on the time dams and diestrus virgins spent in the open arms of an elevated plus-maze on day 7 postpartum. Dams in the OT infusion experiment were separated from pups 4 h before testing to increase anxiety, which was significantly reduced by 2 ng OT. Significant main effects were found for reproductive state, and significant interactions with infusion type, but are not indicated. *Significantly different from saline controls within the same reproductive state, $p < 0.05$. (*Source*: Figure modified from Lonstein (2007)).

GABA

The inhibitory neurotransmitter, GABA, has long been studied for its involvement in the etiology and treatment of anxiety disorders in humans (Nemeroff, 2003a, b; Roy-Byrne, 2005) and for coordinating anxiety-related responses in animals (Miczek *et al.*, 1995; Blanchard *et al.*, 2003; Millan, 2003; Korff & Harvey, 2006). $GABA_A$ receptors, as opposed to $GABA_B$ receptors, are particularly relevant for the pathophysiology of anxiety (Smith, 2001; Atack, 2003; Lydiard, 2003; Rudolph & Möhler, 2006; Whiting, 2006), and are widespread throughout the central nervous system. They are concentrated in many regions implicated in emotional regulation, including the amygdala, BST, hippocampus, and medial prefrontal cortex (Fénelon & Herbison, 1996; Liu & Glowa, 1999; Vermetten & Bremner, 2002a, b; Shah *et al.*, 2004; Roy-Byrne, 2005; Nelvokov *et al.*, 2006). In fact, reduced expression or low sensitivity of central $GABA_A$ receptors is related to pathological anxiety in humans (Cowley *et al.*, 1993, 1995; Tiihonen *et al.*, 1997; Malizia *et al.*, 1998; Bremner *et al.*, 2000a, b; Goddard *et al.*, 2001) and high anxiety-related behaviors in laboratory animals (Rägo *et al.*, 1991; Concas *et al.*, 1993).

In laboratory rodents, $GABA_A$ receptor agonists are anxiolytic in many behavioral paradigms, including the elevated plus- and T-mazes, light–dark box, Vogel conflict test, and open field (Yasumatsu *et al.*, 1994; Nazar *et al.*, 1997; Sienkiewicz-Jarosz *et al.*, 2003; Bueno *et al.*, 2005; Lippa *et al.*, 2005;). Conversely, peripheral injections of $GABA_A$ receptor antagonists increase anxiety and Fos expression in many regions of the rat brain, including the amygdala, hippocampus, BST, and medial prefrontal cortex (Kurumaji *et al.*, 2003; Singewald *et al.*, 2003; Salchner *et al.*, 2006), suggesting that anxiety behavior could be related to changes in GABA activity in these sites. Specific neural sites where $GABA_A$ agonists and benzodiazepines decrease anxiety, and $GABA_A$ receptor inhibition increases anxiety, include the amygdala, hippocampus, medial prefrontal cortex, and PAG (Shibata *et al.*, 1989; Millan, 2003; Shah & Treit, 2004; Shah *et al.*, 2004; Bueno *et al.*, 2005; Rezayat *et al.*, 2005; Roy-Byrne, 2005; Akirav *et al.*, 2006).

As might be predicted, increased GABAergic neurotransmission is, in part, responsible for

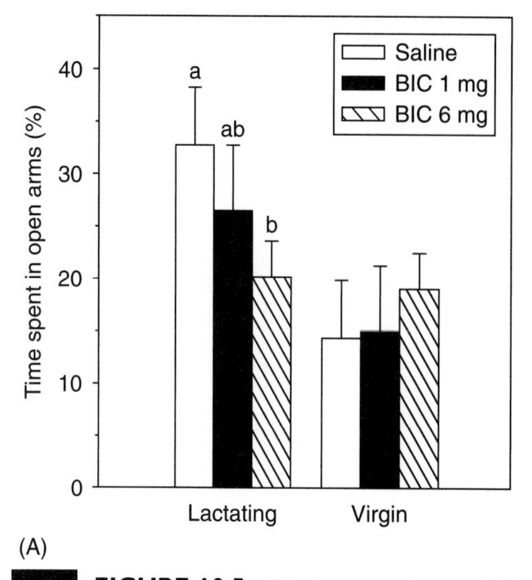

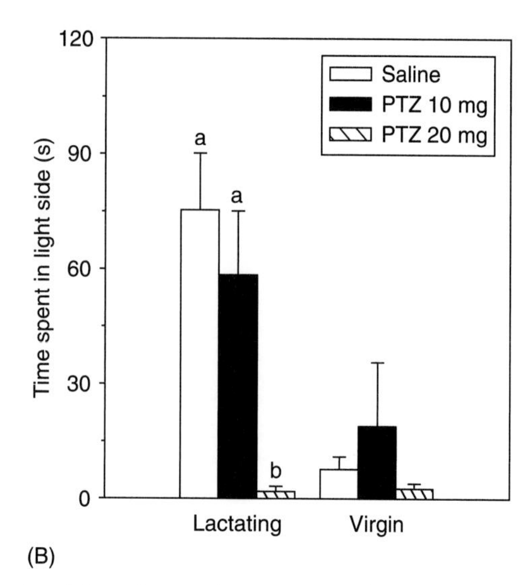

FIGURE 10.5 Effects of peripheral administration of bicuculline on the percentage of time spent in the open arms of an elevated plus-maze (A), and effects of PTZ on time spent in the light side of a light–dark box (B). Different letters above bars indicate significant differences between groups within reproductive state, $p < 0.05$. Significant main effect of reproductive state and interactions with drugs found, but not indicated.

the postpartum reduction in anxiety-related behaviors. Work from the Hansen laboratory demonstrated that lactating rats interacting with pups have significantly higher cerebrospinal fluid (CSF) levels of GABA compared to dams that had been without their pups for 6 h (Qureshi *et al.*, 1987). This drop in CSF levels of GABA could be reversed if mothers were reunited with infants (Qureshi *et al.*, 1987), demonstrating that dams' GABAergic tone depends on relatively recent sensory cues from pups. Further evidence indicating GABA's involvement in this decreased postpartum anxiety is that virgin female rats treated with benzodiazepines show as little freezing in response to a noise burst as do lactating rats, and that peripheral administration of agents that inhibit the $GABA_A$ receptor (FG-7142, pentylenetetrazol, and caffeine) potentiate mother rats' freezing in this paradigm (Hansen *et al.*, 1985). Additional support for an infant- and GABA-mediated decrease in anxiety is that punished drinking is greater in lactating vs. virgin rats, that this difference is significantly magnified when dams have their pups present during testing (Ferreira *et al.*, 1989), and that systemic injection of the $GABA_A$ antagonist pentylenetetrazol decreases dams' punished drinking (Hansen, 1990).

Our laboratory has recently expanded upon these results, and demonstrated that peripheral injection of the $GABA_A$ receptor antagonists bicuculline or pentylenetetrazol, or the benzodiazepine inverse agonist FG-7142, increase dams' anxiety-related behaviors in two widely used and pharmacologically reliable paradigms, the elevated plus-maze and the light–dark box (Miller & Lonstein, 2007). As can be seen in Figure 10.5, bicuculline (6 mg/kg) significantly decreases the percentage of time dams spend in the open arms of an elevated plus-maze, and pentylenetetrazol (20 mg/kg) decreases the duration dams spend in the white side of a light–dark box. These studies support the hypothesis that dams' decreased anxiety in response to infant touch is at least partially due to increased GABA neurotransmission. They do not identify brain regions involved in this phenomenon, though. Identifying such sites could be difficult, because compared to the neural regulation of pain (Hunt & Koltzenburg, 2005), the neurobiological and behavioral consequences of positive, non-noxious touch (not involving suckling) are not well studied. Infant-mediated increases in maternal GABA activity probably occur in many hypothalamic, limbic, or hindbrain sites traditionally implicated in anxiety, and these sites are strongly modulated by projections from somatosensory regions of the cortex and thalamus (e.g., Barone *et al.*, 1994; Shi &

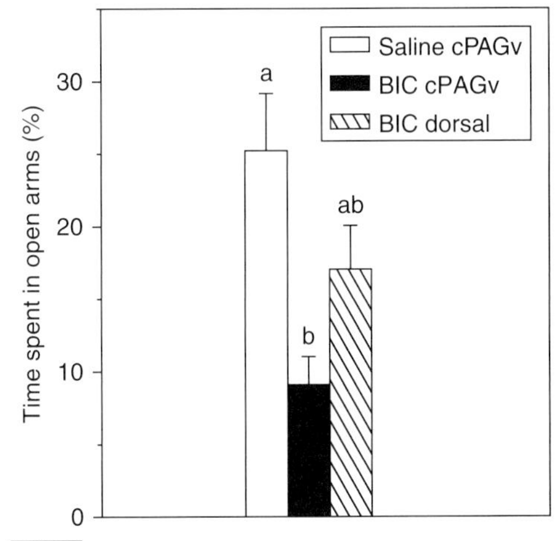

FIGURE 10.6 Effects of cPAGv infusion of bicuculline (2.5 ng/hemisphere) into the cPAGv, or into "missed" locations dorsal to the cPAGv, on the percentage of time spent in the open arms of an elevated plus-maze on day 7 postpartum. Different letters above bars indicate significant differences between groups, $p < 0.05$. (*Source*: Figure modified from Lonstein (2007)).

Cassell, 1998; Veinante & Freund-Mercier, 1998; Gauriau & Bernard, 2004; Gabbott *et al.*, 2005; Vertes *et al.*, 2006). Even so, infusing the $GABA_A$ antagonist bicuculline into the central, lateral, and basolateral amygdaloid nuclei does not significantly affect dams' freezing in response to a sudden acoustic stimulus, indicating that, while the amygdala is important in other models of emotional regulation, it might not be crucial for GABAergic modulation of postpartum responses to fearful or anxiogenic stimuli (Hansen & Ferreira, 1986). Furthermore, there is also no effect of infusing bicuculline into the ventromedial hypothalamus on dams' freezing in response to an acoustic burst (Hansen & Ferreira, 1986).

Our laboratory has instead recently focused on the PAG as an important site of GABAergic action for postpartum anxiety-related behavior. As noted above, the PAG is necessary not only for anxiety-related behavior in other rodent models, but also specifically in postpartum rats, as electrolytic lesions of this region increase dams' anxiety behavior as measured in an elevated plus-maze (Lonstein *et al.*, 1998). In addition, as mentioned above, contact with pups appears to be necessary for reduced anxiety in postpartum rats (Neumann, 2003; Lonstein, 2005). Physical contact with pups with or without suckling increases *c-fos* expression in the cPAGv (Lonstein & Stern, 1997a, b), perhaps indicating that the physical contact from the pups is modulating PAG activity in a manner that affects anxiety, which we believe occurs through GABAergic action in the PAG. In support, our laboratory has found that infusing a $GABA_A$ receptor antagonist into the cPAGv of lactating rats increases their anxiety-related behavior in an elevated plus-maze (Figure 10.6; Miller *et al.*, in preparation).

NOREPINEPHRINE

Central and peripheral noradrenergic activity is positively correlated with anxiety in humans (Post *et al.*, 1978; Ko *et al.*, 1983; Sevy *et al.*, 1989; Lepola *et al.*, 1990; Leckman *et al.*, 1995; Geracioti *et al.*, 2001), and in non-lactating laboratory rats (Tanaka *et al.*, 2000; Neophytou *et al.*, 2001; Dazzi *et al.*, 2002; Fendt *et al.*, 2005; Debiec & LeDoux, 2006). Levels of NE in the CSF of lactating women or laboratory rats have not been examined, but it remains possible that a global suppression of noradrenergic activity during lactation helps maintain a low-anxiety state. It is known that NE release specifically in the PVN drops when dams are with their pups, and increases when pups are removed (Toufexis *et al.*, 1998). Expression of various noradrenergic receptors in the PVN is also modified during lactation (Toufexis *et al.*, 1998), further reducing PVN responsiveness to noradrenergic stimulation (Windle *et al.*, 1997b). This seems inconsistent with the finding that Fos is expressed in TH-immunoreactive cells of the A1 and A6 groups in the brainstem after dams have physical contact with pups (Li *et al.*, 1999), given that the A1 group provides substantial noradrenergic input to the PVN (Swanson & Sawchenko, 1983; Cunningham *et al.*, 1990; Pacak *et al.*, 1993). This Fos expression is observed after dams are separated from pups for 48 h and then reunited for 90 min (Li *et al.*, 1999), however, and may reflect an increase in NE release specifically when dams are separated and then reunited with pups, rather than what occurs during undisturbed interactions.

These changes in noradrenergic activity while mothers interact with pups influence dams' reflexive startle in response to an acoustic burst, a reflex that is modified by emotional state (Koch, 1999; Toufexis *et al.*, 1999). Site-specific injections of noradrenergic drugs are necessary to determine if NE in the PVN or elsewhere in the brain affects dams' startle responding or other emotional behaviors. However, NE turnover in the caudal PVN, which has greater projections throughout the nervous system compared to the rostral PVN, is not increased by suckling (Sawchenko & Swanson, 1982). Furthermore, most parvocellular neurons in the PVN are not excited by NE (Daftary *et al.*, 2000). It is unclear how NE release in the PVN during interactions with pups might directly influence anxiety, and it may instead be more involved in other PVN-mediated effects, such as reduced physiological stress responsiveness when dams are not in the presence of their pups (Lightman *et al.*, 2001; Walker *et al.*, 2001; Deschamps *et al.*, 2003).

We are currently examining the hypothesis that, instead of acting in the hypothalamus, infant contact reduces NE release in the ventral BST (vBST; including parts of the fusiform, dorsomedial, subcommissural, anterolateral, and anteroventral nuclei; Swanson, 1998) to alleviate mothers' anxiety. The BST is involved in the ability to "cope" with potentially anxiogenic stimuli. Large BST lesions exacerbate the effects of stress on restraint stress-induced stomach ulcers and cause male rats to be more behaviorally responsive to fear-associated cues (Henke, 1984). In a conditioned fear inhibition paradigm, presentation of the stimulus that indicates safety and the absence of electric shock increases Fos expression in only a few neural sites – including the vBST (Campeau *et al.*, 1997). The vBST is also activated during anxiogenic social threats in primates (Kalin *et al.*, 2005). The vBST has one of the densest noradrenergic innervations of the rat brain (Woulfe *et al.*, 1990; Fendt *et al.*, 2005), which arises from the A1 and A2 groups in the brainstem (Riche *et al.*, 1990; Woulfe *et al.*, 1990; Roder & Ciriello, 1994). Conditioned or unconditioned anxiogenic stimuli increase NE release in the vBST (Onaka & Yagi, 1998; Fendt *et al.*, 2005), and pharmacological reduction of noradrenergic activity in the vBST reduces anxiety or fear responses (Onaka & Yagi, 1998; Fendt *et al.*, 2005).

Recent data from our lab implicates the vBST in how infant contact modulates anxiety in lactating rats. As discussed above, separating mothers from their infants increases dams' anxiety to levels found in diestrous virgins (Lonstein, 2005). We examined Fos expression in the brains of separated and non-separated dams that were placed in an elevated plus-maze, gently handled, or simply left in their home cages. We found that, although the plus-maze behavior differed between the groups, only a small number of neural sites showed differences in Fos expression between separated and non-separated dams. The greatest difference between them was in the vBST, with greater Fos expression in less-anxious dams that had their pups before exposure to the plus-maze (Smith & Lonstein, in press; Figure 10.7). Control dams with or without pups that were not placed in the plus-maze had no difference in Fos in the vBST, indicating that the display of maternal behavior before testing did not generate this increase in Fos (Numan & Numan, 1994, 1995), and that it was instead in response to the elevated plus-maze.

It was somewhat unexpected that dams with pups before testing, and therefore lower

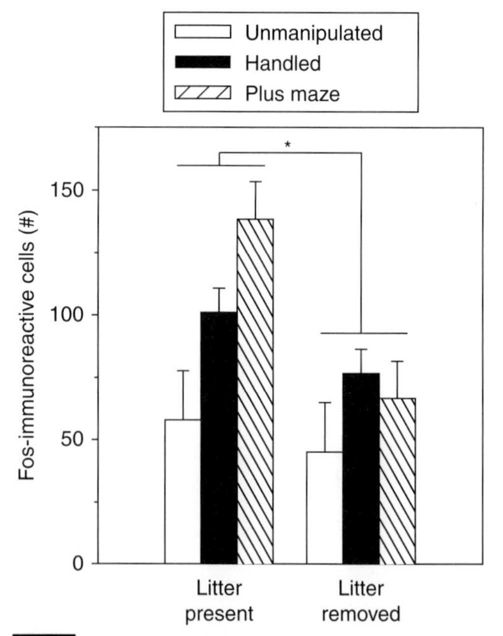

FIGURE 10.7 Number of Fos-immunoreactive cells in the vBST of lactating female rats that were unhandled, handled, or exposed to an elevated plus-maze after having their litters removed or not 4 h before testing. *Significant main effect of litter presence on Fos expression, $p < 0.05$. (*Source*: Figure modified from Lonstein (2007)).

anxiety, had greater Fos in the vBST than more anxious dams. NE in the vBST inhibits glutamate release and cellular activity (Casada & Dafny, 1993; Forray *et al.*, 1999; Egli *et al.*, 2005), however, so removing this inhibition could increase neural activity and Fos expression in the dams' vBST. Recent contact with pups may decrease the amount of NE release in the vBST when dams are exposed to anxiogenic stimuli. Resultant increases in glutamate release and vBST activity could readily inhibit anxiety, as many vBST neurons are GABAergic (Muganini & Oertel, 1985; Stefanova *et al.*, 1997). Activation of inhibitory interneurons, or inhibitory projection neurons from the vBST to many areas of the brain, could suppress anxiety in lactating rats. It may be that the vBST is a primary source of GABAergic input to the cPAGv, and when mothers have been in recent contact with infants, this projection is disinhibited in the face of anxiogenic stimuli to suppress their anxiety-related behaviors. We are currently examining these possibilities, and already have preliminary data indicating that increasing noradrenergic tone with a peripheral injection of yohimbine increases dams' anxiety as measured in an elevated plus-maze to levels found in diestrus virgins (Smith & Lonstein, unpublished data).

CONCLUSIONS

Somatosensory inputs that mothers and infants provide each other are critical regulators of their physiology, psychological state, and behavior. How the early postpartum timing, type, and frequency of infant touch affect lactation, HPA function, maternal bonding, and maternal behaviors in human and non-human mothers has been extensively studied (e.g., Anisfeld *et al.*, 1990; Stern, 1996, 1997; Heinrichs *et al.*, 2002; Charpak *et al.*, 2005; Tu *et al.*, 2005; Britton *et al.*, 2006; Gonzalez-Mariscal, 2007; Moore & Anderson, 2007; Fleming, this volume; Levy, this volume), but how infant touch influences maternal emotional state has been somewhat neglected. Considering that a positive maternal emotional state is requisite for adequate maternal motivation and subsequent care of infants, this is a topic of tremendous importance.

Infants' ability to utilize somatosensation as a means of regulating their mother's emotional state is only one context in which touch contributes to positive mood. Not surprisingly, parturient women also benefit from positive touch coming from another adult (Imura *et al.*, 2006), and the anxiolytic benefits of massage have been reported for children (Hernandez-Reif *et al.*, 1999; Field, 2005; Beider & Moyer, 2007), adolescents (Field *et al.*, 1992, 1998; Diego *et al.*, 2002), and the elderly (Sharpe *et al.*, 2007). Massage reduces anxiety in cardiac (Anderson & Cutshall, 2007) and cancer (Cassileth & Vickers, 2004; Wilkinson *et al.*, 2007) patients, as well as for people with dementia (Viggo-Hansen *et al.*, 2006) and other psychological disorders (Field, 2002; Elkins *et al.*, 2005). Furthermore, the benefits of positive physical contact are not only psychological, but also physiological (Kurosawa *et al.*, 1995; Hernandez-Reif *et al.*, 2004; Field *et al.*, 2005).

Given the importance of infant touch on maternal emotional state, it is possible that natural differences in sensitivity to somatosensory inputs contribute to healthy or unhealthy emotional states in women. Sensitivity of the skin on and surrounding the nipples and breasts, and the somatosensory cortex representation of this region, increases after giving birth (Robinson & Short, 1977; Xerri *et al.*, 1994) and declines as lactation progresses (Drife & Baynham, 1988). The surface area of the areola also increases within days after giving birth (Schaal *et al.*, 2006). These changes in sensitivity are probably a consequence of fluctuating ovarian hormones affecting sensory receptor density and responsiveness (Komisaruk *et al.*, 1972; Bereiter & Barker, 1980; Blacklock & Smith, 2004), and may be specific to this region of the body, because such changes do not occur in other areas, including the very sensitive fingertips (Whipple *et al.*, 1990). In animals, somatosensory thresholds are also altered by parturition and contact with infants, although the direction of change reported has been inconsistent (Gintzler, 1980; Rushen *et al.*, 1993; Cruz *et al.*, 1996). Women that are hyposensitive to suckling and other somatosensory cues from their infants may be somewhat immune to the potentially positive effects of these cues on anxiety. Conversely, women that are hypersensitive may even find such inputs aversive or painful. A woman's level of anxiety might eventually be found to be a cause, effect, or both, of her differential sensitivity to infant tactile cues (Kopp & Gruzelier, 1989; Wilhelm *et al.*, 2001;

Neugebauer *et al.*, 2004; Tang & Gibson, 2005; Fishbain *et al.*, 2006; Borsook *et al.*, 2007). Investigation into these possibilities would greatly contribute to the understanding of how infants regulate maternal emotional state, as well as other processes influenced by somatosensory communication between infants and their mothers, such as the onset or duration of breastfeeding or postpartum amenorrhea (Drife & Baynham, 1988; Kappel *et al.*, 1997).

REFERENCES

Adam, E. K., Gunnar, M. R., and Tanaka, A. (2004). Adult attachment, parent emotion, and observed parenting behavior: Mediator and moderator models. *Child Dev.* 75, 110–122.

Akirav, I., Raizel, H., and Maroun, M. (2006). Enhancement of conditioned fear extinction by infusion of the $GABA_A$ agonist muscimol into the rat prefrontal cortex and amygdala. *Eur. J. Neurosci.* 23, 758–764.

Altshuler, L. L., Hendrick, V., and Cohen, L. S. (2000). An update on mood and anxiety disorders during pregnancy and the postpartum period. *Prim. Care Companion J. Clin. Psychiatr.* 2, 217–222.

Amico, J. A., Mantella, R. C., Vollmer, R. R., and Li,X. (2004). Anxiety and stress responses in female oxytocin deficient mice. *J Neuroendocrinol.* 16, 319–324.

Anderson, P. G., and Cutshall, S. M. (2007). Massage therapy: A comfort intervention for cardiac surgery patients. *Clin. Nurse Spec.* 21, 161–165.

Anisfeld, E., Casper, V., Nozyce, M., and Cunningham, N. (1990). Does infant carrying promote attachment? An experimental study of the effects of increased physical contact on the development of attachment. *Child Dev.* 61, 1617–1627.

Atack, J. R. (2003). Anxioselective compounds acting at the GABA(A) receptor benzodiazepine binding site. *Curr. Drug Targets – CNS Neurol. Disord.* 2, 213–232.

Bale, T. L., Davis, A. M., Auger, A. P., Dorsa, D. M., and McCarthy, M. M. (2001). CNS region-specific oxytocin receptor expression: Importance in regulation of anxiety and sex behavior. *J. Neurosci.* 21, 2546–2552.

Barnett, B., and Parker, G. (1986). Possible determinants, correlates, and consequences of high levels of anxiety in primiparous mothers. *Psychol. Med.* 16, 177–185.

Barone, F. C., Cheng, J. T., and Wayner, M. J. (1994). GABA inhibition of lateral hypothalamic neurons: Role of reticular thalamic afferents. *Brain Res. Bull.* 33, 699–708.

Beider, S., and Moyer, C. A. (2007). Randomized controlled trials of pediatric massage: A review. *Evid. Based Complement Alternat. Med.* 4, 23–34.

Bereiter, D. A., and Barker, D. J. (1980). Hormone-induced enlargement of receptive fields in trigeminal mechanoreceptive neurons. I. Time course, hormone, sex and modality specificity. *Brain Res.* 184, 395–410.

Blacklock, A. D., and Smith, P. G. (2004). Estrogen increases calcitonin gene-related peptide-immunoreactive sensory innervation of rat mammary gland. *J. Neurobiol.* 59, 192–204.

Blanchard, D. C., Griebel, G., and Blanchard, R. J. (2003). The mouse defense test battery: Pharmacological and behavioral assays for anxiety and panic. *Eur. J. Pharmacol.* 463, 97–116.

Borsook, D., Becerra, L., Carlezon, W. A., Shaw, M., Renshaw, P., Elman, I., and Levine, J. (2007). Reward-aversion circuitry in analgesia and pain: Implications for psychiatric disorders. *Eur. J. Pain* 11, 7–20.

Bosch, O. J., Kromer, S. A., Brunton, P. J., and Neumann, I. D. (2004). Release of oxytocin in the hypothalamic paraventricular nucleus, but not central amygdala or lateral septum in lactating residents and virgin intruders during maternal defence. *Neuroscience* 124, 439–448.

Bosch, O. J., Meddle, S. L., Beiderbeck, D. I., Douglas, A. J., and Neumann, I. D. (2005). Brain oxytocin correlates with maternal aggression: Link to anxiety. *J. Neurosci.* 25, 6807–6815.

Brandao, M. L., Troncoso, A. C., de Souza Silva, M. A., and Huston, J. P. (2003). The relevance of neuronal substrates of defense in the midbrain tectum to anxiety and stress: Empirical and conceptual issues. *Eur. J. Pharmacol.* 463, 225–233.

Breitkopf, C. R., Primeau, L. A., Levine, R. E., Olson, G. L., Wu, Z. H., and Berenson, A. B. (2006). Anxiety symptoms during pregnancy and postpartum. *J. Psychosom. Obstet. Gynaecol.* 27, 157–162.

Bremner, J. D., Innis, R. B., Southwick, S. M., Staib, L., Zoghbi, S., and Charney, D. S. (2000a). Decreased benzodiazepine receptor binding in prefrontal cortex in combat-related posttraumatic stress disorder. *Am. J. Psychiatr.* 157, 1120–1126.

Bremner, J. D., Innis, R. B., White, T., Fujita, M., Silbersweig, D., Goddard, A. W., Staib, L., Stern, E., Cappiello, A., Woods, S., Baldwin, R., and Charney, D. S. (2000b). SPECT [I-123]iomazenil measurement of the benzodiazepine receptor in panic disorder. *Biol. Psychiatr.* 47, 96–106.

Bridges, R. S. (1996). Biochemical basis of parental behavior in the rat. In *Parental Care: Evolution, Mechanisms, and Adaptive Significance. Advances in the Study of behavior* (J. S. Rosenblatt and C. T. Snowden, Eds.), Vol. 25, pp. 215–242. Academic, New York.

Britton, J. R., Britton, H. L., and Gronwaldt, V. (2006). Breastfeeding, sensitivity, and attachment. *Pediatrics* 118, 1436–1443.

Bueno, C. H., Zangrossi, H., and Viana, M. B. (2005). The inactivation of the basolateral nucleus of the rat amygdala has an anxiolytic effect in the elevated T-maze and light/dark transition tests. *Braz. J. Med. Biol. Res.* 38, 1697–1701.

Buijs, R. M., De Vries, G. J., Van Leeuwen, F. W., and Swaab, D. R. (1983). Vasopressin and oxytocin: Distribution and putative functions in the brain. *Prog. Brain Res.* 60, 115–122.

Campeau, S., Falls, W. A., Cullinan, W. E., Helmreich, D. L., Davis, M., and Watson, S. J. (1997). Elicitation and reduction of fear: Behavioural and neuroendocrine indices and brain induction of the immediate-early gene *c-fos*. *Neuroscience* 78, 1087–1104.

Casada, J. H., and Dafny, N. (1993). Responses of neurons in bed nucleus of the stria terminalis to microiontophoretically applied morphine, norepinephrine and acetylcholine. *Neuropharmacology* 32, 279–284.

Cassileth, B. R., and Vickers, A. J. (2004). Massage therapy for symptom control: Outcome study at a major cancer center. *J Pain Symptom Manag.* 28, 244–249.

Champagne, F., and Meaney, M. J. (2001). Like mother, like daughter: Evidence for non-genomic transmission of parental behavior and stress responsivity. *Prog. Brain Res.* 133, 287–302.

Champagne, F. A., and Curley, J. P. (2005). How social experiences influence the brain. *Curr. Opin. Neurobiol.* 15, 704–709.

Charpak, N., Ruiz, J. G., Zupan, J., Cattaneo, A., Figueroa, Z., Tessier, R., Cristo, M., Anderson, G., Ludington, S., Mendoza, S., Mokhachane, M., and Worku, B. (2005). Kangaroo mother care: 25 years after. *Acta Paediatr.* 94, 514–522.

Clifford, T. J., Campbell, M. K., Speechley, K. N., and Gorodzinsky, F. (2006). Factors influencing full breastfeeding in a southwestern Ontario community: Assessments at 1 week and at 6 months postpartum. *J. Hum. Lact.* 22, 292–304.

Cogill, S. R., Caplan, H. L., Alexandra, H., Robson, K. M., and Kumar, R. (1986). Impact of maternal postnatal depression on cognitive development in young children. *Br. Med. J.* 292, 1165–1167.

Concas, A., Sanna, E., Cuccheddu, T., Mascia, M. P., Santoro, G., Maciocco, E., and Biggio, G. (1993). Carbon dioxide inhalation, stress and anxiogenic drugs reduce the function of $GABA_A$ receptor complex in the rat brain. *Prog. Neuropsychopharmacol. Biol. Psychiatr.* 17, 651–661.

Cowley, D. S., and Roy-Byrne, P. (1989). Panic disorder during pregnancy. *J. Psychosom. Obstet. Gynecol.* 10, 193–210.

Cowley, D. S., Roy-Byrne, P. P., Greenblatt, D. J., and Hommer, D. W. (1993). Personality and benzodiazepine sensitivity in anxious patients and control subjects. *Psychiatr. Res.* 47, 151–162.

Cowley, D. S., Roy-Byrne, P. P., Radant, A., Ritchie, J. C., Greenblatt, D. J., Nemeroff, C. B., and Hommer, D. W. (1995). Benzodiazepine sensitivity in panic disorder: Effects of chronic alprazolam treatment. *Neuropsychopharmacology* 12, 147–157.

Cruz, Y., Martinez-Gomez, M., Manzo, J., Hudson, R., and Pacheco, P. (1996). Changes in pain threshold during the reproductive cycle of the female rat. *Physiol. Behav.* 59, 543–547.

Cunningham, E. T., Bohn, M. C., and Sawchenko, P. E. (1990). Organization of adrenergic inputs to the paraventricular and supraoptic nuclei of the hypothalamus in the rat. *J. Comp. Neurol.* 292, 651–667.

Daftary, S. S., Boudaba, C., and Tasker, J. G. (2000). Noradrenergic regulation of parvocellular neurons in the rat hypothalamic paraventricular nucleus. *Neuroscience* 96, 743–751.

Dazzi, L., Vignone, V., Seu, E., Ladu, S., Vacca, G., and Biggio, G. (2002). Inhibition by venlafaxine of the increase in norepinephrine output in rat prefrontal cortex elicited by acute stress or by the anxiogenic drug FG 7142. *J. Psychopharmacol.* 16, 125–131.

De Bellis, M. D., Broussard, E. R., Herring, D. J., Wexler, S., Moritz, G., and Benitez, J. G. (2001). Psychiatric co-morbidity in caregivers and children involved in maltreatment: A pilot research study with policy implications. *Child Abuse Neglect* 25, 923–944.

Debiec, J., and LeDoux, L. E. (2006). Noradrenergic signaling in the amygdala contributes to the reconsolidation of fear memory: Treatment implications for PTSD. *Ann. NY Acad. Sci.* 1071, 521–524.

Dennis, C. L. (2006). Identifying predictors of breastfeeding self-efficacy in the immediate postpartum period. *Res. Nurs. Health* 29, 256–268.

Depaulis, A. and Bandler, R. (Eds.) (1991). *The Midbrain Periaqueductal Gray Matter: Functional, Anatomical, and Neurochemical Organization.* Plenum, New York.

Deschamps, S., Woodside, B., and Walker, C. D. (2003). Pups presence eliminates the stress hyporesponsiveness of early lactating females to a psychological stress representing a threat to the pups. *J. Neuroendocrinol.* 15, 486–497.

Diego, M. A., Field, T., Hernandez-Reif, M., Shaw, J. A., Rothe, E. M., Castellanos, D., and Mesner, L. (2002). Aggressive adolescents benefit from massage therapy. *Adolescence* 37, 597–607.

Drife, J. O., and Baynham, K. (1988). Breast sensitivity and lactational amenorrhea. *Br. J. Obstet. Gynaecol.* 95, 824–826.

Dunckley, P., Wise, R. G., Fairhurst, M., Hobden, P., Aziz, Q., Chang, L., and Tracey, I. (2005). A comparison of visceral and somatic pain processing in the human brainstem using functional magnetic resonance imaging. *J. Neurosci.* 25, 7333–7341.

Egli, R. E., Kash, T. L., Choo, K., Savchenko, V., Matthews, R. T., Blakely, R. D., and Winder, D. G. (2005). Norepinephrine modulates glutamatergic transmission in the bed nucleus of the stria terminalis. *Neuropsychopharmacology* 30, 657–668.

Elkins, G., Rajab, M. H., and Marcus, J. (2005). Complementary and alternative medicine use by psychiatric inpatients. *Psychol. Rep.* 96, 163–166.

Engle, P. L., Scrimshaw, S. C., Zambrana, R. E., and Dunkel-Schetter, C. (1990). Prenatal and postnatal anxiety in Mexican women giving birth in Los Angeles. *Health Psychol.* 9, 285–299.

Feijo, L., Hernandez-Reif, M., Field, T., Burns, W., Valley-Gray, S., and Simco, E. (2006). Mothers' depressed mood and anxiety levels are reduced after massaging their preterm infants. *Infant Behav. Dev.* 29, 476–480.

Fendt, M., Siegl, S., and Steiniger-Brach, B. (2005). Noradrenaline transmission within the ventral bed nucleus of the stria terminalis is critical for fear behavior induced by trimethylthiazoline, a component of fox odor. *J. Neurosci.* 25, 5998–6004.

Fénelon, V. S., and Herbison, A. E. (1996). *In vivo* regulation of specific $GABA_A$ receptor subunit messenger RNAs by increased GABA concentrations in rat brain. *Neuroscience* 71, 661–670.

Fenelon, V. S., Poulain, D. A., and Theodosis, D. T. (1993). Oxytocin neuron activation and Fos expression: A quantitative immunocytochemical analysis of the effect of lactation, parturition, osmotic and cardiovascular stimulation. *Neuroscience* 53, 77–89.

Ferreira, A., Hansen, S., Nielsen, M., Archer, T., and Minor, B. G. (1989). Behavior of mother rats in conflict tests sensitive to antianxiety agents. *Behav. Neurosci.* 103, 193–201.

Ferreira, A., Pereira, M., Agrati, D., Uriarte, N., and Fernandez-Guasti, A. (2002). Role of maternal behavior on aggression, fear and anxiety. *Physiol. Behav.* 77, 197–204.

Field, T. (2002). Massage therapy. *Med. Clin. N. Am.* 86, 163–171.

Field, T. (2005). Massage therapy for skin conditions in young children. *Dermatol. Clin.* 23, 717–721.

Field, T., Morrow, C., Valdeon, C., Larson, S., Kuhn, C., and Schanberg, S. (1992). Massage reduces anxiety in child and adolescent psychiatric patients. *J. Am. Acad. Child Adolesc. Psychiatr.* 31, 125–131.

Field, T., Schanberg, S., Kuhn, C., Field, T., Fierro, K., Henteleff, T., Mueller, C., Yando, R., Shaw, S., and Burman, I. (1998). Bulimic adolescents benefit from massage therapy. *Adolescence* 33, 555–563.

Field, T., Hernandez-Reif, M., Diego, M., Schanberg, S., and Kuhn, C. (2005). Cortisol decreases and serotonin and dopamine increase following massage therapy. *Int. J. Neurosci.* 115, 1397–1413.

Figueira, R. J., Peabody, M. F., and Lonstein, J. S. (2008). Oxytocin receptor activity in the ventrocaudal periaqueductal gray modulates anxiety-related behavior in postpartum rats. *Behavioral Neuroscience*, in press.

Fishbain, D. A., Cole, B., Cutler, R. B., Lewis, J., Rosomoff, H. L., and Rosomoff, R. S. (2006). Chronic pain and the measurement of personality: Do states influence traits?. *Pain Med.* 7, 509–529.

Fleming, A. S., and Luebke, C. (1981). Timidity prevents the virgin female rat from being a good mother: Emotionality differences between nulliparous and parturient females. *Physiol. Behav.* 27, 863–868.

Fleming, A. S., Ruble, D. N., Flett, G. L., and Van Wagner, V. (1990). Adjustment in first-time mothers: Changes in mood and mood content during the early postpartum months. *Dev. Psychol.* 26, 137–143.

Fleming, A. S., Morgan, H. D., and Walsh, C. J. (1996). Experiential factors in postpartum regulation of maternal care. In *Advances in the Study of Behavior* (J. S. Rosenblatt and C. T. Snowden, Eds.), Vol. 25, pp. 295–332. Academic, New York.

Forray, M. I., Bustos, G., and Gysling, K. (1999). Noradrenaline inhibits glutamate release in the rat bed nucleus of the stria terminalis: *In vivo* microdialysis studies. *J Neurosci. Res.* 55, 311–320.

Forster, D. A., McLachlan, H. L., and Lumley, J. (2006). Factors associated with breastfeeding at six months postpartum in a group of Australian women. *Int. Breastfeed. J.* 1, 18.

Francis, D. D., Champagne, F. C., and Meaney, M. J. (2000). Variations in maternal behaviour are associated with differences in oxytocin receptor levels in the rat. *J. Neuroendocrinol.* 12, 1145–1148.

Fujita, M., Endoh, Y., Saimon, N., and Yamaguchi, S. (2006). Effect of massaging babies on mothers: Pilot study on the changes in mood states and salivary cortisol level. *Complement Ther. Clin. Pract.* 12, 181–185.

Gabbott, P. L. A., Warner, T. A., Jays, P. R. L., Salway, P., and Busby, S. J. (2005). Prefrontal cortex in the rat: Projections to subcortical autonomic, motor, and limbic centers. *J. Comp. Neurol.* 492, 145–177.

Galler, J. R., Harrison, R. H., Biggs, M. A., Ramsey, F., and Forde, V. (1999). Maternal moods predict breastfeeding in Barbados. *J. Dev. Behav. Pediatr.* 20, 80–87.

Galler, J. R., Harrison, R. H., Ramsey, F., Forde, V., and Butler, S. C. (2000). Maternal depressive symptoms affect infant cognitive development in Barbados. *J. Child Psychol. Psychiatr.* 6, 747–757.

Gauriau, C., and Bernard, J. F. (2004). Posterior triangular thalamic neurons convey nociceptive messages to the secondary somatosensory and insular cortices in the rat. *J. Neurosci.* 24, 752–761.

Geracioti, T. D., Baker, D. G., Ekhator, N. N., West, S. A., Hill, K. K., Bruce, A. B., Schmidt, D., Rounds-Kugler, B., Yehuda, R., Keck, P. E., and Kasckow, J. W. (2001). CSF norepinephrine concentrations in posttraumatic stress disorder. *Am. J. Psychiatr.* 158, 1227–1230.

Gintzler, A. R. (1980). Endorphin-mediated increases in pain threshold during pregnancy. *Science* 210, 193–195.

Goddard, A. W., Mason, G. F., Almai, A., Rothman, D. L., Behar, K. L., Petroff, O. A. C., Charney, D. S., and Krystal, J. H. (2001). Reductions in occipital cortex GABA levels in panic disorder detected with ^{1}H-magnetic resonance spectroscopy. *Arch. Gen. Psychiatr.* 58, 556–561.

Gonzalez-Mariscal, G. (2007). Mother rabbits and their offspring: Timing is everything. *Dev. Psychobiol.* 49, 71–76.

Hansen, S. (1990). Mechanisms involved in the control of punished responding in mother rats. *Horm. Behav.* 24, 186–197.

Hansen, S., and Ferreira, A. (1986). Effects of bicuculline infusions in the ventromedial hypothalamus and amygdaloid complex on food intake and affective behavior in mother rats. *Behav. Neurosci.* 100, 410–415.

Hansen, S., Ferreira, A., and Selart, M. E. (1985). Behavioural similarities between mother rats and benzodiazepine-treated non-maternal animals. *Psychopharmacology* 86, 344–347.

Hard, E., and Hansen, S. (1985). Reduced fearfulness in the lactating rat. *Physiol. Behav.* 35, 641–643.

Heinrichs, M., Meinlschmidt, G., Neumann, I., Wagner, S., Kirschbaum, C., Ehlert, U., and Hellhammer, D. H. (2001). Effects of suckling on hypothalamic–pituitary–adrenal axis responses to psychosocial stress in postpartum lactating women. *J. Clin. Endocrinol. Metab.* 86, 4798–4804.

Heinrichs, M., Neumann, I., and Ehlert, U. (2002). Lactation and stress: Protective effects of breastfeeding in humans. *Stress* 5, 195–203.

Heinrichs, M., Baumgartner, T., Kirschbaum, C., and Ehlert, U. (2003). Social support and oxytocin interact to suppress cortisol and subjective responses to psychosocial stress. *Biol. Psychiatr.* 54, 1389–1398.

Henke, P. G. (1984). The bed nucleus of the stria terminalis and immobilization stress: Unit activity, escape behaviour, and gastric pathology in rats. *Behav. Brain Res.* 11, 35–45.

Hernandez-Reif, M., Field, T., Krasnegor, J., Martinez, E., Schwartzman, M., and Mavunda, K. (1999). Children with cystic fibrosis benefit from massage therapy. *J. Pediatr. Psychol.* 24, 175–181.

Hernandez-Reif, M., Ironson, G., Field, T., Hurley, J., Katz, G., Diego, M., Weiss, S., Fletcher, M. A., Schanberg, S., Kuhn, C., and Burman, I. (2004). Breast cancer patients have improved immune and neuroendocrine functions following massage therapy. *J. Psychosom. Res.* 57, 45–52.

Hunt, S., and Koltzenburg, M. (Eds.) (2005). *The Neurobiology of Pain.* Oxford University Press, Oxford

Imura, M., Misao, H., and Ushijima, H. (2006). The psychological effects of aromatherapy-massage in healthy postpartum mothers. *J Midwifery Womens Health* 51, e21–ee27.

Insel, T. R. (1986). Postpartum increases in brain oxytocin binding. *Neuroendocrinology* 44, 515–518.

Insel, T. R. (1990). Regional changes in brain oxytocin receptors post-partum: Time course and relationship to maternal behaviour. *J. Neuroendocrinol.* 2, 539–545.

Jenck, F., Broekkamp, C. L., and Van Delft, A. M. (1989). Effects of serotonin receptor antagonist on PAG stimulation induced aversion: Different contributions of 5HT1, 5HT2, and 5HT3 receptors. *Psychopharmacology* 97, 489–495.

Jirikowski, G. F., Caldwell, J. D., Pilgrim, C., Stumpf, W. E., and Pedersen, C. A. (1989). Changes in immunostaining for oxytocin in the forebrain of the female rat during late pregnancy, parturition and early lactation. *Cell Tissue Res.* 256, 411–417.

Johnstone, L. E., and Higuchi, T. (2001). Food intake and leptin during pregnancy and lactation. *Prog. Brain Res.* 133, 215–227.

Kalin, N. H., Shelton, S. E., Fox, A. S., Oakes, T. R., and Davidson, R. J. (2005). Brain regions associated with the expression and contextual regulation of anxiety in primates. *Biol. Psychiatr.* 58, 796–804.

Kappel, R. M., Dijkstra, R.., Storm van Leeuwen, J. B., Houpt, P., and Kuyper, M. (1997). Nipple sensitivity and lactation in two methods of breast reduction. *Eur. J. Plast. Surg.* 20, 60–65.

Kendell, R. E., Chalmers, J. C., and Platz, C. (1987). Epidemiology of puerperal psychosis. *Br. J. Psychiatr.* 150, 662–673.

Kendrick, K. M., Keverne, E. B., Baldwin, B. A., and Sharman, D. F. (1986). Cerebrospinal fluid levels of acetylcholinesterase, monoamines and oxytocin during labour, parturition, vaginocervical stimulation, lamb separation and suckling in sheep. *Neuroendocrinology* 44, 149–156.

Kendrick, K. M., Keverne, E. B., Hinton, M. R., and Goode, J. A. (1992). Oxytocin, amino acid and monoamine release in the region of the medial preoptic area and bed nucleus of the stria terminalis of the sheep during parturition and suckling. *Brain Res.* 569, 199–209.

Kim, J. J., Rison, R. A., and Fanselow, M. S. (1993). Effects of amygdale, hippocampus, and periaqueductal gray lesions on short- and long-term contextual fear. *Behav. Neurosci.* 1067, 1093–1098.

Ko, G. N., Elsworth, J. D., Roth, R. H., Rifkin, B. G., Leigh, H., and Redmond, D. E. (1983). Panic-induced elevation of plasma MHPG levels in phobic-anxious patients. Effects of clonidine and imipramine. *Arch. Gen. Psychiatr.* 40, 425–430.

Koch, M. (1999). The neurobiology of startle. *Prog. Neurobiol.* 59, 107–128.

Komisaruk, B. R., Adler, N. T., and Hutchison, J. (1972). Genital sensory field: Enlargement by estrogen treatment in female rats. *Science* 178, 1295–1298.

Kopp, M., and Gruzelier, J. (1989). Electrodermally differentiated subgroups of anxiety patients and controls. II: Relationships with auditory, somatosensory and pain thresholds, agoraphobic fear, depression and cerebral laterality. *Int. J. Psychophysiol.* 7, 65–75.

Korff, S., and Harvey, B. H. (2006). Animal models of obsessive-compulsive disorder: Rationale to understanding psychobiology and pharmacology. *Psychiatr. Clin. N. Am.* 29, 371–390.

Kumar, R., and Robson, M. K. (1984). A prospective study of emotional disorders in childbearing women. *Br. J. Psychiatr.* 144, 35–47.

Kurosawa, M., Lundeberg, T., Agren, G., Lund, I., and Uvnas-Moberg, K. (1995). Massage-like stroking of the abdomen lowers blood pressure in anesthetized rats: Influence of oxytocin. *J. Auton. Nerv. Syst.* 56, 26–30.

Kurumaji, A., Umino, A., Tanami, M., Ito, A., Asakawa, M., and Nishikawa, T. (2003). Distribution of anxiogenic-induced c-Fos in the forebrain regions of developing rats. *J. Neural Transm.* 110, 1161–1168.

Leckman, J. F., Goodman, W. K., Anderson, G. M., Riddle, M. A., Chappell, P. B., McSwiggan-Hardin, M. T., McDougle, C. J., Scahill, L. D., Ort, S. I., Pauls, D. L., Cohen, D. J., and Price, L. H. (1995). Cerebrospinal fluid biogenic amines in obsessive compulsive disorder, Tourette's syndrome, and healthy controls. *Neuropsychopharmacology* 12, 73–86.

Lederman, S. A. (2004). Influence of lactation on body weight regulation. *Nutr. Rev.* 62, S112–sS119.

LeDoux, J. E., Iwata, J., Cicchetti, P., and Reis, D. J. (1988). Different projections of the central amygdaloid nucleus mediate autonomic and behavior correlates of conditioned fear. *J. Neurosci.* 8, 2517–2529.

Lepola, U., Jolkkonen, J., Pitkanen, A., Riekkinen, P., and Rimon, R. (1990). Cerebrospinal fluid monoamine metabolites and neuropeptides in patients with panic disorder. *Ann. Med.* 22, 237–239.

Li, C., Chen, P., and Smith, M. S. (1999). Neural populations in the rat forebrain and brainstem activated by the suckling stimulus as demonstrated by cFos expression. *Neuroscience* 94, 117–129.

Lightman, S. L., Windle, R. J., Wood, S. A., Kershaw, Y. M., Shanks, N., and Ingram, C. D. (2001). Peripartum plasticity within the hypothalamo–pituitary–adrenal axis. *Prog. Brain Res.* 133, 111–129.

Lippa, A., Czobor, P., Stark, J., Beer, B., Kostakis, E., Gravielle, M., Bandyopadhyay, S., Russek, S. J., Gibbs, T. T., Farb, D. H., and Skolnick, P. (2005). Selective anxiolysis produced by ocinaplon, a $GABA_A$ receptor modulator. *Proc. Natl Acad. Sci.* 102, 7380–7385.

Liu, M., and Glowa, J. R. (1999). Alterations of $GABA_A$ receptor subunit mRNA levels associated with increases in punished responding induced by acute alprazolam administration: An *in situ* hybridization study. *Brain Res.* 882, 8–16.

Lonstein, J. S. (2005). Reduced anxiety during lactation requires recent interactions with pups, but not their suckling or peripheral sources of hormones. *Horm. Behav.* 47, 241–255.

Lonstein, J. S. (2007). Regulation of anxiety during the postpartum period. *Front. Neuroendocrinol.* 28, 115–141.

Lonstein, J. S., and Gammie, S. C. (2002). Sensory, hormonal, and neural control of maternal aggression in laboratory rodents. *Neurosci. Biobehav. Rev.* 26, 869–888.

Lonstein, J. S., and Morrell, J. I. (2006). Neuropharmacology and neuroendocrinology of maternal motivation and behavior. In *Handbook of Neurochemistry and Molecular Biology. Vol. 18. – Behavioral Neurobiology* (J. D. Blaustein, Ed.), pp. 195–245. Springer Press, New York.

Lonstein, J. S., Simmons, D. A., and Stern, J. M. (1998). Functions of the caudal periaqueductal gray in lactating rats: Kyphosis, lordosis, maternal aggression, and fearfulness. *Behav. Neurosci.* 112, 1502–1518.

Lonstein, J. S., and Stern, J. M. (1997a). Role of the midbrain periaqueductal gray in maternal nurturance and aggression: *c-Fos* and electrolytic lesion studies in lactating rats. *J. Neurosci.* 17, 3364–3378.

Lonstein, J. S., and Stern, J. M. (1997b). Somatosensory contributions to c-fos activation within the caudal periaqueductal gray of lactating rats: Effects of perioral, rooting, and suckling stimuli from pups. *Horm. Behav.* 32, 155–166.

Lund, I., Ge, Y., Yu, L. C., Uvnas-Moberg, K., Wang, J., Yu, C., Kurosawa, M., Agren, G., Rosen, A., Lekman, M., and Lundeberg, T. (2002). Repeated massage-like stimulation induces long-term effects on nociception: Contribution of oxytocinergic mechanisms. *Eur. J. Neurosci.* 16, 330–338.

Lydiard, R. B. (2003). The role of GABA in anxiety disorders. *J. Clin. Psychiatr.* 64, 21–27.

Malizia, A. L., Cunningham, V. J., Bell, C. J., Liddle, P. F., Jones, T., and Nutt, D. J. (1998). Decreased brain $GABA_A$-benzodiazepine receptor binding in panic disorder. *Arch. Gen. Psychiatr.* 55, 715–720.

Manassis, K., Bradley, S., Goldberg, S., Hood, J., and Swinson, R. P. (1994). Attachment in mothers with anxiety disorders and their children. *J. Am. Acad. Child Adolesc. Psychiatr.* 33, 1106–1113.

Mantella, R. C., Vollmer, R. R., Li, X., and Amico, J. A. (2003). Female oxytocin-deficient mice display enhanced anxiety-related behavior. *Endocrinology* 144, 2291–2296.

Mayer, A. D., Carter, L., Jorge, W. A., Mota, M. J., Tannu, S., and Rosenblatt, J. S. (1987). Mammary stimulation and maternal aggression in rodents: Thelectomy fails to reduce pre- or postpartum aggression in rats. *Horm. Behav.* 21, 501–510.

McCarthy, M. M., McDonald, C. H., Brooks, P. J., and Goldman, D. (1996). An anxiolytic action of oxytocin is enhanced by estrogen in the mouse. *Physiol. Behav.* 60, 1209–1215.

McNeilly, A. S. (2001). Lactational control of reproduction. *Reprod. Fertil. Dev.* 13, 583–590.

Meddle, S. L., Bishop, V. R., Gkoumassi, E., van Leeuwen, F. W., and Douglas, A. J. (2007). Dynamic changes in oxytocin receptor expression and activation at parturition in the rat brain. *Endocrinology* 148, 5095–5104.

Menard, J., and Treit, D. (1999). Effects of centrally administered anxiolytic compounds in animal models of anxiety. *Neurosci. Biobehav. Rev.* 23, 591–613.

Miczek, K. A., Weerts, E. M., Vivian, J. A., and Barros, H. M. (1995). Aggression, anxiety and vocalizations in animals: $GABA_A$ and 5-HT anxiolytics. *Psychopharmacology* 121, 38–56.

Millan, M. J. (2003). The neurobiology and control of anxious states. *Prog. Neurobiol.* 70, 83–244.

Miller, S. M., and Lonstein, J. S. (2007). Effects of anxiogenic drugs on the anxiety behavior of postpartum and virgin rats in the elevated plus maze and light–dark box. *Parental Brain Conf.*, Boston, MA, June 7–10.

Miller, S. M., Peabody, M. F., and Lonstein, J. S. $GABA_A$ receptor antagonists administered peripherally or in the ventrocaudal periaqueductal gray increase anxiety in the anxiety-resistant lactating rat (in preparation).

Moore, P. S., Wiley, S. E., and Sigman, M. (2004). Interactions between mothers and children: impacts of maternal and child anxiety. *J. Abnorm. Psych.* 113, 471–476.

Moore, E. R., and Anderson, G. C. (2007). Randomized controlled trial of very early mother–infant skin-to-skin contact and breastfeeding status. *J. Midwifery Womens Health* 52, 116–125.

Muganini, E., and Oertel, W. H. (1985). An atlas of the distribution of GABAergic neurons and terminals in the rat CNS as revealed by GABA immunohistochemistry. In *Handbook of Chemical Neuroanatomy. Vol. 4. GABA and Neuropeptides in the CNS, Part I* (A. Björklund and T. Hökfelt, Eds.), pp. 436–553. Elsevier Science Publishers, Amsterdam.

Nashold, B. S., Wilson, W. P., and Slaughter, D. G. (1969). Sensations evoked by stimulation of the midbrain of man. *J. Neurosurg.* 30, 14–24.

Nayak, M. B., and Milner, J. S. (1998). Neuropsychological functioning: Comparison of mothers at high- and low-risk for child physical abuse. *Child Abuse Neglect* 22, 687–703.

Nazar, M., Jessa, M., and Pla nik, A. (1997). Benzodiazepine-$GABA_A$ receptor complex ligands in two models of anxiety. *J. Neural Transm.* 104, 733–746.

Nelvokov, A., Areda, T., Innos, J., Kõks, S., and Vasar, E. (2006). Rats displaying distinct exploratory activity also have different expression patterns of γ-aminobutyric acid- and cholecystokinin-related genes in brain regions. *Brain Res.* 1100, 21–31.

Nemeroff, C. B. (2003a). Anxiolytics: Past, present, and future agents. *J. Clin. Psychiatr.* 64, 3–6.

Nemeroff, C. B. (2003b). The role of GABA in the pathophysiology and treatment of anxiety disorders. *Psychopharmacol. Bull.* 37, 133–146.

Neophytou, S. I., Aspley, S., Butler, S., Beckett, S., and Marsden, C. A. (2001). Effects of lesioning noradrenergic neurones in the locus coeruleus on conditioned and unconditioned aversive behaviour in the rat. *Prog. Neuropsychopharmacol. Biol. Psychiatr.* 25, 1307–1321.

Neugebauer, V., Li, W., Bird, G. C., and Han, J. S. (2004). The amygdala and persistent pain. *Neuroscientist* 10, 221–234.

Neumann, I. D. (2001). Alterations in behavioral and neuroendocrine stress coping strategies in pregnant, parturient and lactating rats. *Prog. Brain Res.* 133, 143–152.

Neumann, I. D. (2003). Brain mechanisms underlying emotional alterations in the peripartum period in rats. *Depress. Anxiety* 17, 111–121.

Neumann, I., Ludwig, M., Engelmann, M., Pittman, Q. J., and Landgraf, R. (1993). Simultaneous microdialysis in blood and brain: Oxytocin and vasopressin release in response to central and peripheral osmotic stimulation and suckling in the rat. *Neuroendocrinology* 58, 637–645.

Neumann, I. D., Johnstone, H. A., Hatzinger, M., Liebsch, G., Shipston, M., Russell, J. A., Landgraf, R., and Douglas, A. J. (1998). Attenuated neuroendocrine responses to emotional and physical stressors in pregnant rats involve adenohypophysial changes. *J. Physiol.* 508, 289–300.

Neumann, I. D., Torner, L., and Wigger, A. (2000). Brain oxytocin: Differential inhibition of neuroendocrine stress responses and anxiety-related behaviour in virgin, pregnant and lactating rats. *Neuroscience* 95, 567–575.

Numan, M., and Numan, M. J. (1994). Expression of Fos-like immunoreactivity in the preoptic area of maternally behaving virgin and postpartum rats. *Behav. Neurosci.* 108, 379–394.

Numan, M., and Numan, M. J. (1995). Importance of pup-related sensory inputs and maternal performance for the expression of Fos-like immunoreactivity in the preoptic area and ventral bed nucleus of the stria terminalis of postpartum rats. *Behav. Neurosci.* 109, 135–149.

Numan, M., and Insel, T. R. (2003). *The Neurobiology of Parental Behavior*. Springer, New York.

O'Brien, L. M., Heycock, E. G., Hanna, M., Jones, P. W., and Cox, J. L. (2004). Postnatal depression and faltering growth: A community study. *Pediatrics* 113, 1242–1247.

Ogawa, S., Kow, L. M., and Pfaff, D. W. (1992). Effects of lordosis-relevant neuropeptides on midbrain periaqueductal gray neuronal activity *in vitro*. *Peptides* 13, 965–975.

Olazabal, D. E., and Ferreira, A. (1997). Maternal behavior in rats with kainic acid-induced lesions of the hypothalamic paraventricular nucleus. *Physiol. Behav.* 61, 779–784.

Onaka, T., and Yagi, K. (1998). Role of noradrenergic projections to the bed nucleus of the stria terminalis in neuroendocrine and behavioral responses to fear-related stimuli in rats. *Brain Res.* 788, 287–293.

Pacak, K., Palkovits, M., Kvetnansky, R., Kopin, I. J., and Goldstein, D. S. (1993). Stress-induced norepinephrine release in the paraventricular nucleus of rats with brainstem hemisections: A microdialysis study. *Neuroendocrinology* 58, 196–201.

Pereira, M., Uriarte, N., Agrati, D., Zuluaga, M. J., and Ferreira, A. (2005). Motivational aspects of maternal anxiolysis in lactating rats. *Psychopharmacology* 180, 241–248.

Post, R. M., Lake, C. R., Jimerson, D. C., Bunney, W. E., Wood, J. H., Ziegler, M. G., and Goodwin, F. K. (1978). Cerebrospinal fluid norepinephrine in affective illness. *Am. J. Psychiatr.* 135, 907–912.

Qureshi, G. A., Hansen, S., and Sodersten, P. (1987). Offspring control of cerebrospinal fluid GABA concentrations in lactating rats. *Neurosci. Lett.* 75, 85–88.

Rägo, L., Adojaan, A., Harro, J., and Kiivet, R. A. (1991). Correlation between exploratory activity in an elevated plus-maze and number of central and peripheral benzodiazepine binding sites. *Arch. Pharmacol.* 343, 301–306.

Rezayat, M., Roohbakhsh, A., Zarrindast, M. R., Massoudi, R., and Djahanguiri, B. (2005). Cholecystokinin and GABA interaction in the dorsal hippocampus of rats in the elevated plus-maze test of anxiety. *Physiol. Behav.* 84, 775–782.

Riche, D., De Pommery, J., and Menetrey, D. (1990). Neuropeptides and catecholamines in efferent projections of the nuclei of the solitary tract in the rat. *J. Comp. Neurol.* 293, 399–424.

Ring, R. H., Malberg, J. E., Potestio, L., Ping, J., Boikess, S., Luo, S., Schechter, L. E., Rizzo, S., Rahman, Z., and Rosenzweig-Lipson, S. (2006). Anxiolytic-like activity of oxytocin in male mice: Behavioral and autonomic evidence, therapeutic implications. *Psychopharmacology* 185, 218–225.

Robinson, J. E., and Short, R. V. (1977). Changes in breast sensitivity at puberty, during the menstrual cycle, and at parturition. *Br. Med. J.* 1, 1188–1191.

Roder, S., and Ciriello, J. (1994). Collateral axonal projections to limbic structures from ventrolateral medullary A1 noradrenergic neurons. *Brain Res.* 638, 182–188.

Rosenblatt, J. S., and Siegel, H. I. (1975). Hysterectomy-induced maternal behavior during pregnancy in the rat. *J. Comp. Physiol. Psychol.* 89, 685–700.

Roy-Byrne, P. P. (2005). The GABA-benzodiazepine receptor complex: Structure, function, and role in anxiety. *J. Clin. Psychiatr.* 66(suppl 2), 14–20.

Rudolph, U., and Möhler, H. (2006). GABA-based therapeutic approaches: $GABA_A$ receptor subtype functions. *Curr. Opin. Pharmacol.* 6, 18–23.

Rushen, J., Foxcroft, G., and De Passille, A. M. (1993). Nursing-induced changes in pain sensitivity, prolactin, and somatotropin in the pig. *Physiol. Behav.* 53, 265–270.

Salchner, P., Sartori, S. B., Sinner, C., Wigger, A., Frank, E., Landgraf, R., and Singewald, N. (2006). Airjet and FG-7142-induced Fos expression differs in rats selectively bred for high and low anxiety-related behavior. *Neuropharmacology* 50, 1048–1058.

Sawchenko, P. E., and Swanson, L. W. (1982). Immunohistochemical identification of neurons in the paraventricular nucleus of the hypothalamus that project to the medulla or to the spinal cord in the rat. *J. Comp. Neurol.* 205, 260–272.

Schaal, B., Doucet, S., Sagot, P., Hertling, E., and Soussignan, R. (2006). Human breast areolae as scent organs: Morphological data and possible involvement in maternal–neonatal coadaptation. *Dev. Psychobiol.* 48, 100–110.

Sevy, S., Papadimitriou, G. N., Surmont, D. W., Goldman, S., and Mendlewicz, J. (1989). Noradrenergic function in generalized anxiety disorder, major depressive disorder, and healthy subjects. *Biol. Psychiatr.* 25, 141–152.

Shah, A. A., and Treit, D. (2004). Infusions of midazolam into the medial prefrontal cortex produce anxiolytic effects in the elevated plus-maze and shock-probe burying tests. *Brain Res.* 996, 31–40.

Shah, A. A., Sjovold, T., and Treit, D. (2004). Inactivation of the medial prefrontal cortex with the $GABA_A$ receptor agonist muscimol increases open-arm activity in the elevated plus-maze and attenuates shock-probe burying in rats. *Brain Res.* 1028, 112–115.

Sharpe, P. A., Williams, H. G., Granner, M. L., and Hussey, J. R. (2007). A randomised study of the effects of massage therapy compared to guided relaxation on well-being and stress perception among older adults. *Complement Ther. Med.* 15, 157–163.

Shi, C. J., and Cassell, M. D. (1998). Cascade projections from somatosensory cortex to the rat basolateral amygdala via the parietal insular cortex. *J. Comp. Neurol.* 399, 469–491.

Shibata, S., Yamashita, K., Yamamoto, E., Ozaki, T., and Ueki, S. (1989). Effects of benzodiazepine and GABA antagonists on anticonflict effects of antianxiety drugs injected into the rat amygdala in a water-lick suppression test. *Psychopharmacology* 98, 38–44.

Sienkiewicz-Jarosz, H., Szyndler, J., Czlonkowska, A. I., Siemiątkowski, M., Maciejak, P., Wislowska, A., Zienowicz, M., Lehner, M., Turzyńska, D., Bidziński, A., and Plaźnik, A. (2003). Rat behavior in two models of anxiety and brain [^{3}H]muscimol binding: Pharmacological, correlation, and multifactor analysis. *Behav. Brain Res.* 145, 17–22.

Singewald, N., Salchner, P., and Sharp, T. (2003). Induction of c-Fos expression in specific areas of the fear circuitry in rat forebrain by anxiogenic drugs. *Biol. Psychiatr.* 53, 275–283.

Sjogren, B., Widstrom, A. M., Edman, B., and Unvas-Moberg, K. (2000). Changes in personality pattern during the first pregnancy and lactation. *J. Psychosom. Obstet. Gynaecol.* 21, 31–38.

Sjolin, S., Hofvander, T., and Hillervik, C. (1977). Factors related to early termination of breast feeding. A retrospective study in Sweden. *Acta Paediatr. Scand.* 66, 505–511.

Smith, C. D., and Lonstein, J. S. (2008). Contact with infants influences anxiety-induced *c-fos* activity in the postpartum rat brain. *Behavioral Brain Research*, in press.

Smith, T. A. (2001). Type A gamma-aminobutyric acid ($GABA_A$) receptor subunits and benzodiazepine binding: Significance to clinical syndromes and their treatment. *Br. J. Biomed. Sci.* 58, 111–121.

Smith, M. S., and Grove, K. L. (2002). Integration of the regulation of reproductive function and energy balance: Lactation as a model. *Front. Neuroendocrinol.* 23, 225–256.

Stefanova, N., Bozhilova-Pastirova, A., and Ovtscharoff, W. (1997). Distribution of GABA-immunoreactive nerve cells in the bed nucleus of the stria terminalis in male and female rats. *Eur. J. Histochem.* 41, 23–28.

Stern, J. M. (1996). Somatosensation and maternal care in Norway rats. In *Parental Care: Evolution, Mechanisms, and Adaptive Significance. Advances in the Study of Behavior* (J. S. Rosenblatt and C. T. Snowden, Eds.), Vol. 25, pp. 243–294. Academic, San Diego.

Stern, J. M. (1997). Offspring-induced nurturance: Animal–human parallels. *Dev. Psychobiol.* 31, 19–37.

Stern, J. M., and Siegel, H. I. (1978). Prolactin release in lactating, primiparous and multiparous thelectomized and maternal virgin rats exposed to pup stimuli. *Biol. Reprod.* 19, 177–182.

Stern, J. M., and Kolunie, J. M. (1993). Maternal aggression of rats is impaired by cutaneous anesthesia of the ventral trunk, but not by nipple removal. *Physiol. Behav.* 54, 861–868.

Swanson, L. W. (1998). *Brain Maps: Structure of the Rat Brain*, 2nd edn.. Elsevier Science Publishers B.V., Amsterdam.

Swanson, L. W., and Sawchenko, P. E. (1983). Hypothalamic integration: Organization of the paraventricular and supraoptic nuclei. *Annu. Rev. Neurosci.* 6, 269–324.

Tanaka, M., Yoshida, M., Emoto, H., and Ishii, H. (2000). Noradrenaline systems in the hypothalamus, amygdala and locus coeruleus are involved in the provocation of anxiety: Basic studies. *Eur. J. Pharmacol.* 405, 397–406.

Tang, J., and Gibson, S. J. (2005). A psychophysical evaluation of the relationship between trait anxiety, pain perception, and induced state anxiety. *J. Pain* 6, 612–619.

Tiihonen, J., Kuikka, J., Räsänen, P., Lepola, U., Koponen, H., Liuska, A., Lehmusvaara, A., Vainio, P., Könönen, M., Bergström, K., Yu, M., Kinnunen, I., Akerman, K., and Karhu, J. (1997). Cerebral benzodiazepine receptor binding and distribution in generalized anxiety disorder: A fractal analysis. *Mol. Psychiatr.* 2, 463–471.

Toufexis, D. J., Thrivikraman, K. V., Plotsky, P. M. Morilak, D.A., Huang, N. and Walker, C.D. (1998). Reduced noradrenergic tone to the hypothalamic paraventricular nucleus contributes to the stress hyporesponsiveness of lactation, *J. Neuroendocrinol.* 10, 417–442.

Toufexis, D. J., Rochford, J., and Walker, C. D. (1999). Lactation induced reduction in rats' acoustic startle is associated with changes in noradrenergic neurotransmission. *Behav. Neurosci.* 113, 176–184.

Tu, M. T., Lupien, S. J., and Walker, C. D. (2005). Measuring stress responses in postpartum mothers: Perspectives from studies in human and animal populations. *Stress* 8, 19–34.

Turner, S. M., Beidel, D. C., Roberson-Nay, R., and Tervo, K. (2002). Parenting behaviors in parents with anxiety disorders. *Behav. Res. Ther.* 41, 541–554.

Veinante, P., and Freund-Mercier, M. J. (1998). Intrinsic and extrinsic connections of the rat central extended amygdala: An *in vivo* electrophysiological study of the central amygdaloid nucleus. *Brain Res.* 794, 188–198.

Vermetten, E., and Bremner, J. D. (2002a). Circuits and systems in stress. I. Preclinical studies. *Depress. Anxiety* 15, 126–147.

Vermetten, E., and Bremner, J. D. (2002b). Circuits and systems in stress. II. Applications to neurobiology and treatment in posttraumatic stress disorder. *Depress. Anxiety* 16, 14–38.

Vertes, R. P., Hoover, W. B., Do Valle, A. C., Sherman, A., and Rodriguez, J. J. (2006). Efferent projections of reuniens and rhomboid nuclei of the thalamus in the rat. *J. Comp. Neurol.* 499, 768–796.

Vianna, D. M., Landiera-Fernandez, J., and Brandao, M. L. (2001). Dorsolateral and ventral regions of the periaqueductal gray matter are involved in distinct types of fear. *Neurosci. Biobehav. Rev.* 25, 711–719.

Viggo-Hansen, N., Jorgensen, T., and Ortenblad, L. (2006). Massage and touch for dementia. *Cochrane Database Syst. Rev.* 4, 1–15. CD004989

Wakerley, J. B., Clarke, G., and Summerlee, A. J. S. (1994). Milk ejection and its control, In *The Physiology of Reproduction* (E. Knobil and J. D. Neill, Eds.), Vol. 2, 2nd edn., pp. 1131–1178. Raven, New York.

Walker, C. D., Toufexis, D. J., and Burlet, A. (2001). Hypothalamic and limbic expression of CRF and vasopressin during lactation: Implications for the control of ACTH secretion and stress hyporesponsiveness. *Prog. Brain Res.* 133, 99–110.

Whipple, E. E., and Webster-Stratton, C. (1991). The role of parental stress in physically abusive families. *Child Abuse Neglect* 15, 279–291.

Whipple, B., Josimovich, J. B., and Komisaruk, B. R. (1990). Sensory thresholds during the antepartum, intrapartum and postpartum periods. *Int. J. Nurs. Stud.* 27, 213–221.

Whiting, P. J. (2006). GABA-A receptors: A viable target for novel anxiolytics?. *Curr. Opin. Pharmacol.* 6, 24–29.

Wilhelm, F. H., Kochar, A. S., Roth, W. T., and Gross, J. J. (2001). Social anxiety and response to touch: Incongruence between self-evaluative and physiological reactions. *Biol. Psychol.* 58, 181–202.

Wilkinson, S. M., Love, S. B., Westcombe, A. M., Gambles, M. A., Burgess, C. C., Cargill, A., Young, T., Maher, E. J., and Ramirez, A. J. (2007). Effectiveness of aromatherapy massage in the management of anxiety and depression in patients with cancer: A multicenter randomized controlled trial. *J. Clin. Oncol.* 25, 532–539.

Windle, R. J., Shanks, N., Lightman, S. L., and Ingram, C. D. (1997). Central oxytocin administration reduces stress-induced corticosterone release and anxiety behavior in rats. *Endocrinology* 138, 2829–2834.

Windle, R. J., Brady, M. M., Kunanandam, T., Da Costa, A. P., Wilson, B. C., Harbuz, M., Lightman, S. L., and Ingram, C. D. (1997). Reduced response of the hypothalamo–pituitary–adrenal axis to alpha1-agonist stimulation during lactation. *Endocrinology* 138, 3741–3748.

Windle, R. J., Gamble, L. E., Kershaw, Y. M., Wood, S. A., Lightman, S. L., and Ingram, C. D. (2006). Gonadal steroid modulation of stress-induced hypothalamo–pituitary–adrenal activity and anxiety behavior: Role of central oxytocin. *Endocrinology* 147, 2423–2431.

Woodruff-Borden, J., Morrow, C., Bourland, S., and Cambron, S. (2002). The behavior of anxious parents: Examining mechanisms of transmission of anxiety from parent to child. *J. Clin. Child Adolesc. Psychol.* 31, 364–374.

Woulfe, J. M., Flumerfelt, B. A., and Hrycyshyn, A. W. (1990). Efferent connections of the A1 noradrenergic cell group: A DBH immunohistochemical and PHA-L anterograde tracing study. *Exp. Neurol.* 109, 308–322.

Xerri, C., Stern, J. M., and Merzenich, M. M. (1994). Alterations of the cortical representation of the rat ventrum induced by nursing behavior. *J. Neurosci.* 14, 1710–1721.

Yasumatsu, H., Morimoto, Y., Yamamoto, Y., Takehara, S., Fukuda, T., Nakao, T., and Setoguchi, M. (1994). The pharmacological properties of Y-23684, a benzodiazepine receptor partial agonist. *Br. J. Pharmacol.* 111, 1170–1178.

Yoshimura, R., Kiyama, H., Kimura, T., Araki, T., Maeno, H., Tanizawa, O., and Tohyama, M. (1993). Localization of oxytocin receptor messenger ribonucleic acids in the rat brain. *Endocrinology* 133, 1239–1246.

Young, L. J., Muns, S., Wang, Z., and Insel, T. R. (1997). Changes in oxytocin receptor mRNA in rat brain during pregnancy and the effects of estrogen and interleukin-6. *J. Neuroendocrinol.* 9, 859–865.

Zelkowitz, P., and Papageorgiou, A. (2005). Maternal anxiety: An emerging prognostic factor in neonatology. *Acta Paediatr.* 94, 1704–1705.

11

THE ROLE OF THE BRAIN SEROTONERGIC SYSTEM IN THE ORIGIN AND TRANSMISSION OF ADAPTIVE AND MALADAPTIVE VARIATIONS IN MATERNAL BEHAVIOR IN RHESUS MACAQUES

DARIO MAESTRIPIERI

Department of Comparative Human Development,
The University of Chicago, Chicago, IL 60637, USA

INTRODUCTION

The systematic study of maternal behavior in non-human primates, and particularly in rhesus macaques, began in the late 1950s and was driven by the notion that primate research could elucidate the biological mechanisms and functional significance of parent–child attachment (Bowlby, 1969). One influential line of research begun by Harry Harlow at the University of Wisconsin aimed to understand the nature of mother–infant attachment and the extent of maternal influences on infant development. The main experimental paradigm used in this research involved depriving rhesus macaque infants of their mothers early in life and exposing them to different rearing environments (Harlow, 1959). Early studies focused on the behavioral consequences of maternal deprivation for the infants (e.g., Harlow & Harlow, 1962; Kaufman & Rosenblum, 1967), while subsequent research incorporated hormonal responses to separation and artificial rearing (e.g., Champoux *et al.*, 1989; Levine & Wiener, 1988) as well as assessments of brain structure and function in maternally deprived individuals (e.g., Kraemer *et al.*, 1989; Martin *et al.*, 1991; Ginsberg *et al.*, 1993). In the 1960–1980 period, this line of research was pursued in many laboratories around the world and culminated in the development of a psychobiological theory of attachment, in which primate mothers were viewed as playing a fundamental role in the normative neurological, physiological, and sociobehavioral development of their offspring (Kraemer, 1992). This theory, however, was not based on studies of mothers and their behavior but only on the presumed effects of maternal absence on infant development. Although Harlow's early studies of surrogate mothers made a fundamental contribution to our understanding of infant attachment to a caregiver, later research and its extrapolations have been criticized on both conceptual and empirical grounds (see Insel, 1992; Kagan, 1992; Maestripieri, 2003a; Maestripieri & Wallen, 2003).

A different line of research begun by Robert Hinde at the University of Cambridge emphasized the study of rhesus macaque mothers and infants in naturalistic social environments, the quantification of different aspects of maternal behavior with ethological observational methods, and the investigation of naturally occurring interindividual variation in mothering styles (Hinde & Spencer-Booth, 1968, 1971). Hinde's approach had a pervasive influence on the field such that virtually every observational study of primate maternal behavior conducted in the last four decades has used some of his concepts or empirical measures. This research has enhanced our knowledge of the causes of naturally occurring interindividual variation in mothering style as well as of its consequences for offspring socio-behavioral development (e.g., see Altmann, 1980; Berman, 1984; Fairbanks, 1996, 2003; Maestripieri, 1999).

Studies of macaques, baboons, and vervet monkeys have shown that most variability in mothering style occurs along the two orthogonal dimensions of maternal protectiveness and rejection (Tanaka, 1989; Schino *et al.*, 1995; Fairbanks, 1996; Maestripieri, 1998a). The maternal protectiveness dimension includes variation in the extent to which the mother physically restrains her infant, initiates proximity and contact, and cradles and grooms her infant. The maternal rejection dimension includes the extent to which the mother limits the timing and duration of contact, suckling, or carrying. Although maternal behavior changes as a function of infant age and the mother's own age and experience, individual differences in mothering style tend to be consistent over time and across infants (Hinde & Spencer-Booth, 1971; Fairbanks, 1996).

There is now a great deal of evidence that variation in mothering style is accounted for by a combination of maternal characteristics (e.g., age, parity, dominance rank, temperament), infant characteristics (e.g., sex, age, baseline activity levels), and those of the surrounding environment (e.g., stress and support from other group members, ecological variables) (Maestripieri, 1999; Fairbanks, 2003). There is also evidence that individual differences in mothering style have long-term effects on the offspring's tendency to respond to challenges or explore the environment. For example, infants reared by highly rejecting (or less responsive) mothers generally develop independence at an earlier age (e.g., spend more time out of contact with their mothers, explore the environment more, and play more with their peers) than infants reared by mothers with low rejection levels (Simpson, 1985; Simpson & Simpson, 1985; Simpson *et al.*, 1989; Simpson & Datta, 1990). In contrast, infants reared by more protective mothers appear to be delayed in the acquisition of their independence and are relatively fearful and cautious when faced with challenging situations (Fairbanks & McGuire, 1987, 1988, 1993; Vochteloo *et al.*, 1993). Effects of mothering style on offspring behavior have been shown to extend into adulthood (Fairbanks & McGuire, 1988; 1993; Schino *et al.*, 2001; Bardi *et al.*, 2005; Bardi & Huffman, 2006; Maestripieri *et al.*, 2006b). These effects have also been demonstrated with experimental manipulations of maternal behavior (Vochteloo *et al.*, 1993) and with infant cross-fostering studies (Maestripieri, 2005a, b; Maestripieri *et al.*, 2006b). Mothering style, and particularly its maternal rejection component, has also been shown to affect the offspring's parenting behavior in adulthood, as there are often significant similarities in maternal behavior between mothers and daughters (Fairbanks, 1989; Berman, 1990; Maestripieri, 2005a; Maestripieri *et al.*, 2007).

Research on naturally occurring variation in primate maternal behavior has recently been revitalized by studies on laboratory rodents, in which the neurobiological and molecular mechanisms underlying the cross-generational effects of naturally occurring individual differences in maternal behavior have been elucidated. These studies have shown that variation in rates of maternal grooming/licking results in dramatic and long-lasting differences in the behavior, responsiveness to stress, and reproduction of the offspring, including the transmission of maternal style across generations (e.g., Meaney, 2001; Champagne & Curley, 2008). These cross-generational effects of maternal behavior are mediated by epigenetic modifications of gene expression (through DNA methylation mechanisms) for glucocorticoid and estrogen receptors in particular areas of the brain, which in turn, trigger a cascade of neuroendocrine effects involving the HPA and HPG axis, and a variety of neuropeptide hormones and neurotransmitters (see Champagne & Curley, 2008, for review). Although research on molecular brain mechanisms in non-human primates is constrained by a number of factors, the success of rodent studies in explaining how variation in maternal behavior affects offspring biobehavioral development has prompted primate researchers to investigate whether similar or different mechanisms may operate in primates.

BRAIN SEROTONIN AND NATURALLY OCCURRING VARIATION IN PRIMATE MATERNAL BEHAVIOR

In 1998, I began with my research collaborators a long-term project to investigate the neurobiological and neuroendocrine mechanisms underlying the cross-generational effects of naturally occurring variation in maternal behavior in rhesus macaques. We focused on individual differences in mothering style along the protectiveness and rejection dimension, as well as on one extreme expression of variable parenting behavior: physical abuse of offspring. The project was conducted at the Field Station of the Yerkes National Primate Research Center in Lawrenceville, Georgia, where previous studies had shown that 5–10% of adult females in the macaque population abuse their offspring and that abusive parenting runs in families, being present in some matrilines for more than 6–7 generations and completely absent in others (Maestripieri *et al.*, 1997; Maestripieri, 1998b; Maestripieri & Carroll, 1998a, b). At the Yerkes Field Station, the rhesus macaque population includes over 1,500 individuals, and monkeys are housed in large outdoor corrals, where they live in social groups of naturalistic size and composition. In our research project, the individuals are studied in their own social groups, where they have the opportunity to express naturally occurring variation in behavioral tendencies. The monkeys are trained for capture and handling, so that procedures involving experimental testing and collection of biological samples are generally brief and the subjects are immediately returned to their groups for observation.

The project involved the longitudinal study, from birth to adulthood, of 16 females that were cross-fostered at birth between abusive and non-abusive mothers, along with the study of 43 males and females that were born and raised by their biological mothers, half of which were abusive and half non-abusive. In addition to studying the social development and behavioral reactivity to stress of offspring exposed to variable maternal behavior in infancy, we assessed the development of hypothalamic–pituitary–adrenal (HPA) function and the functionality of brain monoamine systems such as serotonin, dopamine, and norepinephrine. This was accomplished by measuring the plasma concentrations of ACTH and cortisol in a variety of experimental conditions, and by measuring the CSF concentrations of serotonin, dopamine, and norepinephrine metabolites (5-HIAA, HVA, and MHPG, respectively) at 6-month intervals (see Maestripieri *et al.*, (2006a) for details of the experimental procedures). A subset of infants and their mothers were also genotyped for the polymorphism in the serotonin transporter (SERT) gene (Lesch *et al.*, 1996), which has been shown to modulate the effects of early experience on adult behavior and psychopathology in both humans and rhesus macaques. In particular, individuals with the short (*s*) allele of this gene are more likely to develop anxiety disorders and dysregulation of the HPA axis as a result of early adverse experience than individuals with the long (*l*) allele (Lesch *et al.*, 1996; Bennett *et al.*, 2002; Caspi *et al.*, 2002, 2003; Barr *et al.*, 2004a, b). Analyses of the hormonal data are still in progress, therefore the rest of this chapter will focus on the behavioral and the brain monoamine data, with particular emphasis on serotonin.

We found that abusive mothers were significantly more likely to carry the *s* allele of the SERT gene than non-abusive mothers; however, there was no significant difference in the prevalence of the *l* and *s* alleles between the offspring of abusive and non-abusive mothers (McCormack *et al.* unpublished data). Individual differences in the CSF concentrations of 5-HIAA in the offspring measured at 6, 12, 18, 24, 30, and 36 months of age were highly stable over time (see also Higley *et al.*, 1992). Infants heterozygous (*l/s* genotype) or homozygous for the long or the short allele (*l/l* and *s/s* genotype) of the SERT gene did not differ significantly from each other in their CSF concentrations of 5-HIAA. Moreover, we found no significant variation in CSF concentrations of 5-HIAA in relation to infant abuse experienced in the first 3 months of life (abuse is concentrated in the first month and generally ends by the end of the third month) (Maestripieri *et al.*, 2006b). Therefore, we focused our analysis of offspring development on the effects of exposure to variable mothering style in infancy. We found stable individual differences in many measures of maternal behavior in the first 6 postpartum months and, similar to previous studies, we found that these measures clustered around two factors, or mothering style dimensions: protectiveness (making contact, restraining, cradling, and grooming the infant) and rejection (breaking contact, rejecting the infant's attempt to make contact). We obtained composite measures of these two dimensions and classified all mothers in our sample as being high or low in protectiveness and high or low in rejection depending on whether their scores were above or below the median value for the composite measures.

The individuals exposed to high rates of maternal rejection in infancy had significantly lower CSF concentrations of 5-HIAA across their first 3 years of life than the individuals exposed to low rates of maternal rejection (Maestripieri *et al.*, 2006a). Data were analyzed separately for cross-fostered and non-cross-fostered individuals and a similar relation between maternal rejection and CSF 5-HIAA was found in both groups, suggesting that this association was not due to genetic similarities between mothers and offspring. In contrast, there were no differences in CSF 5-HIAA between offspring reared by high and low protectiveness mothers. Long-term effects of early experience on the development of the brain serotonergic system have also been reported in other studies of rhesus macaques (Kraemer *et al.*, 1989; Higley *et al.*, 1992; Shannon *et al.*, 2005) as well as rodents (e.g., Ladd *et al.*, 1996; Gardner *et al.*, 2005).

When we examined the development of various social and non-social behaviors prior to the onset of puberty, we found no differences in relation to early maternal protectiveness, and only one difference in relation to maternal rejection: the offspring of high rejection mothers engaged in more solitary play than those of low rejection mothers (Maestripieri *et al.*, 2006b). Thus, exposure to variable maternal behavior early in life had little impact on the offspring's social interactions with other group members prior to puberty. Consistent with this, we found that the general affiliative and aggressive tendencies of cross-fostered females in their first 2 years of life were more similar to those of their biological mothers than to those of their foster mothers (Maestripieri, 2003b). We also found, however, a significant negative correlation between CSF 5-HIAA and rates of scratching (Maestripieri *et al.*, 2006b), suggesting that individuals with low CSF 5-HIAA were more anxious than those with high 5-HIAA (see Schino *et al.* (1991) and Maestripieri *et al.* (1992) for the relation between scratching and anxiety). Differences in anxiety associated with serotonergic function may have contributed to some of the effects of exposure of variable mothering style we observed after our female subjects reached puberty and gave birth for the first time.

Although there were no similarities between the maternal protectiveness scores of mothers and daughters, the maternal rejection rates of daughters closely resembled those of their mothers (Maestripieri *et al.*, 2007). The resemblance was particularly strong for the cross-fostered females and their foster mothers. This finding is consistent with a previously reported intergenerational correlation of maternal rejection rates in another population of rhesus macaques (Berman, 1990) and suggests that this correlation is the result of early experience and not of genetic similarities between mothers and daughters. Although learning through direct experience with one's own mother and/or observations of maternal interactions with siblings may play a role in the intergenerational transmission of maternal rejection in macaques (Berman, 1990), biological mechanisms are also important. In our study, we found that the cross-fostered females' CSF concentrations of 5-HIAA were negatively correlated with their rates of maternal rejection such that the individuals with lower CSF 5-HIAA exhibited higher rates of rejection with their infants (Maestripieri *et al.*, 2007). Therefore, exposure to variable rates of maternal rejection in infancy may affect the development of the brain serotonergic system, and variation in serotonergic function, in turn, may contribute to the expression of maternal rejection with one's own offspring later in life. Interestingly, a preliminary study by Lindell *et al.* (1997) found that the CSF 5-HIAA concentrations of rhesus macaque mothers were significantly correlated with those of their 9-month-old infants, but this study did not assess whether these correlations had a genetic or environmental nature. Evidence of both genetic and environmental effects on CSF concentrations of 5-HIAA and other monoamine metabolites was provided by Rogers *et al.* (2004) in a study of a large pedigreed population of baboons.

In addition to demonstrating the intergenerational transmission of maternal rejection rates, we also found evidence for the intergenerational transmission of infant abuse. Specifically, about half of the females who were abused by their mothers early in life, whether cross-fostered or non-cross-fostered (all cross-fostered females reared by abusive mothers were also abused by them), exhibited abusive parenting toward their first-born offspring, whereas none of the females reared by non-abusive mothers did (including those born to abusive mothers; Maestripieri, 2005a). Moreover, the abused females, both cross-fostered and non-cross-fostered, who became abusive mothers had lower CSF 5-HIAA concentrations than the abused females who did not become abusive mothers (Maestripieri *et al.*, 2006a). This finding suggests that experience-induced long-term alterations in serotonergic function in females reared by abusive mothers contribute to the manifestation of abusive parenting in adulthood. It is possible that experience-induced reduction in serotonergic function results in elevated anxiety and impaired impulse control, and that high anxiety and impulsivity increase the probability of occurrence of abusive parenting (e.g., Troisi & D'Amato, 1991, 1994), perhaps in conjunction with social learning resulting from direct experience of abuse early in life or observation of abusive parenting displayed by one's own mother with siblings. The intergenerational transmission of infant abuse, however, is likely to be a complex process with multiple determinants and influences. The finding that abusive mothers were more likely to carry the *s* allele of the SERT gene suggests that genetic variation in brain serotonergic function may play a role in the manifestation of abusive parenting and its transmission across generations. In order to understand the complex relationship between serotonin and abusive parenting, one must understand the relation between serotonin and maternal rejection as well as the relationship between maternal rejection and abuse.

SEROTONIN AND MATERNAL BEHAVIOR

The brain serotonergic system is believed to play an important role in impulse control and in reducing the probability that risky, dangerous, or aggressive behaviors will be expressed in response to internal pressures or external stimuli (e.g., Gollan *et al.*, 2005). Consistent with a large body of human research (e.g., Linnoila & Virkkunen, 1992), studies of rhesus macaques and vervet monkeys have shown that, in adult males, low levels of CSF 5-HIAA are associated with high impulsivity, risk-taking behavior, and propensity to engage in severe forms of aggression (see Higley, 2003, for a review). In young males, low levels of CSF 5-HIAA are associated with earlier age of emigration from the natal group (e.g., Mehlman *et al.*, 1995) and with the attainment of high dominance rank in adulthood (Fairbanks *et al.*, 2004). Similar to the adult males, adult monkey females with low CSF 5-HIAA have been reported to be more likely to be wounded, to engage in violent aggression, and to be lower ranking than females with high CSF 5-HIAA (see Higley, 2003 for review; but see Cleveland *et al.*, 2004). Adult females with low CSF 5-HIAA also appear to be less socially oriented, spending more time alone, grooming less, and having fewer conspecifics in close proximity (Cleveland *et al.*, 2004; rhesus macaque abusive mothers fit this behavioral profile quite well; see Maestripieri, 1998b). Westergaard *et al.* (2003) also reported that the infants born to adult rhesus females with low CSF 5-HIAA concentrations are more likely to die within a year after birth than infants born to females with high CSF 5-HIAA concentrations.

Although serotonin is an obvious candidate neurotransmitter for the regulation of maternal care, there is surprisingly little information about the relationship between serotonin and maternal behavior, not only in non-human primates but in rodents as well (Numan & Insel, 2003). A study of rats found that lesions of the median raphe serotonergic neurons disrupted maternal behavior on day 1 of lactation (Barofsky *et al.*, 1983), but the results of this study have been questioned and the effects on maternal behavior have been ascribed to the surgical procedure rather than to the specific loss of a serotonergic pathway (Numan & Insel, 2003). Other studies of rats involving pharmacological manipulations of the brain serotonergic system reported some effects on maternal aggression but few or no effects on pup retrieval or other aspects of maternal care (reviewed by Numan & Insel, 2003; but see Johns *et al.*, 2005). Studies with knockout mice or with mice with a null mutation for a serotonin receptor gene reported higher fearfulness, impulsivity, and hyperactivity in these individuals but also impaired maternal behavior (Brunner *et al.*, 1999; Gingrich & Hen, 2001; Weller *et al.*, 2003; Jacobs & Emeson, 2006).

Studies involving brain lesions or genetic manipulations in primates are problematic, therefore researchers who have investigated the relationship between serotonin and maternal behavior have relied on measures of the serotonin metabolite 5-HIAA in the CSF. Early studies reported that monkey mothers with low CSF 5-HIAA were more protective and restrictive, and that their infants spent more time in contact with them, than mothers with high CSF 5-HIAA (Lindell *et al.*, 1997; Fairbanks *et al.*, 1998). Cleveland *et al.* (2004) found no relationship between CSF 5-HIAA and maternal behavior in the first few postpartum days, but on postpartum days 15 and 20, females with

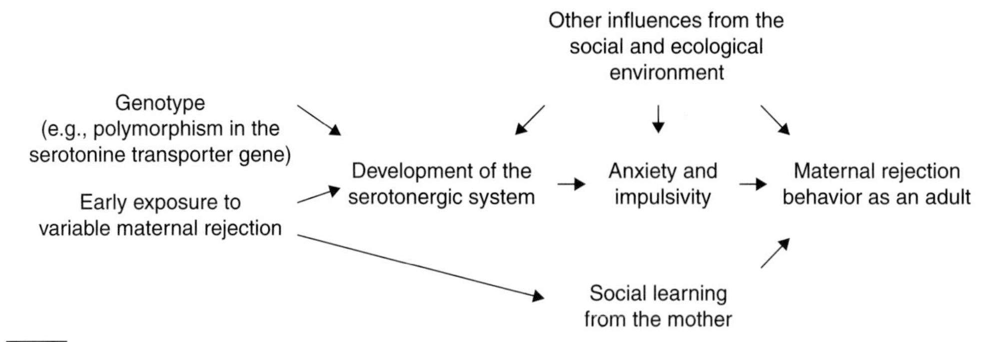

FIGURE 11.1 A schematic representation of the possible role of genetic and experiential factors in the development and expression of maternal rejection behavior by adult female rhesus monkeys, including the brain serotonergic system and anxiety and impulsivity as important mediating mechanisms.

low CSF 5-HIAA broke contact and left their infants less frequently than females with high CSF 5-HIAA. A preliminary study in our laboratory reported a positive correlation between CSF 5-HIAA concentrations measured during pregnancy and maternal rejection behaviors in the first postpartum month in multiparous females (Maestripieri *et al.*, 2005). Our more recent work involving multiple measurements of CSF 5-HIAA during development, however, reported a negative correlation between CSF 5-HIAA and maternal rejection among first-time mothers (Maestripieri *et al.*, 2007). Taken together, these studies support the notion that variation in serotonergic function can contribute to the expression of differences in maternal behavior, although the relationship between serotonin and primate maternal behavior is not yet fully understood (but see Figure 11.1 for a schematic representation of the possible relationship between genes and early experience, serotonergic function, and maternal rejection).

The relative lack of studies of serotonin and maternal behavior in rodents may reflect the belief that the motivational bases of maternal behaviors such as nest-building, crouching over the pups, licking/grooming, and pup retrieval depend on the direct actions of hormones and peptides such as prolactin and oxytocin in specific regions of the brain (e.g., the medial preoptic area of the hypothalamus; Numan & Insel, 2003). Although serotonin might affect maternal motivation through its actions on oxytocin or prolactin release, serotonin and other monoamines are viewed as having aspecific effects on emotionality, motivation, or memory rather than specific effects on parentally motivated behaviors (Insel & Winslow, 1998; Numan & Insel, 2003). Research with rodents has established a strong link between anxiety/impulsivity and maternal aggression (e.g., Lonstein & Gammie, 2002), but the emotional substrate of rodent maternal behavior is not well established (but see Weller *et al.*, 2003; Johns *et al.*, 2005).

Emotions, however, play a fundamental role in the regulation of maternal behavior in non-human primates and humans (Dix, 1991; Pryce, 1992; Maestripieri, 1999). Emotions can be powerful elicitors of maternal behavior and play a crucial role in mediating the impact of the surrounding environment on the mother–infant dyad. For example, Pryce (1992) argued that two emotional systems, the attraction–arousal system and the anxiety system, play a central role in the regulation of primate maternal behavior. The attraction–arousal system involves the activation of positive emotions (e.g., excitement or joy) that elicit nurturing maternal behavior, whereas the anxiety system involves the activation of negative emotions (e.g., anxiety and fear) that elicit protective or rejecting maternal behaviors. Whereas the postpartum period is associated with lower reactivity to stress in rodents (Tu *et al.*, 2006; but see Deschamps *et al.*, 2003), pregnancy and the postpartum period in non-human primates and humans are characterized by high emotional instability and reactivity. For example, high cortisol levels and high arousability in the early postpartum period have been associated with greater sensitivity to infant cues and greater maternal responsiveness in humans (Fleming *et al.*, 1987, 1997; see also Maestripieri *et al.*, 2008 under review, for rhesus macaques). Interestingly, etiological theories of postpartum psychosis based on estrogen's interaction with serotonin systems have been proposed (Fink & Sumner, 1996). For example, it has been shown that variation in the SERT genotype affects susceptibility to bipolar affective puerperal psychosis (Coyle *et al.*, 2000).

Motherhood is a psychologically stressful condition in human and non-human primates. In rhesus macaques, the first few months of an infant's life result into a number of anxiety-eliciting situations for the mother (Maestripieri, 1993a). There are marked individual differences in anxiety among rhesus mothers, and such differences translate into differences in maternal style (Maestripieri, 1993b). Maternal anxiety has also been implicated in the etiology of infant abuse (Troisi & D'Amato, 1984, 1991, 1994). Although the role of emotionality, and particularly of impulsivity, in primate maternal behavior is still poorly understood, it is possible that impulsivity affects how primate mothers interact with their infants, and that high impulsivity is expressed as high rejection rates as well as, as other studies suggest, greater maternal protectiveness. Our recent findings suggest that variation in impulsivity and maternal rejection originates, at least in part, from early experience and that there may be causal relationships between these two variables, such that high rates of maternal rejection result in low serotonergic function, which in turn results in high rates of maternal rejection later in life.

Maternal rejection also has a complex relation with infant abuse, perhaps not dissimilar from the relationship between child neglect and abuse in humans. Although abusive parenting in monkeys is probably maladaptive (Maestripieri, 1998b), maternal rejection is a behavior that belongs to the normal maternal repertoire and is used by mothers to limit the amount of time spent by infants in bodily and nipple contact, thus encouraging the infant's social and nutritional independence (e.g., Simpson & Simpson, 1985). Abusive parenting in rhesus macaques co-occurs with high rates of maternal rejection. Abusive mothers begin rejecting their infants shortly after birth (rejection normally begins after 3–4 weeks) and continue to do so at much higher rates than non-abusive mothers (Maestripieri, 1998b; McCormack *et al.*, 2006). Although we found no direct effects of infant abuse on CSF 5-HIAA, the observed significant effects of maternal rejection on CSF 5-HIAA were likely driven by abused infants, who were exposed to much higher levels of rejection than non-abused infants. Rejection occurs more frequently than abuse and although it does not cause physical harm to the infants, it may be even more psychologically traumatic than abuse. Interestingly, human studies have found that child neglect tends to have stronger and more consistent effects on brain structure and function in maltreatment victims than physical abuse does, although both are transmitted across generations (e.g., Glaser, 2000; DeBellis, 2005). Although social learning probably plays an important role in the intergenerational transmission of both maternal rejection and abuse in monkeys, our results suggest that rejection is more likely than abuse to cause long-term alterations in neuroendocrine and emotional functioning, and that these alterations may contribute to the expression of both rejecting and abusive parenting later in life.

ACKNOWLEDGMENTS

The research reviewed in this chapter was supported by NIH and involved the participation of many collaborators and assistants including Richelle Fulks, Anne Graff, Dee Higley, Stephen Lindell, Kai McCormack, Nancy Megna, Timothy Newman, and Mar Sanchez.

REFERENCES

Altmann, J. (1980). *Baboons Mothers and Infants.* Harvard University Press, Cambridge, MA.

Bardi, M., and Huffman, M. A. (2006). Maternal behavior and maternal stress are associated with infant behavioral development in macaques. *Dev. Psychobiol.* 48, 1–9.

Bardi, M., Bode, A. E., Ramirez, S. M., and Brent, L. Y. (2005). Maternal care and the development of the stress response. *Am. J. Primatol.* 66, 263–278.

Barofsky, A. L., Taylor, J., Tizabi, Y., and Jones-Quartey, K. (1983). Specific neurotoxin lesions of median raphe serotonergic neurons disrupt maternal behavior in the lactating rat. *Endocrinology* 113, 1884–1893.

Barr, C. S., Newman, T. K., Shannon, C., Parker, C., Dvoskin, R. L., Becker, M. L., Schwandt, M., Champoux, M., Lesch, K. P., Goldman, D., Suomi, S. J., and Higley, J. D. (2004a). Rearing condition and rh5-HTTLPR interact to influence LHPA-axis response to stress in infant macaques. *Biol. Psychiatr.* 55, 733–738.

Barr, C. S., Newman, T. K., Lindell, S., Shannon, C., Champoux, M., Lesch, K. P., Suomi, S. J., and Higley, J. D. (2004b). Interaction between serotonin transporter gene variation and rearing condition in alcohol preference and consumption in female primates. *Arch. Gen. Psychiatr.* 61, 1146–1152.

Bennett, A. J., Lesch, K. P., Heils, A., Long, J. C., Lorenz, J. G., Shoaf, S. E., Champoux, M., Suomi, S. J., Linnoila, M. V., and Higley, J. D. (2002). Early experience and serotonin transporter gene variation interact to influence primate CNS function. *Mol. Psychiatr.* 7, 118–122.

Berman, C. M. (1984). Variation in mother–infant relationships: Traditional and nontraditional factors. In *Female Primates: Studies by Women Primatologists* (M. F. Small, Ed.), pp. 17–36. Alan Liss, New York.

Berman, C. M. (1990). Intergenerational transmission of maternal rejection rates among free-ranging rhesus monkeys. *Anim. Behav.* 39, 329–337.

Bowlby, J. (1969). . Basic Books, New York.

Brunner, D., Buhot, M. C., Hen, R., and Hofer, M. (1999). Anxiety, motor activation, and maternal–infant interactions in 5HT1b knockout mice. *Behav. Neurosci.* 113, 587–601.

Caspi, A., McClay, J., Moffitt, T. E., Mill, J., Martin, J., Craig, I. W., Taylor, A., and Poulton, R. (2002). Role of genotype in the cycle of violence in maltreated children. *Science* 297, 851–854.

Caspi, A., Sugden, K., Moffitt, T. E., Taylor, A., Craig, I. W., Harrington, H., McClay, J., Mill, J., Martin, J., Braithwaite, A., and Poulton, R. (2003). Influence of life stress on depression: Moderation by a polymorphism in the 5-HTT gene. *Science* 301, 386–389.

Champagne, F. A., and Curley, J. P. (2008). The trans-generational influence of maternal care on offspring gene expression and behavior in rodents. In *Maternal Effects in Mammals* (D. Maestripieri and J. M. Mateo, Eds.). The University of Chicago Press, Chicago. in press

Champoux, M., Coe, C. L., Schanberg, S. M., Kuhn, C. M., and Suomi, S. J. (1989). Hormonal effects of early rearing conditions in the infant rhesus monkey. *Am. J. Primatol.* 19, 111–117.

Cleveland, A., Westergaard, G. C., Trenkle, M. K., and Higley, J. D. (2004). Physiological predictors of reproductive outcome and mother–infant behaviors in captive rhesus macaque females (*Macaca mulatta*). *Neuropsychopharmacology* 29, 901–910.

Coyle, N., Jones, I., Robertson, E., Lendon, C., and Craddock, N. (2000). Variation at the serotonin transporter gene influences susceptibility to bipolar affective puerperal psychosis. *Lancet* 356, 1490–1491.

DeBellis, M. D. (2005). The psychobiology of neglect. *Child Maltreatment* 10, 150–172.

Deschamps, S., Woodside, B., and Walker, C. D. (2003). Pups' presence eliminates the stress hyporesponsiveness of early lactating females to a psychological stress representing a threat to the pups. *J. Neuroendocrinol.* 15, 486–497.

Dix, T. (1991). The affective organization of parenting: Adaptive and maladaptive processes. *Psych. Bull.* 110, 3–25.

Fairbanks, L. A. (1989). Early experience and cross-generational continuity of mother–infant contact in vervet monkeys. *Dev. Psychobiol.* 22, 669–681.

Fairbanks, L. A. (1996). Individual differences in maternal styles: Causes and consequences for mothers and offspring. *Adv. Study Behav.* 25, 579–611.

Fairbanks, L. A. (2003). Parenting. In *Primate Psychology* (D. Maestripieri, Ed.), pp. 144–170. Harvard University Press, Cambridge, MA.

Fairbanks, L. A., and McGuire, M. T. (1987). Mother–infant relationships in vervet monkeys: Response to new adult males. *Int. J. Primatol.* 8, 351–366.

Fairbanks, L. A., and McGuire, M. T. (1988). Long-term effects of early mothering behavior on responsiveness to the environment in vervet monkeys. *Dev. Psychobiol.* 21, 711–724.

Fairbanks, L. A., and McGuire, M. T. (1993). Maternal protectiveness and response to the unfamiliar in vervet monkeys. *Am. J. Primatol.* 30, 119–129.

Fairbanks, L. A., Melega, W. P., and McGuire, M. T. (1998). CSF 5-HIAA is associated with individual differences in maternal protectiveness in vervet monkeys. *Am. J. Primatol.* 45, 179–180. (abstract)

Fairbanks, L. A., Jorgensen, M. J., Huff, A., Blau, K., Hung, Y., and Mann, J. J. (2004). Adolescent impulsivity predicts adult dominance attainment in male vervet monkeys. *Am. J. Primatol.* 64, 1–17.

Fink, G., and Sumner, B. E. H. (1996). Estrogen and mental state. *Nature* 383, 306.

Fleming, A. S., Steiner, M., and Anderson, V. (1987). Hormonal and attitudinal correlates of maternal behavior during the early postpartum period in first-time mothers. *J. Reprod. Infant Psychol.* 5, 193–205.

Fleming, A. S., Steiner, M., and Corter, C. (1997). Cortisol, hedonics, and maternal responsiveness in human mothers. *Horm. Behav.* 32, 85–98.

Gardner, K. L., Thrivikraman, K. V., Lightman, S. L., Plotsky, P. M., and Lowry, C. A. (2005). Early life experience alters behavior during defeat: Focus on serotonergic systems. *Neuroscience* 136, 181–191.

Gingrich, J. A., and Hen, R. (2001). Dissecting the role of the serotonin system in neuropsychiatric disorders using knockout mice. *Psychopharmacology* 155, 1–10.

Ginsberg, S. D., Hof, P. R., McKinney, W. T., and Morrison, J. H. (1993). The noradrenergic innervation density of the monkey paraventricular nucleus is not altered by early social deprivation. *Neurosci. Lett.* 158, 130–134.

Glaser, D. (2000). Child abuse and neglect and the brain: A review. *J. Child Psychol. Psychiatr.* 41, 97–116.

Gollan, J. K., Lee, R., and Coccaro, E. F. (2005). Developmental psychopathology and neurobiology of aggression. *Dev. Psychopathol.* 17, 1151–1171.

Harlow, H. F. (1959). Affectional response in the infant monkey. *Science* 130, 421–432.

Harlow, H. F., and Harlow, M. K. (1962). The effect of rearing conditions on behavior. *Bull. Menninger Clin.* 26, 213–224.

Higley, J. D. (2003). Aggression. In *Primate Psychology* (D. Maestripieri, Ed.), pp. 17–40. Harvard University Press, Cambridge.

Higley, J. D., Suomi, S. J., and Linnoila, M. (1992). A longitudinal study of CSF monoamine metabolite and plasma cortisol concentrations in young rhesus monkeys: Effects of early experience, age, sex and stress on continuity of interindividual differences. *Biol. Psychiatr.* 32, 127–145.

Hinde, R. A., and Spencer-Booth, Y. (1968). The study of mother–infant interaction in captive group-living rhesus monkeys. *Proc. R. Soc. Lond. B* 169, 177–201.

Hinde, R. A., and Spencer-Booth, Y. (1971). Towards understanding individual differences in rhesus mother–infant interaction. *Anim. Behav.* 19, 165–173.

Insel, T. R. (1992). Oxytocin and the neurobiology of attachment. *Behav. Brain Sci.* 15, 515–516.

Insel, T. R., and Winslow, J. T. (1998). Serotonin and neuropeptides in affiliative behaviors. *Biol. Psychiatr.* 44, 207–219.

Jacobs, M. M., and Emeson, R. B. (2006). Modulation of maternal care by editing of the serotonin 2C receptor (5-HT2CR). *J. Pharm. Sci.* 101, 80. (abstract)

Johns, J. M., Joyner, P. W., McMurray, M. S., Elliott, D. L., Hofler, V. E., Middleton, C. L., Knupp, K., Greenhill, K. W., Lomas, L. M., and Walker, C. H. (2005). The effects of dopaminergic/serotonergic reuptake inhibition on maternal behavior, maternal aggression, and oxytocin in rats. *Pharmacol. Biochem. Behav.* 81, 769–785.

Kagan, J. (1992). The meanings of attachment. *Behav. Brain Sci.* 15, 517–518.

Kaufman, I. C., and Rosenblum, L. A. (1967). The reaction to separation in monkeys: Anaclitic depression and conservation-withdrawal. *Psychosom. Med.* 29, 649–675.

Kraemer, G. W. (1992). A psychobiological theory of attachment. *Behav. Brain Sci.* 15, 493–541.

Kraemer, G. W., Ebert, M. H., Schmidt, D. E., and McKinney, W. T. (1989). A longitudinal study of the effect of different social rearing conditions on cerebrospinal fluid norepinephrine and biogenic amine metabolites in rhesus monkeys. *Neuropsychopharmacology* 2, 175–189.

Ladd, C. O., Owens, M. J., and Nemeroff, C. B. (1996). Persistent changes in corticotropin-releasing factor neuronal systems induced by maternal deprivation. *Endocrinology* 137, 1212–1218.

Lesch, K. P., Bengel, D., Heils, A., Sabol, S. Z., Greenberg, B. D., Petri, S., Benjamin, J., Muller, C., Hamer, D., and Murphy, D. (1996). Association of anxiety-related traits with a polymorphism in the serotonin transporter gene regulatory region. *Science* 274, 1527–1531.

Levine, S., and Wiener, S. G. (1988). Psychoendocrine aspects of mother–infant relationships in nonhuman primates. *Psychoneuroendocrinology* 13, 143–154.

Lindell, S. G., Higley, J. D., Shannon, C., and Linnoila, M. (1997). Low levels of CSF 5-HIAA in female rhesus macaques predict mother–infant interaction patterns and mother's CSF 5-HIAA correlates with infant's CSF 5-HIAA. *Am. J. Primatol.* 42, 129. (abstract)

Linnoila, V. M., and Virkkunen, M. (1992). Aggression, suicidality, and serotonin. *J. Clin. Psychiatr.* 53, 46–51.

Lonstein, J. S., and Gammie, S. C. (2002). Sensory, hormonal, and neural control of maternal aggression in laboratory rodents. *Neurosci. Biobehav. Rev.* 26, 869–888.

Maestripieri, D. (1993a). Maternal anxiety in rhesus macaques (*Macaca mulatta*). I. Measurement of anxiety and identification of anxiety-eliciting situations. *Ethology* 95, 19–31.

Maestripieri, D. (1993b). Maternal anxiety in rhesus macaques (*Macaca mulatta*). II. Emotional bases of individual differences in mothering style. *Ethology* 95, 32–42.

Maestripieri, D. (1998a). Social and demographic influences on mothering style in pigtail macaques. *Ethology* 104, 379–385.

Maestripieri, D. (1998b). Parenting styles of abusive mothers in group-living rhesus macaques. *Anim. Behav.* 55, 1–11.

Maestripieri, D. (1999). The biology of human parenting: Insights from nonhuman primates. *Neurosci. Biobehav. Rev.* 23, 411–422.

Maestripieri, D. (2003a). Attachment. In *Primate Psychology* (D. Maestripieri, Ed.), pp. 108–143. Harvard University Press, Cambridge, MA.

Maestripieri, D. (2003b). Similarities in affiliation and aggression between cross-fostered rhesus macaque females and their biological mothers. *Dev. Psychobiol.* 43, 321–327.

Maestripieri, D. (2005a). Early experience affects the intergenerational transmission of infant abuse in rhesus monkeys. *Proc. Natl Acad. Sci. USA* 102, 9726–9729.

Maestripieri, D. (2005b). Effects of early experience on female behavioural and reproductive development in rhesus macaques. *Proc. R. Soc. Lond. B* 272, 1243–1248.

Maestripieri, D., and Carroll, K. A. (1998a). Child abuse and neglect: Usefulness of the animal data. *Psych. Bull.* 123, 211–223.

Maestripieri, D., and Carroll, K. A. (1998b). Risk factors for infant abuse and neglect in group-living rhesus monkeys. *Psychol. Sci.* 9, 65–67.

Maestripieri, D., and Wallen, K. (2003). Nonhuman primate models of developmental psychopathology: Problems and prospects. In *Neurodevelopmental Mechanisms in Psychopathology* (D. Cicchetti and E. Walker, Eds.), pp. 187–214. Cambridge University Press, Cambridge.

Maestripieri, D., Schino, G., Aureli, F., and Troisi, A. (1992). A modest proposal: Displacement activities as an indicator of emotions in primates. *Anim. Behav.* 44, 967–979.

Maestripieri, D., Wallen, K., and Carroll, K. A. (1997). Infant abuse runs in families of group-living pigtail macaques. *Child Abuse Neglect* 21, 465–471.

Maestripieri, D., Lindell, S. G., Ayala, A., Gold, P. W., and Higley, J. D. (2005). Neurobiological characteristics of rhesus macaque abusive mothers and their relation to social and maternal behavior. *Neurosci. Biobehav. Rev.* 29, 51–57.

Maestripieri, D., Higley, J. D., Lindell, S. G., Newman, T. K., McCormack, K., and Sanchez, M. M. (2006a). Early maternal rejection affects the development of monoaminergic systems and adult abusive parenting in rhesus macaques. *Behav. Neurosci.* 120, 1017–1024.

Maestripieri, D., McCormack, K., Lindell, S. G., Higley, J. D., and Sanchez, M. M. (2006b). Influence of parenting style on the offspring's behavior and CSF monoamine metabolites levels in crossfostered and noncrossfostered female rhesus macaques. *Behav. Brain Res.* 175, 90–95.

Maestripieri, D., Lindell, S. G., and Higley, J. D. (2007). Intergenerational transmission of maternal behavior in rhesus monkeys and its underlying mechanisms. *Dev. Psychobiol.* 49, 165–171.

Maestripieri, D., Hoffman, C. L., Fulks, R., and Gerald, M. S. (2008). Plasma cortisol responses to stress in lactating and nonlactating female rhesus macaques. *Horm. Behav.* 53, 170–176.

Martin, L. J., Spicer, D. M., Lewis, M. H., Gluck, J. P., and Cork, L. C. (1991). Social deprivation of infant rhesus monkeys alters the chemoarchitecture of the brain: I. Subcortical regions. *J. Neurosci.* 11, 3344–3358.

McCormack, K. M., Sanchez, M. M., Bardi, M., and Maestripieri, D. (2006). Maternal care patterns and behavioral development of rhesus macaque abused infants in the first 6 months of life. *Dev. Psychobiol.* 48, 537–550.

Meaney, M. J. (2001). Maternal care, gene expression, and the transmission of individual differences in stress reactivity across generations. *Ann. Rev. Neurosci.* 24, 1161–1192.

Mehlman, P. T., Higley, J. D., Faucher, I., Lilly, A. A., Taub, D. M., Vickers, J., Suomi, S. J., and Linnoila, M. (1995). Correlation of CSF 5-HIAA concentration with sociality and the timing of emigration in free-ranging primates. *Am. J. Psychiatr.* 152, 907–913.

Numan, M., and Insel, T. R. (2003). *The Neurobiology of Parental Behavior*. Springer, New York.

Pryce, C. R. (1992). A comparative systems model of the regulation of maternal motivation in mammals. *Anim. Behav.* 43, 417–441.

Rogers, J., Martin, L. J., Comuzzie, A. G., Mann, J. J., Manuck, S. B., Leland, M., and Kaplan, J. R. (2004). Genetics of monoamine metabolites in baboons: Overlapping sets of genes influence levels of 5-hydroxyindolacetic acid, 3-hydroxy-4-methoxyphenylglycol, and homovanillic acid. *Biol. Psychiatr.* 55, 739–744.

Schino, G., Troisi, A., Perretta, G., and Monaco, V. (1991). Measuring anxiety in nonhuman primates: Effect of lorazepam on macaque scratching. *Pharm. Biochem. Behav.* 38, 889–891.

Schino, G., D'Amato, F. R., and Troisi, A. (1995). Mother–infant relationships in Japanese macaques: Sources of interindividual variation. *Anim. Behav.* 49, 151–158.

Schino, G., Speranza, L., and Troisi, A. (2001). Early maternal rejection and later social anxiety in juvenile and adult Japanese macaques. *Dev. Psychobiol.* 38, 186–190.

Shannon, C., Schwandt, M. L., Champoux, M., Shoaf, S. E., Suomi, S. J., Linnoila, M., and Higley, J. D. (2005). Maternal absence and stability of individual differences in CSF5-HIAA concentrations in rhesus monkey infants. *Am. J. Psychiatr.* 162, 1658–1664.

Simpson, M. J. A. (1985). Effects of early experience on the behaviour of yearling rhesus monkeys (*Macaca mulatta*) in the presence of a strange object: Classification and correlation approaches. *Primates* 26, 57–72.

Simpson, A. E., and Simpson, M. J. A. (1985). Short-term consequences of different breeding histories for captive rhesus macaque mothers and young. *Behav. Ecol. Sociobiol.* 18, 83–89.

Simpson, M. J. A., and Datta, S. B. (1990). Predicting infant enterprise from early relationships in rhesus macaques. *Behaviour* 116, 42–63.

Simpson, M. J. A., Gore, M. A., Janus, M., and Rayment, F. D. G. (1989). Prior experience of risk and individual differences in enterprise shown by rhesus monkey infants in the second half of their first year. *Primates* 30, 493–509.

Tanaka, I. (1989). Variability in the development of mother–infant relationships among free-ranging Japanese macaques. *Primates* 30, 477–491.

Troisi, A., and D'Amato, F. R. (1984). Ambivalence in monkey mothering: Infant abuse combined with maternal possessiveness. *J. Nerv. Ment. Dis.* 172, 105–108.

Troisi, A., and D'Amato, F. R. (1991). Anxiety in the pathogenesis of primate infant abuse: A pharmacological study. *Psychopharmacology* 103, 571–572.

Troisi, A., and D'Amato, F. R. (1994). Mechanisms of primate infant abuse: The maternal anxiety hypothesis. In *Infanticide and Parental Care* (S. Parmigiani and F.v. Saal, Eds.), pp. 199–210. Harwood, London.

Tu, M. T., Lupien, S. J., and Walker, C.-D. (2006). Measuring stress in postpartum mothers: Perspectives from studies in human and animal populations. *Stress* 8, 19–34.

Vochteloo, J. D., Timmermans, P. J. A., Duijghuisen, J. A. H., and Vossen, J. M. H. (1993). Effects of reducing the mother's radius of action on the development of mother–infant relationships in longtailed macaques. *Anim. Behav.* 45, 603–612.

Weller, A., Leguisamo, A. C., Towns, L., Ramboz, S., Bagiella, E., Hofer, M., Hen, R., and Brunner, D. (2003). Maternal effects in infant and adult phenotypes of 5HT(1A) and 5HT(1B) receptor knockout mice. *Dev. Psychobiol.* 42, 194–205.

Westergaard, G. C., Cleveland, A., Trenkle, M. K., Lussier, I. D., and Higley, J. D. (2003). CSF 5-HIAA concentrations as an early screening tool for predicting significant life history outcomes in female specific-pathogen-free (SPF) rhesus macaques (*Macaca mulatta*) maintained in captive breeding groups. *J. Med. Primatol.* 32, 95–104.

12 POSTPARTUM DEPRESSION: THE CLINICAL DISORDER AND APPLICATION OF PET IMAGING RESEARCH METHODS

EYDIE L. MOSES-KOLKO[1], CAROLYN C. MELTZER[2], SARAH L. BERGA[3] AND KATHERINE L. WISNER[4]

[1] *University of Pittsburgh, School of Medicine, Department of Psychiatry, Pittsburgh, PA, USA*
[2] *Emory University School of Medicine, Departments of Radiology, Neurology, and Psychiatry and Behavioral Sciences, Atlanta, GA, USA. University of Pittsburgh, School of Medicine, Departments of Psychiatry and Neurology, Pittsburgh, PA, USA*
[3] *Emory University School of Medicine, Departments of Gynecology and Obstetrics, Psychiatry and Behavioral Sciences, Atlanta, GA, USA*
[4] *University of Pittsburgh, School of Medicine, Departments of Psychiatry, Obstetrics and Gynecology, and Epidemiology, Pittsburgh, PA, USA*

PART I: THE CLINICAL DISORDER

INTRODUCTION

Postpartum depression (PPD) has drawn tremendous public attention in the last decade. Most deeply etched in our minds are the heart-wrenching stories of infanticidal mothers who suffer from severe postpartum disorders (Spinelli, 2004). Also in our awareness are public figures who have struggled with and have ultimately triumphed over PPD (Osmond *et al.*, 2001; Shields, 2005). PPD has finally come to the attention of United States law makers, far behind the United Kingdom and other countries that have for decades devoted resources to sustain the mental health of new mothers. In the state of New Jersey, new legislation requires screening of all recently delivered women for PPD. Similar legislation (The MOTHERS Act: Mom's Opportunity to Access Health, Education, Research, and Support for Postpartum Depression) is under review as national law (http://thomas.loc.gov/cgi-bin/query/z?c109:S.3529). Mandatory screening for PPD offers the promise of greater recognition and treatment of PPD.

DEFINITION OF PPD

While the term "PPD" has widespread recognition value, it is becoming a less accurate descriptor as we learn more about the breadth of puerperal disorders. According to the Diagnostic and Statistical Manual-edition IV (Association, 1994), PPD is an episode of major depressive disorder (MDD) that begins within 4 weeks of childbirth. Evolving research, however, suggests that this time specifier needs to be reassessed. Thirty to sixty percent of postpartum mood disorders begin during pregnancy (Stowe *et al.*, 2005) (personal communication K. L. Wisner) and the incidence of major depression clearly extends beyond 1 month postpartum

(Kendell *et al.*, 1976, 1987; Gaynes *et al.*, 2005; Munk-Olsen *et al.*, 2006). "Perinatal depression" has therefore taken the place of the term "PPD" in clinical and research parlance. In addition, puerperal disorders extend beyond the boundaries of perinatal depression to include anxiety (e.g., panic disorder, obsessive compulsive disorder), manic, and psychotic disorders (Moses-Kolko & Feintuch, 2002), which have received far less research attention.

According to the Diagnostic and Statistical Manual-edition IV (Association, 1994), an episode of major depression is defined by a period of 2 weeks in which there is persistent low/sad mood (dysphoria) and/or loss of interest (anhedonia). A total of five symptoms (including dysphoria and/or anhedonia) must be present in the majority of the 2-week period and can include disturbances in appetite, sleep, energy, psychomotor activity, concentration, and the symptoms of guilt/worthlessness, hopelessness, and suicidal thoughts. Depression must be associated with significant distress or functional impairment and cannot occur solely during bereavement or be the direct physiological effect of a substance or medical condition.

THE ANTENATAL WINDOW OF OPPORTUNITY

The recognition and management of mental illness in the antenatal period is critical because of potential adverse effects of depression, anxiety, and stress on fetal development. Severe stress, such as the death of a child or spouse during early pregnancy, is associated with an increased risk of fetal birth defects (Hansen *et al.*, 2000). As reviewed by Bonari *et al.*, (2004), maternal anxiety, depression symptoms, and stress are associated with greater obstetric complications [miscarriage, vaginal bleeding, preterm birth, operative deliveries, intensive neonatal care admission complications, and preeclampsia (Kurki *et al.*, 2000)] as well as increased rates of low birth weight (adjusted for gestational age) and small head circumference in offspring. Potential mechanisms of intrauterine growth restriction and preterm birth include anxiety/stress effects to increase uterine artery resistance (Teixeira *et al.*, 1999) and serum corticotrophin releasing hormone (CRH) concentration (Wadhwa *et al.*, 2004). Maternal depression and stress may also modify fetal autonomic nervous and hypothalamic–pituitary–adrenal (HPA) axis system regulation (Monk *et al.*, 2003) to increase offspring susceptibility for psychiatric disturbance.

Screening for and management of stress and psychiatric disorders during the antenatal period has the potential to improve pregnancy outcomes. Pregnant women can be highly motivated to modify behaviors that have potential to negatively impact the fetus, such as smoking, using illicit drugs, and drinking alcohol, all of which are common in mental illness (Wisner *et al.*, 2007). Heightened awareness of adverse effects of maternal obesity on fetal development is a strong motivator for some women to pursue exercise programs and healthier diets preconceptionally and throughout pregnancy. Several research and clinical programs harness this unique time in women's lives to assist women achieve their goals for behavioral change (Armstrong *et al.*, 2003).

PSYCHOSOCIAL CONTEXT OF PREGNANCY

Becoming a mother is a significant role change that requires substantial intrapsychic and interpersonal adjustment, especially for primiparous women. Positive experiences with children, with one's own parents, and higher levels of social support, including marital harmony are factors that mitigate risk for depression (O'Hara & Swain, 1996). Because the majority of pregnancies are unplanned, the effectiveness of coping with a pregnancy and planning for the arrival of a newborn depends on the structure and level of commitment of the antenatal parental/family unit. Women of lower socioeconomic status and in minority groups are at increased risk for perinatal depression (Hobfoll *et al.*, 1995; Beeghly *et al.*, 2003). Potentially complicating the perinatal course are other serious psychosocial stressors, including intimate partner violence, single parenthood, housing problems, adolescent pregnancy, and pre-existing roles as caregivers for older children and parents (Scholle & Kelleher, 2003). Perinatal programs that provide access to social services have shown preliminary benefit for pregnancy outcomes and depression prevention (Zlotnick *et al.*, 2006; Wisner *et al.*, 2007).

PREGNANCY DOES NOT PROTECT WOMEN FROM DEPRESSION

In Research Diagnostic Criteria and DSM-IV-based studies, the incidence of antenatal minor and major depression ranged from 7% to 26% (O'Hara *et al.*, 1984, 1990; Hobfoll *et al.*, 1995). Despite high rates of antenatal depression, psychiatric hospitalization was not increased antenatally (Munk-Olsen *et al.*, 2006). Because MDD encompasses somatic symptoms that are pervasive in pregnancy (i.e., fatigue, appetite changes, sleep abnormalities), the diagnosis of MDD may be missed (Klein & Essex, 1995). In cases of suspected antenatal MDD, a score of 14 or higher on the 10-item Edinburgh Postnatal Scale for Depression (EPDS) (Cox *et al.*, 1987), is highly suggestive of MDD (Murray & Cox, 1990).

The work of Cohen *et al.* (2006) instructs us further that pregnancy is not protective against depression. Among women with an MDD history who discontinued antidepressant medications periconceptionally, nearly 70% experienced antenatal depressive relapse compared to 26% of women who maintained their medications during pregnancy (a 5-fold increased risk for relapse; Figure 12.1). Even temporary antenatal discontinuation of antidepressants was associated with increased risk for depressive relapse. Demographic risk factors for pregnancy-associated depression relapse included age 32 or younger, depression duration greater than 5 years, and a history of greater than 4 prior depressive episodes.

POSTPARTUM BLUES

Baby blues is a ubiquitous syndrome that affects 50–80% of new mothers and consists of tearfulness, mood lability, feeling overwhelmed, loss of appetite, insomnia, and also joy and elation. Blues generally do not come to the attention of a physician because the syndrome is mild, short-lived, the mother lacks functional impairment, and mother–infant attachment is unaffected. Because this condition raises the risk for PPD 3-fold (Henshaw *et al.*, 2004), longitudinal monitoring of women with blues may be reasonable in women with personal and family history of depression. Whether the commonly observed elation and euphoria of the early puerperium is best captured within the blues spectrum or as a subsyndromal manifestation of bipolar disorder (Heron *et al.*, 2005) deserves further study.

PPD

PPD that meets criteria for minor or major depressive disorder affects 7–26% of women after they deliver a baby (O'Hara *et al.*, 1984, 1990; Hobfoll *et al.*, 1995). Using the PPD

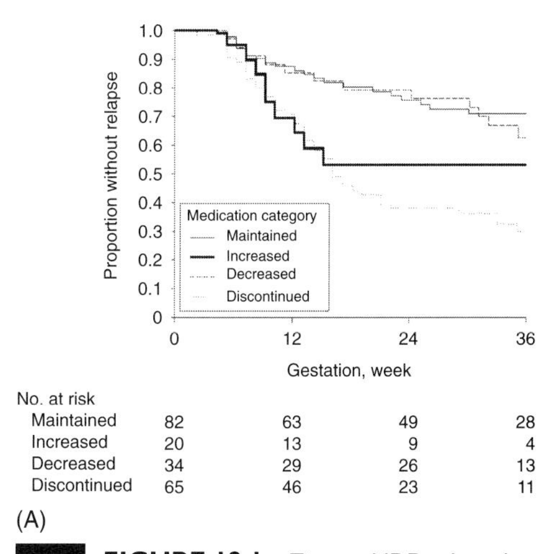

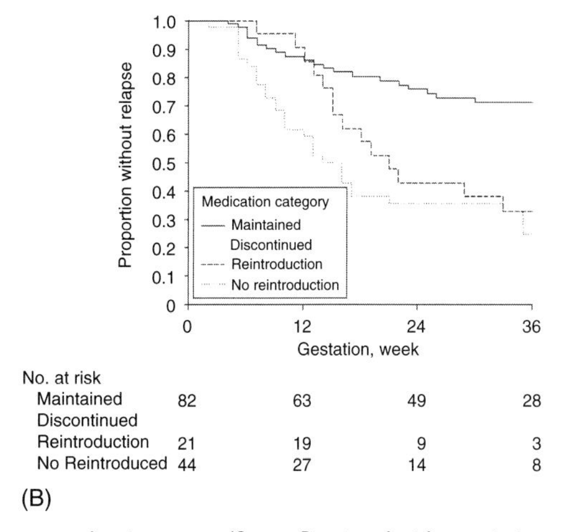

FIGURE 12.1 Time to MDD relapse by pregnancy medication status. (*Source*: Reprinted with permission Cohen *et al.*, 2006.)

prevalence rate of 14.5% from the most recent meta-analysis of minor and major depression onset in the first 3 months postpartum (Gaynes *et al.*, 2005), nearly 600,000 new mothers in the United States will experience PPD every year. PPD is more common in primiparous women (Munk-Olsen *et al.*, 2006) and in women with a personal and family history of mood disorder (O'Hara & Swain, 1996). In a recent obstetric-psychiatric database linkage study, Munk-Olsen *et al.* (2006) reported relative risks of 2.0–3.5 for first-time psychiatric hospitalization in the first 5 months postpartum for women with unipolar depression (relative to months 6–11 postpartum) (Figures 12.2 and 12.3). They also reported that outpatient psychiatric contacts increase 1.6–2.6-fold in the first month postpartum. These researchers found no increased risk for psychiatric hospitalization or outpatient contact in fathers throughout the perinatal period, a finding that gives further credence to the maternal-specific risk for postpartum psychopathology. In their comparison of parents and non-parents, Munk-Olsen *et al.* (2006) found an age-by-parental status interaction, whereby younger parents (ages 18–27) and older non-parents (ages 27–38) had increased relative risk of psychiatric hospitalization.

Similar to depression that occurs at other stages of the life span, PPD is phenomenologically heterogeneous. A large proportion of women with PPD have co-morbid anxiety symptoms (Hendrick *et al.*, 2000), anxiety disorders, or chronic depression that preceded the postpartum period. PPD can also follow a mild and insidious course which makes it less apparent to affected women and their families. Women with PPD may minimize symptoms and fail to recognize the need for treatment because of ongoing stigmatization of mental illness.

Untreated PPD poses unfortunate public health consequences. Maternal depression is associated with adverse developmental effects in offspring, including slower neurocognitive development (Grace *et al.*, 2003), and higher risk

Diagnosis	Pregnancy†	0–30 d	31–60 d	3–5 mo	6–11 mo
Unipolar depressive disorders					
No. of Cases	56	**38**	**48**	**113**	81
Rate per 1,000 person-years	0.116	**0.748**	**0.949**	**0.563**	0.274
*RR (95% CI)	0.44 (0.31–0.62)	**2.79 (1.90–4.11)**	**3.53 (2.47–5.05)**	**2.08 (1.57–2.77)**	1.00

FIGURE 12.2 Diagnosis-specific risks of first-time hospital admission in mothers 0–11 months postpartum time since birth of first live-born child. (*Source:* Adapted from Munk-Olsen *et al.*, 2006.) † Pregnancy was defined as beginning 270 days before birth of child and ending on day of birth. * Relative risk is adjusted for age and calendar time (5-year groups).

Diagnosis	Pregnancy†	0-30 d	31–60 d	3–5 mo	6–11 mo
Bipolar affective disorders					
No. of Cases	2	**26**	**7**	11‡	
Rate per 1,000 person-years	0.004	**0.512**	**0.138**	0.022‡	
*RR (95% CI)	0.19 (0.04–0.86)	**23.33 (11.52–47.24)**	**6.30 (2.44–16.25)**	1.00‡	

FIGURE 12.3 Diagnosis-specific risks of first-time hospital admission in mothers 0–11 months postpartum time since birth of first live-born child. (*Source:* Adapted from Munk-Olsen *et al.*, 2006.) † Pregnancy was defined as beginning 270 days before birth of child and ending on day of birth. * Relative risk is adjusted for age and calendar time (5-year groups).

of infant attachment disorders (Murray, 1992) and psychopathology in childhood (Goodman & Gotlib, 1999; Misri *et al.*, 2006). Conversely, treatment of maternal depression can improve psychiatric outcomes in children (Weissman *et al.*, 2006). Breastfeeding may have a role in decreasing anxiety and improving well-being in mothers (Altemus *et al.*, 1995; Mezzacappa *et al.*, 2000; Mezzacappa & Katkin, 2002); however, because antenatal depression is associated with lack of confidence and intention to breastfeed (D. Bogen unpublished), depressed women may not avail themselves of this strategy and infants may be deprived of the health benefits of breast milk.

Women who present with PPD may have a primary diagnosis of bipolar disorder. Bipolarity poses a 23-fold increase in risk for psychiatric hospitalization in the first postpartum month (Figure 12.3) (Munk-Olsen *et al.*, 2006). The diagnosis of bipolar disorder requires the presence of major depressive episodes as well as hypomania, mania, or mixed depression-manic episodes. Hypomania is defined as a period of four or more days in which there is persistently elevated, expansive, or irritable mood along with at least three of the following symptoms: inflated self-esteem, decreased need for sleep, increased talkativeness, racing thoughts, distractibility, increase in goal-directed activity, and excessive involvement in pleasurable activities with high potential for painful consequences. Whereas hypomania is associated with lack of functional impairment (or a potential functional improvement), mania is a functionally impairing state which lasts at least 1 week (American Psychiatric Association, 1994). During mixed mood states, full criteria for both mania and major depression must be present. Lifetime bipolar illness is categorized as bipolar I disorder if there has ever been a manic or mixed episode while bipolar II disorder requires the presence of one hypomanic episode.

While there are no reported gender differences in the incidence of bipolar I (1%) and bipolar spectrum (4–8%) disorders, women's lifetime pattern of the illness is characterized by a higher incidence of depressive, rapid cycling (≥4 episodes per year), and mixed episodes relative to men (Leibenluft, 2000). In the NIMH Bipolar Genetics Initiative substudy related to illness in women (Blehar *et al.*, 1998), 50% of mothers reported postpartum illness episodes, of which all were depressive in polarity. Bipolar depression is commonly associated with atypical features that include hyperphagia, hypersomnia, and diurnal variation (where mood and energy vary between morning and night), and is often treatment refractory (Hlastala *et al.*, 1997). Because bipolar disorder is frequently associated with substance abuse (Kessler *et al.*, 1997; O'Brien *et al.*, 2004), chaotic lifestyles, and postpartum psychosis (PPP) (see below), clinicians must be vigilant for the effects of this disorder not only on the identified patients, but also on their families.

POSTPARTUM ANXIETY DISORDERS

Postpartum anxiety disorders include obsessive compulsive disorder (OCD), panic disorder, and generalized anxiety disorder (GAD). OCD can be exacerbated antepartum as women prepare for the baby's arrival. "Nesting" is an evolutionarily adaptive behavior described in many species, but can reach maladaptive proportions clinically (Altemus, 2001). Postpartum obsessions commonly focus on the baby's safety and are associated with compensatory compulsive behaviors that include checking of the baby, handwashing, and cleaning. OCD or obsessional thoughts that coexist with MDD include distressing aggressive obsessions to harm the infant (Wisner *et al.*, 1999), without aggressive intent. Clinicians must differentiate aggressive obsessions from a psychotic disorder in which aggressive or infanticidal thoughts are perceived by the mother as appropriate and necessary to spare the child a life in a bad world, for example (altruistic delusion; see PPP below).

Panic disorder can also worsen perinatally. Expansion of the pelvic organs into the abdominal space and consequent contraction of the respiratory space may alter the oxygen–carbon dioxide balance to increase the likelihood of panic attacks (March & Yonkers, 2001). Panic disorder can be associated with agoraphobia and interfere with obstetric and pediatric medical visits. GAD and chronic stress may not come to the attention of a psychiatrist until there is a co-morbid depressive disorder. A common presentation of GAD is repeated office calls and visits for reassurance regarding the health status of the pregnancy or infant.

POSTPARTUM PSYCHOSIS

PPP is considered a psychiatric emergency because it seriously impedes judgment and function. Infanticide and suicide are devastating outcomes of PPP (Spinelli, 2004). Women with PPP can appear delirious, with marked disorientation, confusion, and disruption of thought and speech processes (Brockington *et al.*, 1981; Wisner *et al.*, 1994). Neurovegetative changes in energy, psychomotor activity, appetite, and sleep are also marked. Psychotic symptoms can range from milder perceptual disturbance (illusions) to nihilistic delusions that are accompanied by thoughts and acts of suicide and infanticide. Infanticide occurred at a rate of 9 in 100,000 in the United States between 1988 and 1991, with PPP suspected to be the most common precipitant (Overpeck, 1998).

PPP is a disorder that usually emerges in the first postpartum month and is therefore considered a psychiatric disorder linked to disruptions that commence at delivery (i.e., hormonal, circadian). The risk for a primiparous woman to develop PPP in the first postpartum month is 35–38-fold greater relative to her risk in pregnancy (Kendell *et al.*, 1987) (Figure 12.4) or in the 6–11 months after childbirth (Munk-Olsen *et al.*, 2006). While PPP affects 1–2 in 1,000 women, the incidence is likely much higher among women with bipolar disorder (Sit *et al.*, 2006). The likelihood of PPP recurrence is 60% after a single episode (Robertson *et al.*, 2005) and 100% after two prior episodes (Sichel *et al.*, 1995). Susceptibility genes for PPP are under investigation after the discovery of 85% concordance of PPP among parous sisters (Jones & Craddock, 2001), and may include genes in chromosomal regions 16p13 and 8q24 (Jones *et al.*, 2007).

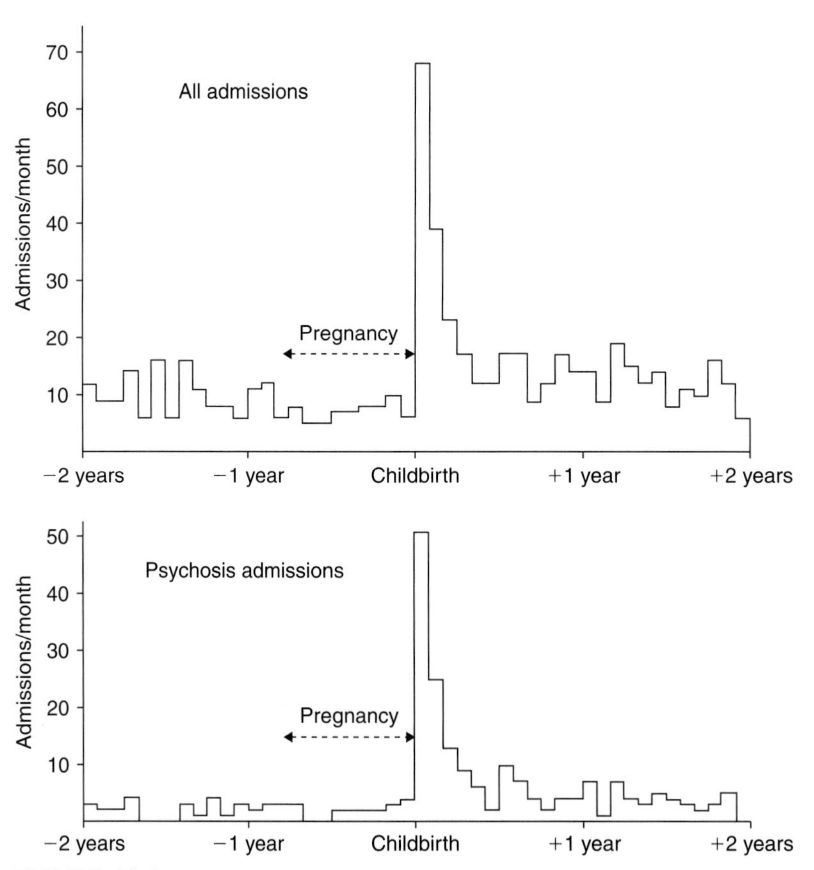

FIGURE 12.4 Temporal relationship between psychiatric hospital inpatient admission and childbirth. (*Source*: Reprinted with permission Kendell *et al.*, 1987.)

SCREENING FOR PERINATAL DEPRESSION

Screening for antenatal and PPD can be easily implemented through the use of the EPDS (Cox *et al.*, 1987). This is a 10-item, self-report scale for which a cut-off score of ≥ 10 has the greatest sensitivity and specificity for measuring major and minor depression (Hanusa *et al.*, in press; Peindl *et al.*, 2004). This scale was designed to emphasize cognitive and affective facets of mood and de-emphasize somatic and anergia symptoms, which are so common in perinatal women regardless of depression status. The EPDS has been validated against gold-standard diagnostic interviews to establish major depression, such as the Schedule for Affective Disorders and Schizophrenia and the Structured Clinical Interview for DSM-IV (First *et al.*, 1998) and has been translated into numerous languages.

TREATMENT OF PERINATAL DEPRESSION

Risk-Benefit Decision-Making Process

Despite the emergence of new, controversial contributions to this field, what remains constant is that the clinical approach remains grounded in a thorough risk-benefit decision-making process between physician, mother, and (often) her partner (Wisner *et al.*, 2000). In this manner, treatment selection occurs in the context of a detailed discussion. The discussion begins with a thorough review of maternal diagnosis, treatment history, and response. Next, a detailed review of risk to fetus/infant posed by effective medications for the disorder and by the disorder itself is provided with respect to six stages of fetal/child development: (1) fetal viability, (2) first-trimester-associated birth defects, (3) growth and structural development, (4) neonatal toxicity/withdrawal, (5) lactation, and (6) neurobehavioral teratogenicity. Next information is gathered from the mother (and partner) about what her values are as she considers her options. Finally the physician reflects back to the mother what the mother's choice is. This process is iterative over time: as symptoms evolve, pregnancy progresses, and new clinical information emerges, potentially new clinical information and risk considerations are brought to bear upon decision-making.

Non-pharmacologic Strategies

Because there are limitations in our knowledge of potential fetal/infant risks conferred by gestational and lactational psychotropic exposure, non-pharmacologic strategies are important treatment modalities. Psychotherapy is an effective treatment for perinatal depression (Appleby *et al.*, 1997; O'Hara *et al.*, 2000; Spinelli, 2001; Spinelli & Endicott, 2003). Preliminary efficacy has been established for bright light therapy (Epperson *et al.*, 2004), omega-3-fatty acids (for review see (Freeman, 2006), and acupuncture (Manber *et al.*, 2004) in the treatment of perinatal depression. Aerobic exercise (30 min aerobic activity 3–5 times/week) was effective for the treatment of mild to moderate depression (Dunn, 2005) and remains to be tested in perinatal women. Non-pharmacological treatment approaches also have a role in prevention of perinatal depression in women at risk (Zlotnick *et al.*, 2006) or as adjuncts to psychotropics in women with moderate to severe depressive disorders.

Antidepressant Medications

Antidepressant use in pregnancy is on the rise. According to data from a 2003 medicaid cohort, approximately 500,000 pregnant women (13.4% of pregnancies) filled a prescription for an antidepressant medication (Cooper *et al.*, 2007). While early reports indicated that first-trimester use of serotonin reuptake inhibitors (SRIs) was not associated with increased rates of birth defects (Chambers *et al.*, 1996; Goldstein *et al.*, 1997; Nulman *et al.*, 1997), recent studies indicated increased risk. Exposure to paroxetine was associated with 2–3% cardiovascular defect risk relative to 1% baseline risk in several recent reports (Berard *et al.*, 2006; Bar-Oz *et al.*, 2007; Louik *et al.*, 2007), but not others (Alwan *et al.*, 2007). Post week 20 of gestation use of SRIs was associated with persistent pulmonary hypertension of the newborn (PPHN) risk in 12 cases per 1,000 relative to the baseline risk of 1–2 cases per 1,000 live births (Chambers *et al.*, 2006). The low absolute risk of these birth defects underscores the importance of accurate communication of potential risk due both to the psychotropic and to the disorder itself. Other classes of antidepressants including the tricyclic antidepressants (McElhatton *et al.*,

1996; Ericson *et al.*, 1999) and buproprion (Chun-Fai-Chan *et al.*, 2005) (http://pregnancyregistry.gsk.com/bupropio.html) have not been associated with increased risk of birth defects compared to the general population risk of 3–5%.

The latter part of pregnancy is a period of rapid fetal growth as well as ongoing central nervous system (CNS) development. Both maternal selective serotonin-reuptake inhibitor (SSRI) use (Simon *et al.*, 2002; Suri *et al.*, 2007) and maternal depression (Bonari *et al.*, 2004; Oberlander *et al.*, 2006) have been associated with preterm delivery and reduced infant birth weight. Additional studies are needed that parse out antidepressant-specific vs. maternal depression risks on infant outcomes (Chambers *et al.*, 2007; Kelly *et al.*, 2007).

Once the umbilical cord is severed, the infant is removed from the metabolic enzymes of the mother and also from the continued maternal–fetal transfer of the psychotropic agent. Thirty percent of infants with late gestational exposure to SRI antidepressants will experience poor neonatal adaptation (Chambers *et al.*, 1996; Moses-Kolko *et al.*, 2005b), which most commonly consists of tremor, rigidity, irritability, and feeding, sleeping, and respiratory disturbances. Less than 1% of cases meet criteria for a severe syndrome (Moses-Kolko *et al.*, 2005b). Neonatal syndrome related to *in utero* SRI exposure is generally mild and short-lived, lasting no more than 1 month, and does not appear to negatively impact child development (Oberlander *et al.*, 2004). The syndrome may be more common after exposure to antidepressants with long half-lives and active metabolites (i.e., fluoxetine) or anticholinergic effects (i.e., paroxetine and tricyclic antidepressants) (Moses-Kolko *et al.*, 2005b). Antidepressant dose reduction in late pregnancy may mitigate the syndrome but must be balanced against maternal risk for recurrence (Wisner & Perel, 1988).

Lactational antidepressant transfer is much lower than cross-placenta antidepressant transfer. Negligible drug gets into the circulation of nursing infants except when the infant has combined lactational and *in utero* antidepressant exposure (particularly for long-acting antidepressants). Favored antidepressants during lactation include sertraline, paroxetine, and nortriptyline because of their short half-lives, undetectability in infant serum, and absence of associated adverse events (Weissman *et al.*, 2004). Adequately powered neurobehavioral studies revealed absence of *in utero* antidepressant effects on later child development (Nulman *et al.*, 1997, 2002). Conversely, inadequately treated maternal depression remains the most compelling risk factor for behavioral and developmental difficulties in offspring (Murray, 1992; Goodman & Gotlib, 1999; Misri *et al.*, 2006).

Mood Stabilizer Medications

Selection of a mood stabilizer medication is a clinical challenge because of known teratogenicity conferred by most agents. Lithium is a preferred mood stabilizer because of its lower association with teratogencitiy compared to antiepileptic drugs (AED). *In utero* lithium exposure is associated with a 1 in 1,000 to 1 in 2,000 risk of Ebstein's Anomaly (downward displacement of the tricuspid valve) relative to a 1 in 20,000 risk in the general population (Cohen *et al.*, 1994). Other risks of lithium exposure during pregnancy include increased infant birth weight and polyhydramnios. Neonatal effects described after *in utero* lithium exposure include cyanosis, hypotonia, hypothyroidism, diabetes insipidus, a decrease in bone calcium, and T-wave changes on EKG. Because lithium equilibrates across the placenta, dose reduction prior to delivery or at the onset of labor is recommended (Newport *et al.*, 2005). Although lactation while taking lithium was discouraged previously by the American Academy of Pediatrics, upon reevaluation of such risk in a small sample of well-educated mothers, low serum lithium concentrations in the range of 0.09–0.25 meq/L were detected in nursing infants and no persistent medical concerns arose in the closely monitored infants (Viguera *et al.*, 2007). Neurobehavioral data reveal lack of developmental effects in lithium-exposed children relative to their non-exposed siblings (Schou, 1976).

AED are commonly used for the treatment of bipolar disorder. The majority of pregnancy data for AEDs are obtained from populations of women with seizure disorders, which themselves pose increased risk for adverse obstetrical outcomes. Fetal death was increased (2.9–3.6%) after first-trimester pregnancy exposure to valproate and carbamazepine, but not lamotrigine (Meador *et al.*, 2006). First-trimester AED use was also associated with increased rates of malformations in cardiac, orofacial, urologic, skeletal, and neural tube systems. The rank order of AED-associated congenital malformations was: valproate (17.4%), carbamazepine (4.5%), lamotrigine (1.0%) (Meador *et al.*, 2006). Lamotrigine was not associated with increased rates of birth defects relative to the general populations in most studies (Vajda *et al.*, 2004; Cunnington & Tennis (2005); Meador *et al.*, 2006; Cunnington *et al.*, 2007). A possible relationship between early pregnancy lamotrigine exposure and oral clefts (8.9 cases per 1,000 vs. 0.16 cases per 1,000 in the general population) (Holmes *et al.*, 2006) requires replication.

Late pregnancy use of valproate and carbamazepine is associated with facial dysmorphism, cardiac and genitourinary effects, digital and nail hypoplasia, low birth weight, small head circumference, developmental delay, and mental retardation. Neonatal effects after valproate and carbamazepine exposure include hepatic toxicity, specifically hyperbilirubinemia. Vitamin K-dependent clotting deficits due to vitamin K deficiency related to these drugs necessitate maternal vitamin K administration at 36-weeks gestation and neonatal vitamin K administration at birth. Both valproate and carbamazepine are absorbed into breastmilk; however, the drug dose to nursing infants is less than 10% of the maternal dose (Bar-Oz *et al.*, 2000; Piontek *et al.*, 2000). An isolated report of thrombocytopenia in a 3-month old infant with valproate exposure and several reports of hepatotoxicity and poor feeding with carbamazepine exposure signal the importance for infant monitoring. Little is known about later pregnancy effects of lamotrigine. Infants exposed to lamotrigine through breastmilk could develop therapeutic levels of the drug (Ohman *et al.*, 2000) concurrent with poor capacity for lamotrigine metabolism; this indicates the importance of careful monitoring for skin rash and hepatotoxicity in lactationally exposed infants.

The antipsychotic medication class has largely been effective in the management of mania and the atypical antipsychotics may have a role in the maintenance phase of bipolar disorder. Because the phenothiazine antipsychotic perphenazine was found to be highly efficacious in a large clinical trial of subjects with schizophrenia (Stroup *et al.*, 2003), it is an agent worthy of consideration in bipolar disorder as well. As a mid-potency neuroleptic, the likelihood of extrapyramidal motor symptoms and sedation is low with perphenazine, which improves tolerability. A small increase (2.4% relative to 2.0%) in the rate of birth defects was observed with the phenothiazines relative to non-teratogenic controls which was attributed to factors associated with the underlying illness (Altshuler *et al.*, 1996).

The main limitation to atypical antipsychotic agents is that relatively little data is available regarding use in pregnancy and lactation. In samples of 96 and 68 women who took olanzapine (McKenna *et al.*, 2003) (Lilly database) and risperidone (Coppola *et al.*, 2007), respectively, in pregnancy, neither atypical antipsychotic agent was associated with increased birth defect rates relative to non-teratogens. Potential metabolic effects associated with the atypical antipsychotic class of drugs, including weight gain, hyperlipidemia, and insulin resistance (Newcomer, 2007), increase the risk of obstetric and fetal complications (Watkins *et al.*, 2003). The potential for transmission of metabolic defects from mother to offspring via *in utero* and lactational exposure to atypical antipsychotics also must be factored into the decision-making process. Olanzapine has been associated with a 20% rate of adverse effects in a small sample of lactationally exposed infants (Lilly database). Clozapine possesses the same metabolic concerns as olanzapine. In addition, adverse birth outcomes and the potential for agranulocytosis with clozapine use detract from the benefits of this medication in highly refractory patients. The atypical agent quetiapine has not been shown to be teratogenic, however small sample sizes limit the conclusions (Gentile, 2004). Only case report data exist for aripiprazole and quetiapine.

CONCLUSION

In summary, the perinatal period is replete with intrapsychic, interpersonal, and biological adjustments that render many women vulnerable to mental illness onset and episode recurrence. Increased dissemination of information regarding these illnesses stands to improve recognition and treatment. Ongoing research efforts are needed to fine tune the risk-benefit decision-making process for the use of psychotropics perinatally. In the absence of the ideal information database, sound clinical evaluation to select an efficacious drug (based on patient history and the literature) and a thorough risk-benefit discussion with patients are the mainstays of good perinatal psychiatric management.

PART II: NEUROBIOLOGICAL MECHANISMS OF POSTPARTUM DEPRESSION

OVERVIEW

The perinatal maternal brain undergoes dramatic changes as described in depth in other book chapters. The neuroendocrine adaptations and variations that occur perinatally no doubt represent endocrinology at its finest. Perinatal neuroplasticity, neurogenesis, and amplification of hormones and peptides are associated with evolutionarily beneficial adaptations in cognitive (i.e., spatial navigation; Kinsley & Lambert, 2006), emotional (fear reduction; Neumann, 2003), and maternal caregiving function (Numan *et al.*, 2006) in preclinical studies (putative model shown in Figure 12.5).

The nature of human maternal neuroplasticity and whether such neuroplasticity is possible in perinatal women with mental illness (or susceptibility to mental illness) is an important area for future investigation. Because healthy maternal adaptation confers benefits to offspring at the level of emotional, cognitive, and physical development, insights into mental illness-related aberrant maternal brain changes will facilitate the development of treatments to foster healthier adaptation to the maternal role. In the section that follows, we (1) summarize what is known about human maternal neurobiological adaptation and underscore the paucity of CNS research, (2) present a neuroendocrine model of PPD, (3) provide the rationale, methods, and risks for conducting positron emission tomography (PET) studies of central serotonin and dopamine (DA) receptors, and (4) present preliminary PET findings. Methodological considerations in the selection of postpartum samples, including issues of mood disorder polarity and endocrine context, postpartum timing, lactation, and menstruation are discussed.

SYNTHESIS OF EXTANT CNS STUDIES OF PPD

The majority of early neurobiological PPD research focused on comparisons of basal, peripheral monoamine, and hormone concentrations between postpartum women with and without depression or baby blues [for summary see (Moses-Kolko & Feintuch, 2002)]. With the exception of differences reported in allopregnanolone (Nappi *et al.*, 2001) and tryptophan availability (Bailara *et al.*, 2006) in women with postpartum blues relative to those without blues, no robust findings emerged. However, this research was limited in several ways. Peripheral measures are often disparate from CNS measures. Likewise, resting state hormone collection is likely less sensitive than neuroendocrine system challenge paradigms to discriminate depressed from non-depressed groups. The assessment of lactation and inclusion of lactation as a covariate was not routinely done. This research led to the conclusion that it was not absolute peripheral hormonal concentrations that mediated PPD, but rather a specific neurobiological alteration in susceptible women in

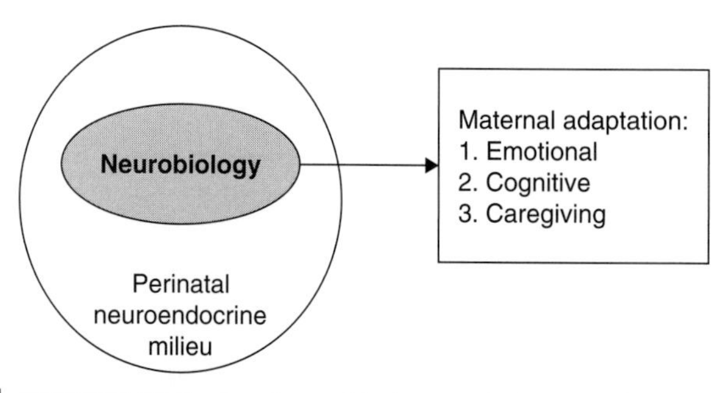

FIGURE 12.5 Putative model of maternal adaptation. Perinatal neurobiological changes directly affect all domains of maternal function.

response to a uniform hormonal change experienced by all childbearing women. This conclusion has gained momentum with the report that pregnancy simulation (continuous leuprolide-induced chemical gonadectomy over the 5-month study course, with add-back estradiol and progesterone for 8 weeks, followed by hormone withdrawal for 12 weeks) was associated with depressive symptom recurrence in 60% of women with a history of puerperal depression, but in none of the healthy controls (Bloch *et al.*, 2000). The fact that add-back estradiol and progesterone during the pregnancy simulation phase was also marginally associated with recurrence of depressive symptoms provides evidence of important endocrine-related effects on brain systems relevant to mood in the antenatal as well as postnatal period.

A NEUROENDOCRINE MODEL OF PPD

We posit that the perinatal hormonal context potently and rapidly activates a neural circuitry that promotes maternal adaptation, which includes the domains of cognition, emotion, and nurturance (Figure 12.5). In our version of the model proposed by Numan & Insel (2003) and Numan *et al.*, 2006 (see Figure 12.6), human maternal adaptation behaviors are determined by the balance of approach and fear system activation as modulated by inputs from hypothalamus, prefrontal cortex, hippocampus, and brainstem monoamines. PPD may result from unmodulated corticotrophin releasing factor (CRF) (from paraventricular hypothalamic nucleus and central nucleus of the amygdala) activation of the fear system (Numan & Insel, 2003). This hypothesis is supported by longer lasting and more severe blunting of ACTH responses to ovine CRF in depressed compared to euthymic postpartum women (Magiakou *et al.*, 1996). In addition, insufficient ventral tegmental area (VTA) to nucleus accumbens (NAc) dopaminergic signal would weaken the maternal approach system in PPD. Higher apomorphine-stimulated growth hormone concentrations in women who went on to develop PPD relative to those who remained healthy (Wieck *et al.*, 1991) suggested maladaptation in DA–hypothalamic circuits in PPD. Direct and indirect (via hypothalamus) cortical afferents into approach and fear systems are also suspected of being altered in PPD given evidence of such in MDD studies (Phillips *et al.*, 2003).

Emerging studies that employ functional magnetic resonance imaging (fMRI) in healthy, non-depressed, postpartum women, and mothers lend support for our model. Activation in amygdala, rostral anterior cingulate cortices,

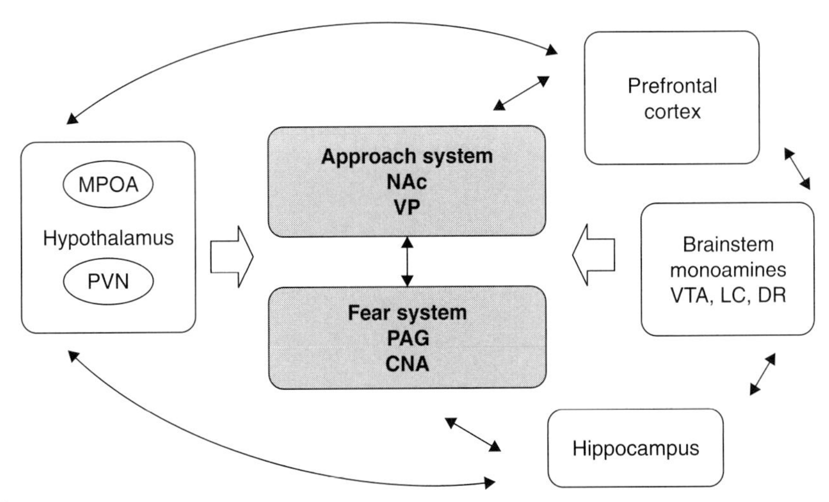

FIGURE 12.6 Neuroendocrine model of PPD. CNA, central nucleus of the amygdala; DR, dorsal raphe; LC, locus coeruleus; MPOA, medical preoptic area; NAc, nucleus accumbens; PAG, periaqueductal gray; PVN, paraventricular nucleus; VP, ventral pallidum; VTA, ventral tegmental area.

and striatum was commonly detected in contrasts of infant-laden vs. neutral auditory and visual cues and contrasts of own vs. other child (Lorberbaum *et al.*, 2002; Leibenluft *et al.*, 2004; Ranote *et al.*, 2004; Strathearn *et al.*, 2007). Using proton magnetic spectroscopy, Epperson *et al.* (2006) detected a reduction in occipital cortical GABA in postpartum women regardless of depressive status, which was interpreted as an index of puerperal cortical excitation. We have reported 5HT1A receptor decreases in PPD relative to postpartum controls in mesiotemporal cortex (amygdala and hippocampus) and anterior cingulate cortex (Moses-Kolko *et al.*, 2007, in press), which could be related to increased activation of fear systems (Fisher *et al.*, 2006) and/or HPA axis dysregulation (Lopez *et al.*, 1999). Our findings of D2 receptor alterations in puerperal relative to non-puerperal women provide preliminary evidence for modifications of the approach system, which may be uniquely modulated in depression (Moses-Kolko *et al.*, 2006). Below, we provide additional rationale and results from our PET studies of PPD.

SEROTONIN-1A RECEPTOR SYSTEM

The serotonergic (5HT) system has been widely implicated in mood disorders since the initial discovery of reduced cerebrospinal fluid (CSF) 5HIAA (primary 5HT metabolite) in major depression (Garlow *et al.*, 2000). Evidence of a specific association between the 5HT system and depressive disorders in reproductive-aged women includes the robust response of depressive disorders in reproductive-aged women to SSRIs [MDD (Kornstein *et al.*, 2000), PMDD (Wikander *et al.*, 1998; Freeman *et al.*, 1999), and PPD (Stowe *et al.*, 1995; Wisner *et al.*, 2006)]. Because female gonadal steroids have extensive effects on the 5HT system (Biegon, 1983; Sumner *et al.*, 1995; Cyr *et al.*, 1998; Pecins-Thompson *et al.*, 1998; Bethea *et al.*, 2000), the perinatal reproductive transition may pose a destabilizing challenge to the serotonergic system of susceptible women.

The *in vivo* study of human 5HT function is made possible through PET of molecular sites of 5HT action. We have focused on postsynaptic 5HT1A receptors located in the nonbrainstem limbic regions of our model because of well-known modulation of the 5HT1A receptor system by steroid hormones. In preclinical hormonal manipulation experiments, estradiol administration was associated with decreased 5HT1A receptor mRNA (Osterlund *et al.*, 1998) and receptor protein (Osterlund *et al.*, 2000), which was augmented by progesterone (Pecins-Thompson & Bethea, 1999). It is possible that following the massive estrogen and progesterone discharge of pregnancy and withdrawal of the postpartum, women develop PPD as a result of inadequate re-equilibration of 5HT1A receptor function. An absence of 5HT1A receptor plasticity across menstrual cycle phases was reported in the reproductive mood disorder, PMDD (Nordstrom *et al.*, 2004). Like estradiol, glucocorticoids, and paradigms of chronic unconditioned stress in animals also reduce 5HT1A receptor mRNA expression and receptor protein (Lopez *et al.*, 1999). Because hypercortisolism is also a dominant neuroendocrine state of childbearing, it is conceivable that suppressed concentration of 5HT1A receptors in pregnancy (by cortisol, estradiol, and progesterone) may be maintained in vulnerable women following the natural hypercortisolism of the puerperium.

5HT1A receptor dysfunction has also been repeatedly shown in MDD. Deficits in 5HT1A function and reductions in density and binding were demonstrated with neuroendocrine challenge studies (Lesch *et al.*, 1990), postmortem studies (Bowen *et al.*, 1989; Lopez *et al.*, 1998), and *in vivo* PET studies (Drevets *et al.*, 1999; Sargent *et al.*, 2000; Bhagwagar *et al.*, 2004), although there are some inconsistencies (Stockmeier *et al.*, 1998; Parsey *et al.*, 2006).

In our published PET studies (Moses-Kolko *et al.*, 2007), we evaluated brain 5HT1A receptor binding in postpartum depressed relative to control subjects with the selective 5HT1A receptor radioligand, carbonyl-labeled [^{11}C]WAY-100635 ([^{11}C]WAY). We hypothesized that PPD would be accompanied by a reduction of 5HT1A receptor binding relative to postpartum control women due to combined vulnerabilities of depressive illness, hypercortisolemia, and long-term hyperestrogenemia during pregnancy, all of which are associated with 5HT1A receptor reductions. Nine unmedicated PPD (4 bipolar; 5 unipolar) and seven postpartum control subjects underwent 5HT1A-PET procedures as described below. PPD was defined as prevalent depression, as six of nine PPD subjects experienced depression onset antenatally. Brain regions of interest included raphe nucleus and mesiotemporal, left lateral orbitofrontal, and subgenual cingulate cortices, because these regions are implicated in mood regulation (Phillips *et al.*, 2003) and have been associated with depression-related 5HT1A receptor decreases in prior studies (Drevets *et al.*, 1999). We found that postsynaptic cortical, but not presynpatic raphe nucleus 5HT1A receptor binding was significantly reduced 20–28% in PPD relative to postpartum controls (Figure 12.7). There was no difference in 5HT1A receptor binding on the basis of unipolar or bipolar disorder diagnosis (Figure 12.7). Further research is underway to compare 5HT1A receptor changes in PPD relative to MDD that occurs at other times of the lifespan.

DOPAMINE-2 RECEPTOR SYSTEM

DA system dysfunction has long been hypothesized to have an important role in MDD (Drevets *et al.*, 2001; Willner *et al.*, 2005; Nestler & Carlezon, 2006; Dunlop & Nemeroff, 2007). Because primary rewards are associated with striatal DA increases, striatal DA deficits are proposed to underlie the amotivation and loss of pleasure from rewarding stimuli (anhedonia) in MDD. Depression improvement with DA agonists and worsening with DA antagonists or DA-depleting agents provide face validity for this hypothesis. CSF DA, DA metabolite (Sher *et al.*, 2006), postmortem (Bowden *et al.*, 1997; Klimek *et al.*, 2002), and nuclear imaging studies (D'haenen & Bossuyt, 1994; Ebert *et al.*, 1996; Klimke *et al.*, 1999; Allard & Norlen, 2001; Parsey *et al.*, 2001; Meyer *et al.*, 2006) of D2 receptor changes in MDD further support the DA hypothesis of MDD; however, discrepant findings across imaging studies signal the need for further research. Gender (Pohjalainen *et al.*, 1998; Munro *et al.*, 2006; Riccardi *et al.*, 2006) and estradiol (Lammers *et al.*, 1999) effects on striatal DA systems may contribute to contradictory D2 receptor binding findings in MDD.

In addition to a putative gender-specific role of striatal DA in MDD, striatal DA function is uniquely regulated in the puerperium. Higher striatal DA concentrations were detected both at 4 days postpartum compared to estrous controls [decreased homovanillic acid/DA ratios; (Glaser *et al.*, 1990)] and in parous compared to nulliparous rodents (Byrnes *et al.*, 2001). Apomorphine-induced disruption of prepulse inhibition and greater stereotypy (mesolimbic behaviors) in parous compared to nulliparous rodents (Byrnes *et al.*, 2001) suggested either

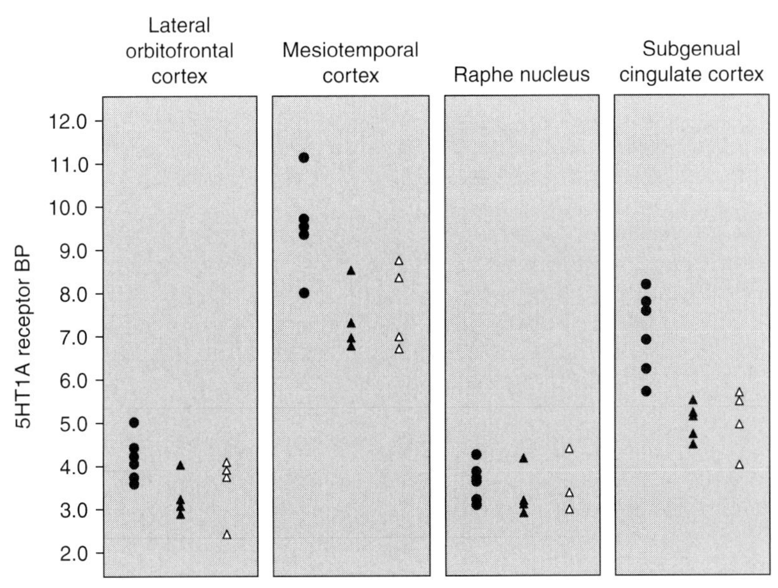

FIGURE 12.7 Postsynaptic 5HT1A receptors are reduced in postpartum unipolar (▲) and bipolar (△) depressed relative to postpartum controls (●).

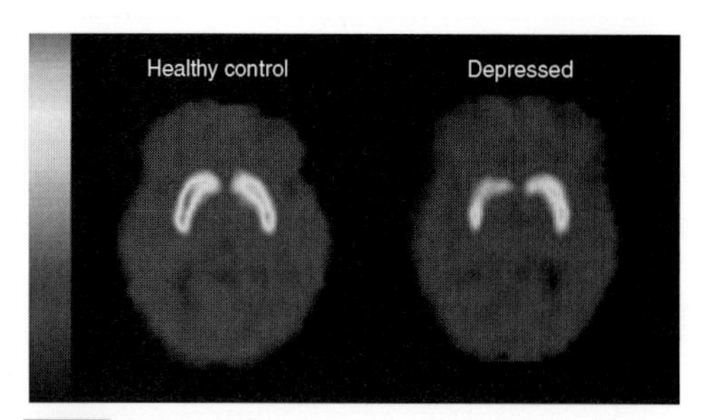

FIGURE 12.8 Reduced D2 receptor binding in depression. (See Color Plate)

increased D2 receptor density or increased DA release after reproductive experience. These postpartum DA system changes are associated with healthy maternal behavior, as maternal–pup care giving is associated with increased extracellular striatal DA (Champagne *et al.*, 2004) and D2 receptor blockade or DA system neurotoxin injection blocks maternal pup retrieval (Hansen *et al.*, 1991). Human studies of perinatal DA function are also suggestive of increased DA-receptor affinity and DA concentration (Gregoire *et al.*, 1990) changes in healthy controls (Petraglia *et al.*, 1987).

We used PET with the D2 receptor-specific radioligand [^{11}C]raclopride to examine the effect of depression and postpartum status on striatal D2 receptor binding (Moses-Kolko *et al.*, 2006). We hypothesized that DA activity would be dampened in pregnancy on the basis of high circulating estradiol levels, and that DA activity would rebound to adaptive levels in the puerperium. For women with susceptibility to MDD, we posited failure of DA system re-equilibration postpartum, leading either to hypo or hyperdopaminergic activity in mesolimbic circuits. Because [^{11}C]raclopride is displaced from D2 receptors by endogenous DA, we hypothesized increased D2 receptor binding (lower extracellular DA) in MDD and reduced D2 receptor binding (increased extracellular DA) in postpartum women.

Subjects were recruited into four groups as follows: non-postpartum control (n = 18), postpartum control (n = 7), non-postpartum MDD (n = 4), and PPD (n = 5). D2 receptor imaging was performed as previously described (Drevets *et al.*, 2001). The brain region of primary interest was the anteroventral striatum which encompasses the NAc. D2 receptor binding potential (BP) was reduced 10% in postpartum relative to non-postpartum women in anteroventral striatum, as hypothesized. Contrary to our hypothesis, but consistent with chronic stress models of depression (Grant *et al.*, 1998), D2 receptor binding was 16% lower in the MDD relative to the healthy control group in anteroventral striatum (Moses-Kolko *et al.*, 2006, see Figure 12.8). Further sample accrual is underway to adequately power analyses of depression-by-postpartum interaction effects on D2 receptor binding.

PET METHODS

PET is a unique imaging modality that allows for the *in vivo* assessment of molecular-level processes. Components of the PET imaging process are described elsewhere (Phelps, 2004). PET measures of specific receptor binding include distribution volume (V_T, ml/ml) and BP, (unitless). Both measures are related to B_{max}/K_D, where B_{max} (nM) is the concentration of receptors available for binding (not bound to endogenous ligand) and K_D (nM) is the equilibrium dissociation constant that is an index of the affinity of the PET tracer for the binding site.

5HT1A Receptor System

High specific activity [^{11}C]WAY is a robust 5HT1A receptor radioligand (Pike *et al.*, 1996), whose binding follows the regional rank order of 5HT1A receptor density observed in human postmortem tissue (Hall *et al.*, 1997). [^{11}C]WAY is insensitive to endogenous 5HT and can also bind to high and low affinity (internalized) receptor states. Tracer kinetic methods for [^{11}C]WAY have been extensively evaluated and refined (Parsey *et al.*, 2000, 2005; Price *et al.*, 2002a). Greatest accuracy of the BP parameter requires acquisition of the arterial input function, whereas high reliability of the BP parameter is achieved with a non-arterial-based reference tissue model (Parsey *et al.*, 2000).

D2 Receptor System

[^{11}C]raclopride provides robust *in vivo* measures of striatal D2 receptor binding, has been carefully validated, and is widely used by the field (Laruelle *et al.*, 2003). Because [^{11}C]raclopride is vulnerable to endogenous DA competition, it can be used in functional challenge studies to yield an index of intrasynaptic DA release (Laruelle, 2000). The simplified reference tissue model is a valid method for measurement of [^{11}C]raclopride BP and has largely taken the place of arterial-based graphical and compartmental models (Mawlawi *et al.*, 2001). Improved PET scanner resolution in the last decade offers the potential to resolve functionally distinct striatal subregions (Mawlawi *et al.*, 2001). The anteroventral striatum, which encompasses the NAc, is a region of particular interest in the study of PPD because NAc DA function has been specifically implicated in maternal behavior (Champagne *et al.*, 2004) and mood (Drevets *et al.*, 2001).

CONSIDERATION OF RISKS TO MOTHER AND INFANT DURING PET IMAGING

The risks to consider in the informed consent process for PPD subjects enrolled in PET studies include exposure to radioactivity, complications of arterial cannulation, and delay to treatment.

Radioactivity Exposure

Radioligand selection dictates the amount of radioactivity exposure in a PET study. One Sievert (Sv) per year is the limit identified for radiation protection of the general population. A single radioligand study generally involves less than 5 mSv of radioactivity exposure. Multiple radiotracer studies and longitudinal studies that involve serial scan acquisitions must comply with a maximum radiation exposure of 50 mSv per year, which is the annual whole body radiation exposure permitted to radiation workers by federal regulations. Pregnancy is an exclusion for PET studies due to teratogenic effects of radiation. Lactating women can be safely included in PET studies on the basis of our report that negligible radioactivity (0.6–2.7 μSv) was present in expressed breast milk post-PET scan (Moses-Kolko *et al.*, 2005a). While high dose radiation exposure is a well-established risk factor for cancer induction, whether low radiation doses (0–100 mSv) pose cancer risk remains controversial (Semelka *et al.*, 2007). As the numerical risk of mortality from radiation-induced cancer is 5% per Sv (The international commission on radiation protection: Radiation dose to patients from radiopharmaceuticals: Addendum 6 to icrp publication 53, 2002), the estimated mortality from a single radioligand PET study is negligible (2.5 per 10,000 individuals) compared to the risk of cancer overall (42 per 100 individuals) (Semelka *et al.*, 2007).

Arterial Cannulation

The PET data modeling approach differs on the basis of the radioligand of interest. For some radioligands, the arterial input function radioactivity (over the course of the scan) remains essential to derive the most accurate estimate of BP. Potential complications of radial arterial cannulation include bleeding, infection, thrombosis, or nerve damage. Individuals with clotting disorders (i.e., Factor V Leiden mutation) should not undergo research-related arterial cannulation due to increased complication risk. The Allen's test is routinely used to confirm adequate collateral ulnar artery circulation to the hand prior to radial artery cannulation. Bleeding is minimized by instructing subjects to minimize repetitive wrist movements for 36 h after removal of the arterial catheter.

Delay to Treatment

Due to the potential confound of medication effects on brain binding measures, it is desirable for depressed subjects to delay the start of medication until completion of scan procedures. A delay to treatment exposes subjects to the risk of continued depression and possible worsening of symptoms including feeling suicidal or homicidal, and need for hospitalization. Such risks can be managed through regular contact with depressed individuals and/or their designated emergency contacts. In addition, arrangements are made at the time of study enrollment so that antidepressant treatment can be initiated immediately post-scan.

SAMPLE SELECTION AND SAMPLING ISSUES

Psychiatric Factors

Because many women with PPD have lifetime mood disorder phenomenology that is consistent with bipolar disorder, careful screening for previously undetected bipolar disorder is an important component of PPD research.

Furthermore, identification of neurophysiologic differences between bipolar and unipolar women with PPD may provide important clues about these disorders more generally.

Although PPD is defined by the Diagnostic and Statistical Manual of Mental Disorders-IV (American Psychiatric Association, 1994) as a major depressive episode that begins within 4 weeks of delivery, PPD definitions frequently vary according to the study hypotheses. Because women with PPD often experience depressive episode onset during actual (Stowe *et al.*, 2005) and simulated (Bloch *et al.*, 2000) pregnancy, antenatal depression onset may be an acceptable definition for some studies.

Inclusion of subjects with primary depression (rather than depression that occurs secondary to a medical illness, bereavement, or another psychiatric disorder) provides greater likelihood for detection of depression-specific neural circuits in PPD. Likewise, selection of subjects with a moderate level of depression enhances the likelihood of detecting depression-related neurobiological alterations.

Age and Endocrine Factors

Careful matching for age heightens the likelihood of detecting neurobiological differences across groups in neuroimaging studies because healthy aging has been associated with reductions in neuroreceptors (Volkow *et al.*, 1996; Meltzer *et al.*, 2001). The definition of "reproductive age" has expanded as the mean age of menarche decreases and as artificial reproductive technologies increase. Because the adolescent brain differs from the adult brain (Yurgelun-Todd, 2007), a lower age threshold of 18 may be appropriate for some studies of PPD. An upper age limit of 35 or 40 might be considered because healthy ovarian function begins to decline at age ≥ 35 (Guzick, 1996). Premenopausal status can be confirmed with FSH < 30 IU/L in all subjects.

Because ovarian hormones have been associated with changes in neural function (Moses *et al.*, 2000; Protopopescu *et al.*, 2005; Dreher *et al.*, 2007), studies in reproductive-aged women are routinely scheduled in the early follicular phase of the menstrual cycle (Smith & Zubieta, 2001), when circulating concentrations of estradiol and progesterone are low. Exogenous reproductive hormone exposures are commonly exclusionary for these studies. Postpartum amenorrhea is an ideal time to schedule studies. The hormonal fluctuations of the early postpartum period should optimally be avoided as should the gradual, unpredictable return of postpartum ovarian function, which may occur in non-lactating women by 6 weeks postpartum (Gray *et al.*, 1987; WHO, 1998). In the absence of burdensome, daily hormonal assessments, menstrual history, lactational history, and measurement of peripheral hormone concentrations offer a skeletal approach to characterization of endocrine milieu.

Comparison Group

Inclusion of a non-postpartum or nulliparous group can help to ascertain whether PPD has distinct neurobiological features from non-postpartum major depression. Non-postpartum women are ideally selected based on menstrual period regularity (i.e., 25–32 days) and ovulatory cycles (day 21 progesterone levels > 25 nmol/L), as irregular and non-ovulatory menstrual cycles could indicate heightened stress-sensitive neural circuits that regulate hypothalamic–pituitary–ovarian (HPO) axis function (Williams *et al.*, 2007).

SUMMARY

Although PPD is a disorder with devastating effects on women, children, and families, it is generally a highly treatable illness once recognized. With ever-increasing attention to PPD in the public health sphere, new mothers in the 21st century have a greater chance for PPD detection. Increased clinical research of PPD has yielded a wide armamentarium of available treatments during lactation that was not available one decade ago. Study of birth outcomes after *in utero* SSRI exposure has broadened the understanding of risks and benefits of available treatments in perinatal depression. Further advances in the treatment and prevention of PPD will be made possible through mechanistic research into specific neurobiological factors that underlie PPD. The application of varied neurophysiologic methods to carefully selected clinical populations has great potential to further enhance maternal mental health.

REFERENCES

Allard, P., and Norlen, M. (2001). Caudate nucleus dopamine d(2) receptors in depressed suicide victims. *Neuropsychobiology* 44(2), 70–73.

Altemus, M. (2001). Obsessive-compulsive disorder during pregnancy and postpartum. In *Management of Psychiatric Disorders in Pregnancy* (K. Yonkers and B. Little, Eds.), pp. 149–163. Arnold, London.

Altemus, M., Deuster, P., Galliven, E., Carter, S., and Gold, P. W. (1995). Suppression of hypothalamic–pituitary–adrenal axis responses to stress in lactating women. *J. Clin. Endocrinol. Metabol.* 80, 2954–2959.

Altshuler, L., Cohen, L., Szuba, M., Burt, V., Gitlin, M., and Mintz, J. (1996). Pharmacologic management of psychiatric illness during pregnancy: Dilemmas and guidelines. *Am. J. Psychol.* 153, 592–606.

Alwan, S., Reefhuis, J., Rasmussen, S. A., Olney, R. S., and Friedman, J. M. (2007). Use of selective serotonin-reuptake inhibitors in pregnancy and the risk of birth defects. *New Engl. J. Med.* 356(26), 2684–2692.

American Psychiatric Association (1994). *Diagnostic and Statistical Manual of Mental Disorders*, 4th edn. American Psychiatric Association, Washington, DC.

Appleby, L., Warner, R., Whitton, A., and Faragher, B. (1997). A controlled study of fluoxetone and cognitive-behavioral counselling in the treatment of postnatal depression. *Br. Med. J.* 314(7085), 932–936.

Armstrong, M. A., Osejo, V. G., Lieberman, L., Carpenter, D. M., Pantoja, P. M., and Escobar, G. J. (2003). Perinatal substance abuse intervention in obstetric clinics decreases adverse neonatal outcomes. *J. Perinatal.* 23, 3–9.

Association, A. P. (1994). *Diagnostic and statistical manual of mental disorders*, 4th edn. Washington, DC.

Bailara, K., Henry, C., Lestage, J., Launay, J., Parrot, F., Swendsen, J., Sutter, A., Roux, D., Dallay, D., and Demotes-Mainard, J. (2006). Decreased brain tryptophan availability as a partial determinant of post-partum blues. *Psychoneuroendocrinology* 31, 407–413.

Bar-Oz, B., Nulman, I., Koren, G., and Ito, S. (2000). Anticonvulsants and breast feeding: A critical review. *Paediatr. Drugs* 2(2), 113–126. Mar–Apr.

Bar-Oz, B., Einarson, T., Einarson, A., Boskovic, R., O'Brien, L., Malm, H., Berard, A., and Koren, G. (2007). Paroxetine and congenital malformations: Meta-analysis and consideration of potential confounding factors. *Clin. Therapeut.* 29(5), 918–926.

Beeghly, M., Olson, K., Weinberg, M., Pierre, S., Downey, N., and Tronick, E. (2003). Prevalence, stability, and socio-demographic correlates of depressive symptoms in black mothers during the first 18 months postpartum. *Matern. Child Health J.* 7(3), 157–168.

Berard, A., Ramos, E., Rey, E., Blais, L., St.-Andre, M., and Oraichi, D. (2006). First trimester exposure to paroxetine and risk of cardiac malformations in infants: The importance of dosage. *Birth Defects Research (Part B)* 80, 18–27.

Bethea, C. L., Mirkes, S. J., Shively, C. A., and Adams, M. R. (2000). Steroid regulation of tryptophan hydroxylase protein in the dorsal raphe of macaques. *Biological Psychiatry.* 47, 562–576.

Bhagwagar, Z., Rabiner, E. A., Sargent, P. A., Grasby, P. M., and Cowen, P. J. (2004). Persistent reduction in brain serotonin1a receptor binding in recovered depressed men measured by positron emission tomography with [11c]way-100635. *Mol. Psychiatr.* 9, 386–392.

Blehar, M. C., DePaulo, J. R., Jr., Gershon, E. S., Reich, T., Simpson, S. G., and Nurnberger, J. I. Jr. (1998). Women with bipolar disorder: Findings from the nimh genetics initiative sample. *Psychopharmacol. Bull.* 34(3), 239–243.

Bloch, M., Schmidt, P., Danaceau, M., Murphy, J., Neiman, L., and Rubinow, D. (2000). Effects of gonadal steroids in women with a history of postpartum depression. *Am. J. Psychiatr.* 157, 924–930.

Biegon, A., Reches, A., Snyder, L., and McEwen, B. S. (1983). Serotonergic and noradrenergic receptors in the rat brain: Modulation by chronic exposure to ovarian hormones. *Life Sci.* 32, 2015–2021.

Bonari, L., Pinto, N., Ahn, E., Einarson, A., Steiner, M., and Koren, G. (2004). Perinatal risks of untreated depression during pregnancy. *Can. J. Psychiatr.* 49(11), 726–734.

Bowden, C., Theodorou, A. E., Cheetham, S. C., Lowther, S., Katona, C. L. E., Crompton, M. R., and Horton, R. W. (1997). Dopamine d1 and d2 receptor binding sites in brain samples from depressed suicides and controls. *Brain Res.* 752, 227–233.

Bowen, D. M., Najlerahim, A., Procter, A. W., Francis, P. T., and Murphy, E. (1989). Circumscribed changes of the cerebral cortex in neuropsychiatric disorders of late life. *Proc. Natl Acad. Sci.* 86, 9504–9508.

Bogen, D. L. (2006). The role of depression on breastfeeding intention and initiation. *Archives of Women's Mental Helath* 10(1), III–IV.

Brockington, I., Cernik, A., Schofield, E. *et al.* (1981). Puerperal psychosis, phenomena and diagnosis. *Arch. Gen. Psychiatr.* 38, 829–833.

Byrnes, E. M., Byrnes, J. J., and Bridges, R. S. (2001). Increased sensitivity of dopamine systems following reproductive experience in rats. *Pharmacol., Biochem. Behav.* 68(3), 481–489.

Chambers, C., Johnson, K., Dick, L., Felix, R., and Jones, K. (1996). Birth outcomes in pregnant women taking fluoxetine. *New Engl. J. Med.* 335, 1010–1015.

Chambers, C., Moses-Kolko, E., and Wisner, K. L. (2007). Antidepressant use in pregnancy: New concerns, old dilemmas. *Expert Review of Neurotherapeutics* 7(7), 761–764.

Chambers, C. D., Hernandez-Diaz, S., Van Marter, L. J., Werler, M. M., Louik, C., Jones, K. L., and Mitchell, A. A. (2006). Selective serotonin-reuptake inhibitors and risk of persistent pulmonary hypertension of the newborn.[see comment]. *New Engl. J. Med.* 354(6), 579–587.

Champagne, F. A., Chretien, P., Stevenson, C. W., Zhang, T. Y., Gratton, A., and Meaney, M. J. (2004). Variations in nucleus accumbens dopamine associated with individual differences in maternal behavior in the rat. *J. Neurosci.* 24(17), 4113–4123.

Chun-Fai-Chan, B., Koren, G., Fayez, I., Kalra, S., Voyer-Lavigne, S., Boshier, A., Shakir, S., and Einarson, A. (2005). Pregnancy outcome of women exposed to bupropion during pregnancy: A prospective comparative study. *Am. J. Obstet. Gynecol.* 192, 932–936.

Cohen, L., Freidman, J., Jefferson, J., Johnson, E., and Weiner, M. (1994). A reevaluation of risk of *in utero* exposure to lithium. *J. Am. Med. Assoc.* 271, 146–150.

Cohen, L. S., Altshuler, L. L., Harlow, B. L., Nonacs, R., Newport, D. J., Viguera, A. C., Suri, R., Burt, V. K., Hendrick, V., Reminick, A. M., Loughead, A., Vitonis, A. F., and Stowe, Z. N. (2006). Relapse of major depression during pregnancy in women who maintain or discontinue antidepressant treatment. *JAMA* 295(5), 499–507.

Cooper, W. O., Willy, M. E., Pont, S. J., and Ray, W. A. (2007). Increasing use of antidepressants in pregnancy. *Am. J. Obstet. Gynecol.* 196. 544.e541-e544.e545.

Coppola, D., Russo, L. J., Kwarta, R. F., Varughese, R., and Schmider, J. (2007). Evaluating the postmarketing experience of risperidone use during pregnancy. *Drug Saf.* 30(3), 247–264.

Cox, J., Holden, J., and Sagovsky, R. (1987). Detection of postnatal depression: Development of the 10-item edinburgh postnatal depression scale. *Br. J. Psychiatr.* 150, 782–786.

Cunnington, M., Tennis, P., and International Lamotrigine Pregnancy Registry Scientific Advisory, C. (2005). Lamotrigine and the risk of malformations in pregnancy.[see comment]. *Neurology*, 64(6), 955–960.

Cunnington, M., Ferber, S., Quartey, G., and International Lamotrigine Pregnancy Registry Scientific Advisory, C. (2007). Effect of dose on the frequency of major birth defects following fetal exposure to lamotrigine monotherapy in an international observational study. *Epilepsia*, 48(6), 1207–1210.

Cunnington, M., and Tennis, P., and the International Lamotrigine Pregnancy Registry Scientific Advisory Committee (2005). Lamotrigine and the risk of malformations in pregnancy. *Neurology* 64(6), 955–960.

Cyr, M., Bosse, R., and Di Paolo, T. (1998). Gonadal hormones modulate 5-ht2a receptors: Emphasis on the rat frontal cortex. *Neuroscience.* 83(3), 829–836.

D'haenen, H. A., and Bossuyt, A. (1994). Dopamine d2 receptors in depression measured with single photon emission computed tomography. *Biol. Psychiatr.* 35, 128–132.

Dreher, J. C., Schmidt, P. J., Kohn, P., Furman, D., Rubinow, D. R., and Berman, K. F. (2007). Menstrual cycle phase modulates reward-related neural function in women. *PNAS* 104(7), 2465–2470.

Drevets, W. C., Frank, E., Price, J. C., Kupfer, D. J., Holt, D., Greer, P. J., Huang, Y., Guatier, C., and Mathis, C. (1999). Pet imaging of serotonin 1a receptor binding in depression. *Biol. Psychiatr.* 46, 1375–1387.

Drevets, W. C., Gautier, C., Price, J. C., Kupfer, D. J., Kinahan, P. E., Grace, A. A., Price, J. L., and Mathis, C. A. (2001). Amphetamine-induced dopamine release in human ventral striatum correlates with euphoria. *Biol. Psychiatr.* 49, 81–96.

Dunlop, B. W., and Nemeroff, C. B. (2007). The role of dopamine in the pathophysiology of depression. *Arch Gen. Psychiatr.* 64, 327–337.

Dunn, A., Trivedi, M., Kampert, J., Clark, C., and Chambliss, H. (2005). Exercise treatment for depression: Efficacy and dose response. *Am J Prev Med.* 28(1), 1–8.

Ebert, D., Feistel, H., Loew, T., and Pirner, A. (1996). Dopamine and depression-striatal dopamine d2 receptor spect before and after antidepressant therapy. *Psychopharmacology* 126, 91–94.

Epperson, C. N., Terman, M., Terman, J. S., Hanusa, B. H., Oren, D. A., Peindl, K. S., and Wisner, K. L. (2004). Randomized clinical trial of bright light therapy for antepartum depression: Preliminary findings. *J. Clin. Psychiatr.* 65(3), 421–425.

Epperson, C. N., Gueorguieva, R., Czarkowski, K. A., Stiklus, S., Sellers, E., Krystal, J. H., Rothman, D. L., and Mason, G. F. (2006). Preliminary evidence of reduced occipital gaba concentrations in puerperal women: A 1h-mrs study. *Psychopharmacology* 186, 425–433.

Ericson, A., Kallen, B., and Wiholm, B. (1999). Delivery outcome after the use of antidepressants in early pregnancy. *Eur. J. Clin. Pharmacol.* 55, 503–508.

First, M. B., Spitzer, R. L., Gibbon, M., and Williams, J. B. W. (1998). *Structured Clinical Interview for dsm-iv Axis i Disorders-Patient Edition.* New York State Psychiatric Institute, Biometrics Research Department, New York.

Fisher, P., Meltzer, C., Ziolko, S., Price, J., Moses-Kolko, E., Serga, S., and Hariri, A. (2006). Capacity for 5-ht1a-mediated autoregulation predicts amygdala reactivity. *Nat. Neurosci.* 9(11), 1362–1363.

Freeman, E. W., Rickels, K., Sondheimer, S. J., and Polansky, M. (1999). Differerential response to antidepressants in women with premenstrual syndrome/premenstrual dysphoric disorder. *Arch. Gen. Psychiatr.* 56, 932–939.

Freeman, M. P. (2006). Omega-3 fatty acids and perinatal depression: A review of the literature and recommendations for future research. *Prostaglandins Leukotrienes and Essential Fatty Acids* 75(4–5), 291–297.

Garlow, S. J., Musselman, D. L., and Nemeroff, C. B. (2000). The neurochemistry of mood disorders: Clinical studies (Chapter 27). In *Neurobiology of Mental Illness* (D. S. Charney, E. J. Nestler, and B. J. Bunney, Eds.), pp. 348–364. Oxford University Press, New York.

Gaynes, B. N., Gavin, N., Meltzer-Brody, S., Lohr, K. N., Swinson, T., Gartlehner, G., Brody, S., and Miller, W. C. (2005). Perinatal depression: Prevalence, screening accuracy, and screening outcomes. *Evid. Rep. Tech. Assess* 119, 1–8.

Gentile, S. (2004). Clinical utilization of atypical antipsychotics in pregnancy and lacation. *Ann. Pharmacother.* 38, 1265–1271.

Glaser, J., Russell, V. A., de Villiers, A. S., Searson, J. A., and Taljaard, J. J. F. (1990). Rat monoamine and serotonin s2 receptor changes during pregnancy. *Neurochem. Res.* 15(10), 949–956.

Goldstein, D. J., Corbin, L. A., and Sundell, K. L. (1997). Effects of first trimester fluoxetine exposure on the newborn. *Obstet. Gynecol.* 89, 713–718.

Goodman, S. H., and Gotlib, I. H. (1999). Risk for psychopathology in the children of depressed mothers: A developmental model for understanding mechanisms of transmission. *Psychol. Rev.* 106(3), 458–490.

Grace, S. L., Evindar, A., and Stewart, D. E. (2003). The effect of postpartum depression on child cognitive development and behavior: A review and critical analysis of the literature. *Arch. Wom. Ment. Health* 6(4), 263–274.

Grant, K. A., Shively, C. A., Nader, M. A., Ehrenkaufer, R. L., Line, S. W., Morton, T. E., Gage, H. D., and Mach, R. H. (1998). Effect of social status on striatal dopamine d2 receptor binding characteristics in cynomologous monkeys assessed with positron emission tomography. *Synapse* 29, 80–83.

Gray, R. H., Campbell, O. M., Zacur, H. A., Labbok, M. H., and Mac Rae, S. L. (1987). Postpartum return of ovarian activity in non-breastfeeding women monitored by urinary assays. *J. Endocrinol. Metabol.* 64, 645–650.

Gregoire, I., el Esper, N., Gondry, J., Boitte, F., Fievet, P., Makdassi, R., Westeel, P. F., Lalau, J. D., Favre, H., de Bold, A. *et al.* (1990). Plasma atrial natriuretic factor and urinary excretion of a ouabain displacing factor and dopamine in normotensive pregnant women before and after delivery. *Am. J. Obstet. Gynecol.* 162(1), 71–76.

Guzick, D. S. (1996). Human infertility: An introduction. In *Rreproductive Endocrinology, Surgery, and Technology* (E. Y. Adashi, J. A. Rock, and Z. Rosenwaks, Eds.), Vol. 2, pp. 1897–1913. Lippincott-Raven, Philadelphia.

Hall, H., Lundkvist, C., Halldin, C., Farde, F., Pike, V. W., McCarron, J. A., Fletcher, A., Cliffe, I. A., Barf, T., Wikstrom, H., and Sedvall, G. (1997). Autoradiographic localization of 5-ht1a receptors in the post-mortem human brain using [3h]way-100635 and [11c]way-100635. *Brain Res.* 745, 96–108.

Hansen, D., Lou, H., and Olsen, J. (2000). Serious life events and congenital malformations: A national study with complete follow-up. *Lancet* 356, 875–880.

Hansen, S., Harthon, C., Wallin, E., Lofberg, L., and Svensson, K. (1991). Mesotelencephalic dopamine system and reproductive behavior in the female rat: Effects of ventral tegmental 6-hydroxydopamine lesions on maternal and sexual responsiveness. *Behav. Neurosci.* 105(4), 588–598.

Hanusa, B. H., Scholle, S. H., Haskett, R. F., Spadaro, K., and Wisner, K. L. (2008). Comparison of three instruments to screen for postpartum depression. *J. Wom. Health.* 17(4).

Hendrick, V., Altshuler, L., Strouse, T., and Grosser, S. (2000). Postpartum and nonpostpartum depression: Differences in presentation and response to pharmacologic treatment. *Depress. Anxiety* 11, 66–72.

Henshaw, C., Foreman, D., and Cox, J. (2004). Postnatal blues: A risk factor for postnatal depression. *J. Psychosom. Obstet. Gynecol.* 25(3–4), 267–272.

Heron, J., Craddock, N., and Jones, I. (2005). Postnatal euphoria: Are 'the highs' an indicator of bipolarity?. *Bipolar Disorders* 7(2), 103–110.

Hlastala, S. A., Frank, E., Mallinger, A. G., Thase, M. E., Ritenour, A. M., and Kupfer, D. J. (1997).

Bipolar depression: An underestimated treatment challenge. *Depress. Anxiety* 5(2), 73–83.

Hobfoll, S. E., Ritter, C., Lavin, J., Hulsizer, M. R., and Cameron, R. P. (1995). Depression prevalence and incidence among inner-city pregnant and postpartum women. *J. Consult. Clin. Psychol.* 63(3), 445–453.

Holmes, L., Wyszynski, D., Baldwin, E., Habecker, E., Glassman, L., and Smith, C. (2006). Increased risk for non-syndromic cleft palate among infants exposed to lamotrigine during pregnancy, *Teratology Society-46th Annual Meeting*. Tuscon, AZ.

The international commission on radiation protection: Radiation dose to patients from radiopharmaceuticals: Addendum 6 to icrp publication 53. (2002). New York, NY: Pergamon Press.

Jones, I., and Craddock, N. (2001). Familiality of the puerperal trigger in bipolar disorder: Results of a family study. *Am. J. Psychiatr.* 158(6), 913–917.

Jones, I., Hamshere, M., Nangle, J. M., Bennett, P., Green, E., Heron, J., Segurado, R., Lambert, D., Holmans, P., Corvin, A., Owen, M., Jones, L., Gill, M., and Craddock, N. (2007). Bipolar affective puerperal psychosis: Genome-wide significant evidence for linkage to chromosome 16.[see comment]. *Am. J. Psychiatr.* 164(7), 1099–1104.

Kelly, M. B., Wisner, K. L., and Cornelius, M. D. (2007). Ssris and birth defects.[comment]. *Epidemiology* 18(3), 411–412. author reply 412–413.

Kendell, R. E., Wainwright, S., and Hailey, A. (1976). The influence of childbirth on psychiatric morbidity. *Psychol. Med.* 6, 297–302.

Kendell, R. E., Chalmers, J. C., and Platz, C. (1987). Epidemiology of puerperal psychoses. *Br. J. Psychiatr.* 150, 662–673.

Kessler, R. C., Rubinow, D. R., Holmes, C., Abelson, J. M., and Zhao, S. (1997). The epidemiology of dsm-iii-r bipolar i disorder in a general population survey. *Psychol. Med.* 27(5), 1079–1089.

Kinsley, C. H., and Lambert, K. G. (January 2006). The maternal brain. *Sci. Am.*, 72–79.

Klein, M., and Essex, M. (1995). Pregnant or depressed? The effect of overlap between symptoms of depression and somatic complaints of pregnancy on rates of major depression in the second trimester. *Depression* 2, 308–314.

Klimke, A., Larisch, R., Janz, A., Vosberg, H., Muller-Gartner, H. W., and Gaebel, W. (1999). Dopamine d2 receptor binding before and after treatment of major depression measured by [123i]ibzm spect. *Psychiatr. Res.* 90(2), 91–101.

Klimek, V., Schenck, J. E., Han, H., Stockmeier, C. A., and Ordway, G. A. (2002). Dopaminergic abnormalities in amygdaloid nuclei in major depression: A postmortem study. *Biol. Psychiatr.* 52, 740–748.

Kornstein, S. G., Schatzberg, A. F., Thase, M. E., Yonkers, K. A., McCullough, J. P., Keitner, G. I., Gelenberg, A. J., Davis, S. M., Harrison, W. M., and Keller, M. B. (2000). Gender differences in treatment response to sertraline versus imipramine in chronic depression. *Am. J. Psychiatr.* 157(9), 1445–1452.

Kurki, T., Hiilesmaa, V., Raitasalo, R., Mattila, H., and Ylikorkala, O. (2000). Depression and anxiety in early pregnancy and risk for preeclampsia. *Obstet. Gynecol.* 95, 487–490.

Lammers, C.-H., D'Souza, U., Qin, Z.-H., Lee, S.-H., Yajima, S., and Mouradian, M. M. (1999). Regulation of striatal dopamine receptors by estrogen. *Synapse* 34, 222–227.

Laruelle, M. (2000). Imaging synaptic neurotransmission with *in vivo* binding copetition techniques: A critical review. *J. Cerebr. Blood Flow Metabol.* 20, 423–451.

Laruelle, M., Slifstein, M., and Huang, Y. (2003). Relationships between radiotracer properties and image quality in molecular imaging of the brain with positron emission tomography. *Mol. Imag. Biol.* 5(6), 363–375.

Leibenluft, E. (2000). Women and bipolar disorder: An update. *Bull. Menninger Clin.* 64, 5–17.

Leibenluft, E., Gobbini, M. I., Harrison, T., and Haxby, J. V. (2004). Mothers' neural activation in reponse to pictures of their children and other children. *Biol. Psychiatr.* 56, 225–232.

Lesch, K., Mayer, S., and Disselkamo-Tietze, J. (1990). 5ht1a receptor responsivity in unipolar depression. Evaluation of ipsapirone-induced acth and cortisol secretion in patients and controls. *Biol. Psychiatr.* 28, 620–628.

Lopez, J. F., Chalmers, D. T., Little, K. Y., and Watson, S. J. (1998). Regulation of serotonin 1a, glucocorticoid, and mineralocorticoid receptor in rat and human hippocampus: Implications for the neurobiology of depression. *Biol. Psychiatr.* 43, 547–573.

Lopez, J. F., Akil, H., and Watson, S. J. (1999). Neural circuits mediating stress. *Biol. Psychiatr.* 46, 1461–1471.

Lorberbaum, J. P., Newman, J. D., Horwitz, A. R., Dubno, J. R., Lydiard, R. B., Hammer, M. B., Bohning, D. E., and George, M. S. (2002). A potential role for thalamocingulate circuitry in human maternal behavior. *Biol. Psychiatr.* 51, 431–445.

Louik, C., Lin, A. E., Werler, M. M., Hernandez-Diaz, S., and Mitchell, A. A. (2007). First-trimester use of selective serotonin-reuptake inhibitors and the risk of birth defects. *New Engl. J. Med.* 356(26), 2675–2683.

Magiakou, M. A., Mastorakos, G., Rabin, D., Dubbert, B., Gold, P. W., and Chrousos, G. P. (1996). Hypothalamic corticotropin-releasing hormone suppression during the postpartum period: Implications for the increase in psychiatric manifestations at this time. *J. Clin. Endocrinol. Metabol.* 81, 1912–1917.

Manber, R., Schnyer, R. N., Allen, J. J., Rush, A. J., and Blasey, C. M. (2004). Acupuncture: A promising treatment for depression during pregnancy. *J. Affect. Disord.* 83(1), 89–95.

March, D., and Yonkers, K. A. (2001). Panic disorder. In *Management of Psychiatric Disorders in Pregnancy* (K. A. Yonkers and B. Little, Eds.), pp. 134–148. Arnold Publishers, London.

Mawlawi, O., Martinez, D., Slifstein, M., Broft, A., Chatterjee, R., Hwang, D. R., Huang, Y., Simpson, N., Ngo, K., Van Heertum, R., and Laruelle, M. (2001). Imaging human mesolimbic dopamine transmission with positron emission tomography. Part i: Accuracy and precision of d(2) receptor parameter measurements in ventral striatum. *J. Cerebr. Blood Flow Metabol.* 21, 1034–1057.

McElhatton, P., Garbis, H., Elefant, E., Vial, T., Bellemin, B., Mastroiacovo, P., Arnon, J., Rodriguez-Pinilla, E., Schaefer, C., Pexieder, T., Merlob, P., and Dal Verme,, S. (1996). The outcome of pregnancy in 689 women exposed to therapeutic doeses of antidepressants. A collaborative study of the european network of teratology information services (entis). *Reproduct. Toxicol.* 10(4), 285–294.

McKenna, K., Levinson, A., Einarson, A., Diav-Citrin, O., Zipursky, R., and Koren, G. (2003). Pregnancy outcome in women recieving atypical antipsychotic drugs: A prospective, multicentre, comparative study. *Birth Defects Res. Part A Clin. Mol. Teratol.* 67(5), 391.

Meador, K., Baker, G., Finnell, R., Smith, J., and Wolff, M. (2006). *In utero* antiepileptic drug exposure: Fetal death and malformations. *Neurology* 67, 407–412.

Meltzer, C. C., Drevets, W. C. *et al.* (2001). Gender-specific aging effects on the serotonin 1a receptor. *Brain Res.* 895(1–2), 9–17.

Meyer, J. H., McNeely, H. E., Sagrati, S., Boovariwala, A., Martin, K., Verhoeff, N. P., Wilson, A. A., and Houle, S. (2006). Elevated putamen d(2) receptor binding potential in major depression with motor retardation: An [11c]raclopride positron emission tomography study. *Am. J. Psychiatr.* 163(9), 1594–1602.

Mezzacappa, E., and Katkin, E. (2002). Breast-feeding is associated with reductions in perceived stress and negative mood in mothers. *Health Psychol.* 21, 187–193.

Mezzacappa, E. S., Guethlein, W., Vaz, N., and Bagiella, E. (2000). A preliminary study of breast-feeding and maternal symptomatology. *Ann. Behav. Med.* 22, 71–79.

Misri, S., Reebye, P., Kendrick, K., Carter, D., Ryan, D., Grunau, R. E., and Oberlander, T. F. (2006). Internalizing behaviors in 4-year-old children exposed *in utero* to psychotropic medications.[see comment]. *Am. J. Psychiatr.* 163(6), 1026–1032.

Monk, C., Myers, M. M., Sloan, R. P., Ellman, L. M., and Fifer, W. P. (2003). The effects of women's stress–elicited physiological activity and chronic anxiety on fetal heart rate. *Dev. Behav. Pediatr.* 24(1), 32–38.

Moses, E. L., Drevets, W. C., Smith, G., Mathis, C. A., Kalro, B. N., Butters, M. A., Leondires, M. P., Greer, P. J., Lopresti, B., Loucks, T. L., and Berga, S. L. (2000). Effects of estradiol and progesterone administration on human serotonin 2a receptor binding: A pet study. *Biol. Psychiatr.* 48, 854–860.

Moses-Kolko, E. L., and Feintuch, M. (2002). Perinatal psychiatric disorders: A clinical review. *Curr. Probl. Obstet. Gynecol. Fertil.* 25(3), 61–112.

Moses-Kolko, E., Meltzer, C. C., Helsel, J. C., Sheetz, M., Mathis, C., Ruszkiewicz, J., Bogen, D., Confer, A. L., and Wisner, K. L. (2005a). No interruption of lactation is needed after [11c]way 100635 or [11c]raclopride pet. *J. Nucl. Med.* 46(10), 1765.

Moses-Kolko, E. L., Bogen, D., Perel, J., Bregar, A., Uhl, K., Levin, B., and Wisner, K. L. (2005b). Neonatal signs after late *in utero* exposure to serotonin reuptake inhibitors: Literature review and implications for clinical applications. *JAMA* 293(19), 2372–2383.

Moses-Kolko, E. L., Wisner, K. L., Lanza di Scalea, T., Kaye, W. H., Price, J. C., Grace, A. A., Berga, S. L., Mathis, C. A., Drevets, W. C., Hanusa, B. H., Becker, C., and Meltzer, C. C. (2006). Reduced ventral striatal d2/d3 receptor binding in depressed reproductive-aged women: A [11c]raclopride positron emission tomography study. *Neuropsychopharmacology* 31(suppl 1s), S87–S88.

Moses-Kolko, E. L., Wisner, K. L., Price, J. C., Berga, S. L., Mathis, C. E., Drevets, W. C., Confer, A. L., Hanusa, B. H., Loucks, T. L., Becker, C., and Meltzer, C. C. (2008). Serotonin 1a receptor reductions in postpartum depression: A positron emission tomography study. *Fertil. Steril.* 89, 685–692.

Munk-Olsen, T., Laursen, T. M., Pedersen, C. B., Mors, O., and Mortensen, P. B. (2006). New parents and mental disorders: A population-based register study.[see comment]. *JAMA* 296(21), 2582–2589.

Munro, C. A., McCaul, M. E., Wong, D. F., Oswald, L. M., Zhou, Y., Brasic, J., Kuwabara, H., Kumar, A., Alexander, M., Ye, W., and Wand, G. S. (2006). Sex differences in striatal dopamine release in healthy adults. *Biol. Psychiatr.* 59(10), 966–974.

Murray, D., and Cox, J. (1990). Screening for depression during pregnancy with ediburgh depression scale (epds). *J. Repro. Infant Psychol.* 8, 99–107.

Murray, L. (1992). The impact of postnatal depression on infant development. *J. Child Psychol. Psychiatr.* 33, 543–561.

Nappi, R., Petraglia, F., Luisi, S., Polatti, F., Farina, C., and Genazzani, A. (2001). Serum allopregnanolone in women with postpartum "blues". *Obstet. Gynecol.* 97(1), 77–80.

Nestler, E. J., and Carlezon, W. A. (2006). The mesolimbic dopamine reward circuit in depression. *Biol. Psychiatr.* 59(12), 1151–1159.

Neumann, I. D. (2003). Brain mechanisms underlying emotional alterations in the peripartum period in rats. *Depress. Anxiety* 17, 111–121.

Newcomer, J. W. (2007). Metabolic considerations in the use of antipsychotic medications: A review of recent evidence. *J. Clin. Psychiatr.* 68(suppl 1), 20–27.

Newport, J., Viguera, A., Beach, A., Ritchie, J., Cohen, L., and Stowe, Z. (2005). Lithium placental passage and obstetrical outcome: Implications for clinical management during late pregnancy. *Am. J. Psychiatr.* 162(11), 2162–2170.

Nordstrom, A. L., Jovanovic, H., Cerin, A., Karlsson, P., and Halldin, C. (2004). Pet study of 5-ht1a receptors at different phases of the menstrual cycle in premenstrual dysphoria. *Neuropsychopharmacology.* Abstract from Annual Meeting.

Nulman, I., Rovet, J., Stewart, D., Wolpin, J., Gardner, H., Theis, J., Kulin, N., and Koren, G. (1997). Neurodevelopment of children exposed *in utero* to antidepressant drugs. *NEJM* 336, 258–262.

Nulman, I., Rovet, J., Stewart, D. E., Wolpin, J., Pace-Asciak, P., Shuhaiber, S., and Koren, G. (2002). Child development following exposure to tricyclic antidepressants or fluoxetine throughout fetal life: A prospective, controlled study. *Am. J. Psychiatr.* 159(11), 1889–1895.

Numan, M., and Insel, T. R. (2003). Human implications. In *The Neurobiology of Parental Behavior* (M. Numan and T. R. Insel, Eds.), pp. 316–342. Springer-Verlag, New York.

Numan, M., Fleming, A. S., and Levy, F. (2006). Maternal behavior. In *Knobil and Neill's Physiology of Reproduction* (J. D. Neill, Ed.) St. Louis, Mo; Elsevier.

O'Brien, C. P., Charney, D. S., Lewis, L., Cornish, J. W., Post, R. M., Woody, G. E., Zubieta, J. K., Anthony, J. C., Blaine, J. D., Bowden, C. L., Calabrese, J. R., Carroll, K., Kosten, T., Rounsaville, B., Childress, A. R., Oslin, D. W., Pettinati, H. M., Davis, M. A., Demartino, R., Drake, R. E., Fleming, M. F., Fricks, L., Glassman, A. H., Levin, F. R., Nunes, E. V., Johnson, R. L., Jordan, C., Kessler, R. C., Laden, S. K., Regier, D. A., Renner, J. A. Jr., Ries, R. K., Sklar-Blake, T., and Weisner, C. (2004). Priority actions to improve the care of persons with co-occurring substance abuse and other mental disorders: A call to action. *Biol. Psychiatr.* 56(10), 703–713.

O'Hara, M., Neunaber, D., and Zekoski, E. (1984). Prospective study of postpartum depression: Prevalence, course, and predictive factors. *J. Abnorm. Psychol.* 93, 158–171.

O'Hara, M. W., and Swain, A. M. (1996). Rates and risks of postpartum depression-a meta-analysis. *Int. Rev. Psychiatr.* 8(1), 37–54.

O'Hara, M. W., Zekoski, E. M., Phillips, L. H., and Wright, E. J. (1990). Controlled prospective study of postpartum mood disorders: Comparison of childbearing and nonchildbearing women. *J. Abnorm. Psychol.* 99(1), 3–15.

O'Hara, M. W., Stuart, S., Gorman, L. L., and Wenzel, A. (2000). Efficacy of interpersonal psychotherapy for postpartum depression. *Arch. Gen. Psychiatr.* 57(11), 1039–1045.

Oberlander, T. F., Misri, S., Fitzgerald, C. E., Kostaras, X., Rurak, D., and Riggs, W. (2004). Pharmacologic factors associated with transient neonatal symptoms following prenatal psychotropic medication exposure. *J. Clin. Psychiatr.* 65(2), 230–237.

Oberlander, T. F., Warburton, W., Misri, S., Aghajanian, J., and Hertzman, C. (2006). Neonatal outcomes after prenatal exposure to selective serotonin reuptake inhibitor antidepressants and maternal depression using population-based linked health data. *Arch. Gen. Psychiatr.* 63(8), 898–906.

Osmond, M., Wilkie, M., and Moore, J. (2001). *Behind the Smile: My Journey Out of Postpartum Depression.* Warner Books, Inc, New York.

Osterlund, M. K., and Hurd, Y. L. (1998). Acute 17b-estradiol treatment downregulates serotonin 5ht1a receptor mrna expression in the limbic system of female rats. *Molecular Brain Research.* 55, 169–172.

Osterlund, M. K., Halldin, C., and Hurd, Y. L. (2000). Effects of chronic 17b-estradiol treatment on the serotonin 5-ht1a receptor mrna and binding levels in the rat brain. *Synapse.* 35, 39–44.

Overpeck, M., Brenner, R., and Trumble, A. (1998). Risk factors for infant homicide in the united states. *N. Engl J Med.* 339, 1211–1216.

Ohman, I., Vitols, S., and Tomson, T. (2000). Lamotrigine in pregnancy: Pharmacokinetics during delivery, in the neonate, and during loctation. *Epilepsia.* 41(6), 709–713.

Parsey, R. V., Slifstein, M., Hwang, D. R., Abi-Dargham, A., Simpson, N., Mawlawi, O., Guo, N. N., Van Hertum, R. V., Mann, J. J., and Laruelle, M. (2000). Validation and reproducibility of measurement of 5ht1a receptor parameters with [carbonyl-11c]way-100635 in human: Comparison of arterial and reference tissue imput functions. *J. Cerebr. Blood Flow Metabol.* 20, 1111–1133.

Parsey, R. V., Oquendo, M. A., Zea-Ponce, Y., Rodenhiser, J., Kegeles, L. S., Pratap, M., Cooper, T. B., Van Heertum, R., Mann, J. J., and Laruelle, M. (2001). Dopamine d(2) receptor availability and amphetamine-induced dopamine release in unipolar depression. *Biol. Psychiatr.* 50(5), 313–322.

Parsey, R. V., Arango, V., Olvet, D. M., Oquendo, M. A., Van Heertum, R. L., and Mann, J. J. (2005). Regional heterogeneity of 5-ht1a receptors in human cerebellum as assessed by positron emission tomography. *J. Cerebr. Blood Flow Metabol.* 25, 785–793.

Parsey, R. V., Oquendo, M. A., Ogden, R. T., Olvet, D. M., Simpson, N., Huang, Y., Van Heertum, R., Arango, V., and Mann, J. J. (2006). Altered serotonin 1a binding in major depression: A [carbonyl-c-11]way 100635 positron emission tomography study. *Biol. Psychiatr.* 59(2), 106–113.

Peindl, K., Wisner, K., and Hanusa, B. (2004). Identifying depression in the first postpartum year: Guidelines for office-based screening and referral. *J. Affect. Disord.*

Pechins-Thompson, M., Brown, N. A., and Bethea, C. L. (1998). Regulation of serotonin re-uptake transporter mrna expression by ovarian steroids in rhesus macaques. *Molecular Brain Research* 53, 120–129.

Pecins-Thompson, M., and Bethea, C. L. (1999). Ovarian steroid regulation of serotonin-1a autoreceptor messenger rna expression in the dorsal raphe of rhesus macaques. *Neuroscience* 89(1), 267–277.

Petraglia, F., De Leo, V., Sardelli, S., Mazzullo, G., Gioffre, W. R., Genazzani, A. R., and D'Antona, N. (1987). Prolactin changes after administration of agonist and antagonist dopaminergic drugs in puerperal women. *Gynecol. Obstet. Investig.* 23(2), 103–109.

Phelps, M. E. (2004). *Pet: Molecular Imaging and Its Biological Applications*. Verlag, Germany, Springer.

Phillips, M. L., Drevets, W. C., Rauch, S. L., and Lane, R. (2003). Neurobiology of emotion perception ii: Implications for major psychiatric disorders. *Biol. Psychiatr.* 54, 515–528.

Pike, V. W., McCarron, J. A., Lemmertsma, A. A., Osman, S., Hume, S. P., Sargent, P. A., Bench, C. J., Cliffe, I. A., Fletcher, A., and Grasby, P. M. (1996). Exquisite deliniation of 5ht1a receptors in human brain with pet and [caronyl-11c]way 100635. *Eur. J. Pharmacol.* 301, R5–R7.

Piontek, C., Baab, S., Peindl, K., and Wisner, K. (2000). Serum valproate levels in 6 breastfeeding mother–infant pairs. *J. Clin. Psychiatr.* 61, 170–172.

Pohjalainen, T., Rinne, J. O., Nagren, K., Syvalahti, E., and Hietala, J. (1998). Sex differences in the striatal doapmine d2 receptor binding characteristics *in vivo*. *Am. J. Psychol.* 155(6), 768–773.

Price, J. C., Xu, L., Mazumdar, S., Meltzer, C. C., Drevets, W. C., Mathis, C. A., Kelley, D. E., Ryan, C. M., and Reynolds, C. F. (2002a). Impact of graphical analysis bias on group comparisons of regional [carbonyl-11c]way binding potential measures. *Neuroimage* 16(3), S72.

Protopopescu, X., Pan, H., Altemus, M., Tuescher, O., Polanecsky, M., McEwen, B., Silbersweig, D., and Stern, E. (2005). Orbitofrontal cortex activity related to emotional processing changes across the menstrual cycle. *Proc. Natl Acad. Sci. USA* 102(44), 16060–16065.

Ranote, S., Elliott, R., Abel, K. M., Mitchell, R., Deakin, J. F. W., and Appleby, L. (2004). The neural basis of maternal responsiveness to infants: An fmri study. *NeuroReport* 15, 1825–1829.

Riccardi, P., Zald, D., Li, R., Park, S., Ansari, M. S., Dawant, B., Anderson, S., Woodward, N., Schmidt, D., Baldwin, R., and Kessler, R. (2006). Sex differences in amphetamine-induced displacement of [18f]fallypride in striatal and extrastriatal regions: A pet study. *Am. J. Psychol.* 163, 1639–1641.

Robertson, E., Jones, I., Haque, S., Holder, R., and Craddock, N. (2005). Risk of puerperal and nonpuerperal recurrence of illness following bipolar affective puerperal (post-partum) psychosis.[see comment]. *Br. J. Psychiatr.* 186, 258–259.

Sargent, P. A., Kjaer, K. H., Bench, C. J., Rabiner, E. A., Messa, C., Meyer, J., Gunn, R. N., Grasby, P. M., and Cowan, P. J. (2000). Brain serotonin 1a receptor binding measured by positron emission tomography with [11c]way-100635: Effects of depression and antidepressant treatment. *Arch. Gen. Psychiatr.* 57, 174–180.

Scholle, S. H., and Kelleher, K. J. (2003). Assessing primary care among young, low-income women. *Women Health* 37(1), 15–30.

Schou, M. (1976). What happened later to the lithium babies? A follow-up study of children born without malformations. *Acta Neurol. Scand.* 54(3), 193–197.

Semelka, R. C., Armao, D. M., Elias, J., Jr., and Huda, W. (2007). Imaging strategies to reduce the risk of radiation in ct studies, including selective substitution with mri. *J. Magn. Reson. Imag.* 25(5), 900–909.

Sher, L., Mann, J. J., Traskman-Bendz, L., Winchel, R., Huang, Y. Y., Fertuck, E., and Stanley, B. H. (2006). Lower cerebrospinal fluid homovanillic acid levels in depressed suicide attempters. *J. Affect. Disord.* 90, 83–89.

Shields, B. (2005). *Down Came the Rain: My Journey Through Postpartum Depression*. Christa Incorporated, New York, NY.

Sichel, D. A., Cohen, L. S., Robertson, L. M. *et al.* (1995). Prophylactic estrogen in recurrent postpartum affective disorder. *Biol. Psychiatr.* 38, 814–818.

Simon, G. E., Cunningham, M. L., and Davis, R. L. (2002). Outcomes of prenatal antidepressant exposure.[see comment]. *Am. J. Psychiatr.* 159(12), 2055–2061.

Sit, D., Rothschild, A., and Wisner, K. (2006). A review of postpartum psychosis. *J. Wom. Health* 15(4), 352–368.

Smith, Y. R., and Zubieta, J. (2001). Neuroimaging of aging and estrogen effects on central nervous system physiology. *Fertil. Steril.* 76(4), 651–659.

Sumner, B. E., and Fink, G. (1995). Estrogen increases the density of 5-ht2a receptors in cerebral cortex and nucleus accumbens in the female rat. *J Steroid Biochem Molec Biol* 54(1/2), 15–20.

Spinelli, M. (2001). Interpersonal psychotherapy for depressed antepartum women. In *Management of Psychiatric Disorders in Pregnancy* (K. Yonkers and B. Little, Eds.), pp. 105–121. Arnold, London.

Spinelli, M. G. (2004). Maternal infanticide associated with mental illness: Prevention and the promise of saved lives.[see comment]. *Am. J. Psychiatr.* 161(9), 1548–1557.

Spinelli, M. G., and Endicott, J. (2003). Controlled clinical trial of interpersonal psychotherapy versus parenting education program for depressed pregnant women. *Am. J. Psychiatr.* 160(3), 555–562.

Stockmeier, C. A., Shapiro, L. A., Dilley, G. E., Kolli, T. N., Friedman, L., and Rajkowska, G. (1998). Increase in serotonin-1a autoreceptors in the midbrain of suicide victims with major depression-postmortem evidence for decreased serotonin activity. *J. Neurosci.* 18(18), 7394–7401.

Stowe, Z. N., Casarella, J., Landry, J., and Nemeroff, C. B. (1995). Sertraline in the treatment of women with postpartum major depression. *Depression* 3, 49–55.

Stowe, Z. N., Hostetter, A. L., and Newport, D. J. (2005). The onset of postpartum depression: Implications for clinical screening in obstetrical and primary care. *Am. J. Obstet. Gynecol.* 192, 522–526.

Strathearn, L., Jian, L., Fonagy, P., and Montague, P. R. (2007). *Infant affect modulates maternal brain reward activation.* Paper presented at the Parental Brain, Boston, MA.

Stroup, T., J. M., and Swartz, M. (2003). The national institute of mental health clinical antipsychotic trials of intervention effectiveness (catie) project: Schizophrenia trial design and protocol development. *Schizophr Bull.* 29, 15–31.

Suri, R., Altshuler, L., Hellemann, G., Burt, V. K., Aquino, A., and Mintz, J. (2007). Effects of antenatal depression and antidepressant treatment on gestational age at birth and risk of preterm birth. *Am. J. Psychiatr.* 164(8), 1206–1213.

Sumner, B. E., and Fink, G. (1995). Estrogen increases the density of 5-ht2a receptors in cerebral cortex and nucleus accumbens in the female rat. *J Steroid Biochem Molec Biol.* 54(1/2), 15–20.

Teixeira, J. M., Fisk, N. M., and Glover, V. (1999). Association between maternal anxiety in pregnancy and increased uterine artery resistance index: A cohort based study. *Br. Med. J.* 16, 153–157.

Vajda, F., Lander, C., O'Brien, T., Hitchcock, A., Graham, J., Solinas, C., Eadie, M., and Cook, M. (2004). Australian pregnancy registry of women taking antiepileptic drugs. *Epilepsia* 45(11), 1466.

Viguera, A. C., Newport, D. J., Ritchie, J., Stowe, Z., Whitfield, T., Mogielnicki, J., Baldessarini, R. J., Zurick, A., and Cohen, L. S. (2007). Lithium in breast milk and nursing infants: Clinical implications. *Am. J. Psychiatr.* 164(2), 342–345.

Volkow, N. D., Wang, G.-J., Fowler, J. S., Logan, J., Gatley, S. J., MacGregor, R. R., Schlyer, D. J., Hitzemann, R., and Wolf, A. P. (1996). Measuring ag-related changes in da d2 receptors with [11c]raclopride and with [18f]n-methylspiroperidol. *Psychiatr. Res.* 67, 11–16.

Wadhwa, P., Garite, T., Porto, M., Glynn, L., Chicz-DeMet, A., Dunkel-Schetter, C., and Sandman, C. (2004). Placental crh, spontaneous preterm birth and fetal growth restriction: A prospective investigation. *Am. J. Obstet. Gynecol.* 191(4), 1063–1069.

Watkins, M. L., Rasmussen, S. A., Honein, M. A., Botto, L. D., and Moore, C. A. (2003). Maternal obsesity and risk for birth defects. *Pediatrics* 111, 1152–1158.

Weissman, A., Levy, B., Hartz, A., Bentler, S., Donohue, M., Ellingrod, V., and Wisner, K. (2004). Pooled analysis of antidepressant levels in lactating mothers, breast milk, and nursing infants. *Am. J. Psychiatr.* 161, 1066–1078.

Weissman, M. M., Pilowsky, D. J., Wickramaratne, P. J., Talati, A., Wisniewski, S. R., Fava, M., Hughes, C. W., Garber, J., Malloy, E., King, C. A., Cerda, G., Sood, A. B., Alpert, J. E., Trivedi, M. H., Rush, A. J., and Team, S. T. D.-C. (2006). Remissions in maternal depression and child psychopathology: A star*d-child report. *JAMA* 295(12), 1389–1398.

WHO (1998). World health organization task force on methods for the natural regulation of fertility. The world health organization multinational study of breast-feeding and lactational amenorrhea. Ii. Factors associated with the length of amenorrhea. *Fertil. Steril.* 70(3), 461–471.

Wieck, A., Kumar, R., Hirst, A. D., Marks, M. N., Campbell, I. C., and Checkley, S. A. (1991).

Increased sensitivity of dopamine receptors and recurrence of affective psychosis after childbirth. *Br. Med. J.* 303(6803), 613–616.

Wikander, I., Sundblad, C., Andersch, B., Dagnell, I., Zylberstein, D., Bengtsson, F., and Eriksson, E. (1998). Citalopram in premenstrual dysphoria: Is intermittent treatment during luteal phases more effective than continuous medication throughout the menstrual cycle?. *J. Clin. Psychopharmacol.* 18(5), 390–398.

Williams, N. I., Berga, S. L., and Cameron, J. L. (2007). Synergism between psychosocial and metabolic stressors: Impact on reproductive function in cynomolgus monkeys. *Am. J. Physiol. Endocrinol. Metabol.* 293(1), E270–E276.

Willner, P., Hale, A. S., and Argyropoulos, S. (2005). Dopaminergic mechanism of antidepressant action in depressed patients. *J. Affect. Disord.* 86, 37–45.

Wisner, K., Peindl, K., and Hanusa, B. (1994). Symptomatology of affective and psychotic illnesses related to childbearing. *J. Affect. Disord.* 30, 77–87.

Wisner, K., Peindl, K., Gigliotti, T., and Hanusa, B. (1999). Obsessions and compulsions in women with postpartum depression. *J. Clin. Psychiatr.* 60, 176–180.

Wisner, K., Zarin, D., Holmboe, E., Appelbaum, P., Gelenberg, A., Leonard, H., and Frank, E. (2000). Risk-benefit decision making for treatment of depression during pregnancy. *Am. J. Psychiatr.* 157, 1933–1940.

Wisner, K. L., and Perel, J. M. (1988). Psychopharmacologic agents and electroconvulsive therapy during pregnancy and the puerperium. In *Psychiatric Consultation in Childbirth Settings* (R. L. Cohen, Ed.), pp. 165–206. Plenum Press, New York.

Wisner, K. L., Hanusa, B. H., Perel, J. M., Peindl, K. S., Piontek, C. M., Sit, D. K. Y., Findling, R. L., and Moses-Kolko, E. L. (2006). Postpartum depression: A randomized trial of sertraline vs. Nortriptyline. *J. Clin. Psychopharmacol.* 26, 353–360.

Wisner, K. L., Sit, D. K. Y., Reynolds, S. K., Bogen, D. L., Sunder, K. R., Altemus, M., Misra, D., and JM, P. (2007). Psychiatric disorders, in obstetrics. In *Normal and Problem Pregnancies* (S. G. Gabbe, J. R. Niebyl, J. L. Simpson, H. Galan, L. Goetzl, E. R. M. Jauniaux, and M. Landon, Eds.), 5th edn., pp. 1249–1288. Elsevier.

Yurgelun-Todd, D. (2007). Emotional and cognitive changes during adolescence. *Curr. Opin. Neurobiol.* 17(2), 251–257.

Zlotnick, C., Miller, I. W., Pearlstein, T., Howard, M., and Sweeney, P. (2006). A preventive intervention for pregnant women on public assistance at risk for postpartum depression. *Am. J. Psychiatr.* 163(8), 1443–1445.

NEUROENDOCRINE ADAPTATIONS OF PARENTING: PREGNANCY, LACTATION, AND OFFSPRING

13 BRINGING FORTH THE NEXT GENERATION ... AND THE NEXT

JOHN A. RUSSELL AND PAULA J. BRUNTON

Laboratory of Neuroendocrinology, Centre for Integrative Physiology, School of Biomedical Sciences, College of Medicine and Veterinary Medicine, University of Edinburgh, Hugh Robson Building, George Square, Edinburgh EH8 9XD, UK

INTRODUCTION

This chapter considers some issues that affect the outcome of pregnancy in terms of parental behavior, including emotionality and desires, and stress-coping. It is well known that the pre- and post-natal development of an individual is shaped by the intrauterine environment and by postnatal experience, interacting with the individual's given genome. Perhaps inevitably, attention has been on effects of adversity, but with a view to optimizing development through understanding how environment can interact with developing systems to permanently and adversely program their functioning to increase vulnerability to mental and physical health in adulthood (Barker *et al.*, 1993; Fall, 2006; Meaney *et al.*, 2007). There are many examples of the impact of pre- and post-natal experiences that can have such consequences. These include intrauterine growth restriction, and low birth weight, which are associated with multiple pregnancies, pre-term birth, maternal protein deprivation, or stress. In particular, low birth weight predicts increased susceptibility in later life to hypertension and cardiovascular disease, type 2 diabetes mellitus, obesity, and depression (Barker *et al.*, 1993; Barker, 2004). An effect on blood pressure is found in early adulthood in some (Moore *et al.*, 1999), but not other studies (Williams & Poulton, 2002), with adult weight as a complicating factor (Huxley *et al.*, 2002; Phillips, 2006; Eriksson *et al.*, 2007).

A predominant unifying, convergent hypothesis is that many of these effects occur through epigenetic programming of the fetal, or neonatal, hypothalamo–pituitary–adrenal (HPA) axis (Meaney *et al.*, 2007) that are a consequence of exposure to excess glucocorticoid during development (Seckl & Holmes, 2007). In particular, altered methylation of promoter regions of the glucocorticoid receptor gene, and consequent altered setting of expression, has been proposed as a key mechanism of programming (Weaver *et al.*, 2004; Meaney *et al.*, 2007). This hypothesis assumes transferral of maternally experienced stress to the fetuses by the maternal glucocorticoid system. Indeed, there is strong evidence that treatment of late pregnant rodents or primates with the synthetic glucocorticoid dexamethasone results in such programming of the fetuses (Andrews & Matthews, 2004; de Vries *et al.*, 2007). However, there are other possibilities that might effect programming, including sympathetic responses to stress, which compromise blood supply to the placenta and hence oxygen and nutrient supply to the fetus(es) (Gitau *et al.*, 2001a, b). The fetus may also be a source of excess glucocorticoid as a result of such stress *in utero* (Cohen & Guillon, 1985). It is clear that *in utero* fetal blood sampling or transfusion in human pregnancy can activate the fetal HPA axis if the procedure involves needling of the intrahepatic vein (which involves fetal skin penetration), rather than a vessel at the uninnervated placental cord insertion, as evidenced by increased fetal blood levels of cortisol, ACTH, and β-endorphin (a POMC derivative, like ACTH, secreted from anterior pituitary corticotrophs) without increased maternal cortisol or β-endorphin secretion (Gitau *et al.*, 2001a, b, 2004).

A consequence of fetal programming is proposed to be enhanced activity or stress-responsiveness of the offspring HPA axis (Meaney *et al.*, 2007). This hypothesis also provides an explanation for early postnatal programming by environmental factors, in particular by the quality of attentive maternal care. Thus, offspring of rat mothers that are more attentive ("high licking/grooming") show increased glucocorticoid receptor expression in the hippocampus, enhancing negative feedback and reducing HPA axis stress responses. The altered glucocorticoid receptor expression here is a consequence of the cutaneous stimulation provided by the mother (Meaney *et al.*, 2007). Furthermore, this type of programming can be imposed on the next generation through the less attentive type of maternal care that postnatally programmed females give to their offspring. Moreover, intrauterine programming by maternal stress, which may be mediated via glucocorticoid action *in utero*, can also be transmitted through the offspring to the next generation without further treatment (Jarvis *et al.*, 2006; Seckl & Holmes, 2007), and even through the male line (Drake & Walker, 2004; Drake *et al.*, 2005; Meaney *et al.*, 2007).

The concept that early life adverse programming involves stable alterations in phenotype leaves open the possibility to overwrite programming if the mechanisms can be understood (Weaver *et al.*, 2007). However, it can also be argued that the transfer to the offspring (*in utero* or neonatally) of the registration of experience of an adverse environment by the mother via glucocorticoid "messengers" is adaptive, and might confer some survival and reproductive advantage on the offspring should they face the same adversity in later life (Coall & Chisholm, 2003; Seckl & Holmes, 2007). Nonetheless, bigger babies (from otherwise normal pregnancies) have better health in adulthood (Gunnarsdottir *et al.*, 2004; Huxley *et al.*, 2007).

Anxiety is an emotional accompaniment of stress, and in women, anxiety and stress during pregnancy, related for example to the workplace, is linked to pre-term birth, low birth weight, and neurobehavioral developmental and behavioral regulation problems for the offspring (Van den Bergh *et al.*, 2005). There are clear physiological or behavioral (detected with ultrasound scanning) responses in fetuses of mothers exposed to stress, generally in late pregnancy. These include increased arousal of fetuses of mothers experiencing high stress or anxiety levels, and altered fetal heart rate variability (Van den Bergh *et al.*, 2005). Offspring of mothers who experienced stress or anxiety during pregnancy have difficulty in interacting with their mother in infancy, impaired formally assessed development (Field *et al.*, 1985; Brouwers *et al.*, 2001; Huizink *et al.*, 2003; King & Laplante, 2005) and behavioral problems which persist into adolescence (Van den Bergh *et al.*, 2005). The timing of maternal stress during pregnancy as a factor in determining vulnerability to programming effects on the fetus seems to be variable between studies (Van den Bergh *et al.*, 2005). From these studies it seems that stress-reduction programs for pregnant women may be desirable, to supplement the natural protective mechanisms discussed below (Facchinetti *et al.*, 2004; Urizar *et al.*, 2004).

In animal models perinatal programming by prenatal stress or maternal separation in lactation, alters trait anxiety in the offspring. Stress in late pregnancy programs increased anxiety-like behavior in offspring, while postnatal handling has the opposite effect (Vallee *et al.*, 1997). However, the direction of the programming effect is evidently determined by genetic factors, including sex (Bowman *et al.*, 2004). Moreover, perinatal stress that is a consequence of maternal stress or quality of maternal behavior increases anxiety in a low-anxiety behavior rat line, and decreases anxiety in a high-anxiety behavior rat line: that is, perinatal stress tends to normalize the phenotype for anxious behavior against the trait predicted from the genotype (Neumann *et al.*, 2005a, b; Bosch *et al.*, 2006).

While the above considerations are relevant to human health, there are also implications of early life programming by intrauterine and early postnatal environment, associated with husbandry practices, on the welfare of domestic, farmed, and captive animals (Jarvis *et al.*, 2006). Our focus in this chapter is on perinatal interactions between mother and offspring, particularly when the mother is stressed in late pregnancy.

NEUROENDOCRINE STRESS RESPONSES

Without pregnancy, exposure to stressors results in the urgent activation of the sympathetic innervation of the adrenal medulla and consequent stimulation of adrenaline secretion as well as activation of the sympathetic innervation of the heart and blood vessels. Together, these responses increase energy availability and blood flow to skeletal muscle that are appropriate for "flight or fight" (Goldstein & Kopin, 2007). Subsequent activation of the HPA axis results in a more prolonged stimulation of glucocorticoid secretion, and consequent increase in energy availability and immune system modulation, as well as feedback actions of glucocorticoid on the brain (Herbert *et al.*, 2006; Goldstein & Kopin, 2007). Stressors are broadly classified as emotional/psychogenic (i.e., are perceived as potentially or actually threatening to homeostasis or survival), or physical (i.e., are actually painful or damaging, disrupting homeostasis). These stressors are primarily processed by rostral, limbic brain networks, or by caudal networks in the medulla oblongata, respectively (Buller, 2003; Herman *et al.*, 2005).

Two mechanisms protect the fetus(es) from adverse programming effects of exposure to excess glucocorticoid as a result of maternal stress. The first is a placental barrier that prevents free transfer of glucocorticoid from mother to fetus, and the second is reduced maternal responses to stressors in late pregnancy.

PLACENTAL BARRIER

The primary protective mechanism is the expression of the enzyme 11β-hydroxysteroid dehydrogenase type 2 (11β-HSD2) in the placenta which inactivates the active glucocorticoid (cortisol in humans, corticosterone in rodents) by conversion to inactive 11-keto metabolites (human, cortisone; rat, 11-dehydrocorticosterone). However, since the placental barrier, 11β-HSD2, is not entirely effective, glucocorticoid transfer from mother to fetus is not completely prevented (Benediktsson *et al.*, 1997; Venihaki *et al.*, 2000). In humans the levels of cortisol in fetal blood and amniotic fluid by mid-pregnancy correlate with the maternal blood cortisol level, suggesting cortisol transfer from mother to fetus. Fetal blood cortisol levels are only about 10% of the maternal levels, indicating that maternal cortisol crosses the placenta to a limited extent (Gitau *et al.*, 2001a, b; Sarkar *et al.*, 2007). Nonetheless, the transfer from mother to fetus of only 10% of a large increase in maternal cortisol secretion during gestation will lead to a much greater proportional increase in fetal circulating cortisol level (Gitau *et al.*, 2001a, b).

Impairment of the placental 11β-HSD2 barrier mechanism increases glucocorticoid transfer from the maternal to the fetal circulation, and evidently reduces birth weight (Stewart *et al.*, 1995). Homozygous 11β-HSD2$^{-/-}$ knockout mice borne of heterozygous (11β-HSD2$^{+/-}$) mothers exhibit programming as offspring; since these 11β-HSD2$^{-/-}$ mice lack the placental enzyme, these findings provide support for the protective role of placental 11β-HSD2 (Holmes *et al.*, 2006). Reduction of placental 11β-HSD2 activity by hypoxia or maternal protein restriction may likewise be a mechanism that allows increased glucocorticoid transfer from mother to fetus in these stressful circumstances (Langley-Evans *et al.*, 1996; Hardy & Yang, 2002). Furthermore, while acute maternal stress increases placental 11β-HSD2 activity in late pregnancy, chronic stress prevents this putative protective response (Welberg *et al.*, 2005) and repeated stress greatly reduces 11β-HSD2 mRNA expression, with less effect on 11β-HSD2 activity (Mairesse *et al.*, 2007). Reduced plasma ACTH levels and adrenal gland weight in the fetuses of stressed mothers suggest that HPA activity in these fetuses is suppressed by corticosterone crossing the placenta from the mother, despite similar terminal levels of plasma corticosterone in stressed and unstressed fetuses (Mairesse *et al.*, 2007).

So, the placental 11β-HSD2 barrier affords some protection against maternal glucocorticoid passage across the placenta. In rats and mice placental 11β-HSD2 expression decreases near the end of pregnancy (Waddell *et al.*, 1998), at which time corticosterone can effectively cross the placenta from mother to fetus, at least at the normal circadian peak levels in the maternal circulation (Cohen & Guillon, 1985; Venihaki *et al.*, 2000). Near the end of pregnancy in these species an alternative robust protective mechanism against programming by glucocorticoid emerges as the maternal HPA axis becomes less responsive to stress.

MATERNAL HPA AXIS HYPORESPONSIVENESS TO STRESS IN LATE PREGNANCY

The HPA axis becomes less responsive to both psychological and physical stressors during late pregnancy in rats (Neumann *et al.*, 1998; Brunton *et al.*, 2005), mice (Douglas *et al.*, 2003), and women (de Weerth & Buitelaar, 2005). This has been proposed as a mechanism for minimizing exposure of the fetus(es) to glucocorticoids. In rats the attenuation of HPA axis responses to stressors becomes evident from around day 15 of gestation, and responses are progressively reduced as term approaches (Neumann *et al.*, 1998); this hyporesponsiveness persists through parturition (Wigger *et al.*, 1999) and lactation (Windle *et al.*, 1997), until weaning. In late pregnant rats HPA axis hyporesponsiveness is reflected by reduced ACTH and corticosterone secretory responses to a range of stressors, including exposure to the elevated plus maze (Neumann *et al.*, 1998), forced swimming (Douglas *et al.*, 1998; Neumann *et al.*, 1998), restraint (Brunton *et al.*, 2000), social stress (Figure 13.1) (Brunton *et al.*, 2003), immune challenge (Brunton *et al.*, 2005), and metabolic peptides signaling a lack of energy (Brunton & Russell, 2003; Brunton *et al.*, 2006). Late pregnant mice (day 17.5–18.5) also exhibit reduced ACTH secretion in response to psychological (novel environment) and physical (forced swimming) stressors (Douglas *et al.*, 2003). In both rats and mice these effects are a consequence of reduced drive from the corticotropin releasing hormone (CRH) (and AVP) neurons in the parvocellular division of the paraventricular nucleus (PVN) at the end of pregnancy. Following stress exposure, upregulation of transcription for CRH, arginine vasopressin (AVP), or immediate early genes is reduced in late pregnancy, compared with virgin animals (da Costa *et al.*, 1996; Brunton & Russell, 2003; Douglas *et al.*, 2003; Brunton *et al.*, 2005, 2006), indicating that these neurons are stimulated less strongly by stressors.

In general, HPA axis responses to stressors are reduced in pregnant women (de Weerth & Buitelaar, 2005). For example, late pregnant women fail to show an increase in ACTH or corticosterone secretion in response to exogenously

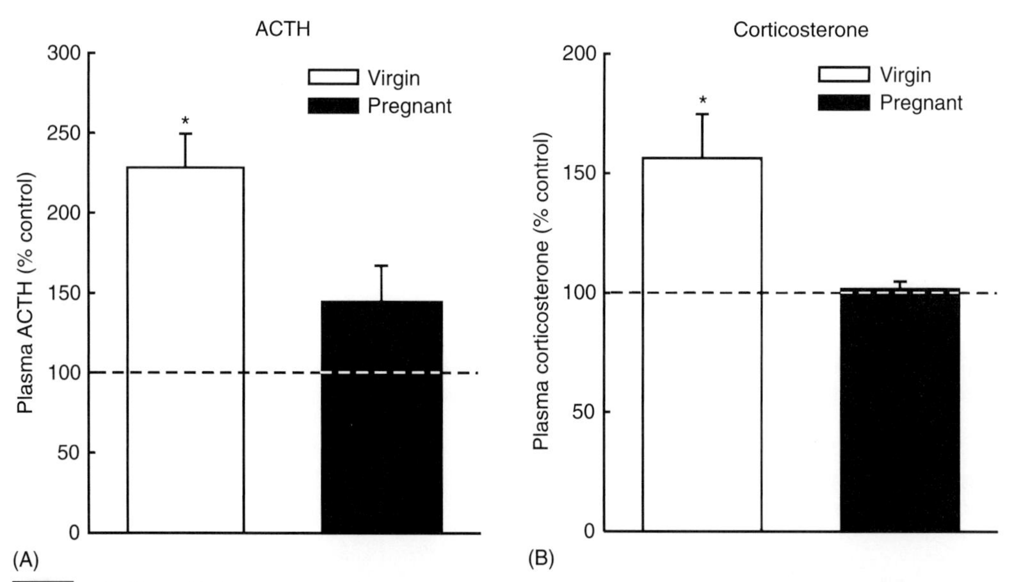

FIGURE 13.1 *Effect of a single 30-min period of social stress on stress hormone secretion: lack of response in late pregnancy.* (A) ACTH and (B) corticosterone measured in trunk blood plasma (expressed as % of respective control) in virgin ($n = 7$) and late (day 21; term = 22.5 days) pregnant ($n = 5$) rats ("intruders") collected immediately following exposure to an aggressive lactating (on day 5–6 of lactation) rat ("resident") for 30 min. Control virgin and pregnant rats were left undisturbed in their home cage. Data are presented as group means ± SEM. Dashed line indicates mean plasma ACTH/corticosterone concentrations in control rats. $^*p < 0.04$ vs. pregnant group; Student's *t*-test (Brunton *et al.*, 2003).

administered CRH (Schulte *et al.*, 1990) and display a suppressed salivary cortisol response (presumably reflecting reduced maternal ACTH secretion) to a physical stressor (cold pressor test)(Kammerer *et al.*, 2002) compared with non-pregnant women. However, the salivary cortisol response to a standard social stress test is undiminished in mid and late pregnancy (Nierop *et al.*, 2006). Furthermore, cortisol content in hair (evidently reflecting chronic cortisol secretion) is greater in pregnant women with higher perceived stress scores (Kalra *et al.*, 2007). These findings may be relevant to the potential impact of emotional stress on fetal programming by maternal glucocorticoids.

The mechanisms underpinning suppressed HPA axis responses to stress in late pregnancy have been studied extensively in the rat. Reduced activation of the CRH/AVP neurons in the PVN is not a result of enhanced negative feedback actions of glucocorticoids (Johnstone *et al.*, 2000), but instead is attributed to reduced excitatory afferent drive to these neurons. Brainstem noradrenergic neurons are activated by stressful stimuli and play an important role in mediating HPA axis responses to stress. In contrast to virgin rats, forced swimming (Douglas *et al.*, 2005) and immune challenge (systemic interleukin-1β; IL-1β) (Brunton *et al.*, 2005) fail to evoke noradrenaline release in the PVN in late pregnancy. Endogenous opioids are involved since HPA axis responses to both of these stressors can be restored in pregnant rats by blocking opioid action with systemic naloxone treatment (Douglas *et al.*, 1998; Brunton *et al.*, 2005). Furthermore, opioid and opioid receptor expression is increased in late pregnancy in brain regions known to influence HPA activity (Douglas *et al.*, 2002; Brunton *et al.*, 2005). Endogenous opioids evidently act presynaptically on noradrenergic terminals in the PVN to inhibit stress-induced noradrenaline release at the end of pregnancy, since naloxone retrodialysed into the PVN reinstates the noradrenaline response to systemic IL-1β (Brunton *et al.*, 2005). In contrast, enhanced endogenous opioid inhibition does not explain reduced HPA axis responses to stress during late pregnancy in the mouse (Douglas *et al.*, 2003), hence the underlying mechanisms involved in this species remain unclear.

Induction of inhibitory opioid tone over HPA axis responses to stress in pregnancy in the rat is not signaled directly to the brain by the increased levels of circulating estrogen or progesterone in pregnancy (Douglas *et al.*, 2000). However, the progesterone metabolite, allopregnanolone, is involved. Allopregnanolone concentrations in the blood and brain are increased in pregnancy, reaching a peak 2 days before term (Concas *et al.*, 1998). Blocking allopregnanolone production restores HPA axis responses to systemic IL-1β in late pregnant rats, while treating virgin rats with allopregnanolone (to mimic pregnancy) suppresses these responses (see Brunton & Russell). Moreover allopregnanolone induces opioid tone over HPA axis responses to systemic IL-1β (see Brunton & Russell, 2008). To date, a role for allopregnanolone in suppressed HPA axis responses to stress during late pregnancy in other species has not been investigated.

MATERNAL SYMPATHETIC AND ADRENOMEDULLARY RESPONSES TO STRESS IN PREGNANCY

Few studies have compared the responses of the sympathetic nervous system or adrenomedullary system to stress in pregnant and non-pregnant subjects. Acute thermal stress (short-term exposure to a heat chamber at 70°C) increases plasma noradrenaline and adrenaline concentrations in non-pregnant women, however in pregnant women this stressor only increases noradrenaline release and has no effect on adrenaline secretion (Vaha-Eskeli *et al.*, 1992). Accordingly, virgin and late pregnant rats display similar noradrenaline secretory responses to air-puff startle, however the adrenaline response is markedly reduced in late pregnancy (Russell *et al.*, in press). The mechanisms responsible for suppressed adrenomedullary responses to stress and indeed whether adrenomedullary responses to other stressors are also attenuated in late pregnancy remain to be elucidated.

DESIRE FOR A FAMILY

A human family is recognized in the Universal Declaration of Human Rights as the "*natural and fundamental group unit of society*" and as such is "*entitled to protection by society and the State.*" The Declaration also establishes that "*men and women of full age, without any limitation due to race, nationality or religion, have the right to marry and to found a family.*" (United Nations Universal Declaration of Human Rights, 1948). The Declaration may be taken to imply that society and states have a responsibility to enable couples to fulfill a wish to found a family, and to optimize the outcome. Not least, this is in the interests of the success and development of each state. The national provision of maternity health services, with antenatal, obstetric, and postnatal care and education about parenthood has enabled safe and healthy pregnancy and fostered the safe birth of healthy babies.

In recent years assisted reproduction technology (ART; *in vitro* fertilization and embryo transfer) has enabled many couples to found a family, where this technology can be afforded by couples or national health services. However, a consequence of ART has been a large increase in the incidence of multiple births, consequent on the practice of transferring multiple embryos to try to ensure successful implantation (El-Toukhy *et al.*, 2006). As reviewed by El-Toukhy *et al.* (2006), the twin birth rate in 2001 in the USA was 59% higher than in 1980 and the triplet and above birth rate was 4-fold greater. Of the >40,000 infants born after ART procedures, 54% were from multiple pregnancies (twins: 46%, triplets and more: 8%). In the USA in 2002 ART accounted for 17% of all multiple births, including 44% of triplets, but only 1.1% of all live births (Reynolds *et al.*, 2003). These data reflect the persistence in ART clinical practice of transferring more than one embryo, despite awareness of the problems of multiple pregnancy. In the last few years the number of embryos transferred has tended to decrease, moving toward official guidelines (Reynolds & Schieve, 2006). The situation in Europe has been similar to that in USA (Blondel & Kaminski, 2002).

In the Peoples Republic of China the one-child family policy, introduced in 1979, has likely contributed with the support for contraception and abortion, to a further decline in fertility rate (Hesketh *et al.*, 2005), but with an increase in the male:female ratio of delivered or surviving babies, indicative of son-preference, for socio-economic reasons. This phenomenon of "missing millions of females," or excess of males, has been aided by the availability of ultrasound technology to determine fetal sex, for the purpose of choosing to abort females, and affects several Asian countries despite government efforts to prohibit this (Hesketh & Xing, 2006). A recent survey has revealed that the optimal desired number of children among women in China is one or two (Hesketh *et al.*, 2005; Ding & Hesketh, 2006). Evidently, a minority (7.5% in rural areas) want three or more (Hesketh *et al.*, 2005), and it seems that some resort to using readily available self-treatment with clomiphene to achieve super-ovulation and a multiple pregnancy, thus circumventing the restrictions (BBC news, 2007). Such self-medication to stimulate super-ovulation and multiple pregnancy is not restricted to China. The desire for multiple pregnancy in the USA increases with age in females with a fertility problem (Gleicher *et al.*, 1995).

The point of these considerations is that increased numbers of multiple births have arisen from the use of ART in developed countries, and perhaps from the one-child family policy in China, with likely increased strain and stress on the parents to cope with the demands of parenthood. The problems include low birth weight, perinatal bereavement due to increased mortality of multiple newborns, greater incidence of postnatal depression, economic issues, strain on the relationship with the partner, and feeling inadequate as a mother, with an excessive demand on time available to care for the infants (Dill, 2006; Garel *et al.*, 2006). Subsequently, the mothers of young twins and triplets have a greater risk of being depressed (Thorpe *et al.*, 1991; Garel *et al.*, 1997; Fisher, 2006). The consequences of the extra multiple births that have arisen from the natural desire of couples to found a family aided by ART on the long-term health of the offspring as adults have yet to emerge. Low birth weight children are expected to have increased HPA axis activity, with some of the consequences discussed above (Clark *et al.*, 1996; Phillips *et al.*, 1998; Phillips *et al.*, 2000), including raised systolic blood pressure (McNeill *et al.*, 2004; Bergvall *et al.*, 2007).

DRUG ABUSE

Cigarettes and Alcohol

It is well established that drug abuse in pregnancy has deleterious effects on the offspring. Alcohol consumption, apart from in excess leading to the fetal alcohol syndrome (Lazzaroni *et al.*, 1993) is associated with low birth weight (Zhang *et al.*, 2005; Jaddoe *et al.*, 2007).

Adult rat offspring of mothers ingesting ethanol show HPA hyper-responsiveness to stressors, with altered feedback regulation (Glavas *et al.*, 2007). Cigarette smoking is also associated with lower birth weight (Bernstein *et al.*, 2005). Fetuses of cigarette smoking mothers have increased ACTH levels in umbilical arterial blood, compared to fetuses of non-smoking mothers, indicating a possible programming effect of exposure to tobacco smoking in pregnancy (McDonald *et al.*, 2006), and the long-term consequences outlined above (Williams & Poulton, 1999). Alcohol consumption and cigarette smoking in combination have a greater effect in reducing birth weight than either alone (Okah *et al.*, 2005).

Opiates and Cocaine

Abuse of opiates such as heroin leads to addiction of the fetus, and hence a withdrawal syndrome in the newborn, and substantially increases the risk of low birth weight (Finnegan, 1985; Hulse *et al.*, 1997). Cocaine abuse in women in pregnancy produces intrauterine growth retardation and reduces birth weight (Bada *et al.*, 2002). Acute cocaine administration stimulates the fetal and maternal HPA axis in sheep (Owiny *et al.*, 1991) and juvenile offspring of rats treated with cocaine prenatally show exaggerated corticosterone responses to stress (Huber *et al.*, 2001). Maternal behavior is highly rewarding, and this aspect of maternity reflects activation of the mesolimbic circuitry that is also involved in hedonic appetite and drug-seeking behavior, involving opiate and dopaminergic mechanisms (Numan, 2007). Indeed, the rewarding aspects of maternal behavior are demonstrated by experiments on rats in which a choice has to be made between access to pups and access to cocaine (which is strongly rewarding; cocaine is preferred by about 50% of lactating rats) (Mattson *et al.*, 2003; see also Chapter 4). These studies indicate competition between cocaine seeking and maternal behavior, although maternal behavior was evidently not compromised. However, the offspring of mothers given cocaine show deficient maternal behavior (Johns *et al.*, 2005).

Cannabis

Compared with the "hard" drugs, the recreational use of cannabis is often regarded as relatively innocuous, and cannabis is commonly abused by pregnant women (Day *et al.*, 1994). Powerful deleterious actions of cannabinoids, which can cross the placenta (Hutchings *et al.*, 1989), have been described on the development of GABAergic connections in the fetal cerebral cortex (Berghuis *et al.*, 2007), in which endocannabinoids are involved (Harkany *et al.*, 2007). Hence cannabis use during pregnancy is likely to interfere with normal brain development. Exogenous cannabinoids act in the brain to activate the HPA axis (Weidenfeld *et al.*, 1994), although it is not clear whether this action occurs in pregnancy, and if it does then this would contribute to any programming actions of cannabinoids on the fetus. It is relevant in the context of programming of the fetal HPA axis to note that endocannabinoids have been shown to have a key role in the negative feedback regulation by glucocorticoid of the activity of CRH neurons in the PVN (Di *et al.*, 2003). The enduring impact of exposing the developing HPA axis prenatally to exogenous cannabinoids is indicated by reduced circulating corticosterone level in male offspring (del Arco *et al.*, 2000). There are also altered HPA axis responses to novelty in rat offspring of mothers treated with cannabinoid in pregnancy; these effects are sexually dimorphic such that female offspring are more responsive and males less so (Navarro *et al.*, 1995). In terms of anxiety, there is a U-shaped relationship between the greater time spent by adult offspring on the open arms of the elevated plus maze and doses of cannabinoid given to the mother in pregnancy, indicating reduced anxiety with the middle cannabinoid doses (Navarro *et al.*, 1995).

Cannabis abuse by pregnant women evidently reduces birth weight, suggesting a programming scenario (Zuckerman *et al.*, 1989; Fergusson *et al.*, 2002). Ten-year-old children from pregnancies during which the mothers smoked marijuana have more behavioral problems than children from drug-free pregnancies (Goldschmidt *et al.*, 2000). Any prenatal effects of cannabis exposure through the mother's abuse of the drug is likely to be conferred on the next generation since teenage offspring of mothers who used cannabis in pregnancy are more likely to be cannabis users (Day *et al.*, 2006).

SOCIAL STRESS

Domestic violence, including psychological, physical, and sexual abuse by the husband, spouse, or partner against women (intimate partner abuse) in pregnancy is common worldwide, independent of nationality, ethnicity, culture, and socio-economic status (Collado-Pena & Villanueva-Egan, 2007). Addressing the problem is a target of the United Nations Millennium Development Goals (United Nations, 2000; World Health Organization, 2005). There is an adverse impact of domestic violence in pregnancy on the well-being of both the mother and child, and potentially the next generation. Recent studies indicate prevalence of domestic violence in pregnancy ranging from 4.3% in China (Guo *et al.*, 2004) to 4.5% in Malaysia (Jahanfar *et al.*, 2007), 3–6% in the UK (Bacchus *et al.*, 2004a, b), 6–9% in the USA (Neggers *et al.*, 2004; Yost *et al.*, 2005; Koenig *et al.*, 2006), 13% in Mexico (Cuevas *et al.*, 2006), 15–18% in India (Peedicayil *et al.*, 2004; Ahmed *et al.*, 2006; Varma *et al.*, 2007), 23% in Pakistan (Fikree *et al.*, 2006), 28% in Uganda (Kaye *et al.*, 2006), >33% in Nicaragua (Valladares *et al.*, 2005), and 37% in Nigeria (Efetie & Salami, 2007). Low socio-economic status is an important risk factor for domestic violence in pregnancy (Koenig *et al.*, 2006), which is cause of maternal mortality (Campero *et al.*, 2006), accounting for 20% of maternal deaths in US cities (Campbell, 2002), and perinatal infant mortality (Yost *et al.*, 2005; Ahmed *et al.*, 2006).

Women experiencing domestic violence in pregnancy see themselves more negatively as mothers than women not abused (Huth-Bocks *et al.*, 2004). Babies born of mothers experiencing domestic violence in pregnancy are more likely to be born early (Boy & Salihu, 2004; Neggers *et al.*, 2004; Schoeman *et al.*, 2005) of lower birth weight, but perhaps not as a result of physical abuse (Murphy *et al.*, 2001; Nunez-Rivas *et al.*, 2003; Boy & Salihu, 2004; Yost *et al.*, 2005; Kaye *et al.*, 2006; Collado-Pena & Villanueva-Egan, 2007), indicating a likelihood of adverse prenatal programming for life. Children of women experiencing domestic violence during pregnancy show greater behavioral problems (Tan & Gregor, 2006). Whether association between raised salivary cortisol level in children and an unstable parental environment (Flinn & England, 1997) or with low socio-economic status and mother's depressed mood (Lupien *et al.*, 2000) relates to domestic violence as a prenatal stressor needs evaluation.

Recognition of the problem of domestic violence inflicted on pregnant women has led to introduction of domestic violence screening and support programs in antenatal clinics (Bacchus *et al.*, 2004b; World Health Organization, 2005; Duncan *et al.*, 2006; Escriba-Aguir *et al.*, 2007), and calls for paternal educational classes for spouses (Efetie & Salami, 2007). A particular effort is being mounted in Moldova (World Health Organization Europe, 2005).

MODELING PRENATAL SOCIAL STRESS

Recently, pigs and rats have been used to model the effects of aggression as a social stressor during pregnancy on the possible programming of offspring. These studies have relevance not only to humans, as discussed above, but also to the welfare of livestock. In European Union countries legislation about housing conditions for farmed animals, including pigs (European Commission Directive, 2001) has led to sows being housed in groups. This mixing of pigs that are not familiar with each other leads to aggressive behavior associated with the establishment of dominance in the group and involves some superficial physical injury, as well as HPA axis activation (Tsuma *et al.*, 1996; D'Eath & Lawrence, 2004; O'Connell *et al.*, 2004). This stressful situation provides a model to evaluate outcomes for the offspring (Jarvis *et al.*, 2005). To date, no studies have compared HPA axis responses to stress in late pregnant and non-pregnant pigs. It is, however, interesting to note that early- and mid-pregnancy pigs display a transient increase in plasma cortisol concentrations following a meal. This does not occur in late pregnant sows (Hay *et al.*, 2000), suggesting the HPA axis may be less responsive to satiety factors at this time.

Our comparable studies on rats are based on the fact that female rats display aggressive behavior during lactation. Lactating rats are highly aggressive toward female rats (pregnant or not) that intrude into the territory they establish around their nest, clearly serving to protect their young (Lonstein & Gammie, 2002; Bosch *et al.*, 2004; Gammie, 2005). Hence, introducing a pregnant female rat into the home cage of a lactating rat provides a reliable means of exposing the intruder to social stress (Bosch. *et al.*, 2004).

Pig Model of Prenatal Social Stress

Mixing a pair of younger primiparous pregnant sows with a pair of older multiparous

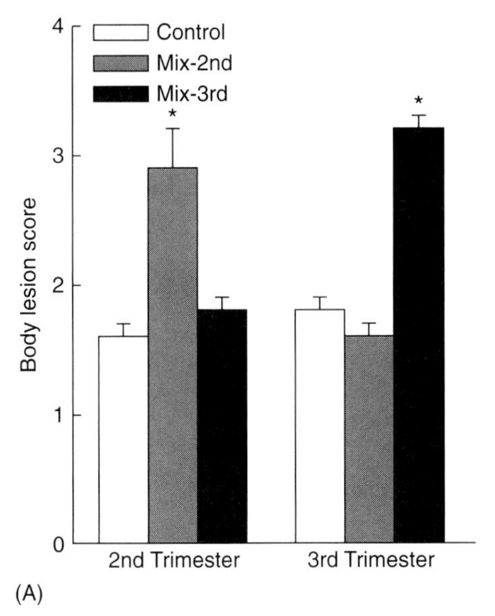

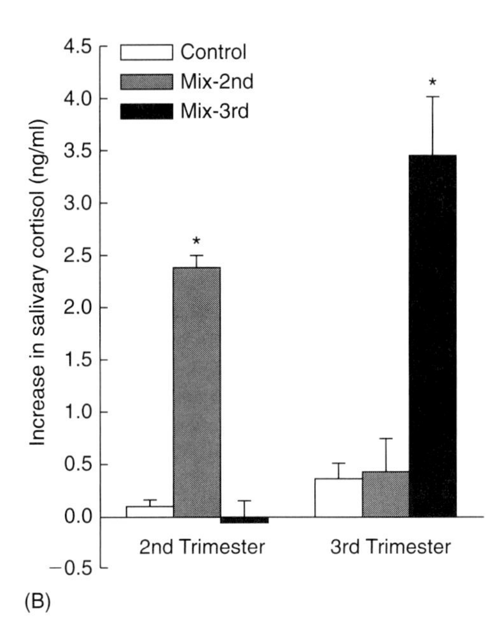

FIGURE 13.2 *Effects of social stress on body lesions and cortisol secretion in pregnant pigs.* Primiparous pregnant pigs were either left undisturbed in their home pen (control, $n = 8$), or exposed to social mixing (housed with two older unfamiliar multiparous sows). One group was exposed to social mixing for two 7-day periods (days 39–45 and 59–65) only during the 2nd trimester (mix-2nd, $n = 8$), and the other group was exposed to social mixing for two 7-day periods (days 77–83 and 97–103) only during the 3rd trimester (mix-3rd, $n = 8$). Gestation in the pig is 114 days. (A) Body lesion score (scratches/lesions scored on a scale of 1–5, 5 being the most severe) was assessed in pregnant pigs during the 2nd and 3rd trimesters on the first 3 days of social mixing to assess aggression by the two older sows or at an equivalent time in unmixed pigs. (B) Increase in salivary cortisol level was calculated by subtracting concentrations measured 3.5 h following the onset of the first period of social mixing in the 2nd (mix-2nd) or 3rd (mix-3rd) trimester, or at an equivalent time in unmixed pigs, from basal concentrations measured the day before. Data are group means ±SEM. $*p < 0.05$ vs. other groups at the same time; one-way ANOVA. (*Source*: Data from Jarvis *et al.* (2006).) *Note*: Although some pigs were not mixed until the 3rd trimester (mix-3rd), their body lesion scores and salivary cortisol level were assessed in the 2nd trimester (before any social mixing). Similarly, pigs mixed only in the 2nd trimester (mix-2nd) were also assessed in the 3rd trimester for body lesions and salivary cortisol.

sows for two episodes of 1 week, either in the 2nd or 3rd trimester, results in the older sows being aggressive toward the younger sows, and establishing social dominance (at the trough) (Jarvis *et al.*, 2006). In this model the younger sows show more body lesions (Figure 13.2) from aggressive interactions than when left in a same-age group, and salivary cortisol measurements in the young sows show substantial increases as a result of mixing with older sows. Hence, the younger sows are clearly subjected to social stress by being housed with older sows (see Figure 13.2; Jarvis *et al.*, 2006). Nonetheless, in this study there were no differences in measures of immediate pregnancy outcome, including piglet birth weight, or perinatal maternal behaviors between the prenatally stressed sows and their controls (Jarvis *et al.*, 2006). However, the female offspring (only females were studied) of prenatally stressed mothers showed differences from controls as they grew and matured in their growth rate, HPA axis function, and in their maternal behaviors when they produced litters. Thus, these females were prenatally programmed by the social stress experienced by their mothers. The programmed offspring gained weight more slowly, showed slower behavioral adjustments toward littermates at weaning, and showed greater and more prolonged salivary cortisol responses when later exposed to social stress (Figure 13.3(C)). Basal CRH mRNA expression in the PVN and amygdala of the programmed females was greater than that in

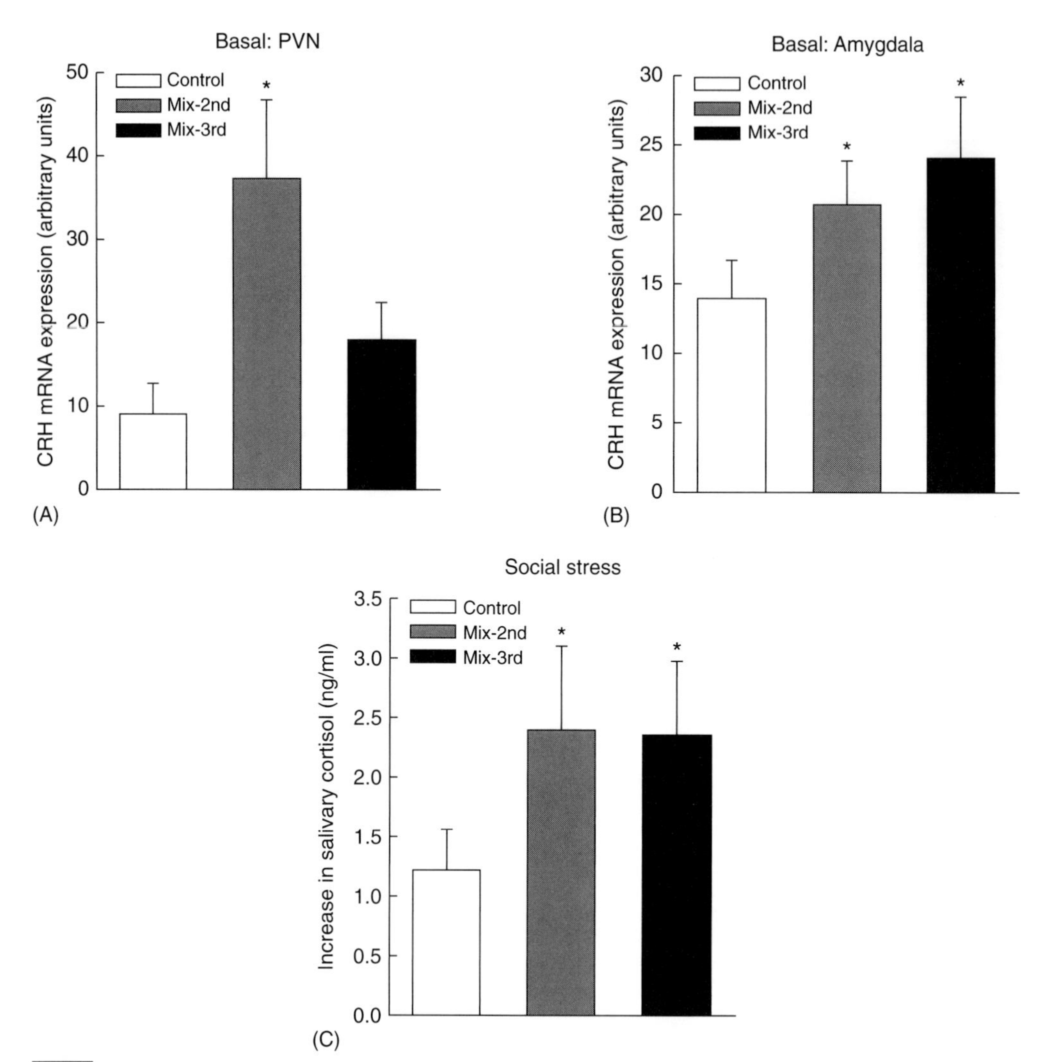

FIGURE 13.3 *Effects of prenatal social stress in pigs on basal CRH mRNA expression in the brain and stress-induced cortisol secretion in the female offspring.* Female offspring born to undisturbed pigs (control) and to those exposed to social mixing during the 2nd (mix-2nd) or 3rd trimester (mix-3rd) were killed at 60 days old. Brains were removed, sectioned and processed to detect CRH mRNA by *in situ* hybridization to determine basal levels of expression in the (A) PVN and (B) amygdala. (C) Salivary cortisol concentration was measured 1 h before (basal) and 1 h after the onset of social mixing in female offspring (age = 67 days) from undisturbed pigs and from those exposed to social mixing during pregnancy. Data are expressed as the increase in salivary cortisol concentration from basal levels. In each case, data are presented as group means ± SEM and group numbers = 8 pigs/group. *$p < 0.05$ vs. controls; two-way ANOVA. (*Source*: Data from Jarvis *et al.* (2006).)

controls (Figures 13.3(A),(B)). Remarkably, of the adult offspring that were inseminated and gave birth, the sows that experienced prenatal stress in the 2nd trimester showed more aggressive behavior toward their piglets (Jarvis *et al.*, 2006). Whether this poor maternal behavior in the prenatally stressed offspring relates to the greater CRH mRNA expression in the amygdala requires further study. For example, CRH action in the amygdala is considered to be important in anxiogenesis (Kanitz *et al.*, 2004; Shekhar *et al.*, 2005), and the use of an appropriately scaled-up elevated plus maze has potential to assess anxiety in pigs (Andersen *et al.*, 2000a, b). Overall, the studies on pigs indicate that prenatal social stress, involving repeated aggressive behavior toward the pregnant sow, has long-lasting programming effects on at least the female offspring, which leads to impaired maternal behavior when these programmed offspring become mothers. These results have welfare implications for considering husbandry practices in relation to the social context of housing farmed pigs, and for the consequences of domestic violence against women in pregnancy for the potential of their female offspring to be good mothers.

Rat Model of Prenatal Social Stress

Commonly, rodent models of social stress utilize the resident–intruder paradigm. In males this involves the intruder being "defeated" by a dominant male (the resident), whereas in females social defeat models exploit maternal aggression exhibited by a lactating resident toward a female intruder approaching the nest. These models reliably and robustly activate the HPA axis (Wotjak *et al.*, 1996; Neumann *et al.*, 2001; Martinez *et al.*, 2002) as well as sympathetic (noradrenaline) and adrenomedullary (adrenaline) responses in the intruder rats (Sgoifo *et al.*, 1996). Divergent behaviors are displayed by the resident and intruder animals. Not surprisingly, resident rats are dominant and exhibit more aggressive behavior, whereas intruder rats are submissive and display defensive behavior (Neumann *et al.*, 2001).

We have exposed virgin and pregnant rats to a single 30-min period of social defeat (by a lactating rat) on day 21 of pregnancy. We found that as is the case with other stressors, pregnant rats display attenuated ACTH and corticosterone secretory responses compared with non-pregnant female intruders (Brunton *et al.*, 2003) (Figure 13.1), despite experiencing similar levels of aggression (number of attacks and latency to attack) displayed by the lactating resident rat. Again, these reduced responses are the result of reduced activation of the CRH/AVP neurons in the parvocellular PVN, reflected by a failure to upregulate expression of mRNA for the immediate early gene, nerve growth factor inducible gene-B (NGFI-B). Whether this social stress paradigm induces sympathetic or adrenomedullary responses in pregnant rats is not known.

Typically, social stress experienced by women in pregnancy, for example through domestic violence, is chronic in nature. To model this in rodents we have adapted the ethologically relevant social stress model described above, and exposed pregnant rats to an unfamiliar aggressive lactating rat for 10 min each day on five consecutive days (days 16–20 of gestation). While this social stress increased ACTH and corticosterone secretion in virgin rats, it did not alter ACTH secretion in the pregnant rats on each of the days tested (Figure 13.4(A)). We did, however, observe a significant corticosterone response in the pregnant rats on days 16–19 (in contrast to rats exposed to a single 30-min social stress on day 21 of pregnancy as shown in Figure 13.1), though this was markedly attenuated (by ~70%), compared with the virgin group (Figure 13.4(B)). The discordance between ACTH and corticosterone levels in the pregnant rats may be attributable to the action of estrogen increasing adrenal sensitivity to ACTH (Figueiredo *et al.*, 2007).

Exposure to social stress during pregnancy does not appear to alter subsequent maternal behavior (Neumann *et al.*, 2005a, b); however, detrimental effects on some aspects of maternal care have been reported in the female offspring (F1 generation) of rats exposed to gestational stress (Bosch *et al.*, 2007). Nevertheless, our preliminary studies indicate that the adult offspring of these rats exposed to chronic social stress during late pregnancy exhibit exaggerated HPA axis and behavioral responses (i.e., increased anxiety) to stress, consistent with the reported findings in prenatally stressed (repeated restraint during late pregnancy) (McCormick *et al.*, 1995; Vallee *et al.*, 1997) or glucocorticoid-exposed offspring (Welberg *et al.*, 2001; Banjanin *et al.*, 2004; Shoener *et al.*, 2006). The results suggest that this repeated social stress paradigm induces programming effects on the offspring, despite suppressed maternal HPA axis responses, and gives rise to the question of whether the reduced maternal corticosterone response to stress in pregnancy, which is no greater than the conserved diurnal corticosterone increase (Atkinson & Waddell, 1995), is sufficient to explain programming, and whether maternal sympathetic and/or adrenomedullary responses to stress contribute to prenatal programming of the offspring. Indeed, elevated levels of catecholamines in the maternal circulation constrict placental blood vessels, decreasing the supply of oxygen and glucose to the fetus(es) and activating fetal sympathetic and adrenomedullary responses (Gu & Jones, 1986). Moreover, up to 12% of maternal noradrenaline is not metabolized by placental monoamine oxidase

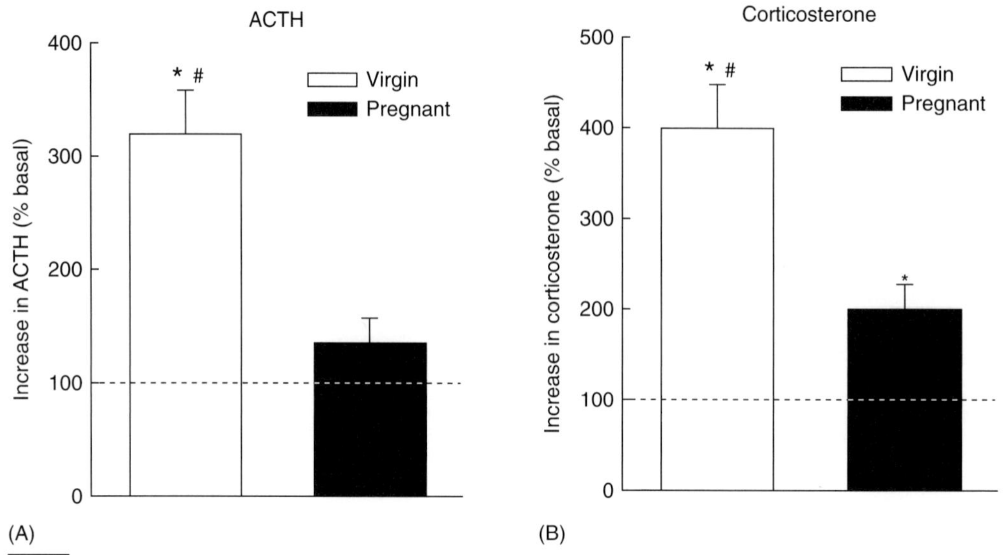

FIGURE 13.4 *Reduced effects of repeated social stress on HPA axis responses in late pregnant rats.* Mean increase in (A) ACTH and (B) corticosterone plasma concentrations (expressed as % of basal concentration) across 5 days, in virgin ($n = 6$–8 rats/day) and pregnant (days 16–20 of pregnancy; $n = 7$–9 rats/day) rats immediately following 10 min social stress (exposure to an aggressive lactating rat, between days 2–8 of lactation) per day for 5 consecutive days. Data are presented as group means ± SEM. Dashed line indicates basal concentrations. $*p < 0.01$ vs. basal; $\#p < 0.001$ vs. pregnant group; Student's *t*-test (Brunton *et al.*, unpublished).

and catechol-O-methyltransferase (Chen *et al.*, 1974, 1976) and is transferred to the fetal compartment (Sandler *et al.*, 1963; Morgan *et al.*, 1972; Saarikoski, 1974; Sodha *et al.*, 1984). Thus, retained noradrenaline responses to stress in late pregnancy (as described above) may also adversely affect the fetus(es). There is a need to investigate whether fetal stress responses (HPA, sympathetic and/or adrenomedullary) are involved in mediating prenatal fetal programming effects.

CONCLUSION

We have drawn attention to some consequences of the strong desire of humans to found a family, which is a basic human right, and the resort to ART with a high risk of a multiple pregnancy. The extensive use of ART (resulting in birth of more than 110,000 live infants in the USA between 1997 and 2000 (Reynolds *et al.*, 2003)) increased the number of triplet births 4-fold (more than 13,000 triplet, or more, infants born between 1997 and 2000 (Reynolds *et al.*, 2003)). In contrast, given the birth of 15 million babies between 1997 and 2000 in the USA, it can be estimated conservatively that at least 900,000 (6%) pregnant women in this period will have been subjected to domestic violence (Neggers *et al.*, 2004). Both multiple pregnancy and domestic violence during pregnancy are expected, for quite different reasons, to be emotionally disturbing or stressful for the pregnant woman. Drawing on human and animal studies, we outlined the concept of adverse lifelong fetal programming through exposure of the mother and fetus to stress and the evidence that excess glucocorticoid causes this programming, breaching the two defense mechanisms that protect the fetus to some extent, namely the placental 11β-HSD barrier (Seckl & Holmes, 2007) and the reduced responsiveness of the mother's neuroendocrine mechanisms to stress (Brunton *et al.*, 2005)(Figure 13.5). It seems that studies are required that are designed to assess the programming impact of their prenatal experiences on the offspring of mothers exposed to domestic violence in pregnancy. We reviewed data from animal studies that model the latter through imposing repeated aggression by conspecifics in pregnancy, and demonstrated that the mothers show reduced

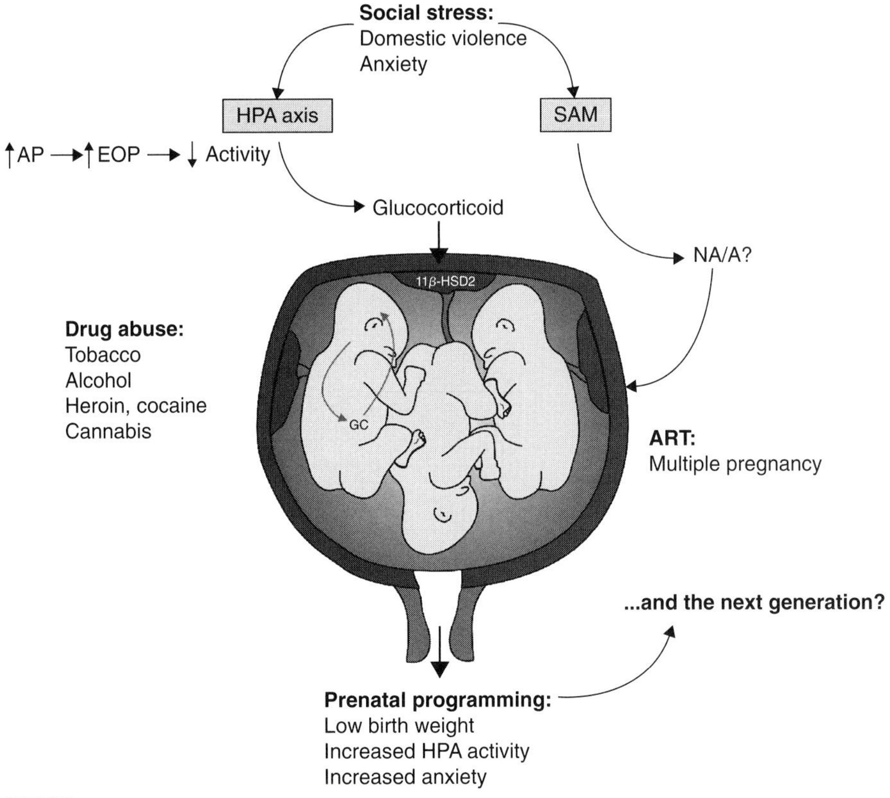

FIGURE 13.5 *Summary diagram.* Preventable factors that contribute to prenatal fetal programming and low birth weight, as discussed in this chapter, are illustrated. The natural protective mechanisms through hyporesponsiveness of the maternal hypothalamo–hypophysial axis and placental 11-β hydroxysteroid dehydrogenase are indicated. A, adrenaline; AP, allopregnanolone; ART, assisted reproduction technology; EOP, endogenous opioid peptide; HPA, hypothalamo–pituitary–adrenal axis; GC, glucocorticoid (cortisol in human, pig; corticosterone in rodents); NA, noradrenaline; SAM: sympathetic and adrenomedullary systems.

HPA axis responses to this stressor. Moreover, the adult offspring have a hyperactive HPA axis and increased anxiety behavior. In the pig model, the female offspring also show deficient maternal behavior when they reproduce (Jarvis *et al.*, 2006). The animal studies have relevance to animal welfare issues, but also point up the possible similar programming consequences for offspring of women exposed to domestic violence in pregnancy.

ACKNOWLEDGMENTS

The Biotechnology and Biological Research Council (BBSRC) provides financial support for the authors' research on prenatal stress and programming.

ABBREVIATIONS

11β-HSD2	11beta-hydroxysteroid dehydrogenase type 2
ACTH	adrenocorticotropic hormone
ART	assisted reproduction technology
AVP	arginine vasopressin
CRH	corticotropin releasing hormone
HPA	hypothalamo–pituitary–adrenal
NGFI-B	nerve growth factor inducible gene-B
POMC	pro-opiomelanocortin
PVN	paraventricular nucleus

REFERENCES

Ahmed, S., Koenig, M. A., and Stephenson, R. (2006). Effects of domestic violence on perinatal and early-childhood mortality: Evidence from north India. *Am. J. Public Health* 96, 1423–1428.

Andersen, I. L., Boe, K. E., Foerevik, G., Janczak, A. M., and Bakken, M. (2000a). Behavioural evaluation of methods for assessing fear responses in weaned pigs. *Appl. Anim. Behav. Sci.* 69, 227–240.

Andersen, I. L., Faerevik, G., Boe, K. E., Janczak, A. M., and Bakken, M. (2000b). Effects of diazepam on the behaviour of weaned pigs in three putative models of anxiety. *Appl. Anim. Behav. Sci.* 68, 121–130.

Andrews, M. H., and Matthews, S. G. (2004). Programming of the hypothalamo–pituitary–adrenal axis: Serotonergic involvement. *Stress* 7, 15–27.

Atkinson, H. C., and Waddell, B. J. (1995). The hypothalamic–pituitary–adrenal axis in rat pregnancy and lactation: Circadian variation and interrelationship of plasma adrenocorticotropin and corticosterone. *Endocrinology* 136, 512–520.

Bacchus, L., Mezey, G., and Bewley, S. (2004a). Domestic violence: Prevalence in pregnant women and associations with physical and psychological health. *Eur. J. Obstet. Gynecol. Reprod. Biol.* 113, 6–11.

Bacchus, L., Mezey, G., Bewley, S., and Haworth, A. (2004b). Prevalence of domestic violence when midwives routinely enquire in pregnancy. *Br. J. Obstet. Gynaecol.* 111, 441–445.

Bada, H. S., Das, A., Bauer, C. R., Shankaran, S., Lester, B., Wright, L. L., Verter, J., Smeriglio, V. L., Finnegan, L. P., and Maza, P. L. (2002). Gestational cocaine exposure and intrauterine growth: Maternal lifestyle study. *Obstet. Gynecol.* 100, 916–924.

Banjanin, S., Kapoor, A., and Matthews, S. G. (2004). Prenatal glucocorticoid exposure alters hypothalamic–pituitary–adrenal function and blood pressure in mature male guinea pigs. *J. Physiol.* 558, 305–318.

Barker, D. J. (2004). The developmental origins of adult disease. *J. Am. Coll. Nutr.* 23, 588S–595S.

Barker, D. J., Hales, C. N., Fall, C. H., Osmond, C., Phipps, K., and Clark, P. M. (1993). Type 2 (non-insulin-dependent) diabetes mellitus, hypertension and hyperlipidaemia (syndrome X): Relation to reduced fetal growth. *Diabetologia* 36, 62–67.

BBC news (2007). Chinese one-child policy <http://www.news.bbc.co.uk/2/hi/asia-pacific/6694135.stm>.

Benediktsson, R., Calder, A. A., Edwards, C. R., and Seckl, J. R. (1997). Placental 11 beta-hydroxysteroid dehydrogenase: A key regulator of fetal glucocorticoid exposure. *Clin. Endocrinol.* 46, 161–166.

Berghuis, P., Rajnicek, A. M., Morozov, Y. M., Ross, R. A., Mulder, J., Urban, G. M., Monory, K., Marsicano, G., Matteoli, M., Canty, A., Irving, A. J., Katona, I., Yanagawa, Y., Rakic, P., Lutz, B., Mackie, K., and Harkany, T. (2007). Hardwiring the brain: Endocannabinoids shape neuronal connectivity. *Science* 316, 1212–1216.

Bergvall, N., Iliadou, A., Johansson, S., de Faire, U., Kramer, M. S., Pawitan, Y., Pedersen, N. L., Lichtenstein, P., and Cnattingius, S. (2007). Genetic and shared environmental factors do not confound the association between birth weight and hypertension: A study among Swedish twins. *Circulation* 115, 2931–2938.

Bernstein, I. M., Mongeon, J. A., Badger, G. J., Solomon, L., Heil, S. H., and Higgins, S. T. (2005). Maternal smoking and its association with birth weight. *Obstet. Gynecol.* 106, 986–991.

Blondel, B., and Kaminski, M. (2002). Trends in the occurrence, determinants, and consequences of multiple births. *Semin. Perinatol.* 26, 239–249.

Bosch, O. J., Kromer, S. A., Brunton, P. J., and Neumann, I. D. (2004). Release of oxytocin in the hypothalamic paraventricular nucleus, but not central amygdala or lateral septum in lactating residents and virgin intruders during maternal defence. *Neuroscience* 124, 439–448.

Bosch, O. J., Kromer, S. A., and Neumann, I. D. (2006). Prenatal stress: Opposite effects on anxiety and hypothalamic expression of vasopressin and corticotropin-releasing hormone in rats selectively bred for high and low anxiety. *Eur. J. Neurosci.* 23, 541–551.

Bosch, O. J., Musch, W., Bredewold, R., Slattery, D. A., and Neumann, I. D. (2007). Prenatal stress increases HPA axis activity and impairs maternal care in lactating female offspring: Implications for postpartum mood disorder. *Psychoneuroendocrinology* 32, 267–278.

Bowman, R. E., MacLusky, N. J., Sarmiento, Y., Frankfurt, M., Gordon, M., and Luine, V. N. (2004). Sexually dimorphic effects of prenatal stress on cognition, hormonal responses, and central neurotransmitters. *Endocrinology* 145, 3778–3787.

Boy, A., and Salihu, H. M. (2004). Intimate partner violence and birth outcomes: A systematic review. *Int. J. Fertil. Womens Med.* 49, 159–164.

Brouwers, E. P. M., Van Baar, A. L., and Pop, V. J. M. (2001). Maternal anxiety during pregnancy and subsequent infant development. *Infant Behav. Dev.* 24, 95–106.

Brunton, P. J., and Russell, J. A. (2003). Hypothalamic–pituitary–adrenal responses to centrally administered orexin-A are suppressed in pregnant rats. *J. Neuroendocrinol.* 15, 633–637.

Brunton, P. J., and Russell, J. A. (2008). The expecant brain: adapting for motherhood. *Nat. Rev. Neurosci.* 9, 11–25.

Brunton, P. J., Ma, S., Shipston, M. J., Wigger, A., Neumann, I. D., Douglas, A. J., and Russell, J. A. (2000). Central mechanisms underlying reduced ACTH stress responses in pregnant rats: Attenuated acute gene activation in the parvocellular paraventricular nucleus (pPVN). *Eur. J. Neurosci.* 12, 184.17.

Brunton, P. J., Meddle, S. L., Krömer, S., Neumann, I. D., and Russell, J. A. (2003). Hypothalamo–pituitary–adrenal responses to social defeat are reduced in pregnant rats. *J. Physiol.* 548P, O32.

Brunton, P. J., Meddle, S. L., Ma, S., Ochedalski, T., Douglas, A. J., and Russell, J. A. (2005). Endogenous opioids and attenuated hypothalamic–pituitary–adrenal axis responses to immune challenge in pregnant rats. *J. Neurosci.* 25, 5117–5126.

Brunton, P. J., Bales, J., and Russell, J. A. (2006). Neuroendocrine stress but not feeding responses to centrally administered neuropeptide Y are suppressed in pregnant rats. *Endocrinology* 147, 3737–3745.

Buller, K. M. (2003). Neuroimmune stress responses: Reciprocal connections between the hypothalamus and the brainstem. *Stress* 6, 11–17.

Campbell, J. C. (2002). Health consequences of intimate partner violence. *Lancet* 359, 1331–1336.

Campero, L., Walker, D., Hernandez, B., Espinoza, H., Reynoso, S., and Langer, A. (2006). The contribution of violence to maternal mortality in Morelos, Mexico. *Salud. Publica Mexico* 48, S297–S306.

Chen, C. H., Klein, D. C., and Robinson, J. C. (1974). Catechol-O-methyltransferase in rat placenta, human placenta and choriocarcinoma grown in culture. *J. Reprod. Fertil.* 39, 407–410.

Chen, C. H., Klein, D. C., and Robinson, J. C. (1976). Monoamine oxidase in rat placenta, human placenta, and cultured choriocarcinoma. *J. Reprod. Fertil.* 46, 477–479.

Clark, P. M., Hindmarsh, P. C., Shiell, A. W., Law, C. M., Honour, J. W., and Barker, D. J. (1996). Size at birth and adrenocortical function in childhood. *Clin. Endocrinol.* 45, 721–726.

Coall, D. A., and Chisholm, J. S. (2003). Evolutionary perspectives on pregnancy: Maternal age at menarche and infant birth weight. *Soc. Sci. Med.* 57, 1771–1781.

Cohen, A., and Guillon, Y. (1985). Effect of fetal decapitation on unbound rat plasma corticosterone concentration at the end of pregnancy. *Biol. Neonate* 47, 163–169.

Collado-Pena, S. P., and Villanueva-Egan, L. A. (2007). Relationship between familial violence during pregnancy and risk of low weight in the newborn. *Ginecol. Obstet. Mexico* 75, 259–267.

Concas, A., Mostallino, M. C., Porcu, P., Follesa, P., Barbaccia, M. L., Trabucchi, M., Purdy, R. H., Grisenti, P., and Biggio, G. (1998). Role of brain allopregnanolone in the plasticity of gamma-aminobutyric acid type A receptor in rat brain during pregnancy and after delivery. *Proc. Natl Acad. Sci. USA* 95, 13284–13289.

Cuevas, S., Blanco, J., Juarez, C., Palma, O., and Valdez-Santiago, R. (2006). Violence and pregnancy in female users of Ministry of Health care services in highly deprived states in Mexico. *Salud. Publica Mexico* 48, S239–S249.

da Costa, A. P. C., Wood, S., Ingram, C. D., and Lightman, S. L. (1996). Region-specific reduction in stress-induced c-fos mRNA expression during pregnancy and lactation. *Brain Res.* 742, 177–184.

Day, N. L., Richardson, G. A., Goldschmidt, L., Robles, N., Taylor, P. M., Stoffer, D. S., Cornelius, M. D., and Geva, D. (1994). Effect of prenatal marijuana exposure on the cognitive development of offspring at age three. *Neurotoxicol. Teratol.* 16, 169–175.

Day, N. L., Goldschmidt, L., and Thomas, C. A. (2006). Prenatal marijuana exposure contributes to the prediction of marijuana use at age 14. *Addiction* 101, 1313–1322.

D'Eath, R. B., and Lawrence, A. B. (2004). *Anim. Behav.* 67, 501–509.

de Vries, A., Holmes, M. C., Heijnis, A., Seier, J. V., Heerden, J., Louw, J., Wolfe-Coote, S., Meaney, M. J., Levitt, N. S., and Seckl, J. R. (2007). Prenatal dexamethasone exposure induces changes in non-human primate offspring cardiometabolic and hypothalamic–pituitary–adrenal axis function. *J. Clin. Invest.* 117, 1058–1067.

de Weerth, C., and Buitelaar, J. K. (2005). Physiological stress reactivity in human pregnancy: A review. *Neurosci. Biobehav. Rev.* 29, 295–312.

del Arco, I., Munoz, R., Rodriguez De Fonseca, F., Escudero, L., Martin-Calderon, J. L., Navarro, M., and Villanua, M. A. (2000). Maternal exposure to the synthetic cannabinoid HU-210: Effects on the endocrine and immune systems of the adult male offspring. *Neuroimmunomodulation* 7, 16–26.

Di, S., Malcher-Lopes, R., Halmos, K. C., and Tasker, J. G. (2003). Nongenomic glucocorticoid inhibition via endocannabinoid release in the hypothalamus: A fast feedback mechanism. *J. Neurosci.* 23, 4850–4857.

Dill, S. (2006). Economic and social implications of multiple gestation: Introduction to theme. *Aust. NZ J. Obstet. Gynaecol.* 46, S29–S30.

Ding, Q. J., and Hesketh, T. (2006). Family size, fertility preferences, and sex ratio in China in the era of the one child family policy: Results from national family planning and reproductive health survey. *Br. Med . J.* 333, 371–373.

Douglas, A. J., Johnstone, H. A., Wigger, A., Landgraf, R., and Neumann, I. D. (1998). The role of endogenous opioids in neurohypophysial and hypothalamo–pituitary–adrenal axis hormone secretory responses to stress in pregnant rats. *J. Endocrinol.* 158, 285–293.

Douglas, A. J., Johnstone, H., Brunton, P., and Russell, J. A. (2000). Sex-steroid induction of endogenous opioid inhibition on oxytocin secretory responses to stress. *J. Neuroendocrinol.* 12, 343–350.

Douglas, A. J., Bicknell, R. J., Leng, G., Russell, J. A., and Meddle, S. L. (2002). Beta-endorphin cells in the arcuate nucleus: Projections to the supraoptic nucleus and changes in expression during pregnancy and parturition. *J. Neuroendocrinol.* 14, 768–777.

Douglas, A. J., Brunton, P. J., Bosch, O. J., Russell, J. A., and Neumann, I. D. (2003). Neuroendocrine responses to stress in mice: Hyporesponsiveness in pregnancy and parturition. *Endocrinology* 144, 5268–5276.

Douglas, A. J., Meddle, S. L., Toschi, N., Bosch, O. J., and Neumann, I. D. (2005). Reduced activity of the noradrenergic system in the paraventricular nucleus at the end of pregnancy: Implications for stress hyporesponsiveness. *J. Neuroendocrinol.* 17, 40–48.

Drake, A. J., and Walker, B. R. (2004). The intergenerational effects of fetal programming: Non-genomic mechanisms for the inheritance of low birth weight and cardiovascular risk. *J. Endocrinol.* 180, 1–16.

Drake, A. J., Walker, B. R., and Seckl, J. R. (2005). Intergenerational consequences of fetal programming by *in utero* exposure to glucocorticoids in rats. *Am. J. Physiol. Regul. Integr. Comp. Physiol.* 288, R34–rR38.

Duncan, M. M., McIntosh, P. A., Stayton, C. D., and Hall, C. B. (2006). Individualized performance feedback to increase prenatal domestic violence screening. *Matern. Child Health J.* 10, 443–449.

Efetie, E. R., and Salami, H. A. (2007). Domestic violence on pregnant women in Abuja, Nigeria. *J. Obstet. Gynaecol.* 27, 379–382.

El-Toukhy, T., Khalaf, Y., and Braude, P. (2006). IVF results: Optimize not maximize. *Am. J. Obstet. Gynecol.* 194, 322–331.

Eriksson, J. G., Forsen, T. J., Kajantie, E., Osmond, C., and Barker, D. J. (2007). Childhood growth and hypertension in later life. *Hypertension* 49, 1415–1421.

Escriba-Aguir, V., Ruiz-Perez, I., and Saurel-Cubizolles, M. J. (2007). Screening for domestic violence during pregnancy. *J. Psychosom. Obstet. Gynaecol.* 28, 133–134.

European Commission Directive (2001). 2001/88/EC <http://ec.europa.eu/food/animal/welfare/farm/pigs_en.htm>.

Facchinetti, F., Tarabusi, M., and Volpe, A. (2004). Cognitive-behavioral treatment decreases cardiovascular and neuroendocrine reaction to stress in women waiting for assisted reproduction. *Psychoneuroendocrinology* 29, 162–173.

Fall, C. H. D. (2006). Developmental origins of cardiovascular disease, type 2 diabetes and obesity in humans. *Adv. Exp. Med. Biol.* 573, 8–28.

Fergusson, D. M., Horwood, L. J., and Northstone, K. (2002). Maternal use of cannabis and pregnancy outcome. *Br. J. Obstet. Gynaecol.* 109, 21–27.

Field, T., Sandberg, D., Garcia, R., Vega-Lahr, N., Goldstein, S., and Guy, L. (1985). Pregnancy problems, postpartum depression, and early mother–infant interactions. *Dev. Psychol.* 21, 1152–1156.

Figueiredo, H. F., Ulrich-Lai, Y. M., Choi, D. C., and Herman, J. P. (2007). Estrogen potentiates adrenocortical responses to stress in female rats. *Am. J. Physiol. Endocrinol. Metab.* 292, E1173–eE1182.

Fikree, F. F., Jafarey, S. N., Korejo, R., Afshan, A., and Durocher, J. M. (2006). Intimate partner violence before and during pregnancy: Experiences of postpartum women in Karachi, Pakistan. *J. Pak. Med. Assoc.* 56, 252–257.

Finnegan, L. P. (1985). Effects of maternal opiate abuse on the newborn. *Fed. Proc.* 44, 2314–2317.

Fisher, J. (2006). Psychological and social implications of multiple gestation and birth. *Aust. N Z J. Obstet. Gynaecol.* 46, S34–sS37.

Flinn, M. V., and England, B. G. (1997). Social economics of childhood glucocorticoid stress response and health. *Am. J. Phys. Anthropol.* 102, 33–53.

Gammie, S. C. (2005). Current models and future directions for understanding the neural circuitries of maternal behaviors in rodents. *Behav. Cogn. Neurosci. Rev.* 4, 119–135.

Garel, M., Salobir, C., and Blondel, B. (1997). Psychological consequences of having triplets: A 4-year follow-up study. *Fertil. Steril.* 67, 1162–1165.

Garel, M., Charlemaine, E., and Blondel, B. (2006). Psychological consequences of multiple births. *Gynecol. Obstet. Fertil.* 34, 1058–1063.

Gitau, R., Fisk, N. M., and Glover, V. (2001). Maternal stress in pregnancy and its effect on the human foetus: An overview of research findings. *Stress* 4, 195–203.

Gitau, R., Fisk, N. M., Teixeira, J. M., Cameron, A., and Glover, V. (2001). Fetal hypothalamic–pituitary–adrenal stress responses to invasive procedures are independent of maternal responses. *J. Clin. Endocrinol. Metab.* 86, 104–109.

Gitau, R., Fisk, N. M., and Glover, V. (2004). Human fetal and maternal corticotrophin releasing hormone responses to acute stress. *Arch. Dis. Child Fetal Neonatal Ed.* 89, F29–fF32.

Glavas, M. M., Ellis, L., Yu, W. K., and Weinberg, J. (2007). Effects of prenatal ethanol exposure on basal limbic–hypothalamic–pituitary–adrenal regulation: Role of corticosterone. *Alcohol Clin. Exp. Res.* 31, 1598–1610.

Gleicher, N., Campbell, D. P., Chan, C. L., Karande, V., Rao, R., Balin, M., and Pratt, D. (1995). The desire for multiple births in couples with infertility problems contradicts present practice patterns. *Hum. Reprod.* 10, 1079–1084.

Goldschmidt, L., Day, N. L., and Richardson, G. A. (2000). Effects of prenatal marijuana exposure on child behavior problems at age 10. *Neurotoxicol. Teratol.* 22, 325–336.

Goldstein, D. S., and Kopin, I. J. (2007). Evolution of concepts of stress. *Stress* 10, 109–120.

Gu, W., and Jones, C. T. (1986). The effect of elevation of maternal plasma catecholamines on the fetus and placenta of the pregnant sheep. *J. Dev. Physiol.* 8, 173–186.

Gunnarsdottir, I., Birgisdottir, B. E., Benediktsson, R., Gudnason, V., and Thorsdottir, I. (2004). Association between size at birth, truncal fat and obesity in adult life and its contribution to blood pressure and coronary heart disease; study in a high birth weight population. *Eur. J. Clin. Nutr.* 58, 812–818.

Guo, S. F., Wu, J. L., Qu, C. Y., and Yan, R. Y. (2004). Domestic abuse on women in China before, during, and after pregnancy. *Chin. Med. J.* 117, 331–336.

Hardy, D. B., and Yang, K. (2002). The expression of 11 beta-hydroxysteroid dehydrogenase type 2 is induced during trophoblast differentiation: Effects of hypoxia. *J. Clin. Endocrinol. Metab.* 87, 3696–3701.

Harkany, T., Guzman, M., Galve-Roperh, I., Berghuis, P., Devi, L. A., and Mackie, K. (2007). The emerging functions of endocannabinoid signaling during CNS development. *Trends Pharmacol. Sci.* 28, 83–92.

Hay, M., Meunier-Salaun, M. C., Brulaud, F., Monnier, M., and Mormede, P. (2000). Assessment of hypothalamic–pituitary–adrenal axis and sympathetic nervous system activity in pregnant sows through the measurement of glucocorticoids and catecholamines in urine. *J. Anim. Sci.* 78, 420–428.

Herbert, J., Goodyer, I. M., Grossman, A. B., Hastings, M. H., de Kloet, E. R., Lightman, S. L., Lupien, S. J., Roozendaal, B., and Seckl, J. R. (2006). Do corticosteroids damage the brain? *J. Neuroendocrinol.* 18, 393–411.

Herman, J. P., Ostrander, M. M., Mueller, N. K., and Figueiredo, H. (2005). Limbic system mechanisms of stress regulation: Hypothalamo–pituitary–adrenocortical axis. *Prog. Neuropsychopharmacol. Biol. Psychiatr.* 29, 1201–1213.

Hesketh, T., and Xing, Z. W. (2006). Abnormal sex ratios in human populations: Causes and consequences. *Proc. Natl Acad. Sci. USA* 103, 13271–13275.

Hesketh, T., Lu, L., and Xing, Z. W. (2005). The effect of China's one-child family policy after 25 years. *N. Engl. J. Med.* 353, 1171–1176.

Holmes, M. C., Abrahamsen, C. T., French, K. L., Paterson, J. M., Mullins, J. J., and Seckl, J. R. (2006). The mother or the fetus? 11beta-hydroxysteroid dehydrogenase type 2 null mice provide evidence for direct fetal programming of behavior by endogenous glucocorticoids. *J. Neurosci.* 26, 3840–3844.

Huber, J., Darling, S., Park, K., and Soliman, K. F. (2001). Altered responsiveness to stress and NMDA following prenatal exposure to cocaine. *Physiol. Behav.* 72, 181–188.

Huizink, A. C., Robles de Medina, P. G., Mulder, E. J., Visser, G. H., and Buitelaar, J. K. (2003). Stress during pregnancy is associated with developmental outcome in infancy. *J. Child Psychol. Psychiatr.* 44, 810–818.

Hulse, G. K., Milne, E., English, D. R., and Holman, C. D. (1997). The relationship between maternal use of heroin and methadone and infant birth weight. *Addiction* 92, 1571–1579.

Hutchings, D. E., Martin, B. R., Gamagaris, Z., Miller, N., and Fico, T. (1989). Plasma concentrations of delta-9-tetrahydrocannabinol in dams and fetuses following acute or multiple prenatal dosing in rats. *Life Sci.* 44, 697–701.

Huth-Bocks, A. C., Levendosky, A. A., Theran, S. A., and Bogat, G. A. (2004). The impact of domestic violence on mothers' prenatal representations of their infants. *Infant Ment. Health J.* 25, 79–98.

Huxley, R., Neil, A., and Collins, R. (2002). Unravelling the fetal origins hypothesis: Is there really an inverse association between birthweight and subsequent blood pressure? *Lancet* 360, 659–665.

Huxley, R., Owen, C. G., Whincup, P. H., Cook, D. G., Rich-Edwards, J., Smith, G. D., and Collins, R. (2007). Is birth weight a risk factor for ischemic heart disease in later life? *Am. J. Clin. Nutr* 85, 1244–1250.

Jaddoe, V. W., Bakker, R., Hofman, A., Mackenbach, J. P., Moll, H. A., Steegers, E. A., and Witteman, J. C. (2007). Moderate alcohol consumption during pregnancy and the risk of low birth weight and preterm birth. The Generation R Study. *Ann. Epidemiol.* 17, 834–840.

Jahanfar, S., Kamarudin, E. B., Sarpin, M. A., Zakaria, N. B., Abdul Rahman, R. B., and Samsuddin, R. D. (2007). The prevalence of domestic violence against pregnant women in Perak, Malaysia. *Arch. Iran Med.* 10, 376–378.

Jarvis, S., Moinard, C., Robson, S. K., Baxter, E., Ormandy, E., Douglas, A. J., Seckl, J. R., Russell, J. A., and Lawrence, A. B. (2006). Programming the offspring of the pig by prenatal social stress: Neuroendocrine activity and behaviour. *Horm. Behav.* 49, 68–80.

Johns, J. M., Elliott, D. L., Hofler, V. E., Joyner, P. W., McMurray, M. S., Jarrett, T. M., Haslup, A. M., Middleton, C. L., Elliott, J. C., and Walker, C. H. (2005). Cocaine treatment and prenatal environment interact to disrupt intergenerational maternal behavior in rats. *Behav. Neurosci.* 119, 1605–1618.

Johnstone, H. A., Wigger, A., Douglas, A. J., Neumann, I. D., Landgraf, R., Seckl, J. R., and Russell, J. A. (2000). Attenuation of hypothalamic–pituitary–adrenal axis stress responses in late pregnancy: Changes in feedforward and feedback mechanisms. *J. Neuroendocrinol.* 12, 811–822.

Kalra, S., Einarson, A., Karaskov, T., Van Uum, S., and Koren, G. (2007). The relationship between stress and hair cortisol in healthy pregnant women. *Clin. Invest. Med.* 30, E103–E107.

Kammerer, M., Adams, D., Castelberg Bv, B., and Glover, V. (2002). Pregnant women become insensitive to cold stress. *BMC Pregnancy Childbirth* 2, 8.

Kanitz, E., Tuchscherer, M., Puppe, B., Tuchscherer, A., and Stabenow, B. (2004). Consequences of repeated early isolation in domestic piglets (*Sus scrofa*) on their behavioural, neuroendocrine, and immunological responses. *Brain Behav. Immun.* 18, 35–45.

Kaye, D. K., Mirembe, F. M., Bantebya, G., Johansson, A., and Ekstrom, A. M. (2006). Domestic violence during pregnancy and risk of low birthweight and maternal complications: A prospective cohort study at Mulago Hospital, Uganda. *Trop. Med. Int. Health* 11, 1576–1584.

King, S., and Laplante, D. P. (2005). The effects of prenatal maternal stress on children's cognitive development: Project Ice Storm. *Stress* 8, 35–45.

Koenig, L. J., Whitaker, D. J., Royce, R. A., Wilson, T. E., Ethier, K., and Fernandez, M. I. (2006). Physical and sexual violence during pregnancy and after delivery: A prospective multistate study of women with or at risk for HIV infection. *Am. J. Public Health* 96, 1052–1059.

Langley-Evans, S. C., Phillips, G. J., Benediktsson, R., Gardner, D. S., Edwards, C. R., Jackson, A. A., and Seckl, J. R. (1996). Protein intake in pregnancy, placental glucocorticoid metabolism and the programming of hypertension in the rat. *Placenta* 17, 169–172.

Lazzaroni, F., Bonassi, S., Magnani, M., Calvi, A., Repetto, E., Serra, F., Podesta, F., and Pearce, N. (1993). Moderate maternal drinking and outcome of pregnancy. *Eur. J. Epidemiol.* 9, 599–606.

Lonstein, J. S., and Gammie, S. C. (2002). Sensory, hormonal, and neural control of maternal aggression in laboratory rodents. *Neurosci. Biobehav. Rev.* 26, 869–888.

Lupien, S. J., King, S., Meaney, M. J., and McEwen, B. S. (2000). Child's stress hormone levels correlate with mother's socioeconomic status and depressive state. *Biol. Psychiatr.* 48, 976–980.

Mairesse, J., Lesage, J., Breton, C., Breant, B., Hahn, T., Darnaudery, M., Dickson, S. L., Seckl, J., Blondeau, B., Vieau, D., Maccari, S., and Viltart, O. (2007). Maternal stress alters endocrine function of the feto-placental unit in rats. *Am. J. Physiol. Endocrinol. Metab.* 292, E1526–E1533.

Martinez, M., Calvo-Torrent, A., and Herbert, J. (2002). Mapping brain repsonse to social stress in rodents with c-fos expression: A review. *Stress* 5, 3–13.

Mattson, B. J., Williams, S. E., Rosenblatt, J. S., and Morrell, J. I. (2003). Preferences for cocaine- or pup-associated chambers differentiates otherwise behaviorally identical postpartum maternal rats. *Psychopharmacology* 167, 1–8.

McCormick, C. M., Smythe, J. W., Sharma, S., and Meaney, M. J. (1995). Sex-specific effects of prenatal stress on hypothalamic–pituitary–adrenal responses to stress and brain glucocorticoid receptor density in adult rats. *Dev. Brain Res.* 84, 55–61.

McDonald, S. D., Walker, M., Perkins, S. L., Beyene, J., Murphy, K., Gibb, W., and Ohlsson, A. (2006). The effect of tobacco exposure on the fetal hypothalamic–pituitary–adrenal axis. *Br. J. Obstet. Gynaecol.* 113, 1289–1295.

McNeill, G., Tuya, C., and Smith, W. C. (2004). The role of genetic and environmental factors in the association between birthweight and blood pressure: Evidence from meta-analysis of twin studies. *Int. J. Epidemiol.* 33, 995–1001.

Meaney, M. J., Szyf, M., and Seckl, J. R. (2007). Epigenetic mechanisms of perinatal programming of hypothalamic–pituitary–adrenal function and health. *Trends Mol. Med.* 13, 269–277.

Moore, V. M., Cockington, R. A., Ryan, P., and Robinson, J. S. (1999). The relationship between birth weight and blood pressure amplifies from childhood to adulthood. *J. Hypertens.* 17, 883–888.

Morgan, C. D., Sandler, M., and Panigel, M. (1972). Placental transfer of catecholamines *in vitro* and *in vivo*. *Am. J. Obstet. Gynecol.* 112, 1068–1075.

Murphy, C. C., Schei, B., Myhr, T. L., and Du Mont, J. (2001). Abuse: A risk factor for low birth weight? A systematic review and meta-analysis. *Can. Med. Assoc. J.* 164, 1567–1572.

Navarro, M., Rubio, P., and de Fonseca, F. R. (1995). Behavioural consequences of maternal exposure to natural cannabinoids in rats. *Psychopharmacology* 122, 1–14.

Neggers, Y., Goldenberg, R., Cliver, S., and Hauth, J. (2004). Effects of domestic violence on preterm birth and low birth weight. *Acta Obstet. Gynecol. Scand.* 83, 455–460.

Neumann, I. D., Johnstone, H. A., Hatzinger, M., Liebsch, G., Shipston, M., Russell, J. A., Landgraf, R., and Douglas, A. J. (1998). Attenuated neuroendocrine responses to emotional and physical stressors in pregnant rats involve adenohypophysial changes. *J. Physiol.* 508, 289–300.

Neumann, I. D., Toschi, N., Ohl, F., Torner, L., and Kromer, S. A. (2001). Maternal defence as an emotional stressor in female rats: Correlation of neuroendocrine and behavioural parameters and involvement of brain oxytocin. *Eur. J. Neurosci.* 13, 1016–1024.

Neumann, I. D., Kromer, S. A., and Bosch, O. J. (2005a). Effects of psycho-social stress during pregnancy on neuroendocrine and behavioural parameters in lactation depend on the genetically determined stress vulnerability. *Psychoneuroendocrinology* 30, 791–806.

Neumann, I. D., Wigger, A., Kromer, S., Frank, E., Landgraf, R., and Bosch, O. J. (2005b). Differential effects of periodic maternal separation on adult stress coping in a rat model of extremes in trait anxiety. *Neuroscience* 132, 867–877.

Nierop, A., Bratsikas, A., Klinkenberg, A., Nater, U. M., Zimmermann, R., and Ehlert, U. (2006). Prolonged salivary cortisol recovery in second-trimester pregnant women and attenuated salivary alpha-amylase responses to psychosocial stress in human pregnancy. *J. Clin. Endocrinol. Metab.* 91, 1329–1335.

Numan, M. (2007). Motivational systems and the neural circuitry of maternal behavior in the rat. *Dev. Psychobiol.* 49, 12–21.

Nunez-Rivas, H. P., Monge-Rojas, R., Grios-Davila, C., Elizondo-Urena, A. M., and Rojas-Chavarria, A. (2003). Physical, psychological, emotional, and sexual violence during pregnancy as a reproductive-risk predictor of low birthweight in Costa Rica. *Rev. Panam. Salud. Publica* 14, 75–83.

O'Connell, N. E., Beattie, V. E., and Moss, B. W. (2004). Influence of replacement rate on the welfare of sows introduced to a large dynamic group. *Appl. Anim. Behav. Sci.* 85, 43–56.

Okah, F. A., Cai, J., and Hoff, G. L. (2005). Term-gestation low birth weight and health-compromising behaviors during pregnancy. *Obstet. Gynecol.* 105, 543–550.

Owiny, J. R., Jones, M. T., Sadowsky, D., Myers, T., Massman, A., and Nathanielsz, P. W. (1991). Cocaine in pregnancy: The effect of maternal administration of cocaine on the maternal and fetal pituitary–adrenal axes. *Am. J. Obstet. Gynecol.* 164, 658–663.

Peedicayil, A., Sadowski, L. S., Jeyaseelan, L., Shankar, V., Jain, D., Suresh, S., and Bangdiwala, S. I. (2004). Spousal physical violence against women during pregnancy. *Br. J. Obstet. Gynaecol.* 111, 682–687.

Phillips, D. I. W. (2006). Birth weight and adulthood disease and the controversies. *Fetal Maternal Med. Rev.* 17, 205–227.

Phillips, D. I. W., Barker, D. J. P., Fall, C. H. D., Seckl, J. R., Whorwood, C. B., Wood, P. J., and Walker, B. R. (1998). Elevated plasma cortisol concentrations: An explanation for the relationship between low birth weight and adult cardiovascular risk factors. *J. Clin. Endocrinol. Metab.* 83, 757–760.

Phillips, D. I., Walker, B. R., Reynolds, R. M., Flanagan, D. E., Wood, P. J., Osmond, C., Barker, D. J., and Whorwood, C. B. (2000). Low birth weight predicts elevated plasma cortisol concentrations in adults from 3 populations. *Hypertension* 35, 1301–1306.

Reynolds, M. A., and Schieve, L. A. (2006). Trends in embryo transfer practices and multiple gestation for IVF procedures in the USA, 1996–2002. *Hum. Reprod.* 21, 694–700.

Reynolds, M. A., Schieve, L. A., Martin, J. A., Jeng, G., and Macaluso, M. (2003). Trends in multiple births conceived using assisted reproductive technology, United States, 1997–2000. *Pediatrics* 111, 1159–1162.

Russell, J. A., Douglas, A. J., and Brunton, P. J. (2008). Reduced hypothalamo-pituitary adrenal axis stress reponses in late pregnancy: central opioid inhibition and noradrenergic mechanisms. *Ann. N. Y. Acad. Sci. in press.*

Saarikoski, S. (1974). Fate of noradrenaline in the human foetoplacental unit. *Acta Physiol. Scand.* 421, 1–82.

Sandler, M., Ruthven, C. R., Contrator, S. F., Wood, C., Booth, R. T., and Pinkerton, J. H. (1963). Transmission of noradrenaline across the human placenta. *Nature* 197, 598.

Sarkar, P., Bergman, K., Fisk, N. M., O'Connor, T. G., and Glover, V. (2007). Ontogeny of foetal exposure to maternal cortisol using midtrimester amniotic fluid as a biomarker. *Clin. Endocrinol.* 66, 636–640.

Schoeman, J., Grove, D. V., and Odendaal, H. J. (2005). Are domestic violence and the excessive use of alcohol risk factors for preterm birth? *J. Trop. Pediatr.* 51, 49–50.

Schulte, H. M., Weisner, D., and Allolio, B. (1990). The corticotrophin releasing hormone test in late pregnancy: Lack of adrenocorticotrophin and cortisol response. *Clin. Endocrinol.* 33, 99–106.

Seckl, J. R., and Holmes, M. C. (2007). Mechanisms of disease: Glucocorticoids, their placental metabolism and fetal "programming" of adult pathophysiology. *Nat. Clin. Pract. Endocrinol. Metab.* 3, 479–488.

Sgoifo, A., de Boer, S. F., Haller, J., and Koolhaas, J. M. (1996). Individual differences in plasma catecholamine and corticosterone stress responses of wild-type rats: Relationship with aggression. *Physiol. Behav.* 60, 1403–1407.

Shekhar, A., Truitt, W., Rainnie, D., and Sajdyk, T. (2005). Role of stress, corticotrophin releasing factor (CRF) and amygdala plasticity in chronic anxiety. *Stress* 8, 209–219.

Shoener, J. A., Baig, R., and Page, K. C. (2006). Prenatal exposure to dexamethasone alters hippocampal drive on hypothalamic–pituitary–adrenal axis activity in adult male rats. *Am. J. Physiol. Regul. Integr. Comp. Physiol.* 290, R1366–R1373.

Sodha, R. J., Proegler, M., and Schneider, H. (1984). Transfer and metabolism of norepinephrine studied from maternal-to-fetal and fetal-to-maternal sides in the *in vitro* perfused human placental lobe. *Am. J. Obstet. Gynecol.* 148, 474–481.

Stewart, P. M., Rogerson, F. M., and Mason, J. I. (1995). Type 2 11 beta-hydroxysteroid dehydrogenase messenger ribonucleic acid and activity in human placenta and fetal membranes: Its relationship to birth weight and putative role in fetal adrenal steroidogenesis. *J. Clin. Endocrinol. Metab.* 80, 885–890.

Tan, J. C., and Gregor, K. V. (2006). Violence against pregnant women in northwestern Ontario. *Ann. NY Acad. Sci.* 1087, 320–338.

Thorpe, K., Golding, J., MacGillivray, I., and Greenwood, R. (1991). Comparison of prevalence of depression in mothers of twins and mothers of singletons. *Br. Med. J.* 302875–302878.

Tsuma, V. T., Einarsson, S., Madej, A., Kindahl, H., Lundeheim, N., and Rojkittikhun, T. (1996). Endocrine changes during group housing of primiparous sows in early pregnancy. *Acta Vet. Scand.* 37, 481–489.

United Nations (2000). Resolution A/RES/55/2. The United Nations Millennium Declaration, <http://www.un.org/millennium/declaration/ares552e.htm>.

United Nations Universal Declaration of Human Rights (1948). Article 16, <http://www.un.org/Overview/rights.html>.

Urizar, G. G. J., Milazzo, M., Le, H. N., Delucchi, K., Sotelo, R., and Munoz, R. F. (2004). Impact of stress reduction instructions on stress and cortisol levels during pregnancy. *Biol. Psychol.* 67, 275–282.

Vaha-Eskeli, K. K., Erkkola, R. U., Scheinin, M., and Seppanen, A. (1992). Effects of short-term thermal stress on plasma catecholamine concentrations and plasma renin activity in pregnant and nonpregnant women. *Am. J. Obstet. Gynecol.* 167, 785–789.

Valladares, E., Pena, R., Persson, L. A., and Hogberg, U. (2005). Violence against pregnant women: Prevalence and characteristics. A population-based study in Nicaragua. *Br. J. Obstet. Gynaecol.* 112, 1243–1248.

Vallee, M., Mayo, W., Dellu, F., Le Moal, M., Simon, H., and Maccari, S. (1997). Prenatal stress induces high anxiety and postnatal handling induces low anxiety in adult offspring: Correlation with stress-induced corticosterone secretion. *J. Neurosci.* 17, 2626–2636.

Van den Bergh, B. R., Mulder, E. J., Mennes, M., and Glover, V. (2005). Antenatal maternal anxiety and stress and the neurobehavioural development of the fetus and child: Links and possible mechanisms. A review. *Neurosci. Biobehav. Rev.* 29, 237–258.

Varma, D., Chandra, P. S., Thomas, T., and Carey, M. P. (2007). Intimate partner violence and sexual coercion among pregnant women in India: Relationship with depression and post-traumatic stress disorder. *J. Affect. Disord.* 102, 227–235.

Venihaki, M., Carrigan, A., Dikkes, P., and Majzoub, J. A. (2000). Circadian rise in maternal glucocorticoid prevents pulmonary dysplasia in fetal mice with adrenal insufficiency. *Proc. Natl Acad. Sci. USA* 97, 7336–7341.

Waddell, B. J., Benediktsson, R., Brown, R. W., and Seckl, J. R. (1998). Tissue-specific messenger ribonucleic acid expression of 11beta-hydroxysteroid dehydrogenase types 1 and 2 and the glucocorticoid receptor within rat placenta suggests exquisite local control of glucocorticoid action. *Endocrinology* 139, 1517–1523.

Weaver, I. C., Cervoni, N., Champagne, F. A., D'Alessio, A. C., Sharma, S., Seckl, J. R., Dymov, S., Szyf, M., and Meaney, M. J. (2004). Epigenetic programming by maternal behavior. *Nat. Neurosci.* 7, 847–854.

Weaver, I. C., D'Alessio, A. C., Brown, S. E., Hellstrom, I. C., Dymov, S., Sharma, S., Szyf, M., and Meaney, M. J. (2007). The transcription factor nerve growth factor-inducible protein A mediates epigenetic programming: Altering epigenetic marks by immediate-early genes. *J. Neurosci.* 27, 1756–1768.

Weidenfeld, J., Feldman, S., and Mechoulam, R. (1994). Effect of the brain constituent anandamide, a cannabinoid receptor agonist, on the hypothalamo–pituitary–adrenal axis in the rat. *Neuroendocrinology* 59, 110–112.

Welberg, L. A. M., Seckl, J. R., and Holmes, M. C. (2001). Prenatal glucocorticoid programming of brain corticosteroid receptors and corticotrophin-releasing hormone: Possible implications for behaviour. *Neuroscience* 104, 71–79.

Welberg, L. A., Thrivikraman, K. V., and Plotsky, P. M. (2005). Chronic maternal stress inhibits the capacity to up-regulate placental 11beta-hydroxysteroid dehydrogenase type 2 activity. *J. Endocrinol.* 186, R7–R12.

Wigger, A., Lorscher, P., Oehler, I., Keck, M. E., Naruo, T., and Neumann, I. D. (1999). Nonresponsiveness of the rat hypothalamo–pituitary–adrenocortical axis to parturition-related events: Inhibitory action of endogenous opioids. *Endocrinology* 140, 2843–2849.

Williams, S., and Poulton, R. (1999). Twins and maternal smoking: Ordeals for the fetal origins hypothesis? A cohort study. *Br. Med. J.* 318, 897–900.

Williams, S., and Poulton, R. (2002). Birth size, growth, and blood pressure between the ages of 7 and 26 years: Failure to support the fetal origins hypothesis. *Am. J. Epidemiol.* 155, 849–852.

Windle, R. J., Wood, S., Shanks, N., Perks, P., Conde, G. L., da Costa, A. P. C., Ingram, C. D., and Lightman, S. L. (1997). Endocrine and behavioural responses to noise stress: Comparison of virgin and lactating female rats during non-disrupted maternal activity. *J. Neuroendocrinol.* 9, 407–414.

World Health Organization (2005). Addressing violence against women and achieving the millennium development goals. <http://www.who.int/gender/documents/MDGs&VAWSept05.pdf>.

World Health Organization Europe (2005). Making pregnancy safer & gender mainstreaming, Republic of Moldova <http://www.euro.who.int/document/MPS/mps_gem_mda_new.pdf>.

Wotjak, C. T., Kubota, M., Liebsch, G., Montkowski, A., Holsboer, F., Neumann, I., and Landgraf, R. (1996). Release of vasopressin within the rat paraventricular nucleus in response to emotional stress: A novel mechanism of regulating adrenocorticotropic hormone secretion? *J. Neurosci.* 16, 7725–7732.

Yost, N. P., Bloom, S. L., McIntire, D. D., and Leveno, K. J. (2005). A prospective observational study of domestic violence during pregnancy. *Obstet. Gynecol.* 106, 61–65.

Zhang, X., Sliwowska, J. H., and Weinberg, J. (2005). Prenatal alcohol exposure and fetal programming: Effects on neuroendocrine and immune function. *Exp. Biol. Med.* 230, 376–388.

Zuckerman, B., Frank, D. A., Hingson, R., Amaro, H., Levenson, S. M., Kayne, H., Parker, S., Vinci, R., Aboagye, K., Fried, L. E., Cabral, H., Timperi, R., and Bauchner, H. (1989). Effects of maternal marijuana and cocaine use on fetal growth. *N. Engl. J. Med.* 320, 762–768.

14

FAST DELIVERY: A CENTRAL ROLE FOR OXYTOCIN

ALISON J. DOUGLAS AND SIMONE L. MEDDLE

Centre for Integrative Physiology, College of Medicine and Veterinary Medicine, University of Edinburgh, Hugh Robson Building, George Square, Edinburgh EH8 9XD, United Kingdom

INTRODUCTION

For many of the decades since its discovery by Dale in 1906, oxytocin was assumed to facilitate birth primarily by acting on the myometrium to cause intermittent contractions to propel the fetus through the birth canal. Due to this early identified role it was named oxytocin, translated from Greek to mean "fast delivery." Today oxytocin is often used clinically to facilitate labor in women. If it is administered to mimic its intermittent secretory profile observed during birth (Higuchi *et al.*, 1986; Summerlee, 1989), oxytocin can advance labor and accelerate established parturition (Antonijevic *et al.*, 1995; Douglas *et al.*, 2001). Furthermore, recent evidence has shown that oxytocin plays a vital role in the actual timing of delivery since transgenic mice lacking oxytocin give birth at random times of day following shifts in the light–dark cycle (Roizen *et al.*, 2007). This peripheral action is not the only perinatal role for oxytocin as it has been established that oxytocin also acts within the brain. This phenomenon was first suggested by De Weid in 1965 and since then several significant roles for central oxytocin have emerged, most notably in mouse null mutation models whereby a lack of oxytocin or its receptor leads to social deficits which have many behavioral consequences (Takayanagi *et al.*, 2005; Billings *et al.*, 2006; Vollmer *et al.*, 2006; Jin *et al.*, 2007). On the basis of this and other extensive evidence, central oxytocin has been allocated various roles in social, sexual, and maternal behaviors.

At birth oxytocin neurons become activated to secrete oxytocin both peripherally and centrally. This not only mediates birth but also underlies specific maternal behaviors including lactation, appropriate maternal care and protection for the offspring (for overview see Kendrick, 2000; Russell *et al.*, 2003; Lonstein & Morrell, 2007). While there is an established role for central oxytocin in parturition and maternal behavior onset, the mechanism of oxytocin neuron activation and availability in the brain perinatally are not well understood. Here we aim to give an overview of the literature supporting oxytocin action in the brain perinatally. We will describe the release, action, and receptor mechanisms underlying brain oxytocin action at labor, and the mechanisms coordinating the multiple roles for oxytocin that facilitate "fast delivery" of offspring and appropriate maternal behavior.

OXYTOCIN: PERINATAL RELEASE AND ACTION IN THE BRAIN

Sources of Oxytocin

Oxytocin release in the brain may come from two main sources: the centrally projecting parvocellular paraventricular nucleus (PVN) neurons whose axons extend to the median eminence and other extra-hypothalamic regions such as the brainstem and spinal cord, and also from the extensive dendrites of magnocellular oxytocin neurons located in the supraoptic nucleus (SON) and a subregion of the PVN that release oxytocin from vesicles into the extracellular space. These effectively enable autocrine and paracrine functions. Additionally, there is a small population of poorly characterized parvocellular oxytocinergic neurons in the median preoptic nucleus, adjacent to the anterior PVN and contained within the medial preoptic area (MPOA). The SON and PVN are integral to birth processes, while the PVN and MPOA (and possibly the SON) are the main hypothalamic regions mediating maternal behaviors.

Perinatal Oxytocin Release Patterns

Extracellular oxytocin concentration within the SON, PVN, and MPOA alters dynamically during parturition in rats and sheep. Elegant experimental approaches using microdialysis or push–pull perfusion in precise hypothalamic areas in combination with very sensitive oxytocin radioimmunoassay have revealed that oxytocin concentration within the SON and PVN increases during birth compared to late pregnancy (Landgraf *et al.*, 1992; Neumann *et al.*, 1993; da Costa *et al.*, 1996). This finding is very specific to oxytocin since the release of the similar nonapeptide, vasopressin, does not increase in these regions during birth (Neumann *et al.*, 1993), suggesting a particular role for central oxytocin at this time. Despite the sensitive assays used to measure oxytocin, the sample times required for detection of changes in extracellular oxytocin remain relatively long, typically spanning the birth of three or more pups in the rat (30 min). This has limited the interpretation of whether intermittent release accompanies the birth of individual pups or whether release is slow and sustained. However, in sheep 15-min samples reveal transiently increased oxytocin in the SON during parturition (da Costa *et al.*, 1996), suggesting that oxytocin release occurs only during the passage of the neonate through the birth canal, decreasing again within less than 1 h of the lamb being born.

Whereas release within the SON and PVN is likely to be primarily dendritic, Ludwig and Leng (2006) have argued that this is dissociated from the sporadic, synchronous burst-like firing pattern that generates the separate and intermittent peripheral secretory profile. It is suggested that dendritic release of oxytocin from magnocellular neurons is probably sustained over a period of time relevant to the appropriate stimuli and is likely to be associated with expression of immediate early genes such as c-*fos*, even inducing its expression. Fos expression in SON and PVN oxytocin neurons increases duringbirth compared to time-matched pre-parturientrats and mice (Luckman, 1995; Douglas *et al.*, 2002). Moreover, preliminary evidence indicates that oxytocin microinjection into the SON in virgin female rats increases Fos compared to vehicle injection (Mike Ludwig, personal communication). Together this evidence indicates a potentially important regulatory role for oxytocin within the SON and PVN itself.

During birth oxytocin release also increases within other brain areas. There is a perinatal rise in oxytocin in the cerebrospinal fluid (CSF) in sheep (Kendrick, 2000) and women (Takagi *et al.*, 1985; Altemus *et al.*, 2004), suggesting that very large amounts are released centrally to diffuse into the CSF whereby they are transported to reach many brain regions. Additionally, in all other brain regions analyzed so far during birth in the rat and sheep, such as in other nuclei of the hypothalamus and in the limbic system (i.e., MPOA, mediobasal hypothalamus, substantia nigra, bed nucleus of the stria terminalis (BnST), lateral septum, hippocampus, and olfactory bulbs) there is increased extracellular oxytocin. Similar findings are observed with vagino-cervical stimulation that mimics fetal passage through the birth canal (for overview Kendrick, 2000). Indeed in some regions, such as the MPOA, oxytocin release increases prior to birth, during labor, although this increase is not detectable 3 h before delivery itself. In other regions like the BnST and olfactory bulbs, oxytocin release only increases during fetus expulsion (Kendrick *et al.*, 1988, 1992). Since limbic regions and the olfactory bulbs positively facilitate maternal behaviors, the increase only during delivery indicates a role in these regions for oxytocin in the mothers' immediate responses to offspring. We might speculate that, as in the SON and PVN (da Costa *et al.*, 1999), oxytocin may activate its target neurons within these other brain regions as well. It is not yet known whether oxytocin is released during parturition in other brain areas important for maternal behaviors such as the ventral tegmental area, the ventral pallidum, or the amygdala. Nonetheless, since oxytocin concentration is elevated perinatally in multiple brain sites, extracellular oxytocin concentration is likely to be increased globally.

ENDOGENOUS OXYTOCIN EFFECTS IN THE BRAIN

Whether the central release of oxytocin within these brain areas causes immediate or permissive effects requires further investigation. Convincing evidence for either of these scenarios is patchy, although slowly accumulating. Proof depends on three aspects: (a) whether endogenous oxytocin has proven perinatal behavioral effects; (b) the distribution and density of oxytocin receptors (OTR) at the time of birth; and (c) whether oxytocin plays a direct role in activating target neurons. For example, do oxytocin sensitive cells become activated at the time oxytocin is proposed to mediate a perinatal effect?

Oxytocin: Perinatal Action in the Hypothalamus

While the facilitatory effects of intracerebral oxytocin on birth and maternal behavior in many species have been extensively reported and reviewed (e.g., Kendrick, 2000; Numan & Insel, 2003; Russell *et al.*, 2003; Lonstein & Morrell, 2007), the precise targets and roles for oxytocin at birth await discovery. Oxytocin antagonist studies and data from oxytocin null mice have established a role for endogenous oxytocin in the onset and progress of maternal behavior (Kendrick, 2000; Numan & Insel, 2003; Takayanagi *et al.*, 2005 Roizen *et al.*, 2007). Additionally, central administration of selective oxytocin agonists validate a role for oxytocin in accelerating birth and the onset of maternal behavior (Neumann *et al.*, 1996; Kendrick, 2000), although the latter depends on sex steroids, previous exposure to young, and the particular species studied.

Reports about precise actions of intra-hypothalamic oxytocin are limited to the SON and PVN, which are known to be oxytocin sensitive (reviewed in Kombian *et al.*, 2002). Around the time of birth it appears that oxytocin release within the SON has direct auto-positive feedback effects. Oxytocin neurons express OTR (Young *et al.*, 1997; Meddle *et al.*, 2007) and specific OTR antagonist retrodialysis within the SON not only attenuates further SON oxytocin release but also slows birth progression (Neumann *et al.*, 1996). These findings reveal a perinatal physiological autoregulatory effect. Such autoregulation was previously suspected because oxytocin induced an increase in intracellular calcium in oxytocin neurons *in vitro* (Lambert *et al.*, 1993) and also autoregulated oxytocin neuron firing rate contributing to the pattern of burst firing and synchronization of cells during the milk rejection reflex (Wakerley & Ingram, 1993; Russell & Leng, 2000), an effect similar to that observed during birth (Summerlee, 1989). Oxytocin may also have an indirect or permissive role since oxytocin agonist administration in the PVN increases noradrenaline release and inhibits glutamate release (da Costa *et al.*, 1999), suggesting a complex regulatory intra-PVN mechanism, although whether endogenous oxytocin acts similarly in the PVN at parturition is unkown. However, there is evidence for increased noradrenaline and glutamate release within the SON during the establishment of birth (Herbison *et al.*, 1997). Also, oxytocin can attenuate inhibitory GABAergic signaling in oxytocin cells in vitro (*refs as in text*), even though extracellular GABA concentration does not alter (Fenelon & Herbison, 2000), so it is possible that intra-SON oxytocin may exert its action via these classical transmitters, potentially presynaptically, during birth.

If we look outside the hypothalamus, evidence suggests that oxytocin-mediated activity in the limbic system, specifically in the BnST and lateral septum, has an important facilitatory effect in the peripartum period for neuronal responses mediating both birth and maternal behavior (Ingram *et al.*, 1990; Housham *et al.*, 1997; Wakerley *et al.*, 1998; Kendrick, 2000). Many of these regions also express OTR (Young *et al.*, 1997; Kendrick, 2000; Meddle *et al.*, 2007), suggesting a direct action. However, Kendrick (2000) has speculated that oxytocin could act indirectly to presynaptically modulate glutamate or GABA release which, during birth in the rat and sheep, is released in a similar pattern to that observed for oxytocin. To our present knowledge it remains to be seen whether oxytocin antagonist administration can block the release, and therefore action, of other transmitters like glutamate and GABA in these target regions perinatally.

AVAILABILITY OF EXTRACELLULAR OXYTOCIN IN THE BRAIN

The action of oxytocin in the brain, as in the periphery, is not only dependent on release dynamics such as location, timing, patterns, or magnitude but also depends on the rate of degradation of oxytocin. Like many neurotransmitters and hormones, including those in the brain, oxytocin availability is controlled by extracellular mechanisms that regulate biological activity. We have known for many years that oxytocin is degraded in peripheral tissue by the oxytocinase enzyme, leucine or cysteine aminopeptidase. The enzyme is a membrane-spanning zinc metalloproteinase that degrades oxytocin preferentially, but oxytocinase is also known to degrade other peptides such as vasopressin and met-enkephalin (reviewed in Claybaugh & Uyehara, 1993). The peripheral activity of this enzyme is particularly striking during pregnancy when activity substantially increases (Mitchell *et al.*, 1997; Matsumoto & Mori, 1998) presumably to prevent peripheral oxytocin from inappropriately causing myometrial contraction or preterm labor. Immediately prior to birth, oxytocinase activity decreases enabling oxytocin to drive uterine contractions. The major source of oxytocinase is the placenta and this is why it is also known by its other name placental leucine aminopeptidase (P-LAP). Oxytocinase is not placenta specific and it can also be found for example, in the uterus, fetus, kidney, and liver. Furthermore, it is expressed in brain membranes (Matsumoto *et al.*, 2001). As early as 1983 it was recognized that oxytocin could be degraded in the brain (Burbach & Lebouille, 1983; Gibson *et al.*, 1991) and that inhibition of aminopeptidases (using amastatin) prolongs behavioral effects of intracerebroventricular oxytocin administration (Meisenberg & Simmons, 1984). The enzyme can also be "converted" into a soluble form (not found in isolated neuronal preparations) or "secreted" so it is available in the extracellular space or plasma (Matsumoto *et al.*, 2001).

Surprisingly very few studies have reported oxytocinase expression or action associated with the major oxytocin neuron location and release sites, that is the SON and PVN; or even in the hypothalamus as a whole. Some evidence indicates that by increasing hypothalamic oxytocin with oxytocinase inhibitors *in vitro*, excitatory neurotransmission in the SON is inhibited (Kombian *et al.*, 1997; Hirasawa *et al.*, 2001). Interestingly, amastatin was found to have no effect on oxytocin-induced glutamate release in isolated SON cultures (Curras-Collazo *et al.*, 2003). Both these papers call into question the role of oxytocinase within the SON, but neither study used a physiological model in which endogenous oxytocin release is known or has a precise function. Since we know that oxytocin is released to act within the PVN and SON perinatally, we have investigated by fluorescence immunocytochemistry whether the SON and PVN express the oxytocinase enzyme. Using an antibody against P-LAP (i.e., a specific antibody for oxytocinase) we have for the first time described specific immunolabeling for P-LAP in neurons of the PVN and SON (Figure 14.1(A),(D)). Furthermore, P-LAP is found to be co-localized in all oxytocin neurons (shown in the PVN, Figure 14.1(B),(C)). P-LAP is also present in other neurons within the PVN, the phenotype of which, by their location within the PVN, may be vasopressinergic or contain corticotrophin-releasing hormone. This recent finding suggests that oxytocin-synthesizing neurons themselves have the capability to degrade oxytocin, and therefore to control extracellular oxytocin availability. The obvious question of whether oxytocinase activity within the SON and PVN plays a role in oxytocin availability perinatally has yet to be answered. Nonetheless aminopeptidase activity that degrades other hypothalamic neuropeptides does appear to change during the estrous cycle and pregnancy (de Gandarias *et al.*, 1993).

PERINATAL DISTRIBUTION, DENSITY AND ACTIVATION OF OXYTOCIN RECEPTOR

In the final hours of pregnancy the sensitivity of the uterus to oxytocin significantly increases and this can be explained by the dramatic increase of oxytocin receptor mRNA expression and oxytocin binding in the myometrium. Coordination of neuroendocrine changes at birth is critical for normal delivery and the sensory information from the uterus is relayed back to the hypothalamic oxytocin system via noradrenergic pathways from the brainstem (Meddle *et al.*, 2000; Douglas *et al.*, 2001). This established positive feedback loop has been termed the Ferguson reflex after Ferguson who first described it in 1941 (Ferguson, 1941).

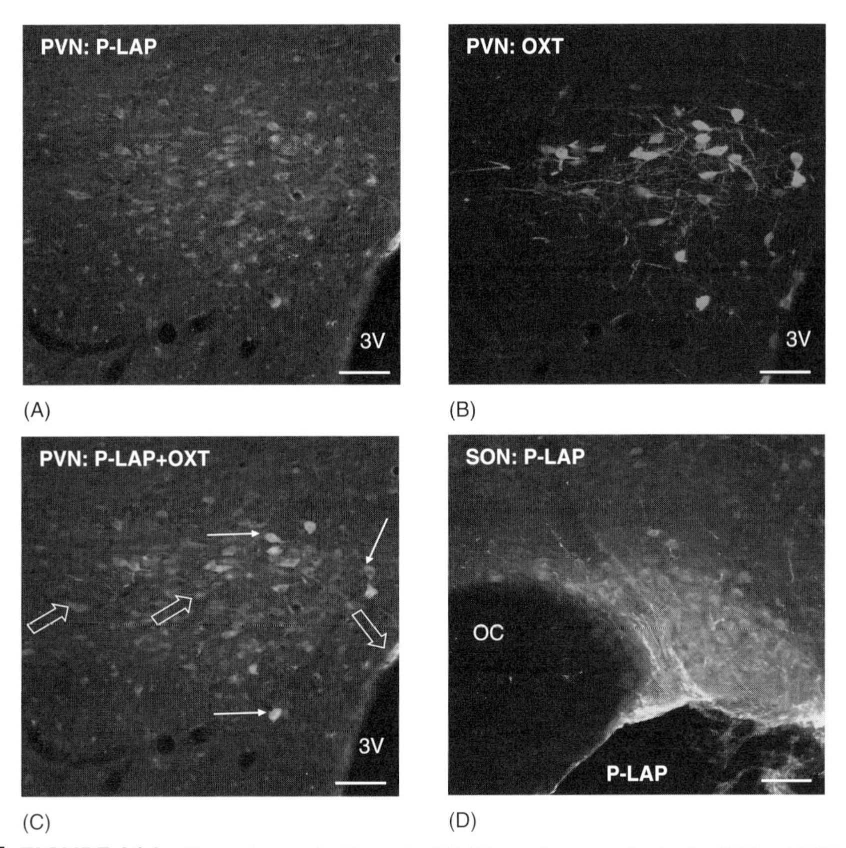

FIGURE 14.1 Photomicrographs illustrating P-LAP protein expression in the PVN and SON. Brain sections from a virgin female rat containing PVN (A)–(C) and SON (D) were processed by fluorescence immunocytochemistry for P-LAP using a rabbit anti-rat antibody (1:1,000 for 48 h) and Streptavidin Alexa Fluor 488 secondary antibody (green). PVN sections were also processed with oxytocin antibody (1:5,000 for 48 h) and Alexa Fluor 568 secondary antibody (red; (B) and (C)). All sections were viewed by confocal microscopy. Examples of double-labeled cells are indicated with filled arrows, unfilled arrows illustrate P-LAP-only labeling. Scale bars = 80 μm. OC = optic chiasm, 3V = third ventricle. (See Color Plate)

As described above, many brain regions are sensitive to oxytocin. OTR mRNA expression and oxytocin binding occur in regions including the brainstem, hypothalamus, limbic system, hippocampus, and olfactory bulbs (e.g., Bealer *et al.*, 2006; Meddle *et al.*, 2007). In the SON of the hypothalamus we can reveal, using triple immunofluorescence and confocal microscopy, that oxytocin receptor is clearly co-localized within all oxytocin neurons, and is also found within most vasopressin neurons (Figure 14.2). OTR binding in the SON and PVN increases significantly during pregnancy (Bealer *et al.*, 2006) and we have recently reported dynamic changes in OTR mRNA expression in the rat brain perinatally, using *in situ* hybridization techniques, that is upregulated in a brain region dependent way (Meddle *et al.*, 2007). We discovered that nuclei important for oxytocin release during birth such as the SON and brainstem regions such as the nucleus tractus solitarii (NTS, A2/C2) and ventrolateral medulla (VLM, A1/C1) showed a significant increase in OTR mRNA around the time of birth. Together these studies provide evidence that OTRs are upregulated during late gestation in brain circuits that regulate oxytocin release at parturition. One functional explanation for this dynamic change is that oxytocin itself is likely to regulate its own release which at birth contributes to the oxytocin burst firing patterns so vital for normal delivery. Together with the fact that oxytocin release increases at birth and is reliant upon intra-oxytocin action

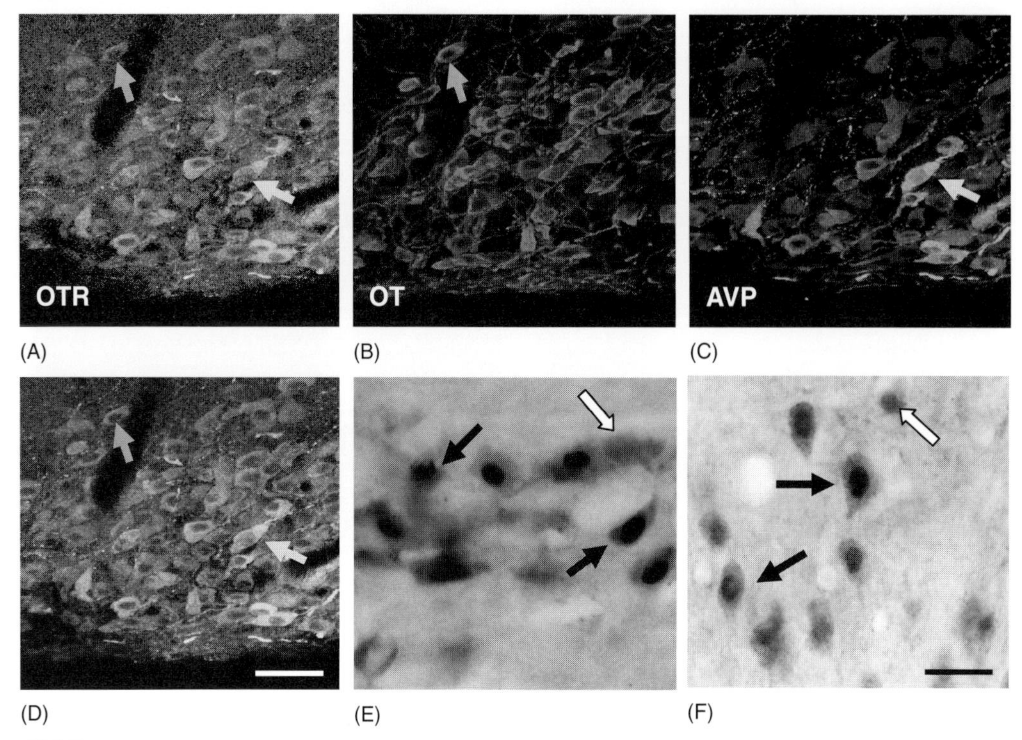

FIGURE 14.2 Confocal photomicrograph of oxytocin receptor (OTR) immunolabeling in the SON (A) (1/1,000 dilution, visualized with goat anti-rabbit Alexa Fluor 488). The same section was immunolabeled for oxytocin (B) (antibody gift from Hal Gainer; 1/4,000 dilution, visualized with goat anti-rabbit Alexa Fluor 647) and vasopressin (C) (AVP; antibody raised in guinea pig from Peninsula labs; diluted 1/1,000, visualized with Alexa Fluor 568). Most oxytocin cells and vasopressin cells appear to express OTR. OTR cells that co-localize with oxytocin appear pale blue in (D) (example shown by pale blue arrow). OTR cells co-localizing with vasopressin appear yellow in (D) (example shown by yellow arrow). Photomicrograph examples of activated OTR-expressing neurons in the (E) SON and (F) NTS of parturient rats. Brains were double immunolabeled for OTR (brown) and Fos (black nuclei) expression. Examples of double-labeled cells for Fos and OTR are indicated with filled arrows, unfilled arrows illustrate cells only labeled for OTR. Scale bars all = 50 μm. (See Color Plate)

(Neumann *et al.*, 1996), this hypothesis is further supported by the finding that at parturition OTR-immunoreactive cells in the SON and magnocellular PVN become activated as confirmed by an increase in Fos expression (Figures 14.2 and 14.3).

Other components of the Ferguson reflex such as the feed forward input to the oxytocin neurons in the SON are also probably regulated by oxytocin. Using fluorescent retrograde tracing studies we have established that populations of activated SON projecting neurons in the NTS and VLM are immunoreactive for OTR (Meddle *et al.*, 2007). As it is thought that SON neurons are excited by noradrenaline at parturition (Douglas *et al.*, 2001), these data suggest that oxytocin may act presynaptically on OTR located on nerve terminals of noradrenergic inputs in the SON and PVN to facilitate noradrenaline release (Onaka *et al.*, 2003). Moreover, oxytocin sensitivity in the brainstem is enhanced at birth as OTR mRNA expression increases at parturition (Meddle *et al.*, 2007). Oxytocin, perhaps originating from paraventricular-spinal projections originating in the pPVN (Buijs, 1978; Rinaman, 1998), may act directly in the brainstem and spinal cord at this time to regulate uterine activity (Benoussaidh *et al.*, 2004), and also to play a role in antinociceptive effects such as regulating pain sensitivity and heart rate.

Successful reproduction not only relies on successful delivery but also on proper maternal care of the offspring. For a rat this includes lactation

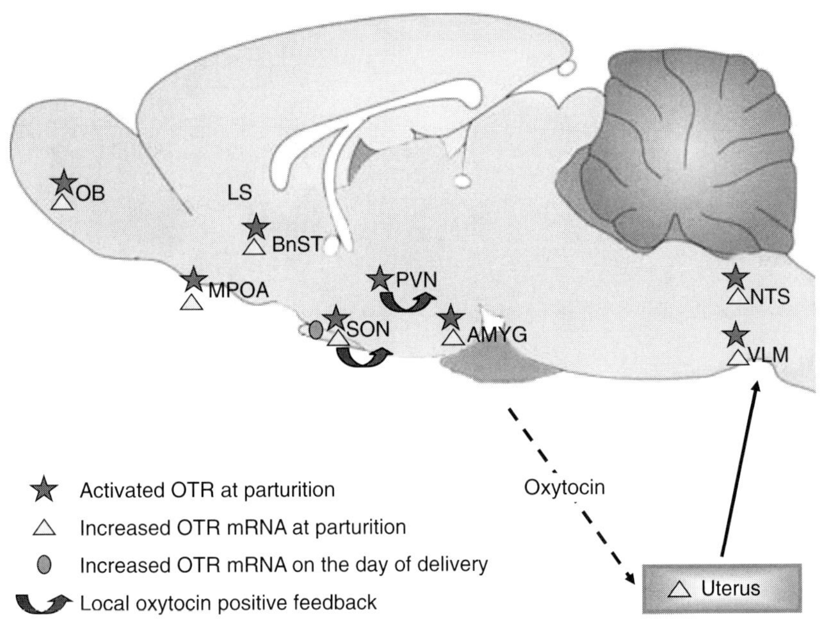

FIGURE 14.3 Important sites of oxytocin action at birth in a sagittal view of the rat brain. At parturition oxytocin is released from the posterior pituitary into the blood stream to act on OTR in the myometrium, but it also acts centrally to exert its action in the brain. At parturition OTR cells become activated (identified by OTR cells expressing Fos protein) in brain regions indicated by *stars* (★). OTR mRNA expression increases in the SON on the expected day of delivery (day 22, △) and at parturition (as indicated by ⬮). Increased OTR mRNA expression is higher in the hypothalamus (SON and PVN) and brainstem (NTS and VLM) but also in brain regions important for maternal behavior such as the olfactory bulbs, MPOA, BnST, and amygdala. Moreover during birth, OTR-expressing neurons in these brain regions express Fos. Oxytocin itself is likely to regulate its own release in the SON and magnocellular region of the PVN which at birth contributes to the oxytocin burst firing patterns so vital for normal delivery. SON, supraoptic nucleus; PVN, paraventricular nucleus; OB, olfactory bulb; LS, lateral septum; BnST, bed nucleus of the stria terminalis; MPOA, medial preoptic area; AMYG, amygdala; NTS, nucleus tractus solitarii; VLM, ventrolateral medulla. (See Color Plate)

in an arched back nursing position, licking and grooming of pups, maternal defense and nest building (e.g., Lonstein & Morrell, 2007). Virgin rats are fearful of newborn pups, but during pregnancy, in anticipation of the pup arrival, their behavior changes. Oxytocin is critical for the induction of maternal behavior (Pedersen, 1997) and the neural circuits that are involved in regulating this multifaceted behavior are oxytocin sensitive. Thus, it is no surprise to find that OTR is located in brain regions important for maternal behavior such as the olfactory bulbs, MPOA, BnST, and amygdala. In these brain regions, during birth, OTR mRNA expression increases and OTR-immunoreactive cells are activated (Figure 14.3; Meddle *et al.*, 2007). In fact in the olfactory bulbs, BnST and medial amygdala expression significantly increases only during birth, after the appearance of the first pup, coinciding with transient oxytocin release in these regions. This extends the previous findings for the central function oxytocin plays by acting in these brain regions to mediate maternal care during the perinatal period. Also at birth, oxytocin may be acting in some of these brain regions such as the BnST to facilitate burst firing of magnocellular oxytocin neurons themselves (e.g., Housham *et al.*, 1997), particularly since it is established that BnST sensitivity to oxytocin varies during the peripartum period (Ingram & Wakerley, 1993). Oxytocin also has a vital role during lactation for milk let down and remarkably in the 4–12 h following birth OTR mRNA expression in the hypothalamus, brainstem, olfactory bulbs, and limbic system is back to levels observed in virgin animals. These dynamic perinatal changes in OTR mRNA expression and binding

(Young *et al.*, 1997; Bealer *et al.*, 2006) may also be related to generation of the milk ejection reflex. Since OTR distribution patterns in the brain determine the quality of reproductive behaviors (Young, 1999), the perinatal transient changes in OTR density must play a vital role in the coordination of perinatal behavior and reproduction success.

CONCLUSION

Dynamic and transient changes occur within the oxytocin-synthesizing, secreting, and target nuclei, and underlie the perinatal role of oxytocin. Not only is oxytocin release increased but also OTR expression is enhanced in selected regions, changing the pattern and potential sites of oxytocin action at birth. Furthermore, the main oxytocin-synthesizing and dendritic release regions, the SON and PVN, express the protein for the oxytocinase enzyme which, if also dynamically regulated perinatally as in the uterus/placenta, indicates a tightly controlled system for the availability and action of extracellular oxytocin within the brain. Together these coordinated regulatory systems are expected to synergize to facilitate and synchronize the central brain functions of oxytocin in labor, birth, and maternal behavior in the immediate perinatal period.

ACKNOWLEDGMENTS

We thank Vicky Tobin and Val Bishop for their immunocytochemistry expertise. We thank the following for their kind and generous donations: Fred Van Leeuwen for oxytocin receptor antibody, Peter Burbach for oxytocin receptor cDNA, Hal Gainer for the oxytocin antibody, and Masafumi Tsujimoto (Riken, Japan) for P-LAP antibody. AJD and SLM were funded by The Wellcome Trust and the BBSRC.

REFERENCES

Altemus, M., Fong, J., Yang, R. R., Damast, S., Luine, V., and Ferguson, D. (2004). Changes in cerebrospinal fluid neurochemistry during pregnancy. *Biol. Psychiatr.* 56, 386–392.

Antonijevic, I. A., Leng, G., Luckman, S. M., Douglas, A. J., Bicknell, R. J., and Russell, J. A. (1995). Induction of uterine activity with oxytocin in late pregnant rats replicates the expression of c-fos in neuroendocrine and brainstem neurons as seen during parturition. *Endocrinology* 136, 154–163.

Bealer, S. L., Lipschitz, D. L., Ramoz, G., and Crowley, W. R. (2006). Oxytocin receptor binding in the hypothalamus during gestation in rats. *Am. J. Physiol.* 291, R53–R58.

Benoussaidh, A., Maurin, Y., and Rampin, O. (2004). Spinal effects of oxytocin on uterine motility in anesthetized rats. *Am. J. Physiol.* 287, R446–R453.

Billings, L. B., Spero, J. A., Vollmer, R. R., and Amico, J. A. (2006). Oxytocin null mice ingest enhanced amounts of sweet solutions during light and dark cycles and during repeated shaker stress. *Behav. Brain Res.* 171, 134–141.

Buijs, R. M. (1978). Intra-hypothalamic and extrahypothalamic vasopressin and oxytocin pathways in rat-pathways to limbic system, medulla-oblongata and spinal-cord. *Cell Tissue Res.* 192, 423–435.

Burbach, J. P., and Lebouille, J. L. (1983). Proteolytic conversion of arginine-vasopressin and oxytocin by brain synaptic membranes. Characterization of formed peptides and mechanisms of proteolysis. *J. Biol. Chem.* 258, 1487–1494.

Claybaugh, J. R., and Uyehara, C. F. (1993). Metabolism of neurohypophysial hormones. *Ann. NY Acad. Sci.* 689, 250–268.

Curras-Collazo, M. C., Gillard, E. R., Jin, J., and Pandika, J. (2003). Vasopressin and oxytocin decrease excitatory amino acid release in adult rat supraoptic nucleus. *J. Neuroendocrinol.* 15, 182–190.

da Costa, A. P., Guevara-Guzman, R. G., Ohkura, S., Goode, J. A., and Kendrick, K. M. (1996). The role of oxytocin release in the paraventricular nucleus in the control of maternal behaviour in the sheep. *J. Neuroendocrinol.* 8, 163–177.

da Costa, A. P. C., de la Riva, C., Guevara-Guzman, R., and Kendrick, K. M. (1999). C-fos and c-jun in the paraventricular nucleus play a role in regulating peptide gene expression, oxytocin and glutamate release, and maternal behaviour. *Eur. J. Neurosci.* 11, 2199–2210.

de Gandarias, J. M., Irazusta, J., Echavarria, E., and Casis, L. (1993). Aspartate-aminopeptidase activity during the estrous cycle and the pregnancy in rat brain and pituitary gland. *Exp. Clin. Endocrinol.* 101, 156–160.

De Wied, D. (1965). The influence of the posterior and intermediate lobe of the pituitary and pituitary peptides on the maintenance of a conditioned avoidance response in rats. *Int. J. Neuropharmacol.* 4, 157–167.

Douglas, A. J., Scullion, S., Antonijevic, I. A., Brown, D., Russell, J. A., and Leng, G. (2001). Uterine contractile activity stimulates supraoptic neurons in term pregnant rats via a noradrenergic pathway. *Endocrinology* 142, 633–644.

Douglas, A. J., Leng, G., and Russell, J. A. (2002). The importance of oxytocin mechanisms in the control of mouse parturition. *Reproduction* 123, 543–552.

Fenelon, V. A., and Herbison, A. E. (2000). Progesterone regulation of GABA(A) receptor plasticity in adult rat supraoptic nucleus. *Eur. J. Neurosci.* 12, 1617–1623.

Ferguson, J. K. W. (1941). A study of the motility of the intact uterus at term. *Surg. Gynecol. Obstet.* 73, 359–366.

Gibson, A. M., Biggins, J. A., Lauffart, B., Mantle, D., and McDermott, J. R. (1991). Human brain leucyl aminopeptidase: Isolation, characterization and specificity against some neuropeptides. *Neuropeptides* 19, 163–168.

Herbison, A. E., Voisin, D. L., Douglas, A. J., and Chapman, C. (1997). Profile of monoamine and excitatory amino acid release in rat supraoptic nucleus over parturition. *Endocrinology* 138, 33–40.

Higuchi, T., Tadokoro, Y., Honda, K., and Negoro, H. (1986). Detailed analysis of blood oxytocin levels during suckling and parturition in the rat. *J. Endocrinol.* 110, 251–256.

Hirasawa, M., Kombian, S. B., and Pittman, Q. J. (2001). Oxytocin retrogradely inhibits evoked, but not miniature, EPSCs in the rat supraoptic nucleus: Role of N- and P/Q-type calcium channels. *J. Physiol.* 532, 595–607.

Housham, S. J., Terenzi, M. G., and Ingram, C. D. (1997). Changing pattern of oxytocin-induced excitation of neurons in the bed nuclei of the stria terminalis and ventrolateral septum in the peripartum period. *Neuroscience* 81, 479–488.

Ingram, C. D., and Wakerley, J. B. (1993). Post-partum increase in oxytocin-induced excitation of neurones in the bed nuclei of the stria terminalis *in vitro*. *Brain Res.* 602, 325–330.

Ingram, C. D., Moos, F., Wakerley, J. B., Guerne, Y., and Richard, P. (1990). Microinjections of oxytocin (OT) into bed nucleus of the stria terminalis (BnST) and lateral septum (LS) facilitate the milk-ejection reflex in anaesthetized, suckled rats. *J. Physiologie* 427.

Jin, D., Liu, H. X., Hirai, H., Torashima, T., Nagai, T., Lopatina, O., Shnayder, N. A., Yamada, K., Noda, M., Seike, T., Fujita, K., Takasawa, S., Yokoyama, S., Koizumi, K., Shiraishi, Y., Tanaka, S., Hashii, M., Yoshihara, T., Higashida, K., Islam, M. S., Yamada, N., Hayashi, K., Noguchi, N., Kato, I., Okamoto, H., Matsushima, A., Salmina, A., Munesue, T., Shimizu, N., Mochida, S., Asano, M., and Higashida, H. (2007). CD38 is critical for social behaviour by regulating oxytocin secretion. *Nature* 446, 41–45.

Kendrick, K. M. (2000). Oxytocin, motherhood and bonding. *Exp. Physiol.* 85, 111S–124S.

Kendrick, K. M., Keverne, E. B., Chapman, C., and Baldwin, B. A. (1988). Intracranial dialysis measurement of oxytocin, monoamine and uric acid release from the olfactory bulb and substantia nigra of sheep during parturition, suckling, separation from lambs and eating. *Brain Res.* 439, 1–10.

Kendrick, K. M., Keverne, E. B., Hinton, M. R., and Goode, J. A. (1992). Oxytocin, amino acid and monoamine release in the region of the medial preoptic area and bed nucleus of the stria terminalis of the sheep during parturition and suckling. *Brain Res.* 569, 199–209.

Koksma, J. J., Van Kesteren, R. E., Rosahl, T. W., Zwart, R., Smit, A. B., Luddens, H., and Brussaard, A. B. (2003). Oxytocin regulates neurosteroid modulation of GABA(A) receptors in supraoptic nucleus around parturition. *J. Neurosci.* 23, 788–797.

Kombian, S. B., Mouginot, D., and Pittman, Q. J. (1997). Dendritically released peptides act as retrograde modulators of afferent excitation in the supraoptic nucleus *in vitro*. *Neuron* 19, 903–912.

Kombian, S. B., Hirasawa, M., Mouginot, D., and Pittman, Q. J. (2002). Modulation of synaptic transmission by oxytocin and vasopressin in the supraoptic nucleus. *Vasopr. Oxy. Gene. Clin. Appl.* 139, 235–246.

Lambert, R. C., Moos, F. C., and Richard, P. (1993). Action of endogenous oxytocin within the paraventricular and supraoptic nuclei: A powerful link in the regulation of the bursting pattern of oxytocin neurons during the milk-ejection reflex in rats. *Neuroscience* 57, 1027–1038.

Landgraf, R., Neumann, I., Russell, J. A., and Pittman, Q. J. (1992). Push–pull perfusion and microdialysis studies of central oxytocin and vasopressin release in freely moving rats during pregnancy, parturition, and lactation. *Ann. NY Acad. Sci.* 652, 326–339.

Lonstein, J. S., and Morrell, J. I. (2007). Neuropharmacology and neuroendocrinology of maternal behavior and motivation. In *Handbook of Neuro chemistry and Molecular Biology* (J. D. Blaustein, Ed.), pp. 195–245. Kluwer Press.

Luckman, S. M. (1995). Fos expression within regions of the preoptic area, hypothalamus and brainstem during pregnancy and parturition. *Brain Res.* 669, 115–124.

Ludwig, M., and Leng, G. (2006). Dendritic peptide release and peptide-dependent behaviours. *Nat. Rev. Neurosci.* 7, 126–136.

Matsumoto, H., and Mori, T. (1998). Changes in cysteine aminopeptidase (oxytocinase) activity in mouse serum, placenta, uterus and liver during pregnancy or after steroid hormone treatments. *Zoo. Sci.* 15, 111–115.

Matsumoto, H., Nagasaka, T., Hattori, A., Rogi, T., Tsuruoka, N., Mizutani, S., and Tsujimoto, M.

(2001). Expression of placental leucine aminopeptidase/oxytocinase in neuronal cells and its action on neuronal peptides. *Eur. J. Biochem.* 268, 3259–3266.

Meddle, S. L., Leng, G., Selvarajah, J., Bicknell, R. J., and Russell, J. A. (2000). Direct pathways to the supraoptic nucleus from the brainstem and the main olfactory bulb are activated during parturition in the rat. *Neuroscience* 101, 1013–1021.

Meddle, S. L., Bishop, V. R., Gkoumassi, E., van Leeuwen, F. W., and Douglas, A. J. (2007). Dynamic changes in oxytocin receptor expression and activation at parturition in the rat brain. *Endocrinology* 148, 5095–5104.

Meisenberg, G., and Simmons, W. H. (1984). Amastatin potentiates the behavioral effects of vasopressin and oxytocin in mice. *Peptides* 5, 535–539.

Mitchell, B. F., Fang, X., and Wong, S. (1997). Metabolism of oxytocin in rat uterus and placenta in late gestation. *Biol. Reprod.* 57, 807–812.

Neumann, I., Russell, J. A., and Landgraf, R. (1993). Oxytocin and vasopressin release within the supraoptic and paraventricular nuclei of pregnant, parturient and lactating rats: A microdialysis study. *Neuroscience* 53, 65–75.

Neumann, I., Douglas, A. J., Pittman, Q. J., Russell, J. A., and Landgraf, R. (1996). Oxytocin released within the supraoptic nucleus of the rat brain by positive feedback action is involved in parturition-related events. *J. Neuroendocrinol.* 8, 227–233.

Numan, M., and Insel, T. R. (2003). *The Neurobiology of Parental Behavior*. Springer, New York.

Onaka, T., Ikeda, K., Yamashita, T., and Honda, K. (2003). Facilitative role of endogenous oxytocin in noradrenaline release in the rat supraoptic nucleus. *Eur. J. Neurosci.* 18, 3018–3026.

Pedersen, C. A. (1997). Oxytocin control of maternal behavior. Regulation by sex steroids and offspring stimuli. *Ann. NY Acad. Sci.* 807, 126–145.

Rinaman, L. (1998). Oxytocinergic inputs to the nucleus of the solitary tract and dorsal motor nucleus of the vagus in neonatal rats. *J. Comp. Neurol.* 399, 101–109.

Roizen, J., Luedke, C. E., Herzog, E. D., and Muglia, L. J. (2007). Oxytocin in the circadian timing of birth. *PLOSOne* 2, e922.

Russell, J. A., and Leng, G. (2000). Veni, vidi, vici: The neurohypophysis in the twentieth century. *Exp. Physiol.* 85, 1S–6S.

Russell, J. A., Leng, G., and Douglas, A. J. (2003). The magnocellular oxytocin system, the fount of maternity: Adaptations in pregnancy. *Front. Neuroendocrinol.* 24, 27–61.

Summerlee, A. J. S. (1989). Relaxin, opioids and the timing of birth in rats. In *Brain Opioid Systems in Reproduction* (R. G. Dyer and R. J. Bicknell, Eds.), pp. 257–270. Oxford University Press, Oxford.

Takagi, T., Tanizawa, O., Otsuki, Y., Sugita, N., Haruta, M., and Yamagi, K. (1985). Oxytocin in the cerebrospinal fluid and plasma of pregnant and nonpregnant subjects. *Horm. Metab. Res.* 17, 308–310.

Takayanagi, Y., Yoshida, M., Bielsky, I. F., Ross, H. E., Kawamata, M., Onaka, T., Yanagisawa, T., Kimura, T., Matzuk, M. M., Young, L. J., and Nishimori, K. (2005). Pervasive social deficits, but normal parturition, in oxytocin receptor-deficient mice. *Proc. Natl Acad. Sci. USA* 102, 16096–16101.

Theodosis, D. T., Koksma, J. J., Trailin, A., Langle, S. L., Piet, R., Lodder, J. C., Timmerman, J., Mansvelder, H., Poulain, D. A., Oliet, S. H. R., and Brussaard, A. B. (2006). Oxytocin and estrogen promote rapid formation of functional GABA synapses in the adult supraoptic nucleus. *Mol. Cell. Neurosci.* 31, 785–794.

Vollmer, R. R., Li, X., Karam, J. R., and Amico, J. A. (2006). Sodium ingestion in oxytocin knockout mice. *Exp. Neurol.* 202, 441–448.

Wakerley, J. B., and Ingram, C. D. (1993). Synchronisation of bursting in hypothalamic oxytocin neurones: Possible co-ordinating mechanisms. *NiPS* 8, 129–133.

Wakerley, J. B., Terenzi, M. G., Housham, S. J., Jiang, Q. B., and Ingram, C. D. (1998). Electro–physiological effects of oxytocin within the bed nuclei of the stria terminalis: Influence of reproductive stage and ovarian steroids. *Prog. Brain Res.* 119, 321–334.

Young, L. J. (1999). Oxytocin and vasopressin receptors and species-typical social behaviors. *Horm. Behav.* 36, 212–221.

Young, L. J., Muns, S., Wang, Z. X., and Insel, T. R. (1997). Changes in oxytocin receptor mRNA in rat brain during pregnancy and the effects of estrogen and interleukin-6. *J. Neuroendocrinol.* 9, 859–865.

15

BIOLOGICAL AND MATHEMATICAL MODELING APPROACHES TO DEFINING THE ROLE OF OXYTOCIN AND DOPAMINE IN THE CONTROL OF MATING-INDUCED PRL SECRETION*

MARC E. FREEMAN[1,2], DE'NISE T. MCKEE[1,2], MARCEL EGLI[4] AND RICHARD BERTRAM[2,3]

[1]*Department of Biological Science,* [2]*Program in Neuroscience,* [3]*Department of Mathematics and Program in Molecular Biophysics, Florida State University, Tallahassee, Florida, USA*
[4]*Space Biology Group, Swiss Federal Institute of Technology Zurich (ETHZ), Technoparkstrasse 1, ETH-Technopark, Zurich, Switzerland, UK*

INTRODUCTION

Any review of *Parenting and the Brain* must actually consider the earlier period during which the organism prepares for parenting. This period, pregnancy, begins with fertilization of the oocyte that has been newly released from the ovarian follicle, implantation of the fertilized oocyte in the wall of the uterus, and development of the placenta which assumes the humoral role of the ovaries and pituitary gland during the mid to latter stages of pregnancy.

All aspects of early pregnancy in all mammals requires the secretion of progesterone from the corpus luteum to provide the appropriate uterine environment for successful development of the conceptus (Stouffer, 2006). In rats, unlike most mammals, the corpus luteum is an ephemeral structure (Stouffer, 2006), secreting small amounts of progesterone for less than 2 days after its formation (Smith, *et al.*, 1975) with the major progestin being its metabolite, 20 α-hydroxyprogesterone (Freeman, 2006; Stouffer, 2006). This latter steroid will not support the appropriate uterine environment to maintain pregnancy. Thus, rodents have short estrous cycles because they have short, subfunctional luteal phases. If, on the other hand, a fertile mating occurs in the rat, this initiates the secretion of sufficient prolactin (PRL) from the anterior pituitary gland to act as a "luteotropin" and "rescue" the corpus luteum and maintain its ability to secrete progesterone for at least 10–12 days (Freeman, *et al.*, 2000; Freeman, 2006).

*Studies from the authors' laboratories were supported by DA-19356, DK-43200, and HD-11669.

PRL exerts this effect by blocking the activity of the enzyme, 20 α-hydroxysteroid dehydrogenase, which converts progesterone to 20 α-hydroxyprogesterone (Freeman, 2006; Stouffer, 2006). The subsequent prolonged maintenance of progesterone secretion for the 20–22 days of pregnancy occurs in response to the placenta subsuming the role of the now quiescent pituitary gland and secreting a family of placental lactogens (Soares, *et al.*, 1991, 1998, 2007a, b; Sahgal, *et al.*, 2000; Dai, *et al.*, 2000; Soares, 2004). The placental lactogens inhibit pituitary PRL secretion by acting at the hypothalamus (Voogt *et al.*, 1982, 1996; Tonkowicz *et al.*, 1983; Tonkowicz & Voogt, 1983a, b, 1984, 1985; Voogt, 1984; Voogt & de Greef, 1989; Arbogast *et al.*, 1992; Tomogane *et al.*, 1992, 1993; Lee & Voogt, 1999) or directly on the pituitary gland (Gorospe & Freeman, 1985a, b).

In order to study the control of mating-induced PRL secretion independent of placental influence, female rats that copulate with a vasectomized male or receive an artificial copulomimetic stimulus on the afternoon of proestrus also results in the "rescue" of the corpus luteum which, in the absence of the placenta, persists for a 12–14 day interval known as pseudopregnancy (Smith *et al.*, 1975, 1976). At the end of this period, the rat ovulates and normal 4–5 day ovarian cycles eventuate (Freeman, 2006). During early pregnancy and the entire pseudopregnancy, PRL is secreted in a unique pattern characterized by two daily "surges" of secretion (Butcher *et al.*, 1972; Freeman, *et al.*, 1974; Smith & Neill, 1976). The first surge reaches its greatest concentration at 03:00–05:00 h. We have designated this the "nocturnal surge" (Freeman *et al.*, 1974). The second, designated "diurnal surge," approaches somewhat lesser peak concentrations at 17:00–19:00 h. This pattern of PRL secretion recurs daily in the absence of reinforcement of the stimulus with the difference being that the last day of the secretory pattern is day 10 post-coitum in pregnancy (Smith *et al.*, 1976) and day 12 after cervical stimulation (CS) during pseudopregnancy (Freeman *et al.*, 1974).

NEURAL SITES CONTROLLING MATING-INDUCED PRL SECRETION

The response of the PRL secretory apparatus to the mating stimulus has been termed a "unique neuroendocrine response" (Gunnet & Freeman, 1983) since, unlike the classical neuroendocrine response to suckling, mating eventuates in this unique, repetitive pattern of PRL secretion even in the absence of reinforcement of the stimulus. There is little doubt that this response must involve the central nervous system. The pelvic nerve is the primary sensory efferent pathway from the uterine cervix to the brainstem (Reiner *et al.*, 1981). The central neural areas responsive to stimulation of the uterine cervix include the lateral reticular nucleus, the gigantocellular reticular nucleus, and the dorsal raphe nucleus (Kawakami & Kubo, 1971; Hornby & Rose, 1976; Allen *et al.*, 1981) as well as the medial amygdala and preoptic areas (Erskine, 1993; Rowe & Erskine, 1993; Polston & Erskine, 2001; Cameron *et al.*, 2004). Lesions of the medial preoptic area cause repetitive pseudopregnancies (Clemens *et al.*, 1976) characterized by the early morning but not the early evening phase of PRL secretion (Freeman & Banks, 1980). Moreover, this lesion will block the evening surge induced by CS (Freeman & Banks, 1980). These studies suggested that the medial preoptic area posses two types of neurons: those tonically inhibitory to the early morning surges and those potentially stimulatory to the early evening surges. The mating stimulus acts by inhibiting the former and stimulating the latter (Gunnet & Freeman, 1984). Another area implicated in mating-induced PRL secretion is the dorsomedial–ventromedial area of the hypothalamus. Electrical stimulation of this area induces a pseudopregnancy characterized by nocturnal and diurnal surges of PRL (Freeman & Banks, 1980; Gunnet *et al.*, 1981). Lesion of this area does not interfere with estrous cyclicity, the ability to mate and support a pregnancy to term, or secrete a nocturnal surge of PRL (Yokoyama & Ota, 1959; Chateau, *et al.*, 1981; Gunnet *et al.*, 1981). Such a lesion only interferes with the diurnal surge of PRL (Gunnet *et al.*, 1981). In fact an intact dorsomedial–ventromedial hypothalamus is required for the stimulatory but not the inhibitory control of the PRL surges by the medial preoptic area (Gunnet & Freeman, 1985). Finally, the exquisitely timed surges of PRL secreted in response to mating are controlled by the suprachiasmatic nucleus (SCN) of the hypothalamus, the so-called master clock. When lesioned, both surges are abolished (Bethea & Neill, 1980). It is therefore likely that mating-induced PRL secretion is a circadian rhythm (Bethea & Neill, 1979).

NEUROENDOCRINE CONTROL OF MATING-INDUCED PRL SECRETION

There is little doubt that tuberoinfundibular neurons in the arcuate nucleus of the hypothalamus deliver dopamine (DA) to the hypophyseal portal vessels that bathe the anterior pituitary gland and thus inhibits PRL secretion (Ben-Jonathan & Hnasko, 2001). It would therefore be anticipated that there is an inverse relationship between DA levels in hypophyseal portal blood and PRL secretion into peripheral plasma in cervically stimulated rats. As expected, there is an inverse relationship between the levels of DA in hypophyseal portal blood and the surges of PRL in peripheral plasma. That is, during the surges, DA levels were 36% lower in cervically stimulated than control ovariectomized rats (deGreef & Neill, 1979). In addition, the amount of DA arriving at the anterior pituitary gland is also inversely proportional and in antiphase to the amount of PRL released into the peripheral plasma of cervically stimulated rats (McKee *et al.*, 2007). However, mimicry of the 36% decrease of DA levels was not sufficient to provoke the same magnitude of PRL release as that seen as a result of CS (deGreef & Neill, 1979). This would argue for the existence of an additional PRL inhibiting factor or possibly a heretofore uncharacterized PRL releasing factor of hypothalamic origin.

The description of an endogenous stimulatory rhythm regulating PRL secretion would be compatible with this notion (Arey *et al.*, 1989). In non-mated rats, pharmacologic blockade of DA receptors at various times of day leads to greatest PRL secretory responses at times of day corresponding to the early morning and early evening surges of PRL. One interpretation of these data is that CS may lower dopaminergic tone sufficiently to allow the PRL releasing rhythm to be expressed. Oxytocin (OT) appears to be the neurohormone which acts directly on the lactotrope to stimulate PRL secretion. Pharmacologic blockade of OT receptors will block both surges of PRL secretion (Arey & Freeman, 1989, 1990; McKee *et al.*, 2007). Moreover, in ovariectomized, non-cervically stimulated rats OT neurons in the paraventricular nucleus express the marker of neuronal activity, c-Fos, in greatest abundance during the early morning and early evening (Arey & Freeman, 1992). OT appears to stimulate PRL secretion by a calcium-dependent mechanism (Egli *et al.*, 2004). In mated animals, the diminution of activity of the neuroendocrine dopaminergic neurons just before initiation of PRL secretion (Lerant *et al.*, 1996; McKee *et al.*, 2007) appears to be due to the activity of vasoactive inhibitory polypeptide (VIP) originating in neurons of the SCN (Egli *et al.*, 2004; Bertram *et al.*, 2006). These dopaminergic neurons have VIP receptors and placement of VIP antisense oligonucleotides into the SCN prevents the fall in DA that precedes the release of PRL. Similarly, placement of the VIP antisense oligonucleotides in the SCN also abolished the afternoon increase in OT and PRL in cervically stimulated rats (Egli *et al.*, 2004).

We have also taken a mathematical modeling approach to define the roles of DA, OT, and VIP in the control of mating-induced PRL secretion.

THE MATHEMATICAL MODEL

We have developed a mathematical model that describes a likely mechanism for the circadian PRL rhythm induced by CS (Bertram *et al.*, 2006). This is a mean field model that uses a single variable to describe the activity level of each cell population. With this model, the circadian PRL rhythm is due to the interactions between four cell populations, as shown in Figure 15.1. Lactotroph cells reside in the pituitary gland and their activity level is denoted by PRL. The other cell populations are located in the hypothalamus. Dopaminergic neurons, with activity level DA, are located in the arcuate and periventricular nuclei. These release DA into portal blood vessels at the median eminence, from where it travels down to the pituitary and affects lactotrophs. DA is known to bind to D2 receptors on the lactotroph plasma membrane, leading to inhibition of lactotroph activity and secretion (Ben-Jonathan & Hnasko, 2001). The inhibitory influence is denoted with an open arrow from DA to PRL in Figure 15.1.

PRL from lactotrophs also feeds back onto DA neurons, in a stimulatory manner (DeMaria *et al.*, 1999). The stimulatory effects of PRL are due to increased gene expression (Freeman *et al.*, 2000) and upregulation of tyrosine hydroxylase (TH) activity (Ma *et al.*, 2005), which synthesizes l-DOPA from tyrosine. The delay in the PRL effect due to gene expression is expected to be several hours, while the delay

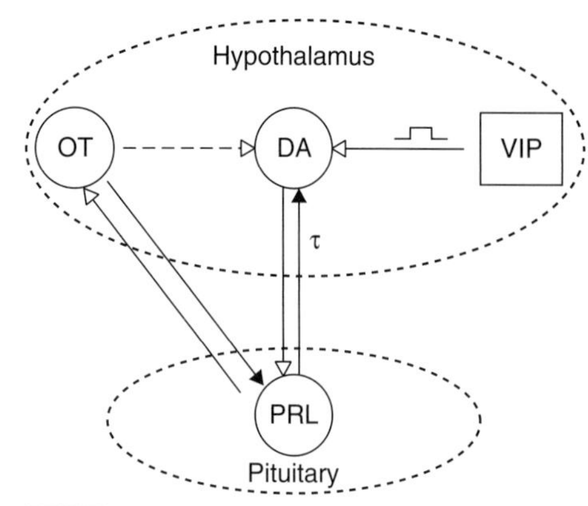

FIGURE 15.1 Illustration of the cell populations included in the mathematical model. Closed arrows represent stimulatory, and open arrows inhibitory, influences. The arrow with adjacent τ indicates a delayed influence. The arrow with a square pulse indicates a square pulse of VIP activity in the early morning, driven by the circadian clock in the SCN.

due to TH upregulation is as short as an hour (Ma *et al.*, 2005). We incorporate a time delay into the model, represented by τ. The delayed stimulatory effect of PRL on DA neurons is represented in Figure 15.1 as a closed arrow with adjacent τ. With the two-way interaction between DA neurons and lactotrophs, one direction stimulatory while the other is inhibitory, there is essentially a delayed negative feedback of PRL on lactotroph activity. This feedback loop has the right characteristics for an oscillation.

Another hypothalamic cell population that influences PRL secretion in the model is the population of oxytocinergic neurons of the paraventricular nucleus. These neurons, whose activity level we represent by OT, send axons to the neural lobe of the pituitary, where OT is released into short portal vessels that project into the anterior lobe of the pituitary, where lactotrophs are located. As noted previously, this OT has been shown to stimulate lactotroph activity (Egli *et al.*, 2004, 2006). PRL in turn affects the activity of OT neurons; initial data suggest that PRL inhibits OT neuron activity (Kokay, *et al.*, 2006). Thus, we assume a stimulatory influence of OT on lactotroph activity, and an inhibitory influence of PRL on OT neuron activity (Figure 15.1). Unlike the PRL–DA interaction, there is no delay here, so the PRL–OT feedback loop is not appropriate for the generation of oscillations.

TABLE 15.1 Parameter values used in the mathematical model

$T_p = 1$	$k_d = 1$	$q = 0.5$	$T_d = 10$
$k_p = 0.3$	$\tau = 3\,\text{h}$	$r_v = 2$	$v_o = 1$
$p_{\text{inj}} = 1{,}000$	$k_o = 1$	$T_o = 3$	

The final cell population in the model is a population of VIP neurons located in the SCN. We assume that the activity of these neurons is driven by the circadian clock within the SCN and is not influenced by OT, DA, or PRL. The VIP neurons innervate DA neurons of the arcuate nucleus (Gerhold *et al.*, 2001), and VIP inhibits DA neuron activity (Gerhold *et al.*, 2002). Under normal lighting conditions, VIP neurons have elevated activity in the early morning (Shinohara *et al.*, 1993), which we simulate as a square pulse of activity of 3-h duration. This is represented as an open arrow with adjacent square pulse in Figure 15.1.

The PRL, DA, and OT variables change over time according to ordinary differential equations. The PRL equation is:

$$\frac{\text{dPRL}}{\text{d}t} = \frac{T_p + v_o\text{OT}}{1 + k_d\text{DA}^2} - q\text{PRL} \tag{15.1}$$

The first term on the right hand side of the equation, T_p, is constant stimulatory drive that reflects the spontaneous activity of many lactotrophs in the absence of DA inhibition. The second term, v_oOT, reflects the stimulatory influence of OT on lactotrophs. The inhibitory influence of DA is simulated by placing DA in the denominator. Finally, the last term, $-q$PRL, represents first-order decay of lactotroph activity. The values of the parameters T_p, v_o, k_d, as well as other model parameters, are given in Table 15.1.

The delay-differential equation for DA is:

$$\frac{\text{dDA}}{\text{d}t} = T_d + k_p PRL_\tau^2 - q\text{DA} - r_v\text{VIP} \times \text{DA} \tag{15.2}$$

The first term on the right hand side, T_d, is tonic drive, reflecting the spontaneous activity of DA neurons. The second term, $k_p\text{PRL}_\tau^2$, reflects the stimulatory action of PRL, with a time delay of $\tau = 3\text{h}$. The third term, $-q$DA, represents first-order decay of DA neuron activity, assuming the same rate constant q as for decay of

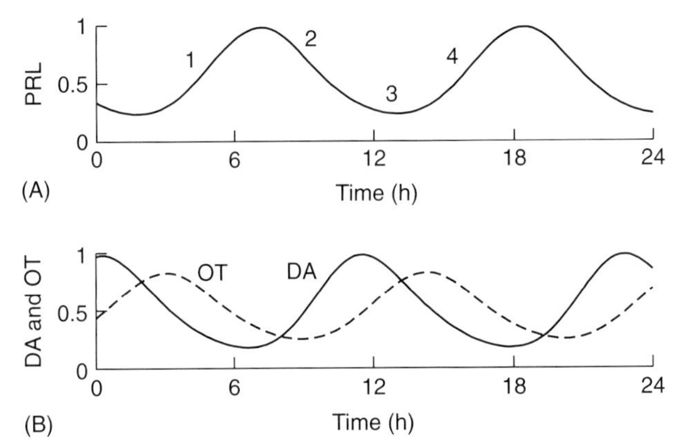

FIGURE 15.2 (A) A circadian PRL rhythm is generated due to interaction with DA and OT neurons. There are two PRL surges per day. The effects of VIP pulses are not included ($r_v = 0$). (B) Rhythmic DA and OT neuron activity accompany the PRL rhythm. The DA (solid) and OT (dashed) oscillations are out of phase with each other, and out of phase with PRL. To initiate and maintain the oscillations we set $T_d = 0$.

lactotroph activity. The final term, $-r_v\text{VIP} \times \text{DA}$, reflects the inhibitory effect of VIP.

The differential equation for OT is:

$$\frac{d\text{OT}}{dt} = \frac{T_o}{1 + k_o\text{PRL}^2} - q\text{OT} \qquad (15.3)$$

The first term on the right hand side, $T/(1 + k_o\text{PRL}^2)$, reflects tonic drive and the inhibitory effect of PRL on OT neuron activity. Notice that there is no time delay here, since recent data show that the inhibitory effect of PRL on OT is rapid (Kokay *et al.*, 2006). The second term, $-q\text{OT}$, is first-order decay of OT neuron activity.

We describe VIP neuron activity as a square pulse that is elevated for 3 h during the morning (VIP = 2) and is 0 for the rest of the day. The VIP variable is inhibitory to DA neuron activity (Eq. 15.2).

The differential equations were solved numerically using the fourth-order Runge–Kutta method implemented in the XPPAUT software package (Ermentrout, 2002). The variables are in arbitrary units (except time, in hours), so curves are presented in normalized form. The model was previously developed in (McKee *et al.*, 2007).

GENESIS OF THE PRL RHYTHM

In the mathematical model, the circadian PRL rhythm is generated by the interaction of PRL, DA, and OT. Figure 15.2 shows a computer simulation of the rhythm during 1 day. The effect of VIP pulses is not included in this figure (we eliminate the effect of VIP by setting $r_v = 0$). At the time point marked as "1" in Figure 15.2(A) the OT level is high and the DA level is declining (Figure 15.2(B)), both of which lead to an increase in PRL. The rise in PRL in turn inhibits its OT neurons and, after a delay, stimulates DA neurons. As a result, the PRL level begins to decline (label "2"). The decline in PRL (label "3") removes inhibition from the OT neurons and removes the delayed stimulation of DA neurons, so that OT is elevated and DA declines. As a result, the PRL level rises again (label "4"). Thus, the interaction between variables results in two PRL surges per day, with later surges in DA and even later surges in OT. That is, the PRL, DA, and OT oscillations are out of phase with one another.

DAILY VIP PULSE SETS THE PHASE OF THE PRL RHYTHM

While the PRL rhythm shown in Figure 15.2 looks correct in some respects, there are two major shortcomings. One is that there is nothing to set the phase of the rhythm. That is, there is nothing to determine when the first PRL surge of the day will occur, while the data shows that the nocturnal surge should occur in the early morning (at about 03:00 h). The other shortcoming is that the magnitudes of the PRL

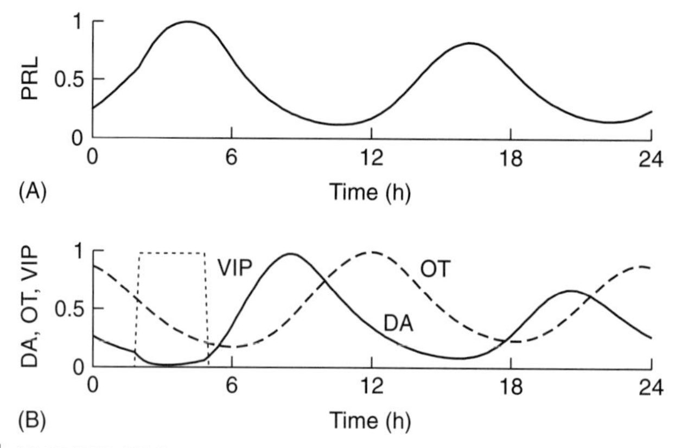

FIGURE 15.3 (A) When the effects of the daily VIP pulse are included in the model ($r_v = 2$), the phase of the PRL rhythm is set so that the nocturnal surge occurs in the early morning and the diurnal surge in the afternoon. The nocturnal surge is also now larger than the diurnal surge. (B) In the model, a VIP pulse (dotted) occurs for 3 h in the early morning (between 2:00 and 5:00 h). This inhibits DA neurons so that the nocturnal PRL surge occurs during the VIP pulse. To initiate and maintain the oscillations we set $T_d = 0$.

surges in Figure 15.2 are the same. However, the data show that the nocturnal PRL surge is typically larger than the diurnal surge (Butcher *et al.*, 1972; Gunnet & Freeman, 1983.

Both shortcomings are resolved when the effect of a daily VIP pulse is added in to the DA differential equation (Eq. 15.2). That is, we now set $r_v = 2$, rather than $r_v = 0$ as in Figure 15.2. The VIP pulse, which occurs early in the morning (Gerhold *et al.*, 2001), inhibits DA neurons, causing DA to prematurely decline (Figure 15.3(B)). The decline in DA results in an increase in PRL, so the first PRL surge of the day (the nocturnal surge) occurs during the VIP pulse (Figure 15.3(A)). The second surge (the diurnal surge) occurs in the afternoon, at about 16:00 h. Thus, the early morning VIP pulse sets the phase of the PRL rhythm. Since this rhythm is due to interaction with DA and OT neurons, the VIP also sets the phases of the oscillations in these variables.

In addition to setting the phase of the PRL rhythm, the morning VIP pulse also amplifies the nocturnal PRL surge (by inhibiting the DA neurons), so that now the nocturnal surge is larger than the diurnal surge (Figure 15.3(A)), as is typically observed experimentally.

In addition to the PRL time course, the model does at good job of reproducing the time courses of DA and OT following CS. The DA content in the anterior lobe of the pituitary gland was shown to exhibit a surge at about 12:00 h (Egli *et al.*, 2006), while the OT plasma concentration had a surge at about 13:00 h (Egli *et al.*, 2006). Thus the two surges occurred between the PRL surges, and the OT surge followed the DA surge, as in Figure 15.3.

MATHEMATICAL MODELING SUGGESTS AN EXPLANATION FOR THE OT-INDUCED PRL RHYTHM

We have shown that a single injection of OT initiates a rhythm of PRL secretion similar to that generated by mating (Egli *et al.*, 2006). Mathematical modeling can be used to help understand the mechanism through which OT injection can initiate a sustained circadian PRL rhythm that is similar to the rhythm induced by CS. We begin by writing a differential equation for the OT that is injected into the system. This is separate from the OT released by OT neurons in the PVN. To distinguish between the two we define OT_{inj} and OT_{PVN} as injected OT and OT from the PVN, respectively. Then $OT = OT_{inj} + OT_{PVN}$. The OT_{inj} differential equation is:

$$\frac{dOT_{inj}}{dt} = p_{inj} I - qOT_{inj} \tag{15.4}$$

where $I = 1$ for 2 h following the beginning of injection, otherwise $I = 0$. The parameter p_{inj} represents the magnitude of the injection, while q is the rate of the OT removal from the

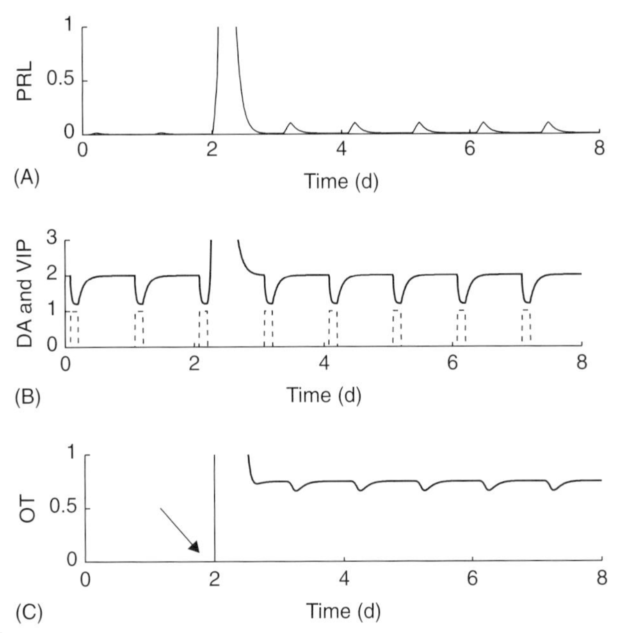

FIGURE 15.4 Simulation of one scenario for the effects of OT injection. In this scenario, the OT activates OT neurons of the PVN, but has no effect on DA neurons. Parameter T_o is initially set to 0, but then increased to 3 at the time of injection (arrow). No circadian PRL rhythm is produced. (A) There is a single PRL surge following injection. (B) A single DA surge is produced by the PRL surge. There are also small declines due to daily VIP pulses (dashed). (C) OT from the injection is cleared from the system before day 3. The OT level subsequently remains at an elevated level, providing stimulatory input to the lactotrophs.

system. The OT_{PVN} is described by Eq. (15.3), with OT replaced by OT_{PVN}. That is:

$$\frac{dOT_{PVN}}{dt} = \frac{T_o}{1 + k_o PRL^2} - qOT_{PVN} \quad (15.5)$$

One potential mechanism through which OT injection triggers a sustained PRL rhythm is through activation of the OT neurons in the PVN, which in turn provide stimulatory input to the lactotrophs. We initially set $T_o = 0$ in Eq. (15.5), so that OT neuron activity is initially depressed. Then to simulate OT injection we activate the neurons by setting $T_o = 3$. The results of this simulation are shown in Figure 15.4. Prior to OT injection (arrow in panel C), DA is elevated, with daily reductions due to the early morning VIP pulse. The overall elevated level of DA keeps PRL pinned at a low level, with very small bumps resulting from the VIP-induced reductions in DA. The OT injection on day 2 results in a spike in PRL, since OT stimulates lactotrophs. This spike in PRL causes a delayed spike in DA, since PRL stimulates DA neurons with a delay. Once the injected OT has cleared the system, OT settles at a plateau since the OT neurons are now activated. This adds additional stimulatory drive to lactotrophs, so that now during daily declines in DA the resulting PRL bump is somewhat amplified. Each PRL bump results in a small decline in OT, since PRL is inhibitory to OT neurons. Most importantly, the OT injection and subsequent activation of OT neurons did not initiate a two-pulse per day PRL rhythm. Even if OT neurons are stimulated at higher (or lower) levels, no PRL rhythm is initiated. Thus, activation of OT neurons alone cannot explain data showing that OT injection induces such a rhythm.

Another possibility is that the OT injection acts directly on lactotrophs to initiate the sustained PRL rhythm. In order for the rhythm to be sustained, one must assume that the stimulatory effect of OT on lactotrophs is somehow sustained long after the OT levels have returned to normal. Even in this case, however, simulations with the model show that no PRL

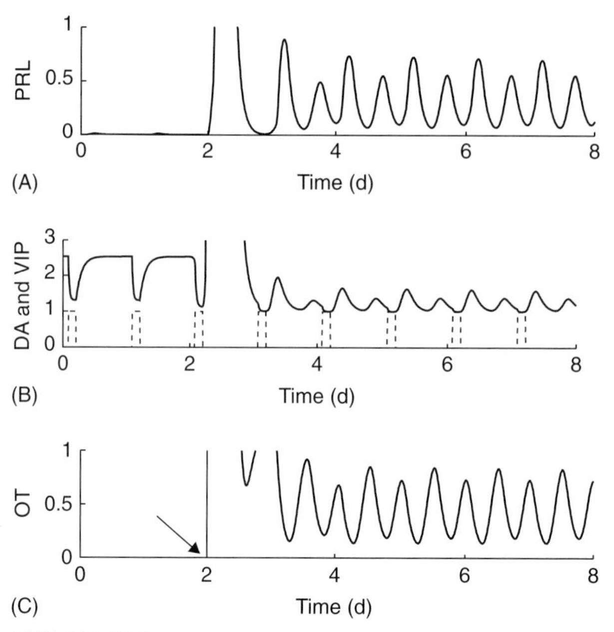

FIGURE 15.5 Simulation of a scenario for the effects of OT injection, in which DA neurons are partially inhibited ($T_d = 0$) and OT neurons are stimulated ($T_o = 3$) in response to the injection. (A) Following an initial PRL surge in direct response to the OT injection, a circadian pattern of PRL is established, with two surges per day. (B) Following an initial DA surge in response to the PRL surge, the overall DA level declines to a range where the PRL rhythm can be initiated and maintained. VIP is represented as a dashed curve. (C) The OT injection (arrow) results in an initial large increase in OT, followed by oscillations due to inhibition from lactotrophs.

rhythm will be initiated. In fact, this scenario is similar to the scenario above, where activated OT neurons provide sustained stimulatory drive to lactotrophs. In both cases, the effect of direct lactotroph stimulation is to just amplify the daily VIP-induced PRL bumps.

A third possibility is that the OT injection partially inhibits DA neurons. This inhibition is not likely to be direct, since OT receptors have not been detected in tuberoinfundibular dopamine (TIDA) neurons (Gimpl & Fahrenholz, 2001). However, it is possible that OT activates another neural population, and that these neurons project to and inhibit DA neurons. If these neurons are bi-stable, then OT injection may switch them from an "off" to an "on" state that would persist until the neurons are reset to the "off" state (Bertram *et al.*, 2006). In Figure 15.5 we show a simulation of this scenario, where DA neurons are partially inhibited (T_d is set to 0) and OT neurons are activated (T_o is set to 3) in response to the OT injection (arrow). As in Figure 15.4 there is a transient spike in PRL and DA in response to the OT injection. Also as before, the OT level is elevated. Unlike before, DA is now lowered in response to the injection. This is crucial, since now DA is in a range where a circadian rhythm can be established through mutual feedback with the lactotrophs, as in Figure 15.2. The pattern in PRL shown in Figure 15.5(A) clearly has the right properties: two surges per day, with a larger nocturnal surge. Thus, it appears that partial inhibition of DA neurons is effective in initiating a two-pulse per day PRL rhythm, while stimulation of OT neurons or lactotrophs is ineffective at this.

MATHEMATICAL MODELING OF THE EFFECTS OF AN OT ANTAGONIST

We have recently found that, in the presence of an OT antagonist (OTA), CS does not immediately initiate surges of PRL but the surges begin 2 days later when the antagonist is cleared (McKee *et al.*, 2007). An important feature of

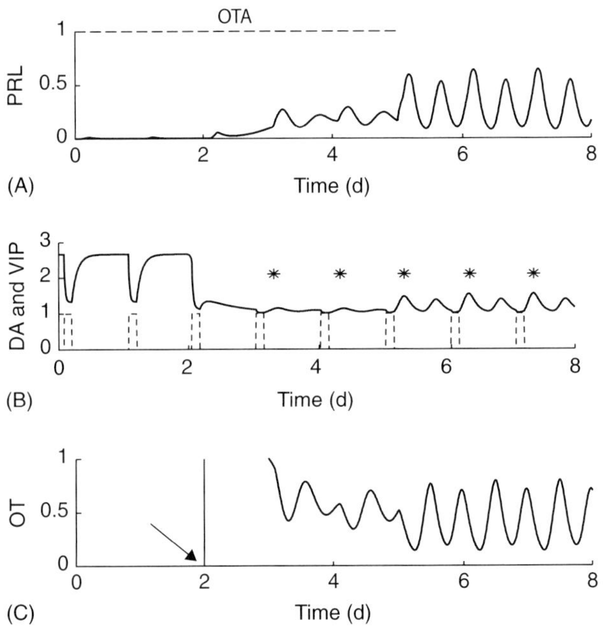

FIGURE 15.6 Simulation of the response to CS (at arrow) in the presence of an OTA (simulated by setting $v_o = 0$). The OTA is removed on day 5 ($v_o = 1$). (A) The PRL level increases somewhat following CS, but a circadian rhythm is not started until the OTA is removed. (B) The DA surge that normally occurs around noon following CS is eliminated by OTA. Once the OTA is removed, DA surges begin. The timing of the dominant surge corresponds to the timing in the control case (no OTA application, asterisks). VIP is shown as a dashed curve. (C) CS is simulated with an OT pulse, which decays within a day. This activates OT neurons, but the effect of these neurons on lactotrophs is antagonized by the OTA.

the OTA that we used in experiments is that it does not cross the blood–brain barrier. Thus, it antagonizes the effects of OT in the pituitary, but not in the hypothalamus. We incorporate the effects of the OTA into the model by setting $v_o = 0$ in the PRL differential equation (Eq. 15.1). Thus, we remove the direct stimulatory effects of OT on the model lactotrophs. If CS induces a surge in OT, as we believe from our experiments with OT injection, then the ability of this surge to partially inhibit DA neurons and activate OT neurons should be unimpeded by the OTA, which does not cross into the hypothalamus. The activation of OT neurons will be of no consequence, since the OT receptors on the lactotrophs are antagonized. However, the inhibition of DA will put DA at the right level to potentially initiate a circadian PRL rhythm.

Figure 15.6 shows a simulation of CS in the presence of an OTA. CS is applied at the arrow, inducing a surge of OT and a decline in DA. The decline in DA leads to a rise in PRL, but the circadian PRL rhythm is not established as long as OTA is present. The reason is that the stimulatory effect of OT from the PVN neurons on lactotrophs is antagonized. Without this, there is not enough stimulatory drive to the lactotrophs to initiate the PRL rhythm. However, although the PRL rhythm is not produced while OTA is present, the "memory" of the CS has been established, in that DA neurons are partially inhibited and OT neuron activity is increased. Thus, when the OTA has been cleared from the system (by day 5 in the figure) a circadian PRL rhythm is quickly established. In addition, the phase of the rhythm is set so that the nocturnal surge occurs in the early morning. The asterisks in Figure 15.6(B) show the timing of the primary DA surge on each day following CS in the absence of OTA (the control). When OTA is present there are no DA surges. However, when OTA is cleared the DA surges occur at the same time as they did in the control. The reason for this is that the daily pulsing of VIP is unaffected by the OTA, and it is this that sets the phase of the PRL rhythm. Our recent biological studies with the OTA have confirmed the predictions of the model (McKee *et al.*, 2007).

CONCLUSIONS

Taking two independent approaches, mathematical modeling and biological experimentation, we have arrived upon the same conclusions. Both approaches agreed that (1) a copulomimetic stimulus at the uterine cervix incites the release of OT and the unique biphasic secretion of PRL, (2) a single injection of OT induces the unique pattern of PRL secretion, (3) a copulomimetic stimulus induces a pattern of DA arriving in the anterior pituitary gland that is in antiphase with the secretion of PRL and is a direct response to PRL acting at the dopaminergic neuron, and (4) OT antagonism during the copulomimetic stimulus prevents induction of the PRL surges which appear 2 days after OTA withdrawal; suggestive of a "memory."

Our use of mathematical modeling is reflective of the potential use of modeling in the study of the parental brain. In particular, modeling can be used to interpret complex biological data, and to suggest new experiments. In our work it has played both roles. The successful use of modeling as a research tool in the study of the parental brain requires some knowledge of the neural circuitry and hormonal pathways involved in the behavior under investigation, and there should be a relatively high degree of quantification in the data. Models tend to be most useful when they are simple enough to be readily understood; the utility of a model often declines as its complexity is increased past some point. The mean field models that we use here are typical of the class of models that would be useful in studying the parental brain. In this class of model one uses a single variable to describe the average behavior of an entire subpopulation of cells. This differs from neural network models, where a set of variables is used to describe each neuron in the population. The mean field models involve many fewer variables, and are appropriate when single-cell resolution is not needed or practical. They are also simple enough to be effective tools for interpreting experimental data and designing new experiments.

REFERENCES

Allen, T. O., Adler, N. T., Greenberg, J. H., and Reivich, M. (1981). Vaginocervical stimulation selectively increases metabolic activity in the rat brain. *Science* 211, 1068–1072.

Arbogast, L. A., Soares, M. J., Tomogane, H., and Voogt, J. L. (1992). A trophoblast-specific factor(s) suppresses circulating prolactin levels and increases tyrosine hydroxylase activity in tuberoinfundibular dopaminergic neurons. *Endocrinology* 131, 105–113.

Arey, B. J., Averill, R. L., and Freeman, M. E. (1989). A sex-specific endogenous stimulatory rhythm regulating prolactin secretion. *Endocrinology* 124, 119–123.

Arey, B. J., and Freeman, M. E. (1989). Hypothalamic factors involved in the endogenous stimulatory rhythm regulating prolactin secretion. *Endocrinology* 124, 878–883.

Arey, B. J., and Freeman, M. E. (1990). Oxytocin, vasoactive intestinal peptide and serotonin regulate the mating-induced surges of prolactin secretion in the rat. *Endocrinology* 126, 279–284.

Arey, B. J., and Freeman, M. E. (1992). Activity of oxytocinergic neurons in the paraventricular nucleus mirrors the periodicity of the endogenous stimulatory rhythm regulating prolactin secretion. *Endocrinology* 130, 126–132.

Ben-Jonathan, N., and Hnasko, R. (2001). Dopamine as a prolactin (PRL) inhibitor. *Endocr. Rev.* 22, 724–763.

Bertram, R., Egli, M., Toporikova, N., and Freeman, M. E. (2006). A mathematical model for the mating-induced prolactin rhythm of female rats. *Am. J. Physiol. Endocrinol. Metabol.* 290, E573–E582.

Bethea, C. L., and Neill, J. D. (1979). Prolactin secretion after cervical stimulation of rats maintained in constant dark or constant light. *Endocrinology* 104, 870–876.

Bethea, C. L., and Neill, J. D. (1980). Lesions of the suprachiasmatic nuclei abolish the cervically stimulated prolactin surges in the rat. *Endocrinology* 107, 1–5.

Butcher, R. L., Fugo, N. W., and Collins, W. C. (1972). Semicircadian rhythm in plasma levels of prolactin during early gestation in the rat. *Endocrinology* 90, 1125–1127.

Cameron, N. M., Carey, P., and Erskine, M. S. (2004). Medullary noradrenergic neurons release norepinephrine in the medial amygdala in females in response to mating stimulation sufficient for pseudopregnancy. *Brain Res.* 1022, 137–147.

Chateau, D., Plas-Roser, S., and Aron, C. L. (1981). Follicular growth, ovulatory phenomena and

ventromedial nucleus lesions in the rat. *Endokrinologie* 77, 257–268.
Clemens, J. A., Smalstig, E. B., and Sawyer, B. D. (1976). Studies on the role of the preoptic area in the control of reproductive function in the rat. *Endocrinology* 99, 728–735.
Dai, G., Wang, D., Liu, B., Kasik, J. W., Müller, H., White, R. A., Hummel, G. S., and Soares, M. J. (2000). Three novel paralogs of the rodent prolactin gene family. *Journal of Endocrinology* 166, 63–75.
deGreef, W. J., and Neill, J. D. (1979). Dopamine levels in hypophysial stalk plasma of the rat during surges of prolactin induced by cervical stimulation. *Endocrinology* 105, 1093–1099.
DeMaria, J. E., Lerant, A. A., and Freeman, M. E. (1999). Prolactin activates all three populations of hypothalamic neuroendocrine dopaminergic neurons in ovariectomized rats. *Brain Res.* 837, 236–241.
Egli, M., Bertram, R., Sellix, M. T., and Freeman, M. E. (2004). Rhythmic secretion of prolactin in rats: Action of oxytocin coordinated by vasoactive intestinal polypeptide of suprachiasmatic nucleus origin. *Endocrinology* 145, 3386–3394.
Egli, M., Bertram, R., Toporikova, N., Sellix, M. T., Blanco, W., and Freeman, M. E. (2006). Prolactin secretory rhythm of mated rats induced by a single injection of oxytocin. *Am. J. Physiol. Endocrinol. Metabol.* 290, E566–E572.
Ermentrout, G. B. (2002). *Simulating, Analyzing and Animating Dynamical Systems: A Guide to XPPAUT for Researchers and Students.* SIAM, Philadelphia.
Erskine, M. S. (1993). Mating-induced increases in FOS protein in preoptic area and medial amygdala of cycling female rats. *Brain Res. Bull.* 32, 447–451.
Freeman, M. E. (2006). Neuroendocrine control of the ovarian cycle of the rat. In *Knobil and Neill's Physiology of Reproduction* (J. D. Neill, Ed.), pp. 2327–2388. Academic Press, San Diego.
Freeman, M. E., and Banks, J. A. (1980). Hypothalamic sites which control the surges of prolactin secretion induced by cervical stimulation. *Endocrinology* 106, 668–673.
Freeman, M. E., Kanyicska, B., Lerant, A., and Nagy, G. M. (2000). Prolactin: structure, function, and regulation of secretion. *Physiological Reviews* 80, 1523–1631.
Freeman, M. E., Smith, M. S., Nazian, S. J., and Neill, J. D. (1974). Ovarian and hypothalamic control of the daily surges of prolactin secretion during pseudopregnancy in the rat. *Endocrinology* 94, 875–882.
Gerhold, L. M., Horvath, T. L., and Freeman, M. E. (2001). Vasoactive intestinal peptide fibers innervate neuroendocrine dopaminergic neurons. *Brain Res.* 919, 48–56.
Gerhold, L. M., Sellix, M. T., and Freeman, M. E. (2002). Antagonism of vasoactive intestinal peptide mRNA in the suprachiasmatic nucleus disrupts the rhythm of FRAs expression in neuroendocrine dopaminergic neurons. *J. Comp. Neurol.* 450, 135–143.
Gimpl, G., and Fahrenholz, F. (2001). The oxytocin receptor system: structure, function, and regulation. *Physiol Rev.* 81, 629–683.
Gorospe, W. C., and Freeman, M. E. (1985a). Detection of prolactin inhibitory activity in uterine epithelial cell secretions and rat serum. *Endocrinology* 116, 1559–1564.
Gorospe, W. C., and Freeman, M. E. (1985b). Effects of placenta and maternal serum on prolactin secretion *in vitro*. *Biol. Reprod.* 32, 279–283.
Gunnet, J. W., and Freeman, M. E. (1983). The mating-induced release of prolactin: a unique neuroendocrine response. *Endocr. Rev.* 4, 44–60.
Gunnet, J. W., and Freeman, M. E. (1984). Hypothalamic regulation of mating-induced prolactin release: Effect of electrical stimulation of the medial preoptic area in conscious femals. *Neuroendocrinology* 38, 12–16.
Gunnet, J.W., Freeman, M.E., (1985). The interaction of the medial preoptic area and the dorsomedial-ventromedial nuclei of the hypothalamus in the regulation of the mating-induced release of prolactin. *Neuroendocrinology* 40, 232–237.
Gunnet, J. W., Mick, C., and Freeman, M. E. (1981). The role of the dorsomedial-ventromedial area of the hypothalamus in control of prolactin secretion induced by cervical stimulation. *Endocrinology* 109, 1846–1850.
Hornby, J. B., and Rose, J. D. (1976). Responses of caudal brainstem neurons to vaginal and somatosensory stimulation in the rat and evidence of genital–nociceptive interactions. *Exp. Neurol.* 51, 363–376.
Kawakami, M., and Kubo, K. (1971). Neuro-correlate of limbic-hypothalamo-pituitary-gonadal axis in the rat: change in limbic-hypothalamic unit activity induced by vaginal and electrical stimulation. *Neuroendocrinology* 7, 65–89.
Kokay, I. C., Bull, P. M., Davis, R. L., Ludwig, M., and Grattan, D. R. (2006). Expression of the long form of the prolactin receptor in magnocellular oxytocin neurons is associated with specific prolactin regulation of oxytocin neurons. *Am. J. Physiol. Regul. Integr. Comp. Physiol.* 290, R1216–R1225.
Lee, Y. S., and Voogt, J. L. (1999). Feedback effects of placental lactogens on prolactin levels and Fos-related antigen immunoreactivity of tuberoinfundibular dopaminergic neurons in the arcuate nucleus during pregnancy in the rat. *Endocrinology* 140, 2159–2166.

Lerant, A., Herman, M. E., and Freeman, M. E. (1996). Dopaminergic neurons of periventricular and arcuate nuclei of pseudopregnant rats: Semicircadian rhythm in fos-related antigens immunoreactivities and in dopamine concentration. *Endocrinology* 137, 3621–3628.

Ma, F. Y., Grattan, D. R., Goffin, V., and Bunn, S. J. (2005). Prolactin-regulated tyrosine hydroxylase activity and messenger ribonucleic acid expression in mediobasal hypothalamic cultures: The differential role of specific protein kinases. *Endocrinology* 146, 93–102.

Mckee, D. T., Poletini, M. O., Bertram, R., and Freeman, M. E. (2007). Oxytocin action at the lactotroph is required for prolactin surges in cervically stimulated ovariectomized rats. *Endocrinology* 48, 4649–4657.

Polston, E. K., and Erskine, M. S. (2001). Excitotoxic lesions of the medial amygdala differentially disrupt prolactin secretory responses in cycling and mated female rats. *J. Neuroendocrinol.* 13, 13–21.

Reiner, P., Woolsey, J., Adler, N. T., and Morrison, A. (1981). A gross anatomical study of the peripheral nerves associated with reproductive function in the female albino rat. In *Neuroendocrinology* of Reproduction (N. T. Adler, Ed.), pp. 545–549. Plenum Press, New York.

Rowe, D. W., and Erskine, M. S. (1993). c-Fos proto-oncogene activity induced by mating in the preoptic area, hypothalamus and amygdala in the female rat: Role of afferent input via the pelvic nerve. *Brain Res.* 621, 25–34.

Sahgal, N., Knipp, G. T., Liu, B., Chapman, B. M., Dai, G., and Soares, M. J. (2000). Identification of two new nonclassical members of the rat prolactin family. *J. Mol. Endocrinol.* 24, 95–108.

Shinohara, K., Tominaga, K., Isobe, Y., and Inouye, S.-I. T. (1993). Photic regulation of peptides located in the ventrolateral subdivision of the suprachiasmatic nucleus of the rat: Daily variations of vasoactive intestinal polypeptide, gastrin- releasing peptide, and neuropeptide Y. *J. Neuroscience* 13, 793–800.

Smith, M. S., Freeman, M. E., and Neill, J. D. (1975). The control of progesterone secretion during the estrous cycle and early pseudopregnancy in the rat: Prolactin, gonadotropin and steroid levels associated with rescue of the corpus luteum of pseudopregnancy. *Endocrinology* 96, 219–226.

Smith, M. S., McLean, B., and Neill, J. D. (1976). Prolactin: The initial luteotropic stimulus of pseudopregnancy in the rat. *Endocrinology* 98, 1370–1377.

Smith, M. S., and Neill, J. D. (1976). Termination at midpregnancy of the two daily surges of plasma prolactin initiated by mating in the rat. *Endocrinology* 98, 696–706.

Soares, M. J. (2004). The prolactin and growth hormone families: pregnancy-specific hormones/cytokines at the maternal-fetal interface. *Reprod. Biol. Endocrinol.* 2, 51.

Soares, M. J., Alam, S. M. K., Duckworth, M. L., Horseman, N. D., Konno, T., Linzer, D. I. H., Maltais, L. J., Nilsen-Hamilton, M., Shiota, K., Smith, J. R., and Wallis, M. (2007a). A standardized nomenclature for the mouse and rat prolactin superfamilies. *Mamm. Genome* 18, 154–156.

Soares, M. J., Faria, T. N., Roby, K. F., and Deb, S. (1991). Pregnancy and the prolactin family of hormones: Coordination of anterior pituitary, uterine, and placental expression. *Endocr. Rev.* 12, 402–423.

Soares, M. J., Konno, T., and Alam, S. M. K. (2007b). The prolactin family: Effectors of pregnancy-dependent adaptations. *Trends. Endocrinol. Metabol.* 18, 114–121.

Soares, M. J., Müller, H., Orwig, K. E., Peters, T. J., and Dai, G. L. (1998). The uteroplacental prolactin family and pregnancy. *Biol. Reprod.* 58, 273–284.

Stouffer, R. L. (2006). Structure, function and regulation of the corpus luteum. In *Physiology of Reproduction* (J. D. Neill, Eds.), pp. 475–526. Academic Press, San Diego.

Tomogane, H., Arbogast, L. A., Soares, M. J., Robertson, M. C., and Voogt, J. L. (1993). A factor(s) from a rat trophoblast cell line inhibits prolactin secretion *in vitro* and *in vivo*. *Biol. Reprod.* 48, 325–332.

Tomogane, H., Mistry, A. M., and Voogt, J. L. (1992). Late pregnancy and rat choriocarcinoma cells inhibit nocturnal prolactin surges and serotonin-induced prolactin release. *Endocrinology* 130, 23–28.

Tonkowicz, P., Robertson, M., and Voogt, J. (1983). Secretion of rat placental lactogen by the fetal placenta and its inhibitory effect on prolactin surges. *Biol. Reprod.* 28, 707–716.

Tonkowicz, P. A., and Voogt, J. L. (1983a). Examination of rat placental lactogen and prolactin at 6-hr intervals during midpregnancy. *Proc. Soc. Exp. Biol. Med.* 173.

Tonkowicz, P. A., and Voogt, J. L. (1983b). Termination of prolactin surges with development of placental lactogen secretion in the pregnant rat. *Endocrinology* 113, 1314–1318.

Tonkowicz, P. A., and Voogt, J. L. (1984). Ovarian and fetal control of rat placental lactogen and prolactin secretion at midpregnancy. *Endocrinology* 114, 254–259.

Tonkowicz, P. A., and Voogt, J. L. (1985). Effect of conceptus number, hysterectomy, and **progesterone** on prolactin surges in rats. *Am. J. Physiol.* 248, E269–E273.

Voogt, J., and de Greef, W. J. (1989). Inhibition of nocturnal prolactin surges in the pregnant rat by incubation medium containing placental lactogen. *Proc. Soc. Exp. Biol. Med.* 191, 403–407.

Voogt, J., Robertson, M., and Friesen, H. (1982). Inverse relationship of prolactin and rat placental lactogen during pregnancy. *Biol.Reprod.* 26, 800–805.

Voogt, J. L. (1984). Evidence for an inhibitory influence of rat placental lactogen on prolactin release *In vitro. Biol. Reprod.* 31, 141–147.

Voogt, J. L., Soares, M. J., Robertson, M. C., and Arbogast, L. A. (1996). Rat placental lactogen-I abolishes nocturnal prolactin surges in the pregnant rat. *Endocrine* 4, 233–238.

Yokoyama, A., and Ota, K. (1959). Effect of oxytocin replacement on lactation in rats bearing hypothalamic lesions. *Endocrinologica Japonica* 6, 268–276.

16 ROLE OF PROLACTIN IN THE METABOLIC ADAPTATIONS TO PREGNANCY AND LACTATION

BARBARA WOODSIDE[1], RACHEL. A. AUGUSTINE[2], SHARON R. LADYMAN[1], LINDSAY NAEF[3] AND DAVID R. GRATTAN[2]

[1] *Center for Studies in Behavioral Neurobiology and Department of Psychology, Concordia University, Montréal, Canada*
[2] *Centre for Neuroendocrinology and Department of Anatomy and Structural Biology, University of Otago, Dunedin, New Zealand*
[3] *Douglas Hospital Research Centre, McGill University, Montréal, Canada*

INTRODUCTION

The successful completion of pregnancy and lactation depends on coordinated adaptation in neuroendocrine, physiological and behavioral mechanisms in the mother that have to be finely tuned to the needs of the developing young (Russell *et al.*, 2001). In the absence of sensory neural input that could signal the pregnant state to the brain, these adaptive changes are initiated by the hormonal changes associated with this state. In mammals, maternal commitment to reproduction continues beyond pregnancy and parturition, with further adaptive changes required for parental behavior and lactation. Thus, signals that persist through pregnancy and lactation may also be important for inducing and maintaining the adaptations of the maternal brain. The lactogenic hormones, including prolactin from the maternal pituitary gland and placental lactogen, are well suited to provide such afferent information, as collectively, these hormones are present at high levels throughout pregnancy and lactation. Accumulating evidence suggests that these hormones exert multiple effects in the maternal brain, perhaps coordinating a range of adaptive changes (Grattan, 2002).

Energy homeostasis represents one critical physiological function that requires obligatory modification to allow successful reproduction. Recent evidence demonstrating the long-term negative effects of perinatal under- or overnutrition (Vickers *et al.*, 2000; Gluckman *et al.*, 2007; Xiao *et al.*, 2007) emphasize the fact that ensuring the appropriate level of nutrient availability to maintain optimal growth at all stages of development is a major facet of successful rearing. This task is complicated by the dramatic changes in the nutrient requirements of the offspring over time. By the end of the first trimester the human fetus weighs only about an ounce, at 6 months about 1.75 lbs and the average weight of a newborn is 7.5 lbs. These changes in growth are reflected in the estimated energetic cost of 375 kJ/day for the first trimester and 1,200 and 1,950 kJ/day for the second and third trimesters, respectively (Butte & King, 2005). The same pattern of accelerated fetal growth and energetic cost over gestation is seen in rats. In mid pregnancy (day 11 postconception) each placental-fetal unit weighs about 0.25 g but 11 days later the new born rat weighs 5–6 g (Naismith, 1969). The task of meeting these changing demands is complicated further by the physiological burden placed on the mother by some of the adaptations to pregnancy. For example, the increase in basal metabolic rate observed in pregnant women results from increased energy expenditure associated with greater blood volume and respiration as well as the increase in tissue mass (Lof *et al.*, 2005). The energetic cost of lactation far outweighs that of pregnancy.

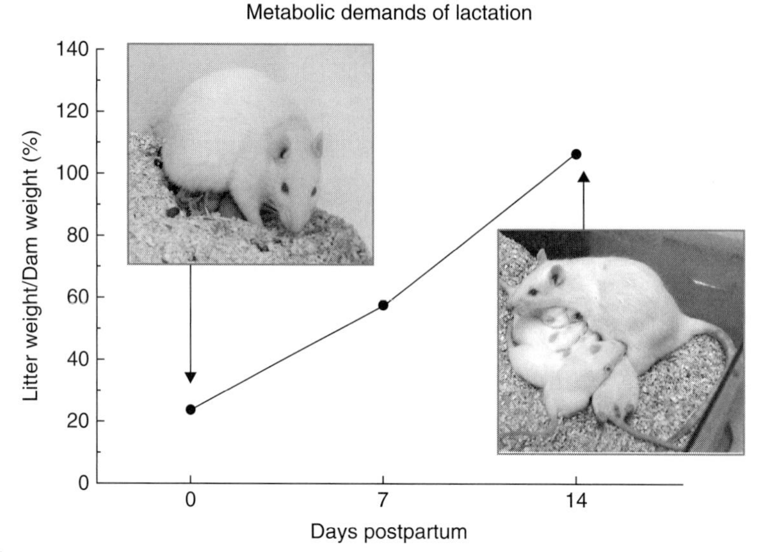

FIGURE 16.1 Growth of litters of eight pups across the first 14 days postpartum expressed as percentage of maternal body weight.

The estimated daily cost of full breastfeeding in women is 2.62 MJ/day and it has been estimated that in well-fed populations about 0.72 MJ/ day maybe mobilized from maternal tissues (Butte & King, 2005). As Figure 16.1 illustrates, the energetic demands of lactation are also high in rats. At peak lactation, rats suckling litters of eight pups produce about 70 ml of milk/day (Brommage, 1989) associated with a 2-fold increase in metabolic rate (Denckla & Bilder, 1975). To meet these rising energy demands, female mammals utilize a variety of adaptive mechanisms during pregnancy and lactation, chief among which are changes in ingestive behavior.

In this chapter we describe some of the metabolic adaptations to pregnancy and lactation, and discuss the neuroendocrine mechanisms underlying these changes. We have focused on research completed in laboratory rodents, but where possible, have drawn comparisons with data from human mothers. Our goal is to examine the hypothesis that the changing pattern of central prolactin receptor activation associated with different stages of pregnancy and lactation is a key signal through which peripheral stimuli modulate energy homeostasis in mothers and thus facilitate the optimal division of energetic resources between mother and young.

METABOLIC ADAPTATIONS TO PREGNANCY AND LACTATION

The mechanisms brought into play to enable female mammals to meet the energetic demands of the developing young include reductions in the expenditure of energy through both locomotor activity and thermogenesis, increased assimilation and storage of nutrients, preferential partitioning of resources from the mother to the young and, most dramatically, marked changes in maternal ingestive behavior. These adaptations can occur in the absence of any obvious metabolic costs and in many cases are driven by the hormonal changes associated with pregnancy and lactation.

Reductions in Energy Expenditure

One of the earliest behavioral signs of pregnancy in a newly mated rat is the absence of the peak in activity usually associated with behavioral estrus (Slonaker, 1924a, b). This reduction in activity is evident by about the fourth day of pregnancy prior to implantation, which in this species, occurs on days 5–6 postconception. Since it is induced by a change in hormonal condition is also seen in rats that are rendered pseudopregnant after an infertile mating (Slonaker,

1924a). Activity remains at diestrous levels for most of pregnancy until there is a further decrease just prior to parturition (Slonaker, 1924a). Low levels of locomotor activity are also seen in lactating rats, particularly in early lactation when they spend a great deal of time in the nest nursing (Grota & Ader, 1974). Whether there is a similar decrease in physical activity level across pregnancy in women is a matter of some debate. Although self-report data from a cohort of British mothers suggest that there is a decrease in locomotor activity in pregnancy, in particular that associated with work and leisure rather than in domestic activity (Clarke *et al.*, 2005), a study of Swedish women suggest that changes in metabolism seen at the end of pregnancy do not reflect a decrease in physical activity (Lof & Forsum, 2006).

By mid pregnancy, the reduction in locomotor activity is accompanied by a reduction in the thermogenic capacity of brown adipose tissue (BAT). The mitochondrial content of this tissue in rats on days 11 and 13 of pregnancy is only half that of nonpregnant rats, and is accompanied by a decrease in β3-adenoreceptors, although uncoupling protein (UCP)1 levels are not changed (Frontera *et al.*, 2005). More profound changes in thermogenesis are seen by the end of pregnancy, when both diet- and cold-induced thermogenesis are suppressed (Abelenda & Puerta, 1987; Imai-Matsumara *et al.*, 1990). The inhibition of BAT thermogenesis persists throughout lactation in most rodent species (Trayhurn, 1989) and is associated with a decrease in GDP binding on mitochondria as well as a reduction in both UCP1 and UCP3 (Xiao *et al.*, 2004a) which is also seen in skeletal muscle (Xiao *et al.*, 2004b). Interestingly the degree of suppression of thermogenesis is related to the energetic cost of lactation since it is positively correlated with litter size (Isler *et al.*, 1984).

Increased Assimilation and Storage of Nutrients

To meet the energetic costs of reproduction, the provision of specific nutrients as well as calories to the young is facilitated by alterations in the processes of absorption. Perhaps the most general of these is an increase in the absorptive capacity of the gut. These changes, which are first seen in mid pregnancy as an increase in the height of the intestinal villi, progress to include hyperplasia of all layers of the intestinal wall (Cripps & Williams, 1975). By the end of lactation this hyperplasia extends to all the intestinal tract (Boyne *et al.*, 1966). In both pregnancy and lactation there are also increases in the content of some enzymes within the walls of the gut (Rolls, 1975; Burdett & Reek, 1979).

There is a 2-fold increase in both calcium and phosphorus absorption across the wall of the small intestine in late pregnancy and throughout lactation. Some of this increase is due to elevated parathyroid hormone secretion that results in increased levels of 1,25-dihydroxycholecalciferol (Robinson *et al.*, 1982). Vitamin D deficient pregnant and lactating rats also show enhanced calcium absorption, however, suggesting that some proportion of the pregnancy-induced increase in calcium absorption is not vitamin D-dependent (Boass *et al.*, 1981; Brommage *et al.*, 1990). As a result of the increased calcium absorption coupled with increased dietary calcium intake, there is an increase in calcium content of the maternal skeleton over pregnancy that is rapidly lost during the ensuing lactation (Miller *et al.*, 1986). In fact, rats can lose between 15% and 40% of their skeletal calcium to milk production (Brommage, 1989). Lactating women may also lose skeletal calcium. One study found a reduction of about 4% in bone mineral density between 2 weeks and 3 months postpartum in a group of fully lactating North American women (Krebs *et al.*, 1997) in spite of increased dietary calcium intake as well as use of calcium supplements. Bone loss was restored on weaning, but the rate of restoration decreased with parity.

This pattern of metabolic regulation, with accumulation of nutrient stores during one phase of the reproductive episode followed by their use at another, is also seen with respect to protein and fat. Rat dams increase lean body mass during the initial phases of pregnancy (Naismith & Morgan, 1976), which has been attributed to decreases in enzymes that control amino acid oxidation and urea synthesis (Naismith & Walker, 1988). In spite of the increase in lean body mass in early pregnancy, there in no net gain of protein when this is assessed at parturition and it has been suggested that the early protein gains may be utilized to support the rapid growth of the fetuses in late pregnancy (Naismith & Morgan, 1976). Although there is evidence from studies of nitrogen excretion in well-nourished healthy women that the efficiency of dietary protein use may increase in late pregnancy, whether there is a similar pattern of protein utilization for fetal growth is controversial (Naismith & Emery, 1988; Mojtahedi *et al.*, 2002).

Despite the energy invested in the developing fetuses, both women and female rats typically gain fat across pregnancy. Butte *et al.* (2003) found an increase of about 5 kg in fat mass in women at 36 weeks postconception. The estimates of fat gain in rats vary considerably but increases of 40–60% have been reported (Naismith *et al.*, 1982). Most of this fat is accrued after the first week of gestation (Naismith *et al.*, 1982; Ladyman & Grattan, 2005). In contrast to protein utilization, fat stores are maintained throughout pregnancy and utilized during lactation. Immediately after parturition rats have higher levels of body fat than age-matched virgins. By the end of the second week postpartum, however, fat levels in the postpartum rats are well below those of cycling controls (Naismith *et al.*, 1982; Woodside *et al.*, 2000).

There is less clear evidence for a similar loss of fat mass during lactation in women. Some studies have reported faster weight loss in breastfeeding mothers than in those who bottlefeed (Dewey *et al.*, 1993) whereas others have found no difference (Haiek *et al.*, 2001). Conversely, others have reported a faster rate of loss of whole body fat mass during the early postpartum period in women who formula fed their babies (Wosje & Kalkwarf, 2004).

Changes in Partitioning of Nutrients

In addition to the increase in efficiency of nutrient processing, pregnancy and lactation are also associated with changes in the partitioning of nutrients between maternal tissues as well as with mechanisms that result in the preferential use of nutrients by the fetus or in milk. The change in partitioning of resources between mother and young that occurs during pregnancy is accomplished in part by changes in insulin secretion and development of peripheral insulin resistance. Early pregnancy is typically associated with normal (Butte, 2000) or increased insulin secretion and sensitivity (Gonzalez *et al.*, 2002; Ramos *et al.*, 2003). As pregnancy progresses, however, there is a gradual decrease in insulin sensitivity in maternal fat and muscle tissue but not in mammary gland (Flint *et al.*, 1979; Flint, 1985; Ramos *et al.*, 2003). Increases in islet cell number and size as well as a lower threshold for glucose-stimulated insulin secretion (Sorenson & Brelje, 1997) induce elevated insulin secretion to compensate for the lack of sensitivity to insulin and hence protect against gestational diabetes. Overall, the result of these changes is a decrease in maternal glucose utilization in late pregnancy. Because neither uterine glucose uptake nor placental glucose transport is changed, there is a diversion of available glucose from the mother to the fetus (Hay *et al.*, 1984), which supports its rapid growth in the last trimester of pregnancy.

Insulin resistance in white adipose tissue persists into lactation and is associated with a reduction in glucose utilization and lipogenesis and increased lipolysis that persists throughout lactation (Vernon & Pond, 1997). These changes are accompanied by decreased glucose utilization in skeletal muscle. The mammary gland remains highly sensitive to insulin, however, and there is increased glucose utilization and lipogenesis in the liver of lactating rats (Vernon & Pond, 1997). This pattern of change of insulin sensitivity during lactation might be expected to divert available nutrients from storage in the mother into milk for the young.

Ingestive Behavior in Pregnancy and Lactation

The adaptations described above facilitate the efficient use of nutrients during pregnancy and lactation and help to buffer the mother from the energetic demands placed on her by the rapid growth of her young during the last trimester of pregnancy as well as during lactation. By far the greatest contribution to meeting the energetic costs of reproduction, however, is made by alterations in the pattern of maternal ingestive behavior. As Figure 16.2 shows, there is a dramatic increase in the total amount of food consumed during pregnancy and especially lactation, and changes in the pattern of food intake as well as in diet selection have been documented.

Increases in food intake appear prior to implantation in the rat and by day 4 postconception female rats are eating significantly more than virgins (Ladyman & Grattan, 2004). The rate of rise in food intake increases rapidly in the second half of pregnancy to levels 50–60%

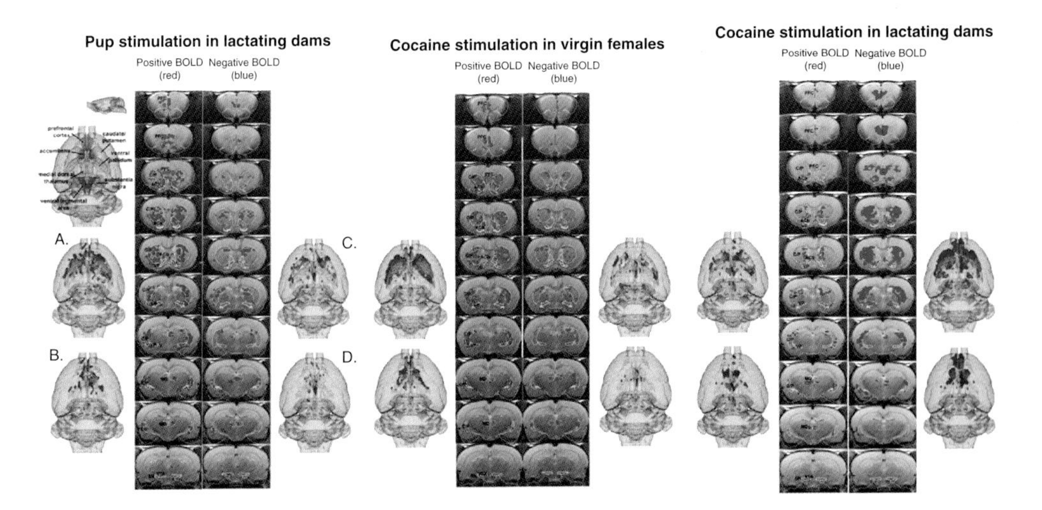

PLATE 4.3 Composite brain maps of positive and negative BOLD responses to suckling pups and cocaine. Shown are data for virgin rats and PPD 4–8 dams given cocaine or dams exposed to suckling pups. The three columns of coronal anatomical scans surrounded by 3D glass brains indicate areas of the mesocorticolimbic and nigrostriatal dopamine systems that showed increases in BOLD responses (red-positive BOLD) or decreases (blue-negative BOLD) for the three conditions. The 3D brain in the upper far left corner highlights the anatomical subregions that were studied. (*Source*: Adapted from Ferris *et al.*, 2005.)

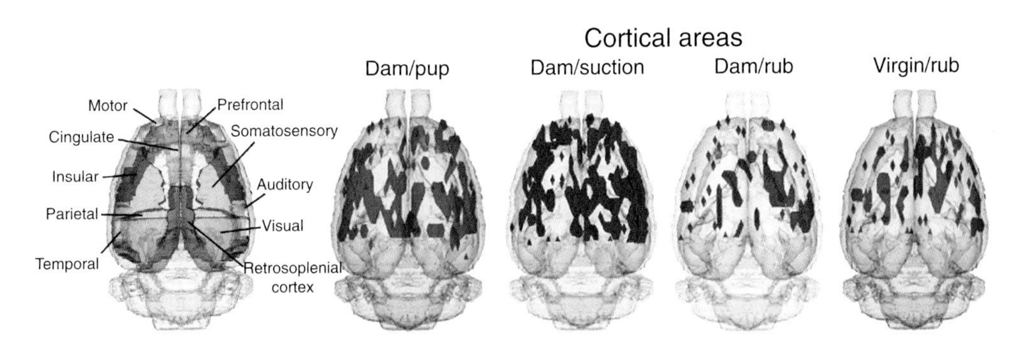

PLATE 4.4 Cortical activation maps showing positive BOLD responses to pup and artificial suckling in PPD 4–8 dams or ventrum rubbing in dams and virgin rats. The 3D glass brains indicate areas of the cortical mantle that showed increases in BOLD responses (red-positive BOLD). The 3D brain in the far left highlights the various major subdivisions of the rat cortex. (*Source*: Adapted from Febo *et al.*, 2008, in press.)

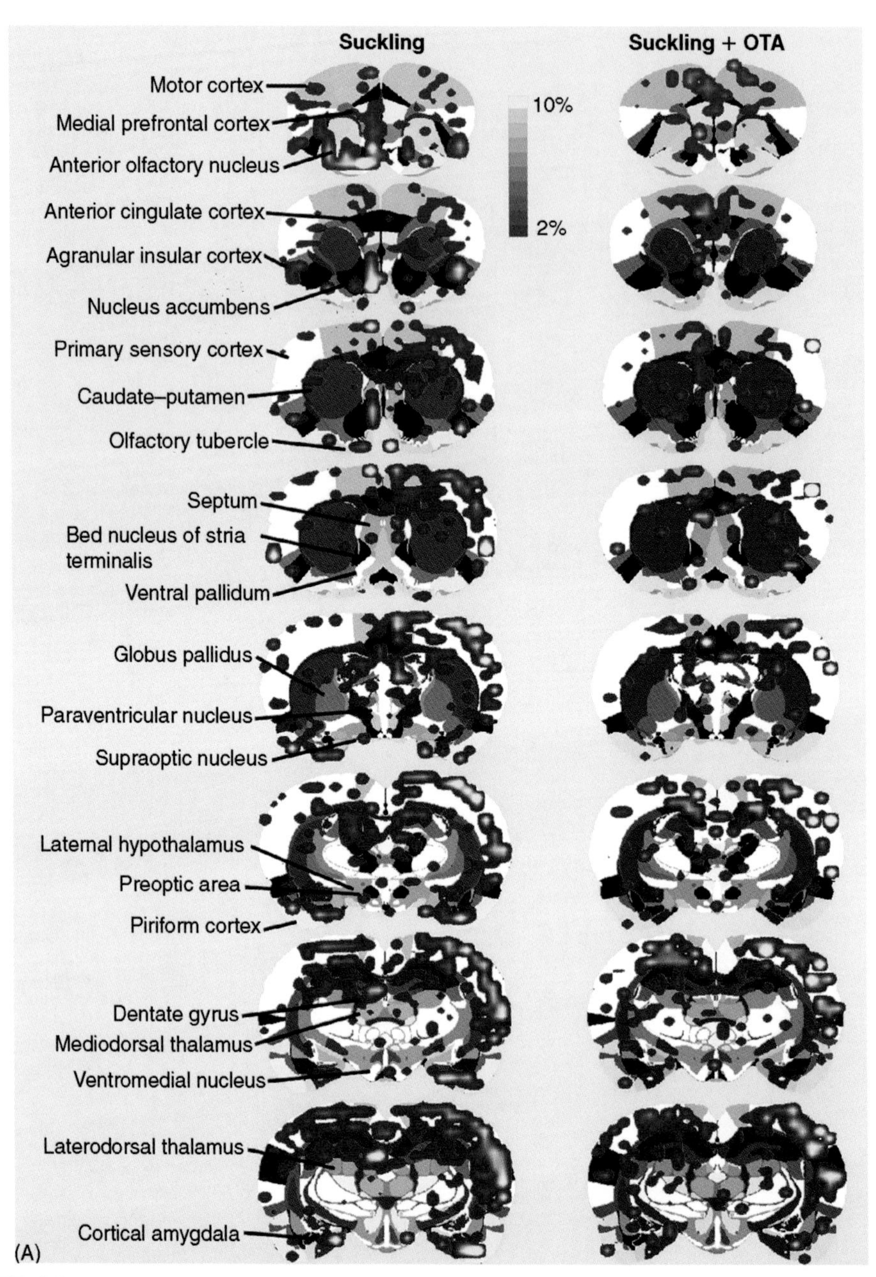

PLATE 4.5 Composite brain maps of positive BOLD activity in response to suckling, suckling after OT receptor blockade (A) and in response to OT administration (B) in PPD 4–8 dams. Colored areas in (A) indicate increases in BOLD activity. BOLD activation maps in (B) were compared to anatomical maps of OT receptor binding. (*Source*: Adapted from Febo *et al.*, 2005b.)

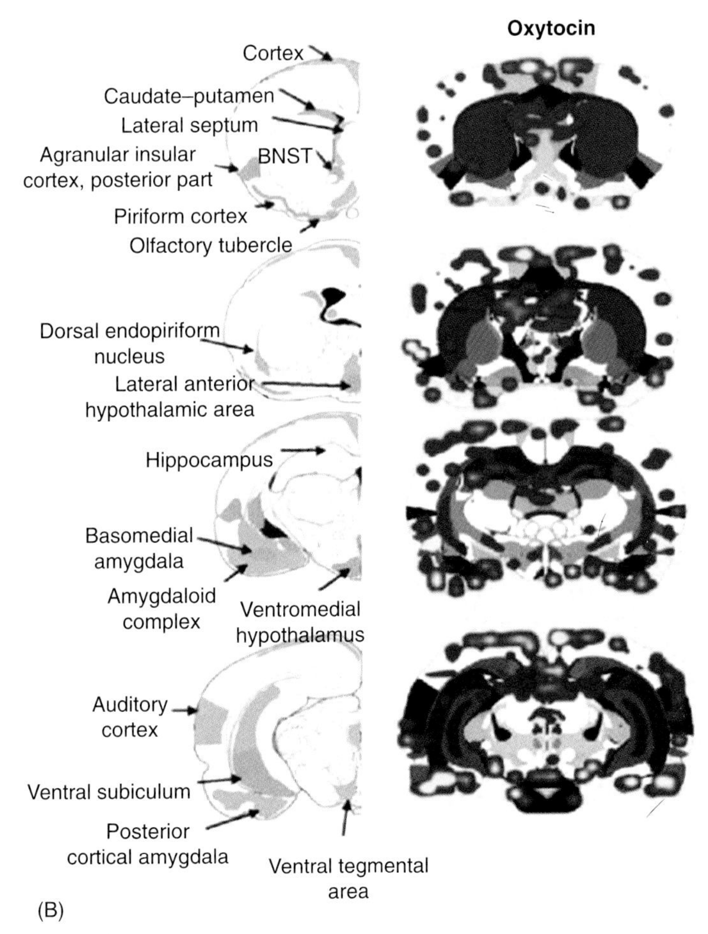

PLATE 4.5 (continued)

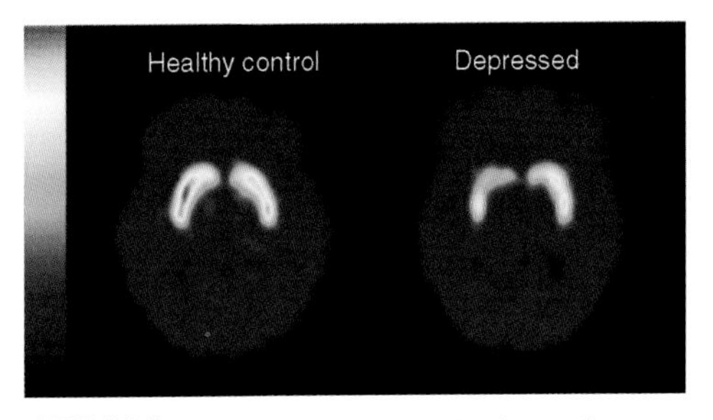

PLATE 12.8 Reduced D2 receptor binding in depression.

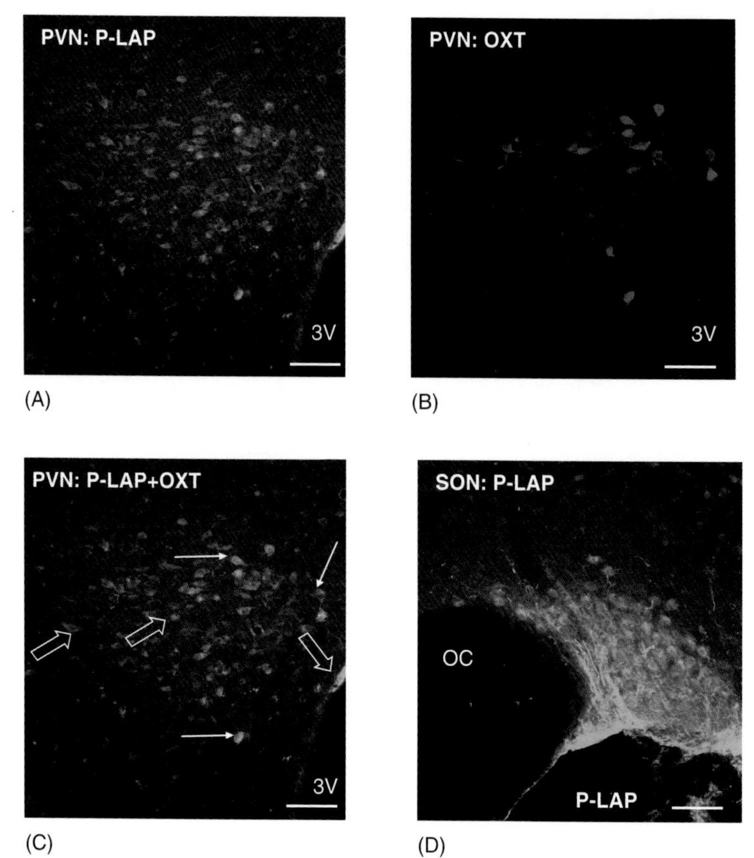

PLATE 14.1 Photomicrographs illustrating P-LAP protein expression in the PVN and SON. Brain sections from a virgin female rat containing PVN (A)–(C) and SON (D) were processed by fluorescence immunocytochemistry for P-LAP using a rabbit anti-rat antibody (1:1,000 for 48 h) and Streptavidin Alexa Fluor 488 secondary antibody (green). PVN sections were also processed with oxytocin antibody (1:5,000 for 48 h) and Alexa Fluor 568 secondary antibody (red; (B) and (C)). All sections were viewed by confocal microscopy. Examples of double-labeled cells are indicated with filled arrows, unfilled arrows illustrate P-LAP-only labeling. Scale bars = 80 μm. OC = optic chiasm, 3V = third ventricle.

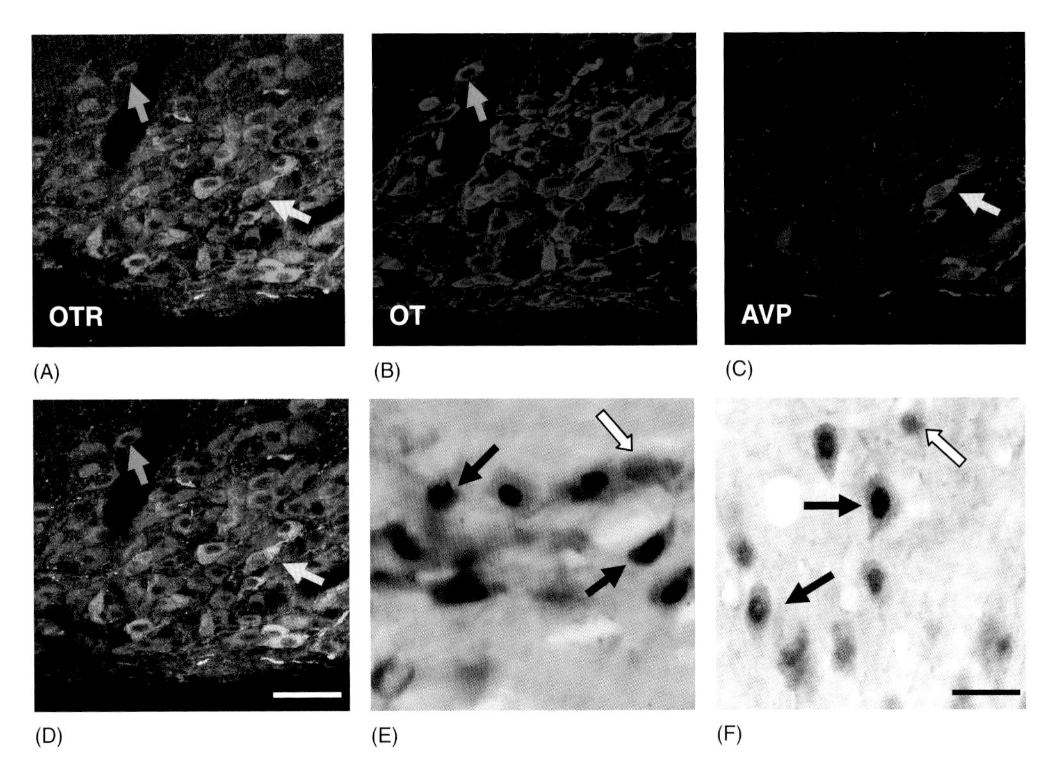

PLATE 14.2 Confocal photomicrograph of oxytocin receptor (OTR) immunolabeling in the SON (A) (1/1,000 dilution, visualized with goat anti-rabbit Alexa Fluor 488). The same section was immunolabeled for oxytocin (B) (antibody gift from Hal Gainer; 1/4,000 dilution, visualized with goat anti-rabbit Alexa Fluor 647) and vasopressin (C) (AVP; antibody raised in guinea pig from Peninsula labs; diluted 1/1,000, visualized with Alexa Fluor 568). Most oxytocin cells and vasopressin cells appear to express OTR. OTR cells that co-localize with oxytocin appear pale blue in (D) (example shown by pale blue arrow). OTR cells co-localizing with vasopressin appear yellow in (D) (example shown by yellow arrow). Photomicrograph examples of activated OTR-expressing neurons in the (E) SON and (F) NTS of parturient rats. Brains were double immunolabeled for OTR (brown) and Fos (black nuclei) expression. Examples of double-labeled cells for Fos and OTR are indicated with filled arrows, unfilled arrows illustrate cells only labeled for OTR. Scale bars all = 50 μm.

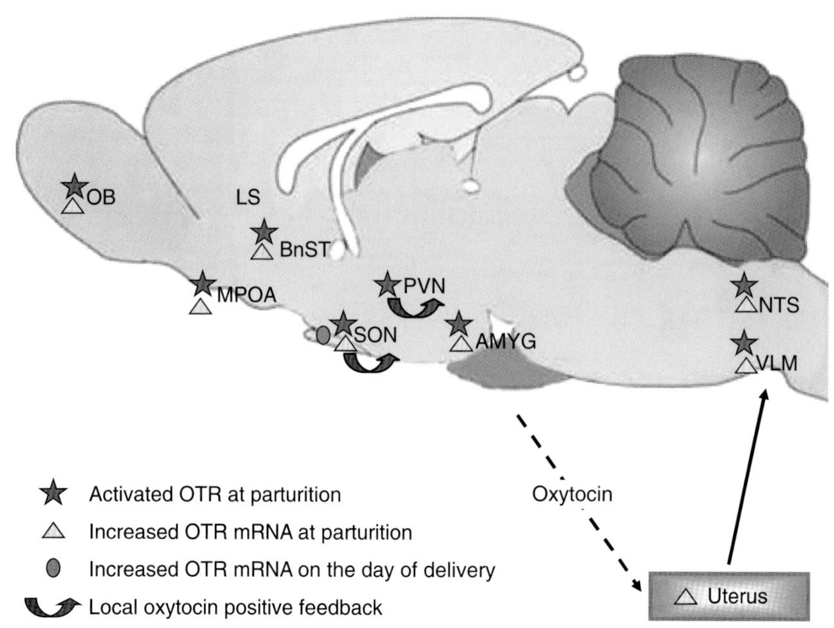

PLATE 14.3 Important sites of oxytocin action at birth in a sagittal view of the rat brain. At parturition oxytocin is released from the posterior pituitary into the blood stream to act on OTR in the myometrium, but it also acts centrally to exert its action in the brain. At parturition OTR cells become activated (identified by OTR cells expressing Fos protein) in brain regions indicated by *stars* (★). OTR mRNA expression increases in the SON on the expected day of delivery (day 22, △) and at parturition (as indicated by ●). Increased OTR mRNA expression is higher in the hypothalamus (SON and PVN) and brainstem (NTS and VLM) but also in brain regions important for maternal behavior such as the olfactory bulbs, MPOA, BnST, and amygdala. Moreover during birth, OTR-expressing neurons in these brain regions express Fos. Oxytocin itself is likely to regulate its own release in the SON and magnocellular region of the PVN which at birth contributes to the oxytocin burst firing patterns so vital for normal delivery. SON, supraoptic nucleus; PVN, paraventricular nucleus; OB, olfactory bulb; LS, lateral septum; BnST, bed nucleus of the stria terminalis; MPOA, medial preoptic area; AMYG, amygdala; NTS, nucleus tractus solitarii; VLM, ventrolateral medulla.

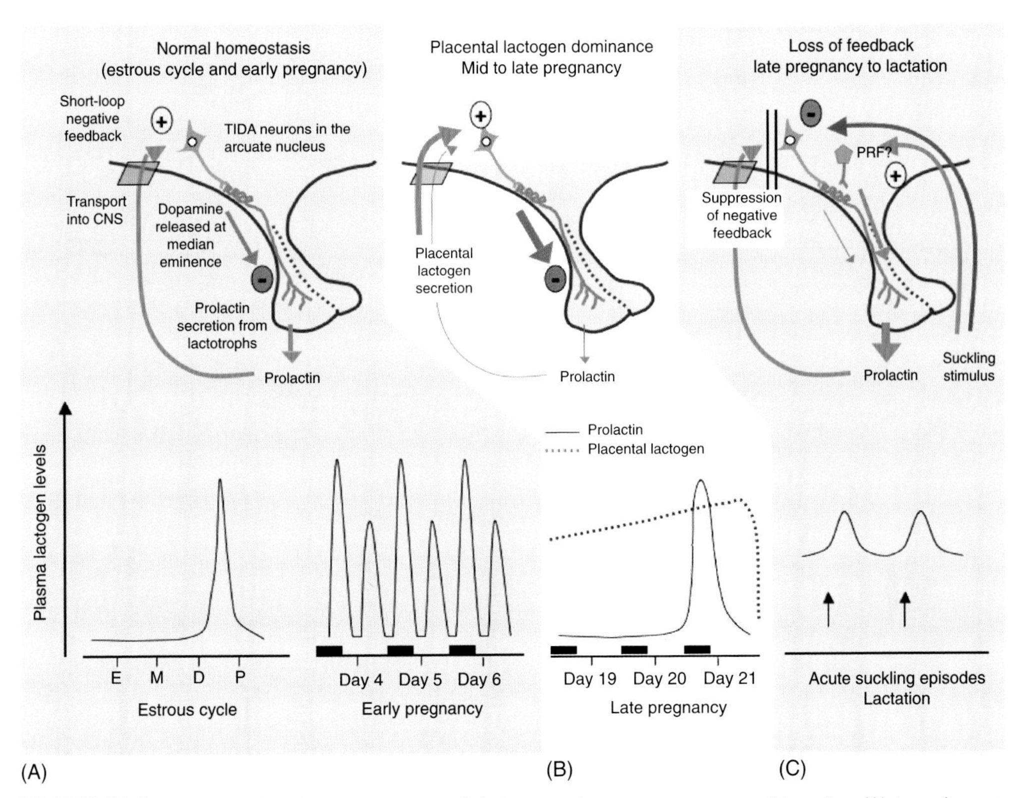

PLATE 16.5 Patterns of prolactin secretion and their control across pregnancy and lactation: (A) in cycling rats and in early pregnancy, (B) in mid to late pregnancy and (C) in late pregnancy and lactation.

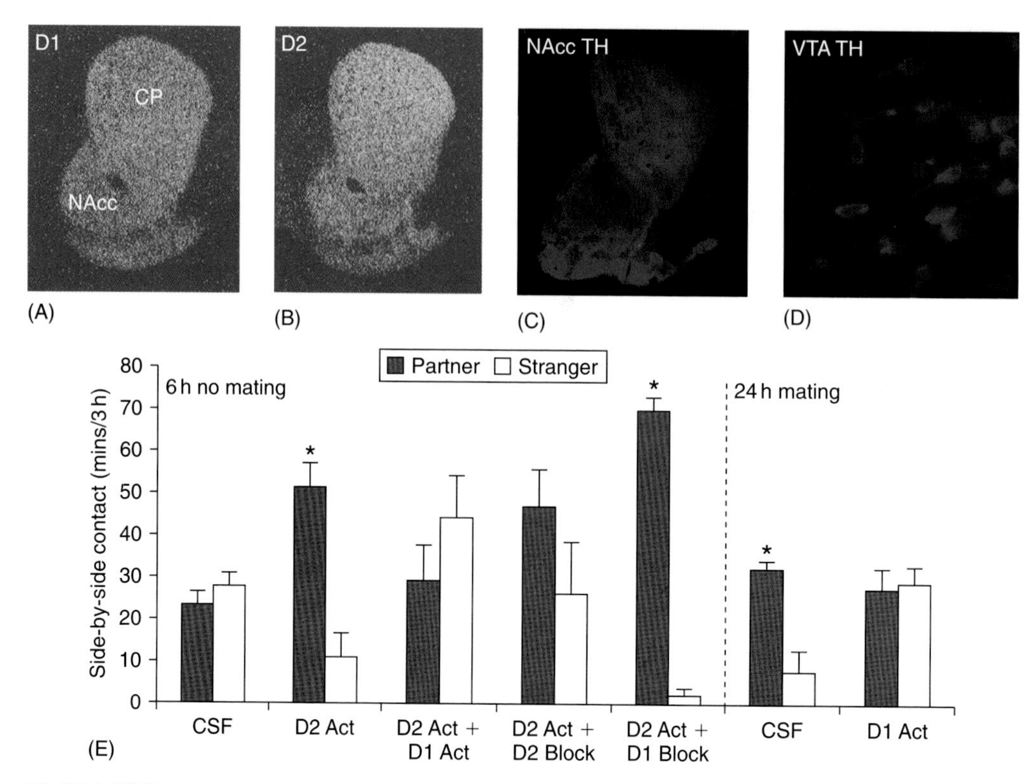

PLATE 22.2 Photo images showing binding for D1-type (A) and D2-type (B) DA receptors in the NAcc and CP as well as fluorescent immunoreactive staining for TH fibers in the NAcc (C) and cell bodies in the VTA (D). (E) Partner preference formation of male prairie voles as a function of NAcc DA pharmacology. Six hours of non-sexual cohabitation does not induce partner preference in control males that received intra-NAcc CSF injections. Activation (Act) of D2 receptors induces partner preference, which can be diminished by activation of D1 receptors or blockade (Block) of D2 receptors. Activation of D2 receptors, together with blockade of D1 receptors, induces robust partner preference formation. Finally, 24 h of mating induces partner preference formation in control males that received intra-NAcc CSF injections, and this behavior is blocked by activation of D1 receptors. $*p < 0.05$. Error bars indicate SEM.

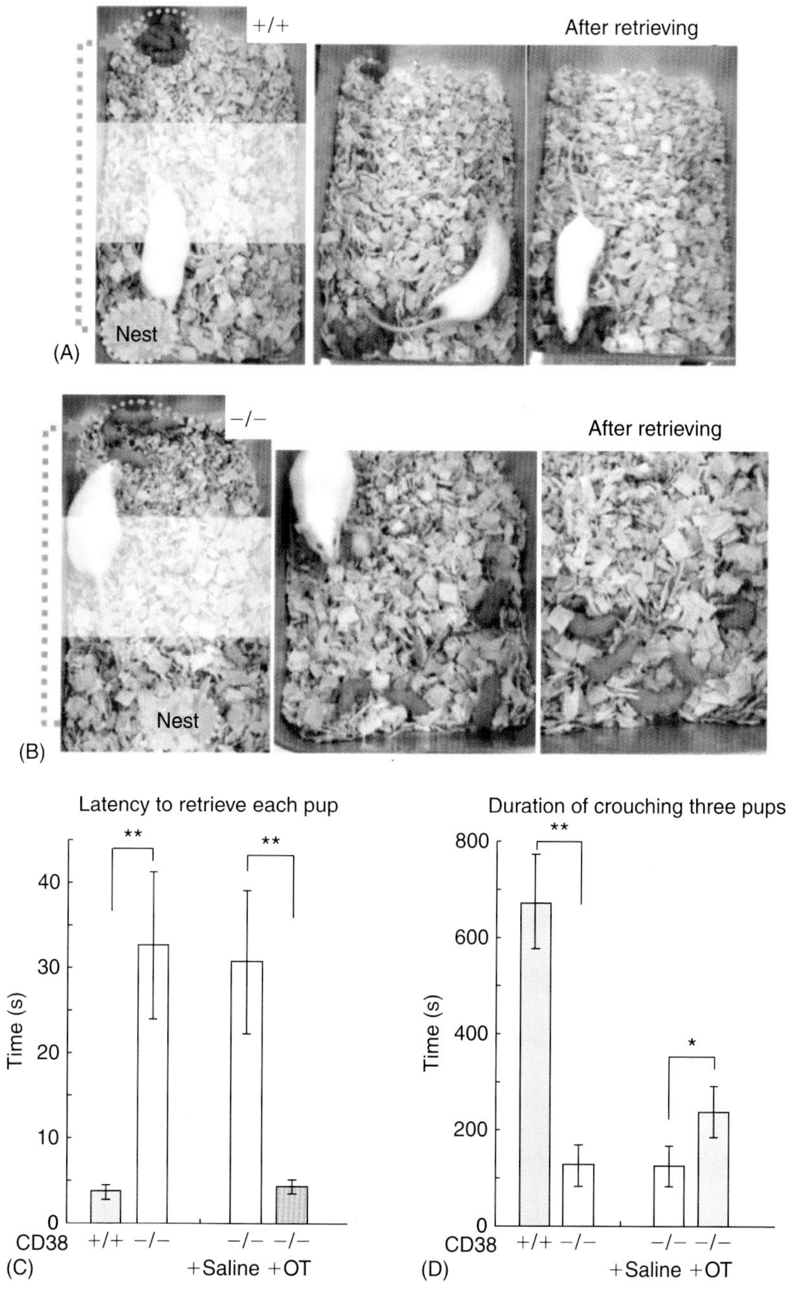

PLATE 23.1 *Retrieving behavior is impaired in CD38$^{-/-}$ mothers.* (A, B) Typical pup retrieval behavior in mothers' home cages. Pups of a CD38$^{+/+}$ (A) or CD38$^{-/-}$ (B) dam were first removed, and five selected pups were placed on the opposite side of the home cage (encircled by dotted green line) from the nest (Nest). Note that the wild-type dam picked pups up in her mouth and transported them to the original nest quickly. The CD38$^{-/-}$ dam picked up pups but dropped and stopped to retrieve furthermore to the original nest. The pups were therefore scattered in the home cage and neglected. (C) Latency to retrieve each pup by mother CD38$^{+/+}$ or CD38$^{-/-}$ mice in separate cages (14), and by mother CD38$^{-/-}$ mice treated with subcutaneous injection of 0.4 mL saline or 3 ng oxytocin (OT)/kg body weight. (D) Time spent crouching over all three pups in the nest by the same set of mothers. $*p < 0.05$; $**p < 0.01$; $N = 6$ (wild type) and $N = 10$ (mutants).

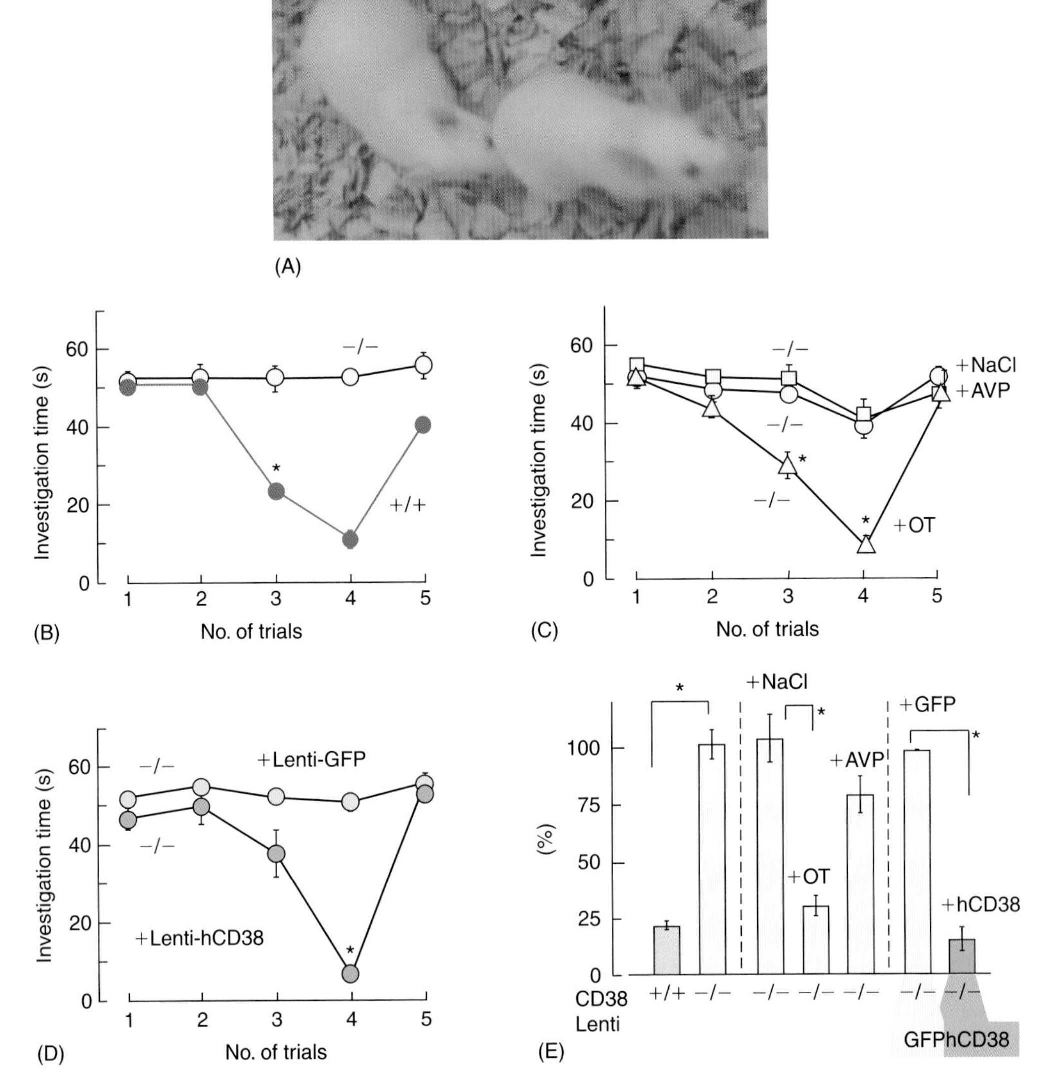

PLATE 23.2 *Social behavior deficit in CD38$^{-/-}$ male mice.* (A) Olfactory investigation in CD38$^{-/-}$ mice used for social recognition test. Social memory by male mice was measured as a difference in ano-genital investigation. (B) Investigation time allocated to investigation of CD38$^{-/-}$ (yellow circle) and CD38$^{+/+}$ (blue circle) male mice. *A significant decrease between each trial compared with the first trial, at $p < 0.01$. (C) Olfactory investigation of CD38$^{-/-}$ males as in (B). A single subcutaneous shot of 0.3 mL saline (NaCl, circle), OT (10 ng/kg; triangle), or vasopressin (AVP; 10 ng/kg; square) to CD38$^{-/-}$ males was administered 10 min before the first pairing. *A significant decrease between each trial compared with the first trial, at $p < 0.01$. (D) The social recognition deficit and its rescue by lenti-GFP (green) and lenti-hCD38 (orange) infection in male CD38$^{-/-}$ mice. *$p < 0.01$ compared to controls. $N = 6$–10 mice in each experiment.

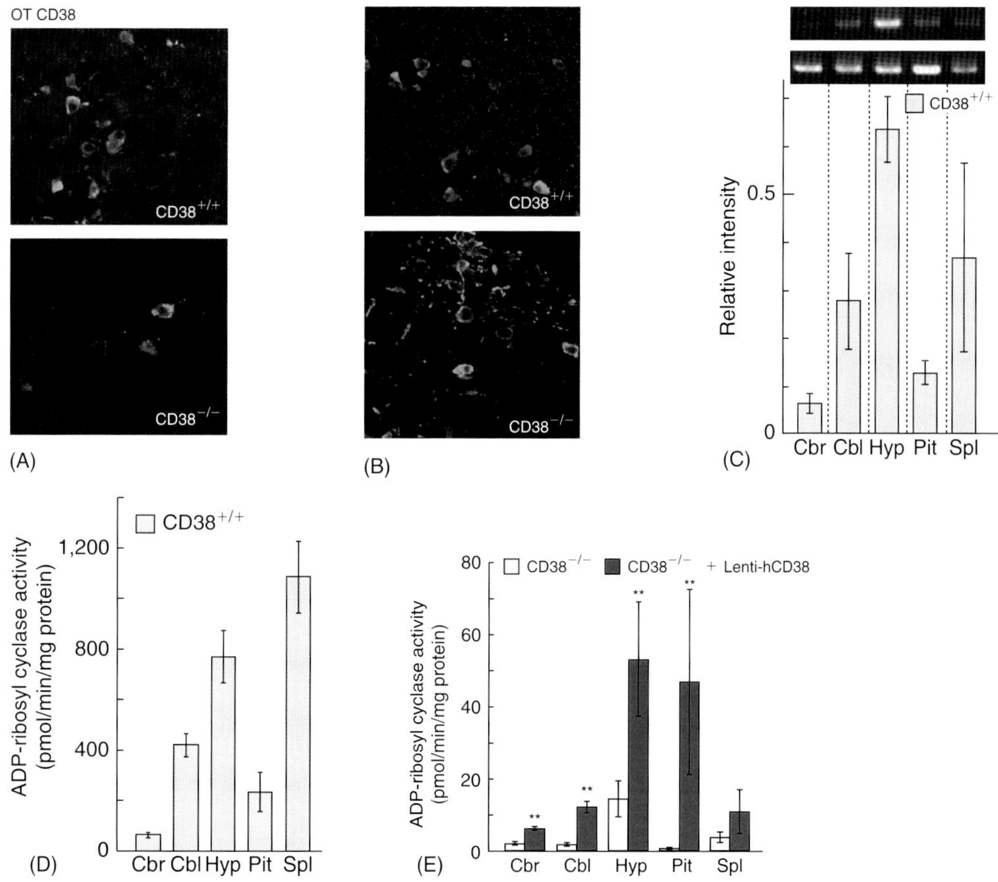

PLATE 23.6 *CD38 expression and ADP-ribosyl cyclase activity.* (A, B) Immunohistochemical analysis of mouse CD38 of hypothalamus in $CD38^{+/+}$ (*upper*) or $CD38^{-/-}$ (*lower*) mice. CD38 stained with anti-mouse CD38 monoclonal antibody (red) and anti-murine OT (A) or AVP (B) antibodies (green) (magnification = $\times$ 630). (C) RT-PCR products of CD38 (*top*) and TATA binding protein (*middle*) of mRNA isolated from indicated tissues in $CD38^{+/+}$ mice and mean relative intensities of the two (bars); $N = 3$. (D) ADP-ribosyl cyclase activity, measured as the rate of cyclic GDP-ribose formation by whole cell homogenates isolated from various tissues of $CD38^{+/+}$ mice; $N = 4$–12. (E) Cyclic GDP-ribose formation by whole cell homogenates isolated from various tissues of $CD38^{-/-}$ mice or $CD38^{-/-}$ mice 2 weeks after receiving lenti-hCD38; $N = 3$–8. Cbr, cerebrum; Cbl, cerebellum; Hyp, hypothalamus; Pit, posterior pituitary; Spl, spleen. $*p < 0.05$; $**p < 0.01$.

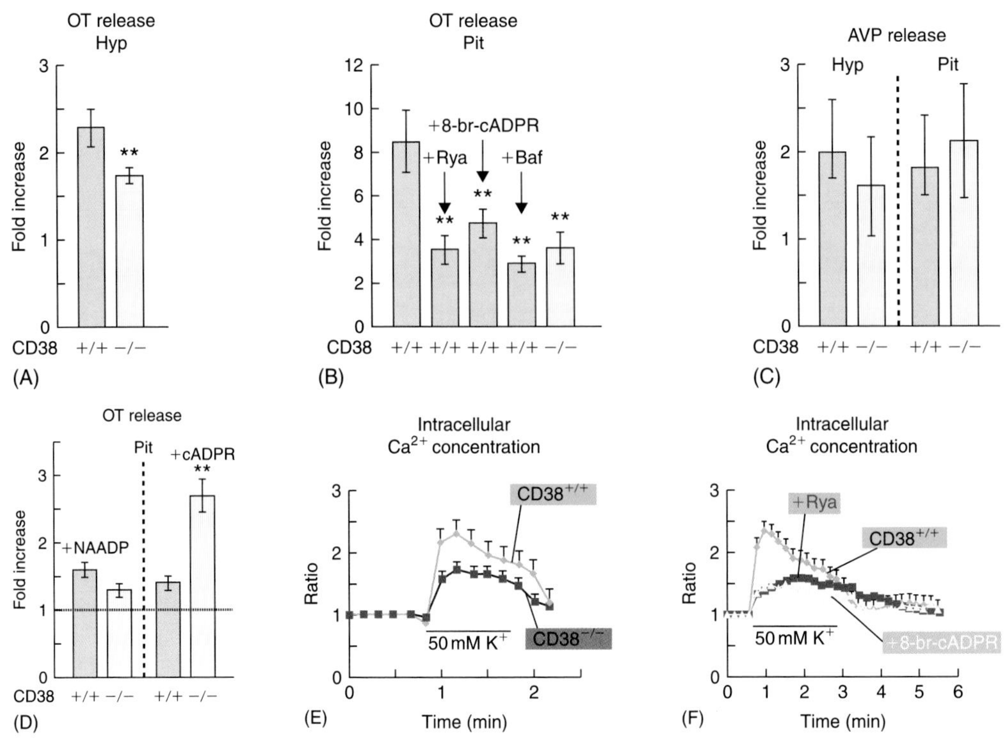

PLATE 23.7 *Oxytocin secretion and calcium signaling.* Enzyme-linked immunoassay of OT (A, B, D) or AVP (C) secreted from hypothalamic cells (Hyp) or nerve endings (Pit) in 35 mm culture dishes stimulated by 70 mM KCl for 2 min. Hypothalamic neurons and nerve endings were acutely dissociated from adult $CD38^{+/+}$ or $CD38^{-/-}$ mice, cultured for 6 h and perfused with Locke solution for at least 45 min. The OT release was measured in the presence or absence of 200 μM ryanodine (Rya), 100 μM 8-bromo-cADPR (8-br-cADPR), or 2 μM bafilomycin (Baf). Values represent the ratio of the concentrations in the perfusate after 70 mM KCl to that in control (5 mM KCl) solution; $N = 4$–12. $*p < 0.05$ and $**p < 0.01$, respectively, compared to control $CD38^{+/+}$. (D) OT release from nerve endings isolated from posterior pituitary of $CD38^{+/+}$ or $CD38^{-/-}$ mice in the presence of 100 nM NAADP or 10 μM cADPR after treatment with 1 μM digitonin to permeabilize the cell membranes for 5 min. Values represent the ratio of OT release in the presence and absence of the effectors; $N = 4$–6. (E) Time course of the increase in $[Ca^{2+}]_i$ stimulated with 50 mM KCl for the indicated period in OT-ergic nerve endings isolated from $CD38^{+/+}$ or $CD38^{-/-}$ mice posterior pituitary; $N = 4$. (F) Time course of the increase in $[Ca^{2+}]_i$ stimulated with 50 mM KCl for the indicated period in the presence or absence of 200 μM ryanodine or 100 μM 8-bromo-cADPR in OT-ergic nerve endings isolated from $CD38^{+/+}$ mice; $N = 3$–4.

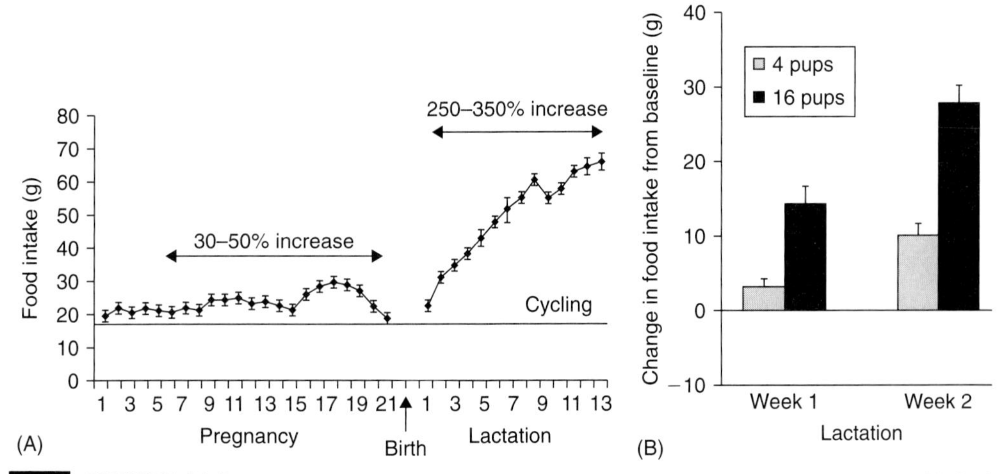

FIGURE 16.2 Food intake in pregnancy and lactation compared to that seen in cycling rats. (A) Daily food intake across pregnancy and lactation in females nursing eight pups. (B) Changes in food intake as a function of number of pups nursed (means ± S.E. are shown).

greater than those seen in cycling controls and then falls dramatically in the 1–2 days prior to parturition (see Figure 16.2). During lactation, there is an even greater increase in food intake to 250–300% of baseline. The elevated food intake of the first week postpartum is associated with larger meal sizes rather than more frequent meals, whereas meal frequency is increased in the second and third weeks postpartum with no further change in meal size (Strubbe & Gorissen, 1980). The change in meal frequency seen in mid and late lactation is associated with more meals being taken toward the end of the light phase (Strubbe & Gorissen, 1980). Food intake in lactation peaks around the time that pups themselves start eating solid food (days 14–16 postpartum), and coincides with peak milk production. Although the amount of food eaten by the dam increases with the number of young nursed, this is not sufficient to maintain similar growth rates across litters of different sizes (Leon & Woodside, 1983).

When given the opportunity to select among diets differing in their macronutrient content, both pregnant and lactating rats show an increase in the proportion of calories eaten as protein (Richter & Barelare, 1938; Leshner *et al.*, 1972; Cohen & Woodside, 1989). Pregnant rats appear to form an aversion to low protein diets (Wilson, 1987) and in the last trimester of pregnancy rats reduce their carbohydrate intake relative to intake of fat and protein. Diet preferences are not limited to macronutrients, there is also evidence for an increase in calcium appetite in lactating rats (Richter & Barelare, 1938; Woodside & Millelire, 1987; Millelire & Woodside, 1989) as well as in salt appetite (Richter & Barelare, 1938; Steinberg & Bindra, 1962).

Food cravings and aversions are commonly reported in pregnant women (Bayley *et al.*, 2002), but it is difficult to relate these to specific macro- or micro-nutrient preferences rather than to sociocultural influences. Empirical studies of pregnant women, however, have reported changes in taste sensitivity as well as in the hedonic value of pure tastants (Bhatia & Puri, 1991; Kolble *et al.*, 2001). Changes in response to the odor of foods are often cited as the underlying cause for food or beverage aversions during pregnancy and Nordin *et al.* (2004) reported that 67% of women experienced increased sensitivity to smell during early pregnancy.

Adaptations of Neural Pathways Controlling Food Intake

A homeostatic model of energy balance might predict that increased food intake of pregnancy and lactation is simply a response to the energy drain represented by the developing young. Although these costs must play some role, the increases in ingestive behavior seen in pregnant and lactating rats do not simply follow

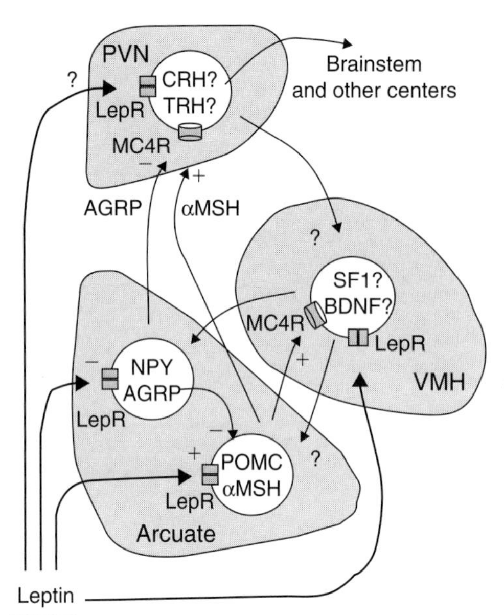

FIGURE 16.3 Hypothalamic pathways through which leptin modulates energy balance. Thick arrows depict direct leptin action on NPY and POMC neurons in the arcuate nucleus as well as undefined cell types in the paraventricular nucleus (PVN) and ventromedial nucleus (VMH). Other arrows depict known and putative interactions between these sites. LepR, Leptin receptor; MC4R, melanocortin 4 receptor: CRH, corticotropin releasing hormone; TRH, thyrotropin releasing hormone; POMC, proopiomelanocortin, αMSH, α-melanocyte stimulating hormone; NPY, neuropeptide Y; AGRP, agouti-related peptide; BDNF, brain-derived neurotrophic factor; SF1, steroidogenic factor 1.

the laws of supply and demand. The initial increases in food intake and body weight in pregnancy, like the reduction in locomotor activity, occur prior to implantation and thus before any direct investment in embryonic development. Moreover, the substantial increase in food intake seen in rats during the second and third weeks of pregnancy occurs against a background of positive energy balance, as reflected in the increase in fat stores and higher circulating levels of the protein hormone leptin (Ladyman & Grattan, 2005).

Leptin is the protein product of the obese gene and is expressed in white adipose tissue from which it enters the circulation and thence gains access to the brain. Leptin acts at multiple hypothalamic sites to suppress food intake (see Figure 16.3). One of the key sites of action is the arcuate nucleus of the hypothalamus. Here leptin stimulates the activity of proopiomelanocortin (POMC) neurons, to produce the anorectic neuropeptide α-melanocyte stimulating hormone (αMSH) and inhibits activity of neurons producing the orexigenic peptides, neuropeptide Y (NPY) and agouti-related peptide (AGRP). Despite elevated levels of circulating leptin during the latter part of pregnancy, however, NPY (Garcia *et al.*, 2003; Rocha *et al.*, 2003) and AGRP (Rocha *et al.*, 2003) mRNA within the arcuate nucleus are not reduced and may even increase while POMC mRNA levels remain relatively stable during pregnancy (Mann *et al.*, 1997; Douglas *et al.*, 2002; Rocha *et al.*, 2003) and the amount of αMSH in the medial basal hypothalamus is decreased (Khorram *et al.*, 1984). The increase in orexigenic and decrease in anorexigenic peptides are consistent with the hormone-induced increase in food intake during pregnancy. The fact that these changes occur when circulating leptin levels are high suggests the presence of a potent orexigenic stimulus and/or a reduction in sensitivity to anorectic signals such as leptin.

In fact, there is an increase in plasma leptin-binding activity in pregnant rats that might contribute to leptin insensitivity at this time by interfering with leptin transport into the brain (Seeber *et al.*, 2002). Moreover, by day 14 of pregnancy, rats do not show the reduction in food intake in response to *centrally* administered leptin that was evident in nonpregnant rats given the same treatment (Ladyman & Grattan, 2004). A particularly interesting aspect of these data is that pregnant rats given an identical leptin treatment on day 7 of pregnancy showed an equivalent response to that of nonpregnant females, suggesting that the hormonal state of the second half of pregnancy or signals from the fetus/placenta are key to inducing leptin resistance.

Despite the apparent leptin insensitivity in arcuate neuronal populations involved in regulation of food intake, there is no evidence of downregulation of leptin receptor mRNA levels in this nucleus during pregnancy (Ladyman & Grattan, 2005), suggesting that there may be changes in leptin-sensitive signal transduction pathways downstream of the receptor. Leptin binds to dimers of receptor molecules, inducing the receptor-associated janus kinase 2 (JAK2) to autophosphorylate and then cross-phosphorylate the other receptor molecule.

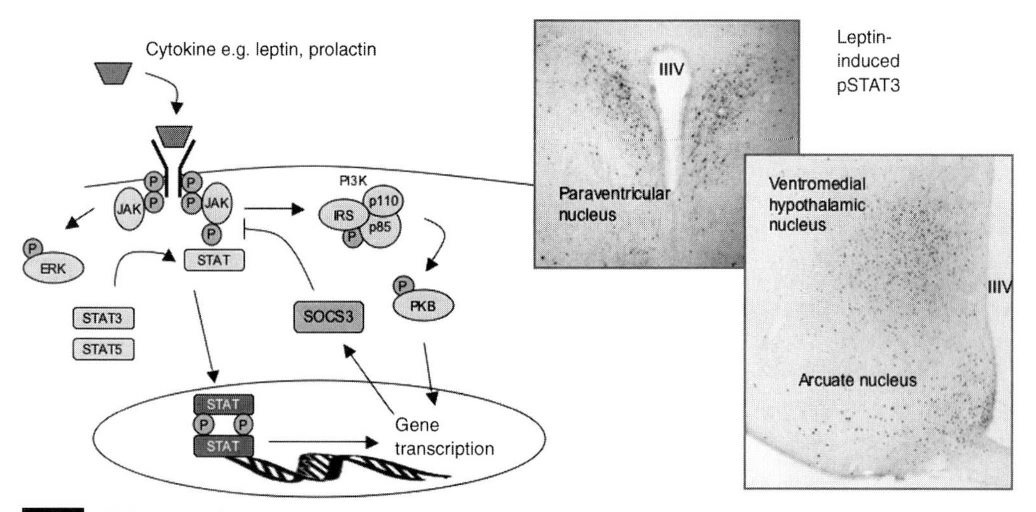

FIGURE 16.4 Cytokine signaling pathways and example of leptin-induced pSTAT3 within the paraventricular, arcuate and ventromedial hypothalamic nuclei. P = phosphorylation of specific proteins.

This leads to activation of a number of different signal transduction pathways (see Figure 16.4). The major pathway involved in leptin's action to suppress food intake is the JAK/signal transducer and activator of transcription (STAT) pathway. The phosphorylated receptor recruits the latent cytoplasmic molecule STAT3, and possibly also STAT5, allowing the STAT molecules to be phosphorylated by JAK2. Phosphorylated STATs then dimerize and translocate to the nucleus, where they bind to regulatory elements on specific genes, influencing gene transcription. Ladyman and Grattan (2004) used the phosphorylation of STAT3 (pSTAT3) as a marker to evaluate leptin responses in the hypothalamus during pregnancy. In the arcuate nucleus, there was a decrease in the total amount of pSTAT3 compared with nonpregnant animals (Ladyman & Grattan, 2005), although the number of cells expressing pSTAT3 did not change (Ladyman & Grattan, 2005). Recent data suggest that leptin action in these arcuate neurons may also involve additional pathways, including the phosphoinositol-3 kinase/protein kinase B (PI3K/PKB) and mitogen-activated protein kinase/extracellular regulated protein kinase (MAPK/ERK) pathways (Bates *et al.*, 2005). Leptin action on NPY and AGRP may be independent of STAT3 (Bates *et al.*, 2003), and appears to require the activity of PI3K/PKB (Morrison *et al.*, 2005). Whether or not these alternative pathways are affected by pregnancy has not been examined.

In contrast to the lack of change in leptin receptor mRNA levels identified in the arcuate nucleus, in the ventromedial hypothalamus (VMH) there was a significant decrease in leptin receptor mRNA levels throughout pregnancy, together with a marked reduction in leptin-induced pSTAT3 (Ladyman & Grattan, 2004, 2005). These data suggest that, during pregnancy, different mechanisms of leptin resistance operate at the hypothalamic sites depicted in Figure 16.3. One of these is seen in leptin-sensitive neurons of the VMH and involves a reduction in leptin receptor expression and leptin-induced activation of STAT3. The other is seen in the arcuate nucleus and occurs in the presence of unchanged leptin receptor levels and relatively normal leptin-induced activation of STAT3.

The energetic costs of lactation are greater than those of pregnancy. Although rats begin lactation in positive energy balance, by the second week postpartum, fat stores are beginning to be depleted despite the dramatic increases in caloric intake. Thus, cues associated with negative energy balance such as a fall in leptin levels are likely to contribute to the increased food intake seen at this time. Nevertheless, there is evidence that in lactation, as in pregnancy, there are modifications in the neural pathways controlling food intake that can be observed, even in the absence of negative energy balance. This is best illustrated by studies in which the energetic demand of lactation is removed by surgical transection of the galactophores, the tubes that carry milk from the mammary gland to the nipple. This model was first described by Cotes and Cross in 1954 (Cotes & Cross, 1954). Rat pups will continue to suckle at the nipples, but no milk delivery is possible and, hence, milk production ceases. By keeping galactophore-cut postpartum rats with healthy foster pups, the suckling stimulus can be maintained without the energetic demands of milk production. Because the hormonal profile of lactation is dependent on suckling stimulation, rather than on milk delivery itself, this model has proved useful in teasing apart the relative amount that cues associated with negative energy balance and those associated with hormonal state contribute to the changes in ingestive behavior and reproductive function seen in lactating rats. Consistent with Cotes and Cross (Cotes & Cross, 1954), we have shown that galactophore-cut rats suckling litters of eight pups eat more and show a greater weight gain than cycling rats (Woodside *et al.*, 2000). Moreover, they do so in spite of increased fat stores and leptin levels, suggesting a level of hypothalamic leptin resistance, similar to that occurring in pregnancy (Woodside *et al.*, 2000). Interestingly, galactophore-cut rats remain sensitive to changes in levels of suckling stimulation, such that just as in intact lactating rats, both food intake and duration of lactational infertility increases with litter size (Woodside & S., 2002).

Other researchers have utilized a model of pup removal followed by acute resuckling to evaluate changes in hypothalamic neuropeptides induced by stimulation from the pups, independent of loss of nutrients through the milk (Brogan *et al.*, 1999). Using this model, Smith and colleagues (Chen & Smith, 2003) have shown that expression of NPY within both the arcuate nucleus and the dorsomedial hypothalamus is increased following acute resuckling, suggesting that signals arising from pup stimulation are sufficient to activate orexigenic pathways.

These data suggest that during both pregnancy and lactation, the neuropeptides controlling energy balance are regulated so that there is an increase in food intake even in the presence of positive energy balance. These changes appear to be mediated by the hormonal changes associated with pregnancy and lactation. Moreover, in the latter part of pregnancy, pregnancy hormones may exert some of their effects by inducing a state of central leptin resistance through a variety of mechanisms.

ROLE OF PROLACTIN IN THE METABOLIC ADAPTATIONS OF PREGNANCY AND LACTATION

Recent data from our laboratories have independently implicated prolactin as a major factor regulating metabolic processes during pregnancy and lactation (Woodside, 2007; Augustine and Grattan, in press). Prolactin is a peptide hormone synthesized and released from lactotrophes in the anterior pituitary. The primary control of prolactin secretion comes from the tuberoinfundibular dopamine neurons (TIDA) in the arcuate nucleus that provide tonic inhibition of prolactin release. Activation of prolactin receptors on these neurons increases dopamine release in the pituitary and provides a short-loop negative feedback on prolactin secretion, and under most conditions, this feedback regulation maintains relatively low levels of prolactin in blood. During pregnancy and lactation, however, there are marked changes in the control of prolactin secretion, resulting in hyperprolactinemia. The changing patterns of prolactin secretion in different stages of the reproductive process, from the reproductive cycle through pregnancy and into lactation, which are illustrated in Figure 16.5, provide a mechanism to differentially influence metabolic processes.

Patterns of Prolactin Secretion in Pregnant and Lactating Rats

Prolactin secretion is relatively low during the estrous cycle in rats, except for an estrogen-induced surge during the afternoon of proestrus (Freeman *et al.*, 1972) with some labs reporting a secondary surge during the afternoon of estrus (Szawka *et al.*, 2005, 2007). The vaginocervical stimulation that female rats experience during mating provides the afferent input of a unique neuroendocrine reflex resulting in the initiation of twice-daily surges of prolactin (Gunnet & Freeman, 1983). Without additional stimulatory input, these prolactin surges continue for over a week, providing the essential luteotropic support to rescue and maintain the corpora lutea and hence establish progesterone secretion in the first half of pregnancy (Smith *et al.*, 1975). Because prolactin surges are induced by vaginocervical stimulation, they do not depend on a fertile mating and can be induced by appropriate artificial stimulation. The prolactin surges induced in the absence of fertilized ova

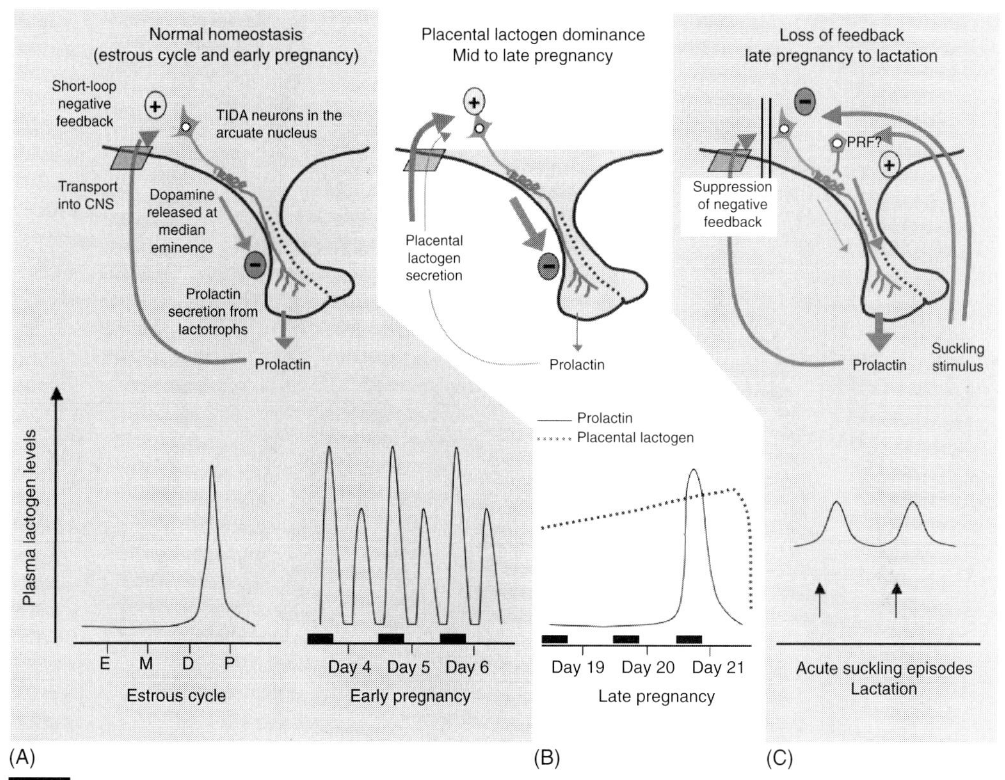

FIGURE 16.5 Patterns of prolactin secretion and their control across pregnancy and lactation: (A) in cycling rats and in early pregnancy, (B) in mid to late pregnancy and (C) in late pregnancy and lactation. (See Color Plate)

result in a state called pseudopregnancy, with hormonal changes that are essentially identical to early pregnancy (Freeman *et al.*, 1974). Prolactin surges persist for about 10 days in pregnant rats and slightly longer (about 12 days) in pseudopregnant rats. In pregnant rats, the prolactin surges are terminated in response to the increased secretion of placental lactogens from the developing conceptus (Tonkowicz & Voogt, 1983; Voogt & de Greef, 1989). The placental lactogens are structurally similar to prolactin and act in an identical manner, activating prolactin receptors on TIDA neurons and thereby inhibiting the secretion of prolactin from the maternal pituitary (Arbogast *et al.*, 1992). Because placental hormone secretion is not under the inhibitory control of the hypothalamus, this placental lactogen secretion essentially bypasses the maternal regulatory feedback pathways to provide high levels of prolactin receptor activation throughout the second half of pregnancy. Thus, at day 10 of pregnancy, there is a change in the pattern of prolactin receptor activation, from a phasic pattern of activation provided by twice-daily prolactin surges during early pregnancy to the chronic stimulation induced by placental lactogen (see Figure 16.5). This chronic secretion of placental lactogen continues until term.

Despite the continued presence of placental lactogens, there is a decrease in TIDA neuronal activity during late pregnancy (Andrews *et al.*, 2001) associated with a surge in pituitary prolactin secretion immediately before parturition (Grattan & Averill, 1990; Andrews *et al.*, 2001). It appears that the TIDA neurons become unresponsive to prolactin or placental lactogen at this time, rendering the short-loop negative feedback system functionally inactive (Grattan & Averill, 1995). There is no evidence that this loss of sensitivity is mediated by a downregulation of prolactin receptors on the TIDA neurons (Kokay & Grattan, 2005). Prolactin-induced activation of the transcription factor, STAT5b, however, which is essential for prolactin-mediated stimulation of dopamine synthesis and release (Grattan *et al.*, 2001), is suppressed during lactation (Anderson *et al.*, 2006a, b). Recent data suggest that this suppression of prolactin-induced STAT signaling might be mediated by an upregulation of a family of endogenous inhibitors of STAT signaling, the suppressors of cytokine signaling (SOCS) proteins (Anderson *et al.*, 2006a, b). This adaptation persists into lactation, meaning that the high levels of pituitary prolactin secretion, which are required for lactation and maternal behavior, can be maintained throughout late pregnancy and lactation unencumbered by a regulatory feedback system.

After parturition, prolactin secretion from the maternal pituitary again becomes the major source of lactogenic hormones in the maternal circulation. Suckling stimulation from the young directly induces maternal prolactin release and thus ensures continued stimulation of milk production. The high levels of suckling-induced prolactin secretion in lactation are associated with a suppression of TIDA neuron activity (Selmanoff & Wise, 1981; Demarest *et al.*, 1983b; Selmanoff & Gregerson, 1985). Despite the hyperprolactinemia induced by suckling, activity of the TIDA neurons remains low (Ben-Jonathan *et al.*, 1980; Demarest *et al.*, 1983a), because the neurons remain less responsive to prolactin (Demarest *et al.*, 1983a; Arbogast & Voogt, 1996). This appears to be due to the continued high expression of SOCS proteins within the arcuate nucleus (Anderson *et al.*, 2006a). In addition to the loss of the inhibitory dopaminergic regulation, there is also some evidence that a stimulatory factor may be involved in the suckling-induced release of prolactin. The identity of this physiological prolactin-releasing factor, if it exists, is unknown at present.

Patterns of Prolactin Secretion in Pregnant and Lactating Women

There has been relatively little research documenting changes in prolactin and placental lactogen secretion during a normal human pregnancy. Both prolactin (Tyson *et al.*, 1972) and human placental lactogen (hPL) (Braunstein *et al.*, 1980) are present in the blood during pregnancy, with increasing amounts detected with advancing gestation. The concurrent rises in both prolactin and placental lactogen suggest that placental lactogen is ineffective at activating the short-loop feedback system to suppress prolactin secretion. It is possible that the TIDA neurons become insensitive to prolactin (and placental lactogen) relatively early in human gestation, although this has not been studied. The human decidua also produces significant amounts of a protein that is identical to human pituitary prolactin. Decidual prolactin is present in amniotic fluid by week 9 of pregnancy increasing to peak levels around mid gestation. Decidual prolactin does not get into the maternal circulation in significant concentrations (Riddick *et al.*, 1979), however, and hence its contribution to lactogenic effects in the mother is minimal. As in other species, suckling induces a rapid increase in plasma prolactin, that is proportional to the duration and intensity of the suckling episode (Diaz *et al.*, 1989).

Thus, in humans and other mammalian species, there are at least three adaptations to ensure high levels of lactogenic hormone activity during pregnancy and lactation. First, the production of placental lactogen bypasses the inhibitory regulation of the maternal pituitary gland, providing constantly elevated levels of lactogenic hormones throughout pregnancy. Second, the maternal TIDA neurons become unresponsive to prolactin (or placental lactogen), allowing TIDA activity to remain low during late pregnancy and lactation despite elevated prolactin. Finally, maternal behavior and the introduction of the suckling stimulus provide the most powerful stimulus to prolactin secretion that is known in mammals. While these elevated levels of lactogenic hormones are clearly required for development of the mammary gland during pregnancy and for milk production during lactation, prolactin and placental lactogen are also able to act in the maternal brain to facilitate the adaptive responses to these changing reproductive states.

Prolactin Receptors are Expressed in the Brain

There are at least two isoforms of the prolactin receptor, a long form and a short form, produced by alternative splicing of the prolactin receptor gene. The different isoforms of the receptor protein have not been specifically identified within brain tissue, but several studies have examined the distribution of the two forms of prolactin receptor mRNA (Bakowska & Morrell, 1997, 2003; Pi & Grattan, 1998a; Augustine *et al.*, 2003;). Both forms are highly

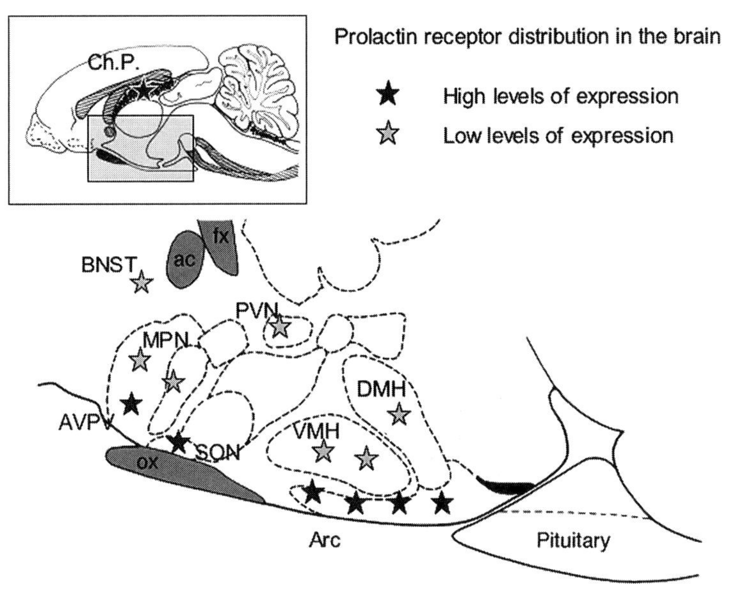

FIGURE 16.6 The distribution of prolactin receptor in the hypothalamus – MPN: medial preoptic nucleus, DMH: dorsomedial hypothalamus, AVPV: anteroventral periventricular area, SON: supraoptic nucleus, BNST: bed nucleus of the stria terminalis.

expressed in the choroid plexus, but the long form is predominant in the hypothalamus. The two forms of the prolactin receptor have identical extracellular portions, and hence are both able to bind prolactin or placental lactogen, but they differ in their ability to activate intracellular signaling pathways. Like the leptin receptor, discussed above, the prolactin receptor is a member of the cytokine family of receptors, and the two hormones have very similar signal transduction pathways. Binding of prolactin to its receptor induces dimerization of the receptor molecules, and then activation of multiple intracellular signaling proteins (Bole-Feysot *et al.*, 1998; Freeman *et al.*, 2000). Ligand interaction with the long form of the receptor leads to phosphorylation of JAK2 and subsequent activation of STAT (Leonard & O'Shea, 1998). Several STAT molecules may be involved in prolactin signal transduction, including STAT1, 3, 5a and 5b (Bole-Feysot *et al.*, 1998). In the TIDA neurons, prolactin action requires phosphorylation of STAT5b (Grattan *et al.*, 2001; Ma *et al.*, 2005). As only the long form of the receptor can activate the JAK–STAT pathway, these data suggest that this isoform is the critical protein mediating prolactin action, at least in these neurons. While the short form of the receptor cannot activate the JAK–STAT pathway, it can mediate some actions of prolactin through the MAPK pathway (Das & Vonderhaar, 1995), and may also play a role in regulating neuronal function.

Central prolactin binding sites were first identified in the choroid plexus (Walsh *et al.*, 1978), and the presence of prolactin receptors in the choroid plexus has since been confirmed in numerous studies (Chiu & Wise, 1994; Pi & Grattan, 1998a, b; Augustine *et al.*, 2003). Prolactin transport into the brain takes place via a saturable mechanism, and it has been suggested that prolactin receptors in the choroid plexus might mediate this transport (Walsh *et al.*, 1987). It is also possible that prolactin is able to enter the brain at circumventricular areas, for example the median eminence, where the blood–brain barrier is compromised. In addition to the ability of prolactin of pituitary origin to enter the brain, several studies over a number of years have shown that the mRNA for prolactin can be detected in the brain (reviewed in Dutt *et al.* (1994)). In particular, prolactin may be expressed in the paraventricular nucleus (PVN) of the hypothalamus, with increased expression during lactation (Torner *et al.*, 2004; also see Chapter 9).

Specific prolactin receptors have also been described in the hypothalamus using binding studies (Muccioli *et al.*, 1991; Muccioli & Di Carlo, 1994), *in vitro* autoradiography (Crumeyrolle-Arias *et al.*, 1993) and immunohistochemistry (Pi & Grattan, 1998b, 1999a).

Similarly, the distribution of prolactin receptor mRNA has been characterized using *in situ* hybridization (Chiu & Wise, 1994; Bakowska & Morrell, 1997, 2003). The distribution of prolactin receptor in the hypothalamus is shown in Figure 16.6. Prolactin receptor expression is observed in arcuate, ventromedial, periventricular, paraventricular, supraoptic and medial preoptic nuclei of the hypothalamus (Pi & Grattan, 1999b). Most of the prolactin receptor containing neurons in the arcuate nucleus are tyrosine hydroxylase positive (Lerant & Freeman, 1998; Kokay *et al.*, 2006), representing the TIDA neurons involved in negative feedback regulation of prolactin secretion. We have also observed prolactin receptor mRNA in enkephalin cells in the arcuate and ventromedial hypothalamic nucleus (Kokay & Grattan, unpublished data). In the supraoptic nuclei, oxytocin neurons express prolactin receptor mRNA, and prolactin acutely inhibits the firing of those neurons (Kokay *et al.*, 2006). Similarly, in the PVN, both magnocellular and parvocellular oxytocin neurons contain prolactin receptor mRNA (Kokay *et al.*, 2006). In other regions, the neurochemical identity of prolactin receptor containing cells is not known. Outside the hypothalamus, prolactin receptor mRNA expression has also been identified in the bed nucleus of the stria terminalis as well as in the medial amygdala.

A dramatic increase in prolactin receptor mRNA expression in the brain during pregnancy has been reported (Sugiyama *et al.*, 1994). This was subsequently shown to primarily reflect changes in the choroid plexus (Sugiyama *et al.*, 1996). Both long and short form prolactin receptors increase in the choroid plexus during pregnancy, and then are present at very high levels throughout lactation, before returning to basal levels following weaning (Augustine *et al.*, 2003). There have been relatively few studies showing the cellular distribution of prolactin receptor protein or mRNA during pregnancy. Using *in situ* hybridization, levels of the long form prolactin receptor mRNA in the medial preoptic nucleus have been demonstrated to increase significantly during late pregnancy compared with early pregnancy (Bakowska & Morrell, 1997). These observations are consistent with changes we have observed during lactation, where levels of prolactin receptor immunoreactivity are significantly elevated compared with nonpregnant females (Pi & Grattan, 1999b). In the arcuate nucleus, levels of prolactin receptor mRNA in TIDA neurons do not change during pregnancy and lactation (Kokay & Grattan, 2005).

PROLACTIN AND THE METABOLIC ADAPTATIONS OF PREGNANCY AND LACTATION

As described above, distinct patterns of prolactin and/or placental lactogen secretion occur during pregnancy and lactation, each under the control of different inputs (see Figure 16.5). Thus, unique patterns of prolactin receptor activation are associated with the different stages of a reproductive episode. It seems likely that these differential patterns of prolactin receptor activation provide a mechanism to convey information about the different phases of reproduction, thus allowing adaptive responses in metabolic pathways and ensuring appropriate changes in behavior and physiology. Evidence suggests that prolactin may act both in peripheral tissues to facilitate metabolic adaptations and in the brain to directly stimulate food intake.

Peripheral Actions of Prolactin on Metabolism

Prolactin acts at the ovary to maintain the corpora lutea and hence the patterns of gonadal steroid levels typical of pregnancy and lactation (Smith *et al.*, 1976). Both estrogen (Nance & Gorski, 1978) and progesterone (Hervey & Hervey, 1967) can exert powerful effects on metabolic pathways and may mediate some of the actions of prolactin. For example, administration of exogenous prolactin to cycling female rats results in the downregulation of liver enzymes that control amino acid oxidation and urea synthesis (Naismith & Walker, 1988), consistent with the hypothesis that prolactin may contribute to the alterations in protein metabolism seen in pregnant and lactating rats. The precise mechanism through which prolactin produces these effects is not clear, although it has been suggested that prolactin-induced increases in serum progesterone concentration are involved (Naismith & Walker, 1988).

Other adaptations reflect a direct action of prolactin on peripheral tissues. The change in islet cell number and sensitivity seen in mid to late pregnancy that helps offset insulin resistance in fat and muscle has been attributed to the effects of placental lactogens acting directly on the islet cells (Brelje *et al.*, 1994). Prolactin receptors are expressed on islet cells (Galsgaard *et al.*, 1999), and prolactin exerts an anti-apoptotic action in this tissue (Fujinaka *et al.*, 2007). Prolactin receptors are also present in the duodenum, and there is evidence from both *in vitro* and *in vivo* studies that prolactin acting on these receptors stimulates the active transport of calcium and hence contributes to the increased calcium absorption seen during pregnancy and lactation (Charoenphandhu *et al.*, 2001, 2006).

Prolactin Stimulates Food Intake

A number of studies have documented the ability of systemic prolactin administration to increase food intake in a variety of species (Buntin & Figge, 1988; Gerardo-Gettens *et al.*, 1989; Noel & Woodside, 1993). In rats, this treatment results not only in increased food intake and body weight gain but also in a suppression of cyclic estrogen release and increases in plasma progesterone concentrations (Smith *et al.*, 1976; Noel & Woodside, 1993). The anorectic effects of estrogen are well established (Nance & Gorski, 1978) and there is also evidence that progesterone treatment stimulates weight gain and fat deposition together with food intake (Hervey & Hervey, 1967). The increase in food intake in female rats following prolactin treatment, therefore, could be mediated through effects on gonadal steroids.

However, intracerebroventricular (icv) administration of prolactin increases food intake without influencing estrous cyclicity whether given by bi-daily injections or via chronic infusions using an osmotic minipump (Noel & Woodside, 1993; Sauve & Woodside, 1996; Naef & Woodside, 2007). Moreover, icv administration of prolactin also increases food intake in ovariectomized rats. Together these data argue for the ability of central prolactin receptor activation to increase food intake independent of steroid hormone levels.

Thus, the phasic prolactin secretion of early pregnancy is likely to contribute to the rapid increase in food intake seen in pregnant rats, both indirectly, by promoting progesterone secretion, and directly, through actions in the hypothalamus. Prolactin receptors are found in many of the nuclei involved in the homeostatic regulation of food intake, including the arcuate, ventromedial and paraventricular nuclei, and hence these nuclei form likely targets for prolactin action during pregnancy and lactation. However, prolactin receptors do not appear to be expressed in the NPY and POMC neurons (Li *et al.*, 1999; Chen & Smith, 2004; Kokay & Grattan, 2005) that regulate appetite. Hence, it seems likely that prolactin acts downstream of the arcuate neurons, such as at the PVN. Consistent with this hypothesis, localized injections of prolactin directly into the PVN stimulate food intake in a dose-dependent manner in female rats (Sauve & Woodside, 2000).

Our recent evidence suggests that one route through which prolactin increases food intake is by inducing a state of leptin resistance. In female rats, chronic icv infusion of prolactin prevents the reduction in body weight or food intake normally seen in response to a central injection of leptin (Naef & Woodside, 2007). This prolactin-induced leptin resistance occurs whether the leptin is administered to satiated rats or after a 24-h fast (Naef & Woodside, 2007). The lack of a behavioral response to leptin was accompanied by a reduction in the ability of leptin to induce Fos expression and pSTAT3 in both the paraventricular and ventromedial hypothalamic nucleus nuclei (Naef & Woodside, 2007). Consistent with these data, we have shown that prolactin infusions induce leptin resistance in pseudopregnant rats (Augustine & Grattan, 2008). Interestingly, the pseudopregnant rats, which are hyperphagic in response to early pregnancy-like changes in prolactin, estrogen and progesterone, remain fully responsive to leptin. It is only after chronic prolactin infusion, mimicking the chronic elevation in placental lactogen secretion characteristic of mid pregnancy, that they become leptin resistant like pregnant animals. Taken together, these data suggest that the prolactin surges during early pregnancy stimulate an initial orexigenic response through direct and indirect actions. The resultant hyperphagia is then maintained, in the face of rising leptin levels and a positive energy balance, by placental lactogen-induced leptin resistance.

We have also shown (Woodside, 2007) that prolactin, acting on central prolactin receptors, contributes to the hyperphagia of lactation. Prolactin is the major lactogenic hormone in the rat, thus examining the effects of prolactin on food intake by pharmacological manipulation in postpartum rats is confounded due to the energetic demands of increased milk production causing an increase in appetite,

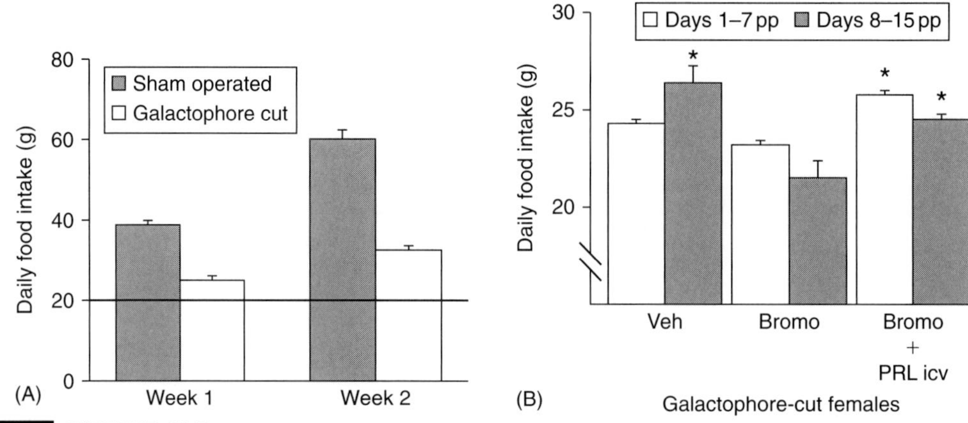

FIGURE 16.7 Food intake in galactophore-cut rats. (A) Food intake of galactophore-cut rats compared with that seen in sham-operated and cycling females. (B) The effects of suppression of prolactin release and its replacement on food intake in galactophore-cut suckled rats.

independent of any direct effects of prolactin. Because galactophore-cut postpartum rats do not deliver milk but have similar hormonal profiles to those of intact lactating rats, we have been able to use the galactophore-cut model to study the specific contributions of suckling-induced prolactin to the hyperphagia seen in galactophore-cut females. When prolactin levels are suppressed in galactophore-cut rats using the dopamine D_2 agonist, bromocriptine, food intake and body weight are decreased. Interestingly, there is also a decrease in plasma leptin concentrations and a small but significant decrease in adiposity, suggesting that in this model, lower leptin levels are actually associated with a *reduction* in food intake. As is shown in Figure 16.7, replacement of prolactin in bromocriptine-treated galactophore-cut rats, by chronic intraventricular infusion, restores food intake without changing the length of lactational infertility suggesting that the orexigenic effects of prolactin acting on prolactin receptors within the brain make a major contribution to the hyperphagia of galactophore-cut rats (Woodside, 2007).

Studies by Smith and her colleagues have also implicated prolactin in stimulating the changes in neuropeptide expression observed in lactating rats (Li *et al.*, 1999). They have reported the presence of prolactin receptor on NPY neurons in the dorsomedial hypothalamus and found that prolactin stimulates NPY expression in this area (Chen & Smith, 2004). They suggest that the upregulation of NPY in this part of the hypothalamus, which is also seen in genetic models of obesity, augments the orexigenic effects of other NPY inputs to the PVN (Li *et al.*, 1999). Whether the prolactin-induced changes in neuropeptides and ingestive behavior in suckled rats are also accompanied by leptin resistance remains to be seen.

CONCLUSION

The studies reviewed here highlight the multiplicity of metabolic adaptations that occur during pregnancy and lactation to enable female mammals to successfully rear their young. Many of these adaptations occur in anticipation of subsequent energetic costs. Thus, pregnancy is associated with adaptive changes in energy balance in preparation for the subsequent metabolic demands of fetal development and lactation. These adaptive changes are to a large extent driven by the hormonal state of the female and in particular by changing patterns of prolactin and placental lactogen release. These peptides act on receptors in the brain to modulate energy balance pathways so as to facilitate the increase in food intake required to provide sufficient nutrients for the young. At least one route through which prolactin receptor activation increases food intake is by inducing a state of central leptin resistance. Because this effect is neither dependent on changes in transport of leptin across the blood–brain barrier nor increases in circulating leptin levels, it represents a novel pathway through which the anorectic effects of leptin are attenuated.

The goal of this chapter has been to specifically elucidate the role of prolactin and placental lactogens in the metabolic changes that accompany pregnancy and lactation. Hyperprolactinemia is also associated with other pathophysiological conditions; it may well be that prolactin receptor activation contributes to the metabolic changes seen in these and a range of physiological conditions that can affect the maternal brain and mental health.

REFERENCES

Abelenda, M., and Puerta, M. L. (1987). Inhibition of diet-induced thermogenesis during pregnancy in the rat. *Pflugers Arch.* 409, 314–317.

Anderson, G. M., Beijer, P., Bang, A. S., Fenwick, M. A., Bunn, S. J., and Grattan, D. R. (2006a). Suppression of prolactin-induced signal transducer and activator of transcription 5b signaling and induction of suppressors of cytokine signaling messenger ribonucleic acid in the hypothalamic arcuate nucleus of the rat during late pregnancy and lactation. *Endocrinology* 147, 4996–5005.

Anderson, S. T., Barclay, J. L., Fanning, K. J., Kusters, D. H., Waters, M. J., and Curlewis, J. D. (2006b). Mechanisms underlying the diminished sensitivity to prolactin negative feedback during lactation: Reduced STAT5 signaling and up-regulation of cytokine-inducible SH2 domain-containing protein (CIS) expression in tuberoinfundibular dopaminergic neurons. *Endocrinology* 147, 1195–1202.

Andrews, Z. B., Kokay, I. C., and Grattan, D. R. (2001). Dissociation of prolactin secretion from tuberoinfundibular dopamine activity in late pregnant rats. *Endocrinology* 142, 2719–2724.

Arbogast, L. A., and Voogt, J. L. (1996). The responsiveness of tuberoinfundibular dopaminergic neurons to prolactin feedback is diminished between early lactation and midlactation in the rat. *Endocrinology* 137, 47–54.

Arbogast, L. A., Soares, M. J., Tomogane, H., and Voogt, J. L. (1992). A trophoblast-specific factor(s) suppresses circulating prolactin levels and increases tyrosine hydroxylase activity in tuberoinfundibular dopaminergic neurons. *Endocrinology* 131, 105–113.

Augustine, R. A., Kokay, I. C., Andrews, Z. B., Ladyman, S. R., and Grattan, D. R. (2003). Quantitation of prolactin receptor mRNA in the maternal rat brain during pregnancy and lactation. *J. Mol. Endocrinol.* 31, 221–232.

Augustine, R. A., and Grattan, D. R. (2008). Induction of central leptin resistance in hyperphagic pseudopregnant rats by chronic prolactin infusion. *Endocrinology* 149, 1049–1055.

Bakowska, J. C., and Morrell, J. I. (1997). Atlas of the neurons that express mRNA for the long form of the prolactin receptor in the forebrain of the female rat. *J. Comp. Neurol.* 386, 161–177.

Bakowska, J. C., and Morrell, J. I. (2003). The distribution of mRNA for the short form of the prolactin receptor in the forebrain of the female rat. *Brain Res. Mol. Brain Res.* 116, 50–58.

Bates, S. H., Stearns, W. H., Dundon, T. A., Schubert, M., Tso, A. W., Wang, Y., Banks, A. S., Lavery, H. J., Haq, A. K., Maratos-Flier, E., Neel, B. G., Schwartz, M. W., and Myers, M. G., Jr. (2003). STAT3 signalling is required for leptin regulation of energy balance but not reproduction. *Nature* 421, 856–859.

Bates, S. H., Kulkarni, R. N., Seifert, M., and Myers, M. G., Jr. (2005). Roles for leptin receptor/STAT3-dependent and independent signals in the regulation of glucose homeostasis. *Cell Metab.* 1, 169–178.

Bayley, T. M., Dye, L., Jones, S., DeBono, M., and Hill, A. J. (2002). Food cravings and aversions during pregnancy: Relationships with nausea and vomiting. *Appetite* 38, 45–51.

Ben-Jonathan, N., Neill, M. A., Arbogast, L. A., Peters, L. L., and Hoefer, M. T. (1980). Dopamine in hypophysial portal blood: Relationship to circulating prolactin in pregnant and lactating rats. *Endocrinology* 106, 690–696.

Bhatia, S., and Puri, R. (1991). Taste sensitivity in pregnancy. *Indian J. Physiol. Pharmacol.* 35, 121–124.

Boass, A., Toverud, S. U., Pike, J. W., and Haussler, M. R. (1981). Calcium metabolism during lactation: Enhanced intestinal calcium absorption in vitamin D-deprived, hypocalcemic rats. *Endocrinology* 109, 900–907.

Bole-Feysot, C., Goffin, V., Edery, M., Binart, N., and Kelly, P. A. (1998). Prolactin (PRL) and its receptor: Actions, signal transduction pathways and phenotypes observed in PRL receptor knockout mice. *Endocr. Rev.* 19, 225–268.

Boyne, R., Fell, B. F., and Robb, I. (1966). The surface area of the intestinal mucosa in the lactating rat. *J. Physiol.* 183, 570–575.

Braunstein, G. D., Rasor, J. L., Engvall, E., and Wade, M. E. (1980). Interrelationships of human chorionic gonadotropin, human placental lactogen, and pregnancy-specific beta 1-glycoprotein throughout normal human gestation. *Am. J. Obstet. Gynecol.* 138, 1205–1213.

Brelje, T. C., Parsons, J. A., and Sorenson, R. L. (1994). Regulation of islet beta-cell proliferation by prolactin in rat islets. *Diabetes* 43, 263–273.

Brogan, R. S., Mitchell, S. E., Trayhurn, P., and Smith, M. S. (1999). Suppression of leptin during lactation: Contribution of the suckling stimulus versus milk production. *Endocrinology* 140, 2621–2627.

Brommage, R. (1989). Measurement of calcium and phosphorus fluxes during lactation in the rat. *J. Nutr.* 119, 428–438.

Brommage, R., Baxter, D. C., and Gierke, L. W. (1990). Vitamin D-independent intestinal calcium and phosphorus absorption during reproduction. *Am. J. Physiol.* 259, G631–G638.

Buntin, J. D., and Figge, G. R. (1988). Prolactin and growth hormone stimulate food intake in ring doves. *Pharmacol. Biochem. Behav.* 31, 533–540.

Burdett, K., and Reek, C. (1979). Adaptation of the small intestine during pregnancy and lactation in the rat. *Biochem. J.* 184, 245–251.

Butte, N. F. (2000). Carbohydrate and lipid metabolism in pregnancy: Normal compared with gestational diabetes mellitus. *Am. J. Clin. Nutr.* 71, 1256S–1261S.

Butte, N. F., and King, J. C. (2005). Energy requirements during pregnancy and lactation. *Public Health Nutr.* 8, 1010–1027.

Butte, N. F., Ellis, K. J., Wong, W. W., Hopkinson, J. M., and Smith, E. O. (2003). Composition of gestational weight gain impacts maternal fat retention and infant birth weight. *Am. J. Obstet. Gynecol.* 189, 1423–1432.

Charoenphandhu, N., Limlomwongse, L., and Krishnamra, N. (2001). Prolactin directly stimulates transcellular active calcium transport in the duodenum of female rats. *Can. J. Physiol. Pharmacol.* 79, 430–438.

Charoenphandhu, N., Limlomwongse, L., and Krishnamra, N. (2006). Prolactin directly enhanced Na+/K+- and Ca2+-ATPase activities in the duodenum of female rats. *Can. J. Physiol. Pharmacol.* 84, 555–563.

Chen, P., and Smith, M. S. (2003). Suckling-induced activation of neuronal input to the dorsomedial nucleus of the hypothalamus: Possible candidates for mediating the activation of DMH neuropeptide Y neurons during lactation. *Brain Res.* 984, 11–20.

Chen, P., and Smith, M. S. (2004). Regulation of hypothalamic neuropeptide Y messenger ribonucleic acid expression during lactation: Role of prolactin. *Endocrinology* 145, 823–829.

Chiu, S., and Wise, P. M. (1994). Prolactin receptor mRNA localization in the hypothalamus by *in situ* hybridization. *J. Neuroendocrinol.* 6, 191–199.

Clarke, P. E., Rousham, E. K., Gross, H., Halligan, A. W., and Bosio, P. (2005). Activity patterns and time allocation during pregnancy: A longitudinal study of British women. *Ann. Hum. Biol.* 32, 247–258.

Cohen, L. R., and Woodside, B. C. (1989). Self-selection of protein during pregnancy and lactation in rats. *Appetite* 12, 119–136.

Cotes, P. M., and Cross, B. A. (1954). The influence of suckling on food intake and growth of adult female rats. *J. Endocrinol.* 10, 363–367.

Cripps, A. W., and Williams, V. J. (1975). The effect of pregnancy and lactation on food intake, gastrointestinal anatomy and the absorptive capacity of the small intestine in the albino rat. *Br. J. Nutr.* 33, 17–32.

Crumeyrolle-Arias, M., Latouche, J., Jammes, H., Djiane, J., Kelly, P. A., Reymond, M. J., and Haour, F. (1993). Prolactin receptors in the rat hypothalamus: Autoradiographic localization and characterization. *Neuroendocrinology* 57, 457–466.

Das, R., and Vonderhaar, B. K. (1995). Transduction of prolactin's (PRL) growth signal through both long and short forms of the PRL receptor. *Mol. Endocrinol.* 9, 1750–1759.

Demarest, K. T., McKay, D. W., Riegle, G. D., and Moore, K. E. (1983a). Biochemical indices of tuberoinfundibular dopaminergic neuronal activity during lactation: A lack of response to prolactin. *Neuroendocrinology* 36, 130–137.

Demarest, K. T., Moore, K. E., and Riegle, G. D. (1983b). Role of prolactin feedback in the semicircadian rhythm of tuberoinfundibular dopaminergic neuronal activity during early pregnancy in the rat. *Neuroendocrinology* 36, 371–375.

Denckla, W. D., and Bilder, G. E. (1975). Investigations into the hypermetabolism of pregnancy, lactation and cold-acclimation. *Life Sci.* 16, 403–414.

Dewey, K. G., Heinig, M. J., and Nommsen, L. A. (1993). Maternal weight-loss patterns during prolonged lactation. *Am. J. Clin. Nutr.* 58, 162–166.

Diaz, S., Seron-Ferre, M., Cardenas, H., Schiappacasse, V., Brandeis, A., and Croxatto, H. B. (1989). Circadian variation of basal plasma prolactin, prolactin response to suckling, and length of amenorrhea in nursing women. *J. Clin. Endocrinol. Metab.* 68, 946–955.

Douglas, A. J., Bicknell, R. J., Leng, G., Russell, J. A., and Meddle, S. L. (2002). Beta-endorphin cells in the arcuate nucleus: Projections to the supraoptic nucleus and changes in expression during pregnancy and parturition. *J. Neuroendocrinol.* 14, 768–777.

Dutt, A., Kaplitt, M. G., Kow, L. M., and Pfaff, D. W. (1994). Prolactin, central nervous system and behavior: A critical review. *Neuroendocrinology* 59, 413–419.

Flint, D. J. (1985). Role of insulin and the insulin receptor in nutrient partitioning between the mammary gland and adipose tissue. *Biochem. Soc. Trans.* 13, 828–829.

Flint, D. J., Sinnett-Smith, P. A., Clegg, R. A., and Vernon, R. G. (1979). Role of insulin receptors in the changing metabolism of adipose tissue during pregnancy and lactation in the rat. *Biochem. J.* 182, 421–427.

Freeman, M. E., Reichert, L. E., Jr., and Neill, J. D. (1972). Regulation of the proestrus surge of prolactin secretion by gonadotropin and estrogens in the rat. *Endocrinology* 90, 232–238.

Freeman, M. E., Smith, M. S., Nazian, S. J., and Neill, J. D. (1974). Ovarian and hypothalamic control of the daily surges of prolactin secretion during pseudopregnancy in the rat. *Endocrinology* 94, 875–882.

Freeman, M. E., Kanyicska, B., Lerant, A., and Nagy, G. (2000). Prolactin: Structure, function, and regulation of secretion. *Physiol. Rev.* 80, 1523–1631.

Frontera, M., Pujol, E., Rodriguez-Cuenca, S., Catala-Niell, A., Roca, P., Garcia-Palmer, F. J., and Gianotti, M. (2005). Rat brown adipose tissue thermogenic features are altered during mid-pregnancy. *Cell. Physiol. Biochem.* 15, 203–210.

Fujinaka, Y., Takane, K., Yamashita, H., and Vasavada, R. C. (2007). Lactogens promote beta cell survival through JAK2/STAT5 activation and BCL-XL upregulation. *J. Biol. Chem.* 282, 30707–30717.

Galsgaard, E. D., Nielsen, J. H., and Moldrup, A. (1999). Regulation of prolactin receptor (PRLR) gene expression in insulin-producing cells. Prolactin and growth hormone activate one of the rat prlr gene promoters via STAT5a and STAT5b. *J. Biol. Chem.* 274, 18686–18692.

Garcia, M. C., Lopez, M., Gualillo, O., Seoane, L. M., Dieguez, C., and Senaris, R. M. (2003). Hypothalamic levels of NPY, MCH, and prepro-orexin mRNA during pregnancy and lactation in the rat: Role of prolactin. *FASEB J.* 17, 1392–1400.

Gerardo-Gettens, T., Moore, B. J., Stern, J. S., and Horwitz, B. A. (1989). Prolactin stimulates food intake in a dose-dependent manner. *Am. J. Physiol.* 256, R276–R280.

Gluckman, P. D., Lillycrop, K. A., Vickers, M. H., Pleasants, A. B., Phillips, E. S., Beedle, A. S., Burdge, G. C., and Hanson, M. A. (2007). Metabolic plasticity during mammalian development is directionally dependent on early nutritional status. *Proc. Natl Acad. Sci. USA* 104, 12796–12800.

Gonzalez, C. G., Alonso, A., Balbin, M., Diaz, F., Fernandez, S., and Patterson, A. M. (2002). Effects of pregnancy on insulin receptor in liver, skeletal muscle and adipose tissue of rats. *Gynecol. Endocrinol.* 16, 193–205.

Grattan, D. R. (2002). Behavioural significance of prolactin signalling in the central nervous system during pregnancy and lactation. *Reproduction* 123, 497–506.

Grattan, D. R., and Averill, R. L. (1990). Effect of ovarian steroids on a nocturnal surge of prolactin secretion that precedes parturition in the rat. *Endocrinology* 126, 1199–1205.

Grattan, D. R., and Averill, R. L. (1995). Absence of short-loop autoregulation of prolactin during late pregnancy in the rat. *Brain Res. Bull.* 36, 413–416.

Grattan, D. R., Xu, J., McLachlan, M. J., Kokay, I. C., Bunn, S. J., Hovey, R. C., and Davey, H. W. (2001). Feedback regulation of PRL secretion is mediated by the transcription factor, signal transducer, and activator of transcription 5b. *Endocrinology* 142, 3935–3940.

Grota, L. J., and Ader, R. (1974). Behavior of lactating rats in a dual-chambered maternity cage. *Horm. Behav.* 5, 275–282.

Gunnet, J. W., and Freeman, M. E. (1983). The mating-induced release of prolactin: A unique neuroendocrine response. *Endocr. Rev.* 4, 44–61.

Haiek, L. N., Kramer, M. S., Ciampi, A., and Tirado, R. (2001). Postpartum weight loss and infant feeding. *J. Am. Board Fam. Pract.* 14, 85–94.

Hay, W. W., Jr., Sparks, J. W., Wilkening, R. B., Battaglia, F. C., and Meschia, G. (1984). Fetal glucose uptake and utilization as functions of maternal glucose concentration. *Am. J. Physiol.* 246, E237–E242.

Hervey, E., and Hervey, G. R. (1967). The effects of progesterone on body weight and composition in the rat. *J. Endocrinol.* 37, 361–381.

Imai-Matsumara, K., Matsumura, K., Morimoto, A., and Nakayama, T. (1990). Suppression of cold-induced thermogenesis in full-term pregnant rats. *J. Physiol.* 425, 271–281.

Isler, D., Trayhurn, P., and Lunn, P. G. (1984). Brown adipose tissue metabolism in lactating rats: The effect of litter size. *Ann. Nutr. Metab.* 28, 101–109.

Khorram, O., DePalatis, L. R., and McCann, S. M. (1984). Changes in hypothalamic and pituitary content of immunoreactive alpha-melanocyte-stimulating hormone during the gestational and postpartum periods in the rat. *Proc. Soc. Exp. Biol. Med.* 177, 318–326.

Kokay, I. C., and Grattan, D. R. (2005). Expression of mRNA for prolactin receptor (long form) in dopamine and pro-opiomelanocortin neurones in the arcuate nucleus of non-pregnant and lactating rats. *J. Neuroendocrinol.* 17, 827–835.

Kokay, I. C., Bull, P. M., Davis, R. L., Ludwig, M., and Grattan, D. R. (2006). Expression of the long form of the prolactin receptor in magnocellular oxytocin neurons is associated with specific prolactin regulation of oxytocin neurons. *Am. J. Physiol. Regul. Integr. Comp. Physiol.* 290, R1216–R1225.

Kolble, N., Hummel, T., von Mering, R., Huch, A., and Huch, R. (2001). Gustatory and olfactory function in the first trimester of pregnancy. *Eur. J. Obstet. Gynecol. Reprod. Biol.* 99, 179–183.

Krebs, N. F., Reidinger, C. J., Robertson, A. D., and Brenner, M. (1997). Bone mineral density changes during lactation: Maternal, dietary, and biochemical correlates. *Am. J. Clin. Nutr.* 65, 1738–1746.

Ladyman, S. R., and Grattan, D. R. (2004). Region-specific reduction in leptin-induced phosphorylation of signal transducer and activator of transcription-3 (STAT3) in the rat hypothalamus is associated with leptin resistance during pregnancy. *Endocrinology* 145, 3704–3711.

Ladyman, S. R., and Grattan, D. R. (2005). Suppression of leptin receptor messenger ribonucleic acid and leptin responsiveness in the ventromedial nucleus of the hypothalamus during pregnancy in the rat. *Endocrinology* 146, 3868–3874.

Leon, M., and Woodside, B. (1983). Energetic limits on reproduction: Maternal food intake. *Physiol. Behav.* 30, 945–957.

Leonard, W. J., and O'Shea, J. J. (1998). Jaks and STATs: Biological implications. *Ann. Rev. Immunol.* 16, 293–322.

Lerant, A., and Freeman, M. E. (1998). Ovarian steroids differentially regulate the expression of PRL-R in neuroendocrine dopaminergic neuron populations: A double label confocal microscopic study. *Brain Res.* 802, 141–154.

Leshner, A. I., Siegel, H. I., and Collier, G. (1972). Dietary self-selection by pregnant and lactating rats. *Physiol. Behav.* 8, 151–154.

Li, C., Chen, P., and Smith, M. S. (1999). Neuropeptide Y and tuberoinfundibular dopamine activities are altered during lactation: Role of prolactin. *Endocrinology* 140, 118–123.

Lof, M., and Forsum, E. (2006). Activity pattern and energy expenditure due to physical activity before and during pregnancy in healthy Swedish women. *Br. J. Nutr.* 95, 296–302.

Lof, M., Olausson, H., Bostrom, K., Janerot-Sjoberg, B., Sohlstrom, A., and Forsum, E. (2005). Changes in basal metabolic rate during pregnancy in relation to changes in body weight and composition, cardiac output, insulin-like growth factor I, and thyroid hormones and in relation to fetal growth. *Am. J. Clin. Nutr.* 81, 678–685.

Ma, F. Y., Anderson, G. M., Gunn, T. D., Goffin, V., Grattan, D. R., and Bunn, S. J. (2005). Prolactin specifically activates STAT5b in neuroendocrine dopaminergic neurons. *Endocrinology* 146, 5112–5119.

Mann, P. E., Rubin, R. S., and Bridges, R. S. (1997). Differential proopiomelanocortin gene expression in the medial basal hypothalamus of rats during pregnancy and lactation. *Brain Res. Mol. Brain Res.* 46, 9–16.

Millelire, L., and Woodside, B. (1989). Factors influencing the self-selection of calcium in lactating rats. *Physiol. Behav.* 46, 429–434.

Miller, S. C., Shupe, J. G., Redd, E. H., Miller, M. A., and Omura, T. H. (1986). Changes in bone mineral and bone formation rates during pregnancy and lactation in rats. *Bone* 7, 283–287.

Mojtahedi, M., de Groot, L. C., Boekholt, H. A., and van Raaij, J. M. (2002). Nitrogen balance of healthy Dutch women before and during pregnancy. *Am. J. Clin. Nutr.* 75, 1078–1083.

Morrison, C. D., Morton, G. J., Niswender, K. D., Gelling, R. W., and Schwartz, M. W. (2005). Leptin inhibits hypothalamic Npy and Agrp gene expression via a mechanism that requires phosphatidylinositol 3-OH-kinase signaling. *Am. J. Physiol. Endocrinol. Metab.* 289, E1051–E1057.

Muccioli, G., and Di Carlo, R. (1994). Modulation of prolactin receptors in the rat hypothalamus in response to changes in serum concentration of endogenous prolactin or to ovine prolactin administration. *Brain Res.* 663, 244–250.

Muccioli, G., Ghe, C., and Di Carlo, R. (1991). Distribution and characterization of prolactin binding sites in the male and female rat brain: Effects of hypophysectomy and ovariectomy. *Neuroendocrinology* 53, 47–53.

Naef, L., and Woodside, B. (2007). Prolactin/leptin interactions in the control of food intake in rats. *Endocrinology* 148, 5977–5983.

Naismith, D. J. (1969). The foetus as a parasite. *Proc. Nutr. Soc.* 28, 25–31.

Naismith, D. J., and Morgan, B. L. (1976). The biphasic nature of protein metabolism during pregnancy in the rat. *Br. J. Nutr.* 36, 563–566.

Naismith, D. J., and Emery, P. W. (1988). Excretion of 3-methylhistidine by pregnant women: Evidence for a biphasic system of protein metabolism in human pregnancy. *Eur. J. Clin. Nutr.* 42, 483–489.

Naismith, D. J., and Walker, S. P. (1988). Effect of prolactin on amino acid catabolism in the female rat. *Ann. Nutr. Metab.* 32, 305–311.

Naismith, D. J., Richardson, D. P., and Pritchard, A. E. (1982). The utilization of protein and energy during lactation in the rat, with particular regard to the use of fat accumulated in pregnancy. *Br. J. Nutr.* 48, 433–441.

Nance, D. M., and Gorski, R. A. (1978). Similar effects of estrogen and lateral hypothalamic lesions on feeding behavior of female rats. *Brain Res. Bull.* 3, 549–553.

Noel, M. B., and Woodside, B. (1993). Effects of systemic and central prolactin injections on food intake, weight gain, and estrous cyclicity in female rats. *Physiol. Behav.* 54, 151–154.

Nordin, S., Broman, D. A., Olofsson, J. K., and Wulff, M. (2004). A longitudinal descriptive study of self-reported abnormal smell and taste perception in pregnant women. *Chem. Senses* 29, 391–402.

Pi, X. J., and Grattan, D. R. (1998a). Differential expression of the two forms of prolactin receptor mRNA within microdissected hypothalamic nuclei of the rat. *Brain Res. Mol. Brain Res.* 59, 1–12.

Pi, X. J., and Grattan, D. R. (1998b). Distribution of prolactin receptor immunoreactivity in the brain of estrogen-treated, ovariectomized rats. *J. Comp. Neurol.* 394, 462–474.

Pi, X. J., and Grattan, D. R. (1999a). Expression of prolactin receptor mRNA is increased in the preoptic area of lactating rats. *Endocrine* 11, 91–98.

Pi, X. J., and Grattan, D. R. (1999b). Increased expression of both short and long forms of prolactin receptor mRNA in hypothalamic nuclei of lactating rats. *J. Mol. Endocrinol.* 23, 13–22.

Ramos, M. P., Crespo-Solans, M. D., del Campo, S., Cacho, J., and Herrera, E. (2003). Fat accumulation in the rat during early pregnancy is modulated by enhanced insulin responsiveness. *Am. J. Physiol. Endocrinol. Metab.* 285, E318–E328.

Richter, C. P., and Barelare, B. (1938). Nutritional requirements of pregnant and lactating rats studied by the self-selection method. *Endocrinology* 23, 15–24.

Riddick, D. H., Luciano, A. A., Kusmik, W. F., and Maslar, I. A. (1979). Evidence for a nonpituitary source of amniotic fluid prolactin. *Fertil. Steril.* 31, 35–39.

Robinson, C. J., Spanos, E., James, M. F., Pike, J. W., Haussler, M. R., Makeen, A. M., Hillyard, C. J., and MacIntyre, I. (1982). Role of prolactin in vitamin D metabolism and calcium absorption during lactation in the rat. *J. Endocrinol.* 94, 443–453.

Rocha, M., Bing, C., Williams, G., and Puerta, M. (2003). Pregnancy-induced hyperphagia is associated with increased gene expression of hypothalamic agouti-related peptide in rats. *Regul. Pept.* 114, 159–165.

Rolls, B. A. (1975). Dipeptidase activity in the small intestinal mucosa during pregnancy and lactation in the rat. *Br. J. Nutr.* 33, 1–9.

Russell, J. A., Douglas, A. J., Windle, R. J., and Ingram, C. D. (2001). The maternal brain: Neurobiological and neuroendocrine adaptation and disorders in pregnancy and postpartum. In *Progress in Brain Research*, Vol. 133. Elsevier, Amsterdam.

Sauve, D., and Woodside, B. (1996). The effect of central administration of prolactin on food intake in virgin female rats is dose-dependent, occurs in the absence of ovarian hormones and the latency to onset varies with feeding regimen. *Brain Res.* 729, 75–81.

Sauve, D., and Woodside, B. (2000). Neuroanatomical specificity of prolactin-induced hyperphagia in virgin female rats. *Brain Res.* 868, 306–314.

Seeber, R. M., Smith, J. T., and Waddell, B. J. (2002). Plasma leptin-binding activity and hypothalamic leptin receptor expression during pregnancy and lactation in the rat. *Biol. Reprod.* 66, 1762–1767.

Selmanoff, M., and Wise, P. M. (1981). Decreased dopamine turnover in the median eminence in response to suckling in the lactating rat. *Brain Res.* 212, 101–115.

Selmanoff, M., and Gregerson, K. A. (1985). Suckling decreases dopamine turnover in both medial and lateral aspects of the median eminence in the rat. *Neurosci. Lett.* 57, 25–30.

Slonaker, J. R. (1924a). The effect of copulation, pregnancy, pseudopregnancy and lactation on the voluntary activity and food consumption of the albino rat. *Am. J. Physiol.* 71, 362–394.

Slonaker, J. R. (1924b). The effect of pubescence, oestruation and menopause on the voluntary activity in the albino rat. *Am. J. Physiol.* 68, 294–315.

Smith, M. S., Freeman, M. E., and Neill, J. D. (1975). The control of progesterone secretion during the estrous cycle and early pseudopregnancy in the rat: Prolactin, gonadotropin and steroid levels associated with rescue of the corpus luteum of pseudopregnancy. *Endocrinology* 96, 219–226.

Smith, M. S., McLean, B. K., and Neill, J. D. (1976). Prolactin: The initial luteotropic stimulus of pseudopregnancy in the rat. *Endocrinology* 98, 1370–1377.

Sorenson, R. L., and Brelje, T. C. (1997). Adaptation of islets of Langerhans to pregnancy: Beta-cell growth, enhanced insulin secretion and the role of lactogenic hormones. *Horm. Metab. Res.* 29, 301–307.

Steinberg, J., and Bindra, D. (1962). Effects of pregnancy and salt-intake on genital licking. *J. Comp. Physiol. Psychol.* 55, 103–106.

Strubbe, J. H., and Gorissen, J. (1980). Meal patterning in the lactating rat. *Physiol. Behav.* 25, 775–777.

Sugiyama, T., Minoura, H., Kawabe, N., Tanaka, M., and Nakashima, K. (1994). Preferential expression of long form prolactin receptor mRNA in the rat brain during the oestrous cycle, pregnancy and lactation: Hormones involved in its gene expression. *J. Endocrinol.* 141, 325–333.

Sugiyama, T., Minoura, H., Toyoda, N., Sakaguchi, K., Tanaka, M., Sudo, S., and Nakashima, K. (1996). Pup contact induces the expression of long form prolactin receptor mRNA in the brain of female rats: Effects of ovariectomy and hypophysectomy on receptor gene expression. *J. Endocrinol.* 149, 335–340.

Szawka, R. E., Helena, C. V., Rodovalho, G. V., Monteiro, P. M., Franci, C. R., and Anselmo-Franci, J. A. (2005). Locus coeruleus norepinephrine regulates the surge of prolactin during oestrus. *J. Neuroendocrinol.* 17, 639–648.

Szawka, R. E., Rodovalho, G. V., Helena, C. V., Franci, C. R., and Anselmo-Franci, J. A. (2007). Prolactin secretory surge during estrus coincides with increased dopamine activity in the hypothalamus and preoptic area and is not altered by ovariectomy on proestrus. *Brain Res. Bull.* 73, 127–134.

Tonkowicz, P. A., and Voogt, J. L. (1983). Termination of prolactin surges with development of placental lactogen secretion in the pregnant rat. *Endocrinology* 113, 1314–1318.

Torner, L., Maloumby, R., Nava, G., Aranda, J., Clapp, C., and Neumann, I. D. (2004). *In vivo* release and gene upregulation of brain prolactin in response to physiological stimuli. *Eur. J. Neurosci.* 19, 1601–1608.

Trayhurn, P. (1989). Thermogenesis and the energetics of pregnancy and lactation. *Can. J. Physiol. Pharmacol.* 67, 370–375.

Tyson, J. E., Hwang, P., Guyda, H., and Friesen, H. G. (1972). Studies of prolactin secretion in human pregnancy. *Am. J. Obstet. Gynecol.* 113, 14–20.

Vernon, R. G., and Pond, C. M. (1997). Adaptations of maternal adipose tissue to lactation. *J. Mammary Gland Biol. Neoplasia* 2, 231–241.

Vickers, M. H., Breier, B. H., Cutfield, W. S., Hofman, P. L., and Gluckman, P. D. (2000). Fetal origins of hyperphagia, obesity, and hypertension and postnatal amplification by hypercaloric nutrition. *Am. J. Physiol. Endocrinol. Metab.* 279, E83–E87.

Voogt, J., and de Greef, W. J. (1989). Inhibition of nocturnal prolactin surges in the pregnant rat by incubation medium containing placental lactogen. *Proc. Soc. Exp. Biol. Med.* 191, 403–407.

Walsh, R. J., Posner, B. I., Kopriwa, B. M., and Brawer, J. R. (1978). Prolactin binding sites in the rat brain. *Science* 201, 1041–1043.

Walsh, R. J., Slaby, F. J., and Posner, B. I. (1987). A receptor-mediated mechanism for the transport of prolactin from blood to cerebrospinal fluid. *Endocrinology* 120, 1846–1850.

Wilson, J. F. (1987). Severe reduction in food intake by pregnant rats resembles a learned food aversion. *Physiol. Behav.* 41, 291–295.

Woodside, B. (2007). Prolactin and the hyperphagia of lactation. *Physiol. Behav.* 91, 375–382.

Woodside, B., and Millelire, L. (1987). Self-selection of calcium during pregnancy and lactation in rats. *Physiol. Behav.* 39, 291–295.

Woodside, B., and Yorozu, S. (2002). Contributions of milk delivery and hormones to the changes energy balance that accompany lactation in rats. *International Congress on Obesity,* Sao Paulo, Brazil.

Woodside, B., Abizaid, A., and Walker, C.-D. (2000). Changes in leptin levels during lactation: Implications for lactational hyperphagia and anovulation. *Horm. Behav.* 37, 353–365.

Wosje, K. S., and Kalkwarf, H. J. (2004). Lactation, weaning, and calcium supplementation: Effects on body composition in postpartum women. *Am. J. Clin. Nutr.* 80, 423–429.

Xiao, X. Q., Grove, K. L., Grayson, B. E., and Smith, M. S. (2004a). Inhibition of uncoupling protein expression during lactation: Role of leptin. *Endocrinology* 145, 830–838.

Xiao, X. Q., Grove, K. L., and Smith, M. S. (2004b). Metabolic adaptations in skeletal muscle during lactation: Complementary deoxyribonucleic acid microarray and real-time polymerase chain reaction analysis of gene expression. *Endocrinology* 145, 5344–5354.

Xiao, X. Q., Williams, S. M., Grayson, B. E., Glavas, M. M., Cowley, M. A., Smith, M. S., and Grove, K. L. (2007). Excess weight gain during the early postnatal period is associated with permanent reprogramming of brown adipose tissue adaptive thermogenesis. *Endocrinology* 148, 4150–4159.

17

THE ENERGETICS OF PARENTING IN AN AVIAN MODEL: HORMONAL AND NEUROCHEMICAL REGULATION OF PARENTAL PROVISIONING IN DOVES

JOHN D. BUNTIN[1], APRIL D. STRADER[2] AND SELVAKUMAR RAMAKRISHNAN[1]

[1] *Department of Biological Sciences, University of Wisconsin – Milwaukee, Milwaukee, WI, USA*
[2] *Department of Physiology, Southern Illinois University School of Medicine, University of Wisconsin – Milwaukee, Milwaukee, WI, USA*

INTRODUCTION

Parenting is an energetically costly activity in both birds and mammals, but differences in avian and mammalian modes of reproduction profoundly shape the types and temporal patterns of energy expended in the parental effort. For example, viviparity allows mammalian parents to restrict most of their parental behavior expression to the post-partum period. In contrast, birds have an oviparous mode of reproduction that requires parental behavior to be expressed both before and after the young appear if the reproductive effort is to be successful. During incubation, parent birds must actively defend the nest, keep the eggs warm, and turn the eggs periodically to ensure continued embryonic development. After hatching, avian parents typically protect the nest and young from intruders, engage in brooding behavior to provide heat for the nestlings, and provide food for the growing young. Because parents cannot forage for food while they are incubating, and because parents must find enough food after hatching to meet their own nutritional needs and those of their rapidly growing young, most parent birds incur energetic costs that rival those associated with pregnancy and lactation in mammals. These costs vary with clutch size, maturity of the young at hatching, environmental constraints (food availability, diet, temperature, habitat, etc.), and the degree to which males and females share parental duties. In contrast to mammals, which overwhelmingly display exclusive maternal care, over 80% of avian subfamilies show some sort of biparental care, which may serve to decrease energetic costs to individual parents (Silver *et al.*, 1985). On the other hand, this advantage is partially offset by the fact that over 80% of these avian subfamilies also raise altricial young, which are essentially helpless at hatching and must be intensively brooded and fed by the parents over an extended period.

Pigeons and doves in the order Columbiformes are particularly interesting to study from the perspective of parental energetics because they, like flamingos (Lang, 1963) and emperor penguins (Prevost & Vilter, 1963), have evolved an unusual and specialized form of parental care that bears a functional resemblance to mammalian lactation. Plasma levels of prolactin are elevated in pigeon and dove parents of both sexes while they sit on their eggs (Goldsmith *et al.*, 1981; Buntin *et al.*, 1996). In response to this increase, the mucosal epithelial cells of the crop sac, an outpocketing of the esophagus that normally serves as a seed storage organ, begin to proliferate and accumulate large amounts of lipids and proteins during incubation, resulting in a 10-fold increase in thickness (Patel, 1936; Horseman & Buntin, 1995). At the time of hatching, large numbers of nutrient-rich epithelial cells slough off from the crop sac wall to form crop "milk," a cheese-like mass that is regurgitated to the young. Crop "milk" forms the exclusive source of nourishment for the young for the first 3 days (Vandeputte-Poma, 1980), but as plasma prolactin levels gradually decline and the crop sac begins to regress during the second week after hatching (Goldsmith *et al.*, 1981; Buntin *et al.*, 1996), young squabs are fed progressively less crop milk and progressively more grain and other foods obtained through increased parental foraging. (Murton *et al.*, 1963; Vandeputte-Poma, 1980).

Despite obvious differences, there are striking similarities in parental care strategies of Columbiform birds and mammals that result in several shared energetic characteristics. Although young pigeons and doves depend exclusively on regurgitated crop "milk" for a shorter period than their young mammalian counterparts which depend on mammary gland milk, the ability to manufacture these substances confers an advantage in both instances because it partially buffers nutrient delivery to the young from variations in parental food supply (Roberts *et al.*, 1985; Mondloch & Timberlake, 1991). Pigeon and dove parents also resemble lactating mammals in showing increased food intake to meet the energetic challenges associated with accelerated milk production after the young appear (Lea *et al.*, 1992; Mondloch & Timberlake, 1991). Finally, in a manner similar to the provisioning of weanling young in some mammalian taxa (e.g., most carnivores and some primate species; Rapaport, 2006), pigeon and dove parents increase their foraging significantly to provide an additional source of food that can be regurgitated to the young as crop milk production declines. How the best studied Columbiform species, the ring dove (*Streptopelia risoria*), meets these energetic challenges and how these adaptations are regulated physiologically are the principal themes of this chapter.

Energetics of Incubation

The energetic costs of incubation can be tremendous in some avian species. Male emperor penguins, for example, must incubate their eggs continuously and without relief for over 2 months during the winter while their breeding partners are foraging at sea (Groscolas & Robin, 2001). This incubation fast is preceded by a 40–50 day period of food deprivation that accompanies the journey to the breeding site. Not surprisingly, males lose over 40% of their body mass during this combined 4-month period of fasting. Weight loss is also common in avian species in which one sex incubates the eggs without nest relief or provisioning by their mate (Mrosovsky & Sherry, 1980). Birds that adopt this parenting strategy include galliform species such as chickens, grouse, and quail, and anseriform species such as ducks, geese, and swans. In the red junglefowl (*Gallus gallus*), the feral species from which the domestic chicken is derived, incubating hens sit on the eggs for over 23.5 h/day and consumes 80% less food than non-incubating hens. This results in a 20% loss of body mass over the 20-day period of incubation (Sherry *et al.*, 1980). Interestingly, the decrease in food consumption is not due to the fact that incubation and foraging are competing activities, since incubating hens do not increase their food consumption even when given the opportunity to ingest food while simultaneously sitting in the nest. These studies and the results of food deprivation and re-feeding experiments in incubating birds collectively suggest that junglefowl hens continue to regulate their body weight during incubation but do so at a lower set point (Sherry *et al.*, 1980).

The energetic costs of incubation are typically reduced in species in which the incubating breeding partner is provisioned by its mate or in species in which males and females share in the sitting duties. Male and female ring doves are typical of most other socially monogamous species in that they share extensively in all aspects of parental care. During the 14-day incubation period, male and female parents show a daily rhythm of nest occupation that is sex-specific, with females sitting during the evening, night, and early morning hours and males sitting for a 7–9 h period from mid-morning to late afternoon or evening (Ramos & Silver, 1992). This sharing of incubation duties provides opportunities for parents to forage during periods off the nest, and as a result, body weights and food intake do not change appreciably across the incubation period (Ramakrishnan *et al.*, 2007). However, these observations were made in breeding birds kept in small cages in the laboratory with food available *ad libitum*. The pattern of changes in body weight and food consumption during incubation may be different in free-living doves breeding in their natural habitat where foraging may require more effort. In addition, there are species differences in body mass and egg size among different Columbiform species that could influence energetics and food procurement during incubation. For example, small eggs lose heat more rapidly than larger eggs because of their higher surface to volume ratio, which in turn leads to corresponding differences in the amount of heat that parents must transfer to eggs (Ricklefs, 1974). These relationships could explain why pigeons consume approximately 25% less food during incubation than during the pre-laying period when allowed to breed in small laboratory enclosures with food freely available (Mondloch & Timberlake, 1991), while incubating ring dove parents, which incubate smaller eggs, show no decrease in food intake under similar housing conditions (Ramakrishnan *et al.*, 2007).

Energetics of Nestling Care

As with incubation, the energetic costs of caring for young in birds vary considerably from species to species. At one end of the spectrum are brood parasites such as cuckoos and cowbirds, which lay eggs in the nests of their hosts and display no parental care. Similarly, newly hatched mallee fowl young are extremely precocial and require no care from their parents (Frith, 1956). In the vast majority of birds, however, some parental care is required, and the type of care provided and the amount of energy expended varies with the capabilities of the young. Because their foraging and locomotor abilities are well developed at hatching, the vast majority of precocial offspring do not need to be fed by their parents. However, their parents do protect them and lead them to food (Richard-Yris *et al.*, 1983). Parents with precocial young also brood their newly hatched chicks, but this behavior declines rapidly as the thermoregulatory abilities of the young improve (Sherry, 1981; Lea *et al.*, 1982).

Because altricial young emerge at hatching with poor thermoregulatory and locomotor abilities, the parental effort associated with brooding and provisioning these young tends to be higher than that expended by parents rearing precocial young. Based on calculations from five species of birds that rear altricial young (see Drent & Daan, 1980 for review), maximal parental energy expenditure approaches 4 times the basal metabolic rate (BMR), which is comparable to that expended during peak lactation in mammals (Gittleman & Thompson, 1988). More recent calculations have revised the maximum parental energy expenditure upward to over 5 times BMR; however, there also appears to be considerable variation in metabolic energy expenditure associated with the breeding effort among different avian species (Weathers & Sullivan, 1989; Peterson *et al.*, 1990).

In pigeons and doves, constraints imposed by crop sac size and the rate and magnitude of crop sac mucosal proliferation determines the yield of crop milk produced and effectively limits the number of offspring that can be adequately fed by parents to two (Burley, 1980; Westmoreland & Best, 1987). Average daily energy expenditure by parent doves rearing young is substantial, and is estimated at 3.5 times BMR for parents raising two young in 1,000 ft^3 enclosures with food available *ad libitum* (Brisbin, 1969; Drent & Daan, 1980) .

Not surprisingly, pigeon and dove parents must increase their foraging activity significantly in order to meet the energetic demands of crop milk production and to provision their rapidly growing young with seeds. In contrast to the decline in food consumption that occurs during the last few days of incubation, parental food intake begins to increase soon after hatching in both pigeons (Mondloch & Timberlake, 1991) and ring doves (Lea *et al.*, 1992; Koch *et al.*, 2002). According to Mondloch & Timberlake (1991), daily parental food intake in pigeon pairs housed in small enclosures stabilizes at about 50% above pre-laying values by post-hatch day 8–10 and remains at this level for at least 15 days. Ring dove pairs housed under similar conditions show a similar but more pronounced increase, with parents raising 3-week old young consuming over twice as much food each day as pairs sampled at earlier breeding stages (Koch *et al.*, 2002; Strader & Buntin, 2003; Ramakrishnan *et al.*, 2007; see Figure 17.1). Despite these substantial elevations in food intake, parent pigeons and doves show a 5–10% decrease in body mass while provisioning their young, which reflects the fact that a substantial proportion of

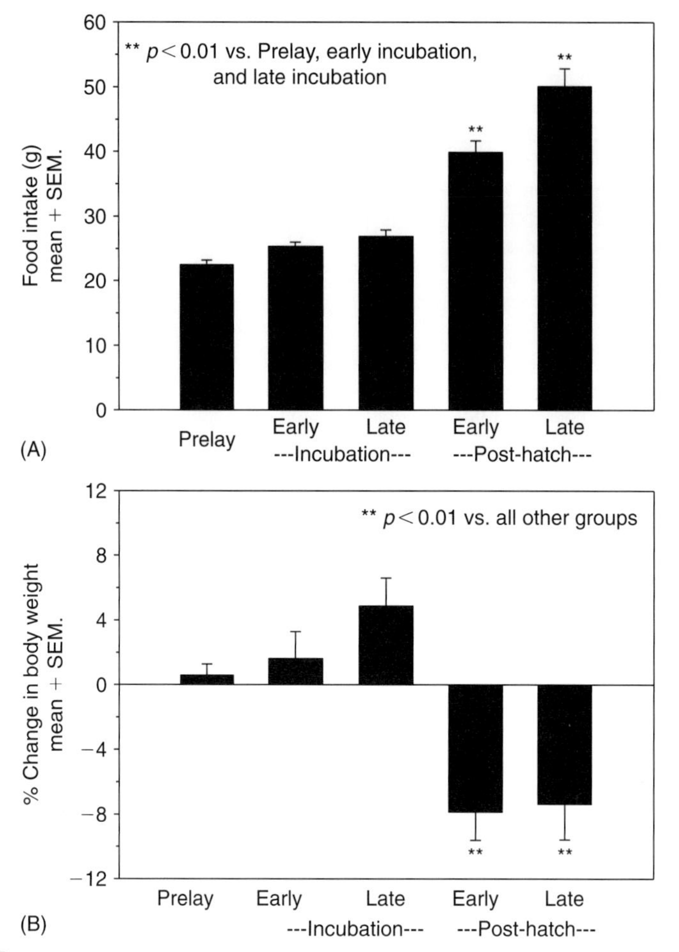

FIGURE 17.1 Changes in food intake (A) and body mass (B) in ring dove parents during the incubation and post-hatching periods of the breeding cycle. Prelay = 4–5 days after pairing; early incubation = 4–5 days after egg laying; late incubation = 11–12 days after egg laying; early post-hatch = 11–12 days after hatching; late post-hatch = 21–22 days after hatching. Food-intake values for each pair were based on the average daily food consumption on the 5 days closest to the breeding cycle days indicated above. Body weight data were based on weights recorded for each bird on the sampling date indicated, and are expressed as a percentage of body weight recorded prior to pairing. (*Source*: Data from Ramakrishnan *et al.* (2007)).

the food acquired during foraging is transferred to the young by regurgitation (Figure 17.1).

Altricial nestlings signal their nutritional needs to their parents by begging. There is ample evidence that hungry nestlings show more begging activity than recently fed young, and that parents respond to this increase in stimulation by increasing the amount or altering the type of food provided to the young (Bengtsson & Ryden, 1983; Whittingham & Robertson, 1993; Price & Ydenberg, 1995; Ottosson *et al.*, 1997; Saino *et al.*, 2000). Evidence for these relationships also exists for Columbiform species. Mondloch (1995) found that food-deprived pigeon squabs spend more time begging and receive more regurgitations from their parents than satiated young. Similarly, Brisbin (1969) reported that ring dove parents ingest more food when provisioning two squabs than when provisioning only one. The foraging adjustments of ring dove parents are closely geared to changes in the energetic requirements of the young, as measured by the frequency of parental regurgitation feeding activity. In addition, such adjustments can occur rapidly. In a recent study on male parents tested with their young

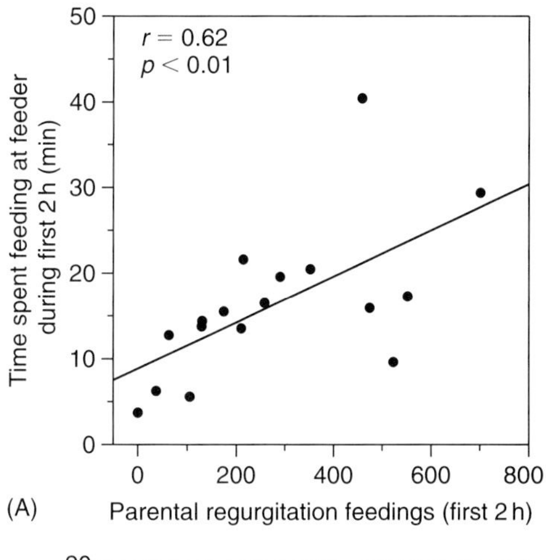

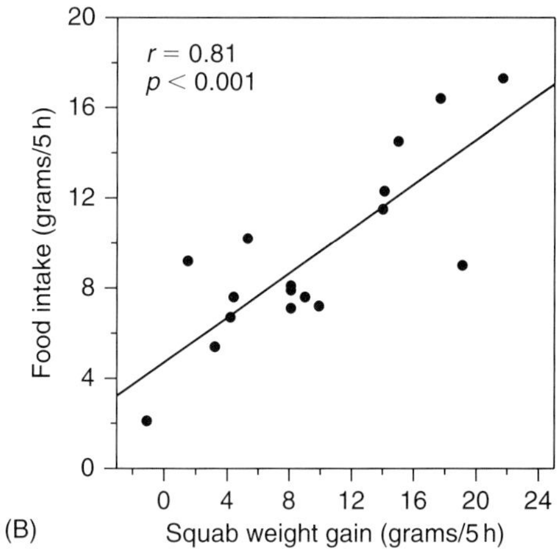

FIGURE 17.2 The relationship between parental provisioning of young and seed consumption in male ring dove parents that were reunited with their 10-day old young after an overnight separation. (A) Parental regurgitation feeding frequency and time spent ingesting seeds at a feeder during the first 2 h after reintroduction. (B) Parental food consumption and squab weight gain during the first 5 h after reintroduction.

after an overnight separation, time spent foraging on seed was positively correlated with the number of regurgitation feedings delivered to 10-day old squabs when assessed at 1 h and 2 h after parents and young were reunited. As a result, the amount of weight gained by the squabs was strongly related to parental food intake when assessed at 5 h after the young were re-introduced (Figure 17.2).

Role of Prolactin in Parental Provisioning

Studies of wild bird populations indicate that parents that rear altricial young typically exhibit elevated levels of plasma prolactin when they are provisioning their offspring (see Buntin, 1996 for review). For example, prolactin levels in blood correlate positively with the frequency of parental feeding visits to the nest in male house finches (Badyaev & Duckworth, 2005), and female red-eyed vireos show higher levels of plasma prolactin, and more parental provisioning, than their male partners (Van Roo *et al.*, 2003). Interestingly, a close link between prolactin and parental provisioning is not restricted to breeding birds, since plasma prolactin levels have also been reported to be elevated in non-breeding "helpers" that assist parents in provisioning young in three cooperatively breeding species (Harris' hawk: Vleck *et al.*, 1991; Florida scrub jay: Schoech *et al.*, 1996; Mexican jay: Brown & Vleck, 1998).

A particularly compelling case for a role of prolactin in parental provisioning in wild bird populations comes from work on male house finches by Badyaev & Duckworth (2005) (Figure 17.3). In the population of birds that they studied, parental investment by the male is strongly correlated with plumage color. Specifically, males with dull yellow plumage provision their nestlings and their incubating mates frequently while males with bright red plumage spend very little time on these activities (Badyaev & Hill, 2002). These behavioral differences are associated with differences in circulating prolactin levels during these breeding stages, with the more parental males exhibiting higher prolactin concentrations. Remarkably, parental feeding visits to the nest increased several fold when non-parental red males were treated with pellets containing the prolactin-releasing peptide vasoactive intestinal peptide (VIP). Conversely, feeding visits to the nest ceased when parental yellow males were treated with pellets containing the dopamine agonist bromocryptine, which reduced plasma prolactin to non-detectable levels (Badyaev & Duckworth, 2005).

In ring doves, prolactin is implicated in every aspect of the parental provisioning process. The progressive rise in plasma prolactin levels that begins at mid-incubation and peaks during the early post-hatching period (Goldsmith *et al.*, 1981; Buntin *et al.*, 1996)

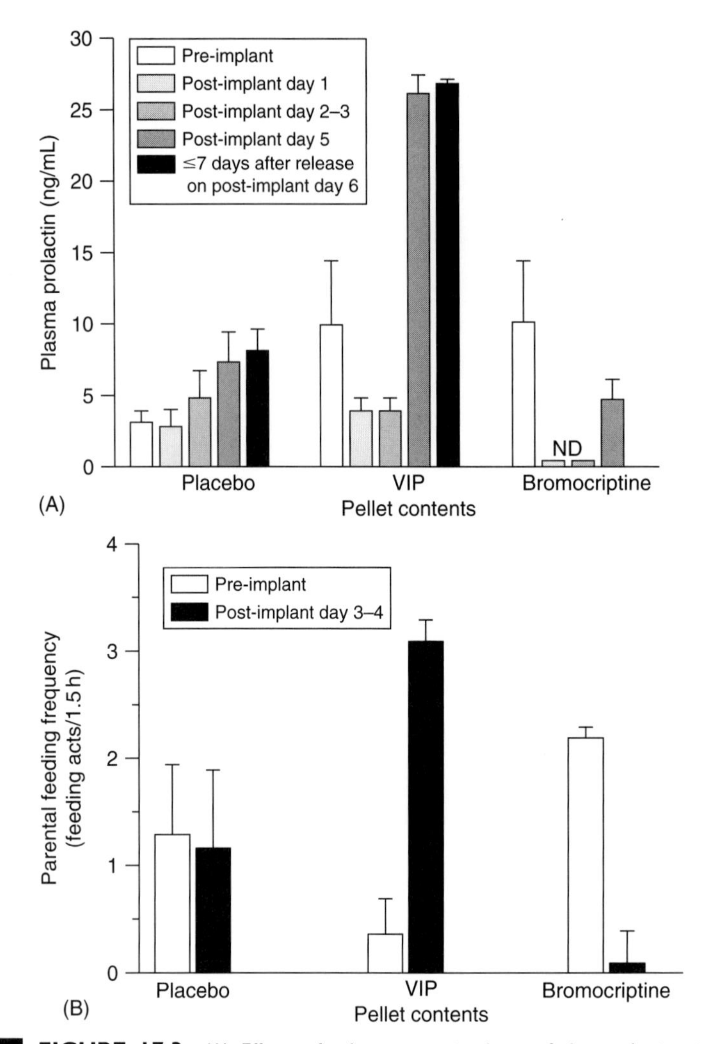

FIGURE 17.3 (A) Effects of subcutaneous implants of the prolactin-stimulating peptide VIP (14-day release) and the prolactin inhibiting drug bromocriptine (21-day release) on changes in plasma prolactin in free-living male house finches that were captured and housed in holding cages during the 5-day treatment period. VIP-treated birds were released from captivity on day 6 and recaptured within 7 days for a final blood sample. (B) Effects of VIP and bromocriptine pellets on parental feeding frequency by male house finch parents. Pellets were implanted 3–5 days after hatching and behavior tests were conducted 7–9 days after hatching. VIP was administered to males with bright red plumage that typically exhibit very little parental provisioning of young. Bromocriptine was administered to males with dull yellow plumage that typically show high levels of parental provisioning of young. ND = not detectable. (*Source*: Data from Badyaev & Duckworth (2005)).

stimulates crop sac development and later crop "milk" formation (Patel, 1936; Horseman & Buntin, 1995), which guarantees that a nutritious source of food will be available to the young when they hatch. Secondly, prolactin stimulates the regurgitation behavior that is necessary to transfer food from the parent's crop sac to the young (Lehrman, 1955; Buntin *et al.*, 1991; Wang & Buntin, 1999). Finally, prolactin stimulates hyperphagia (Buntin & Tesch, 1985; Buntin & Figge, 1988; Buntin, 1989), which results in engorgement of the crop sac with seeds. These seeds that are transferred by parental regurgitation gradually replace crop milk as the principal food source for the rapidly growing young.

Sites of Prolactin Action in Promoting Parental Provisioning in Doves

Crop Milk Formation and Parental Regurgitation

In addition to stimulating crop milk formation in dove parents of both sexes, the prolactin that is secreted into the blood during incubation and the post-hatching period is likely to act at both central and peripheral target sites to stimulate the transfer of milk to the nestlings by regurgitation. Crop sac development leading to crop milk formation is mediated by high affinity receptors for prolactin in crop sac mucosal epithelial cells, which undergo rapid proliferation in response to prolactin stimulation (Horseman & Buntin, 1995). The resulting engorgement of the crop sac with milk may also contribute to the stimulatory effects of prolactin on parental regurgitation activity and resulting transfer of crop milk to the young. In early studies on non-breeding doves with previous breeding experience, Lehrman (1955) found that daily intradermal injections of prolactin stimulated parental regurgitation behavior toward young in 80% of the birds tested. Because this treatment also resulted in extensive crop sac development, and because the incidence of parental regurgitation was reduced to 33% when the crop sac was anesthetized prior to testing, Lehrman hypothesized that some of the effects of prolactin on regurgitation may result from somatosensory or proprioceptive cues generated by the engorgement and distension of the crop sac (or possibly the overlying skin) that accompanies prolactin-induced crop growth and milk production. Subsequent studies have confirmed that crop sac development enhances the incidence and frequency of prolactin-induced parental regurgitation to young in non-breeding doves with previous breeding experience (Buntin *et al.*, 1991). However, it is equally clear that a well developed crop sac is not essential for the expression of this behavior, since incubating dove parents attempt to feed young that are prematurely introduced into their nests before any measurable crop sac growth has occurred (Klinghammer & Hess, 1964; Hansen, 1971). In addition, prolactin can increase the incidence of parental regurgitation behavior and attempts to feed young (parental invitations) in non-breeding doves with previous breeding experience when given directly to the brain by intracerebroventricular (ICV) injection at doses that are too low to stimulate crop sac development and engorgement (Buntin *et al.*, 1991). Together, these data suggest that while prolactin can act directly on the brain to promote both the appetitive and the consummatory components of parental regurgitation, full expression of the behavior requires an additional indirect action mediated by peripheral stimuli associated with crop sac development and distension. At present, it is not clear exactly how and where the prolactin-induced signals that are generated in the central nervous system (CNS) are integrated with these somatosensory or proprioceptive cues to promote parental regurgitation and responsiveness to young. However, immediate early gene expression studies on parents that are interacting with and feeding their young have identified sites of neuronal activation in the dove brain that could be involved in such integration, including the preoptic area (POA), lateral hypothalamus, and lateral septum (Buntin *et al.*, 2006).

The ability of ICV injections of prolactin to induce significant changes in parental regurgitation (Buntin *et al.*, 1991) implies the existence of prolactin receptors in the dove CNS. High affinity binding sites for prolactin are present in dove brain homogenates (Buntin & Ruzycki, 1987) and have been mapped to discrete regions of the dove hypothalamus, POA, and other components of the limbic system by *in vitro* autoradiography using radiolabeled ovine prolactin (Fechner & Buntin, 1989; Buntin *et al.*, 1993). As in mammalian species (Walsh *et al.*, 1987), prolactin binding sites are also found in the choroid plexus in doves (Buntin & Walsh, 1988). In rats, these cells have been implicated in the uptake and transport of prolactin from blood to cerebrospinal fluid, from which the hormone could conceivably access target sites in the brain (Walsh *et al.*, 1987). Brain uptake of blood-borne prolactin has also been demonstrated in doves (Buntin & Walsh, 1988; Buntin, *et al.*, 1993), but whether the prolactin receptors in dove choroid plexus are part of this uptake process remains to be determined. In addition to evidence that prolactin receptors in the brain are accessible to blood-borne prolactin from the anterior pituitary, there is evidence in birds (Ramesh *et al.*, 2000) and in mammals (Torner and Neumann, 2002) for prolactin-like molecules synthesized in the brain that could also interact with these CNS prolactin receptors. A major challenge in future studies will be to determine how these two prolactinergic systems interact to promote changes in brain circuits underlying behavioral expression.

Although prolactin-sensitive areas of the dove brain have not been systematically tested to determine which populations mediate prolactin-induced parental regurgitation, there is strong evidence from lesion studies that the integrity of the POA is essential. In tests with non-breeding male doves with previous breeding experience, Slawski & Buntin (1995) found that axon-sparing lesions of the POA severely disrupted the ability of prolactin to stimulate parental regurgitation and parental feeding invitations toward young. Notably, the magnitude of these behavioral deficits varied directly with the extent of POA damage.

Parental Foraging and Hyperphagia

Two lines of evidence suggest that prolactin plays an important role in stimulating the increase in food consumption that parent doves exhibit while provisioning their young. First, parental hyperphagia begins during the early post-hatching period (Lea *et al.*, 1992; Koch *et al.*, 2002) when peak levels of plasma prolactin are attained (Goldsmith *et al.*, 1981; Buntin *et al.*, 1996). Secondly, prolactin increases food intake in non-breeding pigeons and doves when given systemically (Schooley *et al.*, 1941; Bates *et al.*, 1962; Buntin & Figge, 1988) or by ICV injection (Buntin & Tesch, 1985; Buntin & Figge, 1988; Buntin, 1989). The food-intake response to prolactin is also dose-dependent and sex-specific, with males consuming twice as much food as females (Buntin & Tesch, 1985; Buntin & Figge, 1988). Although the physiological basis of this sex difference remains to be established, the stronger orexigenic response of the male could be adaptive in light of the division of parental responsibilities that develops in many Columbiform birds when breeding conditions are favorable. Under such conditions, the female typically lays a new clutch of eggs before the squabs are old enough to forage on their own. With the female occupied for most of the day with incubating the new clutch, the responsibility for provisioning older offspring is left primarily to the male partner.

Daily food intake is increased by over 100% in male doves given ICV injections of prolactin in amounts that are too low to have any detectable effects on peripheral target organs such as the crop sac (Buntin & Figge, 1988; Figure 17.4). This provides compelling evidence that prolactin can act directly on the brain to exert profound effects on feeding activity. Later microinjection studies by Foreman, *et al.* (1990) and Hnasko & Buntin (1993) further identified specific brain sites of prolactin action in promoting this response. Using doses that are too low to be effective when injected ICV, Hnasko & Buntin (1993) observed increased food intake when prolactin was microinjected into the POA, the ventromedial nucleus (VMN) of the hypothalamus, and the tuberal hypothalamic area, but not when microinjected into four other prolactin-sensitive brain regions (Figure 17.4). These orexigenic effects are likely to be mediated by prolactin receptors on target cells at these loci since Li *et al.* (1995) found that the hyperphagia induced by prolactin injections into the ventromedial hypothalamus was significantly attenuated when antibodies generated against the rat liver prolactin receptor were injected into the same site at 1 h prior to prolactin administration.

Appetitive and Consummatory Aspects of Parental Provisioning Behavior

The behavior that parents exhibit while provisioning their altricial young is similar in some respects to the feeding activity that parents engage in to meet their own nutritional needs. However, unlike feeding behavior, which typically has relatively simple appetitive (food seeking) and consummatory (ingestive) phases, parental provisioning is a complex activity with several appetitive and consummatory components. Initially, parents with altricial young must engage in appetitive food seeking or foraging activity that culminates in food procurement. In some species, such as the Columbiformes, parents actually ingest the food that they procure for the young, while in others, such as raptors and insectivorous passerine species, the food is not eaten by the parent. In either case, parents must bring the procured food back to the nest and then, in a final consummatory act, transfer the food to the young. Whether these different components of parental provisioning are products of the same physiological mechanism has not been systematically explored, but it is likely, based on studies of other motivated behaviors, that they are not (Jewett *et al.*, 1992; Ikemoto & Panksepp, 1996). Furthermore, the physiological regulators involved may influence

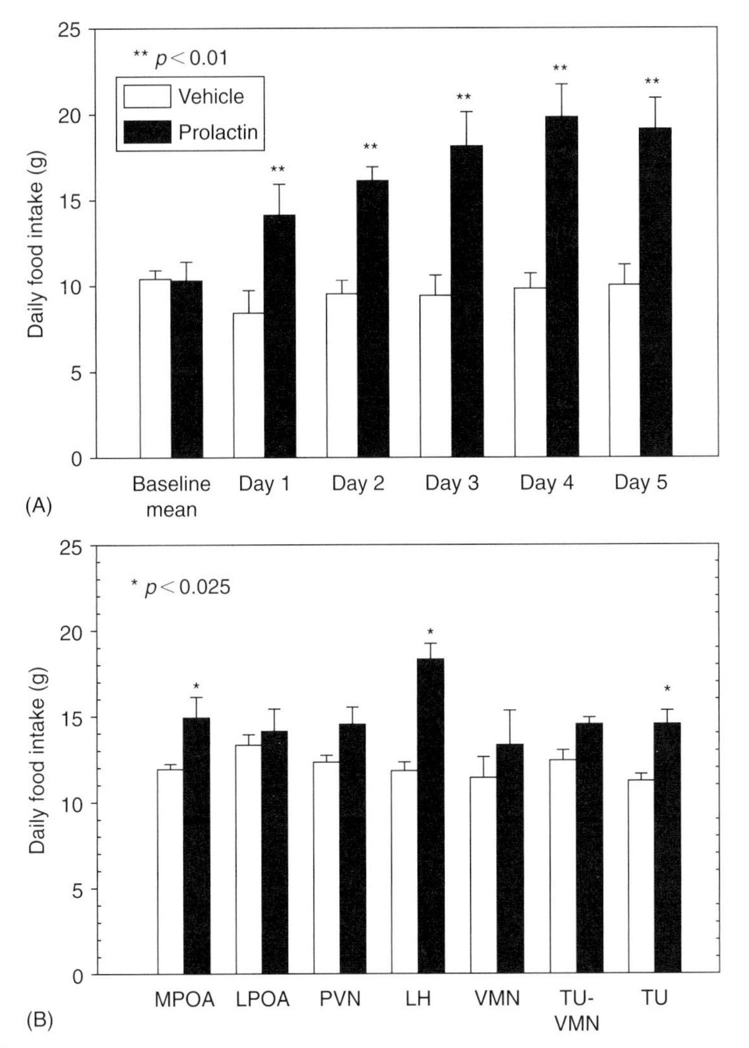

FIGURE 17.4 (A) Effects of 5 days of twice-daily ICV injections of ovine prolactin (1 μg/2 μL) or vehicle on food intake in non-breeding male ring doves. (B) Changes in average daily food intake following 4 days of twice-daily microinjections of ovine prolactin (50 ng/10 nL) or vehicle into seven different prolactin-sensitive brain sites in non-breeding male ring doves. Values are mean ± SEM. MPOA, medial preoptic area; LPOA, lateral preoptic area; PVN, paraventricular nucleus of the hypothalamus; LH, lateral hypothalamus; VMN, ventromedial nucleus of the hypothalamus; TU, tuberal hypothalamus; TU-VMN, medial hypothalamic region between the TU and VMN. (*Source*: Site-specific microinjection data from Hnasko & Buntin (1993)).

these components in complex ways. In several species, for example, corticosterone levels in plasma tend to rise in parents when energetic demands are increased by parental provisioning of large broods (Silverin, 1982), by food shortages (Kitaysky *et al.*, 1999; Groscolas & Robin, 2001), or following the loss of a mate that shares parental duties (Silverin, 1982). Such elevations in plasma corticosterone can result in increased parental foraging effort and food procurement, which can benefit the young by increasing the amount of food delivered (Silverin, 1990; Kitaysky *et al.*, 2001). However, when corticosterone levels rise further, increased parental foraging can become uncoupled from parental provisioning or nest attentiveness, and parents may abandon the parental effort altogether (Silverin, 1982, 1990;

Groscolas & Robin, 2001). It has been suggested that parental outcomes may depend not only on the effects of these energetic challenges on corticosterone secretion, but also on prolactin secretion, which would be expected to promote parental attachment to the nest and young (Groscolas & Robin, 2001; Criscuolo *et al.*, 2006; O'Dwyer *et al.*, 2006).

Experiments in ring doves have shed some light on the neural and hormonal determinants of individual parental provisioning components, but much remains to be characterized. In a recent immunocytochemical mapping study using antibodies to a synthetic fragment of the chicken Fos protein (Buntin *et al.*, 2006), neuronal activation patterns in parent doves that were allowed unrestricted access to their young following a period of overnight separation were compared with those in parents that were deprived of physical contact but exposed to visual, auditory, and olfactory stimuli from their reintroduced young, which were separated from their parents by a wire mesh partition that bisected the cage into two equal sized compartments (Figure 17.5). Parents that received physical contact with their hungry, reintroduced young engaged in vigorous and frequent parental regurgitation activity during the 1.5-h test. In contrast, parents that were physically separated from their young showed very little regurgitation.

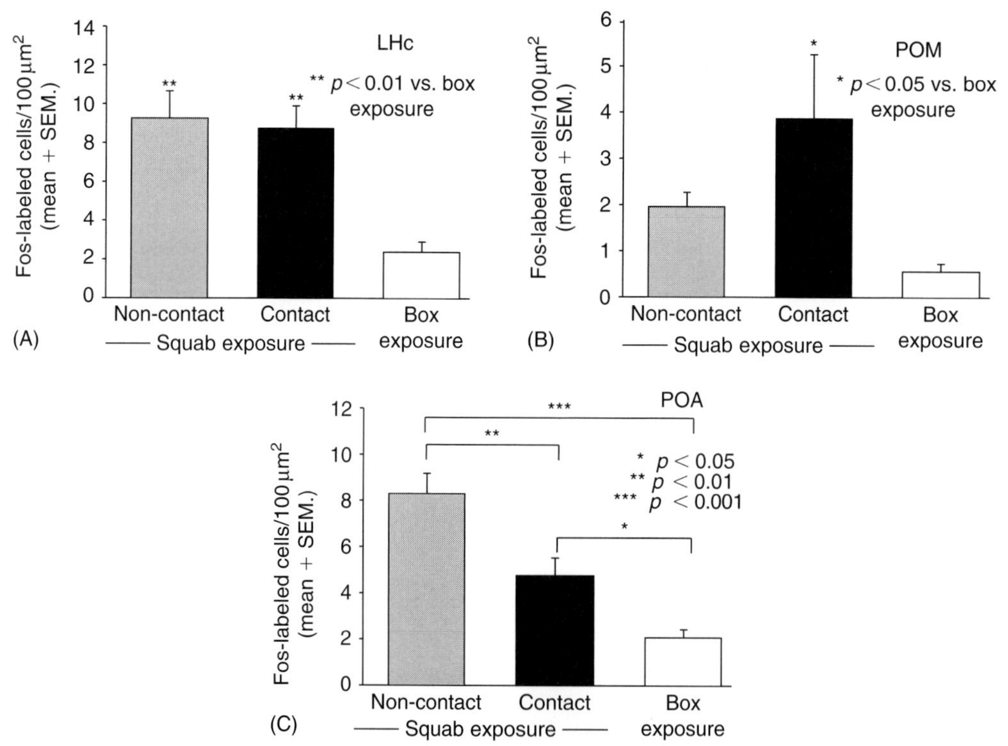

FIGURE 17.5 Fos-immunoreactive cell densities in different brain regions of ring dove parents given 80 min of exposure to their 10-day old young after a period of overnight separation. Individual parents were allowed free access to their young (contact squab exposure), non-tactile exposure to their young from across a wire partition (non-contact squab exposure), or exposure to a neutral stimulus (box) from across a wire mesh partition. The response pattern seen in the caudal portion of the lateral hypothalamus (LHc) was also observed in lateral septum and the rostral portion of the lateral hypothalamus, and was characterized by increased numbers of Fos-staining cells in both squab exposure groups relative to control values. The response pattern observed in the POM was also seen in the bed nucleus of the stria terminalis, and was characterized by elevated numbers of Fos-staining cells in parents allowed unrestricted access to young but not in parents given only non-tactile exposure. The preoptic area (POA) was unique in showing higher numbers of Fos-staining cells in parents given non-tactile exposure to young than in those given free access to squabs. (*Source*: Data from Buntin *et al.* (2006)).

Instead, these birds spent lengthy periods in close proximity to the partition and engaged in frequent appetitive activities such as grasping the wire mesh partition with their feet or flying up the side of the partition in an attempt to gain access to the nest and squab on the other side. Parents in both squab exposure conditions showed a larger number of Fos-labeled cells in the rostral POA, lateral hypothalamus, and lateral septum than did parents in the control group, which received exposure to a cardboard box placed on the opposite side of the wire mesh partition. Parents given the opportunity to physically interact with and feed their young also showed more Fos-staining cells in the medial preoptic nucleus (POM) and the bed nucleus of the pallial commissure than did box-exposed controls, but parents that were barred from physical contact with their young did not. This implies that the POM and bed nucleus of the pallial commissure participate in the expression of the consummatory aspects of parent–young feeding interactions. In contrast to this pattern, parents given only non-tactile exposure to squabs had more Fos-staining cells in the rostral POA than did controls or parents allowed unrestricted access to their young. This suggests a link between neuronal activation in this brain region and the expression of appetitive components of parenting, such as parental approach to the nest and young.

Although the neural substrates underlying appetitive and consummatory components of parental foraging and food procurement in doves have not been experimentally investigated, the possible hormonal mechanisms that regulate these components have been explored. As discussed earlier, prolactin exerts potent effects on the actual consumption of food in non-breeding doves, which presumably mimic similar effects on parental hyperphagia during the post-hatching period. However, the question of whether prolactin also enhances the appetitive, food seeking activity necessary to procure the food to be consumed is difficult to test under laboratory conditions where doves are housed in small enclosures with food available *ad libitum* from dispensers located less than two feet away. To test this question more directly, Gamoke *et al.* (2000) trained non-breeding doves to peck a key for food reward and then used key pecking performance on demanding schedules of reinforcement to assess the degree to which the birds were motivated to work for food under different hormonal treatment and feeding conditions. Not surprisingly, none of the birds given ICV injections of prolactin or vehicle learned the key pecking response when given unrestricted access to food. However, prolactin-treated birds did learn the operant response if the hyperphagia that develops in response to prolactin treatment under free-feeding conditions was prevented by restricting daily food consumption to the level seen under the same feeding regimen during the baseline (BL) period prior to prolactin administration. This supports the idea that prolactin stimulates appetitive components of feeding. In addition, it suggests that prolactin raises the metabolic set point around which food consumption is regulated. Consistent with the idea that metabolic homeostasis persists in prolactin-treated birds are observations that daily food consumption stabilizes at approximately 50% and 100% above BL in female and male doves, respectively, when birds are given several days of ICV prolactin treatment under *ad libitum* feeding conditions (Buntin & Tesch, 1985).

To compare the strength of feeding motivation in prolactin-treated doves with that in chronically food-restricted doves, Gamoke *et al.* (2000) used a progressive ratio schedule of reinforcement that required birds to peck the key 10 additional times for each successive food reward. Birds maintained at their pre-treatment *ad libitum* food-intake values while given ICV prolactin injections continued to work for food until they were required to peck the key 40–60 times between food reinforcements. By comparison, chronically food-restricted doves maintained at 80% of their free-feeding weight exhibited progressive ratio "break points" that were 3–4 times higher than these values. Overall, these results suggest that prolactin enhances motivation to feed when administered ICV at a dose that is sufficient to induce maximal hyperphagia, although the strength of the appetitive behavior induced is less than that induced when food availability is restricted enough to lower body weight by 20%.

The Neurochemistry and Neuroendocrinology of Parental Hyperphagia in Doves

The fact that food intake in breeding doves increases by over 100% when parents are provisioning their young suggests that the neural mechanisms that regulate appetite are profoundly altered during this period. Two neuropeptides with potent and well documented appetite-stimulating properties that could mediate these changes are neuropeptide Y (NPY) and the endogenous melanocortin receptor antagonist agouti-related peptide (AgRP). (Morton & Schwartz, 2001). These neuropeptides are extensively co-localized in neurons of the arcuate nucleus of the hypothalamus in mammals (Hahn *et al.*, 1998) and in the avian homolog of the arcuate nucleus, the tuberal hypothalamus, in Japanese quail (Boswell *et al.*, 2002). Although co-localization of these peptides has not been investigated in other birds, the tuberal hypothalamus is rich in NPY and/or AgRP-immunoreactive cells in several other avian species, including the ring dove (Kuenzel & McMurtry, 1988; Kameda *et al.*, 2001; Strader & Buntin, 2001, 2003; denBoer-Visser & Dubbeldam, 2002; Strader *et al.*, 2003; Mirabella *et al.*, 2004).

Two lines of evidence support the hypothesis that changes in NPY and AgRP activity in the tuberal hypothalamus are involved in the expression of parental hyperphagia in doves. First, NPY and AgRP both stimulate feeding when given by ICV injection to non-breeding male doves, *albeit* with different response latencies and durations (Strader & Buntin, 2001; Strader *et al.*, 2003). Secondly, NPY- and AgRP-immunoreactive neurons in the tuberal region of the ring dove hypothalamus increase in number at the time of hatching and remain elevated throughout the period of parental provisioning (Strader & Buntin, 2003; Ramakrishnan *et al.*, 2007; Figure 17.6). In the case of NPY, the post-hatching increase in number of cells immunostaining for NPY in the mediobasal hypothalamus is likely to reflect increased synthesis of the neuropeptide, since it is accompanied by an increase in NPY mRNA in the same region (Ramakrishnan *et al.*, 2007; Figure 17.6). Although extracellular NPY concentrations in dove brain have yet to be measured, it is likely that this increase in NPY synthesis supports increased NPY release, which could contribute to the increase in food consumption associated with parental provisioning of young. Since AgRP and NPY show virtually identical patterns of changes in immunostaining in the tuberal hypothalamus at this stage, and since AgRP and NPY are likely to be co-expressed in the same neurons in the tuberal hypothalamus (Boswell *et al.*, 2002), it is likely that any orexigenic effects of increased NPY activity during the post-hatching are reinforced by parallel changes in AgRP. However, direct measurements of AgRP gene expression and release will be required to provide convincing evidence for such changes.

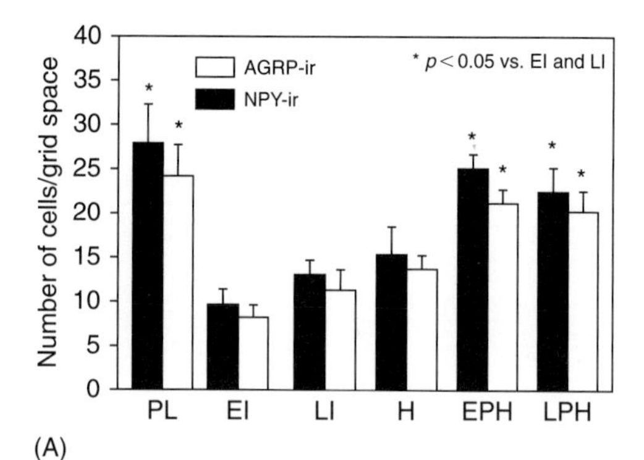

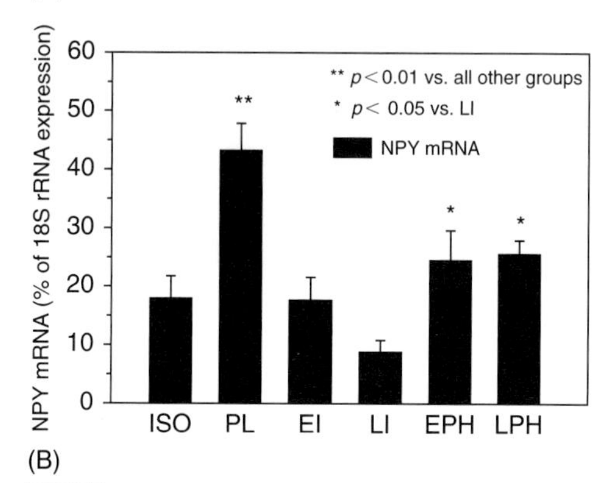

FIGURE 17.6 (A) Changes in numbers of cells immunostaining for NPY (NPY-ir) and AgRP (AGRP-ir) in the tuberal hypothalamus of male and female ring doves as a function of breeding cycle stage. (B) Changes in NPY mRNA in the mediobasal hypothalamus (including the tuberal hypothalamus) of male and female ring doves during six breeding stages. Estimates were obtained using semi-quantitative RT-PCR using 18S rRNA for normalization. Values are mean ± SEM. ISO, non-breeding birds housed in visual isolation from conspecifics; PL, pre-laying (4–5 days after pairing); EI, early incubation (4–5 days after egg laying); LI, late incubation (11–12 days after egg laying); H, hatching date; EPH, early post-hatching (11–12 days after hatching); late post-hatching (21–22 days after hatching). (*Source*: AgRP data from Strader & Buntin (2003); NPY data from Ramakrishnan *et al.* (2007)).

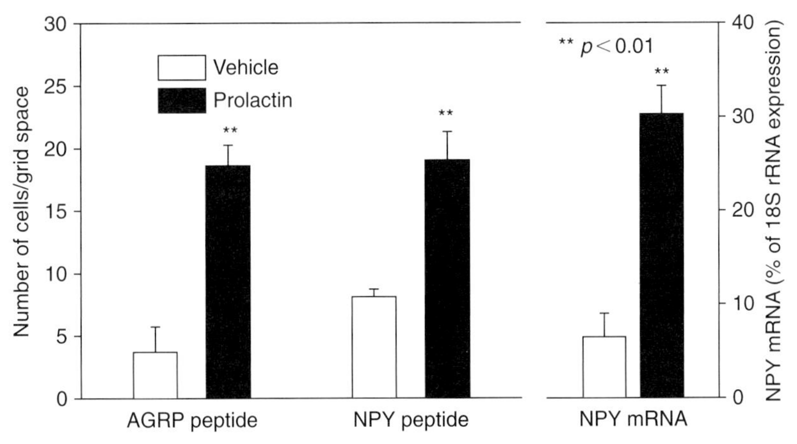

FIGURE 17.7 Effects of ICV injections of ovine prolactin or vehicle on the number of NPY-immunostaining and AgRP-immunostaining cells in the tuberal hypothalamus and the amount of NPY mRNA in the mediobasal hypothalamus of non-breeding male ring doves as assessed by semi-quantitative RT-PCR. Birds used for peptide analysis received 5 days of twice-daily ICV injections of either prolactin (1 μg/2 μL) or vehicle. Birds used for mRNA analysis were given 3 ICV injections of ovine prolactin (1 μg/2 μL) over a 30-h period. Values are mean ± SEM. (*Source*: Data from Strader & Buntin (2001, 2003) and Ramakrishnan *et al.* (2007)).

Onset of Parental Hyperphagia

Male and female dove parents begin to increase their food intake during the early post-hatching period when prolactin has attained its peak concentrations in plasma (Goldsmith *et al.*, 1981; Buntin, *et al.*, 1996; Koch *et al.*, 2002). This temporal correspondence, together with the strong orexigenic effects of prolactin in non-breeding doves (Buntin & Tesch, 1985; Buntin & Figge, 1988; Buntin, 1989), suggests that elevated levels of prolactin at hatching could contribute to the onset of parental hyperphagia. One avenue by which prolactin could exert these effects is by regulating the activity of neuropeptides in the hypothalamus that influence appetite. For several reasons, NPY and AgRP neurons in the dove tuberal hypothalamus are promising candidates as targets of prolactin action in this regard. First, both of these peptides are orexigenic in doves and appear to be up-regulated in the tuberal hypothalamus during the period of parental provisioning (see previous section). In addition, prolactin receptors are concentrated in the tuberal hypothalamus where these neurons are located (Fechner & Buntin, 1989; Buntin *et al.*, 1993) and the tuberal hypothalamus is a site at which prolactin acts to promote feeding activity (Hnasko & Buntin, 1993), Third, non-breeding doves given ICV injections of prolactin show the same increase in NPY-and AgRP-immunoreactivity in the tuberal hypothalamus that parent doves show during the early post-hatching period when plasma prolactin levels are elevated (Strader & Buntin, 2001, 2003; Figure 17.7). Finally, in ring dove parents and in non-breeding doves given ICV prolactin injections, the increase in NPY staining in the tuberal hypothalamus is accompanied by increased numbers of cells that express NPY mRNA in this region (Ramakrishnan *et al.*, 2007; Figure 17.7). Collectively, these findings support the idea that parental hyperphagia is initiated, at least in part, by prolactin-stimulated elevations in the synthesis and release of hypothalamic NPY and AgRP during the early post-hatching period.

Maintenance of Parental Hyperphagia

Plasma prolactin remains at peak levels in parent doves from the time of hatching until the young reach 3–4 days of age. Thereafter, levels gradually decline until they reach pre-incubation BL concentrations between the second and third week post-hatching (Goldsmith *et al.*, 1981; Lea *et al.*, 1986; Buntin *et al.*, 1996). This suggests that while prolactin may be responsible for initiating parental hyperphagia, it is unlikely to contribute to the sustained elevation in food intake that parents show while provisioning older young.

Despite a doubling of daily food intake, parental body mass decreases by 6–8% as a result of the frequent transfer of crop contents to the young during parental regurgitation (Koch *et al.*, 2002; Ramakrishnan *et al.*, 2007). This negative energy state is likely to stimulate parental foraging and hyperphagia during later stages of the post-hatching period, since non-breeding doves that enter a similar negative energy state as a result of food restriction expend a considerable amount of energy engaging in appetitive behaviors that culminate in food reward when tested under progressive ratio schedules in an operant conditioning paradigm (Gamoke *et al.*, 2000). One possible candidate for mediating the effects of negative energy state on feeding activity at this stage is corticosterone. The energetic challenges associated with food restriction and food deprivation are consistently associated with activation of the hypothalamo–pituitary–adrenal axis and increased plasma corticosterone levels in birds (Lea *et al.*, 1992; Pravosudov *et al.*, 2001; Lynn *et al.*, 2003; Rajman *et al.*, 2006). In addition, corticosterone administration has been reported to stimulate foraging activity and food consumption in several avian species (Nagra *et al.*, 1963; Siegel & Van Kampen, 1984; Pravosudov, 2003).

Although the role of corticosterone in regulating parental provisioning by male dove parents remains to be clarified, evidence collected in female parents suggests that energy deficits that develop during the post-hatching period activate the hypothalamo–pituitary–adrenal axis, which in turn results in elevations in plasma corticosterone secretion that help sustain parental foraging and hyperphagia. Plasma levels of corticosterone in breeding females increase by 60% at hatching and remain elevated throughout the post-hatching period of parental provisioning (Koch *et al.*, 2002). In addition, non-breeding female doves given subcutaneous implants of corticosterone increase their food consumption by over 50% (McNeil *et al.*, 2002) and non-breeding doves of both sexes show similar elevations following systemic or intracranial administration of dexamethasone, a synthetic glucocorticoid (Koch *et al.*, 2002). Because the feeding response induced in doves by ICV dexamethasone can be completely blocked by the mammalian glucocorticoid receptor antagonist RU486 (Koch *et al.*, 2002; Figure 17.8), the orexigenic action of corticosterone is likely to be mediated in large part by mammalian-type glucocorticoid receptors resembling those characterized in house sparrow brain by Breuner & Orchinik (2001).

In contrast to females, plasma corticosterone levels in male parents remain unchanged after hatching (Lea *et al.*, 1992; Koch *et al.*, 2002), which raises questions about the functional significance of the robust feeding responses that non-breeding males display in response to subcutaneous or ICV administration of glucocorticoids (Koch *et al.*, 2002). Although these results could reflect a sex difference in hormonal requirements for parental provisioning, males may also show changes in tissue sensitivity to corticosterone after hatching, which could increase the biological potency of corticosterone in the face of unchanging plasma corticosterone titers. Alternatively, male parents may exhibit a decrease in the fraction of total circulating corticosterone bound to plasma proteins such as corticosteroid binding globulin (CBG) and an increase in the fraction circulating in the free form, which has higher bioactivity. The fact that similar sex-specific and breeding stage-specific changes in CBG binding capacity have been reported in other birds lends support to this argument (Silverin, 1986).

Interestingly, the orexigenic peptides NPY and AgRP that are implicated in prolactin-induced feeding may also play a role in corticosterone-induced feeding activity. Numbers of AgRP-and NPY-immunoreactive cells in the dove tuberal hypothalamus are increased by 50–500% during negative energy states such as food deprivation and chronic food restriction that would be expected to elevate plasma corticosterone levels through hypothalamo–pituitary–adrenal axis activation (Strader & Buntin, 2001, 2003; Strader *et al.*, 2003). Moreover, 48-h food deprivation results in a 3-fold increase in the numbers of tuberal hypothalamic cells that exhibit NPY mRNA in male doves (Ramakrishnan *et al.*, 2007). In addition to increasing AgRP-immunoreactivity in the tuberal hypothalamus, corticosterone administration also regulates the activity of NPY neurons in this region, albeit in complex ways (McNeil *et al.*, 2002). While corticosterone administration significantly increased the number of NPY-ir cells in the dorsolateral tuberal hypothalamus, it decreased the number of NPY-ir cells in the ventromedial tuberal hypothalamus. Although interpretation of these results is complicated by the lack of information on changes in NPY gene expression and peptide release in these neurons, it is clear from these

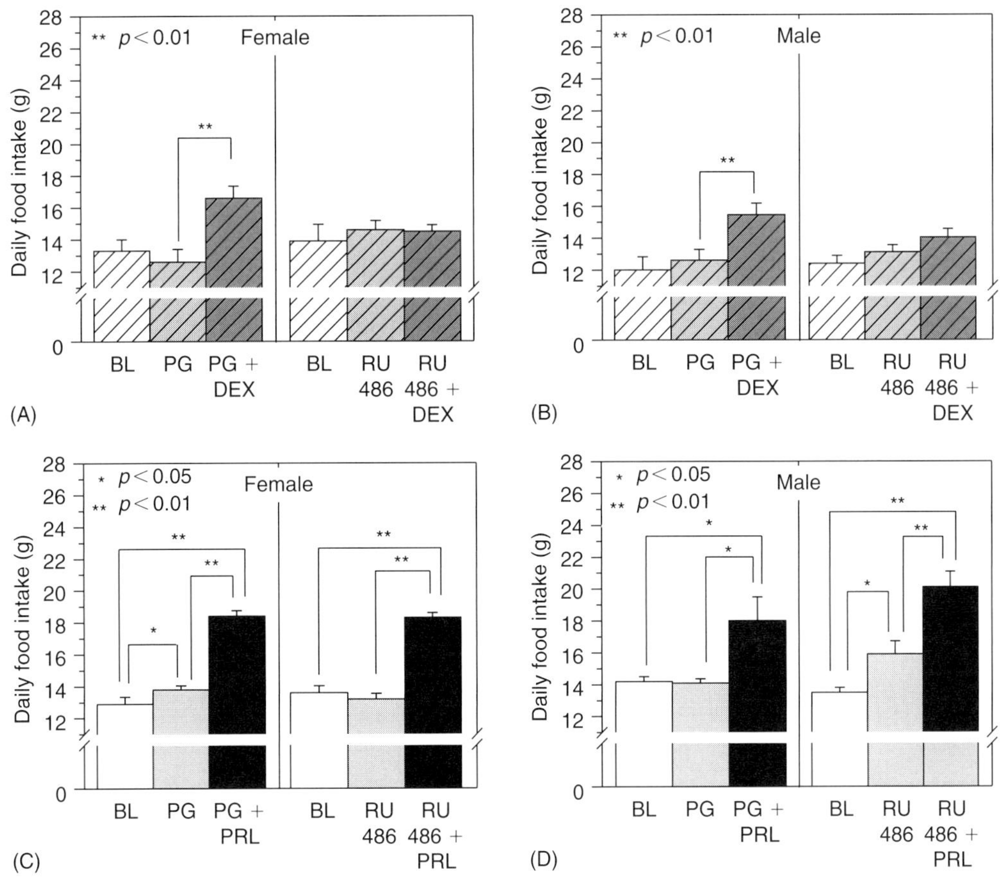

FIGURE 17.8 Effects of the glucocorticoid antagonist RU486 on dexamethasone-induced and prolactin-induced food intake in non-breeding male and female ring doves. Food intake was recorded daily during a 7-day pre-treatment or BL period and during a subsequent 7-day period of chronic ICV infusion of RU486 (100 μg/day) or propylene glycol (PG) vehicle. Birds were then given 5 days of twice-daily ICV injections of dexamethasone (A, B; 1 μg/2 μL) or ovine prolactin (C, D; 1 μg/2 μL) while ICV infusions of RU486 or PG continued. Values are mean ± SEM. (*Source*: Data from Koch *et al.* (2002, 2004)).

results that the orexigenic action of corticosterone is accompanied by significant alterations in the activity of NPY and AgRP neurons in the tuberal hypothalamus.

Do Corticosterone and Prolactin Act Independently to Stimulate Feeding in Doves?

While the negative energy state that develops in parents as a result of provisioning their young may be an important stimulus for maintaining elevated plasma corticosterone levels during later stages of the post-hatching period, other signals may be responsible for the initial increase in plasma corticosterone that occurs in female dove parents soon after hatching (Lea *et al.*, 1992; Koch *et al.*, 2002). The high levels of prolactin that are circulating in blood during the early post-hatching period could play a role in this regard, since Koch *et al.* (2004) found that plasma levels of corticosterone more than doubled when non-breeding doves were given ICV injections of prolactin (Figure 17.9). This idea is supported by evidence for similar effects of ICV administration of prolactin on plasma corticosterone levels in rodents (de Greef *et al.*, 1995) and by evidence from other mammalian studies for stimulatory effects of prolactin on corticotropin-releasing

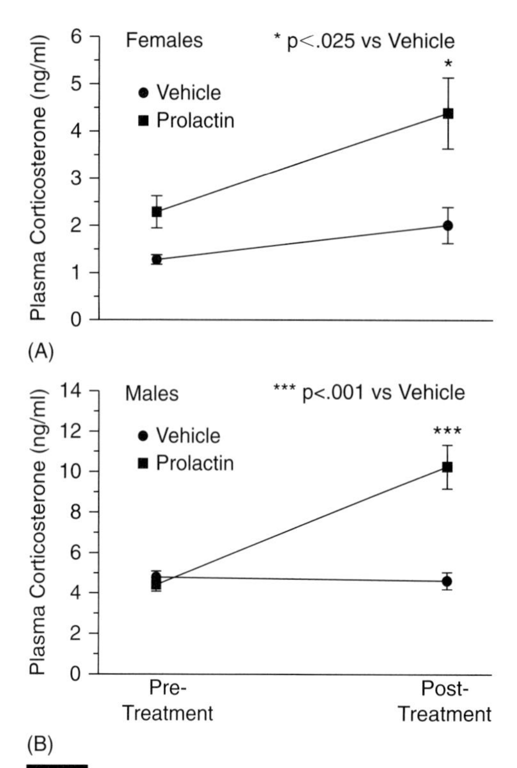

FIGURE 17.9 Stimulatory effects of ICV prolactin injections on plasma corticosterone levels in non-breeding male and female ring doves. Pre-treatment blood samples were obtained at 16:00 h and 04:00 h on a LD14:10 photoperiod with lights on at 06:30 h. Birds were then given twice-daily injections of ovine prolactin (1 μg/2 μL) or vehicle for six consecutive days, followed by blood sampling at 16:00 h and 04:00 h. The pre- and post-treatment corticosterone values shown here are averages of those obtained at the two sampling times. Values are mean ± SEM. (*Source*: Data from Koch *et al.* (2004)).

hormone (CRH) and adrenocorticotropic hormone (ACTH) secretion *in vivo* and *in vitro* (Kooy *et al.*, 1990; Weber & Calogero, 1991). In addition to influencing adrenal glucocorticoid secretion by acting at the level of the hypothalamus, prolactin could conceivably act directly on the adrenal cortex to potentiate the effects of ACTH on steroidogenesis, as has been demonstrated in mammals (Eldridge & Lymangrover, 1984; Albertson *et al.*, 1987; O'Connell *et al.*, 1994).

Based on evidence that prolactin elevates corticosterone levels in plasma, and evidence that both of these peptides exert their orexigenic effects in part by modulating the activity of NPY and AgRP in the tuberal hypothalamus, it could be argued that corticosterone-induced glucocorticoid signaling is responsible for the stimulatory effects of prolactin on food consumption in female dove parents (and possibly males as well; see above) during the early post-hatching period. Koch *et al.* (2004) tested this hypothesis by examining the effects of continuous ICV infusion of the glucocorticoid receptor blocker RU486 on the ability of ICV-injected prolactin to increase food consumption in non-breeding doves of both sexes. The results of this experiment (Figure 17.8) clearly indicated that prolactin-induced feeding is not diminished by a dose of RU486 that was previously shown to completely abolish the orexigenic effects of the synthetic glucocorticoid dexamethasone (Koch *et al.*, 2002). These results suggest that prolactin could stimulate feeding, at least in part, by interacting with target sites in the dove brain to increase hypothalamo–pituitary–adrenocortical axis activation and corticosterone secretion. Nevertheless, even though signaling in the brain via a mammalian-type glucocorticoid receptor is an effective pathway by which glucocorticoids such as corticosterone could elevate feeding in doves, it is not essential for prolactin-induced hyperphagia.

A Model for the Neuroendocrine Control of Parental Hyperphagia in Doves

Figure 17.10 summarizes our current understanding of the hormonal and neurochemical mechanisms involved in the hyperphagia exhibited by parent doves while provisioning their young. The model that is presented is restricted to female doves because the current evidence for glucocorticoid involvement in parental hyperphagia in stronger in females than in males. However, as discussed above, a sex difference in adrenal involvement may, with additional investigation, turn out to be more apparent than real, with females exhibiting post-hatching changes in plasma corticosterone concentrations and males exhibiting post-hatching changes in corticosterone bioactivity.

The model posits that there is a shift in hormonal regulation of parental provisioning of young as the post-hatching period proceeds.

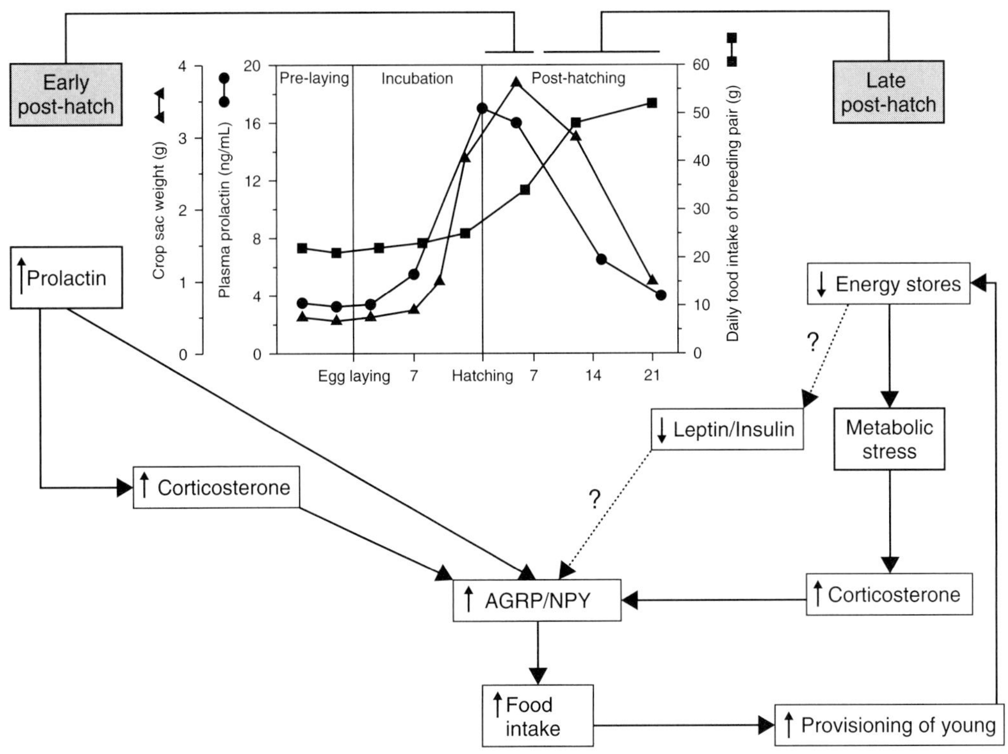

FIGURE 17.10 A model of the neuroendocrine control of parental hyperphagia in female ring doves. Post-hatching elevations in parental food intake appear to be initiated by the high levels of circulating prolactin that are attained during the early post-hatching period and later maintained by elevations in circulating corticosterone that result from energy depletion brought on by intensive parental provisioning activity. Reductions in the secretion of appetite-suppressing hormones such as insulin and leptin could conceivably contribute to parental hyperphagia, and other appetite-regulating neuropeptide systems in the brain other than NPY/AgRP may be involved in mediating hormonal effects. However, these questions remain to explored. See text for further details.

The early post-hatching period is dominated by high levels of prolactin in the blood, which are likely to initiate parental hyperphagia, at least in part, by increasing the activity of NPY- and AgRP-secreting neurons in the tuberal hypothalamus. In females, some of the orexigenic actions of prolactin may well be mediated by increases in plasma corticosterone levels. However, prolactin-induced feeding persists when glucocorticoid signaling is inhibited, which suggests multiple mechanisms of prolactin action in promoting the feeding response.

Despite substantial elevations in parental food intake, the frequent transfer of seeds to the young by parental regurgitation eventually leads to negative energy balance in dove parents. Presumably, this could generate metabolic stress signals that activate the hypothalamo–pituitary–adrenal axis and result in increased secretion of corticosterone. As plasma concentrations of prolactin decline during the second half of the post-hatching period, the sustained elevation in BL plasma corticosterone that results from depletion of energy stores could be a major stimulus for maintaining parental hyperphagia during the remainder of the parental provisioning period. Although the possibility has yet to be experimentally investigated, the hyperphagic actions of corticosterone could be reinforced by reductions in the secretion or signaling potency of appetite-suppressing hormones such as leptin and insulin in parent doves, which would be expected to accompany energy depletion.

Summary and Conclusions

Birds, like mammals, face formidable energetic challenges in raising their offspring. Prior to birth or hatching of their young, the patterns of energy expenditure that parent birds and mammals exhibit, such as those associated with egg production and incubation behavior in birds and those associated with sustaining placental support of embryonic growth and development during pregnancy in mammals, are unique to the oviparous and viviparous modes of reproduction that are characteristic of these two groups. After birth or hatching, avian and mammalian parents are faced with the same energetic challenges of providing adequate food for their rapidly growing offspring, despite the substantial differences in the way in which the food is delivered. A comparison of the physiological mechanisms that birds and mammals enlist in meeting these challenges is facilitated by the fact that the avian group in which these mechanisms have been best characterized, the Columbiformes, is also the group that has evolved a mode of parental provisioning that most closely resembles the mammalian pattern of lactation.

Comparisons based on data from the best studied avian species (the ring dove) and the best studied mammalian species (the rat) suggest either strong evolutionary conservation or remarkable evolutionary convergence in the hormonal and neurochemical mechanisms involved in meeting the energetic demands associated with provisioning young in these two vertebrate taxa. In both species, parents must increase their food intake dramatically in order to meet these demands. Elevated food intake is necessary to support lactogenesis in rats, while in doves, it serves the dual purpose of initially supporting crop milk production and later providing food that is directly transferred to the young by parental regurgitation. Prolactin is a critical agent in stimulating and integrating the physiological and the behavioral events associated with these provisioning efforts in both species. Physiologically, prolactin plays a central role in stimulating mammary gland growth and lactogenesis in rats and crop sac growth and crop milk production in doves. In addition, intracranial injections or infusions of prolactin have been shown to elevate food intake substantially when administered directly to the brain in both doves (see above) and rats (Sauve & Woodside, 1996, 2000). A specific linkage between prolactin and lactational hyperphagia is also supported by evidence that elevated food intake is blunted or abolished in lactating female rat dams given the dopamine agonist and prolactin secretion inhibitor bromocryptine, an effect that can be reversed when bromocryptine treatment is combined with intracranial administration of prolactin (Woodside, 2007).

Similarities in hormonal determinants of lactational hyperphagia in rats and parental hyperphagia in dove are paralleled by similarities in the neurochemical mechanisms that mediate these effects. For several reasons, the neuropeptides NPY and AgRP are likely to play a central role in mediating parental hyperphagia in both species. In addition to being strongly orexigenic in both doves (see above) and rats (see Morton & Schwartz, 2001 for review), the NPY/AgRP neuronal system is upregulated during the period of parental provisioning in specific regions of the rat and dove hypothalamus that are implicated in appetite regulation (dove: see above: rat: Malabu *et al.*, 1994; Pickavance *et al.*, 1996; Li *et al.*, 1999; Garcia *et al.*, 2003; Chen & Smith, 2004; Xiao *et al.*, 2005). In addition, there is direct evidence that prolactin increases NPY expression in the hypothalamus of both species. In lactating rats, Chen & Smith (2004) found that the increase in NPY mRNA expression that occurs in the dorsomedial nucleus of the hypothalamus following suckling stimulation was greatly attenuated when prolactin levels in blood were reduced by administration of the dopamine agonist bromocryptine. Because this suppression was reversible by co-injection of prolactin, and because prolactin receptor and NPY are co-localized in neurons in this region, it is likely that prolactin potentiates suckling induced increases in NPY in this nucleus, which in turn could result in increased food intake via its projection to appetite-regulating neurons in the paraventricular nucleus (PVN) of the hypothalamus (Li *et al.*, 1998). Although prolactin has similar effects on NPY expression in the dove tuberal hypothalamus, NPY neurons in the homologous region of the rat brain, the arcuate nucleus, did not respond to prolactin treatment. Although this suggests that some anatomical differences exist in the neural substrates involved, these findings nevertheless point to remarkable similarities between the two species in the hormonal and neurochemical mechanisms regulating parental hyperphagia.

ACKNOWLEDGMENTS

We wish to thank Dr. A. F. Parlow, NIDDK, for providing the purified ovine prolactin that was used in these studies. We also wish to thank Michelle Ranic and Leah Eisenberg for assistance with data collection. This work was supported by PHS GrantMH41447 from the NIMH.

REFERENCES

Albertson, B. D., Siekiewicz, M. L., Kimball, D., Munabi, A. K., Cassorla, F., and Loriaux, D. L. (1987). New evidence for a direct effect of prolactin on rat adrenal steroidogenesis. *Endocr. Res.* 13, 317–333.

Badyaev, A. V., and Hill, G. E. (2002). Paternal care as a conditional strategy: Distinct reproductive tactics associated with elaboration of plumage ornamentation in the house finch. *Behav. Ecol.* 13, 591–597.

Badyaev, A. V., and Duckworth, R. A. (2005). Evolution of plasticity in hormonally integrated parental tactics. In *Functional Avian Endocrinology* (A. Dawson and P. J. Sharp, Eds.), pp. 375–386. Narosa Publishing House, New Delhi.

Bates, R. W., Miller, R. A., and Garrison, M. M. (1962). Evidence in the hypophysectomized pigeon of a synergism among prolactin, growth hormone, thyroxine, and prednisone upon weight of the body digestive tract, kidney, and fat stores. *Endocrinology* 71, 260–345.

Bengtsson, H., and Ryden, O. (1983). Parental feeding rate in relation to begging behavior in asynchronously hatched broods of the great tit. *Parus major*. *Behav. Ecol. Sociobiol.* 12, 243–251.

Boswell, T., Li, Q., and Takeuchi, S. (2002). Neurons expressing neuropeptide Y mRNA in the infundibular hypothalamus of Japanese quail are activated by fasting and co-express agouti-related protein mRNA. *Brain Res. Mol. Brain Res.* 100, 31–42.

Breuner, C. W., and Orchinik, M. (2001). Seasonal regulation of membrane and intracellular corticosterone receptors in the house sparrow brain. *J. Neuroendocrinol.* 13, 412–420.

Brisbin, I. L. (1969). Bioenergetics of the breeding cycle of the ring dove. *Auk* 86, 54–74.

Brown, J. L., and Vleck, C. M. (1998). Prolactin and helping in birds: has natural selection strengthened helping behavior?. *Behav. Ecol.* 9, 541–545.

Buntin, J. D. (1989). Time course and response specificity of prolactin-induced hyperphagia in ring doves. *Physiol. Behav.* 45, 903–909.

Buntin, J. D. (1996). Neural and hormonal control of parental behavior in birds. *Adv. Study Behav.* 25, 161–213.

Buntin, J. D., and Tesch, D. (1985). Effects of intracranial prolactin administration on maintenance of incubation readiness, ingestive behavior, and gonadal condition in ring doves. *Horm. Behav.* 19, 188–203.

Buntin, J. D., and Ruzycki, E. (1987). Characterization of prolactin binding sites in the brain of the ring dove (*Streptopelia risoria*). *Gen. Comp. Endocrinol.* 65, 243–253.

Buntin, J. D., and Figge, G. R. (1988). Prolactin and growth hormone stimulate food intake in ring doves. *Pharmacol. Biochem. Behav.* 31, 533–540.

Buntin, J. D., and Walsh, R. J. (1988). *In vivo* autoradiographic analysis of prolactin binding sites in brain and choroid plexus of the domestic ring dove. *Cell Tissue Res.* 251, 105–109.

Buntin, J. D., Ruzycki, E., and Witebsky, J. (1993). Prolactin receptors in dove brain: Autoradiographic analysis of binding characteristics in discrete brain regions and accessibility to blood-borne prolactin. *Neuroendocrinology* 57, 738–750.

Buntin, J. D., Becker, G. M., and Ruzycki, E. (1991). Facilitation of parental behavior in ring doves by systemic and intracranial injections of prolactin. *Horm. Behav.* 25, 424–444.

Buntin, J. D., Hnasko, R. M., Zuzick, P. H., Valentine, D. L., and Scammell, J. G. (1996). Changes in bioactive prolactin-like activity in plasma and its relationship to incubation behavior in breeding ring doves. *Gen. Comp. Endocrinol.* 102, 221–232.

Buntin, L., Berghman, L. R., and Buntin, J. D. (2006). Patterns of fos-like immunoreactivity in the brains of parent ring doves (*Streptopelia risoria*) given tactile and non-tactile exposure to their young. *Behav. Neurosci.* 120, 651–664.

Burley, N. (1980). Clutch overlap and clutch size: Alternative and complementary reproductive tactics. *Am. Nat.* 115, 223–246.

Chen, P., and Smith, M. S. (2004). Regulation of hypothalamic neuropeptide Y messenger ribonucleic acid expression during lactation:Rrole of prolactin. *Endocrinology* 145, 823–829.

Criscuolo, F., Bertile, F., Durant, J. M., Raclot, T., Gabrielson, G. W., Massemin, S., and Chastel, O. (2006). Body mass and clutch size may modulate prolactin and corticosterone levels in eiders. *Physiol. Biochem. Zool.* 79, 514–521.

de Greef, W. J., Ooms, M. P., Vreeberg, J. T. M. and Weber, R. F. A. (1995). Plasma levels of luteinizing hormone during hyperprolactinemia: response to central administration of antagonists of corticotropin-releasing factor. *Neuroendocrinology* 61, 19–26.

denBoer-Visser, A. M., and Dubbeldam, J. L. (2002). The distribution of dopamine, substance P, vasoactive intestinal polypeptide and neuropeptide

Y immunoreactivity in the brain of the collared dove, *Streptopelia decaocto*. *J. Chem. Neuroanat.* 23, 1–27.

Drent, R. H., and Daan, S. (1980). The prudent parent: Energetic adjustments in avian breeding. *Ardea* 68, 225–252.

Eldridge, J. C., and Lymangrover, J. R. (1984). Prolactin stimulates and potentiates adrenal steroid secretion *in vitro*. *Horm. Res.* 20, 252–260.

Fechner, J. H., Jr., and Buntin, J. D. (1989). Localization of prolactin binding sites in ring dove brain by quantitative autoradiography. *Brain Res.* 487, 245–254.

Foreman, K. T., Lea, R. W., and Buntin, J. D. (1990). Changes in feeding activity, plasma LH, and testes weight in ring doves following hypothalamic injections of prolactin. *J. Neuroendocrinol.* 2, 667–673.

Frith, H. J. (1956). Breeding habits in the family Megapodiidae. *Ibis* 98, 620–640.

Gamoke, C., Moore, J. C., and Buntin, J. D. (2000). Motivational influences underlying prolactin-induced feeding in doves (*Streptopelia risoria*). *Behav. Neurosci.* 114, 963–971.

Garcia, M. C., López, M., Gualillo, O., Seoane, LM., Diéguez, C., and Senaris, R. M. (2003). Hypothalamic levels of NPY, MCH, and prepro-orexin mRNA during pregnancy and lactation in the rat: Role of prolactin. *FASEB J.* 17, 1392–1400.

Gittleman, J. L., and Thompson, S. J. (1988). Energy allocation in mammalian reproduction. *Am. Zool.* 28, 863–875.

Goldsmith, A. R., Edwards, C., Koprucu, M., and Silver, R. (1981). Concentrations of prolactin and luteinizing hormone in plasma of doves in relation to incubation and development of the crop gland. *J. Endocrinol.* 103, 251–256.

Groscolas, R., and Robin, J.-P. (2001). Long-term fasting and re-feeding in penguins. *Comp. Biochem. Physiol. A.* 128, 645–655.

Hahn, T., Breininger, J. F., Baskin, D. G., and Schwartz, M. W. (1998). Coexpression of AgRP and NPY in fasting activated hypothalamic neurons. *Nature Neurosci.* 1, 271–272.

Hansen, E. W. (1971). Squab-induced crop growth in experienced and inexperienced ring dove (*Streptopelia risoria*) foster parents. *J. Comp. Physiol. Psychol.* 77, 375–381.

Hnasko, R. M., and Buntin, J. D. (1993). Functional mapping of neural sites mediating prolactin-induced hyperphagia in doves. *Brain Res.* 623, 257–266.

Horseman, N. D., and Buntin, J. D. (1995). Regulation of pigeon milk crop milk secretion and parental behaviors by prolactin. *Annu. Rev. Nutr.* 15, 213–238.

Ikemoto, S., and Panksepp, J. (1996). Dissociations between appetitive and consummatory responses by pharmacological manipulations of reward-relevant b rain regions. *Behav. Neurosci.* 110, 331–345.

Jewett, D. C., Cleary, J., Levine, A. S., Schaal, D. W., and Thompson, T. (1992). Effects of neuropeptide Y on food-reinforced behavior in satiated rats. *Pharm. Biochem. Behav.* 42, 207–212.

Jewett, D. C., Cleary, J., Levine, A. S., Schall, D. W., and Thompson, T. (1995). Effects of neuropeptide Y, insulin, 2-deoxyglucose, and food deprivation on food-motivated behavior. *Psychopharmacology* 120, 267–271.

Kameda, Y., Miura, M., and Nishimaki, T. (2001). Localization of neuropeptide Y mRNA and peptide in chicken hypothalamus and their alterations after food deprivation, dehydration, and castration. *J. Comp. Neurol.* 436, 376–388.

Kitaysky, A. S., Wingfield, J. C., and Piatt, J. F. (1999). Dynamics of food availability, body condition and physiological stress response in breeding black-legged kittiwakes. *Funct. Ecol.* 13, 577–584.

Kitaysky, A. S., Wingfield, J. C., and Piatt, J. F. (2001). Corticosterone facilitates begging and affects resource allocation in the black-legged kittiwake. *Behav. Ecol.* 12, 619–625.

Klinghammer, E., and Hess, E. H. (1964). Parental feeding in ring doves (*Streptopelia risoria*): innate or learned? *Z. Tierpsychol.* 21, 338–347.

Koch, K. A., Wingfield, J:C., and Buntin, J. D. (2002). Glucocorticoids and parental hyperphagia in ring doves (*Streptopelia risoria*). *Horm. Behav.* 41, 9–21.

Koch, K. A., Wingfield, J. C., and Buntin, J. D. (2004). Prolactin-induced parental hyperphagia in ring doves: Are glucocorticoids involved? *Horm. Behav.* 46, 498–505.

Kooy, A., de Greef, W. J., Vreeburg, J. T., Hackeng, W. H., Ooms, M. P., Lamberts, S. W., and Weber, R. F. (1990). Evidence for the involvement of corticotropin-releasing factor in the inhibition of gonadotropin release induced by hyperprolactinemia. *Neuroendocrinology* 51, 261–266.

Kuenzel, W. J., and McMurtry, J. (1988). Neuropeptide Y. Brain localization and cental effects on plasma levels in chicks. *Physiol. Behav.* 44, 669–678.

Lang, E. M. (1963). Flamingoes raise their young on a liquid containing blood. *Experientia* 19, 532–533.

Lea, R. W., Sharp, P. J., and Chadwick, A. (1982). Daily variations in the concentration of plasma prolactin in broody bantams. *Gen. Comp. Endocrinol.* 48, 275–284.

Lea, R. W., Vowles, D. M., and Dick, H. R. (1986). Factors affecting prolactin secretion during the breeding cycle of the ring dove (*Streptopelia risoria*) and its possible role in incubation. *J. Endocrinol.* 110, 447–458.

Lea, R. W., Klandorf, H., Harvey, S., and Hall, T. R. (1992). Thyroid and adrenal function in the ring dove (*Streptopelia risoria*) during food deprivation and a breeding cycle. *Gen. Comp. Endocrinol.* 86, 136–146.

Lehrman, D. S. (1955). The physiological basis of parental feeding behavior in the ring dove (*Streptopelia risoria*). *Behaviour* 7, 241–286.

Li, C., Kelly, P. A., and Buntin, J. D. (1995). Inhibitory effects of anti-prolactin receptor antibodies on prolactin binding in brain and prolactin-induced feeding behavior in ring doves. *Neuroendocrinology* 61, 125–135.

Li, C., Chen, P., and Smith, M. S. (1998). Neuropeptide Y (NPY) neurons in the arcuate nucleus (ARH) and dorsomedial nucleus (DMH), areas activated during lactation, project to the paraventricular nucleus of the hypothalamus (PVN). *Regul. Pept.* 75(75), 93–100.

Li, C., Chen, P., and Smith, M. S. (1999). Neuropeptide Y and tuberoinfundibular dopamine activities are altered during lacation: Role of prolactin. *Endocrinology* 140, 118–123.

Lynn, S. E., Breuner, C. W., and Wingfield, J. C. (2003). Short-term fasting affects locomotor activity, corticosterone, and corticosterone binding globulin in a migratory songbird. *Horm. Behav.* 43, 150–157.

Malabu, U. H., Kilpatrick, A., Ware, M., Vernon, R. G., and Williams, G. (1994). Increased neuropeptide Y concentrations in specific hypothalamic regions of lactating rats: Possible relationship to hyperphagia and adaptive changes in energy balance. *Peptides* 15, 83–87.

McNeil, A. M., Strader, A. D., and Buntin, J. D. (2002). Corticosterone influences energy state and orexigenic neuropeptides in female ring doves. *Horm. Behav.* 41, 480. (abstract)

Mirabella, N., Esposito, V., Squillacioti, C., DeLuca, A., and Paino, G. (2004). Expression of agouti-related protein (AgRP) in the hypothalamus and adrenal gland of the duck (*Anas platyrhynchos*). *Anat. Embryol. (Berl.)* 209, 137–141.

Mondloch, C. (1995). Chick hunger and begging affect parental allocation of feedings in pigeons. *Anim. Behav.* 49, 601–613.

Mondloch, C. J., and Timberlake, W. (1991). The effect of parental food supply on parental feeding and squab growth in pigeons, *Columba livia*. *Ethology* 88, 236–248.

Morton, G. J., and Schwartz, M. W. (2001). The NPY/AgRP neuron and energy homeostasis. *Int. J. Obes. Relat. Metab. Disord.* 25(suppl 5), S56–S62.

Mrosovsky, N., and Sherry, D. F. (1980). Animal anorexias. *Science* 22, 837–842.

Murton, R. K., Isaacson, A. J., and Westwood, N. J. (1963). The food and growth of nestling wood-pigeons in relation to the breeding season. *Proc. Zool. Soc. London* 141, 747–781.

Nagra, C. L., Breitenbach, R. P., and Meyer, R. K. (1963). Influence of hormones on food intake and lipid deposition in castrated pheasants. *Poultry Sci.* 42, 77–775.

O'Connell, Y., McKenna, T. J., and Cunningham, S. K. (1994). The effect of prolactin, human chorionic gonadotropin, insulin, and insulin-like growth factor 1 on adrenal steroidogenesis in isolated guinea pig adrenal cells. *J. Steroid Biochem. Mol. Biol.* 48, 235–240.

O'Dwyer, T. W., Buttemer, W. A., Priddel, D. M., and Downing, J. A. (2006). Prolactin, body condition, and the cost of good parenting: An interyear study in a long-lived seabird, Gould's Petrel (*Pterodroma leucoptera*). *Funct. Ecol.* 20, 806–811.

Ottosson, U., Backman, J., and Smith, H. (1997). Begging affects parental effort in the pied flycatcher, *Ficedula hypoleuca*. *Behav. Ecol. Sociobiol.* 41, 381–384.

Patel, M. D. (1936). The physiology of the formation of "pigeon's milk". *Physiol. Zool.* 9, 129–152.

Peterson, C. C., Nagy, K. A., and Diamond, J. (1990). Sustained metabolic scope. *Proc. Natl. Acad. Sci. USA* 87, 2324–2328.

Pickavance, L., Dryden, S., Hopkins, D., Bing, C., Frankish, H., Wang, Q., Vernon, R. G., and Williams, G. (1996). Relationships between hypothalamic neuropeptide Y and food intake in the lactating rat. *Peptides* 17, 577–582.

Pravosudov, V. V. (2003). Long-term moderate elevation of corticosterone facilitates avian food-caching behaviour and enhances spatial memory. *Proc. Roy. Soc. Lond. B.* 270, 2599–2604.

Pravosudov, V. V., Kitaysky, A. S., Wingfield, J. C., and Clayton, N. S. (2001). Long-term unpredictable foraging conditions and physiological stress response in mountain chickadees (*Poecile gambeli*). *Gen. Comp. Endocrinol.* 123, 324–331.

Prevost, J., and Vilter, V. (1963). Histologie de la secretion oesophagienne du Manchot empereur. *Proc. 13th Int. Ornithol.Cong.*, 1085–1094.

Price, K., and Ydenberg, R. (1995). Begging and provisioning in broods of asynchronously hatched yellow-headed blackbird nestlings. *Behav. Ecol. Sociobiol.* 37, 201–208.

Rajman, M., Juráni, M., Lamosová, D., Mácajová, Sedlacková, M., Kost'ál, L., Jezová, D., and Vyboh, P. (2006). The effect of feed restriction on plasma biochemistry in growing meat-type chickens (*Gallus gallus*). *Comp. Biochem. Physiol. A. Mol. Integr. Physiol.* 145, 363–371.

Ramakrishnan, S., Strader, A. D., Wimpee, B., Chen, P., Smith, M. S., and Buntin, J. D. (2007). Evidence for increased neuropeptide Y synthesis in mediobasal hypothalamus in relation to parental hyperphagia and gonadal activation in breeding ring doves. *J. Neuroendocrinol.* 19, 163–171.

Ramesh, R., Kuenzel, W. J., Buntin, J. D., and Proudman, J. A. (2000). Identification of growth-hormone- and prolactin-containing neurons within the avian brain. *Cell Tissue Res.* 299, 371–383.

Ramos, C., and Silver, R. (1992). Gonadal hormones determine sex differences in timing of incubation by doves. *Horm. Behav.* 26, 586–601.

Rapaport, L. (2006). Provisioning in wild golden lion tamarins (*Leontopithicus rosalia*): Benefits to omnivorous young. *Behav. Ecol.* 17, 212–221.

Richard-Yris, M.-A., Garnier, D. H., and Leboucher, G. (1983). Induction of maternal behavior and some hormonal and physiological correlates in the domestic hen. *Horm. Behav.* 17, 345–355.

Ricklefs, R. E. (1974). Energetics of reproduction in birds. In *Avian Energetics* (R. A. Paynter, Ed.), pp. 152–297. Nuttall Ornithological Club, Cambridge MA.

Roberts, S. B., Cole, T. J., and Coward, W. A. (1985). Lactational performance in relation to energy intake in the baboon. *Am. J. Clin. Nutr.* 41, 1270–1276.

Saino, N., Ninni, P., Incagli, M., Calza, S., Sacchi, R., and Møller, A. P. (2000). Begging and parental care in relation to offspring need and condition in the barn swallow (*Hirundo rustica*). *Am. Nat.* 156, 637–649.

Sauve, D., and Woodside, B. (1996). The effect of central administration of prolactin on food intake in virgin female rats is dose-dependent, occurs in the absence of ovarian hormones, and the latency to onset varies with feeding regimen. *Brain Res.* 729, 75–81.

Sauve, D., and Woodside, B. (2000). Neuroanatomical specificity of prolactin-induced hyperphagia in virgin female rats. *Brain Res.* 868, 306–314.

Schoech, S. J., Mumme, R. L., and Wingfield, J. C. (1996). Prolactin and helping behaviour in the cooperatively breeding Florida scrub-jay, *Aphelocoma c. coerulescens. Anim. Behav.* 52, 445–456.

Schooley, J. P., Riddle, O., and Bates, R. W. (1941). Replacement therapy in hypophysectomized juvenile pigeons. *Am. J. Anat.* 69, 124–154.

Sherry, D. F. (1981). Parental care and the development of thermoregulation in red junglefowl. *Behaviour* 76, 240–279.

Sherry, D. F., Mrosovsky, N., and Hogan, J. A. (1980). Weight loss and anorexia during incubation in birds. *J. Comp. Physiol. Psychol.* 94, 89–98.

Siegel, H. M., and Van Kampen, M. (1984). Energy relationships in growing chickens given daily injections of corticosterone. *Br. Poultry Sci.* 25, 477–485.

Silver, R., Andrews, H., and Ball, G. F. (1985). Parental care in an ecological perspective: a quantitative analysis of avian subfamilies. *Am. Zool.* 25, 823–840.

Silverin, B. (1982). Endocrine correlates of brood size in adult pied flycatchers, *Ficedula hypoleuca. Gen. Comp. Endocrinol.* 47, 18–23.

Silverin, B. (1986). Corticosterone-binding proteins and behavioral effects of high plasma levels of corticosterone during the breeding period in the pied flycatcher. *Gen. Comp. Endocrinol.* 64, 67–74.

Silverin, B. (1990). Testosterone and corticosterone and their relation to territoriality and parental behavior in the pied flycatcher. In *Hormones, Brain, and Behavior in Vertebrates. 2. Behavioral Activation in Males and Females-Social Interaction and Reproductive Endocrinology* (J. Balthazart, Ed.), pp. 129–144. S. Karger, Basel.

Slawski, B. A., and Buntin, J. D. (1995). Preoptic area lesions disrupt prolactin-induced parental feeding behavior in ring doves. *Horm. Behav.* 29, 248–266.

Strader, A. D., and Buntin, J. D. (2001). Neuropeptide Y: A possible mediator of prolactin-induced feeding and regulator of energy balance in the ring dove (*Streptopelia risoria*). *J. Neuroendocrinol.* 13, 386–392.

Strader, A. D., and Buntin, J. D. (2003). Changes in agouti-related peptide during the ring dove breeding cycle in relation to prolactin and parental hyperphagia. *J. Neuroendcrinol.* 15, 1046–1053.

Strader, A. D., Schiöth, H. G., and Buntin, J. D. (2003). The role of the melanocortin system and the melanocortin-4 receptor in ring dove (*Streptopelia risoria*) feeding behavior. *Brain Res.* 960, 112–121.

Torner, L., and Neumann, I. D. (2002). The brain prolactin system: Involvement in stress response adaptations in lactation. *Stress* 5, 249–257.

Van Roo, B. L., Ketterson, E. D., and Sharp, P. J. (2003). Testosterone and prolactin in two songbirds that differ in paternal care: The blue-headed vireo and the red-eyed vireo. *Horm. Behav.* 44, 435–441.

Vandeputte-Poma, J. (1980). Feeding, growth, and metabolism of the pigeon, *Columba livia domestica*: duration and role of crop milk feeding. *J. Comp. Physiol.* 135, 97–99.

Vleck, C. M., Mays, N. A., Dawson, J. W., and Goldsmith, A. R. (1991). Hormonal correlates of parental and helping behavior in cooperatively breeding Harris' hawks (*Parabuteo unicinctus*). *Auk* 108, 638–648.

Walsh, R. J., Slaby, F. J., and Posner, B. I. (1987). A receptor-mediated mechanism for the transport of prolactin from blood to cerebrospinal fluid. *Endocrinology* 120, 1846–1850.

Wang, Q., and Buntin, J. D. (1999). The roles of stimuli from young, previous breeding experience, and prolactin in regulating parental behavior in ring doves (*Streptopelia risoria*). *Horm. Behav.* 35, 241–253.

Weathers, W. W., and Sullivan, K. A. (1989). Juvenile foraging proficiency, parental effort, and avian reproductive success. *Ecol. Monogr.* 59, 223–246.

Weber, R. F., and Calogero, A. E. (1991). Prolactin stimulates rat hypothalamic corticotropin-releasing hormone and pituitary adrenocorticotropin secretion *in vitro*. *Neuroendocrinology* 54, 248–253.

Westmoreland, D., and Best, L. B. (1987). What limits mourning doves to a clutch of two eggs? *Condor* 89, 486–493.

Whittingham, L. A., and Robertson, R. J. (1993). Nestling hunger and parental care in red-winged blackbirds. *Auk* 110, 240–246.

Woodside, B. (2007). Prolactin and the hyperphagia of lactation. *Physiol. Behav.* 91, 375–382.

Xiao, X. Q., Grove, K. L., Lau, S. Y., McWeeney, S., and Smith, M. S. (2005). Deoxyribonucleic acid microarray analysis of gene expression pattern in the arcuate nucleus/ventromedial nucleus of the hypothalamus during lactation. *Endocrinology* 146, 4391–4396.

18 MATERNAL FAT INTAKE AND OFFSPRING BRAIN DEVELOPMENT: FOCUS ON THE MESOCORTICOLIMBIC DOPAMINERGIC SYSTEM

CLAIRE-DOMINIQUE WALKER, LINDSAY NAEF, ESTERINA D'ASTI, HONG LONG AND ZHIFANG XU

Department of Psychiatry, McGill University, Douglas Mental Health University Institute, 6875 LaSalle Boulevard, Montreal, Canada, H4H 1R3

INTRODUCTION

The role of the early pre and postnatal environment in shaping development of the offspring is well recognized. Maternal stress, dietary changes, and exposure to environmental factors such as drugs or toxins can have a profound effect on the phenotype of the offspring, altering brain growth, neurotransmission, and behavioral responses with an increased predisposition to pathologies in the adult (Walker & Plotsky, 2005, Kapoor *et al.*, 2006; Phillips, 2007). In particular, modifications in the hypothalamic–pituitary–adrenal (HPA) axis or the dopamine (DA) system have been demonstrated in the adult offspring using a number of different perinatal procedures including handling and maternal separation, natural variations in maternal care, and perinatal hypoxia. These manipulations have long lasting effects on physiology and behavior (Brake *et al.*, 2000, 2004; Fortier *et al.*, 2004; Flores *et al.*, 2005; Meaney & Szyf, 2005).

Maternal nutrition is a fundamental variable in the early environment affecting the developmental "blueprint" of the infant. Indeed nutrition exhibits a wide range of effects on brain development, not only during the brain growth spurt period, but also during early organizational processes such as cellular migration and differentiation, neurogenesis, synaptogenesis, and maturation of neurotransmitter pathways (Georgieff & Innis, 2005). Strong evidence that accumulated in the last decade shows that under- or over-nutrition during development is an important risk factor for later development of obesity and associated metabolic diseases (Godfrey & Barker, 2001; Plagemann, 2006; Gluckman *et al.*, 2007a). In an elegant set of studies, Gluckman *et al.* have extended the hypothesis developed by Barker (2002) suggesting that the mismatch between availability of caloric resources postnatally and that of relative undernutrition and low birth weight prenatally causes many of the genetic and epigenetic changes leading to metabolic and cardiovascular disorders in the adult (Gluckman *et al.*, 2007b; Godfrey *et al.*, 2007).

In addition, numerous reports have demonstrated the importance of intake of specific nutrients including proteins (Morgane *et al.*, 2002; Gallagher *et al.*, 2005) and specific fatty acids (Innis *et al.*, 2000b; Wainwright, 2002; Innis, 2007) in the maternal diet to optimize the offspring's development (Walker, 2005). For instance, specific protein deficiency during gestation leads to deficits in prepulse inhibition (PPI) and changes in glutamate and dopaminergic receptors in the hippocampus and striatum

(Palmer *et al.*, 2004). Prenatal protein malnutrition was also shown to perturb inhibitory activity in the hippocampal formation by increasing GABA release from interneurons (Morgane *et al.*, 2002; Chang *et al.*, 2003). The maternal intake of lipids is also crucial both pre and postnatally (Innis, 2004) for both infant growth, cell membrane formation, and myelinization, processes that can determine visual and mental functions later in childhood (Crawford, 1993; Carlson *et al.*, 1994; Koletzko *et al.*, 2001; Helland *et al.*, 2003). In this chapter, we examine how maternal fat intake and the intake of specific lipids can directly and indirectly, through metabolic effectors, regulate developmental brain processes in the offspring. These dietary effects are certainly not limited to those pathways typically implicated in the control of ingestive behavior and metabolism. Indeed, we have gathered exciting new data to suggest for instance that, in rats, perinatal high fat feeding can have a significant impact on dopaminergic transmission in the adult offspring (Naef, *et al.*, 2008). The mesocortical and mesolimbic DA circuits are central in linking the motivational and rewarding aspects of food to the hypothalamic centers controlling energy balance (Berridge, 1996; Berthoud, 2004; Kelley *et al.*, 2005). Since the mesocorticolimbic DA system is developing for a large part postnatally (Kalsbeek *et al.*, 1988; Antonopoulos *et al.*, 2002), it might thus be very susceptible to the "organizational effects" of early exposure to a high fat diet. Previous studies including our own (Proulx *et al.*, 2001, 2002; Walker *et al.*, 2007) suggesting that leptin can act in the neonate to shape neuronal circuits (Bouret *et al.*, 2004) raise the possibility that organizational effects of a high fat diet are mediated in part through hormonal modulators such as leptin. In this way the early nutritional environment may be able to "program" mesocortical and mesolimbic DA circuits. For example, one of the direct consequences of these "organizational" modifications might be to change the rewarding properties of drugs and/or specific food types.

This chapter will highlight some of the critical aspects of maternal fat intake for the development and long-term function of neurotransmitter systems and in particular the mesocortical and mesolimbic DA system in the offspring. We will first discuss salient developmental aspects of the dopaminergic system and examine the consequences of perinatal high fat feeding on adult DA function. Given that the precise mechanisms through which perinatal high fat feeding might affect the dopaminergic system are still unclear, we will examine the possible role of (1) metabolic hormones (i.e., leptin and insulin) that are increased by the high fat diet and (2) fatty acids that can directly modulate membrane coupled receptors, intracellular signaling systems, activity of ion channels, etc. on the development of the dopaminergic system (see summary provided in Figure 18.1).

DEVELOPMENT OF THE MESOCORTICOLIMBIC DA SYSTEM

The mesocortical and mesolimbic DA system represents an important neuronal substrate for environmental programming since its development is not complete until well into adolescence. Although birth and migration of DA neurons in the ventral tegmental area (VTA) and substantia nigra (SNc) is accomplished mostly by the second half of gestation in rodents (Voorn *et al.*, 1988; Park *et al.*, 2000; McArthur *et al.*, 2005), development of the dopaminergic innervation from the VTA neurons to the striatum and the PFC is an active process that continues well past the third week of postnatal life in rodents (Kalsbeek *et al.*, 1988; Antonopoulos *et al.*, 2002). There is a distinct ontogenetic pattern of DA fiber innervation in the NAc/striatum and the PFC regions in the rat. The majority of DA projections to the striatum/NAc appear to be already established at birth, with a peak in the total density of synapses present between postnatal day (PND) 7 and 14 (Andrews *et al.*, 2006). Pruning of synapses in the NAc region occurs after the second week of life, since the total density of synapses is lower by PND 21 (Antonopoulos *et al.*, 2002). However, it is not clear whether the reduction in synapses involves a reduction in axodendritic synapses, the major synaptic type present in this region, or a reduction in the dendritic arborization of the neurons harboring these synapses. In the PFC, DA innervation is typically a postnatal process (Kalsbeek *et al.*, 1988; Dawirs *et al.*, 1993; Rosenberg & Lewis, 1995). In the rat, DA fibers innervating the PFC appear on PND 4 and the density of fibers is reported to increase until PND 60 when it reaches adult levels (Kalsbeek *et al.*, 1988). This long postnatal period of maturation makes those

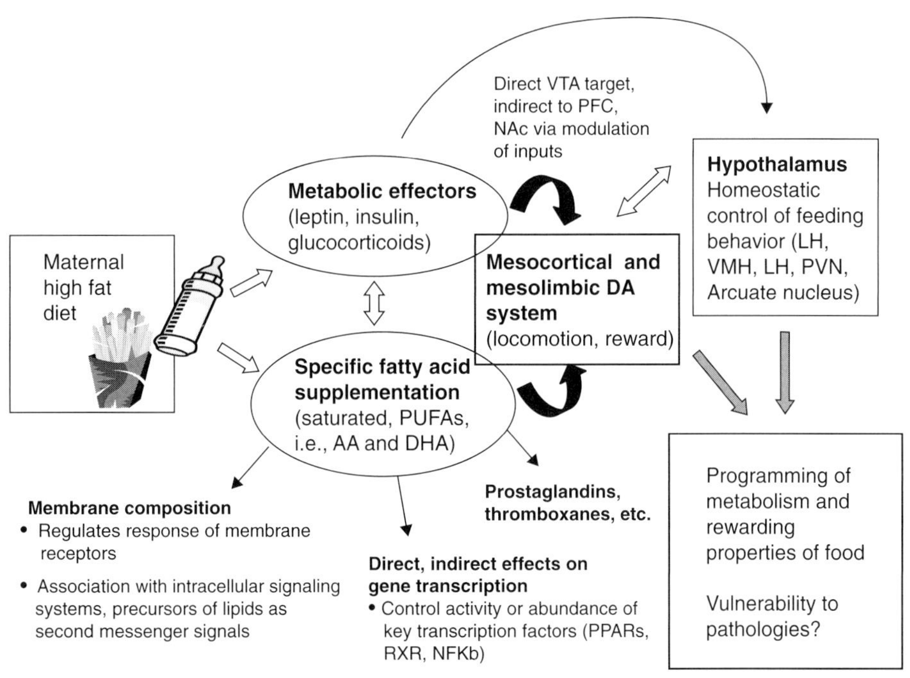

FIGURE 18.1 Potential mechanisms implicated in the effects of perinatal high fat exposure on dopamine neurotransmission in the adult offspring. Maternal high fat diet results in an increase in circulating leptin levels in the offspring and changes the secretion of insulin and glucocorticoids. These metabolic effectors have both direct and indirect effects on the mesocorticolimbic DA system by regulating activity of VTA neurons and DA release in the NAc. The intake of specific polyunsaturated fatty acids in the high fat diet can bring further changes to the DA system via modifications in the biophysical properties of membranes, gene expression or production of signaling molecules. Multiple interactions between the DA system and hypothalamic centers regulating metabolism and energy balance during a critical period of brain development converge to "re-program" the established metabolic set point and rewarding properties of food in the offspring. Such "re-programming" might predispose the offspring to develop several pathologies in the adult.

projections particularly susceptible to any environmental modulation, including those resulting from exposure to a high fat diet perinatally. In agreement with other studies demonstrating that the critical window for functional remediation of DA function after fatty acid deficiency is between birth and the second week of life (Kodas *et al.*, 2002), we speculate that the first 2–3 weeks of life in the rat represent a sensitive time window to affect the development of DA projections from the VTA to the PFC and NAc.

CONSEQUENCES OF PERINATAL MATERNAL FAT INTAKE ON DA FUNCTION IN THE ADULT OFFSPRING

Since many of the anti-psychotic drugs that are used to treat mental pathologies are often associated with significant increases in food intake and body weight gain (Newcomer, 2007), a large research area has developed to investigate the role of DA acting compounds on mechanisms regulating food intake. Fewer studies on the other hand, have investigated how specific nutrients could modify the activity of the DA system in rodents, and most of these studies have been concerned with sucrose ingestion (Rada *et al.*, 2005; Vacca *et al.*, 2007). Recently, attention has focused on fat intake given that it was shown that sham feeding of corn oil in adult rats produces increases in DA release in the NAc that are similar to those obtained with licking on sucrose solutions (Liang *et al.*, 2006). Furthermore, adult mice made obese by exposure to a high fat diet (diet-induced obesity model) exhibit higher D2 receptor mRNA expression (Huang *et al.*, 2005) and dopamine transporter (DAT) density (Huang *et al.*, 2006) in the NAc as well as increases in tyrosine hydroxylase (TH) mRNA in the VTA compared to mice of the same strain that are obesity resistant (Huang *et al.*, 2005). Taken together, these observations suggest that acute fat intake can stimulate DA release in the NAc and that chronic exposure to a high fat diet leads to significant modifications in the mesolimbic DA system in adult rodents.

The question arises whether a similar effect of high fat intake on mesocorticolimbic DA function can be detected in young animals and whether exposure to high fat during a critical period of development of this dopaminergic system could permanently alter this system or "re-program" its functional set point. In our most recent studies (Naef *et al.*, 2007), we found that adult male offspring of mothers fed a high fat diet (30% fat, HF, compared to 5% in controls) during the last week of gestation and throughout lactation displayed significant reductions in locomotor activation following a single acute injection of amphetamine (AMPH) as illustrated for a dose of AMPH of 2mg/kg bw in Figure 18.2. Amphetamine is a psychostimulant drug acting on the DA system to stimulate locomotor activity. In fact, locomotor activation following the administration of AMPH correlates well with the time course of change and amount of DA released in the NAccore (Sharp *et al.*, 1987). Interestingly, in these experiments rats were exposed to the HF regimen only during the perinatal period and placed on either a control diet (CD, 5% fat) or a choice of macronutrients after weaning. The postweaning diet did not appear to modify the reduced locomotor responses of HF offspring. In addition to blunted responses to acute AMPH, we found that behavioral sensitization to AMPH was also reduced in HF compared to CD offspring (Naef *et al.*, 2007). Behavioral sensitization is represented by increases in the strength of a response to a stimulus induced by past experiences with the same or related stimuli. This process recruits several of the mechanisms involved in synaptic plasticity (Ujike *et al.*, 2002). One of the key mechanisms associated with behavioral sensitization is an enhancement of DA release in the NAc following administration of AMPH for instance, and significant changes in the activity of DA neurons in the VTA (Stewart & Badiani, 1993; Vezina, 2004). Thus, it appears that exposure of the mother to a high fat diet during the end of gestation and the postpartum period results in a significant change in brain DA activity in the offspring.

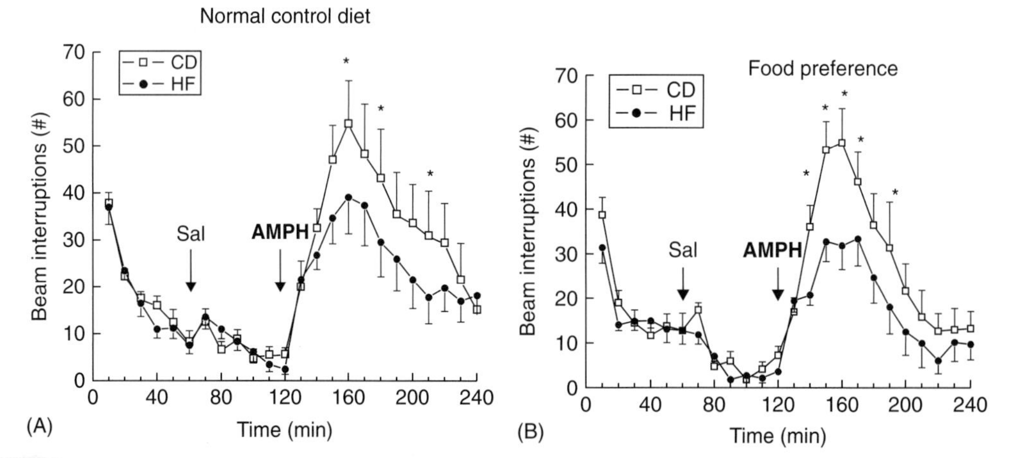

FIGURE 18.2 Reduction in AMPH-induced locomotor activity (# beam crossed/10min) in adult offspring from mothers maintained on a high fat diet (HF, 30% fat and 24% carbohydrates) compared to mothers on a CD (5% fat and 60% carbohydrates) during the last week of gestation and lactation. Adult male offspring (n = 9–10/group) were tested on PNDs 59 through 62 after being weaned either on a CD (panel A) or on a choice of pure macronutrients (lard, sucrose, and protein mix with balanced vitamins, panel B). ANOVA demonstrated a significant effect of neonatal diet ($p = 0.049$) and time ($p < 0.001$) on the amphetamine response (120–240min) as well as a significant neonatal diet × time interaction. There was no difference in basal (0–60min) or saline-induced locomotion (60–120min).*$p<0.05$ compared to HF offspring (*Post-hoc* HSD Tukey test).

We have documented that the diet-induced behavioral changes that we observed in our experiments are not related to differential fat accumulation and differences in availability of plasma and brain AMPH between diet groups. Behavioral effects are associated with significant changes in DA content and synthesis in the VTA, PFC, and NAc of adult rats. It is particularly notable that these effects occurred even in rats maintained on a CD since weaning, suggesting that perinatal dietary changes might have programmed some aspects of DA function in the long term. What are the mechanisms that are responsible for the perinatal effects of fat on the developing DA system? We can speculate on the role of metabolic effectors that are modified during the HF regimen and on the effect of elevated fatty acids present in the milk of HF-fed mothers as described below.

METABOLIC FACTORS AFFECTING DA FUNCTION

Homeostatic circulating factors like insulin, glucocorticoids, and leptin constitute important signals informing the brain and the hypothalamus in particular, about the size and state of peripheral metabolic stores as well as the energy state (Woods *et al.*, 2004). These hormones are greatly influenced, and generally increased by exposure to high fat high caloric diets in adulthood. Leptin and insulin demonstrate some degree of synergy in their actions at the intracellular level. Both leptin and insulin receptors are found throughout the mesocorticolimbic DA system and in particular on VTA neurons that are TH-positive (Figlewicz *et al.*, 2003). These hormones have significant effects on the adult DA system as they reduce the reward effectiveness of perifornical hypothalamic (PFH) stimulation (Shalev *et al.*, 2001; Shizgal *et al.*, 2001; Fulton *et al.*, 2004; Fruhbeck, 2006) and eliminate the positive valence associated with HF diet in a conditioned place preference task (Figlewicz *et al.*, 2004). Recent reports have confirmed that DA and GABA neurons in the VTA can be activated following leptin exposure (Fulton *et al.*, 2006; Hommel *et al.*, 2006). How these two hormones affect the DA system in *adults* is currently being unraveled. For instance, intracerebroventricular leptin administration reduces basal and feeding-evoked DA release in the NAc (Krugel *et al.*, 2003) and DA in the hippocampus of adult rats (Dagon *et al.*, 2005). Similarly, insulin increases expression and synaptic activity of DAT, resulting in a reduction in DA signaling (Figlewicz, 2003) in the VTA and SN. Both hormones appear to inactivate neurons through hyperpolarization, regulation of ATP-dependent K^+ channels, and activation of phosphatidyl inositol-3 kinase (PI3K) intracellular pathway in adults (Plum *et al.*, 2005). Nothing is known of the developmental effects of leptin and insulin on DA transmission, however.

In mothers exposed to a high fat diet, these hormones will be transferred to the pups through the milk (Bonnet *et al.*, 2002; Miralles *et al.*, 2006). Indeed, in our experimental model where mothers are fed a high fat (30% fat compared to 5% in controls) diet during the last week of gestation and throughout lactation, we found that by mid-lactation (PND 8–10), milk from HF-feeding mothers

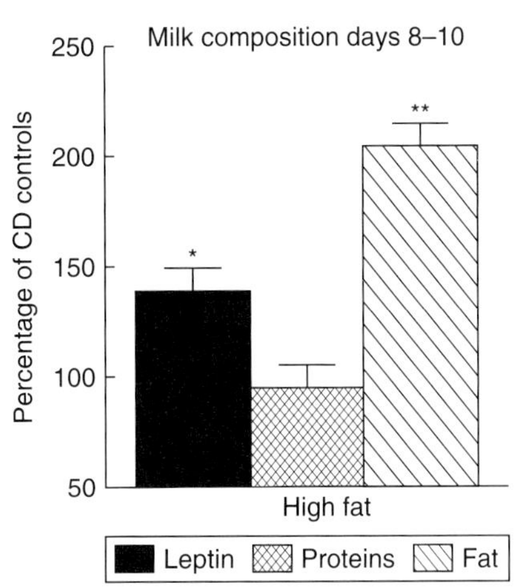

FIGURE 18.3 Feeding on a high fat (30% fat, HF) diet significantly increases milk concentrations of total fat and leptin in mothers on lactation days 8–10 compared to mothers remaining on a CD (5% fat). Milk was obtained from the mother under isoflurane anesthesia and by gentle hand stroking of the nipples after administration of oxytocin (1.5 IU/rat, i.p.). Total fat content was determined by a colorimetry using the sulfuric acid-vanillin reaction (Koski *et al.*, 1990), total protein levels using the Bradford assay and milk leptin concentrations were measured by radioimmunoassay (Proulx, 2002). Values are expressed as percentage of the values observed in the milk of CD-fed mothers. They represent the mean ± SEM of five samples/diet group (CD or HF diet). $^{**}p < 0.01$ and $^{*}p < 0.05$ (Student *t*-test).

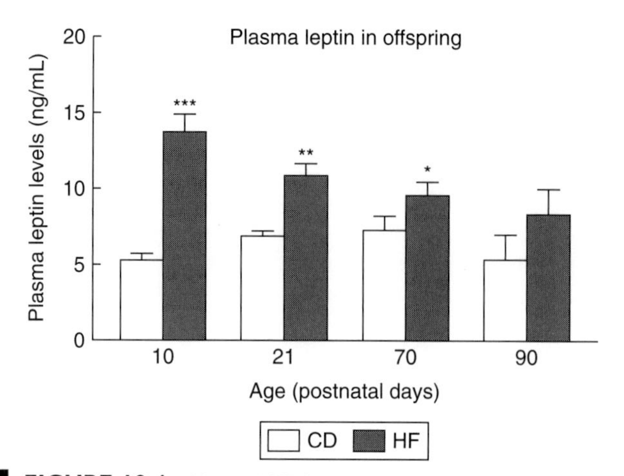

FIGURE 18.4 Maternal HF feeding between day 14 of gestation through PND 14 increases circulating plasma leptin levels in offspring postnatally. All measures were taken in the AM. Plasma leptin concentrations were measured by radioimmunoassay (Proulx *et al.*, 2002). Values are mean ± SEM of 6–11 determinations per age group. $^*p < 0.05$; $^{**}p < 0.01$; $^{***}p < 0.001$ (one- or two-tailed *t*-test).

contained significantly higher concentrations of fat and leptin compared to milk from mothers maintained on a CD (Figure 18.3). The net consequence of increased fat and leptin transfer to the pups via maternal milk is that of tonic increases in circulating plasma levels of leptin in the offspring (Figure 18.4). In addition to milk leptin transfer, increased plasma leptin in pups could be a direct result of the transfer of fatty acids to the pups and activation of leptin production in tissues such as brown adipose tissue and stomach (Oliver *et al.*, 2001; Pico *et al.*, 2002). Interestingly, small, but significant increases in plasma leptin levels are maintained in adult offspring, long after weaning off the HF diet on PND 22. This observation suggests that feeding on high fat milk during the postnatal period maintains a state of positive energy balance in adulthood with increased leptin production and potential increased fat deposition, even when animals are feeding on a normal CD. Tonically elevated leptin secretion in HF offspring might have several consequences on the development of an obesity-like phenotype, but also might interfere with the development of neurotransmitter systems that are targeted by leptin and in particular the mesocorticolimbic DA system. Given the multiple actions of leptin on the DA system, it is therefore plausible that elevated levels of leptin during the perinatal and postnatal periods might affect the activity of VTA neurons at a time when they are still maturing. In addition to leptin and insulin, recent reports also suggest that some of the long-term consequences of chronic fat intake on CNS functions in the adult might be mediated by a reduction in BDNF, CREB, and synapsin I production (Molteni *et al.*, 2002, 2004), compromising neuroplasticity, cognitive functions, and the outcome of traumatic brain injury, for instance (Wu *et al.*, 2003, 2004). Thus, we could hypothesize that the reduced behavioral sensitization observed in adult offspring of the HF-fed mothers (Naef, *et al.*, 2007) might be due to early changes in BDNF CREB and other synaptic proteins and growth factors (i.e., bFGF) that are implicated in sensitization processes (Flores & Stewart, 2000). In this way the early nutritional environment may be able to "program" mesocortical and mesolimbic DA circuits which are central in linking the motivational and rewarding aspects of food to the hypothalamic centers controlling energy balance. For example, one of the direct consequences of these "organizational" modifications might be to change the rewarding properties of drugs and/or specific food types.

SPECIFIC FATTY ACIDS INTAKE AND MODULATION OF NEUROTRANSMITTER FUNCTIONS

The brain is a target for metabolic hormones produced under conditions of high fat intake, but the various fatty acids contained in the diet are also potent regulators of a vast array of functions within the CNS. Polyunsaturated fatty acids (PUFAs) are major components of the phospholipids that comprise the membranes of all cells and neurons and their specific insertion into phospholipids can determine the physicochemical properties of membranes (Murphy, 1990; Jump, 2002; Horrocks & Farooqui, 2004). Membrane composition will directly influence the function of important membrane proteins, such as G-protein-coupled receptors and signaling systems, ion channels, enzymes, etc. Specific long-chain PUFAs such as arachidonic acid (AA, $n-6$) and docosahexaenoic acid (DHA, $n-3$) obtained from the diet act as precursors of several classes of signaling molecules (i.e., prostaglandins, thromboxanes, endocannabinoids, etc.), and as regulators of various transcription factors (i.e., PPARs, NFkappaB, etc.), thereby modulating neuronal functions (Bazan, 2003). Of particular relevance to our discussion of the perinatal effects of dietary fat intake is the observation that most of the brain AA and DHA accumulates during the late prenatal and early postnatal period in rodents and humans, making this period particularly vulnerable to maternal dietary insufficiency of the DHA precursor, alpha-linolenic acid.

In recent years, much attention has focused on the role of DHA and AA in brain development since earlier studies in rodents and primates have unequivocally demonstrated that a dietary deficiency of alpha-linolenic acid produces severe deficits in visual and cognitive functions such as problem solving, recognition memory, etc. (Carlson *et al.*, 1994; Innis, 2000a, b). These observations are consistent with the selective high accumulation of DHA and AA in brain and retina. During pregnancy and lactation, the maternal dietary fatty acid composition influences maternal and fetal plasma concentrations of PUFA as well as tissue composition (Amusquivar & Herrera, 2003). Thus, a maternal diet rich in PUFAs has been shown, for instance, to increase growth cone membranes at birth and DA concentrations in the brain of offsprings (Innis & de La Presa Owens, 2001), while restriction in dietary $n-3$ fatty acids during gestation reduces DHA concentrations in growth cones (Auestad & Innis, 2000). The multiple effects of DHA supplementation on gene expression (Kitajka *et al.*, 2002) and neurotransmission generally indicate a beneficial influence of this fatty acid (see review, Walker, 2005). Dietary fat intake and the subsequent changes in brain membrane fatty acid composition have also been implicated in the development and susceptibility to some mental illnesses (Greenwood & Young, 2001; Naliwaiko *et al.*, 2004).

Several reports have now indicated an important effect of AA on DA release (L'Hirondel *et al.*, 1995) and DHA deficiency or supplementation on DA function in rodents. In adult rodents, DHA deficiency increases AMPH-induced locomotion (Levant *et al.*, 2004) and sensitization of locomotor activity in mice (McNamara *et al.*, 2007) with a negative correlation between locomotor responses and DHA content in the NAc (McNamara *et al.*, 2007). A Deficient DHA diet given to mothers during gestation and lactation increases the expression of DA, glutamate, and acetylcholine receptors in several brain areas in the adult offspring, D2 receptors being over-expressed in particular in the VTA, striatum, amygdale, and NAc (Kuperstein *et al.*, 2005). Interestingly, DHA deficiency appears to be associated with clear hyperactivity of the mesolimbic and hypofunction of mesocortical dopaminergic system since there is a reduction in DA-containing vesicles, vesicular monoamine transporter (VMAT2), and D2 receptor concentrations in this region, and an opposite regulation of DA content and D2 receptors by supplementation with fish oil, a source of DHA in the diet (Chalon *et al.*, 1998, 2001). In agreement with the suggestion of hypofrontality, DA release in the frontal cortex was shown to be markedly reduced in offspring from DHA-deficient mothers (Kodas *et al.*, 2002; Zimmer *et al.*, 2002). Remediation of DHA deficiency during the neonatal period appears to re-establish normal DA secretion in the frontal cortex and NAc, but only if DHA replacement is provided before the second week of life. This observation suggests that there is a critical window of susceptibility to the effects of DHA on the mesocorticolimbic DA system during the first 2 weeks of life in rodents.

Taken together, results from our experiments and those presented with DHA-deficient models or supplementation of the maternal diet with DHA-rich fish oil tend to suggest that specific fatty acids present in the diet during the early developmental period in rodents play a critical role in modulating DA function in the adult offspring. Supplementation of the diet with increased fatty acids and specifically DHA might help prevent the development of dopaminergic dysfunctions that are observed in some mental pathologies and food disorders. Further investigation is clearly required to determine the precise role of early supplementation with specific types of fat and the long-term consequences of such dietary changes on the dopaminergic system of adults.

CONCLUSIONS

We have demonstrated that increased dietary fat intake of the mother could have long-term consequences on the function of the mesocorticolimbic DA system of the offspring, conferring some kind of resistance to the effects of repeated exposure to AMPH, for instance. Neonates naturally feed on a high fat diet during the first weeks of postnatal life and it is therefore tempting to speculate that this naturally elevated intake of fat and of PUFA precursors, as well as high exposure to leptin in pups via the maternal milk are critical components required to ensure adequate brain function and optimal development of dopaminergic function. Maternal regulatory influence on brain neurotransmission of the offspring is expressed at least in part through dietary fat intake, protecting the brain during a critical window of development. Since DA mesocorticolimbic pathways are also tightly associated to structures controlling food intake and energy balance (i.e., arcuate nucleus, VMH, LH, PVN), it is probable that the effects of fatty acids, leptin, and insulin are also exerted on these structures to regulate and integrate the metabolic and motivational aspects of energy intake and expenditure. We have yet to discover the precise mechanisms that mediate the early "programming" of DA function by high fat exposure and understand the role of specific lipids in morphological and functional changes in the dopaminergic pathways. Understanding those early nutritional "programming" events leading to changes in motivational pathways toward food and other rewarding stimuli (i.e., drugs of abuse) will be critical to our ongoing struggle to limit drug intake and to curb the progression of the obesity epidemics. As well, these results will contribute to our view that the early developmental period in most species is determinant for the expression of the adult phenotype and vulnerability to a vast range of pathologies. It will be of particular interest to determine how this early dietary programming may affect the offspring's parental behavior as adults, given the crucial role for the dopaminergic systems in maternal care and the neuroendocrine processes of pregnancy and lactation.

ACKNOWLEDGMENTS

This work was supported by a grant from the Canadian Institutes for Health Research (CIHR) grant #FRN 53350 to CDW and by a fellowship from the Fonds de Recherche en Santé du Quebec (FRSQ) to LN.

ABBREVIATIONS

AMPH	amphetamine
DA	dopamine
DAT	dopamine transporter
DHA	docosahexaenoic acid
LH	lateral hypothalamus
NAc	nucleus accumbens
PFC	prefrontal cortex
PVN	paraventricular nucleus
SNc	substantia nigra
TH	tyrosine hydroxylase
VMAT	vesicular monoamine transporter
VMH	ventromedial hypothalamus
VTA	ventral tegmental area

REFERENCES

Amusquivar, E., and Herrera, E. (2003). Influence of changes in dietary fatty acids during pregnancy on placental and fetal fatty acid profile in the rat. *Biol. Neonate* 83, 136–145.

Andrews, Z. B., Rivera, A., Elsworth, J. D., Roth, R. H., Agnati, L., Gago, B., Abizaid, A., Schwartz, M., Fuxe, K., and Horvath, T. L. (2006). Uncoupling protein-2 promotes nigrostriatal dopamine neuronal function. *Eur. J. Neurosci.* 24, 32–36.

Antonopoulos, J., Dori, I., Dinopoulos, A., Chiotelli, M., and Parnavelas, J. G. (2002). Postnatal development of the dopaminergic system of the striatum in the rat. *Neuroscience* 110, 245–256.

Auestad, N., and Innis, S. M. (2000). Dietary $n-3$ fatty acid restriction during gestation in rats: Neuronal cell body and growth-cone fatty acids. *Am. J. Clin. Nutr.* 71, 312S–314S.

Barker, D. J. (2002). Fetal programming of coronary heart disease. *Trends Endocrinol. Metab.* 13, 364–368.

Bazan, N. G. (2003). Synaptic lipid signaling: Significance of polyunsaturated fatty acids and platelet-activating factor. *J. Lipid Res.* 44, 2221–2233.

Berridge, K. C. (1996). Food reward: Brain substrates of wanting and liking. *Neurosci. Biobehav. Rev.* 20, 1–25.

Berthoud, H. R. (2004). Mind versus metabolism in the control of food intake and energy balance. *Physiol. Behav.* 81, 781–793.

Bonnet, M., Delavaud, C., Laud, K., Gourdou, I., Leroux, C., Djiane, J., and Chilliard, Y. (2002). Mammary leptin synthesis, milk leptin and their putative physiological roles. *Reprod. Nutr. Dev.* 42, 399–413.

Bouret, S. G., Draper, S. J., and Simerly, R. B. (2004). Trophic action of leptin on hypothalamic neurons that regulate feeding. *Science* 304, 108–110.

Brake, W. G., Sullivan, R. M., and Gratton, A. (2000). Perinatal distress leads to lateralized medial prefrontal cortical dopamine hypofunction in adult rats. *J. Neurosci.* 20, 5538–5543.

Brake, W. G., Zhang, T. Y., Diorio, J., Meaney, M. J., and Gratton, A. (2004). Influence of early postnatal rearing conditions on mesocorticolimbic dopamine and behavioural responses to psychostimulants and stressors in adult rats. *Eur. J. Neurosci.* 19, 1863–1874.

Carlson, S. E., Werkman, S. H., Peeples, J. M., and Wilson, W. M. (1994). Long-chain fatty acids and early visual and cognitive development of preterm infants. *Eur. J. Clin. Nutr.* 48(suppl 2), S27–S30.

Chalon, S., Delion-Vancassel, S., Belzung, C., Guilloteau, D., Leguisquet, A. M., Besnard, J. C., and Durand, G. (1998). Dietary fish oil affects monoaminergic neurotransmission and behavior in rats. *J. Nutr.* 128, 2512–2519.

Chalon, S., Vancassel, S., Zimmer, L., Guilloteau, D., and Durand, G. (2001). Polyunsaturated fatty acids and cerebral function: Focus on monoaminergic neurotransmission. *Lipids* 36, 937–944.

Chang, Y. M., Galler, J. R., and Luebke, J. I. (2003). Prenatal protein malnutrition results in increased frequency of miniature inhibitory postsynaptic currents in rat CA3 interneurons. *Nutr. Neurosci.* 6, 263–267.

Crawford, M. A. (1993). The role of essential fatty acids in neural development: Implications for perinatal nutrition. *Am. J. Clin. Nutr.* 57(5 suppl), 703S–710S.

Dagon, Y., Avraham, Y., Magen, I., Gertler, A., Ben-Hur, T., and Berry, E. M. (2005). Nutritional status, cognition, and survival: A new role for leptin and AMP kinase. *J. Biol. Chem.* 280, 42142–42148.

Dawirs, R. R., Teuchert-Noodt, G., and Czaniera, R. (1993). Maturation of the dopamine innervation during postnatal development of the prefrontal cortex in gerbils (*Meriones unguiculatus*). A quantitative immunocytochemical study. *J. Hirnforsch.* 34, 281–290.

Figlewicz, D,P. (2003). Adiposity signals and food reward: Expanding the CNS roles of insulin and leptin. *Am. J. Physiol. Regul. Integr. Comp. Physiol.* 284, R882–R892.

Figlewicz, D. P., Evans, S. B., Murphy, J., Hoen, M., and Baskin, D. G. (2003). Expression of receptors for insulin and leptin in the ventral tegmental area/substantia nigra (VTA/SN) of the rat. *Brain Res.* 964, 107–115.

Figlewicz, D. P., Bennett, J., Evans, S. B., Kaiyala, K., Sipols, A. J., and Benoit, S. C. (2004). Intraventricular insulin and leptin reverse place preference conditioned with high-fat diet in rats. *Behav. Neurosci.* 118, 479–487.

Flores, C., and Stewart, J. (2000). Changes in astrocytic basic fibroblast growth factor expression during and after prolonged exposure to escalating doses of amphetamine. *Neuroscience* 98, 287–293.

Flores, G., Alquicer, G., Silva-Gomez, A. B., Zaldivar, G., Stewart, J., Quirion, R., and Srivastava, L. K. (2005). Alterations in dendritic morphology of prefrontal cortical and nucleus accumbens neurons in post-pubertal rats after neonatal excitotoxic lesions of the ventral hippocampus. *Neuroscience* 133, 463–470.

Fortier, M. E., Joober, R., Luheshi, G. N., and Boksa, P. (2004). Maternal exposure to bacterial endotoxin during pregnancy enhances amphetamine-induced locomotion and startle responses in adult rat offspring. *J. Psychiatr. Res.* 38, 335–345.

Fruhbeck, G. (2006). Intracellular signalling pathways activated by leptin. *Biochem. J.* 393, 7–20.

Fulton, S., Richard, D., Woodside, B., and Shizgal, P. (2004). Food restriction and leptin impact brain reward circuitry in lean and obese Zucker rats. *Behav. Brain Res.* 155, 319–329.

Fulton, S., Pissios, P., Manchon, R. P., Stiles, L., Frank, L., Pothos, E. N., Maratos-Flier, E., and Flier, J. S. (2006). Leptin regulation of the mesoaccumbens dopamine pathway. *Neuron* 51, 811–822.

Gallagher, E. A., Newman, J. P., Green, L. R., and Hanson, M. A. (2005). The effect of low protein diet in pregnancy on the development of brain metabolism in rat offspring. *J. Physiol.* 568, 553–558.

Georgieff, M. K., and Innis, S. M. (2005). Controversial nutrients that potentially affect preterm neurodevelopment: Essential fatty acids and iron. *Pediatr. Res.* 57, 99R–103R.

Gluckman, P. D., Lillycrop, K. A., Vickers, M. H., Pleasants, A. B., Phillips, E. S., Beedle, A. S., Burdge, G. C., and Hanson, M. A. (2007a). Metabolic plasticity during mammalian development is directionally dependent on early nutritional status. *Proc. Natl Acad. Sci. USA* 104, 12796–12800.

Gluckman, P. D., Hanson, M. A., and Beedle, A. S. (2007b). Non-genomic transgenerational inheritance of disease risk. *Bioessays* 29, 145–154.

Godfrey, K. M., and Barker, D. J. (2001). Fetal programming and adult health. *Public Health Nutr.* 4, 611–624.

Godfrey, K. M., Lillycrop, K. A., Burdge, G. C., Gluckman, P. D., and Hanson, M. A. (2007). Epigenetic mechanisms and the mismatch concept of the developmental origins of health and disease. *Pediatr. Res.* 61, 5R–10R.

Greenwood, C.., and Young, S. N. (2001). Dietary fat intake and the brain: A developing frontier in biological psychiatry. *J. Psychiatr. Neurosci.* 26, 182–184.

Helland, I. B., Smith, L., Saarem, K., Saugstad, O. D., and Drevon, C. A. (2003). Maternal supplementation with very-long-chain $n - 3$ fatty acids during pregnancy and lactation augments children's IQ at 4 years of age. *Pediatrics* 111, 39–44.

Hommel, J. D., Trinko, R., Sears, R. M., Georgescu, D., Liu, Z. W., Gao, X. B., Thurmon, J. J., Marinelli, M., and DiLeone, R. J. (2006). Leptin receptor signaling in midbrain dopamine neurons regulates feeding. *Neuron* 51, 801–810.

Horrocks, L. A., and Farooqui, A. A. (2004). Docosahexaenoic acid in the diet: Its importance in maintenance and restoration of neural membrane function. *Prostag. Leukot. Essent. Fatty Acids* 70, 361–372.

Huang, X. F., Yu, Y., Zavitsanou, K., Han, M., and Storlien, L. (2005). Differential expression of dopamine D2 and D4 receptor and tyrosine hydroxylase mRNA in mice prone, or resistant, to chronic high-fat diet-induced obesity. *Mol. Brain Res.* 135, 150–161.

Huang, X. F., Zavitsanou, K., Huang, X., Yu, Y., Wang, H., Chen, F., Lawrence, A. J., and Deng, C. (2006). Dopamine transporter and D2 receptor binding densities in mice prone or resistant to chronic high fat diet-induced obesity. *Behav. Brain Res.* 175, 415–419.

Innis, S. M. (2000a). Essential fatty acids in infant nutrition: Lessons and limitations from animal studies in relation to studies on infant fatty acid requirements. *Am. J. Clin. Nutr.* 71, 238S–244S.

Innis, S. M. (2000b). The role of dietary $n - 6$ and $n - 3$ fatty acids in the developing brain. *Dev. Neurosci.* 22, 474–480.

Innis, S. M. (2004). Polyunsaturated fatty acids in human milk: An essential role in infant development. *Adv. Exp. Med. Biol.* 554, 27–43.

Innis, S. M. (2007). Dietary ($n - 3$) fatty acids and brain development. *J. Nutr.* 137, 855–859.

Innis, S. M., and de La Presa Owens, S. (2001). Dietary fatty acid composition in pregnancy alters neurite membrane fatty acids and dopamine in newborn rat brain. *J. Nutr.* 131, 118–122.

Jump, D. B. (2002). The biochemistry of $n - 3$ polyunsaturated fatty acids. *J. Biol. Chem.* 15(277), 8755–8758.

Kalsbeek, A., Voorn, P., Buijs, R. M., Pool, C. W., and Uylings, H. B. (1988). Development of the dopaminergic innervation in the prefrontal cortex of the rat. *J. Comp. Neurol.* 269, 58–72.

Kapoor, A., Dunn, E., Kostaki, A., Andrews, M. H., and Matthews, S. G. (2006). Fetal programming of hypothalamo–pituitary–adrenal function: Prenatal stress and glucocorticoids. *J. Physiol.* 572, 31–44.

Kelley, A. E., Baldo, B. A., Pratt, W. E., and Will, M. J. (2005). Corticostriatal–hypothalamic circuitry and food motivation: Integration of energy, action and reward. *Physiol. Behav.* 86, 773–795.

Kitajka, K., Puskas, L. G., Zvara, A., Hackler, L., Jr., Barcelo-Coblijn, G., Yeo, Y. K., and Farkas, T. (2002). The role of $n - 3$ polyunsaturated fatty acids in brain: Modulation of rat brain gene expression by dietary $n - 3$ fatty acids. *Proc. Natl. Acad. Sci. USA* 99, 2619–2624.

Kodas, E., Vancassel, S., Lejeune, B., Guilloteau, D., and Chalon, S. (2002). Reversibility of $n - 3$ fatty acid deficiency-induced changes in dopaminergic neurotransmission in rats: Critical role of developmental stage. *J. Lipid Res.* 43, 1209–1219.

Koletzko, B., Rodriguez-Palmero, M., Demmelmair, H., Fidler, N., Jensen, R., and Sauerwald, T. (2001). Physiological aspects of human milk lipids. *Early Hum. Dev.* 65, S3–S18.

Koski, K. G., Hill, F. W., and Lonnerdal, B. (1990). Altered lactational performance in rats fed low carbohydrate diets and its effect on growth of neonatal rat pups. *J. Nutr.* 120, 1028–1036.

Krugel, U., Schraft, T., Kittner, H., Kiess, W., and Illes, P. (2003). Basal and feeding-evoked dopamine release in the rat nucleus accumbens is depressed by leptin. *Eur. J. Pharmacol.* 482, 185–187.

Kuperstein, F., Yakubov, E., Dinerman, P., Gil, S., Eylam, R., Salem, N., Jr., and Yavin, E. (2005). Overexpression of dopamine receptor genes and their products in the postnatal rat brain following maternal $n - 3$ fatty acid dietary deficiency. *J. Neurochem.* 95, 1550–1562.

Levant, B., Radel, J. D., and Carlson, S. E. (2004). Decreased brain docosahexaenoic acid during development alters dopamine-related behaviors in adult rats that are differentially affected by dietary remediation. *Behav. Brain Res.* 152, 49–57.

L'hirondel, M., Cheramy, A., Godeheu, G., and Glowinski, J. (1995). Effects of arachidonic acid on dopamine synthesis, spontaneous release, and uptake in striatal synaptosomes from the rat. *J. Neurochem.* 64, 1406–1409.

Liang, N. C., Hajnal, A., and Norgren, R. (2006). Sham feeding corn oil increases accumbens dopamine in the rat. *Am. J. Physiol. Regul. Integr. Comp. Physiol.* 291, R1236–R1239.

McArthur, S., McHale, E., Dalley, J. W., Buckingham, J. C., and Gillies, G. E. (2005). Altered mesencephalic dopaminergic populations in adulthood as a consequence of brief perinatal glucocorticoid exposure. *J. Neuroendocrinol.* 17, 475–482.

McNamara, R. K., Sullivan, J., and Richtand, N. M. (2008). Omega-3 fatty acid deficiency augments amphetamine-induced behavioral sensitization in adult mice: Prevention by chronic lithium treatment. *J. Psychiatr. Res.* 42(6), 458–468.

Meaney, M. J., and Szyf, M. (2005). Environmental programming of stress responses through DNA methylation: Life at the interface between a dynamic environment and a fixed genome. *Dialogues Clin. Neurosci.* 7, 103–123.

Miralles, O., Sanchez, J., Palou, A., and Pico, C. (2006). A physiological role of breast milk leptin in body weight control in developing infants. *Obesity (Silver Spring)* 14, 1371–1377.

Molteni, R., Barnard, R. J., Ying, Z., Roberts, C. K., and Gomez-Pinilla, F. (2002). A high-fat, refined sugar diet reduces hippocampal brain-derived neurotrophic factor, neuronal plasticity, and learning. *Neuroscience* 112, 803–814.

Molteni, R., Wu, A., Vaynman, S., Ying, Z., Barnard, R. J., and Gomez-Pinilla, F. (2004). Exercise reverses the harmful effects of consumption of a high-fat diet on synaptic and behavioral plasticity associated to the action of brain-derived neurotrophic factor. *Neuroscience* 123, 429–440.

Morgane, P. J., Mokler, D. J., and Galler, J. R. (2002). Effects of prenatal protein malnutrition on the hippocampal formation. *Neurosci. Biobehav. Rev.* 26, 471–483.

Murphy, M. G. (1990). Dietary fatty acids and membrane protein function. *J. Nutr. Biochem.* 1, 68–79.

Naef, L., Srivastava, L., Gratton, A., Hendrickson, H., Owens, S. M., and Walker, C-D. (2008). Maternal high fat diet during the perinatal period alters mesocorticolimbic dopamine in the adult rat offspring: Reduction in the behavioral responses to repeated amphetamine administration. *Psychopharmacology* (Berl) 197, 83–94.

Naliwaiko, K., Araujo, R. L., da Fonseca, R. V., Castilho, J. C., Andreatini, R., Bellissimo, M. I., Oliveira, B. H., Martins, E. F., Curi, R., Fernandes, L. C., and Ferraz, A. C. (2004). Effects of fish oil on the central nervous system: A new potential antidepressant?. *Nutr. Neurosci.* 7, 91–99.

Newcomer, J. W. (2007). Metabolic considerations in the use of antipsychotic medications: A review of recent evidence. *J .Clin. Psychiatr.* 68(suppl 1), 20–27.

Oliver, P., Pico, C., and Palou, A. (2001). Ontogenesis of leptin expression in different adipose tissue depots in the rat. *Pflugers Arch.* 442, 383–390.

Palmer, A. A., Printz, D. J., Butler, P. D., Dulawa, S. C., and Printz, M. P. (2004). Prenatal protein deprivation in rats induces changes in prepulse inhibition and NMDA receptor binding. *Brain Res.* 996, 193–201.

Park, M., Kitahama, K., Geffard, M., and Maeda, T. (2000). Postnatal development of the dopaminergic neurons in the rat mesencephalon. *Brain Dev.* 22(suppl 1), S38–S44.

Phillips, D. I. (2007). Programming of the stress response: A fundamental mechanism underlying the long-term effects of the fetal environment?. *J. Intern. Med.* 261, 453–460.

Pico, C., Sanchez, J., Oliver, P., and Palou, A. (2002). Leptin production by the stomach is up-regulated in obese (fa/fa) Zucker rats. *Obes. Res.* 10, 932–938.

Plagemann, A. (2006). Perinatal nutrition and hormone-dependent programming of food intake. *Horm. Res.* 65(suppl 3), 83–89.

Plum, L., Schubert, M., and Bruning, J. C. (2005). The role of insulin receptor signaling in the brain. *Trends Endocrinol. Metab.* 16, 59–65.

Proulx, K., Clavel, S., Nault, G., Richard, D., and Walker, C. D. (2001). High neonatal leptin exposure enhances brain GR expression and feedback efficacy on the adrenocortical axis of developing rats. *Endocrinology* 142, 4607–4616.

Proulx, K., Richard, D., and Walker, C. D. (2002). Leptin regulates appetite-related neuropeptides in the hypothalamus of developing rats without affecting food intake. *Endocrinology* 143 (12), 4683–4692.

Rada, P., Avena, N. M., and Hoebel, B. G. (2005). Daily bingeing on sugar repeatedly releases dopamine in the accumbens shell. *Neuroscience* 134, 737–744.

Rosenberg, D. R., and Lewis, D. A. (1995). Postnatal maturation of the dopaminergic innervation of monkey prefrontal and motor cortices: A tyrosine hydroxylase immunohistochemical analysis. *J. Comp. Neurol.* 358, 383–400.

Shalev, U., Yap, J., and Shaham, Y. (2001). Leptin attenuates acute food deprivation-induced relapse to heroin seeking. *J. Neurosci.* 21, RC129.

Sharp, T., Zetterstrom, T., Ljungberg, T., and Ungerstedt, U. (1987). A direct comparison of amphetamine-induced behaviours and regional brain dopamine release in the rat using intracerebral dialysis. *Brain Res.* 401, 322–330.

Shizgal, P., Fulton, S., and Woodside, B. (2001). Brain reward circuitry and the regulation of energy balance. *Int. J. Obes. Relat. Metab. Disord.* 25(suppl 5), S17–S21.

Stewart, J., and Badiani, A. (1993). Tolerance and sensitization to the behavioral effects of drugs. *Behav. Pharmacol.* 4, 289–312.

Ujike, H., Takaki, M., Kodama, M., and Kuroda, S. (2002). Gene expression related to synaptogenesis, neuritogenesis, and MAP kinase in behavioral sensitization to psychostimulants. *Ann. NY Acad. Sci.* 965, 55–67.

Vacca, G., Ahn, S., and Phillips, A. G. (2007). Effects of short-term abstinence from escalating doses of d-amphetamine on drug and sucrose-evoked dopamine efflux in the rat nucleus accumbens. *Neuropsychopharmacology* 32, 932–939.

Vezina, P. (2004). Sensitization of midbrain dopamine neuron reactivity and the self-administration of psychomotor drugs. *Neurosci. Biobehav. Rev.* 27, 827–839.

Voorn, P., Kalsbeek, A., Jorritsma-Byham, B., and Groenewegen, H. J. (1988). The pre- and postnatal development of the dopaminergic cell groups in the ventral mesencephalon and the dopaminergic innervation of the striatum of the rat. *Neuroscience* 25, 857–887.

Wainwright, P. E. (2002). Dietary essential fatty acids and brain function: A developmental perspective on mechanisms. *Proc. Nutr. Soc.* 61, 61–69.

Walker, C.-D. (2005). Nutritional aspects modulating brain development and the responses to stress in early neonatal life. *Prog. Neuropsychopharmacol. Biol. Psychiatr.* 29, 1249–1263.

Walker, C.-D., and Plotsky, P. M. (2002). Glucocorticoids, stress and development. In *Hormones, Brain and Behavior* (D. Pfaff, A. Arnold, A. Etgen, S. Farbach, and R. Rubin, Eds.), pp. 487–534. Academic Press, New York.

Walker, C.-D., Long, H., Williams, S., and Richard, D. (2007). Long-lasting effects of neonatal leptin on hippocampal function, synaptic proteins and NMDA receptor subunits in the rat. *J. Neurosci. Res.* 85, 816–828.

Woods, S. C., D'Alessio, D. A., Tso, P., Rushing, P. A., Clegg, D. J., Benoit, S. C., Gotoh, K., Liu, M., and Seeley, R. J. (2004). Consumption of a high-fat diet alters the homeostatic regulation of energy balance. *Physiol. Behav.* 83, 573–578.

Wu, A., Molteni, R., Ying, Z., and Gomez-Pinilla, F. (2003). A saturated-fat diet aggravates the outcome of traumatic brain injury on hippocampal plasticity and cognitive function by reducing brain-derived neurotrophic factor. *Neuroscience* 119, 365–375.

Wu, A., Ying, Z., and Gomez-Pinilla, F. (2004). The interplay between oxidative stress and brain-derived neurotrophic factor modulates the outcome of a saturated fat diet on synaptic plasticity and cognition. *Eur. J. Neurosci.* 19, 1699–1707.

Zimmer, L., Vancassel, S., Cantagrel, S., Breton, P., Delamanche, S., Guilloteau, D., Durand, G., and Chalon, S. (2002). The dopamine mesocorticolimbic pathway is affected by deficiency in $n-3$ polyunsaturated fatty acids. *Am. J. Clin. Nutr.* 75, 662–667.

MATERNAL CARE: FROM GENES TO ENVIRONMENT

19

MATERNAL INFLUENCE ON OFFSPRING REPRODUCTIVE BEHAVIOR: IMPLICATIONS FOR TRANSGENERATIONAL EFFECTS

FRANCES A. CHAMPAGNE

Department of Psychology, Columbia University, New York, NY 10027, USA

Recent evidence suggests that the non-genomic inheritance of variations in maternal care may be mediated by epigenetic regulation of hypothalamic steroid receptors. Unlike germ-line epigenetic transmission, this mode of inheritance is dependent on the quality of maternal care received in infancy. This mechanism of inheritance allows the reproductive behavior of one generation to influence that of future generations. Empirical investigation of trans-generational effects in mammals suggests that throughout the preconceptual, prenatal, and postnatal periods, maternal influence can stably alter the reproductive behavior of female offspring. In this chapter, evidence for the generational transmission of maternal effects on reproduction and the possible mechanisms that mediate this transmission will be reviewed. These studies present a novel approach to understanding the origin of similarities in phenotype between parents and offspring.

Traditional views regarding the mechanisms mediating the inheritance of characteristics by offspring from parental generations have been limited to the transmission of gene sequences that influence physiology and behavior. However, in a number of species, cross-fostering studies have demonstrated that similarities in phenotype between parents and offspring can be mediated by non-genomic mechanisms such as maternal effects (Mousseau & Fox, 1998). These environmentally mediated effects have been demonstrated to induce profound alterations in offspring development that can be equivalent or even greater in magnitude to those attributed to genetic effects (Kruuk & Hadfield, 2007). Amongst mammals there is an extended period of mother–infant interaction throughout the prenatal and postnatal periods which would allow maternal effects to influence multiple aspects of phenotype. During gestation, the developing fetus is dependent on the transfer of resources across the placenta and can be influenced by maternal endocrine changes (Gluckman *et al.*, 2005; Weinstock, 2005). During the postnatal period, mammalian mother–infant interactions typically involve intense physical contact which serves to maintain offspring body temperature and allow offspring to suckle. In addition to meeting the physical needs of offspring, this post-partum care is critical in shaping offspring neuroendocrine systems involved in the regulation of stress responsivity and social behavior. As a consequence, offspring reproductive and social behavior can be influenced by the quality of the postpartum care received in infancy (Meaney, 2001). Evidence for this developmental effect comes primarily from rodent and primate studies, though there is certainly correlational support for these effects in humans. In this chapter, the role of maternal effects in shaping offspring reproductive behavior in mammals will be explored and the implications of these effects for subsequent generations will be discussed. The role of epigenetic mechanisms in mediating these maternal effects will also be explored.

Neurobiology of the Parental Brain

MATERNAL INFLUENCE ON THE DEVELOPING EMBRYO

Maternal influence during early embryonic development is often overlooked as a potential mechanism influencing offspring behavior. However, oocyte maternal mRNAs serve a critical role in the early stages of development that occur immediately after fertilization. Prior to maturation, levels of transcription and translation in the oocyte are extremely high which serves to "stock-pile" proteins and mRNA that will be needed in early zygotic development (Wassarman & Kinloch, 1992). Zygotic genes are silenced through chromatin mediated suppression of transcription during the post-fertilization phase and thus all protein synthesis is mediated by maternal mRNAs and enzymes until zygotic gene activation during the two-cell embryo stage (Nothias *et al.*, 1995; De Sousa *et al.*, 1998). Moreover, it is the gradients of maternal mRNA present in the zygote that will serve to drive segmentation of the embryo and generate anterior–posterior polarity of the nervous system (Gavis & Lehmann, 1992; Scott, 2000).

Though there is evidence that paternal mRNA can also be transferred to the oocyte environment during fertilization and possibly influence development (Boerke *et al.*, 2007), it is clear that the maternal contribution to the developing embryonic nervous system is an essential feature of the successful progression from a fertilized zygote to a maturing embryo and fetus. Perturbation to this maternal environment during the preconception period can thus have sustained effects on offspring. The duration of this preconception period can be particularly lengthy for the oocyte as maturation of the female gametes can occur over the course of decades (Bukovsky *et al.*, 2005). This is an extensive period of time during which environmental conditions can act to alter maternal mRNA. Oocyte gene expression has been found to be altered as a function of maternal age (Janny & Menezo, 1996; Eichenlaub-Ritter, 1998) and nutrition (Symonds *et al.*, 2005). In the case of *in vitro* fertilization (IVF), the quality of the culture medium in which oocytes are maintained during the pre- and post-fertilization periods has implications for zygote and embryo survival rates and development associated with changes in oocyte gene expression (Corcoran *et al.*, 2005, 2007). Alterations in gene expression and transcription factors within the zygote have been suggested to lead to impairments in genomic imprinting amongst embryos derived from assisted reproductive technologies (Fernandez-Gonzalez *et al.*, 2004, 2007). Thus, even prior to conception, the quality of the maternal environment can have sustained effects on offspring phenotype.

MATERNAL INFLUENCE ON THE DEVELOPING FETUS

Though it is likely that preconceptional maternal factors can alter development directly, there is also considerable interaction between preconceptional and gestational maternal condition on offspring development. In mammals, gestation can be characterized as a period when many demands are made on maternal physiology to support the energetic needs of the developing fetus (Weissgerber & Wolfe, 2006). When fetal demands exceed the capacity of maternal physiology, there are increased incidences of maternal hypertension and diabetes (Haig, 1993; Kaaja & Greer, 2005) that can influence offspring risk of cardiovascular disease and glucose intolerance (Rocha *et al.*, 2005; Malcolm *et al.*, 2006; McLean *et al.*, 2006). Low birth weight, particularly as a consequence of reduced maternal food intake during pregnancy, is associated with a multitude of long-term metabolic outcomes, including cardiovascular disease, diabetes, elevated blood pressure, glucose intolerance, and obesity (Roseboom *et al.*, 2000, 2006; Painter *et al.*, 2005). These outcomes are explained in terms of the "developmental origins" hypothesis, which argues that when the fetus detects low energy supplies from the mother, changes in gene expression occur that prepare the organism for a nutritionally poor postnatal environment (Barker, 2004; McMillen & Robinson, 2005). These adaptive developmental changes allow for the storage of energy as fat in adulthood, which leads to obesity and other metabolic health consequences for offspring.

Exposure to toxins during sensitive periods of brain development can also lead to pathological outcomes. Maternal alcohol consumption produces fetal alcohol syndrome (Olegard *et al.*, 1979; Hoyseth & Jones, 1989); increased smoking is associated with failure to thrive

(Higgins, 2002; Bernstein *et al.*, 2005); and prenatal exposure to cocaine and heroin are strong risk factors for pregnancy complications, infant mortality, infant drug dependency, and learning difficulties (Fulroth *et al.*, 1989; Wagner *et al.*, 1998). Exposure to severe psychosocial stressors has also been associated with low birth weight, reduced infant survival, and modified sex ratio (Zorn *et al.*, 2002; Rondo *et al.*, 2003; Catalano *et al.*, 2005). Stressful life events, such as death in the family, starting a new job, or chronic exposure to daily hassles have been found to alter maternal and fetal physiology (Tambyrajia & Mongelli, 2000).

Psychosocial stress activates the maternal hypothalamic–pituitary–adrenal (HPA) axis resulting in the release of glucocorticoids (de Weerth *et al.*, 2007). Though enzymes within the placenta, such as 11-β-hydroxysteroid dehydrogenase-2 (11-βHSD-2), can inactivate glucocorticoids and thus buffer the developing fetus from these steroid hormones, severe stress may overwhelm the capacity of this enzymatic conversion (Edwards *et al.*, 1996; Seckl *et al.*, 1999). Moreover, in cases of severe maternal undernutrition the levels of expression of 11-βHSD-2 are decreased, thus exposing the fetus to an increased risk of both metabolic and stress-related disorders (Lesage *et al.*, 2001). Investigation of the neurobiological impact on offspring of prenatal stress has been primarily based on a rodent model and demonstrates a number of long-term changes that can be observed in adulthood. Prenatally stressed offspring exhibit elevated plasma corticosterone (Stohr *et al.*, 1998), increased corticotrophin releasing hormone (CRH) mRNA in the amygdala (Cratty *et al.*, 1995), and reduced monoamine and catecholamine turnover (Peters, 1982; Takahashi *et al.*, 1992). Behaviorally, these offspring are more hyperactive, inhibited of novelty, and impaired on cognitive tasks (Weinstock *et al.*, 1988). Prenatally stressed offspring display impairments in social behavior, such as reduced frequency of play and physical interaction with siblings associated with alterations in oxytocinergic pathways (Braadstad, 1998; Lee *et al.*, 2007). Reproductive success of prenatally stressed females is impaired accompanied by deficits in sexual and maternal behaviors (Herrenkohl, 1979). Prenatally stressed females engage less frequently in lordosis, exhibit reduced paced mating behavior (Frye & Orecki, 2002), display longer latencies to retrieve pups (Kinsley & Bridges, 1988), and provide less frequent tactile stimulation toward pups during the postpartum period (Champagne & Meaney, 2006). These effects may in part be attributed to a masculinization of the female brain in response to high levels of prenatal androgen exposure (Kaiser *et al.*, 2003; Kaiser & Sachser, 2005).

POSTNATAL MATERNAL INFLUENCE ON DEVELOPING OFFSPRING

Following parturition, mammalian offspring continue to be dependent on mother–infant interactions for growth and survival. Resource exchange during this period is typically mediated by lactation and the provisioning of heat from the maternal ventrum. As is the case for prenatal nutrition, the dietary intake of lactating females can have sustained effects on offspring development. Maternal diet during lactation has been demonstrated to alter gene expression of milk proteins within the mammary gland (Zubieta & Lonnerdal, 2006). Placing lactating females on a high fat diet during the perinatal period results in increased levels of leptin protein in milk with implications for metabolism, food preference, and stress responsivity of suckled offspring (Trottier *et al.*, 1998; Bayol *et al.*, 2007). Postpartum dietary restriction, much like prenatal dietary restriction, has consequences for growth and metabolism (Zambrano *et al.*, 2005) as well as altering onset of puberty in female offspring (Curley *et al.*, 2004; da Silva Faria *et al.*, 2004).

Though the nutritional aspects of the mother–infant interaction provide a necessary component of maternal care there is clearly a developmental role for other features of the postpartum contact between mother and infant as demonstrated by the maternal deprivation studies of Harlow in the 1950s and 1960s (Arling & Harlow, 1967; Harlow & Suomi, 1971, 1974). Correlational studies of maternal deprivation in humans (Beckett *et al.*, 2002; Parker & Nelson, 2005; Gunnar & van Dulmen, 2007) and experimental evidence from studies of the effect of deprivation (Hall, 1975; West, 1993; Gonzalez *et al.*, 2001; Lovic & Fleming, 2004) or prolonged maternal separation (Lehmann *et al.*, 1999, 2000;

Lippmann *et al.*, 2007) in rodents support the finding of these early studies and suggest that disruption of the mother–infant relationship has profound consequences for offspring neurobiology and behavior. Variations in the quality of mother–child interactions have been studied as a mediator of individual differences in development. Elevated fear responses, increased negative affect, and decreased frontal EEG asymmetry have been found amongst infants of mothers exhibiting low levels of maternal sensitivity (Hane & Fox, 2006). Parental bonding, defined as the level of care and overprotection displayed by mothers, can be used to predict offspring risk of depression (Parker, 1983; Parker, 1990; Hane & Fox, 2006) and stress-induced dopamine release in the ventral striatum (Pruessner *et al.*, 2004). Likewise, disorganized attachment relationships between mother and infant is associated with elevated cortisol in infancy (Hertsgaard *et al.*, 1995) and an increased risk of psychopathology in adulthood (Sroufe, 2005).

Amongst primates and rodents there are also substantial individual differences in postpartum maternal care. Vervet monkeys engage in variations in frequency of contact with infants (Fairbanks & McGuire, 1988; Fairbanks, 1989), whereas rhesus macaques have been demonstrated to vary in maternal rejection rates (Hinde & Simpson, 1975; Berman, 1990), maternal overprotection (Bardi & Huffman, 2002), and frequency of infant abuse (Maestripieri, 1998; Maestripieri *et al.*, 1999). These patterns of maternal care are likewise associated with variations in the development and behavior of offspring (Bardi & Huffman, 2002, 2006; McCormack *et al.*, 2006). In particular, frequency of maternal rejection and abuse has pervasive effects, including increased behavioral indices of distress, increased withdrawal from social contact, and disruptions to serotonergic functioning (Maestripieri, 2005; Maestripieri *et al.*, 2005, 2006).

Study of the long-term consequences of maternal care is challenging in humans and primates due to the lengthy interval between infancy and adulthood and the limitations inherent in studying neurobiological change. However, this empirical approach can also be used in a rodent model as both rats and mice display a considerable degree of natural variation in maternal behavior (Champagne *et al.*, 2003a, 2007). In Long-Evans rats, offspring reared by dams that exhibit high levels of pup licking/grooming (LG) have an attenuated HPA response to stress and are more exploratory in a novel environment associated with elevated hippocampal glucocorticoid receptor (GR) mRNA, decreased hypothalamic CRH mRNA, and increased density of benzodiazepine receptors in the amygdala (Liu *et al.*, 1997, 2000; Caldji *et al.*, 1998; Francis *et al.*, 1999b). These offspring exhibit enhanced performance on tests of spatial leaning and memory, elevated hippocampal brain derived neurotrophic factor mRNA, and increased synaptophysin (Liu *et al.*, 2000). Neuronal survival is increased and apoptosis decreased within the hippocampus of high LG offspring (Weaver *et al.*, 2002; Bredy *et al.*, 2003). Dopamine release in the medial prefrontal cortex associated with stress responsivity in males (Zhang *et al.*, 2005) and nucleus accumbens dopamine release in response to reward in females are also altered as a function of LG (Champagne *et al.*, 2004). Both sexual and maternal behaviors of female offspring are altered by maternal LG with offspring of high LG dams displaying reduced lordosis, longer intervals between intromissions, and high levels of LG toward their own offspring (Francis *et al.*, 1999a; Champagne *et al.*, 2003a; Cameron & Meaney, 2006; Uriarte *et al.*, 2007). The increases in maternal care of high LG offspring are associated with elevated hypothalamic oxytocin receptor binding (Francis *et al.*, 2000; Champagne *et al.*, 2001), increased sensitivity to estrogen (Champagne *et al.*, 2003b), and increased expression of estrogen receptor alpha (ERα) in the medial preoptic area (MPOA; Champagne *et al.*, 2003b). Conversely, the decreased sexual receptivity of high LG offspring is associated with decreased ERα expression in the ventral medial hypothalamus (Cameron & Meaney, 2006).

TRANSMISSION OF MATERNAL INFLUENCE ACROSS GENERATIONS IN MAMMALS

The maternal effects described in previous sections range in origin from the preconceptual to postpartum period and alter multiple aspects of offspring phenotype. However, when aspects of social and reproductive behavior are altered as a consequence of mother–infant interactions, these effects may extend beyond the offspring generation to grand-offspring. Prenatal exposure to endocrine disrupters has been found to alter reproductive success of males through decreased sperm count and viability (Anway *et al.*, 2005) as well as altering the attractiveness of these males to females in a mate preference test (Crews *et al.*, 2007). Interestingly, these effects are observed 2–4 generations beyond the initial exposure. Prenatally stressed female offspring exhibit reduced hypothalamic oxytocin receptor binding and decreased maternal LG and these effects are transmitted to the grand-offspring of gestationally stressed dams (Champagne & Meaney, 2006). Denenberg (Denenberg & Rosenberg, 1967) illustrated that response to novelty could be transmitted across multiple generations. Postpartum maternal dietary protein restriction alters offspring growth and metabolism and grand-offspring likewise have altered glucose and insulin levels (Zambrano *et al.*, 2005). Mother–infant attachment classifications are similar across generations of female offspring resulting in a transgenerational inheritance of risk or resilience depending on the quality of that attachment (Benoit & Parker, 1994; Sroufe, 2005). Frequency of postpartum maternal behavior is also transmitted across rhesus and pigtail macaques matrilines as are rates of maternal rejection and infant abuse (Berman, 1990; Maestripieri *et al.*, 1997, 1999). Maternal LG can be transmitted to female offspring and grand-offspring such that under stable environmental conditions, the frequency of LG received in infancy reliably predicts frequency of LG exhibited in adulthood. Thus the offspring of high LG dams are themselves high LG and this aspect postpartum care alters the LG behavior of subsequent generations. This same pattern is observed in the low LG matriline. Thus, maternal effects occurring during both the prenatal and postnatal periods can be transmitted across generations in mammals, particularly through matrilines in which aspects of reproductive behavior are altered.

MECHANISMS OF INTERGENERATIONAL TRANSMISSION

Maternal effects can exert stable, long-term effects on offspring phenotype both within and across generations. Though this pattern of inheritance resembles what would be observed if a genetically determined trait were to be passed from parent to offspring, evidence has emerged that supports an environmentally mediated mechanism of inheritance. Cross-fostering studies have demonstrated that non-stressed rat pups reared by a gestationally stressed dam exhibit behavioral hyperactivity (Moore & Power, 1986). Rhesus macaques infants born to abusive mothers who are then fostered to non-abusive mothers do not display abusive behavior (Maestripieri, 2005). Conversely, infants taken from non-abusive mothers and fostered to abusive females are themselves abusive as adults. Amongst offspring of Long-Evans rats, LG behavior of foster mothers rather than biological mothers is associated with offspring stress response, maternal care, and GR and ERα expression (Liu *et al.*, 1997; Francis *et al.*, 1999a; Champagne *et al.*, 2003a, 2006). As such, this generational transmission appears to be behaviorally mediated. However, the adult phenotype persists long after the cessation of the intense period of mother–infant interaction believed to be the critical signal to development. Exploration of the mechanisms mediating this stable inheritance implicates epigenetic regulation of gene transcription in hypothalamic brain regions implicated in the expression of maternal behavior.

Investigation of the mechanisms through which maternal care can influence subsequent generations has focused primarily on the transmission of maternal LG in Long-Evans rats. Differences in the expression of ERα within the MPOA are thought to account for the differences in estrogen sensitivity and oxytocin receptor binding amongst high and low LG mothers and offspring (Champagne *et al.*, 2003b). These variations in gene expression emerge in infancy and are maintained into adulthood when females are rearing their own litters. One of the molecular mechanisms through which long-term maintenance of patterns of gene expression can be achieved is an epigenetic modification known as DNA methylation (Razin, 1998; Turner, 2001). The regulation of transcription is mediated by several factors within the cell nucleus which ultimately control RNA polymerase binding to the transcription start site. Access to DNA is normally limited as it is wrapped tightly around a complex of histone proteins. Moreover, chemical alterations to DNA can occur at specific cytosine nucleotides within the promotor sequence of the DNA whereby a methyl group is added to the cytosine forming 5-methylcytosine. This modification does not alter the gene sequence but does inhibit gene expression by preventing transcription factor binding and attracting enzymes which further serve to condense chromatin (Strathdee & Brown, 2002). Moreover, DNA methylation is a very stable modification to DNA and the patterns of methylation within the nucleus can be maintained following mitosis. Thus DNA methylation is referred to as an epigenetic mechanism of gene regulation.

There is emerging evidence for the role of DNA methylation and associated enzymes in maintaining many of the maternal effects discussed in previous sections of this chapter. Abnormal development of embryos conceived through IVF is associated with disruption of the patterns of DNA methylation of imprinted genes (Gosden *et al.*, 2003; Allegrucci *et al.*, 2005; Li *et al.*, 2005). These particular genes do not exhibit biallelic expression and only the mother's or father's gene copy is typically expressed in the developing organism. The parental allele that is not expressed is silenced by methylation of the gene sequence. When oocyte or sperm are stored *in vitro*, it is possible that the enzymatic machinery responsible for genomic imprinting is disrupted leading to growth abnormalities and increases in syndromes associated with imprinted genes. Dietary restriction during gestation can also alter offspring epigenetic patterns. Choline is an essential nutrient within the diet and provides methyl groups needed for DNA methylation (Blusztajn & Wurtman, 1981). When maternal dietary restriction during pregnancy reduces choline consumption there are altered patterns of methylation and gene expression in offspring tissue, including effects on insulin-like growth factor 2 (Kovacheva *et al.*, 2007). Augmenting the amount of methyl donors in maternal diet has been found to alter aspects of offspring phenotype associated with methylation status of the agouti and axin fused gene (Wolff *et al.*, 1998; Waterland *et al.*, 2006). Likewise, the transgenerational effects on growth and metabolism of prenatal protein restriction is associated with decreased methylation of the GR amongst offspring and grand-offspring (Zambrano *et al.*, 2005). Prenatal exposure to endocrine disruptors alters DNA methylation patterns in sperm that are transmitted across multiple generations of offspring and influence reproductive success (Anway *et al.*, 2005; Crews *et al.*, 2007). Moreover, supplementing the maternal diet with methyl donors can reverse the damage caused by early exposure to endrocrine disruptors (Dolinoy *et al.*, 2007).

Epigenetic regulation of ERα has been studied extensively in the context of the molecular pathways in tumor progression in breast cancer. ERα is highly expressed in breast cancer cells and may mediate estrogen-induced increases in cell proliferation (Murphy *et al.*, 2003). Hypermethylation of the ERα promotor has been demonstrated to decrease ERα expression and modify sensitivity of tumor cells to estrogen (Yan *et al.*, 2001). Dietary intake of methyl donors such as follate has been found to alter ERα promotor methylation and ERα gene expression (Davis & Uthus, 2004). The role of postpartum maternal care in altering DNA methylation patterns within the ERα promotor has been investigated in the offspring of high and low LG dams. Elevated levels of methylation at several sites within the ERα promotor of MPOA tissue are characteristic of the offspring of low LG dams whereas low levels of methylation are detected in the ERα promotor of the offspring of high LG dams (Champagne *et al.*, 2006). This differential methylation has implications for the binding of transcription factors at the signal transducer and activator of transcription 5 response element within the promotor and is thought to produce the variations in gene expression and behavior observed in adult female offspring. Thus, the downstream target of postpartum maternal care may be similar to that of other maternally mediated effects such as germ cell condition, toxin exposure, and nutrition (Figure 19.1).

CONCLUSION

Maternal effects can occur across development and influence many aspects of offspring phenotype. When these effects alter reproductive behavior there can be consequences for subsequent generations of offspring. There is evidence for reproductive changes in offspring as a consequence of preconceptual, prenatal, and postnatal maternal environment. Traditional views on the mechanisms through which these effects occur have been limited to genetic germline perturbations resulting in transgenerational inheritance to subsequent generations. However,

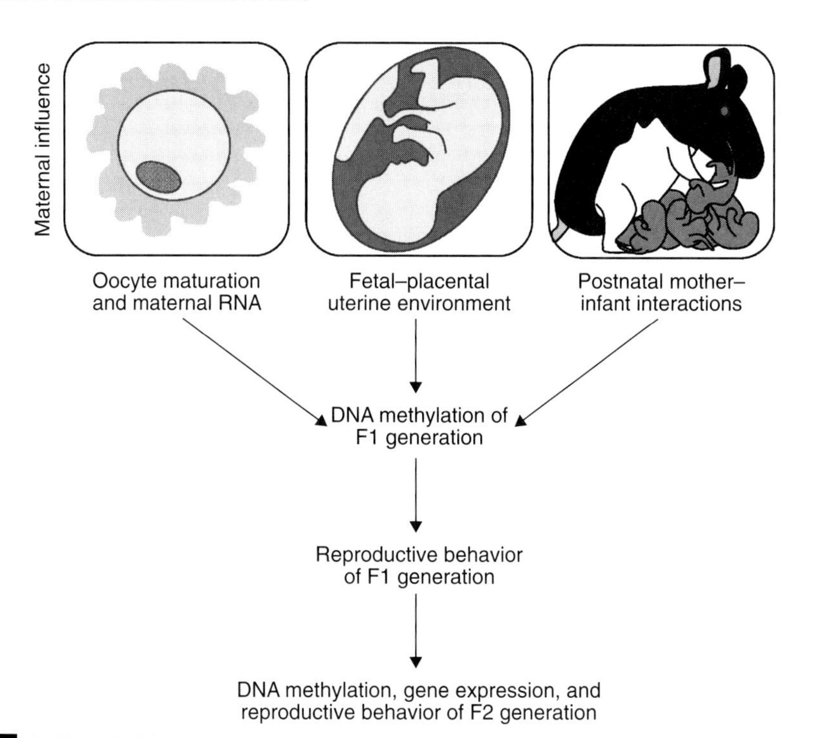

FIGURE 19.1 Maternal influence on offspring development can originate within the preconceptual, prenatal, and postnatal periods. Post-fertilization progression from zygote to embryo and patterning of the embryonic nervous system is dependent on oocyte mRNA levels. During the prenatal period, fetal development is influenced by maternal stress, nutrient intake, and endocrine activation. Postnatally, lactation and maternal care influence growth and behavior of infants. These maternal influences are thought to alter offspring (F1) gene expression and behavior through mechanisms such as DNA methylation. When the epigenetically altered genes influence reproductive behavior, these F1 modifications can be transmitted to the next generation of offspring (F2). Thus, through maternal influence, it is proposed that non-genomic transmission of phenotypes can be mediated.

increasing understanding of the stable regulation of gene expression through DNA methylation in response to maternal cues presents a novel approach to the study of the non-genomic inheritance. From an evolutionary perspective, these effects may represent an adaptive response to changes in environmental conditions to which offspring will be exposed in adulthood. Environments may decrease fertility, sexual receptivity, and maternal care under conditions of severe threat or disruption such that females delay investment in reproduction until more favorable conditions arise. Conversely, when resources are plentiful and threat is low, mothers and offspring may increase the quantity and quality of reproductive investment. The finding that the offspring of low LG dams display high levels of sexual receptivity and low levels of maternal care suggests a third approach to reproductive behavior. Natural variations in maternal care may induce a reproductive strategy in which trade-offs between offspring sexual and maternal are maintained by region specific changes in hypothalamic steroid receptor expression. The pervasiveness of these environmental effects allows a developing organism to display developmental plasticity and prepare future generations to be adapted to the ecological niche in which reproduction will occur.

REFERENCES

Allegrucci, C., Thurston, A., Lucas, E., and Young, L. (2005). Epigenetics and the germline. *Reproduction* 129, 137–149.

Anway, M. D., Cupp, A. S., Uzumcu, M., and Skinner, M. K. (2005). Epigenetic transgenerational actions of endocrine disruptors and male fertility. *Science* 308, 1466–1469.

Arling, G. L., and Harlow, H. F. (1967). Effects of social deprivation on maternal behavior of rhesus monkeys. *J. Comp. Physiol. Psychol.* 64, 371–377.

Bardi, M., and Huffman, M. A. (2002). Effects of maternal style on infant behavior in Japanese macaques (*Macaca fuscata*). *Dev. Psychobiol.* 41, 364–372.

Bardi, M., and Huffman, M. A. (2006). Maternal behavior and maternal stress are associated with infant behavioral development in macaques. *Dev Psychobiol.* 48, 1–9.

Barker, D. J. (2004). The developmental origins of adult disease. *J. Am. Coll. Nutr.* 23, 588S–595S.

Bayol, S. A., Farrington, S. J., and Stickland, N. C. (2007). A maternal "junk food" diet in pregnancy and lactation promotes an exacerbated taste for "junk food" and a greater propensity for obesity in rat offspring. *Br. J. Nutr.* 98, 843–851.

Beckett, C., Bredenkamp, D., Castle, J., Groothues, C., O'Connor, T. G., and Rutter, M. (2002). Behavior patterns associated with institutional deprivation: A study of children adopted from Romania. *J. Dev. Behav. Pediatr.* 23, 297–303.

Benoit, D., and Parker, K. C. (1994). Stability and transmission of attachment across three generations. *Child Dev.* 65, 1444–1456.

Berman, C. (1990). Intergenerational transmission of maternal rejection rates among free-ranging rheus monkeys on Cayo Santiago. *Anim. Behav.* 44, 247–258.

Bernstein, I. M., Mongeon, J. A., Badger, G. J., Solomon, L., Heil, S. H., and Higgins, S. T. (2005). Maternal smoking and its association with birth weight. *Obstet. Gynecol.* 106, 986–991.

Blusztajn, J. K., and Wurtman, R. J. (1981). Choline biosynthesis by a preparation enriched in synaptosomes from rat brain. *Nature* 290, 417–418.

Boerke, A., Dieleman, S. J., and Gadella, B. M. (2007). A possible role for sperm RNA in early embryo development. *Theriogenology* 68(suppl 1), S147–S155.

Braadstad, B. O. (1998). Effects of prenatal stress on behavior of offspring of laboratory and farmed mammals. *Appl. Anim. Behav. Sci.* 61, 159–180.

Bredy, T. W., Grant, R. J., Champagne, D. L., and Meaney, M. J. (2003). Maternal care influences neuronal survival in the hippocampus of the rat. *Eur. J. Neurosci.* 18, 2903–2909.

Bukovsky, A., Caudle, M. R., Svetlikova, M., Wimalasena, J., Ayala, M. E., and Dominguez, R. (2005). Oogenesis in adult mammals, including humans: A review. *Endocrine* 26, 301–316.

Caldji, C., Tannenbaum, B., Sharma, S., Francis, D., Plotsky, P. M., and Meaney, M. J. (1998). Maternal care during infancy regulates the development of neural systems mediating the expression of fearfulness in the rat. *Proc. Natl Acad. Sci. USA* 95, 5335–5340.

Cameron, N. M., and Meaney, M. J. (2006). Effect of maternal care and ovariectomy on sexual behavior and c-Fos activity in the female rat. *Front. Neuroendocrinol.* 27, 144.

Catalano, R., Bruckner, T., Hartig, T., and Ong, M. (2005). Population stress and the Swedish sex ratio. *Paediatr. Perinat. Epidemiol.* 19, 413–420.

Champagne, F., Diorio, J., Sharma, S., and Meaney, M. J. (2001). Naturally occurring variations in maternal behavior in the rat are associated with differences in estrogen-inducible central oxytocin receptors. *Proc. Natl Acad. Sci. USA* 98, 12736–12741.

Champagne, F. A., and Meaney, M. J. (2006). Stress during gestation alters postpartum maternal care and the development of the offspring in a rodent model. *Biol. Psychiatr.* 59, 1227–1235.

Champagne, F. A., Francis, D. D., Mar, A., and Meaney, M. J. (2003a). Variations in maternal care in the rat as a mediating influence for the effects of environment on development. *Physiol. Behav.* 79, 359–371.

Champagne, F. A., Weaver, I. C., Diorio, J., Sharma, S., and Meaney, M. J. (2003b). Natural variations in maternal care are associated with estrogen receptor alpha expression and estrogen sensitivity in the medial preoptic area. *Endocrinology* 144, 4720–4724.

Champagne, F. A., Chretien, P., Stevenson, C. W., Zhang, T. Y., Gratton, A., and Meaney, M. J. (2004). Variations in nucleus accumbens dopamine associated with individual differences in maternal behavior in the rat. *J. Neurosci.* 24, 4113–4123.

Champagne, F. A., Weaver, I. C., Diorio, J., Dymov, S., Szyf, M., and Meaney, M. J. (2006). Maternal care associated with methylation of the estrogen receptor-alpha1b promoter and estrogen receptor-alpha expression in the medial preoptic area of female offspring. *Endocrinology* 147, 2909–2915.

Champagne, F. A., Curley, J. P., Keverne, E. B., and Bateson, P. P. (2007). Natural variations in postpartum maternal care in inbred and outbred mice. *Physiol. Behav.* 91, 325–334.

Corcoran, D., Fair, T., and Lonergan, P. (2005). Predicting embryo quality: mRNA expression and the preimplantation embryo. *Reprod. Biomed. Online* 11, 340–348.

Corcoran, D., Rizos, D., Fair, T., Evans, A. C., and Lonergan, P. (2007). Temporal expression of transcripts related to embryo quality in bovine embryos cultured from the two-cell to blastocyst stage *in vitro* or *in vivo*. *Mol. Reprod. Dev.* 74, 972–977.

Cratty, M. S., Ward, H. E., Johnson, E. A., Azzaro, A. J., and Birkle, D. L. (1995). Prenatal stress increases corticotropin-releasing factor (CRF) content and release in rat amygdala minces. *Brain Res.* 675, 297–302.

Crews, D., Gore, A. C., Hsu, T. S., Dangleben, N. L., Spinetta, M., Schallert, T., Anway, M. D., and Skinner, M. K. (2007). Transgenerational epigenetic imprints on mate preference. *Proc. Natl Acad. Sci. USA* 104, 5942–5946.

Curley, J. P., Barton, S., Surani, A., and Keverne, E. B. (2004). Coadaptation in mother and infant regulated by a paternally expressed imprinted gene. *Proc. Biol. Sci.* 271, 1303–1309.

da Silva Faria, T., da Fonte Ramos, C., and Sampaio, F. J. (2004). Puberty onset in the female offspring of rats submitted to protein or energy restricted diet during lactation. *J. Nutr. Biochem.* 15, 123–127.

Davis, C. D., and Uthus, E. O. (2004). DNA methylation, cancer susceptibility, and nutrient interactions. *Exp. Biol. Med. (Maywood)* 229, 988–995.

De Sousa, P. A., Caveney, A., Westhusin, M. E., and Watson, A. J. (1998). Temporal patterns of embryonic gene expression and their dependence on oogenetic factors. *Theriogenology* 49, 115–128.

de Weerth, C., Wied, C. C., Jansen, L. M., and Buitelaar, J. K. (2007). Cardiovascular and cortisol responses to a psychological stressor during pregnancy. *Acta Obstet. Gynecol. Scand.* 86, 1181–1192.

Denenberg, V. H., and Rosenberg, K. M. (1967). Nongenetic transmission of information. *Nature* 216, 549–550.

Dolinoy, D. C., Huang, D., and Jirtle, R. L. (2007). Maternal nutrient supplementation counteracts bisphenol A-induced DNA hypomethylation in early development. *Proc. Natl Acad. Sci. USA* 104, 13056–13061.

Edwards, C. R., Benediktsson, R., Lindsay, R. S., and Seckl, J. R. (1996). 11 beta-Hydroxysteroid dehydrogenases: Key enzymes in determining tissue-specific glucocorticoid effects. *Steroids* 61, 263–269.

Eichenlaub-Ritter, U. (1998). Genetics of oocyte ageing. *Maturitas* 30, 143–169.

Fairbanks, L. A. (1989). Early experience and cross-generational continuity of mother–infant contact in vervet monkeys. *Dev. Psychobiol.* 22, 669–681.

Fairbanks, L. A., and McGuire, M. T. (1988). Long-term effects of early mothering behavior on responsiveness to the environment in vervet monkeys. *Dev. Psychobiol.* 21, 711–724.

Fernandez-Gonzalez, R., Moreira, P., Bilbao, A., Jimenez, A., Perez-Crespo, M., Ramirez, M. A., Rodriguez De Fonseca, F., Pintado, B., and Gutierrez-Adan, A. (2004). Long-term effect of *in vitro* culture of mouse embryos with serum on mRNA expression of imprinting genes, development, and behavior. *Proc. Natl Acad. Sci. USA* 101, 5880–5885.

Fernandez-Gonzalez, R., Ramirez, M. A., Bilbao, A., De Fonseca, F. R., and Gutierrez-Adan, A. (2007). Suboptimal *in vitro* culture conditions: An epigenetic origin of long-term health effects. *Mol. Reprod. Dev.* 74, 1149–1156.

Francis, D., Diorio, J., Liu, D., and Meaney, M. J. (1999a). Nongenomic transmission across generations of maternal behavior and stress responses in the rat. *Science* 286, 1155–1158.

Francis, D. D., Champagne, F. A., Liu, D., and Meaney, M. J. (1999b). Maternal care, gene expression, and the development of individual differences in stress reactivity. *Ann. NY Acad. Sci.* 896, 66–84.

Francis, D. D., Champagne, F. C., and Meaney, M. J. (2000). Variations in maternal behaviour are associated with differences in oxytocin receptor levels in the rat. *J. Neuroendocrinol.* 12, 1145–1148.

Frye, C. A., and Orecki, Z. A. (2002). Prenatal stress alters reproductive responses of rats in behavioral estrus and paced mating of hormone-primed rats. *Horm. Behav.* 42, 472–483.

Fulroth, R., Phillips, B., and Durand, D. J. (1989). Perinatal outcome of infants exposed to cocaine and/or heroin *in utero*. *Am. J. Dis. Child.* 143, 905–910.

Gavis, E. R., and Lehmann, R. (1992). Localization of nanos RNA controls embryonic polarity. *Cell* 71, 301–313.

Gluckman, P. D., Hanson, M. A., and Pinal, C. (2005). The developmental origins of adult disease. *Matern. Child Nutr.* 1, 130–141.

Gonzalez, A., Lovic, V., Ward, G. R., Wainwright, P. E., and Fleming, A. S. (2001). Intergenerational effects of complete maternal deprivation and replacement stimulation on maternal behavior and emotionality in female rats. *Dev. Psychobiol.* 38, 11–32.

Gosden, R., Trasler, J., Lucifero, D., and Faddy, M. (2003). Rare congenital disorders, imprinted genes, and assisted reproductive technology. *Lancet* 361, 1975–1977.

Gunnar, M. R., and van Dulmen, M. H. (2007). Behavior problems in postinstitutionalized internationally adopted children. *Dev. Psychopathol.* 19, 129–148.

Haig, D. (1993). Genetic conflicts in human pregnancy. *Q. Rev. Biol.* 68, 495–532.

Hall, W. G. (1975). Weaning and growth of artificially reared rats. *Science* 190, 1313–1315.

Hane, A. A., and Fox, N. A. (2006). Ordinary variations in maternal caregiving influence human infants' stress reactivity. *Psychol. Sci.* 17, 550–556.

Harlow, H. F., and Suomi, S. J. (1971). Social recovery by isolation-reared monkeys. *Proc. Natl Acad. Sci. USA* 68, 1534–1538.

Harlow, H. F., and Suomi, S. J. (1974). Induced depression in monkeys. *Behav. Biol.* 12, 273–296.

Hertsgaard, L., Gunnar, M., Erickson, M. F., and Nachmias, M. (1995). Adrenocortical responses to the strange situation in infants with disorganized/disoriented attachment relationships. *Child Dev.* 66, 1100–1106.

Herrenkohl, L. R. (1979). Prenatal stress reduces fertility and fecundity in female offspring. *Science* 208, 1097–1099.

Higgins, S. (2002). Smoking in pregnancy. *Curr. Opin. Obstet. Gynecol.* 14, 145–151.

Hinde, R. A., and Simpson, M. J. (1975). Qualities of mother–infant relationships in monkeys. *Ciba Found. Symp.* 39–67.

Hoyseth, K. S., and Jones, P. J. (1989). Ethanol induced teratogenesis: Characterization, mechanisms and diagnostic approaches. *Life Sci.* 44, 643–649.

Janny, L., and Menezo, Y. J. (1996). Maternal age effect on early human embryonic development and blastocyst formation. *Mol. Reprod. Dev.* 45, 31–37.

Kaaja, R. J., and Greer, I. A. (2005). Manifestations of chronic disease during pregnancy. *JAMA* 294, 2751–2757.

Kaiser, S., and Sachser, N. (2005). The effects of prenatal social stress on behaviour: Mechanisms and function. *Neurosci. Biobehav. Rev.* 29, 283–294.

Kaiser, S., Kruijver, F. P., Swaab, D. F., and Sachser, N. (2003). Early social stress in female guinea pigs induces a masculinization of adult behavior and corresponding changes in brain and neuroendocrine function. *Behav. Brain Res.* 144, 199–210.

Kinsley, C. H., and Bridges, R. S. (1988). Prenatal stress and maternal behavior in intact virgin rats: Response latencies are decreased in males and increased in females. *Horm. Behav.* 22, 76–89.

Kovacheva, V. P., Mellott, T. J., Davison, J. M., Wagner, N., Lopez-Coviella, I., Schnitzler, A. C., and Blusztajn, J. K. (2007). Gestational choline deficiency causes global- and Igf2 gene- DNA hypermethylation by upregulation of Dnmt1 expression. *J. Biol. Chem.* 282, 31777–31788.

Kruuk, L. E., and Hadfield, J. D. (2007). How to separate genetic and environmental causes of similarity between relatives. *J. Evol. Biol.* 20, 1890–1903.

Lee, P. R., Brady, D. L., Shapiro, R. A., Dorsa, D. M., and Koenig, J. I. (2007). Prenatal stress generates deficits in rat social behavior: Reversal by oxytocin. *Brain Res.* 1156, 152–167.

Lehmann, J., Pryce, C. R., Bettschen, D., and Feldon, J. (1999). The maternal separation paradigm and adult emotionality and cognition in male and female Wistar rats. *Pharmacol. Biochem. Behav.* 64, 705–715.

Lehmann, J., Stohr, T., and Feldon, J. (2000). Long-term effects of prenatal stress experiences and postnatal maternal separation on emotionality and attentional processes. *Behav. Brain Res.* 107, 133–144.

Lesage, J., Blondeau, B., Grino, M., Breant, B., and Dupouy, J. P. (2001). Maternal undernutrition during late gestation induces fetal overexposure to glucocorticoids and intrauterine growth retardation, and disturbs the hypothalamo–pituitary adrenal axis in the newborn rat. *Endocrinology* 142, 1692–1702.

Li, T., Vu, T. H., Ulaner, G. A., Littman, E., Ling, J. Q., Chen, H. L., Hu, J. F., Behr, B., Giudice, L., and Hoffman, A. R. (2005). IVF results in *de novo* DNA methylation and histone methylation at an Igf2-H19 imprinting epigenetic switch. *Mol. Hum. Reprod.* 11, 631–640.

Lippmann, M., Bress, A., Nemeroff, C. B., Plotsky, P. M., and Monteggia, L. M. (2007). Long-term behavioural and molecular alterations associated with maternal separation in rats. *Eur. J. Neurosci.* 25, 3091–3098.

Liu, D., Diorio, J., Tannenbaum, B., Caldji, C., Francis, D., Freedman, A., Sharma, S., Pearson, D., Plotsky, P. M., and Meaney, M. J. (1997). Maternal care, hippocampal glucocorticoid receptors, and hypothalamic–pituitary–adrenal responses to stress. *Science* 277, 1659–1662.

Liu, D., Diorio, J., Day, J. C., Francis, D. D., and Meaney, M. J. (2000). Maternal care, hippocampal synaptogenesis and cognitive development in rats. *Nat. Neurosci.* 3, 799–806.

Lovic, V., and Fleming, A. S. (2004). Artificially-reared female rats show reduced prepulse inhibition and deficits in the attentional set shifting task – reversal of effects with maternal-like licking stimulation. *Behav. Brain Res.* 148, 209–219.

Maestripieri, D. (1998). Parenting styles of abusive mothers in group-living rhesus macaques. *Anim. Behav.* 55, 1–11.

Maestripieri, D. (2005). Early experience affects the intergenerational transmission of infant abuse in rhesus monkeys. *Proc. Natl Acad. Sci. USA* 102, 9726–9729.

Maestripieri, D., Wallen, K., and Carroll, K. A. (1997). Infant abuse runs in families of group-living pigtail macaques. *Child Abuse Neglect* 21, 465–471.

Maestripieri, D., Tomaszycki, M., and Carroll, K. A. (1999). Consistency and change in the behavior of rhesus macaque abusive mothers with successive infants. *Dev. Psychobiol.* 34, 29–35.

Maestripieri, D., Lindell, S. G., Ayala, A., Gold, P. W., and Higley, J. D. (2005). Neurobiological characteristics of rhesus macaque abusive mothers and their relation to social and maternal behavior. *Neurosci. Biobehav. Rev.* 29, 51–57.

Maestripieri, D., Higley, J. D., Lindell, S. G., Newman, T. K., McCormack, K. M., and Sanchez, M. M. (2006). Early maternal rejection affects the development of monoaminergic systems and adult abusive parenting in rhesus macaques (*Macaca mulatta*). *Behav. Neurosci.* 120, 1017–1024.

Malcolm, J. C., Lawson, M. L., Gaboury, I., Lough, G., and Keely, E. (2006). Glucose tolerance of offspring of mother with gestational diabetes mellitus in a low-risk population. *Diabet. Med.* 23, 565–570.

McCormack, K., Sanchez, M. M., Bardi, M., and Maestripieri, D. (2006). Maternal care patterns and behavioral development of rhesus macaque abused infants in the first 6 months of life. *Dev. Psychobiol.* 48, 537–550.

McLean, M., Chipps, D., and Cheung, N. W. (2006). Mother to child transmission of diabetes mellitus: Does gestational diabetes program Type 2 diabetes in the next generation?. *Diabet. Med.* 23, 1213–1215.

McMillen, I. C., and Robinson, J. S. (2005). Developmental origins of the metabolic syndrome: Prediction, plasticity, and programming. *Physiol. Rev.* 85, 571–633.

Meaney, M. J. (2001). Maternal care, gene expression, and the transmission of individual differences in stress reactivity across generations. *Annu. Rev. Neurosci.* 24, 1161–1192.

Moore, C. L., and Power, K. L. (1986). Prenatal stress affects mother–infant interaction in Norway rats. *Dev. Psychobiol.* 19, 235–245.

Mousseau, T. A., and Fox, C. W. (1998). *Maternal Effects as Adaptations.* Oxford University Press, New York.

Murphy, L., Cherlet, T., Lewis, A., Banu, Y., and Watson, P. (2003). New insights into estrogen receptor function in human breast cancer. *Ann. Med.* 35, 614–631.

Nothias, J. Y., Majumder, S., Kaneko, K. J., and DePamphilis, M. L. (1995). Regulation of gene expression at the beginning of mammalian development. *J. Biol. Chem.* 270, 22077–22080.

Olegard, R., Sabel, K. G., Aronsson, M., Sandin, B., Johansson, P. R., Carlsson, C., Kyllerman, M., Iversen, K., and Hrbek, A. (1979). Effects on the child of alcohol abuse during pregnancy. Retrospective and prospective studies. *Acta Paediatr. Scand. Suppl.* 275, 112–121.

Painter, R. C., Roseboom, T. J., and Bleker, O. P. (2005). Prenatal exposure to the Dutch famine and disease in later life: An overview. *Reprod. Toxicol.* 20, 345–352.

Parker, G. (1983). Parental "affectionless control" as an antecedent to adult depression. A risk factor delineated. *Arch. Gen. Psychiatr.* 40, 956–960.

Parker, G. (1990). The Parental Bonding Instrument: A decade of research. *Soc. Psychiatr. Psychiatr. Epidemiol.* 25, 281–282.

Parker, S. W., and Nelson, C. A. (2005). The impact of early institutional rearing on the ability to discriminate facial expressions of emotion: An event-related potential study. *Child Dev.* 76, 54–72.

Peters, D. A. (1982). Prenatal stress: Effects on brain biogenic amine and plasma corticosterone levels. *Pharmacol. Biochem. Behav.* 17, 721–725.

Pruessner, J. C., Champagne, F., Meaney, M. J., and Dagher, A. (2004). Dopamine release in response to a psychological stress in humans and its relationship to early life maternal care: A positron emission tomography study using [^{11}C]raclopride. *J. Neurosci.* 24, 2825–2831.

Razin, A. (1998). CpG methylation, chromatin structure and gene silencing – a three-way connection. *EMBO J.* 17, 4905–4908.

Rocha, S. O., Gomes, G. N., Forti, A. L., do Carmo Pinho Franco, M., Fortes, Z. B., de Fatima Cavanal, M., and Gil, F. Z. (2005). Long-term effects of maternal diabetes on vascular reactivity and renal function in rat male offspring. *Pediatr. Res.* 58, 1274–1279.

Rondo, P. H., Ferreira, R. F., Nogueira, F., Ribeiro, M. C., Lobert, H., and Artes, R. (2003). Maternal psychological stress and distress as predictors of low birth weight, prematurity and intrauterine growth retardation. *Eur. J. Clin. Nutr.* 57, 266–272.

Roseboom, T., de Rooij, S., and Painter, R. (2006). The Dutch famine and its long-term consequences for adult health. *Early Hum. Dev.* 82, 485–491.

Roseboom, T. J., van der Meulen, J. H., Osmond, C., Barker, D. J., Ravelli, A. C., Schroeder-Tanka, J. M., van Montfrans, G. A., Michels, R. P., and Bleker, O. P. (2000). Coronary heart disease after prenatal exposure to the Dutch famine, 1944–45. *Heart* 84, 595–598.

Scott, L. A. (2000). Oocyte and embryo polarity. *Semin. Reprod. Med.* 18, 171–183.

Seckl, J. R., Nyirenda, M. J., Walker, B. R., and Chapman, K. E. (1999). Glucocorticoids and fetal programming. *Biochem. Soc. Trans.* 27, 74–78.

Sroufe, L. A. (2005). Attachment and development: A prospective, longitudinal study from birth to adulthood. *Attach. Hum. Dev.* 7, 349–567.

Stohr, T., Schulte Wermeling, D., Szuran, T., Pliska, V., Domeney, A., Welzl, H., Weiner, I., and Feldon, J. (1998). Differential effects of prenatal stress in two inbred strains of rats. *Pharmacol. Biochem. Behav.* 59, 799–805.

Strathdee, G., and Brown, R. (2002). Aberrant DNA methylation in cancer: Potential clinical interventions. *Expert Rev. Mol. Med.* 4, 1–17.

Symonds, M. E., Budge, H., Stephenson, T., and Gardner, D. S. (2005). Experimental evidence for long-term programming effects of early diet. *Adv. Exp. Med. Biol.* 569, 24–32.

Takahashi, L. K., Turner, J. G., and Kalin, N. H. (1992). Prenatal stress alters brain catecholaminergic activity and potentiates stress-induced behavior in adult rats. *Brain Res.* 574, 131–137.

Tambyrajia, R. L., and Mongelli, M. (2000). Sociobiological variables and pregnancy outcome. *Int. J. Gynaecol. Obstet.* 70, 105–112.

Trottier, G., Koski, K. G., Brun, T., Toufexis, D. J., Richard, D., and Walker, C. D. (1998). Increased fat intake during lactation modifies hypothalamic–pituitary–adrenal responsiveness in developing rat pups: A possible role for leptin. *Endocrinology* 139, 3704–3711.

Turner, B. (2001). *Chromatin and Gene Regulation.* Blackwell Science Ltd, Oxford.

Uriarte, N., Breigeiron, M. K., Benetti, F., Rosa, X. F., and Lucion, A. B. (2007). Effects of maternal care on the development, emotionality, and reproductive functions in male and female rats. *Dev. Psychobiol.* 49, 451–462.

Wagner, C. L., Katikaneni, L. D., Cox, T. H., and Ryan, R. M. (1998). The impact of prenatal drug exposure on the neonate. *Obstet. Gynecol. Clin. North Am.* 25, 169–194.

Wassarman, P. M., and Kinloch, R. A. (1992). Gene expression during oogenesis in mice. *Mutat. Res.* 296, 3–15.

Waterland, R. A., Dolinoy, D. C., Lin, J. R., Smith, C. A., Shi, X., and Tahiliani, K. G. (2006). Maternal methyl supplements increase offspring DNA methylation at axin fused. *Genesis* 44, 401–406.

Weaver, I. C., Grant, R. J., and Meaney, M. J. (2002). Maternal behavior regulates long-term hippocampal expression of BAX and apoptosis in the offspring. *J. Neurochem.* 82, 998–1002.

Weinstock, M. (2005). The potential influence of maternal stress hormones on development and mental health of the offspring. *Brain Behav. Immun.* 19, 296–308.

Weinstock, M., Fride, E., and Hertzberg, R. (1988). Prenatal stress effects on functional development of the offspring. *Prog. Brain Res.* 73, 319–331.

West, J. R. (1993). Use of pup in a cup model to study brain development. *J. Nutr.* 123, 382–385.

Weissgerber, T. L., and Wolfe, L. A. (2006). Physiological adaptation in early human pregnancy: adaptation to balance maternal-fetal demands. *Appl. Physiol. Nutr. Metab.* 31, 1–11.

Wolff, G. L., Kodell, R. L., Moore, S. R., and Cooney, C. A. (1998). Maternal epigenetics and methyl supplements affect agouti gene expression in Avy/a mice. *FASEB J.* 12, 949–957.

Yan, L., Yang, X., and Davidson, N. E. (2001). Role of DNA methylation and histone acetylation in steroid receptor expression in breast cancer. *J. Mammary Gland Biol. Neoplasia* 6, 183–192.

Zambrano, E., Martinez-Samayoa, P. M., Bautista, C. J., Deas, M., Guillen, L., Rodriguez-Gonzalez, G. L., Guzman, C., Larrea, F., and Nathanielsz, P. W. (2005). Sex differences in transgenerational alterations of growth and metabolism in progeny (F2) of female offspring (F1) of rats fed a low protein diet during pregnancy and lactation. *J. Physiol.* 566, 225–236.

Zhang, T. Y., Chretien, P., Meaney, M. J., and Gratton, A. (2005). Influence of naturally occurring variations in maternal care on prepulse inhibition of acoustic startle and the medial prefrontal cortical dopamine response to stress in adult rats. *J. Neurosci.* 25, 1493–1502.

Zorn, B., Sucur, V., Stare, J., and Meden-Vrtovec, H. (2002). Decline in sex ratio at birth after 10-day war in Slovenia: Brief communication. *Hum. Reprod.* 17, 3173–3177.

Zubieta, A. C., and Lonnerdal, B. (2006). Effect of suboptimal nutrition during lactation on milk protein gene expression in the rat. *J. Nutr. Biochem.* 17, 604–610.

20 PARENT-OF-ORIGIN EFFECTS ON PARENTAL BEHAVIOR

JAMES P. CURLEY[1,2]

[1] *Sub-Department of Animal Behaviour, University of Cambridge, Cambridge, UK.*
[2] *Department of Psychology, Columbia University, New York, USA.*

Mothers and fathers have a differential impact on their offspring's parental care. Such parent-of-origin effects can be demonstrated through the use of reciprocal hybrids. One source of parent-of-origin effects is genomic imprinting. There is increasing evidence that paternally expressed genes, those that are only ever expressed when inherited from the father, play an important role in the regulation of mammalian reproduction. In addition, variations in the quality of maternal or paternal care received by offspring early in life can lead to altered development with consequences for adult neuroendocrinology and parental behavior. There is also evidence that genes inherited on sex chromosomes can differentially regulate parental behavior. This chapter will discuss the evidence for these potential mediators of parent-of-origin effects and discuss evolutionary explanations for the occurrence of differential maternal and paternal influences on offspring behavior.

RECIPROCAL HYBRIDS

One strategy for investigating phenotypes that may be subject to parent-of-origin effects is to mate a mother of one species with a father of another, and then to perform the reciprocal mating. The resulting interspecific hybrids, which can occasionally be found in the field and irregularly in captivity, have long held a fascination for biologists including Charles Darwin and Alfred Wallace (Darwin, 1956). These hybrids have been found to differ greatly from both their parents and one another in several aspects of their phenotype. For instance, mules (horse mother and donkey father) are generally larger than hinnies (donkey mother and horse father), and resemble donkeys in their mane, ears, head shape, coat color and temperament more so than hinnies (Gray, 1971, 1972). Another relatively common interspecific hybrid is that between lions and tigers. Ligers (lion father and tiger mother) are much bigger than tigons (tiger father and lion mother) and are also more lion-like in their coloration, though tigons exhibit the more pronounced striping characteristic of tigers. In addition, the cubs of both crosses have mixed vocalizations and altered social behavior (Gray, 1971, 1972). Many other hybrids sporting a variety of exotic portmanteau names have been reported to show reciprocal differences in growth, body form, coloration and behavior (Gray, 1972). However, due to the rarity of such crosses, only anecdotal references exist to any changes in maternal care in interspecific hybrids.

In a laboratory setting, the behavior of reciprocal hybrids generated between strains or sub-species of mice can be useful in researching parent-of-origin effects. This strategy has been successfully employed for the investigation of emotional behavior in reciprocal hybrids generated between inbred BalbC and C57BL/6J (B6) mice. These two inbred strains differ significantly in many aspects of behavior and physiology with B6 mice typically being more active and exploratory than BalbC mice (Trullas & Skolnick, 1993; Calatayud & Belzung, 2001). B6C mice (B6 mother and BalbC father) show decreased reactivity and increased exploration in a novel environment compared to CB6 mice

(BalbC mother and B6 father), demonstrating a maternal inheritance of the B6-typical response to novelty. Hybrid F1 mice with a B6 mother and DBA/2J (D2) father consume more ethanol in a forced intake paradigm than F1 mice with D2 mothers and B6 fathers, again demonstrating a maternal inheritance of the B6-type reward phenotype (Gabriel & Cunningham, 2007). Parent-of-origin analysis of spontaneous infanticidal behavior toward newborn pups has been studied in virgin male and female reciprocal hybrids generated between the CF-1 strain and a wild-stock strain of mice (Perrigo *et al.*, 1993). Female hybrids exhibit infanticidal behavior similar to the CF-1 background regardless of whether their mother or father was a CF-1, however, male hybrids of both crosses inherit their maternal phenotype, suggesting that some aspects of parental motivation may be subject to parent-of-origin effects.

Recently, the postpartum care of reciprocal hybrid dams has been investigated more fully in the hybrid offspring of B6, BalbC and 129Sv mice strains. These particular strains differ in postpartum levels of nursing and pup licking/grooming (LG; Figure 20.1) (Shoji & Kato, 2006; Champagne *et al.*, 2007). 129Sv dams have a passive style of postpartum care exhibiting high levels of nursing and low levels of tactile stimulation in the form of LG whereas B6 dams have a more active style of care associated with lower levels of nursing but higher levels of LG (Champagne *et al.*, 2007). Other studies have consistently reported that BalbC dams also differ significantly in their maternal care from B6 dams, exhibiting both lower nursing and LG levels than B6 (Carola *et al.*, 2006; Shoji & Kato, 2006). An initial study investigating the maternal care of CB6 and B6C hybrid dams did not find any difference between each hybrid in levels of nursing or LG over either days 1–10 or days 11–20 of the postpartum period (Calatayud *et al.*, 2004), though CB6 dams were found to decrease the frequency of both nursing and LG displayed between the first and last 10-days postpartum whereas B6C dams did not. However, the observation periods in this study may have been too brief to detect behavioral differences (10 min/day). In a study with prolonged maternal observations (4 h/day), CB6 dams were found to engage in LG less frequently over the first week postpartum compared to B6C dams (Carola *et al.*, 2006), though no difference was observed in nursing frequencies between the hybrids. The failure to detect parent-of-origin effects on nursing behavior may be because the majority of observations (75%) occurred during the light phase of the light cycle when the dams are least active and passive nursing is highest (Shoji & Kato, 2006). Nevertheless, these data are consistent with a maternal inheritance of LG patterns and are supported by preliminary data from our research group indicating the influence of maternal strain (B6 or 129Sv) on hybrid offspring maternal LG and nursing behavior.

Thus, evidence from studies of reciprocal mouse hybrids suggests that several behaviors including infanticide, postpartum nursing and LG are differentially influenced by each parent. Moreover, there is intriguing evidence that the distribution of hormone receptors known to regulate maternal care may also be subject to such parent-of-origin effects. Female reciprocal hybrids between two sub-populations of prairie voles (Kansas and Illinois voles) possess estrogen receptor alpha distributions in the

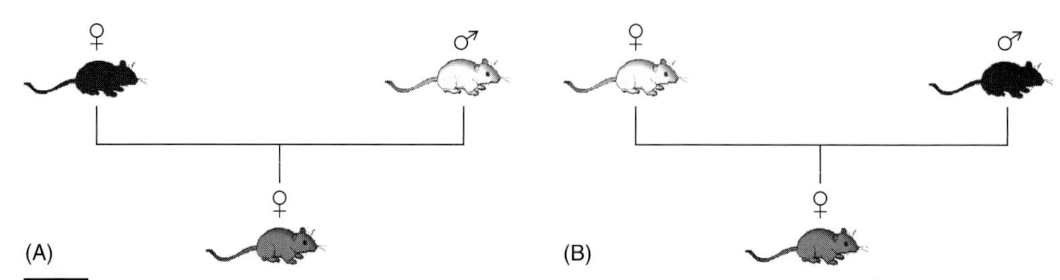

FIGURE 20.1 Parent-of-origin effects revealed by reciprocal hybrids. Reciprocal hybrid mice are produced by mating the father of one strain or sub-species with the mother of another strain or sub-species (A) and vice-versa (B). Phenotyping of the offspring of both crosses is a useful tool to revealing behaviors such as infanticide, nursing and LG that are subject to parent-of-origin effects.

bed nucleus of the stria terminalis (BNST) and medial amygdala typical of the father's sub-population (Kramer *et al.*, 2006). There are many possible mechanisms through which these neural and behavioral parent-of-origin effects can occur. In this chapter I shall discuss evidence for several of these mechanisms, including (1) genomic imprinting, (2) maternal effects, (3) paternal effects and (4) the influence of sex chromosomes.

GENOMIC IMPRINTING

In the early 1980s, there were attempts to create viable mammalian embryos that consisted of either two maternal haploid (parthenogenetic-PG) or two paternal haploid (androgenetic-AG) genomes (McGrath & Solter, 1984; Surani *et al.*, 1984). These attempts were unsuccessful however as embryos failed to reach term, dying a few days post-implantation. The PG conceptus possessed rudimentary placentae and survived only until the 25 somite stage, whereas the AG conceptus had extensive placental tissue but only survived until the 6–8 somite stage. These observations that maternal and paternal autosomal genomes contribute unequally to mammalian development led to the discovery of imprinted genes (McGrath & Solter, 1984; Surani *et al.*, 1984; Cattanach & Kirk, 1985; McLaughlin *et al.*, 1996). These genes are only ever expressed in male and female offspring when inherited from one parent with the other parental gene copy silenced. Thus, the expression of an imprinted gene is dependent on the sex of the individual *passing on* the gene rather than the sex of the individual *inheriting* the gene. This is achieved mechanistically through the erasure and re-establishment of epigenetic modifications (imprints) such as CpG methylation at imprinting domains during germ cell development (Hajkova *et al.*, 2002). For the majority of autosomal genes, both parents pass on a functionally equivalent gene to their offspring, however, approximately 100 imprinted genes and related transcripts do not follow such classical Mendelian inheritance (Reik & Walter, 2001). One parent actively silences their own gene copy meaning that there is parent-of-origin specific monoallelic expression in the offspring.

Though PG and AG mice do not survive beyond implantation, chimeric mice that contain a mixture of either PG or AG cells and at least 60% normal wild-type cells (that contain both maternal and paternal genomes) are viable (Keverne *et al.*, 1996). Labeling of PG and AG cells with a LacZ reporter gene reveals a startling dichotomy in the proliferation of these cells within the developing brain. PG cells contribute significantly to the developing neocortex, striatum and hippocampus, but are excluded from brain areas such as the hypothalamus, BNST, septum and preoptic area (Allen *et al.*, 1995). Interestingly, these areas are associated with the hormonal regulation of maternal care and other appetitive behaviors (Albert & Walsh, 1984; Numan, 2006). Even at the earliest stages of embryonic brain development (E9 and E10), PG cells are absent from the basal forebrain plate, whereas at this stage AG cells are found in all neural tissue (Allen *et al.*, 1995). As development proceeds, AG cells become localized to the mediobasal forebrain and are almost completely absent from telencephalic structures thereby implicating genes of paternal origin in the regulation of reproductive behaviors including maternal care (Keverne *et al.*, 1996).

The role of paternally expressed genes (those expressed only from the father and silenced when inherited from the mother) in the regulation of maternal behavior has been investigated in mice that lack either a Peg1 (Paternally Expressed Gene 1) or Peg3 (Paternally Expressed Gene 3) gene (Lefebvre *et al.*, 1998; Li *et al.*, 1999). Peg3 encodes a large zinc finger protein with 11 widely spaced C2H2 motifs (Kuroiwa *et al.*, 1996), and is involved in the regulation of apoptotic pathways through translocation of Bax proteins from cytosol to mitochondria (Deng & Wu, 2000; Johnson *et al.*, 2002). The Peg3 protein through its involvement in the P53 and TNF signaling pathways has the capacity to shape the development of those tissues such as the placenta and hypothalamus in which it is highly expressed (Li *et al.*, 1999). Indeed, pregnant and postparturient female Peg3 mutant mice exhibit a number of behavioral and physiological deficits associated with dysfunction of the hypothalamus (Li *et al.*, 1999; Curley *et al.*, 2004). All of these impairments have been observed in mutant females mated with wild-type males, and who were therefore raising wild-type pups that expressed a normal copy of the Peg3 gene (see Figure 20.2).

During gestation, Peg3 mutant females do not increase their food intake as much as control females and carry over fewer fat reserves into the

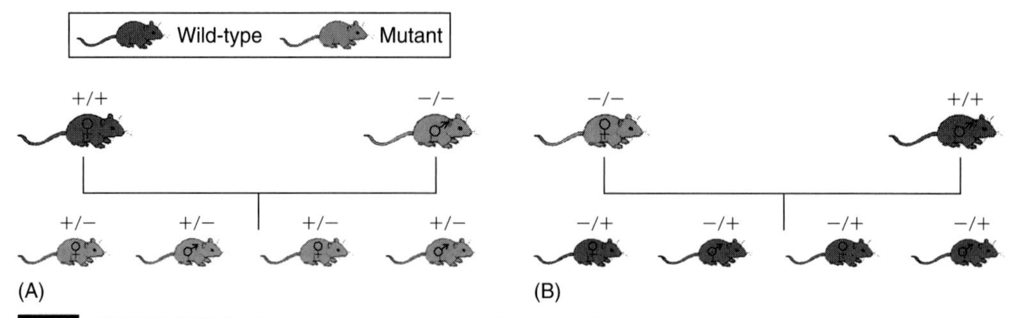

FIGURE 20.2 Co-adaptation in mother and infant regulated by paternally expressed genes. When a male carrying a dysfunctional copy of either the Peg3 or Peg1 genes is mated with a wild-type female (A), all their offspring express the mutation during embryogenesis, in the placenta and the hypothalamus. Such offspring are growth retarded *in utero*, do not prime mothers to behave as maternally during gestation, are impaired in suckling postnatally and have delayed onset of thermogenesis and puberty. When a female carrying a dysfunctional copy of these genes is mated with a wild-type male, all of her offspring are wild-type (B). As a consequence of the presence of this mutation in the hypothalamus this female will exhibit a number of maternal care deficits. Therefore, through expression in tissues in both mother and offspring, paternally expressed genes ensure the successful development of offspring.

postpartum period (Curley *et al.*, 2004). Mutant mothers also provide fewer resources during pregnancy to their wild-type offspring evidenced by smaller pups and decreased litter size. These pups continue to be growth retarded throughout lactation and even lose weight between the first and second days postpartum as mutant females are slower to let-down milk from their mammary glands (Curley *et al.*, 2004). Consistent with these findings, pup mortality is also higher in the offspring of Peg3 mutant mothers, both during the pre-and postnatal periods (Curley *et al.*, 2004). Behaviorally, Peg3 mutant females are impaired in their ability to retrieve pups during the early lactation period, being slower to return the pups to the nest in a standard retrieval test (Li *et al.*, 1999; Curley *et al.*, 2004). During this test, mutant mothers are also slower to build a nest and crouch over pups and therefore provide less warmth for their altricial pups. These deficits are not related to either the pup genotype or *in utero* experiences, as Peg3 mutant females remain up to 11 times slower than wild-types to retrieve cross-fostered pups (Li *et al.*, 1999). Moreover, similar deficits are observed in non-parturient virgin females despite repeated pup exposure (Li *et al.*, 1999). Preliminary data suggest that these mutant females also engage in less frequent postpartum nursing and LG during the first week postpartum (Champagne *et al.*, 2005).

These patterns of maternal care appear to be related to alterations in the regulation of the hypothalamic oxytocinergic system. In rodents, synthesis of oxytocin and oxytocin receptors is increased in late gestation in preparation for parturition (Caldwell *et al.*, 1987; Bale *et al.*, 1995; Larcher *et al.*, 1995). Central infusion of oxytocin accelerates the onset of maternal care (Pedersen & Prange, 1979; Pedersen *et al.*, 1982) whereas antagonists to oxytocin or oxytocin receptors inhibits maternal responding (Fahrbach *et al.*, 1985; Pedersen *et al.*, 1985). Postparturient mutant Peg3 females have relatively fewer oxytonergic positive neurons in the paraventricular nucleus (PVN) of the hypothalamus, supraoptic nucleus (SON) and medial preoptic area (MPOA) (Li *et al.*, 1999). Lack of the Peg3 protein may lead to these changes in two non-mutually exclusive ways. As a transcription factor, Peg3 may serve to co-ordinate oxytocin and oxytocin receptor mRNA synthesis. Amongst Peg3 mutant females, the absence of this transcription factor may lead to reduced levels of these proteins which are critical for the expression of maternal care. However, a more probable explanation is that the lack of Peg3 expression during hypothalamic development results in disruptions to apoptotic signaling and cell proliferation in the MPOA, PVN and SON resulting in a lack of organization of maternal oxytocinergic neural networks.

A second paternally expressed gene, Peg1 (or MEST, mesoderm-specific transcript), is also critical for the expression of maternal care

(Lefebvre *et al.*, 1998). This gene is expressed throughout development and in the adult CNS, particularly in the ventral forebrain, hypothalamus, amygdala, hippocampus and main and accessory olfactory bulbs (Lefebvre *et al.*, 1998). Initial work with Peg1 mutant mice produced on a 129Sv background indicated higher mortality rates and smaller pup size amongst wild-type pups born to mutant mothers. Postparturient mutant dams are less likely to engage in placentophagia, slower to engage in suckling and are slower to retrieve pups, nest build and crouch in a retrieval test (Lefebvre *et al.*, 1998). As with Peg3 mutant dams, Peg1 mutants do not differ in latency to sniff pups in this test or to find a hidden food pellet, indicating that the behavioral differences observed are not due to gross olfactory deficits. Recent data indicate that Peg1 dams produced on a mixed hybrid 129Sv/B6 background also engage in lower levels of LG during the first week postpartum (unpublished data). It has not yet been determined how these behavioral differences manifest, though alterations in oxytocinergic signaling is a strong possibility. It is interesting to note the similarities in behavioral phenotype between mutant Peg1 and Peg3 dams despite the differences in function of these two genes. The exact cellular function of Peg1 is still unknown, but it appears to encode a polypeptide similar to the α/β hydrolase fold enzymes, (Lefebvre *et al.*, 1997). It remains to be seen if other paternally expressed genes or even maternally expressed genes are similarly involved in the regulation of maternal behavior. Evidence from studies of Peg1 and Peg3 mutants implicates the maternal epigenetic silencing of imprinted genes early in development as being significant in the development of several physiological and behavioral aspects of parental care. Interestingly, recent studies have implicated imprinted genes in both mate choice (Isles *et al.*, 2001, 2002) and male sexual behavior (Swaney *et al.*, 2007) suggesting a wide-ranging role for genomic imprinting in the regulation of male and female reproductive behavior.

MATERNAL EFFECTS

Variations in offspring development in response to variations in maternal environmental cues, referred to as maternal effects, have been well studied in evolutionary biology across a wide diversity of taxa (Mousseau & Fox, 1998). Maternal effects are extremely important in mammals and there is an extensive intimate relationship between a mother and her offspring (Reinhold, 2002). Effects of maternal environment on offspring behavioral development may occur from gametogenesis through to the pre-conceptual period, fertilization and implantation, but predominate during gestation and the postpartum lactation period when neural development accelerates (Fleming *et al.*, 2004; Champagne & Curley, 2005). Maternal effects are considered to be adaptive in that they enable offspring to exhibit development plastically in response to cues provided by the mother, thus maximizing their suitability to the future adult environment (Gluckman *et al.*, 2005b; Bateson, 2007).

In rodents, subjecting pregnant dams to pharmacological, physiological, hormonal and nutritional manipulations can have long-term consequences for offspring behavior in adulthood; particularly stress responsivity, cognitive ability, and response to reward (Weinstock *et al.*, 1988; Weinstock, 2005; Gluckman *et al.*, 2005a, b; Bateson, 2007). In addition, the consequences of these manipulations highlight the importance of *in utero* experiences of females on their subsequent parental care. Exposure in late gestation to oxaxepam (a benzodiazepine derivative that leads to an increase in binding of GABA at $GABA_A$ receptors) leads to higher levels of maternal aggression by females toward a male intruder, and alters postpartum nursing patterns (Bignami *et al.*, 1992). Likewise, mice exposed prenatally to methamphetamine have been shown to passively nurse their own offspring more frequently and to retrieve pups more quickly (Slamberova *et al.*, 2007). Another approach to studying prenatal influences on development is restraint stress. Female offspring of rat dams that are subjected to physical restraint during gestation display decreased levels of nursing associated with elevated corticotropin-releasing hormone (CRH) mRNA and vasopressin in the parvocellular region of the PVN (Bosch *et al.*, 2007). Other studies have shown that prenatal restraint stress results in decreased postpartum LG by dams and female offspring associated with reductions in hypothalamic oxytocin receptor binding densities (Champagne & Meaney, 2006). However, in these studies it is difficult to dissociate the effects of the gestational stressor directly on the offspring's development from indirect effects that occur due to the effects of gestational stress on postpartum maternal behavior. For instance, it has been demonstrated that gestational restraint stress can alter the nursing and licking of pups by dams, which may then lead to differences in the offspring behavior (Moore & Power, 1986; Smith *et al.*, 2004; Patin *et al.*, 2005).

To understand the role of the prenatal environment as a potential mediator of parent-of-origin effects, one strategy is to conduct embryo transfer between inbred strains of mice. Using this technique it would be possible to evaluate whether prenatal factors are critical in the establishment of strain-specific maternal behaviors. Though this approach has not yet been applied to study maternal care, it has been successfully utilized to illustrate that long-term changes in strain-specific adult behaviors such as anxiety, learning, memory and motor activity can indeed originate through strain-specific prenatal environments (Denenberg *et al.*, 1996, 1998, 2001; Francis *et al.*, 2003; Rose *et al.*, 2006). These studies raise the question of what particular prenatal factors can lead to long-term changes in behavior and how these effects are maintained. Undoubtedly, given the organizing effects of hormones, steroids, proteins and other metabolites it is likely that variation in the circulating levels of these factors *in utero* is critical for shaping brain development and leads to individual differences in adult behavior and maternal care (Kaiser & Sachser, 1998; Seckl & Meaney, 2004; Kemme *et al.*, 2007).

The majority of research on maternal effects in mammals has focused on how alterations in the postnatal mother–infant relationship can modify offspring brain development and behavior (Champagne & Curley, 2005). This research has demonstrated that mother–infant interactions early in development can mediate the inheritance of styles of maternal care. In humans, it has been shown that individuals who are abused as infants are more likely to abuse their own children (Chapman & Scott, 2001). However, the inheritance of variations in care is not restricted to negative parenting. Natural variations in patterns of mother–infant attachment have also been shown to be inherited transgenerationally. Women who experienced a secure attachment to their mother are more likely to have a secure relationship with their own child, and similarly those women who had an insecure relationship with their mother are more likely to have an insecure relationship with their own child (Benoit & Parker, 1994). Patterns of maternal care have also been shown to be transgenerationally inherited down matrilines in primate groups (Berman, 1990). In vervet monkeys, the degree of mother–infant contact is positively correlated with the amount of contact displayed by female offspring toward their own infants (Fairbanks, 1989), whereas amongst rhesus monkeys the degree of infant grooming has been found to be transmitted through the matriline (Maestripieri *et al.*, 2007). Furthermore, both rhesus monkey and macaque mothers are more likely to abuse or reject their infants if they themselves were abused or rejected by foster mothers demonstrating that the rearing environment is critical for the inheritance of this maternal behavior (Berman, 1990; Maestripieri *et al.*, 1997, 2007; Maestripieri, 2005). In rhesus monkeys this effect appears to be related to maternal modulation of serotonergic systems, with high levels of maternal rejection related to decreased levels of serotonin and its metabolites (Maestripieri *et al.*, 2006, 2007). These studies provide an elegant illustration of the role of maternal behavior it self in mediating the intergenerational transmission of maternal styles.

In rodents, study of the transgenerational inheritance of maternal care has utilized several experimental approaches. Maternal separation, the removal of pups from the mother for daily repeated sessions of up to 5 h, has been used to demonstrate that disruptions to the mother–infant relationship can lead to behavioral and neuroendocrine alterations in offspring, including changes in the expression and distribution of hormones and neuropeptides that co-ordinate maternal care (Sanchez *et al.*, 2001). Maternally separated animals are not only subjected to decreased levels of maternal somatosensory stimulation, but may also be subjected to inappropriate maternal care following periods of separation. Fleming and colleagues have conducted a series of studies in which rat pups received either no maternal separation, 3 h of separation daily (days 2–9 postnatal), 5 h of separation daily (days 4–20 postnatal) or complete maternal separation (24 h/day for days 4–20; these offspring were artificially reared (AR) in plastic cups, being fed via a pump and stimulated manually) (Gonzalez *et al.*, 2001; Lovic *et al.*, 2001; Fleming *et al.*, 2002). Those females that had experienced the briefest levels of maternal separation did not exhibit any differences in maternal care. However, females that had been separated for 5 h/day showed decreased levels of LG and nursing of pups (Fleming *et al.*, 2002). Those females that were AR displayed very low levels of LG and nursing of offspring over the first week, as well as deficits in pup retrieval (Gonzalez *et al.*, 2001; Gonzalez & Fleming, 2002; Melo *et al.*, 2006). AR females found pups to be a less rewarding stimulus, were less responsive to pups after a period of separation (Melo *et al.*, 2006), and had decreased expression of c-Fos in the MPOA in response to pup exposure (Gonzalez & Fleming, 2002). Moreover, following estrogen priming, virgin AR females exhibit reduced LG and crouching compared to maternally separated rats (Novakov & Fleming, 2005).

Though the use of artificial separations demonstrates the potential significance of the role of maternal care in shaping offspring behavior, more recent work by Michael Meaney and colleagues has taken advantage of the occurrence of natural variations in levels of maternal care to illustrate the intergenerational transfer of maternal phenotypes (Francis *et al.*, 1999; Champagne & Meaney, 2001; Meaney, 2001; Champagne *et al.*, 2003a). Long-Evans rat dams exhibit consistent levels of LG toward pups over successive litters, indicating that LG behavior is a stable trait of individual dams (Champagne *et al.*, 2003a). When large cohorts of dams are observed extensively during the postpartum period, a normal distribution of maternal LG is observed, with some dams licking their pups with a low frequency (if they are −1 standard deviation of the mean) and other dams exhibiting high levels of LG (if they are +1 standard deviation of the mean). Due to the stability of LG, pups can be cross-fostered between high and low LG dams to determine whether postnatal LG is critical in mediating variations in offspring behavior. Through the use of such cross-fostering it has been shown that the degree of LG received by pups during the first week postpartum is critical in predicting how female offspring will behave toward their own pups. Levels of maternal LG are associated with hypothalamic oxytocin receptor binding density and levels of estrogen receptor alpha (ERα) in the MPOA of offspring (Champagne *et al.*, 2001, 2003b). Thus, high levels of maternal LG received by females during infancy are correlated with increased levels of gene expression in brain regions that are critical for the expression of maternal care. Furthermore, the mechanism for this transmission is epigenetic. High levels of LG received by pups during the first week postpartum lead to lower levels of methylation of the ERα gene promoter in the MPOA, corresponding to increased expression of ERα and elevated levels of maternal care exhibited by these females in adulthood (Champagne *et al.*, 2006). This series of experiments provides insight into the mechanisms regulating maternal effects and suggests that epigenetic regulation of gene expression and development may mediate maternal influence on offspring behavior.

An additional approach to studying maternal effects is to manipulate the mother–infant relationship in such a way as to shift levels of nursing and LG. For example, if rat dams are made anosmic through olfactory bulbectomy, they display reduced LG toward pups during the first week postpartum compared to sham operated controls (Fleming *et al.*, 1979). Interestingly, those female offspring reared by bulbectomized dams were found to engage in less LG toward their own offspring (Fleming *et al.*, 1979). In this study control sham operated dams exhibited lower levels of LG toward pups when they had larger litters and females raised in such large litters displayed reduced LG in adulthood. In a more recent study, rat dams that were exposed to a predator (cat) odor following parturition displayed increased frequencies of LG and nursing of pups during the first week postpartum (McLeod *et al.*, 2007). This increased maternal care was found to be inherited by cross-fostered female offspring (i.e., pups that had never experienced the predator odor but were reared by predator-exposed females). These behavioral effects were associated with an increase in the expression of estrogen receptors (both alpha and beta) in the MPOA of female offspring. Thus, the quality of maternal care can alter neuroendocrine systems regulating maternal behavior and thus mediate transgenerational effects in matrilines.

Tactile stimulation received during early development seems to be a critical feature of the mother–infant interaction and mediate the observed changes in offspring development. For example, the maternal deficits associated with artificial rearing may be ameliorated if pups are provided with high levels of tactile stimulation in the form of manual stroking, or if pups are paired with a sibling and exposed to maternal odors throughout their first weeks of life (Fleming *et al.*, 2002; Melo *et al.*, 2006). Another strategy for increasing tactile stimulation of pups in early development is to implement a communal rearing paradigm. Lactating female mice will readily form a communal nest if group housed thus providing pups with higher levels of mother–infant stimulation and elevated pup–pup interaction (Sayler & Salmon, 1969). Females reared in such a manner display higher levels of nursing and LG during the first week postpartum, are quicker to retrieve pups, and are less aggressive and more subordinate toward a male intruder in an aggression test (Curley *et al. unpublished*). Tactile stimulation may thus trigger physiological systems that stimulate epigenetic factors which in turn promote up-regulation of the neuroendocrine modulators of maternal care. These maternal effects lead to stable long-term changes in reproductive behavior and promote the transgenerational cycle of maternal care (see Figure 20.3).

PATERNAL EFFECTS

In mammals, the mother is the sole caregiver of her offspring from fertilization through to parturition. Moreover, in the majority of mammalian species, the mother is the only parent to provide postpartum care and as such has the opportunity to shape offspring behavioral development. However, about 8–9% of mammalian species exhibit monogamy, a mode of social organization where males and females form long-term social or sexual partnerships and where males routinely take part in infant care (Kleiman, 1977). In these species, variations in the quality of paternal care as well as maternal care may lead to variation in the brain and behavioral development of offspring. Amongst monogamous Californian deer mice, males exhibit several parental behaviors including grooming, retrieval and huddling (Trainor & Marler, 2001; Frazier *et al.*, 2006). Interestingly, variations in levels of several of these paternal behaviors are correlated with behavioral changes in the offspring. For

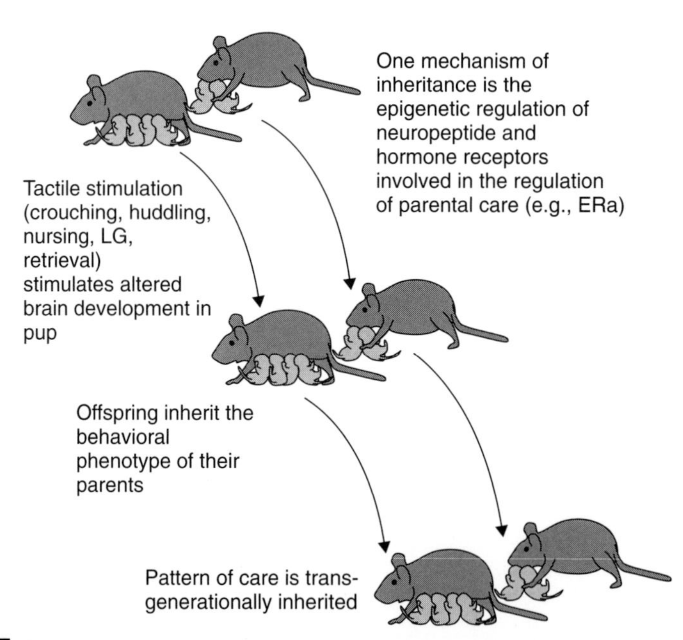

FIGURE 20.3 Transgenerational inheritance of parental behaviors. Evidence suggests that postpartum behaviors that have a component of tactile stimulation of offspring may be transgenerationally inherited. Variations in *in utero* experiences also impact the parental behavior of offspring and may be passed to future generations via the matriline. Recent work suggests that this inheritance may be mediated through epigenetic regulation of gene expression of receptors that are involved in regulating parental care.

instance, lower paternal grooming is related to poorer performance of offspring on a novel object recognition task (Bredy *et al.*, 2004). Offspring receiving less paternal grooming also have increased vasopressin in the PVN and higher levels of circulating corticosterone and progesterone suggesting that paternal behavior may modulate offspring stress-responses (Frazier *et al.*, 2006). Male Californian mice that are stimulated to retrieve and grab their pups have offspring that are more aggressive and have increased vasopressin in the dorsal BNST and decreased vasopressin in the ventral BNST (Frazier *et al.*, 2006). Moreover, when cross-fostered to the less paternal white-footed mouse, male Californian mice experience fewer retrievals when young and, as adults, are less aggressive in resident-intruder tests and have reduced vasopressin in the BNST and SON (Bester-Meredith & Marler, 2001). Significantly, there also appears to be a paternal transmission of parental care in this species, as Californian mice reared by white-footed mice exhibit less retrieval behavior toward their own offspring (Bester-Meredith & Marler, 2003). This transmission of paternal effects may be related to levels of vasopressin in the BNST as individual variations in paternal retrieval are correlated with levels of this peptide (Bester-Meredith & Marler, 2003). Therefore, as with postnatal maternal effects, the tactile stimulation received via different paternal behaviors is related to the development of specific offspring behaviors including parental care, and these appear to be related to a reorganization of hypothalamic vasopressinergic systems. From an evolutionary perspective, it would be anticipated that fathers would be as good an indicator of future environmental quality as mothers and would therefore have potential to exert a strong influence on offspring behavioral plasticity.

SEX CHROMOSOMES

There is increasing evidence that genes located on sex chromosomes may be expressed in the brain and can play a significant role in the regulation of social, aggressive and reproductive behaviors, independent from the effects of SRY (testes-determining factor) on gonadal development (De Vries *et al.*, 2002; Arnold, 2003; Arnold *et al.*, 2004). One strategy to investigate the relationship between sex chromosomes and behavioral phenotype has been to dissociate sex chromosome complement and gonad determination by producing transgenic mice that (1) have a deleted SRY gene, (2) contain genes from the X or Y chromosome on autosomes or (3) express X or Y genes on the opposite sex chromosome (Arnold *et al.*, 2004). For example, one transgenic mouse model took advantage of a translocation of the SRY gene and a small number of other genes from the Y chromosome to the paternal X chromosome during meiosis, thereby producing XX^{SRY} males. If these males were found to be indistinguishable from XY males in behavior then it could be suggested that SRY contains all the information required to masculinize the brain through its role in hormonal regulation. However, if the behavior of XX^{SRY} and XY males differs, this would indicate a role for other genes on the Y chromosome in generating variations in phenotype. When tested in a pup sensitization and retrieval test, XX^{SRY} males were found to be significantly less infanticidal and much better at retrieving pups than XY males (Reisert *et al.*, 2002). This finding suggests that genes operating outside of that translocated region of the Y chromosome may be expressed in the male rodent brain and decrease parental care and increase pup aggression.

Another approach to studying the role of sex chromosomes in mediating offspring behavior is to delete the SRY gene from the Y chromosome and re-insert into an autosome. Female XX and XY^- (that completely lack the SRY gene) mice differ significantly in pup retrieval behavior, with XY^- females being slower to retrieve and retrieving fewer pups than normal XX females (Gatewood *et al.*, 2006). XY^- females also have an increased density of vasopressin fibers in the lateral septum than XX females, suggesting a potential mechanism for these behavioral effects (De Vries *et al.*, 2002; Gatewood *et al.*, 2006). These brain and behavioral differences cannot be attributed to the gonadal determining effects of the SRY gene as both groups lacked this gene, but appear to involve differences in gene expression on either the X or Y chromosome, and thus relate to brain-specific expression of sex chromosome genes (Arnold *et al.*, 2004; Dewing *et al.*, 2006). Therefore, there is increasing evidence that the sex chromosomes inherited by an individual from their mother or father could play a role in the regulation of parental behaviors, though the exact mechanism through which this occurs is yet unknown.

CONCLUSION

Few studies have provided a systematic evaluation of the influence of parent-of-origin effects on parental behavior or the mechanisms through which these effects are achieved. However, it is clear from the study of reciprocal hybrids that some aspects of maternal behavior and the underlying neuroendocrine regulators of maternal care differ in a parent-dependent manner. In future studies, it would be useful to extend this research to other strains and species that differ in a broad range of maternal behaviors and in multiple hormone and receptor systems related to reproduction. The efficacy of this strategy would certainly benefit from the use of embryo transfer which would permit the dissociation of classical maternal effects from other sources of parent-of-origin influence (Isles *et al.*, 2001). In addition to the sources of maternal and paternal influence discussed in this chapter there is emerging evidence for the role of maternal mRNAs (Schier, 2007) and the inheritance of environmentally induced epigenetic modifications to the paternal germline (Anway *et al.*, 2005) as potential mediators of offspring development. Taken together, these studies illustrate that gene expression and behavior vary in ways that are a departure from traditional Mendelian laws of inheritance. There are a number of fascinating parent-of-origin systems for inheriting differences in gene expression, brain development and behavior, thus providing exciting avenues for future research.

ACKNOWLEDGMENTS

The author would like to thank Bob Bridges and the organizing committee of the Parental Brain Meeting for the opportunity to present a paper on this topic, to Frances Champagne for providing a critical reading of the manuscript, and to also thank Darwin College, Cambridge, and The Leverhulme Trust for financial assistance.

REFERENCES

Albert, D. J., and Walsh, M. L. (1984). Neural systems and the inhibitory modulation of agonistic behavior: A comparison of mammalian species. *Neurosci. Biobehav. Rev.* 8, 5–24.

Allen, N. D., Logan, K., Lally, G., Drage, D. J., Norris, M. L., and Keverne, E. B. (1995). Distribution of parthenogenetic cells in the mouse brain and their influence on brain development and behavior. *Proc. Natl Acad. Sci. USA* 92, 10782–10786.

Anway, M. D., Cupp, A. S., Uzumcu, M., and Skinner, M. K. (2005). Epigenetic transgenerational actions of endocrine disruptors and male fertility. *Science* 308, 1466–1469.

Arnold, A. P. (2003). The gender of the voice within: The neural origin of sex differences in the brain. *Curr. Opin. Neurobiol.* 13, 759–764.

Arnold, A. P., Xu, J., Grisham, W., Chen, X., Kim, Y. H., and Itoh, Y. (2004). Minireview: Sex chromosomes and brain sexual differentiation. *Endocrinology* 145, 1057–1062.

Bale, T. L., Pedersen, C. A., and Dorsa, D. M. (1995). CNS oxytocin receptor mRNA expression and regulation by gonadal steroids. *Adv. Exp. Med. Biol.* 395, 269–280.

Bateson, P. (2007). Developmental plasticity and evolutionary biology. *J. Nutr.* 137, 1060–1062.

Benoit, D., and Parker, K. C. (1994). Stability and transmission of attachment across three generations. *Child Dev.* 65, 1444–1456.

Berman, C. (1990). Intergenerational transmission of maternal rejection rates among free-ranging rheus monkeys on Cayo Santiago. *Anim. Behav.* 44, 247–258.

Bester-Meredith, J. K., and Marler, C. A. (2001). Vasopressin and aggression in cross-fostered California mice (Peromyscus californicus) and white-footed mice (Peromyscus leucopus). *Horm. Behav.* 40, 51–64.

Bester-Meredith, J. K., and Marler, C. A. (2003). Vasopressin and the transmission of paternal behavior across generations in mated, cross-fostered Peromyscus mice. *Behav. Neurosci.* 117, 455–463.

Bignami, G., Laviola, G., Alleva, E., Cagiano, R., Lacomba, C., and Cuomo, V. (1992). Developmental aspects of neurobehavioural toxicity. *Toxicol. Lett.* 64–65. Spec No 231–237

Bosch, O. J., Musch, W., Bredewold, R., Slattery, D. A., and Neumann, I. D. (2007). Prenatal stress increases HPA axis activity and impairs maternal care in lactating female offspring: Implications for postpartum mood disorder. *Psychoneuroendocrinology* 32, 267–278.

Bredy, T. W., Zhang, T. Y., Grant, R. J., Diorio, J., and Meaney, M. J. (2004). Peripubertal environmental enrichment reverses the effects of maternal care on hippocampal development and glutamate receptor subunit expression. *Eur. J. Neurosci.* 20, 1355–1362.

Calatayud, F., and Belzung, C. (2001). Emotional reactivity in mice, a case of nongenetic heredity? *Physiol. Behav.* 74, 355–362.

Calatayud, F., Coubard, S., and Belzung, C. (2004). Emotional reactivity in mice may not be inherited but influenced by parents. *Physiol. Behav.* 80, 465–474.

Caldwell, J. D., Greer, E. R., Johnson, M. F., Prange, A. J., Jr., and Pedersen, C. A. (1987). Oxytocin and vasopressin immunoreactivity in hypothalamic and extrahypothalamic sites in late pregnant and postpartum rats. *Neuroendocrinology* 46, 39–47.

Carola, V., Frazzetto, G., and Gross, C. (2006). Identifying interactions between genes and early environment in the mouse. *Gene. Brain Behav.* 5, 189–199.

Cattanach, B. M., and Kirk, M. (1985). Differential activity of maternally and paternally derived chromosome regions in mice. *Nature* 315, 496–498.

Champagne, F. A., and Meaney, M. J. (2001). Like mother, like daughter: Evidence for non-genomic transmission of parental behavior and stress responsivity. *Prog. Brain Res.* 133, 287–302.

Champagne, F. A., and Curley, J. P. (2005). How social experiences influence the brain. *Curr. Opin. Neurobiol.* 15, 704–709.

Champagne, F. A., Curley, J. P., Swaney, W. T. and Keverne, E. B. (2005). Regulation of olfaction, anxiety, and maternal behavior in mice by paternally expressed genes, *Horm. Behav.* 48, 92–93.

Champagne, F. A., and Meaney, M. J. (2006). Stress during gestation alters postpartum maternal care and the development of the offspring in a rodent model. *Biol. Psychiatr.* 59, 1227–1235.

Champagne, F. A., Diorio, J., Sharma, S., and Meaney, M. J. (2001). Naturally occurring variations in maternal behavior in the rat are associated with differences in estrogen-inducible central oxytocin receptors. *Proc. Natl Acad. Sci. USA* 98, 12736–12741.

Champagne, F. A., Francis, D. D., Mar, A., and Meaney, M. J. (2003a). Variations in maternal care in the rat as a mediating influence for the effects of environment on development. *Physiol. Behav.* 79, 359–371.

Champagne, F. A., Weaver, I. C., Diorio, J., Sharma, S., and Meaney, M. J. (2003b). Natural variations in maternal care are associated with estrogen receptor alpha expression and estrogen sensitivity in the medial preoptic area. *Endocrinology* 144, 4720–4724.

Champagne, F. A., Weaver, I. C., Diorio, J., Dymov, S., Szyf, M., and Meaney, M. J. (2006). Maternal care associated with methylation of the estrogen receptor-alpha1b promoter and estrogen receptor-alpha expression in the medial preoptic area of female offspring. *Endocrinology* 147, 2909–2915.

Champagne, F. A., Curley, J. P., Keverne, E. B., and Bateson, P. P. (2007). Natural variations in postpartum maternal care in inbred and outbred mice. *Physiol. Behav.* 91, 325–334.

Chapman, D., and Scott, K. (2001). The impact of maternal intergenerational risk factors on adverse developmental outcomes. *Dev. Rev.* 21, 305–325.

Curley, J. P., Barton, S., Surani, A. and Keverne, E. B. (2004). Coadaptation in mother and infant regulated by a paternally expressed imprinted gene, *Proc. Roy. Soc. B.* 271, 1303–1309.

Darwin, C. R. (1956). The Darwin-Wallace essays of 1858. In *The Darwin Reader* (M. Bates, and P. S. Humphrey, Eds.), pp. 99–114. Charles Scribner's Sons., New York.

De Vries, G. J., Rissman, E. F., Simerly, R. B., Yang, L. Y., Scordalakes, E. M., Auger, C. J., Swain, A., Lovell-Badge, R., Burgoyne, P. S., and Arnold, A. P. (2002). A model system for study of sex chromosome effects on sexually dimorphic neural and behavioral traits. *J. Neurosci.* 22, 9005–9014.

Denenberg, V. H., Sherman, G., Schrott, L. M., Waters, N. S., Boehm, G. W., Galaburda, A. M., and Mobraaten, L. E. (1996). Effects of embryo transfer and cortical ectopias upon the behavior of BXSB-Yaa and BXSB-Yaa+mice. *Dev. Brain Res.* 93, 100–108.

Denenberg, V. H., Hoplight, B. J., and Mobraaten, L. E. (1998). The uterine environment enhances cognitive competence. *NeuroReport* 9, 1667–1671.

Denenberg, V. H., Hoplight, B., Sherman, G. F., and Mobraaten, L. E. (2001). Effects of the uterine environment and neocortical ectopias upon behavior of BXSB-Yaa+mice. *Dev. Psychobiol.* 38, 154–163.

Deng, Y., and Wu, X. (2000). Peg3/Pw1 promotes p53-mediated apoptosis by inducing Bax translocation from cytosol to mitochondria. *Proc. Nat. Acad. Sci. USA* 97, 12050–12055.

Dewing, P., Chiang, C. W., Sinchak, K., Sim, H., Fernagut, P. O., Kelly, S., Chesselet, M. F., Micevych, P. E., Albrecht, K. H., Harley, V. R., and Vilain, E. (2006). Direct regulation of adult brain function by the male-specific factor SRY. *Curr. Biol.* 16, 415–420.

Fahrbach, S. E., Morrell, J. I., and Pfaff, D. W. (1985). Possible role for endogenous oxytocin in estrogen-facilitated maternal behavior in rats. *Neuroendocrinology* 40, 526–532.

Fairbanks, L. A. (1989). Early experience and cross-generational continuity of mother–infant contact in vervet monkeys. *Dev. Psychobiol.* 22, 669–681.

Fleming, A., Vaccarino, F., Tambosso, L., and Chee, P. (1979). Vomeronasal and olfactory system modulation of maternal behavior in the rat. *Science* 203, 372–374.

Fleming, A., Bonebrake, R., Istwan, N., Rhea, D., Coleman, S., and Stanziano, G. (2004). Pregnancy and economic outcomes in patients treated for recurrent preterm labor. *J. Perinatol.* 24, 223–227.

Fleming, A. S., Kraemer, G. W., Gonzalez, A., Lovic, V., Rees, S., and Melo, A. (2002). Mothering begets mothering: The transmission of behavior and its neurobiology across generations. *Pharmacol. Biochem. Behav.* 73, 61–75.

Francis, D., Diorio, J., Liu, D., and Meaney, M. J. (1999). Nongenomic transmission across generations of maternal behavior and stress responses in the rat. *Science* 286, 1155–1158.

Francis, D. D., Szegda, K., Campbell, G., Martin, W. D., and Insel, T. R. (2003). Epigenetic sources of behavioral differences in mice. *Nat. Neurosci.* 6, 445–446.

Frazier, C. R., Trainor, B. C., Cravens, C. J., Whitney, T. K., and Marler, C. A. (2006). Paternal behavior influences development of aggression and vasopressin expression in male California mouse offspring. *Horm. Behav.* 50, 699–707.

Gabriel, K. I., and Cunningham, C. L. (2007). Effects of maternal strain on ethanol responses in reciprocal F1 C57BL/6J and DBA/2J hybrid mice. *Gene. Brain Behav.* .

Gatewood, J. D., Wills, A., Shetty, S., Xu, J., Arnold, A. P., Burgoyne, P. S., and Rissman, E. F. (2006). Sex chromosome complement and gonadal sex influence aggressive and parental behaviors in mice. *J. Neurosci.* 26, 2335–2342.

Gluckman, P. D., Hanson, M. A., and Pinal, C. (2005a). The developmental origins of adult disease. *Matern. Child Nutr.* 1, 130–141.

Gluckman, P. D., Hanson, M. A., Spencer, H. G., and Bateson, P. (2005b). Environmental influences during development and their later consequences for health and disease: Implications for the interpretation of empirical studies. *Proc. Biol. Sci.* 272, 671–677.

Gonzalez, A., and Fleming, A. S. (2002). Artificial rearing causes changes in maternal behavior and c-fos expression in juvenile female rats. *Behav. Neurosci.* 116, 999–1013.

Gonzalez, A., Lovic, V., Ward, G. R., Wainwright, P. E., and Fleming, A. S. (2001). Intergenerational effects of complete maternal deprivation and replacement stimulation on maternal behavior and emotionality in female rats. *Dev. Psychobiol.* 38, 11–32.

Gray, A. P. (1971). *Mammalian hybrids*. Commonwealth Agricultural Bureau, London.

Gray, A. P. (1972). *Mammalian hybrids. A check-list with bibliography*. Commonwealth Agricultural Bureaux Farnham Royal, London.

Hajkova, P., Erhardt, S., Lane, N., Haaf, T., El-Maarri, O., Reik, W., Walter, J., and Surani, M. A. (2002). Epigenetic reprogramming in mouse primordial germ cells. *Mech. Dev.* 117, 15–23.

Isles, A. R., Baum, M. J., Ma, D., Keverne, E. B., and Allen, N. D. (2001). Urinary odour preferences in mice. *Nature* 409, 783–784.

Isles, A. R., Baum, M. J., Ma, D., Szeto, A., Keverne, E. B., and Allen, N. D. (2002). A possible role for imprinted genes in inbreeding avoidance and dispersal from the natal area in mice. *Proc. Biol. Sci.* 269, 665–670.

Johnson, M. D., Wu, X., Aithmitti, N., and Morrison, R. S. (2002). Peg3/Pw1 is a mediator between p53 and Bax in DNA damage-induced neuronal death. *J. Biol. Chem.* 277, 23000–23007.

Kaiser, S., and Sachser, N. (1998). The social environment during pregnancy and lactation affects the female offsprings' endocrine status and behaviour in guinea pigs. *Physiol. Behav.* 63, 361–366.

Kemme, K., Kaiser, S., and Sachser, N. (2007). Prenatal maternal programming determines testosterone response during social challenge. *Horm. Behav.* 51, 387–394.

Keverne, E. B., Fundele, R., Narasimha, M., Barton, S. C., and Surani, M. A. (1996). Genomic imprinting and the differential roles of parental genomes in brain development. *Dev. Brain Res.* 92, 91–100.

Kleiman, D. G. (1977). Monogamy in mammals. *Q. Rev. Biol.* 52, 39–69.

Kramer, K. M., Carr, M. S., Schmidt, J. V., and Cushing, B. S. (2006). Parental regulation of central patterns of estrogen receptor alpha. *Neuroscience* 142, 165–173.

Kuroiwa, Y., Kaneko-Ishino, T., Kagitani, F., Kohda, T., Li, L. L., Tada, M., Suzuki, R., Yokoyama, M., Shiroishi, T., Wakana, S., Barton, S. C., Ishino, F., and Surani, M. A. (1996). Peg3 imprinted gene on proximal chromosome 7 encodes for a zinc finger protein. *Nat. Genet.* 12, 186–190.

Larcher, A., Neculcea, J., Breton, C., Arslan, A., Rozen, F., Russo, C., and Zingg, H. H. (1995). Oxytocin receptor gene expression in the rat uterus during pregnancy and the estrous cycle and in response to gonadal steroid treatment. *Endocrinology* 136, 5350–5356.

Lefebvre, L., Viville, S., Barton, S. C., Ishino, F., and Surani, M. A. (1997). Genomic structure and parent-of-origin-specific methylation of Peg1. *Hum. Mol. Genet.* 6, 1907–1915.

Lefebvre, L., Viville, S., Barton, S. C., Ishino, F., Keverne, E. B., and Surani, M. A. (1998). Abnormal maternal behaviour and growth retardation associated with loss of the imprinted gene Mest. *Nat. Genet.* 20, 163–169.

Li, L., Keverne, E. B., Aparicio, S. A., Ishino, F., Barton, S. C., and Surani, M. A. (1999). Regulation of maternal behavior and offspring growth by paternally expressed Peg3. *Science* 284, 330–333.

Lovic, V., Gonzalez, A., and Fleming, A. S. (2001). Maternally separated rats show deficits in maternal care in adulthood. *Dev. Psychobiol.* 39, 19–33.

Maestripieri, D. (2005). Early experience affects the intergenerational transmission of infant abuse in rhesus monkeys. *Proc. Natl Acad. Sci. USA* 102, 9726–9729.

Maestripieri, D., Wallen, K., and Carroll, K. A. (1997). Infant abuse runs in families of group-living pigtail macaques. *Child Abuse Negl.* 21, 465–471.

Maestripieri, D., Higley, J. D., Lindell, S. G., Newman, T. K., McCormack, K. M., and Sanchez, M. M. (2006). Early maternal rejection affects the development of monoaminergic systems and adult abusive parenting in rhesus macaques (*Macaca mulatta*). *Behav. Neurosci.* 120, 1017–1024.

Maestripieri, D., Lindell, S. G., and Higley, J. D. (2007). Intergenerational transmission of maternal behavior in rhesus macaques and its underlying mechanisms. *Dev. Psychobiol.* 49, 165–171.

McGrath, J., and Solter, D. (1984). Completion of mouse embryogenesis requires both the maternal and paternal genomes. *Cell* 37, 179–183.

McLaughlin, K. J., Szabo, P., Haegel, H., and Mann, J. R. (1996). Mouse embryos with paternal duplication of an imprinted chromosome 7 region die at midgestation and lack placental spongiotrophoblast. *Development* 122, 265–270.

McLeod, J., Sinal, C. J., and Perrot-Sinal, T. S. (2007). Evidence for non-genomic transmission of ecological information via maternal behavior in female rats. *Gene. Brain Behav.* 6, 19–29.

Meaney, M. J. (2001). Maternal care, gene expression, and the transmission of individual differences in stress reactivity across generations. *Annu. Rev. Neurosci.* 24, 1161–1192.

Melo, A. I., Lovic, V., Gonzalez, A., Madden, M., Sinopoli, K., and Fleming, A. S. (2006). Maternal and littermate deprivation disrupts maternal behavior and social-learning of food preference in adulthood: tactile stimulation, nest odor, and social rearing prevent these effects. *Dev. Psychobiol.* 48, 209–219.

Moore, C. L., and Power, K. L. (1986). Prenatal stress affects mother–infant interaction in Norway rats. *Dev. Psychobiol.* 19, 235–245.

Mousseau, T. A., and Fox, C. W. (1998). *Maternal effects as adaptations.* Oxford University Press.

Novakov, M., and Fleming, A. S. (2005). The effects of early rearing environment on the hormonal induction of maternal behavior in virgin rats. *Horm. Behav.* 48, 528–536.

Numan, M. (2006). Hypothalamic neural circuits regulating maternal responsiveness toward infants. *Behav. Cognit. Neurosci. Rev.* 5, 163–190.

Patin, V., Lordi, B., Vincent, A., and Caston, J. (2005). Effects of prenatal stress on anxiety and social interactions in adult rats. *Brain Res. Dev. Brain Res.* 160, 265–274.

Pedersen, C. A., and Prange, A. J. Jr. (1979). Induction of maternal behavior in virgin rats after intracerebroventricular administration of oxytocin. *Proc. Natl Acad. Sci. USA* 76, 6661–6665.

Pedersen, C. A., Ascher, J. A., Monroe, Y. L., and Prange, A. J. Jr. (1982). Oxytocin induces maternal behavior in virgin female rats. *Science* 216, 648–650.

Pedersen, C. A., Caldwell, J. D., Johnson, M. F., Fort, S. A., and Prange, A. J. Jr. (1985). Oxytocin antiserum delays onset of ovarian steroid-induced maternal behavior. *Neuropeptides* 6, 175–182.

Perrigo, G., Belvin, L., Quindry, P., Kadir, T., Becker, J., van Look, C., Niewoehner, J., and vom Saal, F. S. (1993). Genetic mediation of infanticide and parental behavior in male and female domestic and wild stock house mice. *Behav. Genet.* 23, 525–531.

Reik, W., and Walter, J. (2001). Genomic imprinting: parental influence on the genome. *Nat. Rev. Genet.* 2, 21–32.

Reinhold, K. (2002). Maternal effects and the evolution of behavioral and morphological characters: A literature review indicates the importance of extended maternal care. *J. Hered.* 93, 400–405.

Reisert, I., Karolczak, M., Beyer, C., Just, W., Maxson, S. C., and Ehret, G. (2002). Sry does not fully sex-reverse female into male behavior towards pups. *Behav. Genet.* 32, 103–111.

Rose, C., Schwegler, H., Hanke, J., Rohl, F. W., and Yilmazer-Hanke, D. M. (2006). Differential effects of embryo transfer and maternal factors on anxiety-related behavior and numbers of neuropeptide Y (NPY) and parvalbumin (PARV) containing neurons in the amygdala of inbred C3H/HeN and DBA/2J mice. *Behav. Brain Res.* 173, 163–168.

Sanchez, M. M., Ladd, C. O., and Plotsky, P. M. (2001). Early adverse experience as a developmental risk factor for later psychopathology: Evidence from rodent and primate models. *Dev. Psychopathol.* 13, 419–449.

Sayler, A., and Salmon, M. (1969). Communal nursing in mice: Influence of multiple mothers on the growth of the young. *Science* 164, 1309–1310.

Schier, A. F. (2007). The maternal–zygotic transition: Death and birth of RNAs. *Science* 316, 406–407.

Seckl, J. R., and Meaney, M. J. (2004). Glucocorticoid programming. *Ann. NY Acad. Sci.* 1032, 63–84.

Shoji, H., and Kato, K. (2006). Maternal behavior of primiparous females in inbred strains of mice: A detailed descriptive analysis. *Physiol. Behav.* 89, 320–328.

Slamberova, R., Pometlova, M., and Rokyta, R. (2007). Effect of methamphetamine exposure during prenatal and preweaning periods lasts for generations in rats. *Dev. Psychobiol.* 49, 312–322.

Smith, J. W., Seckl, J. R., Evans, A. T., Costall, B., and Smythe, J. W. (2004). Gestational stress induces post-partum depression-like behaviour and alters maternal care in rats. *Psychoneuroendocrinology* 29, 227–244.

Surani, M. A., Barton, S. C., and Norris, M. L. (1984). Development of reconstituted mouse eggs suggests imprinting of the genome during gametogenesis. *Nature* 308, 548–550.

Swaney, W. T., Curley, J. P., Champagne, F. A., and Keverne, E. B. (2007). Genomic imprinting mediates sexual experience-dependent olfactory learning in male mice. *Proc. Natl Acad. Sci. USA* 104, 6084–6089.

Trainor, B. C., and Marler, C. A. (2001). Testosterone, paternal behavior, and aggression in the monogamous California mouse (Peromyscus californicus). *Horm. Behav.* 40, 32–42.

Trullas, R., and Skolnick, P. (1993). Differences in fear motivated behaviors among inbred mouse strains. *Psychopharmacology (Berl)* 111, 323–331.

Weinstock, M. (2005). The potential influence of maternal stress hormones on development and mental health of the offspring. *Brain Behav. Immun.* 19, 296–308.

Weinstock, M., Fride, E., and Hertzberg, R. (1988). Prenatal stress effects on functional development of the offspring. *Prog. Brain Res.* 73, 319–331.

21 OXYTOCIN AND INDIVIDUAL VARIATION IN PARENTAL CARE IN PRAIRIE VOLES

DANIEL E. OLAZÁBAL[1] AND LARRY J. YOUNG[2]

[1] *Department of Physiology, School of Medicine, Udelar, Montevideo, 11800, Uruguay.*
[2] *Department of Psychiatry and Behavioral Sciences, Center for Behavioral Neuroscience, Yerkes National Primate Research Center, Emory University School of Medicine, Atlanta, GA 30322, USA*

PARENTAL BEHAVIOR IN PRAIRIE VOLES

Prairie voles (*Microtus ochrogaster*) are socially monogamous rodents that display high levels of affiliative behavior and biparental care (Thomas & Birney, 1979; Getz *et al.*, 1981; McGuire & Novak, 1984; Salo *et al.*, 1993; DeVries *et al.*, 1997; Olazábal & Young, 2005). Both males and females contribute to the rearing of their offspring with minimal or no difference in their response to pups. Sexually naive prairie voles of both sexes also display spontaneous parental responses when exposed to pups in nonreproductive contexts (Roberts *et al.*, 1998; Olazábal & Young, 2005). This spontaneous parental response includes licking, grooming, and hovering over the pups immediately after the first exposure, as well as subsequent nest building, and retrieval. However, there is significant variability in parental responsiveness both within and across populations. According to our studies and others using different populations of prairie voles, most males (70–80%) display spontaneous parental care, and a small percentage (10–20%) generally ignore or neglect the pups. Infanticidal responses in males are not uncommon but occur in only ~10% of the cases. Sexually naïve female prairie voles, by contrast, are much more variable in their response to pups than males. About ~50–60% of female prairie voles show spontaneous maternal responses when first exposed to pups, while about 40–50% either ignore/neglect the pups or display infanticidal behavior when exposed to them (Roberts *et al.*, 1998; Lonstein & DeVries, 2000, 2001; Olazábal & Young, 2005).

We have found that variation in the response to pups in naïve females is associated with variation in measures of emotionality and nonmaternal affiliative behavior. Adult females that display maternal behavior spend less time immobile, and make more crosses through the center of an open field arena than females that attack pups (Figure 21.1(A),(B)). This observation suggests that females that display less anxiety-like behavior and/or increased exploratory behavior are more likely to interact positively with novel pups (Olazábal & Young, 2005). Interestingly, juvenile prairie voles, which display high levels of spontaneous maternal behavior, exhibit lower levels of anxiety-like behavior in the open field than adults (Olazábal & Young, 2005). Spontaneously maternal adult females also display increased affiliative behavior toward same-sex conspecifics. Generalized affiliative behavior in adult females was assessed by quantifying social interactions in a three-chambered arena in which female conspecifics were tethered to two of the chambers, with the third neutral chamber remaining empty, or neutral. Spontaneously maternal adult female prairie voles spent more time in social contact and less time isolated in a neutral cage than adults that attacked the pups (Figure 21.1(C)) (Olazábal & Young, 2005).

Neurobiology of the Parental Brain

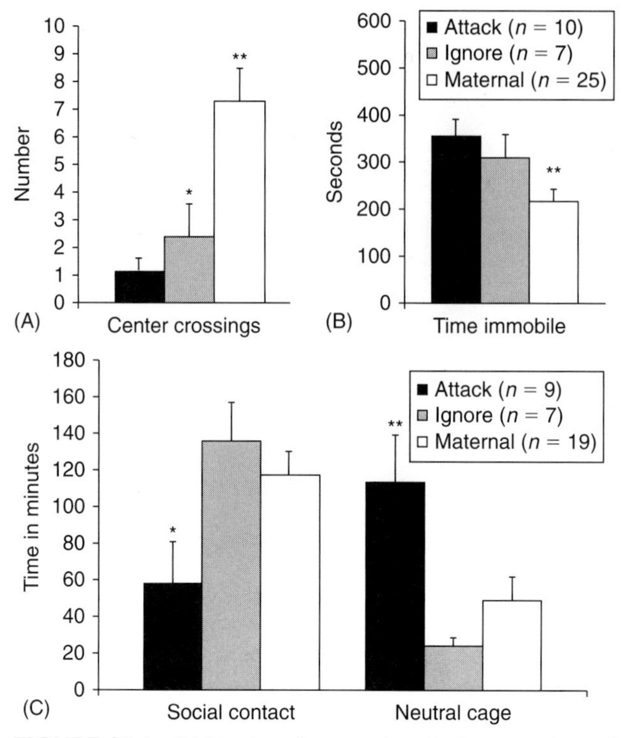

FIGURE 21.1 (A) Number of crosses through the center during the first 10 min in the open field test. Maternal animals made more crosses through the center than animals that attacked the pups. (B) Time spent immobile during the first 10 min of the open field test. Maternal animals spent less time immobile than animals that attacked the pups. (C) Time spent in social contact or in the neutral cage by animals that were maternal, that ignored or attacked the pups. Animals that attacked the pups spent more time in the neutral cage and less time in social contact with conspecifics compared to animals that ignored the pups or showed maternal behavior. Data expressed as mean ± S.E. $^{**}p < 0.01$; $^{*}p \leq 0.05$.

This suggests that increased anxiety-like behavior, and decreased affiliative behavior are negatively associated with maternal responsiveness in adult female prairie voles.

Significant variability in parental responsiveness to pups has also been found across different laboratories, and even within laboratories over time (Roberts *et al.*, 1998; Lonstein & DeVries, 2000; Olazábal & Young, 2005). The possible factors underlying this variability include genetic differences between laboratory colonies, genetic drift due to inbreeding, or differences in the handling, management, or care of the colonies (Bales *et al.*, 2007). Regardless of the origin of this variability in parental responsiveness, both within a lab colony and across different labs, it provides an excellent opportunity to investigate the neurobiological mechanisms that underlie natural variation in parental behavior. Our studies, which include both comparisons across species, correlational analyses within prairie voles, and direct pharmacological manipulations, suggest that variations in the oxytocinergic systems regulating social affiliation and attachment also contribute to variation in parental responsiveness in prairie voles.

OXYTOCIN AND MATERNAL CARE

Oxytocin (OT) is a peptide that has been widely implicated in the regulation of social cognition, affiliative behavior, and maternal behavior in several mammalian species, including humans (Pedersen & Prange, 1979, 1985; Fahrbach *et al.*, 1985; Peterson *et al.*, 1991; Insel, 1992; Keverne & Kendrick, 1992; McCarthy *et al.*, 1992; Uvnas-Moberg, 1998; Gimpl & Fahrenholz, 2001; Carter, 2003; Kosfeld *et al.*, 2005; Burbach *et al.*, 2006; Takayanagi *et al.*, 2006). Unlike prairie voles, sexually naïve adult female rats typically do not display spontaneous maternal care unless repeatedly exposed to pups. However, intracerebroventricular infusion of OT accelerates the display of maternal responsiveness in virgin female rats (Pedersen & Prange, 1979). In sheep, OT also plays a role in both the induction of maternal responsiveness and the selectivity of the ewe for a lamb (Kendrick *et al.*, 1997). Thus the OT system is a logical candidate for producing variation in maternal responsiveness.

Despite the seemingly conserved role of the OT system in the regulation of affiliative behavior and maternal care, there are remarkable species differences in the distribution of OTR in the brain which may contribute to species differences in behavior (Insel & Shapiro, 1992; Barberis & Tribollet, 1996; Young, 1999). For example, OTR in the nucleus accumbens (NA) of female prairie voles plays a critical role in partner preference formation following cohabitation and mating (Young *et al.*, 2001; Young & Wang, 2004). While prairie voles have high densities of OTR in the NA, non-monogamous meadow and montane voles do not (Young & Wang, 2004). OTR is present in several brain regions involved in the regulation of maternal behavior, including the medial preoptic area (MPOA; Numan *et al.*, 1977; Rosenblatt & Ceus, 1998; Numan & Insel, 2003), ventral tegmental area (VTA, Gaffori & Le Moal, 1979; Hansen *et al.*, 1991), olfactory bulb (Fleming & Rosenblatt, 1974a, b), NA (Hansen *et al.*, 1993; Lonstein *et al.*, 1998; Keer & Stern, 1999; Li & Fleming, 2003a, b; Numan & Insel, 2003; Champagne *et al.*, 2004), caudate putamen (CP; Felicio *et al.*, 1996), and lateral septum (LS; Fleischer & Slotnick, 1978; Flannelly *et al.*, 1986). While each of these brain regions has been implicated in various aspects of maternal behavior, the OT system has only been previously implicated in the MPOA, VTA, and olfactory bulb with regard to the facilitation of maternal behavior (Pedersen *et al.*, 1994; Yu *et al.*, 1996; Kendrick *et al.*, 1997).

Parental behavior is not exclusive to adults, and it is present in weanling animals from several species (Mayer & Rosenblatt, 1979). Although this behavior is sometimes called alloparental, all the behavioral components that define parental care in adult rodents are found at the age of 20 days in prairie voles. This provides a great opportunity to investigate differences in the response to pups, because not all rodents display parental care at this age. While female prairie voles (20 days of age) generally show an immediate maternal response toward pups (~85%), rats of the same age require 1–3 days of pup exposure before expressing maternal behavior (Gandelman, 1973; Bridges *et al.*, 1974; Mayer & Rosenblatt, 1979; Roberts *et al.*, 1998; Lonstein & DeVries, 2000; Olazábal & Morrell, 2005; Olazábal & Young, 2005). By contrast, juvenile mice or meadow voles in our laboratory fail to show any maternal responses toward pups (data unpublished). As rats mature, there is a developmental decline in responsiveness to pups (Bridges *et al.*, 1974; Mayer & Rosenblatt, 1979), which is paralleled by a decline in OTR binding in the NA (Shapiro & Insel, 1989). In addition, central administration of OT to 20-day-old rats increased the time spent licking and in contact with pups (Peterson *et al.*, 1991). This evidence led us to examine the potential relationship between OTR density in the NA and variation in maternal care.

OT RECEPTOR DENSITY IN THE NA AND LS AND VARIABILITY IN MATERNAL CARE

The role of the NA in maternal behavior is still unclear, however there is growing evidence that the NA is involved in promoting maternal responses and the processing of pup-related stimuli (Hansen *et al.*, 1993; Lonstein *et al.*, 1998; Keer & Stern, 1999; Li & Fleming, 2003a, b; Numan & Insel, 2003; Champagne *et al.*, 2004; Olazábal & Morrell, 2005). For example, agonists of the receptor of dopamine D1 into the NA facilitate maternal behavior in the rat (Stolzenberg *et al.*, 2007). In addition, injections of dopamine receptor antagonists or 6-hydroxydopamine into the NA, which eliminates dopaminergic input into the NA from the VTA, disrupt active components of maternal behavior (Hansen *et al.*, 1991; Numan *et al.*, 2005). It has also been proposed that ventral stimulation and suckling induced inhibition of dopamine may be required for kyphosis and quiescent nursing. The NA (shell subregion) projections to the ventrolateral part of the periaqueductal gray region (Groenewegen *et al.*, 1996; Stern & Lonstein 2001) have been implicated in the mediation of these postures in the rat.

The LS also plays an important role in the regulation of maternal behavior. There is increased c-Fos expression in this brain region when rats are exposed to pups (Stack & Numan, 2000) and lesions of the LS block maternal behavior in several species including rats, mice, and rabbits (Cruz & Beyer, 1972; Fleischer & Slotnick, 1978). Considering the possibility that OTR in the NA and other brain regions, that is LS and MPOA, might be associated with variation in parental responsiveness, we investigated whether differences in the expression of the OTR in these brain regions had a significant impact on the response to pups across and within species. We expected that a comparative approach would give us insight into the neurobiology underlying individual differences in behavior.

Comparisons of receptor distributions among prairie voles, rats, mice, and meadow voles showed that prairie voles have higher levels of OTR in the NA than rats, mice, and meadow voles, while the last two species have higher levels of OTR in the LS (Insel, 1992; Tribollet *et al.* 1992; Young, 1999; Olazábal & Young, 2006a) (Figures 21.2 and 21.3). OTR binding density in the brain correlated positively, in the case of NA, and negatively, in the case of LS, with the disposition of prepubertal animals of these species to display maternal care after exposure to pups. Prairie voles, the only species in which juvenile females show spontaneous maternal response, has higher OTR binding in the NA and lower binding in the LS compared to mice and meadow voles which do not show any maternal response as juveniles. Rats, which show parental behavior after a few days of pup exposure, have lower OTR density in the NA compared to prairie voles, but higher values than meadow voles and mice (Figures 21.2 and 21.3; Olazábal & Young, 2006a). No differences across species were found in the density of OTR in the MPOA.

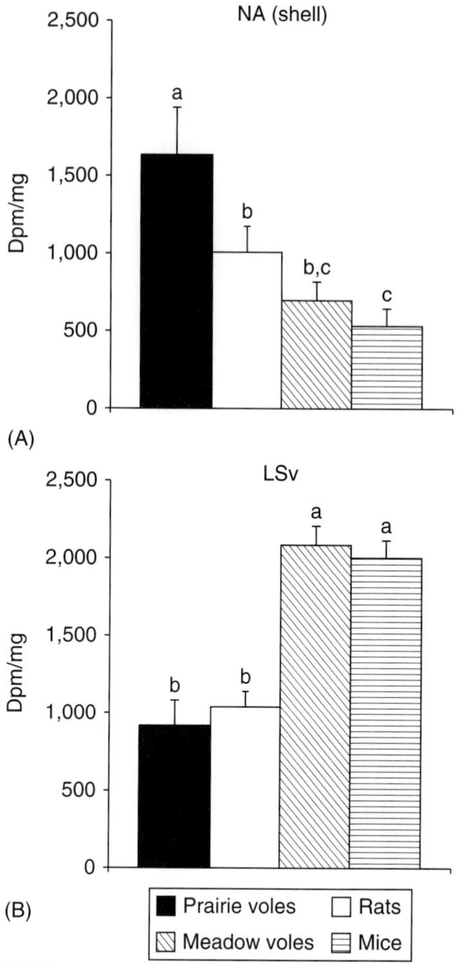

FIGURE 21.2 OTR binding density in the NA (shell subregion), and the LS (ventral subregion) in prairie voles, rats, meadow voles, and mice. Data are expressed as mean ± SE (groups not sharing a letter differ significantly from each other; $p < 0.05$).

Since the goal of the comparative study was to identify potential mechanisms underlying individual variation in behavior, we then correlated OTR density in these brain regions with the quality of parental behavior displayed by juvenile female prairie voles. Females were tested for maternal behavior and later sacrificed to remove the brain and measure OTR density. Similar to what was previously found in the cross-species comparisons, higher quality of juvenile parental behavior, indicated by more time spent crouching over the pups, was positively correlated to OTR density in the NA (Figure 21.4, Olazábal & Young, 2006a). Juveniles that did not show maternal care had higher density of OTR in the LS than maternal juveniles (4501 dpm/mg ± 225; 3251 ± 178, respectively). No correlation was found between density of OTR in the MPOA and maternal care.

Following this experiment we decided to investigate whether the relationship between OTR in the NA and maternal behavior across species and within juvenile prairie voles, was also present in adult prairie voles. Interestingly, we found that adult females that displayed

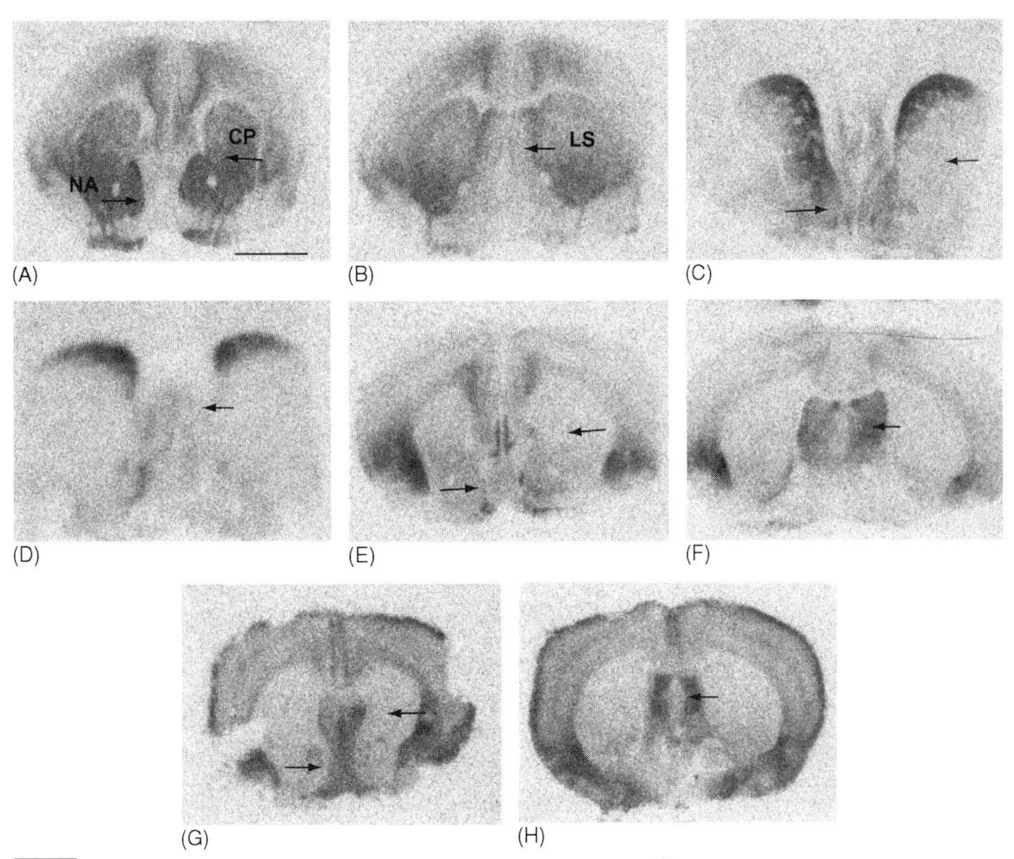

FIGURE 21.3 Autoradiograms of brain sections illustrating I^{125} OTA binding representative of prairie voles (A, B), rats (C, D), meadow voles (E, F), and mice (G, H). Note that the OTR density is higher in the LS (see arrow) of meadow voles and mice (F, H) than in prairie voles and rats (B, D). In the shell subregion of the NA (see arrow) binding is lower in meadow voles and mice (E, G), intermediate in rats (C), and high in prairie voles (A). Scale bar = 2 mm.

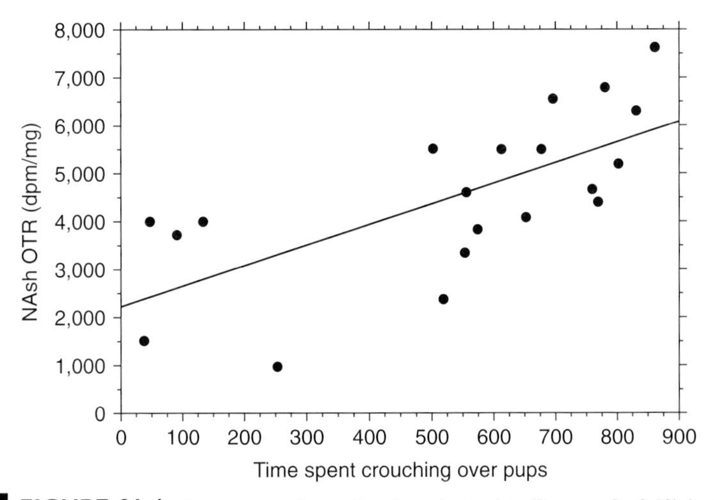

FIGURE 21.4 Scattergram illustrating the relationship (Pearson $R=0.69$) between OTR binding in the nucleus accumbens (shell; NAsh) and the time spent adopting crouching posture.

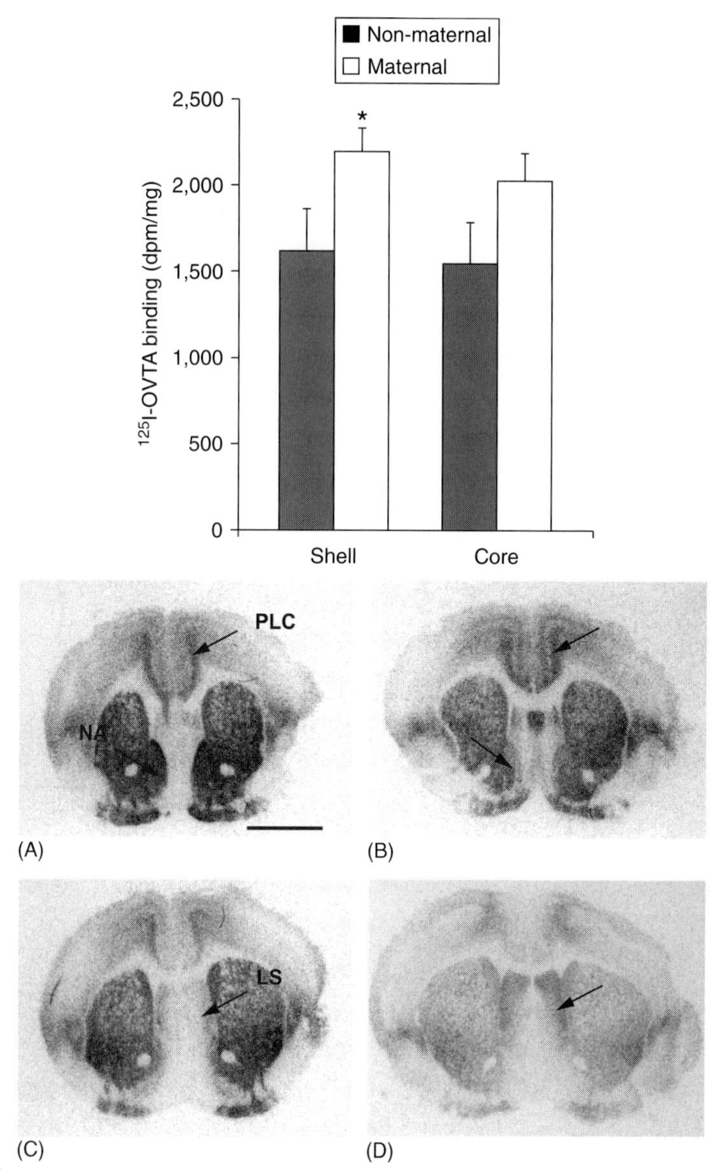

FIGURE 21.5 *Upper panel*: OTR binding in maternal and non-maternal animals. OTR binding in the shell and core subdivisions of the NA was higher in maternal (white bars) than in non-maternal females (black bars). Data are expressed as means ± SE ($*p < 0.05$). *Lower panel*: Autoradiographs of brain sections showing the signal for I^{125} OTA for two animals representative of the maternal (A, C) and non-maternal female (B, D) groups. Note that while OTR binding is clearly higher in the NA of the maternal animal, no differences are seen in the prelimbic cortex (PLC) between maternal and non-maternal females. In addition, the LS binding is higher in non-maternal females demonstrating that the difference is not due to overall decrease in OTR binding or technical artifacts (C, D). Scale bar = 2 mm.

spontaneous maternal behavior also had higher density of OTR in the NA (shell subregion) than non-maternal females (Figure 21.5). However, no differences were found in the density of OTR in the MPOA and LS between maternal and non-maternal adult females.

Finally, we directly tested the role of OTR in the regulation of spontaneous maternal behavior

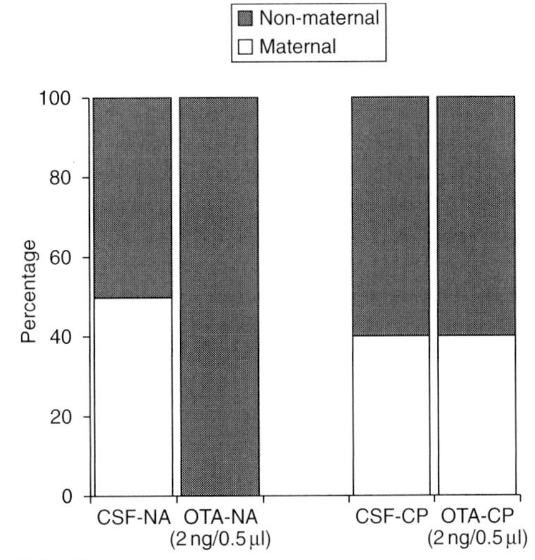

FIGURE 21.6 Oxytocin antagonist (OTA) infusions into the NA, but not the CP, block maternal responsiveness. Bars show percentage of maternal (white) and non-maternal (black) animals. A higher number of CSF-NA infused controls displayed maternal responses than did OTA-NA infused animals. No difference was found between OTA-CP and CSF-CP infused animals.

by infusing an OTR antagonist into the NA. Ten animals were injected bilaterally either with the OTR antagonist d(CH2)5,[Tyr(Me)2,Thr4,Orn8, Tyr9-NH2]-vasotocin (OTA) 2 ng/0.5 µL in CSF or vehicle (CSF). Two hours later these animals were exposed to pups for 15 min and tested for maternal behavior. Our findings demonstrate that when OTR in the NA are blocked by OTA, females fail to display maternal behavior, while similar infusions into the CP have no effect on maternal behavior (Figure 21.6; Olazábal & Young, 2006b).

OT FACILITATES MATERNAL RESPONSES IN PRAIRIE VOLES

These set of studies (Olazábal & Young, 2006b; 2006b) were the first, to our knowledge, in which species differences in brain–behavior relationships have been used to identify potential neural mechanisms underlying individual variation in social behavior. We also found for the first time that OT, extensively involved in the regulation of maternal behavior in other species, was also modulating maternal responses in prairie voles. Furthermore, this is the first evidence that OTR in the NA are involved in the regulation of parental responsiveness in a non-reproductive context. OT has long been proposed to facilitate the process of bringing conspecifics into close proximity (affiliation) for the formation of social attachments (Insel, 1992; Pedersen *et al.*, 1994; Carter, 2003).

The mechanisms underlying OT facilitation on maternal behavior are unknown and may be complex. These behavioral effects might be mediated by OT/dopamine interactions in the NA, as suggested previously (Kovacs *et al.*, 1990; Liu & Wang, 2003). OT may increase the attractive value and the reinforcement properties of pup-related stimuli, and/or disinhibit the approach to the novel stimuli (pups), reducing neophobia and the reactivity to them (Mayer & Rosenblatt, 1975; Fleming & Anderson, 1987). This is especially interesting considering in light of the finding that less anxious and neophobic female prairie voles were also more maternal and affiliative.

Stern and coworkers (Stern & Taylor, 1991; Keer & Stern, 1999) have also proposed that OT interacts with dopamine in the NA to reduce locomotor activity during female–pup interaction. Blockage of the dopaminergic activity in the NA facilitates the passive component of maternal behavior, thereby increasing the time dams spend nursing over pups. The D1/D2 antagonist *cis*-flupenthixol infused into the NA enhanced kyphotic nursing (Keer & Stern, 1999), and low dosages of haloperidol resulted in a more rapid onset and longer duration of nursing (Stern & Keer, 1999). The NA (shell subregion) sends projections to the ventrolateral part of the periaqueductal gray region (Groenewegen *et al.*, 1996) that has been implicated in the regulation of crouching posture in rats (Stern & Lonstein, 2001). Ventral stimulation in naïve female prairie voles pups may also trigger this pathway as has already been proposed for males (Lonstein, 2002).

Our finding also agrees with evidence showing that OT in the NA modulates social bonding, specifically pair bonding in prairie voles (Young, 1999; Young *et al.*, 2001). Central infusions of OT also enhance social interactions in male and female prairie voles (Witt *et al.*, 1990; Cho *et al.*, 1999), and OT antagonists into the ventricular system or into the NA reduce the time a female prairie vole spend with her male partner (Insel *et al.*, 1995; Young, 1999). Although maternal behavior and pair bonding are a consequence of different types of social stimulation, and have different biological significance, both require a non-aggressive approach, followed by a positive stimulation that drives the animal to form a preference for staying close to the source of stimulus. OT action within the NA might functionally modify the valence of the stimulus, facilitating the salience of attractive features from the pups to induce a more positive and rewarding experience, and subsequent parental responses. Thus, maternal responsiveness and pair bonding may share common neurobiological mechanisms. This observation has interesting evolutionary implications in that it indicates that the evolution of pair bonding may emerge through the modulation of preexisting circuits and neurochemistry that are involved in regulating mother–infant relationships.

Previous studies also support the relationship between individual differences in maternal responses and OTR binding in the brain (Francis *et al.*, 2000, 2002; Champagne *et al.*, 2001). Increased OTR binding in the MPOA, the LS, the central nucleus of the amygdala, paraventricular nuclei of the hypothalamus, and the bed nucleus of the stria terminalis have been associated with increased levels of licking and grooming in lactating female rats (Champagne *et al.*, 2001). Although, the MPOA is a principal component of the neural circuit that regulates maternal behavior (Numan *et al.*, 1977; Rosenblatt & Ceus, 1998; Numan & Insel, 2003), variability in parental care is not related to the density of OTR in this brain region, as also shown in adult sensitized rats (Francis *et al.*, 2000, 2002).

In prairie voles we found a negative (juveniles) or no (adult) relationship between LS OTR binding and maternal behavior (Olazábal & Young, 2006a; 2000b). Prior studies in prairie voles also failed to find a positive relationship between OTR binding in the LS and maternal behavior (Insel & Shapiro, 1992; Wang *et al.*, 2000; Olazábal & Young, 2006a). Together, these findings suggest that the NA and LS may have different functions in the neural circuit that supports juvenile female maternal responsiveness. OT in the LS, for example, might play a significant role in pregnant females to reduce defensive responses and anxiety at the time of parturition. The difference between previous findings in rats (Champagne *et al.*, 2001) and these in prairie voles might be due to species differences in the mechanisms by which OT stimulates maternal behavior.

WHAT MECHANISMS GENERATE DIVERSITY IN OTR AND PARENTAL RESPONSIVENESS?

We have yet to explore mechanisms that determine or contribute to the variability in the expression of OTR in the brains of the different species or individuals. Both genetic and epigenetic mechanisms could contribute to the variation in OTR expression and hence parental behavior. For example, Hammock and Young (2005) reported that genetic variation in the promoter of the vasopressin receptor contributes to individual variation in vasopressin receptor expression in the brain, partner preference formation, and some aspects of paternal behavior. Subtle species differences in potential regulatory elements in the prairie and montane vole OTR gene have been observed, but to date there is no evidence that variations in these elements contribute to species differences or individual variation in OTR expression (Young *et al.*, 1996). Early environmental experiences and/or organizational effects of steroid or peptide exposure could also contribute to the variation in behavior, as studied by several laboratories (Young *et al.*, 1997; Francis *et al.*, 2002; Pedersen & Boccia, 2002; Young & Gainer, 2003).

Understanding the factors that give rise to these differences in parental responsiveness is critical from a developmental viewpoint. The expression of OTR in the brain of prairie voles increases during the first two weeks of life and remains fairly unchanged thereafter (Figure 21.7). It is possible that during the early postnatal period OT may be critical in shaping developmental changes in the brain that are critical for the juvenile and adult behavior. When those differences in the expression of OTR between individuals originate is unknown. However, differences in the expression of OTR in the brain might be a consequence of events that occur after birth, at birth, at the intrauterine environment, or be a product of the interaction of multiple events that modify the regulation of the genetic information that codify for this system. Further research on these topics will help clarify OT's role within this context.

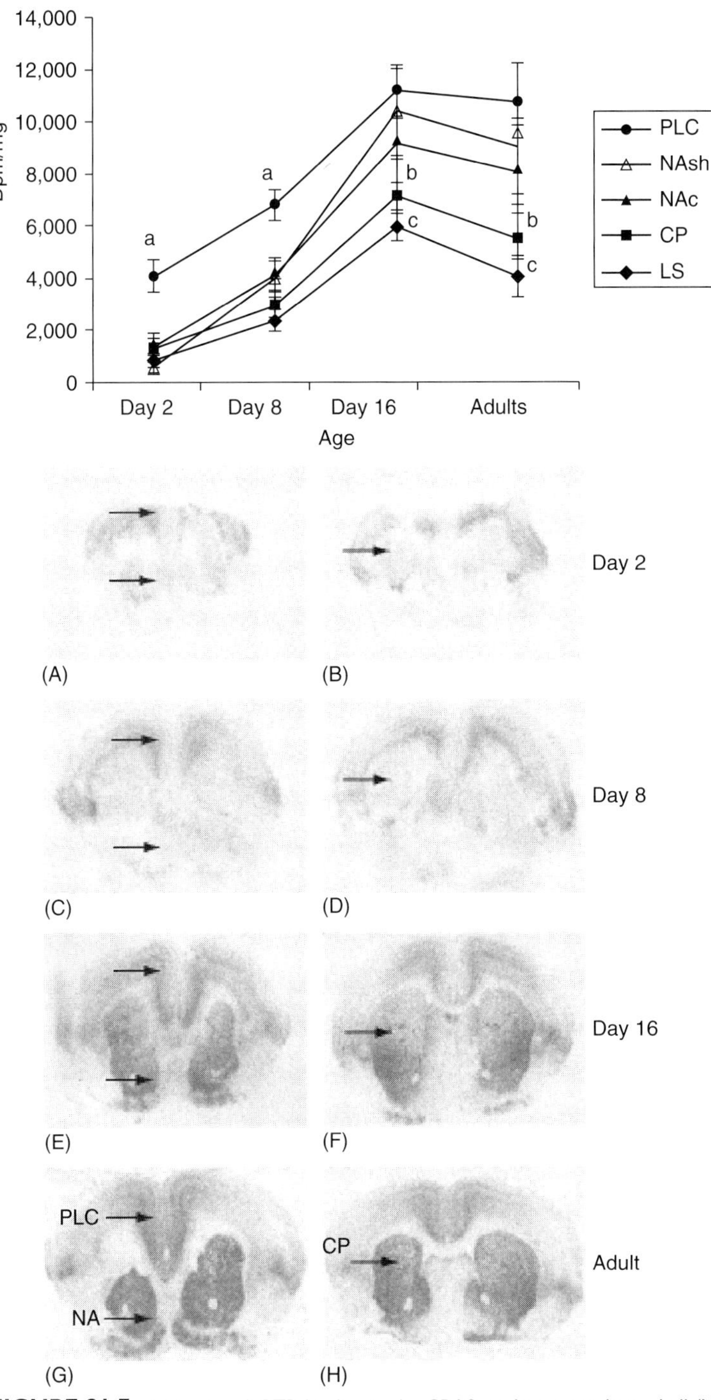

FIGURE 21.7 *Upper panel*: OTR binding in the CP, LS, nucleus accumbens shell (NAsh) and core (NAc), and PLC at ages 2, 8, 16, 60–90 days. Data are expressed as means ± S.E. a: statistically significant different from NA, CP, and LS; b: statistically significant different from PLC; c: statistically significant different from PLC and NA (shell). *Lower panel*: Autoradiographs of brain sections showing the signal for I^{125} OTA for four prairie voles at ages 2 days (A, B), 8 days (C, D), 16 days (E, F), and adults (G, H). Note that while OTR binding in the NA and CP is almost absent at ages 2 and 8 days, there is already significant OTR binding in the PLC at these ages. Scale bar = 2 mm.

In this chapter we have presented our recent studies exploring the variability in maternal care in prairie voles and its possible regulation by the oxytocinergic system. These investigations reveal the value of combining comparative and developmental approaches in generating hypothesis regarding neural mechanisms underlying variability in behavior. We found that, in prairie voles, there are emotional components that significantly influence the adult response to pups and their affiliative behavior. A role of the NA OTR system in regulating parental responsiveness through mechanisms that might involve changes in the attractive value of pups or novel stimuli and OT's role in the maternal responses of lactating prairie voles are topics of future investigations. Our findings do suggest that variability in the expression of receptors, such as OTR, might be a common mechanism underlying the evolutionary or developmental acquisition of select sociobehavioral traits such as spontaneous parental behavior and pair bonding. Understanding the mechanisms generating the variation in OTR expression across or within species will be critical to understanding the variability in the response to pups found in nature.

REFERENCES

Bales, K. L., Lewis-Reese, A. D., Pfeifer, L. A., Kramer, K. M., and Carter, C. S. (2007). Early experience affects the traits of monogamy in a sexually dimorphic manner. *Dev. Psychobiol.* 49, 335–342.

Barberis, C., and Tribollet, E. (1996). Vasopressin and oxytocin receptors in the central nervous system. *Crit. Rev. Neurobiol.* 10, 119–154.

Bridges, R. S., Zarrow, M. X., Goldman, B. D., and Denenberg, V. H. (1974). A developmental study of maternal responsiveness in the rat. *Physiol. Behav.* 12, 149–151.

Burbach, P., Young, L. J., and Russell, J. (2006). Oxytocin: Synthesis, secretion and reproductive functions. In *Knobil and Neill's Physiology of Reproduction* (J. D. Neill, Ed.), Boston, 3rd edn., pp. 3055–3127. Elsevier.

Carter, C. S. (2003). Developmental consequences of oxytocin. *Physiol. Behav.* 79, 383–397.

Cruz, M. L., and Beyer, C. (1972). Effects of septal lesions on maternal behavior and lactation in the rabbit. *Physiol. Behav.* 9, 361–365.

Champagne, F., Diorio, J., Sharma, S. H., and Meaney, M. J. (2001). Naturally occurring variations in maternal behavior in the rat are associated with differences in estrogen-inducible central oxytocin receptors. *Proc. Natl. Acad. Sci. USA* 98, 12736–12741.

Champagne, F. A., Chretien, P., Stevenson, C. W., Zhang, T. Y., Gratton, A., and Meaney, M. J. (2004). Variations in nucleus accumbens dopamine associated with individual differences in maternal behavior in the rat. *J. Neurosci.* 24, 4113–4123.

Cho, M. M., DeVries, A. C., Williams, J. R., and Carter, C. S. (1999). The effects of oxytocin and vasopressin on partner preferences in male and female prairie voles (*Microtus ochrogaster*). *Behav. Neurosci.* 113, 1071–1080.

DeVries, A. C., Johnson, C. L., and Carter, C. S. (1997). Familiarity and gender influence social preferences in prairie voles (*Microtus ochrogaster*). *Can. J. Zool.* 75, 295–301.

Fahrbach, S. E., Morrell, J. I., and Pfaff, D. W. (1985). Role of oxytocin in the onset of estrogen-facilitated maternal behavior. In *Oxytocin: Clinical and Laboratory Studies* (J. A. Amico and A. G. Robinson, Eds.), pp. 372–387. Elsevier.

Felicio, L. F., Florio, J. C., Sider, L. H., Cruz-Casallas, P. E., and Bridges, R. S. (1996). Reproductive experience increases striatal and hypothalamic dopamine levels in pregnant rats. *Brain Res. Bull.* 40(4), 253–256.

Flannelly, K. J., Kemble, E. D., Blanchard, D. C., and Blanchard, R. J. (1986). Effects of septal-forebrain lesions on maternal aggression and maternal care. *Behav. Neural Biol.* 45, 17–30.

Fleischer, S., and Slotnick, B. M. (1978). Disruption of maternal behavior in rats with lesions of the septal area. *Physiol. Behav.* 21, 189–200.

Fleming, A. S., and Rosenblatt, J. S. (1974a). Olfactory regulation of maternal behavior in rats. I. Effects of olfactory bulb removal in experienced and inexperienced lactating and cycling females. *J. Comp. Physiol. Psychol.* 86, 221–232.

Fleming, A. S., and Rosenblatt, J. S. (1974b). Olfactory regulation of maternal behavior in rats. II. Effects of peripherally induced anosmia and lesions of the lateral olfactory tract in pup-induced virgins. *J. Comp. Physiol. Psychol.* 86, 233–246.

Fleming, A. S., and Anderson, V. (1987). Affect and nurturance: Mechanisms mediating maternal behavior in two female mammals. *Prog. Neuropsychopharmacol. Biol. Psychiatr.* 11, 121–127.

Francis, D. D., Champagne, F. C., and Meaney, M. J. (2000). Variations in maternal behavior are associated with differences in oxytocin receptor levels in the rat. *J. Neuroendocrinol.* 12, 1145–1148.

Francis, D. D., Young, L. J., Meaney, M. J., and Insel, T. R. (2002). Naturally occurring differences in maternal care are associated with the expression of oxytocin and vasopressin (V1a) receptors: Gender Differences. *J. Neuroendocrinol.* 14, 349–353.

Gaffori, O., and Le Moal, M. (1979). Disruption of maternal behavior and appearance of cannibalism after ventral mesencephalic tegmentum lesions. *Physiol. Behav.* 23, 317–323.

Gandelman, R. (1973). The ontogeny of maternal responsiveness in female Rockland-Swiss albino mice. *Horm. Behav.* 4, 257–268.

Getz, L. L., Carter, C. S., and Gavish, L. (1981). The mating system of the prairie vole, *Microtus ochrogaster*: Field and laboratory evidence for pair-bonding. *Behav. Ecol. Sociobiol.* 8, 189–194.

Gimpl, G., and Fahrenholz, F. (2001). The oxytocin receptor system: Structure, function, and regulation. *Physiol. Rev.* 81, 629–683.

Groenewegen, H. J., Wright, C. I., and Beijer, A. V. (1996). The nucleus accumbens: Gateway for limbic structures to reach the motor system? *Prog. Brain. Res.* 107, 485–511.

Hammock, E. A., and Young, L. J. (2005). Microsatellite instability generates diversity in brain and sociobehavioral traits. *Science* 308, 1630–1634.

Hansen, S., Harthon, C., Wallin, E., Lofberg, L., and Svensson, K. (1991). Mesotelencephalic dopamine system and reproductive behavior in the female rat: Effects of ventral tegmental 6-hydroxydopamine lesions on maternal and sexual responsiveness. *Behav. Neurosci.* 105, 588–598.

Hansen, S., Bergvall, A. H., and Nyiredi, S. (1993). Interaction with pups enhances dopamine release in the ventral striatum of maternal rats: A microdialysis study. *Pharmacol. Biochem. Behav.* 45, 673–676.

Insel, T. R. (1992). Oxytocin-A neuropeptide for affiliation: Evidence from behavioral, receptor autoradiographic, and comparative studies. *Psychoneuroendocrinology* 17, 3–35.

Insel, T. R., and Shapiro, L. E. (1992). Oxytocin receptor distribution reflects social organization in monogamous and polygamous voles. *Proc. Natl Acad. Sci. USA* 89, 5981–5985.

Insel, T. R., Winslow, J. T., Wang, Z. X., Young, L., and Hulihan, T. J. (1995). Oxytocin and the molecular basis of monogamy. *Adv. Exp. Med. Biol.* 395, 227–234.

Keer, S. E., and Stern, J. M. (1999). Dopamine receptor blockade in the nucleus accumbens inhibits maternal retrieval and licking, but enhances nursing behavior in lactating rats. *Physiol. Behav.* 67, 659–669.

Kendrick, K. M., Costa, A. P., Broad, K. D., Ohkura, S., Guevara, R., Levy, F., and Keverne, E. B. (1997). Neural control of maternal behavior and olfactory recognition of offspring. *Brain Res. Bull.* 44, 383–395.

Keverne, E. B., and Kendrick, K. M. (1992). Oxytocin facilitation of maternal behavior in sheep. *Ann. NY Acad. Sci.* 652, 83–101.

Kosfeld, M., Heinrichs, M., Zak, P. J., Fischbacher, U., and Fehr, E. (2005). Oxytocin increases trust in humans. *Nature* 435, 673–676.

Kovacs, G. L., Sarnyai, Z., Babarczi, E., Szabo, G., and Telegdy, G. (1990). The role of oxytocin–dopamine interactions in cocaine-induced locomotor hyperactivity. *Neuropharmacology* 29, 365–368.

Li, M., and Fleming, A. S. (2003a). The nucleus accumbens shell is critical for normal expression of pup-retrieval in postpartum female rats. *Behav. Brain Res.* 145, 99–111.

Li, M., and Fleming, A. S. (2003b). Differential involvement of nucleus accumbens shell and core subregions in maternal memory in postpartum female rats. *Behav. Neurosci.* 117, 426–445.

Liu, Y., and Wang, Z. X. (2003). Nucleus accumbens oxytocin and dopamine interact to regulate pair bond formation in female prairie voles. *Neuroscience* 121, 537–544.

Lonstein, J. S. (2002). Effects of dopamine receptor antagonism with haloperidol on nurturing behavior in the biparental prairie vole. *Pharmacol. Biochem. Behav.* 74, 11–19.

Lonstein, J. S., and DeVries, G. J. (2000). Sex differences in the parental behavior of rodents. *Neurosci. Biobehav. Rev.* 24, 669–686.

Lonstein, J. S., and DeVries, G. J. (2001). Social influences on parental and nonparental responses toward pups in virgin female prairie voles (*Microtus ochrogaster*). *J. Comp. Psychol.* 115, 53–61.

Lonstein, J. S., Simmons, D. A., Swann, J. M., and Stern, J. M. (1998). Forebrain expression of c-fos due to active maternal behaviour in lactating rats. *Neuroscience* 82, 267–281.

Mayer, A. D., and Rosenblatt, J. S. (1975). Olfactory basis for the delayed onset of maternal behavior in virgin female rats: Experimental effects. *J. Comp. Physiol. Psychol.* 89, 701–710.

Mayer, A. D., and Rosenblatt, J. S. (1979). Ontogeny of maternal behavior in the laboratory rat: Early origins in 18- to 27-day-old young. *Dev. Psychobiol.* 12, 407–424.

McCarthy, M. M., Kow, L. M., and Pfaff, D. W. (1992). Speculations concerning the physiological significance of central oxytocin in maternal behavior. *Ann. NY Acad. Sci.* 652, 70–82.

McGuire, B., and Novak, M. (1984). A comparison of maternal behaviour in the meadow vole (*Microtus Pennsylvanicus*), prairie vole (*M. Ochrogaster*) and pine voles (*M. Pinetorum*). *Anim. Behav.* 32, 1132–1141.

Numan, M., and Insel, T. R. (2003). *The Neurobiology of Parental Behavior.* Springer-Verlag, New York.

Numan, M., Rosenblatt, J. S., and Komisaruk, B. R. (1977). Medial preoptic area and onset of maternal behavior in the rat. *J. Comp. Physiol. Psychol.* 91, 146–164.

Numan, M., Numan, M. J., Pliakou, N., Stolzenberg, D. S. Mullins, O. J., Murphy, J. M., and Smith, C. D. (2005). The effects of D1 or D2 dopamine receptor antagonism in the medial preoptic area, ventral pallidum, or nucleus accumbens on the maternal retrieval response and other aspects of maternal behavior in rats. *Behav. Neurosci.* 119, 1588–1604.

Olazábal, D. E., and Morrell, J. I. (2005). Juvenile rats show immature neuronal patterns of c-Fos expression to first pup exposure. *Behav. Neurosci.* 119, 1097–1110.

Olazábal, D. E., and Young, L. J. (2005). Variability in "spontaneous" maternal behavior is associated with anxiety-like behavior and affiliation in naïve juvenile and adult female prairie voles (*Microtus ochrogaster*). *Dev. Psychobiol.* 47, 166–178.

Olazábal, D. E., and Young, L. J. (2006a). Species and individual differences in juvenile female alloparental care are associated with oxytocin receptor density in the striatum and the lateral septum. *Horm. Behav.* 49, 681–687.

Olazábal, D. E., and Young, L. J. (2006b). Oxytocin receptors in the nucleus accumbens facilitate "spontaneous" maternal behavior in adult female prairie voles. *Neuroscience* 141, 559–568.

Pedersen, C. A., and Boccia, M. A. (2002). Oxytocin links mothering received, mothering bestowed and adult stress responses. *Stress* 5, 259–267.

Pedersen, C. A., and Prange, A. J. (1979). Induction of maternal behavior in virgin rats after intracerebroventricular administration of oxytocin. *Proc. Natl Acad. Sci. USA* 76, 6661–6665.

Pedersen, C. A., and Prange, A. J. (1985). Oxytocin and mothering behavior in the rat. *Pharmacol. Ther.* 28, 287–302.

Pedersen, C. A., Caldwell, J. D., Walker, Ch., Ayers, G., and Mason, G. A. (1994). Oxytocin activates the postpartum onset of rat maternal behavior in the ventral tegmental and medial preoptic areas. *Behav. Neurosci.* 108, 1163–1171.

Peterson, G., Mason, G. A., Barakat, A. S., and Pedersen, C. A. (1991). Oxytocin selectively increases holding and licking of neonates in preweanling but not postweanling juvenile rats. *Behav. Neurosci.* 105, 470–477.

Roberts, R. L., Miller, A. K., Taymans, S. E., and Carter, C. S. (1998). Role of social and endocrine factors in alloparental behavior of prairie voles (*Microtus ochrogaster*). *Can. J. Zool.* 76, 1862–1868.

Rosenblatt, J. S., and Ceus, K. (1998). Estrogen implants in the medial preoptic area stimulate maternal behavior in male rats. *Horm. Behav.* 33, 23–30.

Salo, A. L., Shapiro, L. E., and Dewsbury, D. A. (1993). Comparisons of nipple attachment and incisor growth among four species of voles (*Microtus*). *Dev. Psychobiol.* 27, 317–330.

Shapiro, L. E., and Insel, T. R. (1989). Ontogeny of oxytocin receptors in rat forebrain: A quantitative study. *Synapse* 4, 259–266.

Stack, E. C., and Numan, M. (2000). The temporal course of expression of c-Fos and Fos B within the medial preoptic area and other brain regions of postpartum female rats during prolonged mother–young interactions. *Behav. Neurosci.* 114(3), 609–622.

Stern, J. M., and Taylor, L. A. (1991). Haloperidol inhibits maternal retrieval and licking, but enhances nursing behavior and litter weight gains in lactating rats. *J. Neuroendocrinol.* 3, 591–596.

Stern, J. M., and Keer, S. E. (1999). Maternal motivation of lactating rats is disrupted by low dosages of haloperidol. *Behav. Brain Res.* 99, 231–239.

Stern, J. M., and Lonstein, J. S. (2001). Neural mediation of nursing and related maternal behaviors. *Prog. Brain Res.* 133, 263–278.

Stolzenberg, D. S., McKenna, J. B., Keough, S., Hancock, R., Numan, M. J., and Numan, M. (2007). Dopamine D1 receptor stimulation of the nucleus accumbens or the medial preoptic area promotes the onset of maternal behavior in pregnancy-terminated rats. *Behav. Neurosci.* 121, 907–919.

Takayanagi, Y., Yoshida, M., Bielsky, I. F., Ross, H. R., Kawamata, M., Onaka, T., Yanagisawa, T., Kimura, T., Matzuk, M. M., Young, L. J., and Nishimori, K. (2006). Pervasive social deficits but normal parturition in oxytocin receptor-deficient mice. *Proc. Natl Acad. Sci. USA* 102, 16096–16101.

Thomas, J. A., and Birney, E. C. (1979). Parental care and mating system of the prairie vole, *Microtus ochrogaster*. *Behav. Ecol. Sociobiol.* 5, 171–186.

Tribollet, E., Dubois-Dauphin, M., Dreifuss, J. J., Barberis, C., and Jard, S. (1992). Oxytocin receptors in the central nervous system. Distribution, development, and species differences. *Ann. NY Acad. Sci.* 652, 29–38.

Uvnas-Moberg, K. (1998). Oxytocin may mediate the benefits of positive social interaction and emotions. *Psychoneuroendocrinology* 23, 819–835.

Wang, Z. X., Liu, Y., Young, L. J., and Insel, T. R. (2000). Hypothalamic vasopressin gene expression increases in both males and females postpartum in a biparental rodent. *J. Neuroendocrinol.* 12, 111–120.

Witt, D. M., Carter, C. S., and Walton, D. (1990). Central and peripheral effects of oxytocin administration in prairie voles (*Microtus ochrogaster*). *Pharmacol. Biochem. Behav.* 37, 63–69.

Young, L. J. (1999). Frank A. Beach Award. Oxytocin and vasopressin receptors and species-typical social behaviors. *Horm. Behav.* 36, 212–221.

Young, L. J., and Wang, Z. (2004). The neurobiology of pair bonding. *Nat. Neurosci.* 7, 1048–1054.

Young, L. J., Huot, B., Nilsen, R., Wang, Z., and Insel, T. R. (1996). Species differences in central oxytocin receptor gene expression: Comparative analysis of promoter sequences. *J. Neuroendocrinol.* 10, 777–783.

Young, L. J., Winslow, J. T., Wang, Z., Gingrich, B., Guo, Q., Matzuk, M. M., and Insel, T. R. (1997). Gene targeting approaches to neuroendocrinology: Oxytocin, maternal behavior, and affiliation. *Horm. Behav.* 31, 221–231.

Young, L. J., Lim, M. M., Gingrich, B., and Insel, T. R. (2001). Cellular mechanisms of social attachment. *Horm. Behav.* 40(2), 133–138.

Young, W. S., III, and Gainer, H. (2003). Transgenesis and the study of expression, cellular targeting and function of oxytocin, vasopressin and their receptors. *Neuroendocrinology* 78, 185–203.

Yu, G-Z., Kaba, H., Okutani, S., Takahashi, S., and Higuchi, T. (1996). The olfactory bulb: A critical site of action for oxytocin on the induction of maternal behaviour. *Neuroscience* 72, 1083–1088.

22 DOPAMINE REGULATION OF PAIR BONDING IN MONOGAMOUS PRAIRIE VOLES

KYLE L. GOBROGGE, YAN LIU AND ZUOXIN WANG

Department of Psychology and Program in Neuroscience, Florida State University, Tallahassee, FL, USA

INTRODUCTION

Historically, the field of behavioral neuroendocrinology has primarily focused on using traditional laboratory rodents. Data from these studies have provided valuable information for our better understanding of the neuroanatomical, neurochemical, hormonal, and molecular genetic mechanisms underlying several types of classic social behaviors, including aggression, mating, and maternal care. It has been shown that species differences exist in animal's social behaviors and underlying mechanisms, and thus caution needs to be taken when data are extrapolated from one species to another. Further, traditional laboratory rodents do not readily display certain types of social behaviors, and thus are not appropriate for some investigations. For example, laboratory rats and mice do not display strong social bonds between mates, and males usually do not display paternal behavior. Therefore, efforts have been put forth to develop new animal models for research in behavioral neuroendocrinology. Because the prairie vole (*Microtus ochrogaster*) naturally displays mating-induced pair bonding and bi-parental care, this rodent species has been utilized as an unique model to uncover the underlying neuroendocrine mechanisms controlling specific social behaviors associated with a monogamous life strategy (Carter *et al.*, 1995; Insel *et al.*, 1998; Young *et al.*, 1998; Wang & Aragona, 2004; Young & Wang, 2004).

It is not surprising that complex social behaviors, such as pair bonding, are regulated by multiple neurotransmitter systems in specific neural circuits, as will be illustrated later in this chapter. On the other hand, it is also not surprising that the same neurotransmitter is involved in multiple physiological and behavioral functions. For example, in this chapter we will discuss the data that illustrate mating induces dopamine (DA) release, and released DA acts in a region- and receptor-specific manner to regulate different aspects of pair-bonding behavior in prairie voles. Interestingly, previous studies on the neurobiology of maternal behavior in rats have shown that pup exposure also induces DA release (Hansen *et al.*, 1993; Febo & Ferris, 2007; Lavi-Avnon *et al.*, 2007). Released DA, then, acts on specific types of DA receptors (Numan, 2007) in particular brain regions (Miller & Lonstein, 2005; Numan *et al.*, 2005; Stolzenberg *et al.*, 2007) to regulate different aspects of maternal behavior (Miller & Lonstein, 2005; Numan *et al.*, 2005). These striking similarities and overlaps found in the neurochemical regulation of social behavior indicate a highly conserved neural circuit controlling natural motivation. Therefore, understanding the neurobiology of one type of social behavior, such as pair bonding between adults, may likely provide useful information for a better understanding of the neurobiology of other types of social behavior, such as bonding between parents and offspring. Given the fact that the prairie vole displays both types of bonding behaviors (Lonstein & De Vries, 1999),

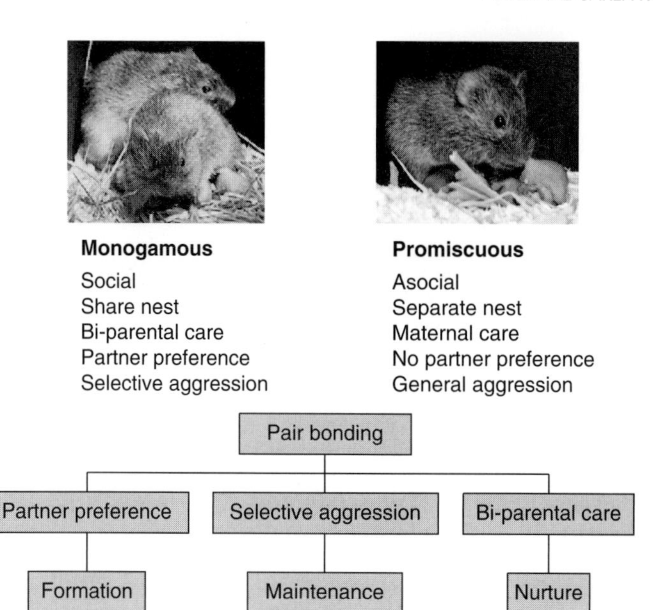

FIGURE 22.1 Photo images illustrating a pair of prairie voles displaying bi-parental care (left) and a female meadow vole exhibiting maternal behavior (right). Species differences in social behavior are compared. The functional significance of pair bonding is further subdivided by behavioral phenotype and function.

vole research may provide unique opportunities for investigation of the neuronal and hormonal mechanisms underlying pair-bonding behaviors critical to parental investment.

THE VOLE MODEL FOR COMPARATIVE STUDIES

Voles are a group of microtine rodent species that are genetically and taxonomically similar, yet show remarkable differences in their life strategy and social behavior (Young & Wang, 2004). These animals, therefore, have provided excellent opportunities for comparative studies examining their social behaviors associated with different life strategies. For example, prairie and pine (*M. pinetorum)* voles are highly social and monogamous, whereas meadow (*M. pennsylvanicus*) and montane (*M. montanus*) voles are asocial and promiscuous (Jannett, 1982; Dewsbury, 1987; Insel & Hulihan, 1995) (Figure 22.1). In the laboratory prairie and pine voles are bi-parental, with both parents participating in caring for their young. Meadow and montane voles are primarily maternal and males do not stay in the maternal nest (McGuire & Novak, 1984, 1986; Oliveras & Novak, 1986). Following mating, prairie voles develop pair bonds between mates and males even display aggression toward conspecific strangers-behaviors that are not exhibited by promiscuous meadow or montane voles (see Figure 22.1; Insel *et al.*, 1995; Lim *et al.*, 2004). Interestingly, these vole species do not differ in their non-social behaviors including grooming, locomotor/exploratory behavior, tunnel and runway construction, nest building, and swimming (Tamarin, 1985), further indicating associations between species-specific social behavior and life strategy.

Similar comparative approaches have also been utilized in studies examining neuroendocrine mechanisms of social behavior in voles. For example, by studying the central vasopressin (arginine vasopressin, AVP) system, monogamous and promiscuous voles are found to differ in the distribution pattern (Insel *et al.*, 1994; Smeltzer *et al.*, 2006), 5′ promotor region sequence (Young *et al.*, 1997), and microsatellite polymorphisms (Hammock & Young, 2005) of central AVP V1a receptors. In addition, these vole species also differ in central AVP activity during mating and reproduction (Bamshad *et al.*, 1993; Wang *et al.*, 1994). Similarly, monogamous and promiscuous voles differ in brain distribution patterns of oxytocin (OT) receptors (Insel & Shapiro, 1992),

DA neuroanatomy (Northcutt *et al.*, 2007), and DA receptors (Aragona *et al.*, 2006). It is important to note that in some cases when multiple species were compared, clear differences in AVP and OT receptor distributions were found between species with different life strategy and social behavior (Insel & Shapiro, 1992). It should also be noted that no species differences are found in other central systems, such as benzodiazepine and opioid receptors (Insel & Shapiro, 1992), indicating neurochemical specificity of species differences in the AVP, OT, and DA systems associated with different social behavior.

THE PRAIRIE VOLE AND PAIR BONDING

Prairie voles have provided excellent opportunities for studies of bi-parental care and pair bonding (Wang & Aragona, 2004; Young & Wang, 2004). In this chapter, we will focus primarily on pair bonding. Field data indicate that the same male and female prairie vole pairs are often found together, throughout life, even after migrating to different locations (Getz *et al.*, 1981). In the lab, pair bonding can be reliably measured by two behavioral indices, classified as partner preference formation and selective aggression (Getz & Hofmann, 1986; Dewsbury, 1987; Carter & Getz, 1993).

An animal's partner preference is tested in a three-chamber apparatus. Chambers are connected by plastic hollow tubes. In our lab, photo beams that are mounted on the tubes are linked to a computer to record the number of cage entries and time spent in each cage by the subject. During a 3 h test, a familiar partner, an animal that has social and/or sexual experience with the subject, and a conspecific stranger, an animal that has never encountered the subject, are tethered in their separate cages whereas the subject can move freely throughout the apparatus. A partner preference is defined by experimental animals spending significantly more time in side-by-side contact with their familiar partner vs. a stranger. It has been shown that 24 h of social cohabitation with *ad libitum* mating reliably induces partner preference formation in both male and female prairie voles (Winslow *et al.*, 1993; Insel *et al.*, 1995) (Figure 22.2(E)). It has also been shown that partner preference, once formed, is enduring. In male prairie voles, for example, partner preference still exists 2 weeks after mating even in the absence of continued exposure to a familiar partner (Insel *et al.*, 1995). Therefore, a 24 h mating paradigm has been used to examine if manipulation of central systems, such as blockade of a particular neurotransmitter receptor, can block mating-induced pair bonding. Finally, in both male and female prairie voles, 6 h of social cohabitation without mating does not induce partner preference formation (Insel *et al.*, 1995) (Figure 22.2(E)). This behavioral paradigm has been employed to examine if activation of a central system is sufficient to induce pair bonding.

Another behavioral index of pair bonding is selective aggression, which is more prominently studied in male than in female prairie voles. Selective aggression is often tested using a resident-intruder paradigm (Winslow *et al.*, 1993; Wang *et al.*, 1997). A conspecific intruder, either a male or a female, is introduced into the male resident cage and their behavioral interactions are observed for 6–10 min. Data have shown that sexually naïve male prairie voles are highly affiliative when they are exposed to a conspecific male or female (Winslow *et al.*, 1993). However, 24 h of mating completely changes their behavior. Besides displaying partner preference, these males also show high levels of aggressive behavior selectively toward unfamiliar conspecifics, but not their familiar mate (Winslow *et al.*, 1993; Wang *et al.*, 1997; Aragona *et al.*, 2006; Gobrogge *et al.*, 2007) (Figure 22.3(A)). Like partner preference formation, selective aggression is also enduring (Insel *et al.*, 1995; Aragona *et al.*, 2006; Gobrogge *et al.*, 2007). Interestingly, pair-bonded males even display selective aggression toward an unfamiliar female that is sexually receptive and could serve as his potential mate (Gobrogge *et al.*, 2007). It has been suggested that because animals actively reject both potential competitors and mates, selective aggression plays an important role for the maintenance of already established pair bonds and for the development of healthy offspring (Carter *et al.*, 1995; Lonstein & De Vries, 1999; Aragona *et al.*, 2006; Gobrogge *et al.*, 2007). Finally, although aggression has been much less studied in female voles, these animals are found to exhibit similar aggressive behavior as males, and this behavior is influenced by female's social and sexual experience (Bowler *et al.*, 2002).

Together, these reliably expressed and measurable behavioral characteristics make the prairie vole an excellent model for investigation

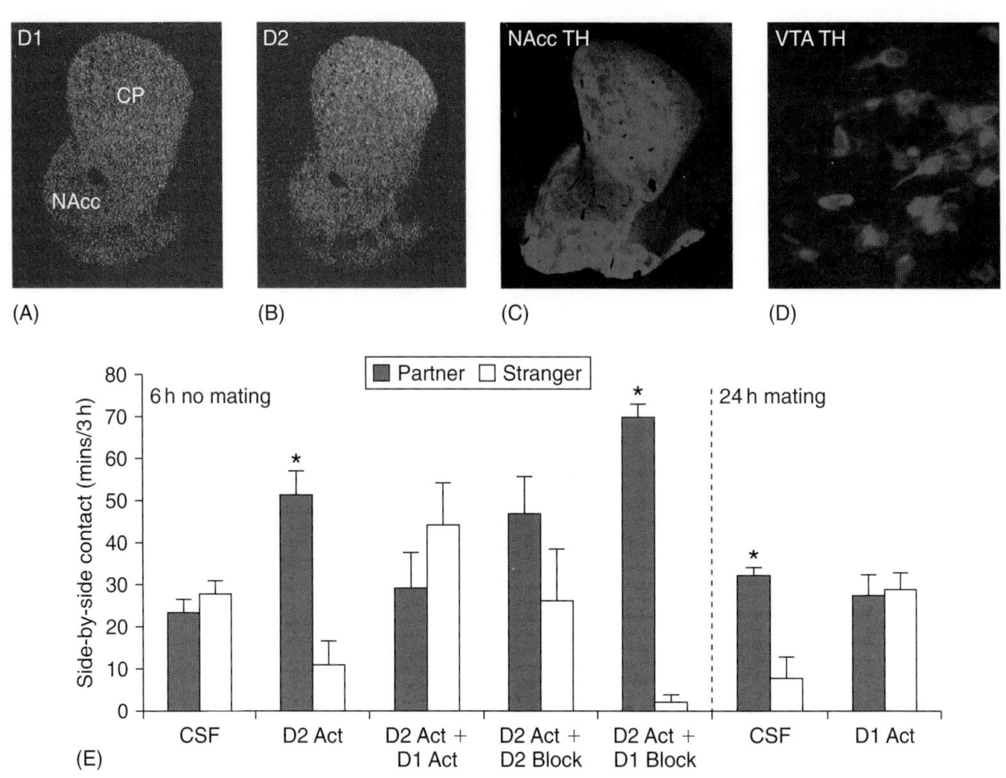

FIGURE 22.2 Photo images showing binding for D1-type (A) and D2-type (B) DA receptors in the NAcc and CP as well as fluorescent immunoreactive staining for TH fibers in the NAcc (C) and cell bodies in the VTA (D). (E) Partner preference formation of male prairie voles as a function of NAcc DA pharmacology. Six hours of non-sexual cohabitation does not induce partner preference in control males that received intra-NAcc CSF injections. Activation (Act) of D2 receptors induces partner preference, which can be diminished by activation of D1 receptors or blockade (Block) of D2 receptors. Activation of D2 receptors, together with blockade of D1 receptors, induces robust partner preference formation. Finally, 24 h of mating induces partner preference formation in control males that received intra-NAcc CSF injections, and this behavior is blocked by activation of D1 receptors. *$p < 0.05$. Error bars indicate SEM. (See Color Plate)

of the neuronal and hormonal mechanisms underlying pair-bonding behaviors critical to a monogamous life strategy.

DA NEUROANATOMY AND INTRACELLULAR SIGNALING

A detailed description of DA neuroanatomy can be found in earlier reports (Zeiss, 2005). Here, we will briefly mention the aspects of the central DA system related to DA regulation of pair bonding. Anatomically, central DA is segregated into three distinct pathways: nigrostriatal, incertohypothalamic, and mesocorticolimbic. DA cell bodies projecting, pre-synaptically, from the substantia nigra (A9) synapse onto the dorsal striatum and comprise the nigrostriatal path (Swanson, 1982). Incertohypothalamic paths extend from DA cell bodies of the A12-14 cell groups and project to the medial preoptic area (MPOA) and paraventricular nucleus (Cheung *et al.*, 1998). The mesocorticolimbic path represents DA cell bodies originating in the ventral tegmental area (VTA) (A10) projecting to the medial prefrontal cortex (mPFC) and nucleus accumbens (NAcc) (Swanson, 1982; Carr & Sesack, 2000). In addition, DA cells in the anterior hypothalamus (AH), including the lateral division, also project to forebrain areas including the striatum, lateral septum (LS), NAcc, and mPFC (Lindvall & Stenevi, 1978; Maeda & Mogenson, 1980).

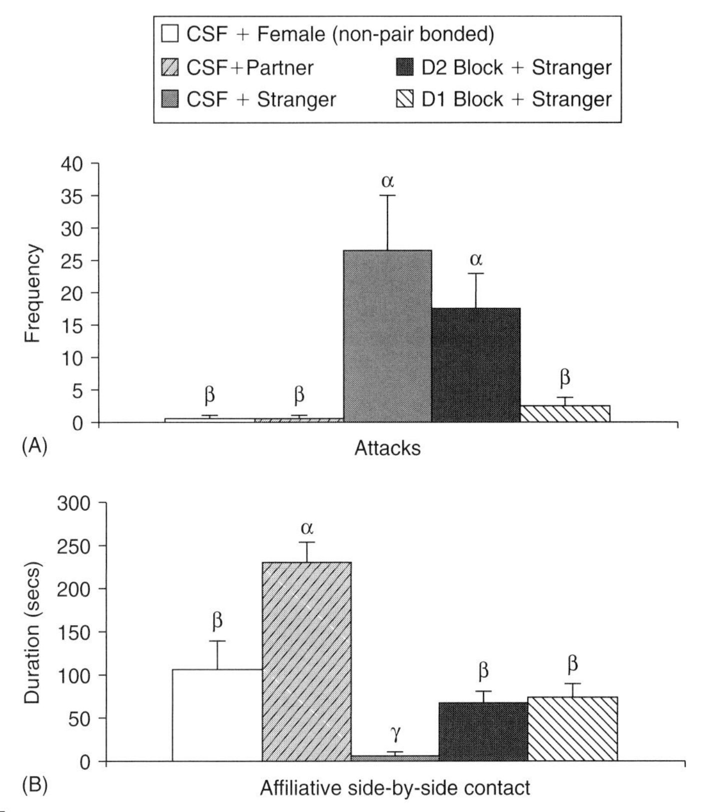

FIGURE 22.3 (A) Selective aggression in male prairie voles. Sexually naïve males (non-pair bonded) do not show selective aggression toward a conspecific female. Males that are pair bonded for 2 weeks (remaining 4 groups) display selective aggression toward a conspecific female, but not their familiar mate. This selective aggression is abolished by blockade of D1, but not D2, receptors in the NAcc. (B) Social affiliation of male prairie voles with stimulus animals. Males display high levels of affiliative behavior toward their familiar partner but not toward a conspecific female stranger. Blockade of either D1 or D2 receptors in the NAcc increases affiliative behavior to the level comparable to that displayed by non-pair bonded males. Group differences are expressed as $p < 0.05$ and bars with different Greek letters differed significantly from each other. Error bars indicate SEM.

DA preferentially binds to two families of receptors: D1-like and D2-like. Both types of DA receptors are found in the mPFC, NAcc, LS, and amygdala (AMY) (Boyson *et al.*, 1986), and many of these brain regions are important in social behavior (Newman, 1999). DA D1-like receptors consist of both D1 and D5 subtypes whereas D2-like DA receptors consist of D2, D3, and D4 subtypes. DA receptors are part of the large family of G-protein coupled receptors having seven transmembrane domains spanning intracellular and extracellular cell membrane surfaces. D1-like receptors are directly coupled to both stimulatory (s) Gα and Gαolf proteins (Neve *et al.*, 2004). Stimulation of these G-proteins leads to activation of adenylate cyclase, cAMP, and protein phosphatase-1 inhibitor DARP-32 (Neve *et al.*, 2004). Conversely, D2-like receptors couple to inhibitory (i) Gα proteins and, when activated, down-regulate adenylate cyclase, cAMP, and protein phosphatase-1 inhibitor DARP-32 (Neve *et al.*, 2004). Phosphorylation of cAMP, with concurrent inhibition of protein phosphatase-1, stimulates other intracellular receptors, ion channels, and nuclear transcription factors (Neve *et al.*, 2004). Further, D1-receptor stimulation has been found to play a critical role in calcium influx. Particularly, calcium entry via the L-type calcium channel which has been shown to be important for cellular long-term potentiation leading to learning and memory in the aging brain (Deyo *et al.*, 1989; Yamada *et al.*, 1996), has direct influences on gene expression (Murphy *et al.*, 1991). Because the focus of this chapter is not on DA intracellular mechanisms, a detailed description of the DA intracellular signaling pathways can be found in other excellent reviews (Calon *et al.*, 2000; Callier *et al.*, 2003; Neve *et al.*, 2004; Pollack, 2004; Surmeier *et al.*, 2007).

DA INVOLVEMENT IN PAIR BONDING

Because pair-bonding behavior represents a form of reward associative learning, in which DA plays a critical role, it is not surprising to find that central DA is involved in pair bonding. In prairie voles, DA producing cells and projections are found in several brain regions important to social behavior and these brain regions include the bed nucleus of the stria terminalis (BNST), MPOA, periventricular nucleus, AH, NAcc (Figure 22.2(C)), VTA (Figure 22.2(D)), striatum, AMY, and LS (Aragona *et al.*, 2003; Lansing & Lonstein, 2006; Gobrogge *et al.*, 2007; Northcutt *et al.*, 2007). Additionally, DA D1-type receptors are found in the NAcc (Figure 22.2(A)), caudate putamen (CP) (Figure 22.2(A)), and AMY with DA D2-type receptors being located in the NAcc (Figure 22.2(B)), CP (Figure 22.2(B)), AMY, substantia nigra, and VTA (Aragona *et al.*, 2003, 2006; Liu *et al.*, unpublished).

The direct evidence of DA involvement in pair bonding came from pharmacological experiments. Peripheral and intracerebroventricular (icv) administration of a non-selective DA receptor antagonist, haloperidol, blocked mating-induced partner preference whereas administration of a non-selective DA receptor agonist, apomorphine, induced this behavior without mating in both male and female prairie voles (Wang *et al.*, 1999; Aragona *et al.*, 2003). In subsequent studies, DA in the NAcc has been shown to be critical in pair bonding. In both male and female prairie voles, mating induces an increase in DA release or DA turnover in the NAcc (Gingrich *et al.*, 2000; Aragona *et al.*, 2003). Blockade of DA receptors in the NAcc blocks partner preference formation induced by mating whereas activation of DA receptors in the NAcc induces this behavior without mating (Wang *et al.*, 1999; Gingrich *et al.*, 2000; Aragona *et al.*, 2003) (Figure 22.2(E)). This DA effect also appears to be region-specific in the NAcc. DA in the rostral NAcc shell, but not caudal shell or core, plays a site-specific role in controlling the development of mating-induced partner preference formation in male prairie voles (Aragona *et al.*, 2003, 2006). Further, data from female prairie voles indicate the importance of NAcc DA in partner preference formation, but not expression, as only blockade of DA receptors prior to mating (mating is essential for partner preference formation), but not after mating, blocked partner preference (Wang *et al.*, 1999).

An interesting finding is that intra-NAcc DA effects on partner preference formation are receptor specific; activation of D2-type, but not D1-type, receptors facilitates partner preference formation, whereas activation of D1-type receptors diminishes partner preference induced either by mating or by D2 receptor activation (Aragona *et al.*, 2006) (Figure 22.2). As described above, D1-type receptors are coupled with Gs proteins leading to activation of cAMP and D2-type receptors are coupled with Gi proteins leading to inhibition of cAMP activity. In a more recent study we found that manipulation of the cAMP pathway in the NAcc has opposing effects on partner preference formation; activation of the cAMP pathway blocks, whereas inhibition of cAMP activity induces partner preference in male prairie voles (Aragona & Wang, 2007). These data provide further evidence highlighting the intracellular mechanisms underlying receptor-specific DA regulation of pair bonding.

This receptor-specific DA regulation of partner preference formation indicates a unique model for neurotransmitter regulation of complex social behaviors. When male prairie voles are exposed to a female, DA is released in the NAcc and activation of D2-type receptors results in the formation of partner preference. Thereafter, 2 weeks of pair-bonding experience induces neuroplastic changes, indicated by a significant increase in D1-type receptors in the NAcc (Aragona *et al.*, 2006). Interestingly, this increase in D1-type receptor density is associated with enhanced selective aggression by male prairie voles toward conspecific strangers (males or females), and intra-NAcc blockade of D1-type receptors blocks selective aggression (Aragona *et al.*, 2006) (Figure 22.3). It is suggested that selective aggression plays an important role in the maintenance of already established pair bonds as males aggressively reject conspecific male or female strangers that could serve as competitors or potential mates (Aragona *et al.*, 2006; Gobrogge *et al.*, 2007).

It should be noted that the NAcc is not the only brain area in which DA regulates pair bonding. Several other brain regions could also be involved. For example, a species difference in DA receptor binding is found in the mPFC. Prairie voles have a higher level of D2-type, but a lower level of D1-type, receptor binding in the mPFC compared to meadow voles (Smeltzer *et al.*, 2006). Manipulation of DA receptors in the mPFC also alters partner preference formation in female prairie voles, indicating a potential role of mPFC DA in the mediation of NAcc–DA regulation of pair bonding (Wommack & Wang, 2007). The VTA, which provides the major source of DA projections to the NAcc and mPFC, is also involved in pair bonding. Glutamate and gamma-amino-butyric- acid (GABA) receptor blockade in the VTA, which alters DA activity in the NAcc, induces partner preference formation in the absence of mating in male prairie voles (Curtis & Wang, 2005b). Finally, we recently discovered a site- and behavior-specific association between increased cellular activity of tyrosine hydroxylase (TH) cells (indicated by TH/fos double labeling) in the AH and increased selective aggression in pair-bonded male prairie voles (Gobrogge *et al.*, 2007), implicating a potential role of AH-DA in pair-bond maintenance.

DA–NEUROCHEMICAL INTERACTIONS AND PAIR-BONDING BEHAVIOR

It is not surprising that complex social behaviors, such as pair bonding, are regulated by multiple neurotransmitter systems. In addition to DA, several other neurochemicals, such as AVP, OT, corticotrophin releasing factor (CRF), GABA, and glutamate, have also been implicated in pair bonding (Williams *et al.*, 1992, 1994; Winslow *et al.*, 1993; Carter *et al.*, 1995; Wang *et al.*, 1998; Liu *et al.*, 2001; Aragona *et al.*, 2003; Liu & Wang, 2003; Lim & Young, 2004; Curtis & Wang, 2005b). As this review is focused on DA, we will discuss interactions between DA and other neurotransmitters in the regulation of pair bonding.

Central OT has been implicated in pair bonding in female prairie voles, as blockade of OT receptors (icv) blocks mating-induced partner preference whereas administration of OT receptor agonists induces this behavior without mating (Williams *et al.*, 1994). OT receptors are expressed extensively in the prairie vole's NAcc but are almost absent in the NAcc of promiscuous meadow and montane voles, further indicating the potential role of NAcc OT in pair bonding in monogamous voles (Insel & Shapiro, 1992). In our study, intra-NAcc administration of an OT receptor antagonist blocked partner preference formation induced by a DA D2-type agonist, whereas blockade of D2-type, not D1-type, DA receptors blocked OT-induced partner preference formation in female prairie voles (Liu & Wang, 2003). These data suggest that concurrent access to both OT and DA D2-type receptors in the NAcc is important in partner preference formation.

Central AVP, particularly AVP in the LS and ventral palladium (VP), has also been implicated in pair bonding in male prairie voles (Winslow *et al.*, 1993; Liu *et al.*, 2001; Pitkow *et al.*, 2001). In a recent study in male meadow voles, that usually do not exhibit pair bonding, viral vector gene transfer was performed to increase AVP V1a receptor expression in the VP. Subsequently, these meadow voles displayed mating-induced partner preference formation (Lim *et al.*, 2004). However, administration of a DA D2-type receptor antagonist blocked this mating-induced partner preference, indicating interactions between DA and AVP in pair-bonding behavior (Lim *et al.*, 2004).

Further, stress and stress hormones (e.g., glucocorticoids), have also been found to interfere with pair bonding in female prairie voles (DeVries *et al.*, 1995). Peripheral administration of glucocorticoid and mineralocorticoid receptor antagonists facilitates partner preference formation without mating, whereas a concurrent central (icv) administration of DA receptor antagonists blocks this behavior in female prairie voles, indicating an interaction between DA and glucocorticoids in partner preference formation (Curtis & Wang, 2005a). Finally, recent data also indicate a presynaptic interaction between DA and both glutamate and GABA receptors in partner preference formation in male prairie voles (Curtis & Wang, 2005b).

It should be noted that previous studies in other rodents have suggested potential mechanisms for DA interaction with other neurotransmitters. For example, the dopaminergic system can regulate central AVP release (Bridges *et al.*, 1976; Buijs *et al.*, 1984; Lindvall *et al.*, 1984) by modulating the reflexive activity of AVP cell bodies (Moos & Richard, 1982). Reciprocally, AVP administration to striatal rat brain slice preparations significantly increases DA release (Tyagi *et al.*, 1998), implicating potential pre-synaptic effects between DA and AVP systems. Further, OT–DA interactions have been observed, in previous work, to regulate several rodent behaviors including penile erections, yawning, and grooming (Drago *et al.*, 1986; Argiolas *et al.*, 1988, 1989; Kovacs *et al.*, 1990). However, the nature of the intracellular mechanisms underlying DA interactions with other neurochemicals in the regulation of pair bonding are virtually unknown and require further investigation.

DA INVOLVEMENT IN NATURAL AND DRUG REWARD

The above-described data have demonstrated that the mesolimbic DA circuit is important in mediating social reward associated with pair bonding in monogamous prairie voles. In fact, central DA has also been implicated in other types of natural reward including food intake (Hernandez & Hoebel, 1989; Clifton & Somerville, 1994; Martel & Fantino, 1996; Volkow *et al.*, 2002; Pecina *et al.*, 2003), body weight homeostasis (Winn & Robbins, 1985; Neill *et al.*, 2002; Wang *et al.*, 2002), mating (Brackett *et al.*, 1986; Pfaus *et al.*, 1990; Meisel *et al.*, 1993; Wersinger & Rissman, 2000; Fisher *et al.*, 2005; Hull & Dominguez, 2006), and parental behavior (Giordano *et al.*, 1990; Keer & Stern, 1999; Lonstein, 2002; Numan, 2007). In rats, for example, food seeking induces DA release in the NAcc, and released DA activates D1-type receptors to facilitate this behavior, illustrating a receptor-specific DA regulation of natural motivation (Eyny & Horvitz, 2003). Mating has also been shown to induce DA release in several brain regions including the NAcc, CP, and MPOA in both male and female rats (Pfaus *et al.*, 1990; Hull *et al.*, 1993; Mas *et al.*, 1995; Triemstra *et al.*, 2005). In a series of elegant studies in male rats, mating was found to induce DA release in the MPOA before and during copulation (Hull *et al.*, 1993). Released DA then acts on post-synaptic DA receptors (Hull *et al.*, 2004) and interacts with other neurochemicals, such as glutamate and GABA (Dominguez *et al.*, 2006), to facilitate genital/motor reflexes (e.g., erection, mounting, and ejaculation) and appetitive processing of the incentive value of potential mates (Hull & Dominguez, 2007). It has been shown that the MPOA has a direct anatomical and neurochemical connection with the mesolimbic DA system (Maeda & Mogenson, 1980) and can act synergistically with the VTA–NAcc circuit to regulate the motivational salience associated with mating (Moses *et al.*, 1995).

The role of mesolimbic DA in mediating actions of drugs of abuse has been well established. DA signaling has been implicated in processing a variety of abused drugs including cocaine, amphetamine, nicotine, alcohol, marijuana, heroin, and morphine (Maldonado, 2003; Nestler & Malenka, 2004; Baler & Volkow, 2006; Marsden, 2006). The mesolimbic DA pathway has been consistently shown to play a significant role in behavioral patterns of drug addiction, including seeking, withdrawal, and relapse (Self *et al.*, 1998; Koob, 2000; Koob & Le Moal, 2001; Yun *et al.*, 2004). A detailed description of DA regulation of drugs of abuse is beyond the scope of the current chapter. However, we would like to point out that in many cases, DA regulates drugs of abuse in a similar way as DA regulates natural reward. For example, activation of D2-type receptors in the rostral NAcc shell facilitates whereas activation of D1-type receptors diminishes cocaine-seeking behavior in rats (Self *et al.*, 1996). These D1 and D2 opposing effects on drug seeking are similar to our findings that NAcc D2 activation promotes whereas NAcc D1 activation prevents pair-bond formation in prairie voles (Aragona *et al.*, 2006). It has been hypothesized that the central DA system has evolved to mediate behavioral processes essential for survival, such as feeding and reproduction (Ikemoto & Panksepp, 1999; Insel, 2003). Drugs of abuse, however, can usurp this neural circuitry mediating associative learning reinforced by natural rewards leaving an animal motivationally paralyzed to exclusively seek artificial drugs while neglecting social affiliation (Yamaguchi & Kandel, 1985; Kaestner, 1995), proper food intake (Ericsson *et al.*, 1996; Chassler, 1997), and parental behavior (Kandel *et al.*, 1994).

IMPLICATIONS OF THE VOLE MODEL FOR HUMAN MENTAL HEALTH

In human mental illness, clinical differences in social behavior underlie several forms of disease sequelae. DA, AVP, and OT have been shown to be involved in several human behaviors including social bonding (Bartels & Zeki, 2004; Gonzaga *et al.*, 2006; Marazziti *et al.*, 2006), autism (Kendrick, 2004), physical aggression (Pedersen, 2004), and parenting (Swain *et al.*, 2007). These neurochemicals work in concert with one-another to control complex social interactions. In prairie voles, these neurochemical circuits have evolved to produce a monogamous life strategy. By understanding the basic neuroendocrinology of pair-bonding behavior in voles, we can hope to better clarify the neural chemistry of mental health deficits associated with impaired social attachment, pathological aggression, and deficits in parental care in humans.

Further, drug addiction has been a tremendous problem for many humans. Drugs have such a powerful control over behavior because they act on highly conserved brain regions, including the mesolimbic DA pathway, that evolved to mediate behavioral processes essential for survival (Panksepp *et al.*, 2002; Insel, 2003). Although the major focus on treatment has been on pharmacological intervention targeted at neural circuitry known to mediate drug reward, the effects of social attachment on drug addiction should be given attention especially since strong social bonds are likely associated with reduced addiction in humans (Havassy *et al.*, 1995; Moos *et al.*, 2002; Duncan *et al.*, 2003). Unfortunately, basic studies examining interactions between attachment and addiction are sparse partially due to the fact that traditional laboratory rats and mice show limited attachment behavior between adults. In vole studies, DA has been found to regulate pair bonding (Aragona *et al.*, 2006), similarly, as it regulates drugs of abuse in other rodents (Self *et al.*, 1996). A behavioral model has recently been established to study drugs of abuse in prairie voles (Aragona *et al.*, 2007). Thus, there is great potential to develop the prairie vole as the premier rodent model for the study of interactions between social- and drug reward, which may have important consequences for addiction prevention.

ACKNOWLEDGMENTS

We are grateful to Dr. Joel Wommack, Ms. Kimberly Young, and Ms. Claudia Lieberwirth for their critical reading of the manuscript. Research reviewed in this chapter from our lab was supported by NIH grants NIH NS Program Training Grant to FSU (T32 NS-7437), MH F31-79600 to KLG, and MHR01-58616, DAR01-19627, and DAK02-23048 to ZW.

REFERENCES

Aragona, B. J., and Wang, Z. (2007). Opposing regulation of pair bond formation by cAMP signaling within the nucleus accumbens shell. *J. Neurosci.* 27, 13352–13356.

Aragona, B. J., Liu, Y., Curtis, J. T., Stephan, F. K., and Wang, Z. (2003). A critical role for nucleus accumbens dopamine in partner-preference formation in male prairie voles. *J. Neurosci.* 23, 3483–3490.

Aragona, B. J., Liu, Y., Yu, Y. J., Curtis, J. T., Detwiler, J. M., Insel, T. R., and Wang, Z. (2006). Nucleus accumbens dopamine differentially mediates the formation and maintenance of monogamous pair bonds. *Nat. Neurosci.* 9, 133–139.

Aragona, B. J., Detwiler, J. M., and Wang, Z. (2007). Amphetamine reward in the monogamous prairie vole. *Neurosci. Lett.* 418, 190–194.

Argiolas, A., Melis, M. R., and Gessa, G. L. (1988). Yawning and penile erection: Central dopamine–oxytocin–adrenocorticotropin connection. *Ann. NY Acad. Sci.* 525, 330–337.

Argiolas, A., Collu, M., D'Aquila, P., Gessa, G. L., Melis, M. R., and Serra, G. (1989). Apomorphine stimulation of male copulatory behavior is prevented by the oxytocin antagonist d(CH2)5 Tyr(Me)-Orn8-vasotocin in rats. *Pharmacol. Biochem. Behav.* 33, 81–83.

Baler, R. D., and Volkow, N. D. (2006). Drug addiction: The neurobiology of disrupted self-control. *Trends Mol. Med.* 12, 559–566.

Bamshad, M., Novak, M. A., and De Vries, G. J. (1993). Sex and species differences in the vasopressin innervation of sexually naive and parental prairie voles, *Microtus ochrogaster* and meadow voles, Microtus pennsylvanicus. *J. Neuroendocrinol.* 5, 247–255.

Bartels, A., and Zeki, S. (2004). The neural correlates of maternal and romantic love. *Neuroimage* 21, 1155–1166.

Bowler, C. M., Cushing, B. S., and Carter, C. S. (2002). Social factors regulate female–female aggression and affiliation in prairie voles. *Physiol. Behav.* 76, 559–566.

Boyson, S. J., McGonigle, P., and Molinoff, P. B. (1986). Quantitative autoradiographic localization of the D1 and D2 subtypes of dopamine receptors in rat brain. *J. Neurosci.* 6, 3177–3188.

Brackett, N. L., Iuvone, P. M., and Edwards, D. A. (1986). Midbrain lesions, dopamine and male sexual behavior. *Behav. Brain Res.* 20, 231–240.

Bridges, T. E., Hillhouse, E. W., and Jones, M. T. (1976). The effect of dopamine on neurohypophysial hormone release *in vivo* and from the rat neural lobe and hypothalamus *in vitro*. *J. Physiol.* 260, 647–666.

Buijs, R. M., Geffard, M., Pool, C. W., and Hoorneman, E. M. (1984). The dopaminergic innervation of the supraoptic and paraventricular nucleus. A light and electron microscopical study. *Brain Res.* 323, 65–72.

Callier, S., Snapyan, M., Le Crom, S., Prou, D., Vincent, J. D., and Vernier, P. (2003). Evolution and cell biology of dopamine receptors in vertebrates. *Biol. Cell* 95, 489–502.

Calon, F., Hadj Tahar, A., Blanchet, P. J., Morissette, M., Grondin, R., Goulet, M., Doucet, J. P., Robertson, G. S., Nestler, E., Di Paolo, T., and Bedard, P. J. (2000). Dopamine-receptor stimulation: Biobehavioral and biochemical consequences. *Trends Neurosci.* 23, S92–S100.

Carr, D. B., and Sesack, S. R. (2000). Projections from the rat prefrontal cortex to the ventral tegmental area: target specificity in the synaptic associations with mesoaccumbens and mesocortical neurons. *J. Neurosci.* 20, 3864–3873.

Carter, C. S., and Getz, L. L. (1993). Monogamy and the prairie vole. *Sci. Am.* 268, 100–106.

Carter, C. S., DeVries, A. C., and Getz, L. L. (1995). Physiological substrates of mammalian monogamy: The prairie vole model. *Neurosci. Biobehav. Rev.* 19, 303–314.

Chassler, L. (1997). Understanding anorexia nervosa and bulimia nervosa from an attachment perspective. *Clin. Soc. Work J.* 25, 407–423.

Cheung, S., Ballew, J. R., Moore, K. E., and Lookingland, K. J. (1998). Contribution of dopamine neurons in the medial zona incerta to the innervation of the central nucleus of the amygdala, horizontal diagonal band of Broca and hypothalamic paraventricular nucleus. *Brain Res.* 808, 174–181.

Clifton, P. G., and Somerville, E. M. (1994). Disturbance of meal patterning following nucleus accumbens lesions in the rat. *Brain Res.* 667, 123–128.

Curtis, J. T., and Wang, Z. (2005a). Glucocorticoid receptor involvement in pair bonding in female prairie voles: The effects of acute blockade and interactions with central dopamine reward systems. *Neuroscience* 134, 369–376.

Curtis, J. T., and Wang, Z. (2005b). Ventral tegmental area involvement in pair bonding in male prairie voles. *Physiol. Behav.* 86, 338–346.

DeVries, A. C., DeVries, M. B., Taymans, S., and Carter, C. S. (1995). Modulation of pair bonding in female prairie voles (*Microtus ochrogaster*) by corticosterone. *Proc. Natl Acad. Sci. USA* 92, 7744–7748.

Dewsbury, D. A. (1987). The comparative psychology of monogamy. *Nebr. Symp. Motiv.* 35, 1–50.

Deyo, R. A., Straube, K. T., and Disterhoft, J. F. (1989). Nimodipine facilitates associative learning in aging rabbits. *Science* 243, 809–811.

Dominguez, J. M., Gil, M., and Hull, E. M. (2006). Preoptic glutamate facilitates male sexual behavior. *J. Neurosci.* 26, 1699–1703.

Drago, F., Caldwell, J. D., Pedersen, C. A., Continella, G., Scapagnini, U., and Prange, A. J., Jr. (1986). Dopamine neurotransmission in the nucleus accumbens may be involved in oxytocin-enhanced grooming behavior of the rat. *Pharmacol. Biochem. Behav.* 24, 1185–1188.

Duncan, G. J., Wilkerson, B., and England, P. (2003). Cleaning up their act: The impact of marriage and cohabitation on licit and illicit drug use. Northwestern University, WP.

Ericsson, M., Poston, W. S., 2nd, and Foreyt, J. P. (1996). Common biological pathways in eating disorders and obesity. *Addict. Behav.* 21, 733–743.

Eyny, Y. S., and Horvitz, J. C. (2003). Opposing roles of D1 and D2 receptors in appetitive conditioning. *J. Neurosci.* 23, 1584–1587.

Febo, M., and Ferris, C. F. (2007). Development of cocaine sensitization before pregnancy affects subsequent maternal retrieval of pups and prefrontal cortical activity during nursing. *Neuroscience* 148, 400–412.

Fisher, H., Aron, A., and Brown, L. L. (2005). Romantic love: An fMRI study of a neural mechanism for mate choice. *J. Comp. Neurol.* 493, 58–62.

Getz, L. L., and Hofmann, J. E. (1986). Social organization in free-living prairie voles, *Microtus ochrogaster*. *Behav. Ecol. Sociobiol.* 18, 275–282.

Getz, L. L., Carter, C. S., and Gavish, L. (1981). The mating system of the prairie vole, *Microtus ochrogaster*. Field and laboratory evidence for pair-bonding. *Behav. Ecol. Sociobiol.* 8, 189–194.

Gingrich, B., Liu, Y., Cascio, C., Wang, Z., and Insel, T. R. (2000). Dopamine D2 receptors in the nucleus accumbens are important for social attachment in female prairie voles (*Microtus ochrogaster*). *Behav. Neurosci.* 114, 173–183.

Giordano, A. L., Johnson, A. E., and Rosenblatt, J. S. (1990). Haloperidol-induced disruption of retrieval behavior and reversal with apomorphine in lactating rats. *Physiol. Behav.* 48, 211–214.

Gobrogge, K. L., Liu, Y., Jia, X., and Wang, Z. (2007). Anterior hypothalamic neural activation and neurochemical associations with aggression in pair-bonded male prairie voles. *J. Comp. Neurol.* 502, 1109–1122.

Gonzaga, G. C., Turner, R. A., Keltner, D., Campos, B., and Altemus, M. (2006). Romantic love and sexual desire in close relationships. *Emotion* 6, 163–179.

Hammock, E. A., and Young, L. J. (2005). Microsatellite instability generates diversity in brain and sociobehavioral traits. *Science* 308, 1630–1634.

Hansen, S., Bergvall, A. H., and Nyiredi, S. (1993). Interaction with pups enhances dopamine release in the ventral striatum of maternal rats: A microdialysis study. *Pharmacol. Biochem. Behav.* 45, 673–676.

Havassy, B. E., Wasserman, D. A., and Hall, S. M. (1995). Social relationships and abstinence from cocaine in an American treatment sample. *Addiction* 90, 699–710.

Hernandez, L., and Hoebel, B. G. (1989). Food intake and lateral hypothalamic self-stimulation covary after medial hypothalamic lesions or ventral midbrain 6-hydroxydopamine injections that cause obesity. *Behav. Neurosci.* 103, 412–422.

Hull, E. M., and Dominguez, J. M. (2006). Getting his act together: Roles of glutamate, nitric oxide, and dopamine in the medial preoptic area. *Brain Res.* 1126, 66–75.

Hull, E. M., and Dominguez, J. M. (2007). Sexual behavior in male rodents. *Horm. Behav.* 52, 45–55.

Hull, E. M., Eaton, R. C., Moses, J., and Lorrain, D. (1993). Copulation increases dopamine activity in the medial preoptic area of male rats. *Life Sci.* 52, 935–940.

Hull, E. M., Muschamp, J. W., and Sato, S. (2004). Dopamine and serotonin: Influences on male sexual behavior. *Physiol. Behav.* 83, 291–307.

Ikemoto, S., and Panksepp, J. (1999). The role of nucleus accumbens dopamine in motivated behavior: A unifying interpretation with special reference to reward-seeking. *Brain Res. Rev.* 31, 6–41.

Insel, T. R. (2003). Is social attachment an addictive disorder? *Physiol. Behav.* 79, 351–357.

Insel, T. R., and Shapiro, L. E. (1992). Oxytocin receptor distribution reflects social organization in monogamous and polygamous voles. *Proc. Natl Acad. Sci. USA* 89, 5981–5985.

Insel, T. R., and Hulihan, T. J. (1995). A gender-specific mechanism for pair bonding: Oxytocin and partner preference formation in monogamous voles. *Behav. Neurosci.* 109, 782–789.

Insel, T. R., Wang, Z. X., and Ferris, C. F. (1994). Patterns of brain vasopressin receptor distribution associated with social organization in microtine rodents. *J. Neurosci.* 14, 5381–5392.

Insel, T. R., Preston, S., and Winslow, J. T. (1995). Mating in the monogamous male: Behavioral consequences. *Physiol. Behav.* 57, 615–627.

Insel, T. R., Winslow, J. T., Wang, Z., and Young, L. J. (1998). Oxytocin, vasopressin, and the neuroendocrine basis of pair bond formation. *Adv. Exp. Med. Biol.* 449, 215–224.

Jannett, F. (1982). Nesting patterns of adult voles, *Microtus montanus*, in field populations. *J. Mammal.* 63, 495–498.

Kaestner, R. (1995). The effects of cocaine and marijuana use on marriage and marital stability. *National Bureau of Economic Research Working Paper No. 5038*.

Kandel, D. B., Rosenbaum, E., and Chen, K. (1994). Impact of maternal drug-use and life experiences on preadolescent children born to teenage mothers. *J. Marriage Fam.* 56, 325–340.

Keer, S. E., and Stern, J. M. (1999). Dopamine receptor blockade in the nucleus accumbens inhibits maternal retrieval and licking, but enhances nursing behavior in lactating rats. *Physiol. Behav.* 67, 659–669.

Kendrick, K. M. (2004). The neurobiology of social bonds. *J. Neuroendocrinol.* 16, 1007–1008.

Koob, G. F. (2000). Neurobiology of addiction. Toward the development of new therapies. *Ann. NY Acad. Sci.* 909, 170–185.

Koob, G. F., and Le Moal, M. (2001). Drug addiction, dysregulation of reward, and allostasis. *Neuropsychopharm.* 24, 97–129.

Kovacs, G. L., Sarnyai, Z., Barbarczi, E., Szabo, G., and Telegdy, G. (1990). The role of oxytocin–dopamine interactions in cocaine-induced locomotor hyperactivity. *Neuropharm.* 29, 365–368.

Lansing, S. W., and Lonstein, J. S. (2006). Tyrosine hydroxylase-synthesizing cells in the hypothalamus of prairie voles (*Microtus ochrogaster*): Sex differences in the anteroventral periventricular preoptic area and effects of adult gonadectomy or neonatal gonadal hormones. *J. Neurobiol.* 66, 197–204.

Lavi-Avnon, Y., Weller, A., Finberg, J. P., Gispan-Herman, I., Kinor, N., Stern, Y. et al. (2007). The reward system and maternal behavior in an animal model of depression: A microdialysis study. *Psychopharmacology (Berl.)* .

Lim, M. M., and Young, L. J. (2004). Vasopressin-dependent neural circuits underlying pair bond formation in the monogamous prairie vole. *Neuroscience* 125, 35–45.

Lim, M. M., Wang, Z., Olazabal, D. E., Ren, X., Terwilliger, E. F., and Young, L. J. (2004). Enhanced partner preference in a promiscuous species by manipulating the expression of a single gene. *Nature* 429, 754–757.

Lindvall, O., and Stenevi, U. (1978). Dopamine and noradrenaline neurons projecting to the septal area in the rat. *Cell. Tissue Res.* 190, 383–407.

Lindvall, O., Bjorklund, A., and Skagerberg, G. (1984). Selective histochemical demonstration of dopamine terminal systems in rat di- and telencephalon: New evidence for dopaminergic innervation of hypothalamic neurosecretory nuclei. *Brain Res.* 306, 19–30.

Liu, Y., and Wang, Z. X. (2003). Nucleus accumbens oxytocin and dopamine interact to regulate pair bond formation in female prairie voles. *Neuroscience* 121, 537–544.

Liu, Y., Curtis, J. T., and Wang, Z. (2001). Vasopressin in the lateral septum regulates pair bond formation in male prairie voles (*Microtus ochrogaster*). *Behav. Neurosci.* 115, 910–919.

Lonstein, J. S. (2002). Effects of dopamine receptor antagonism with haloperidol on nurturing behavior in the biparental prairie vole. *Pharmacol. Biochem. Behav.* 74, 11–19.

Lonstein, J. S., and De Vries, G. J. (1999). Sex differences in the parental behaviour of adult virgin prairie voles: Independence from gonadal hormones and vasopressin. *J. Neuroendocrinol.* 11, 441–449.

Maeda, H., and Mogenson, G. J. (1980). An electrophysiological study of inputs to neurons of the ventral tegmental area from the nucleus accumbens and medial preoptic-anterior hypothalamic areas. *Brain Res.* 197, 365–377.

Maldonado, R. (2003). The neurobiology of addiction. *J. Neural Transm. Suppl.* 66, 1–14.

Marazziti, D., Dell'Osso, B., Baroni, S., Mungai, F., Catena, M., Rucci, P., Albanese, F., Giannaccini, G., Betti, L., Fabbrini, L., Italiani, P., Del Debbio, A., Lucacchini, A., and Dell'Osso, L. (2006). A relationship between oxytocin and anxiety of romantic attachment. *Clin. Pract. Epidemol. Ment. Health* 2, 28.

Marsden, C. A. (2006). Dopamine: The rewarding years. *Br. J. Pharmacol.* 147(Suppl 1), S136–S144.

Martel, P., and Fantino, M. (1996). Influence of the amount of food ingested on mesolimbic dopaminergic system activity: A microdialysis study. *Pharmacol. Biochem. Behav.* 55, 297–302.

Mas, M., Fumero, B., and Gonzalez-Mora, J. L. (1995). Voltammetric and microdialysis monitoring of brain monoamine neurotransmitter release during sociosexual interactions. *Behav. Brain Res.* 71, 69–79.

McGuire, B., and Novak, M. (1984). A comparison of maternal behavior in the meadow vole (*Microtus pennsylvanicus*), prairie vole (*M. ochrogaster*) and pine vole (*M. pinetorum*). *Anim. Behav.* 32, 1132–1141.

McGuire, B., and Novak, M. (1986). Parental care and its relation to social organization in the montane vole. *J. Mammal.* 67, 305–311.

Meisel, R. L., Camp, D. M., and Robinson, T. E. (1993). A microdialysis study of ventral striatal dopamine during sexual behavior in female Syrian hamsters. *Behav. Brain Res.* 55, 151–157.

Miller, S. M., and Lonstein, J. S. (2005). Dopamine d1 and d2 receptor antagonism in the preoptic area produces different effects on maternal behavior in lactating rats. *Behav. Neurosci.* 119, 1072–1083.

Moos, F., and Richard, P. (1982). Excitatory effect of dopamine on oxytocin and vasopressin reflex releases in the rat. *Brain Res.* 241, 249–260.

Moos, R. H., Nichol, A. C., and Moos, B. S. (2002). Risk factors for symptom exacerbation among treated patients with substance use disorders. *Addiction* 97, 75–85.

Moses, J., Loucks, J. A., Watson, H. L., Matuszewich, L., and Hull, E. M. (1995). Dopaminergic drugs in the medial preoptic area and nucleus accumbens: Effects on motor activity, sexual motivation, and sexual performance. *Pharmacol. Biochem. Behav.* 51, 681–686.

Murphy, T. H., Worley, P. F., and Baraban, J. M. (1991). L-type voltage-sensitive calcium channels mediate synaptic activation of immediate early genes. *Neuron* 7, 625–635.

Neill, D. B., Fenton, H., and Justice, J. B., Jr. (2002). Increase in accumbal dopaminergic transmission correlates with response cost not reward of hypothalamic stimulation. *Behav. Brain Res.* 137, 129–138.

Nestler, E. J., and Malenka, R. C. (2004). The addicted brain. *Sci. Am.* 290, 78–85.

Neve, K. A., Seamans, J. K., and Trantham-Davidson, H. (2004). Dopamine receptor signaling. *J. Recept. Signal Transduct. Res.* 24, 165–205.

Newman, S. W. (1999). The medial extended amygdala in male reproductive behavior. A node in the mammalian social behavior network. *Ann. NY Acad. Sci.* 877, 242–257.

Northcutt, K. V., Wang, Z., and Lonstein, J. S. (2007). Sex and species differences in tyrosine hydroxylase-synthesizing cells of the rodent olfactory extended amygdala. *J. Comp. Neurol.* 500, 103–115.

Numan, M. (2007). Motivational systems and the neural circuitry of maternal behavior in the rat. *Dev. Psychobiol.* 49, 12–21.

Numan, M., Numan, M. J., Pliakou, N., Stolzenberg, D. S., Mullins, O. J., Murphy, J. M., and Smith, C. D. (2005). The effects of D1 or D2 dopamine receptor antagonism in the medial preoptic area, ventral pallidum, or nucleus accumbens on the maternal retrieval response and other aspects of maternal behavior in rats. *Behav. Neurosci.* 119, 1588–1604.

Oliveras, D., and Novak, M. (1986). A comparison of paternal behavior in the meadow vole, *Microtus pennsylvanicus*, the pine vole, *Microtus pinetorum*, and prairie vole, *Microtus ochrogaster*. *Anim. Behav.* 34, 519–526.

Panksepp, J., Knutson, B., and Burgdorf, J. (2002). The role of brain emotional systems in addictions: A neuro-evolutionary perspective and new 'self-report' animal model. *Addiction* 97, 459–469.

Pecina, S., Cagniard, B., Berridge, K. C., Aldridge, J. W., and Zhuang, X. (2003). Hyperdopaminergic mutant mice have higher "wanting" but not "liking" for sweet rewards. *J. Neurosci.* 23, 9395–9402.

Pedersen, C. A. (2004). Biological aspects of social bonding and the roots of human violence. *Ann. NY Acad. Sci.* 1036, 106–127.

Pfaus, J. G., Damsma, G., Nomikos, G. G., Wenkstern, D. G., Blaha, C. D., Phillips, A. G., and Fibiger, H. C. (1990). Sexual behavior enhances central dopamine transmission in the male rat. *Brain Res.* 530, 345–348.

Pitkow, L. J., Sharer, C. A., Ren, X., Insel, T. R., Terwilliger, E. F., and Young, L. J. (2001). Facilitation of affiliation and pair-bond formation by vasopressin receptor gene transfer into the ventral forebrain of a monogamous vole. *J. Neurosci.* 21, 7392–7396.

Pollack, A. (2004). Coactivation of D1 and D2 dopamine receptors: In marriage, a case of his, hers, and theirs. *Sci. STKE* . pe50.

Self, D. W., Barnhart, W. J., Lehman, D. A., and Nestler, E. J. (1996). Opposite modulation of cocaine-seeking behavior by D1- and D2-like dopamine receptor agonists. *Science* 271, 1586–1589.

Self, D. W., Genova, L. M., Hope, B. T., Barnhart, W. J., Spencer, J. J., and Nestler, E. J. (1998). Involvement of cAMP-dependent protein kinase in the nucleus accumbens in cocaine self-administration and relapse of cocaine-seeking behavior. *J. Neurosci.* 18, 1848–1859.

Smeltzer, M. D., Curtis, J. T., Aragona, B. J., and Wang, Z. (2006). Dopamine, oxytocin, and vasopressin receptor binding in the medial prefrontal cortex of monogamous and promiscuous voles. *Neurosci. Lett.* 394, 146–151.

Stolzenberg, D. S., McKenna, J. B., Keough, S., Hancock, R., Numan, M. J., and Numan, M. (2007). Dopamine D1 receptor stimulation of the nucleus accumbens or the medial preoptic area promotes the onset of maternal behavior in pregnancy-terminated rats. *Behav. Neurosci.* 121, 907–919.

Surmeier, D. J., Ding, J., Day, M., Wang, Z., and Shen, W. (2007). D1 and D2 dopamine-receptor modulation of striatal glutamatergic signaling in striatal medium spiny neurons. *Trends Neurosci.* 30, 228–235.

Swain, J. E., Lorberbaum, J. P., Kose, S., and Strathearn, L. (2007). Brain basis of early parent–infant interactions: Psychology, physiology, and *in vivo* functional neuroimaging studies. *J. Child Psychol. Psychiatr.* 48, 262–287.

Swanson, L. W. (1982). The projections of the ventral tegmental area and adjacent regions: A combined fluorescent retrograde tracer and immunofluorescence study in the rat. *Brain Res. Bull.* 9, 321–353.

Tamarin, R. (1985). Biology of new world *Microtus*. *Am. Soc. Mamm. Spec.* Pub 8.

Triemstra, J. L., Nagatani, S., and Wood, R. I. (2005). Chemosensory cues are essential for mating-induced dopamine release in MPOA of male Syrian hamsters. *Neuropsychopharm.* 30, 1436–1442.

Tyagi, M. G., Handa, R. K., Stephen, P. M., and Bapna, J. S. (1998). Vasopressin induces dopamine release and cyclic AMP efflux from the brain of water-deprived rats: Inhibitory effect of vasopressin V2 receptor-mediated phosphorylation. *Biol. Signals Recept.* 7, 328–336.

Volkow, N. D., Wang, G. J., Fowler, J. S., Logan, J., Jayne, M., Franceschi, D., Wong, C., Gatley, S. J., Gifford, A. N., Ding, Y. S., and Pappas, N. (2002). "Nonhedonic" food motivation in humans involves dopamine in the dorsal striatum and methylphenidate amplifies this effect. *Synapse* 44, 175–180.

Wang, G. J., Volkow, N. D., and Fowler, J. S. (2002). The role of dopamine in motivation for food in humans: Implications for obesity. *Expert Opin. Ther. Targets* 6, 601–609.

Wang, Z., and Aragona, B. J. (2004). Neurochemical regulation of pair bonding in male prairie voles. *Physiol. Behav.* 83, 319–328.

Wang, Z., Smith, W., Major, D. E., and De Vries, G. J. (1994). Sex and species differences in the effects of cohabitation on vasopressin messenger RNA expression in the bed nucleus of the stria terminalis in prairie voles (*Microtus ochrogaster*) and meadow voles (*Microtus pennsylvanicus*). *Brain Res.* 650, 212–218.

Wang, Z., Hulihan, T. J., and Insel, T. R. (1997). Sexual and social experience is associated with different patterns of behavior and neural activation in male prairie voles. *Brain Res.* 767, 321–332.

Wang, Z., Young, L. J., De Vries, G. J., and Insel, T. R. (1998). Voles and vasopressin: A review of molecular, cellular, and behavioral studies of pair bonding and paternal behaviors. *Prog. Brain Res.* 119, 483–499.

Wang, Z., Yu, G., Cascio, C., Liu, Y., Gingrich, B., and Insel, T. R. (1999). Dopamine D2 receptor-mediated regulation of partner preferences in female prairie voles (*Microtus ochrogaster*): A mechanism for pair bonding? *Behav. Neurosci.* 113, 602–611.

Wersinger, S. R., and Rissman, E. F. (2000). Dopamine activates masculine sexual behavior independent of the estrogen receptor alpha. *J. Neurosci.* 20, 4248–4254.

Williams, J. R., Carter, C. S., and Insel, T. (1992). Partner preference development in female prairie voles is facilitated by mating or the central infusion of oxytocin. *Ann. NY Acad. Sci.* 652, 487–489.

Williams, J. R., Insel, T. R., Harbaugh, C. R., and Carter, C. S. (1994). Oxytocin administered centrally facilitates formation of a partner preference in female prairie voles (*Microtus ochrogaster*). *J. Neuroendocrinol.* 6, 247–250.

Winn, P., and Robbins, T. W. (1985). Comparative effects of infusions of 6-hydroxydopamine into nucleus accumbens and anterolateral hypothalamus induced by 6-hydroxydopamine on the response to dopamine agonists, body weight, locomotor activity and measures of exploration in the rat. *Neuropharm.* 24, 25–31.

Winslow, J. T., Hastings, N., Carter, C. S., Harbaugh, C. R., and Insel, T. R. (1993). A role for central vasopressin in pair bonding in monogamous prairie voles. *Nature* 365, 545–548.

Wommack, J. C., and Wang, Z. (2007). Dopamine in the prefrontal cortex regulates partner preference formation in female prairie voles, Poster P1.28, Presented at the *Society for Behavioral Neuroendocrinology*, Pacific Grove, CA.

Yamada, S., Uchida, S., Ohkura, T., Kimura, R., Yamaguchi, M., Suzuki, M., and Yamamoto, M. (1996). Alterations in calcium antagonist receptors and calcium content in senescent brain and attenuation by nimodipine and nicardipine. *J. Pharmacol. Exp. Ther.* 277, 721–727.

Yamaguchi, K., and Kandel, D. B. (1985). On the resolution of role incompatibility-A life event history analysis of family roles and marijuana use. *Am. J. Sociol.* 90, 1284–1325.

Young, L. J., and Wang, Z. (2004). The neurobiology of pair bonding. *Nat. Neurosci.* 7, 1048–1054.

Young, L. J., Winslow, J. T., Nilsen, R., and Insel, T. R. (1997). Species differences in V1a receptor gene expression in monogamous and nonmonogamous voles: Behavioral consequences. *Behav. Neurosci.* 111, 599–605.

Young, L. J., Wang, Z., and Insel, T. R. (1998). Neuroendocrine bases of monogamy. *Trends Neurosci.* 21, 71–75.

Yun, I. A., Wakabayashi, K. T., Fields, H. L., and Nicola, S. M. (2004). The ventral tegmental area is required for the behavioral and nucleus accumbens neuronal firing responses to incentive cues. *J. Neurosci.* 24, 2923–2933.

Zeiss, C. J. (2005). Neuroanatomical phenotyping in the mouse: The dopaminergic system. *Vet. Pathol.* 42, 753–773.

23

SOCIAL MEMORY, MATERNAL CARE, AND OXYTOCIN SECRETION, BUT NOT VASOPRESSIN RELEASE, REQUIRE CD38 IN MICE

HARUHIRO HIGASHIDA, DUO JIN, HONG-XIANG LIU, OLGA LOPATINA, SHIGERU YOKOYAMA, KEITA KOIZUMI, MINAKO HASHII, MD. SAHARUL ISLAM, KENSHI HAYASHI AND TOSHIO MUNESUE

Kanazawa University, 21st Century COE Program on Innovative Brain Science on Development, Learning and Memory, Kanazawa 920-8640, and Department of Biophysical Genetics, Kanazawa University Graduate School of Medicine, Kanazawa 920-8640, Japan

INTRODUCTION

CD38 has been identified as a transmembrane receptor that triggers proliferation and immune responses in lymphocytes (Lee, 2001; Deaglio *et al.*, 2006; Hunt *et al.*, 2006). CD38, thus, is frequently used as a malignancy or differentiation marker in chronic lymphocytic leukemia or HIV infection. CD38 is present in many tissues, such as the brain and pancreas (Higashida *et al.*, 2001; Lee, 2001; Okamoto & Takasawa, 2002). It can catalyze the formation of cyclic ADP-ribose (cADPR) and nicotinic acid adenine dinucleotide phosphate (NAADP) by ADP-ribosyl cyclase from NAD^+ and NAD phosphate (Howard *et al.*, 1993; Takasawa *et al.*, 1993; Higashida *et al.*, 2001; Lee, 2001, 2005; Okamoto & Takasawa, 2002). cADPR mobilizes Ca^{2+} from ryanodine-sensitive intracellular Ca^{2+} stores in the endoplasmic reticulum and NAADP liberates it from other pools located in lysosomes or secretory granules. The two molecules act as second messengers independent from inositol 1,4,5-trisphosphate (IP_3) (Howard *et al.*, 1993; Takasawa *et al.*, 1993; Kato *et al.*, 1999; Higashida *et al.*, 2001; Lee, 2001; Johnson & Misler, 2002; Okamoto & Takasawa, 2002; Yamasaki *et al.*, 2004; Lee, 2005; Gerasimenko *et al.*, 2006). The CD38-dependent cADPR or NAADP signaling pathway of Ca^{2+} regulation has been demonstrated in cultured anterior pituitary tumor cells (Soares *et al.*, 2005; Jin *et al.*, 2007). Here, we describe effects of these pathways on posterior pituitary (neurohypophysial) hormone function in accordance with our recent results (Jin *et al.*, 2007).

Oxytocin (OT) and arginine vasopressin (AVP) are secreted into the blood stream from endings or into the brain from dendrites of hypothalamic neurons (Hatton, 1990). These peptides were originally identified as regulators of the function of peripheral target organs. They mediate uterine contraction, milk ejection, penile erection, and urine concentration (Hatton, 1990; Russell *et al.*, 2003; Argiolas & Melis, 2005).

But they also exert many behavioral effects via the amygdala and other brain regions (Bartels & Zeki, 2004; Insel & Fernald, 2004; Keverne & Curley, 2004; Young & Wang, 2004; LaBar & Cabeza, 2006). Specifically, OT modulates a wide range of sexual and social behaviors, including social recognition, pair bonding, mate guarding, and parental care (Keverne & Curley, 2004; Young & Wang, 2004; LaBar & Cabeza, 2006). Furthermore, it has been suggested OT may be involved in human love, trust, or fear (Bartels & Zeki, 2004; Fries *et al.*, 2005; Kirsch *et al.*, 2005; Kosfeld *et al.*, 2005). Pharmacological studies indicate that peripheral OT administration facilitates social recognition in rats (Popik *et al.*, 1992), increases feelings of trust in healthy human males (Kosfeld *et al.*, 2005), and reduces repetitive behavior in autistic and Asperger's disorders (Hollander *et al.*, 2003). Mouse mutants for the OT gene (Ferguson *et al.*, 2000) and OT receptor gene (Takayanagi *et al.*, 2005) fail to develop social memory or maternal care, indicating that the OT signaling system functions as part of the neurobiological basis of attachments (Insel & Young, 2001; Insel & Fernald, 2004; Young & Wang, 2004). Moreover, the OT system is thought to have some relationship to the human autistic diseases associated with abnormal social behavior (Lim *et al.*, 2005).

Depolarization-secretion coupling of OT and AVP release has been reported in both somatodendritic and neurohypophysial-terminals of hypothalamic neurons (Belin & Moos, 1986; Hatton, 1990; Ludwig & Leng, 2006). Ca^{2+} release from IP_3- and/or ryanodine-sensitive intracellular Ca^{2+} stores is involved in the stimulated OT release (De Crescenzo *et al.*, 2004; Ludwig & Leng, 2006). However, the intracellular Ca^{2+} signaling mechanisms underlying OT or AVP secretion are not fully understood and the mechanism for OT and AVP may not be the same. To address this issue, we used CD38 gene knockout mice ($CD38^{-/-}$; Kato *et al.*, 1999), and discovered that CD38-dependent cADPR- and NAADP-sensitive intracellular Ca^{2+} mobilization plays a key role in OT release, but not AVP secretion, from soma and axon terminals of hypothalamic neurons, exerting profound actions on social behaviors. Alteration of CD38 function and the resultant disturbance of only OT secretion may be associated with some forms of impaired human behavior in the autism spectrum disorders.

DEFECTS IN MATERNAL NURTURING IN $CD38^{-/-}$ MICE

Whereas $CD38^{-/-}$ mice are apparently healthy and viable (Kato *et al.*, 1999; Jin *et al.*, 2007), little is known about their behavioral states. In an initial study we examined the phenotypic characteristics those mice might express. The behaviors of the $CD38^{-/-}$ mice were first observed in open field and plus maze tests. These tests revealed a significantly greater locomotor activity in $CD38^{-/-}$ mice than in their wild-type ($CD38^{+/+}$) ones (Jin *et al.*, 2007) with no or little abnormalities, anxiety, and fear in $CD38^{-/-}$ mice.

When we monitored the responses of postpartum the $CD38^{-/-}$ mice showed a change in maternal behavior. In this study all 1–3 day-old pups were removed and five pups were returned and placed at the opposite side from the nest in the dam's home cage (Figure 23.1, encircled by dotted line). Wild-type $CD38^{+/+}$ dams typically picked a pup up in her mouth and would transport it to the original nest, repeating this behavior with the other four pups (Figure 23.1(A)). In contrast, $CD38^{-/-}$ dams often would pick up one pup and then drop it more than once *en route* to the nest. Frequently, the $CD38^{-/-}$ mothers did not retrieve pups to the original nest site (Figure 23.1(B)). Sometimes, the dam collected pups in other locations and then made new nests. As a consequence, the pups were often scattered in the home cage and neglected. Latencies to retrieve all five pups were scored as well as retrieval responses to other locations within the test cage.

Animals were subjected to retrieval testing 4 times. Overall, the performance of $CD38^{+/+}$ mice improved over test days. $CD38^{+/+}$ dams took longer to retrieve all times pups in the first test, but the second, third, and fourth tests took much shorter times (Jin *et al.*, 2007). $CD38^{-/-}$ dams took much longer to begin retrieving the first pup and to finish retrieving the fifth pup in the first and second tests. The dam's response was much quicker in the third and fourth tests, but they tended to make new nests to which pups were retrieved. These results demonstrate clear deficits in the

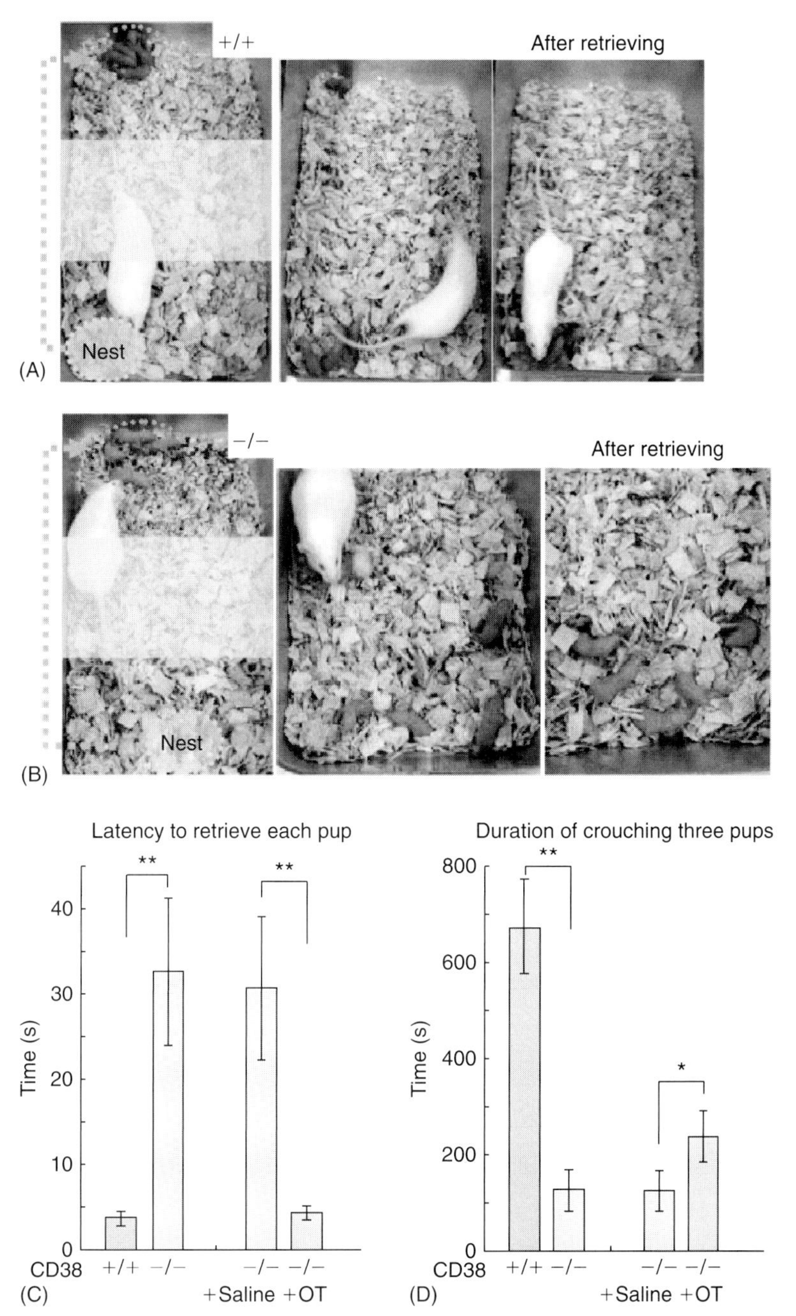

FIGURE 23.1 *Retrieving behavior is impaired in CD38$^{-/-}$ mothers.* (A, B) Typical pup retrieval behavior in mothers' home cages. Pups of a CD38$^{+/+}$ (A) or CD38$^{-/-}$ (B) dam were first removed, and five selected pups were placed on the opposite side of the home cage (encircled by dotted line) from the nest (Nest). Note that the wild-type dam picked pups up in her mouth and transported them to the original nest quickly. The CD38$^{-/-}$ dam picked up pups but dropped and stopped to retrieve furthermore to the original nest. The pups were therefore scattered in the home cage and neglected. (C) Latency to retrieve each pup by mother CD38$^{+/+}$ or CD38$^{-/-}$ mice in separate cages (14), and by mother CD38$^{-/-}$ mice treated with subcutaneous injection of 0.4 mL saline or 3 ng oxytocin (OT)/kg body weight. (D) Time spent crouching over all three pups in the nest by the same set of mothers. $^{*}p < 0.05$; $^{**}p < 0.01$; $N = 6$ (wild type) and $N = 10$ (mutants). (See Color Plate)

mother's pup-care behavior in CD38$^{-/-}$ postpartum mice.

AMNESIA IN SOCIAL MEMORY IN CD38$^{-/-}$ MALE MICE

Maternal nurturing is impaired in OT receptor-deficient mice (Takayanagi *et al.*, 2005), and the OT receptor or OT gene knockout male mice fail to develop social familiarity (Ferguson *et al.*, 2000; Takayanagi *et al.*, 2005). This impairment depends predominantly on olfactory cues which can critically influence reproductive success (Kendrick *et al.*, 1997). In deficient mice there is a consistent decrease OT receptor in olfactory investigation (Figure 23.2(A)) during repeated or prolonged encounters with a conspecific female (Ferguson *et al.*, 2000). We examined this social memory in young adult male mice that experienced repeated pairings with the same female mouse. CD38$^{+/+}$ mice displayed a significant decline in the time spent investigating a female upon subsequent pre-sentations of the same female in trials 3 and 4, as compared to trial 1 (Figure 23.2(B)). This decrease was not due to a general decline in olfactory investigation, because presentation of a novel female during trial 5 resulted in a similar amount of investigation as trial 1 with the original female. In contrast, CD38$^{-/-}$ males showed sustained high levels of investigation at each encounter with the same female and the same level of investigation when presented with a new female at trial 5 (Figure 23.2(B)).

The impairment of social memory did not depend on deficits in main olfactory bulb function, since CD38$^{-/-}$ mice did not have deficits in either olfactory-guided foraging or habituation to a non-social olfactory stimulus as tested by the preference ratio of consumption of isovaleric acid solution in their drinking water (Figure 23.3(A)). It is proposed that the function of CD38 in social behavior could be very specific to the particular neural circuitry involved (Insel & Fernald, 2004; Keverne & Curley, 2004), although to date the data do not exclude a more general cognitive dysfunction. To evaluate these possibilities, mice were trained in a passive avoidance test. The CD38$^{-/-}$ mice could learn the shock as well as wild-type animals (Figure 23.3(B)). Since they perform normally on this learning paradigm, it is concluded that the CD38 mutants do not suffer from global cognitive dysfunction. Rather, it can be said that CD38$^{-/-}$ males with persistent interest during repeated presentations fail to develop their social memory, resembling an amnesia found in OT gene or OT receptor knockout mice (Ferguson *et al.*, 2000; Takayanagi *et al.*, 2005).

PLASMA OT AND VASOPRESSIN LEVELS

Next, we measured plasma OT and AVP levels in CD38 wild-type and mutant mice. Significantly lower concentrations of OT were detected in CD38$^{-/-}$ than in control wild-type mice (Figure 23.4(A)). In sharp contrast, OT concentrations in the hypothalamus and posterior pituitary of CD38$^{-/-}$ mice were increased to a much higher level than in controls (Figure 23.4(B)). In contrast, measurement of AVP, a member of the same posterior pituitary hormone family, revealed little or no decrease in plasma concentrations (Figure 23.4(A)). Moreover, no significant increase in tissue concentrations was evident in CD38$^{-/-}$ mice (data not shown).

Electron microscopic analysis revealed a higher level of neurosecretory hormones in the posterior pituitary. Neurohypophysial axon swellings were filled with secretory dense core vesicles (Figure 23.4(C),(D)) which could be stained with the anti-OT antibody (Figure 23.4(E)). In CD38$^{+/+}$ mice the endings at the neurovascular contact zone contained many microvesicles and vacuoles, but were generally devoid of dense core vesicles (Figure 23.4(C)), evidence of strong secretory activity (Hatton, 1990). In CD38$^{-/-}$ mice, dense core vesicles occupied axon swellings and endings as well (Figure 23.4(D)), indicating overstorage. Thus, our results suggest that OT release, but not the synthesis and vesicular packaging of the hormone, seems to be selectively and severely impaired in CD38$^{-/-}$ mice. Interestingly, AVP secretion is not significantly altered in CD38 mutants.

RESCUE BY INJECTION OF OT OR CD38 RE-EXPRESSION

If low levels of plasma OT are responsible for the abnormal behavior observed in CD38$^{-/-}$

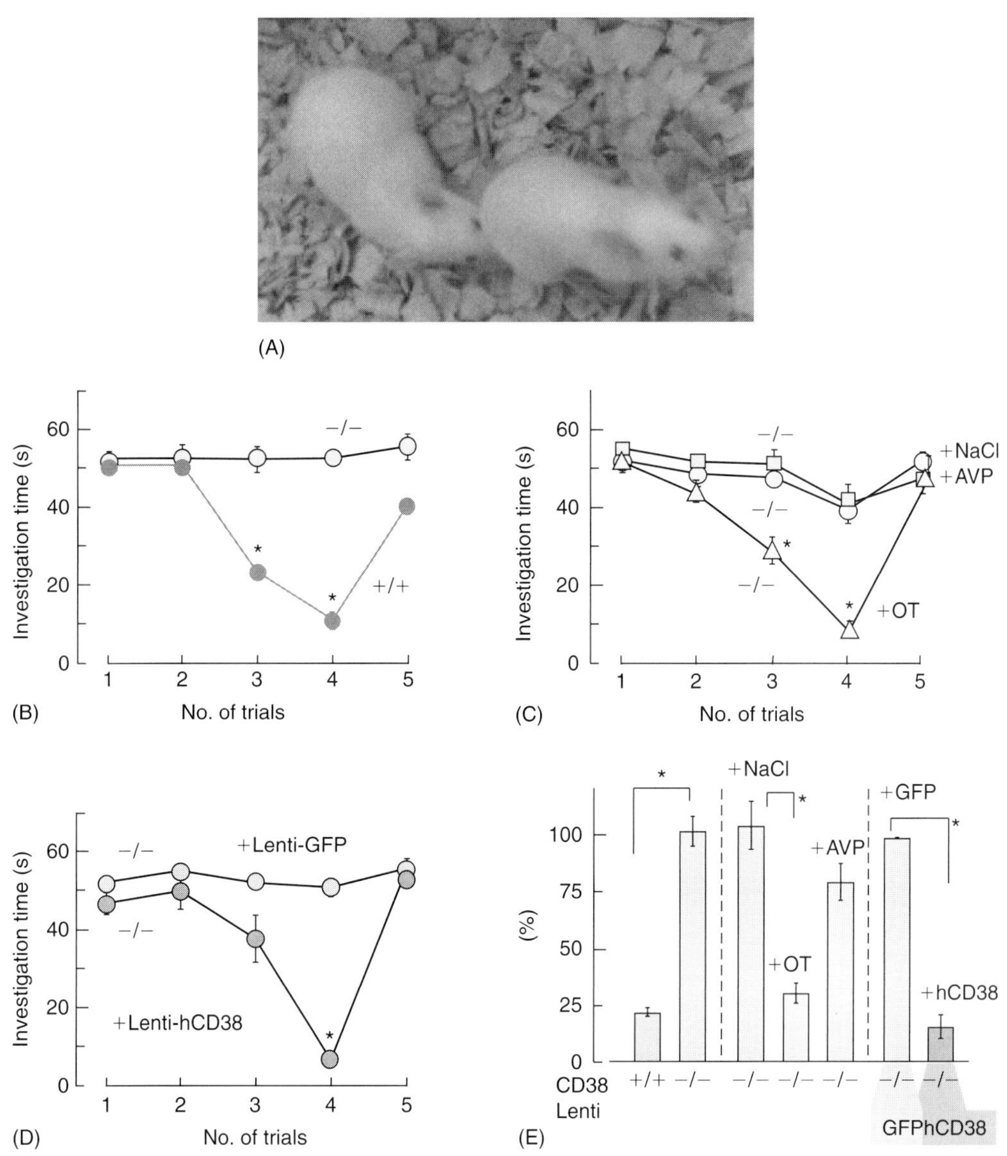

FIGURE 23.2 *Social behavior deficit in CD38$^{-/-}$ male mice.* (A) Olfactory investigation in CD38$^{-/-}$ mice used for social recognition test. Social memory by male mice was measured as a difference in ano-genital investigation. (B) Investigation time allocated to investigation of CD38$^{-/-}$ (light-dark circle) and CD38$^{+/+}$ (filled circle) male mice. *A significant decrease between each trial compared with the first trial, at $p < 0.01$. (C) Olfactory investigation of CD38$^{-/-}$ males as in (B). A single subcutaneous shot of 0.3 mL saline (NaCl, circle), OT (10 ng/kg; triangle), or vasopressin (AVP; 10 ng/kg; square) to CD38$^{-/-}$ males was administered 10 min before the first pairing. *A significant decrease between each trial compared with the first trial, at $p < 0.01$. (D) The social recognition deficit and its rescue by lenti-GFP (light-dark) and lenti-hCD38 (filled) infection in male CD38$^{-/-}$ mice. *$p < 0.01$ compared to controls. $N = 6$–10 mice in each experiment. (See Color Plate)

mice, it is expected that the replacement therapy by OT should result in the rescue of social behaviors. Since a single subcutaneous administration of OT affects memory storage within 10 min in mice (Boccia *et al.*, 1998), we used this same protocol, after having confirmed that OT in the CSF was elevated after subcutaneous injection of OT in both types of mice

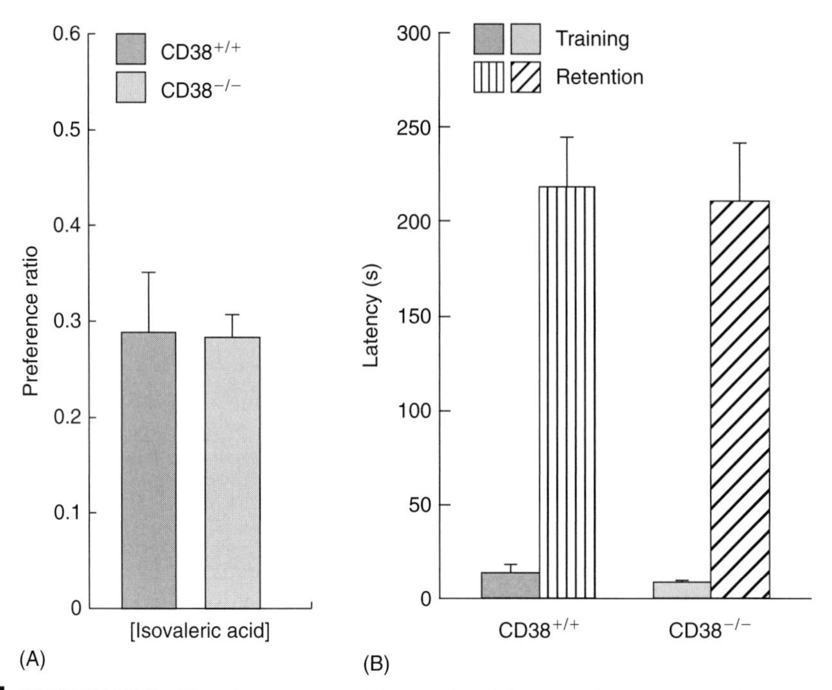

FIGURE 23.3 *The olfactory sense and the step-through latencies of both genotypes.* (A) The olfactory sense of male mice was assayed in an aversive conditioning paradigm in which the mice learn to avoid a distinctive odor of 10 μM isovaleric acid. $CD38^{-/-}$ mice were indistinguishable from wild-type controls in their ability to detect and avoid isovaleric acid (N = 8 or 18), showing that $CD38^{-/-}$ mice are not anosmic. (B) Learning and memory in male mice were measured as step-through latencies in passive avoidance tasks. The step-through latencies in $CD38^{-/-}$ mice were comparable to those in wild-type mice. Each value represents the mean ± SEM (N = 19 for $CD38^{+/+}$ mice and N = 12 for $CD38^{-/-}$ mice).

(Figure 23.5(C),(D)). Acute subcutaneous OT injections substantially rescued the maternal nurturing behaviors of females, measured by the latency to retrieve pups and the duration of crouching over the pups (Figure 23.1(C),(D)). The identical treatment also prevented impaired social recognition in $CD38^{-/-}$ male mice (Figure 23.2(C)). In contrast, AVP had no significant effects in $CD38^{-/-}$ mice under the same conditions (Figure 23.2(C)). Not surprisingly, OT administration had no significant effects on these behavioral endpoints in wild-type mice (data not shown).

A significant improvement following a single injection of OT confirmed that a deficit in OT release is responsible for the social behavior deficits in CD38 mutants. To test this further, we examined the link between the decrease in plasma OT and loss of CD38 expression in hypothalamic neurons by locally inserting the human CD38 gene in the hypothalamic neuronal population. $CD38^{-/-}$ mice received an infusion of lentiviral vectors carrying human CD38 (lenti-hCD38) or green fluorescence protein (GFP) (lenti-GFP) genes into the third ventricle, targeting the hypothalamo–pituitary region (Jin *et al.*, 2007). Two weeks after the injection, mice receiving lenti-hCD38 showed re-expression of human CD38 mRNA and human CD38 immunoreactivity within the hypothalamus and posterior pituitary (Jin *et al.*, 2007). Mice receiving lenti-GFP did not exhibit any human-specific CD38 immunoreactivity.

The low plasma and CSF levels of OT and the high tissue concentrations of OT were reversed in lenti-hCD38-injected $CD38^{-/-}$ mice (Figures 23.4(A),(B) and Figure 23.5(A)), while no comparable change was observed in mice receiving lenti-GFP. Plasma (Figure 23.4(A)) and tissue AVP levels (data not shown) in lenti-hCD38-injected $CD38^{-/-}$ mice were not significantly altered. Further, social memory in $CD38^{-/-}$ male mice receiving lenti-hCD38, but not lenti-GFP, showed the characteristic decline

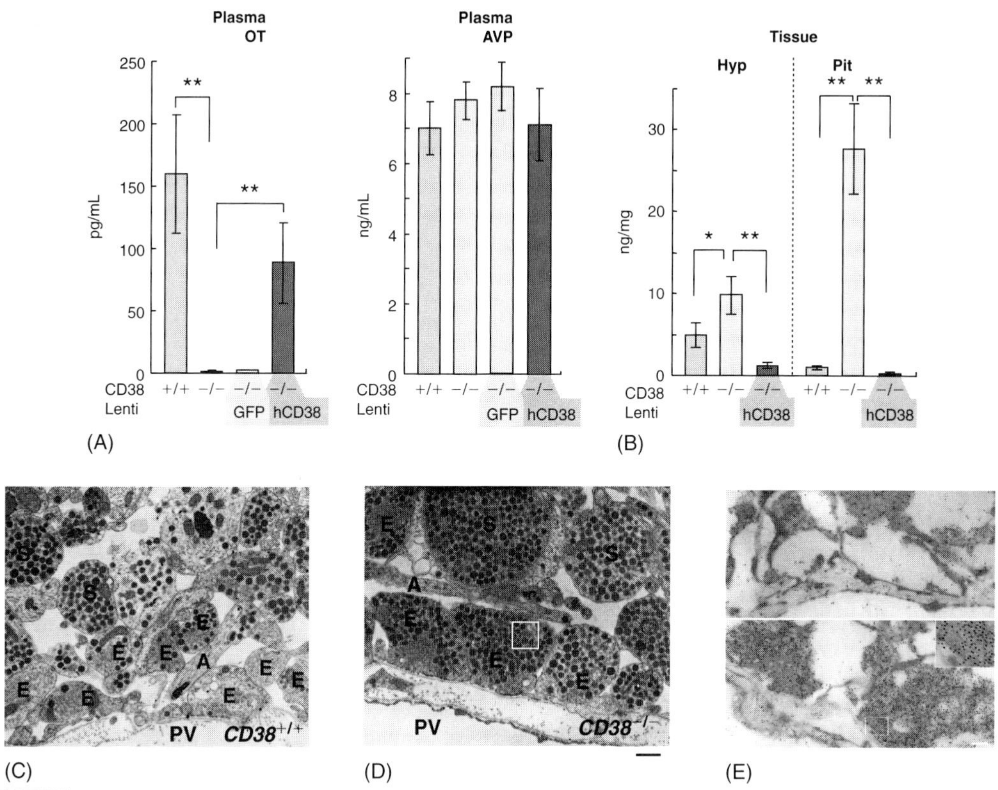

FIGURE 23.4 *Oxytocin and vasopressin concentrations.* Levels of plasma OT or AVP (A) and tissue OT (B) concentrations in $CD38^{+/+}$ and $CD38^{-/-}$ mice. OT levels in $CD38^{-/-}$ mice injected with lenti-GFP or lenti-hCD38. $*p < 0.05$; $**p < 0.001$; $N = 5$–8. (C, D) Electron micrographs of the posterior pituitary gland in $CD38^{+/+}$ (C) and $CD38^{-/-}$ (D) mice. Axon, A, and axonal endings, E, near the perivascular space, PV are empty (C) or full (D) of dense core vesicles. S indicates dense core vesicle-filled neurosecretory axonal swellings (Herring's body). Bar indicates 500 nm (×8,000). (E) Immunogold localization of OT in the posterior pituitary of $CD38^{+/+}$ (*upper*) or $CD38^{-/-}$ (*lower*) mice. The neurohypophysis was processed for immunoelectron microscopy using OT antiserum. OT was localized in axon terminals near the perivascular region to a lesser extent in $CD38^{+/+}$ mice. OT immunogold particles, as shown in the inset, were localized throughout terminals and swellings (not illustrated in this montage) in $CD38^{-/-}$ mice. Bar = 200 nm. Hyp, hypothalamus; Pit, pituitary.

of $CD38^{+/+}$ mice in the time spent investigating a female over trials, with full recovery following the introduction of a new female (Figure 23.2(D)). The recovery rate following lenti-hCD38 re-expression was comparable to that observed after OT administration in $CD38^{-/-}$ mice and in wild-type mice.

CD38 EXPRESSION AND ENZYME ACTIVITY

Along with the loss of CD38 expression, brain tissue from $CD38^{-/-}$ mice shows a reduction of ADP-ribosyl cyclase activity as measured by both cADPR (Kato *et al.*, 1999) and NAADP (Chini *et al.*, 2002) production. We reconfirmed this phenotype by immunohistochemical analysis of hypothalamic periventricular regions and by other measurements (Figure 23.6). CD38 immunoreactivity in wild-type mice was present as punctate dots on or mostly outside OT-ergic and/or AVP-ergic neurons, while staining in $CD38^{-/-}$ mice appeared to be at background levels (Figure 23.6(A),(B)). Surprisingly, wild-type mice showed the highest CD38 mRNA levels in the hypothalamic region among brain regions with low levels in the posterior pituitary gland (Figure 23.6(C)),

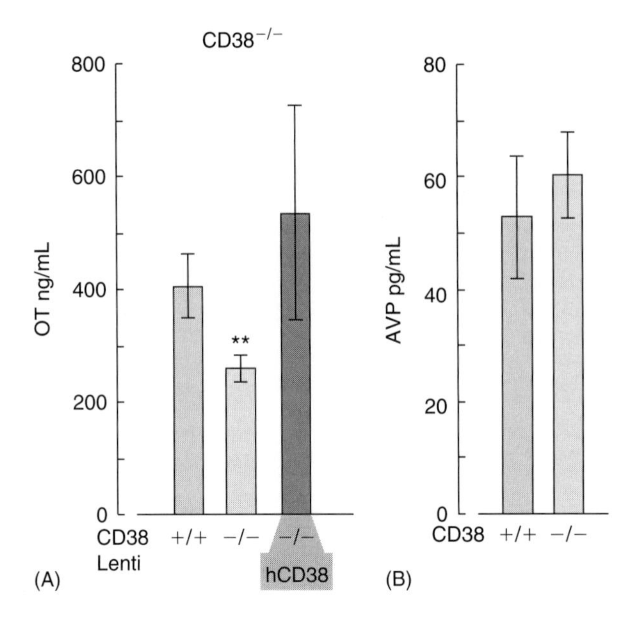

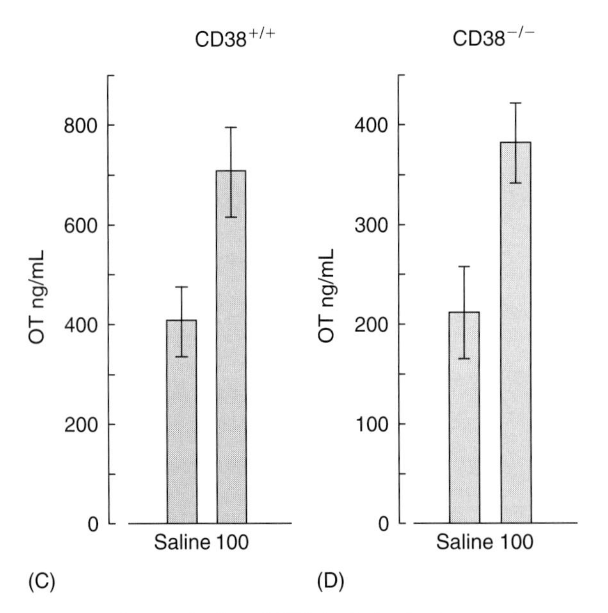

FIGURE 23.5 *Cerebrospinal oxytocin or vasopressin levels.* (A) Cerebrospinal (CSF) OT was significantly higher in the control ($CD38^{+/+}$) vs. knockout ($CD38^{-/-}$) mice. The CSF OT level was rescued in the $CD38^{-/-}$ mice re-expressing hCD38 by injection of a lentiviral vector into the third ventricle ($N = 6$). $^{**}p < 0.01$, from both values ($N = 3$). (B) AVP concentrations in CSF. There were no differences in the AVP concentration between $CD38^{+/+}$ and $CD38^{-/-}$ mice ($N = 3$). (C, D) CSF OT levels measured 10 min after subcutaneous injection of OT at the concentration of 100 ng/mL or saline (0.2 mL) in $CD38^{+/+}$ mice (C) or $CD38^{-/-}$ mice (D). $N = 3$–5. $^{**}p < 0.01$ from saline controls.

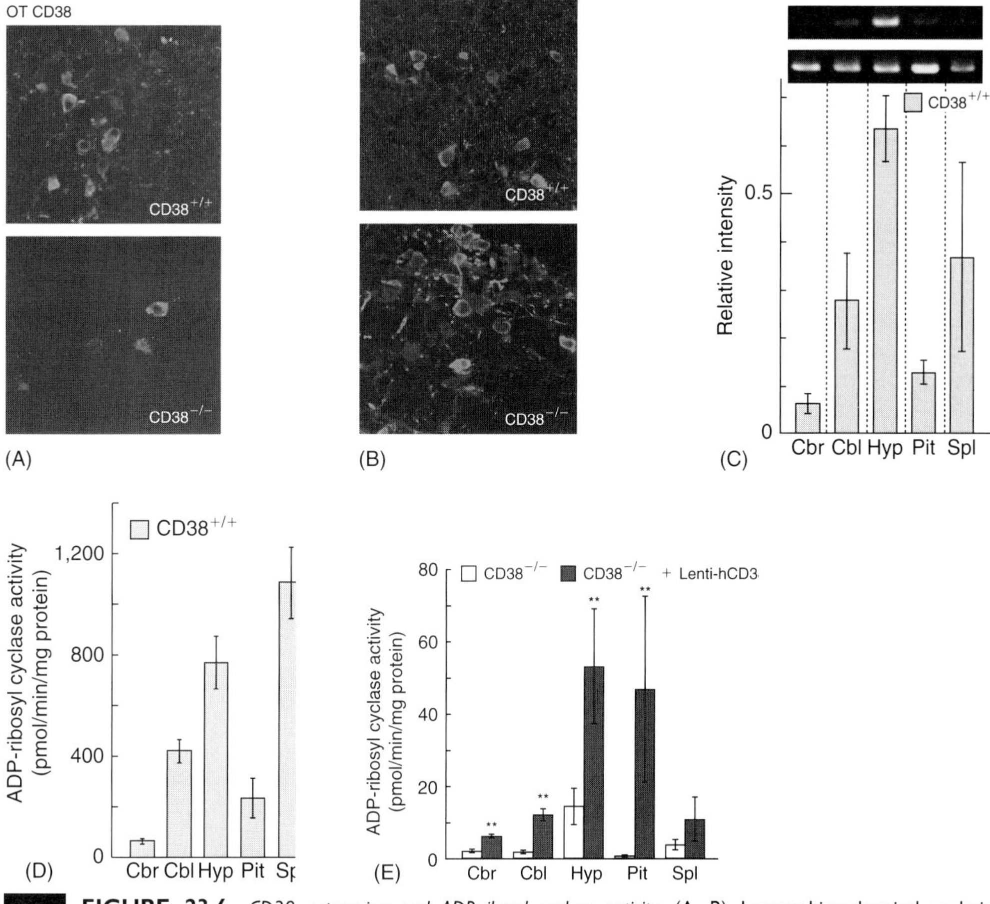

FIGURE 23.6 *CD38 expression and ADP-ribosyl cyclase activity.* (A, B) Immunohistochemical analysis of mouse CD38 of hypothalamus in CD38$^{+/+}$ (*upper*) or CD38$^{-/-}$ (*lower*) mice. CD38 stained with anti-mouse CD38 monoclonal antibody and anti-murine OT (A) or AVP (B) antibodies (magnification = $\times$ 630). (C) RT-PCR of CD38 (*top*) and TATA binding protein (*middle*) of isolated from indicated tissues in CD38$^{+/+}$ mice and mean relative intensities of the two (bars); $N = 3$. (D) ADP-ribosyl cyclase activity, measured as the rate of cyclic GDP-ribose formation by whole cell homogenates isolated from various tissues of CD38$^{+/+}$ mice; $N = 4$–12. (E) Cyclic GDP-ribose formation by whole cell homogenates isolated from various tissues of CD38$^{-/-}$ mice or CD38$^{-/-}$ mice 2 weeks after receiving lenti-hCD38; $N = 3$–8. Cbr, cerebrum; Cbl, cerebellum; Hyp, hypothalamus; Pit, posterior pituitary; Spl, spleen. $*p < 0.05$; $**p < 0.01$. (See Color Plate)

suggesting that CD38 is mainly produced in the hypothalamus. This was accompanied by high levels of ADP-ribosyl cyclase activity in brain hypothalamic homogenates (Figure 23.6(D)), as measured by the production of cyclic guanosine dinucleotide phosphate ribose (cGDPR) from nicotinamide guanine dinucleotide (NGD$^+$) (the most sensitive assay of CD38 (Ceni *et al.*, 2006)). CD38$^{-/-}$ mice had very low ADP-ribosyl cyclase activity in both the hypothalamus and neurohypophysis (Figure 23.6(E)). Of course, the reduced ADP-ribosyl cyclase activity in CD38$^{-/-}$ mice was partly restored following lentivector-mediated re-expression of human CD38.

OT SECRETION AND CA^{2+} TRANSIENTS *IN VITRO*

High (50 mM) potassium-induced depolarization produced more than 2-fold increase in OT secretion from isolated hypothalamic neurons

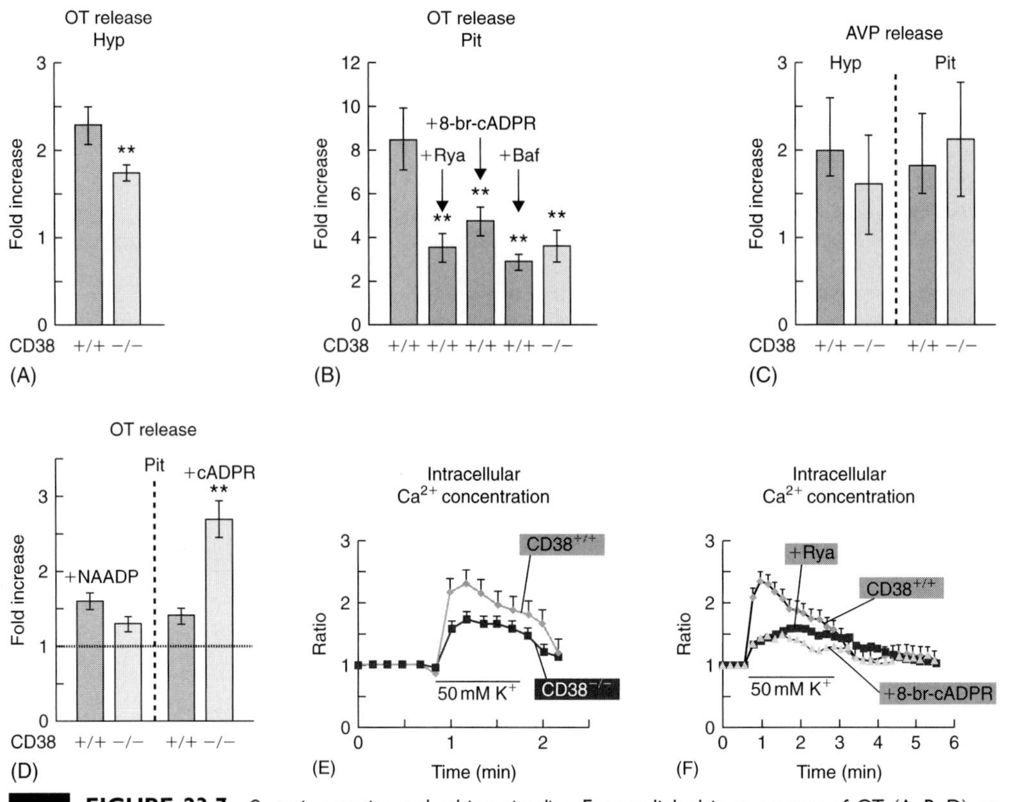

FIGURE 23.7 *Oxytocin secretion and calcium signaling.* Enzyme-linked immunoassay of OT (A, B, D) or AVP (C) secreted from hypothalamic cells (Hyp) or nerve endings (Pit) in 35 mm culture dishes stimulated by 70 mM KCl for 2 min. Hypothalamic neurons and nerve endings were acutely dissociated from adult $CD38^{+/+}$ or $CD38^{-/-}$ mice, cultured for 6 h and perfused with Locke solution for at least 45 min. The OT release was measured in the presence or absence of 200 μM ryanodine (Rya), 100 μM 8-bromo-cADPR (8-br-cADPR), or 2 μM bafilomycin (Baf). Values represent the ratio of the concentrations in the perfusate after 70 mM KCl to that in control (5 mM KCl) solution; $N = 4$–12. $^{*}p < 0.05$ and $^{**}p < 0.01$, respectively, compared to control $CD38^{+/+}$. (D) OT release from nerve endings isolated from posterior pituitary of $CD38^{+/+}$ or $CD38^{-/-}$ mice in the presence of 100 nM NAADP or 10 μM cADPR after treatment with 1 μM digitonin to permeabilize the cell membranes for 5 min. Values represent the ratio of OT release in the presence and absence of the effectors; $N = 4$–6. (E) Time course of the increase in $[Ca^{2+}]_i$ stimulated with 50 mM KCl for the indicated period in OT-ergic nerve endings isolated from $CD38^{+/+}$ or $CD38^{-/-}$ mice posterior pituitary; $N = 4$. (F) Time course of the increase in $[Ca^{2+}]_i$ stimulated with 50 mM KCl for the indicated period in the presence or absence of 200 μM ryanodine or 100 μM 8-bromo-cADPR in OT-ergic nerve endings isolated from $CD38^{+/+}$ mice; $N = 3$–4. (See Color Plate)

in $CD38^{+/+}$ mice (Figure 23.7(A)) and an 8-fold increase in secretion from their axon terminals in the posterior pituitary gland (Figure 23.7(B)). In $CD38^{-/-}$ mice secretion was reduced from both tissues, particularly in the posterior pituitary, while there was no reduction in AVP release from either brain region (Figure 23.7(C)). OT release from pituitary nerve endings was reduced to levels found in $CD38^{-/-}$ mice after treatment with a variety of antagonists: 200 μM ryanodine (a full antagonist of ryanodine receptors at this high dose), 100 μM 8-bromo-cADPR (a cADPR antagonist at the ryanodine receptor binding sites) and 2 μM bafilomycin (a NAADP antagonist) (Figure 23.7(B); Gerasimenko *et al.*, 2006). Furthermore, OT release was enhanced by extracellularly applied cADPR (10 μM) and NAADP (100 nM) from the permeabilized nerve endings of both types of mice already primed by Ca^{2+}. In particular, cADPR showed a 2.5-fold enhancement of OT release in the $CD38^{-/-}$ preparations (Figure 23.7(D)). These findings demonstrate that these metabolites of ADP-ribosyl cyclase are essential for OT secretion and indicate that the deficit in OT secretion and reduced plasma OT levels in $CD38^{-/-}$ mice might reasonably be attributed to the reduced ADP-ribosyl cyclase activity. However, this does not reflect a general defect in neurosecretion, since AVP secretion was unaffected.

OT secretion in hypothalamic cells is triggered by increased intracellular calcium concentrations ($[Ca^{2+}]_i$) due to the Ca^{2+} influx through voltage-dependent Ca^{2+} channels associated with repetitive action potentials (Belin & Moos, 1986; OuYang *et al.*, 2004). The participation of the thapsigargin-sensitive IP_3-mediated $[Ca^{2+}]_i$ increase in the soma and dendrites and miniature Ca^{2+} release from ryanodine receptors (Ca^{2+} syntillas) in the terminals of hypothalamic neurons were previously shown (De Crescenzo *et al.*, 2004; Sasaki *et al.*, 2005; Ludwig & Leng, 2006). Here, we investigated transient $[Ca^{2+}]_i$ increases in isolated nerve endings with the Ca^{2+}-sensing dye, Oregon Green (De Crescenzo *et al.*, 2004; Sasaki *et al.*, 2005). Incubation with 50 mM KCl elevated $[Ca^{2+}]_i$ to 220% of the pre-stimulation level in $CD38^{+/+}$ but to 160% of that in $CD38^{-/-}$ mice (Figure 23.7(E)). The $[Ca^{2+}]_i$ increases in $CD38^{+/+}$ mice was markedly inhibited in the presence of ryanodine or 8-bromo-cADPR (Figure 23.7(F)).

CD38'S ROLE IN OT-MEDIATED SOCIAL RECOGNITION AND MATERNAL BEHAVIOR

In the present study we have demonstrated that social and maternal behaviors are significantly impaired in $CD38^{-/-}$ mice as compared with wild-type controls. The CD38 mutant showed exploratory and investigative behavior toward pups and females, yet seem unable to effectively process pup- and female-induced cues that normally elicit nurturing and social recognition responses (Insel & Fernald, 2004). Social recognition is a unique form of learning and memory that utilizes distinct neural mechanisms that are specific to social cognition (van Wimersma Greidanus & Maigret, 1996; LaBar & Cabeza, 2006). In the current paradigm the effects normally last for an hour or more. These actions seem geared more toward the differentiation of conspecifics over short interactions or durations, and involve the ability to distinguish between two similar, but different, olfactory signatures (Ferguson *et al.*, 2000), comparable to long-term memories in pair bonding in monogamous voles (Insel & Young, 2001; Young & Wang, 2004). Thus, our results for the first time demonstrate a genetic involvement of CD38 in the detection of short-lasting social memory.

In CD38 knockout mice, the plasma OT level was significantly decreased. Replacement of OT, but not AVP, by subcutaneous administration and CD38 gene re-expression in the brain rescued maternal and social recognition behaviors in $CD38^{-/-}$ mice, indicating that the "social brain system" (Insel & Fernald, 2004; Lim *et al.*, 2005) is intact in $CD38^{-/-}$ mice. Since it has been reported that such behaviors are dependent on signaling of either OT or AVP (Ferguson *et al.*, 2000; Keverne & Curley, 2004; Bielsky *et al.*, 2005), our data strongly suggest that the diminished secretion of OT in the $CD38^{-/-}$ mice is the primary cause of the behavioral defects (Figure 23.8). As shown in the histological analysis, the peptides are produced in the hypothalamus and stored in the axon terminals of the neurons in the posterior pituitary. In the *in vitro* experiments, high potassium stimulation-induced secretion of OT and the $[Ca^{2+}]_i$ elevation were significantly attenuated by ryanodine in axon terminals isolated from $CD38^{+/+}$ mice. Thus, attenuation of the intracellular ryanodine-susceptible Ca^{2+} amplification that is necessary for stimulation-secretion coupling for the OT nerve terminals may be the primary cause of the behavioral abnormality (Figures 23.8 and 23.9).

Re-expression of human CD38 by lentiviral vectors in the current $CD38^{-/-}$ mouse studies clearly resulted in a significant recovery in social recognition similar to that seen in OT administration. The ventricular injection of the viral vector (Torashima *et al.*, 2006) may not introduce a single site-specific expression of a single gene. However, judging from ADP-ribosyl cyclase activity found in $CD38^{-/-}$ with lenti-hCD38 mice, CD38 was undoubtedly re-expressed mostly in the hypothalamus and pituitary, and insignificantly in other brain regions. The degree of re-expression in the hypothalamo–hypophysis is probably only partial, since CD38 was not re-expressed in all hypothalamic regions, but most efficiently at

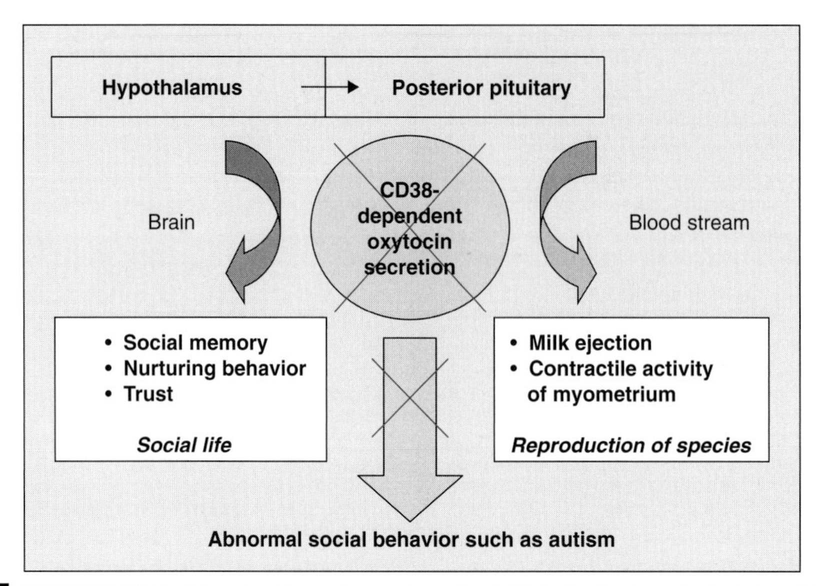

FIGURE 23.8 *Disruption of oxytocin secretion and social behavior.* A schematic depiction of CD38-dependent secretion of OT into the brain and blood stream and its effects on behavioral and reproductive functions in humans. The working hypothesis is that the disruption of CD38-dependent release of OT leads to the abnormal social behavior resembling that in autistic spectrum disorders.

the surface of the hypothalamus close to the ventricular space. Notwithstanding this shortcoming, the current CD38 viral vector provides the first direct evidence for the critical role of CD38 in OT release in the hypothalamus and posterior pituitary.

A significant observation here is that the AVP release is unaffected by the CD38 null mutation. OT and AVP neurons are relatively distinct from each other (Yamashita *et al.*, 2002), although they are both synthetized in the supraoptic and paraventricular nuclei. Thus, only the secretion of OT appears to involve this type of release of Ca^{2+} from ryanodine-sensitive stores (Figure 23.9). The simplest explanation for this is a difference in the targets for the enzymatic products of CD38, cADPR, and NAADP (Lee, 2001, 2005; Yamasaki *et al.*, 2004; Gerasimenko *et al.*, 2006). Hence, even though CD38 is widely distributed throughout the hypothalamus and is not restricted to OT neurons (and hence may exert a paracrine action through extracellular synthesis and cellular uptake of cADPR and NAADP (De Flora *et al.*, 2004)), these will only affect secretion of OT, not AVP.

The finding that CD38 gene disruption results in significantly lower OT secretion suggests that CD38 may have a major role in OT neurons through these metabolites. The results of this study strongly suggest the involvement of cADPR- and NAADP-ryanodine receptor-mediated Ca^{2+} mobilization from intracellular stores on the mechanisms of OT release from neurohypophysial axon endings (Figure 23.9), though the exact mechanism of AVP release remains to be solved. Recently a new mechanism involving acidic lysosomes or granules (Yamasaki *et al.*, 2004; Gerasimenko *et al.*, 2006), which are abundant in the neural lobes of the pituitary (Hatton, 1990), has been proposed as the site of action of NAADP, which subsequently acts as a Ca^{2+}-induced Ca^{2+}-release mediator to prolonged Ca^{2+} transients. However, other possibilities, such as an unknown interaction of CD38 with the neurosecretory machinery, or a channel effect of CD38 to facilitate release of neuropeptides, as shown for CD38's nucleotide transport (De Flora *et al.*, 2004), cannot be excluded.

It is quite interesting that mice lacking the OT or AVP genes (Ferguson *et al.*, 2000; Insel & Young, 2001; Keverne & Curley, 2004; Young & Wang, 2004), OT or AVP receptor genes (Bielsky *et al.*, 2005; Takayanagi *et al.*, 2005), and CD38 gene (Jin *et al.*, 2007) show similar and profound behavioral anomalies that collectively resemble those of human autism, a pervasive neuropsychiatric disorder with marked defects in social interaction.

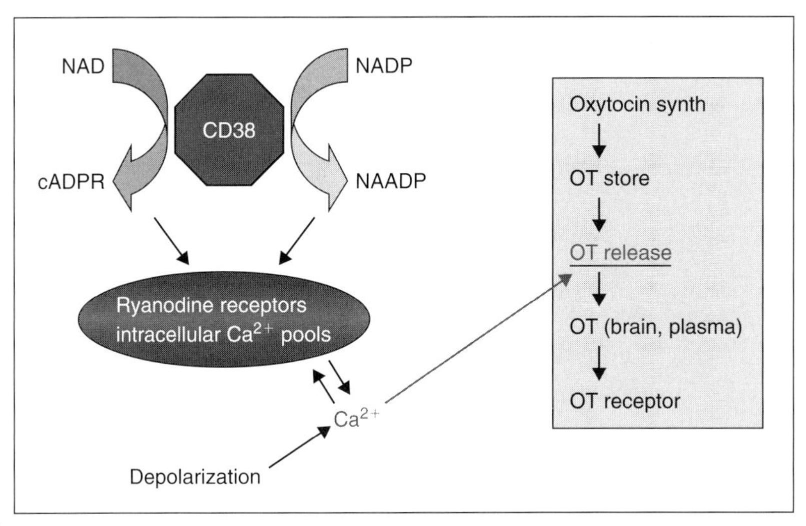

FIGURE 23.9 *Oxytocin secretion due to CD38- and ryanodine-sensitive-pool-dependent calcium signal amplification.* The scheme presents the hypothesis indicating that OT release in the OT system from synthesis to binding to receptors is regulated by depolarization-induced Ca^{2+} influx and cADPR- and NAADP-induced Ca^{2+} mobilization from ryanodine-sensitive Ca^{2+} pools.

Our (and others') results together demonstrate that the behavioral impairment is achieved by disruption of either one of two OT and AVP systems (Kendrick *et al.*, 1997; Keverne & Curley, 2004; Jin *et al.*, 2007). This suggests that one signal which is intact in a given knockout mice is insufficient to produce normal behavior, but rather both signals are required for such behaviors. Thus, we can speculate that CD38-dependent loss of OT function might have an important implication for behavior disruptions (Baron-Cohen & Bellmonte, 2005). It would be interesting to test whether disease-causing variants in the coding or non-coding sequence of human CD38 can be detected in the spectrum of autistic patients, comparable to the CD38 mutations found in a subset of diabetes patients (Yagui *et al.*, 1998); such studies have not yet been done. Furthermore, enhancement of OT levels due to the CD38 expression could potentially be of therapeutic benefit, given that plasma OT levels are decreased in autistic children (Modahl *et al.*, 1998) and that OT infusion induces reduction of repetitive behaviors in adult patients with autistic and Asperger's disorders (Hollander *et al.*, 2003). Similarly, the role of CD38 in postpartum mood disorders also merits consideration based on OT's important role in parental care.

REFERENCES

Argiolas, A., and Melis, M. R. (2005). Central control of penile erection: Role of the paraventricular nucleus of the hypothalamus. *Prog. Neurobiol.* 76, 1–21.

Baron-Cohen, S., and Belmonte, M. K. (2005). Autism: A window onto the development of the social and the analytic brain. *Annu. Rev. Neurosci.* 28, 109–126.

Bartels, A., and Zeki, S. (2004). The neural correlates of maternal and romantic love. *Neuroimage* 21, 1155–1166.

Belin, V., and Moos, F. (1986). Paired recordings from supraoptic and paraventricular oxytocin cells in suckled rats: Recruitment and synchronization. *J. Physiol.* 377, 369–390.

Bielsky, I. F., Hu, S. B., Ren, X., Terwilliger, E. F., and Young, L. J. (2005). The V1a vasopressin receptor is necessary and sufficient for normal social recognition: A gene replacement study. *Neuron* 47, 503–513.

Boccia, M. M., Kopf, S. R., and Baratti, C. M. (1998). Effects of a single administration of oxytocin or vasopressin and their interactions with two selective receptor antagonists on memory storage in mice. *Neurobiol. Learn. Mem.* 69, 136–146.

Ceni, C., Pochon, N., Villaz, M., Muller-Steffner, H., Schuber, F., Baratier, J., De Waard, M., Ronjat, M., and Moutin, M. J. (2006). The CD38-independent

ADP-ribosyl cyclase from mouse brain synaptosomes: A comparative study of neonate and adult brain. *Biochem. J.* 395, 417–426.

Chini, E. N., Chini, C. C., Kato, I., Takasawa, S., and Okamoto, H. (2002). CD38 is the major enzyme responsible for synthesis of nicotinic acid-adenine dinucleotide phosphate in mammalian tissues. *Biochem. J.* 362, 125–130.

De Crescenzo, V., ZhuGe, R., Velazquez-Marrero, C., Lifshitz, L. M., Custer, E., Carmichael, J., Lai, F. A., Tuft, R. A., Fogarty, K. E., Lemos, J. R., and Walsh, J. V., Jr. (2004). Ca^{2+} syntillas, miniature Ca^{2+} release events in terminals of hypothalamic neurons, are increased in frequency by depolarization in the absence of Ca^{2+} influx. *J. Neurosci.* 24, 1226–1235.

De Flora, A., Zocchi, E., Guida, L., Franco, L., and Bruzzone, S. (2004). Autocrine and paracrine calcium signaling by the CD38/NAD^+/cyclic ADP-ribose system. *Ann. NY Acad. Sci.* 1028, 176–191.

Deaglio, S., Vaisitti, T., Aydin, S., Ferrero, E., and Malavasi, F. (2006). In-tandem insight from basic science combined with clinical research: CD38 as both marker and key component of the pathogenetic network underlying chronic lymphocytic leukemia. *Blood* 108, 1135–1144.

Ferguson, J. N., Young, L. J., Hearn, E. F., Matzuk, M. M., Insel, T. R., Winslow, J. T. (2000). Social amnesia in mice lacking the oxytocin gene. *Nat. Genet.* 25, 284–288.

Fries, A. B., Ziegler, T. E., Kurian, J. R., Jacoris, S., and Pollak, S. D. (2005). Early experience in humans is associated with changes in neuropeptides critical for regulating social behavior. *Proc. Natl Acad. Sci. USA* 102, 17237–17240.

Gerasimenko, J. V., Sherwood, M., Tepikin, A. V., Petersen, O. H., and Gerasimenko, O. V. (2006). NAADP, cADPR and IP_3 all release Ca^{2+} from the endoplasmic reticulum and an acidic store in the secretory granule area. *J. Cell Sci.* 119, 226–238.

Hatton, G. I. (1990). Emerging concepts of structure–function dynamics in adult brain: The hypothalamo–neurohypophysial system. *Prog. Neurobiol.* 34, 437–504.

Higashida, H., Hashii, M., Yokoyama, S., Hoshi, N., Chen, X. L., Egorova, A., Noda, M., and Zhang, J. S. (2001). Cyclic ADP-ribose as a second messenger revisited from a new aspect of signal transduction from receptors to ADP-ribosyl cyclase. *Pharmacol. Ther.* 90, 283–296.

Hollander, E., Novotny, S., Hanratty, M., Yaffe, R., DeCaria, C. M., Aronowitz, B. R., and Mosovich, S. (2003). Oxytocin infusion reduces repetitive behaviors in adults with autistic and Asperger's disorders. *Neuropsychopharmacology* 28, 193–198.

Howard, M., Grimaldi, J. C., Bazan, J. F., Lund, F. E., Santos-Argumedo, L., Parkhouse, R. M., Walseth, T. F., and Lee, H. C. (1993). Formation and hydrolysis of cyclic ADP-ribose catalyzed by lymphocyte antigen CD38. *Science* 262, 1056–1059.

Hunt, P. W., Deeks, S. G., Bangsberg, D. R., Moss, A., Sinclair, E., Liegler, T., Bates, M., Tsao, G., Lampiris, H., Hoh, R., and Martin, J. N. (2006). The independent effect of drug resistance on T cell activation in HIV infection. *AIDS* 20, 691–699.

Insel, T. R., and Fernald, R. D. (2004). How the brain processes social information: Searching for the social brain. *Annu. Rev. Neurosci.* 27, 697–722.

Insel, T. R., and Young, L. J. (2001). The neurobiology of attachment. *Nat. Rev. Neurosci.* 2, 129–136.

Jin, D., Liu, H. X., Hirai, H., Torashima, T., Nagai, T., Lopatina, O., Shnayder, N. A., Yamada, K., Noda, M., Seike, T., Fujita, K., Takasawa, S., Yokoyama, S., Koizumi, K., Shiraishi, Y., Tanaka, S., Hashii, M., Yoshihara, T., Higashida, K., Islam, M. S., Yamada, N., Hayashi, K., Noguchi, N., Kato, I., Okamoto, H., Matsushima, A., Salmina, A., Munesue, T., Shimizu, N., Mochida, S., Asano, M., and Higashida, H. (2007). CD38 is critical for social behavior by regulating oxytocin secretion. *Nature* 446, 41–45.

Johnson, J. D., and Misler, S. (2002). Nicotinic acid-adenine dinucleotide phosphate-sensitive calcium stores initiate insulin signaling in human beta cells. *Proc. Natl Acad. Sci. USA* 99, 14566–14571.

Kato, I., Yamamoto, Y., Fujimura, M., Noguchi, N., Takasawa, S., and Okamoto, H. (1999). CD38 disruption impairs glucose-induced increases in cyclic ADP-ribose, $[Ca^{2+}]_i$, and insulin secretion. *J. Biol. Chem.* 274, 1869–1872.

Kendrick, K. M., Da Costa, A. P., Broad, K. D., Ohkura, S., Guevara, R., Levy, F., and Keverne, E. B. (1997). Neural control of maternal behavior and olfactory recognition of offspring. *Brain Res. Bull.* 44, 383–395.

Keverne, E. B., and Curley, J. P. (2004). Vasopressin, oxytocin and social behavior. *Curr. Opin. Neurobiol.* 14, 777–783.

Kirsch, P., Esslinger, C., Chen, Q., Mier, D., Lis, S., Siddhanti, S., Gruppe, H., Mattay, V. S., Gallhofer, B., and Meyer-Lindenberg, A. (2005). Oxytocin modulates neural circuitry for social cognition and fear in humans. *J. Neurosci.* 25, 11489–11493.

Kosfeld, M., Heinrichs, M., Zak, P. J., Fischbacher, U., and Fehr, E. (2005). Oxytocin increases trust in humans. *Nature* 435, 673–676.

LaBar, K. S., and Cabeza, R. (2006). Cognitive neuroscience of emotional memory. *Nat. Rev. Neurosci.* 7, 54–64.

Lee, H. C. (2001). Physiological functions of cyclic ADP-ribose and NAADP as calcium messengers. *Annu. Rev. Pharmacol. Toxicol.* 41, 317–345.

Lee, H. C. (2005). Nicotinic acid adenine dinucleotide phosphate (NAADP)-mediated calcium signaling. *J. Biol. Chem.* 280, 33693–33696.

Lim, M. M., Bielsky, I. F., and Young, L. J. (2005). Neuropeptides and the social brain: Potential rodent models of autism. *Int. J. Dev. Neurosci.* 23, 235–243.

Ludwig, M., and Leng, G. (2006). Dendritic peptide release and peptide-dependent behaviors. *Nat. Rev. Neurosci.* 7, 126–136.

Modahl, C., Green, L., Fein, D., Morris, M., Waterhouse, L., Feinstein, C., and Levin, H. (1998). Plasma oxytocin levels in autistic children. *Biol. Psychiatr.* 15, 270–277.

Okamoto, H., and Takasawa, S. (2002). Recent advances in the Okamoto model: The CD38-cyclic ADP-ribose signal system and the regenerating gene protein (Reg)-Reg receptor system in beta-cells. *Diabetes* 51, S462–S473.

OuYang, W., Wang, G., and Hemmings, H. C., Jr. (2004). Distinct rat neurohypophysial nerve terminal populations identified by size, electrophysiological properties and neuropeptide content. *Brain Res.* 1024, 203–211.

Popik, P., Vetulani, J., and van Ree, J. M. (1992). Low doses of oxytocin facilitate social recognition in rats. *Psychopharmacology* 106, 71–74.

Russell, J. A., Leng, G., and Douglas, A. J. (2003). The magnocellular oxytocin system, the fount of maternity: Adaptations in pregnancy. *Front. Neuroendocrinol.* 24, 27–61.

Sasaki, N., Dayanithi, G., and Shibuya, I. (2005). Ca^{2+} clearance mechanisms in neurohypophysial terminals of the rat. *Cell Calcium* 37, 45–56.

Soares, S. M., Thompson, M., and Chini, E. N. (2005). Role of the second-messenger cyclic-adenosine 5′-diphosphate-ribose on adrenocorticotropin secretion from pituitary cells. *Endocrinology* 146, 2186–2192.

Takasawa, S., Tohgo, A., Noguchi, N., Koguma, T., Nata, K., Sugimoto, T., Yonekura, H., and Okamoto, H. (1993). Synthesis and hydrolysis of cyclic ADP-ribose by human leukocyte antigen CD38 and inhibition of the hydrolysis by ATP. *J. Biol. Chem.* 268, 26052–26054.

Takayanagi, Y., Yoshida, M., Bielsky, I. F., Ross, H. E., Kawamata, M., Onaka, T., Yanagisawa, T., Kimura, T., Matzuk, M. M., Young, L. J., and Nishimori, K. (2005). Pervasive social deficits, but normal parturition, in oxytocin receptor-deficient mice. *Proc. Natl Acad. Sci. USA* 102, 16096–16101.

Torashima, T., Okoyama, S., Nishizaki, T., and Hirai, H. (2006). *In vivo* transduction of murine cerebellar Purkinje cells by HIV-derived lentiviral vectors. *Brain Res.* 1082, 11–22.

van Wimersma Greidanus, T. B., and Maigret, C. (1996). The role of limbic vasopressin and oxytocin in social recognition. *Brain Res.* 713, 153–159.

Yagui, K., Shimada, F., Mimura, M., Hashimoto, N., Suzuki, Y., Tokuyama, Y., Nata, K., Tohgo, A., Ikehata, F., Takasawa, S., Okamoto, H., Makino, H., Saito, Y., and Kanatsuka, A. (1998). A missense mutation in the CD38 gene, a novel factor for insulin secretion: Association with Type II diabetes mellitus in Japanese subjects and evidence of abnormal function when expressed *in vitro*. *Diabetologia*, 41, 1024–1028.

Yamasaki, M., Masgrau, R., Morgan, A. J., Churchill, G. C., Patel, S., Ashcroft, S. J., and Galione, A. (2004). Organelle selection determines agonist-specific Ca^{2+} signals in pancreatic acinar and beta cells. *J. Biol. Chem.* 279, 7234–7240.

Yamashita, M., Glasgow, E., Zhang, B. J., Kusano, K., and Gainer, H. (2002). Identification of cell-specific messenger ribonucleic acids in oxytocinergic and vasopressinergic magnocellular neurons in rat supraoptic nucleus by single-cell differential hybridization. *Endocrinology* 143, 4464–4476.

Young, L. J., and Wang, Z. (2004). The neurobiology of pair bonding. *Nat. Neurosci.* 7, 1048–1054.

24 OXYTOCIN AND MOTHERS' DEVELOPMENTAL EFFECTS ON THEIR DAUGHTERS

CORT A. PEDERSEN[1] AND MARIA L. BOCCIA[2]

[1] *Department of Psychiatry, CB# 7160 The University of North Carolina at Chapel Hill, Chapel Hill, NC, USA.*
[2] *Frank Porter Graham Child Development Institute, CB# 8185 The University of North Carolina at Chapel Hill, Chapel Hill, NC, USA.*

Clinical observations and research in human and non-human primates show clearly that nurturing received during infancy and childhood influences the quality of parental and other social behaviors as well as the ability to cope with stress during adulthood (Parkes *et al.*, 1991; Reite & Boccia, 1994; Van Ijzendoorn, 1995; Kraemer, 1997; Cassidy & Shaver, 1999; George & Solomon, 1999; Suomi, 1999). Analogous effects of early nurturing on adult outcomes have been demonstrated in rodents in recent years. This chapter summarizes the numerous ways in which the neuropeptide, oxytocin (OT), contributes to mothers' developmental effects on their daughters.

INTRODUCTORY BACKGROUND AND CONCEPTS

Pup-licking (PL) and crouched nursing have been identified as the components of rat maternal behavior that exert the greatest influence on the development of offspring responses to acute stressors and some aspects of social behavior. This conclusion has emerged from decades of study of the developmental consequences of daily separation of pups from their mothers during the first several postnatal weeks. In early studies, rats that were subjected to daily brief maternal separations (BMS, often referred to as "handling") early in life were found to be less stress responsive, that is, exhibited lower anxiety and hypothalamic–pituitary–adrenal (HPA) axis activation when subjected to acute stressors such as a novel open field or immobilization (Levine, 1957, 1962; Denenberg, 1964; Levine *et al.*, 1967; Ader & Grota, 1969; Bodnoff *et al.*, 1987; Meaney *et al.*, 1996). Daily long (3–6 h) maternal separations (LMS) in the early postnatal period were subsequently shown to have the opposite developmental effects, elevated anxiety and HPA activation by acute stressors during adulthood (Plotsky & Meaney, 1993; Ogawa *et al.*, 1994; Ladd *et al.*, 1996). Although supporting evidence was published early on (Lee & Williams, 1974), only recently has it been firmly established that BMS and LMS have opposite effects on the maternal behavior of rat mothers undergoing repeated separation from their litters (Liu *et al.*, 1997; Caldji *et al.*, 1998, 2000; Boccia & Pedersen, 2001; Gonzalez *et al.*, 2001; Lovic *et al.*, 2001). BMS increased and LMS decreased maternal pup-grooming and arched-back nursing (PG–ABN). Adult offspring of unmanipulated mothers that displayed frequencies of PG–ABN above and below one standard deviation from the mean of the normal distribution of these maternal behaviors exhibited acute stress responses similar to animals subjected respectively to postnatal BMS and LMS (Liu *et al.*, 1997; Caldji *et al.*, 1998; Francis *et al.*, 1999). The distribution of PG–ABN frequencies was determined by making observations every 3 min over 72-min periods conducted 5 times/day (3 in the light phase, 2 in the dark phase) on postnatal days 1–8 (Champagne *et al.*, 2003). This large number of observations was necessary to clearly demonstrate the difference in mean PG–ABN frequencies between high and low mothers (approximately 40%). BMS and high PG–ABN in comparison to LMS and low PG–ABN have opposite effects on the development of neurochemical systems that regulate anxiety and HPA axis responses to acute stress such as benzodiazepine receptor concentrations and $GABA_A$ receptor complex $_{\gamma}2$ subunit gene expression in the amygdala and glucocorticoid receptor concentrations in the dorsal hippocampus. Mothers' PG–ABN frequencies also determine how much of these components of maternal behavior are exhibited by their adult daughters. Cross-fostering studies between high and low PG–ABN mothers show that these components of daughters' maternal behavior are determined by early nurturing experience and not genetic inheritance (Francis *et al.*, 1999). Based on these findings it has been hypothesized that adult stress reactivity is inversely related to frequencies of PG–ABN received in the early postnatal period and that those frequencies contribute substantially to intergenerational transmission of similar levels of stress responsiveness and maternal behavior.

Most of the studies cited above were conducted in Long Evans rats in the laboratory of Dr. Michael Meaney or his collaborators. Other investigators have obtained different outcomes in Long Evans and other strains of rats. Slotten *et al.* (2006) found that Long Evans rats subjected to LMS exhibited less anxiety in a number of tests, had lower basal corticosterone levels and no difference in restraint stress-induced corticosterone release compared to rats that did not experience early maternal separation. Ogawa *et al.* (1994) reported that adult Sprague Dawley rats subjected to 4.5 h of daily separation during postnatal weeks 1–3 compared to animals that had not been separated released less corticosterone during immobilization stress, did not differ in their hippocampal glucocorticoid receptor concentrations, but did exhibit more inhibited behavior in novel open field testing. Several studies in Wistar rats have produced results that are inconsistent with the findings reported by Meaney and colleagues in the Long Evans strain. Pryce *et al.* (2001) found that BMS and LMS did, respectively, increase and decrease maternal PG–ABN but only during a 1-h observation period immediately following the daily separation, not during four other 1-h time periods evenly spaced across the diurnal cycle. Marmendal *et al.* (2004, 2006) reported less emotional reactivity to novel situations in adult rats that were subjected to daily LMS (4 h) on postnatal days 1–15 compared to rats that had undergone daily BMS (3 or 5 min). In Wistar rat lines that had been selected over many generations for high or low anxiety (HA vs. LA), LMS (3 h) on postnatal days 2–15 compared to no separation decreased anxiety and HPA responses to stressors in adult HA offspring and increased anxiety but had no effect on HPA responses in adult LA offspring (Neumann *et al.*, 2005). Unfortunately, the studies summarized in this paragraph either did not measure or inadequately measured maternal behavior so no information was derived about the relationships among maternal separation, frequencies of maternal behaviors and adult offspring outcomes. While the results of these studies are not consistent with those of Meaney and colleagues, they do suggest that maternal separation (and perhaps related alterations in the frequency of some aspects of maternal care) have significant effects on the development of adult stress responses. The directionality of those effects, however, appears to be influenced by genetic background and baseline emotionality.

What has been referred to as pup-grooming (PG) in the studies summarized above is, in fact, simply PL. Also, arched-back nursing has not been clearly defined and may include several categories of rat nursing behavior described with greater precision by Stern (1996), including high and low crouching and hovering. Stern has used the term kyphotic nursing to incorporate all of these nursing postures in which the mother is upright over the pups and elevated enough that the pups can access her nipple line. Therefore, in the remainder of this chapter, we will use the terms pup-licking (PL) and kyphotic nursing (KN) rather than PG or ABN.

MATERNAL EFFECTS ON MOUSE DAUGHTERS

In the mouse, a much smaller body of evidence suggests that maternal separation and maternal behavior frequencies influence the development of offspring stress responses and related neurochemical systems. In most, but not all studies, BMS and LMS respectively decreased and increased anxiety, depression-like behavior and HPA axis responses to acute stressors in adult offspring (D'Amato *et al.*, 1998; MacQueen *et al.*, 2003; Romeo *et al.*, 2003; Moles *et al.*, 2004; Parfitt *et al.*, 2004; Millstein & Holmes, 2007). Insufficient numbers of observations were made to determine whether repeated separations affected mother or offspring maternal behavior (Romeo *et al.*, 2003; Moles *et al.*, 2004). Mother BALB/cJ mice, which are more stress responsive, have been shown in studies employing large numbers of observations to exhibit significantly less PL and KN than C57BL/6J mothers. BALB/cJ mice cross-fostered to C57BL/6J mothers were significantly less anxious and C57BL/6J mice reared by BALB/cJ mothers had higher basal corticosterone levels and reduced $GABA_A\gamma2$ subunit expression in the central nucleus of the amygdala (Caldji *et al.*, 2004; Priebe *et al.*, 2005).

By making numerous observations during 5-time periods/day on postpartum days (PPD) 2, 4, 6 and 8, we determined the distribution of PL, still crouching and other maternal behavior frequencies in 36 C57BL/J6 mothers (Figure 24.1). Adult offspring of the highest 7 and lowest 7 PL frequency mothers were tested in the elevated plus maze and open field (1 h on 3 successive days) during the first week, open field for 3 h with an ip saline injection after the first hour during the second week and for acoustic startle and prepulse inhibition during the third week. As is summarized in Table 24.1, there were numerous significant differences between high and low PL daughters, but not between high and low PL sons. Low PL

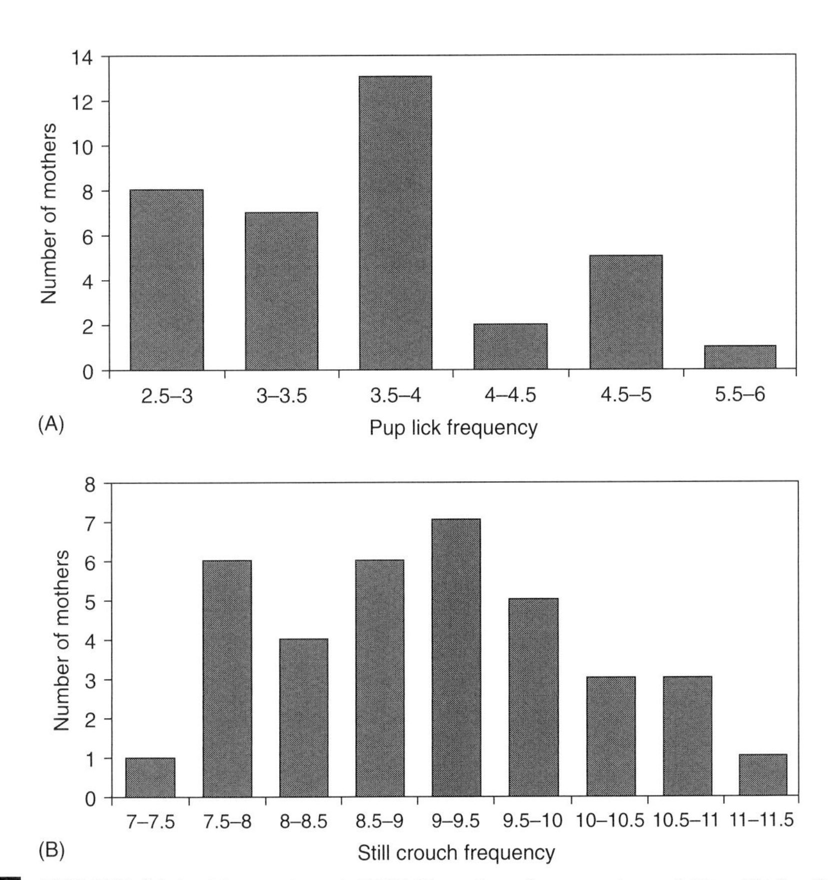

FIGURE 24.1 The number of C57BL/J6 mothers from a cohort of 36 exhibiting PL (A) and still crouched nursing (B) within specific frequency ranges. The frequency for each animal is the mean number of observations (20 total observations/h) during which PL or still crouching was exhibited during each of five 1-h behavior recording sessions (3 during the light phase, 2 during the dark phase) on PPD 2, 4, 6 and 8.

TABLE 24.1 Significant differences between adult female and male C57BL/J6 mice reared by the 7 mothers exhibiting the highest (H) or the 7 mothers exhibiting the lowest (L) pup-licking frequencies from a cohort of 36 mothers on the elevated plus maze (EPM) test and 1-h open field tests conducted on 3 consecutive days during week 1 of testing, a 3-h open field test during week 2 in which an ip saline injection was administered to each animal at the 1-h time point, and acoustic startle testing conducted during week 3 from which prepulse inhibition was calculated. Dist = distance traveled

Tests	Females	Males
EPM	L > H Time in closed arms	No L vs. H difference
Open field: 1 h		
Day 1	L < H Distance traveled (first 5 min)	No L vs. H difference
	Rearing (whole hour)	
	Center time, dist (first 15 min)	
Day 3	L > H Distance traveled	No L vs. H difference
	Center time	
Open field: 3 h		
Overall	L > H Distance traveled	No L vs. H difference
Hour 2	L > H Center time and distance	No L vs. H difference
Post-ip injection	L > H Rearing	No L vs. H difference
	Center distance	
Acoustic startle	L < H Prepulse inhibition	No L vs. H difference

females were more inhibited initially when the open field was novel, failed to habituate upon repeated exposure to this test situation, were more reactive to the acute stress of ip injection and exhibited deficits in prepulse inhibition. Future cross-fostering studies will be necessary to determine how much these variations in outcomes are determined by early experience vs. genetic factors. The results of this study suggest that maternal care may influence the results of behavior phenotyping studies of transgenic or selected strains of mice.

OT SELECTIVELY ENHANCES PL AND KN

Earlier studies found that disruptions of central OT inhibited the onset of all components of maternal behavior in parturient or ovarian steroid-treated female rats but did not appear to affect established mothering behavior (Fahrbach *et al.*, 1985; Pedersen *et al.*, 1985, 1994; Numan & Corodimas, 1985; Van Leengoed *et al.*, 1987; Insel & Harbaugh, 1989). By making numerous observations at short intervals over extended periods of time, we and others have more recently demonstrated that OT selectively enhances frequencies of the PL and KN components of well-established maternal behavior in lactating rats as well as the PL component of spontaneous maternal behavior in nulliparous C57BL/J6 mice (Pedersen & Boccia, 2003; Pedersen *et al.*, 2006).

In the first of our studies, we measured maternal behavior in rat dams during 105-min periods immediately before and starting 2 h after intracerebroventricular (ICV) infusion of 1 μg of a selective oxytocin antagonist (OTA) $d(CH_2)_5$-[Tyr(Me)^2,Thr^4,Tyr-NH$_2^9$] or normal saline vehicle on PPD 2/3 or 6/7. Maternal and other behaviors were scored during each 15-s interval of each 105-min period (420 total observations).

Dams' oral grooming was significantly altered by OTA compared to saline administration. The frequency of PL decreased, although not quite significantly ($p < 0.06$, Figure 24.2(A) top), but the frequency of self-grooming was significantly increased (Figure 24.2(B) bottom). Therefore, PL as a proportion

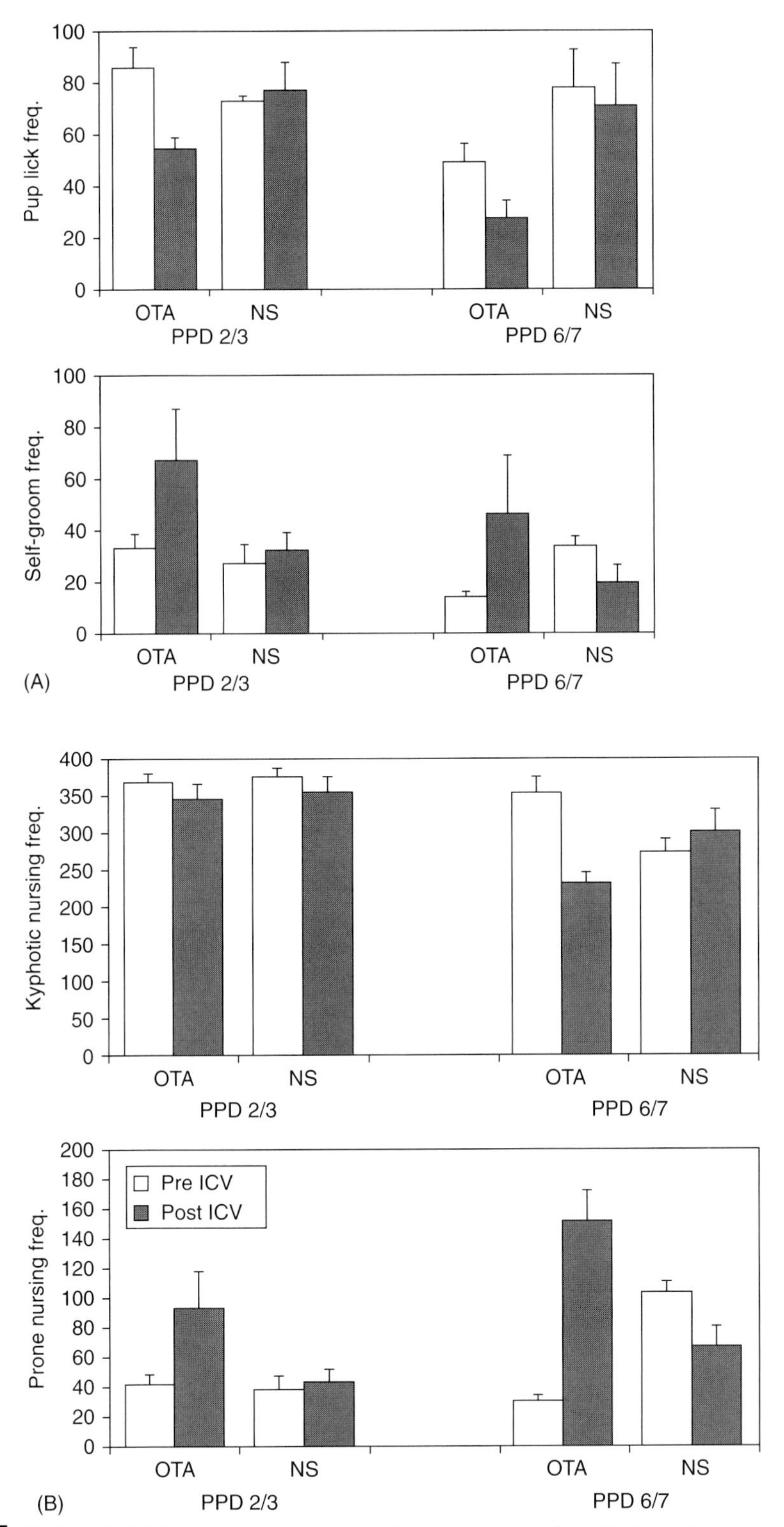

FIGURE 24.2 Modification of mean frequencies (±SEM) of PL (A top), self-grooming (A bottom), kyphotic nursing (B top) and prone nursing (B bottom) by ICV administration of 1 μg oxytocin antagonist (OTA) or normal saline (NS) vehicle. Frequencies for each behavior are the number of 15-s intervals in which the behavior was exhibited during 105-min behavior recording sessions (420 observations/session) immediately preceding and beginning 2 h after ICV injections on PPD 2/3 or 6/7.

of total oral grooming (PL + self-grooming) was significantly decreased by OT antagonism. Furthermore, the frequency of brief PL bouts (<2 s in duration), which were not scored as PL, was increased significantly by blocking OT receptors. In addition, as is illustrated in Figure 24.2(B) (top and bottom), OT antagonism significantly decreased KN frequencies (PPD 6/7 only) and increased the frequencies of prone nursing (i.e., laying ventrum down on pups). Also, the percentage of time in which mothers engaged in sustained quiescent KN bouts >2 min in duration decreased significantly after ICV infusion of OTA. Other components of maternal behavior were not affected by blocking OT and effects on PL and KN were not related to changes in milk ejection.

This study indicates that central OT selectively enhances PL and KN, those components of mothers' behavior that influence the development of their offspring's acute stress responses and their daughters' maternal behavior. Our results suggest that OT directs mothers' oral grooming away from themselves and toward their pups and shifts the balance in upright nursing from prone to kyphotic postures.

In our study, there was a strong trend toward a significant positive correlation between PL frequency before ICV infusion and the magnitude of postinfusion decline in PL in OTA-treated but not saline-treated mothers. Champagne *et al.* (2001) reported the related observation that central administration of an OTA significantly decreased the higher PL frequencies of lactating mothers that had received more maternal licking during infancy but not the lower frequencies of PL of mothers that had received less maternal licking. These findings indicate that the degree to which OT stimulates rat mothers' PL may be determined by the amount of licking received from their mothers.

Nulliparous females in most strains of laboratory mice rapidly exhibit avid maternal behavior toward young pups. We recently published evidence that nulliparous OT gene knockout (OTKO) mice lick newborns significantly less than wild type (WT) mice indicating that OT enhances PL in this species as well as rats (Pedersen *et al.*, 2006, Figure 24.3). This study employed frequent observations over long periods. Our results indicate that, during the maintenance phase of established postpartum maternal behavior in rats, central OT stimulates a substantial fraction of the PL component of spontaneous maternal behavior in laboratory mice. It remains to be determined whether, as has been demonstrated in rats, PL frequencies are related to the amount of maternal behavior received during the postnatal period and if differences in central OT activity account for variations in PL frequencies.

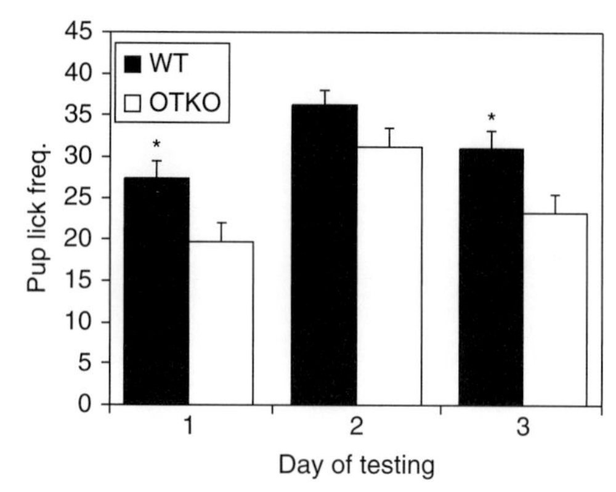

FIGURE 24.3 Comparison of mean (±SEM) PL frequencies between OT knockout (KO) and WT nulliparous mice on test days 1, 2 and 3. Frequencies were the number of 5-s periods at 2-min intervals over 3-h behavior recording sessions during which PL was exhibited (90 observations/session). Over all test days, KOs licked pups at significantly lower frequencies ($F \times [1/25] = 7.573$, $p = 0.011$). $^{*}p < 0.05$, KO < WT on test days 1 and 3.

PL AND KN INFLUENCE OT RECEPTOR EXPRESSION IN THE BRAINS OF ADULT FEMALE OFFSPRING

PL–KN received during the postnatal period influences the expression of OT receptors in several brain regions of adult female rats. Francis *et al.* (2000) and Champagne *et al.* (2001) found that OT binding was significantly higher in the medial preoptic area (MPOA), hypothalamic paraventricular nucleus (PVN), bed nucleus of the stria terminalis (BNST), ventrolateral septum and central nucleus of the amygdala of lactating rats, and the central amygdala of nulliparous females that had received high compared to low levels of PL–KN as young pups. The MPOA is a site in which OT activates postpartum maternal behavior (Pedersen, *et al.*, 1994). OT may also activate maternal behavior in the PVN, BNST and the lateral septum (Kendrick, 2000). The central nucleus of the amygdala is a site in which OT decreases anxiety (Bale *et al.*, 2001). OT release in the PVN diminishes stress activation of the HPA axis (Neumann *et al.*, 2000). Maternal PL–KN may determine daughters' adult acute stress responses and maternal behavior, in part, by altering OT receptor expression in their brains.

OT receptor concentrations rise significantly in the MPOA and probably the BNST, the lateral septum and the PVN in parallel with the upswing in estradiol concentrations during late pregnancy that facilitates parturition and the postpartum onset of maternal behavior (Insel, 1986, 1990; Pedersen *et al.*, 1994; Pedersen, 1997). PL and KN received in infancy may influence late pregnancy hormone-induced up-regulation of OT receptors in brain areas where OT activates maternal behavior and modulates stress responses. Champagne *et al.* (2001) reported that estrogen treatment significantly increased OT receptor binding in the MPOA and lateral septum of nulliparous female offspring of high, but not low, PL–KN mothers. So nurturing received in infancy may broadly influence estrogen regulation of central OT receptor expression.

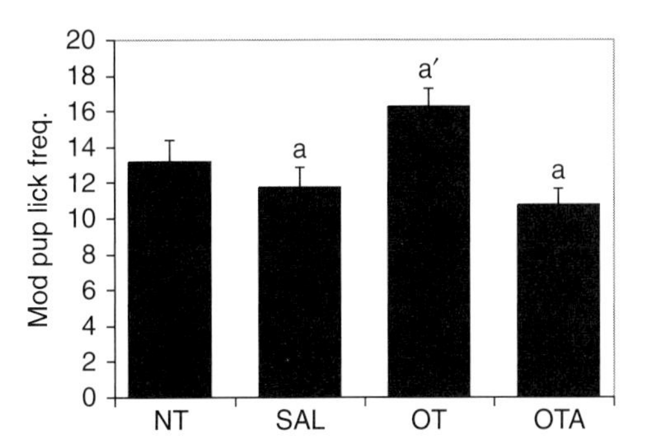

FIGURE 24.4 Comparison of mean (±SEM) PL frequencies among rat mothers that were given daily subcutaneous treatments with oxytocin (OT, 20 μg), OT antagonist (OTA, 20 μg) or normal saline (NS, 0.05 ml) or no treatment (NT) on postnatal days 2–10. Frequencies are based on the number of 10-s observations at 6-min intervals during two 4-h observations/day on PPD 2, 4 and 6 (240 total observations) modified for the frequency of PL received during the early postnatal period (similar scoring system on postnatal days 2, 4 and 6). a′ vs. a indicates $p < 0.05$, OTA > OT and NS.

POSTNATAL OT ACTIVITY INFLUENCES ADULT MATERNAL BEHAVIOR

In rat pups, the distribution of central OT receptors evolves through several phases before the adult pattern of binding emerges around puberty (Shapiro & Insel, 1989; Tribollet *et al.*, 1989). OT receptors arise in a large number of brain sites by postnatal day 5 and then gradually increase in density until postnatal days 10–14. Subsequently, OT binding disappears in many sites. The amygdala is one of the brain structures in which OT receptors appear early and persist into adulthood. The widespread, and in many areas transient, expression of OT receptors in the early postnatal CNS suggests that OT may influence neurochemical development in some brain sites. If OT activity in the young pup brain affects the development of neurochemical systems involved in mothering, manipulation of OT activity during the early postnatal period may alter adult maternal behavior.

To test this hypothesis, we compared the effects of daily subcutaneous administration of OT (20 μg), an OTA (20 μg) or saline or handling alone on postnatal days 2–10 on adult maternal behavior in female rats. This experiment was conducted in experimental litters composed of six female and six male pups each from a different birth litter. Male pups received other treatments and will not be discussed further in this chapter. Between 70 and 100 days of age, female offspring from each treatment group were mated. The behavior of the mothers that reared experimental litters and their postpartum daughters was videotaped for 4 h during the light phase and 4 h during the dark phase on PPD 2, 4 and 6. Maternal and other behaviors were coded during 10-s periods at 6-min intervals on the videotape records (240 total observations). Analysis of covariance with prenatal treatment as a between-subjects variable and the frequency of maternal PL received during the early postnatal period as a covariate was used to examine group differences. PL frequencies in daughters were significantly higher in those that received postnatal OT compared to those that received OTA or saline (Figure 24.4). These results indicate that OT activity in female pups (most likely in their brains, although we have not determined to what extent subcutaneously administered

OT and OTA cross the blood–brain barrier in infant rats) affects how much PL they subsequently exhibit toward their own pups.

MATERNAL BEHAVIOR REGULATION OF OT ACTIVITY IN THE FEMALE PUP BRAIN

Widespread central OT receptor expression occurs during the early postnatal period when daily maternal separation and presumably levels of maternal PL–KN received exert the greatest impact on the development of stress responses and probably maternal behavior (Meaney *et al.*, 1996; Boccia & Pedersen, 2001; Gonzalez *et* al., 2001). Having confirmed that OT activity in young female rat pups influences their adult maternal behavior, we began to examine whether the amount of mothering they receive influences that activity. Previous findings support this hypothesis. Insel & Winslow (1991) reported that central administration of OT suppressed distress vocalizations in rat pups during separations from their mothers. Winslow *et al.* (2000) also observed that OTKO mouse pups vocalized less than WT pups during separations from their mothers. These finding suggest that maternal contact may stimulate OT release in the pup brain and that a decline in that release during maternal separation may contribute to distress vocalizations. In an earlier study (Noonan *et al.*, 1994), we found that OT binding in the dorsal hippocampus and cortex was significantly different on postnatal days 4 and/or 8 in rat pups subjected to brief daily maternal separations which have been reported to increase PL–KN received from mothers (Liu *et al.*, 1997; Caldji *et al.*, 1998; Boccia & Pedersen, 2001). Recently, we have compared OT binding on thin frozen sections through multiple brain areas of 10-day-old female pups reared in litters that received frequencies of maternal licking in the top or bottom 15% of the normal distribution. As is illustrated in Figure 24.5, pups that received greater maternal licking had significantly higher binding in the dorsal hippocampus and the central amygdala but not other brain areas. Future studies involving cross-fostering or experimental litters composed of pups from numerous litters will be necessary to determine whether these OT binding differences are the consequence of variations in maternal behavior or genetic inheritance. The dorsal hippocampus and amygdala are among those brain areas in which BMS vs. LMS and high vs. low frequencies of maternal PL–KN exert the most strikingly contrasting effects on the development of neurochemical systems involved in stress responses (e.g., glucocorticoid receptor concentrations in the dorsal hippocampus, benzodiazepine receptor concentrations and $GABA_{A\ \gamma}2$ subunit gene expression in the amygdala (Caldji *et al.*, 2000)). Perhaps regulation of sensitivity to the

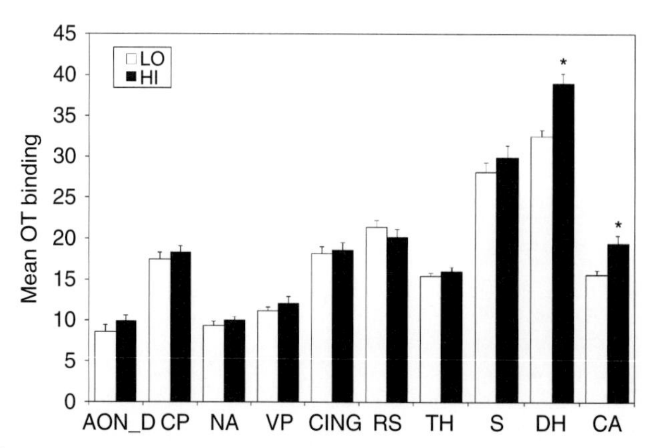

FIGURE 24.5 Comparison of mean (±SEM) OT binding in specific brain areas between 8-day-old female pups reared by mothers exhibiting PL frequencies in the highest (HI) vs. lowest (LO) 15% of a cohort of 92 rat mothers. AON_D, dorsal anterior olfactory nucleus; CP, caudate putamen; NA, nucleus accumbens; VP, ventral pallidum; CING, cingulate cortex; RS, retrosplenial cortex; TH, thalamus; S, subiculum; DH, dorsal hippocampus; AMYG, amygdala. $*p < 0.05$, HI > LO.

putative neurodevelopmental effects of OT in these two brain areas is a mechanism whereby variations in PL (and perhaps KN) affect adult neurochemical outcomes.

Most OT pathways in the brain originate from OT neurons that are located in the PVN. We have begun to examine whether maternal contact, and in particular specific components of maternal behavior, may influence the activation of the PVN of female rat pups. At 8 days of age, 2 female and 2 male rat pups were removed from each experimental litter and placed in individual containers in an incubator (32°C) for 8 h. One female and one male pup from each litter were then returned to their mothers' home cages (SR group). Two hours after return, SR female pups were anesthetized and their brains harvested and fixed in 4% paraformaldehyde solution. Just before return of the SR pups to their mothers, all other pups were removed from the maternal cages. The brains of female pups that had not been separated (NS group) were harvested and fixed. The brains of female pups that had been separated from their mothers but not returned (SN group) were harvested and fixed at the same time as the SN pups. Mothers' behavior was videotaped for 2 h prior to (while non-separated pups were in the nests) and for 2 h after return of the SR pups. Behavior was quantified from the 2-h vid-eotapes by coding all behaviors exhibited at 2-min intervals (60 observations). Fixed pup brains were embedded in paraffin and cut on a microtome into thin (10 μm) coronal sections. Comparable sections through the posterior and mid PVN were immunostained for c-Fos.

The numbers of neurons with c-Fos immunostaining in the left and right PVN on each section were counted blind to the separation condition of the animal. A significantly higher number of c-Fos-immunostaining neurons were found in the PVN of the SR compared to the SN and NS pups (Figure 24.6). Regression analysis confirmed this significant difference among separation groups. However, when the frequencies of PL received during the 2 h prior to sacrifice were added to the regression analysis, the statistical significance of the group differences in numbers of c-Fos-expressing PVN neurons was eliminated. Entering the frequency of other components of maternal behavior received during the 2 h prior to sacrifice did not affect the significance of this comparison. Our results indicate that the greater number of activated neurons in the PVN in the female pups that had been returned to their mothers after an 8-h separation was related specifically to the higher frequency of PL this group received and not other maternal behaviors. OT immunostaining techniques that are reliable in adult rat brains have so far not worked well in discerning which of the PVN neurons in the 8-day-old pup brain express this neuropeptide. Surmounting this technical difficulty is important for determining whether there are significant differences in numbers of activated OT neurons among the SR, SN and NS groups and if they are accounted for by variations in the amount of maternal licking received.

A MODEL OF OT INVOLVEMENT IN MOTHERS' DEVELOPMENTAL EFFECTS ON THEIR DAUGHTERS: CLINICAL IMPLICATIONS

This chapter summarizes evidence that OT is involved in numerous ways in exerting mothers' developmental effects on their daughters' adult

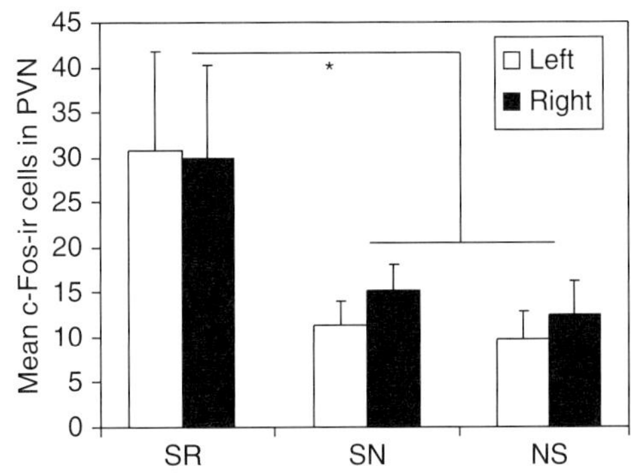

FIGURE 24.6 Comparison of the mean (±SEM) number of c-Fos immunoreactive (c-Fos-ir) cells in the left and right hypothalamic paraventricular nuclei (PVN) among groups of 8-day-old female rat pups ($N = 6$/group). Groups included (1) pups that were separated from their mothers for 8 h and then returned for 2 h before perfusion and brain harvesting (SR), (2) pup that were separated but not returned (SN) and (3) pups that were never separated (NS). Multiple regression of separation condition onto the combined number of c-Fos immunoreactive cells in the right and left PVN, $p = 0.05$. Adding maternal PL received during the 2 h prior to sacrifice to the regression analysis, $p = 0.48$. $^*p < 0.05$, SR (left and right) > SN and NS.

maternal behavior and acute stress responses. Among the strongest findings is that central OT contributes to as much as 40% of mothers' PL and KN frequencies, those components of maternal behavior that exert the greatest developmental effects on their daughters. The amount of PL and KN received during the early postnatal period appears to determine the degree to which OT facilitates these components of maternal behavior in adult daughters possibly by influencing how much OT binding is up-regulated in daughters' brains by rising estrogen levels during late pregnancy. Newer and more tentative evidence indicates that mothers' PL frequency directly influences OT activity in the brains of their young daughters. Specifically the mother's PL frequency is directly related to the concentration of OT receptors in daughters' dorsal hippocampus and central amygdala and may influence the activity of neurons in the PVN, the nucleus containing most of the OT neurons that project to sites outside of the hypothalamus. The widespread distribution and transient expression of OT receptors in numerous brain sites during the early postnatal period suggests that central OT activity may influence the development of a number of neurochemical systems. Control of OT activity in female pup brains may be a mechanism whereby variation in maternal PL and possibly KN frequencies differentially affects development of a number of neurochemical systems, thereby producing contrasting levels of acute stress responses and maternal behavior in adult daughters. Further studies are necessary to confirm whether maternal behavior received by young female pups regulates OT activity in their brains and whether OT activity in the female pup brain influences development of specific neurochemical systems.

OT involvement in the effects of early nurturing on the development of social behavior and stress responses provides a new perspective for understanding the origins of human behavioral and emotional problems that arise from inadequate nurturing and pathological relationships during childhood. Neglect, abuse and deficient emotional attachment and empathic concern tend to be perpetuated within families from one generation to the next. Also, it is not uncommon for children to undergo repeated disruption of attachments during changes in foster care placement. Such early experiences produce persistent difficulties in forming and maintaining relationships and increase vulnerability to depressive, anxiety, addictive and personality disorders (Marcus, 1991; Kessler & Magee, 1993; Atkinson & Zucker, 1997; Bifulco & Moran, 1998). The research summarized above suggests that central OT systems are particularly malleable by nurturing experiences during development and may have lifelong influences on the ability to form close emotional attachments to others especially one's own children. Abuse, neglect and disruptions during early relationships may exert negative effects on the development of central OT systems resulting in deficits in emotional attachment and excessive stress reactivity that lead to abuse and neglect of the next generation of children. These studies could lead to new biological strategies to facilitate recovery from pathological early experiences and to protect children who are caught in these terrible circumstances.

ACKNOWLEDGMENTS

The authors are appreciative of NIH support (MH56243, MH61995, MH066217, HD03110) that made much of this work possible.

REFERENCES

Ader, R., and Grota, L. J. (1969). Effects of early experience on adrenocortical reactivity. *Physiol. Behav.* 4, 303–305.

Atkinson, L., and Zucker, K. J. (1997). *Attachment and Psychopathology*. Guilford Press, New York.

Bale, T. L., Davis, A. M., Auger, A. P., Dorsa, D. M., and McCarthy, M. M. (2001). CNS region-specific oxytocin receptor expression: Importance in regulation of anxiety and sex behavior. *J. Neurosci.* 21, 2546–2552.

Bifulco, A., and Moran, P. (1998). *Wednesday's Child: Research into Women's Experience of Neglect and Abuse in Childhood, and Adult Depression*. Routledge, New York.

Boccia, M. L., and Pedersen, C. A. (2001). Brief vs. long maternal separation in infancy: Contrasting relationships with adult maternal behavior and lactation levels of aggression and anxiety. *Psychoneuroendocrinology* 26, 657–672.

Bodnoff, S. R., Suranyi-Cadotte, B., Quirion, R., and Meaney, M. J. (1987). Postnatal handling reduces novelty-induced fear and increases [^{3}H]-flunitrazepam binding in rat brain. *Eur. J. Pharmacol.* 144, 105–107.

Caldji, C., Tannenbaum, B., Sharma, S., Francis, D., Plotsky, P. M., and Meaney, M. J. (1998). Maternal care during infancy regulates the development of neural systems mediating the expression of fearfulness in the rat. *Proc. Natl Acad. Sci. USA* 95, 5335–5340.

Caldji, C., Francis, D., Sharma, S., Plotsky, P. M., and Meaney, M. J. (2000). The effects of early rearing environment on the development of $GABA_A$ and central benzodiazepine receptor levels and novelty-induced fearfulness in the rat. *Neuropsychopharmacology* 22, 219–229.

Caldji, C., Diorio, J., Anisman, H., and Meaney, M. J. (2004). Maternal behavior regulates benzodiazepine/GABAA receptor subunit expression in brain regions associated with fear in BALB/c and C57BL/6 mice. *Neuropsychopharmacology* 29, 1344–1352.

Cassidy, J., and Shaver, P. R. (1999). *Handbook of Attachment: Theory, Research and Clinical Applications*. The Guilford Press, New York.

Champagne, F. C., Diorio, J., Sharma, S., and Meaney, M. J. (2001). Naturally occurring variations in maternal behavior in the rat are associated with differences in estrogen-inducible central oxytocin receptors. *Proc. Natl Acad. Sci. USA* 98, 12736–12741.

Champagne, F. D., Francis, D. D., Mar, A., and Meaney, M. J. (2003). Variations in maternal care in the rat as a mediating influence for the effects of environment on development. *Physiol. Behav.* 79, 359–371.

D'Amato, F. R., Cabib, S., Ventura, R., and Orsini, C. (1998). Long-term effects of postnatal manipulation on emotionality are prevented by maternal anxiolytic treatment in mice. *Dev. Psychobiol.* 32, 225–234.

Denenberg, V. H. (1964). Critical periods, stimulus input, and emotional reactivity: a theory of infantile stimulation. *Psychol. Rev.* 66, 335–351.

Fahrbach, S. E., Morrell, J. I., and Pfaff, D. W. (1985). Possible role of endogenous oxytocin in estrogen-facilitated maternal behavior in rats. *Neuroendocrinology* 40, 526–532.

Francis, D. D., Diorio, J., Liu, D., and Meaney, M. J. (1999). Nongenomic transmission across generations of maternal behavior and stress responses in the rat. *Science* 286, 1155–1158.

Francis, D. D., Champagne, F. C., and Meaney, M. J. (2000). Variations in maternal behaviour are associated with differences in oxytocin receptor levels in the rat. *J. Neuroendocrinol.* 12, 1145–1148.

George, C., and Solomon, J. (1999). Attachment and caregiving: The caregiving behavioral system. In *Handbook of Attachment: Theory, Research, and Clinical Applications* (J. Cassidy and R. Shaver, Eds.), pp. 649–670. The Guilford Press, New York.

Gonzalez, A., Lovic, V., Ward, G. R., Wainwright, P. E., and Fleming, A. S. (2001). Intergenerational effects of complete maternal deprivation and replacement stimulation on maternal behavior and emotionality in female rats. *Dev. Psychobiol.* 38, 11–32.

Insel, T. R. (1986). Postpartum increases in brain oxytocin binding. *Neuroendocrinology* 44, 515–518.

Insel, T. R. (1990). Regional changes in brain oxytocin receptors postpartum: Time-course and relationship to maternal behaviour. *J. Neuroendocrinol.* 2, 539–545.

Insel, T. R., and Harbaugh, C. R. (1989). Lesions of the hypothalamic paraventricular nucleus disrupt the initiation of maternal behavior. *Physiol. Behav.* 45, 1033–1041.

Insel, T. R., and Winslow, J. T. (1991). Central administration of oxytocin modulates the infant rat's response to social isolation. *Eur. J. Pharmacol.* 203, 149–152.

Kendrick, K. M. (2000). Oxytocin, motherhood and bonding. *Exp. Physiol.* 85S, 11S–124S.

Kessler, R., and Magee, W. J. (1993). Childhood adversities and adult depression: Basic patterns of association in a US national survey. *Psychol. Med.* 23, 679–690.

Kraemer, G. W. (1997). Psychobiology of early social attachment in rhesus monkeys. Clinical implications. *Ann. NY Acad. Sci.* 807, 401–418.

Ladd, C. O., Owens, M. J., and Nemeroff, C. B. (1996). Persistent changes in corticotropin-releasing factor neuronal systems induced by maternal deprivation. *Endocrinology* 137, 1212–1218.

Lee, M. H. S., and Williams, D. I. (1974). Changes in licking behavior of rat mother following handling of young. *Anim. Behav.* 22, 679–681.

Levine, S. (1957). Infantile experience and resistance to physiological stress. *Science* 136, 405–406.

Levine, S. (1962). Plasma-free corticosteroid response to electric shock in rats stimulated in infancy. *Science* 135, 795–796.

Levine, S., Haltmeyer, G. C., Karas, G. G., and Denenberg, V. H. (1967). Physiological and behavioral effects of infantile stimulation. *Physiol. Behav.* 2, 55–59.

Liu, D., Diorio, J., Tannenbaum, B., Caldji, C., Francis, D., Freedman, A., Sharma, S., Pearson, D., Plotsky, P. M., and Meaney, M. J. (1997). Maternal care, hippocampal glucocorticoid receptors, and hypothalamic–pituitary–adrenal responses to stress. *Science* 277, 1659–1662.

Lovic, V., Gonzalez, A., and Fleming, A. S. (2001). Maternally separated rats show deficits in maternal care in adulthood. *Dev. Psychobiol.* 39, 19–33.

Marcus, R. F. (1991). The attachments of children in foster care. *Genet. Soc. Gen. Psychol. Monogr.* 117, 365–394.

MacQueen, G. M., Ramakrishnan, K., Ratnasingan, R., Chen, B., and Young, L. T. (2003). Desipramine treatment reduces the long-term behavioural and neurochemical sequelae of early-life maternal separation. *Int. J. Neuropsychopharm.* 6, 391–396.

Marmendal, M., Roman, E., Eriksson, C. J., Nylander, I., and Fahlke, C. (2004). Maternal separation alters maternal care, but has minor effects on behavior and brain opioid peptides in adult offspring. *Dev. Psychobiol.* 45, 140–152.

Marmendal, M., Eriksson, C. J., and Fahlke, C. (2006). Early deprivation increases exploration and locomotion in adult male Wistar offspring. *Pharmacol. Biochem. Behav.* 85, 535–544.

Meaney, M. J., Diorio, J., Francis, D., Widdowson, J., LaPlante, P., Caldji, C., Sharma, S., Seckl, J. R., and Plotsky, P. M. (1996). Early environmental regulation of forebrain glucocorticoid receptor gene expression: Implications for adrenocortical responses to stress. *Dev. Neurosci.* 18, 49–72.

Millstein, R. A., and Holmes, A. (2007). Effects of repeated maternal separation on anxiety- and depression-related phenotypes in different mouse strains. *Neurosci. Biobehav. Rev.* 31, 3–17.

Moles, A., Rizzi, R., and D'Amato, F. R. (2004). Postnatal stress in mice: Does stressing the mother have the same effect as stressing the pups?. *Dev. Psychobiol.* 44, 230–237.

Neumann, I. D., Wigger, A., Torner, L., Holsboer, F., and Landgraf, R. (2000). Brain oxytocin inhibits basal and stress-induced activity of the hypothalamo–pituitary–adrenal axis in male and female rats: partial action within the paraventricular nucleus. *J. Neuroendocrinol.* 12, 235–243.

Neumann, I. D., Wigger, A., Krömer, S., Frank, E., Landgraf, R., and Bosch, O. J. (2005). Differential effects of periodic maternal separation on adult stress coping in a rat model of extremes in trait anxiety. *Neuroscience* 132, 867–877.

Noonan, L. R., Caldwell, J. D., Li, L., Walker, C. H., Pedersen, C. A., and Mason, G. A. (1994). Neonatal stress transiently alters the development of hippocampal oxytocin receptors. *Dev. Brain Res.* 80, 115–120.

Numan, M., and Corodimas, K. P. (1985). The effects of paraventricular hypothalamic lesions on maternal behavior in rats. *Physiol. Behav.* 35, 417–425.

Ogawa, T., Mikuni, M., Kuroda, Y., Muneoka, K., Mori, K. J., and Takahashi, K. (1994). Periodic maternal deprivation alters stress response in adult offspring, potentiates the negative feedback regulation of restraint stress-induced adrenocortical response and reduces the frequencies of open field-induced behaviors. *Pharm. Biochem. Behav.* 49, 961–967.

Parfitt, D. B., Levin, J. K., Saltstein, K. P., Klayman, A. S., Greer, L. M., and Helmreich, D. L. (2004). Differential early rearing environments can accentuate or attenuate the responses to stress in male C57BL/6 mice. *Brain Res.* 1016, 111–118.

Parkes, C. M., Stevenson-Hinde, J., and Marris, P. (1991). *Attachment Across the Life Cycle.* Routledge, London/NewYork.

Pedersen, C. A. (1997). Oxytocin control of maternal behavior: Regulation by sex steroids and offspring stimuli. *Ann. NY Acad. Sci.* 807, 126–145.

Pedersen, C. A., and Boccia, M. L. (2003). Oxytocin antagonism alters rat dams' oral grooming and upright posturing over pups. *Physiol. Behav.* 80, 233–241.

Pedersen, C. A., Caldwell, J. D., Johnson, M. F., Fort, S. A., and Prange, A. J., Jr. (1985). Oxytocin antiserum delays onset of ovarian steroid-induced maternal behavior. *Neuropeptides* 6, 175–182.

Pedersen, C. A., Caldwell, J. D., Walker, C., Ayers, G., and Mason, G. A. (1994). Oxytocin activates the postpartum onset of rat maternal behavior in the ventral tegmental and medial preoptic areas. *Behav. Neurosci.* 108, 1163–1171.

Pedersen, C. A., Vadlamudi, S. V., Boccia, M. L., and Amico, J. A. (2006). Maternal behavior deficits in nulliparous oxytocin knockout mice. *Genes Brain Behav.* 5, 274–281.

Plotsky, P. M., and Meaney, M. J. (1993). Early, postnatal experience alters hypothalamic corticotropin-releasing factor (CRF) mRNA, median eminence CRF content and stress-induced release in adult rats. *Mol. Brain Res.* 18, 195–200.

Priebe, K., Brake, W. G., Romeo, R. D., Sisti, H. M., Mueller, A., McEwen, B. S., and Francis, D. D. (2005). Maternal influences on adult stress and anxiety-like behavior in C57BL/6J and BALB/cJ mice: A cross-fostering study. *Dev. Psychobiol.* 47, 398–407.

Pryce, C. R., Bettschen, D., and Feldon, J. (2001). Comparison of the effects of early handling and early deprivation on maternal care in the rat. *Dev. Psychobiol.* 38, 239–251.

Reite, M. L., and Boccia, M. L. (1994). Physiological aspects of adult attachment. In *Attachment in Adults: Theory, Assessment, and Treatment* (M. B. Sperling and W. H. Berman, Eds.), pp. 98–127. Guilford Publications, New York.

Romeo, R. D., Mueller, A., Sisti, H. M., Ogawa, S., McEwen, B. S., and Brake, W. G. (2003). Anxiety and fear behaviors in adult male and female C57BL/6 mice are modulated by maternal separation. *Horm. Behav.* 43, 561–567.

Shapiro, L. E., and Insel, T. R. (1989). Ontogeny of oxytocin receptors in rat forebrain: A quantitative study. *Synapse* 4, 259–266.

Slotten, H. A., Kalinichev, M., Hagan, J. J., Marsden, C. A., and Fone, K. C. F. (2006).

Long-lasting changes in behavioural and neuroendocrine indices in the rat following neonatal maternal separation: Gender-dependent effects. *Brain Res.* 1097, 123–132.

Suomi, S. J. (1999). Attachment in rhesus monkeys. In *Handbook of Attachment: Theory, Research, and Clinical Applications* (J. Cassidy and P. R. Shaver, Eds.), pp. 181–197. The Guilford Press, New York.

Stern, J. M. (1996). Somatosensation and maternal care in Norway rats. In *Advances in the Study of Behavior, Parental Care: Evolution, Mechanisms, and Adaptive Significance* (J. S. Rosenblatt and T. Snowdon, Eds.), Vol. 25, pp. 243–294. Academic Press, San Diego.

Tribollet, E., Charpak, S., Schmidt, A., Dubois-Dauphin, M., and Dreifuss, J. J. (1989). Appearance and transient expression of oxytocin receptors in fetal, infant, and peripubertal rat brain studied by autoradiography and electrophysiology. *J. Neurosci.* 9, 1764–1773.

Van Ijzendoorn, M. (1995). Adult attachment representations, parental responsiveness, and infant attachment: A meta-analysis on the predictive validity of the adult attachment interview. *Psychol. Bull.* 117, 387–403.

Van Leengoed, E., Kerker, E., and Swanson, H. H. (1987). Inhibition of postpartum maternal behavior in the rat by injecting an oxytocin antagonist into the cerebral ventricles. *J. Endocrinol.* 112, 275–282.

Winslow, J. T., Hearn, E. F., Ferguson, J., Young, L. J., Matzuk, M. M., and Insel, T. R. (2000). Infant vocalization, adult aggression, and fear behavior of an oxytocin null mutant mouse. *Horm. Behav.* 37, 145–155.

25 STRATEGIES FOR UNDERSTANDING THE MECHANISMS OF MOTHERING AND FATHERING

JODY M. GANIBAN[1], LESLIE D. LEVE[2], GINGER A. MOORE AND JENAE M. NEIDERHISER[3]

[1] *Department of Psychology, The George Washington University, 2125 G Street, NW, Washington, DC 20057, USA*
[2] *Oregon Social Learning Center, 10 Shelton McMurphey Boulevard, Eugene, OR 97401, USA*
[3] *Department of Psychology, Moore Building, The Pennsylvania State University, University Park, PA 16802, USA*

STRATEGIES FOR UNDERSTANDING THE MECHANISMS OF MOTHERING AND FATHERING

It is well established in the human developmental and family literature that parenting has an important impact on the development of children and adolescents. The mechanisms through which parenting influences child development, however, are less clear. The role of genetic influences, via the parents or via the child, were first considered in the early 1980s (Rowe, 1981, 1983), but it took nearly 10 years for researchers to realize the relevance and importance of such influences (e.g., Plomin & Bergeman, 1991). There is now a sizable literature on the roles of genetic and environmental influences on parenting across the lifespan (see reviews in Towers *et al.*, 2001; McGuire, 2003; Ulbricht & Neiderhiser, in press), with findings indicating that both genetic and environmental influences are important for explaining individual differences in parenting. Despite this significant advance in our understanding of the roles of genes and environment in influencing child and adolescent development, few studies have attempted to explicate the mechanisms involved in genetic and environmental influences on parenting. In this chapter, we will first provide a brief overview of studies examining genetic influences on parenting and on genotype–environment correlation (rGE; correlation between genes and environment). We then focus more specifically on factors that can explain genetic influences on parenting (e.g., temperament and personality in parents and in children). We then focus on further advancing our understanding of the mechanisms of environmental influences on parenting. We conclude by considering how advances in our understanding of the mechanisms underlying genetic and environmental influences on parenting can be applied.

Studies Examining Genetic Influences on Parenting

One strategy for understanding the mechanisms involved in parenting is to examine genetic and environmental influences on parenting. As is described below, a number of studies have found that parents' and children's genetic makeup influences parenting and that the patterns of genetic and environmental influences vary in consistent ways based on parenting construct and child age. Considering rGE provides a useful framework for understanding parent- and child-based genetic influences on parenting. Therefore, we next describe rGE, provide a brief overview of studies that have examined genetic and environmental influences on parenting, and provide a brief discussion of developmental patterns of genetic and environmental influences on parenting and how they can be interpreted in regard to rGE.

Genetic Influences on Parenting: rGE

A critical question, especially in regard to genetic influences on parenting, is how to best understand the processes involved. While it makes intuitive sense that the parent's genes influence his/her behaviors and reactions to the child, the pathways through which the child's genes affect the parent is less clear. One of the best explanations for parent- and child-based genetic influences on parenting is rGE. Three types of rGE have been described: passive, active, and evocative (Plomin *et al.*, 1977; Scarr & McCartney, 1983; Rutter & Silberg, 2002). Passive rGE occurs between the parent and the child because they share genes and environments. Passive rGE occurs when genetically influenced characteristics of the parent influence parenting. In this scenario, the child's caregiving environment and genetic makeup are correlated. Evocative rGE, the result of others in the environment responding to genetically influenced characteristics in the child, occurs when genetically influenced characteristics of the child (e.g., temperament) elicit specific responses from the parent. Finally, active rGE occurs when the child actively selects environments that correlate with his/her genetically influenced characteristics. Although active rGE might influence decisions about partner choice or the number of children to have, active rGE is unlikely to directly influence specific parenting behaviors. Thus, we focus in this chapter on passive and evocative rGE. It is worth noting that rGE, as described above, is focused only on genetic influences on parenting and that rGE could operate in studies examining genetic and environmental influences on associations between environmental measures like parenting and child behaviors. An example illustrating how rGE may operate is provided here. A child whose parent has Antisocial Personality Disorder may inherit a genetic vulnerability to externalizing behaviors and be exposed to harsh, inconsistent parenting. In this case, there is not necessarily a causal influence of parenting on the expression of child externalizing behaviors. The two may be associated because of genetic influences shared between parent and child. In the parent, genetic factors contribute to Antisocial Personality Disorder and negative parenting, and in the child, genetic factors contribute to externalizing behaviors.

Figure 25.1 provides a conceptual illustration of the different potential determinants parenting adapted from a paper by Belsky (1984). This figure illustrates three possible determinants of parenting: parent characteristics, child characteristics, and contextual factors. In analyses, such as those described in the next section, where only parenting is examined using a genetically sensitive design, the type of rGE indicated depends on whether the

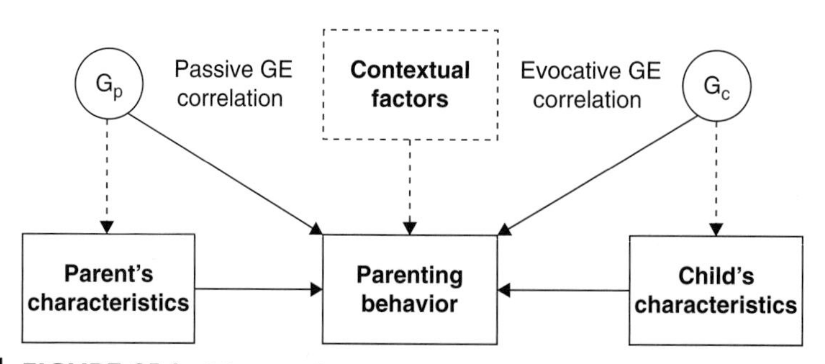

FIGURE 25.1 Schematic illustration of determinants of parenting model adapted from Belsky (1984). According to this figure there are three possible explanations for parenting behavior: parent characteristics, child characteristics, and contextual factors (e.g., marital relationship). The latent variable G_c represents child's genes and the latent variable G_p represents parent's genes.

parents or the children vary in degree of genetic relatedness. Therefore, as illustrated in Figure 25.1, if genetic influences are found on parenting when parents vary in degree of genetic relatedness (e.g., parents' genes are the focus), then passive rGE would be indicated. Genetic influences on parenting when children's genes are the focus suggest evocative rGE. Later we describe studies that have examined parent and child characteristics as possible explanations of genetic and environmental influences on parenting. Those associations are illustrated by the dashed lines in Figure 25.1 with similar interpretation in regard to rGE.

Genetic Influences on Parenting: Child's Genes and Parent's Genes

As is mentioned above, the majority of studies examining genetic influences on parenting have used child-based designs in which the children vary in degree of genetic relatedness to one another. Factor analyses of parenting behaviors assessed using twin and adoption designs (see review in Plomin, 1994) and numerous studies of parenting from a socialization perspective (see Rothbaum & Weisz, 1994) have found two primary orthogonal factors: parental warmth and parental control. In most studies, parental warmth, support, and negativity are influenced by the child's genes and the family-wide environment (i.e., shared environment; Elkins *et al.*, 1997; Lau *et al.*, in press). The construct of parental control, however, shows shared environmental and little to no genetic influence in such studies (Plomin, 1994; Reiss *et al.*, 2000). Taken together, these findings suggest that the child influences parental control and negativity in different ways. For example, parental control might be something that parents work out between each other and then apply to the child in a more general way. However, parental negativity might, in part, be guided by the child's genetically influenced characteristics (which is consistent with evocative rGE and illustrated on the right side of Figure 25.1 by the solid line from G_c to parenting).

There are also parent-based studies of genetic and environmental influences on parenting in which parents vary in degree of genetic relatedness. These designs provide estimates of parent-based genetic contributions to parenting, illustrated by the solid line from G_p to parenting in Figure 25.1. Interestingly, the pattern of findings from parent-based designs differs somewhat from those of child-based designs. Specifically, the findings for warmth, support, and negativity are generally similar in child- and parent-based designs. However, parental control differs across design type, with parent-based designs yielding evidence of genetic influences (Perusse *et al.*, 1992; Kendler, 1996; Losoya *et al.*, 1997; Neiderhiser *et al.*, 2004). By considering child-based and parent-based designs together and examining specific parenting constructs (e.g., warmth, control, and negativity), we can begin to specify the type of rGE that is operating to get a better understanding of the possible mechanisms involved.

Researchers attempting to specify the type of rGE operating for parenting have implemented very different designs, but have converged on similar findings. In two adoption studies, researchers examined associations between biological parent disorders (an estimation of genotype) and adoptive parenting (Ge *et al.*, 1996; O'Connor *et al.*, 1998). The results from both studies suggested that evocative rGE is important, as indicated by the finding that child behavior acted as a mediator of the association between the biological parents' disorders and the adoptive parents' perenting.

Other researchers have compared parent-based and child-based designs. The Twin/Offspring Study in Sweden (TOSS; Reiss *et al.*, 2001) was designed in part as a parent-based complement to the child-based nonshared environment in adolescent development (NEAD; Reiss *et al.*, 2000) study. Analyses comparing these two samples on mothering and fathering have found evidence for passive and evocative rGE for all of the parenting constructs examined, although the patterns of findings tended to vary by reporter, parent, and parenting construct (Neiderhiser *et al.*, 2004; Neiderhiser *et al.*, in press). Generally, the pattern of findings suggested passive rGE for mothers' positivity and monitoring, and evocative rGE for mothers' negativity and control. In other words, the mother may exert more control or demonstrate more negativity *in response* to the child's genetically influenced traits or behaviors. Fathering showed a similar pattern of findings with one exception: evocative rGE was suggested for father's positivity. This suggests that mother and child positivity is shaped by shared genetics, whereas fathers parent more positively *in response* to the child's genetically influenced traits or behaviors. The next step in understanding the mechanisms of rGE is to examine these constructs within the same sample to reduce the likelihood of differences in findings emerging because of sample differences rather than real differences in the mechanisms of effects.

Although the two samples used in the child-based and parent-based comparisons described above used the same measures and included adolescent children of approximately the same ages, the conclusions about rGE relied on two different samples. A more elegant approach is to include the children of twins, thus combining child-based and parent-based designs in a single sample. This design is especially powerful in identifying passive rGE and in distinguishing passive rGE from direct environmental effects transmitted across generations. There have been a number of recent reports using this approach (Silberg & Eaves, 2004; D'Onofrio *et al.*, 2005, 2006; Lynch *et al.*, 2006; Mendle *et al.*, 2006; Harden *et al.*, in press). The most relevant of these for this chapter focused on parenting and considered the effects of harsh parenting on negative outcomes for children (Lynch *et al.*, 2006). Once the shared genetic variance of harsh parenting and negative outcomes for children was accounted for (passive rGE), a direct effect of harsh parenting on negative outcomes for children remained. By extending the children of twins approach to include a child-based sample in the model, it is possible not only to distinguish just between passive rGE and direct environmental effects, but also to identify evocative rGE influences. To date, there is one report using this approach focused on mothering and adolescent internalizing problems (Narusyte *et al.*, 2007). The findings indicate that evocative rGE best explained associations between mothering and adolescent internalizing problems.

Taken together, the findings from studies using child-based and parent-based designs indicate that genetic factors are important and that both passive and evocative rGE are operating. There is evidence that the specific construct of parenting being studied matters a great deal and that "parenting" cannot be treated as a general construct. In addition, there appears to be consistency in mothering and fathering, at least in regard to rGE. Of note, the majority of the published studies examining rGE have used samples of adolescents and their parents. It is possible (even likely, as described below) that genetic and environmental influences on parenting will differ depending on the age of the child being parented.

Genetic and Environmental Influences on Parenting: Developmental Considerations

There is ample evidence that genetic and environmental influences in general change throughout development, and this is true for genetic and environmental influences on parenting. For example, parenting of infants and young children is primarily explained by shared environmental influences with modest genetic influences (Boivin *et al.*, 2005). Genetic influences on parenting tend to increase during middle childhood and adolescence, with shared environmental influences decreasing but typically remaining significant (Rowe, 1981, 1983; Rende *et al.*, 1992; Elkins *et al.*, 1997; Reiss *et al.*, 2000; McGue *et al.*, 2005). Findings for retrospective reports of parental warmth are similar to those for adolescents, with genetic, shared environmental, and nonshared environmental influences explaining nearly equal amounts of variance (Lichtenstein *et al.*, 2003). All of these studies of parenting suggest that the magnitude of shared environmental influences, which in child-based designs might also indicate passive rGE, decreases somewhat from infancy to adolescence and that genetic influences, which most likely indicate evocative rGE, increase in magnitude. It was proposed in the early 1980s that rGE would change throughout the lifespan predictably, with passive rGE effects largest during infancy and evocative rGE becoming more important with the child's age (Scarr & McCartney, 1983). Although not all studies have supported this hypothesis (e.g., Spotts & Neiderhiser, 2003), this is a useful way to understand how genetic and environmental influences on parenting may change throughout child and adolescent development.

Parent-Based Genetic Influences on Parenting

The previous section highlighted studies that identified parent-based genetic influences on parenting. However, it is unclear how these genetic influences are expressed during parent–child interactions. A likely candidate for parent-based genetic effects is the parent's personality, illustrated by the solid line from G_p to parenting in Figure 25.1. Current theories emphasize the biological and genetic origins of basic personality and temperament characteristics (McCrae *et al.*, 2000; Rothbart *et al.*, 2000). Genetic influences on personality characteristics tend to be moderate in magnitude. For example, heritability estimates for neuroticism are around 0.50 for twin samples and around 0.38 for non-twin samples (Bouchard & McGue, 2003). Similar heritability estimates have been reported for extraversion and agreeableness.

Several studies have examined associations between parenting and the Big Five personality characteristics: neuroticism (vs. emotionality stability), extraversion (vs. introversion), agreeableness, conscientiousness, and openness to experience (Belsky & Barends, 2002). In general, neuroticism is associated with less optimal parenting, including less warmth, positive affect, nurturance, responsiveness, and more intrusiveness (Belsky *et al.*, 1995; Kendler *et al.*, 1997; Kochanska *et al.*, 1997; Losoya *et al.*, 1997; Clark *et al.*, 2000; Metsapelto & Pulkkinen, 2003; Spinath & O'Connor, 2003; Kochanska *et al.*, 2004; Prinzie *et al.*, 2004). In other studies, the parent's general negative affectivity and depressive symptoms were also associated with less optimal parenting (Downey & Coyne, 1990; Fish *et al.*, 1991).

Many of these studies also identify associations between agreeableness and positive indices of parenting, including more positive affect and sensitivity and lower levels of detachment (Belsky *et al.*, 1995), more positively toned interactions, responsiveness, and nurturance, (Clark *et al.*, 2000; Kochanska *et al.*, 2004; Metsapelto & Pulkkinen, 2003), and less overreactive parenting (Prinzie *et al.*, 2004). Extraversion has also been associated with more positive affect and sensitivity (Belsky *et al.*, 1995), lower levels of power assertion (Clark *et al.*, 2000), more awareness of child behavior (Kochanska *et al.*, 2004), and nurturance (Metsapelto & Pulkkinen, 2003). Lastly, conscientiousness is correlated with more responsiveness and less power assertion (Clark *et al.*, 2000) and restrictiveness (Metsapelto & Pulkkinen, 2003). Higher levels of parent self-regulation are also associated with more optimal parenting (Brook *et al.*, 1995).

Genetically influenced personality characteristics can affect parenting through several mechanisms, including the following: biasing the recall and evaluation of information, guiding parent's specific responses to the child, and interfering or facilitating sensitivity to the child's emotional cues. In regard to the first mechanism, specific personality characteristics are associated with the tendency to experience negative or positive affect on a daily basis (McCrae *et al.*, 2000). In turn, a proneness to experience negative or positive affect can create biases in how relational experiences are perceived and interpreted, including the recollection of interpersonal experiences (Forgas, 1994; Rusting, 1998). For example, a predisposition to experience negative affect is associated with faster responses to negative vs. positive information and with difficulties shifting attention away from negative information (Derryberry & Reed, 1994, 2002). This same predisposition can also affect information processing. For example, compared to nondepressed/nonanxious individuals, anxious individuals are likely to perceive ambiguous feedback more negatively (Vestre & Caufield, 1986) and to make more negative interpretations when confronted with words or sentences that have multiple meanings (Eysenck *et al.*, 1991). Positive and negative affectivity can also affect autobiographical memory. When asked to recall past experiences, depressed individuals are more likely to recall negative life events, whereas extraverted individuals quickly recall positive life events (Rusting, 1998). It is worth noting that at least one study examining the association between parent–child relationships collected in adulthood and parent–child relationships collected in adolescence found that neither current depressive symptoms nor personality factors appeared to bias recall of parent–child relationships for men or women (Moore *et al.*, 1999). This suggests that, although there is an impact of personality on retrospective recall of some events, retrospective reports of parent–child relationships may be less influenced by such factors.

In regard to the second mechanism, the effects of affectivity can be extended to the social cognitive processes that shape the attributions people make about others and that influence the selection and enactment of behaviors during social interaction (Coie & Dodge, 1998). Consistent with this view, the attributions that the parent makes about the causes of the child's behavior predict parenting behavior (Bugental & Happaney, 2002) and, most likely, are influenced by the parent's personality (Belsky & Barends, 2002).

In regard to the third mechanism, certain personality characteristics may promote or interfere with the parent's sensitivity to the child's cues. Several papers have indicated that a parent who is overwhelmed with anxiety or is depressed tends to be less empathic to his/her child (Downey & Coyne, 1990). The parent's tendency to feel hostile, anxious, or agreeable may be directly expressed through behavior when interacting with the children or may interfere with his/her ability to attend to the child's needs. Other research has pointed toward the importance of the parent's ego control, a form of self-regulation, as an important correlate of parental sensitivity (Brook *et al.*, 1995). A parent who is able to regulate his/her emotions is likely to be more sensitive to the child's needs than a parent who is overwhelmed by anxiety or anger (Belsky & Barends, 2002). There is growing evidence that the basic executive control processes thought to underlie these capacities are genetically influenced (Rothbart & Posner, 2005), as is the case with other personality or temperament characteristics.

Child-Based Genetic Influences on Parenting

The previous section addressed how parent-based genetic influences on parenting are expressed during parent–child interactions. However, child temperament characteristic may also affect parenting and may account for child-based genetic influences on parenting. This is illustrated in Figure 25.1 by the dashed line from G_c to child characteristics and the solid line from G_c to parenting. Numerous phenotypic studies have highlighted associations between negative parenting and child negative emotionality during infancy (Fish & Crockenberg, 1981; Leve *et al.*, 2001) and toddlerhood (Lee & Bates, 1985) and between negative parenting and differential parental treatment during childhood (Jenkins *et al.*, 2003). These associations are detected when researchers rely upon parent self-reports or observational measures (Clark *et al.*, 2000; Kim *et al.*, 2001; Calkins *et al.*, 2004). Furthermore, genetically informed studies have indicated that a child's behavioral characteristics partially explain genetic contributions to parenting. For example, reports from the NEAD study indicate that child aggressiveness primarily accounts for genetic contributions to concurrent parent negativity and to parent negativity over time (Pike *et al.*, 1995; Neiderhiser *et al.*, 1999). Similar findings were reported by Jaffee *et al.* (2004) within the E-risk twin sample, in which child antisocial behavior explained significant variance in the parent's tendency to use corporal punishment. However, it is important to note that, within the Jaffee and colleagues study, the child's genetically influenced characteristics were not predictive of physical maltreatment. Therefore, there are limits to how much child characteristics affect extreme forms of parenting.

In summary, child-based genetic influences on parenting appear to operate through the child's stable behavioral characteristics. Although this might seem straightforward, the underlying mechanism might not be. It is possible that associations between the child's genetically influenced characteristics and parenting reflect passive rGE or evocative rGE. For example, child characteristics and parenting might be correlated because both constructs are influenced by the parent's genes (i.e., passive rGE). If this is the case, then child-based genetic influences on parenting might reflect the indirect effects of the parent's genes. The importance of passive rGE, however, has been difficult to establish because most genetically informed studies cannot estimate parent- and child-based genetic influences on parenting simultaneously. An exception to this limitation, however, is the children of twins design. As more researchers adopt this design, it will be possible to establish the relative importance of passive rGE in understanding parenting.

Another mechanism that has been used to explain child-based genetic influences on parenting is evocative rGE; it is plausible that the child's genetically influenced characteristics influence parenting. For example, a child who is prone to irritability may elicit more negative parental reactions than a child who is not frequently irritable. As is described previously, adoption studies are particularly suited to assess the presence of evocative rGE because adoptive children are raised by nonbiological relatives. Consequently, if child-based genetic influences on parenting are found, they cannot be attributed to passive rGE. Consistent with the evocative rGE mechanism, two adoption studies have reported that adoptees' biological parents' characteristics are correlated with adoptive parents' behavior and that the adoptees' mediated this association (Ge *et al.*, 1996; O'Connor *et al.*, 1998). More recently, Ganiban *et al.* (2007) found that child-based genetic influences on parenting are enhanced amongst children characterized by high negative emotionality. This finding also supports an evocative rGE mechanism, because it suggests that the child exerts more influence on parenting when he/she displays higher levels of anger and distress.

Environmental Influences on Parenting

In addition to the use of genetically informed child- and parent-based designs to further the understanding of genetic and environmental influences on parenting, randomized controlled trial (RCT) designs are an alternative design type that can provide unique information about factors that influence parenting. For example, highly specified environmental RCTs that target specific parenting behaviors and that show change in parenting in the intervention condition (compared to the control condition) can provide evidence of environmental influences on parenting. Further, when such environmentally based parenting interventions lead to change in child behaviors, parenting may operate as an environmental mechanism of change. In this section, we review evidence for environmental influences on parenting, with a focus on an evidence-based intervention program, Multidimensional Treatment Foster Care (MTFC; Chamberlain, 2003), and its efficacy in improving parenting and effecting change in child behavior.

Numerous research groups have shown that parenting behaviors can be modified under controlled intervention conditions to improve parenting practices and child outcomes across development, including infancy (van den Boom, 1994), early childhood (Webster-Stratton & Taylor, 2001; Fisher & Kim, 2007), middle childhood (Martinez & Forgatch, 2001), and adolescence (Eddy & Chamberlain, 2000; Leve *et al.*, 2005). Such interventions (including MTFC) have at their core a focus on teaching parents to use effective parenting practices that are contingent upon the behavior of the child. Below, we review the MTFC program as an illustrative example of how environmental interventions can influence parenting. The MTFC model was originally developed in 1983 in response to a State of Oregon request for proposals for programs developed as community-based alternatives to incarceration and placement in residential care. Since then, clinical teams have run local programs through state and county contracts for youth referred from the mental health, child welfare, and juvenile justice systems. Researchers, in partnership with local agencies (i.e., the State Mental Health Authority and the Oregon Youth Authority), have tested and found support for the efficacy of the MTFC model for boys and girls in preschool, middle childhood, and adolescence.

The MTFC model is centered on helping caregivers implement effective parenting behaviors, including the use of appropriate limit setting and contingent consequences for maladaptive child behavior and the use of positive reinforcement of positive child behavior. The basic model involves placing one (or occasionally two) youth in well-trained and supervised foster homes. Close consultation, training, and support of the foster parents form the cornerstone of the MTFC model. After 20 hours of preservice orientation, foster parents are certified by the state to be foster parents. Case managers with small caseloads (10 families each) maintain daily contact with MTFC parents to collect data on youth adjustment and to provide ongoing consultation, support, and crisis intervention. The basic components of MTFC include the following: (a) daily (M–F) telephone contact with MTFC parents using the Parent Daily Report checklist (PDR; Chamberlain & Reid, 1987); (b) weekly foster parent group meetings focused on supervision, training in parenting practices, and support; (c) an individualized behavior management program implemented daily in the home by the foster parent; (d) individualized skills training/coaching for the youth; (e) family therapy (for the biological/adoptive/relative family of the youth) focused on parent management strategies; (f) close monitoring of school attendance, performance, and homework completion; (g) case management to coordinate the MTFC, family, peer, and school settings; (h) 24-hour-on-call staff availability to MTFC and biological parents; and (i) psychiatric consultation as needed.

The MTFC model has received national attention as a cost-effective alternative to residential care. The results of a series of independent cost–benefit analyses from the Washington State Public Policy group (Aos *et al.*, 1999, 2001) and findings from three randomized trials have led MTFC to be selected as 1 of 10 evidence-based National Model Programs (The Blueprints Programs; Elliott, 1998) by the Office of Juvenile Justice and Delinquency Prevention and as 1 of 9 National Exemplary

Safe, Disciplined, and Drug Free Schools model programs. The MTFC model was also highlighted in two US Surgeon's General reports (US Department of Health and Human Services, 2000a, 2000b) and was selected by the Center for Substance Abuse Prevention and the Office of Juvenile Justice and Delinquency Prevention as an Exemplary I program for Strengthening America's Families (Chamberlain, 1998).

The first RCT of MTFC was conducted in 1986 with youth aged 9–18 who were leaving the Oregon State mental hospital and who were randomly assigned to MTFC or to group care facilities (GC). The results showed that the MTFC youths were placed more quickly, had lower rates of behavioral/emotional problems, and stayed out of the hospital more days in follow-up (Chamberlain & Reid, 1991) than the GC youths. In the second randomized trial (1990–1996), 79 boys referred from juvenile justice who had an average of 14 previous criminal offenses were randomly assigned to MTFC or GC. Results suggested that, compared to the GC boys, the MTFC boys spent less time incarcerated and demonstrated larger decreases in official criminal offenses. Of the MTFC boys, 41% had *no* referrals in the year following exit from placement; of the GC youth, 7% had *no* referrals in the year following exit from placement (Chamberlain & Reid, 1998). The efficacy of the MTFC intervention on reducing girls' criminal offenses and time incarcerated has also been demonstrated (Leve *et al.*, 2005; Chamberlain *et al.*, 2007). These MTFC findings illustrate the potential for parenting interventions to affect child outcomes. Additional analyses that focused on understanding how intervention-driven parenting changes explain group effects on child outcomes showed that the effect of group assignment on boys' subsequent criminal offenses was significantly mediated by a parenting latent construct that was measured during the course of the intervention (supervision, discipline, and positive adult relationships; loadings = 0.79, 0.60, and 0.89, respectively; Eddy & Chamberlain, 2000). That is, intervention-driven parenting changes accounted for the intervention effect on boys' criminal offenses.

A key theme from the MTFC findings is that RCT designs enable conclusions to be drawn about how changes in the environment (implementation of a parenting intervention) influence parenting changes and subsequent changes in child outcomes. The results indicate not only that parenting behaviors such as supervision, discipline, and adult–child relationships can be changed through behavior-based parenting interventions, but that such environmentally mediated changes in parenting are associated with improvements in child outcomes. In a strong test of this parenting-mediated model, Chamberlain *et al.* (in press) implemented a modified version of MTFC in foster and kinship families in the San Diego County Child Welfare System ($N = 700$). The sole focus of this intervention was on directly improving parenting behaviors, and the youth in the study did not directly receive MTFC intervention services or have direct contact with intervention staff members. The results from this RCT indicated that parenting – specifically the ratio of positive reinforcement to discipline – can be improved through the implementation of an environmental manipulation and that such parenting changes mediate reductions in child behavior problems. Given that the intervention did not involve directly treating the youth, these findings provide strong evidence that environmental influences on parenting are associated with changes in child outcomes.

Integrating Theory and Knowledge from RCT Studies of Parenting with Genetic Studies of Parenting

In the earlier sections of this chapter, we reviewed twin, sibling, and adoption research describing genetic influences on parenting and RCTs describing environmental influences on parenting. It is important to note that twin, sibling, and adoption studies generally rely on nonclinical samples, whereas RCTs target families with some degree of clinical symptoms or problems. Patterns of genetic and environmental influences on parenting may be markedly different in the context of family adversity or dysfunction, just as patterns of genetic and environmental influences of child behavior problems differ by severity level (Gjone *et al.*, 1996). Therefore, combining findings from genetically informed studies and RCTs may advance our understanding of the genetic and environmental factors that influence both normative and clinically relevant parenting processes. These two research approaches are not often integrated, and there has yet to be an RCT that uses a genetically sensitive design to test whether parenting can be experimentally manipulated with a resulting change in the expression of genetic influences on behavior. Following the research described above, there are multiple pathways whereby such environmentally mediated effects on genetic influences could manifest.

First, consistent with the parent-based genetic influences on parenting reviewed above, environmental interventions could serve to alter the expressed heritability of a parental personality characteristic that is linked to parenting responses. For example, parental negative affect is a partially heritable personality trait that is linked with difficulty shifting attention away from negative information (Derryberry & Reed, 1994, 2002). A parenting intervention might target improving parenting skill and increasing the frequency of contingent positive responses in response to child positive behavior to shift the balance of negative attention received by the child to more positive feedback for positive behaviors. This focus on teaching the parent to reinforce the child's positive behaviors is a cornerstone of many behavioral interventions, including MTFC. If successful, such parenting interventions might mediate genetic influences on the association between parenting behavior and child behavior problems, thereby reducing the magnitude of passive rGE on the underlying variables targeted by the intervention (e.g. parent negativity) condition as compared to the control condition.

Second, research on parent-based genetic influences on parenting has suggested that the parent's inherited characteristics such as anxiety or anger might interfere with the ability to be appropriately sensitive to the child's needs (Belsky & Barends, 2002). Parenting interventions that focus directly on reducing the expression of the inherited characteristic (i.e., anxiety) in the context of parenting interactions might also serve to mediate the genetic association between parental sensitivity and child adjustment. van den Boom's (1994) sensitivity-based intervention for mothers of irritable infants serves as an example of environmentally mediated mechanisms of change that might operate to reduce the expression of genetic risk on parenting and on subsequent child adjustment as transmitted through parenting. In van den Boom's RCT, mothers were randomly assigned to receive a 3-month intervention aimed at enhancing maternal sensitive responsiveness. The results indicated that, in follow-up assessments, the mothers in the intervention condition were significantly more responsive, stimulating, and visually attentive of their infant's behavior than the mothers in the control group. Further, the infants in the intervention condition were higher on sociability, self-soothing, and exploration and cried less than the infants in the control condition, suggesting a possible environmentally mediated effect between parenting and child outcomes.

Third, an RCT could be used to target child-based genetic influences on parenting. As is reviewed above, a child's inherited behavioral tendencies can affect parenting responses (Ge *et al.*, 1996; O'Connor *et al.*, 1998; Neiderhiser *et al.*, 1999; Jaffee *et al.*, 2004). For example, genetically influenced child antisocial behavior has been shown to elicit harsh parenting (Ge *et al.*, 1996), thus initiating and maintaining a pattern of coercive interactions between the parent and the child. A parenting intervention focused specifically on teaching parents to alter their parenting responses to heritable evocative child behavior might be an effective means of disrupting the coercive process and improving child outcomes. A secondary consequence of such an intervention would be that the rank-order correlations between parenting and inherited child characteristics that were the foci of the intervention would be diminished or eliminated, whereas those that were not the foci of the intervention would remain equivalent between the intervention and control conditions, reflecting the specificity of the genetically influenced processes targeted for intervention. In other words, adverse heritable child traits would not be associated with adverse parenting responses in the intervention condition, whereas such pathways would remain in the control condition (Reiss & Leve, 2007).

Consistent with this hypothesis, it is possible that a successful MTFC intervention case indicates that the intervention helped the parent effectively disregard the child's adverse genetic predispositions, thereby preventing the escalation of risk. Although the MTFC model has demonstrated efficacy in preventing delinquency outcomes and placement disruptions across genders, developmental age ranges, and contexts (Chamberlain & Reid, 1998; Fisher *et al.*, 2005; Leve *et al.*, 2005; Chamberlain *et al.*, 2007; Price *et al.*, in press), some MTFC youth *fail* to make improvements and/or maintain stable placement settings upon completion of treatment. In such cases, the parental monitoring, discipline, and support/warmth behaviors might have been influenced by the child's adverse genetic influences, thereby escalating problem behavior or the child's sensitivity and reactivity to such parenting behavior that might not have ameliorated as a result of the intervention.

However, in each of the examples described above, genetic and environmental contributions to parenting and to change in child adjustment cannot be directly assessed without use of a genetically informed sample; pure psychosocial RCT studies are not able to differentiate whether the mediators of change are genetically influenced. A preventive intervention trial that utilized a prospective genetically informed sample (e.g., an adoption study) would allow associations between specific genetically influenced child characteristics and specific parenting responses to be disentangled and subsequently targeted for intervention at the dyadic or family level. For example, a genetically enhanced intervention (i.e., one that was informed by behavioral genetic studies identifying evocative rGE for a specified child outcome) might augment the standard MTFC intervention by designing components intended to help the parent identify associations between the child's inherited maladaptive characteristics and his/her parenting responses; the parent could then use the role modeling, family therapy, and individual therapy approaches that form the core of the basic MTFC model to focus on breaking the associations between the child's inherited risk behaviors and nonoptimal caregiving responses (for further discussion of this hypothesis, see Reiss & Leve, 2007). Refining our intervention models to consider the pathways whereby child genetic characteristics influence parenting should enable us to develop more precise and effective interventions.

We believe that the field is at an optimal juncture to pursue such translational work and merge knowledge across the two fields. Developmental models that specify the mediating and moderating processes whereby genetic factors and parenting processes jointly produce maladaptive behaviors are becoming well developed and lead naturally to the consideration of how preventive interventions could alter multidetermined maladaptive trajectories (e.g., van Goozen *et al.*, 2007).

REFERENCES

Aos, S., Phipps, P., Barnoski, R., and Leib, R. (1999). *The Comparative Costs and Benefits of Programs to Reduce Crime: A Review of National Research Findings with Implications for Washington State* (No. 99-05-1202). Washington State Institute for Public Policy, Olympia, WA.

Aos, S., Phipps, P., Barnoski, R., and Leib, R. (2001). *The Comparative Costs and Benefits of Programs to Reduce Crime* (No. 01-05-1201). Washington State Institute for Public Policy, Olympia, WA.

Belsky, J. (1984). The determinants of parenting: A process model. *Child Development* 55, 83–96.

Belsky, J., and Barends, M. (2002). Personality and parenting. In *Handbook of Parenting* (M. H. Bornstein, Ed.), Vol. 2, pp. 415–438. Lawrence Erlbaum Associates, Mahwah, NJ.

Belsky, J., Crnic, K., and Woodworth, S. (1995). Personality and parenting: Exploring the mediating role of transient mood and daily hassles. *J. Pers.* 63, 905–929.

Boivin, M., Perusse, D., Dionne, G., Saysset, V., Zoccolillo, M., Tarabulsy, G. M., Tremblay, N., and Tremblay, R. E. (2005). The genetic-environmental etiology of parents' perceptions and self-assessed behaviours toward their 5-month-old infants in a large twin and singleton sample. *J. Child Psychol. Psychiatr.* 46, 612–630.

Bouchard, T. J., and McGue, M. (2003). Genetic and environmental influences on human psychological differences. *J. Neurobiol.* 54, 4–45.

Brook, J. S., Whiteman, M., Balka, E. B., and Cohen, P. (1995). Parent drug use, parent personality, and parenting. *J. Genet. Psychol.* 156, 137–151.

Bugental, D. B., and Happaney, K. (2002). Parental attributions. In *Handbook of Parenting* (M. H. Bornstein, Ed.), Vol. 2, pp. 509–535. Lawrence Erlbaum Associates, Mahwah, NJ.

Calkins, S., Hungerford, A., and Dedmon, S. E. (2004). Mothers' interactions with temperamentally frustrated infants. *Infant Mental Health J.* 25, 219–239.

Chamberlain, P. (1998). Treatment foster care. *Family strengthening series* (NCJ 1734211). U.S. Department of Justice, Washington, DC.

Chamberlain, P. (2003). *Treating Chronic Juvenile Offenders: Advances Made Through the Oregon Multidimensional Treatment Foster Care Model.* American Psychological Association, Washington, DC.

Chamberlain, P., and Reid, J. B. (1987). Parent observation and report of child symptoms. *Behav. Assess.* 9, 97–109.

Chamberlain, P., and Reid, J. B. (1991). Using a specialized foster care community treatment model for children and adolescents leaving the state mental hospital. *J. Community Psychol.* 19, 266–276.

Chamberlain, P., and Reid, J. B. (1998). Comparison of two community alternatives to incarceration for chronic juvenile offenders. *J. Consult. Clin. Psychol.* 6, 624–633.

Chamberlain, P., Leve, L. D., and DeGarmo, D. S. (2007). Multidimensional treatment foster care for girls in the juvenile justice system: 2-year follow-up of a randomized clinical trail. *J. Consult. Clin. Psychol.* 75, 187–193.

Chamberlain, P., Price, J., Leve, L. D., Laurent, H., Landsverk, J., and Reid, J. B. (in press). Prevention of behavior problems for children in foster care: Outcomes and mediation effects. *Prev. Sci.*

Clark, L. A., Kochanska, G., and Ready, R. (2000). Mothers' personality and its interaction with child temperament as predictors of parenting behavior. *J. Pers. Soc. Psychol.* 79, 274–285.

Coie, J. D., and Dodge, K. A. (1998). Aggression and antisocial behavior. In *Handbook of Child Psychology* (N. Eisenberg, Ed.), Vol. 3, pp. 779–862. Wiley & Sons, New York.

Derryberry, D., and Reed, M. A. (1994). Temperament and attention: Orienting toward and away from positive and negative signals. *J. Pers. Soc. Psychol.* 66, 1128–1139.

Derryberry, D., and Reed, M. A. (2002). Anxiety-related attentional biases and their regulation by attentional control. *J. Abnorm. Psychol.* 111, 225–236.

D'Onofrio, B. M., Turkheimer, E., Emery, R. E., Slutske, W. S., Heath, A. C., Madden, P. A., and Martin, N. G. (2005). A genetically informed study of marital instability and its association with offspring psychopathology. *J. Abnorm. Psychol.* 114, 570–586.

D'Onofrio, B. M., Turkheimer, E., Emery, R. E., Slutske, W. S., Heath, A. C., Madden, P. A. F., and Martin, N. G. (2006). A genetically informed study of the processes underlying the association between parental marital instability and offspring adjustment. *Dev. Psychol.* 42, 486–499.

Downey, G., and Coyne, J. C. (1990). Children of depressed parents: An integrative review. *Psychol. Bull.* 108, 50–76.

Eddy, J. M., and Chamberlain, P. (2000). Family management and deviant peer association as mediators of the impact of treatment condition on youth antisocial behavior. *J. Consult. Clin. Psychol.* 68, 857–863.

Elkins, I. J., McGue, M., and Iacono, W. G. (1997). Genetic and environmental influences on parent–son relationships: Evidence for increasing genetic influence during adolescence. *Dev. Psychol.* 33, 351–363.

Elliott, D. S. (1998). *Blueprints for Violence Prevention.* Institute of Behavioral Science, Regents of the University of Colorado, Boulder.

Eysenck, M. W., Mogg, K., May, J., Richards, A., and Mathews, A. (1991). Bias in interpretation of ambiguous sentences related to threat in anxiety. *J. Abnorm. Psychol.* 100, 144–150.

Fish, M., and Crockenberg, S. (1981). Correlates and antecedents of nine-month infant behavior and mother–infant interaction. *Infant Behav. Dev.* 4, 69–81.

Fish, M., Stifter, C. A., and Belsky, J. (1991). Conditions of continuity and discontinuity in infant negative emotionality: Newborn to five months. *Child Dev.* 62, 1525–1537.

Fisher, P. A., and Kim, H. K. (2007). Intervention effects on foster preschoolers' attachment-related behaviors from a randomized trial. *Prev. Sci.* 8, 161–170.

Fisher, P. A., Burraston, B., and Pears, K. (2005). The Early Intervention Foster Care Program: Permanent placement outcomes from a randomized trial. *Child Maltreat.* 10, 61–71.

Forgas, J. P. (1994). Sad and guilty? Affective influences on the explanation of conflict episodes. *J. Pers. Soc. Psychol.* 66, 56–68.

Ganiban, J. M., Ulbricht, J. A., and Neiderhiser, J. M. (2007). Pushing parents' buttons: Understanding child effects on parenting. Poster presented at the Biennial Meeting of the Social for Research on Child Development, Boston, MA.

Ge, X., Conger, R., Cadoret, R., Neiderhiser, J. M., Yates, W., Troughton, E., and Stewart, M. (1996). The developmental interface between nature and nurture: A mutual influence model of child antisocial behavior and parent behaviors. *Dev. Psychol.* 32, 574–589.

Gjone, H., Stevenson, J., Martin Sundet, J., and Eilertsen, D. E. (1996). Changes in heritability across increasing levels of behavior problems in young twins. *Behav. Genet.* 26, 419–426.

Harden, K. P., Turkheimer, E., Emery, R. E., D'Onofrio, B. M., Slutske, W. S., Heath, A. C. et al. (2007). Marital Conflict and Conduct Problems in Children of Twins. *Child Development* 78, 1–18.

Jaffee, S. R., Caspi, A., Moffitt, T. E., Polo-Tomas, M., Price, T. S., and Taylor, A. (2004). The limits of child effects: Evidence for genetically mediated child effects on corporal punishment but not on physical maltreatment. *Dev. Psychol.* 40, 1047–1058.

Jenkins, K., Rasbash, J., and O'Connor, T. G. (2003). The role of the shared family context in differential parenting. *Dev. Psychol.* 39, 99–113.

Kendler, K. S. (1996). Parenting: A genetic-epidemiologic perspective. *Am. J. Psychiatr.* 153, 11–20.

Kendler, K. S., Sham, P. C., and MacLean, C. J. (1997). The determinants of parenting: An epidemiological, multi-informant, retrospective study. *Psychol. Med.* 27, 549–563.

Kim, J. K., Conger, R. D., Lorenz, F. O., and Elder, G. H. (2001). Parent–adolescent reciprocity in negative affect and its relation to early adult social development. *Dev. Psychol.* 37, 775–790.

Kochanska, G., Clark, L. A., and Goldman, M. S. (1997). Implications of mothers' personality for their parenting and their young children's developmental outcomes. *J. Pers.* 65, 389–420.

Kochanska, G., Friesenborg, A. E., Lange, L. A., and Martel, M. M. (2004). Parents' personality and infants' temperament as contributors to their emerging relationship. *J. Pers. Soc. Psychol.* 86, 744–759.

Lau, J. Y. F., Rijsdijk, F. H., and Eley, T. C. (2006). I think, therefore I am: A twin study of attributional style in adolescents. *Journal of Child Psychology and Psychiatry* 47, 696–703.

Lee, C. L., and Bates, J. E. (1985). Mother–child interaction at age two years and perceived difficult temperament. *Child Dev.* 56, 1314–1425.

Leve, L. D., Scaramella, L. V., and Fagot, B. I. (2001). Infant temperament, pleasure in parenting, and marital happiness in adoptive families. *Infant Mental Health J.* 22, 545–558.

Leve, L. D., Chamberlain, P., and Reid, J. B. (2005). Intervention outcomes for girls referred from juvenile justice: Effects on delinquency. *J. Consult. Clin. Psychol.* 73, 1181–1185.

Lichtenstein, P., Ganiban, J., Neiderhiser, J. M., Pedersen, N. L., Hansson, K., Cederblad, M., Elthammar, O., and Reiss, D. (2003). Remembered parental bonding in adult twins: Genetic and environmental influences. *Behav. Genet.* 33, 397–408.

Losoya, S. H., Callor, S., Rowe, D. C., and Goldsmith, H. H. (1997). Origins of familial similarity in parenting: A study of twins and adoptive siblings. *Dev. Psychol.* 33, 1012–1023.

Lynch, S. K., Turkheimer, E., D'Onofrio, B. M., Mendle, J., Emery, R. E., Slutske, W. S., and Martin, N. G. (2006). A genetically informed study of the association between harsh punishment and offspring behavioral problems. *J. Fam. Psychol.* 20, 190–198.

Martinez, C. R., Jr., and Forgatch, M. S. (2001). Preventing problems with boys' noncompliance: Effects of a parent training intervention for divorcing mothers. *J. Consult. Clin. Psychol.* 69, 416–428.

McCrae, R. R., Costa, P. T., Ostendorf, F., Angleitner, A., Hrebickiova, M., Avia, M. D., Sanz, J., and Sanchez-Bernardos, M. L. (2000). Nature over nurture: Temperament, personality, and life span development. *J. Pers. Soc. Psychol.* 78, 173–186.

McGue, M., Elkins, I., Walden, B., and Iacono, W. G. (2005). Perceptions of the parent–adolescent relationship: A longitudinal investigation. *Dev. Psychol.* 41, 971–984.

McGuire, S. (2003). The heritability of parenting. *Parenting: Sci. Pract.* 3, 73–94.

Mendle, J., Turkheimer, E., D'Onofrio, B. M., Lynch, S. K., Emery, R. E., Slutske, W. S., and Martin, N. G. (2006). Family structure and age at menarche: A children-of-twins approach. *Dev. Psychol.* 42, 533–542.

Metsapelto, R., and Pulkkinen, L. (2003). Personality traits and parenting: Neuroticism, extraversion, and openness to experience as discriminative factors. *Eur. J. Pers.* 17, 59–78.

Moore, G. A., Lewinsohn, P., and Cohn, J. F. (1999, June). Stability of recall of parent–child relationships. Presented at the 12th Annual Convention of the American Psychological Society, Miami, FL.

Narusyte, J., Andershed, A. K., Neiderhiser, J. M., and Lichtenstein, P. (2007). Aggression as a mediator of genetic contributions to the association between negative parent-child relationships and adolescent antisocial behavior. *European Child and Adolescent Psychiatry* 16(2), 128–137.

Neiderhiser, J. M., Reiss, D., Hetherington, E. M., and Plomin, R. (1999). Relationships between parenting and adolescent adjustment over time: Genetic and environmental contributions. *Dev. Psychol.* 35, 680–692.

Neiderhiser, J. M., Reiss, D., Pedersen, N. L., Lichtenstein, P., Spotts, E. L., Hansson, K., Cederblad, M., and Ellhammer, O. (2004). Genetic and environmental influences on mothering of adolescents: A comparison of two samples. *Dev. Psychol.* 40, 335–351.

Neiderhiser, J. M., Reiss, D., Lichtenstein, P., Spotts, E. L., and Ganiban, J. (2007). Father adolescent relationships and the role of genotype-environment correlation. *Journal of Family Psychology* 21(4), 560–571.

O'Connor, T. G., Deater-Deckard, K., Fulker, D., Rutter, M., and Plomin, R. (1998). Genotype–environment correlations in late childhood and early adolescence: Antisocial behavioral problems and coercive parenting. *Dev. Psychol.* 34, 970–981.

Perusse, D., Neale, M. C., Heath, A. C., and Eaves, L. J. (1992). Human parental behavior: Evidence for genetic influence and implication for gene-culture transmission (abstract). *Behav. Genet.* 22, 744.

Pike, A., McGuire, S., Hetherington, E. M., Reiss, D., and Plomin, R. (1995). Family environment and adolescent depressive symptoms and antisocial behavior: A multivariate genetic analysis. *Dev. Psychol.* 32, 590–603.

Plomin, R. (1994). *Genetics and Experience. The Interplay Between Nature and Nurture.* Sage Publications, Newbury Park, CA.

Plomin, R., and Bergeman, C. S. (1991). The nature of nurture: Genetic influence on "environmental" measures. *Brain Behav. Sci.* 14, 373–427.

Plomin, R., DeFries, J., and Loehlin, J. C. (1977). Genotype–environment interaction and correlation in the analysis of human behavior. *Psychol. Bull.* 84, 309–322.

Price, J. M., Chamberlain, P., Landsverk, J., Reid, J. B., Leve, L. D., and Laurent, H. (2008). Effects of a foster parent training intervention on placement changes of children in foster care. *Child Maltreatment* 13, 64–75.

Prinzie, P., Onghena, P., Hellinckx, W., Grietens, H., Ghesquiere, P., and Colpin, H. (2004). Parent and child personality characteristics as predictors of negative discipline and externalizing problem behavior in children. *Eur. J. Pers.* 18, 73–102.

Reiss, D., and Leve, L. D. (2007). Genetic expression outside the skin: Clues to mechanisms of Genotype × Environment interaction. *Dev. Psychopathol.* 19, 1005–1027.

Reiss, D., Neiderhiser, J. M., Hetherington, E., and Plomin, R. (2000). *The relationship code: Deciphering genetic and social influences on adolescent development*. Harvard University Press, Cambridge, MA.

Reiss, D., Pedersen, N. L., Cederblad, M., Lichtenstein, P., Hansson, K., Neiderhiser, J. M., and Elthammar, O. (2001). Genetic probes of three theories of maternal adjustment: I. Recent evidence and a model. *Fam. Process* 40, 247–259.

Rende, R., Slomkowski, C. L., Stocker, C., Fulker, D. W., and Plomin, R. (1992). Genetic and environmental influences on maternal and sibling interaction in middle childhood: A sibling adoption study. *Dev. Psychol.* 17, 203–208.

Rothbart, M. K., and Posner, M. I. (2005). Genes and experience the development of executive attention and effortful control. *New Dir. Child Adolesc. Dev.* 109, 101–108.

Rothbart, M. K., Ahadi, S. A., and Evans, D. E. (2000). Temperament and personality: Origins and outcomes. *J. Pers. Soc. Psychol.* 78, 122–135.

Rothbaum, F., and Weisz, J. R. (1994). Parental caregiving and child externalizing behavior in nonclinical samples: A meta-analysis. *Psychol. Bull.* 116, 55–74.

Rowe, D. C. (1981). Environmental and genetic influences on dimensions of perceived parenting: A twin study. *Dev. Psychol.* 17, 203–208.

Rowe, D. C. (1983). A biometrical analysis of perceptions of family environment: A study of twins and singleton sibling kinships. *Child Dev.* 54, 416–423.

Rusting, C. L. (1998). Personality, mood, and cognitive processing of emotional information: Three conceptual frameworks. *Psychol. Bull.* 124, 165–198.

Rutter, M., and Silberg, J. (2002). Gene-environment interplay in relation to emotional and behavioral disturbance. *Annu. Rev. Psychol.* 53, 463–490.

Scarr, S., and McCartney, K. (1983). How people make their own environments: A theory of genotype → environment effects. *Child Dev.* 54, 424–435.

Silberg, J. L., and Eaves, L. J. (2004). Analysing the contributions of genes and parent–child interaction to childhood behavioural and emotional problems: A model for the children of twins. *Psychol. Med.* 34, 347–356.

Spinath, F. M., and O'Connor, T. G. (2003). A behavioral genetic study of the overlap between personality and parenting. *J. Pers.* 71, 785–808.

Spotts, E. L., and Neiderhiser, J. M. (2003). The developmental trajectory of genotype-environment correlation: The increasing role of the individual in selecting environments. In *The Transition to Early Adolescence: Nature and Nurture* (S. A. Petrill, J. K. Hewitt, R. Plomin, and J. C. DeFries, Eds.), pp. 295–309. Oxford University Press, New York.

Towers, H., Spotts, E. L., and Neiderhiser, J. M. (2001). Genetic and environmental influences on parenting and marital relationships: Current findings and future directions. *Marriage Fam. Rev.* 33, 11–29.

Ulbricht, J. A., and Neiderhiser, J. M. (2009). Genotype–environment correlation and family relationships. In *Handbook of Behavioral Genetics* (Y.K. Kim, Ed.), Springer, New York, in press.

US Department of Health and Human Services (2000a). *Child Maltreatment: 2000*. U.S. Department of Health and Human Services, Administration on Children and Families, Washington, DC.

US Department of Health and Human Services. (2000b). Children and mental health. In *Mental Health: A Report of the Surgeon General* (DHHS publication No. DSL 2000-0134-P), pp. 123–220. U.S. Government Printing Office, Washington, DC.

van den Boom, D. C. (1994). The influence of temperament and mothering on attachment and exploration: An experimental manipulation of sensitive responsiveness among lower-class mothers with irritable infants. *Child Dev.* 65, 1457–1477.

van Goozen, S. H. M., Fairchild, G., Snoek, H., and Harold, G. T. (2007). The evidence for a neurobiological model of childhood antisocial behavior. *Psychol. Bull.* 133, 149–182.

Vestre, N. D., and Caufield, B. P. (1986). Perception of neutral personality descriptions by depressed and nondepressed subjects. *Cogn. Ther. Res.* 10, 31–36.

Webster-Stratton, C., and Taylor, T. (2001). Nipping early risk factors in the bud: Preventing substance abuse, delinquency, and violence in adolescence through interventions targeted at young children (0–8). *Prev. Sci.* 2, 165–192.

V

THE NEUROBIOLOGY OF PATERNAL CARE

26

COOPERATIVE BREEDING AND THE PARADOX OF FACULTATIVE FATHERING

SARAH BLAFFER HRDY

Department of Anthropology, University of California Davis, CA, USA

INTRODUCTION: PRIMATE MALE–INFANT INVOLVEMENT

Across the vast majority of the world's 5,400 species of mammals, fathers do remarkably little beyond stake out territories, compete with other males and mate with females – a lot of *sturm* and *drung* punctuated by bellowing, barking, roaring and other spectacular audiovisual displays, but once competitors are routed and the progenitor chosen, its "wham, bam and thank you m'am" and the male is off. In this respect, males belonging to the Order Primates stand out as paragons of paternal responsibility. In most primates, males remain year-round with the females they mated with. There is no unitary explanation, but among the main reasons why primate males tend to stick around after mating are protection against predation (a major source of *adult* mortality) and protecting unweaned young from infanticide by rival male (a major source of *infant* mortality) (Hrdy, 1979; Palombit, 2000; Paul *et al.*, 2000). A population of savanna baboons studied at Moremi in Botswana provides a case in point. Over a 10-year period, 34% or more of infant mortality at Moremi was due to infanticide by adult males (Palombit *et al.*, 1997).

All primates share a deep evolutionary legacy of positive and negative male involvement with infants. Under some conditions, prolonged exposure of males to cues from helpless infants produced opportunities for Natural Selection to favor even more costly and exclusive care directed at infants likely to share genes with them by common descent. In a handful of cases, predispositions to paternal care evolved to the extremes that we see in monogamous titi monkeys (*Callicebus molloch*). As early as the first 3 weeks of life, males carry infants 90% of daytime, more than nursing mothers (Figure 26.1; Wright, 1984). Infants return the favor by actually preferring their fathers over mothers, becoming more agitated (as measured by elevated adrenocortical activity) when separated from their father than from their mother (Hoffman *et al.*, 1995). I know of only one other genus of mammal, the genus *Aotus*, another monogamous South American monkey with biparental care, where fathers rather than mothers are the primary attachment figures (Wolovich *et al.*, 2007).

Given this broad if sporadic primate-wide background of male involvement with infants, as well as the sporadic occurrence of shared care by females-other-than-the mother in many species, it is striking how *little* allomaternal care goes on among our closest primate relations, the Great Apes. Mothers refuse to allow any allomaternal access for the first 6 months or longer, no matter how eager the other group members are to take the infant. Unless the mother is incapacitated, direct male care is rarely observed. Orangutan and chimpanzee fathers, spend little time in the vicinity of mother's and infants. Gorilla silverbacks have

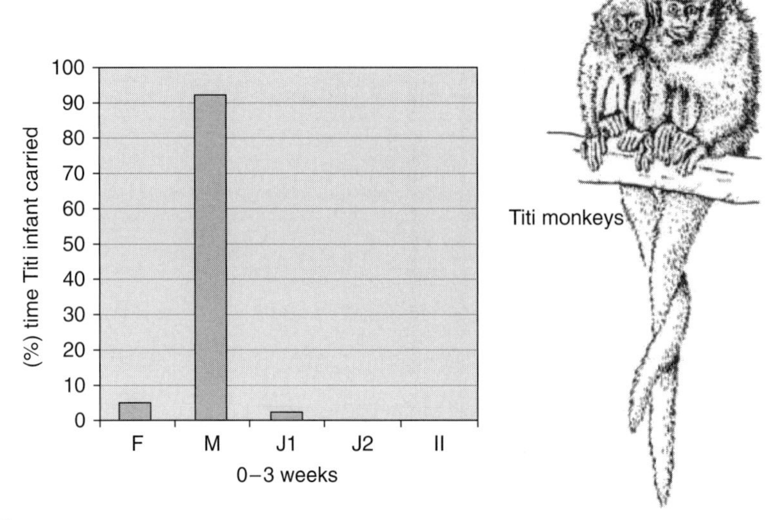

FIGURE 26.1 Percent of time each member of a wild group of titi monkeys (*Callicebus moloch*) in Peru carried the infant during the first 3 weeks after birth. F, adult female; M, adult male; J1, year-old juvenile; J2, 2-year old juvenile; II, independent. (*Source*: Adapted from Wright, 1984: figure 1.)

a high probability of paternity of infants born in their harems, and are unusually protective, so reliably protective that new mothers, and in time, infants themselves, strive to stay close to them. But still, there is no direct care, nothing like carrying or provisioning of infants in any of the Great Apes comparable to what Jeff French sees in his marmosets (see Chapter 30).

THE HUMAN CASE

As in other apes, highly vulnerable, immobile human infants born to nomadic foragers are virtually always held by someone. But there is one important difference: That someone is not necessarily the mother. Among the !Kung San, when these Kalahari people were still living as nomadic hunter-gatherers a newborn might be handed over by a new mother to her mother right after birth, and the grandmother might thereupon massage the newborn and gently shape his skull with her hands (e.g., see Hrdy, 1999; figure 7.9) Among the Hadza, still living as hunters and gatherers in Tanzania, as among the !Kung, "the baby" right from its birth "is likely to be surrounded by relatives, old adult and young... and is carried by them..."(Blurton Jones *et al.*, 1992). According to Marlowe (2005), a Hadza newborn is held by others 85% of the period just after birth.

Among Mbuti pygmies in Central Africa, "the mother emerges and presents the child to the camp..." whereupon "she hands the (baby) to a few of her closest friends and family, not just for them to look at him but for them to hold him close to their bodies..." (Turnbull, 1965; quote from Turnbull, 1978: p. 172). As far away as the Pacific among Phillippine foragers like the Agta, "The infant is eagerly passed from person to person until all in attendance have had an opportunity to snuggle, nuzzle, sniff and admire the newborn... Thereafter he enjoys constant cuddling, carrying, loving, sniffing and affectionate genital stimulation..." (Peterson, 1978; cited in Hewlett, 1991a).

Worldwide, such accounts are typical for hunter-gatherers. When anthropologist Ivey (2000) asked Central African Efé pygmy mothers "who cares for infants?" people immediately replied "We all do" which is literally true since infants are passed around right from the first day of life. Efé infants average 14 different caretakers and by 4 months of age are being carried by allomothers more time than by their own mothers. However, group members are not equally likely to care. When matched by age and sex with other group members, Ivey found that female allomothers are disproportionately older, sisters, aunts and grandmothers. Male allomothers are brothers, cousins and fathers – with grandfathers doing surprisingly little.

In addition to being held and carried by multiple caretakers, long before weaning, as early as 4 months, infants begin to receive treats like honey sweetened saliva, the first installments on several decades of allomaternal supplementation. In all apes breastfeeding goes on for years. Nevertheless, non-human ape infants are nursed for far longer than humans in hunting and gathering societies are. Intervals between births average 60–100 months compared to the 36–48 month intervals typical for nomadic foraging peoples (Knott, 2001). The reason humans can afford much shorter birth intervals, has to do with all the shared care and provisioning that goes on. Whereas other apes, once weaned provision themselves, human children remain dependent and rely on handouts from others for many years to come. As in many other cooperatively breeding birds and some cooperatively breeding mammals, humans have unusually long periods of "post-fledging" or "post-weaning" dependence, what in humans is known as "childhood" (Hrdy, 2005).

Many of us have had occasion to marvel at how much kids today cost. But keep in mind that ever since the Paleolithic and all through the Neolithic, human offspring have always taken a long time and cost a great deal to rear. Whether in foraging or horticultural societies, it takes on the order of 13 million calories to rear a human from birth to maturity, and most foragers will be 19 years or older before they begin to produce as much food as they consume, which means that the mother will bear other infants long before older children are independent. This is why, when Bob Bridges asked me to give a "Father's Day" talk, what came to my mind was this glaring human paradox: men and women mate to produce the costliest and slowest maturing young in all mammaldom, and yet compared to a titi monkey where a male's top priority in all the world is to hold and carry young born to his mate, men's priorities are nothing like so single-minded. Having a father hard-wired to help is not something a human mother can count on.

VARIABILITY IN PATERNAL CARE

Across cultures there is more variation in paternal care in humans than in all 275 species of other primates put together. Routinely, rates of male care are higher in hunter-gatherers than in herding, horticultural/farming or modern post-industrial societies. But even within foraging societies, male involvement ranges from relatively low levels of caretaking among the South African !Kung, to the highest ever reported – infants held by dads 22% of the time – among Central African Aka pygmies. The Aka are net hunters and husbands and wives both participate, taking children with them on expeditions into the forest, where even very young infants remain in ear-shot of their fathers most of the time (Hewlett, 1991a, b).

Contrast this with an average of 50 min per day found among fathers in the United States, up from 20 min a few decades ago (Sayer *et al.*, 2003). Cross-culturally there are also major differences in types of interactions (Lamb *et al.*, 1987; Hewlett & Lamb, 2005). Whereas American and European fathers compress a great deal of hyper-stimulating play into brief periods they spend with infants, in 264 h of systematic observation of father–infant interactions among Aka pygmies, Hewlett reported only one instance of "vigorous" play. Hunter-gatherer style paternal care tends to be soothing, affectionate and custodial and under some ecological conditions, absolutely critical for child survival.

Even more striking than stylistic differences in male care is the variation in paternal commitment between the "Mrs. Doubtfires" of the world who go to great lengths to stay near children (more often found in art than life), and fathers who are more remote, invest little or nothing at all in children at any age. Worldwide, 10–25% of households with children are headed by women. In countries like Botswana, Swaziland, Barbados and Grenada some 40% of households contain children with no father present, while in Zimbabwe, Norway, Germany and the United States, the proportion is closer to 30%. It is difficult to get accurate statistics for men who sire children without knowing or acknowledging it at all. But according to one recent survey in Chile, 42% of children born out of wedlock were receiving no support at all from fathers 6 years later (Engle & Breaux, 1998). In the United States, close to half of all children whose parents divorce lose touch with their fathers shortly after, and by 10 years up to 75% have lost touch (Dominus, 2005). Indeed, according to one survey by the Children's Defense Fund, Americans are 16 times more likely to repay used car loans than pay child support, the delinquency rate for the former being only 3%, compared to the 49% delinquency rate on child support (Associated Press, 1994).

Nor is there is anything particularly modern or evolutionarily novel about absent fathers. When Marlowe (2005) censused Hadza hunter-gatherers still living in the traditional way in southwestern Tanzania, only 36% of children had fathers in their same group. A hemisphere away, among Yanamamo tribes people the chance of a 10-year old child having both a father and a mother still living in the same group were one in three (Chagnon, 1992: p. 177), about the same as among the Aka (Hewlett, 1991a), while among Ongee foragers in the Andaman Islands *none* of the 15-year olds in the sample were still living with either natural parent (Hewlett, 1991a: pp. 19–20).

Even when fathers are around, hunting and fishing are notoriously unreliable ways to stay fed. Hunters may go for days without bringing back meat, in part because hunters prefer big splash prey to small game. Thus specialists on hunter-gatherer ecology and archeology like O'Connell *et al.*, (2002) point out that even though "meat represents a sizable fraction of their families' *annual* caloric intake, it is not acquired reliably enough to satisfy the *daily* nutritional needs of their children." To compensate, women became specialists in tapping more reliable food sources, like nuts, berries, insect grubs and underground tubers, and as among all foraging people, Plio-Pleistocene hominids had to have set up elaborate networks for sharing – a tremendously important part of their lives. Gaps between what children needed and what a mother could supply were potentially met by a range of group members. When he was alive, on hand, and able to provide, the mother's mate could be critically important. Otherwise, child survival often depended on a mother's other male kin or lovers, or post-reproductive helpers. Assistance by older offspring also freed mothers to forage more efficiently. Unrelated group members such as pre-reproductives fostered in from other groups, who like modern *au paires*"pay their keep" by helping, also play roles (Hrdy, 1999: Chapter 11).

THE IMPORTANCE OF ALLOMATERNAL CARE

We've known for a long time that outside of the Great Apes a lot of allomaternal assistance goes on in primates, especially shared care of infants by young females who eagerly practice with borrowed babies among these langurs, leaf monkeys and proboscis monkeys (reviewed in Hrdy, 1999). Furthermore, as females approach the end of their reproductive lives, in female philopatric species where aging females live among close matrilineal kin, old females may become especially dedicated in defending immature kin (Hrdy & Hrdy, 1976; Paul, 2005). Allomaternal defense of infants as well as allomaternal assistance caring for them means that wherever allomothers are available and mothers are willing to give up infants (presumably under conditions where they deem it safe to do so), there is a correlation between allomaternal care and enhanced maternal reproductive success. By the late 1990s, it was becoming apparent that the same correlation between allomaternal assistance and maternal reproductive success that was reported for some other primates, was going to hold for humans as well, especially those in populations with high rates of infant mortality (Hrdy, 1999, 2005 and references therein).

Humans fit the general primate pattern but with an important difference. In the human case, food sharing as well as shared knowledge about resources means that there is so much more that post-menopausal women can do to help kin on a daily basis. The anthropologist Kristen Hawkes first called attention to how hard post-reproductive females worked in hunter-gatherer societies and showed that the presence of older matrilineal kin was correlated with child survival, especially during periods of food shortage. Hawkes *et al.* (1998) have hypothesized that the reason post-menopausal women go on living for decades, rather than for just a year or two after menopause like other primates, is because of what old females can contribute to the survival of matrilineal descendants.

Across a wide range of traditional human societies – African hunter-gatherers, South Asian rice farmers, German peasants and West African horticulturalists, the availability of older matrilineal kin is correlated with faster child growth rates, shorter maternal birth intervals and increased child survival (summarized in Figure 26.2 and references therein; see also Voland *et al.*, 2005). The case of West African Mandinka horticulturalists is especially interesting. Based on a large data set on maternal health and child well-being collected in the middle of the last century, it was known that 40% of infants born (883 of 2,294) died before age five. But no one thought to look at allomaternal

PRESENCE OF OLDER MATRILINEAL KINSWOMEN CORRELATED WITH:

- Faster child growth rates
  - Hadza hunter-gatherers (Hawkes *et al.*, 1998)
  - West African horticulturalists (Sear *et al.*, 2000)
- Shorter maternal birth invervals
  - South Asian swidden agriculturalists (Leonetti, 2002)
- Increased child survival
  - 18th century German peasants (Voland & Beise, 2002)
  - West African horticulturalists (Sear et al., 2002)
- Increased lifetime reproductive success for mother
  - 18th and 19th century Finnish peasants (Lahdenperä *et al.*, 2004)

FIGURE 26.2 Summary of recent studies documenting that the presence of older matrilineal kin is correlated with faster child growth; shorter inter-birth intervals; increased child survival; and/or increased maternal reproductive success over her lifetime.

effects on these high mortality rates until 2000 when Sear and Mace reanalyzed the initial Gambian data set (Sear *et al.*, 2000). Whereas in some societies a father's presence is essential for infant survival (Hill and Hurtado, 1996), in the Gambia data set the father's presence had no detectable impact on infant survival in the first 2 years of life, though if the father died and the mother remarried, the presence of a stepfather turned out to be somewhat detrimental for the survival chances of older children. What did matter very much in this Mandinka case was having a maternal grandmother. Presence of a maternal grandmother was correlated with a halving of the mortality rate from 40% dying before age five, down to 20% (Sear *et al.*, 2002). To my knowledge, these sorts of significant effects from allomothers have only been reported in populations with infant mortality rates in the range of 40% or higher, that is, comparable to those found in wild primates and many foraging societies.

Because hunter-gather sample sizes are almost invariably tiny, the best documentation for this "grandmother effect" comes from archival records kept for pre-industrial European communities such as these 18th and 19th century Finnish farmers where presence of the mother's mother turns out to have been significantly correlated with that woman's enhanced fertility as well as child survival, resulting in significantly higher lifetime reproductive success for these women who have a grandmother on hand to help (Lahdenperä *et al.*, 2004).

Since 2000, there has been an explosion of research documenting the reproductive impact of post-reproductive kin on daughters' reproductive success (reviewed in Voland *et al.*, 2005) – provided kinswomen live long enough, or live close-by, which are noteworthy "ifs." In a sample of Aka pygmy infants studied by anthropologist Courtney Meehan, only one in two Aka infants had either maternal or paternal grandmothers present. In another, comparable sample of Efé infants, only one in four did (Ivey, 2000). Such demographic profiles are consistent with estimates calculated on the basis of archeological from now vanished Paleolithic nomads living under a range of circumstances. The higher the mortality risk, the less either a mother or her slow-maturing children can afford to depend on any specific family composition, and the more important it becomes for other kin, fathers, collateral kin and older siblings, to compensate, relocating if necessary, pitching in as needed.

MATERNAL BET-HEDGING

In environments with high adult mortality and/or extremely unpredictable resources, mothers need to hedge their bets against a shortage of allomaternal assistance. In disparate areas of the world, in parts of North and South America, Africa, Asia and the ancient Near East, one way mothers manage this is by lining up "extra" fathers. Customs and belief systems that function to help mothers pull this off vary tremendously and have become a special interest of mine. I only have time today to talk about the subset of these cases known as "partible paternity" belief systems.

Across a broad swath of forager-horticulturalist Amazonian societies, from the Canela of Brazil in the East, the Matis of Peru in the West, northwards to the Bari of Venezuela or Wayana of French Guyana, southward down to the Takana of Bolivia or the Aché of Paraguay, (also Arawete, Kalina, Kuikuru, Mehinaku,) people subscribe to a convenient folk wisdom about "partible paternity" (Beckerman & Valentine, 2002). A woman believes that semen from each of however many men she had sex with in the preceding 10 months contributed to the growth of the fetus developing inside her body. Fortunately, men believe this as well, and bring gifts of food to women they have mated with during their pregnancy. After birth, such "possible" fathers continue to provision children (e.g., see Hill & Magdalena Hurtado, 1996 for the Aché; Beckerman *et al.*, 1998 for the Bari).

Marriage patterns tend to be quite flexible. Among the Aché for example, the majority of marriages at any given time are monogamous, but these are dynamic unions that fluctuate through time, passing through polygynous or polyandrous phases. Sixty percent (11 of 18) of Aché men spent some time in a polyandrous marriage (one woman, several men) and most women have children with two or more fathers (Hill & Magdalena Hurtado, 1996). Odds are, a woman's official husband *will* be the genetic-father of any child she bears, but not necessarily. This combined with the belief in partible paternity is probably why uncertain paternity does not upset men as much as it might husbands from a more gene-focused society.

"Extra" fathers are socially recognized and may be invited or expected to observe the same dietary restrictions at the time of birth that the mother's official husband is. As a courtesy to the husband, "extra" fathers are expected to be discreet (Pollock, 2002), but in some partible paternity societies it is not only socially acceptable for a husband to permit real brothers or fictive "clan brothers" to have sex with his wife, but polyandrous liaisons during public ceremonies may actually be encouraged as reported for the Canela of Brazil (Crocker & Crocker, 1994).

For several of these societies, we have data on how extra fathers impact child survival (Hill & Magdalena Hurtado, 1996; Beckerman *et al.*, 1998). For example, 80% of 194 Bari children with a secondary possible father in addition to their primary father survived to age 15, compared to only 64% of 628 children without a second father participating (Beckerman *et al.*, 1998). The optimal number of "fathers" turns out to be two. Clearly mothers and children benefit from extra fathers, but because of high levels of child mortality, husbands benefit as well since they have a better than average chance of being the progenitor or at least a relative of the progenitor.

Obviously, women cannot *actually* produce multiple young sired by different fathers as is the case in litters born to mother lions, prairie dogs and wild dogs. Nor did women evolve to produce chimeric young, combining several gene lines within a single individual the way *Calithrix kuhli* can (Ross *et al.*, 2007) so that several males are super-closely related to the young they care for. But humans can rely on these cultural constructs to line up extra fathers, producing in our species the same functional outcomes other animals end up with through conventional evolution.

CIRCUMSTANCES FAVORABLE TO ELICITING MALE CARE

The point here is not that genetic relatedness does not matter, or that sexual jealousy is eliminated. Rather the point is that *eliciting male nurture is more complicated than a man being certain of his paternity.* People like the Aka, who are far more monogamous than the Aché or the Bari, ensure childcare in other ways (Hewlett, 1989). Proximity to infants is a key factor. Men have a lot of leisure time and spend it in camp with prolonged opportunities for intimate interactions (Hewlett, 1992). Over time, cues eliciting nurture in humans have taken on a life of their own, including selection on babies to be irresistible, broadcasting the infantile equivalent of sex appeal (Hrdy, 1999). To understand male care, we need to take into account a range of factors in addition to genetic relatedness. These include belief systems, time in proximity, exposure to infant cues, the man's relationships with the mother and his recent and past experience with children (Fleming *et al.*, 2002; Fleming, 2005). We also need to take into account *residence patterns*. Who else is around can be very important. Recently, in one of the first studies of its kind, Meehan (2005) set out to learn precisely how important.

Like most foragers, Aka pygmies in Meehan's study move around over the course of their lives. It is customary for a husband to come live for a time with his wife and her family where he hunts on behalf of his wife and her kin for a period of years (known as "bride service") until after one or more children have been born. Thereafter, the couple may stay or move with their children back to his people, or to another group altogether. This pattern of remaining near the wife's kin, living "matrilocally," until after children are born means that inexperienced young mothers are likely to be among their own kin when they give birth for the first time, an especially vulnerable time (Hrdy, 1999).

Aka mothers specifically say they prefer to live matrilocally because they have more kin to help. Indeed, mothers residing matrilocally received nearly 5 times more alloparental care than mothers living patrilocally, not counting care by the infant's own older siblings which did not vary with place of residence. Nevertheless, the amount of time the baby was held did not differ in matrilocal and patrilocal settings even though the baby had fewer alloparents eager to help, and even though the mother herself held the baby about the same amount of time. How could this be? The answer was the father, who engaged in 20 times more care in the patrilocal setting, so that fathers'contribution to what Meehan termed "high investment allomaternal care" rose from 2.6% in the matrilocal setting to 62% in the patrilocal setting (Marlowe, 2005 for Hadza foragers; see Figure 26.3; Meehan, 2005).

In other words, foragers are characterized by a highly flexible breeding system that can involve monogamy, polygyny, even polyandry, or any combination thereof, along with proactive and strategic maneuvering by alloparents. Human mothers and their infants are opportunistic and resourceful in eliciting care, and allomothers are flexible about providing it, relocating, adjusting, juggling and compensating in strategic ways.

This flexibility is the hallmark of the human family and provides the key to resolving the paradox of facultative fathering. Paternal care could be less than obligate, and early hominid mothers could overshoot their capabilities to provide because if a father turned out to be an indifferent nurturer, or if he disappeared altogether, his offspring might still pull through, permitting a non-investing "cad" to enjoy his cake and fitness too. Cynical as all this sounds, this model of the family based on cooperative breeding is consistent with a growing body of empirical evidence – most of it collected since the year 2000.

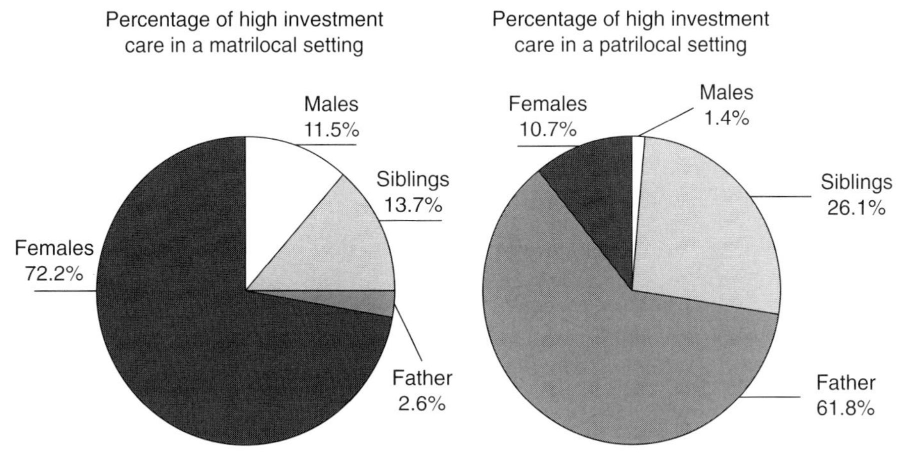

FIGURE 26.3 In Meehan's study of allomaternal childcare among Aka foragers, fathers engaged in more care in patrilocal settings where mothers tended to have fewer matrilineal kinswomen to rely on. (*Source*: Adapted from Meehan, 2005 and from data courtesy of C. Meehan.)

FLEXIBILITY OF HUMAN FAMILIES

Until recently, most anthropological reconstructions of humankind's "Environments of Evolutionary Adaptedness" took "Man the Hunter and his Sex Contract" for granted. It was simply assumed that "monogamous pair-bonding and nuclear families were dominant throughout human history in hunter-gatherer societies..." where monandrous mothers cared for infants in exchange for meat provided by a father certain of his paternity (Lovejoy, 1981). Supposedly, the "most straightforward explanation of the trend toward monogamy (being) that smart female hominids went to work on chimpanzee like hominid-males and – step by step, mate selection by mate selection – shaped them up into loving husbands and fathers with true family values..." (Lawrence & Nohria, 2002: p. 182). No mention was made of what happened when dads failed to adequately provide or of whom else might be involved.

Today however, human behavioral ecologists and sociobiologists attempting to reconstruct the deep history of the human family are increasingly taking the role of alloparents into account. There has been a paradigm shift from models based on a sex contract between man, the hunter and his nurturing mate toward a model based on cooperative breeding (Hrdy, 1999, 2005) that draws on a rich sociobiological literature to explain the evolution of shared care in other cooperatively breeding birds and mammals (Emlen, 1995; Solomon & French, 1997; Koenig & Dickinson, 2004). We continue to assume that our hunter-gatherer ancestors lived in small, intimate family units, but the composition of these families fluctuated through time, and we now take for granted that alloparental assistance was critical for successful childrearing. The highly idealized "nuclear family" (father, mother and their children) was typically just a temporary phase, often a less than optimal phase at that, since by themselves two parents would so rarely have been able to meet the needs of children. In reconstructing the Pleistocene family, the key descriptors I use are "kin-based," child-centered, opportunistic, mobile and very, very flexible. Alloparental safety nets provided the conditions in which highly facultative paternal commitment could evolve.

No one has a machine to travel back through time to observe childcare among African hominids 1.7 million years ago. Reconstructions based on ethnographic evidence from hunter-gatherers and comparative evidence from other primates are admittedly speculative. The best source of information we have about parental brains in early hominids remains the brains of extant humans, along with comparative data on the behavior and neurophysiologies of well-studied cooperatively breeding mammals like marmosets, tamarins, meerkats and voles, creatures who without benefit of a giant neocortex, language or symbolic culture still manage to make highly strategic decisions in regards to allocating care (Emlen, 1995; Bales *et al.*, 2002; Russell *et al.*, 2003; Fite *et al.*, 2005). However, we can not hope to understand the evolution of paternal brains without understanding *alloparental* brains as well. Why do allomothers help? Why does the "donative intent" of grandmothers run so high? Why are human mothers so tolerant of allomothers right after birth, while other apes are not? And how do males in non-cooperatively breeding primates, like chimpanzees, respond to infant cues? Are male responses to infants the same, or different from those Jeff French describes for the cooperatively breeding mammals they study?

A decade ago, in a paper on social and endocrine factors in alloparental behavior among voles, Sue Carter and Lucille Roberts proposed that some cooperative breeders might have "a distinct mechanism for alloparental behavior which is independent of other activational mechanisms for parental behaviors" (Roberts *et al.*, 1998). Even if the evolutionary origins of such behaviors *were* ultimately parental, I think there has been so much selection over time, that Lucille and Sue have to be right, and that in the human case there may be a number of such mechanisms, most of them age- and experience-dependent and highly facultative as Alison Fleming's work with humans is already beginning to suggest. To say then that these are exciting times to be studying the underpinnings of male nurture puts it mildly. There is a vast, nearly infinitely expandable, highly sustainable and as yet untapped human resource out there.

ACKNOWLEDGMENTS

Thanks to Matthew Gibbons for his assistances adapting this lecture for publication here.

REFERENCES

Associated Press (1994). Parents better at car payments than child support, group says. *Sacramento Bee* June 18, A8.

Bales, K., French, J., and Dietz, J. (2002). Explaining variation in maternal care in cooperatively breeding mammals. *Anim. Behav.* 63, 453–461.

Beckerman, S., and Valentine, P. (2002). *Cultures of Multiple Fathers: The Theory and Practice of Partible Paternity in Lowland South America.* University of Florida Press, Gainesville.

Beckerman, S., Lizarralde, R., Ballew, C., Schroeder, S., Fingelton, C., Garrison, A., and Smith, H. (1998). The Bari partible paternity project: Preliminary results. *Curr. Anthropol.* 39, 164–167.

Blurton Jones, N., Smith, L., O' Connell, J., Hawkes, K., and Kamuzora, C. (1992). Demography of the Hadza, an increasing and high density population of savanna foragers. *Am. J. Phys. Anthropol.* 89, 159–181.

Chagnon, N. (1992). *Yanomamo: The Last Days of Eden.* Harcourt Brace & Co., New York.

Crocker, W., and Crocker, J. (1994). *The Canela: Bonding Through Kinship, Ritual and Sex.* Harcourt Brace & Co., Fort Worth, Texas.

Dominus, S. (2005). The father's crusade. *New York Times Magazine* May 8.

Emlen, S. (1995). An evolutionary theory of the family. *Proc. Natl Acad. Sci. USA* 92, 8090–8099.

Engle, P. L., and Breaux, C. (1998). Fathers' involvement with children: Perspectives from developing countries. *Soc. Pol. Report* XII, 1–21.

Fite, J., French, J. A., Patera, K. J., Rukstalis, M., and Hopkins, E. (2005). Opportunistic mothers: Marmoset mothers with elevated and diminished energetic constraints reduce their investment in offspring. *J. Hum. Evol.* 49, 122–142.

Fleming, A. S. (2005). Plasticity of innate behavior: Experiences throughout life affect maternal behavior and its neurobiology. In *Attachment and Bonding: A New Synthesis. Report on 92nd Dahlem Workshop* (C. S. Carter, *et al.*, Ed.), pp. 137–168. The M.I.T. Press, Cambridge.

Fleming, A. S., Kraemer, G. W., Gonzalez, A., Lovic, V., Rees, S., and Melo, A. (2002). Mothering begets mothering: The transmission of behavior and its neurobiology across generations. *Pharmacol. Biochem. Behav.* 73, 61–75.

Hawkes, K., O'Connell, J. F., Blurton Jones, N. G., Alvarez, H., and Charnov, E. L. (1998). Grandmothering, menopause and the evolution of human life histories. *Proc. Natl Acad. Sci. USA* 95, 1336–1339.

Hewlett, B.S. (1989). *Diverse contexts of human infancy*. Prentice Hall: Englewood Cliffs, New Jersey.

Hewlett, B. (1991a). Demography and childcare in preindustrial societies. *J. Anthropol. Res.* 47, 1–37.

Hewlett, B. (1991b). *Intimate Fathers: The Nature and Context of Aka Pygmy Paternal Infant Care.* University of Michigan Press, Ann Arbor.

Hewlett, B. (1992). Husband–Wife reciprocity and the father–infant relationship among Aka pygmies. In *Father–Child Relations: Cultural and Biosocial Contexts* (B. Hewlett, Ed.), pp. 153–176. Aldine De Gruyter, Inc., New York.

Hewlett, B. S., and Lamb, M. eds. (2005). *Hunter-gatherer childhoods.* Piscataway, NJ: Aldine/Transaction.

Hill, K., and Magdalena Hurtado, A. (1996). *Ache Life History: The Ecology and Demography of a Foraging People.* Aldine de Gruyter, Inc., Hawthorne, New York.

Hoffman, K. A., Mendoza, S., Hennessey, M., and Mason, W. (1995). Responses of infant titi monkeys *Callicebus molloch* to removal of one or both parents: Evidence of paternal attachment. *Dev. Psychobiol.* 28, 399–407.

Hrdy, S. B. (1979). Infanticide among animals: A review, classification, and examination of the implications for the reproductive strategies of females. *Ethol. Sociobiol.* 1, 13–40.

Hrdy, S. B. (1999). *Mother Nature: A History of Mothers, Infants and Natural Selection.* Pantheon Inc., New York.

Hrdy, S. B. (2005). Comes the child before man: how cooperative breeding and prolonged postweaning dependence shaped human potential. In *Hunter-Gatherer Childhoods* (B. S. Hewlett and M. E. Lamb, Eds.), pp. 65–91. AldineTransaction, New Brunswick.

Hrdy, S. B., and Hrdy, D. B. (1976). Hierarchical relations among female Hanuman langurs. *Science* 193, 913–915.

Ivey, P. (2000). Cooperative reproduction in Ituri forest hunter-gatherers: Who cares for Efé infants?. *Curr. Anthropol.* 41, 856–866.

Knott, C. (2001). Female reproductive ecology of the apes: Implications for human Evolution. In *Reproductive Ecology and Human Evolution* (P. Ellison, Ed.), pp. 429–463. Aldine de Gruyter, Hawthorne, NY.

Koenig, W. D., and Dickinson, J. L. (2004). *Ecology and Evolution of Cooperative Breeding in Birds.* Cambridge University Press, Cambridge.

Lahdenperä, M., Lummaa, V., Helle, S., Tremblay, M., and Russell, A. F. (2004). Fitness benefits of prolonged post-reproductive lifespan in women. *Nature* 428, 178–181.

Lamb, M., Pleck, J. H., Charnov, E. L., and Levine, J. A. (1987). A biosocial perspective on paternal behavior and involvement. In

Parenting Across the Lifespan (J. B. Lancaster, J. Altmann, A. Rossi, and L. Sherrod, Eds.). Aldine, Hawthorne, NY.

Lawrence, P. R., and Nohria, N. (2002). *Driven: HowHuman Nature Shapes Our Choices*. Jossey-Bass, San Francisco.

Leonetti, D., Nath, D. C., Hemam, N. S., and Neill, D. B. (2002). Cooperative breeding effects among the matrilineal Khasi of N.E. India. *Paper presented at Human Behavior and Evolution Society Meetings*, Rutgers, University, Newark, New Jersey.

Lovejoy, C. O. (1981). The origin of man. *Science* 211, 341–350.

Marlowe, F. (2005). Who tends Hadza children?. In *Hunter-Gatherer Childhoods* (B. S. Hewlett and M. E. Lamb, Eds.), pp. 177–190. AldineTransaction, New Brunswick.

Meehan, C. (2005). The effects of residential locality on parental and alloparental investment among the Aka foragers of the Central African Republic. *Hum. Nat.* 16, 58–80.

O'Connell, J. F., Hawkes, K., Lupo, K. D., and Blurton Jones, N. G. (2002). Male strategies and Plio-Pleistocene archeology. *J. Hum. Evol.* 43, 831–872.

Palombit, R. A. (2000). Infanticide and the evolution of male–female bonds in animals. In *Infanticide by Males and Its Implications* (C. P. Van Schaik and C. R. Janson, Eds.), pp. 239–268. Cambridge University Press, Cambridge.

Palombit, R. A., Seyfarth, R. M., and Cheney, D. L. (1997). The adaptive value of "friendships" to female baboons: Experimental and observational evidence. *Anim. Behav.* 54, 599–614.

Paul, A. (2005). Primate predispositions. In *Grandmotherhood: The Evolutionary Significance of the Second Half of Female Life* (E. Voland, A. Chasiotis, and W. Schiefenhövel, Eds.), pp. 21–37. Rutgers University Press, New Brunswick.

Paul, A., Preuschoft, S., and Van Schaik, C. P. (2000). The other side of the coin: Infanticide and the evolution of affiliative male–infant interactions in old world Primates. In *Infanticide by Males and Its Implications* (C. P. Van Schaik and C. R. Janson, Eds.), pp. 269–292. Cambridge University Press, Cambridge.

Peterson, J. T. (1978). *The Ecology of Social Boundaries: The Agta Foragers of the Phillipines*. University of Illinois Press, Chicago.

Pollock, D. (2002). Partible paternity and multiple paternity among the Kulina. In *Cultures of Multiple Fathers: The Culture and Practice of Partible Parternity in Lowland South America* (S. Beckerman and P. Valentine, Eds.), pp. 42–61. University of Florida Press, Gainesville.

Roberts, R. L., Miller, A. K., Taymans, S. E., and Carter, C. S. (1998). Role of social and endocrine factors in alloparental behavior of prairie voles (*Mircrotus ochrogaster). Can. J. Zool.* 76, 1862–1868.

Ross, C. N., French, J. A., and Orti, G. (2007). Germline chimerism and paternal care in Marmosets (*Callithrix kuhlii*). *Proc. Natl Acad. Sci. USA* 104, 6278–6282.

Russell, A. F., Sharpe, L. L., Brotherton, P. N. M., and Clutton-Brock, T. H. (2003). Cost minimization by helpers in cooperative vertebrates. *Proc. Natl. Acad. Sci. USA* 100, 3333–3338.

Sayer, L. C., Bianchip, S. M., and Robinson, J. P. (2003). Are parents investing less in children? Trends in mothers' and fathers' time with children. *Am. J. Sociol.* 110, 1–43.

Sear, R., Mace, R., and McGregor, I. A. (2000). Maternal grandmothers improve the nutritional status and survival of children in rural Gambia. *Proc. Roy. Soc. Lond. B Biol. Sci.* 267, 461–467.

Sear, R., Steele, F., McGregor, I. A., and Mace, R. (2002). The effects of kin on child mortality in rural Gambia. *Demography* 39, 43–63.

Solomon, N. G., and French, J. A. (1997). The study of mammalian cooperative breeding. In *Cooperative Breeding in Mammals* (N. G. Solomon and J. A. French, Eds.), pp. 1–10. Cambridge University Press, Cambridge.

Turnbull, C. (1965). The Mbuti pygmies: An ethnographic survey. *Anthropological Papers of the American Museum of Natural History* 50(3), 141–282.

Turnbull, C. (1978). The politics of non-aggression. In *Learning Non-Aggression* (A. Montagu, Ed.), pp. 161–221. Oxford University Press, Oxford.

Voland, E., and Beise, J. (2002). Opposite effects of maternal and paternal grandmothers on infant survival in historical Krummhörn. *Behavioral Ecology and Sociobiology* 52, 435–443.

Voland, E., Chasiotis, A., and Schiefenhovel, W. (2005). *Grandmotherhood: The Evolutionary Significance of the Second Half of Female Life.* Rutgers University Press, New Brunswick, NJ.

Wolovich, C. K., Perea-Rodriguez, J. P., and Fernadez-Duque, E. (2007). Food sharing as a form of parental care in wild owl monkeys (*Aotus azarai*). *Abstract. Paper presented at the 77th Annual Meeting of the American Association of Physical Anthropologists*, March 28–31, Philadelphia.

Wright, P. C. (1984). Biparental care in *Aotus trivirgatus* and *Callicebus moloch*. In *Female Primates* (M. Small, Ed.), pp. 59–75. Alan Liss, New York.

27

EARLY EXPERIENCE AND THE DEVELOPMENTAL PROGRAMMING OF OXYTOCIN AND VASOPRESSIN

C. SUE CARTER[1], ERICKA M. BOONE[1] AND KAREN L. BALES[2]

[1] *Department of Psychiatry, Brain Body Center, University of Illinois at Chicago, Chicago, IL 60612, USA*
[2] *Department of Psychology, University of California, Davis, CA 95616, USA*

INTRODUCTION

Genetic and epigenetic factors influence social behaviors across the lifespan. Of particular importance are experiences in early life, especially when the nervous system is rapidly developing. At this time neural and endocrine systems may be vulnerable to modifications, with the capacity to re-tune social behavior in later life. The consequences of early experiences may be mediated by endocrine changes originating in the developing organism itself (including systems that rely on steroids or peptides), or through maternal influences, including the epigenetic effects of postnatal parental stimulation (Kaffman & Meaney, 2007). Compounds of maternal origin also might be transmitted through the mother to the infant, either during the prenatal or postnatal period. In addition, in the postnatal period maternal milk contains biologically active compounds, including prolactin (Grosvenor *et al.*, 1993), cortisol (Glynn *et al.*, 2007), and oxytocin (OT) (Leake *et al.*, 1981). These hormones, in turn, may regulate the endocrine system of the developing neonate, with long-lasting consequences for behavior and physiology.

OT is widely manipulated by medical interventions and potentially by different patterns of infant care. It is particularly common to manipulate OT in women for the purpose of regulating the timing of birth. It has been assumed that exogenous OT (pitocin) given to the mother does not pass through the placenta in amounts sufficient to affect the baby. However, the placental barrier may be disrupted during birth. In addition, exposures to pitocin (used to hasten labor), or treatments such as atosiban (an OT antagonist; OTA used to delay labor) vary in timing, duration, and dose. Treatments given to the mother hold the potential to affect the infant indirectly (e.g., by altering the strength or duration of uterine contractions, relative exposure to hypoxia, or the subsequent ability of the mother to lactate or interact normally with her offspring).

The main focus of the studies described here is how early hormonal experience might influence the offspring. Because of the complexity associated with the birth process, the experimental approaches described focus on direct treatments given to offspring during the immediate postnatal period. We also have centered our initial investigations of the effects of early experiences on social behaviors and neuroendocrine processes that are known to be, in later life, peptide dependent or strongly influenced by OT or arginine vasopressin (AVP), a related neuropeptide. We have focused our chapter on studies in prairie voles, a highly social rodent species in which OT and AVP have been shown to be of particular importance to behavior and physiology.

BACKGROUND ON OXYTOCIN AND VASOPRESSIN

OT and AVP are neuropeptides consisting of a six amino acid ring, with a three amino acid tail.

The molecular structures of OT and AVP differ by two of nine amino acids. OT and AVP are synthesized primarily in the central nervous system. Both peptides are found in particularly high concentrations in the paraventricular (PVN) and the supraoptic nuclei (SON) of the hypothalamus. From the PVN and SON, OT and AVP are carried by axonal transport to the posterior pituitary, where they are released into the blood stream. OT and AVP are also released into the central nervous system. Only one OT receptor (OTR) has been described. The same receptor is present in neural tissue and in other parts of the body including the uterus (Gimpl & Fahrenholz, 2001). Three distinct receptor subtypes have been identified for AVP. Of these, the V1a receptor (V1aR), which is found in the brain, has been associated with parental behavior (Bester-Meredith & Marler, 2003; Bales *et al.*, 2007b) and pair bond formation, especially in males (Winslow *et al.*, 1993). Receptors for both OT and AVP are localized in areas of the nervous system that play a role in reproductive, social, and adaptive behaviors, and in the regulation of the hypothalamic–pituitary–adrenal (HPA) axis and the autonomic nervous system. The AVP V1aR exhibits species and individual differences, due at least in part to variations in the promotor region of the gene responsible for expression of that receptor (Hammock & Young, 2005). The OTR also shows species differences (Witt *et al.*, 1991; Insel & Shapiro, 1992) in expression, although the molecular basis of these differences is less well understood.

AVP peptide synthesis within the central nervous system is sexually dimorphic and influenced by developmental exposure to androgens (De Vries & Simerly, 2002). Specifically, cells that synthesize AVP in the amygdala and bed nucleus of the stria terminalis (BNST) and fibers from these cells that extend into the lateral septum, are more abundant in males. In addition, males seem more responsive than females to AVP (Winslow *et al.*, 1993). These findings, as well as research in knock-out mice made defective for the V1aR, support the general conclusion, with regard to behavioral and emotional regulation, that males are more behaviorally dependent than females on central AVP and AVP receptors (Lim *et al.*, 2004; Bielsky *et al.*, 2005).

Recent studies in humans implicate OT and AVP in the response to stressors, repetitive behaviors and anxiety (reviewed, Carter, 2007). OT may reduce anxiety and reactivity to stressors, as well as repetitive behaviors. Research in animals also implicates OT in maternal behaviors, sexual behavior, social recognition, and social contact, and OT can facilitate pair bonding. AVP, working in conjunction with OT, may be of particular relevance to male social behaviors including parental behavior, pair bond formation, and mate guarding. The effects of AVP probably differ as a function of dose, with lower doses being anxiolytic and higher doses anxiogenic (Carter, 2003).

Sex differences in the AVP system could have broad consequences for behavior and physiology, forming an underlying substrate for processes that are typically attributed to the effects of androgens. It is possible that in males both peptide synthesis and receptor functions are primed by early exposure to androgens, including exposure in the early postpartum period. The capacity of OT manipulations in early life to have long-term effects on neural systems involving AVP would be expected to have sexually dimorphic consequences for many aspects of physiology and behavior. However, in comparison to OT, the possible behavioral functions of AVP in females have received less attention, so these conclusions must remain tentative.

SIMILARITIES OF FUNCTION BETWEEN OT AND AVP

OT and AVP have some overlapping functions, possibly because they can bind to each other's receptors (Barberis & Tribollet, 1996). In some situations the behavioral actions of OT and AVP appear similar, but more often the behavioral effects of OT and AVP seem to be opposing (Carter, 1998). In general OT is associated with increases in sociality, reductions in mobilization, and can induce reductions in the activity of the HPA axis, while AVP may also facilitate certain forms of social behavior, but is often associated with increased arousal and elevations in the activity of the HPA axis. Female reproductive strategies including birth, lactation, and immobile sexual postures may be facilitated by OT, while active coping and overt defensive behaviors may rely on AVP (Porges, 1998). The differential actions of OT and AVP may be of particular importance to understanding sex differences in social behavior.

During critical periods in development, manipulations of OT or AVP may influence the expression of these same peptides, as well as their receptors, with life-long behavioral consequences. However, it is not a simple matter to untangle differential functions of OT and AVP, especially using available pharmacological tools. In addition to the ability of OT and AVP to bind to each other's receptors (Barberis & Tribollet, 1996), the OTA widely used in behavioral studies is not totally selective for the OTR and also affects the AVP V1aR.

EFFECTS OF EARLY HANDLING IN PRAIRIE VOLES

In prairie voles we find that small manipulations, created by different methods of handling the parents or offspring in the postnatal period, can have life-long behavioral consequences (Bales *et al.*, 2007a; see Table 27.1). In a series of experiments, we have examined the effects of differential early behavioral treatments of the infants and family during the neonatal period on later behavioral and neuroendocrine responses in prairie voles. Compared to studies done in rats, which typically involve periods of separation from the mother, the handling manipulations described here in voles were very subtle. During a regular cage change, usually within the first day of life, the parents and offspring were transferred to a fresh cage using a cup, without direct "handling" of the animals (MAN0), or by picking them up by the scruff of the neck for between cage transfers (MAN1). Because pups of this species have milk teeth, they are typically attached to their mother and are transferred with her when she is moved. Observations during the postnatal periods suggest that *mild* disruptions of the family, such as those experienced in the MAN1 treatment, are followed by increases in parental behavior, especially on the part of the mother, but also by the father. In contrast, the MAN0 treatment was associated with lower levels of parental stimulation of the young. Furthermore, excess stimulation in early life, induced by repeated neonatal handling (picked up three times on the first day of life; MAN1 × 3) seemed to distract the parents from showing pup-directed behaviors. Thus, repeated handling, especially in the first days of life, produced immediate and long-term effects that were in some respects similar to those observed in the MAN0 condition (Boone *et al.*, 2006; Tyler, Carter, and Bales, in preparation Tyler *et al.*, 2005).

Behavioral testing as a function of early experience, summarized below, was typically conducted in the postweaning, juvenile period (approximately 21–23 days of age) or in adulthood (at approximately 60–90 days of age). The postweaning period is especially important in prairie voles since the willingness of animals to show social responses within the family, including alloparental behavior (care for

TABLE 27.1 A summary of effects of early experience on prairie voles. Changes are in comparison to MAN1 animals, whose parents are picked up by the scruff of the neck during cage changes. Parents of MAN0 animals are handled with a cup during early cage changes, while parents of MAN1 × 3 animals are picked up by the scruff three times on PND1

	MAN0	MAN1 × 3
Alloparental care	Low in males, normal in females	Low in MAN1 × 3 males
Anxiety (reduced EPM exploration)	High in both sexes	High in males
Pair bonding	Inhibited in females, but not in males	Not measured
OT-ir, PVN	No change from MAN1	Males lower than MAN1
OT-ir, SON	Lower than MAN1	Males lower than MAN1
Baseline corticosterone	No change from MAN1	No change from MAN1
Stress changes in corticosterone	Longer to return to baseline	Higher in MAN1 × 3 males than MAN1 males
Parental care toward own offspring	Lower licking of pups for both males and females	Not measured

offspring that are not their own) can have impact for the development of extended families. In nature increased availability of caretakers is associated with high levels of infant survival (Getz & Carter, 1996).

EARLY EXPERIENCE ALSO ALTERS SUBSEQUENT BEHAVIOR AND MEASURES OF OT AND THE OTRS

As a function of early differential handling, male and female voles both show subsequent changes in behavior; however, the nature of these effects is often sexually dimorphic. For example, animals that received minimal manipulations during the first week of life, here described as "unmanipulated" or MAN0, have been compared to animals exposed to mild stimulation (MAN1). MAN0 males were later less likely to exhibit alloparental care toward pups (see Figure 27.1) and in later life were less exploratory in the open arm of an elevated plus maze (EPM), often used to index anxiety (Bales *et al.*, 2007a). In contrast, MAN0 vs. MAN1 female voles did not differ in alloparental behavior (measured in the postweaning period). However, in adulthood MAN0 females were less likely than MAN1 females to form pair bonds, but were also more anxious, as measured by behavior in the EPM. The first week appears to be a critical period for effects of differential handling, and animals that were first manipulated on postnatal day 8 were similar to MAN0 animals (Bales *et al.*, 2007c).

Among the long-lasting effects of early handling were relative differences in the expression of the OT peptide in the central nervous system, which might account for at least some of the reduced sociality in MAN0 animals that received reduced handling disturbance in early life. In comparison to MAN1 animals, MAN0 voles of both sexes had significantly fewer OT immunoreactive cells in the supraoptic nucleus at 21 days of age (Bales *et al.*, 2007c). Pair bonding in female prairie voles is facilitated by OT and blocked in adulthood by OTA (Williams *et al.*, 1994). Thus, experiences that reduce available OT could produce animals that are less likely to pair bond.

INTERGENERATIONAL EFFECTS OF EARLY EXPERIENCE

There is increasing evidence from rats that early experience, possibly mediated by differential parenting, can be transmitted to subsequent generations of offspring (Francis *et al.*, 1999; Pedersen & Boccia, 2002). Using the handling model described above, infants reared under MAN0 and MAN1 conditions were as adults mated in various combinations (MAN0 female × MAN1 male, MAN0 male × MAN1

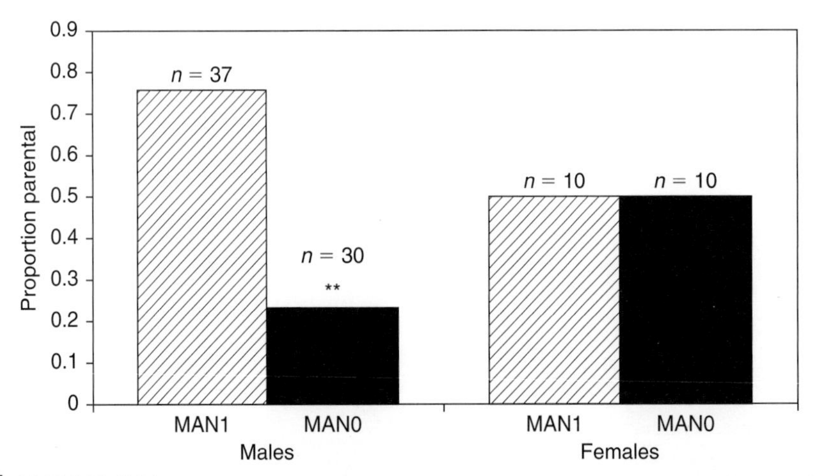

FIGURE 27.1 Proportion of animals that behaved parentally as a function of neonatal manipulation group ($n = 67$, $\chi^2 = 29.16$, $p < 0.0001$). Data for males include result of three replications in two different laboratories. Data for females are from a single experiment carried out at the University of Illinois at Chicago (*Source*: Reproduced from Bales *et al.* (2007a)).

female, MAN1 male × MAN1 female) and examined for possible differences in sexual behavior, reproductive success, and parental behavior toward their own offspring. While most effects on reproduction fell short of statistical significance, pairs containing a MAN0 male took longer to produce their first litter, although the frequencies of sexual activities, such as bouts of mounting and intromission, were actually higher in pairs containing MAN0 versus MAN1 females. Females in pairs with either MAN0 males or MAN0 females licked their pups less than MAN1 × MAN1 pairs (Bales *et al.*, 2007c). Earlier studies in prairie voles have suggested that OT can inhibit male sexual behavior in this species (Mahalati *et al.*, 1991), while OT may enhance parental behavior in both sexes in this species (Bales *et al.*, 2004b, 2007d). In the second generation all animals were treated like the MAN1 condition. The offspring of the second generation, reared by parents with either a MAN0 father or mother, displayed significantly lower frequencies of alloparental behavior than that of pairs reared by MAN1 × MAN1 parents (Bales *et al.*, 2007c). These studies demonstrate that these effects of early handling in prairie voles can be passed to the next generation.

THE CONSEQUENCES OF EARLY EXPERIENCE FOR PEPTIDE RECEPTORS

There is increasing evidence that the long-lasting effects of social experiences may be in part mediated via developmental changes in receptors for OT or AVP. In rats, experience-based changes in both the OT and V1aRs are sexually dimorphic (Champagne *et al.*, 2001; Francis *et al.*, 2002), with more marked changes in OTRs in females and the V1aRs in males. In mice, the consequences of genetically induced deficiencies in the V1aR are also gender specific, having greater consequences in males than females (Lim *et al.*, 2004; Bielsky *et al.*, 2005). In addition, in rats prenatal stress is associated in later life with reductions in the OTR (Champagne & Meaney, 2006).

Our preliminary data indicate that the manipulations of early experience described above also have long-term effects on peptide receptors, particularly OTRs. These effects are not always in the predicted directions and also appear to be sexually dimorphic (Bales *et al.*, 2007c).

DEVELOPMENTAL MANIPULATIONS OF OXYTOCIN

The OT system has exceptional plasticity and may be affected by a variety of developmental factors. OT neurons in adult rats have been described as "immature." These neurons have the capacity to change shape and form new synapses, in part through changes in the glia that normally separate neurons. Both OT and AVP may directly or indirectly influence cellular growth, death or motility, inflammation or differentiation. The potential to remodel the nervous system, especially in early life, offers another process through which OT or AVP may have effects on physiology and behavior. The OTR is also susceptible to epigenetic regulation, for example, by silencing genes via methylation (Kimura *et al.*, 2003; Szyf *et al.*, 2005). The capacity of genes that code for receptors to be silenced or otherwise modified in early life may be particularly relevant to understanding the long-lasting consequences of early experiences, whether originating as behavioral or hormonal experiences.

Possible short-term and long-term consequences for young mammals for exposure to exogenous OT or OTAs have been investigated, primarily in prairie voles (Carter, 2003, 2007). The data collected thus far in animal models have confirmed our original hypothesis that exposure to either OT or an OTA during development can have both immediate and long-lasting consequences for behavior and physiology.

In the experiments described below neonatal prairie voles were treated with either OT (usually 1 mg/kg) or OTA (0.1 mg/kg). Control treatments included either a sterile saline injection or handling. Treatments were given as an intraperitoneal injection on postnatal day 1 (PND1); infants were marked by toe-clipping and then returned to their parents. Litters were comprised of animals of both sexes receiving one experimental treatment (OT or OTA) and one control treatment (saline or handling).

IMMEDIATE OR SHORT-TERM EFFECTS OF NEONATAL OT OR OTA

Neural Activation Following Neonatal Treatment

When neonatal voles were exposed on postnatal day 1 (PND1) to OT or OTA, changes were detected in neural activation, as measured by changes in c-Fos expression within 1 h or less of treatment (Cushing *et al.*, 2003b). Both OTA and OT treatment caused changes in c-Fos expression in the supraoptic area (SON), with OT increasing c-Fos in males while OTA increased c-Fos in females. Additionally OTA decreased c-Fos expression in the medial dorsal thalamic nucleus in females, but not males (Cushing *et al.*, 2003b).

Behavioral and Endocrine Changes in Neonates

In female voles treatment on PND1 with OT or OTA (vs. control procedures) produced changes in corticosterone, and in reactivity to acute social isolation in animals tested on PND8 (Kramer *et al.*, 2003). Females receiving OTA (PND1), relative to controls, had on PND8 elevated basal plasma corticosterone and also vocalized significantly less when isolated from their family. The effect of OTA differed as a function of single vs. repeated injection procedures; females given repeated exposure to OTA (daily injections PND1–7) vocalized significantly more during social isolation than did controls injected daily with saline or handled on PND1–7.

LONG-TERM EFFECTS OF NEONATAL OT AND OTA

Alloparental Behavior, Sociality, and Reproduction in Males

The long-term consequences of transient neonatal exposure to comparatively low doses of OT (1 mg/kg on PND1) tended to be similar to those seen when OT was administered during adulthood. When effects were detected, neonatal OT was associated with increased levels of sociality. For example, males receiving a single neonatal exposure to OT (PND1) formed pair bonds more quickly than animals receiving either no treatment or a saline injection (Bales & Carter, 2003b). The behavioral effects of neonatal OTA (0.1 mg/kg on PND1) often (but not always) were in directions that were opposite to those of OT; in general, reductions in social behavior in later life were seen in animals neonatally exposed to OTA. Males receiving OTA (PND1) were in later life less alloparental and more aggressive toward pups (Bales *et al.*, 2004c). OTA-treated males also tended to be less aggressive toward same-sexed strangers, which in this species may be indicative of increased anxiety (Bales *et al.*, 2004c).

When a higher neonatal dosage of OT (4 mg/kg) was administered, males also formed a pair bond (Figure 27.2). However, males receiving an intermediate dosage of 2 mg/kg did not demonstrate a preference for either the partner or the stranger, suggesting that the effects of OT exposure are not necessarily linear.

We also carried out a preliminary study in males in which we attempted to remediate the deficits in alloparental behavior produced by PND1 exposure to OTA (0.1 mg/kg) (or saline), and using a subsequent treatment on PND8 with OT (1 mg/kg) (or saline) (Bales & Carter, unpublished data). Each male received two injections, including OT, OTA, or SAL (saline, the vehicle control) in the following combinations SAL/SAL, OTA/SAL, or OTA/OT. Alloparental care, measured by huddling over pups on PND21, was low in animals that received OTA/SAL (11%), but was improved to 45% by PND8 exposure to OT (i.e., OTA/OT).

In another study of male prairie voles PND1 exposure to either OTA or OT also was capable of disrupting subsequent sexual behavior and reproductive potential (Bales *et al.*, 2004a). Both OT- and OTA-treated animals, showed in adulthood, indications of deficiencies in sperm transport; OT-treated males also demonstrated temporal differences in mating behavior (Bales *et al.*, 2004a).

Social Behavior in Females

In general females seemed to be less sensitive than males to the effects of PND1 treatments with OTA (0.1 mg/kg) or OT (1 mg/kg). However, females were affected by higher dosages of exogenous OT when given on PND1. OT (2–8 mg/kg) tended to slightly enhance certain aspects of sociality, including postweaning alloparental behavior. As adults these females were also tested under conditions that usually reveal selective social behaviors, including the partner preference behaviors used to index pair bond formation in voles (Williams *et al.*, 1992). In the case of partner preferences, the effects of neonatal exposure to OT were dose dependent and bimodal. Partner preferences were significantly enhanced in the group that

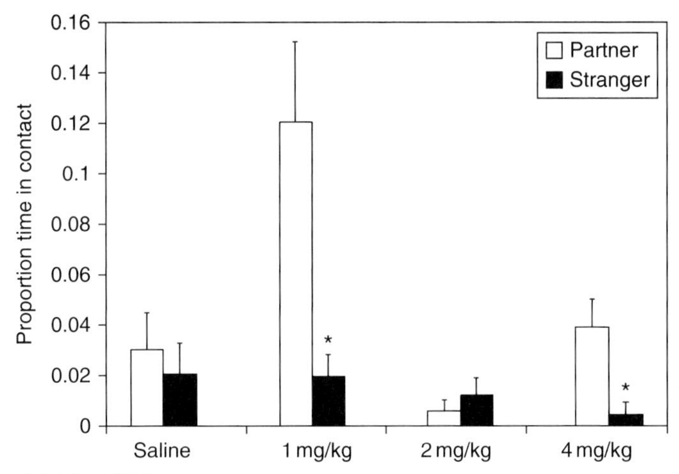

FIGURE 27.2 Effects of early OT treatment in *males* on preference for a partner vs. a stranger measured as a proportion of test time spent in side-to-side contact. Data presented here are from two different studies. 1 mg/kg data are from Bales and Carter (2003b). 2 mg/kg and 4 mg/kg data are previously unpublished, while saline data are combined for the two studies. Both 1 mg/kg ($t(12) = 2.23, p = 0.05$) and 4 mg/kg ($t(10) = 4.29, p = 0.002$) groups showed a statistically significant within-group partner preference.

received neonatal OT at 2 mg/kg, but were disrupted by higher doses of OT (4 mg/kg), and at a very high dose (8 mg/kg) OT-treated females later showed a preference for a stranger rather than the familiar partner. Once more these data reveal the capacity of early peptide exposure to alter later social interactions, but in the case of exposure to the highest dose the familiar animal later was either less preferred or possibly even aversive (Bales *et al.*, 2007d; see Figure 27.3). Alloparenting also varied with neonatal OT dosage, although the optimal dosage for facilitation of alloparenting differed from that for partner preference, suggesting the possibility that neonatal peptides act on these two forms of social behavior through mechanisms that are not identical.

Neonatal OTA in Females

The effects of neonatal OTA (PND1 at 0.1 mg/kg) were generally less disruptive in females than males. We have hypothesized that this comparative insensitivity of females or sensitivity of males to neonatal OTA may be due to OTA-induced disruptions in AVP-dependent processes, which are more critical to male behavior than female behavior. However, in females one interesting effect of neonatal OTA was detected: when female prairie voles were exposed to OTA (0.1 mg/kg on PND1) and then as adults paired with a male, neural activation of the central amygdala was increased (as measured by increased c-Fos expression). In prairie voles the central amygdala is not normally activated after pairing with an individual of the opposite sex, and activation of this area may reflect an increased state of emotional arousal or even fear in the OTA-treated females. No treatment effects were apparent in males in that study (Kramer *et al.*, 2006). The mechanisms for this apparent sex difference remain to be studied, but might in this case be due in females to OTA-induced changes in systems that rely on OT.

EFFECTS ON NEONATAL OT OR OTA ON BRAIN HORMONES AND RECEPTORS

Effects on Brain Peptides

In female prairie voles a single PND1 exposure to either OT (1 mg/kg) or OTA (0.1 mg/kg) was associated with increased central OT, measured at the time of weaning (PND21). In males there were no significant effects of either neonatal OT or OTA on central OT, but changes were detected in the AVP system. In males following neonatal OTA there was an apparent decrease in AVP, also measured on PND21 (Yamamoto *et al.*, 2004). In a related study, effects of early OT (1 mg/kg and lower) on OT

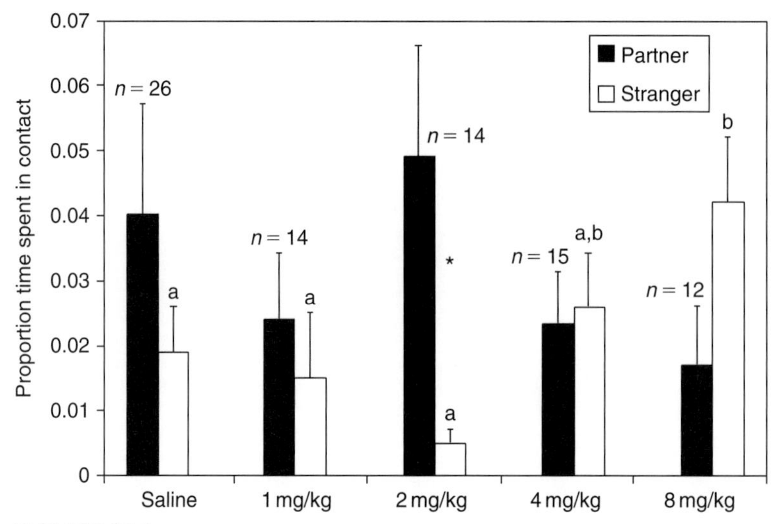

FIGURE 27.3 Effects of early OT treatment in *females* on preference for a partner vs. a stranger, measured as a proportion of test time spent in side-to-side contact. Proportion of test time spent in side-to-side contact with a stranger differs significantly by treatment ($\chi^2(4) = 10.76$, $p = 0.029$), as does the difference between time spent with the partner and time spent with the stranger ($\chi^2(4) = 10.05$, $p = 0.04$) (*Source*: Reproduced from Bales *et al.* (2007d)).

production in the PVN were no longer detectable in adulthood (Kramer *et al.*, 2007).

AVP is important to a variety of male behaviors in this species, including pair bond formation and mate guarding (Winslow *et al.*, 1993), male parental behavior (Wang *et al.*, 1994), and the management of anxiety or reactions to stressful experiences (Carter & Keverne, 2002). Thus, sexually dimorphic consequences of differential availability of either AVP or OT may be explained in part by the capacity of these peptides to be altered by differential peptidergic experiences in early life.

Effects on Peptide Receptors

Brain tissues from animals exposed to these same doses of neonatal OT or OTA were studied using autoradiography for possible changes in the OTR, AVP receptor (V1aR), as well as dopamine receptors (D1 or D2) (Bales *et al.*, 2007b). Effects on the OTR and dopamine receptors of neonatal OT or OTA manipulations were comparatively small, and not statistically significant. However, manipulations of OT on PND1 were associated with a pattern of regional changes in the V1aR system. Sex differences in the distribution of V1aRs in untreated prairie voles were not observed, consistent with earlier reports in this species. However, neonatal manipulations of OT produced long-lasting, sexually dimorphic effects on V1aR levels (Figures 27.4 and 27.5). Particularly striking were changes in the ventral pallidum in OT-treated males – increases in V1aR binding. In contrast, OT-treated females showed decreases in V1aR binding in this region. The ventral pallidum has been implicated in pair bond formation in male prairie voles, and is a region in which AVP V1aRs and dopamine-based "reward" processes may coexist (Young *et al.*, 2005b). Long-lasting increases in the V1aR in the ventral pallidum could help to explain the enhanced sociality of males receiving neonatal OT. Reduced AVP binding following neonatal OTA, especially in males that depend on AVP, might help to explain reductions in later social behavior in OTA exposed males (Carter, 2007).

Males exposed neonatally to OTA also showed reductions in V1aR binding in several other brain regions that have been implicated in both social behavior and emotionality, including the BNST, medial preoptic area, and lateral septum. AVP acting in these same brain regions has been associated later in life with male parental care in several species (Wang *et al.*, 1994; Bester-Meredith & Marler, 2003) and with male–male aggression (Marler *et al.*, 2003), which in pair-bonded prairie voles is

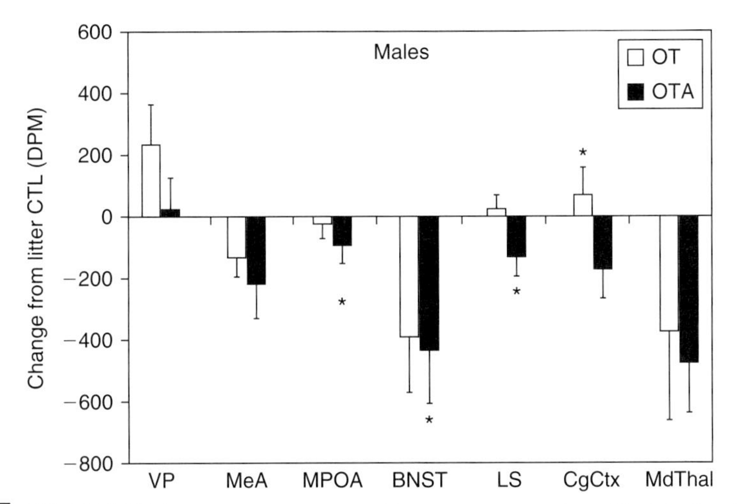

FIGURE 27.4 Effects of early OT or OTA treatment in *males* on V1aR binding. $*p < 0.05$. VP, ventral pallidum; MeA, medial amygdala; MPOA, medial preoptic area; BNST, bed nucleus of the stria terminalis; LS, lateral septum; CgCtx, posterior cingulate cortex; MdThal, mediodorsal thalamus (*Source*: Reproduced from Bales *et al.* (2007b)).

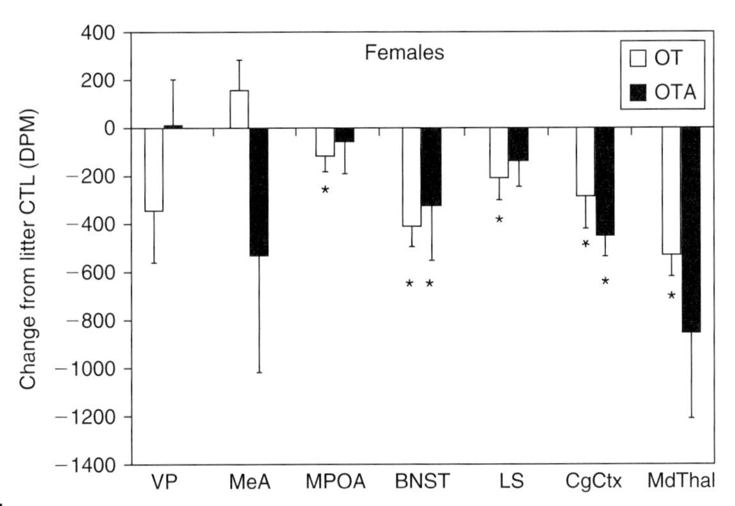

FIGURE 27.5 Effects of early OT or OTA treatment in *females* on V1aR binding. $*p < 0.05$. VP, ventral pallidum; MeA, medial amygdala; MPOA, medial preoptic area; BNST, bed nucleus of the stria terminalis; LS, lateral septum; CgCtx, posterior cingulate cortex; MdThal, mediodorsal thalamus (*Source*: Reproduced from Bales *et al.* (2007b)).

a component of mate guarding (Winslow *et al.*, 1993). We have observed that male, but not female, prairie voles treated neonatally with OTA showed reduced alloparental behavior (Bales *et al.*, 2004c), and tended to show lower levels of same-sex aggression than control males (Bales & Carter, 2003a), supporting a role for AVP and/or the V1aR in these behaviors.

Neonatal Manipulations of AVP also Affect Subsequent Social Behaviors

Earlier studies in prairie voles have revealed that neonatal exposure to AVP (in this case given as daily injections in the first week of life) were associated with a later dose-dependent increase in same sex aggression, especially in males (Stribley & Carter, 1999). In contrast, neonatal exposure to an AVP antagonist was associated in later life with very low levels of aggression. Other aspects of behavior including exploration in an EPM and partner preference formation were not significantly affected by either neonatal AVP or the AVP antagonist. The effects of neonatal AVP manipulations on V1aR binding have not yet been examined in voles. However, because these studies involved repeated daily injections and handling, the effects of AVP cannot be compared directly to those of OT described here, in which treatments were typically given once on the first day of life.

Early Exposure to Gonadal Steroids Facilitates the Response of Adult Male Prairie Voles to Exogenous AVP

In adult male prairie voles AVP plays a major role in pair bond formation (Winslow *et al.*, 1993). However, following neonatal castration, males tested as adults did not form partner preferences in response to centrally administered AVP. Neonatal treatment with testosterone restored the ability of castrated male prairie voles to respond to AVP in adulthood, in this case with the formation of a partner preference. Replacement of testosterone in adulthood did not restore partner preference formation in response to AVP in neonatally castrated males, suggesting once again that the postnatal period is a time when animals are especially sensitive to hormonal manipulations. Interestingly, neonatal castration did not affect the distribution of AVP V1aR, as measured by autoradiography (Cushing *et al.*, 2003a). This conclusion is also supported by the general absence of sex differences in either the OTR or AVP V1aR that have been reported in several studies. The lack of an effect of neonatal castration on the V1aR is in contrast to the significant and sexually dimorphic effects of neonatal manipulations of OT on the V1aR (Bales *et al.*, 2007b). These findings suggest the hypothesis that changes in peptides in early life may be more relevant than changes in gonadal steroids in regulating later peptide receptor distribution.

Estrogen Receptors also Affected by Neonatal OT or OTA

Among the other long-lasting changes that followed neonatal manipulations of OT or OTA were effects on the later expression of estrogen receptors (ERα) (Cushing & Kramer, 2005; Kramer *et al.*, 2007). In female prairie voles, neonatal OT increased the later expression of ERα in the ventromedial hypothalamus, while OTA decreased ERα in the medial preoptic area (Yamamoto *et al.*, 2006). A follow-up study indicated that during the neonatal period OT may affect the expression of ERα by influencing the production of ERα mRNA (Pournajafi-Nazarloo *et al.*, 2007a). On the day of birth female prairie vole pups were treated with OT, OTA, or saline (as above). Within 2 h of treatment, OT significantly increased ERα mRNA expression in the hypothalamus and hippocampus, but not the cortex (measured by RT-PCR). In contrast, neonatal exposure to OTA was associated with a later reduction in the expression of ERα mRNA in the hippocampus. Neonatal treatment did not affect the later expression of the ERα mRNA. Regional specific changes in ERα mRNA expression in females are consistent with studies examining the behavioral and physiological effects of neonatal manipulation of OT in females. Significant effects on ERα of neonatal OT manipulations were not detected initially in males (Cushing & Kramer, 2005). However, in a subsequent study exposure to neonatal OTA was associated with an increase in ERα in the BNST (Kramer *et al.*, 2007). These studies again support the hypothesis that manipulations of OT can have organizational effects, and that the effects of OT are sexually dimorphic. However, in the case of ERα females tended to be more sensitive than males, at least to the effects of neonatal OT. ERαs play a role in a variety of social and reproductive behaviors; alterations in ERαs would be expected to have broad consequences for many systems upon which steroid hormones act both during development and also in adulthood, and are possible mediators of at least some of the effects of neonatal manipulations of peptides.

EFFECTS OF NEONATAL OT OR/AND OTA IN RATS

Reproductive and Endocrine Effects

An extensive analysis of the consequences of neonatal peptides in species other than prairie voles remains to be conducted. However, in rats comparable neonatal manipulations of OT did have long-lasting, sexually dimorphic effects on pituitary levels of OT (Young *et al.*, 2005a). In female rats, treatment with OTA on PND1 significantly decreased pituitary OT levels as adults. In contrast, OTA treatment (PND1) in males resulted in increased pituitary OT levels in adulthood, at least when compared to males treated with OT (PND1). In rats receiving daily OT in the first week of life (PND1–7), subsequent puberty was delayed. This treatment with OT significantly delayed the age at vaginal opening and the age at first estrus (Withuhn *et al.*, 2003).

Cardiovascular and Autonomic Effects of Neonatal OT

In rats early exposure to OT (1 mg/kg) (daily on PND1–14) was followed in later life by increased abundance of $\alpha 2$ adrenergic receptors (Diaz-Cabiale *et al.*, 2004). These changes were most obvious in the hypothalamus and the amygdala, as evaluated by quantitative receptor autoradiography. In offspring from *ad libitum* fed dams, OT treatment significantly increased the density of $\alpha 2$-agonist binding sites in the nucleus tractus solitarius and in the hypothalamus. In addition, in offspring from food-restricted dams, OT treatment also produced a significant increase in indicators of $\alpha 2$-activity, supporting the more general hypothesis that chronic exposure to OT in early life might be capable of reducing sympathomimetic activity and enhancing parasympathetic functions, with broad significance for behavioral and emotional reactivity in later life.

Neonatal exposure to OT in rats also had consequences for gene expression in heart tissue (Pournajafi-Nazarloo *et al.*, 2007b). In this study female and male rats received either OT or saline on PND1. Hearts were collected either on PND1, 1 h following injection, or PND21. At these times the expression of mRNAs (using RT-PCR) was measured for the OTR, atrial natriuretic peptide (ANP), inducible nitric oxide synthase (iNOS), endothelial nitric oxide synthase (eNOS), ERα, and ERβ. OT treatment significantly increased gene expression in the heart for OTR, ANP, and eNOS measured on PND1 in both males and females. As a function of OT treatment, ERα increased only in females. Significant treatment effects were no longer detected in PND21 animals.

SEX DIFFERENCES

The consequences of developmental manipulations of OT are often sexually dimorphic, relying on different neural substrates in males and females (Carter, 2007). One possible reason for the differences between the sexes may be because in males the developmental effects of OT or OTA are mediated, at least in part, through differential effects on AVP (Yamamoto *et al.*, 2004) and/or its receptor (Bales *et al.*, 2007b). In both sexes OT manipulations, either OT (1 mg/kg) or OTA (0.1 mg/kg), did not have a significant effect on the OTR or dopamine receptor (D1 or D2) as measured by autoradiography (Bales *et al.*, 2007b).

Because parental care behavior was a major dependent variable in these studies, we also have analyzed the impact of factors, including peptide administration and stressful experiences, on adult parental behavior. Adult male parental behavior was only eliminated by high dosages of both OTA and a vasopressin V1aR antagonist (Bales *et al.*, 2004b; see Figure 27.6). In one study in adults, prairie voles were either stressed (by a 3-min swim) or not stressed, and then tested with infants. Components of adult male alloparental responses toward infants were increased by the prior swim, while female alloparental behaviors were unaffected (Bales *et al.*, 2006; see Figures 27.7 and 27.8). It is likely that AVP, as well as OT, were released during these stressors. Sex differences in central AVP and sex differences in the physiology of parental behavior (Wang *et al.*, 1998; Bales *et al.*, 2004b), and specifically the reliance of males on AVP, may help to explain the capacity of stressful experiences to enhance male alloparental behavior.

DEVELOPMENTAL SIGNALING CONSEQUENCES OF NEUROPEPTIDES

Maternal OT can act as a signaling mechanism between the mother and fetus. Based on studies in rats, maternal OT, released during birth, also triggers a transient switch in GABA signaling in the fetal brain from excitatory to inhibitory. *In vivo* administration of an OTA before delivery prevented this switch of GABA activity in fetal neurons, and aggravated the severity of anoxic episodes. Thus, it appears that maternal OT inhibits fetal neurons and increases their

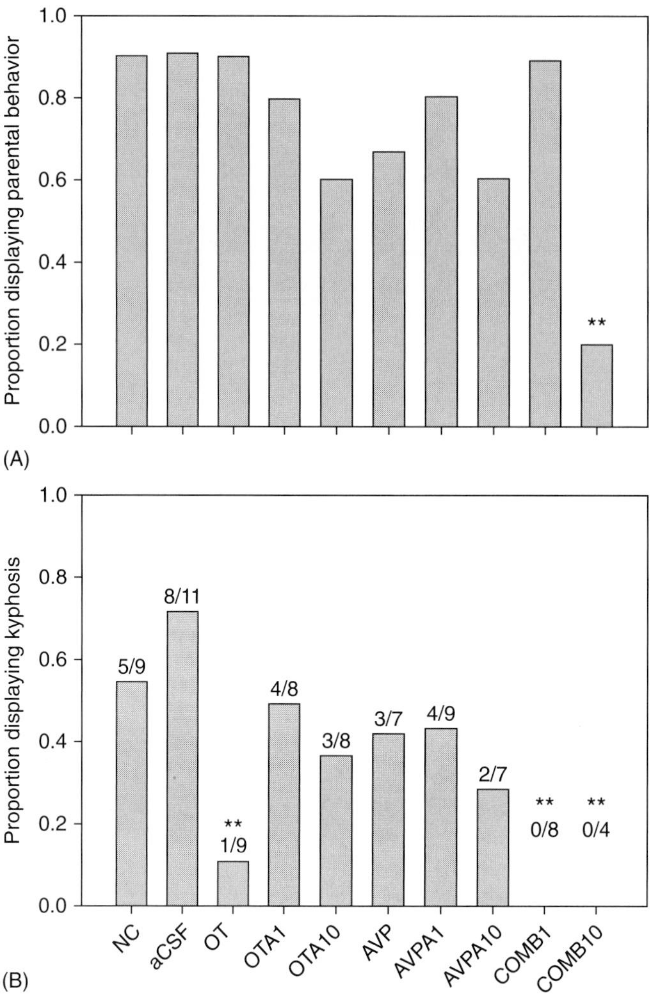

FIGURE 27.6 (A) Proportion of male voles responding parentally (displaying kyphosis, non-kyphotic contact, retrieving, or licking or grooming) toward infants. Significance was assessed in comparison to the aCSF group. Treatments are as follows: control (NC – no cannulation), aCSF (artificial cerebrospinal fluid), OT (oxytocin), OTA1 (1 ng oxytocin antagonist), OTA10 (10 ng OTA), AVP (arginine vasopressin), AVPA1 (1 ng AVP antagonist), AVPA10 (10 ng AVPA), COMB1 (combination treatment, 1 ng AVPA and 1 ng OTA), and COMB10 (combination treatment, 10 ng AVPA and 10 ng OTA). Only the COMB10 group differed significantly from the aCSF control (Fisher's exact probability test, $^{**}p = 0.002$). All group sizes are equal to 9–11 males. (B) Proportion of male voles (out of those not attacking) that displayed kyphosis. Significance was assessed in comparison to the aCSF group. **Different from aCSF group at $p < 0.05$ (*Source*: Reproduced from Bales *et al.* (2004b)).

resistance to hypoxic insult (Tyzio *et al.*, 2006). In addition, the birth-related surge in OT also helps to regulate the synchronization of the fetal hippocampal neurons, possibly allowing the transition from prenatal to postnatal life (Crepel *et al.*, 2007). Such changes would be expected to have consequences for both emotional and cognitive functions.

EARLY EXPERIENCE IN THE CONTEXT OF NATURAL HISTORY

Laboratory experiments have revealed that relatively subtle changes in early experience have long-term and sexually dimorphic consequences for the later social behavior of the offspring.

FIGURE 27.7 Proportion of animals displaying parental behavior after a swim stressor or no swim stressor ($n = 10$ animals/group). Males were more likely to display parental behavior following stress (logistic regression, likelihood ratio $\chi^2 = 3.4522$, $p = 0.032$; one-tailed), while females were not (*Source*: Reproduced from Bales *et al.* (2006)).

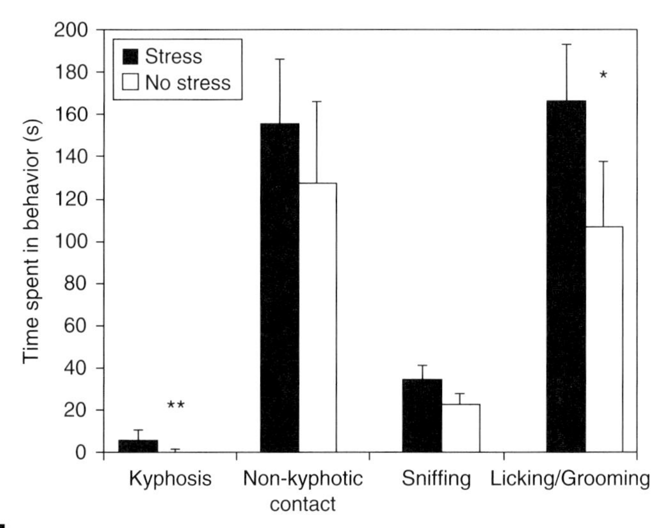

FIGURE 27.8 Time spent (seconds) by males in parental care behaviors, following exposure or no exposure to stress $^{**}p < 0.05$, $^{*}p < 0.1$ (*Source*: Reproduced from Bales *et al.* (2006)).

For example, disruption of the family associated with a single handling in the immediate postnatal period (MAN1) was associated in later life with increased alloparental behavior in male offspring and an increased tendency to form new pair bonds in female offspring. Conversely, reduced handling produced offspring that appeared anxious. Under these conditions female offspring were less likely to form pair bonds. Additional handling or manipulation in the neonatal period did NOT produce further increases in alloparenting, and in fact appeared to inhibit certain aspects of later social behavior. Preliminary data suggest that repeatedly manipulated animals (both parents and offspring) showed increased fear and anxiety. For example, when families were disturbed three times on PND1 (MAN1 × 3), males in particular were later less alloparental compared to MAN1 males. Under some conditions, the more frequently disturbed animals, especially males, showed reduced alloparenting or even attacked pups. Upon an initial disruption most mothers increased their interactions with their pups. However, following repeated disruptions mothers appeared agitated and less attentive to pups (Boone *et al.*, 2006).

These findings may be best interpreted in the context of circumstances that occur under field conditions. In nature prairie voles may live under a variety of social circumstances with consequences for the care of young. A percentage of males and females form pair bonds, live together, and produce their own offspring. However, it has been estimated that approximately 70% of young animals do not form new families, but remain instead as part of a communal breeding group (Getz & Carter, 1996). Under conditions of high mortality or mate abandonment, females may become single parents. However, even if the father is not present, older offspring can help to care for their younger siblings. Variation in the composition of the family would result in naturally occurring differences in the amount of stimulation received by offspring. Such variation could be translated into hormonal messages and epigenetic modifications that would determine whether a young animal would remain in the natal nest, where it would be an alloparent, or alternatively attempt to find a partner and establish a new pair bond and family. Variation in AVP V1aRs has been observed in field caught animals, usually exceeding the variation in animals reared under routine laboratory conditions (Phelps & Young, 2003), which may approximate the MAN1 conditions of our experimental manipulations of early experience.

TRANSLATIONAL IMPLICATIONS OF PERINATAL MANIPULATIONS OF OT

Of particular concern and largely unstudied in humans are the possible consequences of exposure to exogenous peptides, including OT, in the perinatal period. The use of synthetic OT (Pitocin) has grown in prevalence as a method for inducing or augmenting labor. As just one example, Pitocin was used in 93% of the approximately 10,000 births at Northwestern Hospital in Chicago in 2005 (Cynthia Wang, personal communication). Prematurity also continues to be a major medical problem and treatments attempting to delay labor hold the potential to affect the fetus. OTAs, such as Atosiban (currently not approved in the United States, but available in 43 other countries), have been used to delay or prevent premature labor (Husslein, 2002). Other OTAs are being developed as methods for delaying parturition. However, the consequences for human brain and behavior of early OT manipulations, including exogenous OT, or of the use of OTAs remain largely unknown.

The possible effects of early experience or hormonal manipulations also have not been systematically studied in human development. The neural systems that are altered by neonatal peptide manipulations in prairie voles are evolutionarily ancient and have broad behavioral and physiological actions. Our studies suggest that these systems may be accessible to change and thus vulnerable during development to neural changes that could have long-lasting consequences. Results from the present study suggest the need for a deeper understanding of the mechanisms through which manipulations in endogenous or exogenous peptides might affect neuroanatomy, physiology, and behavior. For example, manipulations in OT could have long-lasting consequences for behavioral endophenotypes, including sociality and reactivity to stressors, that are core to personality types and in extreme cases to several psychiatric disorders, including autism (Carter, 2007), anxiety, and depression (Carter & Altemus, 2004), and possibly schizophrenia. Functional sex differences in neuropeptides, including AVP and possibly OT may have particular significance for understanding sexually biased disorders (Carter, 2007). Early manipulations of OT also can program various aspects of the body's management of stressful experiences, including measures of behavior, brain activity and chemistry, stress-related hormones, and even receptors for stress hormones in the heart (Pournajafi-Nazarloo *et al.*, 2007b).

REFERENCES

Bales, K. L., and Carter, C. S. (2003a). Sex differences and developmental effects of oxytocin on aggression and social behavior in prairie voles (*Microtus ochrogaster*). *Horm. Behav.* 44, 178–184.

Bales, K. L., and Carter, C. S. (2003b). Developmental exposure to oxytocin facilitates partner preferences in male prairie voles (*Microtus ochrogaster*). *Behav. Neurosci.* 117, 854–859.

Bales, K. L., Abdelnabi, M., Cushing, B. S., Ottinger, M. A., and Carter, C. S. (2004a). Effects of neonatal oxytocin manipulations on male reproductive potential in prairie voles. *Physiol. Behav.* 81, 519–526.

Bales, K. L., Kim, A. J., Lewis-Reese, A. D., and Carter, C. S. (2004b). Both oxytocin and vasopressin may influence alloparental behavior in male prairie voles. *Horm. Behav.* 45, 354–361.

Bales, K. L., Pfeifer, L. A., and Carter, C. S. (2004c). Sex differences and effects of manipulations of oxytocin on alloparenting and anxiety in prairie voles. *Dev. Psychobiol.* 44, 123–131.

Bales, K. L., Kramer, K. M., Lewis-Reese, A. D., and Carter, C. S. (2006). Effects of stress on parental care are sexually dimorphic in prairie voles. *Physiol. Behav.* 87, 424–429.

Bales, K. L., Lewis-Reese, A. D., Pfeifer, L. A., Kramer, K. M., and Carter, C. S. (2007a). Early experience affects the traits of monogamy in a sexually dimorphic manner. *Dev. Psychobiol.* 49, 335–342.

Bales, K. L., Plotsky, P. M., Young, L. J., Lim, M. M., Grotte, N. D., Ferrer, E., and Carter, C. S. (2007b). Neonatal oxytocin manipulations have long-lasting, sexually dimorphic effects on vasopressin receptors. *Neuroscience* 144, 38–45.

Bales, K. L., Stone, A. I., Boone, E. M., and Carter, C. S. (2007c). Long-term and intergenerational effects of early handling on oxytocin, pair-bonding, and parenting in monogamous prairie voles. *Soc. Neurosci. Abst.* 209.15.

Bales, K. L., Van Westerhuyzen, J. A., Lewis-Reese, A. D., Grotte, N. D., Lanter, J. A., and Carter, C. S. (2007d). Oxytocin has dose-dependent developmental effects on pair-bonding and alloparental care in female prairie voles. *Horm. Behav.* 52, 274–279.

Barberis, C., and Tribollet, E. (1996). Vasopressin and oxytocin receptors in the central nervous system. *Crit. Rev. Neurobiol.* 10, 119–154.

Bester-Meredith, J. K., and Marler, C. A. (2003). Vasopressin and the transmission of paternal behavior across generations in mated, cross-fostered Peromyscus mice. *Behav. Neurosci.* 117, 455–463.

Bielsky, I. F., Hu, S.-B., and Young, L. J. (2005). Sexual dimorphism in the vasopressin system: Lack of an altered behavioral phenotype in female V1a receptor knockout mice. *Behav. Brain Res.* 164, 132–136.

Boone, E., Sanzenbacher, L. L., Carter, C. S., and Bales, K. L. (2006). Sexually dimorphic effects of early experience on alloparental care and adult social behaviors in voles. *Soc. Neurosci. Abst.* 578.6.

Carter, C. S. (1998). Neuroendocrine perspectives on social attachment and love. *Psychoneuroendocrinology* 23, 779–818.

Carter, C. S. (2003). Developmental consequences of oxytocin. *Physiol. Behav.* 79, 383–397.

Carter, C. S. (2007). Sex differences in oxytocin and vasopressin: Implications for autism spectrum disorders?. *Behav. Brain Res.* 176, 170–186.

Carter, C. S., and Altemus, M. (2004). Oxytocin, vasopressin, and depression. In *Current and Future Developments in Psychopharmacology* (J. A. den Boer, M. S. George, and G. J. ter Horst, Eds.), pp. 201–216.

Carter, C. S., and Keverne, E. B. (2002). The neuroendocrinology of social attachment and love. In *Hormones, Brain, and Behavior* (D. Pfaff, Ed.), Vol. 4, pp. 299–337.

Champagne, F., Diorio, J., Sharma, S., and Meaney, M. J. (2001). Naturally occurring variations in maternal behavior in the rat are associated with differences in estrogen-inducible central oxytocin receptors. *Proc. Natl Acad. Sci. USA* 98, 12736–12741.

Champagne, F. A., and Meaney, M. J. (2006). Stress during gestation alters postpartum maternal care and the development of the offspring in a rodent model. *Biol. Psychiatr.* 59, 1227–1235.

Crepel, V., Aronov, D., Jorquera, I., Represa, A., Ben-Ari, Y., and Cossart, R. (2007). A parturition-associated nonsynaptic coherent activity pattern in the developing hippocampus. *Neuron* 54, 105–120.

Cushing, B. S., and Kramer, K. M. (2005). Mechanisms underlying epigenetic effects of early social experience: The role of neuropeptides and steroids. *Neurosci. Biobehav. Rev.* 29, 1085–1105.

Cushing, B. S., Okorie, U., and Young, L. J. (2003a). The effects of neonatal castration on the subsequent behavioural response to centrally administered arginine vasopressin and the expression of V1a receptors in adult male prairie voles. *J. Neuroendocrinol.* 15, 1021–1026.

Cushing, B. S., Yamamoto, Y., Hoffman, G. E., and Carter, C. S. (2003b). Central expression of c-Fos in neonatal male and female prairie voles in response to treatment with oxytocin. *Dev. Brain Res.* 143, 129–136.

De Vries, G. J., and Simerly, R. B. (2002). Anatomy, development, and function of sexually dimorphic neural circuits in the mammalian brain. In *Hormones, Brain, and Behavior* (D. Pfaff, Ed.), Vol. 4, pp. 137–192.

Diaz-Cabiale, Z., Olausson, H., Sohlstrom, A., Agnati, L. F., Narvaez, J. A., Uvnas-Moberg, K., and Fuxe, K. (2004). Long-term modulation by postnatal oxytocin of the alpha(2)-adrenoceptor agonist binding sites in central autonomic regions and the role of prenatal stress. *J. Neuroendocrinol.* 16, 183–190.

Francis, D., Diorio, J., Liu, D., and Meaney, M. J. (1999). Nongenomic transmission across generations of maternal behavior and stress responses in the rat. *Science* 286, 1155–1158.

Francis, D. D., Young, L. J., Meaney, M. J., and Insel, T. R. (2002). Naturally occurring differences in maternal care are associated with the expression of oxytocin and vasopressin (V1a) receptors: Gender differences. *J. Neuroendocrinol.* 14, 349–353.

Getz, L. L., and Carter, C. S. (1996). Prairie vole partnerships. *Am. Sci.* 84, 56–62.

Gimpl, G., and Fahrenholz, F. (2001). The oxytocin receptor system: Structure, function, and regulation. *Physiol. Rev.* 81, 629–683.

Glynn, L. M., Davis, E. P., Schetter, C. D., Chicz-Demet, A., Hobel, C. J., and Sandman, C. J. (2007). Postnatal maternal cortisol levels predict temperament in healthy breastfed infants. *Early Hum. Dev.* 83, 675–681.

Grosvenor, C. E., Picciano, M. T., and Baumrucker, C. R. (1993). Hormones and growth factors in milk. *Endo. Rev.* 14, 710–728.

Hammock, E. A., and Young, L. J. (2005). Microsatellite instability generates diversity in brain and sociobehavioral traits. *Science* 308, 1630–1634.

Husslein, P. (2002). Development and clinical experience with the new evidence-based tocolytic atosiban. *Acta Obstet. Gynecol. Scand.* 81, 633–641.

Insel, T. R., and Shapiro, L. E. (1992). Oxytocin receptor distribution reflects social organization in monogamous and polygamous voles. *Proc. Natl Acad. Sci. USA* 89, 5981–5985.

Kaffman, A., and Meaney, M. J. (2007). Neurodevelopmental sequelae of postnatal maternal care in rodents: Clinical and research implications of molecular insights. *J. Child Psychol. Psychiatr.* 48, 224–244.

Kimura, T., Saji, F., Nishimori, K., Ogita, K., Nakamura, H., Koyama, M., and Murata, Y. (2003). Molecular regulation of the oxytocin receptor in peripheral organs. *J. Mol. Endocrinol.* 30, 109–115.

Kramer, K. M., Cushing, B. S., and Carter, C. S. (2003). Developmental effects of oxytocin on stress response: Single versus repeated exposure. *Physiol. Behav.* 79, 775–782.

Kramer, K. M., Choe, C., Carter, C. S., and Cushing, B. S. (2006). Developmental effects of oxytocin on neural activation and neuropeptide release in response to social stimuli. *Horm. Behav.* 49, 206–214.

Kramer, K. M., Yoshida, S., Papademetriou, E., and Cushing, B. S. (2007). The organizational effects of oxytocin on the central expression of estrogen receptor alpha and oxytocin in adulthood. *BMC Neurosci.* 8, 71.

Leake, R. D., Weitzman, R. E., and Fisher, D. A. (1981). Oxytocin concentrations during the neonatal period. *Biol. Neonate* 39, 127–131.

Lim, M. M., Hammock, E. A. D., and Young, L. J. (2004). The role of vasopressin in the genetic and neural regulation of monogamy. *J. Neuroendocrinol.* 16, 325–332.

Mahalati, K., Okanoya, K., Witt, D. M., and Carter, C. S. (1991). Oxytocin inhibits male sexual behavior in prairie voles. *Pharmacol. Biochem. Behav.* 39, 219–222.

Marler, C. A., Bester-Meredith, J. K., and Trainor, B. C. (2003). Paternal behavior and aggression: Endocrine mechanisms and nongenomic transmission of behavior. In *Advances in the Study of Behavior*, Vol. 32, pp. 263–323.

Pedersen, C. A., and Boccia, M. L. (2002). Oxytocin links mothering received, mothering bestowed, and adult stress responses. *Stress* 5, 267.

Phelps, S. M., and Young, L. J. (2003). Extraordinary diversity in vasopressin (V1a) receptor distributions among wild prairie voles (*Microtus ochrogaster*): Patterns of variation and covariation. *J. Comp. Neurol.* 466, 564–576.

Porges, S. W. (1998). Love: An emergent property of the mammalian autonomic nervous system. *Psychoneuroendocrinology* 23, 837–861.

Pournajafi-Nazarloo, H., Carr, M. S., Papademetriou, E., Schmidt, J. V., and Cushing, B. S. (2007a). Oxytocin increases ER-alpha mRNA expression in the hypothalamus and hippocampus of neonatal female voles. *Neuropeptides.* 41, 39–44.

Pournajafi-Nazarloo, H., Perry, A., Papademetriou, E., Parloo, L., and Carter, C. S. (2007b). Neonatal oxytocin treatment modulates oxytocin receptor, atrial natriuretic peptide, nitric oxide synthase, and estrogen receptor mRNA expression in rat heart. *Peptides* 28, 1170–1177.

Stribley, J. M., and Carter, C. S. (1999). Developmental exposure to vasopressin increases aggression in adult prairie voles. *Proc. Natl Acad. Sci. USA* 96, 12601–12604.

Szyf, M., Weaver, I. C. G., Champagne, F. A., Diorio, J., and Meaney, M. J. (2005). Maternal programming of steroid receptor expression and phenotype through DNA methylation in the rat. *Front. Neuroendocrinol.* 26, 162.

Tyzio, R., Cossart, R., Khalilov, I., Minlebaev, M., Hubner, C. A., Represa, A., Ben-Ari, Y., and Khazipov, R. (2006). Maternal oxytocin triggers a transient inhibitory switch in GABA signaling in the fetal brain during delivery. *Science* 314, 1788–1792.

Tyler, A. N., Michel, G. F., Bales, K. L., and Carter, C. S. (2005). Do brief early disturbances of parents affect parental care in the bi-parental prairie vole (*Microtus ochrogaster*)? *Dev. Psychobiol.* 47, 451.

Wang, Z. X., Ferris, C. F., and De Vries, G. J. (1994). Role of septal vasopressin innervation in paternal behavior in prairie voles (*Microtus ochrogaster*). *Proc. Natl Acad. Sci. USA* 91, 400–404.

Wang, Z. X., Young, L. J., De Vries, G. J., and Insel, T. R. (1998). Voles and vasopressin: A review of molecular, cellular, and behavioral studies of pair bonding and paternal behaviors. *Prog. Brain Res.* 119, 483–499.

Williams, J. R., Catania, K. C., and Carter, C. S. (1992). Development of partner preferences in female prairie voles (*Microtus ochrogaster*): The role of social and sexual experience. *Horm. Behav.* 26, 339–349.

Williams, J. R., Insel, T. R., Harbaugh, C. R., and Carter, C. S. (1994). Oxytocin centrally administered facilitates formation of a partner preference in female prairie voles (*Microtus ochrogaster*). *J. Neuroendocrinol.* 247–250.

Winslow, J. T., Hastings, N., Carter, C. S., Harbaugh, C. R., and Insel, T. R. (1993). A role for central vasopressin in pair bonding in monogamous prairie voles. *Nature.* 365, 545–548.

Withuhn, T. F., Kramer, K. M., and Cushing, B. S. (2003). Early exposure to oxytocin affects the age of vaginal opening and first estrus in female rats. *Physiol. Behav.* 80, 135–138.

Witt, D. M., Carter, C. S., and Insel, T. R. (1991). Oxytocin receptor binding in female prairie voles – Endogenous and exogenous estradiol stimulation. *J. Neuroendocrinol.* 3, 155–161.

Yamamoto, Y., Cushing, B. S., Kramer, K. M., Epperson, P. D., Hoffman, G. E., and Carter, C. S. (2004). Neonatal manipulations of oxytocin alter expression of oxytocin and vasopressin immunoreactive cells in the paraventricular nucleus of the hypothalamus in a gender specific manner. *Neuroscience* 125, 947–955.

Yamamoto, Y., Carter, C. S., and Cushing, B. S. (2006). Neonatal manipulation of oxytocin affects expression of estrogen receptor alpha. *Neuroscience* 137, 157–164.

Young, E., Carter, C. S., Cushing, B. S., and Caldwell, J. D. (2005a). Neonatal manipulations of oxytocin alter oxytocin levels in the pituitary of adult rats. *Horm. Metabol. Res.* 37, 397–401.

Young, L. J., Murphy Young, A. Z., and Hammock, E. A. (2005b). Anatomy and neurochemistry of the pair bond. *J. Comp. Neurol.* 493, 51–57.

28

THE EFFECTS OF PATERNAL BEHAVIOR ON OFFSPRING AGGRESSION AND HORMONES IN THE BIPARENTAL CALIFORNIA MOUSE

CATHERINE A. MARLER[1], BRIAN C. TRAINOR[2], ERIN D. GLEASON[3], JANET K. BESTER-MEREDITH[4] AND ELIZABETH A. BECKER[3]

[1] *University of Wisconsin-Madison, Department of Psychology, Department of Zoology, Madison, WI, USA*
[2] *University of California-Davis, Department of Psychology, Davis, CA, USA*
[3] *University of Wisconsin-Madison, Department of Psychology, Madison, WI, USA*
[4] *Seattle Pacific University, Biology Department, Seattle, WA, USA*

A variety of studies have predicted an effect of parental behavior on postnatal development of offspring aggression. This is an attractive hypothesis, in part, because numerous studies demonstrate an association between parenting style and offspring aggression in humans (review by Marler *et al.*, 2005; recent example: Vitaro *et al.*, 2006), with harsh and punitive parenting being associated with high levels of offspring aggression. Furthermore, animal studies, which allow manipulations of parental behavior, have found causal interactions between stress and variations in maternal behavior. Separation from the mother and the level of maternal grooming/licking behavior significantly alter the stress axis and associated behaviors such as maternal behavior and exploration of novel environments (e.g., Liu *et al.*, 1997; Caldji *et al.*, 1998; Gonzalez *et al.*, 2001; Francis *et al.*, 2003; Weaver *et al.*, 2004). Other examples of intergenerational transmission of maternal behavior include that found in rhesus macaques involving rejection behavior and serotonin (Maestripieri *et al.*, 2007). In contrast, there is scarce empirical evidence for an association between parenting and offspring aggression. A number of studies with inbred mice from a variety of strains have revealed a relatively robust resistance to postnatal parental effects (review by Haug & Pallaud, 1981). The lack of an influence of parental behavior on aggression has been further supported by a cross-fostering study in rats (Lucion *et al.*, 1994) and another in rhesus macaques (Maestripieri, 2003). The majority of studies have tested for maternal effects because the model systems used are typically not biparental species. We describe how specific paternal behaviors in *Peromyscus* mice can influence aggression of offspring and we have begun to identify mechanisms through which this might occur. In an interesting parallel with huddling/grooming behavior (e.g., Meaney, 2001), there is also evidence for transmission of aggression across multiple generations.

Neurobiology of the Parental Brain

ASSOCIATIONS BETWEEN PATERNAL BEHAVIOR AND MALE OFFSPRING AGGRESSION

One of the best examples of an effect of paternal behavior on offspring behavior is seen in the highly territorial California mouse. Males and females of this species do not breed unless they have established a territory (Ribble & Salvioni, 1990). Furthermore, field studies indicate that males and females form exclusive mating pairs (Ribble, 1991) and exhibit high levels of parental care (Gubernick & Alberts, 1987; Bester-Meredith *et al.*, 1999). California mice groom pups and huddle with them to aid in thermoregulation. In addition, male California mice spontaneously retrieve their pups between postpartum days 17 and 21 (Bester-Meredith *et al.*, 1999). Here we focus on retrievals as being particularly important for the effect of paternal behavior on offspring aggression. When a pup is retrieved, the parent grips the pup with its mouth posterior to the shoulder and lifts it off of the cage floor. These retrievals primarily occur when pups are mobile, and often pups are partially retrieved such that the pup is grabbed by the parent, but the pup is then released or escapes before being lifted from the floor (Marler *et al.*, 2003). In these respects, retrieval behavior during later development in the California mouse differs qualitatively and quantitatively from retrieval behavior as it is commonly measured in maternal behavior experiments in rats and mice during early development. These parental behaviors expressed by males are a critical component of this species' life history, because removal of the father under field conditions causes a significant decrease in offspring survival at weaning (Gubernick & Teferi, 2000). In the California mouse, aggressive behavior and parental behavior are inextricably linked, because breeding can not take place unless the parents have a territory.

The link between aggression and parental behavior has been established in a series of correlative and manipulative studies suggesting that parental behavior has critical effects on the development of aggressive behavior. The first signs that offspring behavior might be affected by parental behavior came from cross-fostering studies between California mice and white-footed mice (*P. leucopus*), a species that exhibits low levels of parental care and is less aggressive than California mice in resident-intruder (R-I) and neutral arena aggression tests (Bester-Meredith *et al.*, 1999; Bester-Meredith & Marler, 2001). Male California mice cross-fostered to white-footed mice were less aggressive in R-I tests than California mice raised by parents of the same species, thereby becoming more similar to their foster parents (Bester-Meredith & Marler, 2001). Correlational analyses from the same study suggested that a specific component of the parental environment might play a key role in affecting the development of aggressive behavior. Male white-footed mice huddle with, groom and retrieve pups at very low levels regardless of whether the pups are white-footed mice or California mice (Bester-Meredith & Marler, 2001). However, correlational analyses suggested that retrieval behavior and not huddling and grooming is instrumental in the development of aggression. The number of times pups were retrieved by their fathers was negatively correlated with offspring's attack latencies in R-I tests of the adult offspring, suggesting an interesting link between paternal behavior and offspring aggression (Bester-Meredith & Marler, 2003b).

To test whether male retrievals have a significant effect on offspring aggression, retrieval behavior was experimentally manipulated (Frazier *et al.*, 2006). In the control group, pups were removed from their nest and then returned to their nest. In the retrieval group, pups were removed from their nest and placed on the opposite side of their cage. The retrieval manipulation led to a significant increase in male retrieval and grabbing behavior during postpartum days 15–21 compared to the control group. When tested as adults, male offspring from the retrieval group had significantly shorter attack latencies in R-I aggression tests than male offspring from the control group (Figure 28.1). The effect of retrievals on aggression was limited to the context of the R-I test, as there was no effect of retrievals on aggression in a neutral arena test. One consequence of the retrieval manipulation is that it increased the activity of pups during retrieval tests. The possibility that increased activity as pups, and not retrievals, caused the increases in R-I aggression was examined using partial correlations. A significant negative correlation between male R-I attack latency and male parental retrievals when controlling for pup activity indicated that the effect of retrieval manipulation on aggression is not mediated

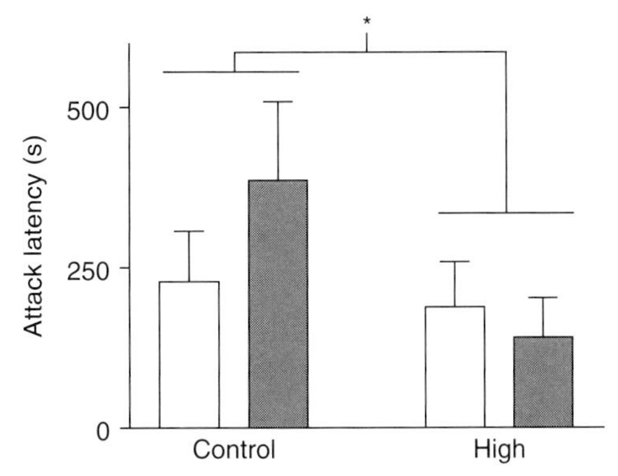

FIGURE 28.1 Male offspring attack latency in R-I aggression tests. Control males experienced sham retrieval manipulations during development; high retrieval males experienced experimentally increased retrievals during development. Open bars, pups raised by intact fathers; black bars, pups raised by castrated fathers. Effect of retrieval manipulation: $*p < 0.05$. (*Source*: From Frazier *et al.*, 2006).

by changes in pup activity. Experimentally elevated retrieval behavior also affected female offspring aggression. Females in the high retrieval group were more aggressive than controls in a R-I test (Frazier *et al.*, 2006). Paternal retrievals therefore appear to influence aggression of both male and female offspring.

Overall, studies in California mice indicate that male parental behavior is an important conduit for the cross-generational transmission of aggressive behavior. In rodents with larger litter sizes, such as rats, it is thought that interactions among juveniles play a critical role in the development of aggressive behavior (Panksepp *et al.*, 1984). Compared to socially housed individuals, juvenile rats that are housed individually from day 22 to day 35 postpartum are more submissive when tested in a R-I test as an intruder (van den Berg *et al.*, 1999). This may be due to reduced social interactions with siblings. Pinning behavior among juveniles (when one rat pins another on its back) during play fighting resembles patterns used by adults in aggressive interactions (Gordon *et al.*, 2002). A difference, however, between rats and California mice is that rats tend to have large litter sizes, thus ensuring social interactions among offspring. On average less than two pups are born to California mouse litters in the field (Gubernick & Teferi, 2000), so that pups frequently will not interact with other siblings. It may be this decreased effect of sibling interactions in California mice that reveals the strong interaction between paternal retrievals and offspring aggression; aggression that is necessary for gaining a territory and acquiring a mate.

DO MALES AND FEMALES PLAY DIFFERENT ROLES IN SHAPING AGGRESSION OF FUTURE GENERATIONS?

Unlike other female rodents, female California mice show high aggressiveness prior to mating that may be related to female-biased dispersal and high levels of mate competition (Ribble, 1992). Perhaps not surprisingly, female California mice and white-footed mice resemble males in that their aggression is shaped by the behavior of their parents during early development. However, parental effects on aggression may differ depending on the sex of the parent and the sex of the offspring.

Behavioral interactions between offspring and their same-sex parent may be particularly important in shaping adult aggression. As is discussed earlier, paternal retrievals have been positively associated with R-I aggression in male offspring (Bester-Meredith & Marler, 2003b; Frazier *et al.*, 2006). In females, a similar association between high R-I aggression and retrievals by the same-sex parent has been identified (Bester-Meredith & Marler, 2007). However, in these cross-fostering studies, paternal retrievals show no association with female offspring aggression and maternal retrievals show no association with male offspring aggression. These findings suggest that in biparental species like California mice, changes in the behavior of one parent might only alter the aggressiveness of same-sex offspring, allowing parents to fine-tune the behavior of offspring in response to environmental challenges. It is important to note, however, that experimental elevations of paternal retrievals did increase aggression of female offspring in the R-I paradigm (Frazier *et al.*, 2006). Manipulations of maternal retrievals will better test the possibility of sex differences in influencing offspring behavior. Other early manipulations of the social environment during development such as social subjugation during puberty also show differential effects on female and male aggression. Whereas males respond to social subjugation with increased aggression toward smaller opponents (Delville *et al.*, 1998), female aggression is not affected by this treatment (Taravosh-Lahn & Delville, 2004).

Other differences between the sexes exist in how parents influence different types of aggression in adult *Peromyscus*. In contrast to the R-I test discussed earlier, neutral arena aggression tests require both animals to be placed simultaneously into a clean testing arena. This simple alteration in the testing paradigm elicits a type of aggression that appears to differ from R-I aggression in its underlying physiological mechanism (e.g., Christie & Barfield, 1979; Teskey & Kavaliers, 1987; Bester-Meredith *et al.*, 2005). In addition, males show more offensive aggression as residents in R-I paradigms than as combatants in neutral arenas but express more defensive aggression in neutral arenas than in R-I paradigms (reviewed in Mast & Marler, unpublished data; Olivier & Mos, 1992). Males and females also may differ in the flexibility of their aggressive responses within these two paradigms. Although cross-fostering to a less aggressive species decreases aggression in both male and female California mice, female California mice show decreased aggression only in the neutral arena test, but not in the R-I paradigm (Bester-Meredith & Marler, 2007). For male California mice, the reverse was apparent: males show flexibility in aggression in response to cross-fostering only in the R-I paradigm, not in the neutral arena (Bester-Meredith & Marler, 2001). These sex differences in how the same social manipulation affects adult aggressiveness may reflect the unusual ecology of this species. The female-biased dispersal pattern of California mice (Ribble, 1992) may lead to females being more responsive to social cues that induce dispersal and/or alter competitiveness prior to territory establishment within neutral areas. Perhaps for males, competition for mates may make flexibility with territorial aggression becoming more important.

In addition to these physiological and behavioral sex differences in association with the R-I and neutral arena aggression tests, the two aggression paradigms appear to differ in how they are linked with the early rearing environment. Paternal nest-building is positively associated with neutral arena aggression in female, but not male *Peromyscus* (Bester-Meredith & Marler, 2003a, 2007). Furthermore, although both males and females show associations between neutral arena aggression and a composite score of maternal behavior, the direction of this effect differs between the sexes. The amount of time that mothers spend huddling, nursing, grooming pups, and residing in the nest is negatively associated with neutral arena aggressiveness in their adult male offspring (Bester-Meredith & Marler, 2003a) but positively associated with this same aggression measure in adult female offspring (Bester-Meredith & Marler, 2007). Although the reason for this sex difference in responsiveness to the early environment is not well understood, clear sex differences exist in the way that the brain responds to other manipulations of grooming behavior during early development. For example, exposure to high licking and grooming during development increases oxytocin receptor binding in females, but not males (Francis *et al.*, 2002). This same manipulation shows the opposite effect on receptor binding for the vasopressin V1a receptor: an increase is found in males, but not females (Francis *et al.*, 2002). In addition, naturally existing sex differences in the brain may be stimulated in part by parents apportioning licking and grooming in a sexually dimorphic way. Rat mothers preferentially groom male pups (Moore & Morelli, 1979) and grooming increases Fos-immunoreactivity in the ventral portion of the medial preoptic area in pups (McCarthy *et al.*, 1997). Other long-term effects of maternal grooming on offspring neural development have been well documented, including cross-generational effects of licking and grooming on the oxytocin and vasopressin receptors that are associated with social behavior (Champagne & Meaney, 2006) and the glucocorticoid receptors associated with stress responsiveness (reviewed by Fish *et al.*, 2004). Cross-fostering paradigms in rats have demonstrated that maternal grooming with unrelated pups induces cross-generational behavioral changes that are associated with alterations in DNA methylation and corresponding changes in gene activity (Weaver *et al.*, 2004). These studies suggest that parental care can alter the behavioral phenotype of offspring because the rearing environment affects gene regulation within the brain.

The sex of the parent providing parental care may also be important in determining how the behavior shapes offspring neural development. Although a composite score of *maternal* behavior is associated with level of neutral arena aggression of offspring of both sexes as described above, *paternal* behavior shows no association with offspring neutral arena aggression in *Peromyscus* mice (Bester-Meredith & Marler, 2003a, 2007). The lack of a link between paternal behavior and offspring neutral arena aggression is further supported using a hormonal manipulation of paternal behavior. By decreasing testosterone to reduce paternal huddling and grooming, it was demonstrated that a reduction in paternal grooming increases offspring adrenal activity, but does not alter aggression in either neutral arena or R-I aggression tests (Frazier *et al.*, 2006). While similar manipulations of maternal behavior have not been conducted, these results suggest that there may be tradeoffs between the parenting styles of mothers and fathers in biparental species that may differentially shape the behavioral development of their offspring. Similarly, in another biparental species, the prairie vole, exposure to swim stress increases paternal but not maternal care toward pups (Bales *et al.*, 2006). Alterations in the behavior of one parent, but not the other parent, under specific environmental conditions may allow the behavior of offspring to be fine-tuned in response to the environment.

IS THERE SPECIES VARIATION IN THE EFFECTS OF MATERNAL AND PATERNAL BEHAVIOR ON OFFSPRING AGGRESSION?

In contrast to the biparental species described above, many rodent species do not show biparental care. Therefore, studies that have examined the effects of environmental perturbation on offspring behavior in rats of the genus *Rattus* and mice of the genus *Mus* have focused primarily on the role of the mother (e.g., Francis *et al.*, 2002; Champagne & Meaney, 2006). Exposure to a predator also increases maternal licking and grooming in female rats (McLeod *et al.*, 2007), suggesting that environmental challenges may produce alterations in maternal behavior that will affect the behavioral phenotype of the offspring in both biparental and uniparental species. In addition, pups from a highly aggressive line of house mice also received higher maternal care in the form of increased nursing (Mendl & Paul, 1990b; Benus & Rondigs, 1996) and maternal grooming (Benus & Rondigs, 1996) in comparison to mice selected to be less aggressive. However, these effects of maternal care on aggression appear to be pup-driven because cross-fostering does not eliminate the increased amount of parental care received by pups of the highly aggressive line (Benus & Rondigs, 1996).

In contrast, the effects of *paternal* care on the behavioral phenotypes of rats of the genus *Rattus* and mice of the genus *Mus* have not been a frequent research focus because males provide little care in these species. In house mice, males may affect the aggressiveness of their offspring through direct genetic effects on aggression (e.g., Carlier *et al.*, 1991) and through indirect effects because paternal, not maternal, genotypes influence litter size in mice (Hager & Johnstone, 2003). Therefore, at least in some strains of mice, the genotype of a father may indirectly alter sibling interactions by influencing the number of siblings present in a litter, which may in turn influence the type and amount of maternal care provided by the mother of the litter (Mendl & Paul, 1990a). More direct effects of paternal behavior in uniparental species have rarely been examined, although the presence of the father during development may increase the aggressiveness of house mice offspring (Mugford & Nowell, 1972). Why the father's presence would alter aggression is not well understood because differences in paternal behavior between non-aggressive and aggressive strains of house mice have not been identified (Mendl & Paul, 1990b).

As described earlier, direct effects of paternal care on aggression have been examined in less parental white-footed mice (e.g., Bester-Meredith & Marler, 2003a). The effects of exposure to increased maternal and paternal care on aggression in white-footed mice via the cross-fostering manipulation contrast sharply with the effects of the cross-fostering manipulation in California mice. In California mice, but not white-footed mice, cross-fostering altered adult aggression in both sexes. Cross-fostering to the less aggressive white-footed mouse led to decreased R-I aggression in male California mice and decreased neutral arena aggression in female California mice (Bester-Meredith & Marler, 2001, 2007). This same cross-fostering manipulation only altered aggression in male, but not female, white-footed mice, and the venue of these changes in aggression also differed between the species (Bester-Meredith & Marler, 2001).

As discussed earlier, some of these sex and species differences may be attributable to differences in ecology. In addition, aggression may develop differently in white-footed mice, a species where females provide most of the care for offspring, than in California mice, a biparental species. The involvement of two parents in day-to-day care of offspring may allow parents to show flexibility in the amount of care that they provide under challenging environmental conditions. When male California mice provide less huddling and grooming to their offspring, their mates partially compensate for this decrease (Trainor & Marler, 2001). Because male and female parental care show differential effects on offspring behavior, this increase in maternal care may influence offspring. In addition, experimentally increasing pup retrievals by fathers increases aggression (Frazier *et al.*, 2006). Biparental care also may influence the development of behavior by altering sibling interactions indirectly because removal of the father can reduce litter size in the wild (Gubernick & Teferi, 2000). A similar mechanism may exist in rats where exposure to a predator can alter the expression of maternal behavior (McLeod *et al.*, 2007). Because the environment may influence the amount and type of parental care provided by each parent, changes in the environment may alter the behavioral phenotype of the offspring. When the amount of parental care that offspring will show as adults is shaped by the behavior of their parents, it is possible that increases or decreases in parental care in response to environmental changes may be transmitted across generations through non-genomic mechanisms (Bester-Meredith & Marler, 2003b).

CAN ALTERATIONS IN PARENTAL HUDDLING AND GROOMING TOWARD OFFSPRING CAUSE CHANGES IN OFFSPRING AGGRESSION AND IS THIS MEDIATED THROUGH VASOPRESSIN?

An important goal in studying the transmission of aggressive behavior across generations is to elucidate both the behavioral and physiological/neurobiological mechanisms through which this occurs. Currently, the most evidence is available for the neuropeptide arginine vasopressin (AVP). AVP and its homolog arginine vasotocin (AVT) have been implicated in the modulation of a wide variety of social behaviors in vertebrates. Olfactory communication, vocal communication, pair bonding, paternal behavior, social recognition, sexual behavior, and offensive aggression have all been shown to be responsive to AVP/AVT (Goodson & Bass, 2001). In a wide variety of species, aggressive behavior is associated with variation in this peptide and can be altered via AVP/AVT manipulations. For example, in non-mammalian species, male zebra finches are more aggressive following infusion of AVT into the septum (Goodson & Adkins-Regan, 1999), as are resident beaugregory damselfish given intramuscular injections of AVT (Santangelo & Bass, 2006). Within mammals, microinjections of AVP into the anterior hypothalamus can decrease the latency for a resident to attack an intruder (Ferris & Potegal, 1988) and intracerebroventricular (ICV) administration of an AVP V1a receptor antagonist shows the opposite effect in California mice (Bester-Meredith *et al.*, 2005). In humans, individuals with personality disorders have cerebral spinal fluid AVP concentrations that are positively correlated with life history of aggression (Coccaro *et al.*, 1998). Thus, AVP is linked to aggressive behavior in a variety of taxa and contexts, and is a logical candidate for transmission of aggression between generations.

Several studies have recently confirmed that AVP can be involved in the transmission of aggressive behavior from parent to offspring. California mice cross-fostered to the less aggressive white-footed mouse show decreased AVP-immunoreactivity (AVP-ir) in the bed nucleus of the stria terminalis (BNST) and the supraoptic nucleus (SON). These neurobiological changes are also behaviorally associated with a decrease in R-I aggression (Bester-Meredith & Marler, 2001). The behavioral conduit for these changes seems to be one specific component of paternal behavior, retrieving behavior, as described earlier. Experimentally increasing retrievals by the California mouse father leads to an increase in AVP-ir staining in the dorsal bed nucleus of the stria terminalis (dBNST), and increased R-I aggression in adult male offspring (Frazier *et al.*, 2006). Importantly, fathers that retrieve their pups also have higher AVP-ir staining in the BNST as compared to non-retrieving fathers (Bester-Meredith & Marler, 2003a). AVP has been shown to enhance paternal behaviors in other species as well; injections of AVP into the lateral septum cause sexually naïve male prairie voles to contact pups more quickly, and spend more time grooming and crouching over the young (Wang *et al.*, 1994). In the less paternal meadow vole, central ICV administration of AVP also significantly increases paternal grooming, huddling, and contact time in previously pup-unresponsive males (Parker & Lee, 2001). In summary, retrieving behavior plays a significant role in the transmission of aggression both within individuals and across generations, and this relationship is most closely associated with AVP.

Manipulations of licking/grooming and huddling behavior of California mouse pups provide an interesting contrast and reveal different connections between paternal behavior and stress (Bester-Meredith and Marler, 2003a; Frazier *et al.* 2006). Castrated California mouse fathers exhibit reduced huddling and grooming behavior of young, but not retrieving behavior; these fathers in turn raise offspring that show normal attack latencies in tests of R-I aggression and neutral arena aggression as adults, and increased AVP-ir in the paraventricular nucleus (PVN), but not the BNST (Trainor & Marler, 2001, 2002; Frazier *et al.*, 2006). Changes in AVP in the PVN have been associated with stress-related disorders such as depression (Rivier & Vale, 1983; Merali *et al.* 2006). Moreover, these offspring also have increased adrenal activity as evidenced by higher progesterone and corticosterone levels. Overall, these data suggest that reduced huddling and grooming of pups and subsequent increases of AVP-ir in the PVN are related to stress, but not to aggression. This work adds to a wide body of evidence in rats demonstrating that animals experiencing low levels of maternal licking/grooming and arched-back nursing during development have heightened hypothalamic–pituitary–adrenal (HPA) activity in response to stress as adults (Caldji *et al.*, 1998).

Other studies also support the idea that changes in AVP-ir in the PVN reflect some aspect of stress. Neurons in the PVN are known to respond to stressors by secreting corticotropin-releasing hormone (CRH) and AVP (Whitnall, 1993) and are thus critical to the endocrinology of the stress response. It is intriguing to note that similar to California mice raised by castrated fathers, rats and mice undergoing a maternal separation (MS) paradigm may experience a comparable reduction in maternal licking and grooming. Interestingly, rats selectively bred for high levels of anxiety-related behavior show higher AVP mRNA expression in the PVN (Bosch *et al.*, 2006). Administration of AVP receptor antagonist in the PVN of high anxiety behavior rats also blocks corticosterone increases in response to a CRH challenge (Keck *et al.*, 2002). Taken together, these studies suggest that stress associated with lower levels of licking and grooming by the parent may be driving changes in adult AVP-ir staining within the PVN.

Even though AVP-ir changes in the PVN of California mice that were exposed to reduced huddling and grooming were not associated with changes in aggression (Frazier *et al.*, 2006), there appears to be wide species variation in this association. Certain social paradigms such as MS or low licking/grooming and huddling during development may also cause changes in stress that impact the PVN and future aggression and/or AVP levels (Veenema *et al.*, 2006, 2007). While changes to AVP in the PVN are consistent across studies, the effects on adult male aggression differ when rats and C57BL/6 mice are compared. In rats, daily MS for the first 14 days of life results in adult males with higher AVP mRNA expression and AVP-ir staining in the PVN and SON, following R-I tests (Veenema *et al.*, 2006) again linking AVP in the PVN and stress as in California mice. MS males are also more aggressive than control animals, such that while groups showed equal time involved in social behaviors as well as equivalent attack latencies and number of attacks, MS males spent a larger percentage of time engaged in aggressive behaviors. In both rats and California mice, adult RI aggression can be elevated by early life events, although this may be caused by different aspects of parental behavior including increased retrievals in the California mouse and the experience of MS in rats. In rats, these changes in R-I aggression are linked to increased AVP in the PVN, rather than in the BNST as in California mice.

Further complications, however, arise when more species comparisons are made. C57BL/6 male mice undergoing an MS paradigm of 3 h of separation daily for the first 14 days of life show an increase in attack latency and more AVP-ir in the PVN (Veenema *et al.*, 2007). This decrease in aggression contrasts with the increased aggression observed in MS rats, and the lack of change in aggression in California mice in response to developmental stressors. This variation occurs despite an increase in AVP-ir in the PVN in all three species. These differences indicate that while stress influences AVP-ir in the PVN, other factors such as social system or coping behavior may influence whether stress translates into a change in aggression.

Overall, it appears that AVP is linked to aggression and can play an important role in transmission of aggression across generations. This role may differ depending on species, and what type of experiences individuals are exposed to during development. At the level of the brain, it appears that AVP-ir in the BNST is tied to aggression, whereas AVP-ir in the PVN reflects some aspect of stress that has the ability to influence aggression. While these species differences are intriguing, they suggest that more work is needed to fully understand the role of vasopressin in the cross-generational transmission of aggression. The variation between behavioral paradigms and their effects on AVP in different parts of the brain may allow us to eventually separate stress-based vs. non-stress-based factors that impact aggression.

IS TESTOSTERONE A MISSING LINK FOR TRANSMISSION OF AGGRESSION ACROSS GENERATIONS?

Changes in an animal's social environment might alter neuropeptide systems through interactions with gonadal hormones. Vasopressin appears to influence several social behaviors, such as aggression and paternal care, and gonadal hormones can alter the functioning of the AVP neuropeptide circuit. As previously mentioned, in the California mouse, fathers can influence the future aggressive behavior of their offspring and changes in both paternal and aggressive behavior may be mediated by AVP (Bester-Meredith & Marler, 2003a; Frazier *et al.*, 2006). The mechanisms underlying these effects, however, remain unclear. Here we examine not only the relationship between AVP and testosterone (T), but also their combined effects and how they may function in the transmission of aggression.

Much attention has been given to understanding how gonadal hormones influence behavior through interactions with neuropeptides. Of most relevance to the California mouse studies is the finding that expression of AVP in the rat is androgen dependent during development (De Vries *et al.*, 1981; De Vries & Miller, 1998). While neonatal castration does not affect the distribution of AVP receptors in the brain, it does reduce AVP fiber density in the lateral septum and cell number in the BNST and medial amygdala, both of which can be restored by testosterone replacement (Cushing *et al.*, 2003). Furthermore, the AVP neurochemical system is influenced by gonadal hormones throughout adulthood, as castration of adult male Brown-Norway rats causes a significant decrease of AVP expression in the extrahypothalamic AVP fiber system (De Vries *et al.*, 1984). This decrease is similar to that found in aging male rats, and peripheral administration of gonadal hormones restored fiber levels in the BNST and medial amygdala to that of young, intact males (Goudsmit *et al.*, 1988). This restoration of innervation is further evidence for the importance of examining peripheral hormones as possible mediators of the organization of the central nervous system.

Studies of social behavior during adulthood also indicate that the relationship between testosterone and AVP remains important at all stages of life. Flank marking in the adult male Syrian hamster (*Mesocricetus auratus*), for example, can be induced by injections of AVP into the medial preoptic area (Ferris *et al.*, 1984). Moreover, studies examining the influence of testosterone on the induction of flank marking by AVP show significant reduction in marking in castrated male hamsters as compared to intact males, demonstrating that testosterone influences the amount of marking produced by the AVP infusions (Albers *et al.*, 1988; Albers & Cooper, 1995).

Testosterone also has been independently linked with the regulation of aggressive behavior in a large number of species including humans (reviewed by Nelson, 2000; e.g., Coccaro *et al.*, 1998; Gould & Ziegler, 2007). One of the classic paradigms demonstrating the effects of testosterone during development involves intrauterine position (IUP), which describes how individuals experience different levels of hormones during prenatal development based on sex of adjacent offspring in utero. In some species, IUP can account for a large amount of behavioral variability within the individuals (Bateson & Young, 1979; Clark *et al.*, 1990; Clark *et al.*, 1993; Hernandez *et al.*, 2006) such that pups become more masculine or feminine physiologically, morphologically, and behaviorally. Male house mice (*Mus musculus*) that are exposed to greater amounts of testosterone *in utero* have been shown to be more aggressive later in life (Ryan & Vandenbergh, 2002). Castration of male marmosets (*Callithrix jacchus*) at infancy decreases future aggression, indicating that secretion of testicular hormones in male marmosets has important effects on the development of aggressive behavior (Dixson, 1993). In California mice, we do not yet know what the developmental effects of testosterone are on aggression. However, testosterone injections in adult male California mice, mimicking natural transient increases following an aggressive encounter, cause the males to behave more aggressively than control animals in an encounter the following day (Trainor *et al.*, 2004). These studies have shown that in many species, administration of testosterone facilitates aggressive behavior.

It is also critical to understand how testosterone interacts with neuropeptides and influences aggressive behaviors. In one such study, castration of male golden hamsters reduced densities of AVP receptor binding in the ventrolateral hypothalamus (VLH), a site that has been previously found to be involved in aggression. Microinjections of AVP into the VLH accelerated the onset of aggression in intact or testosterone treated, and not in castrated males, suggesting that testosterone facilitates aggression via modulation of the vasopressinergic system (Delville *et al.*, 1996). These studies suggest that testosterone is a possible mechanism for mediating the effects of social behavior on the AVP neurochemical system.

Parental care is another social behavior associated with testosterone, but how the interaction between AVP and testosterone influence paternal care has yet to be explored. Early studies examining the effects of testosterone on paternal behavior suggest that androgens, as well as aggressive behavior, are significantly reduced during times of paternal care in a number of species (e.g., Wingfield *et al.*, 1990). In the male Mongolian gerbil (*Meriones unguiculatus*), there exists a tradeoff between paternal care and high levels of testosterone. Gonadectomized males spent significantly more time huddling and grooming their pups than sham and intact males (Clark & Galef, 1999). Lower levels of testosterone have also been associated with paternal care in human fathers (Gray *et al.*, 2007). Furthermore, a number of avian studies have also extensively documented a negative association between paternal care and testosterone (Silverin, 1980; Hegner & Wingfield, 1987; Schoech & Ketterson, 1998; Van Roo, 2004). It has been proposed that there are selection pressures that can cause more variation in the association between testosterone and paternal care depending on factors such as the temporal pattern of aggression and paternal behavior (Marler *et al.*, 2003; Storey *et al.*, 2006). Male Lapland longspurs (*Calcarius lapponicus*) treated with testosterone visited their nests as frequently as controls on postnatal days 4 and 5 (Hunt & Hahn, 1999). Similarly, castration failed to reduce paternal attentiveness in male dwarf hamsters (*Phodopus campbelli*) (Hume & Wynne-Edwards, 2005). This lack of a decrease in testosterone during paternal care has recently been found in a wide range of species (Rodgers *et al.*, 2006). Male bluegill sunfish (*Lepomis macrochirus*), for example, display high pre-spawning testosterone levels that are equally as high at hatching (Kindler & Philipp, 1989; Magee & Neff, 2006). California mice also represent a case where testosterone is differentially associated with paternal behavior such that the two are positively associated (Trainor & Marler, 2001, 2002; Marler *et al.*, 2003; Trainor *et al.*, 2003), possibly because high aggression and paternal behavior can be expressed simultaneously during the postpartum estrus (Gubernick, 1988). This maintenance of androgen levels during periods of parental care suggests that testosterone may be essential to these paternal behaviors. These varying associations between testosterone and paternal behavior suggest that the role of androgens can vary significantly between species.

Because testosterone is linked with both aggression and paternal behavior and can impact the AVP system, we hypothesized that pups will experience a transient increase in testosterone that may serve to facilitate neuronal changes in the extrahypothalamic AVP system leading to the behavioral changes seen in adults experiencing high levels of retrievals (Becker, Moore, Auger & Marler, unpublished data). Preliminary data indicate a rise in testosterone can occur in response to paternal retrievals between days 18 and 21, and because AVP expression is present before birth (De Vries & Villalba, 1997), this suggests that paternal care can directly affect the transient hormone profile of the male offspring during the early stages of sexual differentiation, but prior to adolescence. This is a time point typically viewed as quiescent for testosterone activity. A rise in testosterone during this period would suggest that plasticity of androgens in response to the social environment is important for the regulation of aggression throughout both development and adulthood.

SUMMARY

Our series of experiments with *Peromyscus* combined with studies from other species reveal interesting effects of parental behavior on aggression. Both paternal pup retrievals and some aspect(s) of MS appear to influence aggression of offspring. Paternal retrievals, however, may have a unique effect on offspring aggression that function in part through changes to the AVP neurochemical system associated with the BNST. In comparison, MS and paternal huddling and grooming may function through the PVN as a result of stress, although effects on aggression appear to vary depending on species. The effects of paternal behavior were not uniform such that pup retrievals, but not huddling and grooming appeared to influence aggression. Thus, paternal effects on aggression may operate through mechanisms different than those traditionally associated with stress. This is also supported by preliminary data indicating a change in testosterone levels of pups in response to paternal retrievals. Despite the lack of an effect of changes in paternal huddling and grooming of offspring on aggression, it is possible that paternal separation in California mice may cause changes in aggression similar to those of MS in either rats or house mice.

A careful comparison of aspects of aggression that are influenced by paternal retrievals vs. parental separation could be very illuminating in that we might predict different outcomes for different types of aggression. For example, paternal retrievals and parental separation might affect offensive or defensive components of aggression (i.e., location of attack) differently (Oyegbile & Marler, 2006). Mechanisms associated with both sexual differentiation and stress may impact behavior during this developmental time period between early sexual differentiation and puberty.

We have not tested the effects of maternal retrievals on offspring aggression, but it may have the same effect on offspring aggression as suggested by an association between maternal retrievals and aggression specifically in female offspring. Maternal retrieval manipulations will need to be performed to further test sex differences suggested by the cross-fostering studies. Parents may differentially influence the behavior of their male and female offspring.

A question of interest to those studying both animals and humans is whether changes in offspring behavior and associated underlying mechanisms as a result of changes in paternal behavior can cross multiple generations. It is important to note that male California mice exposed to fewer paternal retrievals in the cross-fostering studies also retrieved their own pups less frequently (Bester-Meredith and Marler, 2003a). Furthermore, some paternal and aggressive behaviors are positively linked within the California mice such that paternal and aggressive behaviors have been positively correlated in mated males (Trainor and Marler, 2001). Based on these studies, we predict that future generations that are exposed to fewer paternal retrievals will also be less aggressive and express fewer paternal retrievals to their own offspring. Thus parental and aggressive behaviors may be positively associated within individuals and perpetuated across generations as has been indicated in humans (Serbin & Karp, 2004). The links between aggressive and paternal behavior that are critical for breeding to occur may also have a major impact in inducing individual variation in behavior that shapes how future generations also respond to their environment.

ACKNOWLEDGMENTS

This research was supported by NRSA Predoctoral Fellowship F31 MH12287, a Vilas Professional Development Award, and a Vilas Graduate Fellowship to J.K.B., NRSA Predoctoral Fellowship F31 MH64328 to B.C.T. and grants IOB-0620042 and IBN-9703309 to C.A.M. We thank Matthew Fuxjager for helpful comments on the manuscript.

REFERENCES

Albers, H. E., and Cooper, T. T. (1995). Effects of testosterone on the behavioral response to arginine vasopressin microinjected into the central gray and septum. *Peptides* 16, 269–273.

Albers, H. E., Liou, S. Y., and Ferris, C. F. (1988). Testosterone alters the behavioral response of the medial preoptic-anterior hypothalamus to microinjection of arginine vasopressin in the hamster. *Brain Res.* 456, 382–386.

Bales, K. L., Kramer, K. K., Lewis-Reese, A. D., and Carter, C. S. (2006). Effects of stress on parental care are sexually dimorphic in prairie voles. *Physiol. Behav.* 87, 424–429.

Bateson, P., and Young, M. (1979). The influence of male kittens on the object play of their female siblings. *Behav.Neural Biol.* 27, 374–378.

Becker, E. A., Moore, B., Auger, C. and Marler, C. A. Paternal care influences offspring testosterone in Peromyscus mice (unpublished data).

Benus, R. F., and Rondigs, M. (1996). Patterns of maternal effort in mouse lines bidirectionally selected for aggression. *Anim. Behav.* 51, 67–75.

Bester-Meredith, J. K., and Marler, C. A. (2001). Vasopressin and aggression in cross-fostered California mice (*Peromyscus californicus*) and white-footed mice (*Peromyscus leucopus*). *Horm. Behav.* 40, 51–64.

Bester-Meredith, J. K., and Marler, C. A. (2003a). Vasopressin and the transmission of paternal behavior across generations in mated cross-fostered *Peromyscus* mice. *Behav. Neurosci.* 117, 455–463.

Bester-Meredith, J. K., and Marler, C. A. (2003b). The association between male offspring aggression and paternal and maternal behavior of *Peromyscus* mice. *Ethology* 109, 797–808.

Bester-Meredith, J. K., and Marler, C. A. (2007). Social experience during development and female offspring aggression in *Peromyscus* mice. *Ethology* 113, 899–900.

Bester-Meredith, J. K., Young, L. J., and Marler, C. A. (1999). Species differences in paternal behavior and aggression in *Peromyscus* and their associations with vasopressin immunoreactivity and receptors. *Horm. Behav.* 36, 25–38.

Bester-Meredith, J. K., Martin, P. A., and Marler, C. A. (2005). Manipulations of vasopressin alter aggression differently across testing conditions in monogamous and nonmonogamous *Peromyscus* mice. *Aggress. Behav.* 31, 189–199.

Bosch, O. J., Krömer, S. A., and Neumann, I. D. (2006). Prenatal stress: Opposite effects on anxiety and hypothalamic expression of vasopressin and corticotropin-releasing hormone in rats selectively bred for high and low anxiety. *Eur. J. Neurosci.* 23, 541–551.

Caldji, C., Tannenbaum, B., Sharma, S., Francis, D., Plotsky, P. M., and Meaney, M. J. (1998). Maternal care during infancy regulates the development of neural systems mediating the expression of fearfulness in the rat. *Proc. Natl Acad. Sci. USA* 95, 5335–5340.

Carlier, M., Roubertoux, P. L., and Pastoret, C. (1991). The Y chromosome effect on intermale aggression in mice depends on the maternal environment. *Genetics* 129, 231–236.

Champagne, F. A., and Meaney, M. J. (2006). Stress during gestation alters postpartum maternal care and the development of the offspring in a rodent model. *Biol. Psychiatr.* 59, 1227–1235.

Christie, M. H., and Barfield, R. J. (1979). Effects of castration and home cage residency on aggressive behavior in rats. *Horm. Behav.* 13, 85–91.

Clark, M. M., and Galef, B. G. (1999). A testosterone-mediated trade-off between parental and sexual effort in male Mongolian gerbils (*Meriones unguiculatus*). *J. Comp. Psychol.* 113, 388–395.

Clark, M. M., Karpiuk, P., and Galef, B. G., Jr. (1993). Hormonally mediated inheritance of acquired characteristics in Mongolian gerbils. *Nature* 364, 712.

Clark, M. M., Malenfant, S. A., Winter, D. A., and Galef, B. G., Jr. (1990). Fetal uterine position affects copulation and scent marking by adult male gerbils. *Physiol. Behav.* 46, 301–305.

Coccaro, E. F., Kavoussi, R. J., Hauger, R. L., Cooper, T. B., and Ferris, C. F. (1998). Cerebrospinal fluid vasopressin levels: Correlates with aggression and serotonin function in personality-disordered subjects. *Arch. Gen. Psychiatr.* 55, 708–714.

Cushing, B. S., Okorie, U., and Young, L. J. (2003). Neonatal castration inhibits adult male response to centrally administered vasopressin but does not alter expression of V1a receptors. *J. Neuroendocrin.* 15, 1021–1026.

Delville, Y., Mansour, K. M., and Ferris, C. F. (1996). Testosterone facilitates aggression by modulating vasopressin receptors in the hypothalamus. *Physiol. Behav.* 60, 25–29.

Delville, Y., Melloni Jr, R. H., and Ferris, C. F. (1998). Behavioral and neurobiological consequences of social subjugation during puberty in golden hamsters. *J. Neurosci.* 18, 2667–2672.

De Vries, G. J., and Miller, M. A. (1998). Anatomy and function of extrahypothalamic vasopressin systems in the brain. *Brain Res.* 119, 3–20.

De Vries, G. J., and Villalba, C. (1997). Brain sexual dimorphism and sex differences in parental and other social behaviors. *Ann. NY. Acad. Sci.* 807, 273–286.

De Vries, G. J., Buijs, R. M., and Swaab, D. F. (1981). Ontogeny of the vasopressinergic neurons of the suprachiasmatic nucleus and their extrahypothalamic projections in the rat brain, presence of a sex difference in the lateral septum. *Brain Res.* 218, 67–68.

De Vries, G. J., Buijs, R. M., and Sluiter, A. A. (1984). Gonadal hormone actions on the morphology of the vasopressinergic innervation of the adult rat brain. *Brain Res.* 298, 141–145.

Dixson, A. (1993). Sexual and aggressive behaviour of adult male marmosets (*Callithrix jacchus*) castrated neonatally, prepubertally, or in adulthood. *Physiol. Behav.* 54, 301–307.

Ferris, C., Albers, H. E., Wesolowski, S. M., Goldman, B. D., and Luman, S. E. (1984). Vasopressin injected into the hypothalamus triggers a stereotypic behavior in golden hamsters. *Science* 224, 521–523.

Ferris, C. F., and Potegal, M. (1988). Vasopressin receptor blockade in the anterior hypothalamus suppresses aggression in hamsters. *Physiol. Behav.* 44, 235–239.

Fish, E. W., Shahrokh, D., Bagot, R., Caldji, C., Bredy, T., Szyf, M., and Meaney, M. J. (2004). Epigenetic programming of stress responses through variations in maternal care. *Ann. NY Acad. Sci.* 1036, 167–180.

Francis, D. D., Young, L. J., Meaney, M. J., and Insel, T. R. (2002). Naturally occurring differences in maternal care are associated with the expression of oxytocin and vasopressin (V1a) receptors: Gender differences. *J. Neuroendocrin.* 14, 349–353.

Francis, D. D., Szegda, K., Campbell, G., Martin, W. D., and Insel, T. R. (2003). Epigenetic sources of behavioral differences in mice. *Nat. Neurosci.* 6, 445–446.

Frazier, C. R. M., Trainor, B. C., Cravens, C. J., Whitney, T. K., and Marler, C. A. (2006). Paternal behavior influences development of aggression and vasopressin expression in male California mouse offspring. *Horm. Behav.* 50, 699–707.

Gonzalez, A., Lovic, V., Ward, G. R., Wainwright, P. E., and Fleming, A. S. (2001). Intergenerational effects of complete maternal deprivation and replacement stimulation on maternal behavior and emotionality in female rats. *Dev. Psychobiol.* 38, 11–32.

Goodson, J. L., and Adkins-Regan, E. (1999). Effect of intraseptal vasotocin and vasoactive intestinal polypeptide infusions on courtship song and aggression in the male zebra finch (*Taeniopygia guttata*). *J. Neuroendocrin.* 11, 19–25.

Goodson, J. L., and Bass, A. H. (2001). Social behavior functions and related anatomical characteristics of vasotocin/vasopressin systems in vertebrates. *Brain Res. Rev.* 35, 246–265.

Gordon, N. S., Kollack-Walker, S., Akil, H., and Panksepp, J. (2002). Expression of c-fos gene activation durin rough and tumble play in juvenile rats. *Brain Res. Bull.* 57, 651–659.

Goudsmit, E., Fliers, E., and Swaab, D. F. (1988). Testosterone supplementation restores vasopressin innervation in the senescent rat brain. *Brain Res.* 473, 306–313.

Gould, L., and Ziegler, T. E. (2007). Variation in fecal testosterone levels, inter-male aggression, dominance rank and age during mating and post-mating periods in wild adult male ring-tailed lemurs (*Lemur catta*). *Am. J. Primatol.* 69, 1325–1339.

Gray, P. B., Parkin, J. C., and Samms-Vaughan, M. E. (2007). Hormonal correlates of human paternal interactions: A hospital-based investigation in urban Jamaica. *Horm. Behav.* 52, 499–507.

Gubernick, D. J. (1988). Reproduction in the California mouse, *Peromyscus californicus. J. Mammal.* 69, 857–860.

Gubernick, D. J., and Alberts, J. R. (1987). The biparental care system of the California mouse, *Peromyscus californicus. J. Comp. Psychol.* 101, 169–177.

Gubernick, D. J., and Teferi, T. (2000). Adaptive significance of male parental care in a monogamous mammal. *Proc. Roy. Soc. Lond. B* 267, 147–150.

Hager, R., and Johnstone, R. A. (2003). The genetic basis of family conflict resolution in mice. *Nature* 421, 533–535.

Haug, M., and Pallaud, B. (1981). Effect of reciprocal cross-fostering on aggression of female mice toward lactating strangers. *Dev. Psychobiol.* 14, 177–180.

Hegner, R. E., and Wingfield, J. C. (1987). Effects of experimental manipulation of testosterone levels on parental investment and breeding success in male house sparrows. *Auk* 104, 462–469.

Hernandez, T. R., Leret, M. L., and Almeida, D. (2006). Effect of intrauterine position on sex differences in the GABAergic system and behavior of rats. *Physiol. Behav.* 30, 625–633.

Hume, J., and Wynne-Edwards, K. E. (2005). Castration reduces male testosterone, estradiol, and territorial aggression, but not paternal behavior in biparental dwarf hamsters (*Phodopus campbelli*). *Horm. Behav.* 48, 303–310.

Hunt, K. E., and Hahn, T. P. (1999). Endocrine influences on parental care during a short breeding season: Testosterone and male parental care in Lapland longspurs (*Calcarius lapponicus*). *Behav. Ecol. Sociobiol.* 45, 360–369.

Keck, M. E., Wigger, A., Welt, T., Muller, M. B., Gesing, A., Reul, J. M., Holsboer, F., Landgraf, R., and Neumann, I. D. (2002). Vasopressin mediates the response of the combined dexamethasone/CRH test in hyper-anxious rats: Implications for pathogenesis of affective disorders. *Neuropsychopharmacology* 26, 94–105.

Kindler, P. M., and Philipp, D. P. (1989). Serum 11-ketotestosterone and testosterone concentrations associated with reproduction in male bluegill (*Lepomis macrochirus*: Centrarchidae). *Gen. Comparat. Endocrinol.* 75, 446–453.

Liu, D., Diorio, J., Tannenbaum, B., Caldji, C., Francis, D., Freedman, A., Sharma, S., Pearson, D., Plotsky, P. M., and Meaney, M. J. (1997). Maternal care, hippocampal glucocorticoid receptors, and hypothalamic–pituitary–adrenal responses to stress. *Science* 277, 1659–1662.

Lucion, A. B., De Almeida, R. M., and De Marques, A. A. (1994). Influence of the mother on development of aggressive behavior in male rats. *Physiol. Behav.* 55, 685–689.

Maestripieri, D. (2003). Similarities in affiliation and aggression between cross-fostered rhesus macaque females and their biological mothers. *Dev. Psychobiol.* 43, 321–327.

Maestripieri, D., Lindell, S. G., and Higley, J. D. (2007). Intergenerational transmission of maternal behavior in rhesus macaques and its underlying mechanisms. *Dev. Psychobiol.* 49, 165–171.

Magee, S., and Neff, B. (2006). Plasma levels of androgens and cortisol in relation to breeding behavior in parental male bluegill sunfish, *Lepomis macrochirus*. *Horm. Behav.* 49, 598–609.

Marler, C. A., Bester-Meredith, J. K., and Trainor, B. C. (2003). Paternal behavior and aggression: Endocrine mechanisms and nongenomic transmission of behavior. *Adv. Stud. Behav.* 32, 263–323.

Marler, C. A., Trainor, B. C., and Davis, E. (2005). Paternal behavior and offspring aggression. *Curr. Direct. Psychol. Sci.* 14, 163–166.

Mast, G. and Marler, C. A. The Winner Effect versus the Resident Advantage: The influence of experience and location on winning aggressive encounters (unpublished data).

McCarthy, M. M., Besmer, H. R., Jacobs, S. C., Keiden, G. M. O., and Gibbs, R. B. (1997). Influence of maternal grooming, sex and age on Fos immunoreactivity in the preoptic area of neonatal rats: Implications for sexual differentiation. *Development. Neurosci.* 19, 488–496.

McLeod, J., Sinai, C. J., and Perrot-Sinal, T. S. (2007). Evidence for non-genomic transmission of ecological information via maternal behavior in female rats. *Genes Brain Behav.* 6, 19–29.

Meaney, M. J. (2001). Maternal care, gene expression, and the transmission of individual differences in stress reactivity across generations. *Ann. Rev. Neurosci.* 24, 1161–1192.

Mendl, M., and Paul, E. S. (1990a). Litter composition affects parental care, offspring growth, and the development of aggressive behaviour in wild house mice. *Behaviour* 116, 90–108.

Mendl, M., and Paul, E. S. (1990b). Parental care, sibling relationships, and the development of aggressive behaviour in two lines of wild house mice. *Behaviour* 116, 11–40.

Merali, Z., Kent, P., Du, L., Hrdina, P., Palkovits, M., Faludi, G., Poulter, M. O., Bedard, T., and Anisman, H. (2006). Corticotropin-releasing hormone, arginine vasopressin, gastrin-releasing peptide, and neuromedin B alterations in stress-relevant brain regions of suicides and control subjects. *Biol. Psychiatr.* 59, 594–602.

Moore, C. L., and Morelli, G. A. (1979). Mother rats interact differently with male and female offspring. *J. Comp. Physiol. Psychol.* 93, 677–684.

Mugford, R., and Nowell, N. (1972). Paternal stimulation during infancy: Effects upon aggression and open-field performance in mice. *J. Comp. Physiol. Psychol.* 79, 30–36.

Nelson, R. (2000). *An Introduction to Behavioral Endocrinology*. Sinauer, Sunderland.

Olivier, B., and Mos, J. (1992). Rodent models of aggressive behavior and serotonergic drugs. *Prog. Neuropsychopharmacol. Biol. Psychiatr.* 16, 847–870.

Oyegbile, T. O. and Marler, C. A. (2006). Weak winner effect in a less aggressive mammal: Correlations with corticosterone but not testosterone. *Physiol Behav.* 89,171–179.

Panksepp, J., Siviy, S., and Normansell, L. (1984). The psychobiology of play: Theoretical and methodological perspectives. *Neurosci. Biobehav. Rev.* 8, 465–492.

Parker, K. J., and Lee, T. M. (2001). Central vasopressin administration regulated the onset of facultative paternal behavior in *Microtus pennsylvanicus* (Meadow voles). *Horm. Behav.* 39, 285–294.

Ribble, D. O. (1991). The monogamous mating system of *Peromyscus californicus* as revealed by DNA fingerprinting. *Behav. Ecol. Sociobiol.* 29, 161–166.

Ribble, D. O. (1992). Dispersal in monogamous rodent, *Peromyscus californicus*. *Ecology* 73, 859–866.

Ribble, D. O., and Salvioni, M. (1990). Social organization and nest coocupancy in *Peromyscus californicus*, a monogamous rodent. *Behav. Ecol. Sociobiol.* 26, 9–15.

Rivier, C., and Vale, W. W. (1983). Interaction of corticotropin-releasing factor and arginine vasopressin on adrenocorticotropin secretion *in vivo*. *Endocrinology* 113, 939–942.

Rodgers, E. W., Earley, R. L., and Grober, M. S. (2006). Elevated 11-ketotestosterone during paternal behavior in the Bluebanded goby *(Lythrypnus dalli)*. *Horm. Behav.* 49, 610–614.

Ryan, B., and Vandenbergh, J. G. (2002). Intrauterine position effects. *Neurosci. Biobehav. Rev.* 26, 665–678.

Santangelo, N., and Bass, A. H. (2006). New insights into the neuropeptide modulation of aggression: Field studies or arginine vasotocin in a territorial tropical damselfish. *Proc. Roy. Soc. Lond. B* 273, 3085–3092.

Schoech, S., and Ketterson, E. D. (1998). The effect of exogenous testosterone on paternal behavior, plasma prolactin, and prolactin binding sites in dark-eyed juncos. *Horm. Behav.* 34, 1–10.

Serbin, L. A., and Karp, J. (2004). The intergenerational transfer of psychosocial risk: Mediators of vulnerability and resilience. *Ann. Rev. Psychol.* 55, 333–363.

Silverin, B. (1980). Effects of prolactin on the gonad and body weight of the male pied flycatcher during the breeding period. *Endokrinologie* 76, 45–50.

Storey, A. E., Delahunty, K., McKay, D. W., Walsh, C. J., and Wilhelm, S. I. (2006). Social and hormonal bases of individual differences in the parental behaviour of birds and mammals. *Can. J. Exp. Psych.* 60, 237–245.

Taravosh-Lahn, K., and Delville, Y. (2004). Aggressive behavior in female golden hamsters: Development and the effect of repeated social stress. *Horm. Behav.* 46, 428–435.

Teskey, G. C., and Kavaliers, M. (1987). Aggression, defeat, and opioid activation in mice: influences of social factors, size and territory. *Behav. Brain. Res.* 23, 77–84.

Trainor, B. C., and Marler, C. A. (2001). Testosterone, paternal behavior, and aggression in the monogamous California mouse (*Peromyscus californicus*). *Horm. Behav.* 40, 32–42.

Trainor, B. C., and Marler, C. A. (2002). Testosterone promotes paternal behavior in a monogamous mammal via conversion to oestrogen. *Proc. Roy. Soc. Lond.* 269, 823–829.

Trainor, B. C., Bird, I. M., Alday, N. A., Schlinger, B. A., and Marler, C. A. (2003). Variation in aromatase activity in the medial preoptic area and plasma progesterone is associated with the onset of paternal behavior. *Neuroendocrinology* 61, 165–171.

Trainor, B. C., Bird, I. M., and Marler, C. A. (2004). Opposing hormonal mechanisms of aggression revealed through short-lived testosterone manipulations and multiple winning experiences. *Horm. Behav.* 45, 115–121.

van den Berg, C. L., Hol, T., van Ree, J. M., Spruijt, B. M., Everts, H., and Koolhaas, J. M. (1999). Play is indispensible for an adequate development of coping with social challenges in the rat. *Dev. Psychobiol.* 34, 129–138.

Van Roo, B. (2004). Exogenous testosterone inhibits several forms of male parental behavior and stimulates song in a monogamous songbird: The blue-headed vireo (*Vireo solitarius*). *Horm. Behav.* 46, 678–683.

Veenema, A. H., Blume, A., Niederle, D., Buwalda, B., and Neumann, I. D. (2006). Effects of early life stress on adult male aggression and hypothalamic vasopressin and serotonin. *Eur. J. Neurosci.* 24, 1711–1720.

Veenema, A. H., Bredewold, R., and Neumann, I. D. (2007). Opposite effects of maternal separation on intermale and maternal aggression in C57BL/6 mice: Link to hypothalamic vasopressin and oxytocin immunoreactivity. *Psychoneuroendocrinology* 32, 437–450.

Vitaro, F., Barker, E. D., Boivin, M., Brendgen, M., and Tremblay, R. E. (2006). Do early difficult temperament and harsh parenting differentially predict reactive and proactive aggression? *J. Abnorm. Child Psychol.* 34, 685–695.

Wang, Z., Ferris, C. F., and De Vries, G. J. (1994). Role of septal vasopressin innervation in paternal behavior in prairie voles (*Microtus ochrogaster*). *Proc. Natl Acad. Sci. USA* 91, 400–404.

Weaver, I. C. G., Cervoni, N., Champagne, F. A., D'Alessio, A. C., Sharma, S., Seckl, J. R., Dymov, S., Szyf, M., and Meaney, M. J. (2004). Epigenetic programming by maternal behavior. *Nature Neurosci.* 7, 847–854.

Wingfield, J. C., Ball, G. F., Dufty, A. M., and Hegner, R. E. (1990). Testosterone and aggression in birds. *Am. Sci.* 75, 602–608.

Whitnall, M. H. (1993). Regulation of the hypothalamic corticotropin-releasing hormone neurosecretory system. *Prog. Neurobiol.* 5, 573–629.

29

FATHERS BEHAVING BADLY: THE ROLE OF PROGESTERONE RECEPTORS

TERESA H. HORTON[1], JOHANNA S. SCHNEIDER[1,2], MARIANA A. JIMENEZ[1] AND JON E. LEVINE[1]

[1] *Department of Neurobiology and Physiology, Northwestern University, 2205 Tech Drive, Rm 2-160 Hogan Hall, Evanston, IL 60208, USA*
[2] *Department of Biochemistry and Molecular Genetics, University of Virginia, Jordan Hall Room 1229, PO Box 800733, Charlottesville, VA 22908, USA*

INTRODUCTION: PROGESTERONE AND ITS RECEPTORS

Progesterone(p) is a steroid hormone that received its name because of its essential role in supporting gestation in female mammals (Allen & Wintersteiner, 1934; Butenandt & Westphal, 1934; Hartmann and Wettstein, 1934; Slotta *et al.*, 1934). Despite its prominent role in female reproduction, until recently few functions had been ascribed to progesterone in males; however, recent work has suggested that P may also influence male reproductive behaviors (Witt *et al.*, 1995; Schneider *et al.*, 2005; Wagner, 2006) as well as male-specific development of the hypothalamus (Quadros *et al.*, 2002; Wagner, 2006). In this chapter we summarize data that suggest progesterone plays an important role in structuring the sexual and parental behavior of male rodents.

Progesterone is both a final product of steroidogenesis and an intermediate in steroidogenic pathways leading to the production of both gonadal and adrenal steroids in males and females (Hadley & Levine, 2007). The synthesis of steroid hormones by the adrenal glands and the gonads is under the control of the pituitary gland. The anterior pituitary secretes adrenocorticotropic hormone (ACTH) which stimulates steroidogenesis by the adrenal cortex leading to increased secretion of glucocorticoids, for which progesterone may serve as a precursor. Steroidogenesis and gametogenesis by the testis and ovary are regulated by luteinizing hormone (LH) and follicle stimulating hormone (FSH). The secretion of LH and FSH is regulated by the pulsatile secretion of gonadotropin releasing hormone (GnRH). The bodies of the GnRH containing cells are located in the preopticohypothalamus, but the axons extend to the median eminence where the terminals release GnRH in the hypothalamo–hypophyseal portal circulation to be carried to the anterior pituitary. The pulsatile pattern of GnRH secretion is regulated at both the level of the cell bodies and at the terminals and by many different factors, including positive and negative feedback effects of gonadal steroids (Hadley & Levine, 2007, pp.122–123).

PROGESTERONE ACTIONS IN THE BRAIN

Progesterone acts on neurons through genomic and non-genomic signaling mechanisms (Mani *et al.*, 1997; Mani, 2006; see Figure 29.1). The classical intracellular progesterone receptors(PR), PR-A and PR-B, are members of the nuclear receptor superfamily. The two isoforms of PR are produced from a single gene, but while structurally related are functionally

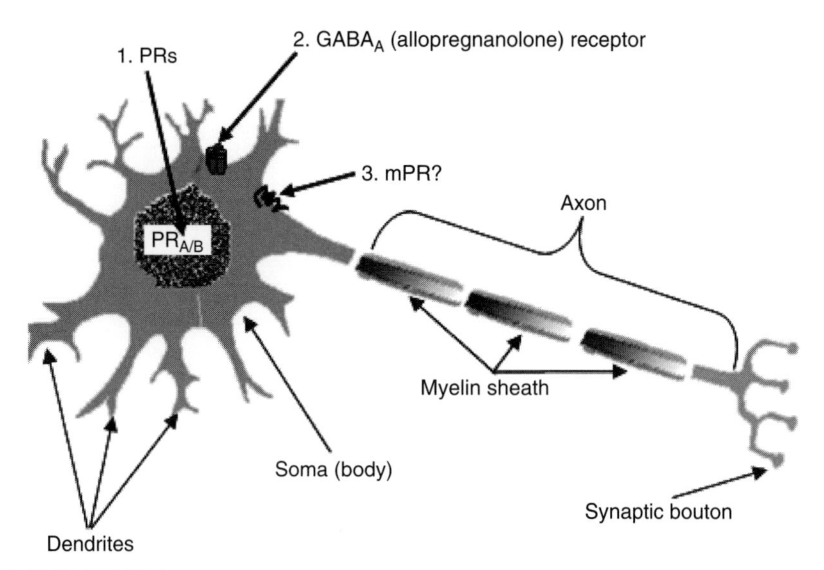

FIGURE 29.1 Cartoon of a neuron illustrating locations of progesterone action. The classical progesterone receptors (1) PR-A and PR-B are members of the nuclear receptor superfamily. Progesterone and its metabolites also work via the GABA$_A$ receptor (2). Recent discoveries suggest a specific membrane progesterone receptor is also located on neurons (3). Progesterone also alters the expression of myelin sheath proteins.

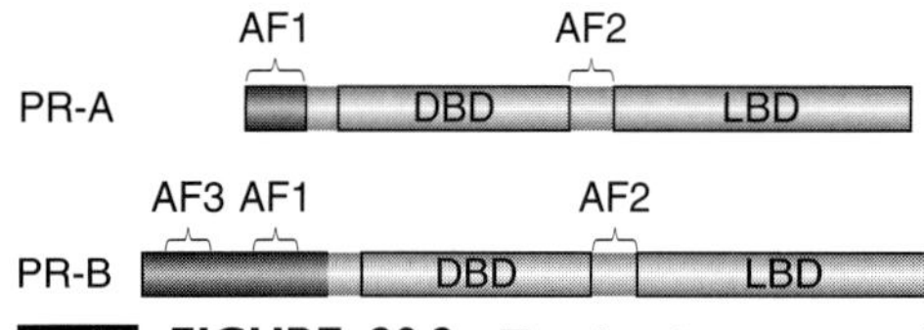

FIGURE 29.2 The domain structure of PR-A and PR-B. The amino terminal region of both isoforms contains a transactivation function (AF1). The DNA binding domain (DBD) of both receptors is associated with dimerization. AF3 of PR-B is located upstream of AF1 in the amino terminal end and results in PR-B activating target genes that cannot be activated by PR-A (*Source*: Redrawn from Conneely and Lydon (2000) and Mani (2006)).

distinct (Conneely & Lydon, 2000; Mani, 2006) (see Figure 29.2). The amino terminal end contains a transactivation function (AF1) that regulates level and promoter specificity of target gene activation (Mani *et al.*, 1997; Conneely & Lydon, 2000). PR-B contains additional amino acids (128–165) in the amino terminal end and a unique activation function (AF3) (Mani, 2006). The presence of AF3 results in PR-B activating target genes that cannot be activated by PR-A (Conneely & Lydon, 2000; Mani, 2006). Each receptor contains a DNA binding domain (DBD) and a ligand binding domain (LBD). The DBD is located downstream of AF-1, interspersed between the DBD and the LBD is another transactivation function (AF-2) which is part of a hypervariable region that also contains nuclear localizations signals (Conneely & Lydon, 2000). The DBD is functionally complex and contains regions that interact with heat-shock proteins, sequences required for dimerization and many other complex, inter- and intramolecular interactions (Conneely & Lydon, 2000; Mani, 2006). Binding of progesterone to its intracellular receptors results in the dissociation of the receptor from heat-shock proteins, enabling the hormone–receptor complex to dimerize, bind to regulatory proteins, and associate with progesterone response elements in the target genes resulting in initiation of gene transcription (Mani, 2006).

Steroid hormones also exert rapid effects on neurons through intracellular signaling pathways that do not involve gene transcription. These effects may involve the interactions of progesterone or its metabolites with intracellular PRs associated with signaling complexes in the cytoplasm or plasma membrane (Mani, 2006; Pluchino *et al.*, 2006). Progesterone and its metabolites can also act as modulators of the γ-aminobutyric acid (GABA$_A$) receptor, oxytocin, the expression of myelin, and the activity of the opioid system (Mani, 2006; Pluchino *et al.*, 2006).

PRs are also activated by substances other than progesterone leading to "ligand-independent" activation (Mani *et al.*, 1997; Auger, 2001). Substances that activate the D_1 subtype of the dopamine receptor can activate PR. The intracellular pathways activated by ligand-dependent and ligand-independent stimulation of PRs may alter gene expression or behavior differentially leading to a rich repertoire of functional outcomes (Auger, 2001).

PRs are expressed in many regions of the brain, including the hypothalamus, basal forebrain, cerebral cortex, cerebellum, and midbrain (MacLusky & McEwen, 1980; Foecking & Levine, 2005). In three regions, the medial preoptic nucleus (mPON), the arcuate-median eminence (ARC-ME), and the ventromedial nucleus (VMN), estrogen induces PR expression (MacLusky & McEwen, 1980; Brown *et al.*, 1987; Kudwa & Rissman, 2003; Foecking & Levine, 2005); however, the induction of PRs in the VMN is greater in females than in males (Brown *et al.*, 1987). This sexual dimorphism in estrogen induction of PRs may account for some of the differences in the actions of progesterone in males and females (Brown *et al.*, 1987; Wagner, 2006). Estrogen also induces expression of both the A and B isoforms of PR from the pituitary of female rats (Szabo *et al.*, 2000). The presence of high levels of progesterone inhibits expression of PR in the pituitary (Turgeon & Waring, 2000). Thus, the interchange between the induction of PR by estrogen and the inhibition of PR by progesterone contributes to conditions leading to the cyclic changes necessary for the female ovulatory cycle.

During the female ovulatory cycle, progesterone plays an important role in modulating behavior as well as changes in hormone secretion. In the early follicular phase (metestrus), progesterone exerts negative feedback on the hypothalamus and pituitary to suppress GnRH and gonadotropin secretion. As the cycle progresses and estradiol secretion increases, progesterone receptors are induced resulting in the conversion of the feedback system to a positive feedback system on proestrus (Levine, 1997). For a short period, in the rat, progesterone will facilitate lordosis; however, progesterone will then desensitize its own receptors resulting in suppression of the behavior and gonadotropin secretion (Levine, 1997; Blaustein, 2008). It is also critical to note that progesterone influences behavior and the secretion of gonadotropins only in the presence of estradiol (Blaustein, 2008); these two hormones appear to work together with their specific effects being determined by their relative concentrations and order of presentation. Progesterone suppresses LH secretion during pregnancy (Al-Gubory *et al.*, 2003), thereby preventing ovulation during pregnancy. Because of its ability to suppress gonadotropin secretion and prevent ovulation, progesterone is used alone or in combination with low doses of estrogen as a contraceptive.

Both progesterone levels and the expression of PRs undergo developmental changes in young animals. Serum progesterone concentrations of male and female rats do not differ prenatally and rise similarly from birth to 20 days of age, after which they tend to increase more rapidly and show increased variability in females, presumably due to the onset of estrous cyclicity (Dohler & Wuttke, 1974; Weisz & Ward, 1980). The messenger RNA for the PRs has been detected as early as 2 days before birth in homogenized tissue extracts containing the hypothalamus and preoptic area of prenatal female rats (Kato & Onouchi, 1983; Kato *et al.*, 1993). Protein for the receptors was first detected in the hypothalamus-preoptic area on the day of birth after which it increased for the first week, then remained stable through days 10 and 28. The pattern of development of the receptors and the sensitivity of the receptors to progesterone appear to differ between males and females (Wagner, 2006). Several reports (reviewed in Wagner, 2006) support the hypothesis that a sex difference exists in the expression of PRs in the developing brain and may contribute to the sexual differentiation of behavior and the control of sex hormone secretion. Given the fact that it will be important to localize differences in expression of receptors to specific nuclei, the use of techniques such as immunohistochemistry and *in situ* hybridization will be needed to reveal the details of these differences.

PROGESTERONE RECEPTOR KNOCKOUT ANIMALS

The advent of genetic engineering technologies has provided new tools with which to explore animal behavior. It is now possible to remove (knockout) or introduce (knockin)

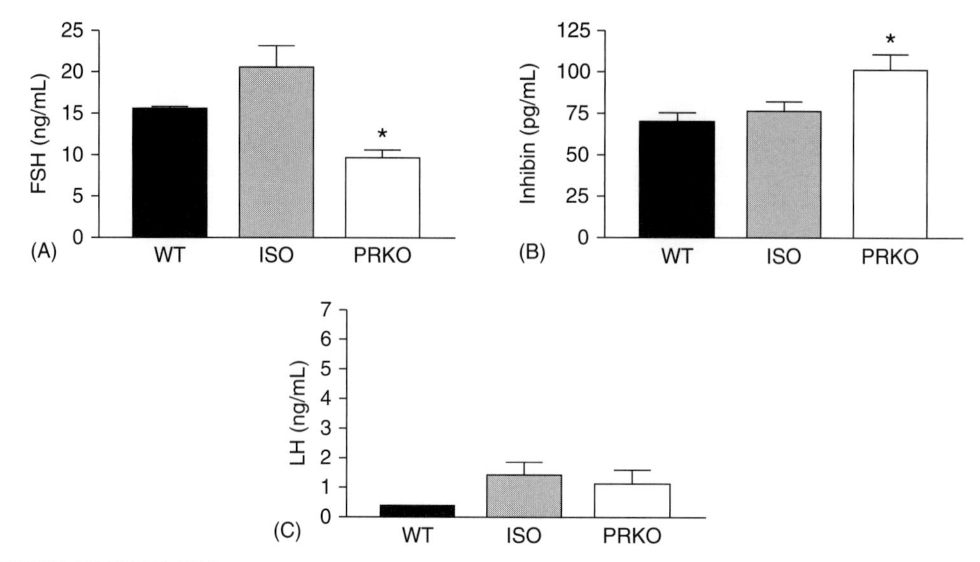

FIGURE 29.3 Plasma FSH, inhibin, and LH levels in male WT C57BL/6, isogenic littermate (ISO), and PRKO mice. (A) Plasma FSH is significantly lower in PRKO males than C57BL/6 and ISO males ($^*p = 0.0001$). (B) Inhibin levels are significantly higher in PRKO males than for WT and ISO males ($^*p = 0.04$). (C) Basal LH levels are not significantly different among WT, male PRKO, and C57BL/6 and ISO mice (*Source*: Redrawn from Schneider *et al.* (2005)).

specific genes with varying degrees of spatial and temporal control. The progesterone receptor knockout (PRKO) mouse was created in the laboratory of Dr Bert O'Malley by John Lydon and colleagues in 1995 (Lydon *et al.*, 1995). Homozygous females of this strain were immediately recognized to be infertile, have abnormal mammary gland development, to exhibit uterine hyperplasia and inflammation, and to lack P-induced sex behavior (Lydon *et al.*, 1995). The obvious next question was to ask what impact the PRKO construct had on reproductive function in males. PRKO males have an altered hypothalamic–pituitary–gonadal axis (Figure 29.3). PRKO males have lower FSH concentrations and elevated inhibin concentrations relative to isogenic wild type (WT) controls and unrelated C57BL6 male mice; however, there are no differences among the three strains in the circulating concentrations of LH (Figure 29.3). The sperm counts of PRKO males are lower than those of WT males at early ages, but by 9–10 weeks of age they have caught up with those of the other strain and are fertile (Schneider *et al.*, 2005).

As with all reproductive steroids, progesterone and its receptors may express both organizational and activational effects on reproduction. The PRKO construct deletes the gene from all tissues at all stages of development, thus it has the potential to disrupt both the organizational and activational functions of progesterone. Any differences between PRKO and WT males are interesting, but additional testing is required to determine whether these differences result from the actions of progesterone and its receptors during prenatal life, early postnatal life, or in adulthood.

PATERNAL BEHAVIOR

Male mammals display a wide range of complex social behaviors toward young that can range from aggression to indifference to parental care (Gibber *et al.*, 1984; Jakubowski & Terkel, 1985; Schneider *et al.*, 2003). Parental care is rare among animals and is estimated to be present in approximately 10% of mammals (Kleiman & Malcolm, 1981) and only 6% of rodent species (Dewsbury, 1981). In addition to a lack of parental care, male mammals may also exhibit extreme aggression toward infants resulting in infanticide. Initially characterized as maladaptive, some studies have concluded that infanticide can increase a male's reproductive success and is therefore an adaptive behavior

(Hrdy, 1979; vom Saal & Howard, 1982). Many species that engage in high levels of parental behavior at one time will, at another stage of life, exhibit aggression toward young. The frequency of infanticide can be influenced by a number of factors including, prenatal hormone exposure, prenatal stress, age, sex, previous sexual experience, previous exposure to young, and previous social experience (Labov, 1980; vom Saal, 1983; Gibber *et al.*, 1984; Gibber & Terkel, 1985; Jakubowski & Terkel, 1985; McCarthy & vom Saal, 1985, 1986; Brown, 1986; McCarthy, 1990; Perrigo *et al.*, 1991; Gubernick *et al.*, 1994). For example, many individuals of the highly parental California mouse (*Peromyscus californicus*) exhibit infanticide when tested prior to mating, but exhibit parental behavior after mating and the birth of their own pups (Gubernick & Nelson, 1989; Gubernick *et al.*, 1994). However, the neural and endocrine mechanisms responsible for these changes in behavior are unknown.

The gonadal steroid testosterone(T) is often the focus of studies of hormones and their role in regulating parental behavior in males. The role of T in parental behavior or infant-directed aggression is highly variable depending on the species studied and the prior sexual experience of the individual. In many studies, T levels have not been found to reliably predict male behavior patterns toward young (Clark & Galef, 1999; Lonstein & De Vries, 1999; Trainor & Marler, 2001, 2002). Human fathers and non-fathers with low T levels are more responsive to infant stimuli (Fleming *et al.*, 2002), and some men experience a decrease in T levels immediately after the birth of their child (Storey *et al.*, 2000; Berg & Wynne-Edwards, 2001). Studies in Mongolian gerbils (Brown *et al.*, 1995) and Djungarian hamsters (Reburn & Wynne-Edwards, 1999) describe an increase in circulating levels of T in males in anticipation of parturition by their mate, followed by a decrease in T in the days after the birth of pups. In cotton-top tamarins, urinary T levels increase during pregnancy and remain elevated after parturition (Ziegler *et al.*, 1996). Urinary T levels were negatively correlated with parental behavior in the marmoset (Nunes *et al.*, 2001), but this effect has not been seen in other non-human primates (Ziegler, 2000). All of the foregoing studies were correlative and did not examine the effects of experimental manipulation of T on parental care or aggression toward young. In one study it was found that exogenous T promotes parental behavior in the highly parental California mouse (Trainor & Marler, 2001) most likely through its aromatization to estrogen in the medial preoptic area (Trainor *et al.*, 2003), a region known to regulate parental behavior in rodents (Sturgis & Bridges, 1997; Numan & Insel, 2003). The role of T in modulating parental behavior thus remains to be fully elucidated.

There are conflicting reports about the effects of castration and T-replacement on parental behavior in many species, including prairie voles (Wang & De Vries, 1993; Lonstein & De Vries, 1999), gerbils (Clark & Galef, 1999), hamsters (Hume & Wynne-Edwards, 2005), and rats (Rosenberg *et al.*, 1971; Quadagno & Rockwell, 1972; Quadagno *et al.*, 1973; Quadagno, 1974; Rosenberg & Sherman, 1975). An important consideration in these studies is that castration also significantly reduces P levels, and it is possible that some of the subsequent behavioral changes could be a consequence of lowered P. For example, in the California mouse virgin males exhibit higher P levels when they are infanticidal and decreased P levels after mating and onset of parental behavior (Trainor *et al.*, 2003). A functional relationship between P- and T-dependent behaviors is further suggested by the observation that the loss of PR expression is correlated with increased androgen receptor expression in the medial preoptic area and bed nucleus of the stria terminalis (Schneider *et al.*, 2005). Additional studies are necessary to elucidate the roles of P and T in regulating both spontaneous and mating-induced parental behaviors by males.

We have used the PRKO male mouse and pharmacological manipulation of WT mice to investigate the role of progesterone in the regulation of parental behavior. PRKO mice are more parental than isogenic WT or unrelated C57BL6 WT males. Using a test of paternal behavior devised by Dr Katherine Wynne-Edwards (Jones & Wynne-Edwards, 2001) we tested virgin males for parental behavior by placing pups in the corner of the males' home cage opposite to the male (Schneider *et al.*, 2003). The index score was calculated by awarding points in the following manner: +1 for contacting pup, +1 for picking up pup, +4 for retrieving pup to nest, +1 for nurturing behaviors such as licking and crouching continuously for at least 2 min after retrieval. A perfect score of +7 indicates the highest level of parental care; a score of 0 indicates no parental behavior. The

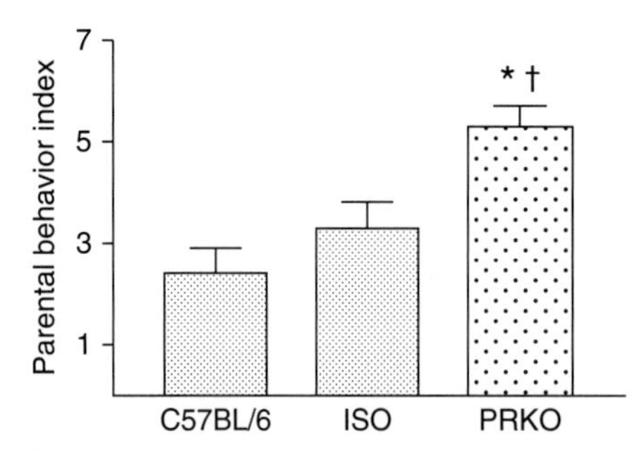

FIGURE 29.4 Virgin PRKO males scored higher on the parental behavior test than males expressing PR. Scores on the paternal behavior index for PRKO were significantly higher than C57BL6 (*$p = 0.001$) and ISO (†$p = 0.01$) strains. Scores on the behavioral index were compared by using Kruskal–Wallis one-way ANOVA for genotype differences (*Source*: Redrawn from Schneider *et al.* (2003)).

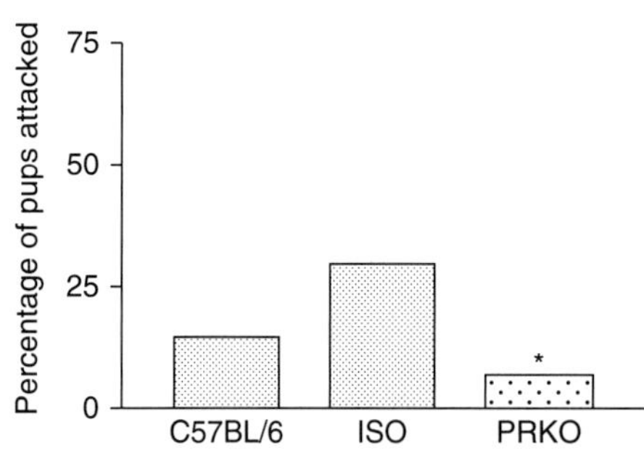

FIGURE 29.5 Virgin PRKO males displayed less infant-directed aggression than males expressing PR. The PR-expressing strains, C57BL6 and ISO, attacked pups during the behavior tests 15% and 29.4% of the times they were presented with pups, respectively. PRKO virgin males attacked pups in only 7.3% of the tests. These differences in frequency to attack were statistically significant (chi-square test, *$p = 0.03$) (*Source*: Redrawn from Schneider *et al.* (2003)).

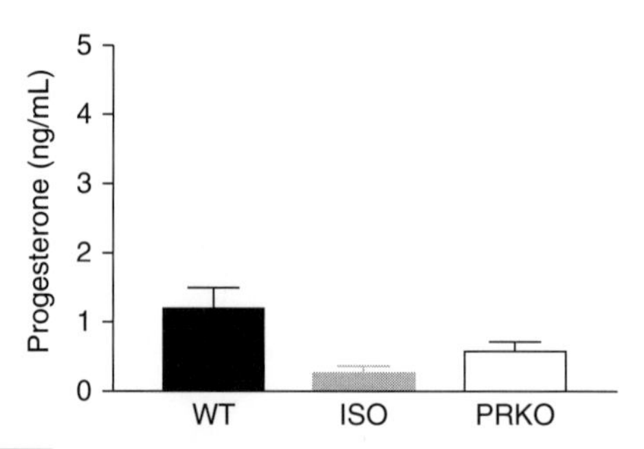

FIGURE 29.6 Plasma concentrations of progesterone in males of three genotypes. Progesterone concentrations did not significantly differ among C57BL6, ISO, or PRKO males (*Source*: Redrawn from Schneider *et al.* (2003)).

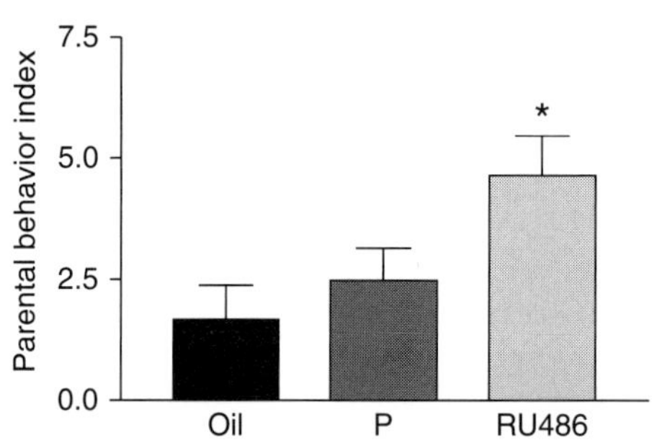

FIGURE 29.7 Treatment with progesterone inhibits and RU486 enhances paternal behavior in virgin C57BL6 mice. Male mice treated with progesterone exhibited similar levels of paternal behavior compared with oil-treated controls. In contrast, RU486 treatment significantly increased paternal behavior in the same strain (Kruskal–Wallis test, *$p = 0.05$). All observers were blind as to the treatment group during testing (*Source*: Redrawn from Schneider *et al.* (2003)).

PRKO males retrieved and nurtured pups more readily than the other genotypes, achieving significantly higher scores on the test (Figure 29.4). The PRKO males are also less aggressive toward the pups than the other males (see Figure 29.5). These differences in behavior occur despite the fact there are no differences in circulating levels of progesterone among the three strains of mice (see Figure 29.6).

To test whether this effect was an organizational or activational effect of the progesterone receptors, adult WT C57BL6 mice were treated with the progesterone antagonist RU486 or progesterone, then tested for parental behavior and aggression toward infants. As seen in the previous experiments, the males treated with RU486 exhibited increased parental behavior index scores (Figure 29.7). In contrast, males treated with progesterone, which elevated their circulating progesterone levels equivalent to those seen in pregnant females, showed increased aggression toward infants (see Figure 29.8).

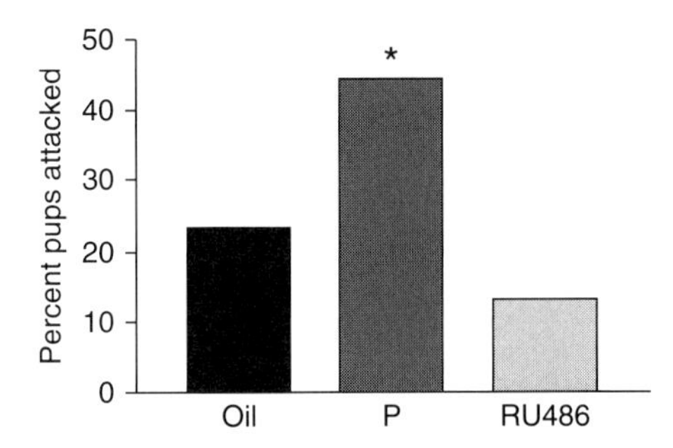

FIGURE 29.8 Treatment with progesterone increases the frequency with which virgin C57BL6 males attacked pups. Progesterone-treated males attacked pups in 44% of the tests compared with 23% of the control-treated animals and only 13% of the RU486-treated males (*chi-square test, $p = 0.037$) (*Source*: Redrawn from Schneider *et al.* (2003)).

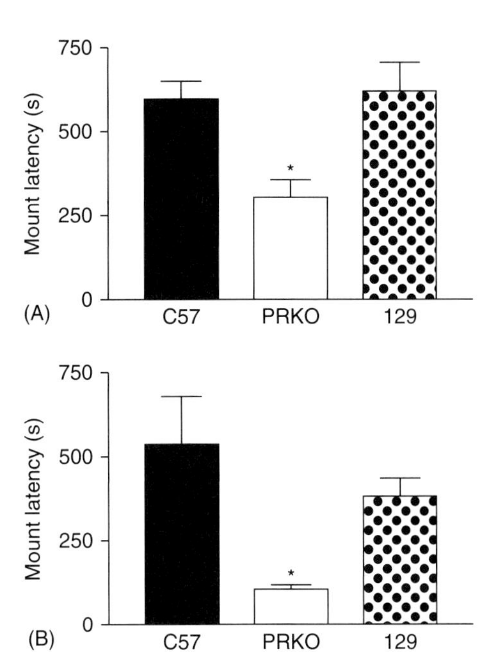

FIGURE 29.9 Latency of male mice to mount a receptive female. (redrawn from Schneider *et al.* 2005). In the initial test with a receptive female (A) the PRKO males exhibited reduced latency to mount, compared with both WT strains (C57 and 129) (*, p = 0.004). In a second test, with a receptive female (B) PRKO males again exhibited reduced latency to mount, compared with both WT strains (*, p < 0.001).

To rule out the possibility that the reduced levels of infant-directed aggression exhibited by the PRKO males was simply the result of an overall reduction in aggression as compared to WT males rather than a specific infant-directed behavior, PRKO and WT males were tested in a resident-intruder test to evaluate their overall levels of aggression (Schneider *et al.*, 2003). There were no significant differences in the latency to attack, the duration of attack, or the number of bouts in which males engaged among the genotypes. These results suggest that activation of PR in adult male mice increases aggression toward young without a commensurate increase in male–male aggression. The results also suggest that the neuronal mechanisms governing one type aggression (infant-directed) are partially separable from other types.

PROGESTERONE, MATING BEHAVIOR, AND ANXIETY

Similarly, it is possible that differences in levels of anxiety or motivation to engage in mating behavior might yield differences in parenting behavior. To test this hypothesis C57Bl6, PRKO and isogenic WT males were tested using common paradigms for measuring latency to mount, anxiety, and mating behavior (Schneider *et al.*, 2005). PRKO males exhibited significantly reduced latencies to mount a female in both their initial and subsequent encounters (Figure 29.9). This rapid initiation of sexual behavior is most likely not due to a reduction of anxiety-related behaviors; PRKO males exhibit an intermediate phenotype on two tests of anxiety-related behavior, the elevated plus maze and the open field test. Thus, the increase in parental behavior exhibited by the PRKO males is accompanied by changes in mating behavior; however, the change is in the direction of increased propensity to engage in sexual activity rather than a decrease which might be expected if the decrease in aggression were due to an overall decrease in motivated behavior.

These results are consistent with observations in primates in which treatment with progesterone inhibits reproductive behavior in male cynomolgus monkeys, but has been observed to increase male-on-male aggression independently of male-on-female aggression while testosterone levels remained unchanged (Zumpe *et al.*, 1991, 2001). In humans the progesterone agonist medroxyprogesterone acetate (MPA) is used clinically to suppress sexual activity in male sex offenders. MPA is reported to reduce pedophilia, incest, and rape (Saleh & Guidry, 2003; Andersen & Tufik, 2006); however, the reported theoretical basis for these treatments relies on the "antiandrogenic" function of progestins rather than other functions of progestins and their receptors. The efficacy of MPA as a treatment for pedophilia, incest, and rape and its numerous side effects have been the subject of many studies, but these studies have not examined the effects on infant-directed behaviors (Cooper, 1986; Saleh & Guidry, 2003; Andersen & Tufik, 2006) Given that high levels of progesterone increase aggression toward infant mice and observations that progesterone is acting via mechanisms other than the antagonism of androgens, perhaps it is time to re-evaluate the use of MPA for the treatment of male sex offenders.

DOES THE ABSENCE OR PHARMACOLOGICAL BLOCKADE OF PRs SIMPLY MIMIC THE PERIPARTUM HORMONAL MILIEU?

In female mammals, the concentration of progesterone changes dramatically at parturition. In the experiments described above when exogenous progesterone has been used to increase aggression toward infants by males, the levels of progesterone produced were similar to those present in pregnant females. Other researchers have observed that progesterone concentrations in the serum of males that were fathers were lower than those of sexually inexperienced males (Trainor *et al.*, 2003). This change in progesterone level correlates with a change from exhibiting infanticidal behavior prior to mating in the California mouse (*Peromyscus californicus*) to exhibiting parental behavior after mating and the birth of their own pups (Gubernick & Nelson, 1989; Gubernick *et al.*, 1994). Although these differences are consistent with our observations in laboratory mice, data from more distantly related rodents, two species of dwarf hamster (*Phodopus*) suggest that the emergence of paternal behavior is not associated with a change in circulating levels of progesterone (Schum & Wynne-Edwards, 2005). These differences illustrate the importance of understanding the phylogeny of species when making analyses; the California mouse and laboratory mice are phylogenetically more closely related to each other than they are to the hamsters.

Even within a single species the expression of parental or infanticidal behavior can be highly variable. For example, in common laboratory strains of mice (*Mus sp.*) the same mouse will alternate between being infanticidal or parental depending on whether he has pups or not. At the time of mating and throughout the pregnancy of their mate, the males remain infanticidal, but coinciding with the periods when males would be likely to encounter their own progeny the tendency to exhibit infanticide is reduced (vom Saal & Howard, 1982; vom Saal, 1985; McCarthy & vom Saal, 1986; Perrigo *et al.*, 1992). Infanticidal behavior will then reemerge spontaneously 50–60 days after mating (McCarthy & vom Saal, 1985, 1986; Perrigo *et al.*, 1992). This is in contrast to behavior observed in rats in which once a male has learned to exhibit infanticidal behavior it will remain infanticidal (Rosenberg & Sherman, 1975). It is thought that the stimulus of ejaculation during mating triggers an unknown neural response that serves to regulate the emergence and inhibition of infanticide in the male often over long time periods (McCarthy & vom Saal, 1985, 1986; Perrigo *et al.*, 1992).

Given that progesterone levels of male California mice that were fathers were lower than those of sexually inexperienced males (Trainor *et al.*, 2003) and that elevated progesterone levels increased aggression toward infants (Schneider *et al.*, 2003), we hypothesized that mating might induce a sequence of events resulting in changes either in progesterone levels or neural sensitivity to progesterone that would mimic the rise and fall of progesterone during pregnancy in the female. We tested this hypothesis by treating males with exogenous progesterone, and then testing whether withdrawal of progesterone would induce parental behavior in male mice. Virgin male C57BL6 mice were implanted with progesterone or empty capsules for 21 days, the capsules were then withdrawn and the males were tested in the pup-retrieval test previously described (Schneider *et al.*, 2003). Contrary to our predictions, withdrawal of progesterone did not result in enhanced parental behavior (Figure 29.10). In fact, the propensity of the males to attack the pups increased (Figure 29.11) forcing us to re-evaluate our hypothesis.

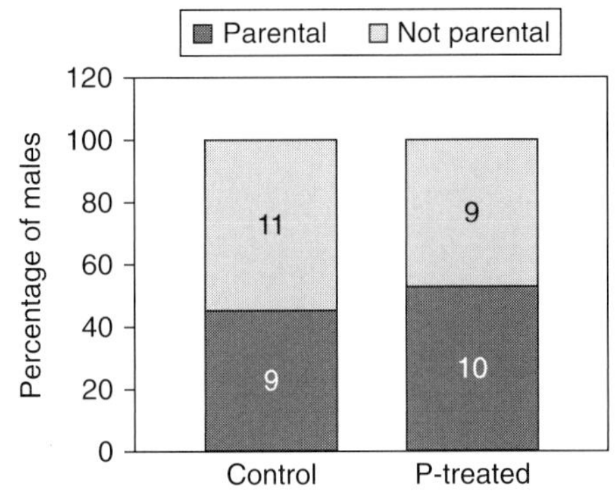

FIGURE 29.10 Proportions of males exhibiting Parental or Nonparental behavior. Males that picked up pups and returned them to their nests during the parental behavior test were categorized as being parental. Approximately fifty percent of adult, virgin, male C57Bl6 mice exhibited parental behavior towards an unfamiliar infant. The propensity to express parental behavior was not altered by prior treatment with progesterone 5 days after removal of the progesterone or control capsule (Chi square test, $p = 0.76$). Data are presented as percentage of each group; numbers inside each bar are the numbers of individuals exhibiting each type of behavior. (Schneider *et al.* unpublished).

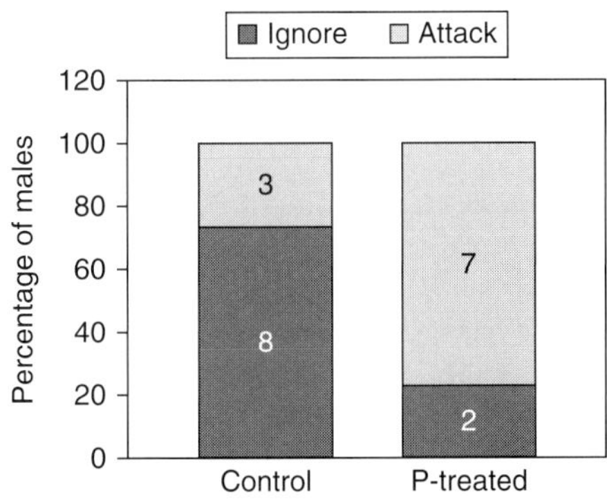

FIGURE 29.11 Percentage of nonparental males that ignored or attacked young. Prior treatment with progesterone significantly increased aggression toward young by the nonparental males. Five days after capsule withdrawal the majority of the subgroup of Control males that were Nonparental ignored the pup (8 of 3); in contrast, the majority of P-treated males that were Nonparental (7 of 9) attacked pups. (Chi square test, $p < 0.02$). (Schneider *et al.* unpublished).

These new results when considered with the data from PRKO mice and mice treated with RU486, suggest that the absence of PRs, or the prolonged pharmacological blockade of PRs, produces an overall behavioral phenotype that is similar to that seen in biparental rodents such as *Peromyscus californicus*, *Microtus ochrogaster*, and *Phodopus campbelli.* These species exhibit other characteristic behaviors, for example the prairie vole, *Microtus ochrogaster,* is socially monogamous and exhibits a high degree of affiliative behavior. In contrast, our results show that the activation of progesterone receptors in laboratory mice by concentrations of progesterone equivalent to those seen by females during pregnancy induces infanticidal behavior toward infants that can persist for at least 5 days following the removal of the steroid.

SUMMARY

Despite its long history as a female hormone, progesterone is emerging as a hormone of importance in male reproductive physiology and behavior. The activation of progesterone receptors either by ligand-dependent or ligand-independent mechanisms may influence organizational and activational aspects of the male reproductive system. These effects include the propensity to mount females, the propensity to care for young, and the propensity to commit acts of aggression toward infants. The deletion or blockade of the progesterone receptor enhances the anticipatory components of sexual activity and promotes parental behavior in male mice (Schneider *et al.*, 2003).

REFERENCES

Al-Gubory, K. H., Hervieu, J., and Fowler, P. A. (2003). Effects of pregnancy on pulsatile secretion of LH and gonadotropin-releasing hormone-induced LH release in sheep: A longitudinal study. *Reproduction* 125, 347–355.

Allen, W. M., and Wintersteiner, O. (1934). Crystalline progestin. *Science* 80, 190–191.

Andersen, M. L., and Tufik, S. (2006). Does male sexual behavior require progesterone? *Brain Res. Rev.* 51, 136–143.

Auger, A. P. (2001). Ligand-independent activation of progestin receptors: Relevance for female sexual behaviour. *Reproduction* 122, 847–855.

Berg, S. J., and Wynne-Edwards, K. E. (2001). Changes in testosterone, cortisol, and estradiol levels in men becoming fathers. *Mayo Clin. Proc.* 76, 582–592.

Blaustein, J. D. (2008). Neuroendocrine regulation of feminine sexual behavior: Lessons from rodent models and thoughts about humans. *Ann. Rev. Psychol.* 59: 93–118.

Brown, R. E. (1986). Social and hormonal factors influencing infanticide and its suppression in adult male Long-Evans rats (*Rattus norvegicus*). *J. Comp. Psychol.* 100, 155–161.

Brown, R. E., Murdoch, T., Murphy, P. R., and Moger, W. H. (1995). Hormonal responses of male gerbils to stimuli from their mate and pups. *Horm. Behav.* 29, 474–491.

Brown, T. J., Clark, A. S., and MacLusky, N. J. (1987). Regional sex differences in progestin receptor induction in the rat hypothalamus: Effects of various doses of estradiol benzoate. *J. Neurosci.* 7, 2529–2536.

Butenandt, A., and Westphal, U. (1934). Zur isolierung und charakterisierung des corpus-luteum-hormons. *Chimishe Berichte* 67, 1440–1445.

Clark, M. M., and Galef, B. G., Jr. (1999). A testosterone-mediated trade-off between parental and sexual effort in male mongolian gerbils (*Meriones unguiculatus*). *J. Comp. Psychol.* 113, 388–395.

Conneely, O. M., and Lydon, J. P. (2000). Progesterone receptors in reproduction: Functional impact of the A and B isoforms. *Steroids* 65, 571–577.

Cooper, A. J. (1986). Progestogens in the treatment of male sex offenders: A review. *Can. J. Psychiatr.* 31, 73–79.

Dewsbury, D. A. (1981). An exercise in the prediction of monogamy in the field from laboratory data on 42 species of muroid rodents. *The Biologist* 63, 138–162.

Dohler, K. D., and Wuttke, W. (1974). Serum LH, FSH, prolactin and progesterone from birth to puberty in female and male rats. *Endocrinology* 94, 1003–1008.

Fleming, A. S., Corter, C., Stallings, J., and Steiner, M. (2002). Testosterone and prolactin are associated with emotional responses to infant cries in new fathers. *Horm. Behav.* 42, 399–413.

Foecking, E. M., and Levine, J. E. (2005). Effects of experimental hyperandrogenemia on the female rat reproductive axis: Suppression of progesterone-receptor messenger RNA expression in the brain and blockade of luteinizing hormone surges. *Gend. Med.* 2, 155–165.

Gibber, J. R., and Terkel, J. (1985). Effect of post-weaning social experience on response of Siberian hamsters (*Phodopus sungorus sungorus*) toward young. *J. Comp. Psychol.* 99, 491–493.

Gibber, J. R., Piontkewitz, Y., and Terkel, J. (1984). Response of male and female Siberian hamsters towards pups. *Behav. Neural Biol.* 42, 177–182.

Gubernick, D. J., and Nelson, R. J. (1989). Prolactin and paternal behavior in the biparental California mouse, (*Peromyscus californicus*). *Horm. Behav.* 23, 203–210.

Gubernick, D. J., Schneider, K. A., and Jeannotte, L. A. (1994). Individual differences in the mechanisms underlying the onset and maintenance of paternal behavior and the inhibition of infanticide in the monogamous biparental California mouse, (*Peromyscus californicus*). *Behav. Ecol. Sociobiol.* 34, 225–231.

Hadley, M., and Levine, J. (2007). *Endocrinology*. Pearson, Prentice Hall, Upper Saddle River, NJ. pp. 340–343.

Hartmann, M., and Wettstein, A. (1934). Ein krystallisiertes hormon aus corpus-luteum. *Helv. Chim. Acta* 17, 878–891.

Hrdy, S. (1979). Infanticide among animals: A review, classification, and examination of the implications for the reproductive strategy of females. *Ethol. Sociobiol.* 1, 13–40.

Hume, J. M., and Wynne-Edwards, K. E. (2005). Castration reduces male testosterone, estradiol, and territorial aggression, but not paternal behavior in biparental dwarf hamsters (*Phodopus campbelli*). *Horm. Behav.* 48, 303–310.

Jakubowski, M., and Terkel, J. (1985). Incidence of pup killing and parental behavior in virgin female and male rats (*Rattus norvegicus*): Differences between Wistar and Sprague-Dawley stocks. *J. Comp. Psychol.* 99, 93–97.

Jones, J. S., and Wynne-Edwards, K. E. (2001). Paternal behaviour in biparental hamsters, *Phodopus campbelli*, does not require contact with the pregnant female. *Anim. Behav.* 62, 453–464.

Kato, J., and Onouchi, T. (1983). Progestin receptors in female rat brain and hypophysis in the development from fetal to postnatal stages. *Endocrinology* 113, 29–36.

Kato, J., Hirata, S., Nozawa, A., and Mouri, N. (1993). The ontogeny of gene expression of progestin receptors in the female rat brain. *J. Steroid Biochem. Mol. Biol.* 47, 173–182.

Kleiman, D. G., and Malcolm, J. R. (1981). The evolution of male parental investment. In *Parental Care in Mammals* (D. J. Gubernick and P. H. Klopfer, Eds.), pp. 347–387. Plenum, New York.

Kudwa, A. E., and Rissman, E. F. (2003). Double oestrogen receptor alpha and beta knockout mice reveal differences in neural oestrogen-mediated progestin receptor induction and female sexual behaviour. *J. Neuroendocrinol.* 15, 978–983.

Labov, J. B. (1980). Factors influencing infanticidal behavior in wild male house mice (*Mus musculus*). *Behav. Ecol. Sociobiol.* 6, 297–303.

Levine, J. E. (1997). New concepts of the neuroendocrine regulation of gonadotropin surges in rats. *Biol. Reprod.* 56, 293–302.
Lonstein, J. S., and De Vries, G. J. (1999). Sex differences in the parental behaviour of adult virgin prairie voles: Independence from gonadal hormones and vasopressin. *J. Neuroendocrinol.* 11, 441–449.
Lydon, J. P., DeMayo, F. J., Funk, C. R., Mani, S. K., Hughes, A. R., Montgomery, C. A., Jr., Shyamala, G., Conneely, O. M., and O'Malley, B. W. (1995). Mice lacking progesterone receptor exhibit pleiotropic reproductive abnormalities. *Genes Dev.* 9, 2266–2278.
MacLusky, N. J., and McEwen, B. S. (1980). Progestin receptors in rat brain: Distribution and properties of cytoplasmic progestin-binding sites. *Endocrinology* 106, 192–202.
Mani, S. K. (2006). Signaling mechanisms in progesterone–neurotransmitter interactions. *Neuroscience* 138, 773–781.
Mani, S. K., Blaustein, J. D., and O'Malley, B. W. (1997). Progesterone receptor function from a behavioral perspective. *Horm. Behav.* 31, 244–255.
McCarthy, M. M. (1990). Short-term early exposure to pups alters infanticide in adulthood in male but not in female wild house mice (*Mus domesticus*). *J. Comp. Psychol.* 104, 195–197.
McCarthy, M. M., and vom Saal, F. S. (1985). The influence of reproductive state on infanticide by wild female house mice (*Mus musculus*). *Physiol. Behav.* 35, 843–849.
McCarthy, M. M., and vom Saal, F. S. (1986). Inhibition of infanticide after mating by wild male house mice. *Physiol. Behav.* 36, 203–209.
Numan, M., and Insel, T. R. (2003). *The Neurobiology of Parental Behavior*. Springer-Verlag, New York.
Nunes, S., Fite, J. E., Patera, K. J., and French, J. A. (2001). Interactions among paternal behavior, steroid hormones, and parental experience in male marmosets (*Callithrix kuhlii*). *Horm. Behav.* 39, 70–82.
Perrigo, G., Belvin, L., and vom Saal, F. S. (1991). Individual variation in the neural timing of infanticide and parental behavior in male house mice. *Physiol. Behav.* 50, 287–296.
Perrigo, G., Belvin, L., and vom Saal, F. S. (1992). Time and sex in the male mouse: Temporal regulation of infanticide and parental behavior. *Chronobiol. Int.* 9, 421–433.
Pluchino, N., Luisi, M., Lenzi, E., Centofanti, M., Begliuomini, S., Freschi, L., Ninni, F., and Genazzani, A. R. (2006). Progesterone and progestins: Effects on brain, allopregnanolone and beta-endorphin. *J. Steroid Biochem. Mol. Biol.* 102, 205–213.
Quadagno, D. M. (1974). Maternal behavior in the rat: Aspects of concaveation and neonatal androgen treatment. *Physiol. Behav.* 12, 1071–1074.
Quadagno, D. M., and Rockwell, J. (1972). The effect of gonadal hormones in infancy on maternal behavior in the adult rat. *Horm. Behav.* 3, 55–62.
Quadagno, D. M., McCullough, J., Ho, G. K., and Spevak, A. M. (1973). Neonatal gonadal hormones: Effect on maternal and sexual behavior in the female rat. *Physiol. Behav.* 11, 251–254.
Quadros, P. S., Goldstein, A. Y., De Vries, G. J., and Wagner, C. K. (2002). Regulation of sex differences in progesterone receptor expression in the medial preoptic nucleus of postnatal rats. *J. Neuroendocrinol.* 14, 761–767.
Reburn, C. J., and Wynne-Edwards, K. E. (1999). Hormonal changes in males of a naturally biparental and a uniparental mammal. *Horm. Behav.* 35, 163–176.
Rosenberg, K. M., and Sherman, G. F. (1975). The role of testosterone in the organization, maintenance and activation of pup-killing behavior in the male rat. *Horm. Behav.* 6, 173–179.
Rosenberg, K. M., Denenberg, V. H., Zarrow, M. X., and Frank, B. L. (1971). Effects of neonatal castration and testosterone on the rat's pup-killing behavior and activity. *Physiol. Behav.* 7, 363–368.
Saleh, F. M., and Guidry, L. L. (2003). Psychosocial and biological treatment considerations for the paraphilic and nonparaphilic sex offender. *J. Am. Acad. Psychiatr. Law* 31, 486–493.
Schneider, J. S., Stone, M. K., Wynne-Edwards, K. E., Horton, T. H., Lydon, J., O'Malley, B., and Levine, J. E. (2003). Progesterone receptors mediate male aggression toward infants. *Proc. Natl Acad. Sci. USA* 100, 2951–2956.
Schneider, J. S., Burgess, C., Sleiter, N. C., DonCarlos, L. L., Lydon, J. P., O'Malley, B., and Levine, J. E. (2005). Enhanced sexual behaviors and androgen receptor immunoreactivity in the male progesterone receptor knockout mouse. *Endocrinology* 146, 4340–4348.
Schum, J. E., and Wynne-Edwards, K. E. (2005). Estradiol and progesterone in paternal and non-paternal hamsters (*Phodopus*) becoming fathers: Conflict with hypothesized roles. *Horm. Behav.* 47, 410–418.
Slotta, K. H., Ruschig, H., and Fels, E. (1934). Reindarstellung der Hormone aus dem Corpusluteum. *Chimishe Berichte* 67, 1270–1271.
Storey, A. E., Walsh, C. J., Quinton, R. L., and Wynne-Edwards, K. E. (2000). Hormonal correlates of paternal responsiveness in new and expectant fathers. *Evol. Hum. Behav.* 21, 79–95.
Sturgis, J. D., and Bridges, R. S. (1997). N-methyl-DL-aspartic acid lesions of the medial preoptic area disrupt ongoing parental behavior in male rats. *Physiol. Behav.* 62, 305–310.
Szabo, M., Kilen, S. M., Nho, S. J., and Schwartz, N. B. (2000). Progesterone receptor A and B messenger

ribonucleic acid levels in the anterior pituitary of rats are regulated by estrogen. *Biol. Reprod.* 62, 95–102.

Trainor, B. C., and Marler, C. A. (2001). Testosterone, paternal behavior, and aggression in the monogamous California mouse (*Peromyscus californicus*). *Horm. Behav.* 40, 32–42.

Trainor, B. C., and Marler, C. A. (2002). Testosterone promotes paternal behaviour in a monogamous mammal via conversion to oestrogen. *Proc. Biol. Sci.* 269, 823–829.

Trainor, B. C., Bird, I. M., Alday, N. A., Schlinger, B. A., and Marler, C. A. (2003). Variation in aromatase activity in the medial preoptic area and plasma progesterone is associated with the onset of paternal behavior. *Neuroendocrinology* 78, 36–44.

Turgeon, J. L., and Waring, D. W. (2000). Progesterone regulation of the progesterone receptor in rat gonadotropes. *Endocrinology* 141, 3422–3429.

vom Saal, F. S. (1983). Variation in infanticide and parental behavior in male mice due to prior intrauterine proximity to female fetuses: Elimination by prenatal stress. *Physiol. Behav.* 30, 675–681.

vom Saal, F. S. (1985). Time-contingent change in infanticide and parental behavior induced by ejaculation in male mice. *Physiol. Behav.* 34, 7–15.

vom Saal, F. S., and Howard, L. S. (1982). The regulation of infanticide and parental behavior: Implications for reproductive success in male mice. *Science* 215, 1270–1272.

Wagner, C. K. (2006). The many faces of progesterone: A role in adult and developing male brain. *Front. Neuroendocrinol.* 27, 340–359.

Wang, Z., and De Vries, G. J. (1993). Testosterone effects on paternal behavior and vasopressin immunoreactive projections in prairie voles (*Microtus ochrogaster*). *Brain Res.* 631, 156–610.

Weisz, J., and Ward, I. L. (1980). Plasma testosterone and progesterone titers of pregnant rats, their male and female fetuses, and neonatal offspring. *Endocrinology* 106, 306–316.

Witt, D. M., Young, L. J., and Crews, D. (1995). Progesterone modulation of androgen-dependent sexual behavior in male rats. *Physiol. Behav.* 57, 307–313.

Ziegler, T. E. (2000). Hormones associated with non-maternal infant care: A review of mammalian and avian studies. *Folia Primatol. (Basel)* 71, 6–21.

Ziegler, T. E., Wegner, F. H., and Snowdon, C. T. (1996). Hormonal responses to parental and non-parental conditions in male cotton-top tamarins, *Saguinus oedipus*, a New World primate. *Horm. Behav.* 30, 287–297.

Zumpe, D., Bonsall, R. W., Kutner, M. H., and Michael, R. P. (1991). Medroxyprogesterone acetate, aggression, and sexual behavior in male cynomolgus monkeys (*Macaca fascicularis*). *Horm. Behav.* 25, 394–409.

Zumpe, D., Clancy, A. N., and Michael, R. P. (2001). Progesterone decreases mating and estradiol uptake in preoptic areas of male monkeys. *Physiol. Behav.* 74, 603–612.

30 FAMILY LIFE IN MARMOSETS: CAUSES AND CONSEQUENCES OF VARIATION IN CAREGIVING

JEFFREY A. FRENCH[1], JEFFREY E. FITE[2] AND CORINNA N. ROSS[3]

[1] *Departments of Psychology and Biology, University of Nebraska at Omaha, Omaha, NE, USA*
[2] *Departments of Psychology, University of Nebraska at Omaha, Omaha, NE, USA*
[3] *Barshop Institute for Longevity and Aging Studies, University of Texas Health Science Center at San Antonio, San Antonio, TX, USA*

INTRODUCTION

In most human societies, child development progresses within the context of an intimate social group that has been described as a "family." There are differences both within and across cultures in the constituents of these family units (e.g., matriarchal kin + offspring vs. a cohabiting man and woman + offspring). Nonetheless, family units share the following common properties: (1) some degree of genetic relatedness among family members, (2) patterns of affiliative and cooperative child-rearing, and (3) co-residence in a common location. The importance of the family environment for development has been recently been highlighted by both professional organizations (e.g., the American Academy of Pediatrics' "Task Force on the Family;" Schor, 2003) and by funding initiatives (the National Institute of Child Health and Development's efforts in the Science and Ecology of Early Development, or SEED).

The impact of early social environments, particularly those features involving maternal care, in shaping the developmental trajectories of developing offspring has been elegantly demonstrated in a variety of animal models (see chapters by Maestripieri, Champagne, and Curley in this book). There is also a growing recognition that variation in the normative functioning of human families can have long-lasting and pervasive effects on a variety of developmental outcomes in individuals that experience these differential early environments. In some cases, poor quality early family life can place children at risk for significant childhood or adult disorders, and higher quality family environments can be protective against these risks. For instance, stress reactivity is modified in children receiving poor quality care early in development (Hane & Fox, 2006; reviews in Gunnar & Quevedo, 2007), and high quality care can overcome the developmental delays in critical brain structures (hippocampus) that derive from low birth weight (Buss *et al.*, 2007). The absence of a father during the early years of a girl's life can lead to earlier menarche (Draper & Harpending, 1982; Belsky *et al.*, 1991), and father-absence, particularly in the first 5 years of life, is associated with a 2-fold increase in early sexual activity and a 3–5 fold increase in teenage pregnancy (Ellis *et al.*, 2003). Even in families with biological fathers present, the quality of the relationship between fathers and daughters, mothers and daughters, and the quality of the marital relationship itself, is associated with variation in the onset of menarche in girls (Ellis *et al.*, 1999).

Given that families represent important environments that profoundly shape developmental trajectories in offspring, and especially given that considerable normal variation exists in the nature of the early experiences encountered by offspring, it is critical that we understand both the *causes* of variation in the quality and quantity of care provided to offspring in a family environment, and the *consequences* of this variation for later biobehavioral profiles in adolescents and adults. In the course of this chapter, we will examine both issues in marmosets, who are mostly monogamous, biparental, and family-living primates from the eastern coastal forests of Brazil. We will highlight both endocrine and genetic influences on variation in maternal and paternal care, and document our preliminary evidence of the pervasive and persistent effects of variation in early care for later somatic, physiological, and behavioral functioning in marmosets.

A PRIMER ON MARMOSET SOCIAL STRUCTURE

Humans share considerable amounts of DNA with the species with which we most recently shared a common ancestor (chimpanzees, *Pan*), and a recent conservative estimate places the overlap at 95% (Britten, 2002). These similarities do not play out, however, in the phenotypic expression of species-specific mating systems. Chimpanzees are known to exhibit mating systems that are characterized by large multimale–multifemale groups, with males mating with multiple females and females mating with multiple males, both within and outside of their social unit, even during a single breeding season (Gagneux *et al.*, 1999). South American primates of the family Callitrichinae (marmosets and tamarins) last shared a common ancestor with *Homo sapiens* somewhere around 35 million years ago (Schrago & Russo, 2003), yet these species exhibit a suite of social and mating characteristics that are remarkably similar to human beings, the details of which are highlighted below.

Marmoset and tamarin social groups are typically extended families made up of 5–15 individuals, and these groups tend to be relatively stable over time (Goldizen, 1987). Social groups are usually comprised of a single breeding adult male and female, perhaps one or more unrelated adults of either sex, and both independent (i.e., juvenile and subadult) and dependent (i.e., infant) offspring of the breeding pair. Strong and persistent social attachments ("pair bonds") develop between adult male and female heterosexual partners making marmosets and tamarins rare among primates (Schaffner *et al.*, 1995; Schaffner & French, 2004). These pair bonds are characterized by high rates of affiliative behavior between males and females, including grooming, huddling, food-sharing, and coordination of general activity. Adult males and females are extremely responsive, both behaviorally and physiologically, to threats to the relationship (e.g., separation from a partner: Shepherd & French, 1999; exposure to same-sex competitor: French, *et al.*, 1995; Schaffner & French, 1997.

Marmosets and tamarins also exhibit a cooperative breeding system typified by three defining characteristics: (1) extended residence of offspring within the family group, (2) breeding activity that is typically limited to a single breeding pair, and (3) alloparental care (i.e., care provided to infants by individuals other than a parent; French, 1997). Unlike most mammals, in which dispersal of offspring from the natal group by one or both sexes occurs around the time of puberty (Chepko-Sade & Halpin, 1987), subadult and adult marmoset offspring remain in the natal group, sometimes for years. It is rare, however, for these adult-aged offspring to engage in breeding attempts. In some cases, adult-aged daughters living in their natal family are endocrinologically suppressed and fail to exhibit normal ovarian cycles (see review in French, 1997), but in other cases, daughters and subordinates are reproductively capable but fail to engage in sexual behavior (Smith *et al.*, 1997; Saltzman *et al.*, 2004). Likewise, subordinate males and sons living in natal groups do not routinely engage in sexual activity, although they are capable of copulating with an unrelated female when given an opportunity, and have levels of testosterone and luteinizing hormone (LH) sufficient to support spermatogenesis and sexual behavior (Baker *et al.*, 1999).

Besides delayed dispersal and cooperative breeding, the most crucial aspect of marmoset reproductive biology from the perspective of this chapter is cooperative infant care. Marmosets typically produce fraternal twins that are cared for by *all* group members, including fathers and older offspring living in the family group (Tardif, 1997). This trait is shared by all species of marmosets and tamarins, although there are subtle differences among species and genera in the distribution and timing of parental effort by mothers, fathers, and alloparents (Santos *et al.*, 1997; Tardif, 1997). Patterns of caregiving provided to infants include licking and grooming, food-sharing and provisioning, play, protection and contact-comfort, and, in the case of breeding females, nursing. But by far the most dramatic and energetically expensive component of offspring care is infant carrying and transport. At birth, the cumulative weight of a twin litter can constitute 15–25% of adult body weight (French, 1997), and by the time weaning is complete at 12 weeks of age, the combined weight of twins approaches 100% of adult body weight. Caregivers that are transporting offspring show reductions in time spent feeding and foraging (Price, 1992) and can lose 10% of their body weight during times of high carrying activity (Sanchez *et al.*, 1999), suggesting that there are real energetic and metabolic costs associated with offspring carrying in marmosets. For mothers, the costs of carrying are compounded by two additional energetic and metabolic demands. First, via lactation mothers provide their rapidly growing offspring with milk that is rich in energy from crude protein (Power *et al.*, 2002). Second, mothers are often pregnant with a new litter (in which they are beginning the investment of 15–25% of their body weight) on the first postpartum ovulation. In our colony, approximately 85% of first postpartum ovulations in marmosets are conceptive ovulations, and these occur 7–14 days postpartum (French *et al.*, 1996).

The Callitrichid Research Center at the University of Nebraska at Omaha has focused on family life in two species of marmosets, the black tufted-ear marmoset (*Callithrix kuhlii*) and more recently, the white-faced marmoset (*C. geoffroyi*). Our animals are housed in large enclosures that contain naturalistic features such as branches and vines, nest boxes, and simulated foraging sites. Most of the work described in this chapter involves behavioral observations of normal, undisturbed parents and offspring, and all of our endocrine work is conducted on urine samples that are collected non-invasively (marmosets are trained to provide first-void urine samples upon arising in the morning; French *et al.*, 1996).

ENDOCRINE CORRELATES OF VARIATION IN PARENTAL CARE

Our approach to evaluating the links between variation in endocrine states and the expression of parental care is strongly shaped by models that integrate behavioral endocrinology and behavioral ecology, particularly trade-off models. These models (e.g., Wingfield *et al.*, 1990; Ketterson & Nolan, 1999; Zera & Harshman, 2001; Ketterson *et al.*, 2005) posit that hormones represent important proximate determinants of life history "decisions," mediating trade-offs between alternative behavioral or reproductive choices. We have focused specifically on the hypothesis that androgenic hormones such as testosterone mediate shifts in investments between parental effort (caring for current offspring) vs. mating effort (time and energy directed toward producing future offspring). There is considerable support for this hypothesis in a wide variety of vertebrate species (review in Ziegler, 2000), but there are also notable exceptions (see contributions by Marler in this book).

In the marmoset, we first addressed the question of how parental labor is divided among group members during different phases of the infants' early life. Marmoset infants are dependent on caregivers for transport for the first 2 months of life, so we recorded rates of carrying effort during this time. The overall effort by mothers, fathers, and juvenile helpers during the 8-week period after birth is shown in Figure 30.1. During the first 2 weeks of postpartum infant life, mothers are the primary offspring caregivers, but show a significant decrease in carrying effort during weeks 3–4 postpartum. This reduction in female care is accompanied by an increase in offspring care by fathers, who exhibit a 40% increase in carrying effort from weeks 1–2 to weeks 3–4. In a typical marmoset family, fathers are therefore the primary caregivers during this period of infant life, and continue to be the primary caregivers throughout the period of infant dependence (although their effort, like other caregivers, decreases as the infants grow older and acquire more independence). Helpers in the group (older siblings of the infants) carry little early in development, but are involved at levels similar to adult female in weeks 3–8. When looking at the pattern of change in parental roles in the social group, then, there are two important transitional elements in parental effort. Males become much *more* engaged in parental care in weeks 3–4, while at the same time, females become much *less* involved in parental care at the same point in time.

To address whether changes in androgens were temporally associated with variation in paternal effort (as predicted by the trade-off

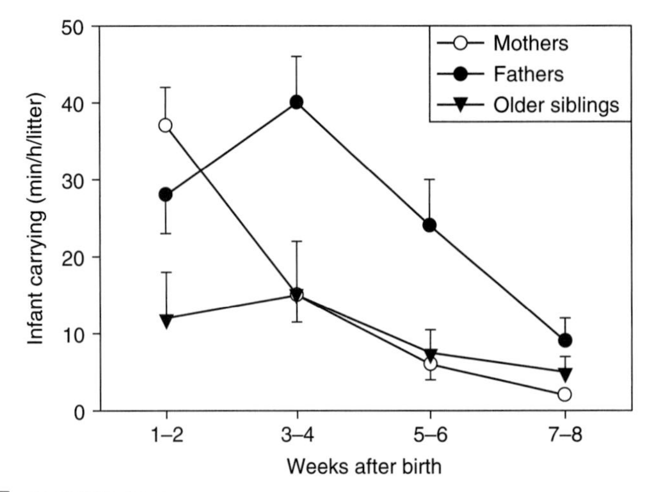

FIGURE 30.1 Rates of infant carrying throughout the first 8 weeks of infant life by the three major classes of marmoset caregivers. Error bars indicate ± SEM.

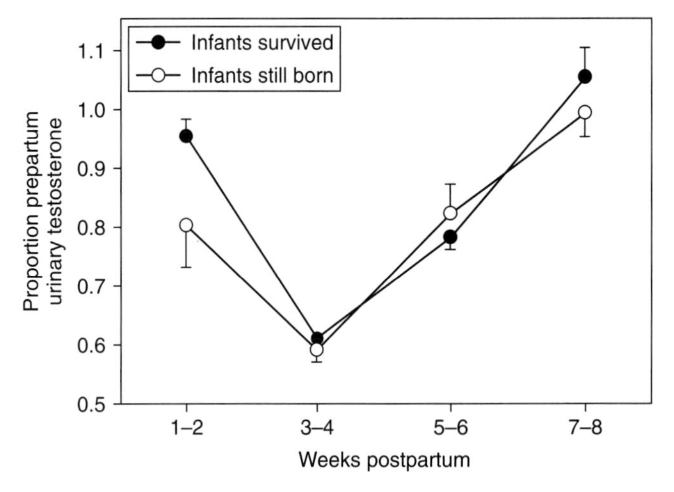

FIGURE 30.2 Changes in urinary testosterone excretion in male marmosets in the first 8 weeks of infant life. Values are expressed as a proportion of prepartum hormone concentrations. (*Source*: Modified from Nunes *et al.*, 2000.)

hypothesis), we monitored changes in testosterone excretion in males across this 8-week period (Nunes *et al.*, 2000). We collected data on a set of males who were actively engaged in parental care with a litter of twins, and on a second group of males whose infants were stillborne or who died of natural causes in the first 2 days of life. These two groups allowed us to determine whether any endocrine changes we observed in males were specifically associated with and potentially triggered by actively engaging in paternal care, or were cued by exposure to cues from pregnant females and/or the events associated with parturition. We found that relative to prepartum baseline testosterone levels, testosterone levels in males dropped slightly during the first 2 weeks postpartum, but then dropped to 60% of baseline levels during the period of maximal male care (weeks 3–4; Figure 30.2). As male carrying effort decreases in weeks 4–8, testosterone levels returned to baseline prepartum concentrations. Similar profiles were noted in males whose infants died within 2 days of birth and in males who reared infants

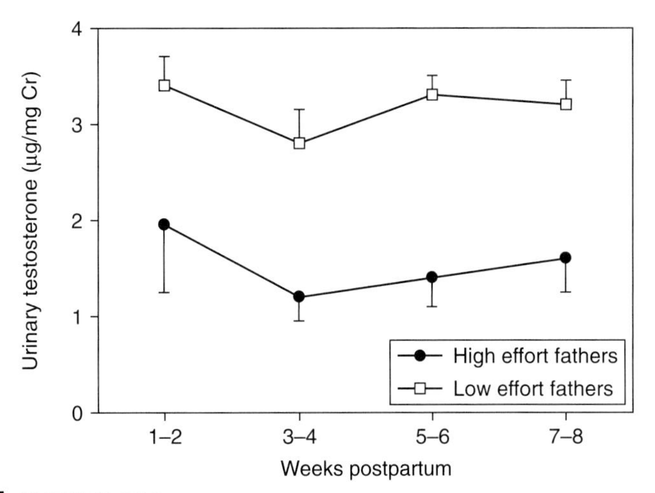

FIGURE 30.3 Urinary testosterone excretion in males that engaged in high levels of paternal care (high effort) vs. males that engaged in lower levels of paternal care (low effort). (*Source*: Modified from Nunes *et al.*, 2001.)

throughout the 8-week period after birth. This suggests that stimuli associated with parturition or the short-term presence of neonates, initiates an endogenous timing mechanism that facilitates later parental care, regardless of the continued presence of live offspring (see similar results in parental penguins: Lormée *et al.*, 1998). In some cases fathers actively consume the placentas associated with the twin litters (personal observations). The consumption of a steroid-rich tissue like the placenta may initiate long-term endocrine changes in males. Alternatively, cues for the change in endocrine status in males may arise from sources other than the infants, including olfactory cues emanating from ovulatory or newly pregnant females (Ziegler *et al.*, 1993), or behaviorally induced changes in endocrine activity associated with the termination of female sexual attractiveness and receptivity.

Our analyses also revealed that previous experience with infants shapes the extent of testosterone suppression in the presence of infants. With age of male, group size, and other potentially confounding variables controlled, males with greater levels of previous infant experience (either as helpers or as fathers) showed significantly lower concentrations of testosterone postpartum than males with less experience. This finding suggests critical interactions between experience as a parent and the regulation of gonadal steroidogenesis, an interaction that may also occur in the relationship between prolactin and parental care in other species of marmosets and tamarins (Ziegler *et al.*, 1996).

Our second study on paternal male marmosets (Nunes *et al.*, 2001) addressed the possibility that individual differences in male paternal effort were associated with variation in gonadal steroid concentrations. We identified a population of males that carried infants at high rates (high effort males; males that carried on average 33 min/h) and a population that carried infants at low rates (low effort; less than 15 min carrying/h), and monitored patterns of paternal care and hormone excretion across two successive litters of twin offspring. While these two populations of males did not differ on a host of demographic and experiential measures (including age, previous experience, size of groups, length of pairing with the female), there were profound differences between males in endocrine profiles during the postpartum period. High effort males had significantly lower concentrations of testosterone than low effort males throughout the period of infant dependence on caregiving (Figure 30.3). The magnitude of the difference between these subsets of males is remarkable – fathers that engage in less paternal care have levels of excreted testosterone that were 2-fold higher than levels in fathers that engage in substantial parental care. It is of interest to note, however, that both populations of males showed reductions in urinary testosterone excretion during weeks 3–4 of infant life, which replicated the effect we had observed earlier (Nunes *et al.*, 2000). Cortisol (CORT) also varied as a function of male parental effort, with low effort males showing dramatically higher levels than high effort males, especially in the first 2 weeks postpartum. We also noted an experience effect on hormones levels in this study, with fathers (both high and low effort) caring for their second set of infants having significantly lower testosterone than they did while caring for their first set. Thus, the timing of changes in testosterone in males across the period of infant dependence (low levels during high paternal effort) and the endocrine differences between males that vary in paternal effort (highly involved fathers have lower testosterone) are observations consistent with the trade-off hypothesis.

Androgens and Variation in Maternal Care

As seen in Figure 30.1, there is a dramatic transition in amount of infant care provided by the mother, beginning in the third week of infant life. Mothers exhibit an almost 4-fold reduction in carrying infants at this time, although they are still actively nursing their offspring during the short period of time they are also carrying infants. We know that the timing and magnitude of this reduction in maternal care is associated with at least one demographic variable and one reproductive variable (Fite, *et al.*, 2005b). First, females reduce their carrying effort more dramatically during this time when there are helpers in the group than when no helpers are available. Second, female marmosets who conceive in the first postpartum ovulation (7–14 days postpartum) immediately reduce their carrying effort in the current set of infants while females who conceive on the second or third postpartum ovulation (30–50 days postpartum) exhibit a much more gradual decline in carrying efforts (Figure 30.4(A)). The reduction in care in the first instance is rapid and dramatic – females reduce infant carrying from ~40 min/h to <10 min/h if they conceive on the first postpartum ovulation. We have suggested that female marmosets are remarkably "opportunistic," reducing maternal effort when they can (additional helpers) or when they have to (in the face of combined lactational and gestational energetic demand. Thus, females appear to facultatively adjust caregiving effort as a function of energetic and social contexts.

There is growing interest in the role of heterotypical gonadal steroid (i.e., androgens in females and estrogens in males) in the regulating of sociosexual behavior, including paternal care (Schum & Wynne-Edwards, 2004; Ketterson *et al.*, 2005). We evaluated the possibility that androgens may play a role in regulating maternal responsiveness, in a manner similar to the pattern suggested by our data on male parental effort. Unpublished data collected on our marmosets by Erin Kinnally revealed that marmoset mothers (experienced and inexperienced) exhibited significant increases in urinary testosterone in weeks 3–4 postpartum, which corresponds exactly to the time at which females exhibit a 4-fold reduction in their efforts in transporting

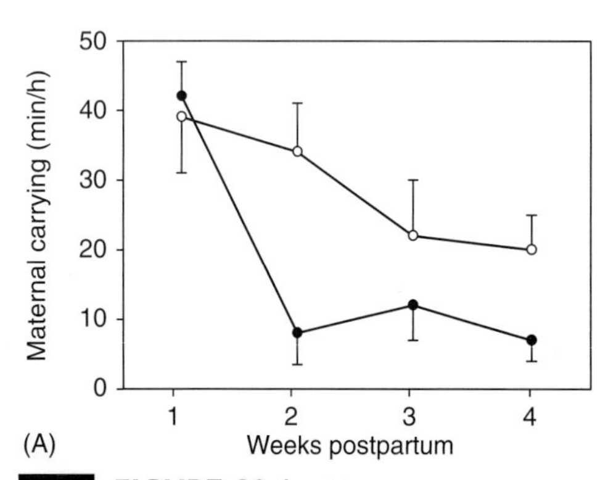

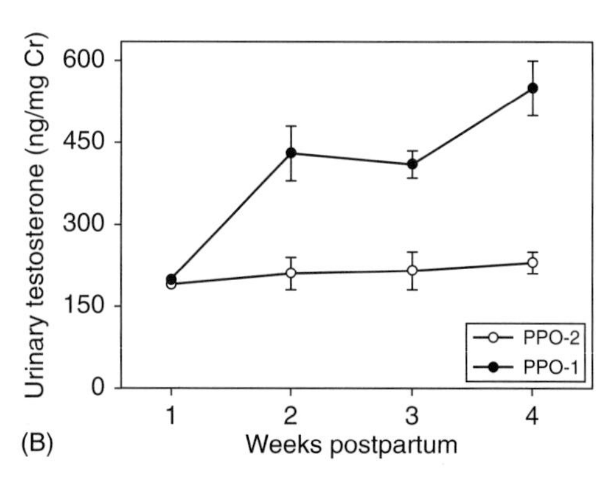

FIGURE 30.4 (A) Maternal care of offspring and (B) patterns of urinary testosterone excretion in female marmosets during the first 4 weeks postpartum. PPO-1, females conceived on their first postpartum ovulation; PPO-2, females conceived on the second or later postpartum ovulation. Cycle length in *Callithrix* is approximately 24 days in length. (*Source*: Modified from French *et al.*, 1996.)

offspring (see Figure 30.1). We hypothesized, based on these data, that elevated androgens may represent a proximate signal that reduces maternal investment in current offspring (at least those features, like offspring transport, that can be shifted to individuals other than the mother), thereby allowing greater investment in the future offspring the female is beginning to gestate. Where do these high levels of androgens come from in females? Previous work in a variety of mammals (dogs, baboons, humans) has shown that one of the most distinctive differences between conceptive vs. non-conceptive ovarian cycles is the rapid and dramatic rise in testosterone associated with conceptive cycles, presumably of ovarian origin (Castracane & Goldzieher, 1983; Concannon & Castracane, 1985; Castracane *et al.*, 1998).

Once again, variation in marmoset biology allowed us to test the androgen trade-off hypothesis in females. Most females conceive on the first postpartum ovulation, but on some occasions they do not. We (Fite, *et al.*, 2005a) identified a set of females who on one occasion had a surviving litter of twins and subsequently conceived on the first postpartum ovulation, while on another occasion produced a set of twins, but did not have a conceptive ovarian cycle until the second or third postpartum ovulation. We monitored ovarian steroids (estrogen and progesterone metabolites), and testosterone levels in the postpartum period, and noted levels of maternal carrying effort. Our data demonstrate that traditional ovarian steroids do not change dramatically at conception, but that urinary androgen concentrations in females rise dramatically in the conceptive, but not in the non-conceptive luteal phase. The results are consistent with our hypothesis – females that conceived on the first postpartum ovulation exhibited elevations in testosterone levels, coincident with a reduction in maternal carrying effort (Figure 30.4(B)). When females did not conceive on the first postpartum ovulation, androgens did not rise and maternal effort remained high. The temporal relationship between elevated androgen levels and reduced maternal care in conceptive mothers is supported by analyses that assess the relationship between maternal androgen and carrying efforts. The correlation between androgen levels and carrying effort is negative and highly significant for all data combined ($p < 0.002$) and also for individual females (i.e., higher androgen levels were associated with reduced carrying rates).

Together, our data on androgens and parental care in marmoset families point to an integrated system of hormone–behavior interactions that helps to regulate the timing of parental effort among adult caregivers in marmoset social groups. Male endocrinology varies in ways that reflect their dual roles as breeding partners and as caregivers. In the immediate postpartum phase, testosterone titers in males remain high, during which time males defer to females with regard to infant care, but actively monitor the female's reproductive state and engage in sexual behavior in the postpartum ovulation. Following the postpartum ovulation and mating, testosterone titers fall and males become more nurturant, and indeed become the primary caregivers for infants within the family group. In females, elevated androgens are associated with a reduction in carrying effort for the current litter, presumably conserving energetic resources for the expensive task of producing a future litter of offspring. These results lend support to the growing appreciation that hormones mediate important life history decisions, mating effort vs. parental effort in males, and current vs. future offspring in females.

GENETIC CHIMERISM AND EVOCATIVE GENETIC EFFECTS ON PARENTAL CARE

It is not uncommon for parents and offspring to have correlated traits, such as temperament. To the extent that these traits display some heritability, traditional quantitative geneticists would suggest that parents and children are similar because of a Mendelian transmission of relevant alleles across generations. A more modern approach, however, suggests that some of the "heritable" covariance may be attributed to evocative effects – the differential solicitation of parental behavior that is evoked by genetically based social and behavior phenotypes in children. For example, when children inherit certain behavioral phenotypes (e.g., distractable or resistive), they tend to evoke a certain parental style (e.g., authoritarian). On the other hand, a cooperative child will evoke a different pattern from the same parents (e.g., laissez-faire: see reviews in Collins *et al.*, 2000; Maccoby, 2000; Reiss & Neiderhiser, 2000; Neiderhiser, this book). Knowledge of the nature of these genetic effects contributes to our understanding of normative gene–environment interactions, but in addition, these evocative effects also speak directly to important health-related outcomes. For instance, adopted children reported to be at genetic risk for antisocial behavior (based on biological mothers' self-reported history of antisocial behavior) elicit higher levels of negative parenting from adoptive parents than children not at genetic risk (Ge *et al.*, 1996; O'Connor *et al.*, 1998).

We have discovered what may, in fact, be an important evocative genetic effect in marmosets, in the form of genetic chimerism. A genetic chimera is a unitary organism that is composed of cells that contain alleles from differing genomic lineages. Naturally occurring mammalian chimeras have been known for almost a century, and the most common example is free-martinism in cattle (Lillie, 1917). Fetal female cows that have a male co-twin are rendered sterile, because the presence of chimeric XY-containing cells in the female fetus masculinizes the female's genitalia (see recent review in Capel & Coveney, 2005). It was recently reported that women who have given birth to one or more children maintain cell lines that express the genotype of their offspring (these cells contain alleles that derive from the child's father as well as the mother, and hence are non-self alleles for the mother; Maloney *et al.*, 1999). These chimeric cells can persist in the maternal circulation for decades, and include CD34(+) and CD38(+) cells and other components of the immune system. Chimerism has been noted at an incidence of approximately 33% of healthy mothers, but the incidence is higher in women with autoimmune disorders (Evans *et al.*, 1999), suggesting that, as in free-martinism, there may be moderate to strong costs associated with chimerism.

In marmosets, in contrast, the entire sequence of placentation and embryonic development appears to be designed specifically to produce genetic chimeras of the fraternal twins that develop *in utero*. Early in embryonic development, the placental anastomoses fuse and form a single chorion (Wislocki, 1932). Fusion of the twins' placentas begins on day 19 and is complete by day 29, forming a single chorion with anastomoses connecting the embryos which are still at a presomite stage in development (Merker *et al.*, 1988; Missler *et al.*, 1992). The fusion of the chorions and a delay in embryonic development at this stage allows the exchange of embryonic stem cells via blood flow between the twins (Benirshke *et al.*, 1962; Benirshke & Brownhill, 1963). As a result, the infants are genetic chimeras with tissues derived from self and sibling embryonic cell lineages. A similar phenomenon occurs with rare frequency in humans. A recent report on dizygous (fraternal) male and female human twins sharing a single placental chorion demonstrated chimerism in blood-derived products, including the presence of XY-bearing cells in the female, and XX-bearing cells in the male (Souter *et al.*, 2003). While there is little doubt that tissues derived from hematopoietic origin are universally chimeric in marmosets (Benirshke *et al.*, 1962; Gengozian *et al.*, 1964), the existence of chimeric cells in non-hematopoietic tissues, including germ-line cells such as sperm and egg is controversial (Gengozian *et al.*, 1980), and has not been systematically evaluated.

Our recent analyses of chimerism in marmosets (Ross, *et al.*, 2007) were the first to use modern molecular genetic tools to examine chimerism in a variety of tissues and with a number of microsatellite DNA primers. Multiple tissue samples were collected from living marmosets and from archived tissues from deceased marmosets in our colony. This colony has been painstakingly pedigreed since its origination in 1991, and there were no ambiguities in paternity in any of the offspring produced in our lab. Samples were blindly coded, DNA was extracted and amplified, and genotypes were determined for five marmoset markers. Majority rule was used to determine "self" genotype (i.e., those alleles that were likely to be inherited vertically from each parent) and "sibling" alleles (those alleles acquired *in utero* from the sibling via horizontal transmission). The ABI scans in Figure 30.5 show typical data for a single marker (CK2). Skin tissue from Twin A is clearly heterozygous at this locus (198/240) and kidney tissue from Twin B is also heterozygous, but with different alleles (216/218). DNA from Twin A's spleen still shows evidence of self alleles at 198/240, but some proportion of the DNA extracted from Twin A's spleen contains alleles expressed by Twin B. Likewise, DNA from Twin B's heart tissue contains a high signal of self alleles (216/218), but the extracted DNA also contains alleles from Twin A.

Although the incidence of chimeric tissue varied among tissue types, every tissue sampled showed evidence of chimerism, perhaps most surprisingly (and importantly) in the germ line. In harvested and purified sperm samples collected from males, 57.1% of the samples

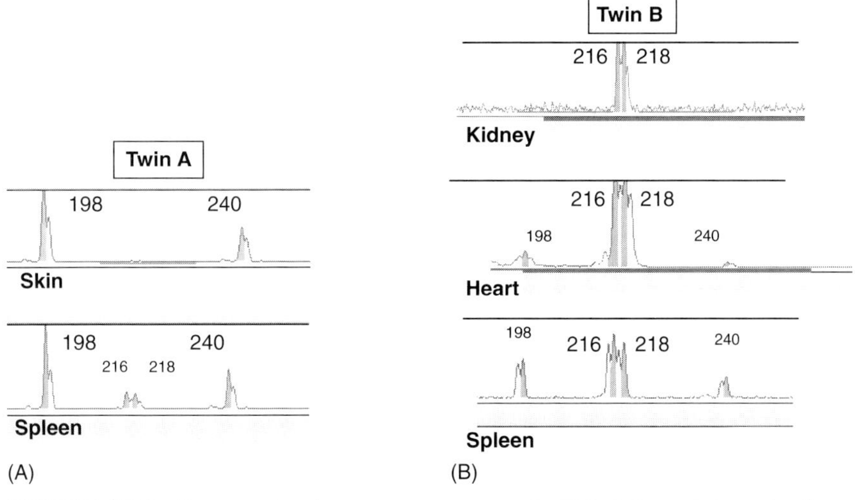

FIGURE 30.5 Genotype data for two marmoset co-twins. Twin A has a designated self genotype of 198/240 for this locus (CK2) as determined by majority rule, and twin B has a designated self genotype of 216/218. This figure also demonstrates that Twin A is chimeric for spleen tissue and nonchimeric for skin, whereas Twin B is chimeric for heart and spleen tissue. (*Source*: Modified and reprinted with permission from Ross *et al.*, 2007.)

contained chimeric sperm (i.e., expressed an allele associated with the intrauterine sibling), and we estimated that 1 out of every 10 sperm cells (~10%) contain chimeric alleles. From a genetic perspective, what is most intriguing is that we have demonstrated that offspring can be sired by males with chimeric sperm that males inherited horizontally from their co-twins. The genealogy in Figure 30.6 demonstrates this point. The parental generation (P) is heterozygous for marker CK2. One of the sons in F1 inherits allele 198 from the father and 240 from the mother. The second son inherits 216 from father and 218 from mother (as determined by majority rule), but also expresses 198 and 240 as alleles contained in sibling cells. This son is paired with a female from another family line who has unique alleles at CK2. Sibling 1 in the F2 generation inherits 216 from father and 232 from mother, but Sibling 2 in F2 inherits the chimeric allele 198 from the father and 220 from the mother. This outcome can only arise if the "father" fertilized the egg with a sperm that was derived from embryonic stem cells that were passed to the male via placental anastomoses with his intrauterine co-twin. Of 34 twin sets evaluated in our analyses, five instances of transmission of chimeric alleles across generations were noted. This phenomenon has clear implications for coefficients of relatedness within marmoset groups.

From a behavioral perspective, it seemed likely that marmoset infants that were chimeric might be treated differently by caregivers, since they may differentially express phenotypic cues regarding relatedness. For example, in generation F1 above, the male marmoset on the far left expresses only one copy of each allele from the parents, while the son in the middle of the genealogy potentially expresses phenotypic products associated with both sets of alleles from each parent. To evaluate the possibility that offspring may be treated differentially, we divided marmoset infants into two categories: those that were chimeric for epithelial tissue (n = 10) and those determined to be nonchimeric for epithelial tissue (n = 20). We compared rates at which each of these classes of infants was carried by mothers vs. fathers during the first 2 weeks of infant life, a period of time during which the mother is typically the primary caregiver (Fite & French, 2000; Nunes *et al.*, 2000, 2001; Fite *et al.* 2005b). Figure 30.7 shows the percentage of carrying offspring by the mothers and fathers, and it is clear that at both 1 and 2 weeks of infant age, fathers carry chimeric infants at 2-fold or higher rates in the first 2 weeks of life than nonchimeric infants. As shown earlier, males are normally not the primary caregivers during this period of time, but if infants possess chimeric alleles in epithelial tissue, fathers appear to be

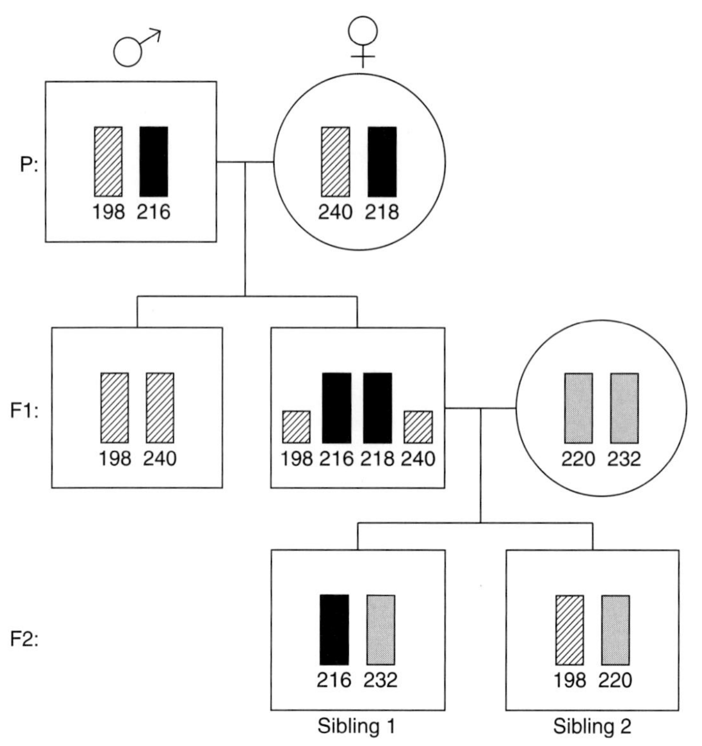

FIGURE 30.6 Example of vertical transmission of chimeric alleles acquired from a sibling co-twin. The parental pair (P) produced two male offspring that had different self genotypes, but one son expressed alleles born by its co-twin (198/240). This son was paired with a female with a different genotype, and Sibling 2 in the F2 generation bears an allele that was acquired by its father (F1) from its male co-twin. (*Source*: Modified and reprinted with permission from Ross *et al.*, 2007.)

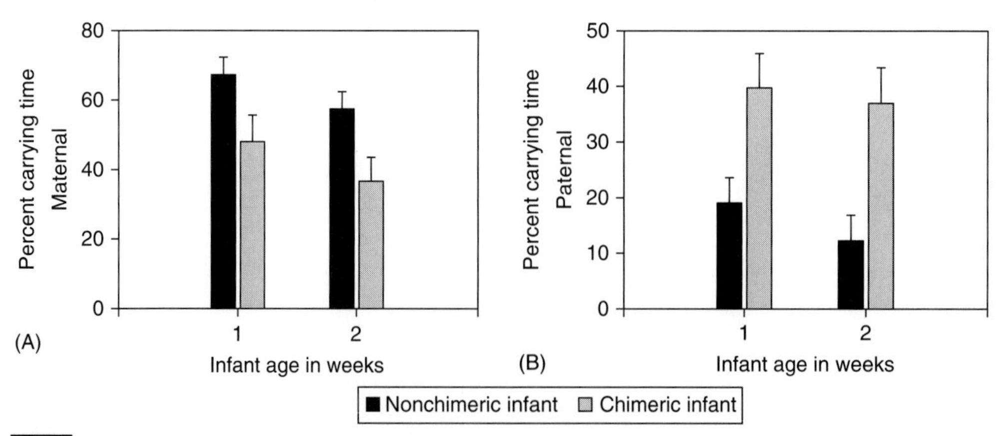

FIGURE 30.7 Patterns of infant care by mothers (A) and fathers (B) for infants whose epithelial tissue was chimeric (i.e., displayed self alleles plus at least one sibling allele) vs. those infants whose epithelial tissue was nonchimeric (contained self alleles only). Mothers with chimeric infants carry significantly less than those with nonchimeric infants ($p = 0.01$), and fathers with chimeric infants carry significantly more than those with nonchimeric infants ($p = 0.002$). (*Source*: Modified and reprinted with permission from Ross *et al.*, 2007.)

particularly attracted to these infants and carry them significantly more than they do nonchimeric infants. Mothers show a slight reduction in the percent of time they carry chimeric infants, but the differences are not as dramatic as the differences displayed by male marmosets toward chimeric vs. nonchimeric offspring. In this dataset, we only recorded carrying effort, and we have no other quantitative information regarding attraction to or aversion from infants on the part of either mothers or fathers. Thus, our data do not allow us to assess whether higher rates of paternal carrying resulted from greater male interest in chimeric infants (e.g., higher rates of transfer and attempted transfer to males, increased sniffing and anogenital licking, shorter latency to retrieve) or, alternatively, the possibility that mothers find chimeric infants less attractive or even aversive and engage in higher rates of rejection, attempted rejections, and infant removals, which would lead to higher male care as a consequence. We are currently addressing this question with more detailed behavioral protocols that will allow us to answer these critical questions regarding the link between chimeric status and the quality, quantity, and distribution of parental caregiving activities within the context of a family system. Differential paternal responsiveness based on the genotype of offspring would appear to be an example of a profound evocative effect in parental care.

CONSEQUENCES OF VARIATION IN EARLY PARENTAL CARE

Impact of Variation in Early Care on Baseline HPA Function and Pubertal Maturation

Adverse or inappropriate maternal care is known to have profound influences on the development of primate offspring. Compelling evidence on this point is available from experimental studies that involve maternal deprivation or separation (Dettling *et al.*, 2002a, b; Sanchez, 2006) or studies of fairly intense maternal abuse and neglect (Maestripieri, 2005; this book). Our approach to addressing the question of early experiences and epigenesist, in contrast, is to examine the impact of *natural* variation in normative early offspring care on subsequent development. Marmoset infants are, on the whole, remarkably well-cared for during the first 2 weeks of life. It is rare to have a set of twin infants that are not carried by one or another caregiver 100% of the time. However, there are significant differences among family groups and even between co-twins in the identity of the caregiver that provides the majority of care during the first 2 weeks of life.

We have recently completed an analysis of the impact of this variation on baseline setpoints in the hypothalamic–pituitary–adrenal (HPA) axis and on the onset of reproductive function (French *et al.*, under review). We studied the development of 12 marmoset infants in family groups where mothers, fathers, and alloparents were present and could provide offspring care. Eleven of the 12 infants were on caregivers 100% of observations, while the 12th was on caregiver 97.3% of observations. Nonetheless, differences in the kinds of experience offspring had during the first 2 weeks of life had profound influences on later physiological markers, in sex-specific ways. First, the variation in the extent to which mothers carried daughters was a significant predictor of age at onset of the first signs of ovarian activity in daughters. The first ovulatory cycle was documented by monitoring patterns of progesterone metabolites in urine samples. Daughters that received more care from mothers *in the first 2 weeks of postnatal life* had earlier pubertal timing than daughters who received less care from mothers ($r = -0.99$, $p < 0.001$). Variation in early caregiving also influenced baseline function in the HPA axis, as 12-month-old daughters who were carried less frequently by mothers in the first 2 weeks of life exhibited higher baseline levels of urinary CORT ($r = -0.84$, $p < 0.03$). When we performed regression analysis on maternal caregiving and CORT levels on pubertal timing, the overall model was significant, explaining 99.2% of variation in pubertal timing. However, removing baseline HPA activity using backward selection did not produce a significant change in R^2, suggesting that the relationship between maternal care and pubertal timing was not mediated by baseline HPA activity at 12 months of age.

Like daughters, sons also received differential care during early life. Also like females, pubertal processes appeared to be influenced by early care – in this case, variation in the amount of care sons received from their fathers. Unlike a discrete event like first ovulation in females, the development of puberty in males is a more continuous process. We estimated degree of pubertal maturation by measuring individual differences in urinary testosterone (T) in males at 12 months of age (the peripubertal stage in marmosets). Sons that were carried by fathers more in the first 2 weeks of life had lower levels of urinary testosterone at 12 months of age than sons that were carried less frequently by the father ($r = 0.83$, $p < 0.03$). This effect appears to be specific to father–son caregiving, since neither variation in caregiving by mothers nor older siblings was correlated with peripubertal testosterone levels in sons. Paternal caregiving was not associated with variation in HPA function at 12 months of age in sons, suggesting that paternal influences on pubertal process in sons is independent of simultaneous activation of the stress system. While we do not have data on early HPA function in infants (i.e., during the first 2 weeks of life while they receive differential caregiving), it may be that differential care may influence pubertal processes via differential early activation of the HPA axis. Infant marmosets are known to be hypercortisolemic (Pryce *et al.*, 2002), and differential early family experiences can alter HPA function (Dettling *et al.*, 2007).

Differential Early Care and Somatic Development in Marmosets

In addition to endocrine function and sexual maturation, we have also been exploring the impact of differential early care on infant and juvenile somatic development. We collect complete somatic measurements on all infants born into our colony on a regular basis, beginning at the age of 2 days and continuing throughout development until adulthood (540 days of age). Our preliminary analyses of somatic growth show that variation in early care influences important components of physical growth. High levels of paternal care early in development have negative consequences for somatic growth. As seen in Figure 30.8, those offspring that receive more paternal care in the first 12 weeks of life weigh less at 300 days of age than those who receive less paternal care, and these differences persist until 480 days of age. Further, several of our measures of physical "robustness" (e.g., chest, abdominal, and thigh circumference) are negatively correlated with early paternal care, even at age

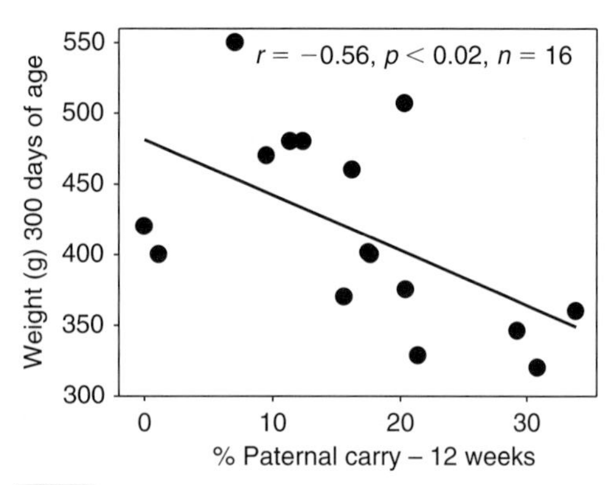

FIGURE 30.8 Weights of juvenile marmosets at 300 days of age varies as a function of the proportion of time they were carried by fathers during the first 12 weeks of life.

points as distant as 540 days. These effects do not appear to be caused by a lack of opportunity to nurse during this early period, since differences in maternal carrying effort (and hence infant ability to gain access to nursing) are not correlated with somatic outcomes later in life.

Differential Early Care and Stress Reactivity in Juvenile Marmosets

We hypothesized that variation in early care in marmosets would differentially program later function in the HPA axis. Our preliminary data show that important parameters of stress reactivity in juvenile marmosets are differentially expressed based on early care (Burrell & French, 2007). We have analyzed stress reactivity at 6 months of age (the juvenile stage) for 14 marmosets (7 male, 7 female) who received differential care early in life. Baseline urine samples were collected, and then juveniles were separated from their natal group and housed in a novel cage in an unfamiliar environment – thus, the stressor constituted both social separation and exposure to environmental novelty. Several aspects of the dynamics of the stress response, assessed by measuring excreted CORT in non-invasively collected urine samples, were affected by differential maternal and paternal care in the first month of life. Variation in paternal care affected stress "reactivity," as shown in Figure 30.9(A), as measured by maximum CORT levels at the

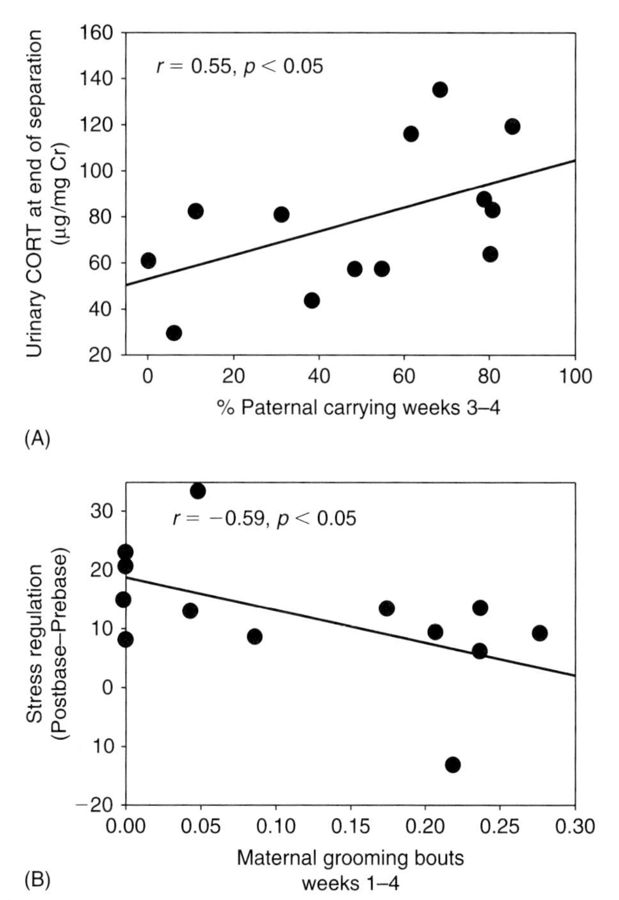

FIGURE 30.9 Differential early family care of marmoset infants alters stress responsivity: (A) 6-month-old marmosets that had received greater paternal carrying during weeks 3–4 of life show higher maximum CORT titers in response to separation and novelty exposure; (B) 6-month-old marmosets that had received more maternal grooming bouts early in life were more likely to return to baseline levels of CORT after a psychosocial stressor than marmosets who received fewer grooming bouts.

end of the 8-h separation. Juveniles that had been carried at high rates by fathers in the second 2-week period in postnatal life had higher CORT at the end of separation, while those juveniles that had been carried less by fathers exhibited lower levels of maximum CORT in response to separation. In contrast, the regulation of the HPA axis, as assessed by the ability of a juvenile to return to baseline HPA function in the absence of a stressor, was influenced by differences in the quality of care provided by the mother. As shown in Figure 30.9(B), 6-month-old marmosets who received high levels of maternal grooming in the first month of life were more likely to have CORT levels that returned to baseline concentrations on the day following the stressor, while those juveniles who received lower levels of maternal grooming had CORT levels that remained elevated on the day following the stressor. Thus, it is clear that although marmoset infants normally receive high quality and consistent care during early life, subtle differences in the amount of care received from potential caregivers in the family differential early care shapes both baseline HPA function (as seen earlier) and the dynamics of HPA responsiveness to stressors. We are currently monitoring stress reactivity and regulation at 12 months (subadult) and 18 months (adult) which will allow us to determine the persistence of these differential early care environments on later stress function.

SUMMARY

Our understanding of social contributions to epigenesis is growing rapidly, particularly those aspects of development that are systematically influenced by variation in early maternal care. In this sense, variation in the nature of dyadic interactions between mothers and offspring can alter developmental trajectories in important ways. The data we've presented in this chapter provide, we believe, a compelling argument that family contexts, with the possibility of triadic and higher order social interactions, also constitute an important component of early epigenesis. The marmoset is an ideal animal model for evaluating these influences, since there are multiple classes of individuals providing care for offspring. Further, the ability to identify important sources of variation in care, and to track the consequences of these differences longitudinally, makes the marmoset particularly useful for providing insight into the ways in which family social environments shape offspring behavioral and physiological phenotypes.

ACKNOWLEDGMENTS

Thanks go to Heather Jensen, Kim Patera, MaLinda Henry, and Denise Hightower for their dedicated service and high standards in maintaining the University of Nebraska at Omaha's Callitrichid Research Center from its inception in 1983. Michael Rukstalis contributed to the data and ideas presented in this chapter, and we thank Angie Burrell for access to her developing dataset on the consequences of variation in parental care in HPA function in marmosets. We gratefully acknowledge the long-standing support of the UNOmaha administration for our efforts in the science and conservation of callitrichid primates. This work presented in this chapter was supported by funds from the National Science Foundation (CRB 90-00094, IBN 92-09528, IBN 97-23842, IBN 00-91030) and the National Institutes of Health (HD 42882).

REFERENCES

Baker, J. V., Abbott, D. H., and Saltzman, W. (1999). Social determinants of reproductive failure in male common marmosets housed with their natal family. *Anim. Behav.* 58, 501–513.

Belsky, J., Steinberg, L., and Draper, P. (1991). Childhood experience, interpersonal development, and reproductive strategy: An evolutionary theory of socialization. *Child Devel.* 62, 647–670.

Benirshke, K., and Brownhill, L. E. (1963). Heterosexual cells in testes of chimeric marmoset monkeys. *Cytogenetics* 2, 331–341.

Benirshke, K., Anderson, J. M., and Brownhill, L. E. (1962). Marrow chimerism in marmosets. *Science* 138, 513–515.

Britten, R. J. (2002). Divergence between samples of chimpanzee and human DNA sequences is 5%, counting indels. *Proc. Natl. Acad. Sci. USA* 99, 13633–13635.

Buss, C., Lord, C., Wasiwalla, M., Hellhammer, D. H., Lupien, S. J., Meaney, M. H., and Pruessner, J. C. (2007). Maternal care modulates the relationship between prenatal risk and hippocampal volume in women but not men. *J. Neurosci.* 27, 2592–2595.

Capel, B., and Coveney, D. (2005). Frank Lillies' freemartin: Illuminating the pathway to 21st century reproductive endocrinology. *J. Exp. Zool. A: Comp. Exp. Biol.* 301, 853–856.

Castracane, V. D., and Goldzieher, J. W. (1983). Plasma androgens during early pregnancy in the baboon (*Papio cynocephalus*). *Fertil. Steril.* 39, 553–559.

Castracane, V. D., Stewart, D. R., Gimpel, T., Overstreet, J. W., and Lasley, B. L. (1998). Maternal serum androgens in human pregnancy: Early increases within the cycle of conception. *Hum. Reprod.* 13, 460–464.

Chepko-Sade, B. D., and Halpin, Z. T. (1987). *Mammalian Dispersal Patterns: The Effects of Social Structure on Population Genetics*. University of Chicago Press, Chicago. 342 pp.

Collins, W. A., Maccoby, E. E., Steinberg, L., Hetherington, E. M., and Bornstein, M. H. (2000). Contemporary research on parenting: The case for nature and nurture. *Am. Psychol.* 55, 218–232.

Concannon, P. W., and Castracane, V. D. (1985). Serum androstenedione and testosterone concentrations during pregnancy and nonpregnant cycles in dogs. *Biol. Reprod.* 33, 1078–1083.

Dettling, A. C., Feldon, J., and Pryce, C. R. (2002a). Early deprivation and behavioral and physiological responses to social separation/novelty in the marmoset. *Pharmacol. Biochem. Behav.* 73, 259–269.

Dettling, A. C., Feldon, J., and Pryce, C. R. (2002b). Repeated parental deprivation in the infant common marmoset (*Callithrix jacchus, primates*) and analysis of its effects on early development. *Biol. Psychiatr.* 52, 1037–1046.

Dettling, A. C., Schnell, C. R., Maier, C., Feldon, J., and Pryce, C. R. (2007). Behavioral and physiological effects of an infant-neglect manipulation in a bi-parental, twinning primate: Impact is dependent on familial factors. *Psychoneuroendocrinology* 32, 331–349.

Draper, P., and Harpending, H. (1982). Father absence and reproductive strategy: An evolutionary perspective. *J. Anthropol. Res.* 38, 255–273.

Ellis, B. J., McFadyen-Ketchum, S., Dodge, K. A., Pettit, G. S., and Bates, J. E. (1999). Quality of early family relationships and individual differences in the timing of pubertal maturation in girls: A longitudinal test of an evolutionary model. *J. Pers. Soc. Psychol.* 77, 387–401.

Ellis, B. J., Bates, J. E., Dodge, K. A., Fergusson, D. M., Horwood, L. J., Pettit, G. S., and Woodward, L. (2003). Does father absence place daughters at special risk for early sexual activity and teenage pregnancy?. *Child Develop.* 74, 801–821.

Evans, P. C., Lambert, N., Maloney, S., Furst, D. E., Moore, J. M., and Nelson, J. L. (1999). Long-term fetal microchimerism in peripheral blood mononuclear cell subsets in healthy women and women with scleroderma. *Blood* 93, 2033–2037.

Fite, J. E., and French, J. A. (2000). Pre- and postpartum sex steroids in female marmosets (*Callithrix kuhlii*): Is there a link with infant survivorship and maternal behavior?. *Horm. Behav.* 38, 1–12.

Fite, J. E., French, J. A., Patera, K. J., Hopkins, E. C., Rukstalis, M., and Ross, C. N. (2005a). Elevated urinary testosterone excretion and decreased maternal caregiving effort in marmosets when conception occurs during the period of infant dependence. *Horm. Behav.* 47, 39–48.

Fite, J. E., Patera, K. J., French, J. A., Rukstalis, M., and Hopkins, E. C. and Ross, C. N. (2005b). Opportunistic mothers: Female Wied's black tufted-ear marmosets (*Callithrix kuhlii*) reduce their investment in offspring when they have to, and when they can. *J. Hum. Evol.* 49, 122–142.

French, J. A. (1997). Regulation of singular breeding in callitrichid primates. In *Cooperative Breeding in Mammals* (N. G. Solomon and J. A. French, Eds.), pp. 34–75. Cambridge University Press, New York.

French, J. A., Schaffner, C. M., Shepherd, R. E., and Miller, M. E. (1995). Familiarity with intruders modulates agonism toward outgroup conspecifics in Wied's black tufted-ear marmoset (*Callithrix kuhli*). *Ethology* 99, 24–38.

French, J. A., Brewer, K. J., Shaffner, C. M., Schalley, J., Hightower-Merritt, D. L., Smith, T. E., and Bell, S. M. (1996). Urinary steroid and gonadotropin excretion across the reproductive cycle in female Wied's black tufted-ear marmosets (*Callithrix kuhli*). *Am. J. Primatol.* 40, 231–245.

Gagneux, P., Boesch, C., and Woodruff, D. S. (1999). Female reproductive strategies, paternity and community structure in wild West African chimpanzees. *Anim. Behav.* 57, 19–32.

Ge, X., Conger, R. D., Dadoret, R. J., Neiderhiser, J. M., Yates, W., Troughton, E., and Stewart, M. A. (1996). The developmental interface between nature and nurture: A mutual influence model of child antisocial behavior and parent behaviors. *Dev. Psychol.* 32, 574–589.

Gengozian, N., Batson, J. S., and Eide, P. (1964). Hematologic and cytogenetic evidence for hematopoietic chimaerism in the marmoset, *Tamarinus nigricollis*. *Cytogenetics* 3, 384–393.

Gengozian, N., Brewen, J. G., Preston, R. J., and Batson, J. S. (1980). Presumptive evidence for the absence of functional germ cell chimerism in the marmoset. *J. Med. Primatol.* 9, 9–27.

Goldizen, A. W. (1987). Tamarins and marmosets: Communal care of offspring. In *Primate Societies* (B. B. Smuts, R. M. Cheney, R. M. Seyfarth, R. W. Wrangham, and T. T. Struhsaker, Eds.), pp. 34–43. University of Chicago Press, Chicago.

Gunnar, M., and Quevedo, K. (2007). The neurobiology of stress and development. *Ann. Rev. Psychol.* 58, 145–173.

Hane, A. A., and Fox, N. A. (2006). Ordinary variations in maternal caregiving influence human infants' stress reactivity. *Psychol. Sci.* 17, 550–556.

Ketterson, E. D., and Nolan, V., Jr. (1999). Adaptation, exaptation, and constraint: A hormonal perspective. *Am. Nat.* 153, S4–S25.

Ketterson, E. D., Nolan, V., Jr., and Sandell, M. (2005). Testosterone in females: Mediator of adaptive traits, constraint on the evolution of sexual dimorphism, or both?. *Am. Nat.* 166, S85–S98.

Lillie, F. R. (1917). The free-martin; a study of the action of sex hormones in the foetal life of cattle. *J. Exp. Zool.* 23, 371–452.

Lormée, H., Jouventin, P., Chastel, O., and Mauget, R. (1998). Endocrine correlates of parental care in an Antarctic winter breeding seabird, the emperor penguin, *Aptenodytes forsteri*. *Horm. Behav.* 35, 9–17.

Maccoby, E. E. (2000). Parenting and its effects on children: On reading and misreading behavior genetics. *Ann. Rev. Psychol.* 51, 1–27.

Maestripieri, D. (2005). Early experience affects the intergenerational transmission of infant abuse in rhesus monkeys. *Proc. Natl Acad. Sci. USA* 102, 9276–9279.

Maloney, S., Smith, A., Furst, D. E., Myerson, D., Rupert, K., Evans, P. C., and Nelson, J. L. (1999). Microchimerism of maternal origin persists into adult life. *J. Clin. Invest.* 104, 41–47.

Missler, M., Wolff, J. R., Rothe, H., Heger, W., Merker, H. J., Treiber, A., Scheid, R., and Crook, G. A. (1992). Developmental biology of the common marmoset: Proposal for a postnatal staging. *J. Med. Prim.* 21, 288–298.

Merker, H-J., Sames, K., Casto, W., Heger, W., and Neubert, D. (1988). The embryology of *Callithrix jacchus*. In *Nonhuman Primates, Developmental Biology and Toxicology* (D. Neubert, H. Merker, and A. Hendrixx, Eds.). Ueberreuta Wissenschaft-Wein, Berlin.

Nunes, S., Fite, J. E., and French, J. A. (2000). Variation in steroid hormones associated with infant-care behaviour and experience in male marmosets (*Callithrix kuhlii*). *Anim. Behav.* 60, 857–865.

Nunes, S., Fite, J. E., Patera, K. J., and French, J. A. (2001). Interactions among paternal behavior, steroid hormones, and parental experience in male marmosets (*Callithrix kuhlii*). *Horm. Behav.* 39, 70–82.

O'Connor, T. G., Deater-Deckard, K., Fulker, D., Rutter, M., and Plomin, R. (1998). Genotype-environment correlations in late childhood and early adolescence: Antisocial behavioral problems and coercive parenting. *Dev. Psychol.* 34, 970–981.

Price, E. C. (1992). The costs of infant carrying in captive cotton-top tamarins. *Am. J. Primatol.* 28, 23–33.

Pryce, C. R., Palme, R., Feldon, J. 2002. Development of pituitary-adrenal endocrine function in the marmoset monkey: infant hypercortisolism is the norm. *J. Clin. Endocrinol. Metab.* 87, 691–699.

Power, M. L., Oftedal, O. T., and Tardif, S. D. (2002). Does the milk of callitrichid monkeys differ from that of larger anthropoids?. *Am. J. Primatol.* 56, 117–127.

Reiss, D., and Neiderhiser, J. M. (2000). The interplay of genetic influences and social processes in developmental theory: Specific mechanisms are coming into view. *Develop. Psychopathol.* 12, 357–374.

Ross, C. N., French, J. A., and Orti, G. (2007). Germ line chimerism and paternal care in marmosets (*Callithrix kuhlii*). *Proc. Natl Acad. Sci. USA* 104, 6278–6282.

Saltzman, W., Pick, R. R., Salper, O. J., Liedl, K. J., and Abbott, D. H. (2004). Onset of plural cooperative breeding in common marmoset families following replacement of the breeding male. *Anim. Behav.* 68, 59–71.

Sanchez, M. M. (2006). The impact of early adverse care on HPA axis development: Nonhuman primate models. *Horm. Behav.* 50, 623–631.

Sanchez, S., Pelaez, F., Gil-Burmann, C., and Kaumanns, W. (1999). Costs of infant-carrying in the cotton-top tamarin (*Saguinus oedipus*). *Am. J. Primatol.* 48, 99–111.

Santos, C. V., French, J. A., and Otta, E. (1997). Infant carrying behavior in callitrichid primates: *Callithrix* and *Leontopithecus*. *Int. J. Primatol.* 18, 889–908.

Schaffner, C. M., and French, J. A. (1997). Group size and aggression: 'Recruitment incentives' in a cooperatively breeding primate. *Anim. Behav.* 54, 171–180.

Schaffner, C. M., and French, J. A. (2004). Social and endocrine responses in male marmosets to the establishment of polyandrous groups. *Int. J. Primatol.* 25, 709–732.

Schaffner, C. M., Shepherd, R. E., Santos, C. V., and French, J. A. (1995). Development of heterosexual social relationships in Wied's black tufted-ear marmoset (*Callithrix kuhli*). *Am. J. Primatol.* 36, 185–200.

Schor, E. L. (2003). American Academy of Pediatrics. Task Force on the Family. Family pediatrics: Report of the Task Force on the Family. *Pediatr.* 111, 1541–1571.

Schrago, C. G., and Russo, C. A. M. (2003). Timing the origin of New World monkeys. *Mol. Biol. Evol.* 20, 1620–1625.

Schum, J. E., and Wynne-Edwards, K. E. (2004). Estradiol and progesterone in paternal and non-paternal hamsters (*Phodopus*) becoming fathers: Conflict with hypothesized roles. *Horm. Behav.* 47, 410–418.

Shepherd, R. E. & French, J. A. (1999). Comparative analysis of sociality in lion tamarins and marmosets: responses to separation from longterm pairmates. *J. Comp. Psychol.* 113, 24–32.

Smith, T. E., Schaffner, C. M., and French, J. A. (1997). Social modulation of reproductive function in female black tufted-ear marmosets (*Callithrix kuhli*). *Horm. Behav.* 31, 159–168.

Souter, V. L., Kapur, R.. P., Nyholt, D.. R., Skogerboe, K., Myerson, D., Ton, C. C., Opheim, K. E., Easterling, . R., Shields, L. E., Montgomery, G. W., and Glass, I. A. (2003). A report of dizygous monochorionic twins. *New Engl. J. Med.* 349, 154–158.

Tardif, S. D. (1997). The bioenergetics of parental behavior and the evolution of alloparental care in marmosets and tamarins. In *Cooperative Breeding in Mammals* (N. G. Solomon and J. A. French, Eds.), pp. 11–33. Cambridge University Press, New York.

Wingfield, J. C., Hegner, R. E., Dufty, A. M., and Ball, G. F. (1990). The "Challenge Hypothesis": Theoretical implications for patterns of testosterone secretion, mating systems, and breeding strategies. *Am. Nat.* 135, 829–846.

Wislocki, G. B. (1932). Placentation in the marmoset (*Oedipomidas geoffroyi*) with remarks on twinning in monkeys. *Anat. Rec.* 52, 381–392.

Zera, A. J., and Harshman, L. G. (2001). Physiology of life history trade-offs in animals. *Ann. Rev. Ecol. Syst.* 32, 95–106.

Ziegler, T. E. (2000). Hormones associated with non-maternal infant care: A review of mammalian and avian studies. *Folia Primatol.* 71, 6–21.

Ziegler, T. E., Epple, G., Snowdon, C. T., Porter, T. A., Belcher, A. M., and Kuederling, I. (1993). Detection of the chemical signals of ovulation in the cotton-top tamarin, (*Saguinus oedipus*). *Anim. Behav.* 45, 313–322.

Ziegler, T. E., Wegner, F. H., and Snowdon, C. T. (1996). Hormonal responses to parental and non-parental conditions in male cotton-top tamarins, *Saguinus oedipus*, a New World primate. *Horm. Behav.* 30, 287–298.

VI

REPRODUCTIVE EXPERIENCE: MODIFICATIONS IN BRAIN AND BEHAVIOR

31 THE NEUROECONOMICS OF MOTHERHOOD: THE COSTS AND BENEFITS OF MATERNAL INVESTMENT

KELLY G. LAMBERT[1] AND CRAIG H. KINSLEY[2]

[1] *Department of Psychology, Randolph-Macon College, Ashland, VA 23005, USA*
[2] *Department of Psychology, Center for Neuroscience, University of Richmond, Richmond, VA 23173, USA*

The terms stimulus and response, cause and effect, action and reaction, costs and benefits are foundations of animal behavior. Indeed, scientists have devoted extensive resources to understanding these guiding principles of animal behavior from a more *proximal* perspective (Glimcher, 2002). Sophisticated procedures currently exist to allow scientists to determine the underlying neurobiological mechanisms accompanying various behavioral responses. In the case of maternal behavior, proximal influences have been observed in the medial preoptic area (mPOA) in the form of Fos protein production (Numan & Numan, 1995; Stafisso-Sandoz *et al.*, 1998), increased perikarya volume, and neuronal complexity (Keyser *et al.*, 2001). Additionally, alterations in dendritic spines in the hippocampus CA1 area have been found to accompany the onset of maternal responsiveness (Pawluski & Galea, 2005; Kinsley *et al.*, 2006a,b). As important as these findings are in determining the underlying neurobiological mechanisms of adaptive maternal responses, further clues may emerge as ultimate mechanisms are considered. In the case of maternal responses, the ultimate consequences of responding to various stimuli differ vastly for a virgin and a lactating female. The new (or even veteran) mother must make a trade-off, a series of cost–benefit calculations, between what is best for her vs. what is best for her offspring, or between safety and security for herself and her pups vs. the danger of leaving the security of a well-hidden nest to go forage in a forbidding environment.

At the most fundamental level, the function of behavior is to incorporate sensory information and existing memories to execute the most adaptive motor responses. According to Paul Glimcher, a pioneer in the emerging field of neuroeconomics, "the goal of behavior is to make the right choices, to choose the course of action that maximizes the survival of the organism's genetic code" (Glimcher, 2003, p.172). Accordingly, neuroeconomics has been defined as the investigation of the neuroarchitecture of the decision-making process via explorations of the integration of sensory and motor systems (Glimcher & Rustichini, 2004; Glimcher *et al.*, 2005). The field of neuroeconomics recognizes the value of economics theory as a tool for defining the problems faced by animals seeking specific goals under varying environmental conditions, as well as the importance of the nervous system in the actual execution of the animal's response (Glimcher, 2003)

Interestingly, research with primates suggests that neural activity in the area of the posterior parietal cortex, specifically the lateral intraparietal area (LIP), is correlated with decision making. Subsequent single cell recording in specific visual tasks led researchers to conclude that cells in this area, once thought to be an association area of the cortex, are carrying out computational comparative analyses to determine the predicted utility of various response options (Dorris & Glimcher, 2004). Recently, it has been suggested that the economic analysis of behavior paradigms may provide valuable

Neurobiology of the Parental Brain

information about how animals solve problems to enhance inclusive fitness (Glimcher, 2002).

The behavioral formula for successfully enhancing the female's inclusive fitness varies greatly between a virgin and mother. In fact, the genetic investment greatly increases as the lactating female now has her own genetic fitness at stake plus 10 or so pups with a significant portion of her genetic code – and, to make her choices even more critical, the lactating wild rat typically is pregnant while she is nursing (Crowcroft, 1973; Gubernick & Klopfer, 1981; Rosenblatt *et al.*, 1985; Sullivan, 2004). Considering these drastic changes in genetic fitness in a neuroeconomics context provides a fascinating foray into new ways of interpreting the summation of neuronal events into an efficient and sharper organism, regulated by the plastic maternal brain. In the case of a mother, the onus is on her to make the most efficient and adaptive decisions. The maternal brain, we believe, has evolved to provide an efficient substrate to quickly and better evaluate the mother's internal state, the local environment, the offspring needs and state, and the exigencies of immediate vs. delayed action. Her investment, the survival of her offspring, costly both metabolically and genetically, lies in the balance.

Over the past decade we have conducted studies investigating spatial memory (foraging) and boldness in the maternal rat, two response modifications that would enhance the mother's efficiency carrying out the more traditional maternal responses. Additionally we have most recently explored motor agility, coordination, and strength in maternal females. As described below, we have found primiparous and multiparous females to exhibit very different response strategies than nulliparous females. Enhanced foraging ability, diminished emotionality and anxiety, and increased motor efficacy represent a shift toward more adaptive responses in these females. The variable nature of the response strategies in the maternal female conveys the potential value of the maternal animal as a model of dynamic neuroeconomics.

FORAGING RESPONSE STRATEGIES

Our initial investigations of the influence of maternal behavior on foraging strategies involved assessing virgins and maternal females in two spatial memory tasks, the radial arm maze and the dry land maze. The results from both our laboratories were unequivocal. Although all groups were placed on comparable food-restriction regimes prior to these memory tests, the nulliparous rats were not as efficient locating the froot loop reward in these early studies. Also interesting was the finding that, in the dry land maze study, the pup-sensitized females also performed better than the nulliparous animals. These initial results suggested that the response strategies were very different between rats with and without some form of maternal experience (Kinsley *et al.*,1999).

Intrigued by the finding that pup-sensitized females performed more efficiently in the dry land maze, we set out to determine the most important aspects of the maternal experience. To do this, the presence of pups was manipulated in both nulliparous and primiparous rats. Hence, virgins with and without pup exposure were used along with primiparous females that were either allowed to keep their pups or had them taken away. Most interesting in this study were the findings in the probe trial, that is, a 5-min trial conducted the day after the traditional dry land maze trials during which no well was baited. We hypothesized that the fully maternal primiparous animals would spend more time in proximity to the previously baited well exhibiting a stronger memory of the task than her less maternal counterparts. The results, however, suggested that the behavioral choices during this trial were even more efficient and adaptive than we anticipated. In addition to spending more time in proximity to the previously baited well, the fully maternal animals also visited more wells (wells that were all baited in the early stages of maze training). Thus, these females seemed to be maximizing the probability of securing the food reward by utilizing different foraging strategies when the "rules" for the task were suddenly changed. This display of dynamic response strategies, observed solely in the fully maternal animals, was indeed surprising and reinforced our earlier observations that the process of decision making in the context of foraging had been altered in significant ways. Of further interest was the observation that these changes persisted after the pups were weaned – these females commenced testing 10 days following the removal of pups. It appeared that the modified response strategies were not directly dependent on the actual maternal experience including pups, hormonal alterations associated with pregnancy and lactation, or some combination of these variables. These results suggested that altered foraging strategies, set in motion during the maternal experience, were long lasting (Lambert *et al.*, 2005).

To confirm the long-lasting nature of these altered foraging strategies, using two different rat strains (Long-Evans and Sprague-Dawley), both of our labs commenced longitudinal studies (Gatewood *et al.*, 2005; Love *et al.*, 2005). In these studies, it became apparent that the altered strategies in decision making persisted throughout the female's life. In one study (Love *et al.*, 2005), we attempted to make the foraging strategy even more ecologically and ethologically relevant by requiring the female to compete with weight-matched females to obtain the froot loop reward. Thus, the nulliparous, primiparous, and multiparous females were placed in the dry land maze following their training and required to negotiate the social novelty so that they could retrieve the single froot loop reward. Interestingly, the females with maternal experience were much more likely to retrieve the froot loop than their virgin counterparts. While the multiparous females retrieved the reward in 60% of the trials, the virgins did so in only 7% of the trials (Love *et al.*, 2005).

Whille methodologically challenging, we have attempted to investigate foraging strategies in a more ethologically relevant context of assessing the females with pups. Although metabolic rates differ among pregnant, lactating, and virgin animals, we were interested in their foraging responses in a task requiring them to leave their safe nesting cage with or without pups to enter an unpredictable environment to find food. In this situation, ultimate consequences of responses, as well as opportunity costs for leaving the nest and pups to venture into unknown territory to secure food could be investigated. Preliminary data indicated that both pregnant and postpartum females were more likely to venture out into the maze to find food and water than the virgin rats (Gatewood *et al.*, 2001). Obviously, the maternal female has more to lose by making the wrong choice in this situation. The opportunity cost for leaving the nest and pups could be the death of her genetic investment should a predator invade the nest; on the other hand, if the mom and pups all starve to death, the cost is even greater.

In another attempt to bridge inherent laboratory artificiality with the "real world" of the lactating female, we tested the predatory behavior of nulliparous and lactating Sprague-Dawley females in an open field. This task requires the rat to make many choices in "real time" to successfully follow a cricket so it can be captured for ingestion. After minimal food restriction and familiarization with the test arena, the females were required to track, capture, and kill a cricket during a 5-min test over 3 days. On average, the mothers captured and killed the prey in ~65 s, compared to ~260 s in the nulliparous females. These data suggest that mothers are better hunters, at least under the restricted conditions with which we examined them in the laboratory. That lactating females may be more motivated due to their enhanced metabolic state was also tested by increasing the length of the food deprivation in the nulliparous females: these related tests showed that "more hunger" did not account for the difference, since being food-restricted for twice as long as the maternal rats did not account for the group differences.

Considering that successfully catching the moving cricket involves efficient sensory, motor, and attentional processes, we were intrigued by the specific mechanisms enabling the animals to perform better at this task. Thus far we have examined the primary sensory systems of the rat, olfaction and audition, as possible mediators but they have not proved to be playing major roles in the enhancement of predation in the mothers. That is, we infused the nares of lactating females with zinc sulfate ($ZnSO_4$) and verified their temporary loss of olfaction; we also tested lactating females in the presence of a white noise generator that obscured the sound of the cricket moving in the arena. Both manipulations failed to meaningfully alter the latency to pursue the crickets (though there was a small and just statistically significant increase in the $ZnSO_4$ group compared to saline). That is, only a small portion of the variance is accounted for by this olfactory effect. We are currently exploring other sensory modalities and combinations of sensory restriction, and our data point to significant contributions of the visual processing system. We have observed larger cell body volumes in the lateral geniculate nucleus of the thalamus in lactating vs. nulliparous females, using nissl (thionin) stained sections (Kinsley *et al.*, 2006a,b,c). In general, the data suggest more efficient processing of attentional/visual cues in the enclosure (Kinsley *et al.*, 2006a,b,c; Sirkin *et al.*, unpublished observations).

Using the Long-Evans rats, we assessed attentional processing in a different task known as the attention set-shifting paradigm. This task requires rats to associate specific odors with small flower pots containing froot loop rewards; subsequently, distracting cues such as different types of bedding and novel odors are introduced and the rat is required to focus solely on the salient cues to discern the flower pot with the actual food reward. In the midst of training, the previous distracting cue of bedding becomes the salient cue requiring the animal to shift response strategies based on current information about "payoffs," while ignoring previously reinforced stimuli (i.e., an extradimensional shift). Thus, a simple discrimination involves learning that the pot with an odor is associated with the reward; a complex discrimination involves learning that the flower pot with vanilla odor is associated with reward

(A)

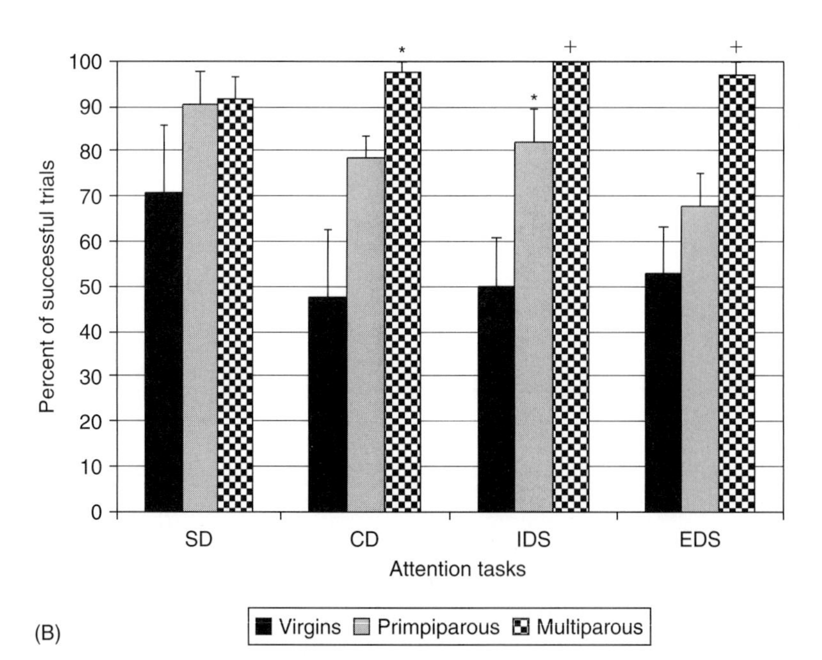

(B)

FIGURE 31.1 *Attention set-shifting paradigm.* Nulliparous rat making error in the extradimensional shift (EDS) trial in the attention set-shifting paradigm (A-left); multiparous female digging for froot loop reward in the correct flower pot choice during the extradimensional shift trials (A-right). Percent of successful trials observed in the simple discrimination (SD), complex discrimination (CD), intradimensional shift (IDS), and extradimensional shift (EDS) trials (B). *Significant difference from nulliparous/virgin animals at $p < 0.05$; +Difference from nulliparous animals at $p < 0.001$; $n = 8$ for each group.

whereas the mango-scented pot is not rewarded; intradimensional shift occurs when the rat subsequently learns that mango, a new stimulus but still within the olfaction sensory dimension, is currently associated with reward; and extradimensional shift requires the animal to learn that a stimulus from a different sensory dimension such as the type of bedding in the pot (yarn vs. confetti, representing a visual/tactile sensory dimension) is the new salient cue. In this case the accuracy of the multiparous animals' choices between the unbaited and baited flower pots ran close to 100% during the complex discrimination, intradimensional shift discrimination, and extradimensional shift discrimination trials, vastly different from the ~50% observed in the nulliparous animals. Whereas all groups showed evidence of learning this task, the multiparous animals made significantly fewer errors during the learning process. In the real world, fewer errors would likely translate into catching the prey before it got away, or finding the food before a competitor retrieved it. This study suggests that attentional processes are enhanced in multiparous animals, likely influencing their response choices in foraging tasks (Higgins *et al.*, 2007a; See Figure 31.1).

RISK ASSESSMENT: BEHAVIORAL VENTURE CAPITALISM

Our initial investigation of emotionality and risk assessment of maternal animals involved placing pregnant and virgin rats into an open field during the later stages of pregnancy. Specifically, primigravid (first pregnancy), multigravid (second pregnancy), and virgin females were exposed to the open field for 30 min. Subsequently, the brains were harvested so that c-Fos immunoreactive tissue could be assessed. The results indicated that, along with showing less behavioral reactivity to the stress of the open field, their brains exhibited less c-Fos immunoreactivity in both the basolateral amygdala and the CA3 region of the hippocampus. In this case, no differences were observed between the primigravid and multigravid groups. These data suggest that the maternal experience leads to a more resilient female; that is, when forced into a stressful situation, these females exhibited less evidence of behavioral and neurobiological stress than their virgin counterparts. A second experiment in this study investigated the effect of restraint stress on nulliparous, primiparous, and multiparous females. Once again, maternal animals exhibited less c-Fos immunoreactivity in the limbic areas mentioned above (Wartella *et al.*, 2003).

The results of these early studies complemented well-established findings that pregnant and parturient rats exhibited diminished activity of the hypothalamic–pituitary–adrenal (HPA) axis (Neumann, 2001; Neumann *et al.*, 2003). As we considered the physiological cost of an excessive stress response in the midst of maintaining a metabolically costly pregnancy and/or lactation, it seemed adaptive for the stress response to be modified for the greatly increased physiological expenditures accompanying the maternal experience. It appeared that the female's ability to avoid allostatic load, requiring the coordination of integrative physiological systems to maintain optimal functioning (McEwen, 2003), would be imperative to the successful maternal rat's ability to continue to provide resources during times of greatly increased demands. In a sense, the maternal animal's stress and vigilance systems appeared to be modified, perhaps fine-tuned, to provide optimal functioning with minimal resources.

To further examine this question, we recently exposed nulliparous, primiparous, and multiparous females to a chronic unpredictable stress paradigm during which the animals were exposed to various ethologically relevant stressors in an unpredictable manner for 16 days. Several behavioral tests were conducted during the chronic stress exposure. In these tests, the multiparous animals exhibited diminished anxiety and efficient responding. Specifically, the multiparous animals exhibited increased exploration in the open field test and, when a small clip was placed on their tail to simulate an insect bite, they responded to this stimulus much faster than the other groups. Additionally, there was a non-significant trend for the multiparous animals to exhibit less conditioned fear when they encountered a stimulus (i.e., block) that had been associated with the forced swim task. Nulliparous and primiparous animals exhibited more defensive behavior toward the block than the multiparous animals. Interestingly, a tell "tail" sign of optimal allostatic functioning was found when the females were placed in a restraint tube so that cardiovascular measures could be obtained. Although no differences in blood pressure were observed among the groups, non-significant trends for the multiparous animals to have a lower heart rate and less blood flow in the tail were observed. It is likely that the diversion of blood from the distal tail to the body's core and muscles would facilitate escape during a stressful experience, representing an efficient allocation of resources. Further evidence of the multiparous female's ability to build resilience against allostatic load was found in the corticosteroid data. Multiparous animals had lower levels than the other groups (specifically, 17% and 63% higher levels were observed for nulliparous and primiparous animals, respectively; Farrell *et al.*, 2007).

We have also been interested in both the emergence and duration of the various stress response alterations observed in the maternal animals. In the longitudinal study described above (Love *et al.*, 2005), animals were exposed to the elevated plus maze every 4 months to assess the amount of time spent in the open arms. Although no differences were observed in the first assessment at 6 months of age, the primiparous animals spent a higher percent of time in the open arm than the virgins at the 10, 14, 18, and 22 month assessments. Additionally, the multiparous animals spent more time in the open arm than their nulliparous counterparts at the 10 and 14 month assessments. These results suggested that risk assessment was altered in repeated exposure to the elevated plus maze. To assess exposure to a novel stimulus during old age, we assessed exploration of novel stimuli at 23 months and found that the multiparous females spent more time in contact with the three novel stimuli than the other groups.

Thus, the various studies we have conducted exploring the stress response in maternal animals suggest that diminished anxiety and stress extend past the stage of pregnancy and weaning, perhaps for the duration of the rat's life (Byrnes & Bridges, 2006). More information is needed to understand the specific parameters of these modifications as well as the consequences for the female. Long past the actual maternal experience, we have found old-age females to exhibit enhanced foraging abilities and, upon examination of their brains, less amyloid precursor protein immunoreactive tissue (Gatewood *et al.*, 2005). The convergence of this longitudinal study with the emotionality studies described in this section suggests that the altered stress response in these females has an additional effect of providing a buffer against the physiological threats of an aging brain. In accordance with this notion, Parsons *et al.* (2004, p. 1099) concluded that "a neuroadaptive mechanism may develop after a first pregnancy that increases the ability to recover from some cognitive deficits after later pregnancies."

STRENGTHENING THE BEHAVIORAL RESPONSE INFRASTRUCTURE

After spending a decade researching the effect of maternal experience on the maternal ancillary behaviors of spatial ability (including foraging and hunting) and emotional resilience, we most recently became interested in modifications of the motor system in parous animals. Considering the importance of motor strength, agility, and coordination to the successful navigation of the maternal rat's environment, we hypothesized that the adaptations of these females would extend to the efficiency of motor responses. Aside from the cognitive and emotional modifications, the successful maternal rat must build a nest, groom and retrieve her pups, return to her nest quickly following a bout of foraging, and deftly defend her pups from conspecifics and predators.

Although little attention has been paid to maternity-induced changes in locomotion and agility, recent research suggests reproductive systems may also play a role in the behavioral choice strategies of the maternal female. For example, researchers have reported the presence of estrogen receptors in the cerebellum, a brain area known for its involvement in coordination and balance (Mitra *et al.*, 2003). Additionally, indirect effects of estrogen on striatal dopaminergic activity have been reported recently (Dluzen, 2005). We were also intrigued by the fascinating results reported by Gregg *et al.* (2007) suggesting that maternal animals exhibited increased levels of the generation of myelin-forming oligodendrocytes and myelinated axons than non-maternal rodents. This study identified prolactin as a potential agent of this myelination modification. Interestingly, this study was conducted as an investigation of anecdotes that women with multiple sclerosis experience remission from their symptoms during pregnancy.

In an attempt to explore further the question of enhanced motor coordination, strength, and agility in the maternal female, we assessed age-matched nulliparous, primiparous, and multiparous females following the weaning of their pups. In this study, all females were assessed for (1) strength via a wire hang test, (2) motor agility and balance via a rod traversal test, and (3) strength, motor learning, and agility via rope-climb training. The results clearly indicated that the animals with maternal experience developed motor modifications to accompany the cognitive and emotional effects observed in our earlier work.

In the wire hang test, females were simply placed on a thick wire so that they were hanging by their paws over a cushioned landing area. The duration of time spent hanging on the wire before releasing was significantly higher for the multiparous animals than the other reproductive groups. Specifically, multiparous animals persisted for 23 s compared to 12 s for the primiparous animals and 14 s for the nulliparous rats. For the rod traversal task, the rats were exposed to three rod surfaces including a sand surface, smooth surface, and smooth surface with hurdles. For each test, traversal time, number of slips, and number of falls were recorded. Focusing on the rough rod test, multiparous animals traversed the rod in 3.7 s compared to 8.2 s and 26.1 s for the primiparous and nulliparous animals, respectively. Additionally, the multiparous animals exhibited fewer slips than the other groups. Whereas a trend for the same pattern of effects was observed in the smooth rod test, the multiparous animals were once again significantly more efficient in the hurdle rod traversal task. Specifically, the multiparous animals traversed the hurdle rod in 6 s whereas it took the primiparous group 18 s and the nulliparous animals 34 s. Considering that this task required memory, boldness, and motor agility, we were encouraged that the effects we had observed on the more specific tasks described earlier in this chapter generalized to the more ethologically relevant tasks requiring the animals to adeptly navigate their threatening environments to protect their genetic investment and strengthen their inclusive fitness (see Figure 31.2).

In the rope-climb test, the animals were shaped to climb the rope by proceeding to knots tied in equidistant increments to retrieve a froot loop reward. During training, we observed the rats to rear up to retrieve the food rewards from the rope early in the training process, but additional training was required for the rats to leave the safety of the climbing platform to engage in what we referred to as a full body climb in which the entire body left the platform. However, at the end of the 3-week training phase, chi-square analyses indicated that a greater number of multiparous animals ventured up the rope to retrieve the reward during the allotted training period. Seventy-five percent of the multiparous animals exhibited the full body climb compared to 13% of the primiparous and 37% of the nulliparous animals (Higgins *et al.*, unpublished observations).

Although we had not planned to do so, we were so intrigued by the multiparous animals' motor skills that we wondered how well they would serve them in the cricket hunting task. We had yet to assess the Long-Evans rats in this task and, although it had been 5 months since the weaning of the respective litters, we were interested in the hunting skills of these females. Testing them in a similar manner as described above, the mean latencies for capturing the crickets were recorded across hunting trials. In this case, the maternal rats continued to exhibit superior performance retrieving the crickets; the multiparous animals caught the crickets in an average of 41 s, the primiparous rats in 71 s, and the nulliparous animals in 184 s. These more successful hunting skills, long after the weaning of her pups, suggest that the brain's response strategies are altered (Higgins *et al.*, 2007b).

THE MOTHERHOOD-INDUCED NEURON-NETWORK EFFECT

In 1982, Robert Metcalfe, the founder of the Ethernet, coined the term *network effect* to convey his observation that the value of a good or service changes in respect to the number of people using it. As the Ethernet became more and more popular with increasing numbers of users, this product was deemed highly important to the consumers. Essentially, the Metcalfe Law of economics posits that "bigger networks are better" (Metcalfe, 2007). In a sense, the maternal brain experiences a network effect as it produces and cares for more and more pups. The convergence of the proximal effects of the maternal experience that have been described over the past several decades – ranging from increased mPOA cell body size to increased numbers of dendritic spines on hippocampal

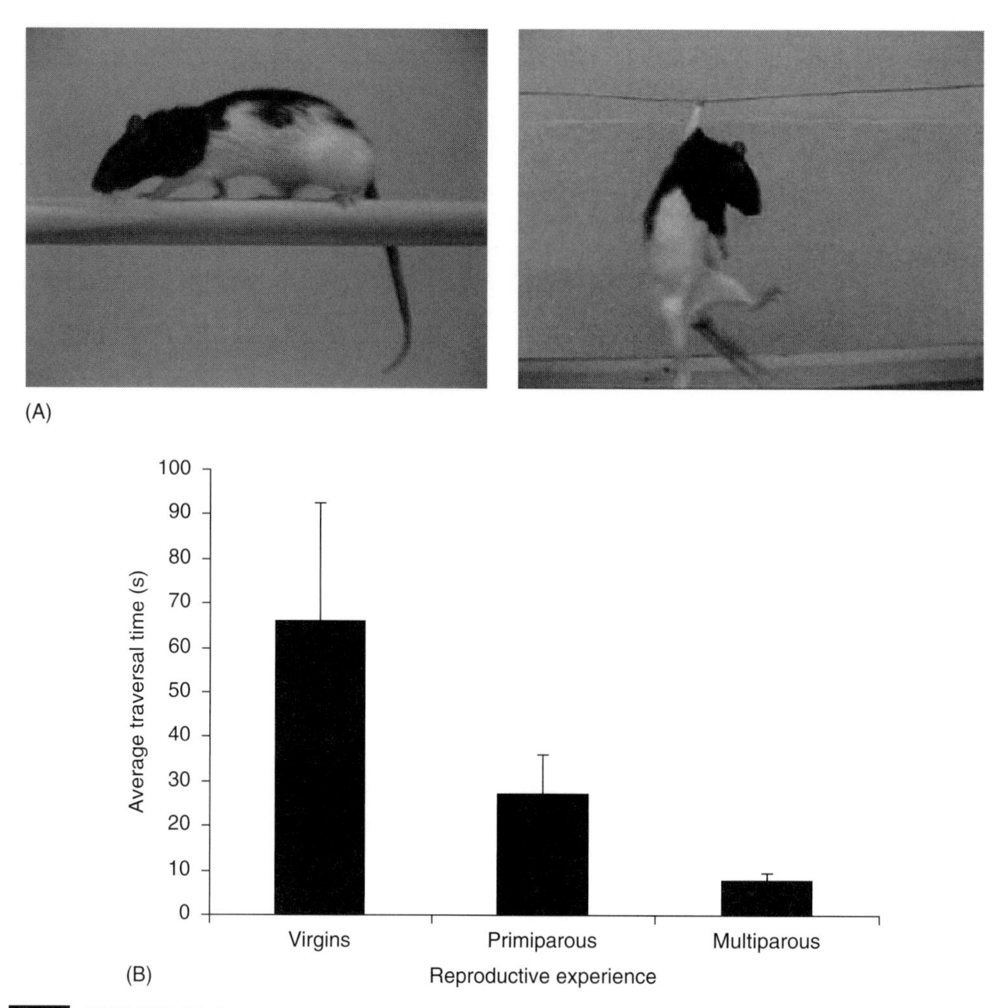

FIGURE 31.2 *Motor efficacy tasks.* Multiparous animal traversing the smooth rod (A-left) and persisting in the hang test (A-right). The average traversal time in the nulliparous, primiparous, and multiparous animals (B); multiparous animals traversed significantly faster than nulliparous/virgin rats ($p < 0.05$); $n = 8$ for each group.

neurons – likely lead to a brain with increased networking potential directed toward caring for her offspring. Recent evidence of increased levels of the neurotrophic factor known as brain-derived neurotrophic factor in the hippocampus provides further evidence of the enhanced neural networking accompanying maternal experience (MacBeth *et al.*, 2007).

According to Glimcher, the 17th century writings of Blaise Pascal emphasized the notion that rational decision making is dependent on the interaction of gain and uncertainty in various response opportunities. As modern economics uses this idea to explore human decision making, it has been argued that game theory lends itself to biological systems more readily than the field of economic behavior for which it was originally designed. In biological systems analyses, inclusive fitness can be measured on a one-dimensional scale (i.e., inclusive fitness); it is difficult, however, to transpose the various outcomes related to financial gains, safety gains, ethical considerations, etc., contributing to human decision making to a single scale (Glimcher, 2002). Thus, the dynamic processes of response strategies observed in the maternal mammal may serve as an interesting model for further explorations of neuroeconomics.

COMPARATIVE MODELS OF PARENTAL NEUROECONOMICS

We have described the interesting intraspecies effects of varying levels of maternal experience in this chapter, but comparative analyses between species also represent a fertile opportunity for response choice explorations. Although the maternal rat spends most of her time in proximity to her pups, the rabbit utilizes a very different strategy. After expending both physical effort and her own fur to build a maternal nest in a burrow, these "working moms" only nurse their litters once a day for a brief duration of about 3 min before sealing the opening of the burrow as she leaves the nest (Gonzalez-Mariscal *et al.*, 2007).

Additionally, factors contributing to the choice strategies made by the maternal brain may be discovered by investigating a different type of parenting brain – that is, the paternal brain. More variability is observed in paternal responses across species with only about 6% of mammalian species exhibiting paternal behavior (Numan & Insel, 2003). Comparing the neurobiological responses of biparental species to uniparental species offers an interesting opportunity to determine key proximal and ultimate mechanisms in closely related congeneric species such as prairie voles and montane voles and California deer mice vs. other more prevalent woodland deer mice. Focusing on a biparental primate species, research has recently shown that vasopressin receptors are more active in the prefrontal cortex of the paternal common marmoset (Kozorovitskiy *et al.*, 2006). Additionally, preliminary findings in our laboratory indicate that vasopressin fibers extend from the paraventricular nucleus of the hypothalamus toward the cortex in males with parenting experience (Everette *et al.*, 2007). Thus, research with paternal models may provide convincing evidence supporting the notion of the parenting-induced "network effect."

In another fascinating exploration of parenting payoffs in males, Velando *et al.* (2006) investigated the *terminal investment hypothesis* that animals should re-direct their energy from themselves and their potential offspring toward the existing offspring as they approach the end of their lifespan. To assess this, both young and old blue-footed booby birds were captured and either injected with *E. coli*, (a molecule that activates the immune system but poses no real threat to their health) or saline 2 weeks prior to the time their mates laid eggs. Compared to controls 18% fewer fledglings were produced in the care of the young dads whereas the older dads turned into superdads as they produced 98% more fledglings than controls. Both age and immune status strongly impact the ultimate consequences of the demanding paternal investment in these birds.

Of course, abandonment or neglect of offspring is not exclusive to male parents. A diverse array of maternal investment in the care of offspring can be found across the animal kingdom. This topic was explored in an interesting article written for *The New York Times* entitled *One thing they aren't: maternal* (Angier, 2006). In addition to the latchkey rabbit pups discussed above, the guinea hen that moves at a pace only the strongest of her brood can keep up with and survive, the panda mom that cares for only one of her twin panda cubs, and the maternal African black eagle that feeds only one of her eaglets while allowing the thriving offspring to peck the starving sibling to death all serve as evidence of the less than nurturing maternal care observed in some mammals. As troubling as these acts of maternal "offspring downsizing" seem, they also provide opportunities to explore the mechanisms underlying such responses. At what point does the rewarding prospect of nurturing one's offspring lead to the decision to neglect, kill, or even eat that genetic investment?

SUMMARY AND CONCLUSIONS

According to Paul MacLean, as maternal mammals entered the evolutionary scene, the unique responses of nursing conjoined with maternal care, audiovocal communication for maintaining mother–offspring contact, and play behavior became staples of the mammalian behavioral repertoire (MacLean, 1990). The consideration of neuroeconomics and neural areas devoted to adaptive choices suggests that another unique aspect, that is, contextually appropriate efficient problem solving, is also linked to the onset of maternal mammalian behavior. Thus, the earliest explorations of economic theories may have been conducted in the brains of these early mammals, providing further support for MacLean's statement that "For more than 180 million years, the female has played the central role in mammalian evolution" (MacLean, 1996, p. 422; Lambert, 2003).

We have attempted to provide convincing evidence that the maternal mammal offers unique opportunities to explore the dynamic mechanisms underlying altered response strategies in specific situations. As the female becomes pregnant, her "genetic stock" increases, which alters the ultimate consequences of her responses. Accordingly, adaptive responses have been observed in females with maternal experience. Although to our knowledge researchers have yet to investigate the effect of maternal experience on specific parietal areas such as the LIP implicated in primate decision making (i.e., the neuroeconomics of choice), it would be interesting to investigate this area in parous animals (Dorris & Glimcher, 2004). Additionally, further investigations of enhanced neural networks accompanying each full maternal experience in the form of neural growth factors, morphological differences, increased synapses, more complex networking of dendritic and/or axonal processes, or even more elaborate scaffolding of glial cells will help to explain the cumulative benefits of each maternal experience. As the gains and costs of mammalian response options are continually calculated, at some point the costs outweigh the gains of providing nurturing responses toward the offspring – learning more about these parenting response thresholds will also offer insight into the neuroeconomics of the parental brain.

REFERENCES

Angier, N. (May 9, 2006). One thing they aren't: Maternal. *The New York Times* .

Byrnes, E. M., and Bridges, R. S. (2006). Reproductive experience alters anxiety-like behavior in the female rat. *Horm. Behav.* 50, 70–76.

Crowcroft, P. (1973). *Mice All Over.* Chicago Zoological Society, Chicago.

Dluzen, D. E. (2005). Unconventional effects of estrogen uncovered. *Trends Pharmacol. Sci.* 26, 485–487.

Dorris, M. C., and Glimcher, P. W. (2004). Activity in posterior parietal cortex is correlated with the relative subjective desirability of action. *Neuron* 44, 365–378.

Everette, A., Fleming, D., Higgins, T., Tu, K., Bardi, M., Kinsley, C. H., and Lambert, K. G. (2007). Paternal experience enhances behavioral and neurobiological responsivity associated with affiliative and nurturing responses. International Behavioral Neurosicence Society, Rio de Janeiro.

Farrell, M., Fleming, D. F., Kinsley, C. H., Bardi, M., and Lambert, K. G. (2007). Maternal experience modifies behavioral, cardiovascular, and endocrinological responses during chronic stress. Unpublished manuscript.

Gatewood, J., Eaton, M., Madonia, L., Babcock, S., Griffin, G., Lambert, K. G., and Kinsley, C. H. (2001). Reproduction-facilitated aging in rats: Parity and pup sensory stimulation may forestall some aspects of senescent memory loss. *Soc. Neurosci. Abstr.* 27, 2001.

Gatewood, J. D., Morgan, M. D., Eaton, M., McNamara, I. M., Stevens, L. F., Macbeth, A. H., Meyer, E. A., Lomas, L. M., Kozub, F. J., Lambert, K. G., and Kinsley, C. H. (2005). Motherhood mitigates aging-related decrements in learning and memory and positively affects brain aging in the rat. *Brain Res. Bull.* 66, 91–98.

Glimcher, P. W. (2002). Decisions, decisions, decisions: Choosing a biological science of choice. *Neuron* 36, 323–332.

Glimcher, P. W. (2003). *Decisions, Uncertainty, and the Brain: The Science of Neuroeconomics.* MIT Press, Cambridge, MA.

Glimcher, P. W., and Rustichini, A. (2004). Neuroeconomics: The consilience of brain and decision. *Science* 306, 447–452.

Glimcher, P. W., Dorris, M. C., and Bayer, H. M. (2005). Physiological utility theory and the neuroeconomics of choice. *Games and Economic Behavior* 52, 213–256.

Gonzalez-Mariscal, G., McNitt, J. I., and Lukefahr, S. D. (2007). Maternal care of rabbits in the lab and on the farm: Endocrine regulation of behavior and productivity. *Horm. Behav.* 52, 86–91.

Gregg, C., Shikar, V., Larsen, P., Mak, G., Chojnacki, A., Yong, V. W., and Weiss, S. (2007). White matter plasticity and enhanced remyelination in the maternal CNS. *J. Neurosci.* 27, 1812–1823.

Gubernick, D., and Klopfer, P. (1981). *Parental Care in Mammals.* Plenum Press, New York.

Higgins, T., Everette, A., Fleming, D., Christon, L., Kinsley, C. H., and Lambert, K. G. (2007a). Maternal experience enhances neurobiological and behavioral responses in an attention set-shifting paradigm. Presentation at The International Behavioral Neuroscience Society, Rio de Janeiro, Brazil.

Higgins, T., Sirkin, M., Drew, M., Chipko, C., Jablow, L., Ferguson, T., Mprlimas. T., Worthington, D., Kinsley, C. H., and Lambert, K. G. (2007b). Maternal experience enhances motor coordination, agility, and strength in Long-Evans rats. Unpublished manuscript.

Keyser, L., Stafisso-Sandoz, G., Gerecke, K., Jasnow, A., Nightingale, L., Lambert, K. G., Gatewood, J., and Kinsley, C. H. (2001). Alterations of medial preoptic area neurons following pregnancy and pregnancy. *Brain Res. Bull.* 55, 737–745.

Kinsley, C. H., and Lambert, K. G. (2006). The maternal brain. *Sci. Am.* 294, 72–79.

Kinsley, C. H., Madonia, L., Gifford, G. W., Tureski, K., Griffin, G. R., Lowry, C., Williams, J., Collins, J., McLearie, H., and Lambert, K. G. (1999). Motherhood improves learning and memory: Neural activity in rats is enhanced by pregnancy and the demands of rearing offspring. *Nature* 402, 137–138.

Kinsley, C. H., Bardi, M., Karelina, K., Rima, B., Christon, L., Friedenberg, J., Sirkin, M., Chipko, C. Victoria, L., Drew, M., Fyfe, C., and Lambert, K. G. (2006a). Track, attack, consume: Pregnancy/parenthood induction of an improved predatory behavioral repertoire and accompanying neural enhancements in the rat. Paper presented at the Society for Neuroscience annual meeting, Atlanta, GA, November, 2006.

Kinsley, C. H., Trainer, R., Stafisso-Sandoz, G., Quadros, P., Keyser-Marcus, L., Hearon, C., Amory-Meyer, E. A., Hester, N., Morgan, M. D., Kozub, F. J., and Lambert, K. G. (2006b). Motherhood and pregnancy hormones modify concentrations of hippocampal neuronal dendritic spines. *Horm. Behav.* 49, 131–142.

Kozorovitskiy, Y., Hughes, M., Lee, K., and Gould, E. (2006). Fatherhood affects dendritic spines and vasopressin V1a receptors in the primate prefrontal cortex. *Nat. Neurosci.* 1094–1095.

Lambert, K. G. (2003). The life and career of Paul MacLean: A journey toward neurobiological and social harmony. *Physiol. Behav.* 79, 343–349.

Lambert, K. G., Berry, A. E., Griffins, G., Amory-Meyers, L., Madonia-Lomas, L., Love, G., and Kinsley, C. H. (2005). Pup exposure differentially enhances foraging ability in primiparous and nulliparous rats. *Physiol. Behav.* 84, 799–806.

Love, G., Torrey, N., McNamara, I., Morgan, M., Banks, M., Hester, N. W., Glasper, E. R., DeVries, A. C., Kinsley, C. H., and Lambert, K. G. (2005). Maternal experience produces long-lasting behavioral modifications in the rat. *Behav. Neurosci.* 119, 1084–1096.

Macbeth, A. H., Scharfman, H. E., MacLusky, N. J., and Luine, V. N. (2008). Multiparity enhances recognition memory: Role of monoaminergic neurotransmitters and brain-derived neurotrophic factor (BDNF). *Horm. Behav.* (in press)

MacLean, P. D. (1990). *The Triune Brain in Evolution: Role in Paleocerebral Functions*. Plenum, New York.

MacLean, P. D. (1996). Women: A more balanced brain. *Zygon* 31, 421–439.

McEwen, B. S. (2003). Glucocorticoids, depression, and mood disorders: Structural remodeling in the brain. *Metabolism* 52, 10–16.

Metcalfe, R. M. (May 7, 2007). It's all in your head. Forbes. http://www.forbes.com/free_forbes/2007/0507/052.html.

Mitra, S. W., Hoskin, E., Yudkovitz, J., Pear, L., Wilkinson, H.a., Hayashi, S., Pfaff, D., Ogawa, S., Rohrer, S. P., Schaeffer, J. M., McEwen, B. S., and Alves, S. E. (2003). Immunolocalization of estrogen receptors in the mouse brain. *Endocrinology* 144, 2055–2067.

Neumann, I. D. (2001). Alterations in behavioral and neuroendocrine stress coping strategies in pregnant, parturient, and lactating rats. *Prog. Brain Res.* 133, 143–152.

Neumann, I. D., Bosch, O. J., Toschi, N., Torner, L., and Douglas, A. J. (2003). No stress responses of the hypothalamo–pituitary–adrenal axis in parturient rats: Lack of involvement of brain oxytocin. *Endocrinology* 144, 2473–2479.

Numan, M., and Insel, T.r. (2003). *The Neurobiology of Parental Behavior*, Springer-Verlag, New York, NY.

Numan, M., and Numan, M. J. (1995). Importance of pup-related sensory inputs and maternal performance for the expression of fos-like immunoreactivity in the preoptic area and ventral bed nucleus of the stria terminalis of postpartum rats. *Behav. Neurosci.* 109, 135–149.

Parsons, T. D., Thompson, E., Buckwalter, D. K., Bluestein, B. W., Stanczyk, F. Z., and Buckwalter, J. G. (2004). Pregnancy history and cognition during and after pregnancy. *Int. J. Neurosci.* 114, 1099–1110.

Pawluski, J. L., and Galea, L. A. (2005). Hippocampal morphology is differentially affected by reproductive experience in the mother. *J. Neurobiol.* 66, 71–81.

Rosenblatt, J. S., Mayer, A. D., and Siegel, H. I. (1985). Maternal behavior among nonprimate mammals. In *Handbook of Behavioral Biology (Vol. 7), Reproduction* (N. Adler, D. Pfaff, and R. W. Goy, Eds.), pp. 229–298. Plenum Press, New York.

Stafisso-Sandoz, G., Polley, G., Holt, E., Lambert, K. G., and Kinsley, C. H. (1998). Opiate disruption of maternal behavior: Morphine reduces, and naloxone restores, c-fos activity I the medial preoptic area of lactating rats. *Brain Res. Bull.* 45, 307–313.

Sullivan, R. (2004). *Rats: Observations on the History and Habitat of the City's Most Unwanted Inhabitants*. Bloomsbury, New York.

Velando, A., Drummond, H., and Torres, R. (2006). Senescent birds redouble reproductive effort when ill: Confirmation of the terminal investment hypothesis. *Proc. Biol. Sci.* 273, 1443–1448.

Wartella, J., Amory, E., Macbeth, A. H., McNamara, I., Stevens, L., Lambert, K. G., and Kinsley, C. H. (2003). Single or multiple reproductive experiences attenuate neurobehavioral stress and fear responses in the female rat. *Physiol. Behav.* 79, 373–381.

32 THE ROLE OF REPRODUCTIVE EXPERIENCE ON HIPPOCAMPAL FUNCTION AND PLASTICITY

JODI L. PAWLUSKI AND LIISA A. M. GALEA

Program in Neuroscience, Department of Psychology and Brain Research Centre, University of British Columbia, Vancouver, BC, Canada.

INTRODUCTION

Approximately 75% of women report short-term memory loss, forgetfulness, disorientation, confusion, lack of concentration, or reading difficulties during late pregnancy and early postpartum (Poser *et al.*, 1986; Parsons & Redman, 1991; Crawley *et al.*, 2003). Anecdotal reports of "baby brain" appear to be substantiated with research findings of decreased verbal recall and spatial ability during late pregnancy and the early postpartum period (Keenan *et al.*, 1998; Buckwalter *et al.*, 1999; Galea *et al.*, 2000; De Groot *et al.*, 2006). To date only one study has aimed to determine the long-term implications of pregnancy and the early postpartum period on cognition in women (Buckwalter *et al.*, 2001). Buckwalter *et al.* (2001) found that at 2 years postpartum, there is an improvement in cognitive abilities compared to late pregnancy and the early postpartum period, suggesting possible enhancing effects of motherhood on memory. Similar findings have been found in animal models; thus *pregnancy* is associated with impaired spatial memory in both primates and rodents (Buckwalter *et al.*, 1999; Galea *et al.*, 2000), while *motherhood* is associated with enhanced spatial memory in rodents (Kinsley *et al.*, 1999).

The effect of the steroid hormones, estradiol and corticosterone, on spatial ability have been well documented (Bowman *et al.*, 2001; Dohanich, 2002), and these hormones are remarkably altered during pregnancy and motherhood (Yoshinaga *et al.*, 1969; Stern *et al.*, 1973; Rosenblatt *et al.*, 1979; Garland *et al.*, 1987). Notably, estradiol increases during pregnancy and decreases across the postpartum period (Yoshinaga *et al.*, 1969; Rosenblatt *et al.*, 1979; Garland *et al.*, 1987; Hapon *et al.*, 2003), while corticosterone increases during late pregnancy and is elevated during the postpartum period (Gala & Westphal, 1965; Stern *et al.*, 1973; Fischer *et al.*, 1995; Atkinson & Waddell, 1995) in the rodent. Interestingly, spatial memory relies in part on the integrity of the hippocampus, an area of the brain that shows remarkable plasticity and is sensitive to the effects of steroid hormones. The hippocampal structure consists of three main regions: cornu ammonis 1 (CA1), cornu ammonis 3 (CA3), and the dentate gyrus (Figure 32.1), and all three regions exhibit marked changes in response to steroid hormones (Galea *et al.*, 2006; Mirescu & Gould, 2006). For example, estradiol increases dendritic spine density in the CA1 region of the hippocampus (Woolley *et al.*, 1990), enhances adult neurogenesis in the dentate gyrus (Tanapat *et al.*, 1999; Ormerod & Galea, 2001a, b), and differentially regulates both spatial working and reference memory (Dohanich, 2002; Holmes *et al.*, 2002). High levels of corticosterone and/or chronic stress decrease dendritic arbors in the CA3 region of the hippocampus (Galea *et al.*, 1997; Magarinos *et al.*, 1998), decrease adult

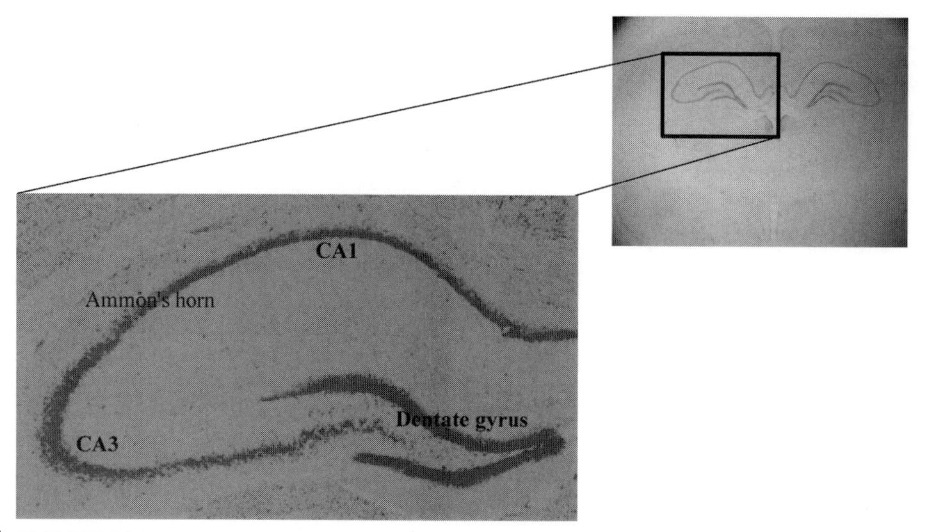

FIGURE 32.1 Representative photo micrograph of the hippocampal structure. The hippocampus consists of three main regions: cornu ammonis 1 (CA1), cornu ammonis 3 (CA3), and the dentate gyrus.

neurogenesis in the dentate gyrus (Cameron & Gould, 1994; Wong & Herbert, 2004), and paradoxically enhance spatial memory in the female rat (Bowman *et al.*, 2001). Given that pregnancy and the postpartum are accompanied by remarkable changes in steroid hormones, including corticosterone (Rosenblatt *et al.*, 1979; Stern *et al.*, 1973), and these hormones have profound effects on structural properties of the hippocampus and behaviors mediated by the hippocampus, it is perhaps not surprising that changes in spatial memory evident with motherhood (Kinsley *et al.*, 1999) may be due to changes in hippocampus plasticity and altered steroid hormone levels.

The first maternal experience (primiparity) has remarkably different effects on the mother than subsequent maternal experiences (Svare & Gandelman, 1976; Kinsley & Bridges, 1988; Bridges *et al.*, 1993). Therefore, when considering the effect of pregnancy and motherhood on the brain and behavior of the dam it is important to take into account the role of reproductive experience. Reproductive experience is defined as the number of times a female has been pregnant and given birth: primiparity refers to being pregnant once, multiparity refers to being pregnant multiple times, and nulliparity refers to never being pregnant. In the rodent, reproductive experience has considerable effects on the behavior, brain, and hormone levels of the mother. Although maternal behavior can be induced in virgin females by continuous exposure to pups, which has been termed *sensitization* (Rosenblatt, 1967), maternal responsiveness (Moltz & Robbins, 1965) and postpartum aggression (Svare & Gandelman, 1976) increase with increasing parity, basal prolactin levels are significantly lower with multiparity compared to primiparity (Bridges *et al.*, 1993), neural sensitivity to opioids is reduced with increased parity in the rat (Kinsley & Bridges, 1988), and the number of astrocytes in the medial preoptic area of the hypothalamus is increased with multiparity compared to primiparity (Featherstone *et al.*, 2000). In addition, primiparity has been suggested to be remarkably different than multiparity as it is a time when "maternal memory," the retention of maternal responsiveness as a consequence of prior experience with pups, is formed (Bridges, 1975; Orpen & Fleming, 1987). Sensitization or pup-exposure given to nulliparous rats also has effects on the brains of these females. For example, sensitized rats have increased c-Fos immunoreactivity in the medial preoptic area of the hypothalamus compared to controls (Kalinichev *et al.*, 2000). Therefore, when considering the effect of pregnancy and motherhood on the brain and behavior of the dam it is important to take into account the role of reproductive experience and whether any effects of motherhood on variables of interest may be due to the effects of pregnancy or pup-exposure alone. This chapter summarizes present findingss on how motherhood and reproductive experience affect hippocampus-dependent learning and memory, hippocampus neurogenesis, hippocampal morphology, and steroid hormone levels in the mother.

THE HIPPOCAMPUS AND MOTHERHOOD

Pregnancy and the postpartum period are a time of heightened neural plasticity. These changes in neural plasticity may be associated with changes in learning and memory performance in the mother. In women, Oatridge *et al.* (2002) have shown that total brain size decreases during pregnancy and returns to preconception size during the postpartum period. In rodents, Galea *et al.* (2000) have shown that hippocampal volume is decreased during pregnancy compared to nulliparous females, and Hamilton *et al.* (1977) have shown that cortical thickness is increased immediately after parturition compared to nulliparous females. In addition, the transition to motherhood is associated with the induction of neural circuitry important for maternal responding (Numan, 2007). This "maternal circuit" involves such areas as the hypothalamus, olfactory bulbs, nucleus accumbens and amygdala, (Numan, 2007) and many of these areas form connections with the hippocampus (Witter, 1993; Vertes, 2006).

The hippocampus has not traditionally been documented as an area important for the onset of motherhood and very little work has focused on this region. One of the only recorded works on the role of the hippocampus in motherhood found that bilateral dorsal hippocampal lesions result in marked alterations in maternal behavior, particularly less time spent nursing, poorer nest building, poorer retrieval of pups, and increased maternal cannibalism than neocortical lesioned dams and sham lesioned controls (Kimble, *et al.*, 1967). In addition, disruption of the main fiber tract projecting to the CA1 and CA3 regions of the hippocampus, via fimbria lesions, results in abnormal nest building and pup retrieval (Terlecki & Sainsbury, 1978). These dams built multiple nest sights and retrieved pups to more than one location (Terlecki & Sainsbury, 1978). Thus, although this early work pointed to a role for the hippocampus in maternal behavior, it was not until recently that the possible relationship between the hippocampus and motherhood was revisited.

HIPPOCAMPUS-DEPENDENT SPATIAL LEARNING AND MEMORY PERFORMANCE IN THE MOTHER

It is widely accepted that spatial memory is dependent on the integrity of the hippocampus (Morris *et al.*, 1982; Moscovitch *et al.*, 2005). Spatial ability is differentially affected by a host of variables including, but not limited to, sex of the subject, steroid hormones, stress, age, or a combination of these variables (Galea *et al.*, 1995; Bowman *et al.*, 2001; Holmes *et al.*, 2002; Montaron *et al.*, 2006). Spatial memory can be compartmentalized into working or reference memory. Working memory is defined as the manipulation and retrieval of trial-unique information that is used to guide prospective action (Olton & Papas, 1979), whereas reference memory is defined as a long-term stable memory (Olton & Papas, 1979). The two common tasks used to test spatial memory in rodents are the radial arm maze and the Morris water maze. Both tasks can be used to determine spatial working and reference memory performance. However the working/reference memory version of the radial arm maze is the only within-subject, within-test design that allows for the examination of both spatial working and reference memory (Olton & Papas, 1979).

As mentioned previously, late pregnancy is associated with impaired spatial working memory in both primates and rodents (Galea *et al.*, 2000; De Groot *et al.*, 2006), whereas motherhood is associated with enhanced spatial working and reference memory in rodents (Kinsley *et al.*, 1999). Kinsley *et al.* (1999) were the first to document that motherhood enhances spatial learning and memory in the dam. They found that after 16 days of lactation or pup-exposure, primiparous and sensitized nulliparous rats had shorter latencies compared to non-sensitized, nulliparous rats when tested on a reference memory version of the dry land version of the water maze (Kinsley *et al.*, 1999). They also found that multiparous dams 14 days after weaning had enhanced working memory performance on the first 6 days of testing compared to nulliparous rats when tested on the working memory version of the radial arm maze (Kinsley *et al.*, 1999). However, this early work did not compare the performance of multiparous, primiparous, and nulliparous females on the same task and dams were tested at different time points during lactation and after weaning (weaning occurs 22–28 days after parturition). Therefore, the effect of reproductive experience and maternal experience (until weaning) on the mother's cognitive performance had yet to be determined.

Recently work from Kinsley's laboratory has demonstrated that parous rats exhibit improvements in spatial memory performance long after weaning (Gatewood *et al.*, 2005; Love *et al.*, 2005). Gatewood *et al.* (2005) found that both multiparous and primiparous rats had shorter latencies to complete a reference and working version of the dry land maze compared to nulliparous rats. In addition, they found that multiparous rats had shorter latencies to complete the task than primiparous rats (Gatewood *et al.*, 2005). These findings were evident when rats were repeatedly tested beginning at 6 months of age (3 weeks after weaning) until 24 months of age. Using a similar task to that of Gatewood *et al.* (2005), Love *et al.* (2005) found that primiparous and multiparous rats had shorter latencies to complete a reference memory version of the dry land maze compared to nulliparous rats when repeatedly tested at 5 and 13 months of age, but not at 9 months of age. However, *after* 13 months of age, there were no differences among multiparous, primiparous, and nulliparous rats on spatial memory performance (Love *et al.*, 2005). This appears to be due to improved performance of nulliparous rats (Love *et al.*, 2005). However, it should be noted that these studies (Kinsley *et al.*, 1999; Gatewood *et al.*, 2005; Love *et al.*, 2005) used latency, and not path length, to determine spatial memory performance. Therefore, it is possible that motor ability, and not memory ability, accounted for the differences observed among nulliparous, primiparous, and multiparous groups.

Recently we have shown that when tested on the working/reference memory version of the radial arm maze at the time of weaning or long after the time of weaning, primiparous, and to a lesser extent multiparous, rats display enhanced spatial reference and working memory performance compared to nulliparous rats (Pawluski *et al.*, 2006a, b; Figure 32.2). This enhanced spatial memory performance was not due to gross differences in motor ability and was not affected by the reproductive state of proestrus (Pawluski *et al.*, 2006 a, b). These findings are consistent with Lemaire *et al.* (2006) who found that primiparous rats exhibited improved reference memory performance compared to nulliparous controls shortly after weaning and up to 16 months later.

Enhanced spatial cognition with motherhood, and especially with primiparity, in rats may be needed for the mother to efficiently locate and retrieve food. In the rodent, food caching activity in the mother is 3 times higher during the first 20 days of mothering (Calhoun, 1963) suggesting that effective and efficient foraging behavior may be required to optimize the time caring for and protecting young. Once reproduction is successful and pups have dispersed from the nest, subsequent maternal experience may be less demanding as the mother has learned the essential maternal behaviors needed for reproductive success. Thus, with subsequent mothering, the induction of maternal behaviors may be rapid, and the mother may not need to spend as much time with her offspring. Hence, the need to efficiently forage is not as great a requirement.

Recent work has begun to investigate the role of hormones, pregnancy, and pup-exposure in mediating changes in spatial learning and memory associated with motherhood (Tomizawa *et al.*, 2003; Lambert *et al.*, 2005; Pawluski *et al.*, 2006a, b; Lemaire *et al.*, 2006). Work by Tomizawa *et al.* (2003) suggests that oxytocin is important for the improvements in reference memory, but not working memory, seen with motherhood. It was shown that nulliparous mice administered oxytocin had improved spatial reference memory compared to nulliparous controls, whereas multiparous mice administered an oxytocin antagonist had impaired reference memory compared to multiparous controls (Tomizawa *et al.*, 2003). Although this work points to a possible role for oxytocin on enhanced spatial reference memory in mothers, it did not compare parous dams with nulliparous females and did not account for a full complement of maternal behaviors as all multiparous mice were tested 3 days after parturition (Tomizawa *et al.*, 2003).

The separate effect of pregnancy and pup-exposure on the improvements in spatial memory with motherhood has also been investigated (Kinsley *et al.*, 1999; Lambert *et al.*, 2005;

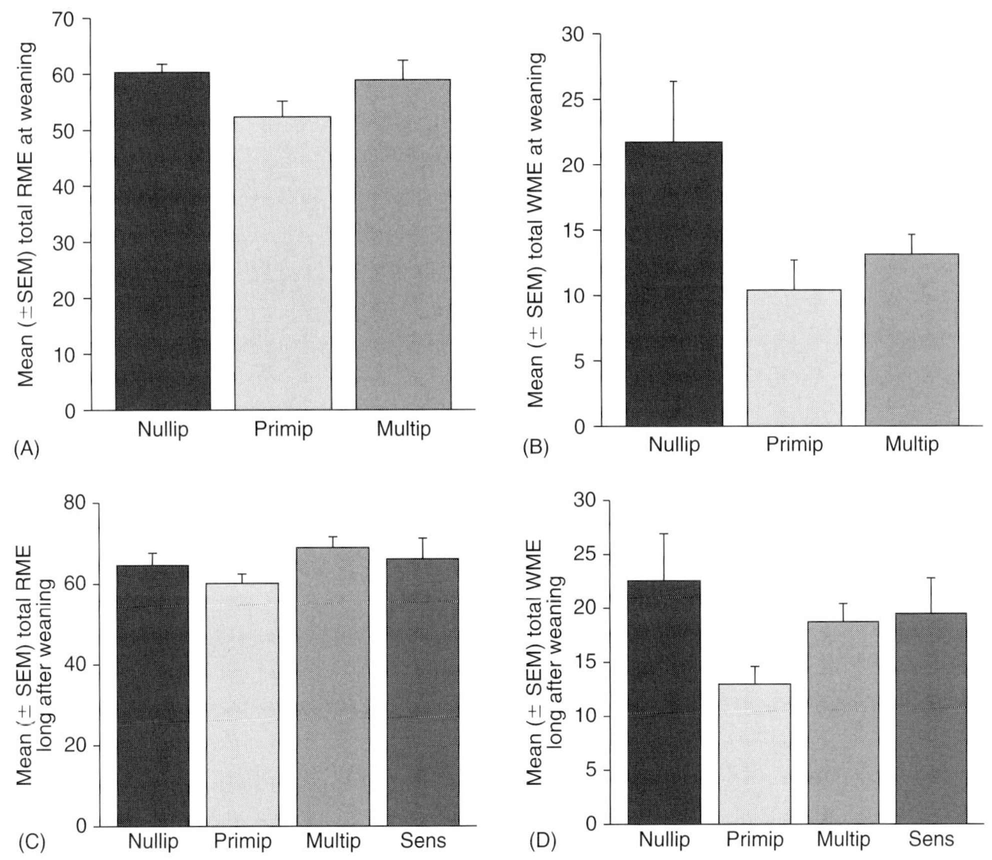

FIGURE 32.2 Mean (±SEM) number of reference memory errors (RME) and working memory errors (WME) with reproductive experience (A, B) at the time of weaning (rats started testing at the time of weaning, testing continued for 30 days after weaning) and (C, D) long after the time of weaning (rats began training on the radial arm maze 60 days after weaning and continued training for another 30 days). Primiparous dams show the greatest improvements in spatial reference and working memory at the time of weaning and long after the time of weaning. Nullip, nulliparous; Primip, primiparous; Multip, multiparous; Sens, sensitized nulliparous. (*Source*: Modified figure from Pawluski *et al.*, 2006a, b, reprinted with permission.)

Pawluski *et al.*, 2006b). As mentioned previously, repeated pup-exposure to nulliparous females, resulting in pup-induced maternal care or sensitization, enhances reference memory performance when compared to nulliparous females (Kinsley *et al.*, 1999). In addition, Lambert *et al.* (2005) have shown that primiparous and nulliparous females exposed to pups for 21 days have some improvements on reference memory performance. These findings suggest that the enhancement in spatial memory with motherhood is due to cues associated with pup-exposure, and/or pup-directed maternal behaviors, and not pregnancy alone. However, the mnemonic benefit from pup-exposure may be short-lived as we have recently shown that the enhancement in spatial memory with 21 days of pup-exposure is not evident long after pup-exposure (Pawluski *et al.*, 2006b). Pawluski *et al.* (2006b) found that when testing *began* 35 days after weaning, sensitized females did not show the same enhancement in learning and memory performance as primiparous dams (Figure 32.2). In addition, parous dams that had pups removed within 24 h of birth failed to complete the task on significantly more days than all other groups (Pawluski *et al.*, 2006b; Figure 32.3). This suggests that the combination of pregnancy, parturition, and active motherhood are all important factors in the alteration in memory performance in the dam long after the time of weaning.

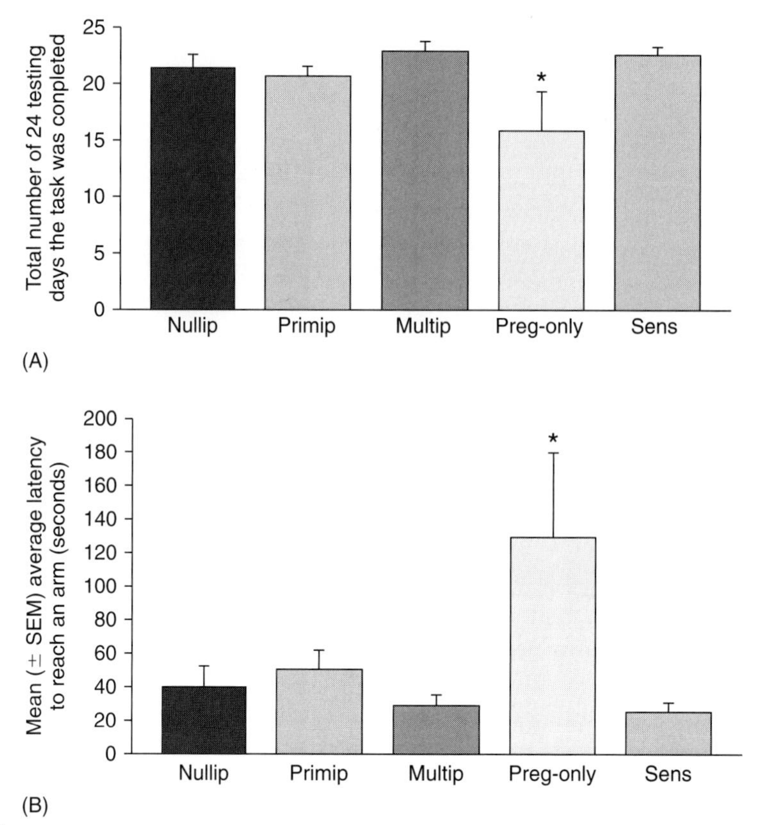

FIGURE 32.3 Mean (±SEM) total (A) number of days the spatial memory task was completed and (B) average latency to enter an arm during testing. Pregnant-only females completed the task on significantly fewer days than all other groups and took significantly longer to enter an arm on the working/reference memory version of the radial arm maze. Nullip, nulliparous; Primip, primiparous; Multip, multiparous; Sens, sensitized; Preg-only, pregnant only (*Source*: Modified figure from Pawluski *et al.*, 2006b, reprinted with permission.)

HIPPOCAMPAL PLASTICITY IN THE MOTHER

As mentioned previously, there is a large degree of neural plasticity with the onset of motherhood (Oatridge *et al.*, 2002; Numan, 2007). Nonetheless, hippocampus plasticity in the mother has only recently been investigated (Tomizawa *et al.*, 2003; Kinsley *et al.*, 2006; Pawluski & Galea, 2006; Lemaire *et al.*, 2006). Tomizawa *et al.* (2003) were the first to demonstrate that electrophysiology in the hippocampus is altered with motherhood. They found that multiparous mice exhibit increased long-lasting long-term potentiation (LTP) along the Schaffer collaterals during the early postpartum period (3 days) compared to nulliparous mice. Interestingly, Tomizawa *et al.* (2003) suggest that oxytocin may be an important mediator of this effect, as oxytocin enhances LTP in the hippocampus of nulliparous mice (Tomizawa *et al.*, 2003). Further research has shown that this increased hippocampal LTP with motherhood exists 2 weeks after weaning and well into aging (Lemaire *et al.*, 2006).

Hippocampus dendritic morphology is also altered with motherhood and reproductive experience (Kinsley *et al.*, 2006; Pawluski & Galea, 2006). Kinsley *et al.* (2006) reported that late pregnant and early lactating dams have increased dendritic spine densities in the CA1 apical region of the hippocampus which could be accounted for by increased levels of estradiol and progesterone during late pregnancy. Other studies have shown that CA1 apical spine density is induced by high levels of estradiol and progesterone in the virgin female rat (Woolley *et al.*, 1990; Woolley & McEwen, 1993).

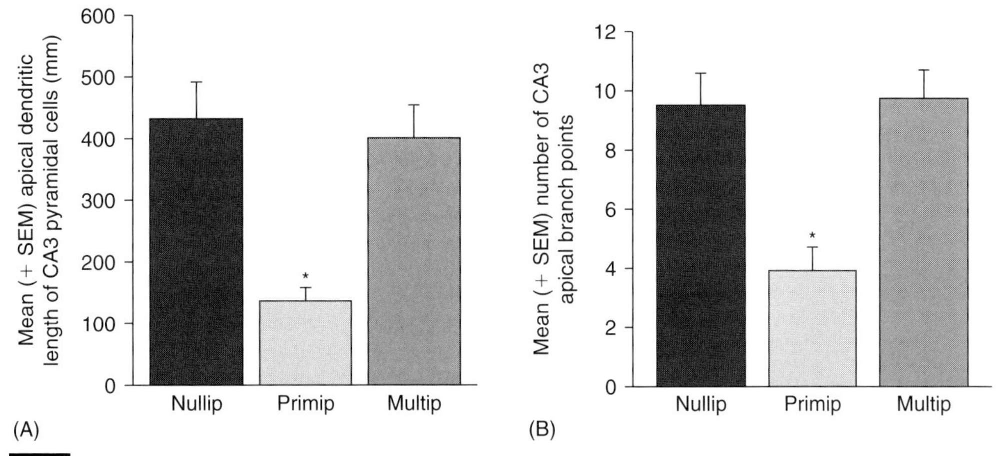

FIGURE 32.4 Mean (± SEM) (A) dendritic length and (B) number of branch points in the apical reagion of CA3 pyramidal neurons in the hippocampus of nulliparous, primiparous, and multiparous rats at the time of weaning. Primiparous rats have shorter dendrites and fewer branch points on CA1 and CA3 pyramidal neurons than multiparous and nulliparous rats. Nullip, nulliparous; Primip, primiparous; Multip, multiparous. (*Source*: Modified figure from Pawluski & Galea, 2006, reprinted with permission; Copyright © 2006 Wiley Periodicals, Inc.)

We have shown that primiparous rats, at weaning, exhibit decreased dendritic lengths and number of branch points in the CA1 and CA3 pyramidal neurons of the hippocampus compared to nulliparous and multiparous rats (Pawluski & Galea, 2006; Figure 32.4). In this same study, multiparous rats were found to have enhanced spine density in the CA1 region of the hippocampus compared to nulliparous and primiparous rats, and spine density correlated with number of male pups in a litter (Pawluski & Galea, 2006). Interestingly, changes in the morphology of CA1 pyramidal neurons with parity are not evident in aged rats, suggesting that these changes are reversible and not permanent (Love *et al.*, 2005).

The dentate gyrus of the hippocampus is one of two areas in the brain where adult neurogenesis occurs at a high rate (Altman, 1962; Cameron & McKay, 2001: the other area being the subventricular zone, SVZ). Adult neurogenesis in the hippocampus was discovered in 1962 (Altman, 1962) and although research is continually investigating the mechanisms responsible for changes in neurogenesis as well as the function of these new neurons, our knowledge of these processes is limited. Neurogenesis consists of at least two processes: *cell proliferation*, defined as the production of new cells, and *cell survival*, defined as new cells that survive to maturity. Factors that affect cell proliferation either suppress or induce mitosis in precursor cells and factors that affect cell survival either promote or prevent the differentiation and/or maturation of labeled cells into mature neurons (Ormerod & Galea, 2001a, b). Therefore, increasing cell proliferation as well as enhancing the survival of new cells can increase the number of new neurons.

The regulation of adult hippocampal neurogenesis is affected by a number of factors including sex of the subject, steroid hormones, stress, age, and environmental enrichment (for review see: Galea, in press; Mirescu & Gould, 2006; Olson *et al.*, 2006). The steroid hormones estradiol and corticosterone differentially affect cell proliferation and cell survival in the adult hippocampus. For example, high estradiol increases cell proliferation (Tanapat *et al.*, 1999), but this effect is time-dependent, such that estradiol increases (within 4 h) and then decreases (within 48 h) cell proliferation in the dentate gyrus (Ormerod & Galea, 2001a, b). Estradiol also enhances cell survival independently of cell proliferation (Ormerod *et al.*, 2004). Interestingly, the mechanism of estradiol's suppression of cell proliferation is, at least, partially dependent on the stimulation of adrenal steroids by estradiol (Ormerod *et al.*, 2003). Both exposure to stress and elevated levels of corticosterone, suppress both cell proliferation and survival (Cameron & Gould, 1994; Wong & Herbert, 2004; Mirescu & Gould, 2006) and luteinizing hormone stimulates

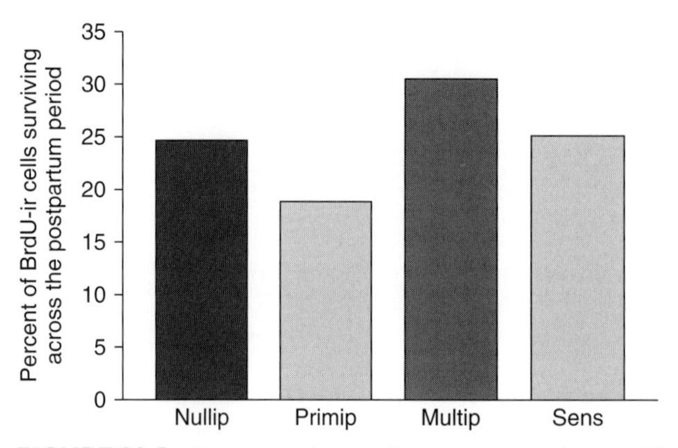

FIGURE 32.5 Percentage of new cells surviving across lactation. This was calculated by dividing the density of cells surviving 21 days after BrdU injection by the density of cells proliferating 24 h after BrdU injection. Multiparous rats had more new cells surviving than nulliparous, primiparous, and sensitized rats. Nullip, nulliparous; Primip, primiparous; Multip, multiparous; Sens, sensitized (*Source*: Modified figure from Pawluski & Galea, 2007, reprinted with permission.)

cell proliferation in the dentate gyrus of female mice (Mak *et al.*, 2007).

Given the dramatic changes in steroid hormones during pregnancy and postpartum together with altered hippocampus-dependent learning and memory in the mother, it is perhaps not surprising that pregnancy and motherhood may alter adult neurogenesis in the hippocampus of the dam. However, very little research has investigated hippocampal adult neurogenesis with reproductive experience. Pregnancy has been suggested to enhance cell survival in the dentate gyrus (Banasr *et al.*, 2001), but has no significant effect on cell proliferation in the dentate gyrus of the dam on gestation days 7 and 21 (Shingo *et al.*, 2003; Furuta & Bridges, 2005). However, increased cell proliferation is evident in the SVZ of the dam during pregnancy and the early postpartum period compared to nulliparous controls (Shingo *et al.*, 2003; Furuta & Bridges, 2005). The postpartum period has recently been associated with decreased cell proliferation in the hippocampus on postpartum day 2 and 8 compared to diestrous, nulliparous controls (Darnaudery *et al.*, 2007; Leuner *et al.*, 2007). These effects were found to be dependent on pup-exposure and adrenal steroid levels (Leuner *et al.*, 2007).

Recently we have investigated the role of reproductive experience and pup-exposure on cell proliferation and cell survival in the hippocampus of the dam during the postpartum period (Pawluski & Galea, 2007). We tested 5 groups of rats: nulliparous, primiparous, multiparous, primiparous with pups removed after 24 h, and nulliparous given pup-exposure to determine whether the decrease in cell proliferation was due to pregnancy or mothering alone. We administered bromodeoxyuridine (BrdU), a DNA synthesis marker which labels dividing cells and their progeny, on day 2 after parturition and perfused rats either 24 h later (cell proliferation) or 21 days later (cell survival). Our results demonstrate that primiparity and multiparity result in decreased hippocampal cell proliferation during the early postpartum period compared to nulliparity, and primiparity with or without pup-exposure results in decreased cell survival across the postpartum period compared to nulliparity, multiparity, and pup-exposure alone (Pawluski & Galea, 2007). Thus, results demonstrate that reproductive experience differentially alters hippocampal neurogenesis in the dam, with more pronounced effects in primiparous dams. Interestingly, compared to initial cell proliferation levels, multiparous dam had the greatest percentage of new neurons surviving in the hippocampus during the postpartum period (Figure 32.5).

The opposing findings of cell proliferation rates during the postpartum period in the dentate gyrus and the SVZ are likely due to the fact that the olfactory bulb (the destination of cells migrating from the SVZ) and the hippocampus mediate very different processes in the dam during the early postpartum period. The olfactory bulbs are crucial for appropriate maternal responding and play a major role in the "maternal circuit" (Fleming and Rosenblatt, 1974; Kinsley & Bridges, 1990; Numan, 2007), while the hippocampus, is required for spatial navigation (Morris *et al.*, 1982; Moser *et al.*, 1995). Furthermore, late gestation and the early postpartum period are associated with the induction of maternal behaviors that are dependent on neural regions of the "maternal circuit" such as the olfactory bulbs and the medial preoptic area but not the hippocampus (Numan, 2007).

ENHANCED SPATIAL MEMORY AND DECREASED HIPPOCAMPAL DENDRITIC MORPHOLOGY IN THE MOTHER

It may seem paradoxical that there is improved spatial memory in primiparous dams (Kinsley *et al.*, 1999; Lemaire *et al.*, 2006; Pawluski *et al.*, 2006a) at a time when there are decreased dendritic arborizations on CA1 and CA3 pyramidal neurons (Pawluski & Galea, 2006). However, a similar pattern of enhanced spatial learning and memory and decreased dendritic arbors in the hippocampus is seen in nulliparous female rats exposed to chronic stress (Galea *et al.*, 1997; Bowman *et al.*, 2001). For example, Galea *et al.* (1997) found that chronically stressed female rats have reduced dendritic arbors on the basal dendrites of CA3 pyramidal cells in the hippocampus. Using the same stress paradigm, Bowman *et al.* (2001) found that chronically stressed female rats exhibit improved spatial memory compared to non-stressed female rats. Therefore, enhanced spatial memory with primiparity (Pawluski *et al.*, 2006a), at a time when there is hippocampal dendritic atrophy (Pawluski & Galea, 2006), is similar to what is seen in chronically stressed females (Galea *et al.*, 1997; Bowman *et al.*, 2001). Furthermore, the effect of chronic stress on hippocampal structure and function in the female is likely due to elevated corticosterone levels and decreased corticosteroid-binding globulin (CBG) (Galea *et al.*, 1997; Magarinos *et al.*, 1998). Because primiparity also results in increased corticosterone and decreased CBG levels during the postpartum period (Pawluski *et al.*, under revision), this suggests a similar mechanism may be behind the enhanced spatial memory and altered hippocampal morphology. It should be noted that the similar pattern of improved hippocampal spatial memory and decreased dendritic arbors in chronically stressed females and primiparous dams (Galea *et al.*, 1997; Bowman *et al.*, 2001; Pawluski & Galea, 2006; Pawluski *et al.*, 2006a) does not indicate that the initial experience of motherhood (primiparity) is analogous to the experience of chronic stress. However, the first maternal experience may be considered a type of stress that is required for the induction of neural and behavioral processes needed for the onset of maternal behavior and offspring survival. In support of this, corticosterone appears to be important for the modulation of maternal memory (Graham *et al.*, 2006).

Hippocampal plasticity and spatial learning and memory are not always linked as multiparous rats do not show the dendritic atrophy in the CA1 and CA3 region of the hippocampus (Pawluski & Galea, 2006), while at the same time multiparous dams exhibit enhanced working memory at weaning compared to nulliparous rats (Pawluski *et al.*, 2006a). Thus, it is possible that different aspects of hippocampal plasticity are altered with reproductive experience and these aspects differentially affect hippocampus-dependent spatial memory. For example, multiparous rats have increased spine density in CA1 pyramidal neurons of the hippocampus at weaning compared to primiparous and nulliparous rats (Pawluski & Galea, 2006). Therefore, increased synaptic transmission and not altered dendritic structure *per se* may be responsible for these changes in learning and memory performance with multiparity. In accordance with this, Tomizawa *et al.* (2003) report greater long-lasting LTP along the Schaffer collateral fibers, which synapse onto the CA1 pyramidal cells, in multiparous mice during the early postpartum period (3–6 days) compared to nulliparous females (unfortunately primiparous mice were not tested). Using a similar paradigm to Tomizawa *et al.* (2003), Lemaire *et al.* (2006) report increased LTP, although to a lesser extent, along the Schaffer collaterals in primiparous rats 2 weeks after weaning and 16 months later compared to nulliparous females. Taken together the findings of Tomizawa *et al.* (2003) and Lemaire *et al.* (2006) demonstrate that LTP is differentially affected at different timepoints during motherhood. However, to date no work has investigated whether there are differences between multiparous, primiparous, and nulliparous rats on measures of LTP during the postpartum period, at the time of weaning, or long past the time of weaning. Thus, it is possible that select aspects of hippocampal plasticity are differentially altered with reproductive experience which in turn may have differential effects on hippocampus-dependent spatial memory.

ENHANCED SPATIAL MEMORY AND DECREASED HIPPOCAMPAL NEUROGENESIS IN THE MOTHER

New neurons produced in the hippocampus of adult rodents are considered functional as they make appropriate synaptic connections and are electrophysiologically active (Zhao *et al.*, 2006). The behavioral function of new neurons in the hippocampus of adult rodents is more controversial, but they appear to play a role in learning and memory (Winocur *et al.*, 2006). Significantly reducing neurogenesis in the dentate gyrus of the hippocampus using the antimitotic agent methylazomethanol acetate (MAM) or focal irradiation results in deficits in performance on hippocampus-dependent tasks (Shors *et al.*, 2002; Winocur *et al.*, 2006). However, it appears that an optimal amount of hippocampal neurogenesis is needed for improved spatial memory. For example, very low or very high levels of hippocampal neurogenesis in rodents, which are evident after cortical and hippocampal lesions (Cameron *et al.*, 1995; Gould & Tanapat, 1997) and seizures (Scarfman, 2004), are associated with reduced, not improved, hippocampus-dependent learning and memory (McNamara *et al.*, 1992; Leung & Shen, 2006).

Decreased cell survival in the hippocampus preceeds increased spatial memory performance in parous dams, which may also seem paradoxical. However, given that there may be an optimal level of neurogenesis for improved spatial memory, it is perhaps not surprising that lower levels of hippocampal neurogenesis in primiparous rats across the postpartum period (Pawluski & Galea, 2007) preceed improved hippocampus-dependent learning and memory at the time of weaning (Pawluski *et al.*, 2006a). In addition, the less extensive decrease in cell survival with multiparity is coincident with slight (and not significant) improvements in spatial memory performance at weaning (Pawluski *et al.*, 2006a; Pawluski & Galea, 2007). This suggests that there may exist an optimal rate of neurogenesis for spatial memory performance in mothers. Additionally, it may be possible that there are differences in adult neurogenesis in the hippocampus of the dam at other time periods during the postpartum period, and these new cells may be differentially involved in spatial learning and memory. Although Leuner *et al.* (2007) demonstrated that there is no significant difference in hippocampal cell proliferation in parous dams at weaning or 2 weeks after weaning compared to nulliparous females, there may be differences in other aspects of hippocampal neurogenesis such as cell survival during this time. Alternatively, neurogenesis may have a minimal role in enhanced spatial memory post-weaning.

It is important to understand the functional role of new neurons in the hippocampus of the mother. For example, primiparous rats with pups removed 24 h after the BrdU injection exhibit the same density of new neurons surviving in the hippocampus across the postpartum period as primiparous rats with pup-exposure (Pawluski & Galea, 2007). However, primiparous rats without pup-exposure were remarkably slower to complete a spatial memory task when tested 8.5 weeks after parturition and pup removal (Pawluski *et al.*, 2006b; Figure 32.3). This illustrates an uncoupling of hippocampal neurogenesis with spatial learning. Furthermore some researchers have suggested that hippocampal neurogenesis is more important for long-term memory and is not important for learning (Snyder *et al.*, 2005); different aspects of memory may be affected by the changes in hippocampal neurogenesis with reproductive experience. Further work is needed to determine the relationship between neurogenesis and hippocampus-dependent spatial learning and memory in the dam.

PERSISTENCE OF IMPROVED SPATIAL MEMORY WITH MOTHERHOOD

Recent findings demonstrate that enhanced spatial memory in motherhood persists past the time of weaning in primiparous rats (Gatewood *et al.*, 2005; Love *et al.*, 2005; Lemaire *et al.*, 2006; Pawluski *et al.*, 2006b). However, the mechanisms underlying this improvement in spatial memory with primiparity are unknown. It is also possible that the enhanced LTP may be important for improved spatial memory in the dam post-weaning. Lemaire *et al.* (2006) found that LTP along the Schaffer collaterals is enhanced in primiparous dams compared to nulliparous females at the same time points that dams have improved spatial reference memory performance compared to nulliparous females (2 weeks and 16 months after weaning).

Furthermore, aging-related deficits may be reduced with reproductive experience as Gatewood *et al.* (2005) have demonstrated that reproductive experience results in decreased amyloid precursor protein immunoreactivity, a marker of neurodegeneration, in the hippocampus of aged (18-month old) primiparous and multiparous dams compared to nulliparous females. Enhanced spatial memory performance was associated with lower amyloid precursor protein immunoreactivity in the hippocampus (Gatewood *et al.*, 2005). Together, these findings suggest that the persistent effect of motherhood on spatial memory is associated with changes in hippocampus structure and plasticity.

ALTERED SPATIAL MEMORY AND HIPPOCAMPAL NEUROGENESIS WITH AND WITHOUT PUP EXPOSURE

The alterations in hippocampal-dependent spatial memory and neural plasticity with motherhood, and primarily primiparity, appear to be due to the combined effect of pregnancy and motherhood, and not pregnancy or pup-exposure alone. Indeed, the effects of pregnancy alone or pup-exposure alone often result in dramatically different responses than those seen with primiparity. For example, pup-exposure to nulliparous rats results in increased cell proliferation and increased cell death, with no significant difference in cell survival, while primiparity results in decreased cell proliferation and decreased cell survival across the postpartum period, regardless of pup-exposure (Pawluski & Galea, 2007). This increased cell proliferation with pup-exposure in nulliparous, but not primiparous, rats suggest that pup-exposure acts as a type of enriched environment for nulliparous females (Pawluski & Galea, 2007). In addition, pregnancy-alone results in dramatically different hippocampus-dependent spatial memory performance, with pregnant-only females failing to complete the task on significantly more of the days than all other groups (Pawluski *et al.*, 2006b; Figure 32.3). Thus, parous females with pups removed within 24 h of birth are affected by the loss of pups more than 60 days later. Furthermore, dams with pups removed for long periods of time throughout the postpartum period exhibit depressive-like behavior compared to dams that did not have pups removed (Boccia *et al.*, 2007). Hence, it appears that parous dams that are given minimal exposure to pups throughout the postpartum period may exhibit a variety of cognitive and behavioral deficits.

POSSIBLE ROLE OF CORTICOSTERONE IN SPATIAL MEMORY AND HIPPOCAMPAL NEUROGENESIS IN THE MOTHER

Pregnancy and motherhood are accompanied by a host of changes in steroid and peptide hormones (Gala & Westphal, 1965; Yoshinaga *et al.*, 1969; Stern *et al.*, 1973; Rosenblatt *et al.*, 1979; Nicholas & Hartmann, 1981) that may play an important role in mediating changes in hippocampal function and plasticity in the mother. Certainly the induction of maternal behavior is dependent on changing levels of estrogens, progesterone, and prolactin during pregnancy (Mann and Bridges, 2001) and reproductive experience may also affect the levels of these hormones. For example, we have found that primiparity results in increased corticosterone levels during the early postpartum period and decreased CBG levels mid-lactation compared to multiparity (Pawluski *et al.*, under revision). Given that there are increased levels of corticosterone and decreased levels of CBG with reproductive experience that precede the cognitive and neural effects of motherhood it is possible that corticosterone may play a role in mediating these effects in the dam. The relationship between increased total corticosterone, decreased CBG, and hippocampus function and plasticity is perhaps not surprising, as previous research have found a similar relationship in chronically stressed female rats. For example, the work of Bowman *et al.* (2001) suggest that prolonged periods of high levels of corticosterone and/or low CBG (Galea *et al.*, 1997) as a result of chronic stress are associated with improved spatial memory in the nulliparous female. This enhanced spatial memory after 21 days of chronic stress in the virgin female may be due to a peak in corticosterone during the early days of the chronic stress, and decreased CBG levels a few days later (Galea *et al.*, 1997; Bowman *et al.*, 2001). This same profile of an initial elevation in corticosterone followed by decreased CBG is evident in primiparous and, to a lesser extent, multiparous rats across the postpartum period (Pawluski *et al.*, under revision). Taken together corticosterone and the availability of corticosterone to target tissues via decreased CBG levels, may play an important role in improved spatial memory at the time of weaning (Pawluski *et al.*, 2006a).

The persistent enhancement in spatial memory after the time of weaning in primiparous dams (Pawluski *et al.*, 2006b) suggests that pregnancy and lactation have permanent effects on the neural circuitry in the dam that influence hippocampal learning as well as maternal behavior. In accordance with this idea, recent work by Bridges & Byrnes (2006) have shown that there are permanent changes in the hypothalamic–pituitary–gonadal (HPG) axis after weaning which result in decreased levels of estradiol at the peak of proestrus in primiparous rats compared to cycling virgin females.

It also seems plausible that increased corticosterone and decreased CBG levels during the postpartum period in the mother are responsible for decreased adult neurogenesis in the hippocampus of the dam. In support of this adrenalectomy and low level of corticosterone replacement eliminates the suppression of cell proliferation during the early postpartum period in the hippocampus of the primiparous dam (Leuner *et al.*, 2007). However, additional research is needed to determine the role of corticosterone on adult hippocampal cell proliferation and cell survival in the adult female.

The action of corticosterone on hippocampal function and plasticity in the mother may also be due to differences in the density and binding capacity of corticosteroid receptors in the hippocampal formation. For example, improved spatial memory performance in chronically stressed nulliparous females is associated with increased glucocorticoid receptor immunoreactivity in CA1 area of the hippocampus and increased mineralocorticoid receptor immunoreactivity in the CA3 area of the hippocampus compared with non-stressed females (Kitraki *et al.*, 2004). In addition, the action of corticosterone on the suppression of cell proliferation in the hippocampus of adult male rats is dependent on both mineralocorticoid and glucocorticoid receptors (Montaron *et al.*, 2003). To date, investigations of similar effects in adult female rats have not been conducted. In lactating rats, however, the binding capacity of glucocorticoid receptors in the hippocampus is decreased (Meaney *et al.*, 1989), suggesting that changes in corticosteroid receptor density and binding capacity in the hippocampus may play a role in corticosterone action on hippocampus-dependent spatial memory and hippocampal neurogenesis in the dam.

POSSIBLE ROLE OF PEPTIDE HORMONES ON THE HIPPOCAMPUS OF THE MOTHER

Recent work has demonstrated that the actions of the peptide hormones, oxytocin and prolactin, like corticosterone, may play an important role in improved hippocampus-dependent spatial memory and decreased hippocampus neurogenesis in the dam. Oxytocin, a hormone which increases in response to milk ejection (Buhimschi, 2004), is associated with enhanced reference memory and increased long-lasting LTP in the hippocampus of the dam shortly after parturition (PD 3–6: Tomizawa *et al.*, 2003). In addition, prolactin, which increases near the end of pregnancy and throughout the postpartum period (Grattan, 2001) and is essential for lactation (Buhimschi, 2004), is thought to mediate increased cell proliferation in the SVZ during the early postpartum period (Shingo *et al.*, 2003). However, prolactin has not been shown to affect cell proliferation in the hippocampus of the nulliparous female (Shingo *et al.*, 2003). To date, investigations are lacking as to the role of oxytocin in adult hippocampal neurogenesis and hippocampal cell morphology as well as the role of prolactin on spatial learning and memory performance. More importantly, how these peptide hormones may interact with basal corticosterone and CBG levels to alter hippocampus-dependent behavior and hippocampal neuroplasticity in the dam is unknown.

CONCLUSIONS

Reproductive experience has dramatic effects on hippocampal function and plasticity. Pregnancy and motherhood exert a marked effect on spatial memory, hippocampal neurogenesis, and hippocampal dendritic structure, effects that appear more pronounced with primiparity. Furthermore changes in corticosterone levels during the postpartum period may play a role in primiparity-induced alterations in hippocampal function and plasticity. In conclusion, the present work demonstrates that the hippocampus, an area not traditionally associated with motherhood, is significantly altered by motherhood and reproductive experience.

REFERENCES

Altman, J. (1962). Are neurons formed in the brains of adult mammals? *Science* 135, 1127–1128.

Atkinson, H. C., and Waddell, B. J. (1995). The hypothalamic–pituitary–adrenal axis in rat pregnancy and lactation: Circadian variation and interrelationship of plasma adrenocorticotropin and corticosterone. *Endocrinology* 136, 512–520.

Banasr, M., Hery, M., Brezun, J. M., and Daszuta, A. (2001). Serotonin mediates oestrogen stimulation of cell proliferation in the adult dentate gyrus. *Eur. J. Neurosci.* 14, 1417–1424.

Boccia, M. L., Razzoli, M., Vadlamudi, S. P., Trumbull, W., Caleffie, C., and Pedersen, C. A. (2007). Repeated long separations from pups produce depression-like behavior in rat mothers. *Psychoneuroendocrinology* 32, 65–71.

Bowman, R. E., Zrull, M. C., and Luine, V. N. (2001). Chronic restraint stress enhances radial arm maze performance in female rats. *Brain Res.* 904(2), 279–289.

Bridges, R. S. (1975). Long-term effects of pregnancy and parturition upon maternal responsiveness in the rat. *Physiol. Behav.* 14, 245–249.

Bridges, R. S., and Byrnes, E. M. (2006). Reproductive experience reduces circulating 17beta-estradiol and prolactin levels during proestrus and alters estrogen sensitivity in female rats. *Endocrinology* 147, 2575–2582.

Bridges, R. S., Felicio, L. F., Pellerin, L. J., Stuer, A. M., and Mann, P. E. (1993). Prior parity reduces post-coital diurnal and nocturnal prolactin surges in rats. *Life Sci.* 53, 439–445.

Buckwalter, J. G., Stanczyk, F. Z., McCleary, C. A., Bluestein, B. W., Buckwalter, D. K., Rankin, K. P., Chang, L., and Goodwin, T. M. (1999). Pregnancy, the postpartum, and steroid hormones: Effects on cognition and mood. *Psychoneuroendocrinology* 24(1), 69–84.

Buckwalter, J. G., Buckwalter, D. K., Bluestein, B. W., and Stanczyk, F. Z. (2001). Pregnancy and post partum: Changes in cognition and mood. *Prog. Brain.Res.* 133, 303–319.

Buhimschi, C. S. (2004). Endocrinology of lactation. *Obstet. Gynecol. Clin. N. Am.* 31, 963–979.

Calhoun, J. B. (Ed.) (1963). The Ecology and Sociology of the Norway Rat. Bethesda, Maryland.

Cameron, H. A., and Gould, E. (1994). Adult neurogenesis is regulated by adrenal steroids in the dentate gyrus. *Neuroscience* 61, 203–209.

Cameron, H. A., and McKay, R. D. (2001). Adult neurogenesis produces a large pool of new granule cells in the dentate gyrus. *J. Comp. Neurol.* 435, 406–417.

Cameron, H. A., McEwen, B. S., and Gould, E. (1995). Regulation of adult neurogenesis by excitatory input and NMDA receptor activation in the dentate gyrus. *J. Neurosci.* 15, 4687–4692.

Crawley, R. A., Dennison, K., and Carter, C. (2003). Cognition in pregnancy and the first year postpartum. *Psychol. Psychother.* 76, 69–84.

de Groot, R. H., Vuurman, E. F., Hornstra, G., and Jolles, J. (2006). Differences in cognitive performance during pregnancy and early motherhood. *Psychol. Med.* 36, 1023–1032.

Darnaudéry, M., Perez-Martin, M., Del Favero, F., Gomez-Roldan, C., Garcia-Segura, L.M., Maccari, S. (2007). Early motherhood in rats is associated with a modification of hippocampal function. *Psychoneuroendocrinology* 32, 803–12.

Dohanich, G. (2002). Gonadal steroids, learning, and memory. In *Hormones, Brain and Behavior* (D. W. Pfaff, Ed.), Vol. 2. Elsevier Science, (USA).

Featherstone, R. E., Fleming, A. S., and Ivy, G. O. (2000). Plasticity in the maternal circuit: Effects of experience and partum condition on brain astrocyte number in female rats. *Behav. Neurosci.* 114, 158–172.

Fischer, D., Patchev, V. K., Hellbach, S., Hassan, A. H., and Almeida, O. F. (1995). Lactation as a model for naturally reversible hypercorticalism plasticity in the mechanisms governing hypothalamo–pituitary–adrenocortical activity in rats. *J. Clin. Invest.* 96, 1208–1215.

Fleming, A.S., Rosenblatt, J.S. (1974). Olfactory regulation of maternal behavior in rats. I. Effects of olfactory bulb removal in experienced and inexperienced lactating and cycling females. *J. Comp. Physiol. Psychol.* 86, 221–232.

Furuta, M., and Bridges, R. S. (2005). Gestation-induced cell proliferation in the rat brain. *Dev. Brain Res.* 156, 61–66.

Gala, R. R., and Westphal, U. (1965). Corticosteroid binding globulin in the rat: Possible role in the initiation of lactation. *Endocrinology* 76, 1079–1088.

Galea, L. A., Kavaliers, M., Ossenkopp, K. P., and Hampson, E. (1995). Gonadal hormone levels and spatial learning performance in the Morris water maze in male and female meadow voles, *Microtus pennsylvanicus*. *Horm. Behav.* 29(1), 106–125.

Galea, L. A., McEwen, B. S., Tanapat, P., Deak, T., Spencer, R. L., and Dhabhar, F. S. (1997). Sex differences in dendritic atrophy of CA3 pyramidal neurons in response to chronic restraint stress. *Neuroscience* 81, 689–697.

Galea, L. A., Ormerod, B., Sampath, S., Kostaras, X., Wilkie, D., and Phelps, M. (2000). Spatial working memory and hippocampal size across pregnancy in rats. *Horm. Behav.* 37, 86–95.

Galea, L. A., Spritzer, M. D., Barker, J. M., and Pawluski, JL. (2006). Gonadal hormone modulation of hippocampal neurogenesis in the adult. *Hippocampus* 16, 225–232.

Galea, L. A. (2008). Gonadal hormone modulation of neurogenesis in the dentate gyrus of adult male and female rodents. *Brain Res Rev.* 57, 332–341.

Garland, H. O., Atherton, J. C., Baylis, C., Morgan, M. R., and Milne, C. M. (1987). Hormone profiles for progesterone, oestradiol, prolactin, plasma renin activity, aldosterone and corticosterone during pregnancy and pseudopregnancy in two strains of rat: Correlation with renal studies. *J. Endocrinol.* 113, 435–444.

Gatewood, J. D., Morgan, M. D., Eaton, M., McNamara, I. M., Stevens, L. F., Macbeth, A. H., Meyer, E. A., Lomas, L. M., Kozub, F. J., Lambert, K. G., and Kinsley, C. H. (2005). Motherhood mitigates aging-related decrements in learning and memory and positively affects brain aging in the rat. *Brain Res. Bull.* 66, 91–98.

Gould, E., and Tanapat, P. (1997). Lesion-induced proliferation of neuronal progenitors in the dentate gyrus of the adult rat. *Neuroscience* 80, 427–436.

Graham, M. D., Rees, S. L., Steiner, M., and Fleming, A. S. (2006). Effects of adrenalectomy and corticosterone replacement on maternal memory in postpartum rats. *Horm. Behav.* 49, 353–361.

Grattan, D. R. (2001). The actions of prolactin in the brain during pregnancy and lactation. *Prog. Brain Res.* 133, 153–171.

Hamilton, W. L., Diamond, M. C., Johnson, R. E., and Ingham, C. A. (1977). Effects of pregnancy and differential environments on rat cerebral cortical depth. *Behav. Biol.* 19, 333–340.

Hapon, M. B., Simoncini, M., Via, G., and Jahn, G. A. (2003). Effect of hypothyroidism on hormone profiles in virgin, pregnant and lactating rats, and on lactation. *Reproduction* 126, 371–382.

Holmes, M. M., Wide, J. K., and Galea, L. A. (2002). Low levels of estradiol facilitate, whereas high levels of estradiol impair, working memory performance on the radial arm maze. *Behav. Neurosci.* 116, 928–934.

Kalinichev, M., Rosenblatt, J. S., Nakabeppu, Y., and Morrell, J. I. (2000). Induction of c-fos-like and fosB-like immunoreactivity reveals forebrain neuronal populations involved differentially in pup-mediated maternal behavior in juvenile and adult rats. *J. Comp. Neurol.* 416, 45–78.

Keenan, P. A., Yaldoo, D. T., Stress, M. E., Fuerst, D. R., and Ginsburg, K. A. (1998). Explicit memory in pregnant women. *Am. J. Obstet. Gynecol.* 179, 731–737.

Kimble, D. P., Rogers, L., and Hendrickson, C. W. (1967). Hippocampal lesions disrupt maternal, not sexual, behavior in the albino rat. *J. Comp. Physiol. Psychol.* 63(3), 401–407.

Kinsley, C. H., and Bridges, R. S. (1988). Parity-associated reductions in behavioral sensitivity to opiates. *Biol. Reprod.* 39, 270–278.

Kinsley, C. H., and Bridges, R. S. (1990). Morphine treatment and reproductive condition alter olfactory preferences for pup and adult odors in female rats. *Develop. Psychobiol.* 23, 331–347.

Kinsley, C. H., Madonia, L., Gifford, G. W., Tureski, K., Griffin, G. R., Lowry, C., Williams, J., Collins, J., McLearie, H., and Lambert, K. G. (1999). Motherhood improves learning and memory. *Nature* 402, 137–138.

Kinsley, C. H., Trainer, R., Stafisso-Sandoz, G., Quadros, P., Marcus, L. K., Hearon, C., Meyer, E. A., Hester, N., Morgan, M., Kozub, F. J., and Lambert, K. G. (2006). Motherhood and the hormones of pregnancy modify concentrations of hippocampal neuronal dendritic spines. *Horm. Behav.* 49, 131–142.

Kitraki, E., Kremmyda, O., Youlatos, D., Alexis, M. N., and Kittas, C. (2004). Gender-dependent alterations in corticosteroid receptor status and spatial performance following 21 days of restraint stress. *Neuroscience* 125(1), 47–55.

Lambert, K. G., Berry, A. E., Griffins, G., Amory-Meyers, E., Madonia-Lomas, L., Love, G., and Kinsley, C. H. (2005). Pup exposure differentially enhances foraging ability in primiparous and nulliparous rats. *Physiol. Behav.* 84, 799–806.

Lemaire, V., Billard, J. M., Dutar, P., George, O., Piazza, P. V., Epelbaum, J., Le Moal, M., and Mayo, W. (2006). Motherhood-induced memory improvement persists across lifespan in rats but is abolished by a gestational stress. *Eur. J. Neurosci.* 23, 3368–3374.

Leuner, B., Mirescu, C., Noiman, L., and Gould, E. (2007). Maternal experience inhibits the production of immature neurons in the hippocampus during the postpartum period through elevations in adrenal steroids. *Hippocampus* 17(6), 434–442.

Leung, L. S., and Shen, B. (2006). Hippocampal CA1 kindling but not long-term potentiation disrupts spatial memory performance. *Learn. Mem.* 13, 18–26.

Love, G., Torrey, N., McNamara, I., Morgan, M., Banks, M., Hester, N. W., Glasper, E. R., Devries, A. C., Kinsley, C. H., and Lambert, K. G. (2005). Maternal experience produces long-lasting behavioral modifications in the rat. *Behav. Neurosci.* 119, 1084–1096.

McNamara, R. K., Kirkby, R. D., dePape, G. E., and Corcoran, M. E. (1992). Limbic seizures, but not kindling, reversibly impair place learning in the Morris water maze. *Behav. Brain Res.* 28, 167–175.

Magarinos, A. M., Orchinik, M., and McEwen, B. S. (1998). Morphological changes in the hippocampal CA3 region induced by non-invasive glucocorticoid administration: A paradox. *Brain Res.* 809, 314–318.

Mak, G. K., Enwere, E. K., Gregg, C., Pakarainen, T., Poutanen, M., Huhtaniemi, I., and Weiss, S. (2007). Male pheromone-stimulated neurogenesis in the adult female brain: Possible role in mating behavior. *Nat. Neurosci.* 10, 1003–1011.

Mann, P. E., and Bridges, R. S. (2001). Lactogenic hormone regulation of maternal behavior. *Prog. Brain Res.* 133, 251–262.

Meaney, M. J., Viau, V., Aitken, D. H., and Bhatnagar, S. (1989). Glucocorticoid receptors in brain and pituitary of the lactating rat. *Physiol. Behav.* 45, 209–212.

Mirescu, C., and Gould, E. (2006). Stress and adult neurogenesis. *Hippocampus* 16, 233–238.

Montaron, M. F., Piazza, P. V., Aurousseau, C., Urani, A., Le Moal, M., and Abrous, D. N. (2003). Implication of corticosteroid receptors in the regulation of hippocampal structural plasticity. *Eur. J. Neurosci.* 18, 3105–3111.

Montaron, M. F., Drapeau, E., Dupret, D., Kitchener, P., Aurousseau, C., Le Moal, M., Piazza, P. V., and Abrous, D. N. (2006). Lifelong corticosterone level determines age-related decline in neurogenesis and memory. *Neurobiol. Aging* 27, 645–654.

Moltz, H., and Robbins, D. (1965). Maternal behavior of primiparous and multiparous rats. *J. Comp. Physiol. Psychol.* 60, 417–421.

Morris, R. G., Garrud, P., Rawlins, J. N., and O'Keefe, J. (1982). Place navigation impaired in rats with hippocampal lesions. *Nature* 24, 681–683.

Moscovitch, M., Rosenbaum, R. S., Gilboa, A., Addis, D. R., Westmacott, R., Grady, C., McAndrews, M. P., Levine, B., Black, S., Winocur, G., and Nadel, L. (2005). Functional neuroanatomy of remote episodic, semantic and spatial memory: A unified account based on multiple trace theory. *J. Anat.* 207(1), 35–66.

Moser, M.B., Moser, E.I., Forrest, E., Andersen, P., Morris, R.G. (1995). Spatial learning with a minislab in the dorsal hippocampus. *Proc. Natl. Acad. Sci. USA* 92, 9697–9701.

Nicholas, K.R., and Hartmann, P. E. (1981). Progressive changes in plasma progesterone, prolactin and corticosteroid levels during late pregnancy and the initiation of lactose synthesis in the rat. *Aust. J. Biol. Sci.* 34, 445–454.

Numan, M. (2007). Motivational systems and the neural circuitry of maternal behavior in the rat. *Dev. Psychobiol.* 49, 12–21.

Oatridge, A., Holdcroft, A., Saeed, N., Hajnal, J. V., Puri, B. K., Fusi, L., and Bydder, G. M. (2002). Change in brain size during and after pregnancy: Study in healthy women and women with preeclampsia. *AJNR Am. J. Neuroradiol.* 23, 19–26.

Olson, A. K., Eadie, B. D., Ernst, C., and Christie, B. R. (2006). Environmental enrichment and voluntary exercise massively increase neurogenesis in the adult hippocampus via dissociable pathways. *Hippocampus* 16, 250–260.

Olton, D. S., and Papas, B. C. (1979). Spatial memory and hippocampal function. *Neuropsychologia* 17, 669–682.

Ormerod, B. K., and Galea, L. A. M. (2001a). Reproductive status influences cell proliferation and cell survival in the dentate gyrus of adult female meadow voles: A possible regulatory role for estradiol. *Neuroscience* 102, 369–379.

Ormerod, B. K., and Galea, L. A. M. (2001b). Mechanisms and function of neurogenesis in the adult. In *Toward a Theory of Neuroplasticity* (C. Shaw and J. McEachern, Eds.), pp. 85–100. Taylor and Francis Publishers, New York.

Ormerod, B. K., Lee, T. T., and Galea, L. A. (2003). Estradiol initially enhances but subsequently suppresses (via adrenal steroids) granule cell proliferation in the dentate gyrus of adult female rats. *J. Neurobiol.* 55, 247–260.

Ormerod, B. K., Lee, T. T., and Galea, L. A. (2004). Estradiol enhances neurogenesis in the dentate gyri of adult male meadow voles by increasing the survival of young granule neurons. *Neurosci.* 128, 645–654.

Orpen, B. G., and Fleming, A. S. (1987). Experience with pups sustains maternal responding in postpartum rats. *Physiol. Behav.* 40, 47–54.

Parsons, C., and Redman, S. (1991). Self-reported cognitive change during pregnancy. *Aust. J. Adv. Nurs.* 9(1), 20–29.

Pawluski, J.L., Charlier, T.D., Lieblich, S.E., Hammond, G.L., and Galea, L.A.M. (under revision). Reproductive experience affects corticosterone and corticosteroid binding globulin during the postpartum period in the dam. *Physiology and Behavior*.

Pawluski, J. L., and Galea, L. A. M. (2007). Reproductive experience alters hippocampal neurogenesis during the postpartum period in the dam. *Neuroscience*. 149, 53–67.

Pawluski, J. L., and Galea, L. A. (2006). Hippocampal morphology is differentially affected by reproductive experience in the mother. *J. Neurobiol.* 66, 71–81.

Pawluski, J. L., Walker, S. K., and Galea, L. A. (2006a). Reproductive experience differentially affects spatial reference and working memory performance in the mother. *Horm. Behav.* 49, 143–149.

Pawluski, J. L., Vanderbyl, B. L., Ragan, K., and Galea, L. A. (2006b). First reproductive experience persistently affects spatial reference and working memory in the mother and these effects are not due to pregnancy or 'mothering' alone. *Behav. Brain Res.* 175, 157–165.

Poser, C. M., Kassirer, M. R., and Peyser, J. M. (1986). Benign encephalopathy of pregnancy. Preliminary clinical observations. *Acta Neurol. Scand.* 73(1), 39–43.

Rosenblatt, J. S. (1967). Nonhormonal basis of maternal behavior in the rat. *Science* 156, 1512–1514.

Rosenblatt, J. S., Siegel, H. I., and Mayer, A. D. (1979). Blood levels of progesterone, estradiol and prolactin in pregnant rats. *Adv. Study Behav.* 10, 225–311.

Scarfman, H. E. (2004). Functional implications of seizure-induced neurogenesis. *Adv. Exp. Med. Biol.* 548, 192–212.

Shingo, T., Gregg, C., Enwere, E., Fujikawa, H., Hassam, R., Geary, C., Cross, J. C., and Weiss, S. (2003). Pregnancy-stimulated neurogenesis in the adult female forebrain mediated by prolactin. *Science* 299, 117–120.

Shors, T. J., Townsend, D. A., Zhao, M., Kozorovitskiy, Y., and Gould, E. (2002). Neurogenesis may relate to some but not all types of hippocampal. *Hippocampus* 12, 578–584.

Snyder, J. S., Hong, N. S., McDonald, R. J., and Wojtowicz, J. M. (2005). A role for adult neurogenesis in spatial long-term memory. *Neuroscience* 130, 843–852.

Stern, J. M., Goldman, L., and Levine, S. (1973). Pituitary–adrenal responsiveness during lactation in rats. *Neuroendocrinology* 12, 179–191.

Svare, B., and Gandelman, R. (1976). Postpartum aggression in mice: The influence of suckling stimulation. *Horm. Behav.* 7, 407–416.

Tanapat, P., Hastings, N. B., Reeves, A. J., and Gould, E. (1999). Estrogen stimulates a transient increase in the number of new neurons in the dentate gyrus of the adult female rat. *J. Neurosci.* 19, 5792–5801.

Terlecki, L. J., and Sainsbury, R. S. (1978). Effects of fimbria lesions on maternal behavior in the rat. *Physiol. Behav.* 21(1), 89–97.

Tomizawa, K., Iga, N., Lu, Y., Moriwaki, A., Matsushita, M., Li, S., Miyamoto, O., Itano, T., and Matsui, H. (2003). Oxytocin improves long-lasting spatial memory during motherhood through MAP kinase cascade. *Nat. Neurosci.* 6, 384–389.

Vertes, R. P. (2006). Interactions among the medial prefrontal cortex, hippocampus and midline thalamus in emotional and cognitive processing in the rat. *Neuroscience* 142, 1–20.

Winocur, G., Wojtowicz, J. M., Sekeres, M., Snyder, J. S., and Wang, S. (2006). Inhibition of neurogenesis interferes with hippocampus-dependent memory function. *Hippocampus* 16, 296–304.

Witter, M. P. (1993). Organization of the entorhinal–hippocampal system: A review of current anatomical data. *Hippocampus* 3, 33–44.

Wong, E. Y., and Herbert, J. (2004). The corticoid environment: A determining factor for neural progenitors' survival in the adult hippocampus. *Eur. J. Neurosci.* 20, 2491–2498.

Woolley, C. S., and McEwen, B. S. (1993). Roles of estradiol and progesterone in regulation of hippocampal dendritic spine density during the estrous cycle in the rat. *J. Comp. Neurol.* 336, 293–306.

Woolley, C. S., Gould, E., Frankfurt, M., and McEwen, B. S. (1990). Naturally occurring fluctuation in dendritic spine density on adult hippocampal pyramidal neurons. *J. Neurosci.* 10, 4035–4039.

Yoshinaga, K., Hawkins, R. A., and Stocker, J. F. (1969). Estrogen secretion by the rat ovary *in vivo* during the estrous cycle and pregnancy. *Endocrinology* 85, 103–112.

Zhao, C., Teng, E. M., Summers, R. G., Jr, Ming, G. L., and Gage, F. H. (2006). Distinct morphological stages of dentate granule neuron maturation in the adult mouse hippocampus. *J. Neurosci.* 26, 3–11.

33

NEUROENDOCRINE AND BEHAVIORAL ADAPTATIONS FOLLOWING REPRODUCTIVE EXPERIENCE IN THE FEMALE RAT

ELIZABETH M. BYRNES, BENJAMIN C. NEPHEW AND ROBERT S. BRIDGES

Tufts University, Cummings School of Veterinary Medicine, Department of Biomedical Sciences, North Grafton, MA, USA.

INTRODUCTION

The combined experiences of pregnancy, parturition and maternal care can induce significant and long-lasting modifications in circulating hormone levels. In women with prior reproductive experience, circulating levels of both prolactin and estrogen are reduced when examined several years after childbirth (Yu *et al.*, 1981; Bernstein *et al.*, 1985; Musey *et al.*, 1987; Wang *et al.*, 1988; Eliassen *et al.*, 2007). The impact of these parity-induced changes in hormone secretion on women's health has largely focused on disease states associated with neuroendocrine factors, such as breast cancer (Wang *et al.*, 1988; Eliassen *et al.*, 2007). Few clinical studies, however, have examined the long-term effects of these hormonal changes on cognitive or emotional processes. Given the growing number of reports demonstrating the influence of hormones on the learning, memory and mood states, the paucity of clinical data regarding the long-term psychological effects of prior reproductive experience is somewhat surprising. For several years we have used the female rat to explore the neural and endocrine consequences of parity. The use of this animal model has provided insight into the influence of reproductive experience on both hormones and behavior. The current chapter will summarize some of our recent findings.

REPRODUCTIVE EXPERIENCE AND CIRCULATING HORMONES

Similar to the changes in parous women, reduced circulating hormone levels are also observed in parous female rats. Moreover, as observed in women, the effect of reproductive experience on hormone secretion was not observed at all sampling times, but rather, was limited by the stage of the reproductive cycle (menstrual cycle in women, estrous cycle in rats). Thus, in primiparous female rats prolactin levels were significantly reduced in on the afternoon of proestrus (Byrnes & Bridges, 2005). These data are shown in Figure 33.1. Interestingly, this effect was only observed in females that were allowed to care for their young postpartum indicating that the reduction

Neurobiology of the Parental Brain

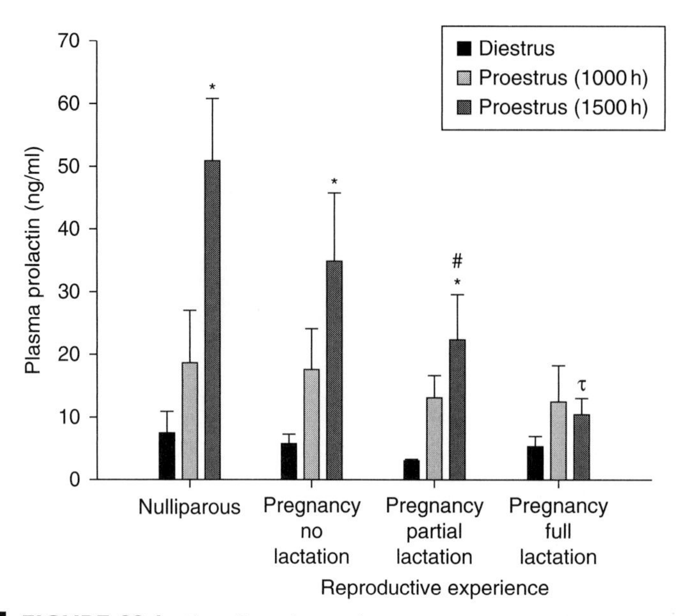

FIGURE 33.1 The effect of reproductive experience on plasma prolactin secretion (mean ± SEM) in cycling female rats. $^{*}p < 0.01$ as compared to diestrus within reproductive experience group. $^{\#}p < 0.01$ as compared to nulliparous females on proestrus (1500 h). $\tau p < 0.04$ as compared to both nulliparous and pregnancy no lactation females on proestrus (1500). (*Source*: Byrnes and Bridges, 2005)

in prolactin secretion following reproductive experience is not solely induced by the hormonal changes of pregnancy. Additional studies also observed a reduction in estradiol levels across the afternoon of proestrus, as well as a significant shift in the sensitivity of parous females to exogenously administered estradiol (Bridges *et al.*, 2005) and prolactin (Anderson *et al.*, 2006). These studies demonstrate that as in women, reproductive experience results in long-term changes in the endocrine system of female rats. Based on these findings our initial behavioral studies examined parity-mediated alterations in anxiety-like behavior, a behavior that is influenced by the hormonal milieu and which is known to vary across the estrous cycle (Mora *et al.*, 1996; Diaz-Veliz *et al.*, 1997; Frye *et al.*, 2000; Marcondes *et al.*, 2001).

REPRODUCTIVE EXPERIENCE AND ANXIETY-LIKE BEHAVIOR

In women, a number of anxiety-related conditions are known to be influenced by the hormonal environment. For example, several studies have demonstrated premenstrual worsening of symptoms for obsessive compulsive disorder (OCD), panic disorder (PD), as well as generalized anxiety disorder (GAD) (Ensom, 2000). Recent epidemiological findings indicate that postpartum onset and/or worsening of OCD and GAD occurs in a significant proportion of women affected by these conditions (Abramowitz *et al.*, 2003; Halbriech, 2005; Wenzel *et al.*, 2005). Moreover, some researchers have suggested that postpartum anxiety disorders may be more prevalent than postpartum depression (Wenzel *et al.*, 2005). Thus, it is clear that hormonal fluctuations can influence anxiety-related disorders. However, very little data exists regarding the long-term impact of reproductive experience on anxiety-related disorders or anxiety symptoms in a sub-clinical population.

In rodents, hormonal fluctuations have also been shown to influence the expression of anxiety-like behavior. For example, during mid- to late-pregnancy rats demonstrate decreased anxiety-like behavior (Faturi *et al.*, 2006) when compared to ovariectomized controls. A similar anxiolytic effect is observed postpartum (Lonstein, 2005). In addition, several studies have reported decreased anxiety-like behaviors during proestrus and/or estrus as compared to diestrus (Mora *et al.*, 1996; Diaz-Veliz *et al.*,

1997; Frye *et al.*, 2000; Marcondes *et al.*, 2001). Finally, more recent studies indicate that both estradiol and prolactin may play a role in the decreased anxiety-like behavior observed in proestrus females (Marcondes *et al.*, 2001; Torner *et al.*, 2001, 2002; Frye & Walf, 2004; Lund *et al.*, 2005). Thus, given the significant changes in prolactin and estradiol secretion observed following reproductive experience, we examined the effect of parity on the expression of anxiety-like behavior in cycling nulliparous and primiparous rats.

All of our reproductive experience studies were conducted in female Sprague-Dawley rats (Crl:CD[SD]BR) purchased from Charles River Laboratories. For all experiments, animals were maintained in temperature- (21–25°C) and light- (14:10 light–dark cycle) controlled rooms and food and water were available *ad libitum*. Animals were maintained in accordance with the National Research Council (NRC) Guide for the Care and Use of Laboratory Animals (© 1996, National Academy of Science), and methods were approved by the Institutional Animals Care and Use Committee of Tufts University, Cummings School of Veterinary Medicine.

Nulliparous and parous females were age-matched to eliminate age as a potential confound. All parous females had their litters weaned at postnatal day 21 and no testing was performed until at least 6 weeks post-weaning. In studies examining the effects of the estrous cycle on anxiety-like behavior, estrous cycles are monitored for at least 8 days prior to testing and only females demonstrating normal 4-day estrous cycles were used. Finally, all behavioral testing was conducted in a quiet behavioral testing room between 13:00 and 15:00 h under standard vivarium lighting.

Anxiety-like behavior was measured using the elevated plus maze (EPM) and open field tasks. All plus maze testing was conducted using a fully automated EPM (Hamilton-Kinder; Poway, CA). Open field activity was monitored with the SmartFrame® Open Field Activity System (Hamilton-Kinder). Each test lasted 5 min. All data were automatically collected and quantified using MotorMonitor® software (Hamilton-Kinder). EPM measures included overall activity (beam breaks) and percent of time spent on the open. Open field measures included overall activity (beam breaks) and percent time spent in the center. Increased time in the open arms and increased time in the center of the open field are interpreted as reduced anxiety-like behavior (Pellow *et al.*, 1985; Lister, 1990).

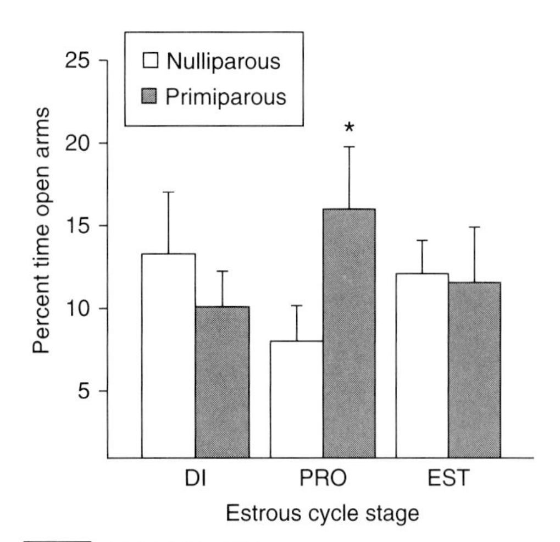

FIGURE 33.2 The effect of reproductive experience on the percent of time (mean ± SEM) spent exploring the open arms of the EPM. DI, diestrus; PRO, proestrus; EST, estrus. $*p < 0.05$ as compared to nulliparous, proestrus females. (Adapted from Byrnes and Bridges, 2006)

In the first study reproductive experience increased the percentage of time females spent on the open arm of the EPM when compared to non-experienced controls (see Figure 33.2). This effect, however, was only observed in proestrus females. Thus, primiparous females displayed less anxiety-like behavior when compared to age-matched, nulliparous females on the afternoon of proestrus. Similar findings were observed in a separate set of subjects tested on the open field task. As illustrated in Figure 33.3, primiparous females spent more time in the center of the open field when compared to nulliparous females tested on proestrus. These shifts in anxiety-like behavior following reproductive experience occur in conjunction with reduced levels of prolactin and estradiol, suggesting that the hormonal regulation of anxiety-like behavior has been modified by parity.

In addition to changes in prolactin levels across the cycle, reproductive experience also modulates the effect of aging on prolactin levels. As the female ages, prolactin levels become significantly higher (Demarest *et al.*, 1982; Console *et al.*, 1997). This rise in prolactin levels is dampened in females with reproductive experience (Wang *et al.*, 1988; Eliassen *et al.*, 2007). Could this dampened prolactin secretion alter anxiety levels in aging primiparous

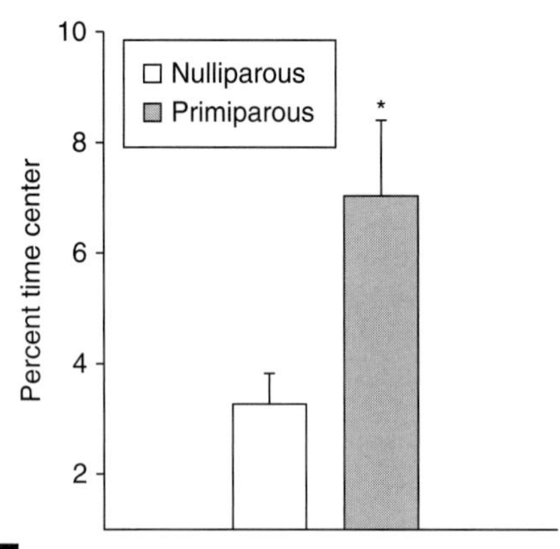

FIGURE 33.3 The effect of reproductive experience on the percent of time (mean ± SEM) spent exploring the center of the open field. All females tested on the afternoon of proestrus. $*p < 0.05$ as compared to nulliparous females. (Adapted from Byrnes and Bridges, 2006)

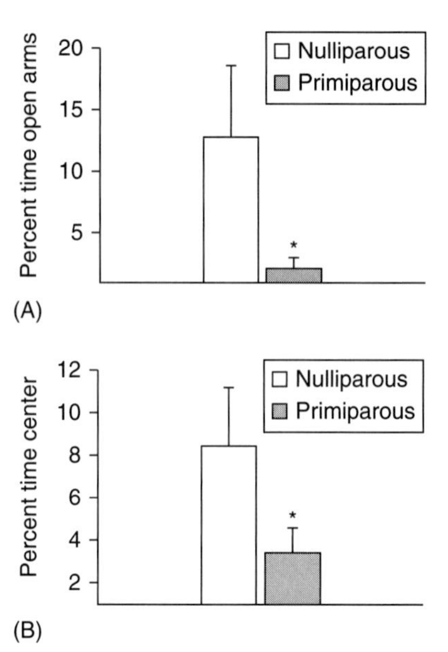

FIGURE 33.4 The effect of reproductive experience on the percent of time (mean ± SEM) spent exploring the open arms of the EPM (A) or the center of the open field (B). All females were tested at 12–13 months of age and all were in constant estrus. $*p < 0.05$ as compared to nulliparous females. (Adapted from Byrnes and Bridges, 2006)

females? Are anxiety-like behaviors in reproductively senescent (i.e., non-cycling) females even altered by reproductive experience? The next set of studies was designed to answer this latter question, that is, does reproductive experience influence anxiety-like behavior in non-cycling, middle-aged females.

REPRODUCTIVE EXPERIENCE, AGING AND ANXIETY-LIKE BEHAVIOR

All behavioral testing was conducted in females between 12 and 13 months of age who were determined to be in constant estrous. As shown in Figure 33.4, both the percent of time in the open arm of the EPM and percent of time spent in the center of the open field were significantly decreased in primiparous females. These findings indicate that unlike young primiparous females, middle-aged primiparous females demonstrated a significant increase in anxiety-like behavior when compared to nulliparous controls. Moreover, in both young and middle-aged primiparous females, ovariectomy eliminated the effect of reproductive experience on anxiety-like behavior (see Figure 33.5), indicating a role for circulating hormones in parity-induced changes in anxiety-like behavior in both young and middle-aged females.

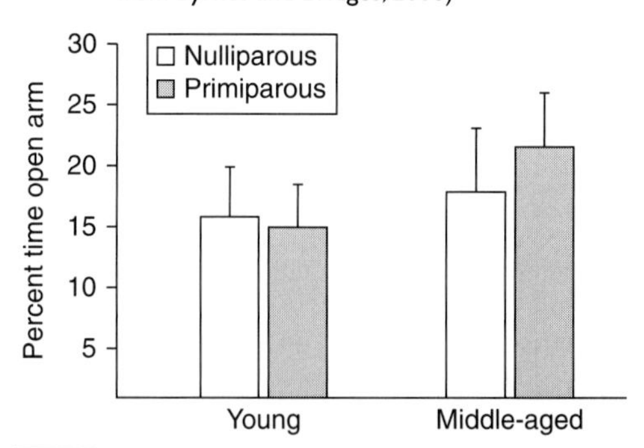

FIGURE 33.5 The effect of reproductive experience on the percent of time (mean ± SEM) spent exploring the open arms of the EPM. Young females were tested at 5–6 months of age, middle-aged females were tested at 12–13 months of age. (Adapted from Byrnes and Bridges, 2006)

In summary, these results indicate that reproductive experience alters anxiety-like behavior. However, these effects depend on the stage of the estrous cycle during which the female is tested and the age of the animal. As ovariectomy can eliminate these effects, circulating hormone levels likely regulate this effect. At this point, however, it is unclear which ovarian hormones are involved in the shift in anxiety-like behavior in primiparous females. It is quite possible that reproductive experience alters the manner in which circulating hormones affect any number of neural systems resulting in behavioral effects that are unique to the experienced females but which are still hormone-dependent.

Having determined that shifts in anxiety-like behavior following reproductive experience can be modulated by the hormonal status of the female, we next wanted to determine whether other factors could influence this behavior in parous females. As studies indicate that pups are critical modulators of anxiety-like behavior in postpartum females, we decided to examine the influence of pups on anxiety-like behavior in females with reproductive experience. Furthermore, to elucidate the influence of pregnancy vs. the influence of maternal care in the absence of the hormonal changes associated with pregnancy, we compared primiparous females to "sensitized" nulliparous females (i.e., females induced to display maternal behavior).

INFLUENCE OF PREGNANCY AND MOTHERING ON PUP-INDUCED SHIFTS IN ANXIETY-LIKE BEHAVIOR

It is well established that nulliparous female rats do not spontaneously display maternal behavior toward foster pups. However, if females are continuously housed with pups they will eventually display many of the components of maternal behavior, including retrieving the pups to a nest site, grouping the pups in the nest and crouching over the pups in a nursing-like posture (Rosenblatt, 1967). In order to control for the amount of maternal care experienced by primiparous and nulliparous females, all subjects in this next study (Scanlan *et al.*, 2006) were allowed 2 days of maternal experience. Thus, primiparous females were housed with their own pups for 2 days after parturition after which time their pups were removed. Sensitized nulliparous females were continuously housed with three donor pups (refreshed daily) until they displayed two consecutive days of full maternal behavior (defined as retrieving, grouping and crouching over three donor pups during a 15 min maternal behavior test). After 2 days of maternal care, both primiparous and sensitized nulliparous females were isolated from young and individually housed for next 55 days. After this time, three donor pups were placed in the subject's home cage and all females

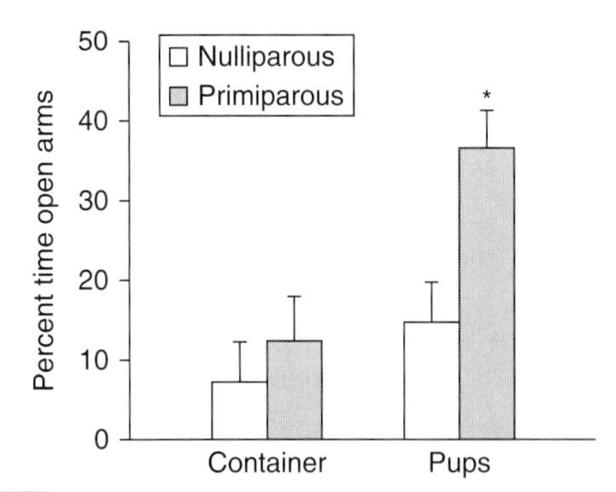

FIGURE 33.6 The effect of pups present in the testing environment on the percent of time (mean + SEM) spent exploring the open arms of the EPM by the maternal nulliparous and primiparous females. *$p < 0.05$ as compared to all other groups. (Adapted from Scanlan *et al.*, 2006)

were monitored daily for the re-induction of maternal care. Once maternal care had been re-established for two consecutive days, females were tested for the expression of anxiety-like behavior on the EPM. In this particular EPM task, however, a container was placed 15 cm from the end of one of the open arms. For half of the subjects the container remained empty, whereas for the others three 2–3 day-old pups were placed in the container. The distance from the end of the open arm precluded direct interaction with the pups, only allowing distal sensory stimulation. At the end of the task subjects were returned to their home cage and 2 h later were perfused and their brains were processed for Fos-immunoreactivity.

EPM data for the re-induced maternal nulliparous and primiparous subjects are shown in Figure 33.6. Overall, primiparous females spent a greater percentage of their time exploring the open arm of the plus maze when compared to sensitized nulliparous females. Moreover, the presence of pups significantly increased the percent of time spent on the open arm in primiparous females, but had no effect on sensitized nulliparous females. It is important to note, however, that primiparous females were not simply spending all of their time at the end of the open arm next to the pup container. In fact, there was no significant difference between the amounts of time spent on the open arm next to the pups when compared to the opposite open arm. Instead, it would appear that the presence of pups increased exploration of both open arms. Thus, prior parity regulates

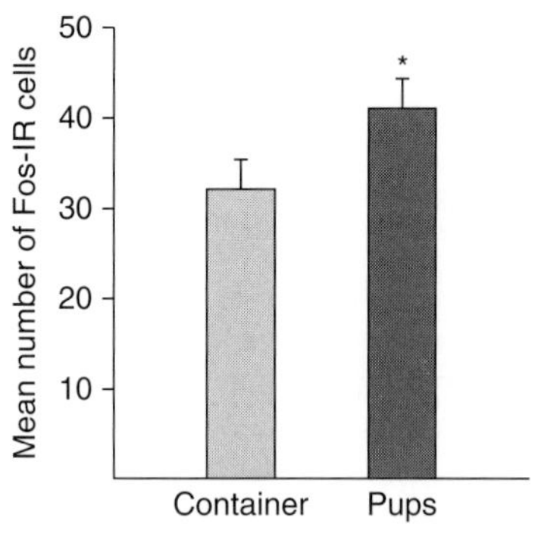

FIGURE 33.7 Mean (+SEM) number of Fos-IR nuclei activated within the nucleus accumbens in females tested on the EPM in the presence or absence of pups. *$p < 0.05$ as compared to females tested in the absence of pups. (Adapted from Scanlan *et al.*, 2006)

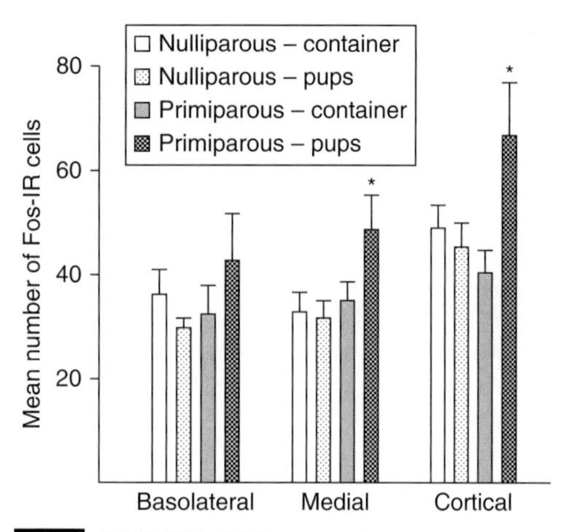

FIGURE 33.8 Mean (+SEM) number of Fos-IR nuclei activated within the amygdala in females tested on the EPM in the presence or absence of pups. *$p < 0.05$ as compared to maternal nulliparous females tested in the presence of pups. (*Source*: Scanlan *et al.*, 2006)

the influence of the pup stimuli on the expression of anxiety-like behavior, and this effect can not be produced simply by inducing nulliparous females to behavior maternally. These data indicate that the physiological changes that occur during pregnancy and parturition are a prerequisite for pup-mediated alterations in anxiety-like behavior.

Analysis of the Fos-immunoreactivity demonstrated significantly higher Fos activation within the shell region of the nucleus accumbens, an area implicated in maternal behavior expression (Numan, 2006) and maternal memory (Li & Fleming, 2003), in all subjects tested in the presence of pups, regardless of their prior experience (see Figure 33.7). Fos activation within the amygdala, however, demonstrated clear differences depending on both the presence of pups and the females prior experience. As illustrated in Figure 33.8, Fos-immunoreactivity in both the medial and cortical amygdala nuclei was increased in primiparous females tested in the presence of pups. Representative sections are shown in Figure 33.9. Thus, primiparous females spent more time on the open arm of the plus maze and had increased Fos-immunoreactivity in the medial and cortical nuclei of the amygdala.

The medial and cortical amygdala nuclei receive direct input from both the main and accessory olfactory bulb (Scalia & Winans, 1975). Moreover, both of these regions are sexually dimorphic and estrogen sensitive (Cooke & Wooley, 2005). In females, these nuclei are involved in numerous functions including maternal behavior (Fleming & Walsh, 1994, Meurisse *et al.*, 2005), sexual behavior (Bennett *et al.*, 2002), stress responsiveness (Trneckova *et al.*, 2006), social learning and memory (Ferguson *et al.*, 2002; Bielsky & Young, 2004) and aggression (Gammie & Nelson, 2001). Previous findings in rats demonstrate that reproductive experience significantly alters performance on learning and memory tasks as well as significantly decreasing stress responsiveness (Wartella *et al.*, 2003; Gatewood *et al.*, 2005). Moreover, in female sheep parity-related changes in maternal care have also been documented (Meurisse *et al.*, 2005) and appear to be regulated within the medial amygdala. Thus, reproductive experience appears to modify the manner in which the medial and cortical amygdala processes incoming sensory stimuli thereby altering behavioral outcomes. We suggest that many of these behavioral adaptations enhance the reproductive success of the female. Based on this hypothesis, we examined the effect of reproductive experience on maternal aggression, a behavior known to be both regulated by the medial amygdala and important for reproductive success.

REPRODUCTIVE EXPERIENCE AND MATERNAL AGGRESSION

Maternal aggression is a well-defined behavioral phenomenon in rodents (Lonstein & Gammie, 2002). When a postpartum female is caring for her young, she will readily attack a conspecific (either a male or a non-lactating female) that

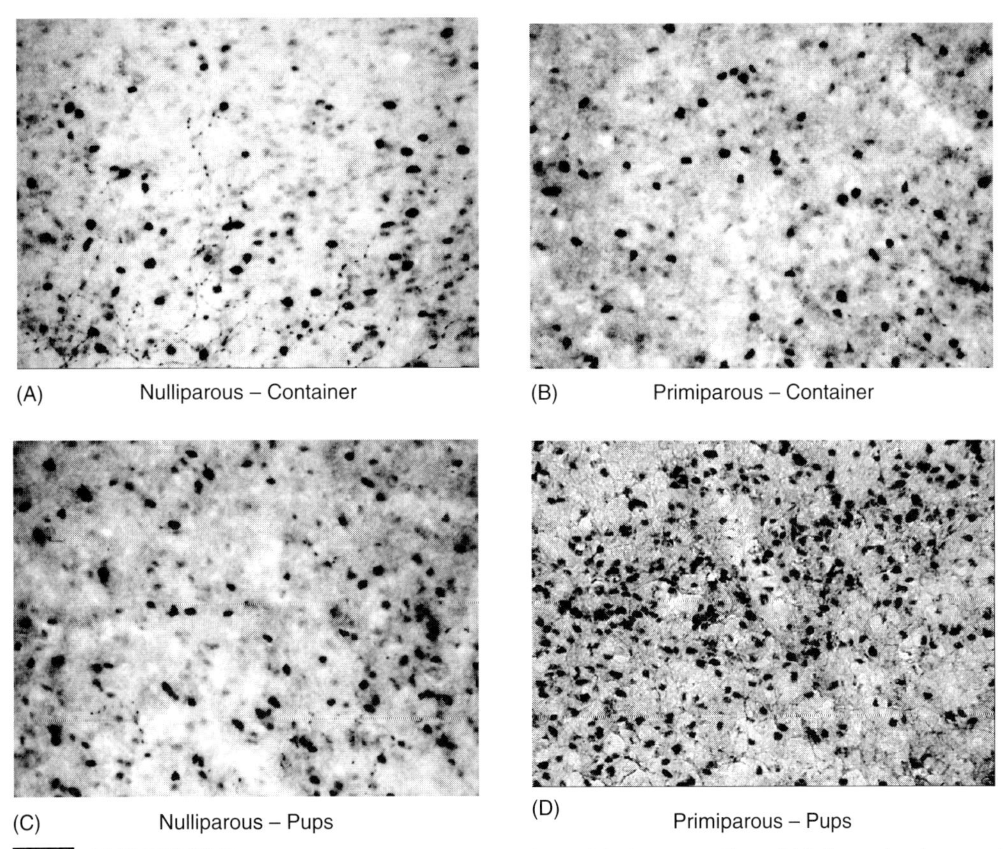

FIGURE 33.9 Photomicrographs of the cortical amygdala depicting c-Fos cell labeling in females tested on the EPM in the presence or absence of pups. Panel A, Nulliparous – Container; Panel B, Primiparous – Container; Panel C, Nulliparous – Pups; Panel D, Primiparous – Pups. (Adapted from Scanlan *et al.*, 2006)

intrudes into her home environment. This response helps protect her young from predation. In the rat, maternal aggression subsides once the mother has been separated from her pups for as little as 2.5 h (Deschamps *et al.*, 2003). In this next study, the effects of repeated pregnancy and lactation/maternal care on the intensity of maternal aggression was measured by comparing the responses of first-time mothers (primiparous) to age-matched, second-time mothers (multiparous). All females were tested for maternal aggression in the presence of a naïve juvenile male on postpartum day 5. As shown in Table 33.1, both the number of attacks and attack duration were significantly increased in multiparous females. These data clearly demonstrate increased maternal aggression in females with prior reproductive experience. The neural changes associated with this increased aggression have yet to be defined; however, we would predict that changes in the medial amygdala may play a role in this effect.

TABLE 33.1 Influence of reproductive experience on maternal aggression

Reproductive Experience	Number of Attacks	Attack Duration (s)
Primiparous	8.5 ± 1.7	8.7 ± 2.8
Multiparous	26.9 ± 3.7*	128.5 ± 36.9*

*$p < 0.01$ as compared to primiparous females.

SUMMARY

Overall, the current findings indicate that reproductive experience can significantly alter behavioral expression. Moreover, these behavioral adaptations are often conditional, depending on the females' reproductive status. Pregnancy and mothering both appear to play a role in the manifestation of these parity effects. Long-term alterations in hormone secretion and a shift in the sensitivity of neuroendocrine parameters following reproductive experience likely underlie the behavioral adaptations. Finally, the medial and cortical nuclei of the amygdala may be critical for the integration of endocrine and sensory stimuli in modifying the behavior of parous females. We propose that these changes enhance the reproductive competency of the female and increase her chances of future reproductive success.

REFERENCES

Abramowitz, J. S., Schwartz, S. A., Moore, K. M., and Leunzmann, K. R. (2003). Obsessive-compulsive symptoms in pregnancy and the puerperium: A review of the literature. *Anxiety Disord.* 17, 461–478.

Anderson, G. M., Grattan, D. R., van den Ancker, W., and Bridges, R. S. (2006). Reproductive experience increases prolactin responsiveness in the medial preoptic area and arcuate nucleus of female rats. *Endocrinology* 147(10), 4688–4694.

Bennett, A. L., Greco, B., Blasberg, M. E., and Blaustein, J. D. (2002). Response to male odours in progestin receptor- and oestrogen receptor-containing cells in female rat brain. *J. Neuroendocrinol.* 14(6), 442–449.

Bernstein, L., Pike, M. C., Ross, R. K., Judd, H. L., Brown, J. B., and Henderson, B. E. (1985). Estrogen and sex hormone-binding globulin levels in nulliparous and parous women. *J. Natl Canc. Inst.* 74(4), 741–745.

Bielsky, I. F., and Young, L. J. (2004). Oxytocin, vasopressin, and social recognition in mammals. *Peptides* 25(9), 1565–1574.

Bridges, R. S., and Byrnes, E. M. (2006). Reproductive experience reduces circulating estradiol-17β and prolactin levels during proestrus and alters estrogen sensitivity in female rats. *Endocrinology* 147, 2575–2582.

Byrnes, E. M., and Bridges, R. S. (2005). Lactation reduces prolactin levels in reproductively experienced female rats. *Horm. Behav.* 49, 278–282.

Byrnes, E. M., and Bridges, R. S. (2006). Reproductive experience alters anxiety-like behavior in the female rat. *Hormones and Behavior* 50, 70–76.

Console, G. M., Gomez Dumm, C. L., Brown, O. A., Ferese, C., and Goya, R. G. (1997). Sexual dimorphism in the age changes of the pituitary lactotrophs in rats. *Mech. Ageing Dev.* 95(3), 157–166.

Cooke, B. M., and Woolley, C. S. (2005). Sexually dimorphic synaptic organization of the medial amygdala. *J. Neurosci.* 25(46), 10759–10767.

Deschamps, S., Woodside, B., and Walker, C. D. (2003). Pups presence eliminates the stress hyporesponsiveness of early lactating females to a psychological stress representing a threat to the pups. *J. Neuroendocrinol.* 15(5), 486–497.

Demarest, K. T., Moore, K. E., and Riegle, G. D. (1982). Dopaminergic neuronal function, anterior pituitary dopamine content, and serum concentrations of prolactin, luteinizing hormone and progesterone in the aged female rat. *Brain Res.* 247(2), 347–354.

Diaz-Veliz, G., Alacron, T., Espinoza, C., Dussaubat, N., and Mora, S. (1997). Ketanserin and anxiety levels: Influence of gender, estrous cycle, ovariectomy and ovarian hormones in female rats. *Pharmcol. Biochem. Behav.* 58, 637–642.

Eliassen, A. H., Tworoger, S. S., and Hankinson, S. E. (2007). Reproductive factors and family history of breast cancer in relation to plasma prolactin levels in premenopausal and postmenopausal women. *Int. J.Canc.* 120(7), 1536–1541.

Ensom, M. H. (2000). Gender-based differences and menstrual cycle-related changes in specific diseases: Implications for pharmacotherapy. *Pharmacotherapy* 20, 523–539.

Faturi, C., Teixeira-Silva, F., and Leite, J. R. (2006). The anxiolytic effect of pregnancy in rats is reversed by finasteride Pharmacology. *Biochem. Behav.* 85(3), 569–574.

Ferguson, J. N., Young, L. J., and Insel, T. R. (2002). The neuroendocrine basis of social recognition. *Front. Neuroendocrinol.* 23(2), 200–224.

Fleming, A. S., and Walsh, C. (1994). Neuropsychology of maternal behavior in the rat: c-fos expression during mother–litter interactions. *Psychoneuroendocrinology* 19(5-7), 429–443.

Frye, C. A., and Walf, A. A. (2004). Estrogen and/or progesterone administered systemically or to the amygdale can have anxiety-, fear-, and pain-reducing effects in ovariectomized rats. *Behav. Neurosci.* 118, 306–313.

Frye, C. A., Petralia, S. M., and Rhodes, M. E. (2000). Estrous cycle and sex differences in performance on anxiety tasks coincide with increases in hippocampal progesterone and 3α, 5α-THP. Pharmacol. Biochem. Behav. 67, 587–596.

Gammie, S. C., and Nelson, R. J. (2001). cFOS and pCREB activation and maternal aggression in mice. *Brain Res.* 898(2), 232–241.

Gatewood, J. D., Morgan, M. D., Eaton, M., McNamara, I. M., Stevens, L. F., Macbeth, A. H., Meyer, E. A., Lomas, L. M., Kozub, F. J., Lambert, K. G., and Kinsley, C. H. (2005). Motherhood mitigates aging-related decrements in learning and memory and positively affects brain aging in the rat. *Brain Res. Bull.* 66, 91–98.

Halbriech, U. (2005). Postpartum disorders: Multiple interacting underlying mechanisms and risk factors. *J Affect. Disord.* 88, 1–7.

Li, M., and Fleming, A. S. (2003). Differential involvement of the nucleus accumbens shell and core subregions in maternal memory in postpartum female rats. *Behav. Neurosci.* 117, 426–445.

Lister, R. G. (1990). Ethologically based animal models of anxiety disorders. *Pharmacol. Therap.* 46, 321–340.

Lonstein, J. S. (2005). Reduce anxiety in postpartum rats requires recent physical interactions with pups, but is independent of suckling and peripheral sources of hormones. *Horm. Behav.* 47, 241–255.

Lonstein, J. S., and Gammie, S. C. (2002). Sensory, hormonal, and neural control of maternal aggression in laboratory rodents. *Neurosci. Biobehav. Rev.* 26(8), 869–888.

Lund, T. D., Rovis, T., Chung, W. C. J., and Handa, R. J. (2005). Novel actions of estrogen receptor-β on anxiety-related behaviors. *Endocrinology* 146, 797–807.

Marcondes, F. K., Miguel, K. J., Melo, L. L., and Spadari-Brattfisch, R. C. (2001). Estrous cycle influences the response of female rats in the elevated plus-maze test. *Physiol. Behav.* 74, 435–440.

Mora, S., Dussaubat, and Diaz-Veliz, G. (1996). Effects of estrous cycle and ovarian hormones on behavioral indices of anxiety in female rats. *Psychoneuroendocrinology* 21, 609–620.

Meurisse, M., Gonzalez, A., Delsol, G., Caba, M., Levy, F., and Poindron, P. (2005). Estradiol receptor-alpha expression in hypothalamic and limbic regions of ewes is influenced by physiological state and maternal experience. *Horm. Behav.* 48(1), 34–43.

Musey, V. C., Collins, D. C., Musey, P. I., Martino-Saltzman, D., and Preedy, J. R. K. (1987). Long-term effect of a first pregnancy on the secretion of prolactin. *New Engl. J. Med.* 316, 229–234.

Numan, M. (2006). Hypothalamic neural circuits regulating maternal responsiveness toward infants. *Behav. Cognit. Neurosci. Rev.* 5, 163–190.

Pellow, S., Chopin, P., File, S. E., and Briley, M. (1985). Validation of open:closed arm entries in an elevated plus-maze as a measure of anxiety in the rat. *J. Neurosci. Meth.* 14, 149–167.

Rosenblatt, J. S. (1967). Nonhormonal basis of maternal behavior in the rat. *Science* 156, 1512–1514.

Scalia, F., and Winans, S. S. (1975). The differential projections of the olfactory bulb and accessory olfactory bulb in mammals. *J. Comp. Neurol.* 161, 31–55.

Scanlan, V. F., Byrnes, E. M., and Bridges, R. S. (2006). Reproductive experience and activation of maternal memory. *Behav. Neurosci.* 120, 676–686.

Torner, L., Toschi, N., Pohlinger, A., Landgraf, R., and Neumann, I. D. (2001). Anxiolytic and anti-stress effects of brain prolactin: Improved efficacy of antisense targeting of the rolactin receptor by molecular modeling. *J. Neurosci.* 21, 3207–3214.

Torner, L., Toschi, N., Nava, G., Clapp, C., and Neumann, I. D. (2002). Increased hypothalamic expression of prolactin in lactation: Involvement in behavioral and neuroendocrine stress responses. *Eur. J. Neurosci.* 14, 1381–1389.

Trneckova, L., Armario, A., Hynie, S., Sida, P., and Klenerova, V. (2006). Differences in the brain expression of c-fos mRNA after restraint stress in Lewis compared to Sprague-Dawley rats. *Brain Res.* 1077(1), 7–15.

Wang, D. Y., De Stavola, B. L., Bulbrook, R. D., Allen, D. S., Kwa, H. G., Verstraeten, A. A., Moore, J. W., Fentiman, I. S., Hayward, J. L., and Gravelles, I. H. (1988). The permanent effect of reproductive events on blood prolactin levels and its relation to breast cancer risk: A population study of postmenopausal women. *Eur. J. Cancer Clin. Oncol.* 24, 1225–1231.

Wartella, J., Amory, E., Lomas, L. M., Macbeth, A., McNamara, I., Stevens, L., Lambert, K. G., and Kinsley, C. H. (2003). Single or multiple reproductive experiences attenuate neurobehavioral stress and fear responses in the female rat. *Physiol. Behav.* 79(3), 373–381.

Wenzel, A., Haugen, E. N., Jackson, L. C., and Brendle, R. J. (2005). Anxiety symptoms and disorders at eight weeks postpartum. *Anxiety Disord.* 19, 295–311.

Yu, M. C., Gerkins, V. R., Henderson, B. E., Brown, J. B., and Pike, M. C. (1981). Elevated levels of prolactin in nulliparous women. *Brit. J. Cancer* 43, 826–831.

34

PLASTICITY IN THE MATERNAL NEURAL CIRCUIT: EXPERIENCE, DOPAMINE, AND MOTHERING

ALISON S. FLEMING, ANDREA GONZALEZ, VERONICA M. AFONSO AND VEDRAN LOVIC

Department of Psychology, University of Toronto, Mississauga, Ont., Canada, L5L1C6

INTRODUCTION

The maternal brain is a complex "place" and the 2007 Boston "Parental Brain Conference" was devoted to its analysis, reflecting just how complex it is, how much we know, and how much we have yet to learn. As the final speaker at the conference, I (Fleming) was in the unenviable position of having to follow the most creative minds in the field and pretend to have something new and original to say. I was in the enviable position to not have to define, introduce, or explain most of the basic concepts or ideas about mothering. I solved the problem by basing my talk and hence the present chapter largely on the work of other people, but framing the story somewhat differently. As well, to have something new to say I contrasted where I could what we know about psychobiological processes in the regulation of rat and human maternal behavior. So as not to improve on our own prose, selected portions of this chapter may also be found in Numan *et al.* (2006), Fleming and Gonzalez (in preparation), and Gonzalez *et al.* (in preparation). Hence, the following!

Over the past 30 years the behavioral processes that are recruited when a new mother becomes maternal, the neural substrates of those processes, and the neurochemical circuitry in the final common pathway for the expression of maternal behavior have been characterized. Overall, four psychobiological systems are intrinsic to the regulation of maternal behavior: (1) maternal affect, (2) maternal experience and memory, (3) maternal hedonics and reward, and (4) maternal attention and sensitivity. This fourth system is relatively understudied and acts in concert with other systems regulating the expression of maternal behavior. Our hypothesis is that the mesocorticolimbic brain dopamine (DA) neurotransmitter system acts as a unifying mechanism that ties the functions of the different systems together.

Below, we briefly describe what we know and have yet to learn about the different psychobiological systems as they relate to the regulation of mothering. For references to older studies, we refer the reader to many recent review chapters or books, in particular, Fleming *et al.* (1999), Fleming and Li (2002), Numan and Insel (2003), Fleming (2005), and Numan *et al.* (2006).

THE WANTING SYSTEM: WHAT MOTIVATES A MOTHER TO MOTHER IN THE FIRST PLACE?

Maternal Affect

The first system, studied initially in the 1970s and 1980s relates to the role of affective changes in the new mother in the onset of her mothering. Our earlier work demonstrated that the new

Neurobiology of the Parental Brain

mother experiences reduced fear and neophobia in general and an enhanced likelihood to approach pups, with the birth of the litter (Numan *et al.*, 2006). It showed further that this change in emotion normally occurs under the influence of the parturitional hormones (estradiol, progesterone, prolactin, and oxytocin) and that enhanced maternal responsiveness can be produced experimentally in the virgin animal by manipulations that reduce ongoing "fear," that is, handling stimulation, olfactory bulbectomy, and anxiolytics (Numan *et al.*, 2006).

Although we do not know if in humans new mothers experience a decrease in generalized fear or in "neophobia" at parturition, it is clear that new mothers do experience increased lability, in the affective domains of well-being (Fleming *et al.*, 1988), depression (Fleming *et al.*, 1988; Cohn *et al.*, 1990; Field *et al.*, 1990), and anxiety (Righetti-Veltema *et al.*, 2003). We have argued that these changes heighten the mothers' overall emotional state which may allow for more intense feelings when first interacting with the infant.

Varied affect states also translate into differential interactions with the infant. There is ample evidence demonstrating that feelings of well-being are significantly associated with more positive attitudes about the infant and motherhood in general (Fleming *et al.*, 1988, 1997a, b) and are associated with higher levels of positive and "sensitive" responding to the infant (Cohn *et al.*, 1990; Field *et al.*, 1990; Righetti-Veltema *et al.*, 2003). In fact, it is likely that because of these positive feelings, mothers are able to withstand considerable duress, and sleep deprivation to remain attentive to and in close proximity to their new infants. However, as indicated above, these positive feelings are neither uniform within mothers, who often experience postpartum lability, or across mothers, where the predominant feelings may initially be intense anxiety and, sometimes, depression (Fleming *et al.*, 1988; Righetti-Veltema *et al.*, 2003). Despite the fact that the majority of mothers experience positive affect with the birth of a new baby, the relation between maternal behavior and affect in humans has primarily been studied in the context of postpartum depression (Beck, 2006). Comparisons between depressed and non-depressed mothers soon after the birth indicate that depressed mothers are irritated, respond less sensitively, less contingently, and are either withdrawn or intrusive during interaction with their infants compared to non-depressed mothers (Fleming *et al.*, 1988; Cohn *et al.*, 1990; Field *et al.*, 1990; Righetti-Veltema *et al.*, 2002; Reck *et al.*, 2004). The effects of postpartum depression on mothers' interactive behaviors are also reflected in depressed mothers' responses to infant cries. We find that in comparison to non-depressed mothers, depressed mothers were particularly prone to feeling anxious when listening to audiotapes of infant cries and differentially responded more negatively to pain as opposed to hunger cries (Gonzalez *et al.*, in preparation), indicating that the problematic interactions seen in depressed mothers may reflect a more general emotional reaction to infant cues.

Maternal Experience and Memory – Parity Effects

The second set of studies explored maternal learning, focusing on behavioral, sensory, neural, and neurochemical mechanisms mediating the maintenance of maternal behavior beyond the period when parturitional hormones facilitate the expression of maternal behavior. In addition, we have explored the effects of multiple parities and associated experiences on mothering (see Bridges, 1975, 1977; Cohen & Bridges, 1981; Fleming & Li, 2002; Li & Fleming, 2003a). Differences in behavior that occur between first-time (primiparous) and multiparous mothers are often subtle or difficult to detect, and they may reflect not only differences in experience, but also differences in the mother's maturity, age, or lactational competence (Beach, 1937; Carlier & Noirot, 1965; Moltz & Robbins, 1965; Swanson & Campbell, 1979; Numan & Insel, 2003). Mothers may exhibit differences in the intensity of particular behaviors, coordination of the different components, or in the speed and efficiency with which behaviors, like retrieval, are expressed (Carlier & Noirot, 1965; Swanson & Campbell, 1979).

In the postpartum mother (when hormonal influences are present) the effects of maternal experience and parity on maternal behavior are most evident when the mother is challenged under specific conditions. Multiparous rats are less disturbed than primiparous mothers by a variety of experimental manipulations including C-section (Moltz *et al.*, 1966), endocrine manipulations (Moltz & Wiener, 1966; Moltz *et al.*, 1969), morphine administration

(Kinsley & Bridges, 1988; Bridges, 1990), and brain lesions (Schlein *et al.*, 1972; Fleming & Rosenblatt, 1974; Franz *et al.*, 1986; Numan, 1994). In addition, maternal experience can affect the threshold of activation of behavior in response to pups and/or their cues. Pup cues which are initially ineffective in eliciting maternal behavior in first-time mothers, may become effective in multiparous animals (Noirot, 1972).

The experiences acquired during multiple births and lactation and the retention of maternal responsiveness as a result of postpartum experience in the first-time mother are likely mediated by the same mechanisms. We know from the maternal experience and maternal memory work that this retention of responsiveness is based on learning that occurs when the mother interacts with her pups at the time of parturition, is enhanced by the hormones of parturition, and is stimulated by odors, touch characteristics and, possibly, vocalizations of the pups (see (Fleming, 2005; Numan *et al.*, 2006). This learning within a species-characteristic context is dependent on multimodal input and on the DA and opioid systems in either the medial preoptic area (MPOA) or accumbens for its acquisition, and on the noradrenergic and DA systems in the accumbens for its consolidation (Moffat *et al.*, 1993; Byrnes *et al.*, 2002; Li & Fleming, 2003a; Li *et al.*, 2004; Parada *et al.*, in press).

In contrast to the rat where more strictly physiological mechanisms are paramount, in humans, prior maternal experience is the single most reliable predictor of positive responsiveness to offspring (Robson & Kumar, 1980; Kaitz *et al.*, 2000). In comparison to primiparous mothers, during the initial postpartum period multiparous mothers are more attracted to the body odors of newborn infants even though they are not better at discriminating their own infants' odors from another infant (Schaal *et al.*, 1980). They respond more rapidly and sympathetically to the cry of their own infant (Bernal, 1972) and more sympathetically to pain than to hunger cries (Stallings *et al.*, 2001). They also show greater heart-rate accelerations to the cry of their own infant (Wisenfeld & Malatesta, 1982) but also show lower arousal responses, suggesting a lower overall arousal state and reduced anxiety (Boukydis & Burgess, 1982).

By 3-months postpartum, when primiparous mothers have become familiar with their infants, parity differences seem to be reversed. Primiparous mothers expressed more affectionate affect, showed higher levels of reciprocal interactions, and provided more infant-directed vocalizations and stimulation than multiparous mothers (Belsky *et al.*, 1984). These relationships between maternal experience, maternal attitudes, and behavior may be mediated by mother's mood, since prior maternal experience is also a strong predictor of postpartum affect (Fleming *et al.*, 1988, Corter & Fleming, 2002)

In conclusion, parity appears to alter the nervous system so that hormones and/or stimuli have an effect that they would not have had otherwise. It is not clear whether parity differences reflect experiences acquired while caring for young or their cues, or other factors like the mothers' age, level of physiological maturation, endocrine/lactational status or other social and psychological developmental factors (Schino *et al.*, 1995). For maternal learning and memory to occur the young must have rewarding properties, which brings us to the third set of studies that we have explored.

Maternal Hedonics and Reward

The third program of work or system directed at understanding the initiation of responsiveness explored perceptual changes that the mother undergoes when she gives birth; where young and their cues (visual, odors, and vocalizations) become more salient and take on a positive valence for the mother via experience.

In the rat, maternal hedonics is reflected in mothers' performance on a variety of preference tests, pairing pup odors with diestrous odors, or the conditioned place preference (CPP) task. As shown in Figure 34.1, lactating females are attracted to maternal nest odors, whereas nulliparous females are not (see also Numan *et al.*, 2006). Likewise, new mothers also orient more to ultrasounds than do nulliparous females (Farrell & Alberts, 2002). These changes in hedonics from pre-pregnancy to the postpartum period ensures that mothers will sustain approach behavior in response to pups and this leads to pup stimuli taking on reinforcing properties, and becoming conditioned stimuli to sustain the maternal learning (Magnusson & Fleming, 1995; Lee *et al.*, 2000; Mattson *et al.*, 2001, 2003). Enhanced pup salience and reward are influenced by the "maternal hormones" and depend

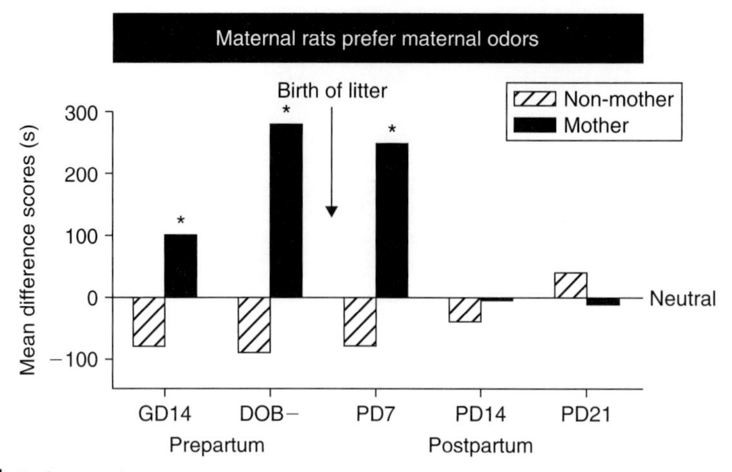

FIGURE 34.1 Time spent by pre- and postpartum (Mother) and nulliparous (virgin, Non-Mother) female rats with the bedding from a lactating dam as opposed to clean wood shavings (scores represent differences between the duration spent sniffing maternal bedding and clean bedding). Before birth (Prepartum, GD14 and day of birth, DOB-) until the first week postpartum (PD7) mother rats are attracted to bedding from maternal animals compared to the non-mothers (*$p < .05$). To address the issue of familiarity, in this same study virgins were exposed to virgin (hence familiar) nest material showed no such preference for their own and familiar odor (not shown). Adapted from Bauer (1983).

on the integrity of both the amygdala and the nucleus accumbens (NAC).

Among humans as well, the young become more salient for the new mother which often involves a shift in their value from a negative to a positive valence. For instance, with little experience with their infants, new human mothers find the infant body odor more attractive than do non-mothers and come to recognize their own infants based on odor (Corter & Fleming, 2002) (Figures 34.2 and 34.3). Visual cues also enhance mothering. When parents view silent videotapes of their infants, the sight of their own baby's crying or smiling causes heart-rate (HR) deceleration then acceleration (Wisenfeld & Klorman, 1978). Mothers responded with HR acceleration when the gaze of an unfamiliar infant was directed toward them, but did not display this arousal when the infant was looking away (Leavitt & Donovan, 1979). At the behavioral level, the infant's gaze evokes mother's gaze and leads to "en face" behavior, which Klaus *et al.* (1975) described as species-typical maternal behavior.

Similarly, human mothers develop appropriate physiological and behavioral responses to infant cries, often experienced by non-parents as aversive. Infants' cries signal need, and mothers often respond by approaching (Stallings *et al.*, 2001). The acoustic and emotion-inducing properties of these signals are well characterized (LaGasse *et al.*, 2005) and are more easily identified and discriminated as mothers gain experience with them.

We explored the subjective, autonomic, and endocrine responses to recorded "pain" and "hunger" cries of unfamiliar infants in non-mothers vs. new mothers. Non-mothers tend to experience emotions that discourage approach responses, whereas new mothers show the opposite. For instance, as shown in Figure 34.4, in comparison to first-time mothers, non-mothers were less sympathetic and less alerted to the intense "pain" cries of infants (but not to "hunger" cries). Further, in comparison to more sympathetic mothers, less sympathetic mothers showed reduced baseline heart rates and salivary stress hormones prior to and while listening to recorded cries (Stallings *et al.*, 2001).

These studies suggest that infant stimuli powerfully elicit maternal behavior and are important as precursors and components of infant–mother interaction. These hedonic effects are related to the amount of experience mothers have with their infants and their hormonal profile during the early postpartum period (Corter & Fleming, 2002). In both rats and humans, hedonically salient infant stimuli orient the mother toward her infant and often involve a shift from withdrawal or aversion to approach and attraction.

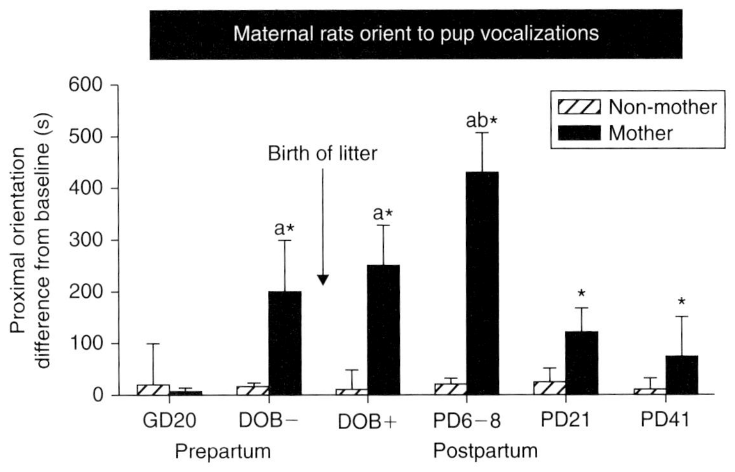

FIGURE 34.2 Behavior of pregnant/lactating dams (Mother) and virgin females (Non-Mother) during exposure to vocalizing pups, as opposed to silent pups (data not shown). Mothers approached and maintained proximal orientation to a vocalizing pup far more than did virgin females. Elevated levels of proximal orientation in pregnant females appeared within hours of birth (DOB-), increasing from birth (DOB+), with significant maximal levels of responding to pup-vocalizations occurring in the 1st week postpartum (postpartum days, PD, 6–8), and, declined by the time of weaning (PD21–41). No change in responding across testing sessions was observed by virgin (cycling) females. Significant differences ($p < .05$) from: [a]GD20; [b]PD21; [c]PD41, and *Non-Mothers at a given time point. Adapted from Farrell and Alberts (2002).

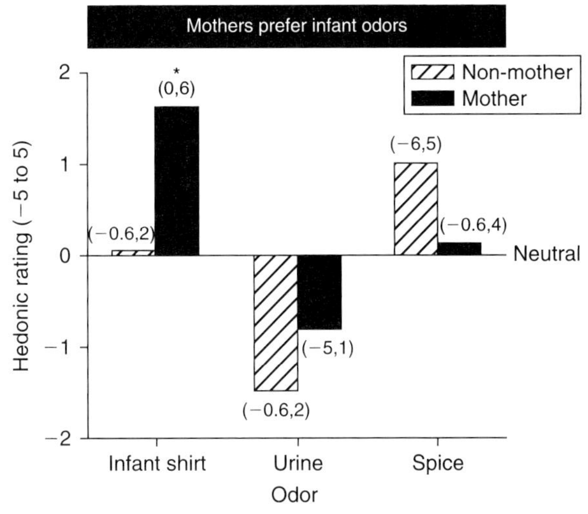

FIGURE 34.3 Hedonics ratings (negative to positive) of different odours (presented on cotton shirts and swabs in Baskin-Robbins containers and concealed by a perforated lid) by first-time postpartum (day 2) mothers and nulliparous non-mothers. As measured by the visual analogue scale (VAS) pleasantness, mothers are more attracted (*$p < 0.05$) to infant body odors than are non-mothers – a result not mirrored with control odours. (*Source*: Adapted from Fleming *et al.* (1993).)

These three lines of biobehavioral evidence converge to increase our understanding of mothers' willingness to engage in maternal behavior at the outset; that is, they support the analyses of aspects of "maternal motivation," or factors that influence mothers' robust responsiveness to their offspring. An excellent review of the analysis of maternal motivation

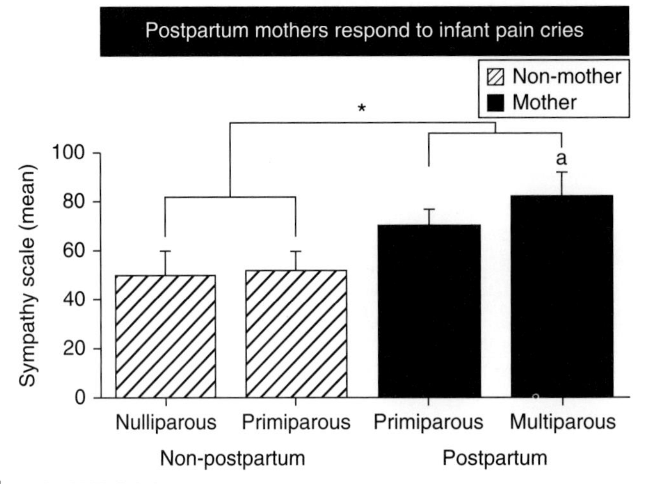

FIGURE 34.4 Sympathy ratings in response to presentation of recorded infant pain cries (not own infant). Postpartum mothers (Mother) are more sympathetic (*$p < .05$) to infant pain cries than are non-postpartum mothers (Non-Mother) despite having had a previous postpartum experience (i.e., primiparous, non-postpartum). However, postpartum mothers that have had a previous postpartum experience (i.e., multiparous, postpartum) demonstrate augmented sympathy ratings compared to the primiparous postpartum mothers ($^{a}p < .05$). Adapted from Stallings, Fleming, Corter, Worthman, & Steiner, (2001).

and its neurobiology may be found in Numan and Insel (2003) and Numan *et al.* (2006).

THE DOING SYSTEM: WHAT DETERMINES THE QUALITY OF MOTHERS' NURTURANT BEHAVIOR?

Maternal Attention and Sensitivity

Once an animal is motivated to mother and expresses all the appropriate consummatory behaviors, under the influence of the hormones of parturition (Numan *et al.*, 2006), a number of other systems contribute to both qualitative and quantitative aspects of its expression. There are large individual differences in mothers' licking, crouching, time in nest and on top of pups, and contingency of responsiveness to pup vocalization and movement cues. This contingent responding is mediated by differences in mothers' attention, both toward pups and in general (Francis *et al.*, 2000; Lovic *et al.*, 2001; Champagne *et al.*, 2003; Lovic & Fleming, 2004; Afonso *et al.*, 2007). As illustrated in Figure 34.5, mother rats who perform better on a variety of attention tasks, especially extradimensional shifting (ED shift) and prepulse inhibition tasks, are less easily distracted, more attentive to their litter, and lick their pups more (Lovic & Fleming, 2004; Lovic *et al.*, in preparation).

Our human research verifies relations between maternal attention and maternal responsiveness. Mothers who respond "sensitively" to their infants exhibit behaviors that are contingent, timely, and appropriate in relation to the infants' behavior. These behaviors are complex and involve cross-situation adaptations. During play, components of sensitivity include structuring interactions, following the infant's lead, adjusting the interaction so as to remain attuned to the infant. Under conditions of infant distress, four response phases include noticing the distress signal and intervening, and the timing and effectiveness of the interventions. If the infant is not soothed by the intervention, the parent must switch interventions. At the same time, the parent must balance interactions with the infant against competing stimuli and demands. These behaviors involve coordination across varied situations in response to multiple signals and fully functioning attentional processes.

Using the Cambridge Neuropsychological Test Automated Battery CANTAB®, we found that mothers who respond with fewer errors on the ED shifting task (hence greater cognitive flexibility), but not on measures of sustained attention, at 2–6 months postpartum, interact

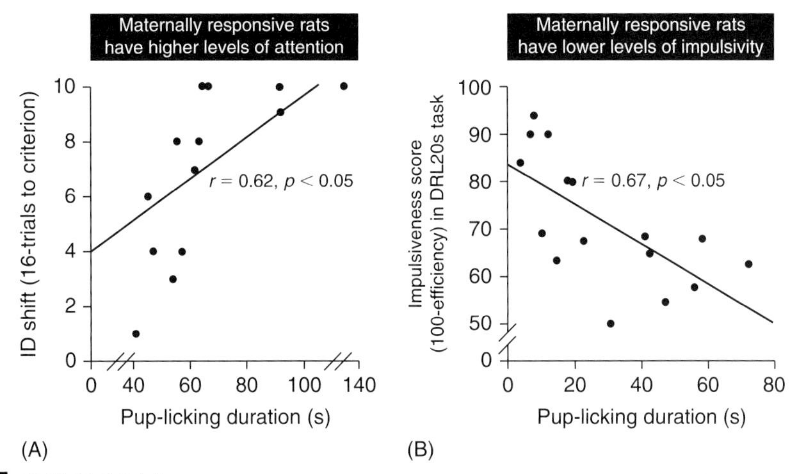

FIGURE 34.5 Correlations between maternal behavior and attention (left) or impulsivity (right). In postpartum rats, higher levels of attention (measured with ID/ED set-shifting task) and lower levels of impulsivity (measured with DLR-20s) are associated with increased pup-licking durations. Adapted from Lovic & Fleming, 2004; Lovic & Fleming, (in prep).

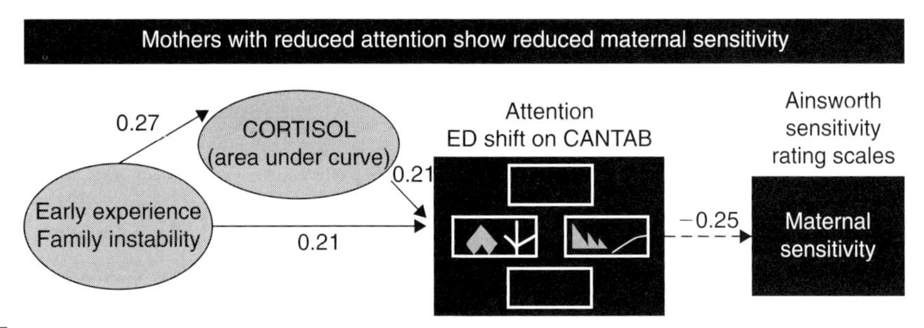

FIGURE 34.6 Mothers (4 months postpartum) with reduced performance on extradimensional set-shifting task (ED shift in CANTAB) show reduced maternal sensitivity, as measured by increased contingent responding to infant cues. Disruptions in early life that may affect cortisol level correlate with reduced attention. Numbers represent β weights (Gonzalez, Jenkins, Steiner, & Fleming, in prep)..

more sensitively with their infants, based on the Ainsworth (1979) attachment scales, and show more contingent responding to infant cues (Gonzalez *et al.*, in preparation). These relations are illustrated in Figure 34.6. In terms of selective attention and mothers' sensitivity to their infants (Atkinson *et al.*, 2005; Atkinson *et al.*, in press) as assessed by the emotional Stroop task (Williams *et al.*, 1996), at 6 and 12 months postpartum mothers who showed greater delays in responding to target stimuli, as compared to neutral stimuli, also showed less sensitivity toward the infant. These results and other results by Atkinson *et al.* (2005) suggest that in humans attentional mechanisms are central to maternal sensitivity.

NEUROANATOMY OF MATERNAL BEHAVIOR

Thanks to the concerted effort of a number of laboratories using convergent technologies (lesions, stimulation, hormone implants, c-Fos and gfap expression, electrophysiology, fMRI, and microdialysis), we now have quite a good understanding of the neural circuitry underlying maternal behavior, most especially the circuitry within the limbic–hypothalamic regions. This work is best described by Michael Numan, the scientist who gave life to the involvement of the MPOA (see Numan & Insel, 2003, Numan *et al.*, 2006) in the maternal behavior neural circuitry.

The maternal circuit involves both excitatory and inhibitory systems. Most work in the area has focused on the final common path for the expression of the behavior, the medial preoptic area/ventral bed nucleus of the stria terminalis (MPOA/vBNST), and its downstream projections into the midbrain (ventral tegmental area, VTA) and hindbrain (periaqueductal gray, PAG), and sensory, limbic, and cortical systems that project to the MPOA/vBNST. The MPOA contains receptors for all the hormones implicated in the activation of maternal behavior, including receptors for estradiol, progesterone, prolactin, oxytocin, vasopressin, and opioids (see Numan *et al.*, 2006). As we have emphasized (Numan *et al.*, 2006), neurons projecting into the MPOA are involved in other behavioral changes, including changes in mothers affect (amygdala, orbitofrontal cortex), stimulus salience (amygdala and striatum/NAC), attention (NAC and medial prefrontal cortex, mPFC), and memory (NAC, mPFC). Some of these sites also contain hormone receptors (amygdala, mPFC) and may be the sites where the periparturitional hormones act to change behavior at the time of parturition (Numan *et al.*, 2006). The relatively complicated neuroanatomy of maternal behavior is based predominantly on work with rats, voles, sheep, and primates (summarized in Numan *et al.*, 2006; see Chapter 1). Taken together, these cross-species studies indicate a striking similarity in the neuroanatomy underlying mothering.

Work on neural bases of maternal behavior in humans is derived primarily from fMRI studies where mothers, non-mothers, and sometimes fathers are presented with either pictures of their own infants or same-aged unfamiliar infants (Bartels & Zeki, 2004; Leibenluft *et al.*, 2004; Nitschke *et al.*, 2004), recorded infant cries (Seifritz *et al.*, 2003), or videotapes of infants (Ranote *et al.*, 2004). These studies in general focus on the effects of these infant signals or cues on brain activation in regions described above. These brain regions are known to mediate positive and negative emotions, stimulus salience, and are involved in reinforcement, cognitive and executive function, memory, experience, and reproductive and affiliative behaviors.

Presently, there are a few studies exploring mothers' responses to infants' cries and/or pictures. Collectively, these studies demonstrate that many of the same hypothalamic, limbic, and cortical sites important for emotional or social (face) processing (Seifritz *et al.*, 2003; Bartels & Zeki, 2004; Leibenluft *et al.*, 2004; Nitschke *et al.*, 2004) or for regulation of maternal behavior in other mammals (Lorberbaum *et al.*, 2002; Seifritz *et al.*, 2003; Bartels & Zeki, 2004; Swain *et al.*, 2007) are implicated. A subset of these also demonstrates that mothers exhibit a different pattern of responses to infant cues than do non-mothers or fathers (Seifritz *et al.*, 2003). Others have illustrated that brain systems underlying maternal motivation include systems that are both specific to maternal behavior and systems that are recruited in a variety of motivational contexts and involve rewarding social stimuli. For instance, the two studies by Bartels and Zeki (2000, 2004) presented either infant-related pictures or pictures of romantic partners to women and found that various brain regions were activated by both sets of stimuli (especially the striatum, middle insula, and anterior cingulate cortex), whereas other regions were more specialized and were activated exclusively by infant stimuli (orbitofrontal cortex (in PFC) and the PAG area) or only by romantic stimuli (dentate gyrus, hippocampus, hypothalamus).

In general, "both studies on maternal and romantic attachment revealed activity that was not only overlapping to a large extent with each other, but also with the reward circuitry of the human brain" (Bartels & Zeki, 2004) and with sites rich in receptors for two neurohormones that have been associated with social affiliation, oxytocin, and vasopressin (Bartels & Zeki, 2004).

DA AND MOTHERING

Mesolimbic System and the Nucleus Accumbens

There is now extensive evidence from animal research that the neurotransmitter DA is intimately involved in the normal expression of maternal behavior, in reinforcement processes, associated learning, and attention (Giordano *et al.*, 1990; Hansen *et al.*, 1991a, b; Di Chiara *et al.*, 1999; Keer & Stern, 1999; Byrnes *et al.*, 2002; Li & Fleming, 2003a, b; Lonstein *et al.*, 2003; Silva *et al.*, 2003; Champagne *et al.*, 2004; Numan *et al.*, 2005; Chudasama & Robbins, 2006; Stefani & Moghaddam, 2006; Numan, 2007). As an example of mesolimbic effects on these functions, lesions to the NAC which receive considerable DA input from the midbrain, retard mothers' initial pup-retrieval responses and block the formation of a maternal

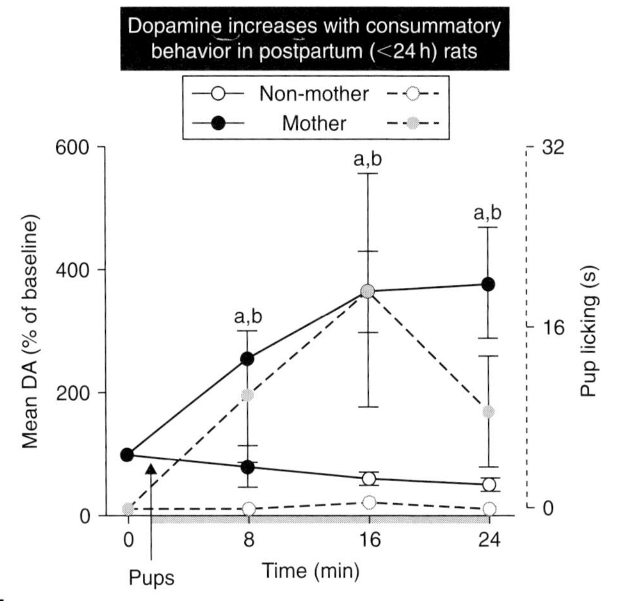

FIGURE 34.7 Dopamine responses to pup-stimuli (as measured from microdialysis samples from the nucleus accumbens shell) in freely moving females. Dopamine (DA) increases in response to pup-stimuli in females with increased sensitivity to pup cues (*Mother*: postpartum lactating dams) relative to females that are less sensitive to these cues (*Non-Mother*: pup-naïve, virgin females). These dopamine responses mirror consummatory behaviors in lactating dams. [a]DA and [b]behavioral differences ($p < .05$) between Mother and Non-Mother groups at a given time point during the presentation of pup-stimuli (gray bar on *x*-axis). Adapted from Afonso, King & Fleming, (2008).

memory, possibly by altering the emotional valence of the conditioned pup stimulus (Keer & Stern, 1999; Lee *et al.*, 2000; Li & Fleming, 2003a, b; Numan *et al.*, 2005). The administration of DA-blockers to the NAC (or, in some cases, the MPOA) have the same basic effect (Keer & Stern, 1999; Li & Fleming, 2003a, b; Miller & Lonstein, 2005; Parada *et al.*, in press), while administration of D1R and D2R agonists into NAC of pregnancy-terminated females enhances the onset of maternal behavior (Numan *et al.*, 2005; Stolzenberg *et al.*, 2007). Finally, simple exposure of maternally experienced animals to pups as opposed to food results in a rapid release of DA from the NAC (Hansen *et al.*, 1993; Champagne *et al.*, 2004; Afonso *et al.*, 2008).

In a series of recent studies of microdialysis we have followed up the seminal work of Hansen and colleagues (Hansen *et al.*, 1991a,b, 1993) exploring the precise conditions within the maternal context that result in DA release in the striatum. Recently, we have observed in postpartum rats (within 24 h after birth) significantly elevated NAC shell DA levels (as measured *in vivo*) in response to pups compared to their basal levels and compared to levels of cycling females, also exposed to pups (Afonso *et al.*, 2008) (Figure 34.7). This DA response was also observed in OVX female rats given maternal hormone-like treatments, and in comparison to levels shown by cholesterol-treated females, or their own basal levels. As shown in Figure 34.8, additionally, we have found that cycling rats with prior maternal experiences with pups through postpartum experiences or females with recent continuous donor pup exposure (5–10 days) had increased DA release associated with initial interactions (Afonso *et al.*, 2007). However, the profile of the response was not as prolonged as seen in animals that are hormonally primed at the time of pup exposure. DA responses in the NAC shell to pup stimuli also correlated significantly with: (a) approach behaviors (i.e., pup sniffing) in non-lactating rats and (b) consummatory behaviors (i.e., pup licking) in lactating and hormone-treated rats.

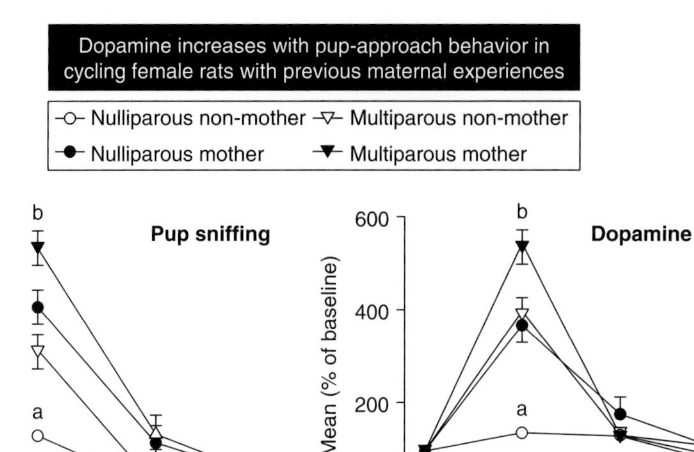

FIGURE 34.8 Dopamine responses to pup-stimuli (as measured from microdialysis samples from the nucleus accumbens shell) in freely moving females. Dopamine (DA) increases initially (only at 8 min) in response to pup-stimuli in females with increased sensitivity to pup cues (*Mother*: nulliparous and multiparous, pup-induced females) relative to females that are less sensitive to these cues (*Non-Mother*: nulliparous and multiparous females with no pup-induced sensitivity, respectively). These dopamine responses mirror approach behaviors in cycling females. [a]Inexperienced, virgin females differed ($p < .05$) from all the experienced groups. [b]Very experienced mothers (prior parities and recent pup-induced sensitization) differed ($p < .05$) from other less experienced females. Pup-stimuli presentation (gray bar on x-axis). Adapted from Afonso *et al.*, (2008).

Mesocortical System and mPFC

Although it has not been intensively studied in rats, there is evidence that although the prefrontal cortex may not be involved in the initiation of maternal responding, it is likely involved in the temporal organization of many of the maternal behaviors and, along with the NAC system, in the positive tone of pup salience (Dietrich & Allen, 1998; Schneider & Koch, 2005; Schultz, 2007). This region is particularly interesting because it receives extensive dopaminergic input (Leonard, 1969; Thierry *et al.*, 1986), contains heavy concentrations of DA receptors (Floresco & Magyar, 2006), is involved in the regulation of attentional/motivational processes known to be dopaminergically mediated (Bassareo *et al.*, 2002), and interacts with (inhibits) DA release in the NAC (Jackson *et al.*, 2001).

In terms of its role of the mPFC in maternal, Stamm (1955) and Slotnick (1967) found that mPFC lesions in female rats resulted in disorganized and persistent retrieving, where animals would seem to perseverate in their retrievals, picking pups up but failing then to release them into the nest (Stamm, 1955; Slotnick, 1967). A more recent report by Afonso *et al.* (2007) found that lesions of the mPFC in females prior to mating disrupted the frequency, duration, and execution of many maternal behavior sequences associated with licking, hovering, and especially retrieval in which postpartum requires the coordination of sensory and motor information in a spatial context (Sutherland *et al.*, 1988; Kolb & Whishaw, 1989; Kolb, 1990; Kolb & Cioe, 2001).

Moreover, lactating mothers exposed to pup-related nest odors show a unique pattern of electrical activity in the mPFC, a pattern not seen in proestrous or estrous rats (Hernandez-Gonzalez *et al.*, 2005). In addition, Walsh *et al.* (1996) report elevated c-Fos activation in the mPFC during their first experience with young, an effect also found for ewes (Kendrick *et al.*, 1997). As shown in Figure 34.9, given its general attentional inhibitory function (Afonso *et al.*, 2007) we also find that mPFC lesioned animals showing retrieval deficits also showed

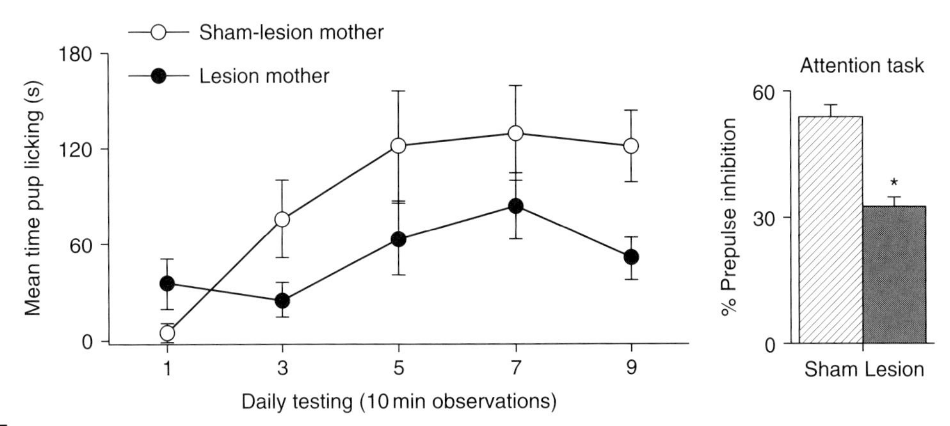

FIGURE 34.9 Rats with medial prefrontal lesions, N-methyl-D-aspartic acid (NMDA) show reduced licking (black filled) across days. Similarly these impaired ($^*p<.05$) mothers have reduced attention, as measured in a prepulse inhibition to the acoustic (72dB) startle reflex task (inset graph). Adapted from Afonso *et al.* (2007).

reduced inhibition in the prepulse inhibition task, an effect of mPFC lesions similar to those previously reported in some (Afonso *et al.*, 2007, Parada *et al.*, in press), although not all studies (see Swerdlow *et al.*, 1995; Lacroix *et al.*, 1998, 2000). The mPFC also appears to play a role in higher levels of sensory processing, especially of emotionally relevant stimuli (Takenouchi *et al.*, 1999; Jodo *et al.*, 2000; Schneider & Koch, 2005).

The pattern of DA release in the mPFC as measured by microdialysis during the onset of maternal behavior and during the ongoing expression of maternal behavior in response to pups merits investigation. Based on research involving the NAC, we predict that if the two sites differentially mediate aspects of maternal responding, then the profile of DA release in the mPFC in response to pups and food reward would differ from that in the NAC shell following maternal hormone and pup-experience manipulations previously used in our *in vivo* studies.

In humans the importance of planning, cognitive flexibility, and selective attention for mothering is considerably greater than that in rodents. For this reason the importance of the prefrontal cortex is likely to be a key as a regulatory mechanism in the organization of mothering. Within the context of the circuitry involved in cognitive flexibility and selective attention (performance on the ID/ED shifting and the Stroop), several imaging studies demonstrate differential activation of neural areas associated with these tasks, including the anterior cingulate and the dorsolateral prefrontal cortex. These same areas are also activated when viewing pictures of own vs. other children and when hearing infant cries (Lorberbaum *et al.*, 2002; Bartels & Zeki, 2004; Leibenluft *et al.*, 2004; Ranote *et al.*, 2004). A schematic of the brain and summary of studies highlighting activation and deactivation in response to attention tasks and infant cues is provided in Figure 34.10.

While it is difficult to extrapolate the meaning of differential activation and deactivation of these areas given that mothers are usually instructed to passively view or listen to the presented stimuli (and are not actively responding, judging or manipulating them), it is clear that these areas are involved in either the emotional or cognitive processing of infant relevant stimuli and are associated with other systems involved in emotion regulation, theory of mind and empathy (see Figure 34.10).

CONCLUSION

Before mothers can respond appropriately and sensitively to their offspring they must "want" to do so. The motivation to mother is multiply regulated by hormonal and experiential

Activation of frontal cortex in response to infant cues and attentional tasks

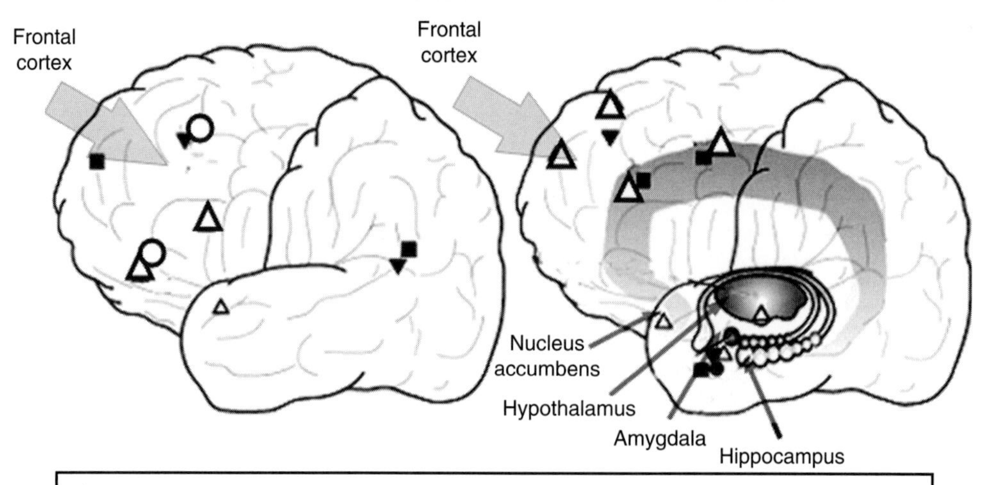

△ **Activation in mothers in response to infant cues (cries pictures).** (Bartels and Zeki, 2004; Lorberbaum *et al.*, 2002; Nitschke *et al.*, 2004; Popeski *et al.*, in prep;Selfritz *et al.*, 2003; Strathearn and McClure, 2002; Swain *et al.*, 2007)

○ **Activation during cognitive and flexibility (ID/ED and WCST) tasks.** (Alvarez and Emory, 2006; Barceló *et al.*,, 2000; Cabeza & Nyberg, 2003; Rogers *et al.*, 2000).

▼ **Deactivation in mothers response to infant cues (cries and pictures).** (see above.)

■ **Activation of empathy** (Völlm *et al.*, 2006).

FIGURE 34.10 Overlapping clusters of activation and deactivation in various neuroimaging studies with mothers' in response to infant cues (cries and pictures), tests of cognitive flexibility (ID/ED shifting and WCST) and in empathy and theory of mind (TOM). Left: lateral surface view. Right: mid-saggital view. Shapes are placed schematically within a region and are not meant to indicate a precise location within a region. Adapted from Gonzalez et al., (submitted).

effects on mothers' emotions, their perceptions of their young, and their ability to gain from experiences with them. These changes normally occur at the time of parturition or after extensive experience with young and are mediated by mesolimbic DA systems interacting with an extensive forebrain maternal neural circuit. The act of mothering and the appropriateness and flexibility in the quality of mothering provided results from these changes in affect, perception, and hedonics. Moreover, it depends on mothers attending adequately to their young, responding contingently to them, and showing flexibility in which, and when, responses occur. This latter aspect of mothering relates more to the "doing" of mothering and depends on higher cortical systems, especially within the prefrontal cortex. Although it has not been adequately studied in either rats or human mothers, we propose that the DA system is likely a major player in the success and appropriateness of prefrontal system functioning within the context of maternal behavior.

REFERENCES

Afonso, V. M., Sison, M., Lovic, V., and Fleming, A. S. (2007). Medial prefrontal cortex lesions in the female rat affect sexual and maternal behavior and their sequential organization. *Behav. Neurosci.* 121, 515–526.

Afonso, V. M., Grella, S. L., Chatterjee, D., and Fleming, A. S. (2008). Maternal experience affects accumbal dopaminergic responses to pups. *Brain Res.*1198, 115–123.

Afonso, V. M., King, S., Novokov, M., Burton, C., and Fleming, A. S. (2008). Hormones and maternal experiences affect accumbal dopaminergic responses. Program No. 103, *Abstract Viewer*. Chicago, IL: Association for Psychological Science. Online.

Ainsworth, M. D. (1979). Infant–mother attachment. *Am. Psychol.* 34, 932–937.

Alvarez, J. A., and Emory, E. (2006). Executive function and the frontal lobes: A meta-analytic review. *Neuropsychol. Rev.* 16, 17–42.

Atkinson, L., Goldberg, S., Raval, V., Pederson, D., Benoit, D., Moran, G., Poulton, L., Myhal, N., Zwiers, M., Gleason, K., and Leung, E. (2005). On the relation between maternal state of mind and sensitivity in the prediction of infant attachment security. *Dev. Psychol.* 41, 42–53.

Atkinson, L., Leung, E., Goldberg, S., Benoit, D., Poulton, L., and Myhal, N. (in press). Attachment and selective attention: Disorganization and emotional Stroop reaction time. *Dev. Psychopathol.*

Barcelo, F., Suwazono, S., and Knight, R. T. (2000). Prefrontal modulation of visual processing in humans. *Nat. Neurosci.* 3, 399–403.

Bartels, A., and Zeki, S. (2000). The neural basis of romantic love. *Neuroreport* 11, 3829–3834.

Bartels, A., and Zeki, S. (2004). The neural correlates of maternal and romantic love. *Neuroimage* 21, 1155–1166.

Bassareo, V., De Luca, M. A., and Di Chiara, G. (2002). Differential expression of motivational stimulus properties by dopamine in nucleus accumbens shell versus core and prefrontal cortex. *J. Neurosci.* 22, 4709–4719.

Bauer, J. H. (1983). Effects of maternal state on the responsiveness to nest odors of hooded rats. *Physiol. Behav.* 30, 229–232.

Beach, F. A. (1937). The neural basis of innate behavior: I. Effects of cortical lesions upon the maternal behavior pattern in the rat. *J. Comp. Physiol. Psychol.* 24, 393–436.

Beck, C. T. (2006). Postpartum depression: It isn't just the blues. *Am. J. Nurs.* 106, 40–50. quiz 50-41

Belsky, J., Rovine, M., and Taylor, D. (1984). The Pennsylvania Infant and Family Development Project, III: The origins of individual differences in infant–mother attachment: Maternal and infant contributions. *Child Dev.* 55, 718–728.

Bernal, J. (1972). Crying during the first 10 days of life and maternal responses. *Dev. Med. Child Neurol.* 14, 362–372.

Boukydis, C. F., and Burgess, R. L. (1982). Adult physiological response to infant cries: Effects of temperament of infant, parental status, and gender. *Child Dev.* 53, 1291–1298.

Bridges, R. S. (1975). Long-term effects of pregnancy and parturition upon maternal responsiveness in the rat. *Physiol. Behav.* 14, 245–249.

Bridges, R. S. (1977). Parturition: Its role in the long term retention of maternal behavior in the rat. *Physiol. Behav.* 18, 487–490.

Bridges, R. S. (1990). Endocrine regulation of parental behavior in rodents. In *Mammalian Parenting: Biochemical, Neurobiological, and Behavioral Determinants* (N. A. Krasnegor and R. S. Bridges, Eds.), pp. 93–117. Oxford University Press, New York, NY.

Byrnes, E. M., Rigero, B. A., and Bridges, R. S. (2002). Dopamine antagonists during parturition disrupt maternal care and the retention of maternal behavior in rats. *Pharmacol. Biochem. Behav.* 73, 869–875.

Cabeza, R., and Nyberg, L. (2003). Functional neuroimaging of memory. *Neuropsychologia* 41, 241–244.

Carlier, C., and Noirot, E. (1965). Effects of previous experience on maternal retrieving by rats. *Anim. Behav.* 13, 423–426.

Champagne, F. A., Francis, D. D., Mar, A., and Meaney, M. J. (2003). Variations in maternal care in the rat as a mediating influence for the effects of environment on development. *Physiol. Behav.* 79, 359–371.

Champagne, F. A., Chretien, P., Stevenson, C. W., Zhang, T. Y., Gratton, A., and Meaney, M. J. (2004). Variations in nucleus accumbens dopamine associated with individual differences in maternal behavior in the rat. *J. Neurosci.* 24, 4113–4123.

Chudasama, Y., and Robbins, T. W. (2006). Functions of frontostriatal systems in cognition: Comparative neuropsychopharmacological studies in rats, monkeys and humans. *Biol. Psychol.* 73, 19–38.

Cohen, J., and Bridges, R. S. (1981). Retention of maternal behavior in nulliparous and primiparous rats: Effects of duration of previous maternal experience. *J. Comp. Physiol. Psychol.* 95, 450–459.

Cohn, J. F., Campbell, S. B., Matias, R., and Hopkins, J. (1990). Face-to-face interactions of postpartum depressed and nondepressed mother–infant pairs at 2 months. *Dev. Psychol.* 35, 119–123.

Corter, C., and Fleming, A. S. (2002). Psychobiology of maternal behavior in human beings. In *Handbook of Parenting: Biology and Ecology of Parenting* (M. H. Bornstein, Ed.), pp. 141–182. Lawrence Erlbaum Associates, New Jersey.

Di Chiara, G., Loddo, P., and Tanda, G. (1999). Reciprocal changes in prefrontal and limbic dopamine responsiveness to aversive and rewarding stimuli after chronic mild stress: Implications for the psychobiology of depression. *Biol. Psychiatr.* 46, 1624–1633.

Dietrich, A., and Allen, J. D. (1998). Functional dissociation of the prefrontal cortex and the hippocampus in timing behavior. *Behav. Neurosci.* 112, 1043–1047.

Farrell, W. J., and Alberts, J. R. (2002). Maternal responsiveness to infant Norway rat (*Rattus norvegicus*) ultrasonic vocalizations during the maternal behavior cycle and after steroid and experiential induction regimens. *J. Comp. Psychol.* 116, 286–296.

Field, T., Healy, B., Goldstein, S., and Guthertz, M. (1990). Behavior-state matching and synchrony in mother–infant interactions of non-depressed versus depressed dyads. *Dev. Psychol.* 26, 7–14.

Fleming, A., and Gonzalez, A. Neurobiology of human maternal care. In *Endocrinology of Social Relationships* (P. Gray P. Ellison, Eds), Harvard University Press, Cambridge, MA (in preparation).

Fleming, A. S. (2005). Plasticity of innate behavior. In *Attachment and Bonding: A New Synthesis* (C. S. Carter, K. E. Grossman, S. B. Hrdy, M. E. Lamb, S. W. Porges, and N. Sachser, Eds.). MIT Press, Cambridge MA.

Fleming, A. S., and Li, M. (2002). Psychobiology of maternal behavior and its early determinants in nonhuman mammals. In *Handbook of Parenting* (M. H. Bornstein, Ed.), Vol.2, pp. 61–97. Lawrence Erlbaum Associates, Mahwah, NJ.

Fleming, A. S., and Rosenblatt, J. S. (1974). Olfactory regulation of maternal behavior in rats. I. Effects of olfactory bulb removal in experienced and inexperienced lactating and cycling females. *J. Comp. Physiol. Psychol.* 86, 221–232.

Fleming, A. S., Ruble, D. N., Flett, G. L., and Shaul, D. (1988). Postpartum adjustment in first-time mothers: Relations between mood, maternal attitudes and mother–infant interactions. *Dev. Psychol.* 24, 71–81.

Fleming, A. S., Corter, C., Franks, P., Surbey, M., Schneider, B., and Steiner, M. (1993). Postpartum factors related to mother's attraction to newborn infant odors. *Dev. Psychobiol.* 26, 115–132.

Fleming, A. S., Ruble, D., Krieger, H., and Wong, P. Y. (1997a). Hormonal and experiential correlates of maternal responsiveness during pregnancy and the puerperium in human mothers. *Horm. Behav.* 31, 145–158.

Fleming, A. S., Steiner, M., and Corter, C. (1997b). Cortisol, hedonics, and maternal responsiveness in human mothers. *Horm. Behav.* 32, 85–98.

Fleming, A. S., O'Day, D. H., and Kraemer, G. W. (1999). Neurobiology of mother–infant interactions: Experience and central nervous system plasticity across development and generations. *Neurosci. Biobehav. Rev.* 23, 673–685.

Floresco, S. B., and Magyar, O. (2006). Mesocortical dopamine modulation of executive functions: Beyond working memory. *Psychopharmacology (Berl)* 188, 567–585.

Francis, D. D., Champagne, F. C., and Meaney, M. J. (2000). Variations in maternal behaviour are associated with differences in oxytocin receptor levels in the rat. *J. Neuroendocrinol.* 12, 1145–1148.

Franz, J. R., Leo, R. J., Steuer, M. A., and Kristal, M. B. (1986). Effects of hypothalamic knife cuts and experience on maternal behavior in the rat. *Physiol. Behav.* 38, 629–640.

Giordano, A. L., Johnson, A. E., and Rosenblatt, J. S. (1990). Haloperidol-induced disruption of retrieval behavior and reversal with apomorphine in lactating rats. *Physiol. Behav.* 48, 211–214.

Gonzalez, A., Atkinson, L., and Fleming, A. S. Attachment and the comparative psychobiology of mothering. In M. de Haan & M. Gunnar (Eds), *Handbook of Developmental Social Neuroscience*. Guilford press.

Gonzalez, A., Steiner, M., and Fleming, A. Depressed mothers show altered physiological responsiveness to infant cries (in preparation).

Hansen, S., Harthon, C., Wallin, E., Lofberg, L., and Svensson, K. (1991a). The effects of 6-OHDA-induced dopamine depletions in the ventral and dorsal striatum on maternal and sexual behavior in the female rat. *Pharmacol. Biochem. Behav.* 39, 71–77.

Hansen, S., Harthon, C., Wallin, E., Lofberg, L., and Svensson, K. (1991b). Mesotelencephalic dopamine system and reproductive behavior in the female rat: Effects of ventral tegmental 6-hydroxydopamine lesions on maternal and sexual responsiveness. *Behav. Neurosci.* 105, 588–598.

Hansen, S., Bergvall, A. H., and Nyiredi, S. (1993). Interaction with pups enhances dopamine release in the ventral striatum of maternal rats: A microdialysis study. *Pharmacol. Biochem. Behav.* 45, 673–676.

Hernandez-Gonzalez, M., Prieto-Beracoechea, C., Navarro-Meza, M., Ramos-Guevara, J. P., Reyes-Cortes, R., and Guevara, M. A. (2005). Prefrontal and tegmental electrical activity during olfactory stimulation in virgin and lactating rats. *Physiol. Behav.* 83, 749–758.

Jackson, M. E., Frost, A. S., and Moghaddam, B. (2001). Stimulation of prefrontal cortex at physiologically relevant frequencies inhibits dopamine release in the nucleus accumbens. *J. Neurochem.* 78, 920–923.

Jodo, E., Suzuki, Y., and Kayama, Y. (2000). Selective responsiveness of medial prefrontal cortex neurons to the meaningful stimulus with a low probability of occurrence in rats. *Brain Res.* 856, 68–74.

Kaitz, M., Chriki, M., Bear-Scharf, L., Nir, T., and Eidelman, A. I. (2000). Effectiveness of primiparae and multiparae at soothing their newborn infants. *J. Gen. Psychol.* 161, 203–215.

Keer, S. E., and Stern, J. M. (1999). Dopamine receptor blockade in the nucleus accumbens inhibits maternal retrieval and licking, but enhances nursing behavior in lactating rats. *Physiol. Behav.* 67, 659–669.

Kendrick, K. M., Da Costa, A. P. C., Broad, K. D., Ohkura, S., Guevara, R., Levy, F., and Keverne, E. B. (1997). Neural control of maternal behaviour and olfactory recognition of offspring. *Brain Res. Bull.* 44, 383–395.

Kinsley, C. H., and Bridges, R. S. (1988). Parity-associated reductions in behavioral sensitivity to opiates. *Biol. Reprod.* 39, 270–278.

Klaus, M. H., Trause, M. A., and Kennell, J. H. (1975). Does human maternal behaviour after delivery show a characteristic pattern?. *Ciba Found. Symp.*, 69–85.

Kolb, B. (1990). Animal models for human PFC-related disorders. *Prog. Brain Res.* 85, 501–519.

Kolb, B., and Cioe, J. (2001). Cryoanethesia on postnatal day 1, but not day 10, affects adult behavior and cortical morphology in rats. *Brain Res. Dev. Brain Res.* 130, 9–14.

Kolb, B., and Whishaw, I. Q. (1989). Plasticity in the neocortex: Mechanisms underlying recovery from early brain damage. *Prog. Neurobiol.* 32, 235–276.

Lacroix, L., Broersen, L. M., Weiner, I., and Feldon, J. (1998). The effects of excitotoxic lesion of the medial prefrontal cortex on latent inhibition, prepulse inhibition, food hoarding, elevated plus maze, active avoidance and locomotor activity in the rat. *Neuroscience* 84, 431–442.

Lacroix, L., Broersen, L. M., Feldon, J., and Weiner, I. (2000). Effects of local infusions of dopaminergic drugs into the medial prefrontal cortex of rats on latent inhibition, prepulse inhibition and amphetamine induced activity. *Behav. Brain Res.* 107, 111–121.

LaGasse, L. L., Neal, A. R., and Lester, B. M. (2005). Assessment of infant cry: Acoustic cry analysis and parental perception. *Ment. Retard. Dev. Disabil. Res. Rev.* 11, 83–93.

Leavitt, L., and Donovan, W. (1979). Perceived infant temperament, focus of control, and maternal physiological response to infant gaze. *J. Res. Pers.* 13, 267–278.

Lee, A., Clancy, S., and Fleming, A. S. (2000). Mother rats bar-press for pups: Effects of lesions of the MPOA and limbic sites on maternal behavior and operant responding for pup-reinforecement. *Behav. Brain Res.* 108, 215–231.

Leibenluft, E., Gobbini, M. I., Harrison, T., and Haxby, J. V. (2004). Mothers' neural activation in response to pictures of their children and other children. *Biol. Psychiatr.* 56, 225–232.

Leonard, C. M. (1969). The prefrontal cortex of the rat: I. Cortical projection of the mediodorsal nucleus. II. Efferent connections. *Brain Res.* 12, 321–343.

Li, M., and Fleming, A. S. (2003a). The nucleus accumbens shell is critical for normal expression of pup-retrieval in postpartum female rats. *Behav. Brain Res.* 145, 99–111.

Li, M., and Fleming, A. S. (2003b). Differential involvement of the nucleus accumbens shell and core subregions in maternal memory in postpartum female rats. *Behav. Neurosci.* 117, 426–445.

Li, M., Davidson, P., Budin, R., Kapur, S., and Fleming, A. S. (2004). Effects of typical and atypical antipsychotic drugs on maternal behavior in postpartum female rats. *Schizophr. Res.* 70, 69–80.

Lonstein, J. S., Dominguez, J. M., Putnam, S. K., DeVries, G. J., and Hull, E. M. (2003). Intracellular preoptic and striatal monoamines in pregnant and lactating rats: Possible role in maternal behavior. *Brain Res.* 970, 149–158.

Lorberbaum, J. P., Newman, J. D., Horwitz, A. R., Dubno, J. R., Lydiard, R. B., Hammer, M. B., Bohning, D. E., and George, M. S. (2002). A potential role for thalamocingulate circuitry in human maternal behavior. *Biol. Psychiatr.* 51, 431–445.

Lovic, V., and Fleming, A. S. (2004). Artificially-reared female rats show reduced prepulse inhibition and deficits in the attentional set shifting task – reversal of effects with maternal-like licking stimulation. *Behav. Brain Res.* 148, 209–219.

Lovic, V., Gonzalez, A., and Fleming, A. S. (2001). Maternally separated rats show deficits in maternal care in adulthood. *Dev. Psychobiol.* 39, 19–33.

Lovic, V., Palombo, D. J., Kraemer, G. W., and Fleming, A. S. Relationship between motor impulsiveness and maternal behaviour (in preparation).

Magnusson, J. E., and Fleming, A. S. (1995). Rat pups are reinforcing to the maternal rat: Role of sensory cues. *Psychobiology* 23, 69–75.

Mattson, B. J., Williams, S., Rosenblatt, J. S., and Morrell, J. I. (2001). Comparison of two positive reinforcing stimuli: Pups and cocaine throughout the postpartum period. *Behav. Neurosci.* 115, 683–694.

Mattson, B. J., Williams, S. E., Rosenblatt, J. S., and Morrell, J. I. (2003). Preferences for cocaine- or pup-associated chambers differentiates otherwise behaviorally identical postpartum maternal rats. *Psychopharmacology (Berl)* 167, 1–8.

Miller, S. M., and Lonstein, J. S. (2005). Dopamine d1 and d2 receptor antagonism in the preoptic area produces different effects on maternal behavior in lactating rats. *Behav. Neurosci.* 119, 1072–1083.

Moffat, S. D., Suh, E. J., and Fleming, A. S. (1993). Noradrenergic involvement in the consolidation of maternal experience in postpartum rats. *Physiol. Behav.* 53, 805–811.

Moltz, H., and Robbins, D. (1965). Maternal behavior of primiparous and multiparous rats. *J. Comp. Physiol. Psychol.* 60, 417–421.

Moltz, H., and Wiener, E. (1966). Effects of ovariectomy on maternal behavior of primiparous and multiparous rats. *J. Comp. Physiol. Psychol.* 61, 455–460.

Moltz, H., Robbins, D., and Parks, M. (1966). Caesarean delivery and maternal behavior of primiparous and multiparous rats. *J. Comp. Physiol. Psychol.* 61, 455–460.

Moltz, H., Levin, R., and Leon, M. (1969). Differential effects of progesterone on the maternal behavior of primiparous and multiparous rats. *J. Comp. Physiol. Psychol.* 67, 36–40.

Nitschke, J. B., Nelson, E. E., Rusch, B. D., Fox, A. S., Oakes, T. R., and Davidson, R. J. (2004). Orbitofrontal cortex tracks positive mood in mothers viewing pictures of their newborn infants. *Neuroimage* 21, 583–592.

Noirot, E. (1972). The onset of maternal behavior in rats, hamsters, and mice. In *Advances in the Study of Behavior* (D. S. Lehrman, *et al.*, Eds.), Vol. 4, pp. 1569–1645. Academic Press, New York.

Numan, M. (1994). A neural circuitry analysis of maternal behavior in the rat. *Acta Paediatr. Suppl.* 397, 19–28.

Numan, M. (2007). Motivational systems and the neural circuitry of maternal behavior in the rat. *Dev. Psychobiol.* 49, 12–21.

Numan, M., and Insel, T. R. (2003). *The Neurobiology of Parental Behavior*. Springer-Verlag, New York.

Numan, M., Numan, M. J., Pliakou, N., Stolzenberg, D. S., Mullins, O. J., Murphy, J. M., and Smith, C. D. (2005). The effects of D1or D2 dopamine receptor antagonism in the medial preoptic area, ventral pallidum, or nucleus accumbens on the maternal retrieval response and other aspects of maternal behavior in rats. *Behav. Neurosci.* 119, 1588–1604.

Numan, M., Fleming, A. S., and Levy, F. (2006). Maternal Behavior. In *Knobil and Neill's Physiology of Reproduction* (J. D. Neill, Ed.), pp. 1921–1993. Elsevier, San Diego.

Parada, M., King, S., Li, M., and Fleming, A. S. (in press). The roles of accumbal dopamine D1 and D2 receptors in maternal memory in rats. *Behav. Neurosci.*

Popeski, N., Scherling, C., Fleming, A. S., Lydon, J., Pruessner, J. C., and Meaney, M. J. Maternal adversity alters patterns of neuronal activation in response to infant cues (in preparation).

Ranote, S., Elliott, R., Abel, K. M., Mitchell, R., Deakin, J. F., and Appleby, L. (2004). The neural basis of maternal responsiveness to infants: An fMRI study. *Neuroreport* 15, 1825–1829.

Reck, C., Hunt, A., Fuchs, T., Weiss, R., Noon, A., Moehler, E., Downing, G., Tronick, E. Z., and Mundt, C. (2004). Interactive regulation of affect in postpartum depressed mothers and their infants: An overview. *Psychopathology* 37, 272–280.

Righetti-Veltema, M., Conne-Perreard, E., Bousquet, A., and Manzano, J. (2002). Postpartum depression and mother–infant relationship at 3 months old. *J. Affect. Disord.* 70, 291–306.

Righetti-Veltema, M., Bousquet, A., and Manzano, J. (2003). Impact of postpartum depressive symptoms on mother and her 18-month-old infant. *Eur. Child Adolesc. Psychiatr.* 12, 75–83.

Robson, K. S., and Kumar, R. (1980). Delayed onset of maternal affection after childbirth. *Br. J. Psychiatr.* 136, 347–353.

Rogers, R. D., Andrews, T. C., Grasby, P. M., Brooks, D. J., and Robbins, T. W. (2000). Contrasting cortical and subcortical activations produced by attentional-set shifting and reversal learning in humans. *J. Cogn. Neurosci.* 12, 142–162.

Schaal, B., Montagner, H., Hertling, E., Bolzoni, D., Moyse, A., and Quichon, R. (1980). [Olfactory stimulation in the relationship between child and mother]. *Reprod. Nutr. Dev.* 20, 843–858.

Schino, G., D'Amato, F. R., and Troisi, A. (1995). Mother–infant relationships in Japanese macaques: Sources of inter-individual variation. *Anim. Behav.* 49, 151–158.

Schlein, P. A., Zarrow, M. X., Cohen, H. A., Denenberg, V. H., and Johnson, N. P. (1972). The differential effect of anosmia on maternal behaviour in the virgin and primiparous rat. *J. Reprod. Fertil.* 30, 139–142.

Schneider, M., and Koch, M. (2005). Behavioral and morphological alterations following neonatal excitotoxic lesions of the medial prefrontal cortex in rats. *Exp. Neurol.* 195, 185–198.

Schultz, W. (2007). Behavioral dopamine signals. *Trends Neurosci.* 30, 203–210.

Seifritz, E., Esposito, F., Neuhoff, J. G., Luthi, A., Mustovic, H., Dammann, G., von Bardeleben, U., Radue, E. W., Cirillo, S., Tedeschi, G., and Di Salle, F. (2003). Differential sex-independent amygdala response to infant crying and laughing in parents versus nonparents. *Biol. Psychiatr.* 54, 1367–1375.

Silva, M. R. P., Bernardi, M. M., Cruz-Casallas, P. E., and Felicio, L. F. (2003). Pimozide injections into the nucleus accumbens disrupt maternal behaviour in lactating rats. *Pharmacol. Toxicol.* 93, 42–47.

Slotnick, B. M. (1967). Disturbances of maternal behavior in the rat following lesions of the cingulate cortex. *Behaviour* 29, 204–236.

Stallings, J., Fleming, A. S., Corter, C., Worthman, C., and Steiner, M. (2001). The effects of infant cries and odors on sympathy, cortisol, and autonomic responses in new mothers and nonpostpartum women. *Parenting Sci. Prac.* 1, 71–100.

Stamm, J. S. (1955). The function of the median cerebral cortex in maternal behavior in rats. *J. Comp. Physiol. Psychol.* 48, 347–356.

Stefani, M. R., and Moghaddam, B. (2006). Rule learning and reward contingency are associated with dissociable patterns of dopamine activation in the rat prefrontal cortex, nucleus accumbens, and dorsal striatum. *J. Neurosci.* 26, 8810–8818.

Stolzenberg, D. S., McKenna, J. B., Keough, S., Hancock, R., Numan, M. J., and Numan, M. (2007). Dopamine D-sub-1 receptor stimulation of the nucleus accumbens or the medial preoptic area promotes the onset of maternal behavior in pregnancy-terminated rats. *Behav. Neurosci.* 121, 907–919.

Strathearn, L., and McClure, S. M. (2002). A functional MRI study of maternal responses to infant facial cues. program no. 517.5.2002 *Abstract viewer/ltinerary planner*. Washington DC: *Soc. Neurosci.* online.

Sutherland, R. J., Whishaw, I. Q., and Kolb, B. (1988). Contributions of cingulate cortex to two forms of spatial learning and memory. *J. Neurosci.* 8, 1863–1872.

Swain, J. E., Lorberbaum, J. P., Kose, S., and Strathearn, L. (2007). Brain basis of early parent–infant interactions: Psychology, physiology, and *in vivo* functional neuroimaging studies. *J. Child Psychol. Psychiatr.* 48, 262–287.

Swanson, L. J., and Campbell, C. S. (1979). Induction of maternal behavior in nulliparous goldon hamsters (*Mesocricetus auratus*). *Behav. Neural Biol.* 26, 364–371.

Swerdlow, N. R., Lipska, B. K., Weinberger, D. R., Braff, D. L., Jaskiw, G. E., and Geyer, M. A. (1995). Increased sensitivity to the sensorimotor gating-disruptive effects of apomorphine after lesions of medial prefrontal cortex or ventral hippocampus in adult rats. *Psychopharmacology (Berl)* 122, 27–34.

Takenouchi, K., Nishijo, H., Uwano, T., Tamura, R., Takigawa, M., and Ono, T. (1999). Emotional and behavioral correlates of the anterior cingulate cortex during associative learning in rats. *Neuroscience* 93, 1271–1287.

Thierry, A. M., Le Douarin, C., Penit, J., Ferron, A., and Glowinski, J. (1986). Variation in the ability of neuroleptics to block the inhibitory influence of dopaminergic neurons on the activity of cells in the rat prefrontal cortex. *Brain Res. Bull.* 16, 155–160.

Vollm, B. A., Taylor, A. N., Richardson, P., Corcoran, R., Stirling, J., McKie, S., Deakin, J. F., and Elliott, R. (2006). Neuronal correlates of theory of mind and empathy: A functional magnetic resonance imaging study in a nonverbal task. *Neuroimage* 29, 90–98.

Walsh, C. J., Fleming, A. S., Lee, A., and Magnusson, J. E. (1996). The effects of olfactory and somatosensory desensitization on fos-like immunoreactivity in the brains of pup-exposed postpartum rats. *Behav. Neurosci.* 110, 134–153.

Williams, J., Mathews, A., and MacLeod, C. (1996). The emotional Stroop task and psycopathology. *Psychol. Bull.* 120, 3–24.

Wisenfeld, A., and Klorman, R. (1978). The mother's psychophysiological reactions to contrasting affective expressions by her own and an unfamiliar infant. *Dev. Psychol.* 14, 294–304.

Wisenfeld, A., and Malatesta, C. Z. (1982). Infant distress: Variable affecting responses of caregivers and others. In *Parenting: Its Causes and Consequences* (L. W. Hoffman, *et al.*, Eds.), pp. 123–139. Erlbaum Associates, Hillsdale, NJ.

INDEX